Year	Name	Country	Contribution
1858	Charles Darwin	Britain	Presents evidence that natural selection guides the evolutionary process.
1858	Alfred R. Wallace	Britain	Independently comes to same conclusions as Darwin.
1865	Louis Pasteur	France	Disproves the theory of spontaneous generation for bacteria; shows that infections are caused by bacteria and develops vaccines against rabies and anthrax.
1866	Gregor Mendel	Austria	Proposes basic laws of genetics based on his experiments with garden peas.
1869	J. Friedrich Miescher	Switzerland	Discovers that the nucleus contains a chemical he called nuclein, now termed DNA.
1878	Joseph Lister	Britain	Devises a method of sterilizing the operating room to prevent infection in surgical patients.
1880	Walther Flemming	Germany	Studies the movement of chromosomes during mitosis.
1882	Robert Koch	Germany	Establishes the germ theory of disease and develops many techniques in bacteriology.
1888	Wilhelm Roux	Germany	Founds the science of embryology by performing experiments on embryos.
1892	August Weismann	Germany	Formulates the germ plasma theory, which states that only germ plasma is passed from generation to generation.
1897	Eduard Büchner	Germany	Extracts enzymes from yeast and uses them to bring about fermentation.
1900	Hugo de Vries, Erich von Tschermak, Karl Correns	Holland, Austria, Germany	Independently rediscover Mendel's laws.
1900	Walter Reed	United States	Discovers that yellow fever virus is transmitted by a mosquito.
1901	Hugo de Vries	Holland	States that mutations account for the presence of variations among members of a species.
1901	Santiago Ramón y Cajal	Spain	Suggests that neurons are separated by synapses.
1902	Walter S. Sutton, Theodor Bovari	United States, Germany	Suggest that genes are on the chromosomes, after noting the similar behavior of genes and chromosomes.
1902	William M. Bayliss, E. H. Starling	Britain	Found the study of endocrinology by demonstrating the action of the hormone secretin.
1903	Karl Landsteiner	Austria	Discovers ABO blood types.
1904	Ivan Pavlov	Russia	Shows that conditioned reflexes affect behavior based on experiments with dogs.

continued on inside back cover

BIOLOGY

BIOLOGY

Third Edition

Sylvia S. Mader

CONTRIBUTORS

Cellular Energy and Botany
W. Dennis Clark, Arizona State University

Genetics
Robert M. Kitchin, The University of Wyoming

Ecology
Thomas C. Emmel, The University of Florida

Critical Thinking Case Studies
Robert D. Allen, Inver Hills Community College

WCB Wm. C. Brown Publishers

Book Team

Editor *Kevin Kane*
Developmental Editor *Carol Mills*
Production Editor *Sherry Padden*
Visuals/Design Consultant *Marilyn Phelps*
Designer *Mark Elliot Christianson*
Art Editor *Donna Slade*
Photo Editor *Michelle Oberhoffer*
Permissions Editor *Mavis M. Oeth*
Visuals Processor *Joseph P. O'Connell*

WCB **Wm. C. Brown Publishers**

President *G. Franklin Lewis*
Vice President, Publisher *George Wm. Bergquist*
Vice President, Publisher *Thomas E. Doran*
Vice President, Operations and Production *Beverly Kolz*
National Sales Manager *Virginia S. Moffat*
Advertising Manager *Ann M. Knepper*
Marketing Manager *Craig S. Marty*
Editor in Chief *Edward G. Jaffe*
Production Editorial Manager *Colleen A. Yonda*
Production Editorial Manager *Julie A. Kennedy*
Publishing Services Manager *Karen J. Slaght*
Manager of Visuals and Design *Faye M. Schilling*

Front cover © Ray Coleman/Allstock, Inc.

Back cover © G. Dimijian/Allstock, Inc.

The credits section for this book begins on page 821, and is considered an extension of the copyright page.

"Selected passages in this book are based on material from *Biology* by Leland G. Johnson (Wm. C. Brown Publishers, 1983)."

Library of Congress Catalog Card Number: 89–50426

ISBN 0–697–05638–4

Printed in the United States of America by Wm. C. Brown Publishers, 2460 Kerper Boulevard, Dubuque, IA 52001

10 9 8 7 6 5 4 3

This book is dedicated to my students, who always showed a keen interest in biology and many of whom warmed their instructor's heart by deciding to major in biology.

BRIEF CONTENTS

CONTENTS

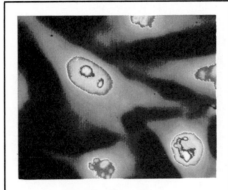

PART 1
The Cell 20

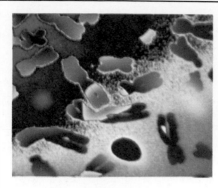

P A R T 2
Genetic Basis of Life 142

P A R T 3
Evolution and Diversity 278

PART 4
Plant Structure and Function 442

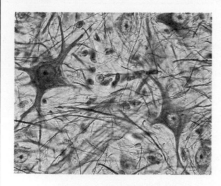

PART 5
Animal Structure and Function 508

PART 6
Behavior and Ecology 698

READINGS

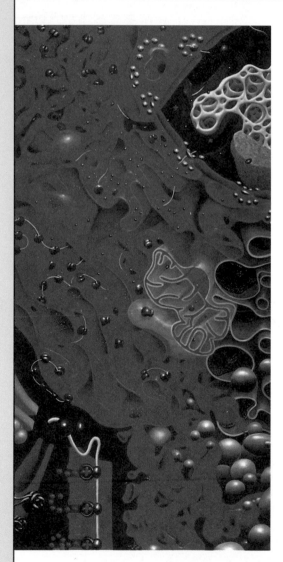

CHAPTER 5

Cell Structure and Function

Your study of this chapter will be complete when you can:

1. State two tenets of the cell theory.
2. List several similarities and differences between prokaryotic and eukaryotic cells.
3. Give several differences between bright-field light microscopy and transmission electron microscopy. Name several other types of microscopes that are available today.
4. Explain why cells are so very small on the basis of cell volume versus cell surface dimensions.
5. Describe the structure of a prokaryotic cell, and give a function for each part mentioned.
6. Describe the structure of the nucleus of a eukaryotic cell, and give a function for each part mentioned.
7. Name the structures that form the endomembrane system, and tell how they are related to one another.
8. Explain the relationship between chloroplasts and mitochondria; describe their structure in terms of compartmentalization of processes.
9. List four evidences for the endosymbiotic theory, which explains the origination of mitochondria and chloroplasts.
10. Name the components of the cytoskeleton; describe the structure and functions of each component.
11. Contrast the structure of prokaryotic, eukaryotic animal, and eukaryotic plant cells.

Artist's representation of a cell's interior illustrates its complexity. After you have studied this chapter, locate the nucleus at the upper right, the mitochondria, Golgi complex, vesicles and rough endoplasmic reticulum (RER). At the lower left, an enlarged portion of a mitochondrion is shown as well as an enlarged portion of rough ER. Look close and notice that protein synthesis is occuring.

55

Behavioral Objectives

Each chapter begins with a list of behavioral objectives designed to help students identify the major concepts of the chapter. Their study of the chapter is complete when they can satisfy these objectives.

Dramatic Visuals Program

Colorful, informative photographs and illustrations enhance the learning program of the text as well as spark interest and discussion of important topics.

Bold-Faced Key Terms

Important terms are boldfaced and defined on first mention.

Text Line Art

Graphic diagrams placed immediately after or within textual passages help clarify difficult concepts and enhance learning.

Concept Summaries

At the ends of major sections within each chapter, concept summaries briefly highlight key concepts in the section, helping students focus their study efforts on the basics.

Figure 1.6
The coral reef, a marine ecosystem in which populations of organisms interact between themselves and the physical environment. The reef consists of the skeletons of stony corals, colonial animals that deposit calcium carbonate without their tissues. Only the outer layer of the reef is alive, the rest forms an inert substratum that provides nooks and crannies where the many different tropical fishes can hide from their predators. Some types of fishes venture forth only at night and feed on plankton, tiny microscopic plants and animals that drift with the current. The coral animals themselves also feed often at night, taking in whatever comes within reach of their extended tentacles. Other fishes are active during the day. These are the grazers that hover about the surface of the coral and snoop around in crevices eating algae or worms, or shrimp and crabs if their jaws are able to crunch up this type of food. The parrot fish even feeds on the stony coral itself. All of the smaller fishes are prey to larger carnivores such as moray eels, jack fishes, and barracudas.

Palau coral reef

Tubastraea coral

red grouper

moray eel

8   Introduction

Glucose Metabolism, an Overview

Most organisms (fig. 9.1) break down carbohydrate molecules (e.g., glucose) in order to acquire a supply of ATP molecules. We can write the overall equation for carbohydrate breakdown in this manner:

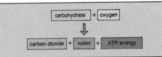
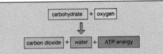

ATP, the energy currency of cells (p. 103), provides the energy that cells need for transport work, mechanical work, and synthetic work. Why do living things convert carbohydrate energy to ATP energy? Why not, for example, utilize glucose energy directly, thereby bypassing the need for mitochondria? The answer is that the energy content of glucose is too great for individual cellular reactions; ATP contains the *right* amount of energy and enzymes are adapted through evolution to couple ATP breakdown to energy-requiring cellular processes (fig. 7.12).

In cells the energy of carbohydrates is converted to that of ATP molecules, the energy currency of cells.

You know from the previous chapters that chloroplasts within plant and algal cells produce the energy-rich carbohydrates that typically undergo cellular respiration in plants and animals (fig. 9.1b). **Cellular respiration,** then, is the link between the energy captured by plant cells and the energy utilized by both plant and animal cells. It is the means by which energy flows from the sun through living things. This is true because, for example, plants produce the food eaten by animals.

Whereas the energy content of food is slowly dissipated, the chemicals, themselves, cycle. Note that when carbohydrate is broken down, the low-energy molecules, carbon dioxide and water, are given off. These are the raw materials for photosynthesis (fig. 9.1b).

The energy content of the organic molecules that are broken down in cells was originally provided by solar energy.

Figure 9.2 gives an overview of glucose metabolism. Glucose is the molecule that most cells use as an energy source in order to produce ATP energy. The first part of glucose metabolism is called glycolysis, a process that occurs in the cytosol and is anaerobic—it does not require oxygen. During **glycolysis** little breakdown occurs as glucose is converted to two molecules of **pyruvate**[1]. Oxidation by removal of hydrogen atoms (e$^-$ + H$^+$) does occur, though, and the energy of this oxidation is used to generate two molecules of ATP.

[1]Pyruvate is the ionized form of pyruvic acid. The terms can be used interchangeably, although at the pH of the cell the ionized form is prevalent.

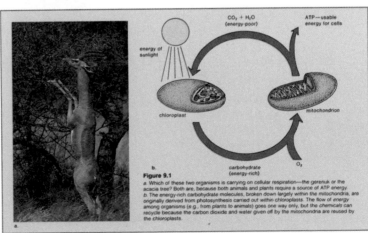

Figure 9.1
a. Which of these two organisms is carrying on cellular respiration—the gerenuk or the acacia tree? Both are, because both animals and plants require a source of ATP energy. b. The energy-rich carbohydrate molecules, broken down largely within the mitochondria, are originally derived from photosynthesis carried out within chloroplasts. The flow of energy among organisms (e.g., from plants to animals) goes one way only, but the chemicals can recycle because the carbon dioxide and water given off by the mitochondria are reused by the chloroplasts.

Glycolysis and Cellular Respiration   **125**

The *Biology* Learning System   **xvii**

2. During transcription, one strand of DNA (the sense strand) serves as a template for the formation of messenger RNA (mRNA). Messenger RNA, therefore, has a series of bases that are complementary to those in DNA.
3. Messenger RNA is processed before it leaves the nucleus during which time the introns are removed.
4. Messenger RNA carries a sequence of codons to the ribosomes, which are composed of ribosomal RNA (rRNA) and proteins.
5. Transfer RNA (tRNA) molecules, each of which is bonded to a particular amino acid, have anticodons that pair complementarily to the codons in mRNA.
6. During translation, the linear sequence of codons of the mRNA determines the order in which the tRNA molecules and their attached amino acids arrive at the ribosomes, and this determines the primary structure (linear sequence of amino acids) of a protein.

Table 16.2 Participants in Protein Synthesis

Name of Molecule	Special Significance	Definition
DNA	Code (cell genotype)	Sequence of DNA bases in threes
mRNA	Codon	Complementary sequence of RNA bases in threes
tRNA	Anticodon	Sequence of three bases complementary to codon
rRNA	Ribosomes	Site of protein synthesis
Amino acids	Building blocks for protein	Transported to ribosomes by tRNAs
Protein	Enzyme (cell phenotype)	Amino acids joined in a predetermined order

Practice Problem

This is a segment of a DNA molecule. What are (1) the messenger RNA codons, (2) the possible tRNA anticodons, and (3) the sequence of amino acids in the polypeptide?

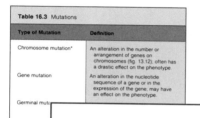

Transcribed strand

*Answers to Practice Problems are given on page 000 of appendix D.

Definition of a Gene and Gene Mutations

Classical (Mendelian) geneticists thought of a gene as a particle on a chromosome. To molecular geneticists, however, a gene is a sequence of DNA bases that code for a product. Most often the gene product is a polypeptide chain. However, there are exceptions because rRNA and tRNA, for example, are gene products themselves. They are involved in protein synthesis, but they do not code for protein.

Gene Mutations

A gene is a sequence of DNA nucleotide bases and a **gene mutation** is a change in that sequence. Table 16.3 lists the different types of mutations so you can become familiar with the terminology that pertains to mutations.

Types of Gene Mutations

Frameshift Mutations **Frameshift mutations** can occur when a single base is inserted into or deleted from a gene. Both of

Table 16.3 Mutations

Type of Mutation	Definition
Chromosome mutation*	An alteration in the number or arrangement of genes on chromosomes (fig. 13.12); often has a drastic effect on the phenotype.
Gene mutation	An alteration in the nucleotide sequence of a gene or in the expression of the gene; may have an effect on the phenotype.
Germinal muta...	
Somatic muta...	

*See page 197.

Tables

Numerous strategically-placed tables list and summarize important information, making it readily accessible for efficient study.

Chapter Summaries

At the end of each chapter is a summary. The style of the summary has been changed for this edition from running text to a listing of important points. This should help students more readily indentify important concepts and better facilitate their learning of chapter content.

Objective Questions

Located at the end of each chapter, these questions require multiple-choice answers that test basic recall of the chapter's key points. The answers to the objective questions are listed in Appendix D.

Practice Problems

Practice problems at the end of the genetics chapters help students master basic gentic quantifications. Answers to these problems can be found in Appendix D.

Readings

Throughout **Biology**, selected readings reinforce major concepts in the book. Most readings are written by the author, but a few are excerpted from popular magazines. A reading may provide insight into the process of science or show how a particular kind of scientific knowledge is applicable to the students' everyday lives.

Cell Biology: Technology and Techniques

The study of cells is dependent upon the use of microscopes and biochemical techniques to detect cellular components and decipher their functions.

Microscopes
In the **bright-field light microscope**, light rays passing through a specimen are brought to a focus by a set of glass lenses, and the resulting image is then viewed by the human eye. In the **transmission electron microscope**, electrons passing through a specimen are brought to a focus by a set of magnetic lenses, and the resulting image is projected onto a fluorescent screen or photographic film. Let's use these statements to examine the differences between the two types of microscopes.

The Means of illumination
Almost everyone knows that an electron microscope magnifies to a greater extent than does a light microscope. A light microscope can magnify objects a few thousand times but an electron microscope can magnify them hundreds of thousands of times. The difference lies in the means of illumination. The paths of the light rays and electrons moving through space are wavelike, but the wavelength of electrons is much shorter than the wavelength of light. This difference in wavelength not only accounts for the ability of the electron microscope to magnify to a greater extent, it also accounts for its greater resolving power. The greater the resolving power, the greater the detail eventually seen. **Resolution** is the *minimum* distance that there can be between two objects before they are simply seen as one larger object. If oil is placed between the sample and the objective lens of the light microscope, the resolving power is increased, and if ultraviolet light is used instead of visible light, it is also increased. But typically, a light microscope can resolve down to 0.2 μm while the electron microscope can resolve down to 0.0001 μm. If the resolving power of the average human eye is set at 1, then that of the typical light microscope is about 500 and that of the electron microscope is 100,000.

The Lenses and Means of Viewing the Object
Light rays can be bent (refracted) and brought to a focus as they pass through glass lenses, but electrons don't pass through glass. Electrons have a charge that allows them to be brought to a focus by magnetic lenses. The human eye utilizes light to see an object and can't utilize electrons for the same purpose. Therefore, electrons leaving the specimen in the electron microscope are directed toward a screen or photographic plate that is sensitive to their presence. Subsequently, humans can view the screen or photograph to see the image depicted there.

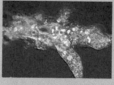

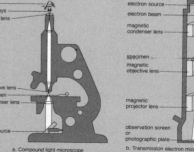
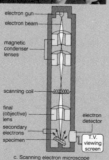

eye
light rays
ocular lens

objective lens
specimen
condenser lens

light source

a. Compound light microscope

electron source
electron beam

magnetic condenser lens

specimen
magnetic objective lens

magnetic projector lens

observation screen or photographic plate

b. Transmission electron microscope

electron gun
electron beam

magnetic condenser lenses

scanning coil

final (objective) lens

secondary electrons
specimen

electron detector

T.V. viewing screen

c. Scanning electron microscope

Figure 1
a. Light micrograph of *Amoeba proteus,* a single-celled organism. A light microscope uses light to view a specimen. The specimen can be living if it is thin enough to allow light to pass through. A light microscope does not magnify or distinguish as much detail as the electron microscope. *b.* Transmission electron micrograph (TEM) of a portion of a pseudopodium, an extension by which *Amoeba proteus* moves. Cytoskeleton (p. 70) elements are visible. The transmission electron microscope uses electrons to "view" the specimen. The specimen is nonliving and must be thin enough to allow electrons to pass through. The magnification and amount of detail seen is far greater than with the light microscope. *c.* Scanning electron micrograph (SEM) of *Amoeba proteus.* Magnification, ×180. The scanning microscope scans the surface of the specimen with an electron beam. The secondary electrons given off are collected and the result is a three-dimensional image of the specimen.

Study Questions

Study Questions

These questions appear at the end of each chapter. They call for specific, short essay answers that truly challenge students' mastery of the chapter's basic concepts.

Thought Questions

Located at the end of each chapter, these questions require narrative responses that challenge students to apply the chapter's basic concepts to related real-life problems or questions.

Selected Key Terms

A selected list of bold-faced, key terms from the chapter appears at the end of each chapter. Each term is accompanied by its phonetic spelling and a page number indicating where the term is introduced and defined in the chapter.

Further Readings

A list of readings at the end of each part suggests references that can be used for further study of topics covered in the chapters of that part. The items listed in this section were carefully chosen for readability and accessibility.

Critical Thinking Case Studies

Each part ends with a case study designed to help students think critically by participating in the process of science. At many institutions, instructors are encouraged to develop the writing skills of their students. In such cases, instructors could require students to write out their answers to the questions in each case study. Suggested answers for each of these questions appear in the Instructor's Manual for the text.

PREFACE

Biology is a college text that covers biological principles in an easy to understand manner. It strives to use other forms of life, in addition to humans, as examples, and makes frequent reference to plants and invertebrates. Also, this text often mentions the contributions of recognized biologists and discusses some of their experiments in detail.

The text is dedicated to my students who have taught me how to best present information so that it is understandable. When we become immersed in a topic we have a tendency to lose sight of what it must be like to approach the topic for the first time. Ever aware of this pitfall, the needs of the beginning student were constantly kept in mind as I prepared the third edition of *Biology*.

Instructors and students alike responded favorably to the first and second editions of *Biology*. The third edition follows essentially the same format but there have been many changes. For example, many chapters have been divided into two chapters so that each one can now be read in one sitting. The illustration program has been extensively revised and many beautiful new full-color illustrations fill the book. The text's pedagogy has been strengthened, especially by the addition of Critical Thinking Case Studies at the end of each part. Each Case Study provides the student with an opportunity to participate in the process of science and to learn to think critically. There is little doubt that instructors feel the need to bolster the ability of students to think critically and we have responded in a way that should be appreciated. These case studies were written by Dr. Robert Allen and myself and are at a level suitable for the incoming freshman. Dr. Allen, who is presently Dean of Instruction at Inver Hills Community College, holds a Ph.D. in Biophysics (from UCLA) and is a frequent presenter of National Science Foundation Chatauqua courses on "Critical Thinking in Science Teaching." Dr. Allen also contributed to the suggested answers for all Case Study questions. These answers appear in the Instructor's Manual for the text.

Organization of the Text

The organization of the text has been revised in two ways. First, it has been divided into more chapters (49 in this edition versus 32 in the previous one) and second, the chapters have been consolidated into an introduction and six parts (versus twelve parts in the previous edition). But the general flow of topics remains the same as in the second edition.

Introduction

As before, chapter 1 discusses the characteristics of life and introduces important biological concepts. Life has both a unity and diversity because all living things have evolved from the first cell or cells, and each one is adapted to a way of life within an ecosystem.

Chapter 2 thoroughly explains the scientific method and gives examples of both experimental and observational biological research. The biological use of the word "theory" is also fully explained.

Part 1 The Cell

These chapters concern cell structure and function and give students a foundation for the rest of the book. An understanding of cells requires a knowledge of their molecular makeup; therefore, we begin the part with the basics of chemistry before considering the cell itself.

In addition, there are now three separate chapters devoted to energetics; instructors have the option of using all three or any one of them. The first of these three chapters gives an overview.

Part 2 Genetic Basis of Life

We are all aware that the twentieth century has seen an explosion in our knowledge concerning the genetic basis of life and that it is important for students to know about this tremendous revolution. In response, this part features expanded coverage of classical and molecular genetics. A new chapter on human genetic disorders presents the very latest that is known in this area.

The biotechnology chapter is at an appropriate level and also is quite modern in its approach. Instructors will appreciate the opportunity to be able to introduce this topic to their students in such a practical and interesting manner.

Part 3 Evolution and Diversity

I have received many favorable comments regarding the logical sequence of these topics: genetics, evolution, and diversity. One topic flows smoothly into the other because genetics is the cornerstone of evolution, and evolution produces diversity. As suggested by adopters, the three evolution chapters have now been placed together in sequence. Again instructors have the option of assigning all three or any one of them. The diversity coverage has been divided into more chapters, expanded slightly, and enriched by the inclusion of the most up-to-date classification system.

Part 4 Plant Structure and Function

Four chapters are devoted to flowering plant anatomy and physiology. There is a new chapter on plant reproduction, making it possible to cover embryonic development and fruits in more detail. While the botany chapters had already been well received, they have been much improved due to the input of many instructors who specialize in this area.

Part 5 Animal Structure and Function

Most of these chapters have been rewritten to include the very latest information on immunity, nutrition, the endocrine system, and development. As before, the chapters contain comparative information about animals and only secondarily focus on human structure and function.

Part 6 Behavior and Ecology

The ecology section was reorganized and it now includes behavior. A new and logical sequence of topics from populations, to communities, to the biosphere is introduced. Environmental concerns are emphasized in the last two chapters. Some instructors may wish to begin the year's work with this unit and that is certainly a workable alternative.

Aids to the Reader

Biology was written so that students might enjoy, appreciate, and come to understand the field of biology. The following text features are especially designed to assist student learning.

Text Introduction

The introductory section (chapters 1 and 2) lays a foundation upon which the rest of the text depends. The first chapter reviews the characteristics of life and presents the fundamental concepts of biology. The second chapter explains, in detail, the scientific method so that students have an appreciation for the methodology used to arrive at conclusions in science. Various other experiments are described throughout the text.

Study Objectives

Each chapter begins with a list of objectives designed to help students identify the major concepts of the chapter. These objectives will also help students prepare for examinations. Their study of the chapter is complete when they can satisfy these objectives.

Key Terms

Knowing the definitions of terms in biology is vital to the learning process. Key terms that are boldfaced in the chapter are defined in context. Especially significant key terms appear in a selected key term list at the end of each chapter. These are page-referenced to where they are introduced and defined.

Readings

Throughout *Biology,* selected readings reinforce major concepts in the book. Most readings are written by the author, but a few are from popular magazines. A reading may provide insight into the process of science or show how scientific knowledge is applicable to everyday concerns.

Drawings, Photographs, and Tables

The drawings, photographs, and tables in *Biology* are designed to help students learn basic biological concepts as well as the specific content of the chapters. Often it is easier to understand a given process by studying illustrations, especially when the illustrations are carefully coordinated with the text, as is the case here. The photographs were selected not only to please the eye but also to emphasize specific points in the text. The tables summarize and list important information, making it readily available for efficient study.

Chapter Summaries

At the end of each chapter is a summary. The style of the summary has been changed from a running text to a listing of important points. This should help students identify important concepts and facilitate their learning of chapter content.

Chapter Questions

Three kinds of questions—objective questions, study questions and thought questions—appear at the close of each chapter. The style of the objective questions has been changed to multiple choice. This allows for more varied questions than before. Answers to these questions appear in Appendix D at the back of the book. The study questions allow students to test their basic understanding of the concepts in the chapter. The thought questions provide another opportunity for students to think critically. They require the students to apply their knowledge in some new and novel manner.

Selected Key Terms

A list of key terms appears at the end of each chapter. Each term is accompanied by its phonetic spelling and a page number indicating where the term is introduced and defined in the chapter.

Further Readings

The list of readings at the end of each part suggests references that can be used for further study of the topics covered in the chapters of that part. The items listed in this section were carefully chosen for readability and accessibility.

Critical Thinking Case Studies

Each part ends with a case study designed to help students think critically by participating in the process of science. At many institutions, instructors are encouraged to develop the writing skills of their students. In such cases, instructors could require students to write out their answers to the questions in each Case Study. Suggested answers for each of these questions appear in the Instructor's Manual for the text.

Glossary

The glossary is composed of the selected key terms that appear at the end of each chapter. Each glossary entry is composed of a definition, a phonetic spelling to aid pronunciation, and is page-referenced to where it was first introduced and defined in the text.

Additional Aids

Instructor's Manual/Test Item File

Prepared by Jay Templin, the Instructor's Manual/Test Item File is designed to assist instructors as they plan and prepare for classes using *Biology*. For each chapter in the text, the Instructor's Manual provides an outline and overview, lecture objectives and suggestions, essay questions with answers, and a listing of selected films. The Test Item File contains approximately 50 objective questions per textbook chapter with answers.

Suggested answers for the Critical Thinking Case Studies that appear at the end of each part in the text are placed at the end of the corresponding parts in the Instructor's Manual.

In addition, there is a conversion table for instructors using the second edition of *Biology,* along with suggested course outlines for quarter, semester, and two-semester courses.

Student Study Guide

The Student Study Guide that accompanies the text was revised by Kathy Thompson. For each text chapter, there is a corresponding Study Guide chapter that includes a list of study objectives, a pretest, study exercises, and a posttest. Answers to all questions appear at the end of each Study Guide chapter, providing the student with immediate feedback.

Micrograph Slides

A boxed set of 50 slides feature photomicrographs and electron micrographs from the textbook.

Laboratory Manual

The Laboratory Manual that accompanies *Biology* was written by Cynthia Handler and me. Its thirty-three exercises provide enough variety to meet the needs of a broad spectrum of class designs. As an aid to the preparation of the labs, each exercise begins with a list of materials. Student aids include a list of learning objectives at the beginning of each exercise and numerous full-color illustrations throughout. In each exercise, ample space is provided for students to record their observations as the lab proceeds. A laboratory review, consisting of a series of questions, ends each exercise. Answers to the review questions are provided in an appendix.

Customized Laboratory Manual

The Laboratory Manual's thirty-three exercises are now available as individual "lab separates," so instructors can custom-tailor the manual to their particular course needs. The separates, which are published in one-color at a greatly reduced price, will be collated and bound by WCB on request.

Laboratory Resource Guide

More extensive information regarding lab preparation can be found in the Laboratory Resource Guide. Developed by Cynthia Handler and me, the Guide is designed to help instructors make the laboratory experience a more meaningful one for the student. It includes suggested sources for materials and supplies, directions for making up solutions and otherwise setting up the laboratory, expected time and results for the exercises, and suggested answers to all questions in the Laboratory Manual.

Transparencies and Lecture Enrichment Kit

A set of 250 transparency acetates also accompanies the text. These feature key illustrations from the text in both two and full color. They are accompanied by a Lecture Enrichment Kit, which is a set of lecture notes featuring a summary of all textual information on the process or element depicted in each transparency, as well as additional high-interest information not presented in the text.

Critical Thinking Case Study Workbook

Written by Robert Allen, this ancillary includes thirty-three additional critical thinking case studies of the type found in the text. Like the text case studies, they are designed to immerse students in the "process of science" and challenge them to solve problems in the same way biologists do. The case studies here are divided into three levels of difficulty (introductory, intermediate, and advanced) to afford instructors greater choice and flexibility.

Videodisc with Hypercard® and Linkway®

Bio Sci II, a general biology videodisc program with Hypercard® and Linkway® interface from Videodiscovery is now available exclusively with this text. The videodisc includes photos and art found in *Biology* as well as a broad assortment of illustrations, photos, charts, diagrams, and motion pictures with narration for use as lecture support, individual student study, or student group activities.

WCB QuizPak

Student computer software programs are available with this edition of *Biology*. WCB QuizPak provides students with true-false and multiple-choice questions for each chapter in the text. Using these programs in the Learning Resource Center will help students to prepare for examinations. They are available on Apple and IBM PC diskettes.

WCB TestPak

A computerized testing service, provides instructors with either a mail-in/call-in testing program or the complete test item file on diskette for use with the Apple and IBM PC computers. WCB TestPak requires no programming experience.

Acknowledgments

Many people have worked diligently to make this book a reality. My editor, Kevin Kane, directed the efforts of all. Carol Mills, my developmental editor, served as a liaison between the editor, myself and many other people. Among them were Thomas and Katherine Brock, Katherine C. Noonan, and Ruth B. Siegel, who carefully read the entire manuscript and provided many valuable suggestions. Sherry Padden was the production editor; Donna Slade the art director; Michelle Oberhoffer the photo researcher; and Mark Christianson the designer. My thanks to each of them for a job well done.

A special word of thanks goes to Carlyn Iverson who was the primary artist for the book. Her work is exquisitely beautiful and breathtaking! Mark Lefkowitz also provided many dramatic renditions that will be thoroughly enjoyed.

The Contributors

With this edition, there were several contributors to various sections of the book. In many cases, they did not just review the manuscript, but provided me with outlines for how they would write particular chapters. With their help, it was easier to make sure the content was sufficient, accurate, and up-to-date. The contributors were:

W. Dennis Clark (Ph.D. University of Texas), professor of botany at Arizona State University. Dr. Clark's expertise in the areas of chemistry, evolution, and physiology of plants was extremely helpful in the revision of the plant chapters.

Thomas C. Emmel (Ph.D. Stanford University), professor of zoology at the University of Florida. He assisted invaluably by providing updated information in ecology, where he has published numerous articles and two books.

Robert M. Kitchin (Ph.D. University of California–Berkeley), professor of genetics at the University of Wyoming. He reviewed the second edition genetics chapters and outlined new information and events taking place in this rapidly changing and expanding field.

The Ancillary Contributors

The ancillary package that accompanies *Biology* utilized the resources of many instructors currently teaching biology.

Critical Thinking Case Study Book: Robert Allen (Ph.D. University of California–Los Angeles) is Dean of Instruction at Inver Hills Community College, whose special interests include research on the Development of Critical Thinking skills in Science. He is a frequent presenter on the subject for the National Science Foundation Chatauqua series.

Instructor's Manual—Test Item File: Jay Templin (Ed.D. Temple University) is an instructor at Widener University who has authored numerous student and instructor's ancillaries for several college biology textbooks. He is also a co-author of the *College Board Achievement Test* in Biology.

Student Study Guide: Kathy Thompson (M.S. Tulane University) is an instructor and General Biology course coordinator for the Department of Zoology and Physiology at Louisiana State University.

Laboratory Manual and Laboratory Resource Guide: Cynthia Handler (Ph.D. The Ohio State University) is a Post Doctoral Fellow in the Division of Medicine and Environmental Toxicology at Thomas Jefferson University and was the General Biology Laboratory Coordinator at the University of Delaware.

QuizPak: Leslie Wiemerslage (Ph.D. University of Pennsylvania) is an instructor at Belleville Area Community College who prepared new questions for this computerized software program.

Lecture Enrichment Kit: Contributors who provided lecture extensions to accompany the transparencies are:

James Averett *Nassau Community College*
Clyde Bottrell *Tarrant County Jr. College*
Lisa Dillman *Navarro College*
Ronald Leavitt *Brigham Young University*
Kenneth D. Mace *Stephen F. Austin State University*
Trudy McKee
Vincent Puglisi *Nassau Community College*
Ralph Rascati *Kennesaw College*
Prentiss Shepherd *University of Lowell*
Gary Smith *Tarrant County Jr. College, N.E.*
Gerald Summers *University of Missouri*
Jay Templin *Widener University*
Thomas Terry *University of Connecticut*
Karl Ulrich *Western Montana College*
Dennis Vrba *North Iowa Area Community College*
Tim Wood *Wright State University*

Reviewers

Many instructors of introductory biology courses around the country reviewed portions of the manuscript. With many thanks, we list their names here.

First Edition

A. Lester Allen *Birgham Young University*
William E. Barstow *University of Georgia*
Lester Bazinet *Community College of Philadelphia*
Eugene C. Bovee *University of Kansas*
Larry C. Brown *Virginia State University*
L. Herbert Bruneau *Oklahoma State University*
Carol B. Crafts *Providence College*
John D. Cunningham *Keene State College*
Dean G. Dillery *Albion College*
H. W. Elmore *Marshall University*
David J. Fox *University of Tennessee*
Larry N. Gleason *Western Kentucky University*
E. Bruce Holmes *Western Illinois University*
Genevieve D. Johnson *University of Iowa*
Malcolm Jollie *Northern Illinois University*
Karen A. Koos *Rio Hondo College*
William H. Leonard *University of Nebraska—Lincoln*
A. David Scarfe *Texas A & M University*
Carl A. Scheel *Central Michigan University*
Donald R. Scoby *North Dakota State University*
John L. Zimmerman *Kansas State University*

Second Edition

David Ashley *Missouri Western State College*
Jack Bennett *Northern Illinois University*
Oscar Carlson *University of Wisconsin–Stout*
Arthur Cohen *Massachusetts Bay Community College*
Rebecca McBride DiLiddo *Suffolk University*
Gary Donnermeyer *St. John's University (Minnesota)*
D. C. Freeman *Wayne State University*
Sally Frost *University of Kansas*
Maura Gage *Palomar College*
Betsy Gulotta *Nassau Community College*
W. M. Hess *Brigham Young University*
Richard J. Hoffmann *Iowa State University*
Steven J. Loring *New Mexico State University*
Trudy McKee
Brian Myres *Cypress College*
John M. Pleasants *Iowa State University*
Jay Templin *Widener University*

Third Edition

Wayne P. Armstrong *Palomar College*
Mark S. Bergland *University of Wisconsin–River Falls*
Richard Blazier *Parkland College*
William F. Burke *University of Hawaii*
Donald L. Collins *Orange Coast College*
Ellen C. Cover *Manatee Community College*
John W. Crane *Washington State University*
Calvin A. Davenport *California State University–Fullerton*
Robert Ebert *Palomar College*
Darrell R. Falk *Pt. Loma Nazarene College*
Jerran T. Flinders. *Brigham Young University*
Sally Frost *University of Kansas*
Elizabeth Gulotta *Nassau Community College*
Madeline M. Hall *Cleveland State University*
James G. Harris *Utah Valley Community College*
Kenneth S. Kilborn *Shasta College*
Donald R. Kirk *Shasta College*
Jon R. Maki *Eastern Kentucky University*
Ric Matthews *Miramar College*
Joyce B. Maxwell *California State University–Northridge*
Leroy McClenaghan *San Diego State University*
Leroy E. Olson *Southwestern College*
Barbara Yohai Pleasants *Iowa State University*
David M. Prescott *University of Colorado*
Robert R. Rinehart *San Diego State University*
Mary Beth Saffo *University of California—Santa Cruz*
Walter H. Sakai *Santa Monica College*
Frederick W. Spiegel *University of Arkansas–Fayetteville*
Gerald Summers *University of Missouri–Columbia*
Marshall Sundberg *Louisiana St. University*
Kathy S. Thompson *Louisiana State University*
Anna J. Wilson *Oklahoma City Community College*
Timothy S. Wood *Wright State University*

C H A P T E R   1

A View of Life

Your study of this chapter will be complete when you can:

1. *List and explain five characteristics of life.*
2. *Describe how populations of organisms interact within ecosystems.*
3. *Explain how evolution accounts for both the unity and diversity of life.*

Not only structure and function but also behavior is subject to the evolutionary process. The open mouths of these blackbird chicks is a sign stimulus that causes the parents to feed them the nourishment they need to carry on the activities of life.

Life. Except for the most desolate and forbidding regions of the polar ice caps, all the earth is teeming with life. Without life, our planet would be nothing but a barren rock hurtling through silent space. The variety of life on earth is staggering. Human beings are part of it. So are giraffes, whales, houseflies, crabgrass, mushrooms, and penguins (fig. 1.1). The variety of living things ranges from unicellular diatoms, much too small to be seen by the naked eye, all the way up to giant redwood trees that can reach heights of 90 meters (300 feet) or more. This variety seems overwhelming—indeed, this entire book would not be large enough to contain even a list of all the separate types of living things scientists continue to identify. There are over 800,000 known types of insects alone. However, despite this extreme variety, biologists have been able to categorize all living things into just a few major groups. This text recognizes five major categories, or **kingdoms,** of life, as described in the reading below.

All these groups of organisms share certain common characteristics. They give us insight into the nature of life and help us distinguish living things from nonliving things.

Characteristics of Life

Living Things Are Organized

One of the fundamental discoveries of biology is that living things as well as nonliving things are composed only of chemicals and obey the same chemical and physical laws. Even so, living things are organized quite differently from nonliving ones. While nonliving things tend to be rather homogeneous inside, living things are made of different parts. The complex organization (fig. 1.2) of living things begins with the **cell,** the smallest, most basic unit of life. Cells can be broken down into smaller

Classification of Organisms

Taxonomy is the discipline of identifying and classifying organisms according to certain rules (taxo—to put into order; nomy—law, rule, custom). When organisms are given accepted scientific names it helps biologists communicate with one another. The scientific name of an organism is a binomial (bi—two; nomen—name) given in Latin. For instance, the name of the garden pea is *Pisum sativum*. The first word of the name is the genus name and the second word is the specific epithet that describes to which species in a genus we are referring. Binomial nomenclature, first popularized by the Swedish taxonomist Linnaeus, has now been accepted by international groups of taxonomists. The accompanying figure gives the scientific names for the organisms shown.

Another aspect of taxonomy is the grouping of species into larger groups according to their shared characteristics. This kind of grouping is done all the time by people to ease communication. For instance, a gardener may refer to some plants in her garden as poisonous plants and others as edible plants, or she may designate them as herbs, shrubs, and trees depending on the circumstances. Taxonomists do the same sort of thing, but they place organisms in a series of set categories.

At each level above species, there are more and more organisms in the category. All the organisms placed in one genus have fairly specific characteristics in common while those that are in the same kingdom have general characteristics in common. Most taxonomists now recognize the grouping of organisms into bigger and bigger groups, defined by more and more general characteristics, is a consequence of the evolutionary history of organisms. Species are defined by recently evolved unique characteristics while the characteristics that apply to all the organisms in a kingdom are anciently evolved and have been passed on ever since.

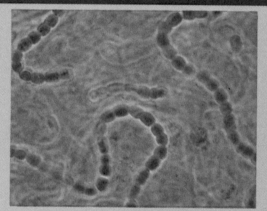
a.

b.

Figure 1

Representatives from the five kingdoms recognized in this text: *a.* kingdom Monera: *Nostoc,* a cyanobacterium; *b.* kingdom Protista: *Euglena,* a single-celled organism; *c.* kingdom

Fungi: *Coprinus comatus,* a shaggy mane mushroom; *d.* kingdom Plantae: *Rosa* hybrid, a flowering plant; *e.* kingdom Animalia: *Carcacal carcacal,* a lynx (short-tailed cat).

Organisms	Human Being	Corn
Kingdom	Animalia	Plantae
Phylum (Division)	Chordata	Anthophyta
Class	Mammalia	Monocotyledons
Order	Primates	Commelinales
Family	Hominidae	Poaceae
Genus	*Homo*	*Zea*
Species	*Homo sapiens*	*Zea mays*

There is much debate among taxonomists as to how many kingdoms should be recognized and what the general characteristics are for each one. That is because recent studies have brought to light new knowledge about the structure and function of organisms. Originally, there were only two kingdoms, the Plantae and Animalia, the plants and animals. Plants were literally things that were planted, immobile, while animals were animated, moved around. The discovery of photosynthesis gave an added characteristic to plants, they make their own food, while animals have to eat. After microscopy began in the seventeenth century, our knowledge of the world of living things was broadened and it became harder and harder to

Figure 1.1

a. Male and female rockhopper penguins court on the rocky shore where they will reproduce. Only two-feet tall, these small penguins are named for their habit of leaping from rock to rock. Penguins are birds, but their feathers are short and their forelimbs are modified into flippers. *b.* Rockhoppers greet another by braying; the sound produced has been likened to the screech of a rusty wheelbarrow. *c.* Yellow feathers above the eyes give a rockhopper a striking pair of eyebrows.

a.

b.

c.

c.

d.

e.

decide what is a plant and what is an animal. What do you call a photosynthetic microorganism that swims around and occasionally eats another microorganism? Problems like this have led to a number of proposals to redefine the kingdom concept since the latter part of the last century. As a result, some taxonomists have developed schemes in which there are as few as two kingdoms while others have proposed over fifteen. This type of disagreement is good for progress in taxonomy, but confusing to beginning biology students; therefore, we have selected a classification system that has wide acceptance today.

In the classification system used in this text, there are five kingdoms, which are listed and described in the accompanying table. You will note that three of these kingdoms contain multicellular forms. These organisms differ in their mode of nutrition: plants photosynthesize and make their own food; animals ingest preformed food; fungi break down preformed food externally by secreting digestive enzymes into the environment.

Kingdoms of Life

Name of Kingdom	Organization	Type of Nutrition	Representative Organisms
Monera	Small, simple single cell (sometimes chains or mats)	Absorb food (some photosynthetic)	Bacteria including cyanobacteria
Protista	Large, complex single cell (sometimes chains or colonies)	All types	Protozoans, algae of various types
Fungi	Multicellular filamentous form with specialized complex cells	Absorb food	Molds and mushrooms
Plantae	Multicellular form with specialized complex cells	Photosynthesize food	Mosses, ferns, woody, and nonwoody flowering plants
Animalia	Multicellular form with specialized complex cells	Ingest food	Sponges, worms, insects, fish, amphibians, reptiles, birds, and mammals

Figure 1.2

All living things are organized. A dragonfly sleeping on a Michigan pine tree's needles in the morning exemplifies the degree to which plants and animals are differently organized.

Figure 1.3

All organisms require an outside source of nutrient materials and energy. Many animals actively seek, kill, and then, like all other animals, ingest their food. *a.* A lioness stalking her prey, a group of zebras, moves closer and closer with muscles taut. *b.* She explodes into her charge joined by other lionesses. The zebras, sensing danger, bolt away. *c.* One zebra of the herd is trapped and cannot escape the claws and jaws of the lionesses.

a.

parts, even down to the atoms making up the molecules found in cells, but these smaller parts are not alive. Just as a whole is often greater than its parts, a living cell possesses the various characteristics of life, whereas its parts do not.

Some organisms are single cells (see fig. 1.4*a*). In many others, however, a large number of similar cells form a tissue—for example, nerve tissue. Tissues make up organs, like a brain or nerves, and then these organs often work together within a system such as a nervous system. Figure 3.1 depicts the levels of organization that are found in organisms.

Living Things Metabolize

The orderliness of your room or office cannot be maintained without an input of energy, and this is true for an organism, too. The organization of the individual is maintained by an intake of materials and energy from its surroundings (i.e., **environment**) since cells are capable of using these to bring about repair and growth of the organism. **Metabolism** is the term used for all the chemical and energy transformations that occur in cells as repair and growth occur.

Animals acquire materials and energy by feeding on other organisms (fig. 1.3). But the ultimate source of energy for all of life on earth comes from the sun. Plants and plantlike organisms are able to capture a small portion of the radiant energy that reaches earth and carry on photosynthesis—the transformation of carbon dioxide and water into energy-rich sugar molecules. Animals that eat plants acquire this energy and pass it on when they themselves become food for other animals. In this way, energy derived from sunlight flows from one organism to another.

Homeostasis

Living things continue to metabolize as long as their internal environment remains fairly constant. Materials enter and leave the organism in various ways, but the internal state must remain suitable for life. This is **homeostasis**—the maintenance of internal conditions within certain boundaries. In animals, body temperature, water balance, and the composition of the blood normally vary only within a narrow range.

Organisms have various mechanisms and structures for maintaining homeostasis. For example, a lizard basks in the sun to raise its internal temperature. However, if the external temperature rises too high, the lizard will scurry for some shade and this will help maintain its normal internal temperature. As another example of homeostasis, consider a biology student who becomes so engrossed in her studies that she forgets to eat lunch. Her liver responds to the emergency by releasing stored sugar into her bloodstream, giving her energy until she eats her next meal.

b.

c.

Living things are highly organized and this organization begins at the cellular level. An intake of materials and energy is needed to maintain an organism's organization. The sun is the ultimate source of energy for life on earth.

Living Things Respond

Because of their organization, living things interact with their environment, responding to it in a variety of ways. Even one-celled organisms respond to outside stimuli, such as changes in chemicals, light, or temperature. The more complex the organism, the more complex the ways in which it can respond to factors in its surroundings.

The ability to respond often results in movement: the leaves of a plant turn toward the sun and animals dart toward safety (fig. 1.3b). The ability to respond helps ensure the survival of the organism and allows it to carry on its daily activities of acquiring energy and reproducing. All together we call these activities the **behavior** of the organism.

Living Things Reproduce

Life comes only from life. Every type of living thing has the ability to **reproduce,** or make a copy similar to itself (fig. 1.4). Some organisms, such as bacteria and protozoa, reproduce by simply dividing in two. However, most multicellular organisms have a more complicated life cycle that involves two parents, one of which produces sperm and the other eggs. The union of a sperm and an egg cell results in an immature individual, which grows and develops through various stages to become the adult.

Heredity

The blueprint for an organism's organization and metabolism are contained within the **genes,** the hereditary factors that are passed on when an organism reproduces. In all organisms, the genes are composed of the complex chemical DNA, which can be copied so that all the cells of a multicellular organism, including sperm and eggs, receive a copy.

Usually the composition of DNA remains the same but on occasion a slight change can occur as it is copied. We call this change a **mutation,** and an offspring that inherits a mutation may have a slightly different organization from its parent or parents. In many instances, this inherited change is detrimental, but on occasion it may prove to be beneficial, so that an offspring has a better chance of surviving and reproducing than its neighbors. For example, a certain moth in England (*Biston betularia*) is normally light colored but a mutated form is dark colored. In polluted areas, the dark form is more prevalent than the light form because birds cannot see the dark form on soot-ladened trees.

Living Things Evolve

The process by which the characteristics of organisms change over time is called **evolution.** These changes occur in groups of interbreeding individuals called **species.** Certain members of a species may inherit a genetic change that causes them to be better adapted or suited to a particular environment. These members can be expected to produce more surviving offspring, which will also have the favorable characteristic. In this way, the attributes of the species members will change over time. For example, the dark-colored moths we just mentioned are better adapted to living in a polluted area than are the light-colored moths of the same species. The dark-colored moths will produce offspring in greater quantity than light-colored moths and these offspring will also be dark colored.

An **adaptation** is any peculiarity of form or function or behavior that promotes the likelihood of a species' continued existence. If the environment should change, and members of a species no longer possess the genetic capability of adapting to the new environment, then extinction can follow.

Living things reproduce, and when they do they pass on genes that determine the form and function of their offspring. Genetic changes account for the ability of a species to evolve and become adapted to a particular environment.

All the various organisms on earth are adapted to particular ways of life and this accounts for the great diversity of life forms. For example, penguins are birds that are adapted to an aquatic existence in the antarctic (fig. 1.1). While the forelimbs of most birds have a structure suitable for flying, the bones in the forelimbs of penguins are shortened and flattened to give a

Figure 1.4

All organisms reproduce. *a.* A single-celled organism such as this *Amoeba proteus* simply divides in two. *b.* Multicellular plants produce seeds, which contain an embryo, an immature form that must develop into the adult. This photograph shows the seed-containing fruit of a flowering dogwood. *c.* Multicellular animals go through developmental stages also. This fertilized queen wasp and workers are building a papery nest. Three stages in wasp development—eggs, larvae, and a pupa in its sealed cell—span the nest. *d.* A mature elephant walks across a plain with her young beside her.

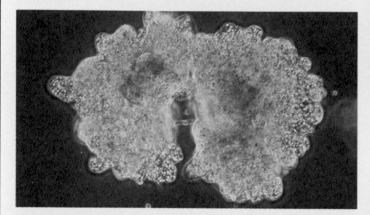
a.

b.

c.

d.

flipper suitable for swimming. Their feet and tails serve as rudders in the water, but the flat feet also allow them to walk on land. Other adaptations govern specifically how the penguin acquires food (materials and energy) and how it reproduces. Rockhopper penguins have a bill adapted to eating small shell fish. Their eggs—one, at most two—are carried on their feet, where they are protected by a pouch of skin. This allows the birds to huddle together for warmth while standing erect and incubating eggs.

Adaptations to different ways of life, including the specific ways in which an organism acquires food and reproduces, explain the great diversity of life forms.

Table 1.1 summarizes the characteristics of living things we have been discussing. The sharing of these same characteristics indicates the unity of all life forms, an idea that we will speak of again on page 10.

Table 1.1 Living Things

Characteristic	Result
1. Are organized	All composed of cells
2. Metabolize	Maintain their organization
3. Respond to stimuli	Interaction with the environment
4. Reproduce	Have offspring
5. Evolve	Become adapted to the environment

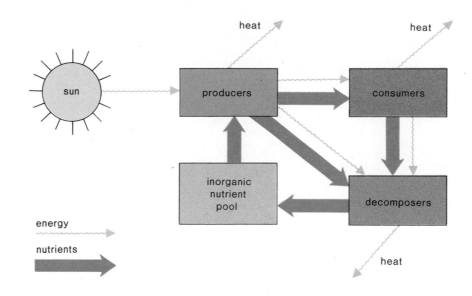

Figure 1.5

The feeding patterns of organisms in an ecosystem can be described in terms of food chains that indicate "who eats whom." The food chain always begins with a photosynthesizing producer population followed by one or more consumer populations. The decomposers are consumer populations that break down dead organic matter and return the chemicals to the producer population. As one population feeds on another, energy flows through the system and is gradually dissipated as heat, but the chemicals (see inorganic nutrient pool) cycle back to the producers once again.

Ecosystems

The organization of life goes beyond separate and individual organisms. All living things on earth are a part of the **biosphere,** an interconnected system that spans the surface of the earth, reaching up into the atmosphere and down into the seas wherever organisms exist. In any portion of the biosphere, such as a particular forest or pond, all members of one species belong to a **population.** The various populations interact with one another and with the physical environment. We call the total of all populations in one natural setting together with their physical environment an **ecosystem** (fig. 1.5).

Populations found within two ecosystems, one terrestrial and the other marine, are described in figures 1.6 and 1.7. The organisms in the terrestrial ecosystem are adapted to a land existence while those in the marine ecosystem are adapted to an aquatic existence. In addition, their specific adaptations allow them to play particular roles in each ecosystem; for example, in the forest, some populations contain plants that photosynthesize while others are animals that feed on plants. The diversity of organisms that we mentioned earlier is best understood in terms of the enormous number of different ways in which organisms carry on their life functions within the ecosystem where they live, acquire energy, and reproduce.

The interactions between populations in an ecosystem, such as a forest or pond, tend to keep the ecosystem in a dynamic balance allowing it to perpetuate itself. Although a forest or a pond will undergo changes, each still tends to be recognizable as such year after year. As in the case of the living organism, it is the organization of an ecosystem that allows it to perpetuate itself. Unlike the individual organism, however, a sufficiently large ecosystem needs no significant outside material support because materials largely cycle as described in figure 1.5. There is only a need for a constant amount of solar energy to keep the system going.

Organisms are members of populations within ecosystems, units of the biosphere. Here their various adaptations allow them to play a role that keeps the system in dynamic balance and assures its continued existence.

The Human Population

The human population tends to modify existing ecosystems for its own purposes. For example, humans clear forests or grasslands in order to grow crops; later they build houses on what was once farmland, and then finally they convert small towns to cities. With each progressive step, fewer and fewer of the original organisms remain and the ecosystem has been completely altered. In the end, there are largely only humans and their domesticated plants and animals where there was once many diverse types of populations.

Unlike the original ecosystem, humans rely on material inputs of various types and they supplement solar energy with fossil fuel (coal and oil) energy in order to maintain farms, towns, and cities. These inputs are eventually converted to outputs such as land pollutants (e.g., trash) and air pollutants (e.g., carbon dioxide).

More and more ecosystems undergo modification as the human population increases. But, we are beginning to learn that the human population is a part of the biosphere and is dependent upon the natural cycles that occur within the biosphere. For example, the air is cleaner over forests than in a city because trees are able to absorb carbon dioxide. As more and more of the biosphere is converted to towns and cities, fewer of the natural cycles are able to function adequately to sustain the human population.

Figure 1.6

The coral reef, a marine ecosystem in which populations of organisms interact between themselves and the physical environment. The reef consists of the skeletons of stony corals, colonial animals that deposit calcium carbonate without their tissues. Only the outer layer of the reef is alive, the rest forms an inert substratum that provides nooks and crannies where the many different tropical fishes can hide from their predators. Some types of fishes venture forth only at night and feed on plankton, tiny microscopic plants and animals that drift with the current. The coral animals themselves also feed often at night, taking in whatever comes within reach of their extended tentacles. Other fishes are active during the day. These are the grazers that hover about the surface of the coral and snoop around in crevices eating algae or worms, or shrimp and crabs if their jaws are able to crunch up this type of food. The parrot fish even feeds on the stony coral itself. All of the smaller fishes are prey to larger carnivores such as moray eels, jack fishes, and barracudas.

Palau coral reef

Tubastraea coral

moray eel

red grouper

lionfish

Figure 1.7

A temperate deciduous forest is a terrestrial ecosystem found in the northern hemisphere where the climate is mild. Various types of insects and worms as well as bacteria and fungi live off the rich organic matter found in the soil. Some birds, like the robin and tufted titmouse feed off soil organisms while others like the red-eyed vireo and woodpeckers prefer the insects that reside in the trees. Various small mammals such as rabbits, squirrels, and white-footed mice live mainly on vegetation, fruits, and insects. Wolves, bobcats, and gray foxes, which feed on these small animals, were more prevalent in days gone by; however, birds of prey like hawks and owls are still common and feed on these animals.

virgin forest at Porcupine Mountains State Park, Michigan

Eastern chipmunk

barn owl

millipede

bobcat

marsh marigolds

The recognition that the human population is not separate from but is a part of the biosphere is one of the most important contributions of recent time. It makes us realize that there are ecological consequences to an ever increasing human population size, and that there is value in unaltered ecosystems.

The human population tends to modify existing ecosystems to suit the needs of humans. We are now beginning to realize that this process can lead to detrimental effects for the human population.

The Unity and Diversity of Life

Earlier we listed and discussed the characteristics seen in all living things. These common characteristics indicate that there is a sameness, or unity, to life. Such a remarkable uniformity can be explained by the descent of organisms from a common ancestor. Evidence from many and varied sources (see chapter 20) tells us that life began as a single cell and that present day organisms evolved from this common origin over a great period of time. In other words, the process of evolution explains the unity we observe in all living things.

We have also witnessed the diversity of life, particularly in two ecosystems—a coral reef and a temperate deciduous forest. Within ecosystems, organisms are adapted to particular ways of life which allow them to carry on their life functions of acquiring energy and reproducing. These specific adaptations explain the diversity of life. In other words, the process of evolution also explains the diversity we observe among living things.

Evolution is the very cornerstone of biology and we will be referring to this process again and again throughout the text.

Evolution from a common ancestor explains the unity of life, and adaptations to different ways of life explains the diversity of life.

Summary

1. Even though living things are diverse, they share certain characteristics in common.
2. All living things are organized; the cell is the smallest unit of life. In many organisms, cells form tissues, tissues form organs, and organs form systems.
3. Living things need an outside source of materials and energy. These are used during metabolism for repair and growth of the organism.
4. Organisms differ as to how they acquire materials and energy. Plants photosynthesize and make their own food; animals either eat plants or other animals. Therefore, there is a flow of energy from plants through all other living things.
5. Living things need a relatively constant internal environment and have various mechanisms and strategies for maintaining homeostasis.
6. Living things respond to external stimuli; taken together, these reactions constitute the behavior of the organism.
7. Living things reproduce and pass on genes, hereditary factors that control organization and metabolism. Changes in the hereditary material allow evolution to occur.
8. Organisms are adapted to various types of environments and to different ways of acquiring energy and reproducing.
9. Members of the same species form a population within an ecosystem. Within the ecosystem, populations interact with each other and with the physical environment.
10. Two examples of ecosystems, the coral reef and the temperate deciduous forest, show that the adaptations of organisms allow them to play a particular role within an ecosystem.
11. The human population is of the biosphere, and the diversity of the biosphere should be preserved in order to ensure the continuance of the human population.
12. The process of evolution can explain both the unity and the diversity of life. Descent from a common ancestor explains why all organisms share the same characteristics, and adaptation to various ways of life explains the diversity of life forms.

Objective Questions

For questions 1 to 5, match the statements in the key with the sentences below.

Key:
   a. Living things are organized.
   b. Living things metabolize.
   c. Living things respond.
   d. Living things reproduce.
   e. Living things evolve.

1. Genes made up of DNA are passed from parent to child.
2. Zebras run away from approaching lions.
3. All organisms are composed of cells.
4. Cells use materials and energy for growth and repair.
5. There are many different kinds of living things.
6. Evolution from the first cell(s) best explains why
   a. ecosystems have populations of organisms.
   b. photosynthesizers produce food.
   c. diverse organisms share common characteristics.
   d. All of these.
7. Adaptation to a way of life best explains why living things
   a. display homeostasis.
   b. are diverse.
   c. began as single cells.
   d. are classified into five kingdoms.
8. Into which kingdom would you place a multicellular land organism that carries on photosynthesis?
   a. kingdom Monera
   b. kingdom Protista
   c. kingdom Plantae
   d. kingdom Animalia

Study Questions

1. What are the five common characteristics of life listed in the text?
2. What evidence can you cite to show that living things are organized?
3. Why do living things require an outside source of materials and energy?
4. What is passed from generation to generation when organisms reproduce? What has to happen to the hereditary material in order for evolution to occur?
5. Choose an organism from a temperate forest or coral reef and tell how it is adapted to its way of life.
6. What is an ecosystem and why should human beings preserve ecosystems?
7. How does evolution explain both the unity and diversity of life?
8. What kingdoms are used in the classification system adopted by this text? What type of organisms are found in each kingdom?

Thought Questions

1. After considering that books store information, explain the phrase that DNA stores information.
2. Some people don't realize that plants are alive. Help enlighten such people by showing that plants exhibit all the characteristics of life.
3. The occurrence of mutations cause variations among the members of a population. Do you suppose that sexual reproduction also contributes to variations? Why?

Selected Key Terms

kingdom (king′dum) 2
cell (sel) 2
environment (en-vi′ron-ment) 4
metabolism (mĕ-tab′o-lizm) 4
homeostasis (ho″me-o-sta′sis) 4

behavior (be-hāv′yor) 5
reproduce (re″pro-dūs′) 5
gene (jēn) 5
mutation (mu-ta′shun) 5
evolution (ev″o-lu′shun) 5

species (spe′shēz) 5
adaptation (ad″ap-ta′shun) 5
biosphere (bi′ŏ-sfēr) 7
population (pop″u-la′shun) 7
ecosystem (ek″o-sis′tem) 7

CHAPTER 2

The Scientific Method

Your study of this chapter will be complete when you can:

1. *Outline the basic steps of the scientific method.*
2. *Explain how inductive and deductive reasoning are used during the steps of the scientific method.*
3. *Give an example of an investigation that follows the steps of the scientific method. Identify the experimental variable and the dependent variable.*
4. *Design a controlled experiment.*
5. *Tell why investigators often report mathematical data and publish their results in scientific journals.*
6. *Show that descriptive research also follows the scientific method.*
7. *Explain what is meant by a "theory" in science and give examples of scientific theories.*

A scientist presents a laboratory flask that contains DNA, the hereditary material of all living things. Biochemical experiments are almost always performed in the laboratory where it is possible to isolate and manipulate the chemicals found within living things.

S cience helps human beings understand the natural world and is concerned solely with information gained by observing and testing that world. Scientists, therefore, ask questions only about events in the natural world and expect that the natural world will in turn provide all the information that is needed to understand that event. It is the aim of science, too, to be objective rather than subjective. It is very difficult to make objective observations and come to objective conclusions since human beings are often influenced by their particular prejudices. Still, anything less than a completely objective observation or conclusion is not considered scientific. Finally, the conclusions of science are subject to change whenever new findings so dictate. Quite often in science, new studies, which might utilize new techniques and equipment, indicate that previous conclusions need to be modified or changed entirely.

Scientists ask questions and carry on investigations that pertain to the natural world. The conclusions of these investigations are tentative and subject to change.

Scientific Method

Scientists, including biologists, employ an approach for gathering information that is known as the **scientific method.** Although the approach is as varied as scientists themselves, there are still certain processes that can be identified as typical of the scientific method. Figure 2.1 outlines the primary steps in the scientific method. On the basis of **data** (factual information) that may have been collected by someone previously, a scientist formulates a tentative statement, called a **hypothesis,** that can be used to guide further observations and experimentation. It's often said that this portion of the scientific method involves **inductive reasoning;** that is, a scientist uses isolated facts to arrive at a general idea that might explain a phenomenon. For example, scientists not only observed the prevalence of dark-colored peppered moths on trees in polluted areas, they also observed the prevalence of light-colored moths on trees in nonpolluted areas. This caused them to formulate the hypothesis that predatory birds were responsible for this unequal distribution of these moths because they feed on the moths they can see (fig. 21.13). Once the hypothesis has been stated, then **deductive reasoning** comes into play. Deductive reasoning begins with a general statement that infers a specific conclusion. It often takes the form of an "if . . . then" statement: If predatory birds are responsible for the unequal distribution of dark- and light-colored moths, then it will be possible to get evidence of birds feeding primarily on light-colored moths in polluted areas and on dark-colored moths in nonpolluted areas. To see if this deduction was correct, the British scientist H.B.D. Kettlewell performed an experiment. He released equal numbers of the two types of moths in the two different areas. He observed that birds captured more dark-colored moths in nonpolluted areas and more light-colored moths in polluted areas. From the unpolluted area, he recaptured 13.7% of the light-colored moths and only 4.7% of the dark form. From the polluted area, he recaptured 27.5% of the dark form and only 13% of the light-colored moths. Mathematical data like these are preferred because they

Figure 2.1
Steps involved in the scientific method. Inductive reasoning is used to formulate a hypothesis from accumulated scientific data, and deductive reasoning is used to decide what observations and experiments would be appropriate in order to test the hypothesis.

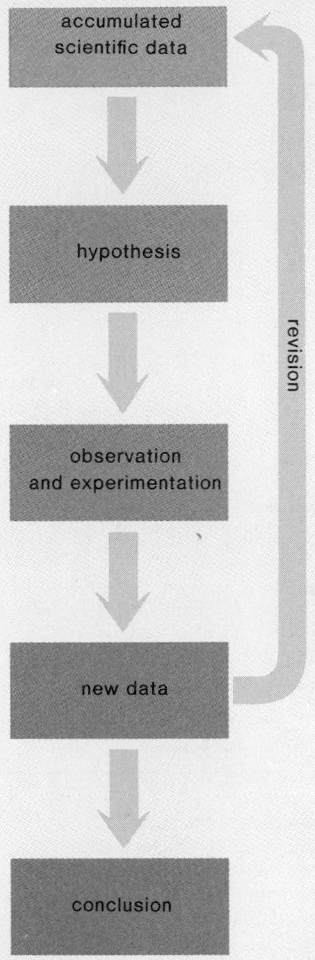

are objective and cannot be influenced by the scientist's subjective feelings. These particular data supported the hypothesis.

Sometimes, as in this case, the data support the hypothesis; however, at other times the data do not support the hypothesis and it has to be rejected. While a hypothesis cannot be proven true, it can be proven false. Because of this characteristic some think of science as what is left after alternative hypotheses have been rejected.

Even though the scientific method is quite variable, it is possible to point out certain steps that characterize the scientific method: making observations, formulating a hypothesis, testing, and coming to a conclusion.

Reporting the Experiment

It is customary to report findings in a **scientific journal** so that the design of the experiment and the results are available to all. This is especially desirable because all experimental results must

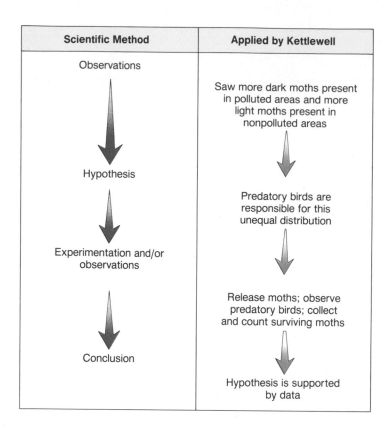

Scientific Method	Applied by Kettlewell
Observations	Saw more dark moths present in polluted areas and more light moths present in nonpolluted areas
Hypothesis	Predatory birds are responsible for this unequal distribution
Experimentation and/or observations	Release moths; observe predatory birds; collect and count surviving moths
Conclusion	Hypothesis is supported by data

Figure 2.2

Design of a controlled experiment. Genetically identical mice are randomly divided into a control group and the experimental groups. All groups are exposed to the same environmental conditions such as housing, temperature, water supply, etc. Only the experimental groups are subjected to the test (experimental variable)—in this case the presence of sweetener S in the food. At the end of the experiment all mice are examined for evidence of bladder cancer.

be repeatable; that is, other scientists using the same procedures are expected to get the same results. If they do not, the original data cannot be considered to have supported the hypothesis.

Often, too, the authors of a report will make suggestions about other experiments that could be done to clarify or broaden our understanding of the matter under study.

The results of observations and experiments are published in a journal where they can be examined by all. These results are expected to be repeatable, that is, to be obtained by anyone doing the exact same procedure.

Controlled Experiment

Kettlewell performed his experiment in the field, but scientists often prefer to work in the laboratory where they can oversee all elements of the experiment.

Elements of an Experiment

All conditions should be held constant except the one being tested. This one element is called the **experimental variable;** in other words, the investigator deliberately varies this portion of the experiment. Then the investigator observes the effects of the experiment; in other words, he observes the **dependent variable:**

Experimental variable:	Dependent variable:
the element of the experiment being tested.	result or change that occurs due to the experimental variable.

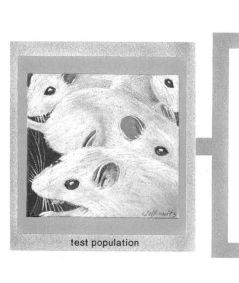

test population

experimental groups

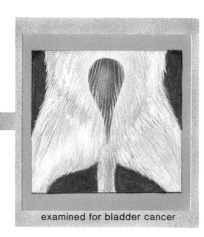

examined for bladder cancer

control group

▦ Constant environmental conditions
☐ Sweetener in food
■ No sweetener in food

In order to be absolutely sure that the results are due to the experimental variable and not due to some unknown factor, it is customary to have a separate sample called the control group. A **control group** goes through all the steps of the experiment except the one being tested.

Design of an Experiment

As an example, suppose physiologists wish to determine if sweetener S is a safe additive for foods. On the basis of available information, they hypothesize that sweetener S would have no effect on health at a low concentration but would cause bladder cancer at a high concentration. They might decide to feed sweetener S to groups of mice at ever greater percentages of the total dietary intake of food:

Group 1: no sweetener S in food (control group)
Group 2: sweetener S is 5% of food
Group 3: sweetener S is 10% of food

Group 11: sweetener S is 50% of food

The experimenters would first place a certain number of randomly chosen inbred (genetically identical) mice into the various groups—say 10 mice per group. If any of the mice are different from one another, the random selection will hopefully distribute them evenly among the groups. The experimenters would also make sure that all conditions such as availability of water, cage conditions, and temperature of surroundings are the same for all groups. The food for each group would be exactly the same except for the amount of sweetener S. (The amount of sweetener S is the experimental variable.) After a designated period of time, the animals would be killed and dissected to examine their bladders for evidence of cancer. (Evidence of bladder cancer is the dependent variable.) Figure 2.2 outlines the design of this experiment.

Results of an Experiment

The data from this experiment would no doubt be presented in the form of a table or graph (fig. 2.3). A statistical test might be run to determine if the difference in the number of cases of bladder cancer between the various groups is significant. After all, if a significant number of mice in the control group developed cancer, the results might be invalid. On the basis of the results, the experimenters will try to come to a recommendation concerning the safety of putting sweetener S in the food of humans. They might determine that any amount over 10% of food intake would be expected to cause a progressively increased incidence of bladder cancer, for example.

Many scientists work in laboratories where they carry out controlled experiments.

Descriptive Research

Scientists don't always gather data by doing experiments. Much of the data that they gather is purely observational, but even so the steps we have described for the scientific method are still applicable: observations are made, a hypothesis is formed, predictions are made on the basis of the hypothesis, and data are collected that may support or disprove the hypothesis. For example, Kenneth E. Glander[1] was interested in determining the feeding pattern of mantled howling monkeys, which are herbivores that feed on plant products (fig. 2.4). He found at least two hypotheses in the scientific literature that could perhaps apply to the monkeys:

1. Most large herbivores try to obtain an optimal mix of nutrients.
2. Mammalian herbivores tend to avoid plant foods that contain toxins.

Glander felt that there was insufficient data to decide if either one or both of these statements was pertinent to the feeding behavior of the howling monkeys. So he and his associates made new observations:

1. They marked off two test sites and identified every type of tree in each site. Altogether they identified 1,699 trees.
2. They observed the feeding behavior of a group of monkeys consisting of two adult males, six adult females, and five juveniles. A total of 1,853 hours of animal observation was accumulated from September 1 of one

[1]Glander, K. E. 1981. Feeding patterns in mantled howling monkeys. In *Foraging behavior: ecological, ethological and psychological approaches*, ed. A. C. Kamil and T. D. Sargent, 231–57. New York: Garland STPM Press.

Figure 2.3

Scientists often acquire mathematical data, which they report in the form of tables or graphs. Mathematical data are more decisive and objective than visual observations. The data in this instance suggest that there is an ever greater chance of bladder cancer if the food contains more than 10% sweetener S. The same experiment will be repeated many times to test these results and statistical analyses will be done to see if the results are significant or simply due to chance alone.

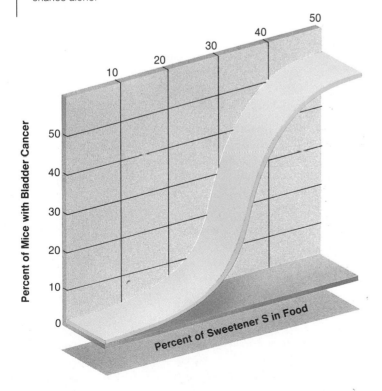

Figure 2.4

Mantled howling monkeys are herbivores that feed on plant products in South American forests. They tend to choose mature leaves and fruit in the wet season and new leaves and flowers in the dry season.

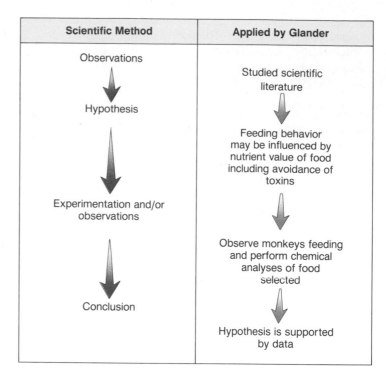

Scientific Method	Applied by Glander
Observations	Studied scientific literature
↓	↓
Hypothesis	Feeding behavior may be influenced by nutrient value of food including avoidance of toxins
↓	↓
Experimentation and/or observations	Observe monkeys feeding and perform chemical analyses of food selected
↓	↓
Conclusion	Hypothesis is supported by data

year to August 23 of the next year. The average observation day lasted 12 hours and 11 minutes.

3. They performed chemical analyses of fresh plant samples from about 50 trees to determine their nutritional (amino acid) content and their toxin content.

Glander found that the study group obtained a majority of their food from 62 relatively rare tree species. During the 12-month study period the monkeys obtained all of their food from 331 different kinds of trees, or 19.5% of all trees available to them. Further, they tended to choose mature leaves and fruits in the wet season and new leaves and flowers in the dry season. The chemical analyses of the plant products showed that the howlers were maximizing their intake of water, total protein, and amino acids, maximizing their intake of digestible protein, and minimizing their intake of fiber, ash, and toxic plant compounds. Glander's conclusion was that the howlers chose those foods that gave the greatest net nutrient return. Glander stated, "A tropical forest should not be viewed as a well-stocked larder waiting to be exploited, but rather as a spatially and temporally changing mosaic of items of varying value and availability."

This proposed model of the nature of the food sources for forest-dwelling herbivores is one of the most important outcomes of Glander's observations. A scientific **model** is a suggested explanation that can help direct future research. Again, we can apply the steps in the scientific method.

A large proportion of scientific information is based on purely observational (descriptive) data, but the steps previously suggested for the scientific method still apply.

Theories and Principles

The ultimate goal of science is to understand the natural world in terms of **theories,** interpretations that take into account the results of many experiments and observations. In a movie a de-

tective might claim to have a theory about the crime. Or you might say that you have a theory about the won-lost record of your favorite baseball team. But in science the word theory is reserved for a conceptual scheme that is supported by a large number of observations and has not yet been found lacking.

Table 2.1 lists some of the unifying theories of biology. You can see that in general the theories pertain to the characteristics of life we discussed in chapter 1. The theory of evolution enables scientists to understand the history of life, the variety of living things, the anatomy, physiology, and development of organisms, and even their behavior. Because the theory of evolution has been supported by so many observations and experiments for over a hundred years, some biologists refer to the "principle of evolution," suggesting that this is the appropriate terminology for theories that are generally accepted as valid by an overwhelming number of scientists.

Theories and Experiments

The scientific method is often presented in the manner of figure 2.1 and scientists do generally follow such a scheme. However, scientists also may make observations and do experiments that they *expect* to have certain results, and that they believe can be explained by a certain theory:

The role of an experiment is illustrative. It allows a scientist to demonstrate the power of his theory, not as a collection of truths, but as a set of ideas. When an experiment succeeds, this shows that a certain way of describing the world has proved itself useful. When an experiment fails it shows that one's concepts were inadequate or confused.[2]

[2]Harre, R. 1981. *Great scientific experiments.* London: Phaidon Press Limited.

Science and Social Responsibility

There are many ways in which science has improved our lives. The most obvious examples are in the field of medicine. The discovery of antibiotics, such as penicillin, and of the vaccines for polio, measles, and mumps have increased our life-spans by decades. Cell-biology research may someday help us understand the mechanisms that cause cancer. Genetic research has produced new strains of agricultural plants that have eased the burden of feeding our burgeoning population.

But science has also fostered technologies of grave concern to some of us. Even though we would not like a cancer patient to be denied the benefits of radiation therapy, we are concerned about the nuclear power industry, and we do not want nuclear wastes stored near our homes. Then, too, there may be undesirable side effects even from those technologies of which we approve. None of us wish to go back to the days in which large numbers of people died of measles or polio. Yet as science has conquered one disease after another, the world's death rate has fallen and the human population has exploded. Few of us are willing to give up technology's gift of the private automobile, but we are concerned about the amount of air pollution that automobiles generate.

All too often we blame science for these side effects, and we say that our lives are infinitely more threatened than they were in "the good old days." We think that scientists should label research good or bad and be able to predict whether any resulting technology would primarily benefit or harm mankind. Yet science, by its very nature, is impartial and simply attempts to study natural phenomena. Science does not make ethical or moral decisions. Once we wish to make value judgments, we must go to other fields of study to find the means to make those judgments. And these judgments must be made by all people. The responsibility for how we use the fruits of science, including a given technology, must rest with people from all walks of life, not upon scientists alone.

Scientists should provide the public with as much information as possible when such issues as nuclear power or recombinant DNA technology are being debated. Then, they, along with other citizens, can help make decisions about the future role of these technologies in our society. All men and women have a responsibility to decide how best to use scientific knowledge in a way that benefits the human species and all living things.

Table 2.1 Unifying Theories of Biology

Name of Theory	Concept	Investigator
Cell	All organisms are composed of cells	Matthias Schleiden 1838 Theodor Schwann 1839
Biogenesis	Life comes only from life	Louis Pasteur 1875
Homeostasis	The internal environment remains within a normal range	Claude Bernard 1851
Evolution	All living things have a common ancestor and are adapted to a particular way of life	Charles Darwin 1858
Gene	Organisms contain coded information that dictates their form, function, and behavior	Gregor Mendel 1860 James Watson and Francis Crick 1953

Figure 2.5

Results of an experiment with nesting mountain bluebirds. The graph shows that resident males are more likely to approach a male model and female mate aggressively before the nest is completed, less likely after the first egg is laid, and least likely after the eggs have hatched. The experimenter gives an evolutionary interpretation to the results: it is more adaptive for a male bluebird to be aggressive before mating in order to be sure the offspring will be his and less adaptive at the other stages mentioned.

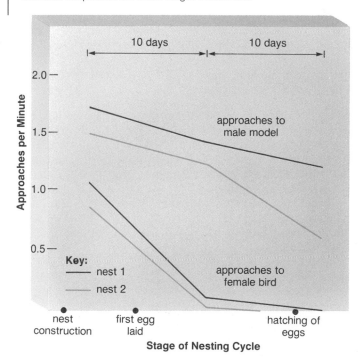

For example, in 1975 David P. Barash[3] conducted a study of two mountain bluebird (fig. 2.5) nests seemingly with the intention of using the theory of evolution to explain their behavior. He posted a model of a male bluebird 1 meter from each nest while the resident male was foraging. Upon returning to the nests, each resident encountered a model in proximity to his female. This was done during three different stages of the nesting cycle. The behavior of the male toward the model and toward his female during the first 10 minutes of his discovery of the model were observed and noted. As a control, a model male robin was also posted four days after the initial presentation of the bluebird model. The robin model elicited no response. The number of aggressive approaches toward the bluebird male model and female mate are given in graph form in figure 2.5. Aggression was most severe when the bluebird male model was presented before the first egg was laid, less severe once the first egg was laid, and least severe after the eggs had hatched.

[3]Barash, D. P. 1976. The male response to apparent female adultery in the mountain bluebird, *Sialia currucoides:* An evolutionary interpretation. *American Naturalist* 110:1097–1101.

In his report of this experiment, Barash explains how evolutionary theory can account for these results. It is adaptive for males to be especially aggressive before the eggs are laid because there are a limited number of nesting sites available to the birds, and he wants to be sure he's the one mating and passing on his genes. It is adaptive for males to be less aggressive after the first egg is laid; the reproductive process has begun and the male is more sure the offspring will be his. Finally, it would be maladaptive for the male to spend time and energy being too aggressive toward a rival and his mate after hatching because his offspring are already present.

Barash's discussion of his results seems to support the suggestion that scientists sometimes do experiments with the intention of explaining their results on the basis of a particular theory. He has shown how his results can be given an evolutionary interpretation. It is adaptive for male bluebirds to be aggressive toward rivals, especially before mating, because it helps ensure that they will be the ones to pass on their genes. Barash shows that this is an observed characteristic among these birds. In doing so, he also shows that in addition to anatomy and physiology, behavior is subject to an evolutionary interpretation.

It is the aim of science to formulate theories and principles that encompass a number of hypotheses concerning the natural world. The power of a theory is evident when it can be used to explain diverse observations and experiments.

Summary

1. Scientific investigations always pertain to the natural world and the conclusions of the investigations are always subject to change.
2. In general, scientists use a methodology called the scientific method. Based on an observation made by themselves or found in the scientific literature, inductive reasoning is used to formulate a hypothesis. Using deductive reasoning, a plan of action is devised and carried out. The results either support or fail to support the hypothesis.
3. Most scientists do controlled experiments. Particularly, in the laboratory we note that (1) the environmental conditions are held constant, (2) there is a control group that goes through all the steps except the one being tested, and (3) there are one or more experimental groups.
4. In a controlled experiment, the experimental variable is that portion of the experiment that is being tested, and the dependent variable is the results of the experiment.
5. Much of biological investigation is descriptive rather than experimental. Descriptive research requires that the investigator make careful observations. Descriptive research still follows the steps outlined for the scientific method.
6. The goal of science is to understand the natural world in terms of theories, which can be used to explain diverse observations and experiments. Some theories are so well supported that many suggest they should be called principles.
7. Investigators sometimes make observations or do experiments with the intended purpose of explaining their results on the basis of a particular theory.

Objective Questions

1. Which of these steps in the scientific method would be the second step?
   a. experimentation
   b. conclusion
   c. researching the scientific literature
   d. forming the hypothesis
2. Which of these steps in the scientific method would be the third step?
   a. experimentation
   b. conclusion
   c. researching the scientific literature
   d. forming the hypothesis
3. With which of these steps in the scientific method would you associate inductive reasoning?
   a. experimentation
   b. conclusion
   c. researching the scientific literature
   d. forming the hypothesis
4. With which of these steps in the scientific method would you associate deductive reasoning?
   a. experimentation
   b. conclusion
   c. researching the scientific literature
   d. forming the hypothesis

5. Which is the experimental variable in this experiment?
   a. Conditions like temperature and housing were the same for all groups.
   b. The amount of sweetener S in food was varied.
   c. Two percent of the group fed 10% sweetener S got bladder cancer and 90% of the group fed 50% sweetener S got bladder cancer.
   d. The data were presented as a graph.
6. Which is the control group in this experiment?
   a. All mice in this group died because a lab assistant forgot to give them water.
   b. All mice in this group received 10% sweetener S in food.
   c. All mice in this group received no sweetener S in food.
   d. Some mice in all groups got bladder cancer; therefore there was no control group.

7. Which would be an example of descriptive research?
   a. Jones put broken egg shells in the nest of some gulls and watched their behavior.
   b. Smith measured the length of the twigs of all trees in the marked off area.
   c. Green put pesticide into one jar of amoeba and not the other jar.
   d. Kettlewell counted how many moths the birds did not eat.
8. Which of these statements is correct?
   a. Because scientists speak of the "theory of evolution" they believe that evolution does not have much merit.
   b. A theory is simply a hypothesis that needs further experimentation and observation.
   c. Theories are hypotheses that have failed to be supported by experimentation and observation.
   d. The term "theory" in science is reserved for those hypotheses that have proven to have the greatest explanatory power.

Study Questions

1. There is no standard scientific method but still it is possible to identify certain steps that are commonly accepted as comprising this method. What are these steps?
2. Apply these steps to Kettlewell's experiment regarding unequal distribution of dark- and light-colored moths.
3. Which of the steps identified in question 1 requires the use of inductive reasoning, which requires the use of deductive reasoning?
4. Why do scientists prefer mathematical data and use charts and graphs to report such data?
5. Give several useful purposes for the existence of scientific journals.
6. Design a controlled experiment using some other example than the one given in the text.
7. What is meant by descriptive research and how does it differ from experimental research?
8. What is the difference between a hypothesis and a theory in science? List five biological theories that pertain to all living things, and name some of the investigators that helped formulate these theories.

Thought Questions

1. In what ways are scientific conclusions different from religious or philosophical beliefs?
2. Evolutionists have been gathering information to support either the graduated hypothesis of evolution (species evolve slowly over a long period of time) or the punctuated hypothesis of evolution (species remain the same for a long period of time and then evolve suddenly). Is this evidence that evolution is "merely a theory?"
3. Criticize the phrase, "proven by scientific investigation." Should phrases like this be used by the advertising media?

Selected Key Terms

scientific method (si″en-tif′ik meth′ud) 13
data (da′tah) 13
hypothesis (hi-poth′ē-sis) 13
inductive reasoning (in-duk′tiv re′zun-ing) 13

deductive reasoning (de-duk′tiv re′zun-ing) 13
scientific journal (si″en-tif′ik jer′nal) 13
experimental variable (ek-sper″i-men′tal va′re-ah-b'l) 14

dependent variable (de-pen′dent va′re-ah-b'l) 14
control group (kon-trōl′ grōōp) 15
model (mod′el) 16
theory (the′o-re) 16

Suggested Readings for Introduction

Baker, J., and Allen, G. 1971. *Hypothesis, prediction, and implication in biology.* Reading, MA: Addison-Wesley.

Barash, D. P. 1976. The male response to apparent female adultery in the mountain bluebird, *Sialia currucoides:* An evolutionary interpretation. *American Naturalist.*

Glander, K. E. 1981. Feeding patterns in mantled howling monkeys. A. C. Kamil and T. D. Sargent, eds. *Foraging behavior, ecological, ethological and psychological approaches.* New York: Garland STPM Press.

Life: Origin and evolution: Readings from Scientific American. Intro. by C. E. Folsome. 1979. San Francisco: W. H. Freeman.

Scientific American Editors. 1978. *Evolution: A Scientific American book.* San Francisco: W. H. Freeman.

Volpe, E. P. 1981. *Understanding evolution.* 4th ed. Dubuque, IA: Wm. C. Brown Publishers.

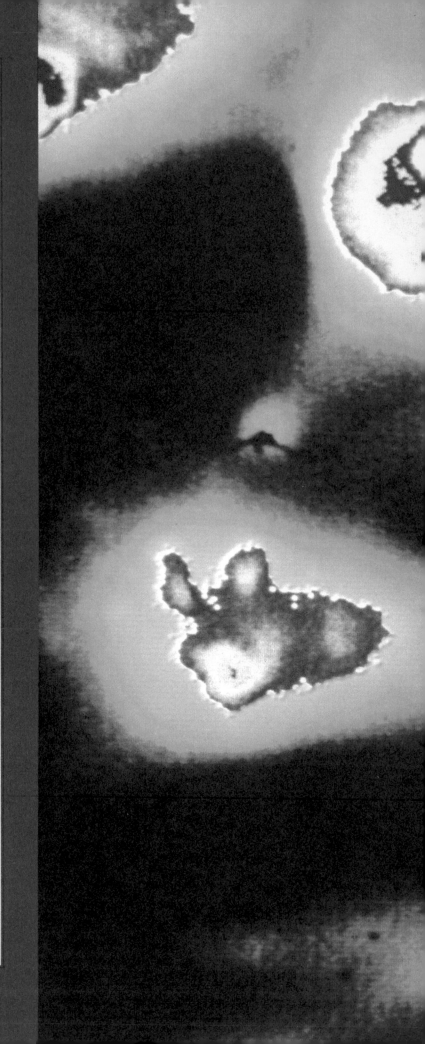

P A R T 1

The Cell

Special microscopy techniques reveal particular features about cells, the fundamental units of life. All living things are composed of cells, and the term multicellular is indeed appropriate because it is cells that form the structure of the organism and carry on its functions. Each cell is a biochemical factory capable of performing all sorts of reactions necessary to maintaining itself.

Cells have a complex structure but their individual parts are not alive, only the cell is alive. They reproduce and each one comes from a preexisting cell.

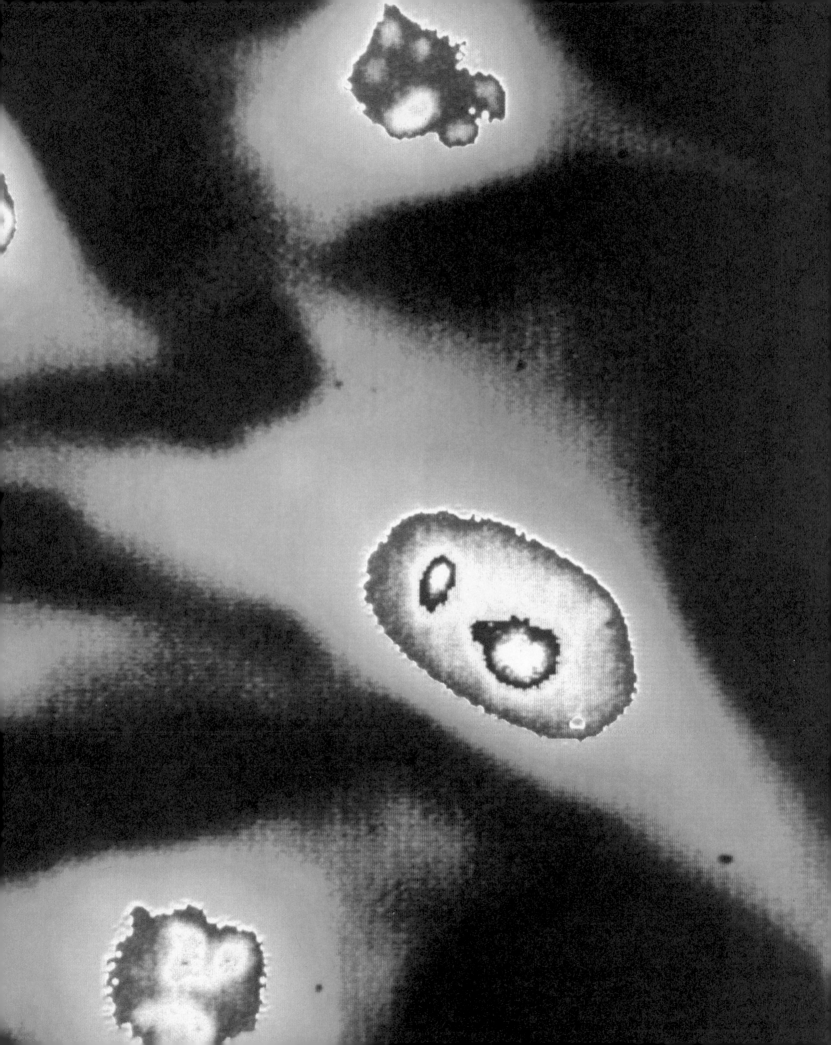

CHAPTER 3

Basic Chemistry

Your study of this chapter will be complete when you can:

1. *Name and describe the subatomic particles of an atom indicating which one accounts for the occurrence of isotopes.*
2. *Describe and discuss the energy levels (electron shells) of an atom including the orbitals of the first two levels.*
3. *Draw a simplified atomic structure of any atom with an atomic number less than 20.*
4. *Distinguish between ionic and covalent reactions and draw representative atomic structures for ionic and covalent molecules.*
5. *Tell which atom has been reduced and which oxidized in a particular reaction.*
6. *Describe the chemical properties of water and explain their importance for living things.*
7. *Define an acid and base; describe the pH scale, and state the significance of buffers.*

Inorganic compounds can, under certain conditions, form beautiful colored patterns. This is an image of sulfide, sodium, and potassium compounds in a color developer.

ometimes it is difficult to realize that life has a chemical basis. Maybe this is because we usually deal with whole objects such as trees, dogs, people and don't often consider their minute composition. If we did consider it, though, we would soon learn that an organism is composed of organ systems having organs that are made up of tissues, that tissues contain cells made up of molecules, and that the molecules themselves are atoms bonded together (fig. 3.1). This shows that fundamentally, living things are made up of chemicals.

The structure and the function of living things are dependent upon chemicals. The success of the cheetah that darts across the plain to capture the gazelle is dependent upon the chemical reactions that occur throughout its body, such as those that allow the eyes to see and the muscles to contract. Also, a bittersweet vine can climb a fence only because of the chemical reactions that permit growth to occur.

As recently as the nineteenth century, scientists believed that only nonliving things, like rocks and metals, consisted of chemicals alone. They felt that living things, such as the cheetah and bittersweet vine, are different and must have a special force, called a vital force. However, scientific investigation has repeatedly shown that both nonliving and living things have only a chemical and physical basis. Therefore, it is proper for us to first study chemical principles as an introduction to our study of life. In this chapter, we consider the basic facts of chemistry, and in the next chapter we will be taking a look at some complex molecules that are especially associated with living things. Then we will discuss how complex molecules are organized into the structures found within cells. In this way, not only the sameness but also the uniqueness of living things will begin to be apparent.

Living things are composed of chemicals organized to provide the structure and to perform the functions necessary to life.

Figure 3.1

Levels of organization from atom to organism. The chemical basis of life becomes apparent when we realize that fundamentally all living things are composed of small units of matter called atoms. Atoms join together to form molecules. Molecules are organized into cells. Cells of the same type form tissues, and different types of tissues make up organs. Several organs are found within organ systems, and organ systems comprise the organism.

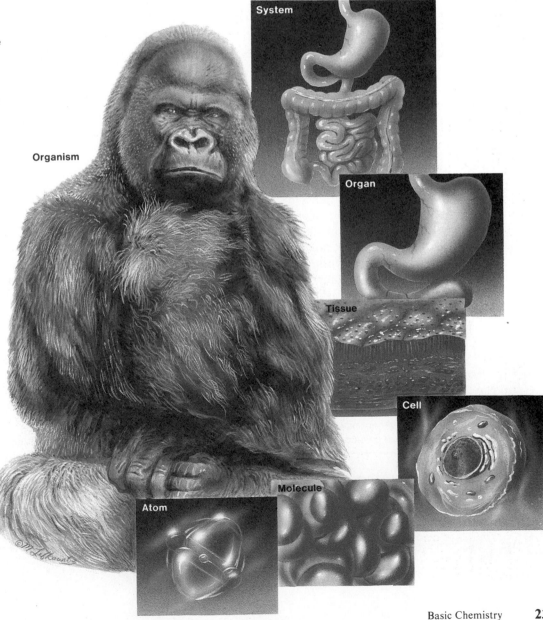

System

Organ

Tissue

Cell

Organism

Molecule

Atom

Table 3.1 Common Elements in Living Things*

Element	Symbol	Atomic Number	Atomic Mass†	Comment
Hydrogen	H	1	1	These elements make up most biological molecules.
Carbon	C	6	12	
Nitrogen	N	7	14	
Oxygen	O	8	16	
Phosphorus	P	15	31	
Sulfur	S	16	32	
Sodium	Na	11	23	These elements occur mainly as dissolved salts.
Magnesium	Mg	12	24	
Chlorine	Cl	17	35	
Potassium	K	19	39	
Calcium	Ca	20	40	
Iron	Fe	26	56	These elements also play vital roles.
Copper	Cu	29	64	
Zinc	Zn	30	65	

*A periodic table of the elements is given in appendix B.
†Average of most common isotopes.
The atomic number gives the number of protons (and electrons in electrically neutral atoms). The number of neutrons is equal to the atomic mass minus the atomic number.

Matter

Matter refers to anything that takes up space and has mass (weight). It's helpful to remember that matter can exist as either a solid, a liquid, or a gas. That way we can realize that not only are we matter but also the water we drink and the air we breathe are matter.

All matter, both nonliving and living, is composed of certain basic substances called **elements.** Most matter contains more than one type of element but some matter contains only one type of element. For example, air contains more than one type of element but the oxygen we respire contains only the element of that name. It's quite remarkable that there are only ninety-two naturally occurring elements (see appendix B). We know they are elements because they can't be broken down to substances that have different properties. A *property* is a chemical or physical characteristic like denseness, smell, taste, and reactivity.

Only six elements—carbon, hydrogen, nitrogen, oxygen, phosphorus, and sulfur—make up most (about 98%) of the body weight of organisms. The acronym CHNOPS helps us remember these six elements. The other elements noted in table 3.1 also play important roles in cells.

All living and nonliving things are matter composed of elements. Six elements in particular are commonly found in living things.

Figure 3.2

Model of a simple atom, helium (He). Atoms contain subatomic particles that are located where shown. Protons and neutrons are found within the nucleus and electrons are outside the nucleus. *a.* The degree of stippling shows the probable location of the electrons in the helium atom. *b.* The average location is sometimes represented by a circle and each electron is represented by a small sphere.

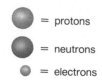 = protons

= neutrons

= electrons

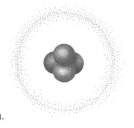

a.

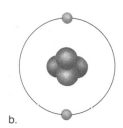

b.

Atoms

In the early 1800s, the English scientist John Dalton proposed that elements actually contain tiny particles called **atoms.** He also deduced that there is only one type of atom in each type of element. You can see why, then, the name assigned to each element is the same as the name assigned to the type of atom it contains. Some of the names we use for the elements (atoms) are derived from English and some are derived from Latin. One or two letters, called the **atomic symbol,** are used to stand for this name. For example, the symbol H stands for a hydrogen atom and the symbol Na (for *natrium* in Latin) stands for a sodium atom. Table 3.1 gives the symbols for the other common elements (atoms) found in living things.

Atoms Have Weight

From our discussion of elements, you would expect that atoms have a certain weight. The weight of an atom is in turn dependent upon the presence of certain subatomic particles. Although physicists have identified a number of subatomic particles, we will consider only the most stable of these: **protons, neutrons,** and **electrons.** Protons and neutrons are located within the nucleus of an atom and the electrons move about the nucleus. Figure 3.2 shows the arrangement of the subatomic particles in helium, an atom that has only two electrons.

Our concept of an atom has changed greatly since Dalton's day. If we could draw an atom the size of a football field, the nucleus would be like a gumball in the center of the field and the electrons would be tiny specks whirling about in the upper stands. Most of an atom is empty space. We should also realize that we can only indicate where the electrons are expected to be most of the time. In our analogy, the electrons may even stray outside the stadium at times.

Atoms contain protons, neutrons, and electrons that are arranged in a definite manner.

Table 3.2 Subatomic Particles

Name	Charge	Mass
Electron	One negative unit	Almost no mass
Proton	One positive unit	One atomic unit
Neutron	No charge	One atomic unit

Figure 3.3
All the atoms of a particular element have the same atomic number, which tells the number of protons. This number is often written as a subscript at the lower left of the atomic symbol. The atomic weight (mass) is often written as a superscript at the upper left of the atomic symbol. For example, the carbon atom can be noted in this way:

atomic weight ——— $^{12}_{6}C$ ——— atomic symbol
atomic number ———

Isotopes are atoms of the same element that differ in weight. The number of protons stays the same but the number of neutrons can vary. The above signifies the most common isotope of carbon, other possible isotopes are:

$^{13}_{6}C$ $^{14}_{6}C$ $^{15}_{6}C$

Figure 3.4
The thyroid gland is a butterfly-shaped organ located in front of the throat region of humans. It produces the hormone thyroxine that contains iodine. Therefore, if a patient is administered radioactive iodine (atomic number 53) a scan of the thyroid viewed sometime later will show the distribution of the iodine throughout the tissue.

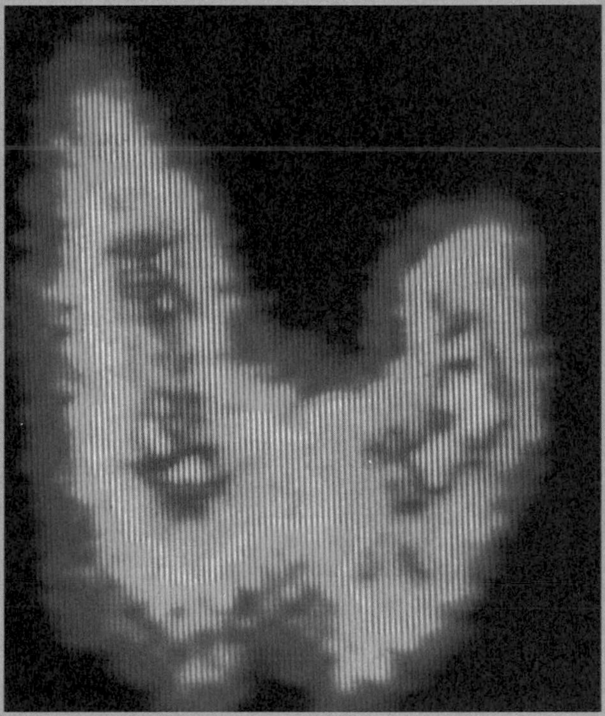

The subatomic particles are so light that their weight is indicated by special units called atomic mass units (table 3.2). Protons and neutrons each weigh only one atomic mass unit. In comparison electrons have almost no weight; an electron weighs about 1/1800 that of a proton or neutron. So, it is customary to disregard the combined weight of the electrons when calculating the total weight, called the **atomic weight,** of an atom.

All atoms of an element have the same number of protons. This is called the atom's **atomic number.** In table 3.1 atoms are listed according to increasing atomic number, as they are in the periodic table of the elements found in appendix B. This serves as an indication that it is the number of protons (i.e., the atomic number) that determines the uniqueness of an atom. The number of protons not only contributes to the physical properties (weight) of an atom, but also *indirectly* determines the chemical properties, which we will be discussing next.

Isotopes

All atoms of an element have the same number of protons but may differ in the number of neutrons. Table 3.1 allows you to calculate only the usual number of neutrons for hydrogen, carbon, and oxygen. But any specific carbon atom, for example, can have a few more or less neutrons than the number indicated. Atoms that have the same atomic number and differ only by the number of neutrons are called **isotopes** (fig. 3.3). Most isotopes are stable, but a few are unstable and tend to break down to more stable forms. They emit radiation as they break down.

These radioactive isotopes can be detected by physical or photographic means and are used by biologists as "labels" in biochemical experiments. For example, researchers used radioactive carbon dioxide ($^{14}CO_2$) to detect the sequential biochemical steps that occur during photosynthesis. Radioactive isotopes are also sometimes used in medical diagnostic procedures. For instance, because the thyroid gland uses iodine, it is possible to administer radioactive iodine and then scan the gland to determine the location of the iodine isotope (fig. 3.4).

^{14}Carbon

Most carbon atoms have six protons and six neutrons (shown as ^{12}C, carbon twelve) and are not radioactive. However, ^{14}C (carbon fourteen) has six protons and eight neutrons and is radioactive. With the release of radioactivity, ^{14}C spontaneously breaks down over a period of time into ^{14}N (nitrogen fourteen). A small amount of ^{14}C is present in all living things. Scientists can determine the age of a fossil because the rate at which ^{14}C breaks down is known. If the fossil still contains organic matter, the relative amounts of ^{14}C and ^{14}N are measured and the age is calculated. However, this method is unreliable for measuring ages in excess of twenty thousand years. The radioactivity becomes so slight that it is difficult to make an accurate determination.

Figure 3.5

Electron energy levels. *a.* Electrons possess stored energy to different degrees. Those with the least amount of energy are found closest to the nucleus and those with more energy are found at a greater distance from the nucleus. These energy levels are also called electron shells. *b.* An electron can absorb energy from an outside source, such as sunlight and then move to a higher energy level. *c.* An electron can later give up the energy absorbed and drop back to its former energy level. This feature of electron behavior is important during the photosynthetic process to be studied in chapter 8.

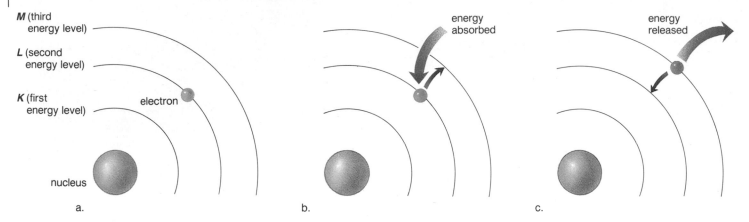

Atoms Have Chemical Properties

Protons and electrons carry a charge; protons have a positive (+) electrical charge and electrons have a negative (−) charge. When an atom is electrically neutral, the number of protons equals the number of electrons. Therefore, in such atoms, the atomic number tells you the number of electrons. For example, a carbon atom has an atomic number of six (table 3.1) and when electrically neutral it has six electrons. Atoms differ in their chemical properties (the way they react with other atoms) because they each have a different number of electrons arranged about the nucleus.

The chemical properties of atoms differ because the number and arrangement of their electrons is different.

Atomic Energy Levels

All electrons have the same weight and charge but they vary in their energy content. **Energy** is defined as the ability to do work. We recognize this in everyday life by suggesting that those who are energetic do more work than those who are not. Electrons differ in the amount of their **potential energy,** that is, stored energy ready to do work. The relative amount of stored energy is designated by placing electrons at ever higher **energy levels,** also called **electron shells,** about the nucleus (fig. 3.5). Electrons with the lowest amount of potential energy are located in the shell closest to the nucleus, called the K shell. Electrons in the next higher shell, called the L shell, have more energy, and so forth as we proceed from shell to shell outside the nucleus (fig. 3.5). An analogy may help you appreciate that the further electrons are from the nucleus, the more potential energy they possess. Falling water has potential energy as witnessed by its use to turn a waterwheel connected to a shaft that transfers the energy to machinery for grinding grain, cutting wood, or weaving cloth for example. The higher the waterfall, the greater the amount of energy released per unit amount of water.

Our analogy is not exact because the potential energy possessed by electrons is not due to gravity; it is due to the attraction between the positively charged protons and the negatively charged electrons. It takes energy to keep an electron further away from the nucleus as opposed to closer to the nucleus. You have often heard that sunlight provides the energy for the photosynthetic process in green plants but it may come as a surprise to learn that when a pigment such as chlorophyll absorbs the energy of the sun, electrons move to a higher energy level about the nucleus. When these electrons return to their previous level, the energy released is used to make food molecules (fig. 3.5*b* and *c*).

Electrons Occupy Orbitals

It's convenient to imagine electrons occupying concentric electron shells about a nucleus, as in figures 3.2 and 3.5. However, such depictions should be regarded as diagrams of convenience. In actuality, it is not possible to determine where rapidly moving electrons are from moment to moment. Still, it is possible to describe the pattern of their motion. These volumes of space where electrons are found most of the time are called **orbitals.**

In the first energy level, there is only a single spherical orbital (fig. 3.6) where at most two electrons will be found about the nucleus. The space is spherical because the most likely location for each electron is a fixed distance in all directions from the nucleus. At the second energy level, there are four orbitals; one of these is spherical but the other three are dumbbell-shaped. Since each orbital can hold two electrons, there are a maximum of eight electrons in this shell. Higher shells can be more complex and contain more orbitals if they are inside another shell. If such a shell is the outer shell, it too has only four orbitals and a maximum number of eight electrons.

Electron Configurations

Although it is a simplification to depict atoms by indicating only electron shells, it is a convenient way to keep track of the number of electrons in the outer shell (fig. 3.7). This is important because the number of electrons in the outer shell determines

Figure 3.6

Electron orbitals. Each electron energy level (see figure 3.5) contains one or more orbitals, a volume of space in which the rapidly moving electrons are most likely found. The nucleus is at the intersection of *x, y,* and *z* axes. *a.* The first energy level (electron shell) contains only one orbital, which has a spherical shape. Two electrons can occupy this orbital. *b.* The second energy level contains one spherical-shaped orbital and also three dumbbell-shaped orbitals at right angles to each other. Since two electrons can occupy each orbital there are a total of eight electrons in the second electron shell. Each orbital is drawn separately here but actually the second spherical orbital surrounds the first spherical orbital and the dumbbell-shaped orbitals pass through the spherical ones.

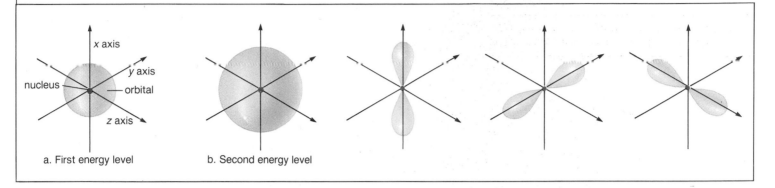

a. First energy level   b. Second energy level

Figure 3.7

Since the manner in which atoms react with one another is dependent upon the number of electrons in the outer shell, it is beneficial to have a convenient way to determine and show this number. These examples of electrically neutral atoms will help you learn to do this.

Hydrogen, the smallest atom, has only one proton and one electron; therefore the outer shell has only one electron. Carbon, on the other hand, has an atomic number of six. It has six protons in the nucleus and six electrons in the shells (two electrons in the first shell and four electrons in the second, or outer shell). Each lower-energy level is filled with electrons before the next higher-energy level contains any electrons. Magnesium, with an atomic number of twelve, has three shells (two electrons in the first shell, eight electrons in the second shell, and two electrons in the third, or outer shell).

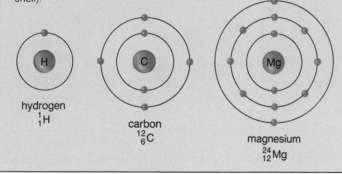

hydrogen
$^{1}_{1}H$

carbon
$^{12}_{6}C$

magnesium
$^{24}_{12}Mg$

whether an atom reacts with another atom and the manner in which the atom reacts. If an atom has only one shell, this outer shell is complete when it has two electrons. Otherwise the **octet rule,** which states that the outer shell is complete when it has eight electrons, holds. It is observed that atoms with eight electrons in the outer shell do not react at all. They are said to be inert. Atoms that have less than eight electrons in the outer shell do react with other atoms in such a way that, after the reaction, each has a completed outer shell. Atoms can give up, accept, or share electrons in order to have a completed outer shell.

The manner in which atoms react with one another is determined by the number of electrons in the outer shell.

Compounds and Molecules

When two or more atoms of different elements combine together, a **compound** results. A **molecule** is the smallest part of a compound that still has the properties of that compound. Molecules can also form when two or more atoms of the same element react with one another.

Chemical Bonds

Electrons possess energy; therefore the bonds that exist between atoms in molecules are energy relationships. Energy is required for a bond to form, and energy is released when a bond is broken. The provision of energy for bond formation and the utilization of energy released when a bond is broken is of top priority to organisms.

Ionic Bonding

Ionic bonds form when electrons are transferred from one atom to another. For example (fig. 3.8), sodium (Na), with only one electron in its third shell, tends to be an electron *donor.* Once it gives up this electron, the second shell, with eight electrons, becomes its outer shell. Chlorine (Cl), on the other hand, tends to be an electron *acceptor.* Its outer shell has seven electrons, so it needs only one more electron to have a completed outer shell. When a sodium atom and a chlorine atom come together, an electron is transferred from the sodium atom to the chlorine atom. Now both atoms have eight electrons in their outer shells (fig. 3.8).

This electron transfer, however, causes a charge imbalance in each atom. The sodium atom has one more proton than it has electrons; therefore, it has a net charge of $+1$ (symbolized by Na^+). The chlorine atom has one more electron than it has protons; therefore, it has a net charge of -1 (symbolized by Cl^-). Such charged atoms are called **ions.** Sodium (Na^+) and chlorine (Cl^-) are not the only biologically important ions. Some, such as potassium (K^+), are formed by the transfer of a single

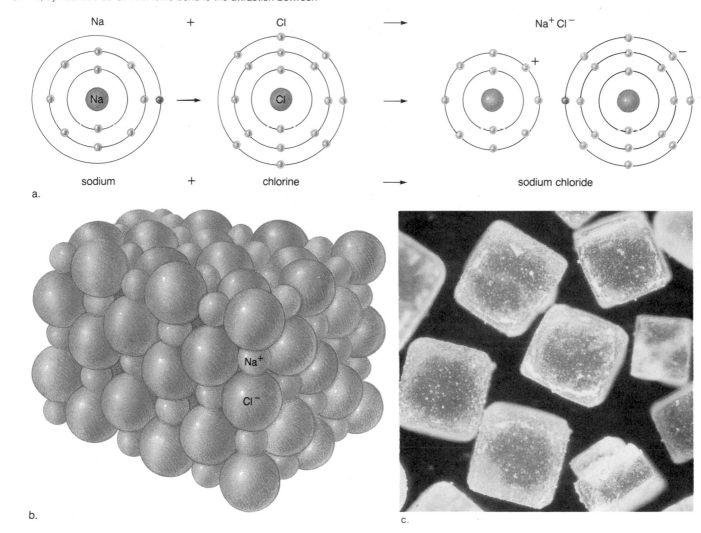

Na + Cl ⟶ Na⁺Cl⁻

sodium + chlorine ⟶ sodium chloride

a.

b.

c.

electron to another atom; others, such as calcium (Ca⁺⁺) and magnesium (Mg⁺⁺), are formed by the transfer of two electrons.

Ionic compounds are held together by an attraction between the charged ions called the **ionic bond.** For example, when sodium reacts with chlorine, an ionic compound called sodium chloride (Na⁺Cl⁻) results, and the reaction is called an ionic reaction. Sodium chloride is a salt commonly known as table salt (fig. 3.8c) because it is used at the table to season our food. Salts can exist as dry solids but when such a compound is placed in water, the ions separate as the salt dissolves. For example, Na⁺Cl⁻ separates into Na⁺ and Cl⁻. Ionic compounds are most commonly found in this dissociated (ionized) form in biological systems because these systems are 70% to 90% water (fig. 3.9).

The transfer of electron(s) between atoms results in ions that are held together by an ionic bond, the attraction of negative and positive charges.

Covalent Bonding

A **covalent bond** results when two atoms *share* electrons in such a way that each atom has a completed outer shell. Let's consider the hydrogen atom, which has one electron in the first shell, a shell that is complete when it contains two electrons. If hydrogen is in the presence of a strong electron acceptor, it gives up its electron to become a hydrogen ion (H⁺), but if this is not possible it can share with another atom and thereby have a completed outer shell. For example, a hydrogen atom can share with another hydrogen atom. In this case, the two orbitals overlap

Figure 3.9

Killer whale emerges from the sea. Seawater and the blood of vertebrates is strikingly similar in the kinds of salts present and in the relative concentrations of these salts. Life is believed to have originated in the sea and the first organisms were adapted to a seawater environment. Since then, life forms have also become adapted to the freshwater and land environment. Even so, the blood of these animals retains something of the same pattern of salts as the sea.

Figure 3.10

Covalently bonded molecules. In a covalent bond, atoms share electrons. Sharing is represented as an overlapping of outer shells and the shared electrons are counted as belonging to each atom. When this is done, each atom has a completed outer shell. *a*. A molecule of hydrogen (H_2) contains two hydrogen atoms sharing a pair of electrons. This single covalent bond can be represented in any of the three ways shown. *b*. A molecule of oxygen (O_2) contains two oxygen atoms sharing two pairs of electrons. This results in a double covalent bond. *c*. A molecule of methane (CH_4) contains one carbon atom bonded to four hydrogen atoms. Each hydrogen atom has a completed outer shell because the first shell only holds two electrons. By sharing with four hydrogens, the carbon atom has a completed outer shell containing eight electrons.

Electron Model	Structural Formula	Molecular Formula
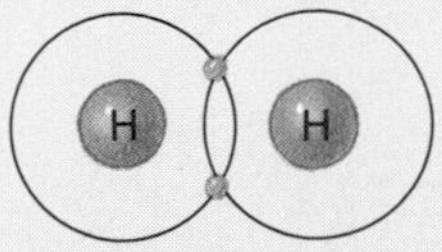 a.	H — H	H_2
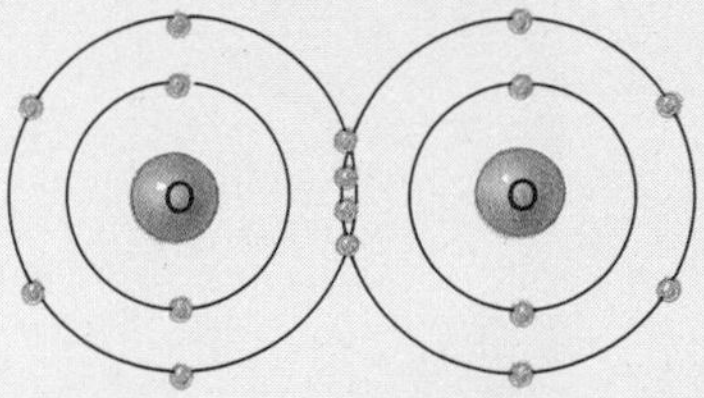 b.	O = O	O_2
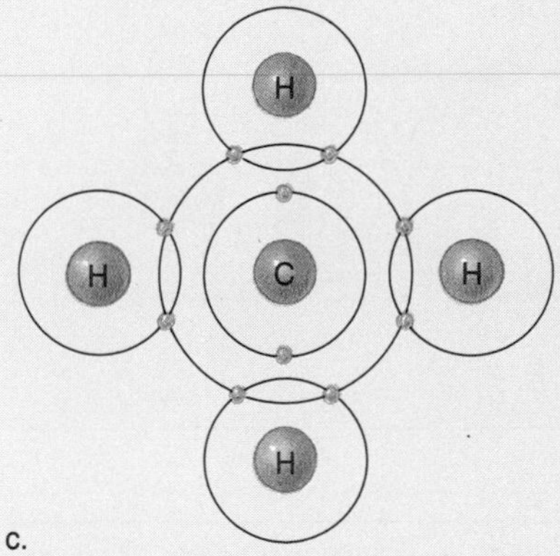 c.	$H - \overset{\displaystyle H}{\underset{\displaystyle H}{C}} - H$	CH_4

and the electrons are shared between them (fig. 3.10*a*). Because they share the electron pair, each atom has a completed outer shell. When a reaction results in a covalent molecule it is called a covalent reaction.

A more common way to symbolize that atoms are sharing electrons is to draw a line between the two atoms as in H — H. Sometimes the line is even omitted and the molecule is simply written as H_2.

Double and Triple Bonds

Besides a single bond, like that between two hydrogen atoms, a double bond sometimes allows two atoms to complete their octets. In a double covalent bond two atoms share two pairs of electrons (fig. 3.10*b*). In order to show that oxygen gas (O_2) contains a double bond, the molecule may be written as O = O.

It is even possible for atoms to form triple covalent bonds as in nitrogen gas (N_2), which may be written as N ≡ N. Single covalent bonds between atoms are quite strong but double and triple bonds are even stronger.

In a covalent molecule, atoms share electrons; not only single bonds but also double and even triple bonds are possible.

Polar Covalent Bonds

Normally, the sharing of electrons between two atoms is fairly equal, and the covalent bond is nonpolar. All the molecules in figure 3.10 including methane (CH_4) (fig. 3.10*c*) are nonpolar. However, in the case of water (H_2O), the sharing of electrons between oxygen and each hydrogen is not completely equal. The larger oxygen atom with the greater number of protons dominates the H_2O association and attracts the electron pair to a greater extent. This causes the oxygen atom to assume a slightly negative charge, showing that it is electronegative in relation to the hydrogens. Each hydrogen atom assumes a slightly positive charge, showing that it is electropositive in relation to oxygen. The unequal sharing of electrons in a covalent bond is called a **polar covalent bond,** and the molecule itself is a polar molecule (fig. 3.11).

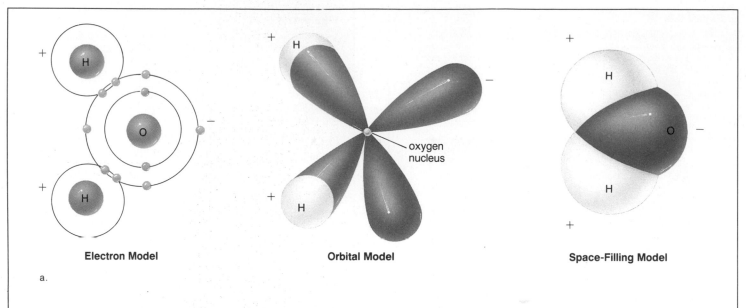

Electron Model   **Orbital Model**   **Space-Filling Model**

oxygen
nucleus

a.

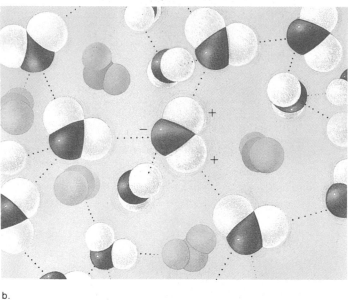

b.

Figure 3.11

Water molecule. *a.* Three models of the structure of water. *Far left,* the two-dimensional model of the structure of water does not indicate how the molecule is orientated in space. *Center,* when an atom with two energy levels (such as oxygen) combines with other atoms, the four orbitals of the outer shell (fig. 3.6*b*) rearrange to give tear-shaped orbitals that point toward the corners of a tetrahedron. However, in water only two of the orbitals are utilized in covalent bonding. *Far right,* the space-filling model shows that the water molecule has a V shape. Water is a polar molecule, and because the oxygen attracts the electrons more strongly than do the hydrogens there is a partial positive charge on each hydrogen and a partial negative charge on the oxygen. *b.* Hydrogen bonding between water molecules. A hydrogen bond is the attraction of a partially positive hydrogen to a partially negative atom in the vicinity. Each water molecule can bond to four other molecules in this manner. When water is in its liquid state, some hydrogen bonds are forming and others are breaking at all times.

The water molecule has an electronegative end and an electropositive end indicating that it is a polar molecule.

Hydrogen Bonding

Polarity within a water molecule causes the hydrogen atoms in one molecule to be attracted to oxygen atoms in other molecules (fig. 3.11*b*). This attractive force creates a weak bond called a **hydrogen bond.** This bond is often represented by a dotted line because a hydrogen bond is easily broken. Hydrogen bonding is not unique to water. It occurs whenever an electropositive hydrogen atom is attracted to an electronegative atom in another molecule or subunits of the same molecule.

Although a hydrogen bond is more easily broken than a covalent bond, multiple hydrogen bonds are strong. Molecules inside cells often have many hydrogen bonds that help them maintain their proper structure and function. We will see that many of the important properties of water and other molecules are due to the presence of hydrogen bonding.

A hydrogen bond occurs between a slightly positive hydrogen atom of one molecule and a slightly negative atom of another molecule.

Chemical Reactions

Chemical reactions are often indicated by writing a chemical equation. The atoms or molecules to the left of an arrow are the *reactants,* the substances that react with one another to give a product(s). The *product(s)* are placed to the right of the arrow. The reactants and products are designated by formulas that give the number of atoms and an indication of how they bond within

Figure 3.12

a. The temperature of water changes slowly and a great deal of heat is needed for vaporization to occur. *b.* Ice is less dense than liquid water and therefore bodies of water freeze from the top down, making ice fishing possible after a hole is drilled through the ice. *c.* The heat needed to vaporize water can help animals in a hot climate to maintain internal temperatures.

a.

b.

c.

the compound or molecule. For example, the equation for the formation of water from hydrogen gas and oxygen gas is written as:

$$2H_2 \;+\; O_2 \longrightarrow 2H_2O$$
hydrogen   oxygen   water

Notice that all atoms present in the reactant(s) are accounted for in the product(s); there are four hydrogen atoms and two oxygen atoms on both sides of the equation. The equation must be balanced in this way because mass can never be created or disappear.

Oxidation-Reduction Reactions

Oxidation-reduction reactions are an important type of reaction in cells but the terminology was derived from studying reactions outside of cells. When oxygen combines with a metal, oxygen receives electrons and becomes negatively charged; the metal loses electrons and becomes positively charged.

Today, the terms **oxidation** and **reduction** are applied to many ionic reactions whether or not oxygen is involved. Very simply, *oxidation refers to the loss of electrons, and reduction refers to the gain of electrons*. In our previous ionic reaction, $Na + Cl \rightarrow Na^+Cl^-$, the sodium has been oxidized (loss of electron) and the chlorine has been reduced (gain of electron).

The terms *oxidation* and *reduction* are also applied to certain covalent reactions. In this case, however, oxidation is the loss of hydrogen atoms, and reduction is the gain of hydrogen atoms. A hydrogen atom contains one proton and one electron; therefore when a molecule loses a hydrogen atom; it has lost an

electron; when a molecule gains a hydrogen atom, it has gained an electron. We will have occasion to refer to this form of oxidation-reduction again in chapter 5.

When oxidation occurs, an atom becomes oxidized (loses electrons). When reduction occurs, an atom becomes reduced (gains electrons). These two processes occur concurrently in oxidation-reduction reactions.

Water

Properties of Water

Life evolved in water, and living things are 70% to 90% water. The chemical properties of water are absolutely essential to the continuance of life (fig. 3.12). Water is a polar molecule—the oxygen end of the molecule is electronegative and the hydrogen end is electropositive. Water molecules are hydrogen-bonded one to the other (fig. 3.11). Hydrogen bonds are much weaker than covalent bonds within a single water molecule, but they still cause water molecules to cling together. Without hydrogen bonding between molecules, water would boil at −80° C and freeze at −100° C, making life as we know it impossible. But because of hydrogen bonding, water is a liquid at temperatures suitable for life. Water has some other important properties.

1. *Water is the universal solvent and facilitates chemical reactions both without and within living systems.* When a salt, such as sodium chloride (Na^+Cl^-), is put into water, the electronegative ends of the water molecules are

attracted to the sodium ions, and the electropositive ends of the water molecules are attracted to the chlorine ions. This causes the sodium ions and the chlorine ions to separate and dissolve in water:

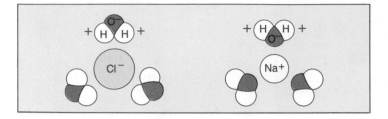

Water is also a solvent for larger molecules that contain ionized atoms or are polar molecules:

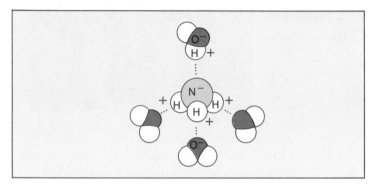

When ions and molecules disperse in water, they move about and collide, allowing reactions to occur.

2. *Water molecules are cohesive.* Water flows freely, yet water molecules do not break apart. They cling together because of hydrogen bonding. Water molecules also adhere to (stick to) surfaces, particularly polar surfaces. Therefore, water can fill a tubular vessel and still flow so that dissolved and suspended molecules are evenly distributed throughout a system. For these reasons, water is an excellent transport system both without and within living organisms. One-celled organisms rely on external water to transport nutrient and waste molecules, but multicellular organisms often contain specialized vessels in which water serves to transport nutrients and wastes.

3. *The temperature of liquid water rises and falls more slowly than that of most other liquids.* It requires a calorie of heat energy to raise the temperature of 1 gram of water 1 degree. This is about twice the amount required for other covalently bonded liquids. The many hydrogen bonds that link water molecules together help water absorb heat without a change in temperature. Because water holds a large amount of heat, its temperature falls more slowly. This property of water is important not only for aquatic organisms but also for all living things. Water protects organisms from rapid temperature changes and helps them maintain their normal internal temperatures.

4. *Water tends to remain a liquid rather than change to ice or steam.* Converting 1 gram of liquid water to ice requires the loss of 80 calories of heat energy. Converting

Figure 3.13
Water is the only common molecule that can be a solid, liquid, or gas according to the environmental temperature. At ordinary temperatures and pressure, water is a liquid and it takes a large input of heat to change it to steam. (When we perspire the heat of our bodies is causing water to vaporize and therefore, sweating causes us to cool off.) In contrast, water gives off heat when it freezes and this heat will keep the environmental temperature higher than expected. Can you see why there are less severe changes in temperature along the coast?

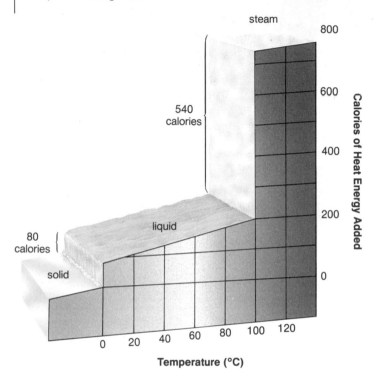

1 gram of water to steam requires an input of 540 calories of heat energy (fig. 3.13). Changing water to steam requires that the hydrogen bonds be broken; this accounts for the very large amount of heat needed for evaporation. This property of water helps moderate the earth's temperature so that it is compatible with the continuance of life. It also gives animals in a hot environment an efficient way to get rid of excess body heat. When an animal sweats, the body heat is used to vaporize the fluid, thus cooling the animal.

5. *Frozen water is less dense than liquid water.* As water cools, the molecules come closer together until they are most dense at 4° C (fig. 3.14), but they are still moving about. Below 4° C, the water molecules cease moving, and hydrogen bonding becomes more rigid but also more open. This makes ice less dense than liquid water; this is why ice floats on liquid water. Bodies of water always freeze from the top down. When a body of water freezes on the surface, the ice acts as an insulator to prevent the water below it from freezing. This protects many aquatic organisms so that they can survive the winter.

Table 3.3 summarizes these five properties of water.

Water has unique properties that allow cellular activities to occur and make life on earth possible.

Figure 3.14

Most substances contract when they solidify but water expands. At 0°C water is least dense and at 4°C it is most dense. As water freezes, the water molecules assume a lattice structure in which the molecules are fixed by hydrogen bonds that cause the molecules to be distant from one another. This very unusual property of water means that ice floats on liquid water. In contrast, when water vaporizes at 100°C, all the hydrogen bonds are broken and the molecules move away from one another.

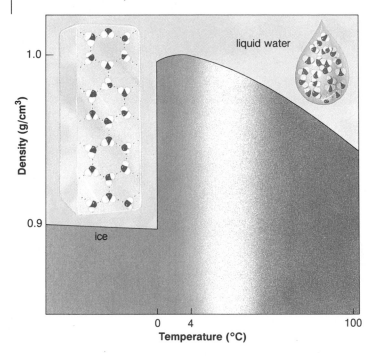

Table 3.3 Water

Properties	Chemistry	Result
Universal solvent	Polarity	Facilitates chemical reactions
Adheres and is cohesive	Polarity; hydrogen bonding	Serves as transport medium
Resists changes in temperature	Hydrogen bonding	Helps keep body temperatures constant
Resists change of state (from liquid to ice and from liquid to steam)	Hydrogen bonding	Moderates earth's temperature Evaporation helps bodies remain cool
Less dense as ice than as liquid water	Hydrogen bonding changes	Ice floats on water

Ionization of Water

Water has another property that has important consequences for living things. It tends to ionize or dissociate, releasing an equal number of hydrogen ions (H^+) and hydroxyl ions (OH^-):

$$H-O-H \rightleftharpoons H^+ + OH^-$$
water   hydrogen   hydroxyl
        ion   ion

This equation is written with a double arrow because the reaction is reversible. Just as water can release hydrogen and hydroxyl ions, these same ions can join together to form water. In fact, only a few water molecules at a time are actually dissociated.

There are many other substances that also dissociate and contribute ions to a given solution, a liquid containing a dissolved substance. **Acids** are molecules that release hydrogen ions when they dissociate. A strong acid, such as hydrochloric acid (HCl), almost completely dissociates when put into water:

$$HCl \longrightarrow H^+ + Cl^-$$
hydrochloric   hydrogen   chloride
acid   ion   ion

In contrast, **bases** are molecules that can take up hydrogen ions. Since the hydroxyl ion can combine with a hydrogen ion to form water, it is a strong base. Sodium hydroxide (NaOH) is termed a strong base because it almost completely dissociates when put into water:

$$NaOH \longrightarrow Na^+ + OH^-$$
sodium   sodium   hydroxyl
hydroxide   ion   ion

pH Scale

Acids and bases affect the relative concentration of hydrogen ions and hydroxyl ions. Biologists use the **pH scale** because it indicates the relative concentrations of H^+ and OH^- in a solution (fig. 3.15). The pH scale range is from 0 to 14. The pH of pure water is 7; this is neutral pH.

pH 7
$$[H^+] = [OH^-]^1$$

Acids lower the pH; any pH below 7 indicates that there are more hydrogen ions than hydroxyl ions.

pH 7 → pH 0
$[OH^-]$ is less than $[H^+]$
$[H^+]$ is greater than $[OH^-]$

Bases increase the pH; any pH above 7 indicates that there are fewer hydrogen ions than hydroxyl ions.

pH 7 → pH 14
$[OH^-]$ is greater than $[H^+]$
$[H^+]$ is less than $[OH^-]$

[1]Square brackets are conventionally used to denote concentrations.

Notice in figure 3.15 that a pH change of 1 unit involves a tenfold change in hydrogen ion concentration. Thus pH 6 has ten times the $[H^+]$ of pH 7, and pH 5 has one hundred times the $[H^+]$ of pH 7.

Buffers

Most organisms maintain a pH of about 7; for example, human blood has a pH of about 7.4. A much higher or lower pH causes illness. Normally, pH stability is possible because organisms have built-in mechanisms to prevent pH changes. Buffers are the most important of these mechanisms. A **buffer** is a chemical or a combination of chemicals that can both take up and release hydrogen ions. Carbonic acid (H_2CO_3) helps buffer human blood because it is a weak acid that does not totally dissociate:

$$H_2CO_3 \rightleftharpoons H^+ + HCO_3^-$$

carbonic acid   hydrogen ion   bicarbonate ion

When excess hydrogen ions are present in blood, the reaction goes to the left and carbonic acid forms to maintain the pH. If hydroxyl ions are added to blood, water will form, effectively

Acid Deposition

Normally rainwater has a pH of about 5.6 because the carbon dioxide in the air combines with water to give a weak solution of carbonic acid. Rain falling in northeastern United States and southeastern Canada has a pH between 5.0 and 4.0. One has to remember that a pH of 4 is ten times more acidic than a pH of 5 to appreciate the increase in acidity this represents.

There is very strong evidence now that this observed increase in rainwater acidity is a result of the burning of fossil fuels like coal and oil, as well as the gasoline derived from oil. When these substances are burned, sulfur oxides and nitrogen oxides are produced and combine with water vapor in the atmosphere to form acids. These acids fall out of the atmosphere as rain or snow, in a process properly called wet deposition but more often called acid rain. Also, dry particles of sulfate and nitrate salts can fall out of the atmosphere and this is called dry deposition. The net result is that the burning of fossil fuels and related products leads to the presence of acids in the air and these return to the earth sometime later.

These air pollutants can be carried in the wind to places far distant from their origin. Acid deposition in Canada and northeastern United States is due to the burning of fossil fuels in factories and power plants located in the Midwest. Similarly, the Scandinavian countries are the recipients of air pollutants from England and northern Europe. Unfortunately, regulations that require the use of tall smokestacks to reduce local air pollution only causes the pollutants to be carried further away.

Acid deposition has created a very serious situation in certain areas of the world. In the United States, vulnerable areas include not only the Northeast but also the Great Smokey Mountains, the lakes of Wisconsin and Minnesota, the Pacific Northwest and the Colorado Rockies. The soil in these areas is thin and lacks limestone (calcium carbonate, $CaCO_3$), which can buffer acid deposition. The first cry of alarm came from Sweden, which reported as early as the 1960s that its lakes were dying. Today dying lakes are found across northeastern North America, Western Europe, and Eastern Europe. The acid rain leaches aluminum from the soil and carries this element

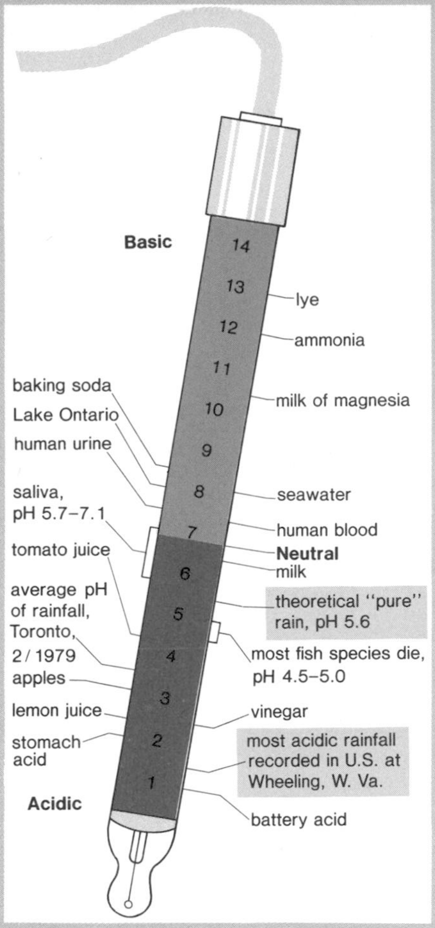

Figure 1
The pH of various solutions including acid rain depicted on the glass electrode of a pH meter, an instrument that automatically measures pH.

into the lakes. The acid also allows the conversion of mercury deposits in lake bottom sediments to soluble methyl mercury. Not only do the lakes become more acid, they also show accumulation of substances that are toxic to living things. The fish die and so do most living things eventually. Sweden now reports 15,000 of its lakes are too acidic to support any higher aquatic life (see figure).

In the early 1980s, evidence began to accumulate that showed that acid precipitation was also damaging to forests. As of mid-1986, some 19 countries in Europe reported damage to their woodlands ranging from roughly 5% to 15% of the forested area in Yugoslavia and Sweden to 50% or more in the Netherlands, Switzerland, and West Germany. More than one-fifth of Europe's forests are now damaged.

These aren't the only effects of acid deposition. Other effects include reduction of agricultural yields, damage to marble and limestone monuments and buildings, and even to the illnesses of humans. Acid deposition is believed to be implicated in the increased incidence of lung cancer and possibly colon cancer along the east coast of the United States. Tom McMillan, Canadian Minister of the Environment, says that acid rain "is destroying our lakes, killing our fish, undermining our tourism, retarding our forests, harming our agriculture, devastating our heritage and threatening our health."

There are of course, things that can be done. We could

a. whenever possible use alternative energy sources such as solar, wind, hydropower, and geothermal energy.
b. use low-sulfur coal or remove the sulfur impurities from coal before it is burned.
c. require the use of scrubbers to remove sulfur from factory and power plant emissions.
d. require people to use mass transit rather than drive their own automobiles.
e. reduce our energy needs through other means of energy conservation.

These measures and possibly others could be taken immediately. It is only necessary for us to determine that they are worthwhile.

This reading is based on these references: Miller, G. Tyler, 1985, *Living in the Environment*. Wadsworth Publishing Co., Belmont, CA and Brown, Lester R., et al., 1988, *State of the World*. W. W. Norton and Co., New York.

Figure 3.15

The pH scale showing the relative concentrations of H^+ and OH^-. Notice that this particular representation should be read in this manner: At pH 7, for every 10,000,000 H^+ ions, there are 10,000,000 OH^- ions.

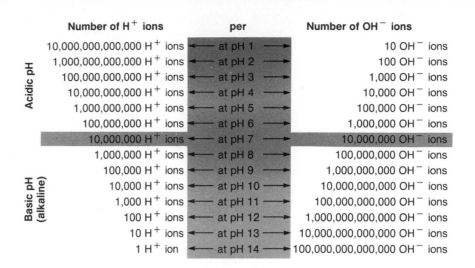

Number of H^+ ions	per	Number of OH^- ions
Acidic pH		
10,000,000,000,000 H^+ ions	at pH 1	10 OH^- ions
1,000,000,000,000 H^+ ions	at pH 2	100 OH^- ions
100,000,000,000 H^+ ions	at pH 3	1,000 OH^- ions
10,000,000,000 H^+ ions	at pH 4	10,000 OH^- ions
1,000,000,000 H^+ ions	at pH 5	100,000 OH^- ions
100,000,000 H^+ ions	at pH 6	1,000,000 OH^- ions
10,000,000 H^+ ions	at pH 7	10,000,000 OH^- ions
1,000,000 H^+ ions	at pH 8	100,000,000 OH^- ions
100,000 H^+ ions	at pH 9	1,000,000,000 OH^- ions
Basic pH (alkaline)		
10,000 H^+ ions	at pH 10	10,000,000,000 OH^- ions
1,000 H^+ ions	at pH 11	100,000,000,000 OH^- ions
100 H^+ ions	at pH 12	1,000,000,000,000 OH^- ions
10 H^+ ions	at pH 13	10,000,000,000,000 OH^- ions
1 H^+ ion	at pH 14	100,000,000,000,000 OH^- ions

removing hydrogen ions. Other types of bases can directly combine with hydrogen ions. When a base removes hydrogen ions the reaction goes to the right and the pH is maintained.

It is possible to overcome an organism's buffering ability but usually buffers keep the pH within normal limits despite many biochemical reactions that either release or take up hydrogen or hydroxyl ions.

Acids have a pH that is less than 7 and bases have a pH that is greater than 7. Organisms contain buffers that help maintain the pH within a normal range.

Summary

1. Both living and nonliving things are composed of matter consisting of elements. Each element contains atoms of just one type. The acronym CHNOPS recalls the most common elements (atoms) found in living things.

2. Atoms contain subatomic particles; protons and neutrons in the nucleus determine the weight of an atom. Electrons are outside the nucleus and have almost no weight.

3. The atomic number indicates the number of protons and the number of electrons in electrically neutral atoms. Protons have a positive charge and electrons have a negative charge and before atoms react the charges are equal.

4. Isotopes are atoms of the same type that differ by the number of neutrons. Radioactive isotopes are used as tracers in biological experiments and medical procedures.

5. Electrons occupy energy levels (electron shells) at discrete distances from the nucleus. When electrons absorb energy from an outside source they move to a higher level and when they drop back they release energy. Certain biological processes are powered by this release of energy.

6. Electron shells contain orbitals. The first shell contains a single spherical-shaped orbital. The second shell contains four orbitals, the first is spherical-shaped and the others are dumbbell-shaped.

7. The number of electrons in the outer shell determines the reactivity of an atom. The first shell is complete when it has two electrons; the second shell is complete when it has eight electrons; other shells can hold more electrons but if they are an outer shell they are also complete with eight electrons. This is the octet rule.

8. Most atoms, including those common to living things, do not have completed outer shells and this causes them to react with one another to form compounds and/or molecules. Following the reaction, the atoms do have completed outer shells.

9. When an ionic reaction occurs, one or more electrons are transferred from one atom to another. The ionic bond is an attraction between the resulting ions.

10. When a covalent reaction occurs, atoms share electrons. A covalent bond is the sharing of these electrons. There are single, double, and triple covalent bonds.

11. In polar covalent bonds, the sharing of electrons is not equal; one end of the bond (molecule) is electronegative, and the other end is electropositive. A hydrogen bond is a weak bond that occurs between an electropositive hydrogen atom and an electronegative atom present in another molecule or another part of the same molecule. Hydrogen bonds help maintain the shape of cellular molecules.

12. Chemical equations are used to symbolize chemical reactions. An atom that has lost electrons (or hydrogen atoms) has been oxidized and an atom that has gained electrons (or hydrogen atoms) has been reduced.

13. Water is a polar molecule and hydrogen bonding occurs between water molecules. These two features account for the unique properties of water, which are summarized in table 3.3. These features allow cellular activities to occur and make life on earth possible.

14. Water dissociates to give an equal number of hydrogen ions and hydroxyl ions. This is termed neutral pH. In acid solutions there are more hydrogen ions than hydroxyl ions; they have a pH less than 7. In basic solutions there are more hydroxyl ions than hydrogen ions; they have a pH greater than 7. Cells are sensitive to pH changes and biological systems tend to have buffers that help keep the pH within a normal range.

Objective Questions

1. Which of the subatomic particles contributes almost no weight to an atom?
   a. protons in the electron shells
   b. electrons in the nucleus
   c. neutrons in the nucleus
   d. electrons at various energy levels
2. An atom that has two electrons in the outer shell would most likely
   a. share to acquire a completed outer shell.
   b. lose these two electrons and become a negatively charged ion.
   c. lose these two electrons and become a positively charged ion.
   d. form hydrogen bonds only.
3. The atomic number tells you
   a. the number of neutrons in the nucleus.
   b. the number of protons in the atom.
   c. the weight of the atom.
   d. to what element the atom belongs.
4. An orbital is
   a. the same volume and charge as an electron shell.

   b. the same space and energy content as the energy level.
   c. the volume of space most likely occupied by an electron.
   d. what causes an atom to react with another atom.
5. A covalent bond is indicated by
   a. plus and minus charges attached to atoms.
   b. dotted lines between hydrogen atoms.
   c. concentric circles about a nucleus.
   d. overlapping electron shells or a straight line between atomic symbols.
6. An atom has been oxidized when
   a. it combines with oxygen.
   b. it gains an electron.
   c. it loses an electron.
   d. Both (a) and (c).
7. In which of these are the electrons shared unequally?
   a. double covalent bond
   b. triple covalent bond
   c. hydrogen bond
   d. polar covalent bond

8. In the molecule

   a. all atoms have eight electrons in the outer shell.
   b. all atoms are sharing electrons.
   c. carbon could accept more hydrogen atoms.
   d. All of these.
9. Which of these properties of water is not due to hydrogen bonding?
   a. stabilizes temperature inside and outside cell
   b. molecules are cohesive
   c. universal solvent
   d. ice floats on water
10. Acids
   a. release hydrogen ions.
   b. have a pH value above 7.
   c. take up hydroxyl ions.
   d. Both (a) and (b).

Study Questions

1. Name the kinds of subatomic particles studied; tell their weight, charges, and locations in an atom. Which of these varies in isotopes?
2. Define energy level and orbital. What is their relationship?
3. Draw a simplified atomic structure for carbon having six protons and six neutrons.

4. Draw an atomic representation for the molecule $Mg^{++}Cl_2^{-}$. Using the octet rule, explain the structure of the compound.
5. Tell whether CO_2 (O — C — O) is an ionic or covalent compound. Why does this arrangement satisfy all atoms involved?
6. Explain why water is a polar molecule. What is the relationship between polarity of the molecule and hydrogen bonding between water molecules?

7. Define oxidation and reduction and tell which has been oxidized and which reduced in this equation:

$$4Fe + 3O_2 \rightarrow 2Fe_2O_3$$

   Satisfy yourself that this equation is balanced.
8. Name five properties of water and relate them to the structure of water including its polarity and hydrogen bonding between molecules.
9. Define an acid and a base. On the pH scale which number(s) indicate an acid, a base, neutral pH?

Thought Questions

1. Discuss the concept that life has a chemical basis.

2. In what ways is life dependent upon the properties of water?

3. Ammonia NH_3 is a polar molecule that can become the ammonium ion NH_4^+. Explain this on the basis of molecular structure.

Selected Key Terms

atom (at′om) 24
atomic weight (ah-tom′ik wāt) 25
atomic number (ah-tom′ik num′ber) 25
isotope (i′so-tōp) 25
orbital (or′bi-tal) 26
compound (kom′pownd) 27

molecule (mol′ĕ-kūl) 27
ionic bond (i-on′ik bond) 28
covalent bond (ko-va′lent bond) 28
polar covalent bond (po′lar ko-va′lent bond) 29
hydrogen bond (hi′dro-jen bond) 30

oxidation (ok″si-da′shun) 31
reduction (re-duk′shun) 31
acid (as′id) 33
base (bās) 33
pH scale (pe āch skāl) 33
buffer (buf′er) 34

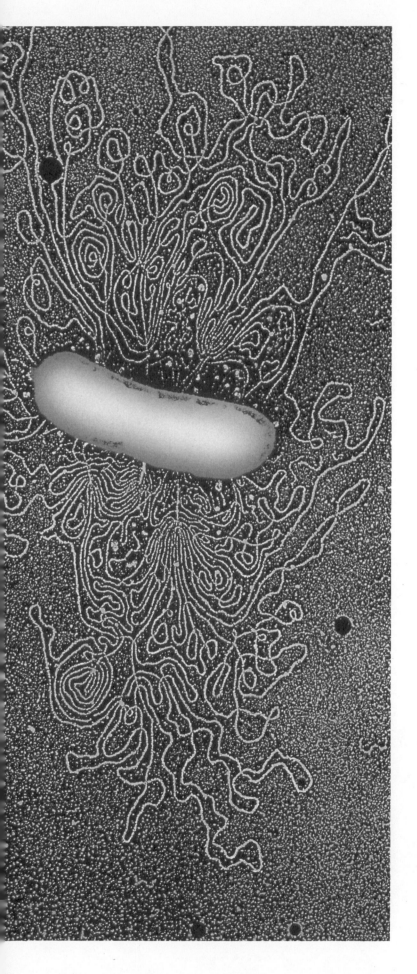

CHAPTER 4

The Chemistry of Life

Your study of this chapter will be complete when you can:

1. Explain and demonstrate the bonding patterns of carbon.
2. Recognize the various functional groups found in cellular molecules.
3. Distinguish between condensation of monomers and hydrolysis of polymers.
4. Give examples of monosaccharides, disaccharides, and polysaccharides, and state their functions.
5. Recognize the molecular and structural formulas for glucose.
6. Give examples of various lipids, and state their functions.
7. Recognize the structural formula for a saturated and unsaturated fatty acid, and the structural formula for a fat.
8. Give examples of proteins, and state their functions.
9. Recognize an amino acid and demonstrate how a peptide bond is formed.
10. Relate the four levels of structure of a protein to the bonding patterns observed at each level.
11. Give examples of nucleotides and nucleic acids, and state their functions.
12. State the components of a nucleotide, and tell how these monomers are joined to form a nucleic acid.
13. Compare the structures of DNA and RNA

Even this small bacterium contains all the chemicals found in living things. Carbohydrates, proteins, lipids, and nucleic acids are the macromolecules that constitute cells. Fine strands of DNA, normally tightly coiled inside, are spilling out of this ruptured single-celled organism.

Living things are highly organized and therefore it is easy for us to differentiate between the plant, animal, and bacterium in figure 4.1. We might even be inclined to think that these organisms are so different that they must contain entirely different types of chemicals. But this isn't the case. Instead they and all living things contain the same types of molecules. But within the sameness there is diversity. For example, the plant, animal, and bacterium all utilize a carbohydrate molecule for structural purposes but the exact carbohydrate is different in each of them.

The most common elements in living things are remembered by the acronym CHNOPS (see table 3.1) and of these—carbon, hydrogen, nitrogen, and oxygen—constitute about 95% of your body weight. But we will see that both the sameness and the diversity of life is dependent upon the chemical characteristics of carbon, an atom whose chemistry is essential to living things.

The bonding of hydrogen, oxygen, nitrogen, and other atoms to carbon creates the molecules known as **organic** molecules. For a molecule to be organic it must contain carbon and hydrogen and this is why carbon dioxide is termed an inorganic molecule. It is the organic molecules that characterize the structure and function of living things like the primrose, blueshell crab, and bacterium in figure 4.1. **Inorganic** molecules constitute nonliving matter but even so, inorganic molecules like salts (e.g., Na^+Cl^-), also play important roles in living things.

> Of the elements most common to living things, the chemistry of carbon allows the formation of varied organic molecules, accounting for both the sameness and diversity of living things.

Chemistry of the Cell and Organism

Carbon has four electrons in its outer shell and this allows it to bond with as many as four other atoms. Moreover, because these bonds are covalent they are quite strong. Usually carbon bonds to hydrogen, oxygen, nitrogen, or another carbon atom. The ability of carbon to bond to itself makes carbon chains of various lengths and shapes possible (fig. 4.2). Long chains containing fifty carbon atoms are not unusual in living systems. Even much longer chains (containing other atoms, such as nitrogen and oxygen) are found. Carbon can also share more than one pair of electrons with another atom, giving a double covalent bond. Carbon-to-carbon bonding sometimes results in ring compounds of biological significance.

Small Organic Molecules

From your examination of figure 4.2 you can see why it is said that carbon chains form the skeleton or backbone of organic molecules. The small organic molecules in living things—simple sugars, fatty acids, amino acids, and nucleotides—all have a carbon backbone but in addition have characteristic *functional groups*. These add to the diversity of organic molecules and they also increase the reactivity of the molecule.

Very often a functional group contributes polarity to an organic molecule, especially when the group can ionize and become charged due to the loss or gain of a hydrogen ion:

hydrocarbon
(nonpolar)

acid in ionized form
(polar)

Figure 4.1

One thing that these organisms (primrose, blueshell crab, and bacterium) have in common is the use of carbohydrate to provide structure. The *primrose* is held erect partly by the incorporation of the polysaccharide cellulose into the cell walls that surround each cell. The shell of the *crab* contains chitin, a different type of polysaccharide. Most *bacterial* cells, like plant cells, are enclosed by a cell wall but in this case the wall is strengthened by another type of polysaccharide known as peptidoglycan. Magnification, ×17,650.

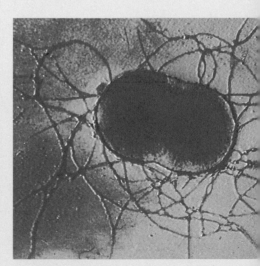

The molecule on the left, being nonpolar, has no tendency to interact with water. It is said to be **hydrophobic.** The molecule on the right with an ionized carboxyl group is polar and does interact with water. It is said to be **hydrophilic.** Other functional groups (OH, CO, and NH_2) are also polar even though they do not ionize (fig. 4.3). Polarity causes molecules to be water soluble. Since cells are 70% to 90% water, ability or inability to interact with water profoundly affects the function of organic molecules in cells.

Another factor contributing to the diversity of organic molecules is the presence of isomers. **Isomers** are molecules that have identical chemical formulae because they contain the same numbers and kinds of atoms, yet they are different molecules because the atoms in each isomer are arranged differently. For example, figure 4.4 shows two compounds with the formula $C_3H_6O_3$. Each of these molecules is assigned its own chemical name because the molecules differ structurally.

Large Organic Molecules

The main chemical components of cells are structures called **macromolecules.** Each of the small organic molecules already mentioned can be a subunit within a particular type of macromolecule. A subunit is called a **monomer,** and the macromolecule is called a **polymer.** Sugars are the monomers within polysaccharides, fatty acids are found in lipids, amino acids join to form proteins, and nucleotides are the basic units of nucleic acids.

Macromolecules	Monomers
polysaccharides	monosaccharides
lipids	glycerol and fatty acids (for fats)
proteins	amino acids
nucleic acids	nucleotides

We will see that macromolecules of the same type can vary because of slight differences in monomer constitution and/or the way they are joined together. Monomers are the backbone of a macromolecule and different types of molecules can be bonded to them. This is why the same type of structural compound can vary in the organisms shown in figure 4.1.

Variety of structure is observed among the simplest and the most complex of organic molecules. Even so, there are only a few different types of macromolecules in living things.

Figure 4.2

Carbon forms the backbone of many different types of molecules. Carbon to carbon bonds bring about (a) straight chains, (b) branched chains, (c) double bonds, and (d) ring compounds. The first formula tells you the number and types of atoms in the compound and the second shows you how they are bonded. The models indicate the shape of the molecules; reactions in cells are often dependent upon the shape of molecules.

Structure	Molecular Formula	Structural Formula	Space-Filling Model
straight chain a.	C_5H_{12}	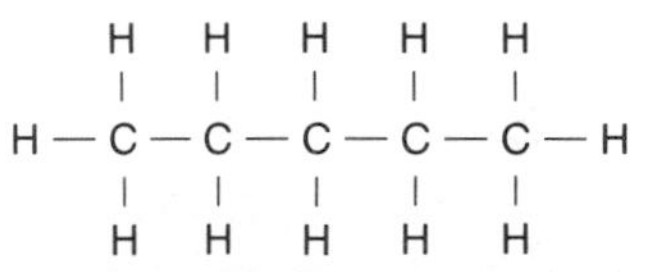	
branched chain b.	C_5H_{12}	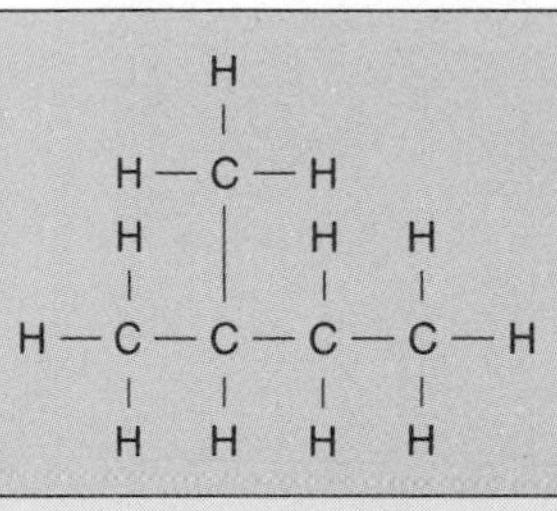	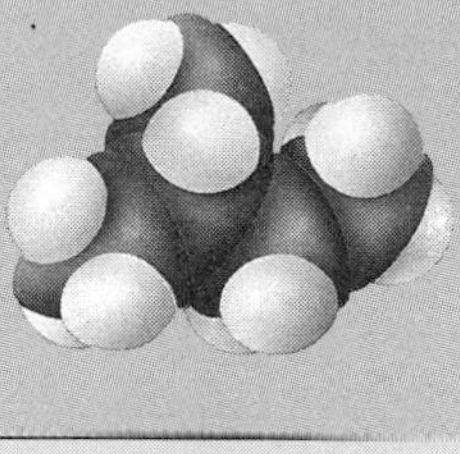
double bond c.	C_2H_4	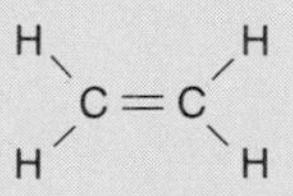	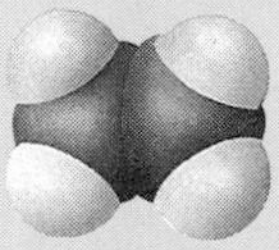
ring d.	C_6H_{12}	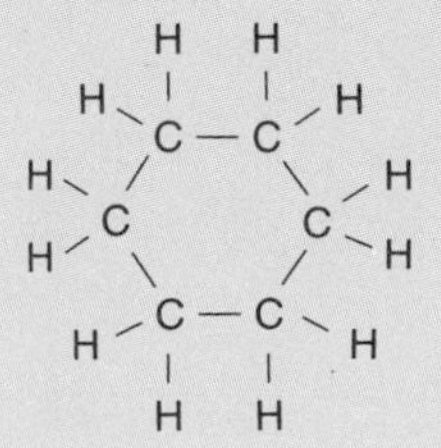	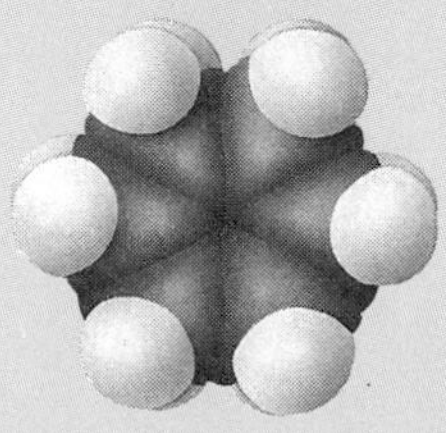

Condensation and Hydrolysis

The manner in which any macromolecule forms within a cell is essentially the same. Two monomers join together when a hydroxyl group (OH) is removed from one monomer and a hydrogen (H) is removed from another (fig. 4.5a). This is a **dehydration synthesis** because the components of water have been removed (*dehydration*) in order to cause the creation (*synthesis*) of a bond. Dehydration synthesis, a condensation reaction, won't take place unless the proper **enzymes** (special substances that speed up chemical reactions in cells) are present and energy is expended. When a bond is formed energy is needed.

Polymers are broken down by **hydrolysis,** a process that is essentially the reverse of condensation: a hydroxyl group from water attaches to one monomer and a hydrogen attaches to the other (fig. 4.5b). This is a hydrolysis reaction because water (*hydro*) is used to break (*lyse*) a bond. When a bond is broken energy is released, or made available.

Macromolecules are formed and broken down in cells as each cell renews its many parts. Also, polysaccharide breakdown makes energy available to cells.

Condensation (the removal of water) joins monomers together to form macromolecules and hydrolysis (the addition of water) breaks macromolecules apart.

Simple Sugars to Polysaccharides

These organic molecules belong to a class of compounds called **carbohydrates.** Carbohydrates usually contain carbon and the components of water—hydrogen and oxygen—in a 1:2:1 ratio. The general formula for any carbohydrate is $(CH_2O)_n$; the n subscript stands for whatever number of these groups there are. For example, the carbohydrate formula $C_6H_{12}O_6 = (CH_2O)_6$.

Figure 4.4

Isomers have the same formula but different configurations. Both of these compounds have the formula $C_3H_6O_3$. *a.* In glyceraldehyde, oxygen is double-bonded to an end carbon. *b.* In dihydroxyacetone, oxygen is double-bonded to the middle carbon.

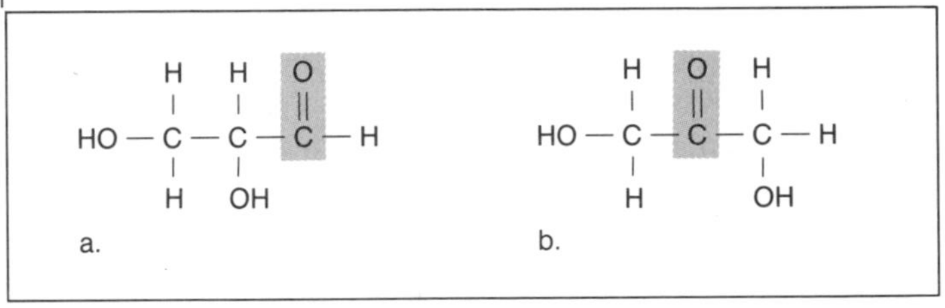

Figure 4.3

Molecules having the exact same type of backbone can still differ according to the type of functional group attached to the backbone. Many of these functional groups are polar, helping to make the molecule soluble in water. In this illustration, the remainder of the molecule (aside from the functional groups) is represented by an R.

Functional Groups		
Name	**Structure**	**Found in**
hydroxyl (alcohol)	R — OH	sugars
carboxyl (acid)	R — C (=O)(OH)	sugars fats amino acids
ketone	R — C(=O) — R	sugars
aldehyde	R — C (=O)(H)	sugars
amine (amino)	R — N (H)(H)	amino acids proteins
sulfhydryl	R — SH	amino acids proteins
phosphate	R — O — P(=O)(OH) — OH	phospholipids nucleotides nucleic acids
R = remainder of molecule		

Figure 4.5

a. In cells, synthesis often occurs when monomers are joined together by condensation (removal of H_2O). *b.* Breakdown occurs when the monomers in a polymer are separated by hydrolysis (the addition of H_2O).

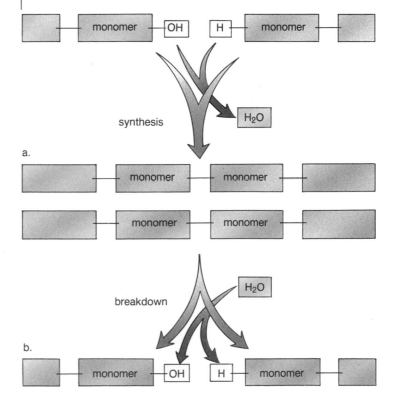

Figure 4.6

a. Glucose is a six-carbon sugar that can exist as an open chain or a ring compound; the ring is more common in cells. The shape of glucose is indicated in the model. Notice that the carbon atoms in glucose are numbered and that the colors allow you to decipher the manner in which the open chain becomes the ring. *b.* Fructose is an isomer of glucose. It too has the molecular formula $C_6H_{12}O_6$. Notice, though, that the atoms are bonded in a slightly different manner in fructose. This causes the open chain to form a slightly different ring and therefore, also contributes to a different shape of the molecule.

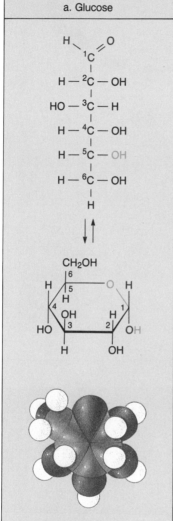

a. Glucose

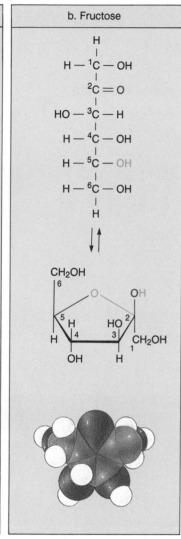

b. Fructose

Simple sugars are **monosaccharides** (one sugar) with a carbon backbone that has from three to seven carbon atoms. The best known sugars are those that have six carbons (*hexoses*). The hexose sugar **glucose** is nearly always used as an immediate energy source in all cells. *Fructose* is a hexose sugar that is frequently found in fruits while *galactose* occurs in milk. These sugars are isomers of one another. They all have the molecular formula $C_6H_{12}O_6$, but they differ in their structures (fig. 4.6). While the structural differences seem very small, actually they cause the molecules to vary in shape. Shape influences the manner in which molecules interact with one another.

There are two five-carbon sugars (*pentoses*) of some significance. They are *ribose* and *deoxyribose* which are found in the nucleic acids, RNA and DNA respectively (p. 50).

A **disaccharide** contains two monosaccharides that have joined together by dehydration synthesis. **Sucrose** is a disaccharide that contains glucose and fructose (fig. 4.7). Sugar is transported within the body of the plant in the form of sucrose and this is the sugar we use at the table to sweeten our food. We acquire this sugar from certain plants, such as sugarcane and beets.

Lactose is a disaccharide found in milk that contains galactose and glucose. *Maltose* (composed of two glucose molecules) is a disaccharide of interest because it is found in our digestive tract as a result of starch digestion.

Polysaccharides

The **polysaccharides** we will be studying are all polymers of glucose.

Starch and Glycogen

The structures of starch and glycogen differ only slightly (fig. 4.8). **Glycogen** is characterized by many side branches, chains of glucose that go off from the main chain. **Starch** has few of these chains.

Plant cells store extra carbohydrates as complex sugars or starches. For instance, when leaf cells are actively producing sugar by photosynthesis, they store some of this sugar within the cell in the form of starch granules (fig. 4.8*b*). Otherwise, sugars are transported to storage organs, especially in roots and modified stems, where they are condensed to starch and stored in granules.

Figure 4.7

Synthesis and hydrolysis of sucrose, a disaccharide containing glucose and fructose units. During synthesis, a bond forms between the glucose and fructose molecules as the components of water are removed. During hydrolysis, the components of water are added as the bond is broken.

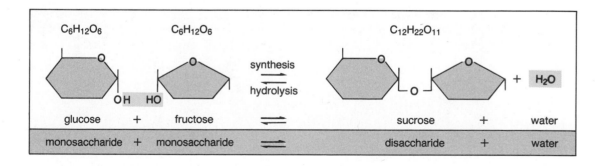

Figure 4.8

a. Starch and glycogen are polymers of glucose. *b.* Starch is the storage form of glucose in plant cells. This is a scanning electron micrograph of starch grains in a plant cell in which starch granules are highly visible. Magnification, ×2,600. *c.* Glycogen is the storage form of glucose in animal cells. This is an electron micrograph of a liver cell in which glycogen granules are visible. Magnification, ×24,000.

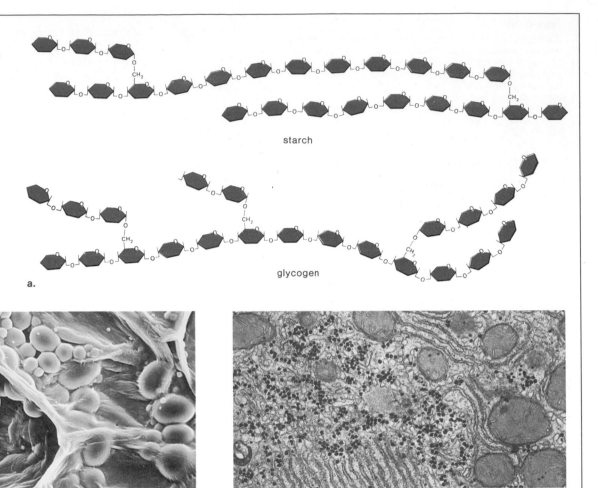

starch

glycogen

a.

b.

c.

Wheat

Humans commonly use only about twelve species of plants as a food source. Three of these are cereals: rice, corn, and wheat. A grain of wheat contains a seed and it is composed of three parts: the embryonic plant, endosperm or stored food, and the seed coat along with other protective layers. The embryonic plant is wheat germ, a substance known to be high in vitamins. The starchy endosperm is used to produce the flour from which our white breads are made. The seed coat is bran, composed mostly of cellulose, which cannot be digested by humans. For this reason, bran is excellent roughage substance, which adds bulk to the diet.

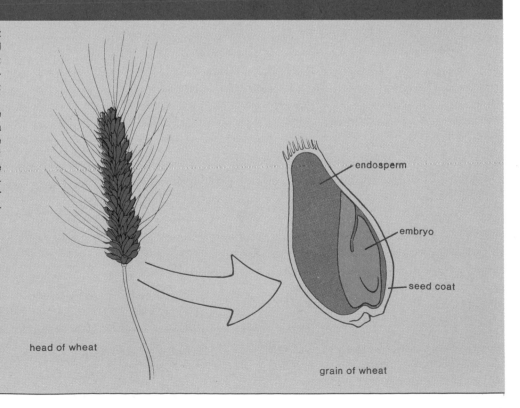

head of wheat

endosperm

embryo

seed coat

grain of wheat

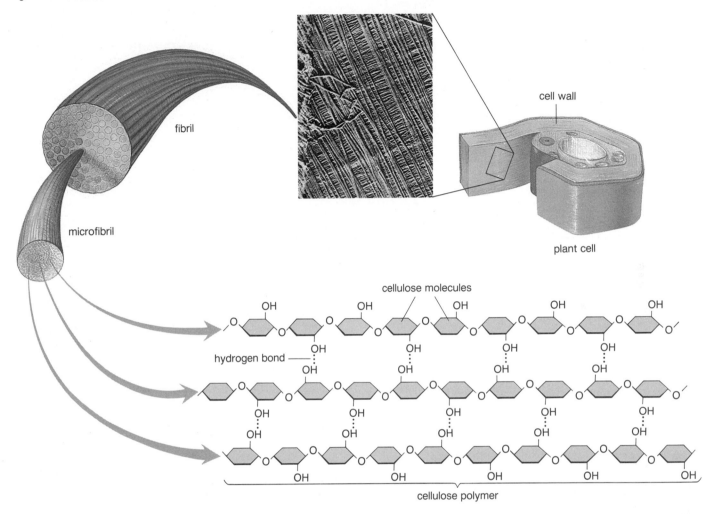

Figure 4.9

Cellulose fibrils are present in plant cell walls. Magnification, ×30,000. Each fibril contains several microfibrils and each microfibril contains many chains of glucose hydrogen-bonded together. Finally, we have a close-up view of three cellulose polymers, each made up of glucose molecules.

fibril

microfibril

cell wall

plant cell

cellulose molecules

hydrogen bond

cellulose polymer

Animal cells store extra carbohydrates as glycogen, sometimes called "animal starch." After a human eats, the liver stores glucose as glycogen (fig. 4.8c). Then, between meals, the liver releases glucose to keep the normal blood concentration of glucose near 0.1%.

Cells use sugars, especially glucose, as an immediate energy source. Glucose is stored as starch in plants and as glycogen in animals.

Cellulose and Chitin

Cellulose contains glucose molecules joined together by a slightly different type of linkage than that found in starch and glycogen. The long chains that result are not branched; many chains are held together by hydrogen bonding to form microfibrils. Several microfibrils, in turn, twist about one another to form fibrils. And these, too, may twist about one another. Layers of cellulose fibrils make up plant cell walls (fig. 4.9). The cellulose fibrils are parallel within each layer, but the layers themselves lie at angles to one another for added strength.

Humans have found lots of uses for cellulose. Cotton fibers are almost pure cellulose and we all wear cotton clothing. Furniture and buildings are made from wood, which contains a high percentage of cellulose. However, most animals, including humans, cannot digest cellulose because the enzymes that digest starch are unable to break the linkage between the glucose molecules in cellulose. Even so, cellulose is a recommended part of our diet because it provides the bulk (also called fiber or roughage) that helps the body maintain regularity of elimination. Grains are rich in both starch and fiber.

Cattle and sheep can receive nutrients from grass only because they have a special stomach chamber, the rumen, where bacteria reside that can digest cellulose. It makes sense, then, that cattle should be range fed rather than grain fed. Unfortunately cattle are often kept in feedlots where they are fed grains. This is ecologically unsound because it requires fossil fuel energy to grow the grain, process it, transport it, and finally feed it to the cattle. Also, since range-fed cattle move about, they produce leaner meat than do feedlot-fed cattle. This is another advantage for humans because of growing evidence that less fatty meat is healthier.

Figure 4.10

A fat molecule consists of glycerol and three fatty acids. These are joined when the hydroxyl groups (OH) of the glycerol react with the carboxyl groups (COOH) of the fatty acids to give a fat and three molecules of water. Stearic, oleic, and palmitic are three common fatty acids in cells; oleic acid is unsaturated.

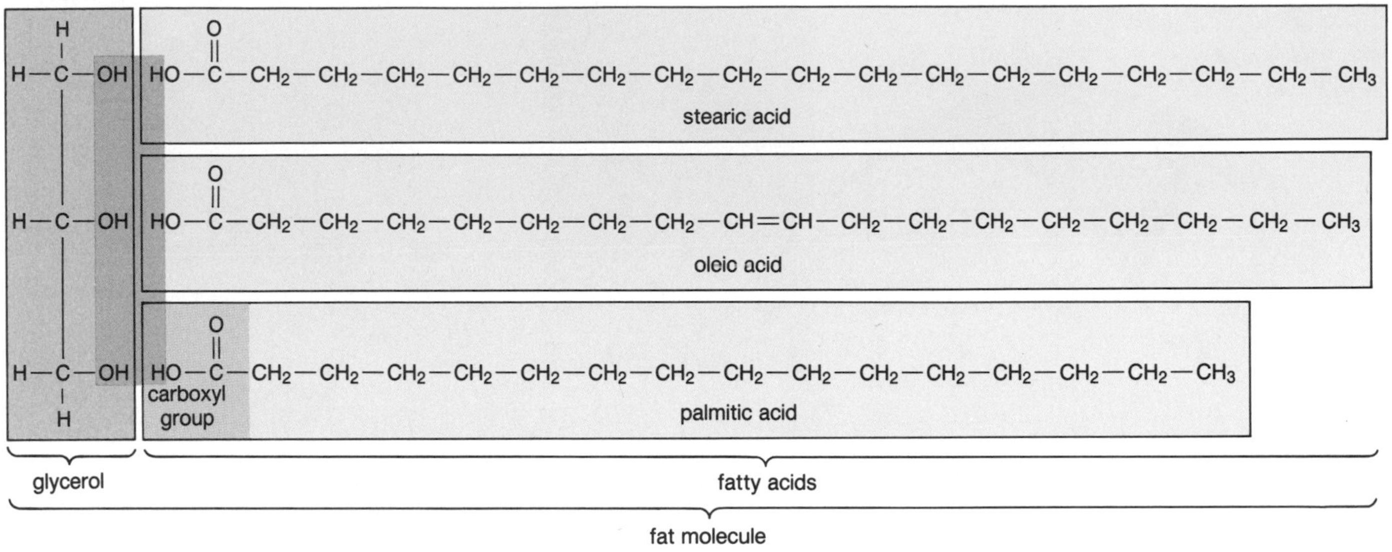

Chitin

Chitin, found in the exoskeleton of crabs and related animals like lobsters and insects, is also a polymer of glucose. However, the glucose is said to be modified because there is an amino group attached to each molecule. The linkage between the glucose molecules is like that found in cellulose; therefore, chitin is not a digestible material. Recently, scientists have discovered how to turn chitin into thread that can be used as suture material. They hope to find other uses for treated chitin. This means that all those discarded crab shells that pile up beside crabmeat processing plants won't have to go to waste.

Fatty Acids to Lipids

A variety of organic compounds are classified as **lipids.** Many of these are insoluble in water because they lack any polarized groups. The most familiar lipids are those found in fats and oils. Organisms use these molecules as long-term energy storage compounds. Phospholipids and steroids are also important lipids found in living things. For example, they are components of plasma membranes.

Lipids are quite varied in structure but they tend to be insoluble in water.

Fats and Oils

Fats and **oils,** sometimes called neutral fats, contain two types of unit molecules: fatty acids and glycerol. Since it takes three fatty acids per one glycerol, these molecules are often called triglycerides (fig. 4.10).

Each **fatty acid** consists of a long hydrocarbon chain with an acidic carboxyl group at one end. Most of the fatty acids in cells contain sixteen to eighteen carbon atoms per molecule, al-

though smaller ones are also found. Fatty acids are either saturated or unsaturated. *Saturated* fatty acids have no double bonds between the carbon atoms. The carbon chain is saturated, so to speak, with all the hydrogens that can be held:

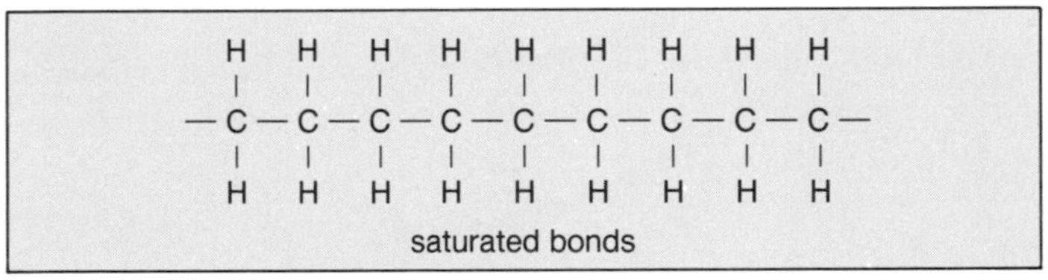

Unsaturated fatty acids have double bonds in the carbon chain wherever the number of hydrogens is less than two per carbon atom:

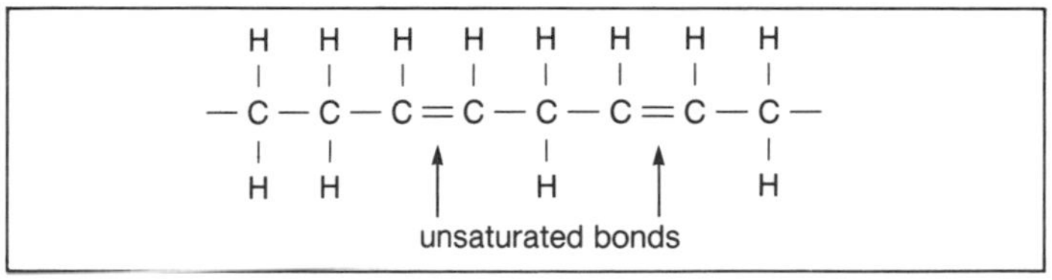

Saturated fatty acids are typically found in animal fat; butter contains saturated fatty acids. Unsaturated fatty acids are most often found in vegetable oils and account for their liquid nature. Vegetable oils, like corn oil and safflower oil, are hydrogenated (hydrogen added) to make margarine but polyunsaturated margarine still contains a large number of unsaturated, or double, bonds.

Glycerol is a compound with three hydroxyl groups (fig. 4.10). When fat is formed, the acid portions of three fatty acids react with these groups so that fat and three molecules of water are formed. Again the larger fat molecule is formed by dehydration synthesis and a fat can be hydrolyzed to its components.

Although a fatty acid does contain a polar group and can mix with water to a degree, a neutral fat has no such groups and is therefore insoluble in water, shown by the fact that you can't mix oils with water. Even after shaking, the oil simply separates out.

Fats and oils are utilized for long-term energy storage because they have mostly hydrogen-carbon bonds, making them a *richer* supply of chemical energy than carbohydrates, which have many C-OH bonds. Molecule per molecule, animal fat contains over twice as much energy as glycogen; gram per gram though, fat stores six times as much energy as glycogen. This is because fat droplets are concentrated and do not contain water. Small birds like the ruby-throated hummingbird (fig. 4.11) store a great deal of fat before they start their long spring and fall migratory flights. About 0.15 gram of fat per gram of body weight is accumulated each day. If the same amount of energy were to be stored as glycogen, a bird would be so heavy it wouldn't be able to fly.

Figure 4.11
A ruby-throated hummingbird, like other small birds that migrate, stores a lot of energy as fat in order to have enough fuel for its long journey. Fat is more energy rich than glycogen and it is found in concentrated droplets; therefore, it is an efficient way to store energy.

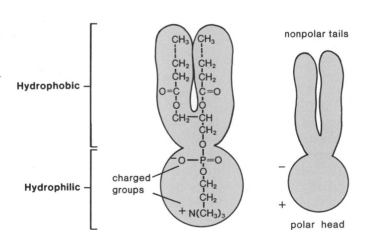

Fats and oils are triglycerides (one glycerol plus three fatty acids) used as long-term storage compounds in plants and animals.

Waxes

In **waxes** a very long-chain fatty acid bonds with a very long-chain alcohol. Waxes are solid and have a high melting point. They are also waterproof and resistant to degradation. They form a protective covering that retards the loss of water for all exposed parts of nonwoody plants. In animals, waxes are involved in skin and fur maintenance. In humans, wax is produced by glands in the outer ear canal. Its function is to trap dust and dirt, preventing it from reaching the eardrum.

Phospholipids

Phospholipids contain a phosphate group and this accounts for their name:

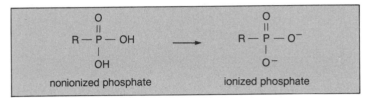

nonionized phosphate   ionized phosphate

Essentially, phospholipids are constructed as neutral fats are, except that in place of the third fatty acid there is a phosphate group or a grouping that contains both phosphate and nitrogen. The group can ionize; therefore, this hydrophilic chain forms what is called the polar head of the molecule while the other two hydrophobic chains are the nonpolar tails (fig. 4.12). When phospholipid molecules are placed in water, they form a sheet in which the polar heads face outward and the nonpolar tails face each other. This property of phospholipids contributes to the structure of membranes.

Phospholipids have nonpolar tails and a polarized head and typically arrange themselves in a double layer in the presence of water.

Figure 4.12
Structurally, phospholipids have a polar head and nonpolar tails (*left and center*), and in water (*right*) the molecules arrange themselves as shown. The polar heads are attracted to water; the nonpolar tails are not.

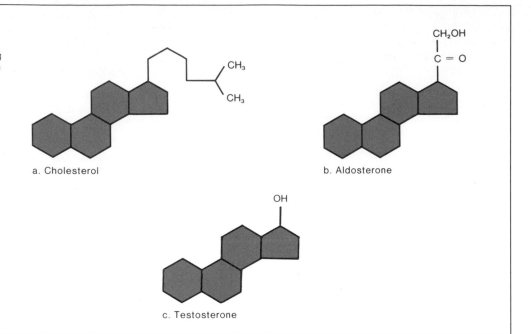

Figure 4.13

Like cholesterol (*a*) steroid molecules have four adjacent rings, but their effects on the body largely depend on the type of grouping attached at the location indicated. The chain in (*b*) is found in aldosterone, which is involved in regulating the sodium and water content in humans, while the chain in (*c*) is found in testosterone, the male sex hormone.

a. Cholesterol

b. Aldosterone

c. Testosterone

Steroids

Steroids are lipids that have entirely different structures than neutral fats. Each steroid has a backbone of four fused carbon rings and varies primarily according to the type of functional group attached to it (fig. 4.13). *Cholesterol* is the precursor of several other steroids such as certain vertebrate hormones including aldosterone, a hormone that helps regulate the sodium content of the blood, and the sex hormones, which help maintain male and female characteristics. Such different functions are due solely to the particular attached groups. Cholesterol is also found in animal cellular membranes.

For many years it has been suggested that a diet high in saturated fats and cholesterol can lead to circulatory disorders due to reduced blood flow caused by the deposit of fatty materials on the linings of blood vessels. This problem is discussed at length in the reading in chapter 33.

Steroids are ring compounds that have a similar backbone but vary according to attached groups. This causes them to have different functions in the body of humans and other animals.

Amino Acids to Proteins

Amino acids are the monomers that undergo condensation to form **proteins,** very large molecules that have both structural and metabolic functions. For example, in animals the proteins myosin and actin are the main components of muscle; insulin is a hormone that regulates the sugar content of the blood; hemoglobin is a transporter of oxygen in blood; and collagen fibers give support to many organs. Proteins are present in the membrane that surrounds each cell and they also exist within the cell. Their other functions are various. Among the most important cell proteins are the **enzymes,** organic catalysts that speed up chemical reactions within cells. Enzymes make "warm chemistry" possible. In a chemical laboratory it is often necessary to heat a reaction flask to bring about a chemical reaction; in only slightly warm cells, the same chemical reactions take place very quickly because a specific enzyme is present for each and every reaction. Enzymes are discussed more fully in chapter 7.

There are about twenty different amino acids in cells, and figure 4.14 gives several examples. All amino acids contain two important functional groups, an acidic carboxyl group (COOH) and an amino group (NH_2), which ionize at normal body pH in this manner:

$$H - N^+ - C - C \overset{O}{\underset{O^-}{\diagup}} \quad \text{or} \quad H_3N^+ - C - COO^-$$

Amino acids differ in the nature of the **R group** (Remainder of the molecule), which ranges in complexity from a single hydrogen to complicated ring compounds. The unique chemical properties of an amino acid depend on those of the R group. For example, some R groups are polar and some are not. Also, the amino acid cysteine contains a sulfhydryl group (SH), which often serves to connect one chain of amino acids to another by a disulfide linkage (S-S).

Peptides

A **peptide** is two or more amino acids joined together, and a **polypeptide** is many amino acids joined to form a long chain of amino acids. Since a protein can contain more than one polypeptide chain, you can see why a protein could have a very large number of amino acids.

Figure 4.14

Representative amino acids; the R groups are in the contrasting color. Some R groups are nonpolar and hydrophobic, some are polar and hydrophilic, and some are ionized and hydrophilic.

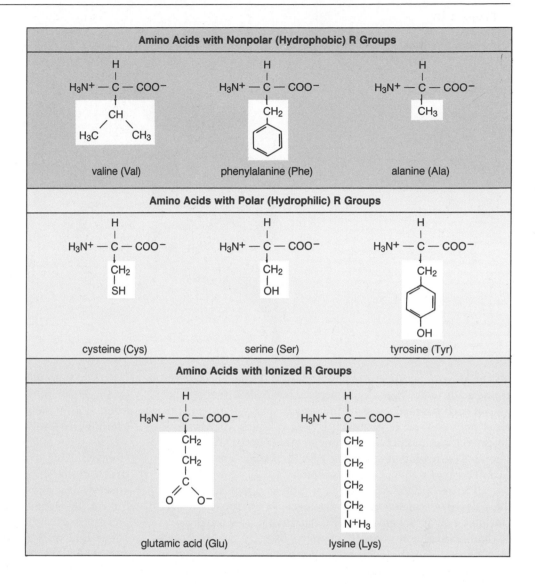

Amino Acids with Nonpolar (Hydrophobic) R Groups

valine (Val) phenylalanine (Phe) alanine (Ala)

Amino Acids with Polar (Hydrophilic) R Groups

cysteine (Cys) serine (Ser) tyrosine (Tyr)

Amino Acids with Ionized R Groups

glutamic acid (Glu) lysine (Lys)

Figure 4.15

Synthesis of a peptide. The peptide bond forms as the components of water are removed from the carboxyl group of one amino acid and the amino group of the other amino acid. Since oxygen is electronegative, there is a partial negative charge on the oxygen and a partial positive charge on the hydrogen within the peptide bond. These charges are not shown.

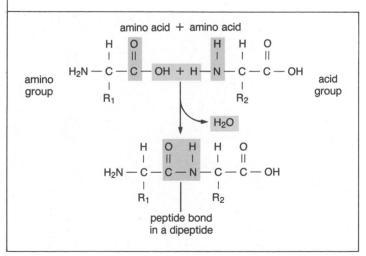

Figure 4.15 shows how two amino acids are joined by a condensation reaction between the carboxyl group of one and the amino group of another. The resulting covalent bond between two amino acids is called a **peptide bond.** The atoms associated with the peptide bond share the electrons unevenly because the oxygen is electroncgative making the hydrogen electropositive. The polarity of the peptide bond means that hydrogen bonding is possible between parts of a polypeptide.

Amino acids are joined by peptide bonds in polypeptides and proteins. Proteins have both structural and metabolic functions in cells. All enzymes, which speed up chemical reactions, are proteins.

Levels of Structure

Each type of polypeptide commonly bends, folds, and twists in a particular way within a protein. An analysis of the final characteristic shape shows that proteins have at least three levels of structure, and some have a fourth level.

Figure 4.16
Secondary structure of polypeptides is dependent upon hydrogen
bonding (dotted lines) between members of the peptide bonds.
a. Alpha (α) helix. *b.* Beta (β) sheet.

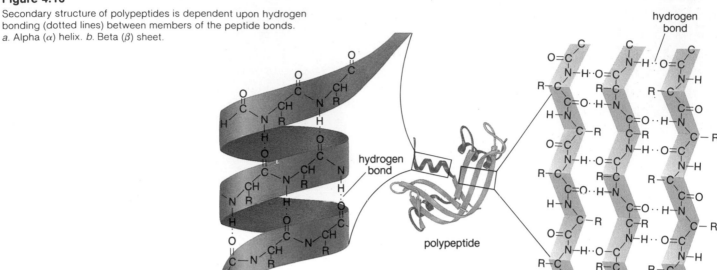

a. Alpha (α) helix   b. Beta (β) sheet

The *primary structure* of a protein is the sequence of the amino acids joined together by peptide bonds. Each protein has its own particular sequence of amino acids and since amino acids differ only by their R groups, it is correct to say that proteins differ from one another by a particular sequence of the R groups. The fact that some of these are polar and some are not influences the final shape of the polypeptide.

The *secondary structure* of a protein comes about when the polypeptide chain takes a particular orientation in space. One common arrangement of the chain is the *alpha (α) helix,* or right-handed coil, with 3.6 amino acids per turn. Hydrogen bonding between the oxygen and nitrogen atoms of different amino acids, in particular, stabilizes the alpha helix (fig. 4.16*a*). Many polypeptides have a secondary structure called a *beta (β) sheet,* or pleated sheet. In this arrangement, the polypeptide chain turns back upon itself and hydrogen bonding occurs between these extended lengths of the polypeptide (fig. 4.16*b*). Some proteins have regions of both alpha helix and beta sheets depending upon the shape of the amino acids in the primary structure.

Now the polypeptide folds and twists into its characteristic shape and this *tertiary structure* is maintained by various types of bonding between the R groups. Indeed, the folding and twisting is determined by those R groups that can bond with one another, giving stability to the shape of the molecule (fig. 4.17). Hydrogen bonds, ionic bonds, and covalent bonding are seen; when cysteines are brought close to one another a disulfide (S-S) bridge occurs between the two amino acids. The hydrophobic R groups, which do not react with other R groups, tend to be pushed together into an inner common locale where they are not exposed to water. (These are called hydrophobic interactions.)

Some proteins have more than one type of polypeptide chain, each with its own primary, secondary, and tertiary structures. Within the protein, these separate chains are arranged to give a fourth level of structure termed the *quaternary structure.* Various types of interactions between the polypeptide chains are observed. Both collagen and hemoglobin have a quaternary structure (fig. 4.18). Collagen is fibrous; the polypeptide chains are organized into long ropelike fibers that make it strong. In contrast, hemoglobin is said to be a globular protein.

A polypeptide chain is folded and twisted to produce a characteristic shape. Four levels of structure have been identified (table 4.1).

Table 4.1 Levels of Protein Structure

Level of Structure	Location of Bond	Type of Bond
Primary	Between amino acids	Covalent (peptide) bond
Secondary	Between members of peptide bond	Hydrogen bond
Tertiary	Between R groups	Hydrogen, ionic, covalent (S-S), hydrophobic interactions*
Quaternary	Between polypeptide chains	Hydrogen, ionic

*Strictly speaking, these are not bonds, but they are very important in creating and stabilizing tertiary structure.

Figure 4.17

Tertiary structure of polypeptides. Disulfide and other types of bonding between the R groups determine the type of folding and twisting of the polypeptide that takes place. *a.* The protein insulin contains two polypeptide chains (A chain and B chain) that are held together by disulfide bonds. The B chain contains both a beta (β) sheet and an alpha (α) helix section. *b.* Ribonuclease is an enzyme whose enzymatic action is dependent upon its final tertiary shape.

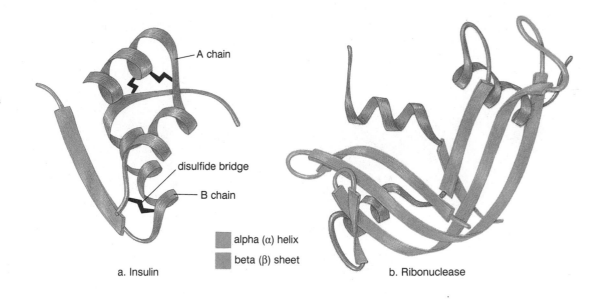

alpha (α) helix
beta (β) sheet

a. Insulin

b. Ribonuclease

Figure 4.18

Quaternary structure of proteins. *a.* Collagen is a fibrous protein. The long polypeptide chains in collagen form a triple helix that is ropelike. Collagen fibers are found in animal connective tissue that supports many organs and constitutes the structure of tendons and ligaments. *b.* Hemoglobin is a globular protein with four polypeptides that assume this position in relation to one another. When hemoglobin carries oxygen, it is attached to iron within the heme group. Heme is not a peptide.

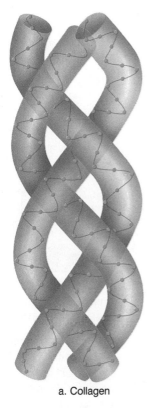

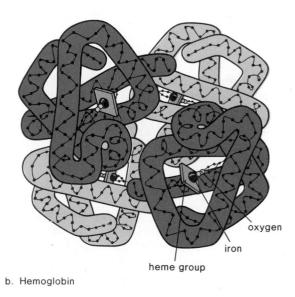

oxygen

iron

heme group

a. Collagen

b. Hemoglobin

Denaturation of Proteins

The final shape of a protein is very important to its function, as will be emphasized when discussing enzyme activity. Both temperature and pH can bring about a change in protein shape. For example, we are all aware that the addition of acid to milk causes curdling; heating causes egg white, a protein called albumin, to congeal or coagulate. When a protein loses its normal configuration, it is said to be denatured. Denaturation occurs because the normal bonding patterns between parts of the molecule have been disturbed. Once a protein loses its normal shape, it is no longer able to perform its usual function.

If the conditions that caused denaturation were not too severe and if these are removed, some proteins will regain their normal shape and biological activity. This shows that the primary structure of a polypeptide forecasts its final shape (fig. 4.19).

The final shape of a protein is dependent upon the primary structure—the sequence of amino acids in the polypeptide(s). The shape of a protein determines how it will interact with other substances in a cell.

Figure 4.19

Denaturation and reactivation of the enzyme ribonuclease. When a protein is denatured it loses its normal shape and activity. If denaturation was gentle and the conditions are removed, some proteins will regain their normal shape. This shows that the normal conformation of the molecule is due to the various interactions between a set sequence of amino acids. Each type of protein has a particular sequence of amino acids.

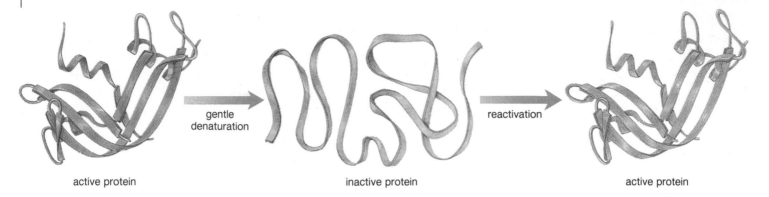

active protein   gentle denaturation   inactive protein   reactivation   active protein

Nucleotides to Nucleic Acids

Every **nucleotide** is a molecular complex of three types of unit molecules: phosphoric acid (phosphate), a pentose sugar, and a nitrogen base.

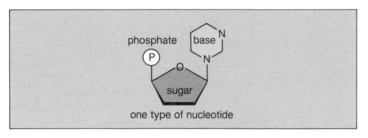

phosphate   base

sugar

one type of nucleotide

Nucleotides have metabolic functions in cells. For example, certain ones are components of coenzymes that perform functions facilitating enzymatic reactions. **ATP** (adenosine triphosphate) is a nucleotide used in cells to supply energy for synthetic reactions and various other energy-requiring processes. Another role of nucleotides is as subunits in nucleic acids.

Nucleic Acids

Nucleic acids are huge polymers of nucleotides with very specific functions in cells; for example, **DNA** (deoxyribonucleic acid) is the genetic material that stores information regarding its own replication and the order in which amino acids are to be joined together to make a protein. Another important nucleic acid, **RNA** (ribonucleic acid), works in conjunction with DNA to bring about protein synthesis.

Both DNA and RNA are polymers of nucleotides. DNA makes up the genes and along with RNA controls protein synthesis within the cell.

In DNA the sugar is deoxyribose and in RNA the sugar is ribose; this difference accounts for their respective names. There are four different types of nucleotides in DNA and RNA. Figure 4.20 shows the types of nucleotides that are present in DNA. The base can be the purines, adenine or guanine, which have a double ring, or the pyrimidines, thymine or cytosine, which have a single ring. These structures are called bases because they have

Figure 4.20

All nucleotides found in DNA contain phosphate, the pentose sugar, deoxyribose, and a base. *a.* The bases adenine and guanine are double-ringed purine bases. *b.* The bases thymine and cytosine are single-ringed pyrimidine bases.

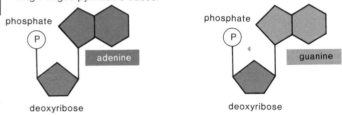

phosphate   adenine   deoxyribose   phosphate   guanine   deoxyribose

a. DNA nucleotides with purine bases

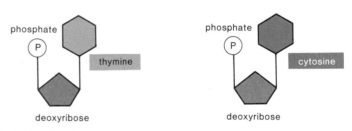

phosphate   thymine   deoxyribose   phosphate   cytosine   deoxyribose

b. DNA nucleotides with pyrimidine bases

chemically basic characteristics that raise the pH of a solution. In RNA the base uracil is used in place of the base thymine.

When nucleotides join together to form DNA or RNA they occur in a definite sequence. The nucleotides form a linear molecule called a strand in which the backbone is made up of phosphate-sugar-phosphate-sugar, with the bases projecting to one side of the backbone. Since the nucleotides occur in a definite order, so do the bases.

RNA is single stranded (fig. 4.21) but DNA is double stranded. The two strands of DNA twist about one another in the form of a double helix (fig. 4.22). The two strands are held together by hydrogen bonds between purine and pyrimidine bases. Thymine (T) in one strand is always paired with adenine (A) in the opposite strand, and guanine (G) is always paired with cytosine (C). This is called complementary base pairing. If we unwind the DNA helix, it resembles a ladder (fig. 4.22c). The sides of the ladder are made entirely of phosphate and sugar molecules, and the rungs of the ladder are made only of the

Figure 4.21

RNA is a single-stranded polymer of nucleotides. When the nucleotides join, the phosphate group of one is bonded to the sugar of the next. The bases project out to the side of the resulting sugar-phosphate backbone.

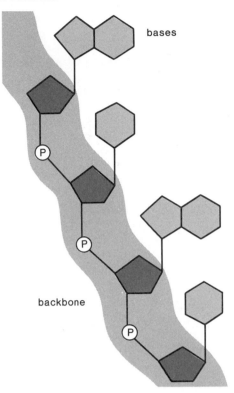

complementary paired bases. The bases can be in any order within a strand but A is always paired with T between strands, and G is always paired with C, and vice versa. Therefore, no matter what the order or the quantity of any particular base pair, the number of purine bases always equals the number of pyrimidine bases.

DNA has a structure like a twisted ladder: sugar-phosphate backbones make up the sides of the ladder; hydrogen-bonded bases make up the rungs of the ladder. RNA differs from DNA in several respects (table 4.2).

The structure and function of nucleic acids are discussed further in chapters 15 and 16 of this text.

Table 4.2 DNA Structure Compared to RNA Structure

	DNA	RNA
Sugar	Deoxyribose	Ribose
Bases	Adenine, guanine, thymine, cytosine	Adenine, guanine, uracil, cytosine
Strands	Double stranded with base pairing	Single stranded
Helix	Yes	No

Figure 4.22

Overview of DNA structure. *a.* Double helix. *b.* Double helix showing complementary base pairing between bases. *c.* When the helix unwinds, the ladder structure of DNA is more evident. Sugar-phosphate molecules make up the sides of the ladder and the bases, held together by hydrogen bonding, make up the rungs. One nucleotide has been boxed in the diagram.

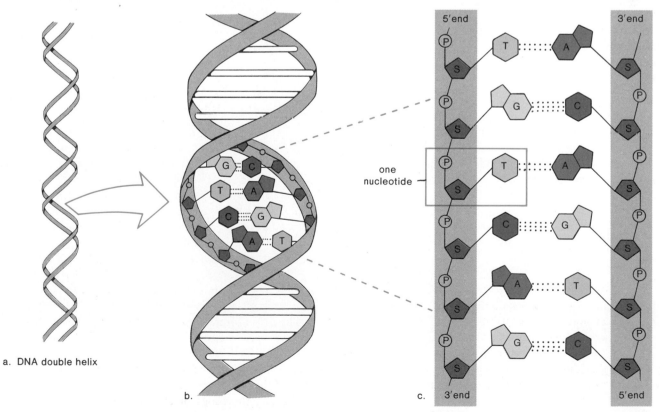

a. DNA double helix

b.

c.

Summary

1. The chemistry of carbon accounts for the diversity of organic molecules found in living things. Carbon can bond with as many as four other atoms and can also bond with itself to form both chains and rings. The latter are called the backbone of the molecule.

2. Each type of small molecule in living things—simple sugars, fatty acids, amino acids, and nucleotides—have a carbon backbone and are characterized by the presence of functional groups. Some functional groups are hydrophobic and some are hydrophilic.

3. Differences in the backbone and attached functional groups cause molecules to have different shapes and shape helps determine how a molecule interacts with other molecules in cells.

4. The large organic molecules in cells are macromolecules, polymers formed by the joining together of monomers. Enzymes carry out both the synthesis (building up) and also hydrolysis (breaking down) of macromolecules. For each bond formed, a molecule of water is removed and for each bond broken, a molecule of water is added.

5. Monosaccharides, disaccharides, and polysaccharides are all carbohydrates. Therefore, the term carbohydrate includes both the monomer (e.g., glucose) and the polymers (e.g., starch, glycogen, and cellulose). Starch and glycogen are energy-storage compounds but cellulose has a structural function in plants.

6. Lipids include a wide variety of compounds that are insoluble in water. Fats and oils allow long-term energy storage and contain one glycerol and three fatty acids. Fats tend to contain saturated fatty acids and oils contain unsaturated fatty acids. In a phospholipid one of the fatty acids contains a phosphate group. Phospholipids in the presence of water form a double layer because the head is polarized and the tails are not. Waxes and steroids are also lipids.

7. Proteins are polymers of amino acids. All enzymes are proteins and proteins also have structural roles in cells and organisms.

8. A polypeptide is a long chain of amino acids joined by peptide bonds. There are twenty different amino acids in cells that differ only by different R groups. Polarity and nonpolarity are an important aspect of the R groups.

9. A polypeptide has three levels of structure; the primary level is the sequence of the amino acids; the secondary level contains alpha (α) helices and beta (β) sheets held in place by hydrogen bonding between peptide bonds; the tertiary level is the final folding and twisting of the polypeptide held in place by bonding and hydrophobic interactions between R groups. Proteins that contain more than one polypeptide have a quaternary level of structure.

10. The shape of a polypeptide influences its biological activity. A polypeptide can be denatured and lose its normal shape and activity and then, if reactivation is possible, it assumes both of these again. This shows that the final shape of a polypeptide is dependent upon the primary structure.

11. Nucleic acids are polymers of nucleotides. Each nucleotide has three components: a sugar, a base, and phosphoric acid (phosphate). DNA, which contains the sugar deoxyribose, is the genetic material that stores information for its own replication and for the order in which amino acids are to be sequenced in proteins. DNA, with the help of RNA, controls protein synthesis.

12. Table 4.3 also summarizes our coverage of organic molecules in cells.

Table 4.3 Organic Molecules in Cells

	Categories	Examples	Functions
Carbohydrates	Monosaccharides six-carbon sugar	Glucose	Immediate energy
	Disaccharides twelve-carbon sugar	Sucrose	Transport sugar in plants
	Polysaccharides polymer of glucose	Starch, glycogen Cellulose	Energy storage Plant cell walls
Lipids	Triglycerides 1 glycerol + 3 fatty acids	Fats, oils	Long-term energy storage
	Waxes fatty acid + alcohol	——	Covering for parts of plants Ear wax
	Phospholipids like triglyceride except one fatty acid has a phosphate group	——	Plasma membrane component
	Steroids backbone of four fused rings	Cholesterol Testosterone	Plasma membrane component Hormones
Proteins	Polypeptides polymer of amino acids have three levels of structure— some proteins have four levels	Enzymes Myosin and actin Insulin Hemoglobin Collagen	Speed up cellular reactions Muscle cell components Regulates sugar content of blood Oxygen carrier in blood Fibrous support of body parts
Nucleic Acids	Nucleic acids polymer of nucleotides	DNA RNA	Genetic material Protein synthesis
	Nucleotides	ATP Coenzymes	Energy carrier Assist enzymes

Objective Questions

For questions 1 to 8, match the items below with those in the key.

Key:
  a. carbohydrate
  b. fats and oils
  c. protein
  d. nucleic acid

D. 1. contains the bases adenine, guanine, cytosine, and thymine
A 2. the six-carbon sugar, glucose
C 3. polymer of amino acids
B 4. glycerol and three fatty acids
C 5. enzymes
B 6. long-term energy storage
D 7. genes
A 8. plant cell walls

9. Which of these is not a characteristic of carbon?
  a. forms four covalent bonds
  b. bonds with itself
  c. is sometimes ionic
  d. forms long chains

10. The functional group COOH is
  a. acidic.
  b. basic.
  c. never ionized.
  d. All of these.

11. A hydrophilic group is
  a. attracted to water.
  b. a polar or ionized group.
  c. found in fatty acids.
  d. All of these.

12. Which of these would be an example of hydrolysis?
  a. amino acid + amino acid → dipeptide + H_2O
  b. dipeptide + H_2O → amino acid + amino acid
  c. Both of these.
  d. Neither of these.

13. Which of these makes cellulose nondigestible?
  a. a polymer of glucose subunits
  b. a fibrous protein
  c. the linkage between the glucose molecules
  d. the peptide linkage between the amino acid molecules

14. A fatty acid is unsaturated if it
  a. contains hydrogen.
  b. contains double bonds.
  c. contains an acidic group.
  d. bonds to glycogen.

15. Which of these is not a lipid?
  a. steroid
  b. fat
  c. polysaccharides
  d. waxes

16. The difference between one amino acid and another is found in the
  a. amino group.
  b. carboxyl group.
  c. R group.
  d. peptide bond.

17. The shape of a polypeptide is
  a. maintained by bonding between parts of the polypeptide.
  b. important to its function.
  c. ultimately dependent upon the primary structure.
  d. All of these.

18. Which of these is the peptide bond?

  a.

  b.   c.

19. Nucleotides
  a. contain a sugar, a base, and a phosphate molecule.
  b. are the monomers for fats and polysaccharides.
  c. join together by covalent bonding between the bases.
  d. All of these.

20. DNA
  a. has a sugar-phosphate backbone.
  b. is single stranded.
  c. has a certain sequence of amino acids.
  d. All of these.

Study Questions

1. How are the chemical characteristics of carbon reflected in the characteristics of organic molecules?
2. Give examples of functional groups and discuss the importance of their being hydrophobic or hydrophilic.
3. What molecules are monomers of the polymers studied in this chapter? How are monomers joined to give polymers and how are polymers broken down to monomers?
4. Name several monosaccharides, disaccharides, and polysaccharides and give a function for each. How are these molecules distinguishable on the basis of structure?
5. Name the different types of lipids and give a function for each type. What is the difference between a saturated and unsaturated fatty acid? Explain the structure of a fat molecule by stating its components and how they are joined together.
6. How does the structure of a phospholipid differ from that of a fat? How do phospholipids form a double layer when in the presence of water?
7. Draw the structure of an amino acid and a dipeptide, pointing out the peptide bond.
8. Discuss the four levels of structure of a protein and relate each level to particular bonding patterns.
9. How is the tertiary structure of a polypeptide related to the primary structure? Mention denaturation as evidence of this relationship.
10. How are nucleotides joined to form nucleic acids? Name several differences between the structure of DNA and RNA.

Thought Questions

1. Silicon is the second member of group IV in the periodic table of the elements but silicon preferentially bonds with oxygen. (Beach sand is SiO_2.) Why does this characteristic eliminate silicon as an alternative to carbon as the basis for organic molecules?

2. Why does meat rather than vegetables give you the best source of dietary organic nitrogen?

3. A whole plant stores energy as starch but many seeds store energy as oil. Can you explain why this is to be expected after reading about the ruby-throated hummingbird on page 45?

Selected Key Terms

organic (or-gan'ik) 38
hydrophobic (hi''dro-fo'bik) 39
hydrophilic (hi''dro-fil'ik) 39
isomer (i'so-mer) 39
dehydration synthesis (de''hi-dra'shun sin'the-sis) 40

hydrolysis (hi-drol'i-sis) 40
carbohydrate (kar''bo-hi'drāt) 40
lipid (lip'id) 44
phospholipid (fos''fo-lip'id) 45
steroid (ste'roid) 46
amino acid (ah-me'no as'id) 46

protein (pro'te-in) 46
enzyme (en'zīm) 46
peptide (pep'tīd) 46
nucleotide (nu'kle-o-tīd) 50
ATP 50
nucleic acid (nu-kle'ik as'id) 50

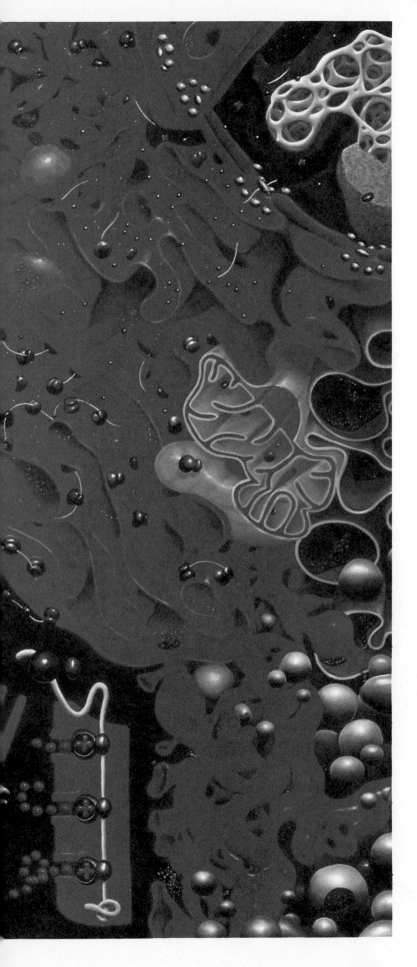

CHAPTER 5

Cell Structure and Function

Your study of this chapter will be complete when you can:

1. State two tenets of the cell theory. PAGE 78

2. List several similarities and differences between prokaryotic and eukaryotic cells.

3. Give several differences between bright-field light microscopy and transmission electron microscopy. Name several other types of microscopes that are available today.

4. Explain why cells are so very small on the basis of cell volume versus cell surface dimensions.

5. Describe the structure of a prokaryotic cell, and give a function for each part mentioned.

6. Describe the structure of the nucleus of a eukaryotic cell, and give a function for each part mentioned.

7. Name the structures that form the endomembrane system, and tell how they are related to one another.

8. Explain the relationship between chloroplasts and mitochondria; describe their structure in terms of compartmentalization of processes.

9. List four evidences for the endosymbiotic theory, which explains the origination of mitochondria and chloroplasts.

10. Name the components of the cytoskeleton; describe the structure and functions of each component.

11. Contrast the structure of prokaryotic, eukaryotic animal, and eukaryotic plant cells.

Artist's representation of a cell's interior illustrates its complexity. After you have studied this chapter, locate the nucleus at the upper right. Also locate the mitochondria, Golgi complex, vesicles, and rough endoplasmic reticulum (RER). At the lower left, an enlarged portion of rough ER is shown. Look close and notice that protein synthesis is occurring.

Cells

A ll the organisms we see about us are made up of cells (fig. 5.1). The atoms and molecules we studied previously are not alive, but the cell is alive. A **cell** is the smallest unit of living matter.

All organisms are made up of cells, and a cell is the smallest unit of living matter.

A few cells, like a hen's or frog's egg, are large enough to be seen by the naked eye but most are not. That's the reason the study of cells did not begin until the improvement of the first microscopes in the seventeenth century. It's hard to know exactly who was the first to see cells, but Antoni van Leeuwenhoek of Holland is famous for observing tiny one-celled living things that no one had seen before. Leeuwenhoek sent his findings to an organization of scientists called the Royal Society in London. Robert Hooke, an Englishman, confirmed Leeuwenhoek's observations, and was first to use the term cell. The tiny chambers he observed in the honeycomb structure of cork reminded him of the rooms or cells in a monastery. Naturally, then, he referred to the boundaries of these chambers as walls.

Although these early microscopists had seen cells, it was more than a hundred years later in the 1830s before Matthias Schleiden published his theory that all plants are composed of cells, and Theodor Schwann published a similar proposal concerning animals. These Germans based their ideas not only on their own work, but on the work of all those who had studied tissues under microscopes. Soon after, another German scientist, Rudolf Virchow, used a microscope to study the life of cells. He came to the conclusion that "every cell comes from a preexisting cell." By the middle of the nineteenth century, biologists clearly recognized that all organisms are composed of self-reproducing elements called cells.

Cells are capable of self-reproduction. Cells come only from preexisting cells.

The previous two statements are often called the **cell theory,** but sometimes are called the **cell doctrine.** Those who use the latter terminology want to make it perfectly clear that there is extensive data to support the cell theory, and that it is universally accepted by biologists.

Cell Structure and Function

Since the early days of cell study, it has become apparent that a cell carries out all those functions we associate with living things, such as those needed for growth and reproduction. Further, particular functions are carried on by certain parts of the cell. Improved microscopy has vastly extended our ability to view the internal structure of cells, and today's standard biochemical techniques have allowed us to determine the function of these cell parts.

Although there are different types of cells, they all have a barrier, called the **plasma membrane,** which separates the contents of the cell from its environment. The structure and function of the plasma membrane is discussed at greater length in

Figure 5.1
All organisms, including both plants and animals, are composed of cells. This is not readily apparent because the techniques of microscopy are needed in order to see the cells. *a.* Corn plant. *b.* Light micrograph of leaf showing many individual cells. *c.* Rabbit. *d.* Light micrograph of the intestinal lining showing that it too is composed of cells. The dark-staining bodies are nuclei.

a.

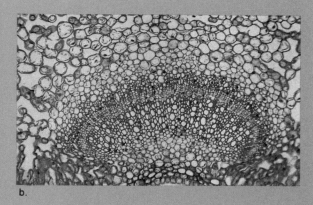
b.

c.

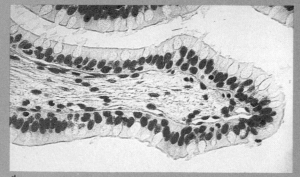
d.

Figure 5.2

Biological measurements. *a.* It takes a microscope to see cells because they are so very small. You can see cells with a light microscope, but not with much detail. It takes an electron microscope to make out most of the organelles in the cells. Note that a bacterium is smaller than the nucleus of a eukaryotic cell. *b.* The units of measurement common to cell biology and their equivalents.

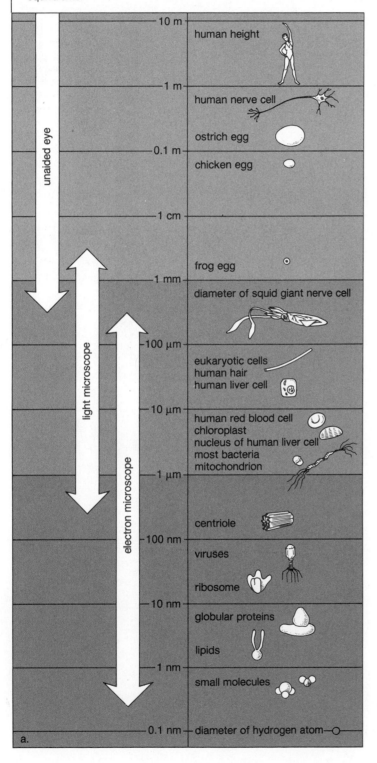

Measurements Used in Cell Biology			
Unit	Symbol	Scale	Seen by
centimeter	cm	0.4 inch	unaided eye
millimeter	mm	0.1 cm	unaided eye
micrometer	μm	0.001 mm	light microscope
nanometer	nm	0.001 μm	electron microscope

b.

chapter 6. For now, we will mention that the plasma membrane regulates the passage of molecules into and out of the cell. The plasma membrane is thin and weak, and alone cannot give much strength to a cell. Some cells, such as those of plant cells and microorganisms (organisms that cannot be seen without a microscope), are strengthened by the addition of a **cell wall** that protects the plasma membrane. Animal cells do not have a cell wall but have other means of support, such as those discussed on pages 70–73.

All cells are surrounded by a plasma membrane that separates them from their environment and is sometimes strengthened by the presence of a cell wall.

Cells are divided into two major groups chiefly on the basis of structure. **Eukaryotic cells** have a true nucleus (*eu* means true, *karyon* means nucleus), a membrane-bound compartment that houses DNA within **chromosomes,** complex threadlike structures. The rest of the cell is also divided by membranes into compartments that perform various functions. Some of these are **organelles,** small membranous bodies whose structure suits their function.

The other type of cell is called a **prokaryotic cell** because the cell lacks a true nucleus (*pro* means before, *karyon* means nucleus). Their DNA is in a single chromosome located within a distinguishable region called the **nucleoid.** Internal membrane does not compartmentalize the cell, and there are few organelles.

The term **cytoplasm** refers to the contents of a cell that lies between the nucleus (or nucleoid) and the plasma membrane. The various structures within the cytoplasm are bathed by a fluid medium known as **cytosol.**

There are two major types of cells: eukaryotic cells have a membrane-bound nucleus, and prokaryotic cells do not have such a nucleus.

Size of Cells

Cells are quite small. Most prokaryotic cells are from 1μm to 10μm in diameter. Eukaryotic cells are much larger, 10μm to 100μm, but still too small to be seen by the naked eye (fig. 5.2). In fact, you could place as many as 10,000 prokaryotic cells and as many as 1,000 eukaryotic cells on the length of this 1 cm line ⌞___⌟.

Cell Biology: Technology and Techniques

The study of cells is dependent upon the use of microscopes and biochemical techniques to detect cellular components and decipher their functions.

Microscopes

In the **bright-field light microscope**, light rays passing through a specimen are brought to a focus by a set of glass lenses, and the resulting image is then viewed by the human eye. In the **transmission electron microscope**, electrons passing through a specimen are brought to a focus by a set of magnetic lenses, and the resulting image is projected onto a fluorescent screen or photographic film. Let's use these statements to examine the differences between the two types of microscopes.

The Means of Illumination

Almost everyone knows that an electron microscope magnifies to a greater extent than does a light microscope. A light microscope can magnify objects a few thousand times but an electron microscope can magnify them hundreds of thousands of times. The difference lies in the means of illumination. The paths of the light rays and electrons moving through space are wavelike, but the wavelength of electrons is much shorter than the wavelength of light. This difference in wavelength not only accounts for the ability of the electron microscope to magnify to a greater extent, it also accounts for its greater resolving power. The greater the resolving power, the greater the detail eventually seen. **Resolution** is the *minimum* distance that there can be between two objects before they are simply seen as one larger object. If oil is placed between the sample and the objective lens of the light microscope, the resolving power is increased, and if ultraviolet light is used instead of visible light, it is also increased. But typically, a light microscope can resolve down to 0.2 μm while the electron microscope can resolve down to 0.0001 μm. If the resolving power of the average human eye is set at 1, then that of the typical light microscope is about 500 and that of the electron microscope is 100,000.

The Lenses and Means of Viewing the Object

Light rays can be bent (refracted) and brought to a focus as they pass through glass lenses, but electrons don't pass through glass. Electrons have a charge that allows them to be brought to a focus by magnetic lenses. The human eye utilizes light to see an object and can't utilize electrons for the same purpose. Therefore, electrons leaving the specimen in the electron microscope are directed toward a screen or photographic plate that is sensitive to their presence. Subsequently, humans can view the screen or photograph to see the image depicted there.

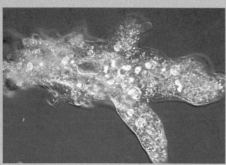

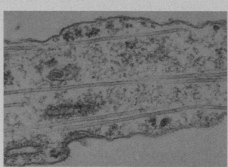

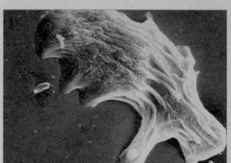

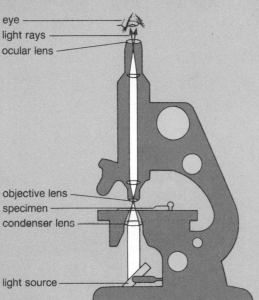

a. Compound light microscope

eye
light rays
ocular lens
objective lens
specimen
condenser lens
light source

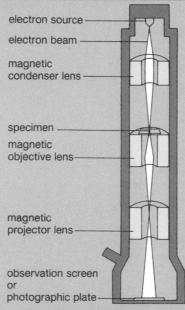

b. Transmission electron microscope

electron source
electron beam
magnetic condenser lens
specimen
magnetic objective lens
magnetic projector lens
observation screen or photographic plate

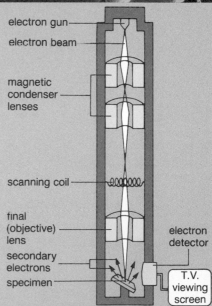

c. Scanning electron microscope

electron gun
electron beam
magnetic condenser lenses
scanning coil
final (objective) lens
secondary electrons
specimen
electron detector
T.V. viewing screen

Figure 1

a. Light micrograph of *Amoeba proteus*, a single-celled organism. A light microscope uses light to view a specimen. The specimen can be living if it is thin enough to allow light to pass through. A light microscope does not magnify or distinguish as much detail as the electron microscope. *b.* Transmission electron micrograph (TEM) of a portion of a pseudopodium, an extension by which *Amoeba proteus* moves. Cytoskeleton (p. 70) elements are visible. The transmission electron microscope uses electrons to "view" the specimen. The specimen is nonliving and must be thin enough to allow electrons to pass through. The magnification and amount of detail seen is far greater than with the light microscope. *c.* Scanning electron micrograph (SEM) of *Amoeba proteus*. Magnification, ×180. The scanning microscope scans the surface of the specimen with an electron beam. The secondary electrons given off are collected and the result is a three-dimensional image of the specimen.

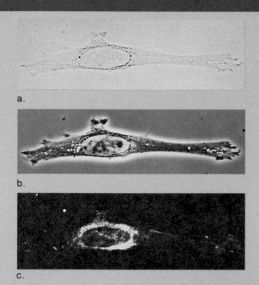

a.

b.

c.

Figure 2
Advanced light microscopes allow the specimen to be viewed using various techniques. *a.* Ordinary bright-field micrograph. *b.* Phase-contrast micrograph. *c.* Dark-field micrograph.

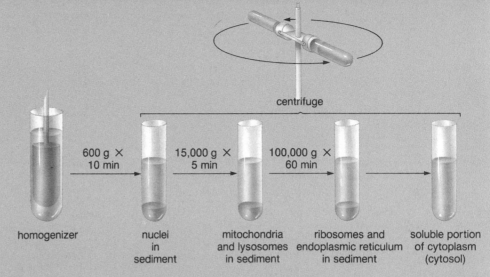

centrifuge

homogenizer → 600 g × 10 min → nuclei in sediment → 15,000 g × 5 min → mitochondria and lysosomes in sediment → 100,000 g × 60 min → ribosomes and endoplasmic reticulum in sediment → soluble portion of cytoplasm (cytosol)

Figure 3
Cell fractionation *a.* Cells are broken open mechanically by the action of the pestle against the sides of the tube. *b.* Contents are centrifuged at ever increasing speed to separate out first larger and then smaller components of the cell.

The Specimen

Organisms that are quite small and thin can be viewed alive by a light microscope but not with an electron microscope. Electrons randomly scatter as they pass through air; therefore, it is necessary to create a vacuum inside an electron microscope. Even with the light microscope, specimens are usually prepared for observation. Cells are killed, fixed so that they do not decompose, and embedded into a matrix. The matrix strengthens the specimen so that it can be thinly sliced. These sections are stained with colored dyes (light microscopy) or with electron-dense metals (electron microscopy). This lengthy treatment is necessary for several reasons: (1) light rays and electrons pass through thin materials only, (2) the dyes create differences in density (e.g., light and dark) that help distinguish detail. Some colored dyes bind preferentially to certain structures and this aids in their identification. The specimen appears colored because white light is composed of many colors of light and the pigments of the dye absorb certain colors and not others. For example, a green object absorbs all the colors except green light, which then remains for us to see.

One of the primary disadvantages of viewing dead, treated cells is that you can never be sure that the living cell is similar or that an artifact (structure not present when the cell is alive) has not been introduced.

Other Microscopes

You'll notice that we made specific reference to the bright-field light microscope and the transmission electron microscope. That's obviously because there are other types of light microscopes and electron microscopes, each with their own particular advantage. The **scanning electron microscope** permits the development of three-dimensional images. A narrow beam of electrons is scanned over the surface of the specimen, which has been coated with a thin metal layer. The metal gives off secondary electrons that are collected to produce a television-type picture of the specimen's surface on a screen.

As mentioned, the light microscope has an advantage over the electron microscope because the untreated specimen can be viewed alive. However, special microscopes other than bright field are needed to permit visualization of detail. **Phase-contrast** and **interference-contrast microscopy** are able to capitalize on the fact that when light passes through a living cell, the phase of each light wave is altered according to the relative density of the cellular structure:

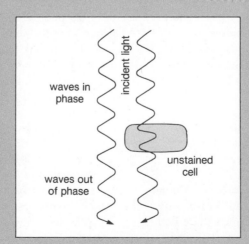

waves in phase

incident light

waves out of phase

unstained cell

Dark-field illumination, as the name suggests, illuminates the object against a dark background. The light has to be directed from the side; only scattered light enters the microscope lens, and as a result, the cell appears vividly lit on a black field. A similar arrangement uses ultraviolet light to detect fluorescence in a specimen.

It should be noted that computers have been increasingly used together with the scanning electron microscope and various types of light microscopes to enhance any detail observed in a picture received on a television screen.

Cell Fractionation

Cell fractionation is a means to separate cell components so that their individual biochemical composition and function can be determined. First, it is necessary to break open a large number of similar type cells. The cells are usually placed in a *homogenizer,* a tube that contains a close-fitting pestle; when the pestle is rotated, the cells are broken. Second, the "freed" contents of the cells are subjected to a spinning action known as *centrifugation.* At a low speed, large particles, like cell nuclei, settle out and are found in the sediment. Smaller particles are still in the fluid (supernatant), which can be poured into a fresh tube and subjected to centrifugation at a higher speed and so forth until the smallest of particles have been separated out.

Now the various cell fractions can be biochemically analyzed. During this analysis, it is sometimes necessary to separate different types of molecules from one another. Some of these techniques are discussed in chapter 18.

An explanation for why cells are so small and why the body of a plant or animal is multicellular can be found by considering surface area and volume relationships. Suppose a cell was 1 mm in each dimension (height, width, depth); its surface area would be

$$6 (1 \times 1) = 6 \text{ mm}^2$$

The volume of the cell would be

$$1 \times 1 \times 1 = 1 \text{ mm}^3$$

If the linear dimensions of the cell are doubled, its surface area would be 24 mm² and its volume would be 8 mm³. Notice that as a result of the cell doubling in size, the volume increased eight times, but the surface increased only four times.

Nutrients enter a cell and wastes exit a cell at the plasma membrane. A large cell requires more nutrients and produces more wastes than does a small cell. Therefore a large cell that is actively metabolizing cannot make do with proportionately less surface area than a small cell. Yet as previously mentioned, as a cell increases in size, surface area decreases dramatically in proportion to volume:

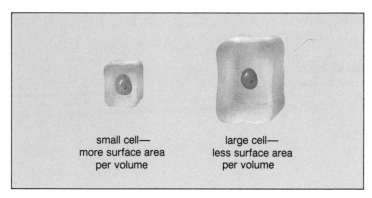

small cell—
more surface area
per volume

large cell—
less surface area
per volume

We would expect, then, that there is a limit of how large an actively metabolizing cell can become. For example, consider that egg cells are among the largest known cells. A chick's egg is several centimeters in diameter, but the egg is not actively metabolizing and contains a large amount of storage material called yolk. Once the egg is fertilized and metabolic activity begins, the egg divides repeatedly. Cell division provides the surface area needed to allow adequate exchange. We can also note that cells, which specialize in absorption, have modifications to greatly increase the surface area per volume of the cell. The columnar cells along the surface of the intestinal villi have microvilli to increase their surface area as do the cells that line the kidney tubules. Modifications are necessary to increase surface area per volume.

A cell needs a surface area (plasma membrane) that can adequately exchange materials with the environment. Surface area to volume relationships require that cells stay small.

There are other factors limiting the size of cells. The nucleus is the control center of the cell, and it can control only a certain amount of cytoplasm. Also, not only must there be an exchange of materials between the cell and external environment, but materials must be made available to various parts of

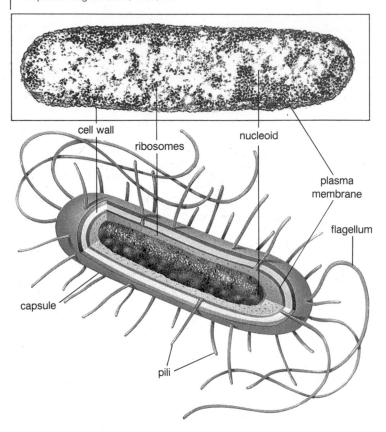

cell wall

ribosomes

nucleoid

plasma membrane

flagellum

capsule

pili

the cytoplasm. A eukaryotic cell is ten times larger than a prokaryotic cell, and we should expect to see modifications for the distribution of molecules inside the cell. The compartmentalization of eukaryotic cells assists their ability to make molecules available where they are needed.

Prokaryotic Cells

Single-celled **bacteria** are the only organisms that are prokaryotic cells. The bacteria are a very diverse group, and some members are photosynthetic. Among these are the **cyanobacteria,** which were formerly called blue-green algae.

Figure 5.3 illustrates the main features of prokaryotic anatomy. There is an exterior **cell wall** containing peptidoglycan, a large complex molecule consisting of polysaccharide polymers cross-linked by short chains of amino acids. In some bacteria, the cell wall is further surrounded by a gelatinous sheath or **capsule.** Motile bacteria usually have long, very thin, appendages called **flagella** that are composed of subunits of a protein called flagellin. The flagella, which rotate like a propeller, move the bacterium rapidly along in a fluid medium. Bacteria also have *pili,* short appendages that help bacteria attach themselves to an appropriate surface.

We have already mentioned that prokaryotic cells lack a defined nucleus. Most of their genes are found within a single loop of DNA that is located within the nucleoid region, but they

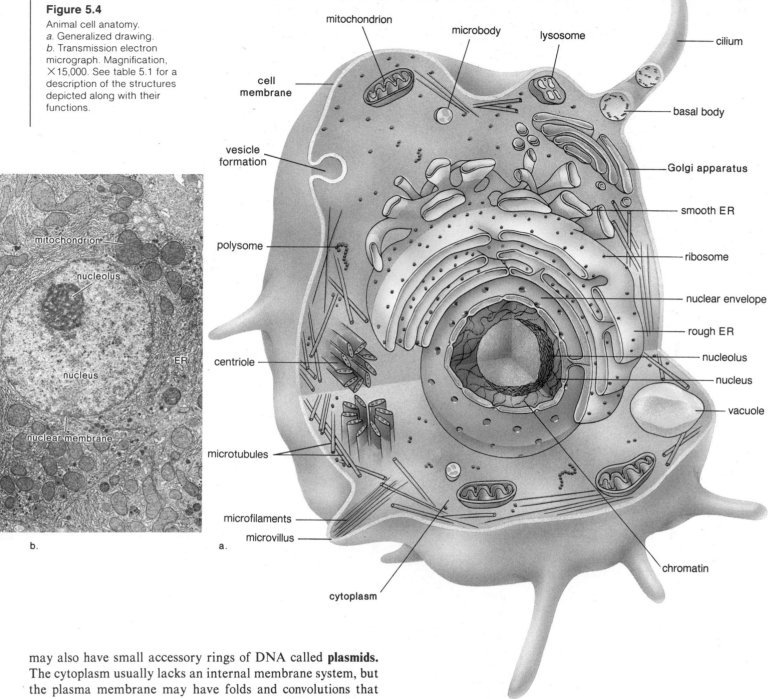

Figure 5.4
Animal cell anatomy.
a. Generalized drawing.
b. Transmission electron micrograph. Magnification, ×15,000. See table 5.1 for a description of the structures depicted along with their functions.

(Labels on micrograph b:) mitochondrion, nucleolus, nucleus, ER, nuclear membrane

b.

(Labels on drawing a:) mitochondrion, microbody, lysosome, cilium, cell membrane, basal body, vesicle formation, Golgi apparatus, polysome, smooth ER, ribosome, centriole, nuclear envelope, rough ER, nucleolus, nucleus, vacuole, microtubules, microfilaments, microvillus, chromatin, cytoplasm

a.

may also have small accessory rings of DNA called **plasmids.** The cytoplasm usually lacks an internal membrane system, but the plasma membrane may have folds and convolutions that extend into the interior of the cell. In addition, the photosynthetic cyanobacteria have light-sensitive pigments, usually within the membrane of flattened disks called **thylakoids.** The cytoplasm contains numerous granules known as **ribosomes** that function in protein synthesis. These are smaller than the ribosomes of eukaryotic cells and are not attached to membranes.

Bacteria are prokaryotic cells with these constant features:	
outer boundary:	cell wall
	plasma membrane
cytoplasm:	ribosomes
	thylakoids (cyanobacteria)
nucleoid:	DNA

Bacteria are so abundant on land, in the sea, and in the air that their total number is believed to be greater than all other organisms combined. We will have an opportunity to mention the importance of bacteria many times in the pages that follow.

Eukaryotic Cells

All cells, aside from bacteria, are eukaryotic. Eukaryotic organisms include the algae, protozoa, and fungi, in addition to plants and animals. In this chapter we are going to concentrate on the structure and function of a typical animal (fig. 5.4) and plant cell (fig. 5.5). Such drawings are composites based on data

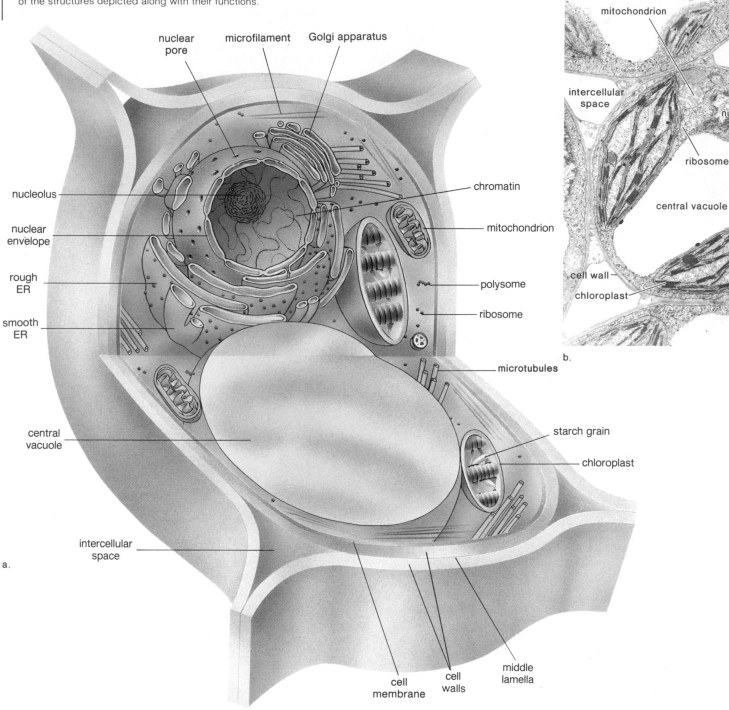

Figure 5.5

Plant cell anatomy. *a.* Generalized drawing. *b.* Transmission electron micrograph. Magnification, ×20,000. See table 5.1 for a description of the structures depicted along with their functions.

nuclear pore

microfilament

Golgi apparatus

nucleolus

nuclear envelope

rough ER

smooth ER

central vacuole

intercellular space

a.

chromatin

mitochondrion

polysome

ribosome

microtubules

starch grain

chloroplast

cell membrane

cell walls

middle lamella

mitochondrion

intercellular space

nucleus

ribosome

central vacuole

cell wall

chloroplast

b.

gathered from microscopic studies. In actuality, there is a great deal of variation among plant and animal cells.

Eukaryotic cells have a nucleus and various organelles (table 5.1), each of which has a structure that suits its function. The cytoplasm is compartmentalized by these organelles and also by other membranous structures.

Nucleus

The nucleus is a prominent structure in the eukaryotic cell. It is usually large enough to be observed through the light microscope, although electron micrographs show much more detail. The **nucleus** is of primary importance in the cell because it is the control center that oversees the metabolic functioning of the cell. The nucleus ultimately determines the cell's characteristics, as experimentation has shown. In one type of green

Table 5.1 Eukaryotic Structures in Animal and Plant Cells

Name	Composition	Function
Cell wall*	Contains cellulose fibrils	Support and protection
Plasma membrane	Bilayer of phospholipid with embedded proteins	Passage of molecules into and out of cell
Nucleus	Nuclear envelope surrounding the nucleoplasm, chromosomes, and nucleoli	Cellular reproduction and control of protein synthesis
Nucleolus	Concentrated area of chromatin, RNA, and proteins	Ribosome formation
Ribosomes	Protein and RNA in two subunits	Protein synthesis
Endoplasmic reticulum	Membranous flattened channels and tubular canals	Synthesis and/or modification of proteins and other substances, and transport by vesicle formation
Rough	Studded with ribosomes	Protein synthesis
Smooth	Having no ribosomes	Various; lipid synthesis in some cells
Golgi complex** (Dictyosome)*	Stack of membranous saccules	Processing and packaging of molecules, e.g., glycoproteins
Vacuoles and vesicles	Membranous sacs	Storage of substances
Lysosome**	Membranous vesicle containing digestive enzymes	Intracellular digestion
Microbodies	Membranous vesicle containing specific enzymes	Various metabolic tasks
Mitochondrion	Inner membrane (cristae) within outer membrane	Cellular respiration
Chloroplast*	Inner membrane (grana) within two outer membranes	Photosynthesis
Cytoskeleton	Microtubules, and microfilaments	Shape of cell, and movement of its parts
Cilia and flagella**	9 + 2 pattern of microtubules	Movement of cell
Centriole**	9 + 0 pattern of microtubules	Forms basal bodies that give rise to microtubules

*Plant cells
**Animal cells

Figure 5.6

The *Acetabularia* are single-cell algae up to 2 inches long that lend themselves to regeneration experiments. *a.* Photo of *Acetabularia mediterranea* commonly called mermaid's wineglass.
b. Experimentation to demonstrate that the nucleus is the control center of the cell. (1) In this experiment, the stalk and cap without a nucleus die, but the base with a nucleus regenerates. (2) In this experiment, the regenerated cap resembles that of the species of the nucleus and not that of the cytoplasm.

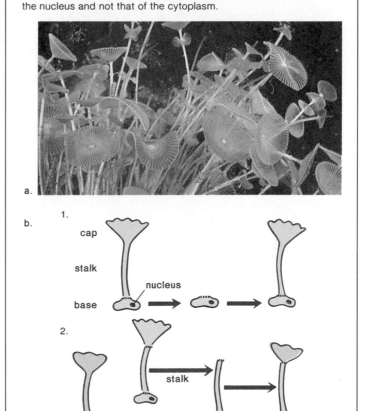

alga, *Acetabularia,* the organism consists of a single cell with a base, stalk, and cap (fig. 5.6). If the stalk and cap are removed from the base (fig. 5.6*b*), the stalk and cap die, but the nucleus-containing base develops into a new organism. The importance of the nucleus is further exemplified when the base is combined with the stalk of a different species. The cap regenerated is like the species of the nucleus rather than like the cytoplasm of the stalk. Similarly, the nucleus of a frog's egg can be removed and the nucleus of another species can be introduced. The tadpole and frog that develops will be like that of the introduced nucleus.

The nucleus is the control center of the cell.

It is clear that the nucleus is the control center because it contains the hereditary material DNA (deoxyribonucleic acid). As mentioned in the previous chapter, DNA carries the instructions for the proper sequencing of amino acids in proteins. DNA always remains within the nucleus, but a special type of RNA, called messenger RNA (mRNA), carries the instructions from DNA out into the cytoplasm where protein synthesis occurs.

When you look at the nucleus even in an electron micrograph, you can't see DNA but you can see chromatin. **Chromatin** looks grainy but actually is a threadlike material that undergoes coiling into rodlike structures called **chromosomes** just before a cell divides. Chemical analysis shows that chromatin, and therefore chromosomes, contain DNA, much protein, and some RNA.

Most likely, too, when you look at an electron micrograph of a nucleus you will see one or two regions that look darker than the rest of the chromatin. These are **nucleoli** where another type of RNA, called ribosomal RNA (rRNA), is being produced. Ribosomes are small bodies in the cytoplasm that contain not only ribosomal RNA, but also contain proteins. These proteins, like all others, are made in the cytoplasm. They migrate into the nucleus and join with ribosomal RNA, forming subunits before passing out into the cytoplasm where they are assembled into ribosomes.

The nucleus is separated from the cytoplasm by a double membrane known as the **nuclear envelope** (fig. 5.7). You might think that the presence of this envelope would prevent materials from passing to and from the nucleus, but electron micrographs show the presence of *pores* in the nuclear envelope. These pores are of sufficient size (100 nm) to permit the passage of ribosomal subunits. High-power electron micrographs show that the pores have a complex structure (fig. 5.7c), so they shouldn't be regarded as simple openings. They have associated proteins that can regulate the passage of materials to and from the nucleus. The fluid portion of the nucleus is called the **nucleoplasm.** The fact that it has a different pH from the cytosol suggests that it has a different composition.

The structural features of the nucleus include:

chromatin (chromosomes):	DNA and proteins
nucleolus:	chromatin and ribosome subunits
nuclear envelope:	double membrane with pores

Figure 5.7

Anatomy of the nucleus. *a.* The fluid portion of the nucleus is the nucleoplasm where chromatin is found. The nucleolus is a special region of DNA where rRNA is produced and joined with proteins from the cytoplasm to form the subunits of ribosomes. The nucleus is bounded by a double membrane that contains pores. *b.* Freeze-fracture (chapter 6) preparation of the nuclear envelope of a rat pancreas cell showing the existence of pores. Magnification, ×145,000. *c.* Each pore contains a pore complex consisting of eight granules and possibly projections that block the opening. These complexes are believed to regulate the passage of materials into and out of the nucleus. Two proposed versions of the pore complex are shown.

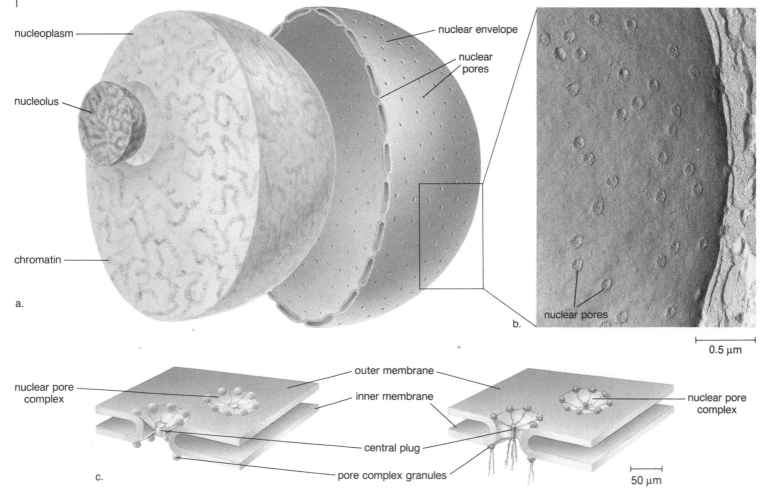

Ribosomes

The ribosomes present in eukaryotic cells are about one-third larger than those found in prokaryotic cells. Both types of ribosomes are composed of two subunits (fig. 5.8), each with its own mix of ribosomal RNA and proteins. We have already mentioned that ribosomal RNA is joined with proteins within the nucleus, but the two subunits are not assembled into one ribosome until they reach the cytoplasm.

Ribosomes function in protein synthesis. As they move along messenger RNA, amino acids are joined in an order originally dictated by DNA. Several ribosomes can be moving along the same messenger RNA at once and the entire complex is called a polysome. Some ribosomes may lie free within the cytosol but most are attached to the **endoplasmic reticulum,** a system of membranous saccules and channels in the cytoplasm. The proteins produced here are often enzymes that either function within the membrane of the reticulum or are exported outside of the cell. Cells that are involved in secreting enzymes, like the glands that produce digestive enzymes for the small intestine, have a large number of ribosomes attached to their endoplasmic reticulum. It is possible for such a cell to contain a few million ribosomes.

Ribosomes are small organelles that are involved in protein synthesis. A polysome contains a group of ribosomes, each one involved in producing a copy of the same protein.

Small-size ribosomes, like those found in bacteria, are also found in mitochondria and chloroplasts.

Endomembrane System

The endomembrane system helps to compartmentalize the cell. Many types of enzymatic reactions occur in cells, and these are facilitated when the reactants for each type reaction are restricted to different locales. Within the endomembrane system

---

Figure 5.8

Rough endoplasmic reticulum. *a.* Electron micrograph of a mouse hepatocyte (liver) cell shows a cross section of many flattened vesicles with ribosomes attached to the side that abuts the cytosol. Magnification, ×75,000. *b.* Drawing that shows the three dimensions of the organelle. *c.* Model of a single ribosome illustrates that each one is actually composed of two subunits. *d.* Method by which the endoplasmic reticulum acts as a transport system.

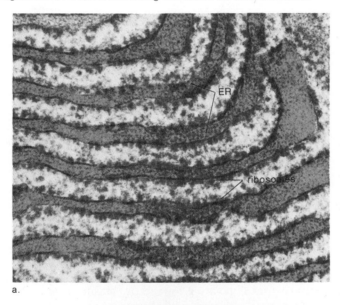

a.

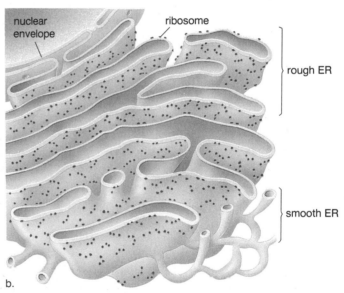

nuclear envelope

ribosome

rough ER

smooth ER

b.

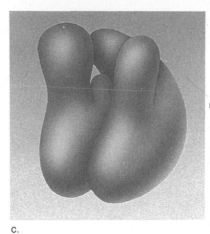

c.

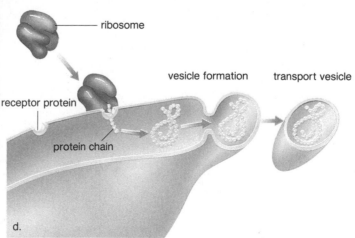

ribosome

vesicle formation

transport vesicle

receptor protein

protein chain

d.

Figure 5.9

The endomembrane system contains the endoplasmic reticulum (ER), which is continuous with the nuclear membrane. Rough ER has attached ribosomes, smooth ER does not. Proteins made at the rough ER move through the lumen of the canals and tubules until they are transported in vesicles from the smooth ER to the Golgi complex. The Golgi complex produces vesicles that move to the plasma membrane and lysosomes that contain digestive enzymes. Macromolecules can enter a cell by vesicle formation and this can fuse with a lysosome, giving a secondary lysosome in which the macromolecules are digested.

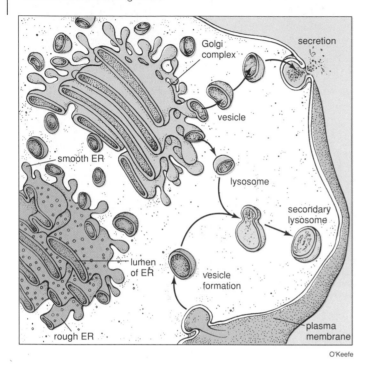

O'Keefe

we find the endoplasmic reticulum, the Golgi complex (dictyosome in plant cells), and various types of vesicles including lysosomes. These are all interrelated and so are presented together in figure 5.9.

Endoplasmic Reticulum

The liver is an organ that performs a lot of different functions in an animal's body; therefore, its cells need many different types of enzymes. It's to be expected then that ribosome-studded endoplasmic reticulum, called **rough ER,** which specializes in protein synthesis, will be extensively developed in these cells. In figure 5.8*b* you can see that the flattened channels of rough ER give way to tubelike canals that lack ribosomes. The latter is called **smooth ER.** Smooth ER has different functions in different cells. Sometimes it specializes in the production of lipids. For example, it is abundant in cells of the testes and adrenal cortex, both of which produce steroid hormones. In the liver it is involved in the detoxification of drugs; in muscle cells it acts as a storage area for calcium ions that are released when contraction occurs.

Rough ER specializes in protein synthesis. Smooth ER has a function suited to the particular type cell; sometimes it specializes in lipid synthesis.

Radioactive tracer studies have shown that the endoplasmic reticulum acts as a transport system for newly produced proteins. First, the proteins enter the lumen (interior space) of the rough ER (fig. 5.8*d*) where they may be combined with other molecules before they proceed to the lumen of smooth ER. Then, a small vesicle pinches off from the smooth ER. Such vesicles are observed to move through the cytoplasm to various locations. For example, they might move to the plasma membrane where they discharge their contents—we call this secretion. More often, vesicles move to and are taken up by the Golgi complex where protein molecules are modified before secretion takes place.

The endoplasmic reticulum is a transport system. Protein molecules move from the lumen of rough ER to that of smooth ER, which sends them enclosed within vesicles usually to the Golgi complex.

Some of the proteins produced by rough ER are never secreted; instead, they become incorporated into the endomembrane system where they carry out enzymatic functions. Also, we might note that new membrane is added to the plasma membrane when a vesicle fuses with it:

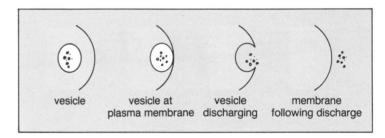

vesicle | vesicle at plasma membrane | vesicle discharging | membrane following discharge

The provision of new membrane is another function of the endoplasmic reticulum.

Golgi Complex (Dictyosome)

In most cells, a **Golgi complex** consists of a central region of large, parallel, curved saccules with smaller, rounded vesicles to the sides (fig. 5.10). The number of curved saccules can vary from about three to twenty. In animal cells, the Golgi complex has a "forming face" nearest the endoplasmic reticulum and a "maturing face" pointing toward the plasma membrane. This terminology is consistent with the observation that vesicles are received from smooth ER at the forming face of the Golgi complex and vesicles leave the complex at the maturing face.

The Golgi complex modifies, or processes, the protein molecules received by way of vesicles from the smooth ER. It activates some of them. For example, the hormone insulin is received by the Golgi in an inactive form, but after certain amino acids are removed, the hormone is active and ready to be secreted. Then it is repackaged within a vesicle that fuses with the plasma membrane. Perhaps the most common type of processing carried on by the Golgi is addition of sugars to proteins, which are then called glycoproteins. If a glycoprotein is also phosphorylated, this seems to be a signal that it is destined to be packaged in a vesicle that will be a lysosome. The Golgi complex produces lysosomes, the organelles discussed next.

Figure 5.10

Golgi complex in an animal cell. The Golgi receives protein-containing vesicles from the smooth ER at its forming face. These proteins are processed while they pass from saccule to saccule of the central region. Finally, they are repackaged in vesicles that leave the maturing face. Some of these vesicles fuse with the plasma membrane allowing secretion to occur, and others are lysosomes that contain digestive enzymes. *a.* Electron micrograph from liver cells of a mammal. Magnification, ×62,000.

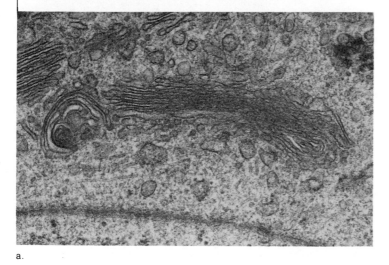

a.

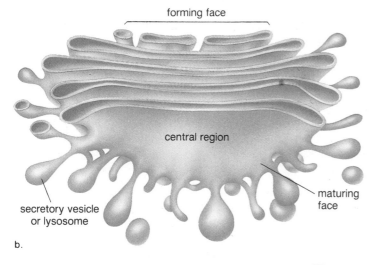

b.

In animal cells, a Golgi complex processes protein molecules, and encloses them in vesicles for secretion. It also produces lysosomes.

In plant cells, Golgi complexes have a different name, orientation, and other functions. They are called **dictyosomes,** which are found in special regions of the cell where their services seem to be needed. For example, the number of dictyosomes increases during cell division because they form vesicles containing polysaccharides needed for a cell plate. The cell plate develops into a cell wall, separating the two new cells. Most likely, dictyosomes also provide vesicles that fuse with newly developing plasma membrane.

In plant cells, dictyosomes provide the material to form new cell walls and new membrane following cell division.

Figure 5.11

Lysosomes in a kidney cell of a mammal. The small dark body is a "primary" lysosome that has not yet fused with an incoming vesicle. The larger bodies are "secondary" lysosomes that have fused with such vesicles. The dark appearance of the lysosomes results from staining for a particular enzyme, acid phosphatase, whose presence is the test for this organelle. Magnification, ×92,000.

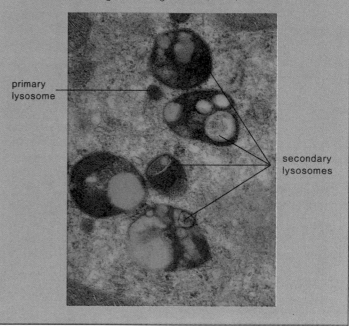

Lysosomes

Lysosomes (fig. 5.11) are membrane-bounded vesicles that contain hydrolytic digestive enzymes. We have already mentioned that lysosomes are produced by the Golgi complex and that the enclosed enzymes are glycoproteins also produced by the Golgi complex. The lysosomal enzymes digest macromolecules. Sometimes macromolecules are brought into a cell by vesicle formation at the plasma membrane (fig. 5.9). A lysosome can fuse with such a vesicle and digest its contents into simpler subunits that then enter the cytoplasm. Some white blood cells defend the body by engulfing bacteria that become enclosed within vesicles. When lysosomes fuse with these vesicles, the bacteria become digested. It should come as no surprise then that even parts of a cell are digested by its own lysosomes (called autodigestion). Normal cell rejuvenation most likely takes place in this manner, but autodigestion is also important during development. For example, when a tadpole becomes a frog, lysosomes are utilized to digest away the cells of the tail. The fingers of a human embryo are at first webbed but the fingers are freed from one another following lysosomal action.

Occasionally, a child is born with a metabolic disorder involving a missing or inactive lysosomal enzyme. In these cases, the lysosomes fill to capacity with a macromolecule that cannot be broken down. The cells become so full of these lysosomes that it brings about the death of the child.

Lysosomes are membrane-bounded vesicles that contain specific enzymes. Lysosomes are produced by Golgi complexes and their hydrolytic enzymes digest macromolecules from various sources.

Microbodies

Microbodies (fig. 5.12) are similar to lysosomes because they are vesicles bounded by a single membrane and contain specific enzymes. In this way, they also function to compartmentalize the cell. Although microbodies are believed to be quite varied, two types are noteworthy: peroxisomes and glyoxysomes.

Peroxisomes are microbodies that contain enzymes that transfer hydrogen atoms to oxygen, forming hydrogen peroxide (H_2O_2), a toxic molecule that is immediately broken down to water by the enzyme catalase. Peroxisomes are abundant in cells that are metabolizing lipids and in liver cells that are metabolizing alcohol. They are believed to help detoxify the alcohol.

Glyoxysomes have been observed in leaves that are carrying on photosynthesis. In that case, they contain enzymes that can metabolize some of the molecules involved in the photosynthetic process. They are also seen in germinating seeds where they are believed to convert oils into sugars to be used as nutrients by the growing plant.

It could be that the endoplasmic reticulum produces microbodies, but it is doubtful because their enzymes are produced by free ribosomes within the cytosol. Afterwards the enzymes enter the microbodies directly from the cytosol.

Vacuoles

A **vacuole** is a large membranous sac; a vesicle is smaller. Animal cells do have vacuoles, but they are much more prominent in plant cells. Typically, plant cells have one or two large vacuoles so filled with a watery fluid that it gives added support to the cell. Most of the central area of the cell is occupied by vacuole, and the rest of the contents of the cell is pushed to the sides (fig. 5.5).

Vacuoles are most often storage areas, but can have specialized functions. Plant vacuoles contain not only water, sugars, and salts, but also pigments and toxic substances. The pigments are responsible for many of the red, blue, or purple colors of flowers and some leaves. The toxic substances help protect a plant from predacious animals. (As long as the substance is contained within the vacuole, it will not be harmful to the plant.) The vacuoles present in protozoans are quite specialized, and include contractile and digestive vacuoles (fig. 24.2).

The endomembrane system includes these organelles:

endoplasmic reticulum: synthesis and/or modification and
　　transport of proteins and other substances
rough ER: protein synthesis
smooth ER: lipid metabolism, in particular
Golgi complex: processing and packaging
lysosomes: intracellular digestion
microbodies: various metabolic tasks
vacuoles: storage areas

Figure 5.9 shows how certain of these organelles are related to each other.

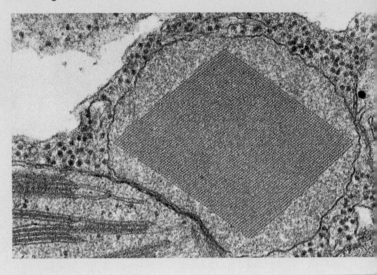

Figure 5.12
Peroxisome in a tobacco leaf. The crystalline, squarelike core of this microbody cross section is believed to consist of the enzyme catalase, which breaks down hydrogen peroxide to water. Microbodies carry out various metabolic reactions and are another means of compartmentalizing enzymatic reactions that occur in cells. Magnification, ×155,000.

Energy-Related Organelles

Life is possible only because of a constant input of energy to maintain the structure of the cell. Chloroplasts and mitochondria are the two eukaryotic membranous organelles that specialize in converting energy into a form that is usable by the cell.

Photosynthesis, which occurs in **chloroplasts,** is the process by which solar energy is converted to chemical-bond energy within carbohydrates. Photosynthesis can be represented by this equation:

$$\text{light energy + carbon dioxide + water} \longrightarrow \text{carbohydrate + oxygen}$$

Here the word "energy" stands for solar energy, the ultimate source of energy for cellular organization. Only plants, algae, and certain bacteria are capable of carrying on photosynthesis. (However, bacteria, being prokaryotic, do not have chloroplasts.)

Cellular respiration, which occurs in **mitochondria,** is the process by which the chemical energy of carbohydrates is converted to that of ATP (adenosine triphosphate), the common carrier of chemical energy in cells. Cellular respiration can be represented by this equation:

$$\text{carbohydrate + oxygen} \longrightarrow \text{carbon dioxide + water + energy}$$

Figure 5.13

Chloroplast structure. *a.* Electron micrograph. Magnification, ×40,000. *b.* Generalized drawing in which the outer and inner membranes have been cut away to reveal the grana.

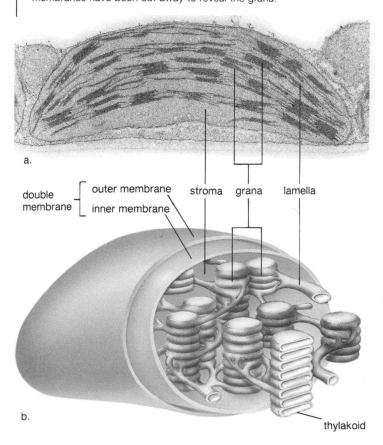

a.

double membrane
outer membrane
inner membrane
stroma   grana   lamella

b.

thylakoid

Figure 5.14

Mitochondrial structure. *a.* Electron micrograph. Magnification, ×70,000. *b.* Generalized drawing in which the outer membrane and portions of the inner membrane have been cut away to reveal the cristae.

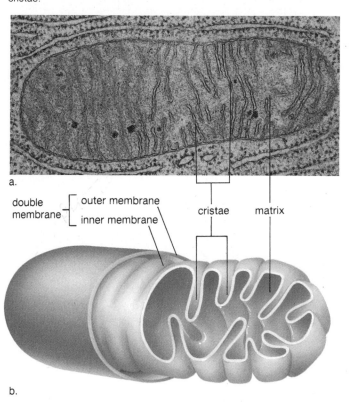

a.

double membrane
outer membrane
inner membrane
cristae   matrix

b.

Here the word "energy" stands for ATP molecules. When a cell needs energy it uses ATP to supply it. ATP energy is used for synthetic reactions, active transport, and all energy-requiring processes in cells. All organisms carry on cellular respiration, and all organisms, except bacteria, have mitochondria. Chloroplasts use the sun's energy to produce carbohydrates, and carbohydrate-derived products are broken down in mitochondria for the purpose of building up ATP molecules.

Chloroplasts and mitochondria supply the eukaryotic cell with usable energy. Energy is needed for various cellular activities, all of them helpful for maintaining the structure of cells.

Chloroplasts

Chloroplasts are about 4 μm to 6 μm in diameter and 1 μm to 5 μm long; they belong to a group of plant organelles known as plastids. Among the plastids are also the *leucoplasts,* which store starch and the *chromoplasts,* which contain red and orange pigments.

A chloroplast is bounded by a double membrane, and inside there is even more membrane organized into flattened sacs called **thylakoids.** The thylakoids are piled up like stacks

of coins, and each stack is called a **granum.** There are membranous connections between the grana called *lamellae* (fig. 5.13). The fluid-filled space about the grana is called the **stroma.**

Photosynthesis requires pigments, which capture solar energy, and enzymes, which synthesize carbohydrate. The green pigments, called chlorophylls, are the major pigments found within the thylakoid membranes of the chloroplasts. Enzymes are found within the stroma. This basically tells you how chloroplasts are organized to carry out photosynthesis. As mentioned previously, the cyanobacteria, which are prokaryotic, usually possess cytoplasmic thylakoids on which photosynthetic pigments are bound.

Chloroplasts have their own DNA and ribosomes, and can produce certain proteins. They also reproduce themselves by division. In other words, a chloroplast has many similarities to a photosynthetic prokaryotic organism, an observation to which we will return.

Mitochondria

Most mitochondria are between 0.5 μm and 1.0 μm in diameter and 7 μm in length, although the size and shape might vary.

Mitochondria, like chloroplasts, are bounded by a double membrane. The inner membrane is folded to form little shelves, called **cristae,** which project into the **matrix,** an inner space filled with a gel-like fluid (fig. 5.14). Analysis reveals that the matrix contains enzymes for breaking down carbohydrate-derived products, while ATP production occurs at the cristae.

Figure 5.15

Endosymbiotic theory. *a*. Mitochondria might be derived from aerobic bacteria that were engulfed by a larger amoebalike eukaryote. *b*. Similarly, chloroplasts may be derived from photosynthetic bacteria that were engulfed by a larger amoebalike eukaryote.

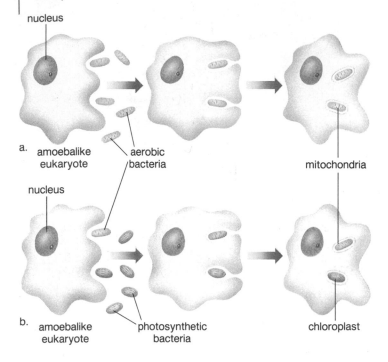

a. amoebalike eukaryote   aerobic bacteria   mitochondria

b. amoebalike eukaryote   photosynthetic bacteria   chloroplast

Like chloroplasts, mitochondria also contain their own DNA and ribosomes, and can produce a few of their own proteins. They also reproduce themselves by division.

Chloroplasts and mitochondria are membranous organelles whose structure is suited to compartmentalizing the processes that occur within them.

Origin of Energy-Related Organelles in Eukaryotic Cells

It's believed that the first eukaryotic cells did not possess energy-related organelles; instead they were acquired through a process of endosymbiosis. The **endosymbiotic theory** states that chloroplasts and mitochondria were originally prokaryotes that came to reside inside a eukaryotic cell, establishing a symbiotic relationship known as mutualism. In other words, today's mitochondria might be derived from an aerobic (oxygen-using) bacterium that was taken up by an eukaryotic cell in exchange for a ready supply of nutrients. Chloroplasts might be derived from a photosynthetic bacterium that permitted its host to carry on photosynthesis (fig. 5.15).

The evidence for the endosymbiotic theory is as follows:

1. Mitochondria and chloroplasts are similar to bacteria in size and structure.
2. Both organelles are bounded by a double membrane—the outer membrane may be derived from the engulfing vesicle, and the inner one may be derived from the plasma membrane of the original prokaryote.
3. Mitochondria and chloroplasts contain a limited amount of genetic material and are capable of self-reproduction. Their DNA is a circular loop like that of bacteria.
4. Although most of the proteins within mitochondria and chloroplasts are now produced by the eukaryotic host, they do possess their own ribosomes, and do produce some proteins. Their ribosomes resemble those of bacteria.

The endosymbiotic theory states that mitochondria and chloroplasts are derived from bacteria that took up residence in (a) eukaryotic cell(s).

The Cytoskeleton, Cell Shape, and Motility

Cells typically have a particular shape; in animals, muscle cells are tubular and nerve cells have long skinny processes. Also, cells are sometimes motile—they can move about from place to place. Some plants and all animals have reproductive sperm cells that move by means of flagella, for example.

Eukaryotic cell shape and motility are dependent upon the **cytoskeleton** an internal framework that contains microtubules, intermediate filaments, and microfilaments that differ as to composition and size.

Microtubules

Microtubules are small cylinders (about 30 nm in diameter) composed of thirteen rows of tubulin dimers, subunits each containing two types of tubulin proteins. When the dimers assemble (polymerize) a microtubule forms; later a microtubule can disassemble (depolymerize) (fig. 5.16). The regulation of microtubule formation is believed to be under the control of a special region in the cell called the *microtubule organizing center* which lies near the nucleus. The center contains various proteins and in animal cells (but not plant cells) the center also contains two centrioles lying at right angles to each other. **Centrioles** are short cylinders, with a 9 + 0 pattern of microtubule triplets, that is, a ring of nine sets of triplets with none in the middle. Before an animal cell divides, the centrioles replicate and the members of each pair are also at right angles to one another. During cell division, the centriole pairs separate so that each new cell receives one pair.[1]

---

[1]For some time it was believed that centrioles were necessary to the formation of the mitotic spindle. However, the spindle apparatus forms normally in plant cells (which do not have centrioles) and in animal cells from which the centrioles have been removed.

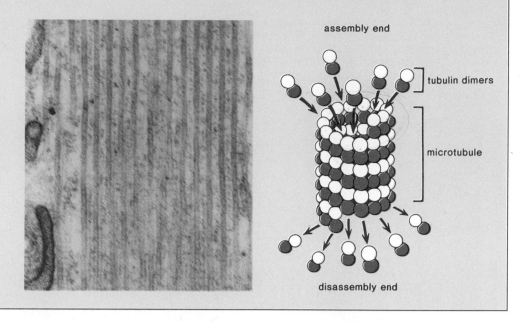

Figure 5.16

a. Electron micrograph of microtubules in a longitudinal section of sperm flagellum.
b. Each microtubule is a cylinder of tubulin dimers. Microtubules are believed to have an assembly end and a disassembly end as shown. This allows them to appear and disappear in the cell.

assembly end

tubulin dimers

microtubule

disassembly end

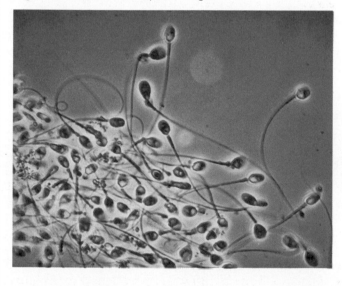

Figure 5.17

Sperm cells use long whiplike flagella to move about. Cilia and flagella have the structure depicted in figure 5.18.

Microtubules that lie free within the cytoplasm are believed to be involved in maintaining the shape of the cell and in directing the movement of cytoplasmic structures like the vesicles formed at the Golgi complex (dictyosome). Microtubules are also arranged in a definite pattern within cilia and flagella, surface structures involved in cell movements.

Cilia and Flagella

Cilia and **flagella** are hairlike projections of cells that can move either in an undulating fashion, like a whip, or stiffly, like an oar. Cells that have these organelles are capable of movement. For example, single-celled paramecia move by means of cilia;

sperm cells move by means of flagella (fig. 5.17). The cells that line our upper respiratory tract are ciliated. The cilia sweep debris trapped within mucus back up into the throat. This action helps keep the lungs clean.

Cilia are much shorter than flagella, but have a similar construction that is quite different from that of prokaryotic flagella (fig. 5.18). They are membrane-bound cylinders enclosing a matrix area. In the matrix are nine microtubule doublets arranged in a circle around two central microtubules (fig. 5.18*b*). This is called the 9 + 2 pattern of microtubules. Each doublet has a pair of arms projecting toward a neighboring doublet and spokes extending toward the central pair of microtubules. Cilia and flagella move when the microtubule doublets slide past one another. The clawlike arms and spokes are involved in causing this sliding action, which requires ATP energy.

Each cilium and flagellum has a **basal body** (fig. 5.18*c*) lying in the cytoplasm at its base. Basal bodies, which are short cylinders with a circular arrangement of nine microtubule triplets called the 9 + 0 pattern, are believed to organize the structure of cilia and flagella and to be derived from centrioles.

Centrioles have a 9 + 0 pattern of microtubules and give rise to basal bodies that organize the 9 + 2 pattern of microtubules in cilia and flagella.

Intermediate Filaments

Intermediate filaments (8 nm to 11 nm in diameter) are intermediate in size between microtubules and microfilaments. They contain bundles of fibers made of protein but the specific type varies according to the tissue. In the skin, where the fibers are made of keratin, the presence of intermediate filaments gives great mechanical strength to cells.

Figure 5.18

Anatomy of cilia and flagella. *a.* Drawing showing that the 9 + 2 pattern of microtubules within a cilium or flagellum is derived (in some unknown way) from the 9 + 0 pattern in a basal body. *b.* Cross-section drawing of the 9 + 2 pattern shows the exact arrangement of microtubules. Notice the clawlike arms of the outer doublets and the spokes that connect them to the central pair. *c.* Electron micrograph of the cross section of a basal body.

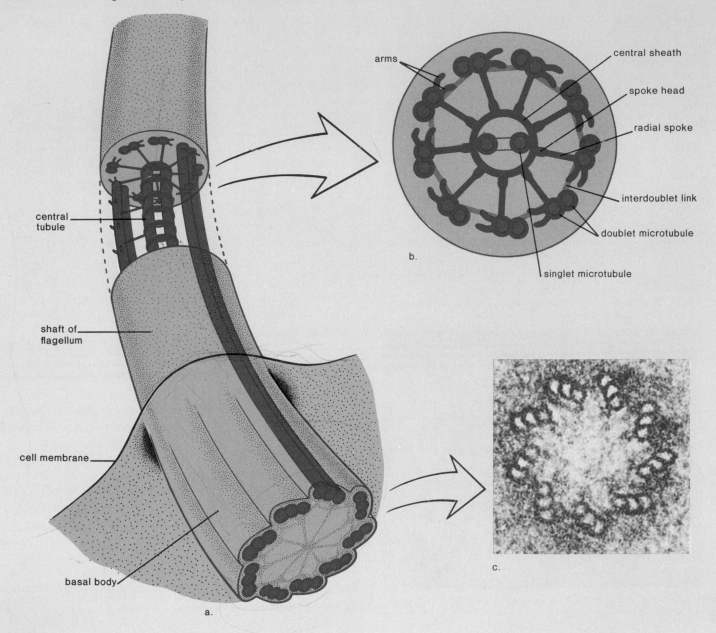

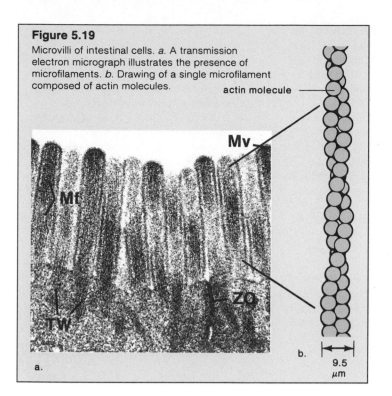

Figure 5.19
Microvilli of intestinal cells. *a.* A transmission electron micrograph illustrates the presence of microfilaments. *b.* Drawing of a single microfilament composed of actin molecules.

actin molecule

Mv

Mf

ZO

TW

a.

b.

9.5 μm

Microfilaments

Microfilaments are long, extremely thin fibers (approximately 7.5 nm in diameter) that usually occur in bundles or networks. In most cells, microfilaments are composed of a protein called actin. Actin microfilaments are well known for their role in muscle contraction where they interact with other protein microfilaments composed of myosin.

Actin microfilaments are seen in various types of cells, especially where movement is occurring. For example, microfilaments are seen in the microvilli that project from intestinal cells. Their presence most likely accounts for the ability of the microvilli to alternately shorten and then extend into the intestine (fig. 5.19). They also form a constricting ring that becomes ever smaller when an animal cell divides into two cells. In plant cells, they seem to be involved in cytoplasmic streaming of cell contents.

Microtubules, intermediate filaments and microfilaments, are components of a cytoskeleton that is involved in cell movements.

Summary

1. All organisms are composed of cells, the smallest units of living matter. Cells are capable of self-reproduction and existing cells come only from preexisting cells.
2. There are two major groups of cells: prokaryotic and eukaryotic. Both types have a plasma membrane and cytoplasm. Eukaryotic cells also have a chromosome-containing nucleus and various organelles. Prokaryotic cells have a DNA-containing nucleoid that is not bounded by a nuclear envelope. They also lack most of the other organelles found in eukaryotic cells. Table 5.2 summarizes the differences between prokaryotic and eukaryotic cells.
3. Cells are very small, and are measured in small units of the metric system. The plasma membrane regulates exchange of materials between the cell and the external environment. Cells must remain small in order to have an adequate amount of plasma membrane per volume of the cell.
4. The nucleus of eukaryotic cells, represented by animal and plant cells, is bounded by a nuclear envelope containing pores, which serve as passageways between the cytoplasm and the nucleoplasm. Within the nucleus, the chromatin undergoes coiling into chromosomes at the time of cell division. The nucleolus is a special region of the

Table 5.2 Comparison of Prokaryotic and Eukaryotic Cells

	Prokaryotic	Eukaryotic Animal	Eukaryotic Plant
Size	*smaller (1–10 μm in diameter)*	*larger (10–100 μm in diameter)*	
Plasma membrane	yes	yes	yes
Cell wall	usually (peptidoglycan)	no	yes (cellulose)
Nuclear envelope	no	yes	yes
Nucleolus	no	yes	yes
DNA	yes (single loop)	yes (chromosomes)	yes (chromosomes)
Mitochondria	no	yes	yes
Chloroplasts	no	no	yes
Endoplasmic reticulum	no	yes	yes
Ribosomes	yes (smaller)	yes	yes
Vacuoles	no	yes (small)	yes (usually large, single vacuole)
Golgi complex	no	yes	yes
Lysosomes	no	always	often
Microbodies	no	usually	usually
Cytoskeleton	no	yes	yes
Centrioles	no	yes	no (in higher plants)
9 + 2 cilia or flagella	no	often	no (in higher plants)

chromatin where ribosomal RNA (rRNA) is produced and where proteins from the cytoplasm gather to form ribosomal subunits. These subunits are joined in the cytoplasm.

5. Ribosomes are organelles that function in protein synthesis. They can be bound to endoplasmic reticulum or exist free within the cytoplasm. Several ribosomes are involved in synthesizing the same type protein. The complex is called a polysome.

6. The endomembrane system includes the endoplasmic reticulum (both rough and smooth), the Golgi complex (dictyosome in plant cells), the lysosomes, and other types of vesicles and vacuoles. The endomembrane system serves to compartmentalize the cell and keep the various biochemical reactions separate from one another. Proteins that are made in the cytoplasm quite often enter the rough ER where they may be modified

before proceeding to the lumen of the smooth ER. The smooth ER has various metabolic functions depending on the cell, but it also forms vesicles that carry proteins and other substances to different locations, particularly to the Golgi complex. The Golgi complex processes proteins and repackages them into lysosomes, which carry out intracellular digestion, or into vesicles that fuse with the plasma membrane. Following fusion, secretion occurs. Also a part of the endomembrane system are microbodies that have special enzymatic functions, and the large single plant vacuole, which not only stores substances but lends support to the plant cell.

7. Cells require a constant input of energy to maintain their structure. Chloroplasts capture the energy of the sun and carry on photosynthesis, which produces carbohydrate. Carbohydrate products are broken down in mitochondria as

ATP is produced. The latter is an oxygen-requiring process called cellular respiration. It is believed that mitochondria and chloroplasts are derived from prokaryotes that took up residence in eukaryotic cells.

8. The cytoskeleton contains microtubules, intermediate filaments, and microfilaments. These are involved in maintaining cell shape and in the movement of cellular components. Microtubules are made up of thirteen rows of tubulin dimers. They are found free within the cytoplasm, but also are present in centrioles and cilia and flagella. Centrioles give rise to basal bodies, which lie at the base of cilia and flagella. Intermediate filaments contain protein fibers and are intermediate in size between microtubules and microfilaments. Microfilaments are usually made up of actin, protein molecules arranged in a helical manner.

Objective Questions

1. The small size of cells is best correlated with
   a. the fact they are self-reproducing.
   b. their prokaryotic versus eukaryotic nature.
   c. an adequate surface area for exchange of materials.
   d. All of these.

2. Which of these is not a true contrast between the light microscope and the transmission electron microscope?

   | Light | Electron |
   a. Uses light to "view" object—uses electrons to "view" object.
   b. Uses glass lenses for focusing—uses magnetic lenses for focusing.
   c. Specimen must be killed and stained—specimen may be alive and nonstained.
   d. Magnification is not as great—magnification is greater.

3. Which of these best distinguishes a prokaryotic cell from a eukaryotic cell?
   a. Prokaryotic cells have a cell wall, but eukaryotic cells never do.
   b. Prokaryotic cells are much larger than eukaryotic cells.
   c. Prokaryotic cells have flagella, but eukaryotic cells do not.
   d. Prokaryotic cells do not have a membrane-bound nucleus, but eukaryotic cells do have such a nucleus.

4. Which of these is not found in the nucleus?
   a. functioning ribosomes
   b. chromatin that condenses to chromosomes
   c. nucleolus that produces ribosomal RNA (rRNA)
   d. nucleoplasm instead of cytoplasm

5. Ribosomal RNA (rRNA)
   a. is made in the cytoplasm.
   b. is made in the nucleus.
   c. is found only on the endoplasmic reticulum.
   d. forms the pores in the nuclear envelope.

6. Vesicles from the smooth ER most likely are on their way to the
   a. rough ER.
   b. lysosomes.
   c. Golgi complex.
   d. plant cell vacuole only.

7. Lysosomes function in
   a. protein synthesis.
   b. processing and packaging.
   c. intracellular digestion.
   d. lipid synthesis.

8. Mitochondria
   a. carry on cellular respiration.
   b. break down ATP to release energy for cells.
   c. contain grana and cristae.
   d. All of these.

9. The endosymbiotic theory explains
   a. the origin of the nucleus.
   b. where the endomembrane system came from.
   c. why prokaryotic cells are different from eukaryotic cells.
   d. the origin of mitochondria and chloroplasts.

10. Which organelle releases oxygen?
   a. ribosomes
   b. Golgi complex
   c. mitochondria
   d. chloroplasts

11. Cilia and flagella contain
   a. microtubules.
   b. microfilaments.
   c. tubulin dimers.
   d. Both (a) and (c).

12. The components of the cytoskeleton
   a. are the same as those of the endoplasmic reticulum.
   b. account for cell shape and subcellular movements.
   c. are found in both prokaryotic and eukaryotic cells.
   d. form the cell wall of plants.

Study Questions

1. What are the two basic tenets of the cell theory?
2. What similar features do both prokaryotic and eukaryotic cells have? What is their major difference?
3. What are the contrasting advantages of light microscopy and electron microscopy?
4. Contrast the size of prokaryotic and eukaryotic cells and tell why cells are so small.
5. Draw a rough sketch of a prokaryotic cell, label its parts, and state a function for each of these.
6. Describe an experiment that supports the concept of the nucleus as the control center of the cell.
7. Distinguish between the nucleolus, ribosomal RNA, and ribosomes.
8. Describe the structure and function of the nuclear envelope and nuclear pores.
9. Trace the path of a protein from rough ER to the plasma membrane.
10. Give the overall equations for photosynthesis and cellular respiration; contrast the two and tell how they are related.
11. Draw a diagram that explains the endosymbiotic theory.
12. What are the three components of the cytoskeleton? What is their structure and function?

Thought Questions

1. Classification schemes are sometimes arbitrary. Can you think of some other way to classify the cells studied in this chapter?
2. If mitochondria were prokaryotes at one time, the cristae came from what part of the cell?
3. What evidence might suggest that centrioles do not give rise to basal bodies?

Selected Key Terms

cell (sel) 56
plasma membrane (plaz′ma mem′brān) 56
cell wall (sel wawl) 57
eukaryotic cell (u″kar-e-ot′ik sel) 57
chromosome (kro′mo-sōm) 57
organelle (or″gan-el′) 57
prokaryotic cell (pro″kar-e-ot′ik sel) 57
nucleoid (nu′kle-oid) 57
cytoplasm (si′to-plazm″) 57

cytosol (si′to-sol) 57
nucleus (nu′kle-us) 62
chromatin (kro′mah-tin) 64
nucleoli (nu-kle′o-li) 64
nuclear envelope (nu′kle-ar en′vĕ-lōp) 64
ribosome (ri′bo-sōm) 65
endoplasmic reticulum (en″do-plas′mik rĕ-tik′u-lum) 65
Golgi complex (gol′je kom′pleks) 66

lysosome (li′so-sōm) 67
microbodies (mi″kro-bod′ez) 68
chloroplast (klo′ro-plast) 68
mitochondrion (mi″to-kon′dre-an) 68
cytoskeleton (si″to-skel′ĕ-ton) 70
microtubule (mi″kro-tu′būl) 70
centriole (sen′tri-ōl) 70
microfilament (mi″kro-fil′ah-ment) 73

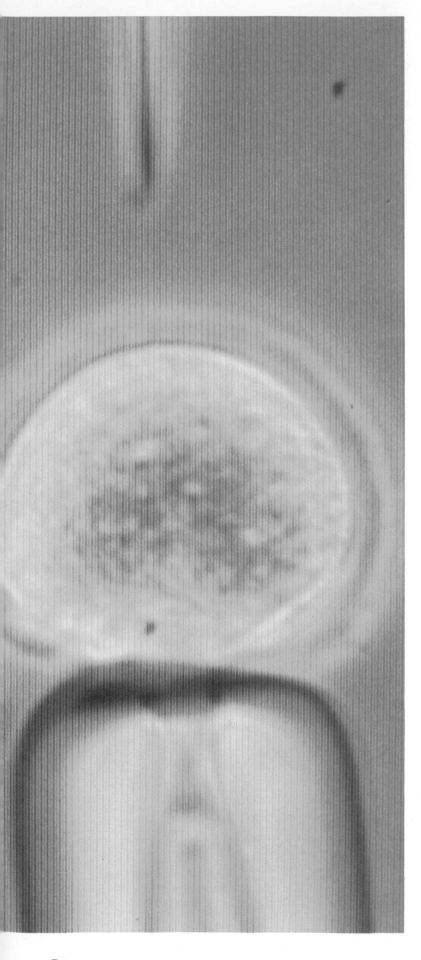

CHAPTER 6

Membrane Structure and Function

Your study of this chapter will be complete when you can:

1. *Contrast the "sandwich" model of protein structure with the "fluid-mosaic" model, and cite the evidence supporting these models.*
2. *Describe the structure of the lipid component, and give a function for each type of lipid in the membrane.*
3. *Describe the protein component of the membrane, and give several functions for these proteins.*
4. *Define diffusion and osmosis, and explain their relevance to cell biology. Describe the appearance of a plant cell and an animal cell in isotonic, hypotonic, and hypertonic solutions.*
5. *Name two types of transport by carrier proteins, and give an example for each.*
6. *Contrast endocytosis and exocytosis. Name three types of endocytosis, and distinguish between them.*
7. *Describe composition of the plant cell wall.*
8. *List and describe three types of junctions that occur between animal cells. Name and describe one type of junction between plant cells.*

This is a microscopic view of a single cell and at the top is a fine probe that will pierce the plasma membrane so that DNA can be removed by the suction tube on the bottom. An intact plasma membrane is necessary to the life of any cell and if it is ruptured the cell cannot continue to exist.

Figure 6.1

When rivers are full of salmon returning to their spawning ground, brown bears catch and eat them. The food that animals eat is beneficial to them only as long as the nutrient molecules enter their cells. It is the cells that must be served by a constant entry of nutrients and exit of metabolic wastes. The plasma membrane regulates the passage of molecules into and out of the cell.

M ulticellular animals take in complex foods (fig. 6.1), but the nutrient molecules they contain are of no value to the organism unless they can cross plasma membranes. The expression "multicellular organism" is appropriate because it is the structure and function of the *cells* that make the organism alive. The composition of the cell is fundamentally different from the medium in which it exists; therefore, it stands to reason that the type and amount of substances that enter and exit cells must be regulated. It is the plasma membrane that sees to this regulation on which life is dependent.

The plasma membrane regulates the passage of molecules into and out of the cell.

Studies of Membrane Structure

At the turn of the century, early investigators noted that lipid-soluble molecules entered cells more rapidly than water-soluble molecules. This prompted them to suggest that lipids might be a component of the plasma membrane. Later, chemical analysis disclosed that the plasma membrane contained phospholipids (fig. 6.2). In 1925, E. Gorter and G. Grendel measured the amount of phospholipid extracted from red blood cells and determined that there was just enough to form a bilayer around the cells.[1] They further suggested that the nonpolar (hydrophobic) tails were directed inward while the polar (hydrophilic) heads were directed outward:

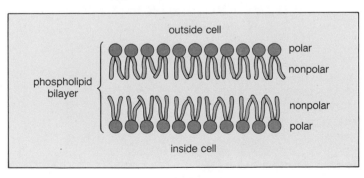

Figure 6.2

The plasma membrane contains phospholipid molecules. Each phospholipid molecule is composed of glycerol bonded to two fatty acid molecules and a grouping that contains phosphate and nitrogen. The molecule arranges itself in such a way that the phosphate and nitrogen form a polar (hydrophilic) head, and the fatty acids form nonpolar (hydrophobic) tails. The presence of a double bond causes a "kink" in the fatty acid tail.

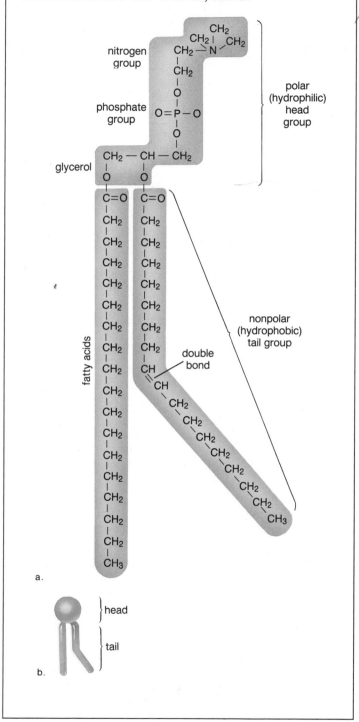

[1] Later investigators have shown that Gorter and Grendel reported too low an amount of lipid, and underestimated the surface area of red blood cells. However, since the two errors cancelled each other, they came to the correct conclusion.

Membrane Structure and Function   **77**

The presence of lipids cannot account for all the properties of the plasma membrane. To account for the permeability of the membrane to certain nonlipid substances, J. Danielli and H. Davson suggested that globular proteins are also a part of the membrane. They proposed the "sandwich" model of membrane structure in which the phospholipid bilayer is a filling between two layers of proteins that were arranged to give channels through which polar substances might pass:

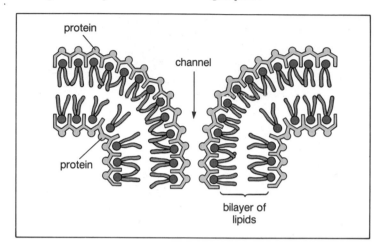

In the late 1950s, electron microscopy had advanced to the point that it was possible to view the plasma membrane and other membranes in the cell. Since the membrane has a sandwichlike appearance (fig. 6.3), it was assumed by J. D. Robertson that the outer dark layer (stained with heavy metals) contained protein plus the hydrophilic heads of the phospholipids. The interior was simply the hydrophobic tails of these molecules. Robertson went on to suggest that all membranes in a cell have basically the same composition and this proposal was called the "unit membrane" model.

This model of membrane structure was accepted for at least ten years, even though investigators began to increasingly doubt its accuracy. For example, not all membranes have the same appearance in electron micrographs, and they certainly don't have the same function. The inner membrane of a mitochondrion is coated with rows of particles, and functions in cellular respiration; it therefore has a far different appearance and function from the plasma membrane. Finally, in 1972, S. Singer and G. Nicolson introduced the **fluid-mosaic model** of membrane structure, which proposes that the membrane is a bilayer of phospholipids, having the consistency of light oil, in which protein molecules are either partially or wholly embedded. The proteins are scattered throughout the membrane, forming a mosaic pattern (fig. 6.4). The fluid-mosaic model of membrane structure is especially supported by electron micrographs of freeze-fracture membranes.

The fluid-mosaic model of membrane structure is widely accepted at this time.

Figure 6.3

Membrane structure. *a.* The red blood cell plasma membrane typically has a three-layered appearance in electron micrographs; there are two outer dark layers and a central light layer. Magnification, ×415,000. *b.* In 1960, Robertson's unit membrane model was widely accepted. This model proposed that the outer dark layers in electron micrographs were made up of protein and the polar heads of phospholipid molecules, while the inner light layer was composed of the nonpolar tails. The Robertson model was challenged, particularly by the Singer and Nicolson fluid-mosaic model, which puts protein molecules within the lipid bilayer. *c.* A new technique, called freeze-fracture, allows an investigator to view the interior of the membrane. Cells are rapidly frozen in liquid nitrogen and then fractured with a knife. The fracture often splits the membrane in two, just in the middle of the lipid bilayer. Platinum and carbon are applied to the fractured surface to produce a faithful replica that is observed by electron microscopy *d.* The electron micrograph is not as smooth as it would be if the unit membrane model was correct. Instead, the micrograph shows the presence of particles as is expected by the fluid-mosaic model.

a. Electron micrograph of red blood plasma membrane

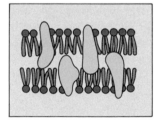

plasma membrane

b. Robertson's unit membrane

Singer and Nicolson fluid-mosaic model

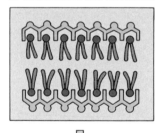

c. Freeze-fracture of membrane

Freeze-fracture of membrane

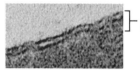

knife

protein

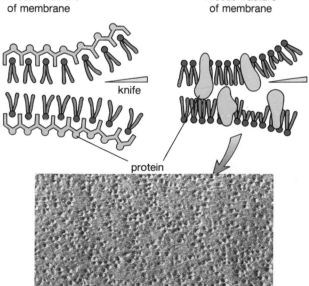

d. Electron micrograph of freeze-fractured membrane shows presence of particles

Figure 6.4

Fluid-mosaic model of the plasma membrane. The membrane is composed of a phospholipid bilayer with embedded proteins. The hydrophilic heads of the phospholipids are at the surfaces of the membrane and the hydrophobic tails make up the interior of the membrane. The carbohydrate chains of glycolipids and glycoproteins are involved in cell to cell recognition. Proteins also serve other functions such as receptors for chemical messengers, passage of molecules through the membrane, and enzymes in metabolic reactions.

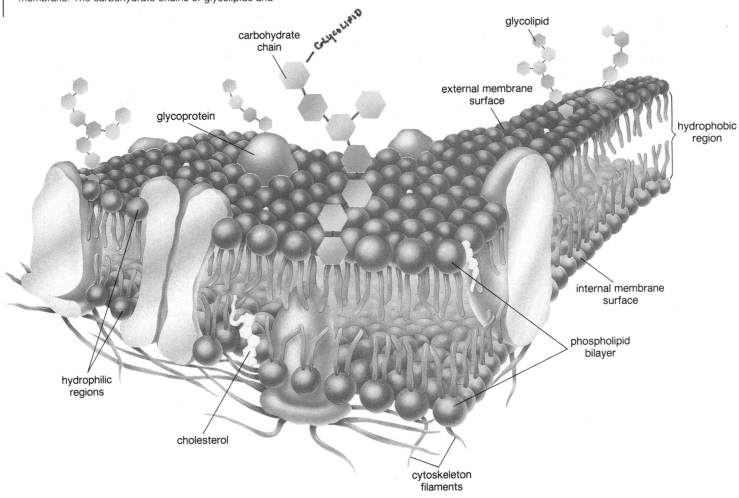

Our perception of the plasma membrane has changed over the years. There have been a series of models, each one developed to suit the evidence available at that time. This illustrates that scientific knowledge is always subject to change, and modifications are made whenever new data is presented. A model is useful because it pulls together the available data and suggests new avenues of research.

Fluid-Mosaic Model of the Plasma Membrane

The fluid-mosaic model of membrane structure (fig. 6.4) has two components, lipids and proteins. The lipids form the matrix of the membrane, and the proteins carry out all of its specific functions.

Lipid Component

Most of the lipids in the plasma membrane are **phospholipids** (fig. 6.2). In a laboratory container, phospholipids spontaneously form a bilayer so that the hydrophilic heads face water on both sides and the hydrophobic tails avoid contact with water.

So that the ends are not exposed to water, the phospholipid bilayer spontaneously forms a sphere—it becomes a barrier just like the plasma membrane:

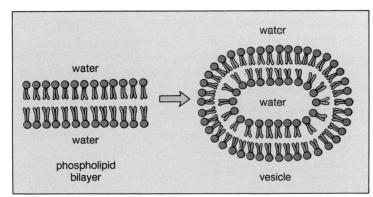

At body temperature, the phospholipid bilayer of the plasma membrane is a liquid; it has the consistency of olive oil. The higher the concentration of unsaturated fatty-acid residues, the more liquid the bilayer. In each monolayer, the hydrocarbon tails wiggle, and the entire molecule can move sideways, even exchanging places with its neighbor. (Phospholipid molecules rarely flip-flop from one layer to the other, because this would require the hydrophilic head to move through the hydrophobic center of the membrane.) The fluidity of a phospholipid bilayer means that cells are pliable. Imagine if they were not—the nerve cells in your neck would crack whenever you nodded your head!

In addition to phospholipids, two other kinds of lipids, **glycolipids** and **cholesterol,** are found in an animal cell plasma membrane. Glycolipids are like phospholipids except that the hydrophilic head is made up of a variety of simple sugars joined to form a straight or branching carbohydrate chain. As figure 6.4 shows, this chain is always directed externally, an observation offers a clue to the function of glycolipids. Investigations suggest that glycolipids might mark a cell as belonging to a particular individual, accounting for such characteristics as specific blood group and why a patient's system sometimes rejects an organ transplant. Glycolipids also regulate the action of plasma membrane proteins involved in the growth of the cell, and in this respect they may have a role in the occurrence of cancer.

In animal cells, cholesterol is a major membrane lipid, equal in amount to phospholipids. Cholesterol has both a hydrophilic and hydrophobic end, and arranges itself in this manner:

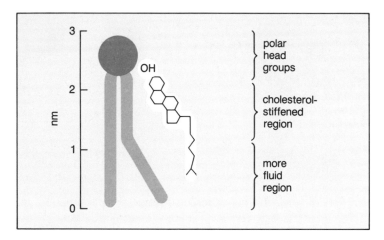

The presence of cholesterol makes the membrane more impermeable to most biological molecules. Small molecules, like amino acids and sugars, do not pass through the membrane easily, because they are polar and the center of the membrane is nonpolar. Passage of molecules through the membrane is dependent upon the protein component.

> The phospholipid bilayer portion of the plasma membrane forms a hydrophobic impermeable barrier that prevents the movement of polar (most biological) molecules through the plasma membrane.

Protein Component

The proteins associated with the plasma membrane are either attached to its inner surface or embedded in the lipid bilayer. Proteins often have hydrophobic and hydrophilic regions; hydrophobic regions occur within the membrane, while the hydrophilic regions project from both surfaces of the bilayer:

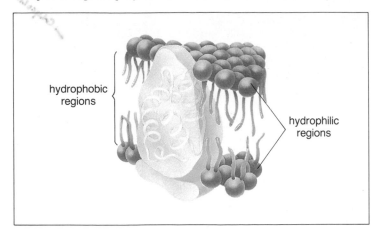

Although proteins are oriented so that a particular hydrophilic region always projects from either the exterior or interior surface, they can move laterally in the fluid lipid bilayer. This has been demonstrated by the experiment explained in figure 6.5.

Some of the proteins associated with the plasma membrane are glycoproteins; they have an attached carbohydrate chain that projects externally from the membrane. Glycoproteins, like glycolipids, make cell-to-cell recognition possible. Their carbohydrate chains are like fingerprints that allow a cell to know if it is in contact with another of the body's cells or a foreign cell. Other glycoproteins have a special configuration that allows them to act as receptors for a chemical messenger like a hormone. When a chemical messenger binds to the protein, a change occurs in its structure, and this initiates an enzymatic reaction within the cytoplasm. Other proteins are themselves enzymes that carry on metabolic reactions. The endomembrane system discussed in chapter 5 certainly contains many such enzymes. As we will discuss in this chapter, certain proteins are involved in the passage of molecules through the membrane. Some of these have a channel through which a polar substance can simply move across the membrane; others are carriers that combine with the substance and help it move across the membrane. Altogether we have mentioned these functions for plasma membrane proteins:

cell-to-cell recognition
receptors for chemical messengers
enzymes for cell reactions
passage of molecules across the membrane

Any particular protein has only one of these functions.

> The movement of molecules through the plasma membrane is dependent upon its protein component. Proteins have other functions as well.

Figure 6.5

a. Fluorescent antibodies are prepared against mouse cell glycoproteins and human cell glycoproteins by the method described. *b.* (1) Mouse and human cells are exposed to their respective antibodies. (2) Proper laboratory treatment causes the cells to fuse. Immediately after fusion, microscopic observation shows that the mouse and human glycoproteins are still separated from one another. (3) Forty minutes after fusion, the glycoproteins are completely intermixed. This shows that the proteins are able to move within the plasma membrane.

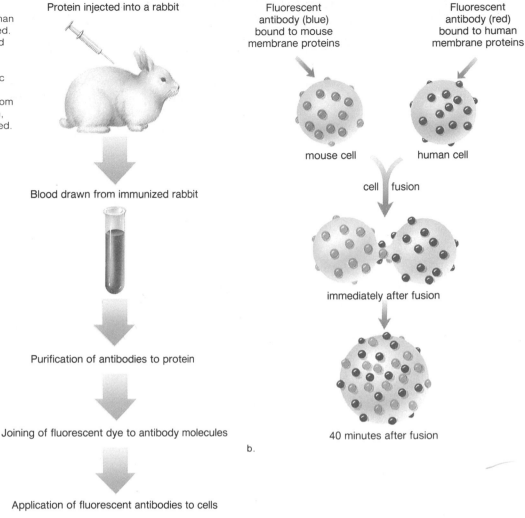

Protein injected into a rabbit

Blood drawn from immunized rabbit

Purification of antibodies to protein

Joining of fluorescent dye to antibody molecules

Application of fluorescent antibodies to cells

a.

Fluorescent antibody (blue) bound to mouse membrane proteins

Fluorescent antibody (red) bound to human membrane proteins

mouse cell

human cell

cell fusion

immediately after fusion

40 minutes after fusion

b.

Movement of Molecules into and out of Cells

If the plasma membrane were completely permeable, all substances would pass through it with no difficulty. However, it is observed that the plasma membrane is semipermeable; some substances are able to pass through and some are not. Proteins are often involved in allowing substances to pass through the membrane. Also, the process of endocytosis and exocytosis brings large molecules into the cell and expels large molecules from the cell. Because the plasma membrane is largely able to regulate which substances will enter and exit the cell, it is said to be **selectively permeable.**

The plasma membrane is selectively permeable—it has special mechanisms to regulate the passage of most molecules into and out of the cell.

Table 6.1 summarizes the various ways that molecules pass into and out of cells.

Diffusion

Ions and molecules in air and water constantly move about, seemingly at random. However, it can be demonstrated that there is a net movement of molecules from a region of higher concentration to a region of lower concentration. For example, when a few crystals of dye are placed in water (fig. 6.6), the dye molecules move in various directions, but their net movement is in a direction away from the concentrated source. The net movement of the water molecules is in the opposite direction. Eventually, the dye is evenly dissolved in the water, resulting in a colored solution. The dye and water molecules continue to move about, but there is no net movement in any one direction.

Diffusion is the movement of molecules from the area of greater concentration to the area of lesser concentration. Diffusion occurs spontaneously, and no energy is required to bring it about. However, diffusion does require a difference in concentrations. A few substances freely diffuse through cell membranes. For example, the respiratory gases, oxygen and carbon dioxide, diffuse into and out of cells. After you have taken in a breath of air, there is a higher concentration of oxygen in the alveoli of your lungs than there is in the blood capillaries surrounding the lungs (fig. 6.7). Oxygen will diffuse down this *concentration gradient* from the lungs into the blood. Water is another molecule that simply diffuses across cell membranes and this has important biological consequences, which we will examine next.

Only a few types of small molecules freely diffuse through the plasma membrane from the area of higher concentration to the area of lower concentration.

Table 6.1 Passage of Molecules into and out of Cells

Name	Direction	Requirements	Examples
Diffusion	Toward lesser concentration	—	Lipid-soluble molecules Water Gases
Transport			
Facilitated	Toward lesser concentration	Carrier	Sugars and amino acids
Active	Toward greater concentration	Carrier plus energy	Sugars, amino acids, and ions
Endocytosis			
Phago-cytosis	Toward inside	Vesicle formation	Cells and subcellular material
Pinocytosis	Toward inside	Vesicle formation	Macromolecules
Receptor-mediated	Toward inside	Vesicle formation	Macromolecules
Exocytosis	Toward outside	Vesicle fuses with plasma membrane	—

Osmosis

The diffusion of water across a selectively permeable membrane has been given a special term: it is called **osmosis.** To illustrate osmosis, a thistle tube covered at one end by a selectively permeable membrane is placed in a beaker of distilled water. Inside the tube is a solution containing both a **solute** (e.g., sugar molecules) and a **solvent** (water) (fig. 6.8). Obviously, the lower solute and higher water concentration is outside the tube, so there will be a net movement of water from outside the tube to inside through the membrane. The solute (sugar molecules) is unable to pass out of the tube because the membrane is impermeable to it; therefore, the level of the solution within the tube rises (fig. 6.8b). The eventual height of the solution in the tube indicates the degree of **osmotic pressure** caused by the flow of water from the area of higher water concentration to the area of lower water concentration.

Notice that in this illustration of osmosis

1. a selectively permeable membrane separates a solution from pure water;
2. a difference in solute and water concentrations exists on the two sides of the membrane;
3. the membrane is impermeable (does not permit passage) to the solute particles;
4. the membrane is permeable (permits passage) to the water, and it moves from the area of higher water (lower solute) concentration to the area of lower water (higher solute) concentration;
5. an osmotic pressure is present; the amount of liquid increases on the side of the membrane with the higher solute concentration.

These considerations will be important as we discuss osmosis in relation to cells placed in different solutions. The plasma membrane is not completely impermeable to solutes such as sugars and salts, but the difference in permeability between water and these solutes is so great that cells do have to cope with the osmotic movement of water.

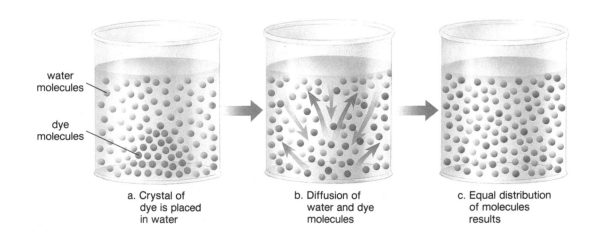

Figure 6.6

Process of diffusion is spontaneous and no energy is required to bring it about.
a. When dye crystals are placed in water they are concentrated in one area.
b. The dye dissolves in the water, and there is a net movement of dye molecules away from the area of concentration. There is a net movement of water molecules in the opposite direction.
c. Eventually, the water and dye molecules are equally distributed throughout the container.

water molecules

dye molecules

a. Crystal of dye is placed in water

b. Diffusion of water and dye molecules

c. Equal distribution of molecules results

Osmosis is the diffusion of water across a selectively permeable membrane. The presence of osmotic pressure is evident when there is an increased amount of water on the side of the membrane that has the higher solute concentration.

Tonicity   Cells can be placed in solutions that are either isotonic, hypotonic, or hypertonic (table 6.2 and figure 6.9). In an **isotonic** solution, a cell neither gains nor loses water because the concentration of solute and water is the same on both sides of the plasma membrane. Most animal cells normally live under isotonic conditions. For example, the tissue fluid that surrounds your cells is normally isotonic to the cells. In a **hypotonic** solution, a cell tends to gain water because the concentration of solute is lower (the prefix hypo means "less than") and the concentration of water is higher than that of the cell. Plants and

algae that live in freshwater ponds live under hypotonic conditions. The cytoplasm and central vacuoles gain water, and the plasma membrane pushes against the rigid cell wall. The resulting pressure, called **turgor pressure,** helps give internal support to the cell. Even plants that live on land are dependent upon turgor pressure to keep them from wilting. Freshwater protozoans are not protected from osmotic swelling by a rigid cell wall, but they have a contractile vacuole that rapidly pumps out excess water (fig. 6.10). Similarly, some organisms, such as marine fishes, live in a **hypertonic** environment in which there is a higher concentration of solute (the prefix hyper means "greater than") and a lower concentration of water than the cells that make up their bodies. Without any protective mechanism, the cells would tend to lose water and the fishes would die from dehydration. However, the gills of marine fish extrude salt from the blood and it remains isotonic to the cells. Our examples illustrate that organisms have quite often evolved mechanisms to cope with the tonicity of their external environment.

When a cell is placed in an isotonic solution, it neither gains nor loses water. When a cell is placed in a hypotonic solution (less solute concentration than isotonic) the cell gains water. When a cell is placed in a hypertonic solution (greater solute concentration than isotonic), the cell loses water and the cytoplasm shrinks.

Figure 6.7
Oxygen (dots) diffuses into the capillaries because there is a higher concentration of oxygen in the alveoli (air sacs) of the lungs than in the capillaries.

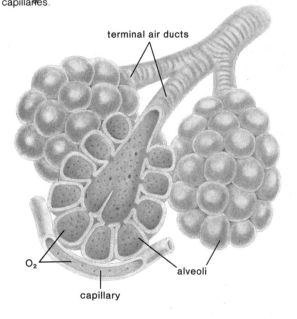

Table 6.2 Effect of Osmosis on Cell

Tonicity of Solution	Concentrations		Net Movement of Water	Effect on Cell
	Solute	Water		
Isotonic	Same as cell	Same as cell	None	None
Hypotonic	Less than cell	More than cell	Cell gains water	Swells, turgor pressure
Hypertonic	More than cell	Less than cell	Cell loses water	Shrinks, plasmolysis

Figure 6.8
Osmosis demonstration. *a.* A thistle tube covered at the broad end by a selectively permeable membrane contains a solute and a solvent. The beaker contains only solvent. *b.* The solute is unable to pass through the membrane, but the solvent passes through in both directions. *c.* There is a net movement of solvent toward the inside of the thistle tube. This causes the solution to rise in the thistle tube until osmotic pressure prevents any further entry.

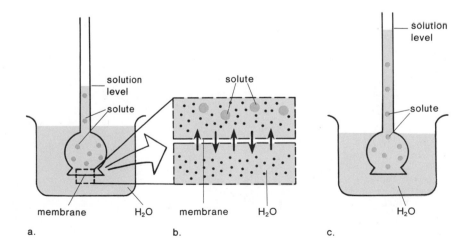

Figure 6.9

Osmosis in plant and animal cells. The arrows indicate the net movement of water. In an isotonic solution, a cell neither gains nor loses water; in a hypotonic solution, a cell gains water; and in a hypertonic solution, a cell loses water.

Plant Cells

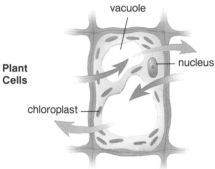
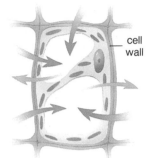
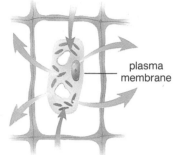

vacuole

nucleus

chloroplast

cell wall

plasma membrane

a. Under isotonic conditions, there is no net movement of water.

b. In a hypotonic environment, vacuoles fill with water, turgor pressure develops, and chloroplasts are seen next to cell wall.

c. In a hypertonic environment, vacuoles lose water, cytoplasm shrinks (called plasmolysis), and chloroplasts are seen in center of cell.

Animal Cells

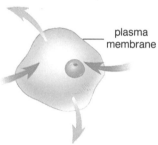

plasma membrane

d. Under isotonic conditions, there is no net movement of water.

e. In a hypotonic environment, water enters cell, which may burst (lyse) due to osmotic pressure.

f. In a hypertonic environment, water leaves the cell, which shrivels (crenation occurs).

Figure 6.10

Many organisms have evolved mechanisms to cope with the tonicity of the surrounding medium. A paramecium (a) lives in fresh water, which is hypotonic to it. It has contractile vacuoles that expel water from the cell. Magnification, ×107. b. Water from the cytoplasm flows into tubules, filling the vacuole. c. The vacuole collapses, appearing to "contract," as water is released through a temporary opening in the plasma membrane.

contractile vacuole

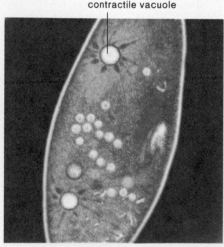
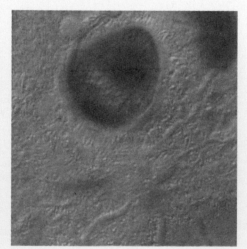

a.

b.

c.

Transport by Carriers

The presence of the plasma membrane impedes the passage of all but a few substances. Yet biologically useful molecules do enter and exit the cell at a rapid rate because there is a transport system in the membrane that delivers them in a timely manner. The proteins that span the plasma membrane act as **carriers** to transport these molecules. Carrier proteins are specific and each can combine with only a certain type of molecule, which is then transported through the membrane. It is not completely understood how carrier proteins function, but after they combine with a substance, they are believed to undergo a *conformational change,* a change in shape that causes the substance to be moved across the membrane.

Some of the proteins in the plasma membrane are carriers that transport biologically useful molecules into and out of the cell.

Facilitated Transport

A model for facilitated transport (fig. 6.11) suggests that after a carrier has assisted the movement of a molecule to one side of the membrane, it is free to assist the passage of another similar molecule. This form of transport is involved in carrying various sugars, amino acids, and nucleotides into or out of a cell according to concentration gradients. Because the molecule is still traveling down its concentration gradient, no significant energy input is required.

Active Transport

Due to **active transport,** molecules or ions accumulate either inside or outside the cell even when this is the region of higher concentration. For example, the plasma membranes of plant root cells are able to extract inorganic ions from the soil due to active transport. In marine fishes we have already mentioned that the cells in the gills are able to extrude salt so that the blood remains isotonic to body cells. In our bodies the thyroid gland accumulates iodine and the kidneys actively reabsorb sodium (Na^+) so that water follows by osmosis. If this did not happen, we would constantly be in danger of dehydration because of water loss by way of the kidneys. As described in the reading (p. 88), the chambered nautilus uses a similar mechanism to achieve buoyancy and stay afloat.

Both a protein carrier and an expenditure of energy are needed for active transport. The energy utilized comes from the breakdown of ATP (adenosine triphosphate). Therefore, it is not surprising that cells primarily involved in active transport, such as kidney cells, have a large number of mitochondria near the membrane where active transport is occurring. Proteins involved in active transport are often called *"pumps"* because just as a water pump uses energy to move water against the force of gravity, proteins use energy to move a substance against its concentration gradient.

One type of pump that is active in all cells, but especially associated with nerve and muscle cells, is the one that moves sodium ions (Na^+) to the outside of the cell and potassium ions

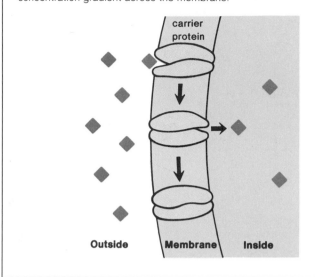

Figure 6.11

Facilitated transport. A carrier protein speeds the rate at which the solute crosses a membrane in the direction of decreasing concentration. Facilitated transport occurs only when there is a concentration gradient across the membrane.

(K^+) to the inside of the cell. These two events are presumed to be linked, and the carrier protein is called a **sodium-potassium pump.** ATP energy released by an ATPase (an enzyme that breaks down ATP) is believed to be necessary to bring about a change in shape of the carrier allowing it to alternately combine with Na^+ and K^+ (fig. 6.12).

Sodium and potassium ions are charged particles; the action of the pump causes the membrane to be positively charged on the outside and negatively charged on the inside. Therefore, the membrane has an electrical gradient; this along with any concentration gradient influences the movement of charged particles. Charged particles tend to move in the direction of their *electrochemical gradient.*

The sodium gradient, established by the sodium-potassium pump, provides the energy that allows an animal cell to take up other types of molecules, such as sugars and amino acids. For example, there are carriers in the plasma membrane of animal cells that combine with both Na^+ and an amino acid. Because there is so much more Na^+ outside the cell than inside the cell, the carrier automatically transports Na^+, and consequently the amino acid as well even if there is already a higher concentration of amino acid inside the cell than outside. Since the sodium-potassium pump establishes a gradient that is utilized to allow the uptake of nutrient molecules, we can understand why it apparently requires one-third of all our energy expenditure!

During facilitated transport (no energy required), small molecules follow their concentration gradients. During active transport (energy required), small molecules go against their concentration gradients.

Figure 6.12

The sodium-potassium pump moves sodium (Na^+) to outside the cell and potassium (K^+) to inside the cell by means of the same type of protein carrier. This establishes not only a concentration gradient for each of these ions, but also an electrical gradient. The inside of a cell is negatively charged while the outside is positively charged.

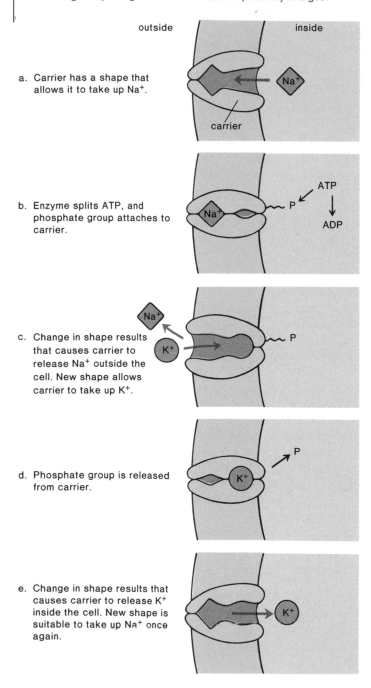

a. Carrier has a shape that allows it to take up Na^+.

b. Enzyme splits ATP, and phosphate group attaches to carrier.

c. Change in shape results that causes carrier to release Na^+ outside the cell. New shape allows carrier to take up K^+.

d. Phosphate group is released from carrier.

e. Change in shape results that causes carrier to release K^+ inside the cell. New shape is suitable to take up Na^+ once again.

Endocytosis and Exocytosis

Some molecules are too large to be transported by protein carriers; instead they are taken into the cell by vesicle formation:

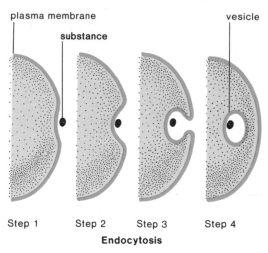

Step 1 Step 2 Step 3 Step 4

Endocytosis

This process, called **endocytosis,** was briefly described in chapter 5. When the material taken in by endocytosis is quite large, the process is called **phagocytosis** (cell eating). Phagocytosis is common to amoeboidtype cells, such as macrophages, which are large phagocytic cells found in the blood of vertebrates. These cells phagocytize bacteria and worn-out red blood cells (fig. 6.13).

Vesicles may also form around large-sized molecules such as proteins; this is called **pinocytosis** (cell drinking). Once formed, these vesicles contain a substance enclosed by membrane. Both phagocytic and pinocytic vesicles can fuse with lysosomes whose enzymes will digest their contents.

The opposite of endocytosis is **exocytosis.** During exocytosis, a vesicle fuses with the membrane, discharging its contents:

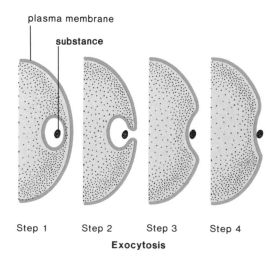

Step 1 Step 2 Step 3 Step 4

Exocytosis

As we noted in chapter 5, vesicles formed at the Golgi complex secrete cell products in this manner.

Figure 6.13

Macrophages are large cells that act as scavengers. These "big eaters" take in (phagocytize) all sorts of debris, including worn out red blood cells, as shown here.

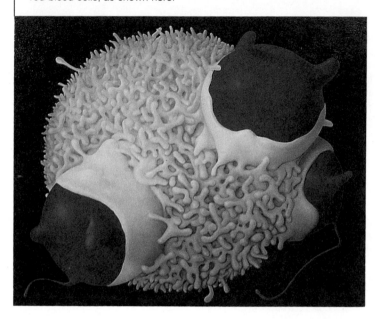

Endocytosis and exocytosis are another way that substances can enter and exit from cells.

It's to be expected that the amount of plasma membrane lost during the process of endocytosis will be replaced by the process of exocytosis, allowing the amount of plasma membrane to remain constant. Investigators have found this to be the case in a process called the receptor-mediated endocytic cycle.

Receptor-Mediated Endocytic Cycle

We have already mentioned that some of the proteins in the plasma membrane are receptors that have the ability to combine with specific substances. During **receptor-mediated endocytosis,** a receptor binds with a specific nutrient molecule, called a ligand, and then these receptors gather at a location where endocytosis is beginning. This location is called a coated pit because there is a layer of fibrous protein, called clathrin, on the cytoplasmic side. When a vesicle forms it is coated, but soon it loses its coat. Eventually, the nutrients enter the cytoplasm from the vesicle, which then fuses with the plasma membrane as exocytosis occurs (fig. 6.14).

Figure 6.14

Receptor-mediated endocytic cycle takes up membrane when endocytosis occurs and returns it when exocytosis occurs. *a.* (1) The receptors in the coated pits combine only with a specific substance, called a ligand. (2) The vesicle that forms is at first coated but soon loses its coat. (3) Ligands leave the vesicle. (4) When exocytosis occurs membrane is returned to the plasma membrane. *b.* Electron micrographs of a coated pit in the process of forming a vesicle.

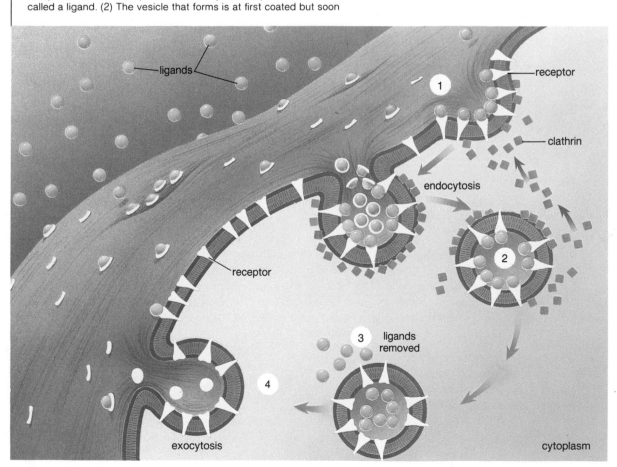
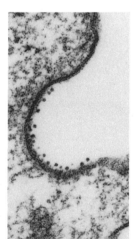
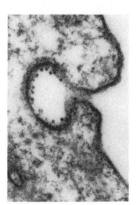

a.

b.

Membrane Structure and Function　**87**

Chambered Nautilus

T he chambered nautilus (*a*), a mollusk that takes its name from its multi-chambered shell, still occurs today in deep tropical oceans but was much more prevalent some 65 million years ago. Along with closely related forms, it dominated the oceans. The advantage these animals must have had was the ability to achieve buoyancy. Thus they were no longer confined to the ocean floor. Recently, scientists discovered that as the nautilus grows, it empties liquid from each newly formed chamber of its spiral shell. This makes it light enough to float in water.

Research has been conducted into the puzzle of how the nautilus empties its chambers. A complex tubular organ, the siphuncle, that contains blood vessels spirals in the same manner as the shell (*b* and *c*). Water from each newly formed chamber makes its way to these blood vessels, but by what means? The blood pressure within these vessels should cause water to *leave* the blood vessels, not enter them!

Electron micrographs have provided a possible explanation. They show that the wall of the siphuncle (*d*) has a folded tissue lining containing many small grooves (*e*). If the cells making up this tissue use active transport to carry solutes from the chamber liquid to the grooves, this would establish a concentration gradient so that water would follow passively by osmosis. This process is called *local osmosis* because it depends on the buildup of solutes by active transport in an isolated area, such as the grooves in question. The cells about these grooves do indeed contain many mitochondria that could supply the energy needed for active transport of the solute into the grooves (*f*) so that water could follow passively by means of osmosis. After being withdrawn, the water moves from the grooves of the tubular organ to collecting channels that funnel it to the blood vessels.

Although the nautilus can empty its chambers, it never attempts to fill them. This observation is consistent with the hypothesis that local osmosis accounts for how the chambered nautilus empties the chambers of its shell.

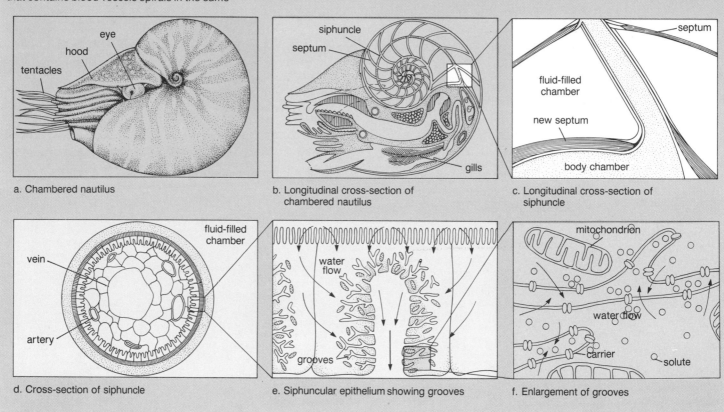

a. Chambered nautilus

b. Longitudinal cross-section of chambered nautilus

c. Longitudinal cross-section of siphuncle

d. Cross-section of siphuncle

e. Siphuncular epithelium showing grooves

f. Enlargement of grooves

Normally, cholesterol bound to a lipoprotein enters the cell by receptor-mediated endocytosis. When individuals lack the particular receptor needed for the process, cholesterol accumulates in the arteries where it contributes to plaque, a buildup of deposits that is implicated in the occurrence of hypertension and heart attack.

Movement   Some animal cells move even though they lack flagella. It has been calculated that in a fibroblast, a type of cell found within humans that contains fibers, there is enough receptor-mediated endocytosis occurring to allow this cell to recirculate its membrane every 50 minutes. A fibroblast can move, it creeps forward about a centimeter a day. Could endocytosis followed by exocytosis be involved? The answer to this question is being investigated, and the answer seems to be yes. Plasma membrane removed by endocytosis at the rear of the cell is added to membrane at the front of the cell, and this causes the cell to move forward.

Receptor-mediated endocytosis brings specific nutrients into the cell, and is followed by exocytosis. This leads to a circulation of the plasma membrane that may contribute to the movement of cells.

Modifications of Cell Surface

The plasma membrane is the outer living boundary of the cell proper, but many cells have an additional component that is formed outside of the membrane. For example, plant cells have a cell wall that is freely permeable and thus does not interfere with the various functions of the plasma membrane.

Figure 6.15

Plant cell wall. All plant cells have a primary cell wall and some have a secondary cell wall. The cell wall, which lends support to the cell, is freely permeable.

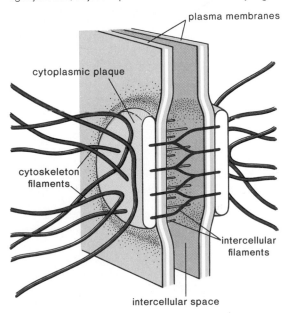

Plant Cell Wall

In addition to a plasma membrane, plant cells are surrounded by a **cell wall** (fig. 6.15) that varies in thickness, depending on the function of the cell. All plant cells have a primary cell wall whose main component is cellulose polymers united into thread-like microfibrils and even larger fibrils (fig. 4.9). Pectin, a sticky substance which occurs in a region called the middle lamella, keeps adjacent plant cells bound together. Some cells in woody plants have a secondary cell wall that forms inside the primary cell wall. The secondary wall has alternating layers of cellulose fibrils reinforced by the addition of lignin, a substance that adds strength. Such secondary walls are important in the structure of woody plants.

The plasma membrane in plants is surrounded by a cell wall, which helps support the cell and does not interfere with the functions of the plasma membrane.

Junctions between Cells

The plasma membrane of cells sometimes interacts with other cells. The junctions that occur between cells are an example of such cellular interaction.

Animal Cells

Three types of junctions are seen between animal cells: spot desmosome, tight junction, and gap junction (fig. 6.16). In a **spot**

Figure 6.16

Cell-to-cell junctions between animal cells. *a.* In a spot desmosome, filaments pass from one cell to the other at a region where plaques of dense material are reinforced by cytoskeleton filaments. *b.* In a tight junction, adjacent proteins are bonded directly together, preventing any passage of materials in the space between the cells. *c.* In gap junctions, donut-shaped proteins from each cell join to form tiny channels. These allow small molecules to pass from cell to cell. These junctions span the "gap" between cells.

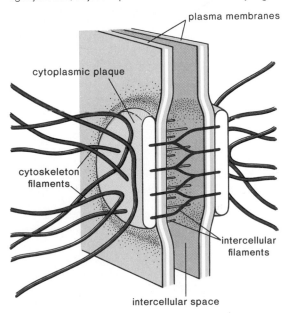

a. Spot desmosome

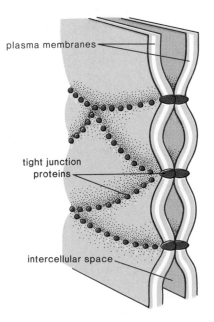

b. Tight junction

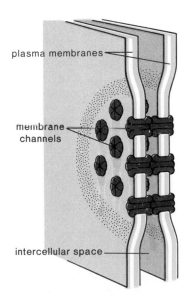

c. Gap junction

desmosome, internal cytoplasmic plaques, firmly attached to the cytoskeleton within each cell, are joined by intercellular filaments. Plasma membrane of adjacent cells is even more closely joined in a **tight junction,** where plasma membrane proteins actually attach to each other, producing a zipperlike fastening. These two junctions lend strength and support to the cells. In addition, there is another type of junction that allows cells to communicate. A **gap junction** is formed when two identical plasma membrane channels join. The channel of each cell is lined by six plasma membrane proteins. A gap junction lends strength to the cells, but it also allows small molecules and ions to pass between them.

All three of these junctions are observed between cells that line the lumen (cavity) of the small intestine. These cells pump the products of digestion out of the intestinal lumen and then deposit them in the blood. The presence of tight junctions, in particular, means that the digestion products must enter the cells. Tight junctions also restrict the lateral movement of proteins in the plasma membrane, allowing the surface facing the lumen of the intestine to maintain a different mix of proteins than does the surface adjacent to the underlying blood vessel. These surfaces carry on different functions depending on the proteins they contain.

In animal cells the plasma membrane is joined by desmosomes and tight junctions. Gap junctions allow small molecules to pass from one cell to the other.

Plant Cells

In plant tissues, the cytoplasm of neighboring cells can be connected by numerous narrow channels that pass through the cell wall. These channels are bounded by plasma membrane and contain cytoplasmic strands, called **plasmodesmata,** plus a tightly constricted tubular portion of endoplasmic reticulum. Plasmodesmata allow direct exchange of materials between neighboring plant tissue cells (fig. 6.17).

In plant cells, plasmodesmata are strands of cytoplasm that allow small molecules to pass from one cell to the other.

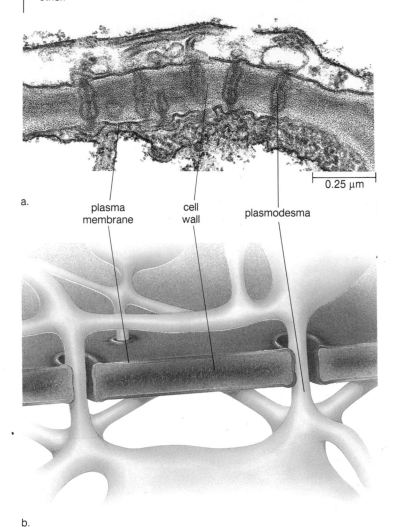

Figure 6.17

Cell-to-cell junction between plant cells. At the plasmodesmata, the plasma membrane of adjacent cells passes through the cell wall to form channels, which allow substances to pass from one cell to the other.

0.25 μm

a.

plasma membrane   cell wall   plasmodesma

b.

Summary

1. The fluid-mosaic model of membrane structure developed by Singer and Nicolson was preceded by several other models. Electron micrographs of freeze-fracture membranes support the fluid-mosaic model and not Robertson's unit membrane concept based on the Danielli and Davson's sandwich model.

2. There are two components of the membrane, lipids and proteins. In the lipid bilayer, phospholipids are arranged with their polar heads at the surfaces and their hydrophobic tails in the interior. The lipid bilayer is fluid, but acts as a barrier to the entrance and exit of most biological molecules. Glycolipids and glycoproteins are involved in marking the cell as belonging to a particular individual.

3. The hydrophobic portion of a protein lies in the lipid bilayer of the plasma membrane, and the hydrophilic portion occurs at the surfaces. Proteins act as receptors, carry on enzymatic reactions, join cells together, form channels or act as carriers to move substances across the membrane.

4. Some molecules (lipid-soluble compounds, water, gases) simply diffuse across the membrane from the area of higher concentration to the area of lower concentration. No energy is required for diffusion to occur.

5. The diffusion of water across a selectively permeable membrane is called osmosis. Water moves across the membrane into the area of lower water (higher solute) content. When cells are in an isotonic solution, they neither gain nor lose water; when they are in a hypotonic solution, they gain water; and when they are in a hypertonic solution, they lose water.

6. Other molecules are transported across the membrane by carrier proteins that span the membrane.
7. During facilitated transport, a protein carrier assists the movement of a molecule down its concentration gradient. No energy is required.
8. During active transport, a protein carrier acts as a pump that allows a substance to move against its concentration gradient. The sodium-potassium pump carries Na^+ to the outside of the cell and K^+ to the inside of the cell. Energy in the form of ATP molecules is required for active transport to occur.

9. Larger substances can enter and exit from a membrane by endocytosis and exocytosis. Endocytosis includes both phagocytosis and pinocytosis. Exocytosis involves secretion.
10. The receptor-mediated endocytic cycle makes use of receptor molecules in the membrane. Once specific substances (e.g., nutrients) bind to their receptors, the entire molecule is drawn into a coated pit, which becomes a coated vesicle. After losing the coat, and after the passage of the nutrients into the cytoplasm, the receptor-containing vesicle fuses with the plasma membrane.

11. Plant cells have a freely permeable cell wall whose main component is cellulose.
12. Junctions between animal cells include spot desmosomes and tight junctions, which help to hold cells together, and gap junctions, which allow passage of small molecules between cells. Plant cells are joined by small channels that span the cell wall and contain plasmodesmata, strands of cytoplasm, which allow materials to pass from one cell to another.

Objective Questions

1. Electron micrographs following freeze-fracture of the plasma membrane indicate that
   a. the membrane is a phospholipid bilayer.
   b. some proteins span the membrane.
   c. protein is found only on the surfaces of the membrane.
   d. glycolipids and glycoproteins are antigenic.
2. A phospholipid molecule has a head and two tails. The tails are found
   a. at the surfaces of the membrane.
   b. in the interior of the membrane.
   c. both at the surfaces and interior of the membrane.
   d. spanning the membrane.
3. Energy is required for
   a. active transport.
   b. diffusion.
   c. facilitated transport.
   d. All of these.
4. When a cell is placed in a hypotonic solution,
   a. solute exits the cell to equalize the concentration on both sides of the membrane.

   b. water exits the cell toward the area of lower concentration.
   c. water enters the cell toward the area of higher solute concentration.
   d. solute exits and water enters the cell.
5. A protozoan's contractile vacuole is more likely to be active
   a. in a hypertonic environment.
   b. in an isotonic environment.
   c. in a hypotonic environment.
   d. when endocytosis is occurring.
6. Active transport
   a. requires a protein carrier.
   b. moves a molecule against its concentration gradient.
   c. requires a supply of energy.
   d. All of these.
7. Glycolipids
   a. are receptors just like glycoproteins.
   b. have a chain of fatty acids attached at the surface of the membrane.
   c. help distinguish one type cell from another type.
   d. have a structure like cholesterol.

8. The sodium-potassium pump
   a. helps establish an electrochemical gradient across the membrane.
   b. concentrates sodium on the outside of the membrane.
   c. utilizes a protein carrier and energy.
   d. All of these.
9. Receptor-mediated endocytosis
   a. is no different from phagocytosis.
   b. brings specific nutrients into the cell.
   c. helps to concentrate proteins in vesicles.
   d. All of these.
10. Plant cells
    a. always have a secondary cell wall, the primary one may disappear.
    b. have channels between cells that allow strands of cytoplasm to pass from cell to cell.
    c. develop turgor pressure when water enters the nucleus.
    d. do not have cell-to-cell junctions like animal cells do.

Study Questions

1. Describe the fluid-mosaic model of membrane structure and the models that preceded it. Cite the evidence that either disproves or supports these models.
2. Tell how the phospholipids are arranged in the plasma membrane. What other lipids are present in the membrane and what functions do they have?

3. Describe how proteins are arranged in the plasma membrane; what are their functions? Describe an experiment that indicates that proteins can move laterally in the membrane.
4. What is diffusion, and what substances can diffuse through a semipermeable membrane?
5. Describe an experiment that measures osmotic pressure.

6. Tell what happens to an animal cell and a plant cell when placed in isotonic, hypotonic, and hypertonic solution.
7. Why do substances have to be assisted through the plasma membrane? Contrast movement by facilitated transport to movement by active transport.

8. Describe how the sodium-potassium pump works, especially mentioning the change in carrier shape that occurs during transport.

9. Contrast endocytosis with exocytosis, and phagocytosis with pinocytosis. What is the receptor-mediated endocytic cycle? How might it be involved in cell movement?

10. Describe the structure of the plant cell wall.

11. What are the three types of cell-to-cell junctions in animal cells? Contrast their structures and functions.

12. Describe the channels through which plasmodesmata connect plant cells.

Thought Questions

1. In the previous chapter we mentioned that microtubules and microfilaments are involved in cell movement. What role might these elements play in receptor-mediated endocytosis?

2. If you are a salt addict, does your blood tend to exert a high or low osmotic pressure? How does this affect blood pressure?

3. What is the significance of using the term "model" when describing membrane structure?

Selected Key Terms

fluid-mosaic model (floo'id mo-za'ik mod'el) 78

phospholipid (fos''fo-lip'id) 79

cholesterol (ko-les'ter-ol) 80

selectively permeable (sĕ-lek'tiv-le per'me-ah-b'l) 81

diffusion (dĭ-fu'zhun) 81

osmosis (oz-mo'sis) 82

osmotic pressure (os-mot'ik presh'ur) 82

isotonic (i''so-ton'ik) 83

hypotonic (hi''po-ton'ik) 83

turgor pressure (tur'gor presh'ur) 83

hypertonic (hi''per-ton'ik) 83

active transport (ak'tiv trans'port) 85

sodium-potassium pump (so'de-um po-tas'e-um pump) 85

endocytosis (en''do-si-to'sis) 86

pinocytosis (pi''no-si-to'sis) 86

exocytosis (ex''o-si-to'sis) 86

CHAPTER 7

Cellular Energy

Your study of this chapter will be complete when you can:

1. *State two energy laws that have important consequences for living things.*
2. *Explain, on the basis of these laws, why living things need an outside source of energy.*
3. *Describe two types of metabolic reactions (pathways) that are found in all cells.*
4. *Describe the structure and function of enzymes and of the factors that affect the yield of enzymatic reactions.*
5. *Describe the regulation of enzymatic reactions (pathways) by the process of inhibition. Contrast competitive inhibition, noncompetitive inhibition, and feedback inhibition.*
6. *Describe the function of the coenzymes NAD$^+$ and FAD, and state their role in the production of ATP by the electron transport system in mitochondria. Contrast this electron transport system to the one in chloroplasts.*
7. *Describe the function of the coenzyme NADP$^+$.*
8. *Explain, with the aid of a diagram, how degradative pathways drive synthetic pathways, emphasizing the role of ATP and NADP$^+$. Tell why an outside source of matter and energy should be included in the diagram.*
9. *Describe the structure and function of ATP and how it is produced by chemiosmotic phosphorylation.*

Organisms use energy in all sorts of ways; for example, most expend a lot of energy on reproduction. A flower uses energy to produce nectar. Insects help bring about cross-fertilization by unintentionally taking pollen from flower to flower.

We can readily see that a butterfly's wing (fig. 7.1) is highly organized but it takes a knowledge of cells to realize that they are also organized. All cells contain the same type of carbon-based molecules and in eukaryotic cells some of these molecules make up the organelles. It is the organelles (fig. 5.4) that show us that a cell is just as organized in its own way as the butterfly's wing. Such organization cannot be maintained without a supply of energy. As an analogy, consider that your room quickly becomes a total mess unless you keep doing work to keep it straight (organized). Similarly, a cell must continually use energy to first form macromolecules from unit molecules, and then arrange the macromolecules into its structural components.

Nature of Energy

Energy is defined as the capacity to do work, that is, to bring about a change. There are several different forms of energy. For example, you use your own mechanical energy to straighten up your room; light energy allows you to see what you are doing; electrical energy makes the motor turn in your vacuum cleaner; and heat energy maintains your room at a comfortable temperature.

Another form of energy, not as obvious as these, is chemical bond energy, the energy that is stored in the bonds that hold atoms together. For example, the chemical bond energy in food provides the energy for your muscles to do mechanical work. Just as chemical bond energy in food can be transformed into the mechanical energy used by your muscles, other forms of energy can be transformed into one another. For example, electrical energy is transformed into light energy within a light bulb, and oil is transformed into electrical energy at a power plant.

A power plant does not create energy, it merely transforms it. Do transformations ever cause energy to be used up? We realize that matter is never created nor destroyed. When we take materials and build a vacuum cleaner, we know that even if we throw the vacuum cleaner away, the materials are still not used up. On page 31 it was mentioned that chemical equations must always be balanced—the same number of atoms are present before and after a reaction—because mass cannot be used up. But what about energy, can it ever be used up?

First Energy Law

As you probably have guessed, not only can energy never be created, it also can never be used up. The same amount of energy has always been present in the universe and will always be present. The first energy law states:

Energy can be converted from one form into another but it cannot be created nor destroyed.

We have what appears to be a paradox, however, because it often seems as if energy has been "lost" when transformations occur. For example, cars need to be refueled and animals have to keep on eating. The answer to the paradox is that neither automobiles nor organisms are isolated. It is said that they are a *system* that has *surroundings*. It is the surroundings that supplies the gasoline (or the food) and when that is used to do work it results in heat that is accepted by the surroundings.

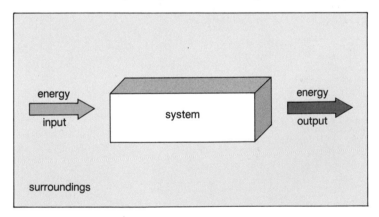

The first energy law can be interpreted to mean that if we put X amount of energy into a system (automobile, organism), then eventually X amount of energy will leave the system even if work has been done.

Second Energy Law

Although energy was not lost in our examples, something seems to have been lost. To determine what it could be, consider that gasoline can be burned to power a car but the heat given off is

Figure 7.1
The complex organization of living things is dramatized by the pattern of color in a butterfly's wing. But the unseen cells that make up the wing are also highly organized, and the nectar the butterfly obtains from the flower is needed to maintain the organization of both the cells and the wing.

no longer able to perform this task. So, the amount of *useful* energy has changed. Similarly, once food energy is used to power muscles, this energy is no longer available because it has been converted to heat. The second energy law states that:

When one form of energy is transformed into another form, some useful energy will always be lost as heat and therefore energy cannot be recycled.

The second energy law implies that only those processes that decrease the amount of useful energy will occur naturally or spontaneously. The word spontaneous does not mean that it has to happen all at once, it simply means that it will happen naturally over time. For example, your room tends to become messy, not neat; water tends to flow downhill, not up; and after death, organisms decay and eventually disintegrate. In other words, the amount of disorder (i.e., entropy) is always increasing in the universe. If this is the case, how do we account for an orderly room or the presence of highly organized structures, such as living cells and organisms? Obviously, these systems must have a continual input of energy and this continual input of energy will eventually increase the entropy of the universe.

While we have been drawing an analogy between non-living systems (a car, your room) and a living system (cell or organism) there is a remarkable difference between the two types of systems. Although living things are continually taking in useful energy and putting out heat just like the nonliving system,

the work performed in between these two events maintains their organization and even allows them to grow. A living organism represents stored energy in the form of chemical compounds:

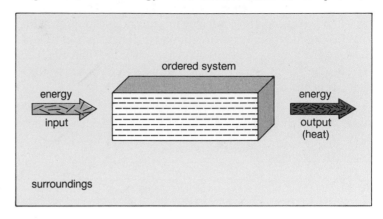

Implication for Living Things

Life does not violate the energy laws discussed earlier, since each living thing merely represents a temporary storage area for useful energy. The useful energy stored in one organism can be used by another to maintain its own organization and because of this, energy flows through a community of organisms (fig. 7.2). As transformations of energy occur, useful energy is lost to the environment in the form of heat until finally it has been completely used up. Since energy cannot recycle there is a need

Figure 7.2
A flow diagram illustrating the loss of useful energy in a community of organisms. Less than 2% of the solar energy reaching the earth is taken up by plants and this energy allows them to make their own food. Herbivores obtain their food by eating plants and carnivores obtain food by eating other animals. Whenever the energy content of food is used by organisms it is eventually converted to heat. With death and decay all the energy temporarily stored in organisms returns as heat to the atmosphere.

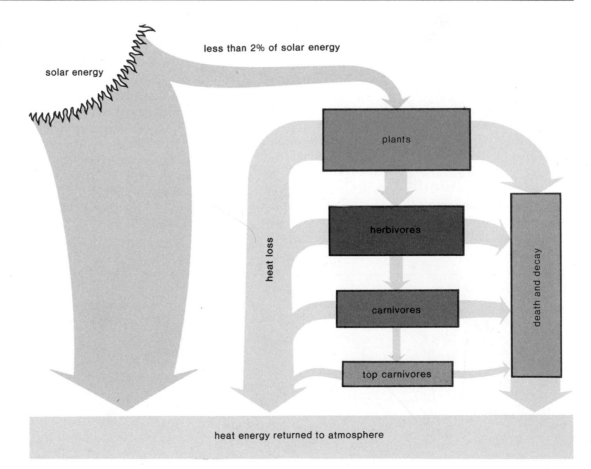

for an ultimate source of energy. This source, which continually supplies all living things with energy, is the sun. The entire universe is tending toward disorder, but in the meantime, solar energy is sustaining all living things. Photosynthesizing organisms like plants and algae are able to capture a small proportion of the solar energy that reaches the earth. They use this energy to make their own organic food, which then becomes the food for all other types of living things also. Without photosynthesizing organisms, life as we know it could not exist.

Living things continually lose useful energy to the environment, but a new supply comes to them in the form of organic food produced by photosynthesizing organisms.

Energy-Balance Sheet

Only 42% of solar energy gets to the earth's surface: the rest is absorbed by or reflected into the atmosphere and becomes heat.

Of this usable portion, only 2% is eventually utilized by plants: the rest becomes heat.

Of this, only 0.1% to 1.6% ever is incorporated into plant material: the rest becomes heat.[1]

Of this, only 20% is eaten by herbivores: a large proportion of the remainder becomes heat.[1]

Of this, only 30% is ever eaten by carnivores: a large proportion becomes heat.[1]

Conclusion: Most of the available energy is never utilized by living things.

---

[1]Plant and animal remains became the fossil fuels that we burn today to provide energy.

Cellular Metabolism

The food an organism eats becomes the nutrient molecules of its cells. Within cells, these molecules take part in a vast array of chemical reactions, collectively termed the **metabolism** of the cell. So far, we have stressed that nutrient molecules can be used as energy sources. In order for this to occur, they have to be broken down and, in fact, they have to be oxidized. You'll recall that oxidation is the removal of an electron either alone or in the form of a hydrogen atom. Oxidative reactions are a *degradative* process that releases energy. In general we can represent them like this:

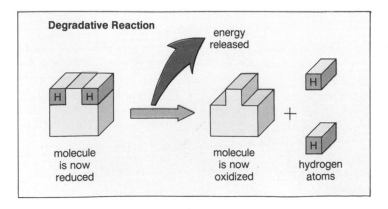

We have already indicated that a cell needs energy to maintain its structure and to grow. In other words, some of the nutrient molecules present in cells are used to form the structure of the cell. For example, amino acids are the unit molecules for proteins, and fatty acids are used to make the lipids found in membranes. There are also *synthetic* reactions in which hydrogen atoms are added to a molecule. These are reduction reactions that can be represented like this:

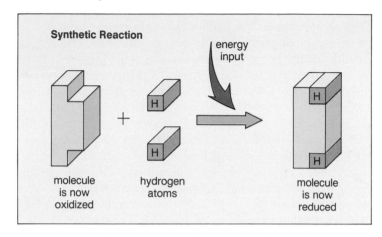

Do you suppose it would be possible to take the energy released by degradative reactions and use it for synthetic reactions? Further, would it be possible to use the hydrogen atoms produced in certain degradative reactions for certain synthetic reactions? The answer to these questions is *yes,* and we are now going to explore how these processes occur inside cells.

Pathways

At any one time there are thousands of reactions occurring in cells. Compartmentalization, which we discussed in detail in chapter 5, helps to segregate different types of reactions but more segregation is needed. Within the different organelles reactions are organized into **metabolic pathways.** Some of these pathways are largely degradative ones that release energy; for example, carbohydrates are broken down in mitochondria (p. 68). Other pathways are largely synthetic ones that require energy input; for example, lipids are synthesized in smooth endoplasmic reticulum.

In general, metabolic pathways can be represented by this diagram:

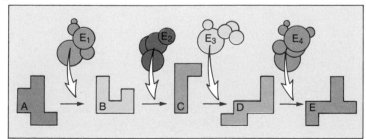

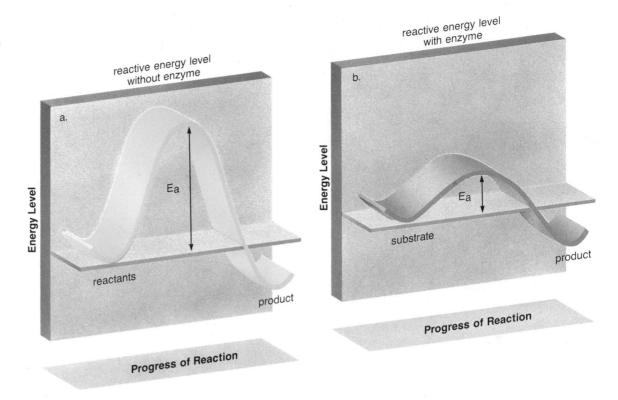

Figure 7.3
Enzymes speed up the rate of chemical reactions because they lower the required energy of activation. *a.* Energy of activation (E_a) that is required for a reaction to occur when an enzyme is not available. *b.* Required energy of activation (E_a) is much lower when an enzyme is available.

reactive energy level without enzyme

Energy Level

E_a

reactants

product

Progress of Reaction

reactive energy level with enzyme

b.

Energy Level

E_a

substrate

product

Progress of Reaction

In this pathway, the letters are products of the previous reaction and reactants of the next reaction. The numbers in the pathway identify different enzymes. The reactant(s) in an enzymatic reaction is called the **substrate(s)** for that enzyme. A is the beginning substrate and E is the end product of the pathway.

This simplified representation of a metabolic pathway does not indicate that the intermediate molecules (B–D) can also be starting points for other pathways. Degradative pathways can even interconnect with synthetic pathways, and in this way degradative pathways sometimes provide not only the energy but also the molecules needed for synthesis of cell components.

Some pathways (reactions) in cells are degradative ones that release energy and some are synthetic ones that require energy input.

Enzymes

As indicated above, every reaction within a metabolic pathway requires an enzyme. **Enzymes** are organic catalysts, globular protein molecules that speed up chemical reactions. Every enzyme is very specific in its action and can speed up only one particular reaction or one type of reaction. Table 7.1 indicates that the name of an enzyme is often formed by adding *ase* to the name of its substrate. Some enzymes are named for the action they perform: for example, a dehydrogenase is an enzyme that removes hydrogen atoms from its substrate.

No reaction can occur in a cell unless its own enzyme is present and active. For example, if enzyme 2 (i.e. E_2) above is missing or not functioning, the pathway will shut down and stop at B. Since enzymes are so necessary in cells, their mechanism of action has been studied extensively.

Table 7.1 Enzymes Named for Their Substrate

Substrate	Enzyme
Lipid	Lipase
Urea	Urease
Maltose	Maltase
Ribonucleic acid	Ribonuclease
Lactose	Lactase

Energy of Activation

Molecules frequently will not react with one another unless they are activated in some way. For example, wood does not burn unless it is heated to a moderately high temperature. In the laboratory, too, activation is very often achieved by heating the reaction flask so that the number of effective collisions between molecules increases, allowing them to react with one another more frequently.

The energy that must be supplied to cause molecules to react with one another is called the **energy of activation.** Figure 7.3 compares the reactive energy level required in the absence of enzyme with the reactive energy level required if enzyme is present. The difference between the nonreactant energy level and the reactive energy level is the necessary energy of activation (E_a). Enzymes lower the necessary energy of activation. For example, the hydrolysis of casein (the protein found in milk)

Figure 7.4
a. The enzyme and substrates before the reaction begins.
b. Formation of the enzyme-substrate complex, holding the substrates together in such a way that they can react. *c.* Following the reaction the enzyme returns to its prior configuration.

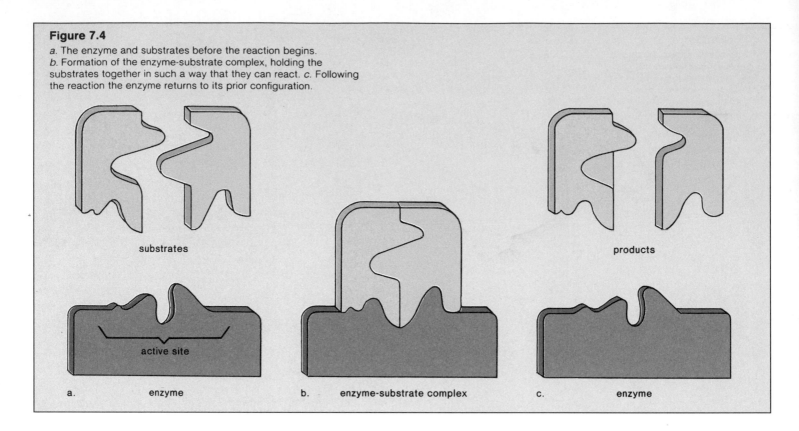

substrates

products

active site

a. enzyme

b. enzyme-substrate complex

c. enzyme

requires an energy of activation of 20,600 Kcal/mole[1] in the absence of an enzyme, but only 12,600 Kcal/mole are required if the appropriate enzyme is present. This example tells us that less energy and therefore less heat is needed to bring about a reaction when an enzyme is present. Enzymes lower the energy of activation by binding with their substrate(s) in such a way that a reaction can occur.

Enzyme-Substrate Complex

The equation, which is pictorially shown in figure 7.4, is often used to indicate that an enzyme forms a complex with its substrate:

$$E + S \rightarrow ES \rightarrow E + P$$

In this equation E = enzyme, S = substrate, ES = enzymes-substrate complex, and P = product.

Notice in figure 7.4 that an enzyme is unaltered by the reaction. Only a small amount of enzyme is actually needed in a cell because enzymes are used over and over again. When the

enzyme-substrate complex forms, the substrates are seemingly attracted to the enzyme because their shapes fit together like a *key fits a lock.* However, it is now thought that the enzyme may very well undergo a slight change in shape in order to more perfectly accommodate the substrates. This is called the **induced-fit model** because as binding occurs the enzyme is induced (undergoes a slight alteration) to achieve maximum fit. The region where the substrate(s) attach is called the **active site,** and it is here that the reaction takes place. After the reaction has been completed, the product(s) is (are) released and the active site returns to its original state.

Some enzymes speed up reactions simply by having adjacent binding sites for substrate molecules. Others do more than that. Figure 7.5 shows a computer-generated model of trypsin, an enzyme that digests protein by breaking peptide bonds. The active site of trypsin contains three amino acids with R groups that actually interact with members of the peptide bond to first break the bond and then introduce the components of water. This illustrates that the formation of the enzyme-substrate complex is very important in speeding up the reaction.

Only (a) certain product(s) can be produced by any particular reactant(s), therefore, an enzyme cannot bring about impossible products. However, the presence of an active enzyme can determine whether a reaction takes place or not. For ex-

[1]Using Kcal (Kcalorie) is a common way to measure heat, and a mole is the molecular weight of a substance expressed in grams.

Figure 7.5

Figure 7.5
Computer generated model of the functional backbone of the enzyme trypsin. Trypsin breaks peptide bonds and it has been found that the three R groups shown in green are in the active site and interact with the members of the peptide bond in order to hydrolyze it.

ample, if substance A can react to form either B or C, then which enzyme is present and active—enzyme 1 or enzyme 2—will determine which product is produced.

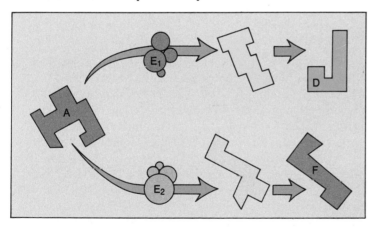

Enzymes are proteins that speed up chemical reactions by lowering the energy of activation. They do this by forming an enyzme-substrate complex.

Conditions Affecting Enzymatic Reactions

Enzymes normally allow reactions to proceed quite rapidly. For example, the breakdown of hydrogen peroxide into water and oxygen can occur 600,000 times a second when the enzyme catalase (an enzyme found in peroxisomes, p. 68) is present. How quickly an enzyme works, however, is affected by certain conditions.

Substrate Concentration

Generally, an enzyme's activity increases with greater substrate concentration because there are more collisions between substrate molecules and the enzyme. More substrate molecules will fill active sites for a larger proportion of the time and so more product will result per unit time. But when the concentration of substrate is so great that the enzyme's active sites are almost continuously filled with substrate, the enzyme's rate of activity cannot increase any more. Maximum velocity has been reached.

Temperature and pH

A higher temperature generally results in an increase in enzyme activity. As the temperature rises, the movement of enzyme and substrate molecules increases and there are more effective contacts between them. If the temperature rises beyond a certain point, however, enzyme activity eventually levels out and then declines rapidly because the enzyme is **denatured** at high temperatures (fig. 7.6*a*). During denaturation an enzyme's shape changes so that its active site can no longer bind substrate molecules efficiently.

A change in pH can also affect enzyme activity (fig. 7.6*b*). Each enzyme has an optimal pH which helps maintain its normal configuration. You will recall that the tertiary structure of a protein is dependent on interactions, such as hydrogen bonding, between R groups (p. 48). A change in pH can alter the ionization of these side chains and disrupt the normal interactions, so that denaturation eventually occurs. Again, the enzyme is then unable to combine efficiently with its substrate.

Adequate substrate concentration, optimum temperature, and pH all promote the speed of enzyme activity.

Regulation of Enzyme Activity

While these factors do affect enzyme activity, the cell also has built-in mechanisms to directly control both enzyme concentration and activity. First of all, cells are able to regulate whether an enzyme is present at all but since this type of regulation involves control of protein synthesis, it is not discussed until chapter 18. Cells also have ways to control the level of activity of enzymes that have already been synthesized and are present in the cytoplasm.

Inhibition   **Inhibition** is a common means by which cells regulate enzyme activity. In *competitive inhibition* another molecule is so close in shape to the enzyme's substrate that it can compete with the true substrate for the active site of the enzyme. This molecule is an inhibitor of the reaction because only the binding of substrate will allow a product to be produced. In *noncompetitive inhibition,* a molecule binds to an enzyme at a site that is not the active site. This site is called the **allosteric site**—from *allo* meaning other and *steric* meaning space or structure. This molecule, too, is an inhibitor because it causes a shift in the three-dimensional structure of the enzyme that prevents the substrate from binding to the active site. In cells, inhibition is usually reversible, that is, the inhibitor is not bound permanently to the enzyme.

Figure 7.6
Rate of an enzymatic reaction is affected by temperature and pH. *a.* At first, as with most chemical reactions, the rate of an enzymatic reaction doubles with every 10-fold rise in temperature. In this graph, which is typical of a mammalian enzyme, the rate is maximum at about 40-fold and then it decreases until it stops altogether, indicating that the enzyme is now denatured. *b.* Pepsin, an enzyme found in the stomach, acts best at a pH of about 2 while trypsin, an enzyme found in the small intestine, prefers a pH of about 8. Other pHs do not properly maintain the shape that enables these proteins to bind with their substrates.

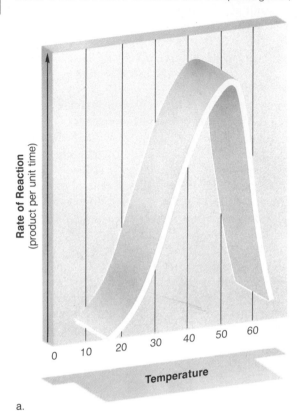

a.

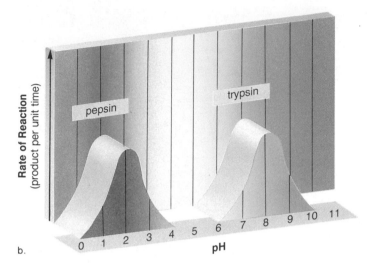

b.

During competitive inhibition, an inhibitor binds to the active site; during noncompetitive inhibition, an inhibitor binds to an allosteric site.

Irreversible inhibition of enzymes does occur but usually is the result of the presence of a poison. For example, penicillin causes the death of bacteria due to irreversible inhibition of an enzyme needed to form the bacterial cell wall. In humans, hydrogen cyanide is an inhibitor of a very important enzyme (cytochrome oxidase) present in all cells, and this accounts for its lethal effect on the human body.

Feedback Inhibition The activity of almost every enzyme in a cell is regulated by feedback inhibition. **Feedback inhibition** is an example of a common biological control mechanism called negative feedback. Just as excessively high temperature can cause a furnace to shut off, so a product produced by an enzyme can inhibit its activity. When the product is in abundance, it binds competitively with its enzyme's active site; as the product is used up, inhibition is reduced and more product can be produced. In this way, the concentration of the product is always kept within a certain range.

Most enzymatic *pathways* are also regulated by feedback inhibition, but in this case the end product of the pathway binds at an allosteric site on the *first* enzyme of the pathway (fig. 7.7). This binding shuts the pathway down and no more product is produced.

In feedback inhibition, the inhibitor is either the product of an allosteric enzyme or the end product of a metabolic pathway that binds to an allosteric site on the first enzyme of the pathway.

Coenzymes

Many enzymes require a nonprotein **cofactor** to assist them in carrying out their functions. Some cofactors are ions; for example, magnesium, Mg^{++}, potassium, K^+, and calcium, Ca^{++}, are very often involved in enzymatic reactions. Some other cofactors, called **coenzymes,** are organic molecules that bind to enzymes and serve as carriers for chemical groups or electrons.

Usually coenzymes participate directly in the reaction. For example, the coenzyme tetrahydrofolate is a carrier for a methyl group (CH_3). Sometimes it works in conjunction with the enzyme thymidylate synthase and donates a methyl group to form thymine, one of the bases found in DNA. In humans, this coenzyme is not present unless our diet includes the vitamin folic acid. **Vitamins** are relatively small organic molecules that are required in trace amounts in our diets, and those of other animals, for health and physical fitness.

The presence of other coenzymes in cells is similarly dependent on our intake of vitamins. For example:

Vitamin	*Coenzyme*
Niacin	NAD^+
B_2 (riboflavin)	FAD
B_1 (thiamine)	Thiamine pyrophosphate
Pantothenic acid	Coenzyme A
B_{12}	Cobamide coenzymes

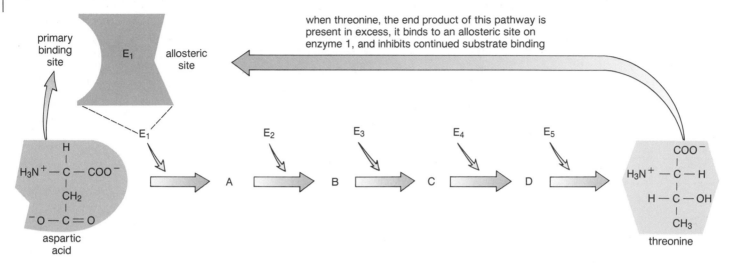

Figure 7.7

The amino acid aspartic acid becomes the amino acid threonine by a sequence of five enzymatic reactions. This pathway is controlled by feedback inhibition.

A deficiency of any one of these vitamins results in a lack of the coenzyme listed and therefore a lack of certain enzymatic actions. In ourselves this eventually results in vitamin-deficiency symptoms. For example, niacin deficiency results in a form of dermatitis called pellagra, and riboflavin deficiency results in fissures at the corners of the mouth.

Coenzymes are nonprotein molecules that assist enzymes in performing their reactions.

NAD⁺ and FAD

NAD^+ (nicotinamide adenine dinucleotide) is a coenzyme that carries electrons and quite often works in conjunction with enzymes called dehydrogenases (fig. 7.8). An enzyme of this type removes two hydrogen atoms ($2e^- + 2H^+$) from its substrate; both electrons are passed to NAD^+ but only one hydrogen ion—NAD^+ becomes NADH:

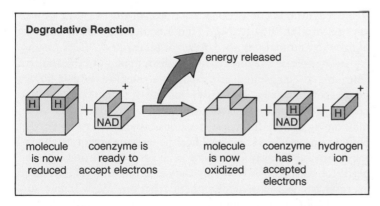

Degradative Reaction

energy released

molecule is now reduced — coenzyme is ready to accept electrons — molecule is now oxidized — coenzyme has accepted electrons — hydrogen ion

This is a degradative oxidation reaction and energy is released, but by comparing this equation to the similar one on page 98, you can see that much less energy has been lost. The energy has been transferred to NAD^+ in the form of high-energy electrons.

Figure 7.8

Secondary structure of an enzyme that degrades alcohol, called alcohol dehydrogenase. NAD^+ is a coenzyme for this enzyme and therefore the enzyme has a binding site not only for its substrate (alcohol) but also for NAD^+. The NAD^+-binding site is in green and yellow; the substrate-binding site is in blue. A bound NAD^+ in purple lies close to the substrate-binding site.

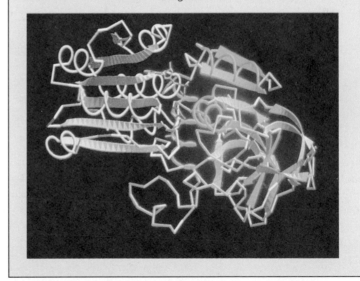

Another coenzyme called **FAD** (flavin adenine dinucleotide) also carries high-energy electrons, but it accepts both hydrogen ions (to become $FADH_2$).

Both NAD^+ and FAD are involved in **cellular respiration,** a metabolic pathway by which carbohydrates are oxidized in a step-by-step manner to carbon dioxide and water. At various junctions along the way these coenzymes accept electrons from certain substrates and carry them to an **electron transport system** consisting of membrane-bound carriers that pass electrons from

Figure 7.9

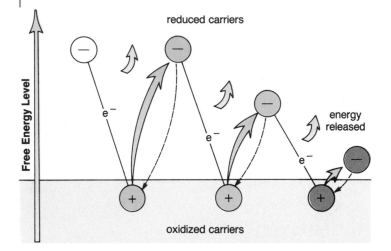

Electron transport system, in which electrons pass from carrier to carrier. With each oxidation reaction some energy is released and a portion of it is trapped for the purpose of producing ATP molecules.

one carrier to another. Every time an electron is transferred oxidation occurs and energy is released; this energy is ultimately used to produce ATP molecules (fig. 7.9).

> NAD^+ and FAD are electron carriers that take electrons to the electron transport system in mitochondria. This system uses the energy of oxidation to produce ATP molecules.

Chloroplasts also have an electron transport system for producing ATP. In this case the electrons that fuel the system were taken from water. Solar energy energized these electrons and as they pass from carrier to carrier, the energy is released and ATP is built up. In the end, a coenzyme called NADP accepts the electrons and becomes reduced to NADPH.

NADP⁺

$NADP^+$ has a structure similar to NAD^+ but it contains a phosphate group that is lacking in NAD^+. This might be a signal to enzymes that its function is slightly different. It does carry electrons and a hydrogen ion just like NAD^+ but its electrons are used to bring about reduction synthesis:

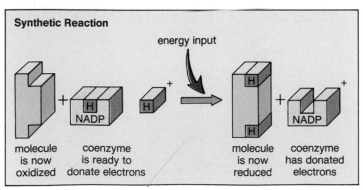

During **photosynthesis** within chloroplasts, carbon dioxide is reduced to a carbohydrate. NADPH supplies the necessary electrons and ATP supplies the necessary energy to bring about this reduction. Both of these are produced within the chloroplast itself.

NADPH also functions outside of chloroplasts. Within the cytosol, plant and animal cells use a special degradative pathway to produce NADPH. This NADPH is used during the synthesis of many necessary molecules in cells. Energy too is required but for this purpose ATP produced by mitochondria is utilized.

> $NADP^+$ is an electron carrier that participates along with ATP in synthetic reactions.

Metabolism Revisited

On page 96, we stated it is possible to harness the energy and hydrogens provided by degradative reactions for synthetic reactions. This is possible *only* because ATP carries energy and NADPH carries electrons between degradative pathways and synthetic pathways (fig. 7.10).

Degradative pathways do run synthetic pathways as long as the cell receives an outside source of energy and matter. The energy laws we discussed on pages 94–95 tell us that life is only possible because organisms (cells) are able to temporarily *store* some of the energy that flows through them.

Of all the organisms, only photosynthesizers such as plants are able to make organic food by utilizing *solar* energy. The green arrows in figure 7.10 illustrate the process of photosynthesis. In chloroplasts, solar energy is used to produce the ATP and NADPH that is used to reduce carbon dioxide to carbohydrates. This and the other macromolecules produced by plants supplies them, and eventually all other living things, with a source of organic food.

Plants and animals use organic food as both an energy source and a source of unit molecules (blue arrows). When used as an energy source, carbohydrates are degraded (oxidized) to carbon dioxide and water, which are excreted. Heat is given off in the process, and some of the energy is used to produce ATP. Degradation also produces a supply of NADPH. Again ATP is a carrier for energy and NADPH is a carrier for electrons utilized in synthetic (reduction) reactions. In this way the flow of energy through the organism (cells) is utilized for the purpose of growth and maintenance of the organism.

The role of enzymes in metabolism must not be forgotten. They make "cold chemistry" possible. Degradative reactions occur spontaneously (naturally) but only if sufficient activation energy is supplied. Enzymes lower the energy of activation making it possible for these reactions to occur very quickly at body temperature. However, since the enzymes involved in synthesis are not the same as those used for degradation, the cell needs the coenzyme $NADP^+$ to carry electrons and ATP to carry energy from degradative to synthetic reactions. We will now turn our attention to exploring the structure and function of ATP in more detail.

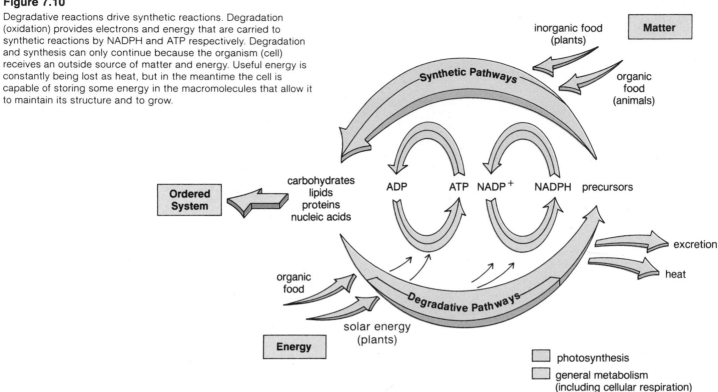

Figure 7.10

Degradative reactions drive synthetic reactions. Degradation (oxidation) provides electrons and energy that are carried to synthetic reactions by NADPH and ATP respectively. Degradation and synthesis can only continue because the organism (cell) receives an outside source of matter and energy. Useful energy is constantly being lost as heat, but in the meantime the cell is capable of storing some energy in the macromolecules that allow it to maintain its structure and to grow.

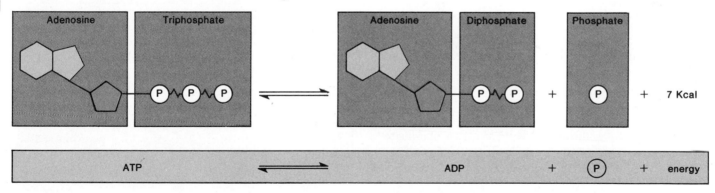

Figure 7.11

ATP, the common energy carrier in cells, is a high-energy molecule as represented by the wavy lines between the phosphate (P) groups. When the last phosphate group is removed, the energy released is used by the cell for various purposes. ATP can be reformed if enough energy is supplied to rejoin ADP and (P).

ATP

ATP (adenosine triphosphate) (fig. 7.11) is the common energy currency of cells; when cells require energy, they "spend" ATP. You might think that this would require our bodies to produce a lot of ATP and indeed it does; the average male needs to produce about 8 kilograms (over 17 lbs) of ATP an hour! Obviously it would be impossible to carry around even a few hours supply of ATP and indeed the body has on hand only about 50 grams (1.8 oz) at any time. The answer to the paradox is that ATP is constantly being recycled from **ADP** and (P) (fig. 7.11). The cell's entire supply of ATP is recycled about once each minute.

Structure of ATP

ATP is a nucleotide composed of the base adenine and the sugar ribose (together called adenosine), and three phosphate groups. ATP is called a "high-energy" compound because a great deal of energy is released when one or two phosphate groups are removed. Under cellular conditions the amount of energy released is about 7.3 kilocalories per mole. The use of the wavy line between the phosphate bonds indicates the high potential energy of these bonds.[2]

[2]The high-energy content of ATP does not reside simply in the phosphate bond; rather it comes from the complex interaction of the atoms within the molecule. The use of the wavy line is for the purpose of convenience only.

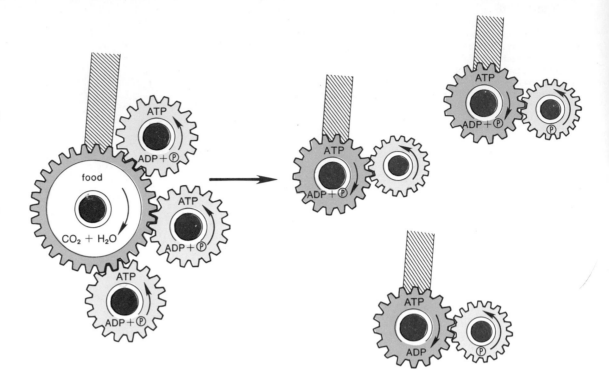

Figure 7.12
The coupling of ATP buildup and breakdown to other reactions. Energy-releasing reactions are shown as gears with arrows going down; energy-requiring reactions are shown as gears with arrows going up. On the far left, degradation of food is coupled to the production of ATP. On the right, breakdown of ATP releases energy to drive energy-requiring reactions. Very often this involves the transfer of a phosphate group from ATP to an intermediate molecule in the reaction, lower right.

The use of ATP as a carrier of energy derived from degradation has some advantages:

1. The energy of degradation can be captured in a step-by-step manner that prevents waste (see fig. 7.9).
2. It provides a common energy currency that can be used in many different types of reactions.
3. When the phosphate bond is broken, the amount of energy released is just about enough for the biological purposes mentioned below and so little energy is wasted.

(1) and (3) are only possible because most reactions (e.g., ATP breakdown and synthetic buildup) are *coupled;* that is, they take place at the same time and at the same place (fig. 7.12) and usually utilize the same enzyme. During the coupling process, a phosphate group is often transferred to an intermediate compound. For example, during active transport a phosphate group is transferred to a membrane-bound protein (fig. 6.12).

Function of ATP

You will recall that at various times we have mentioned at least three uses for ATP:

Chemical work. Supplying the energy needed to synthesize macromolecules that make up the cell.
Transport work. Supplying the energy needed to pump substances across the cell membrane.
Mechanical work. Supplying the energy needed to cause muscles to contract, cilia and flagella to beat, chromosomes to move, and so forth.

We might note here that all organisms, both prokaryotic and eukaryotic, use ATP and that this illustrates the common chemical unity of all living things.

ATP is a carrier of energy in cells. It is the common energy currency because it supplies energy to many different types of reactions.

Formation of ATP

Cells have two ways to make ATP. The first way, called **substrate-level phosphorylation,** occurs in the cytosol. During this process, a high-energy substrate transfers a phosphate group to ADP forming ATP:

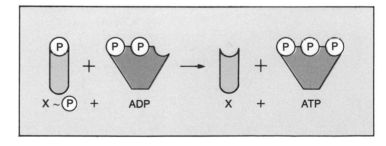

However, substrate-level phosphorylation is not the primary way of forming ATP; most ATP is made within mitochondria and chloroplasts by way of the electron transport system mentioned earlier (fig. 7.9). In this case the phosphorylation that occurs is called **chemiosmotic phosphorylation.**

Figure 7.13

Chemiosmotic theory. *a.* The electron transport system located in the inner membrane deposits hydrogen ions (H⁺) in the intermembrane space between the outer and inner membrane. This establishes an electrochemical gradient that causes H⁺ to flow from the space into the matrix through channels in the F₁ particles. The resultant release of energy allows an enzyme to produce ATP from ADP and Ⓟ. *b.* Similarly, the electron transport system located in the thylakoid membrane deposits hydrogen ions within the thylakoid space. When H⁺ flows out of this space through channels in CF₁ particles, ATP is produced.

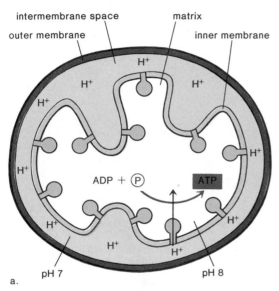

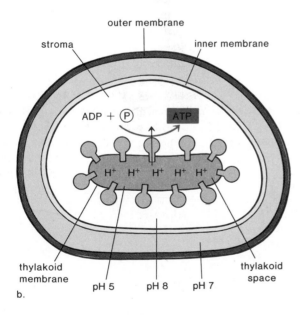

Figure 7.14

Flow diagram for an experiment performed by Andre Jagendorf in 1966 that supports the chemiosmotic theory. An electrochemical gradient was established by first soaking isolated chloroplasts in an acid medium and then abruptly changing the pH to basic. ATP production occurred after ADP and Ⓟ were added. This all occurred in the dark, proving that ATP synthesis is not directly linked to the chloroplast electron transport system, which works only in the light.

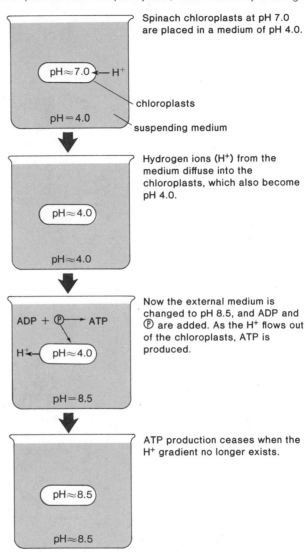

Chemiosmotic Phosphorylation

For many years it was known that ATP synthesis was somehow linked to the electron transport system, but the exact mechanism couldn't be determined. Peter Mitchell, a British biochemist, received the Nobel Prize in 1978 for his **chemiosmotic theory** of ATP production in both mitochondria and chloroplasts.

You'll recall that mitochondria are bounded by a double membrane (fig. 7.13*a*). The membrane-bound carriers of the electron transport system are located along the inner membrane. According to the chemiosmotic theory, hydrogen ions (H⁺) tend to collect in the space between the outer and inner membranes, because they are deposited there by certain carriers. This establishes an electrochemical gradient across the membrane that can be used to provide energy for ATP production. Proteins that span the membrane (called F₁ particles in mitochondria) contain channels that permit H⁺ to pass through the membrane down its electrochemical gradient. The flow of hydrogen ions into the matrix provides the energy that allows an enzyme attached to the particle to produce ATP from ADP + Ⓟ. This process is called chemiosmotic phosphorylation.

The same mechanism is at work in chloroplasts, only here the electron transport system is located in the thylakoid membranes (fig. 7.13b). Certain carriers deposit hydrogen ions within the thylakoid space, and when H$^+$ passes down its electrochemical gradient to the stroma through special channels (called CF$_1$ particles in chloroplasts) ATP is produced by enzymatic action from ADP and Ⓟ. Utilizing chloroplasts, Andre Jagendorf of Cornell University performed a now-famous experiment (fig. 7.14) that lends support to the chemiosmotic theory because it shows that ATP production is indeed tied only to an H$^+$ gradient.

Most of the energy for ATP synthesis is derived from an H$^+$ gradient established across a membrane by an electron transport system.

Summary

1. There are two energy laws that are basic to understanding energy-use patterns in organisms (cells). The first one states that energy cannot be created or destroyed, and the second states that some useful energy is always lost when one form of energy is transformed into another form.
2. In keeping with these laws, living things need an outside source of energy. Plants capture the energy of sunlight to make their own food, and animals either eat plants or other animals to obtain food.
3. Metabolism consists of two kinds of reactions, degradative and synthetic. During degradation, organic molecules derived from food are broken down to release energy; oxidation occurs and hydrogen atoms are removed from substances. During synthesis, an energy-requiring process, molecules derived from food accept hydrogen atoms and are reduced.
4. There are degradative metabolic pathways and synthetic metabolic pathways. Each reaction in a pathway requires its own enzyme. Enzymes speed up chemical reactions because they lower the energy of activation. They can do this because they form a complex with their substrate(s) at a binding site called the active site.
5. Various factors affect the yield of enzymatic reactions, such as the concentration of the substrate(s), the temperature, and the pH. A high temperature or a pH outside the preferred range for that enzyme can lead to denaturation, a change in structure that prevents the enzyme from functioning.
6. The activity of most enzymes is regulated by inhibition. In competitive inhibition, a molecule competes with the substrate for the active site; during noncompetitive inhibition, a molecule binds to an allosteric site. During feedback inhibition, the product of an enzymatic reaction binds to the enzyme; for metabolic pathways, the end product usually binds to an allosteric site on the first enzyme in the pathway.
7. Many enzymes have cofactors or coenzymes that help them carry out a reaction. Coenzymes are nonprotein, organic molecules often derived at least in part from vitamins. NAD$^+$ and FAD are coenzymes that carry high-energy electrons to an electron transport system that is usually located in mitochondria. Here oxidation occurs each time one carrier passes electrons to the next one. At the same time, energy is released and used to produce ATP molecules. Chloroplasts also contain an electron transport system that has energized electrons taken from water.
8. NADPH is also a carrier of electrons but it carries them to various synthetic reactions. These are reduction reactions that require both energy supplied by ATP and hydrogen atoms.
9. Degradative pathways drive synthetic pathways. Degradative pathways supply the ATP and the NADPH that is needed to bring about synthesis. We must never forget, however, that in order for this to occur, the cell needs an outside source of energy and matter. Only a flow of energy through the organism allows it to sustain itself and to grow.
10. ATP is a high-energy molecule that serves as a common carrier of energy in cells. When its last phosphate bond is broken, the reaction supplies the right amount of energy for many reactions. In order to produce ATP, the energy from food breakdown is coupled to ATP production, and in order to pass this energy to synthetic reactions, coupling again occurs.
11. The chemiosmotic theory explains just how the electron transport system produces ATP. The carriers of this system deposit hydrogen ions (H$^+$) on one side of a membrane. When the ions flow down an electrochemical gradient in special channels, an enzyme utilizes the release of energy to make ATP from ADP and Ⓟ.

Objective Questions

1. Which statement is true?
   a. Cells can produce ATP and this shows that they have no need for an outside energy source.
   b. In keeping with the laws of energy, cells always need an outside energy source.
   c. In keeping with the laws of energy, cells can carry on "cold chemistry" and therefore do not give off heat.
   d. Photosynthetic cells do not need ATP molecules.
2. Degradative reactions
   a. are often oxidation reactions.
   b. are the same as synthetic reactions.
   c. require a supply of NADPH molecules.
   d. Both (a) and (c).
3. Enzymes
   a. make it possible for cells to escape the need for energy.
   b. are nonprotein molecules that help coenzymes.
   c. are not affected by a change in pH.
   d. lower the energy of activation.
4. An allosteric site on an enzyme
   a. is the same as the active site.
   b. is where ATP attaches and gives up its energy.
   c. is often involved in feedback inhibition.
   d. All of these.

5. Synthetic reactions
   a. only occur inside chloroplasts.
   b. are needed to maintain the structure of the cell.
   c. supply the ATP needed by degradative reactions.
   d. All of these.
6. A high temperature
   a. can affect the shape of an enzyme.
   b. lowers the energy of activation.
   c. makes cells less susceptible to disease.
   d. Both (a) and (c).
7. Electron transport systems
   a. are found in both mitochondria and chloroplasts.
   b. contain oxidation reactions.

c. are involved in the production of ATP.
   d. All of these.
8. The difference between NAD$^+$ and NADP$^+$ is
   a. only NAD$^+$ production requires niacin in the diet.
   b. one contains high-energy phosphate bonds and the other does not.
   c. one carries electrons to the electron transport system and the other carries them to synthetic reactions.
   d. All of these.
9. ATP
   a. is used only in animal cells and not in plant cells.
   b. carries energy between degradative pathways and synthetic pathways.

c. is needed for chemical work, mechanical work, and transport work.
   d. Both (b) and (c).
10. Chemiosmotic phosphorylation is dependent upon
   a. the diffusion of water across a selectively permeable membrane.
   b. an outside supply of phosphate and other chemicals.
   c. the establishment of an electrochemical H$^+$ gradient.
   d. the ability of ADP to join with (P) even in the absence of a supply of energy.

Study Questions

1. Explain the first energy law using the terms *system* and *surroundings* in your explanation. Explain the second energy law using the term *useful energy* in your explanation.
2. Discuss the importance of solar energy, not only to plants but to all living things.
3. Contrast a degradative reaction with a synthetic reaction.
4. Draw a diagram of a metabolic pathway and use it to explain the terms substrate and product, and the need for various enzymes.

5. In what way do enzymes lower the energy of activation? What does the induced-fit model say about formation of the enzyme-substrate complex?
6. In what way do substrate concentration, temperature, and pH affect the yield of enzymatic reactions?
7. Explain competitive inhibition, noncompetitive inhibition, and feedback inhibition of enzyme activity.

8. Contrast the function of NAD$^+$ and FAD to that of NADP$^+$.
9. In what way do degradative reactions drive synthetic reactions? Why is there still a need for an outside source of energy and matter?
10. Discuss the structure and function of ATP and outline the formation of ATP by chemiosmotic phosphorylation.

Thought Questions

1. Ecologists restate the first energy law as, "There is no free lunch," and the second energy law as, "If you think things are mixed up now, just wait." Explain.

2. Plants perform both photosynthesis and cellular respiration. Why is the latter necessary?

3. Humans are warm blooded. Where does the heat come from?

Selected Key Terms

metabolism (mě-tab′o-lizm) 96
substrate (sub′strāt) 97
enzyme (en′zīm) 97
active site (ak′tiv sīt) 98
denatured (de-na′tūrd) 99
feedback inhibition (fēd′bak in′′hǐ-bish′un) 100
coenzyme (ko-en′zīm) 100

NAD$^+$ (nicotinamide adenine dinucleotide) (nik′′o-tin′ah-mīd ad′ě-nīn di-nu′kle-o-tīd) 101
FAD (flavin adenine dinucleotide) (fla′vin ad′ě-nīn di-nu′kle-o-tīd) 101
cellular respiration (sel′u-lar res′′pǐ-ra′shun) 101
electron transport system (e-lek′tron trans′port sis′tem) 101

NADP$^+$ (nicotinamide-adenine dinucleotide phosphate) (nik′′o-tin′ah-mīd ad′ě-nīn di-nu′kle-o-tīd fos′fāt) 102
photosynthesis (fo′′to-sin′thě-sis) 102
ATP (adenosine triphosphate) (ah-den′o-sēn tri-fos′fāt) 103
ADP (adenosine diphosphate) (ah-den′o-sēn di-fos′fāt) 103

CHAPTER 8

Photosynthesis

Your study of this chapter will be complete when you can:

1. Give at least three examples of the importance of photosynthesis to living things.
2. Relate the visible light range to photosynthesis, and describe the role of chlorophyll.
3. Describe the structure of chloroplasts and give the function of the various parts.
4. Explain the terms light-dependent reactions and light-independent reactions, and describe how the light-dependent reactions drive the light-independent reactions.
5. Trace the noncyclic electron pathway, pointing out significant events as the electrons move.
6. Describe the process of chemiosmotic phosphorylation in chloroplasts.
7. Trace the cyclic electron pathway, pointing out significant events.
8. Describe C_3 photosynthesis and the three stages of the Calvin-Benson cycle.
9. Describe C_4 photosynthesis, and relate the phenomenon of photorespiration to the success of C_4 plants in hot, dry climates.
10. Contrast CAM photosynthesis with C_4 photosynthesis, using the terms ''partitioning in space'' and ''partitioning in time.''

The sun provides the energy that allows photosynthesizers, like trees, to produce their own organic food. The process is extremely wasteful because only 2% of the available energy is ever taken up by plants; yet this amount of energy is sufficient to sustain not only plants but almost all living things. The ultimate source of energy for life on earth is the sun.

H AVE YOU THANKED A GREEN PLANT TODAY? Plants dominate our environment, but most of us spend little time thinking about the various services they perform for us and for all living things. Chief among these is their ability to carry on **photosynthesis,** during which they use sunlight (*photo*) to produce food (*synthesis*):

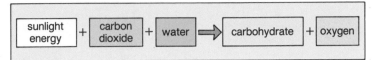

Other organisms, collectively called *algae*,[1] are also major photosynthesizers.

The organic food produced by photosynthesis is not only used by plants themselves, it is also the ultimate source of food for all other living things (fig. 8.1). Also, when organisms break down food to acquire ATP energy they make use of the oxygen

[1]The term algae is used here to mean aquatic, photosynthesizing organisms, including the cyanobacteria, which, although they are prokaryotes, carry out a green-plant type of photosynthesis.

Figure 8.1
This squirrel is a herbivore. It feeds directly on plant material produced by a photosynthesizer. Carnivores, such as a hawk that may feed on this squirrel, are also dependent, although indirectly, on food produced by photosynthesizers.

Photosynthesis—A Historical Perspective

W ith the rise of the scientific method in the seventeenth century, investigators began to perform experiments that eventually led to our present-day knowledge of photosynthesis. The ancient Greeks believed that plants were "soil-eaters" and somehow converted soil into plant material. Apparently to test this hypothesis, a seventeenth-century Dutchman named Jean-Baptiste Van Helmont planted a willow tree weighing 5 pounds in a large pot containing 200 pounds of soil. He watered the tree regularly for five years and then reweighed both the tree and the soil. The tree weighed 170 pounds and the soil weighed only a few ounces less than the original 200 pounds. Van Helmont concluded that the increase in weight of the tree was due primarily to the addition of water. He did not consider the possibility that air had anything to do with it.

Later, a number of other investigators did consider what the role of the air might be. Joseph Priestley (English, 1733–1804) put a plant and candle under a bell jar that had trapped a certain amount of air. He found that after the candle had gone out, the plant could "renew" the air in such a way that the candle would burn if lighted again. The science of chemistry was just emerging at this time and Jan Ingenhousz (Dutch, 1730–1799) identified the gas given off by plants as oxygen. He also observed that the amount of oxygen given off was proportional to the amount of sunlight reaching the plant; in the shade, plants "injured" air and released carbon dioxide. It was Nicholas de Saussure (Swiss, 1767–1845) who found that an increase in the dry weight of a plant was dependent upon the presence of carbon dioxide.

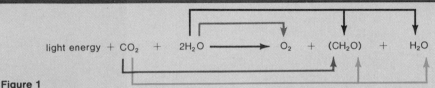

Figure 1
Overall equation for photosynthesis. The arrows show the relationship between the reactants and the products in the equation.

Microscopes were increasingly used during these days and Matthias Schleiden, a German, is credited in 1838 with the conclusion that plants are composed of cells. Julius Sachs (German, 1832–1897) was a plant physiologist who studied plant cells under the microscope. He noted that the green substance named chlorophyll was confined to the chloroplasts. In one experiment with a whole plant, he coated a few leaves with wax and he observed that after exposure to sunlight, only the uncoated ones increased in starch content. (A similar experiment is found in the lab manual that accompanies this text!) Sachs must have been an excellent microscopist because he followed up this experiment by observing that the starch grains in chloroplasts increase in size when a leaf is photosynthesizing.

By the end of the nineteenth century, scientists knew the basic reaction involved in photosynthesis:

carbon dioxide + water

sunlight ↓

carbohydrate + oxygen

In 1905, F. F. Blackman performed experiments in which he steadily increased the amount of light energy while the temperature was held constant; once the rate of photosynthesis was maximum, he increased the temperature. Now the rate of photosynthesis

increased. This led him to conclude that only one part of photosynthesis has to do with light, and that there is a second part that requires the presence of temperature-sensitive enzymes. Today these two parts of photosynthesis are called the light-dependent reaction and the light-independent reaction.

By the 1940s, cell fractionation allowed investigators to study certain parts of cells. Robert Hill isolated chloroplasts from cells and discovered that oxygen was given off in the absence of CO_2 if an electron acceptor was provided. (He used ferric oxalate but the cell uses $NADP^+$.) Scientists realized, therefore, that the oxygen released during photosynthesis came not from carbon dioxide but from water. In 1941, Samuel Ruben and Martin Kamen performed a study that utilized an isotope tracer. A plant was exposed first to carbon dioxide that contained heavy oxygen and then to water that contained heavy oxygen ($^{18}O_2$). Only in the latter instance did the oxygen given off contain this isotope. To explain this finding, it is necessary to show water on both sides of the equation for photosynthesis (*see* figure).

The use of cell fractionation, radioactive tracers, and electron microscopy have continued to add to our knowledge of photosynthesis; some of the most recent of these experiments are covered elsewhere in this chapter.

given off by photosynthesis (fig. 9.1). Nearly all living things are dependent on the oxygen in the atmosphere and all of this oxygen is derived from photosynthesis.

At one time in the distant past, plant and animal matter accumulated without breaking down and then became the fossil fuels (e.g., coal, oil, and gas) that we burn today for energy. This source of energy too was the product of photosynthesis.

Photosynthesis is absolutely essential for the continuance of life, because it is the source of food and oxygen for nearly all living things.

Photosynthesis is an energy transformation in which energy from the sun in the form of light is converted to the energy within carbohydrate molecules. Therefore we will begin our discussion of photosynthesis with the energy source—sunlight.

Sunlight

The radiation coming from the sun can be described in terms of its energy content and its wavelength. The energy comes in discrete packets called **photons.** So, in other words, you can think of radiation as photons that travel in waves:

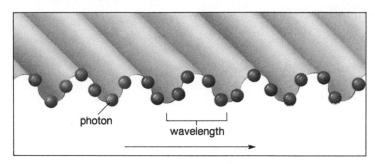

photon wavelength

Figure 8.2 illustrates that solar radiation, or the **electromagnetic spectrum,** can be divided up on the basis of wavelength—gamma rays have the shortest wavelengths and radio waves have

Figure 8.2

The electromagnetic spectrum, the components of visible light, and the absorption spectra of chlorophylls a and b. Visible (white) light is actually made up of a number of different wavelengths of radiation; when it is passed through a prism (or through raindrops) we see these wavelengths as different colors of light. The chlorophylls, like other pigments, absorb only certain wavelengths; for example, the chlorophylls primarily absorb light in the blue-purple range and the

red-orange range. The chlorophylls do not absorb green light to any extent; therefore, light of this color is reflected to our eyes and we see chlorophyll as a green pigment. There are other pigments in plants, such as the carotenoids, that are yellow-orange and are able to absorb in the purple-blue-green range. These pigments become noticeable in the fall when chlorophyll breaks down.

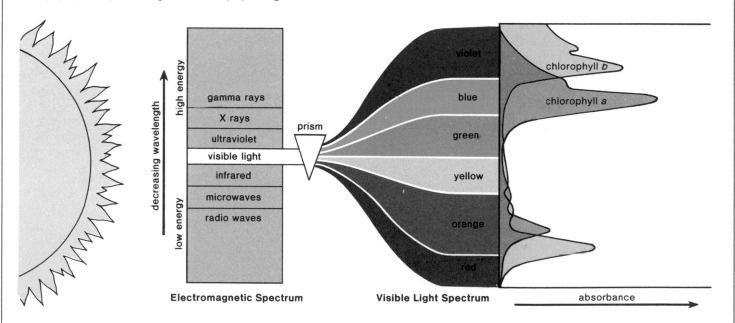

To identify the absorption spectrum of a particular pigment, a purified sample is exposed to different wavelengths of light inside an instrument called a spectrophotometer. A spectrophotometer measures the amount of light that passes through the sample, but from this it can be calculated how much was absorbed. The amount of light absorbed at each wavelength is plotted on a graph and the result is what we call the absorption spectrum, shown at the right side of the figure. How do we know that the peaks shown in

chlorophyll's absorption spectrum indicate the wavelengths used for photosynthesis? Photosynthesis, of course, produces oxygen and therefore we can use the rate of production of this gas as a means to measure the rate of photosynthesis at each wavelength of light. When such data are plotted, the resulting graph (called the action spectrum) is very similar to the absorption spectrum of the chlorophylls. Therefore, we are confident that the light absorbed by the chlorophylls does contribute extensively to photosynthesis.

the longest. The energy content of photons is inversely proportional to the wavelength of the radiation involved; that is, radiation of short wavelength has photons of higher energy level than radiation of long wavelength does. High-energy photons, such as those of short-wavelength, ultraviolet radiation, are dangerous to cells because they can break down organic molecules. Low-energy photons, such as those of infrared radiation, do not damage cells because they only increase the vibrational or rotational energy of molecules; they do not break bonds. But photosynthesis utilizes only a portion of the electromagnetic spectrum known as **visible light.** (It's called visible light because it is that part of the spectrum that allows us to see.) Photons of visible light have just the right amount of energy to promote electrons to a higher electron shell in atoms without harming cells.

We have mentioned previously that only about 42% of solar radiation ever reaches the earth's surface, and most of this radiation is within the visible light range. Higher energy wavelengths are screened out by the ozone layer in the atmosphere and lower energy wavelengths are screened out by water vapor and carbon dioxide before they reach the earth's surface. The conclusion is, then, that both the organic molecules within organisms and certain life processes, such as vision and photosynthesis, are chemically adapted to the radiation that is most prevalent in the environment.

Photosynthesis utilizes a portion of the electromagnetic spectrum (solar radiation) known as visible light.

Chloroplast Structure and Function

The site of photosynthesis is the **chloroplast,** an organelle found in the cells of green plants (fig. 8.3) and certain algae. In the chloroplast a double membrane or envelope surrounds a large central space called the **stroma** that contains an enzyme-rich solution. These enzymes can incorporate CO_2 into organic compounds. The membrane of the stroma forms the **grana,** which looked like piles of seeds (grana means seeds) to early microscopists. The electron microscope reveals that each granum is actually a stack of flattened sacs or disks now called **thylakoids.** The grana are connected to each other by stroma lamellae and therefore it's believed that all the thylakoids in a chloroplast are

Figure 8.3

Leaves are the primary photosynthetic organs of a plant. *Top,* a light microscope shows that a cell from a leaf contains many chloroplasts, the organelles that carry on photosynthesis. *Bottom,* an electron micrograph shows that the stroma within a chloroplast contains grana, each one a stack of flattened membranous sacs called thylakoids.

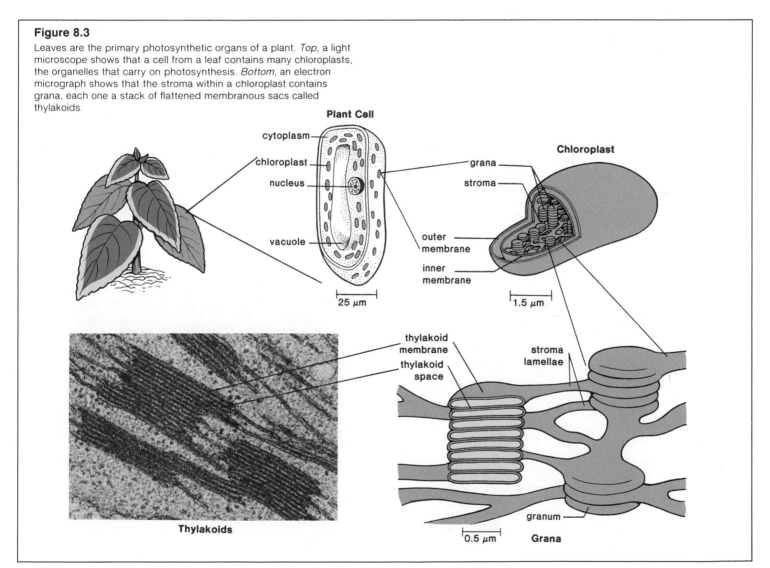

continuous with one another and surround a common interior space. **Chlorophyll** and **carotenoid** pigments, which absorb sunlight energy, are found within the membranes of the thylakoids, making them the energy-generating system of the chloroplasts. Electron transport systems associated with ATP and NADPH production are also present in the thylakoid membranes.

A chloroplast is well organized into two main portions: the stroma and the grana.

Overview of Photosynthesis

Figure 7.10 suggests that photosynthesis consists of two sets of reactions, the degradative reactions and the synthetic ones. The degradative reactions produce the ATP and NADPH needed for the synthetic reactions, during which reduction occurs. The degradative part of photosynthesis consists of the splitting of water with the release of oxygen, and the synthetic portion consists of the reduction of carbon dioxide to carbohydrate.

However, these two sets of reactions are usually called the **light-dependent reactions**—because they require solar energy—and the **light-independent reactions**—because light is not required, that is, they can occur in the dark as well as in light.

From our discussion thus far, it should be obvious that the light-dependent reactions occur in the thylakoids and the light-independent reactions occur in the stroma; furthermore, these are the inputs and outputs of each set:

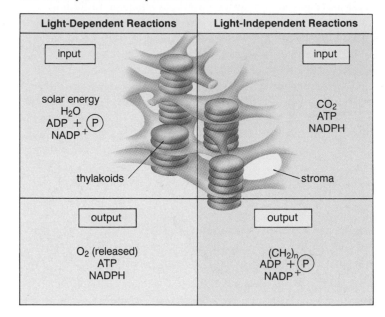

Light-Dependent Reactions	Light-Independent Reactions
input	input
solar energy H_2O ADP + P $NADP^+$	CO_2 ATP NADPH
thylakoids	stroma
output	output
O_2 (released) ATP NADPH	$(CH_2)_n$ ADP + P $NADP^+$

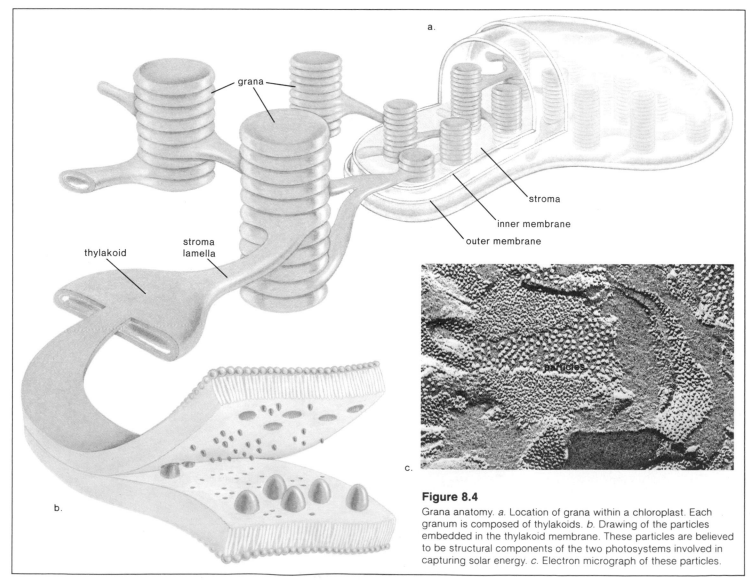

a.

grana

stroma

inner membrane

outer membrane

thylakoid

stroma lamella

particles

b.

c.

Figure 8.4

Grana anatomy. *a.* Location of grana within a chloroplast. Each granum is composed of thylakoids. *b.* Drawing of the particles embedded in the thylakoid membrane. These particles are believed to be structural components of the two photosystems involved in capturing solar energy. *c.* Electron micrograph of these particles.

In the light-dependent reactions, after absorbing solar energy chlorophyll sends energized electrons to an electron transport system that pumps hydrogen ions (H^+) into the thylakoid space. This allows ATP to be produced by chemiosmotic phosphorylation. At the same time, the flow of electrons is used to reduce $NADP^+$ to NADPH. A fresh supply of electrons is derived from water, which splits to release oxygen (O_2).

In the light-independent reactions, the ATP and NADPH made during the light-dependent reactions are used as a source of energy and reducing power, respectively, to drive the synthetic conversion of CO_2 to carbohydrate $(CH_2O)_n$.

The light-dependent reactions of photosynthesis (which take place in the thylakoids) provide the ATP and NADPH needed by the light-independent reactions (which take place in the stroma) to reduce carbon dioxide to carbohydrate.

Light-Dependent Reactions

This portion of photosynthesis requires the participation of two **photosystems,** called Photosystem I and Photosystem II. Electron micrographs of the thylakoid membrane show two different kinds of particles that correspond to these two photosystems (fig. 8.4) and that are bound to the thylakoid membrane. Each photosystem consists of a pigment complex containing several hundred pigment molecules, including chlorophyll and carotenoids, and electron acceptors. These pigment complexes have been termed **light-harvesting antennae** because, just as TV antennae are aimed to pick up signals, so the leaves of a plant turn to allow the pigment complexes to collect as much solar energy as possible.

Photosynthesis begins when the pigment molecules absorb light energy and funnel it to their respective **reaction centers,** that are located within the antennae of both Photosystem I and Photosystem II (fig. 8.5). The antenna of Photosystem I has a reaction-center chlorophyll *a* whose absorption spectrum peaks at a wavelength of around 700 nm and is therefore called **P700** (P stands for pigment). The antenna of Photosystem II has a reaction-center chlorophyll *a* whose absorption spectrum peaks at a slightly shorter wavelength; it is called **P680.** Each reaction-center chlorophyll *a* molecule is associated with an electron acceptor and these two molecules form the essential part of the photosystem. The solar energy received by a reaction center energizes electrons within the chlorophyll molecule so that the electrons can pass from it to an *acceptor molecule.*

In a photosystem, numerous pigment molecules absorb solar energy and funnel it to a reaction-center chlorophyll *a* molecule, which then sends energized electrons on to an acceptor molecule.

Two Electron Pathways

There are two possible pathways that electrons can take during photosynthesis, but only one of these, the **noncyclic electron pathway,** produces both ATP and NADPH. The other, the **cyclic electron pathway,** generates only ATP. Both pathways are dia-

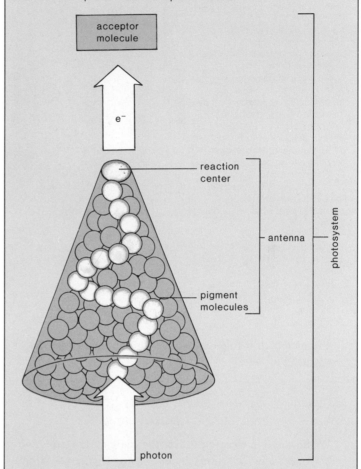

Figure 8.5
Components of a photosystem. An antenna consists of several hundred chlorophyll and carotenoid molecules plus a reaction-center chlorophyll *a* molecule. Light energy is absorbed and then passed from molecule to molecule (shown in yellow) until it reaches a reaction-center chlorophyll *a* molecule. Thereafter, energized electrons are passed to an acceptor molecule.

grammed in figure 8.6. This diagram is often called the Z-scheme, because when turned sideways it resembles that letter. The Z-scheme has the advantage of showing the relative energy levels of the substances involved, but it does not indicate the location of these substances in the thylakoid membrane. Figure 8.7 in the next section does this for you.

The light-dependent reactions consist of a noncyclic electron pathway (producing both ATP and NADPH) and a cyclic electron pathway (producing ATP).

Noncyclic Electron Pathway

From figure 8.6 you can see that each photosystem absorbs sunlight at the same time. However, it is easiest to describe the events as if they occur in a sequential manner, and as if they begin with Photosystem II:

1. As P680 absorbs solar energy (lower left), electrons (e^-) become so highly charged that they can leave the thylakoid-bound chlorophyll molecule. The "hole" left in

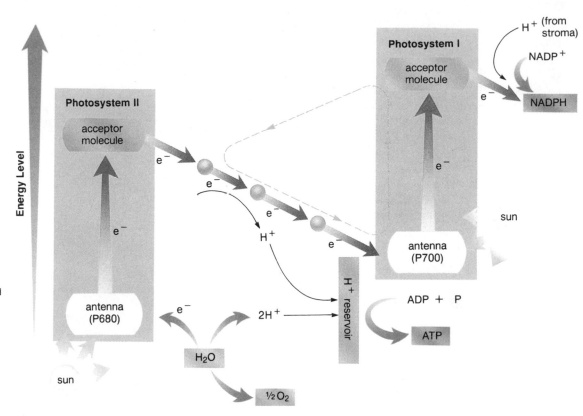

Figure 8.6
The light-dependent reactions contain two electron pathways. Starting at the bottom left, in the noncyclic electron pathway, electrons move from water to P680 and then to an acceptor molecule that passes them down a transport system to P700, which sends them to another acceptor molecule before they finally pass to NADP$^+$. In the cyclic electron pathway (dotted lines), electrons pass from P700 to an acceptor molecule that sends them down the electron transport system before they return to P700. During light-dependent reaction shown here, an H$^+$ reservoir increases in the thylakoid space. This leads to the production of ATP molecules as further described in figure 8.7.

the molecule is filled by electrons that come from the splitting of water:

$$H_2O \longrightarrow 2H^+ + 2e^- + \tfrac{1}{2}O_2$$

While the two electrons enter the photosystem, the two hydrogen ions (H$^+$) stay behind in the thylakoid space, an area that serves as an H$^+$ reservoir. The freed oxygen is the oxygen gas given off during photosynthesis.

2. The electrons that leave P680 are received by an acceptor molecule that sends them down an **electron transport system** consisting of a series of thylakoid-bound carriers, some of which are cytochrome molecules. (For this reason, this electron transport system is sometimes referred to as the **cytochrome system.**) Figure 8.7 shows the location of the parts of this system in the thylakoid membrane. Beginning at the left, hydrogen ions from the stroma are picked up, along with electrons, and at another point these H$^+$ are deposited inside the thylakoid space. Although the passage of two electrons accounts for the contribution of only four H$^+$ to the reservoir, there is an extreme electrochemical gradient across the thylakoid membrane. The concentration of H$^+$ is actually 10,000 times greater inside the thylakoid than in the stroma. Channels (within the CF$_1$ particles) in the thylakoid membrane allow H$^+$ to flow from the thylakoid space into

the stroma. An ATP synthetase enzyme associated with the channel uses this kinetic energy to carry out **chemiosmotic phosphorylation** of ADP to ATP.

3. Electrons leaving the cytochrome system are picked up by P700, the reaction-center chlorophyll *a* molecule of Photosystem I. As this antenna receives solar energy the electrons are boosted to their highest energy level yet and they move from the reaction center to a molecule that passes them on to NADP$^+$. In the end NADPH results:

$$NADP^+ + 2e^- + H^+ \xrightarrow{\text{NADP}^+\text{ reductase}} NADPH$$

Since these hydrogen ions come from the stroma this reaction also serves to increase the H$^+$-electrochemical gradient across the thylakoid membrane.

The preceding events (paragraphs 1–3) are called the process of **noncyclic photophosphorylation,** because (a) it is possible to trace electrons in a one-way (noncyclic) direction from water to NADP$^+$, and (b) the energy from sunlight is used to produce ATP (photophosphorylation).

The noncyclic electron pathway produces both ATP and NADPH, which are used in the stroma to reduce carbon dioxide.

Cyclic Electron Pathway

The other electron pathway in chloroplasts is the cyclic electron pathway. In this pathway, electrons leave P700 and then eventually return to it (dotted line in figure 8.6), instead of reacting

Figure 8.7

Chemiosmotic phosphorylation in a chloroplast. *a.* Entire organelle, showing, in particular, the grana containing the thylakoids and the stroma. *b.* Water-splitting enzymes are located on the inner side of the thylakoid membrane so that the hydrogen ions stay within the thylakoid space. Also, some of these carriers are positioned so that hydrogen ions can be picked up from the stroma and passed across the membrane to be deposited in the thylakoid space. Finally, the enzyme NADP reductase, which reduces $NADP^+$ to NADPH, is probably located on the outer surface of the thylakoid membrane, because $NADP^+$ is located in the stroma. When $NADP^+$ combines with a hydrogen ion, this increases the electrochemical gradient, between thylakoid space and stroma. *c.* The CF_1 particles that allow H^+ to flow down their electrochemical gradient are positioned so that ATP is formed in the stroma.

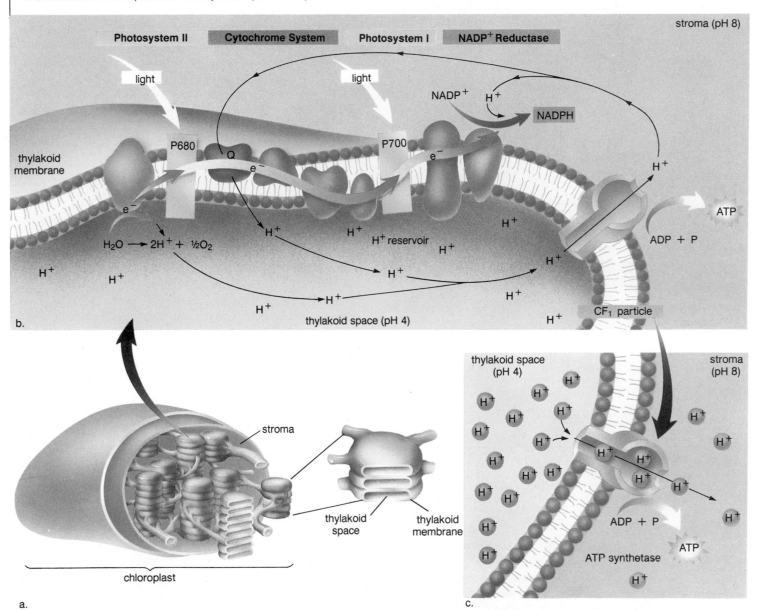

with $NADP^+$. Before they return to P700, they pass down the electron transport system, and ATP is produced by chemiosmotic phosphorylation as previously described. This pathway is sometimes called **cyclic photophosphorylation** because (a) it is possible to trace electrons in a cycle from P700 to P700 and (b) the energy from sunlight is used to produce ATP (photophosphorylation).

It is believed that the cyclic electron pathway, and therefore Photosystem I, evolved very early (before Photosystem II) in the history of the earth. There are some photosynthetic bacteria, even today, that utilize the cyclic electron pathway only.

These bacteria do not produce O_2 because they do not split water. Instead, other sources provide the hydrogen atoms needed to reduce CO_2; for example, the photosynthetic sulfur bacteria take hydrogen atoms from H_2S. With the evolution of Photosystem II, possibly within the cyanobacteria, a much better system became available. Both photosystems working together absorb enough energy to promote the breakdown of water and to produce NADPH, in addition to ATP. It is lucky for us that the breakdown of water also releases oxygen into the atmosphere!

Table 8.1 Characteristics of the Light-Dependent Reactions

Noncyclic Electron Pathway

Participants	Function	Results
Water	Splits to give O$_2$ H$^+$ Electrons	$\boxed{O_2}$ in atmosphere Remains in thylakoid space Pass to Photosystem II
Photosystem II	Absorbs solar energy	Supplies energized electrons
Electron transport system	Buildup of H$^+$	Establishes electrochemical gradient
CF$_1$ particle	Permits flow of H$^+$ down gradient	Chemiosmotic phosphorylation produces $\boxed{ATP}$
Photosystem I	Absorbs solar energy	Supplies energized electrons
NADP$^+$	Final acceptor for electrons	Becomes $\boxed{NADPH}$

Cyclic Electron Pathway

Participants	Function	Results
Photosystem I	Absorbs solar energy	Supplies energized electrons
Electron transport system	Buildup of H$^+$	Establishes electrochemical gradient
CF$_1$ particle	Permits flow of H$^+$ down gradient	Chemiosmotic phosphorylation produces $\boxed{ATP}$

From this discussion, you might think that the cyclic electron pathway is only vestigial and of little use to water-splitting photosynthesizers. However, the cyclic electron pathway may be utilized whenever CO$_2$ is in such limited supply that carbohydrate is not being produced. At these times, all available NADP$^+$ is already NADPH and there is no more NADP$^+$ to receive electrons. Obviously, the energized electrons from Photosystem I have to go somewhere, and under these circumstances, the cyclic electron pathway acts as a sort of safety valve. Finally, the cyclic electron pathway does provide another way to establish a chemiosmotic gradient for ATP production.

The cyclic electron pathway utilizes only Photosystem I and produces only ATP. It probably evolved before Photosystem II.

The characteristics of the light-dependent reactions are summarized in table 8.1.

Light-Independent Reactions

The light-independent reactions are the second stage of photosynthesis. They take their name from the fact that light is not directly required for these reactions to proceed. In this stage of photosynthesis, the NADPH and ATP produced by the noncyclic electron pathway during the first stage are used to reduce carbon dioxide: CO$_2$ becomes CH$_2$O within a carbohydrate molecule. This reduction process is a building-up, or a synthetic, process because it requires the formation of new bonds. Hydrogen atoms (e$^-$ + H$^+$) and energy are needed for reduction synthesis, and these are supplied by NADPH and ATP.

Figure 8.8

a. To discover the reactions involved in carbon dioxide fixation, Calvin and his colleagues used the apparatus shown here. Algae (*Chlorella*) were placed in the flat flask, which was illuminated by two lamps. Radioactive carbon dioxide (^{14}CO$_2$) was added and the algae were killed by transferring them into boiling alcohol (in the beaker below the flat flask). This treatment instantly stopped all chemical reactions in the cells. Carbon-containing compounds were then extracted from the cells and analyzed. *b.* Calvin and his colleagues were able to trace the reaction series, or pathway, through which CO$_2$ was incorporated by tracing the radioactive carbon. They found that the radioactive carbon (shown in boxes) is first incorporated into ribulose bisphosphate (RuBP) and that the resulting molecule immediately splits to form two molecules of phosphoglycerate. By gradually increasing the period of exposure of the algae to ^{14}CO$_2$ before killing, they were able to identify the rest of the molecules in the cycle that is now called the Calvin-Benson cycle (*see* fig. 8.9).

a.

b.

The light-independent reactions use the ATP and NADPH from the light-dependent reactions to reduce carbon dioxide.

Calvin-Benson Cycle—the C₃ Pathway

The reduction of carbon dioxide occurs in the stroma of the chloroplast by means of a series of reactions known as the **Calvin-Benson cycle,** or the **C₃ pathway.** Melvin Calvin received a Nobel Prize for his part in determining these reactions (fig. 8.8), which can be divided into three stages: CO_2 fixation, CO_2 reduction, and regeneration of ribulose bisphosphate.

Carbon Dioxide Fixation

The Calvin-Benson cycle (figs. 8.9 and 8.10) begins when a five-carbon molecule called ribulose bisphosphate (**RuBP**) combines with carbon dioxide. The resulting six-carbon molecule immediately breaks down to form two molecules of phosphoglycerate (PGA). PGA is a three-carbon molecule and that is why the Calvin-Benson cycle is also known as the **C₃ pathway.**

The enzyme that speeds up this first reaction, which is called RuBP carboxylase, or **rubisco,** is responsible for **carbon dioxide fixation**—the attachment of carbon dioxide to an organic compound, in this case, RuBP. Rubisco makes up about 20% to 50% of the protein content in chloroplasts and some say it is the most abundant protein in the world. The reason for its

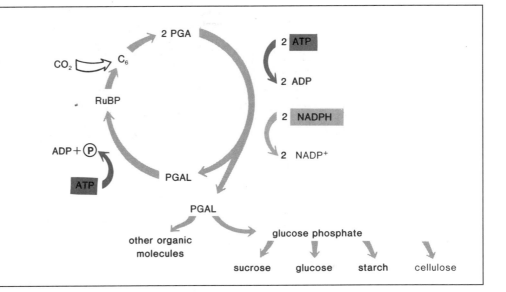

Figure 8.9
Calvin-Benson cycle, simplified. The enzyme rubisco fixes CO_2 to RuBP forming a six-carbon molecule, which immediately breaks down to two PGA molecules. These are then reduced to form two molecules of PGAL. This step uses the NADPH and some of the ATP provided by the light-dependent reactions. PGAL, the end product of photosynthesis, is sometimes used as the starting material for molecules like amino acids or lipids, or it can combine with another molecule of PGAL to form glucose phosphate, a molecule that can be converted to sucrose, glucose, starch, or cellulose. A considerable amount of PGAL is also used to reform RuBP.

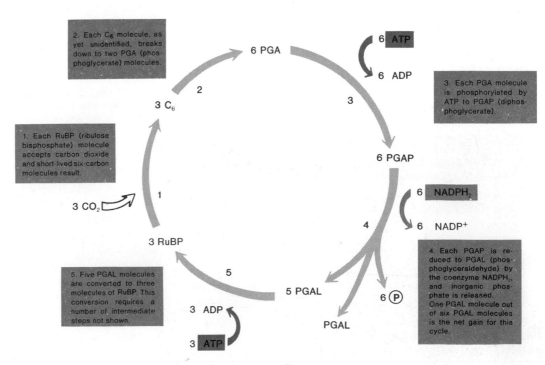

Figure 8.10
Calvin-Benson cycle, detailed. The molecules in the cycle have been multiplied by three because after five molecules of PGAL (total of 15 carbons) are converted to three molecules of RuBP (15 carbons), one molecule of PGAL will remain. Since it takes two PGAL to produce one six-carbon sugar, the numbers shown must be multiplied by two in order for a monosaccharide to be the end product. Compare steps 3 and 4 to steps 5 and 4 in figure 9.5.

1. Each RuBP (ribulose bisphosphate) molecule accepts carbon dioxide and short-lived six-carbon molecules result.

2. Each C₆ molecule, as yet unidentified, breaks down to two PGA (phosphoglycerate) molecules.

3. Each PGA molecule is phosphorylated by ATP to PGAP (diphosphoglycerate).

4. Each PGAP is reduced to PGAL (phosphoglyceraldehyde) by the coenzyme NADPH₂ and inorganic phosphate is released. One PGAL molecule out of six PGAL molecules is the net gain for this cycle.

5. Five PGAL molecules are converted to three molecules of RuBP. This conversion requires a number of intermediate steps not shown.

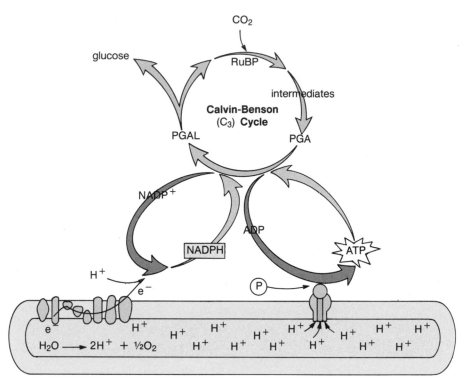

Figure 8.11
Photosynthesis consists of two sets of reactions. The light-independent reactions, which take place in the thylakoid membrane, produce the NADPH and ATP required for CO_2 reduction during the Calvin-Benson cycle, which takes place in the stroma.

abundance may be that it works unusually slow (it processes only about three molecules of substrate per second compared to about 1,000 per second for a typical enzyme) and so there has to be a lot of it to keep this cycle moving.

Reduction of Carbon Dioxide

Each molecule of PGA now undergoes reduction to PGAL (phosphoglyceraldehyde) in two steps:

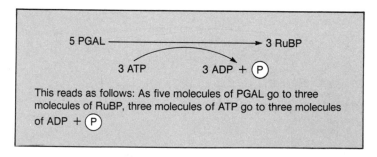

This is read as follows: As PGA goes to PGAL, ATP goes to ADP + (P) and NADPH goes to NADP$^+$.

The reactions that reduce **PGA** to **PGAL** use up the NADPH and some of the ATP formed in the thylakoid membrane during the noncyclic electron pathway (fig. 8.11). These reactions also represent the reduction of carbon dioxide and its conversion to a high-energy molecule. In other words, PGAL contains more hydrogen atoms than does PGA. PGAL is the immediate photosynthetic product of the Calvin-Benson cycle.

For every six turns of the Calvin-Benson cycle (fig. 8.10), there is a net gain of one PGAL molecule. Therefore, *PGAL is sometimes said to be the end product of photosynthesis.* However, since two PGAL can combine to form glucose phosphate, this molecule (or simply glucose) is often called the end product of photosynthesis. Within the leaves of higher plants glucose

molecules are converted to the disaccharide sucrose for transport to other parts of the plant where glucose is used for the synthesis of macromolecules like cellulose and starch, or is broken down to provide energy for ATP formation.

Regeneration of RuBP

For every six turns of the Calvin-Benson cycle, five molecules of PGAL are used to reform three molecules of ribulose bisphosphate (RuBP):

> 5 PGAL ⟶ 3 RuBP
>
> 3 ATP   3 ADP + (P)
>
> This reads as follows: As five molecules of PGAL go to three molecules of RuBP, three molecules of ATP go to three molecules of ADP + (P)

This reaction also utilizes some of the ATP produced by the noncyclic electron pathway. Altogether, the ATP and NADPH consumed in coverting CO_2 to the level of a six-carbon sugar represents about a 30% energy-conversion rate. In other words, 30% of the energy available in ATP and NADPH is finally found in glucose.

During photosynthesis in the C_3 pathway, CO_2 is reduced to PGAL by the Calvin-Benson cycle operating in the stroma. Two PGAL molecules then join to form glucose phosphate, which can be converted to the various organic molecules needed by plants.

Table 8.2 summarizes the light-independent reactions.

Table 8.2 Summary of the Light-Independent Reactions

Participant	Function	Results
RuBP	Takes up CO_2	Carbon dioxide fixation
CO_2	Provides carbon atoms	Reduced to CH_2O
ATP	Provides energy for reduction and generation of RuBP	ADP + P
NADPH	Provides electrons for reduction	$NADP^+$
PGAL	End product of photosynthesis	2 PGAL become glucose phosphate

The C_4 Pathway

Some plants that are adapted to hot, dry environments (fig. 8.12) do not use the Calvin-Benson cycle for carbon dioxide fixation. Instead of using the enzyme RuBP carboxylase (rubisco) to fix carbon dioxide to RuBP, they utilize the enzyme PEP carboxylase (**pepco**) to fix CO_2 to PEP (phosphoenolpyruvate):

$$\text{PEP} + CO_2 \xrightarrow{\text{pepco}} \text{oxaloacetate} \quad (C_4 \text{ plants})$$

$$\text{RuBP} + CO_2 \xrightarrow{\text{rubisco}} \text{2 PGA} \quad (C_3 \text{ plants})$$

These plants are known as C_4 plants, because the product of CO_2-fixation reaction, oxaloacetate, is a C_4, or four-carbon, molecule.

Since we have repeatedly mentioned that there is usually a close correlation between structure and function, you will probably not be surprised to learn that the structure of a leaf from a C_3 plant is different from that of a C_4 plant:

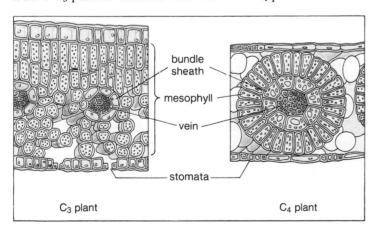

bundle sheath

mesophyll

vein

stomata

C_3 plant

C_4 plant

In a **C_3 plant** the mesophyll cells contain chloroplasts and carry out both the light-dependent and the light-independent reactions (Calvin-Benson cycle). In a **C_4 plant** certain **mesophyll cells** are arranged around the bundle sheath, and only these and the bundle sheath cells contain chloroplasts. The Calvin-Benson cycle occurs only in the **bundle sheath cells,** and not in the mesophyll cells.

Figure 8.12
Percentage of grasses that have C_4 photosynthesis in areas of North America. The higher temperatures in southern locations apparently offer some advantage to these plants because they are more prevalent in these locations.

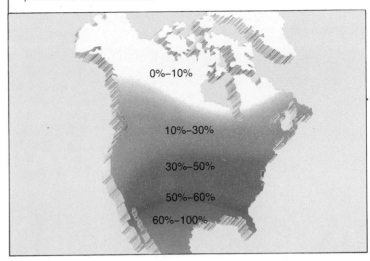

0%–10%

10%–30%

30%–50%

50%–60%

60%–100%

Necessarily, it is the cells that surround the air spaces that take up CO_2. In a C_4 plant, oxaloacetate is formed and reduced in the mesophyll cells, and then is essentially "pumped" into the bundle sheath cells. There it contributes CO_2 to the Calvin-Benson cycle (fig. 8.13), after which pyruvate is returned to the mesophyll cells. It then takes two ATP molecules to regenerate PEP. You would think that these "extra reactions" of the C_4 pathway would represent an energy drain for these plants and yet in hot dry climates the net photosynthetic rate of C_4 plants like sugarcane, corn, and sorghum is about two to three times that of C_3 plants like wheat, rice, and oats.

C_4 photosynthesis utilizes the enzyme pepco to fix carbon dioxide in mesophyll cells. This fixed carbon dioxide then passes to the bundle sheath cells that carry on the Calvin-Benson cycle. C_4 plants have a higher net photosynthetic rate in hot, dry climates than C_3 plants do.

Photorespiration

The explanation for the higher energy efficiency of C_4 plants in hot dry climates compared to C_3 plants lies in the difference between **rubisco** and **pepco,** the respective enzymes for carbon dioxide fixation in C_3 and C_4 plants. O_2 competes with CO_2 for the active site of rubisco, but not for the active site of pepco. If there is an adequate supply of CO_2 inside the leaf, then most of RuBP is converted to two PGA molecules, and the C_3 pathway operates efficiently. But if the supply of CO_2 inside the leaf is inadequate, most of RuBP combines with O_2 giving one molecule of PGA and one molecule of phosphoglycolic acid; the latter rapidly breaks down to release CO_2:

Photorespiration

$$\text{RuBP} + O_2 \longrightarrow \text{PGA} + \text{phosphoglycolic acid} \longrightarrow CO_2$$

Figure 8.13

The C_4 pathway is superimposed on an electron micrograph of the cells of the mesophyll and the bundle sheath in corn. During the C_4 pathway, the enzyme called pepco fixes CO_2 to PEP forming oxaloacetate, which is reduced to a molecule that carries CO_2 to the Calvin-Benson cycle in the bundle sheath cells. Pyruvate then returns to the mesophyll where ATP is used to reform PEP. Magnification, ×8,300.

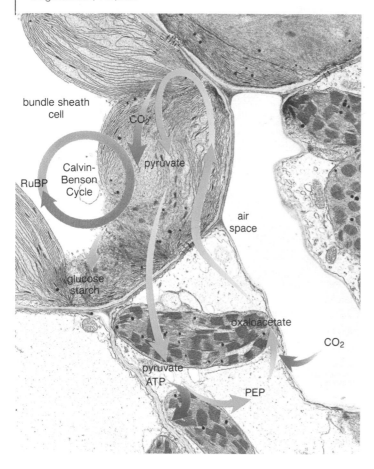

Figure 8.14

Most CAM plants, such as this stonecrop, are succulents that have fleshy stems or leaves. They conserve water by keeping their stomata closed during the day and can live under very arid conditions.

The name for this process, **photorespiration,** is based on the fact that in the presence of light (*photo*), oxygen is taken up and CO_2 is evolved (*respiration*). Obviously, this is just the reverse of the photosynthetic process; photorespiration wastes energy and does nothing to serve the needs of the plant.

You'll notice from page 119 that there are little openings called stomata in the surfaces of leaves through which water can leave and carbon dioxide can enter. If the weather is hot and dry, these openings close in order to conserve water. Therefore, this is also the time when the concentration of CO_2 will be the lowest in the leaf and when photorespiration will predominate in a C_3 plant. C_4 photosynthesis evolved as a way to circumvent this problem, since pepco will never bind to O_2—regardless of the CO_2 concentration in the leaf. The energy-requiring "extra reactions" of C_4 photosynthesis that occur in the mesophyll cells can deliver a constant supply of CO_2 to the Calvin-Benson cycle in the bundle sheath cells. This keeps the CO_2 concentration high in these cells and prevents photorespiration from occurring to any extent, even if the weather is hot and dry.

When the weather is moderate, C_3 plants are not at a disadvantage, but when the weather becomes hot and dry C_4 plants have a distinct advantage, and we can expect them to predominate (fig. 8.11). The same phenomenon can even be observed in the lawns in the cooler parts of this country. In the early summer, C_3 plants such as Kentucky bluegrass and creeping bentgrass predominate, but by mid-summer, crabgrass, a C_4 plant, begins to take over.

C_4 plants have an advantage over C_3 plants when the weather is hot and dry, because their method of CO_2 fixation and their related structure prevent photorespiration from occurring to any extent.

CAM Pathway

CAM plants (fig. 8.14) use pepco (PEP carboxylase) to fix carbon dioxide at *night,* forming malate (malic acid) that is stored in large vacuoles in their mesophyll cells until the next day. CAM stands for crassulacean-acid metabolism. The Crassulaceae are a family of flowering, succulent plants, which includes the stonecrops—low-lying plants adapted to warm, arid regions of the world. CAM was first discovered in these plants, but now it is known to be prevalent among most plants that grow in desert environments, including the cacti.

Whereas C_4 photosynthesis represents partitioning in space—carbon fixation occurs in mesophyll cells and the Calvin-Benson cycle occurs in bundle sheath cells—CAM is an example of partitioning of photosynthesis in time. The malate formed at night is released from the mesophyll vacuoles during the day, and it is taken up by the Calvin-Benson cycle within the *same* cell, which now has NADH and ATP available to it from the light-dependent reactions. The primary reason for this partitioning has to do with the conservation of water again. To conserve water, CAM plants open their stomata only at night, and therefore only at that time is atmospheric CO_2 available to the pepco enzyme. When sunlight reaches the plant, the stomata close, and CO_2 cannot enter the plant.

CAM is usually not as efficient as either C_3 or C_4 photosynthesis, but it does allow plants to live under stressful conditions. Each of the three forms of photosynthesis has its advantages and disadvantages: CAM plants are not efficient photosynthesizers, but they can live under arid conditions; C_4 plants are efficient photosynthesizers, but they do not compete well with C_3 plants below a temperature of 25°C. In other words, they appear to be sensitive to cold conditions.

CAM plants use pepco to carry on carbon dioxide fixation at night when the stomata are open. During the day, the stored carbon dioxide can enter the Calvin-Benson cycle.

Table 8.3 contrasts the features of these three types of plants.

Table 8.3 The Three Pathways of Carbon Dioxide Fixation

Feature	C_3	C_4	CAM
Leaf structure	Bundle sheath cells lack chloroplasts	Bundle sheath cells have chloroplasts	Large vacuoles in mesophyll cells
Enzyme utilized	Rubisco	Pepco	Pepco
Optimum temperature	15°C–25°C	30°C–40°C	35°C
Productivity rate (tons/ hectare/ year)	22 ± 0.3	39 ± 17	Low—variable

Modified with permission from D. K. Northington and J. R. Goodin, *The Botanical World*. St. Louis, 1984, Times Mirror/Mosby College Publishing.

Summary

1. Photosynthesis (1) provides food either directly or indirectly for most living things, (2) produces oxygen, and (3) provided the energy present today in fossil fuels.

2. Photosynthesis uses solar energy in the visible-light range; the photons of this range contain the right amount of energy to energize the electrons of chlorophyll molecules. Specifically, chlorophyll *a* and *b* absorb purple-blue and orange-red wavelengths best. This causes chlorophyll to appear green to us.

3. A chloroplast is bounded by a double membrane and contains two main portions: the liquid stroma and the membranous grana made up of thylakoid sacs. The light-dependent reactions take place in the thylakoids, and the light-independent reactions take place in the stroma.

4. The noncyclic electron pathway of the light-dependent reactions begins when solar energy enters Photosystem II, where it activates the reaction center P680. Energized electrons leave P680 and go to an electron acceptor. The splitting of water replaces these electrons in P680, releases oxygen to the atmosphere, and provides hydrogen ions (H^+) to the thylakoid space. The acceptor molecule passes electrons to Photosystem I by way of an electron transport (cytochrome) system, which pumps hydrogen ions (H^+) into the thylakoid space, contributing to an electrochemical gradient that drives the synthesis of ATP from ADP + Ⓟ. Light energy absorbed by Photosystem I is received by its reaction center P700. The energized electrons that leave this molecule are ultimately received by $NADP^+$, which also combines with H^+ from the stroma to become NADPH.

5. In the cyclic electron pathway, the energized electrons that leave Photosystem I are not sent to $NADP^+$. Instead, they pass down the electron transport system back to Photosystem I again. The cyclic pathway may occur only if there is no free $NADP^+$ to receive the electrons, as a way to prevent an energy overload of Photosystem I. It serves to generate ATP and, in keeping with its name, it evolved first and is the only photosynthetic pathway present in certain bacteria today.

6. Photophosphorylation in chloroplasts is a chemiosmotic process. The buildup of H^+ inside the thylakoid space establishes an electrochemical gradient. When H^+ flows down this gradient through the channel present in CF_1 particles, an ATP synthetase produces ATP from ADP and Ⓟ.

7. The energy yield of the light-dependent reactions is stored in ATP and NADPH. These molecules are now used by the light-independent reactions to reduce carbon dioxide to organic carbon.

8. During C_3 photosynthesis, the enzyme rubisco fixes carbon dioxide to RuBP giving a six-carbon molecule that immediately breaks down to two three-carbon PGA molecules. This is the first stage of the Calvin-Benson cycle. During the second stage, PGA is reduced to PGAL and this step requires the NADPH and some of the ATP from the light-dependent reactions. For every six turns of the cycle, the net gain is one PGAL molecule; the other five PGAL molecules are used to reform three molecules of RuBP. This step also requires ATP for energy.

9. During C_4 photosynthesis, the enzyme pepco fixes carbon dioxide to PEP to form a four-carbon molecule, oxaloacetate, within mesophyll cells. A reduced form of this molecule is pumped into bundle sheath cells where carbon dioxide is released to the Calvin-Benson cycle. The resulting pyruvate returns to mesophyll cells where ATP must be used to regenerate PEP. C_4 plants avoid photorespiration by a partitioning of pathways in space—carbon dioxide fixation in mesophyll cells and Calvin-Benson cycle in bundle sheath cells.

10. During CAM photosynthesis, pepco fixes carbon dioxide to PEP at night to produce malate, which is stored in vacuoles overnight. Carbon dioxide is released the next day and enters the Calvin-Benson cycle within the same cells. This represents a partitioning of pathways in time: carbon dioxide fixation at night and the Calvin-Benson cycle during the day. The plants that carry on CAM photosynthesis are desert plants, in which the stomata only open at night in order to conserve water.

Objective Questions

1. The absorption spectrum of chlorophyll
   a. approximates the action spectrum of photosynthesis.
   b. explains why chlorophyll is a green pigment.
   c. shows that some colors of light are absorbed more than others.
   d. All of these.
2. Chemiosmotic phosphorylation takes place
   a. between the outer and inner membranes of chloroplasts.
   b. in the stroma.
   c. across the thylakoid membrane.
   d. All of these.
3. The final acceptor of electrons during the noncyclic electron pathway is
   a. Photosystem I.
   b. Photosystem II.
   c. ATP.
   d. $NADP^+$.
4. A photosystem contains
   a. pigments, a reaction center, and an electron acceptor.
   b. ADP, Ⓟ, and H^+.
   c. protons, photons, and pigments.
   d. Only (b) and (c).

5. Which of these should not be associated with the electron transport system?
   a. cytochromes
   b. movement of H^+ into the thylakoid space
   c. photophosphorylation
   d. absorption of solar energy
6. The oxygen evolved during photosynthesis comes from
   a. carbon dioxide and the Calvin-Benson cycle.
   b. photosystems and C_4 photosynthesis.
   c. the acceptor molecules.
   d. the breakdown of water.
7. Pepco has an advantage compared to rubisco. The advantage is that
   a. rubisco doesn't function at high temperatures but pepco does.
   b. rubisco fixes carbon dioxide only in C_4 plants, whereas pepco does it in both C_3 and C_4 plants.
   c. rubisco is subject to photorespiration but pepco is not.
   d. Only (b) and (c).

8. The NADPH and ATP from the light-dependent reactions are used to
   a. cause rubisco to fix carbon dioxide.
   b. reform the photosystems.
   c. cause electrons to move along their pathways.
   d. convert PGA to PGAL.
9. CAM photosynthesis
   a. is the same as C_4 photosynthesis.
   b. is an adaptation to cold environments in the southern hemisphere only.
   c. is prevalent in desert plants that close their stomata during the day.
   d. occurs in plants that live in marshy areas.
10. Chemiosmotic phosphorylation depends on
   a. an electrochemical gradient.
   b. a difference in H^+ concentration between the thylakoid space and the stroma.
   c. ATP breaking down to ADP + Ⓟ.
   d. Both (a) and (b).

Study Questions

1. Why is it proper to say that all living things are dependent on solar energy?
2. Contrast the electromagnetic spectrum with the absorption spectrum of chlorophyll. Why is chlorophyll a green pigment?
3. What are the inputs and outputs for the light-dependent reactions and for the light-independent reactions? Where do these reactions take place in the chloroplast?

4. Describe the structure of a photosystem. Explain the terms P700 and P680.
5. Using the Z-scheme on page 114, trace noncyclic and cyclic electron pathways explaining the events as they occur.
6. Contrast the noncyclic and cyclic electron pathways and explain the differences noted.
7. Describe the chemiosmotic process of ATP production mentioning the thylakoid membrane, the electron transport system, the CF_1 particles, and ATP synthetase.

8. Describe the three stages of the Calvin-Benson cycle. What happens during the stage that utilizes ATP and NADPH from the light-dependent reactions? What specific molecules are involved?
9. Explain C_4 photosynthesis, being sure to contrast the actions of rubisco and pepco.
10. Explain CAM photosynthesis, contrasting it to C_4 photosynthesis in terms of partitioning of a pathway.

Thought Questions

1. Some bacterial photosynthesizers don't use water as an electron donor. For example, the green sulfur and purple sulfur bacteria can use hydrogen sulfide (H_2S). Why don't these bacteria emit oxygen to the atmosphere? What do they give off instead?

2. In the previous chapter we mentioned how the process of coupling is used to transfer energy. Is there a direct coupling between the release of energy by the electron transport system and ATP buildup? Explain.

3. Criticize the terminology light-dependent and light-independent reactions. What terminology would you suggest instead?

Selected Key Terms

photon (fo′ton) 110

electromagnetic spectrum (e-lek″tro-mag-net′ik spek′trum) 110

visible light (viz′ĭ-b′l līt) 111

chloroplast (klo′ro-plast) 111

stroma (stro′mah) 111

grana (gra′nah) 111

thylakoid (thi′lah-koid) 111

chlorophyll (klo′ro-fil) 112

light-dependent reaction (līt de-pen′dent re-ak′shun) 112

light-independent reaction (līt in″de-pen′dent re-ak′shun) 112

photosystem (fo″to-sis′tem) 113

chemiosmotic phosphorylation (kem″i-os-mot′ik fos″for-i-la′shun) 114

noncyclic photophosphorylation (non-sik′lik fo″to-fos″for-i-la′shun) 114

cyclic photophosphorylation (sik′lik fo″to-fos″for-i-la′shun) 115

Calvin-Benson cycle (kal′vin ben′sun si′k′l) 117

RuBP 117

carbon dioxide fixation (kar′bon di-ok′sīd fix-sa′shun) 117

C_3 plant 119

C_4 plant 119

CHAPTER 9

Glycolysis and Cellular Respiration

Your study of this chapter will be complete when you can:

1. *Describe the general function of cellular respiration.*
2. *Describe both the anaerobic and aerobic fate of pyruvate.*
3. *Discuss glycolysis, explaining the process of substrate-level phosphorylation and giving the inputs and outputs of the pathway.*
4. *Describe lactate fermentation and alcoholic fermentation and the evolutionary significance of this process.*
5. *Discuss the transition reaction, explaining the role of acetyl CoA and giving the inputs and outputs of the reaction.*
6. *Discuss the Krebs cycle, explaining how it cycles and giving the inputs and outputs of the cycle.*
7. *Discuss the electron transport system, explaining how it contributes to the formation of ATP.*
8. *Describe the arrangement of electron carriers in the inner membrane of mitochondria, and explain the process of chemiosmotic phosphorylation.*
9. *Calculate the yield of ATP molecules per glucose molecule for glycolysis versus cellular respiration.*
10. *Discuss the concept of a metabolic pool and how the breakdown of carbohydrates, proteins, and fats contributes to the pool.*

The energy derived from eating fish powers the muscles of this Eurasian kingfisher so that it can rise from the water in flight. Mitochondria are the organelles that convert the energy of organic compounds to the energy of ATP. Only ATP can cause muscles to contract.

Glucose Metabolism, an Overview

Most organisms (fig. 9.1) break down carbohydrate molecules (e.g., glucose) in order to acquire a supply of ATP molecules. We can write the overall equation for carbohydrate breakdown in this manner:

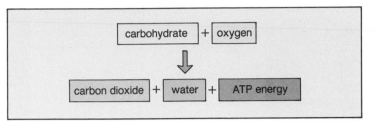

ATP, the energy currency of cells (p. 103), provides the energy that cells need for transport work, mechanical work, and synthetic work. Why do living things convert carbohydrate energy to ATP energy? Why not, for example, utilize glucose energy directly, thereby bypassing the need for mitochondria? The answer is that the energy content of glucose is too great for individual cellular reactions; ATP contains the *right* amount of energy and enzymes are adapted through evolution to couple ATP breakdown to energy-requiring cellular processes (fig. 7.12).

In cells the energy of carbohydrates is converted to that of ATP molecules, the energy currency of cells.

You know from the previous chapters that chloroplasts within plant and algal cells produce the energy-rich carbohydrates that typically undergo cellular respiration in plants and animals (fig. 9.1b). **Cellular respiration,** then, is the link between the energy captured by plant cells and the energy utilized by both plant and animal cells. It is the means by which energy flows from the sun through living things. This is true because, for example, plants produce the food eaten by animals.

Whereas the energy content of food is slowly dissipated, the chemicals, themselves, cycle. Note that when carbohydrate is broken down, the low-energy molecules, carbon dioxide and water, are given off. These are the raw materials for photosynthesis (fig. 9.1b).

The energy content of the organic molecules that are broken down in cells was originally provided by solar energy.

Figure 9.2 gives an overview of glucose metabolism. Glucose is the molecule that most cells use as an energy source in order to produce ATP energy. The first part of glucose metabolism is called glycolysis, a process that occurs in the cytosol and is anaerobic—it does not require oxygen. During **glycolysis** little breakdown occurs as glucose is converted to two molecules of **pyruvate**[1]. Oxidation by removal of hydrogen atoms ($e^- + H^+$) does occur, though, and the energy of this oxidation is used to generate two molecules of ATP.

[1]Pyruvate is the ionized form of pyruvic acid. The terms can be used interchangeably, although at the pH of the cell the ionized form is prevalent.

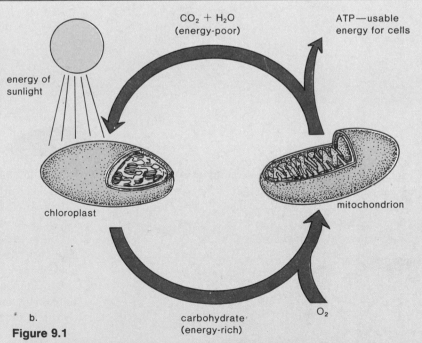

Figure 9.1

a. Which of these two organisms is carrying on cellular respiration—the gerenuk or the acacia tree? Both are, because both animals and plants require a source of ATP energy.
b. The energy-rich carbohydrate molecules, broken down largely within the mitochondria, are originally derived from photosynthesis carried out within chloroplasts. The flow of *energy* among organisms (e.g., from plants to animals) goes one way only, but the *chemicals* can recycle because the carbon dioxide and water given off by the mitochondria are reused by the chloroplasts.

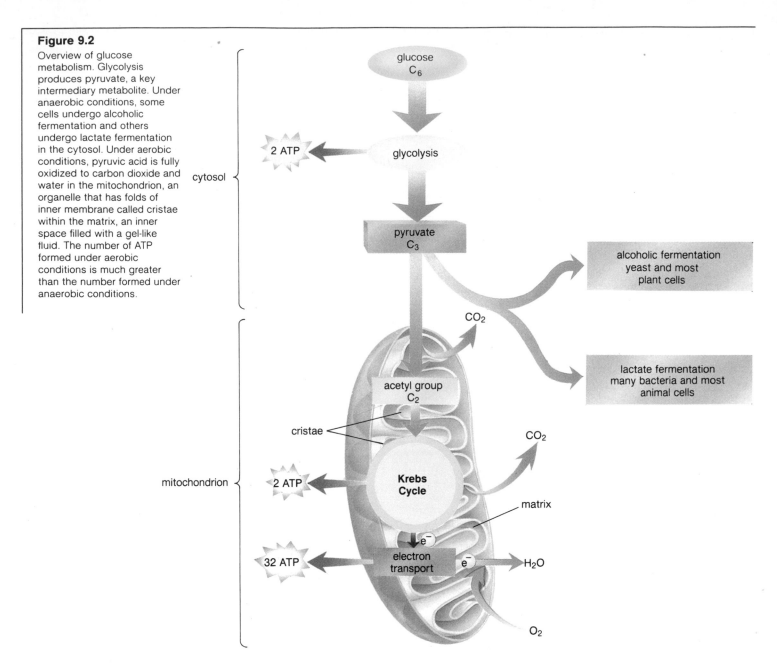

Figure 9.2

Overview of glucose metabolism. Glycolysis produces pyruvate, a key intermediary metabolite. Under anaerobic conditions, some cells undergo alcoholic fermentation and others undergo lactate fermentation in the cytosol. Under aerobic conditions, pyruvic acid is fully oxidized to carbon dioxide and water in the mitochondrion, an organelle that has folds of inner membrane called cristae within the matrix, an inner space filled with a gel-like fluid. The number of ATP formed under aerobic conditions is much greater than the number formed under anaerobic conditions.

glucose
C_6

2 ATP

glycolysis

cytosol

pyruvate
C_3

alcoholic fermentation
yeast and most
plant cells

CO_2

lactate fermentation
many bacteria and most
animal cells

acetyl group
C_2

cristae

CO_2

mitochondrion

2 ATP

**Krebs
Cycle**

matrix

e^-

32 ATP

electron
transport

e^-

H_2O

O_2

What happens next to pyruvate depends on whether or not oxygen is available to the cell. If oxygen is not available, **fermentation** occurs within the cytosol. During fermentation, pyruvate accepts the two molecules of hydrogen that were removed from glucose earlier during glycolysis:

$$\begin{array}{c} 2H \\ \text{pyruvate} \longrightarrow \text{lactate} \end{array}$$

$$\begin{array}{c} 2H \\ \text{pyruvate} \longrightarrow \text{alcohol} + CO_2 \end{array}$$

The reduction of pyruvate to alcohol with the release of CO_2, a process called **alcoholic fermentation,** is utilized by yeast and most plant cells. The reduction of pyruvate to lactate, called

lactate fermentation, is utilized by many bacteria and some animal cells (fig. 9.2). Fermentation, an anaerobic process that will be discussed in more detail later in this chapter, yields only two ATP molecules altogether.

On the other hand, if oxygen is available to the cell, **cellular respiration** occurs, an aerobic process that produces much more energy than fermentation. Now, pyruvate is not reduced, and it enters mitochondria where it is completely oxidized. This time the removal of hydrogen atoms ($e^- + H^+$) is accompanied by the release of carbon dioxide. This gas diffuses out of the mitochondria and the cell, but the hydrogen atoms are sent to an electron transport system located in the membrane of mitochondrial **cristae,** folds of the inner membrane (figs. 5.14 and 9.2). Here, the energy made available as electrons are passed from one carrier to the next is eventually used to generate ATP molecules, and oxygen, the final acceptor for these electrons, is reduced to water.

Figure 9.3

It is interesting to think about how our bodies provide the reactants for aerobic cellular respiration and how they dispose of the products. The air we breathe contains *oxygen* (O_2), and the food we eat contains *glucose*. The oxygen and glucose enter the bloodstream, which carries them to the body's cells, where they diffuse into each and every cell. In mitochondria, glucose products are broken down to carbon dioxide (CO_2) and water as ATP is produced. All three of these then leave the mitochondria. The *ATP* is utilized inside the cell for energy-requiring processes. CO_2 diffuses out of the mitochondria and out of the cell into the bloodstream. The bloodstream takes the *carbon dioxide* to the lungs, where it is exhaled. The *water* molecules produced, called metabolic water, plays little, if any, role in humans. In other organisms, metabolic water helps prevent dehydration.

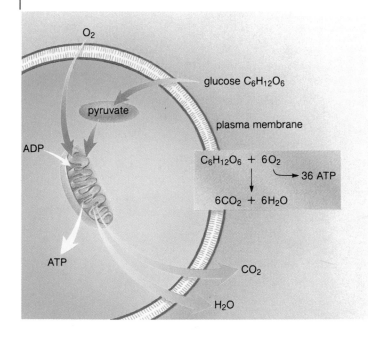

Most of the ATP produced during cellular respiration comes from the electron transport system, and since this system is located within the mitochondria, they are sometimes called the "powerhouse of the cell."

Glycosis, the first part of glucose breakdown, results in the formation of two molecules of pyruvate. Pyruvate can then either undergo fermentation, an anaerobic process, or cellular respiration, an aerobic process. In fermentation, pyruvate is metabolized to end products such as lactate, or alcohol and CO_2, depending on the type of cell. During cellular respiration pyruvate is oxidized completely within mitochondria. They contain an electron transport system which is responsible for most of the ATP production in cells.

Glycolysis and Cellular Respiration

If glycolysis is followed by cellular respiration, it is possible to write an overall reaction for glucose breakdown to carbon dioxide and water as in figure 9.3.

The overall equation indicates that as glucose is broken down, 36 molecules of ATP are formed. The overall reaction does not give a hint that the entire process of glucose breakdown to carbon dioxide and water actually requires many steps.

The burning of an organic material, like wood or fossil fuels can be compared to glucose breakdown. Both of these processes require oxygen and give off carbon dioxide. However, when a material is burned the energy is released all at once, but when glucose is broken down the energy is released bit by bit and stored in ATP molecules. This is possible because the degradative reactions are carried out in metabolic pathways utilizing enzymes that can break down glucose in sequential steps.

Glycolysis (fig. 9.4) is the anaerobic breakdown of glucose to two molecules of pyruvate. The oxidation of glucose by removal of hydrogen atoms provides enough energy for the immediate formation of two ATP molecules. **NAD$^+$**, a coenzyme of dehydrogenases in cells (p. 101), receives the hydrogen atoms in the following reaction:

$$NAD^+ \xrightarrow{2H} NADH + H^+$$

NADH then carries these hydrogens ($e^- + H^+$) to the mitochondria, and pyruvate, the end product of glycolysis, also enters a mitochondrion.

The **transition reaction** connects glycolysis to the Krebs cycle. In this reaction, one carbon atom, in the form of CO_2, is removed from pyruvate and a two-carbon acetyl group remains. The reaction also involves oxidation, and the hydrogen atoms made available are transferred to NAD$^+$, forming more NADH. The **Krebs cycle** is a cyclical series of reactions located within the **matrix,** or inner space, of the mitochondrion, which is filled with a gel-like fluid (figs. 5.14 and 9.2). As the reactions occur, CO_2 is given off and one molecule of ATP is produced directly. As oxidation occurs, some hydrogen atoms are received by NAD$^+$ and others are received by FAD, another coenzyme associated with dehydrogenases:

$$FAD \xrightarrow{2H} FADH_2$$

Both NADH and FADH$_2$ take their hydrogen atoms to the electron transport system located in the cristae of mitochondria. The **electron transport system** (p. 133) is a series of carriers that pass electrons from one to the other until they are finally received by oxygen, which is then reduced to water. This system pumps hydrogen ions into the space between the inner and outer membrane of a mitochondrion, termed the *intermembrane space.* This sets up an electrochemical gradient that leads to the chemiosmotic phosphorylation of ADP to form ATP. The electron transport system produces 32 of the 36 ATP molecules formed during aerobic cellular respiration.

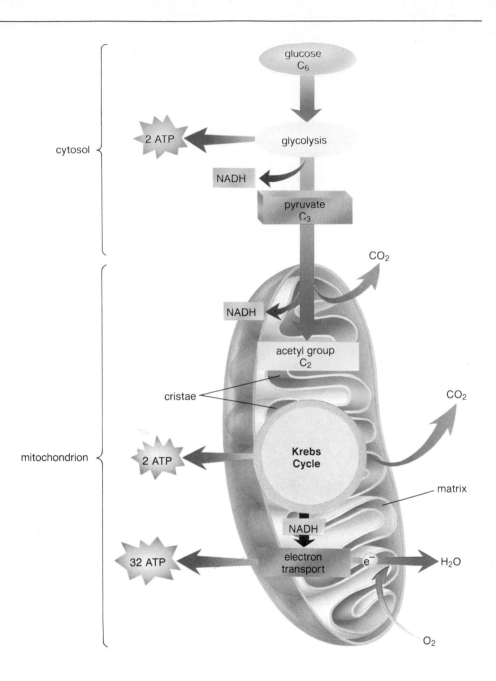

Figure 9.4
Glucose breakdown to carbon dioxide and water utilizes three subpathways (glycolysis, the Krebs cycle, and the electron transport system) plus the transition reaction. NAD$^+$ accepts electrons and becomes NADH and H$^+$ as substrates become oxidized. NADH carries these hydrogen atoms (e$^-$ + H$^+$) to the electron transport system, which uses their energy to produce most of the ATP molecules that result from glucose breakdown.

The complete breakdown of glucose to carbon dioxide and water involves three subpathways and one transition reaction. The pathways are glycolysis, the Krebs cycle, and the electron transport system. Glycolysis occurs in the cytosol, and the other reactions occur in mitochondria, which is where most of the ATP is produced.

Glucose Metabolism outside the Mitochondria

Glycolysis and fermentation are two metabolic pathways that occur outside the mitochondria. They are anaerobic processes that do not require the presence of oxygen.

Glycolysis

The term **glycolysis,** meaning the "splitting of sugar," is appropriate because during the pathway, glucose is split in two. Figure 9.5 gives the main reactions of the pathway. First, two ATP molecules are used to phosphorylate glucose in order to activate it. Later, as hydrogen-carbon bonds are broken, hydrogen atoms are removed from the molecules of the pathway and are picked up by NAD$^+$ which becomes NADH. The energy of oxidation results in high-energy phosphate bonds attached to certain substrates. Now these molecules are used to produce ATP. For example, in figure 9.6 the molecule designated as PGAP gives up a high-energy phosphate to ADP, forming ATP. This process is called **substrate-level phosphorylation** because the substrate has provided the high-energy bond for ATP production.

Figure 9.5

Glycolysis is a metabolic pathway that begins with glucose and ends with pyruvate. Net gain of two ATP molecules can be calculated by subtracting those expended from those produced. Print in boxes explains the reactions.

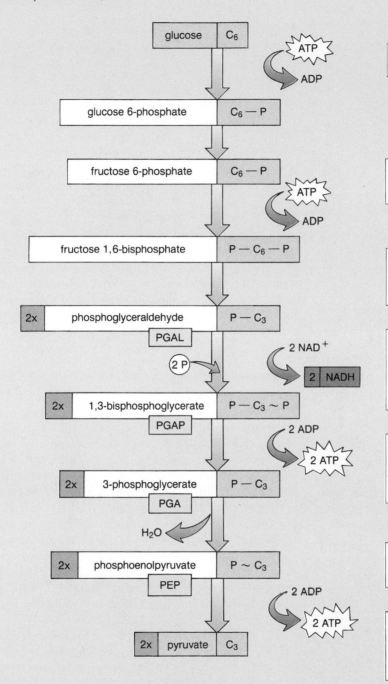

1. Phosphorylation of glucose by ATP produces an activated molecule.

2. Rearrangement, followed by a second ATP phosphorylation, gives fructose bisphosphate.

3. The six-carbon molecule is split into 2 three-carbon PGAL. After this, each reaction has to occur twice per each glucose molecule.

4. Oxidation, followed by phosphorylation, produces 2 NADH molecules and gives 2 PGAP molecules—each with one high-energy phosphate bond, represented by a wavy line.

5. Removal of high-energy phosphate by 2 ADP molecules produces 2 ATP molecules and gives 2 PGA molecules. This pays back the original investment of 2 ATP used in steps 1 and 3.

6. Oxidation by removal of water gives 2 PEP molecules, each with a high-energy phosphate bond.

7. Removal of high-energy phosphate by 2 ADP molecules produces 2 ATP molecules. The net energy yield for the glycolysis pathway is 2 molecules of ATP.

8. Pyruvate is the end product of the glycolytic pathway. If cellular respiration occurs, pyruvate enters the mitochondria for further breakdown.

Figure 9.6

Substrate-level phosphorylation. PGAP transfers its high-energy
phosphate to ADP, forming ATP (step 6 in figure 9.5).

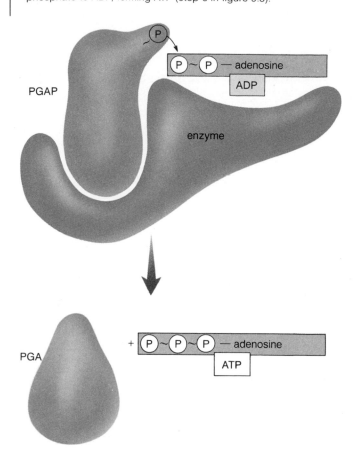

Altogether in the glycolytic pathway (fig. 9.5) four ATP
molecules are produced per glucose molecule but since two ATP
were used to get started, there has been a net gain of only two
ATP.

The inputs and outputs of glycolysis are:

Glycolysis	
inputs	outputs
glucose	2 pyruvates
2 NAD$^+$	2 NADH
2 ADP	2 ATP (net)

Fermentation

In the two most familiar forms of **fermentation,** lactate fermentation and alcoholic fermentation, the end products are
either lactate (in many bacteria and in animal cells) or ethyl
alcohol and carbon dioxide (in yeasts and in most plant cells).
Fermentation allows the glycolytic pathway (fig. 9.5) to keep
operating when oxygen is not available and cellular respiration
cannot occur.

The glycolytic pathway will run as long as it has a supply
of "free" NAD$^+$—that is, NAD$^+$ that can pick up electrons as
it becomes NADH. Ordinarily, NADH passes hydrogen atoms
(e$^-$ + H$^+$) to the electron transport system in mitochondria. If

this system is not working because of a lack of oxygen, NADH
can pass hydrogen atoms to pyruvate in either of the following
reactions:

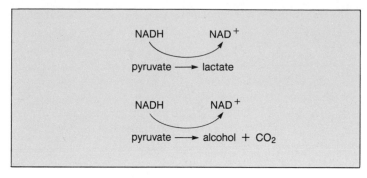

NAD$^+$ is now capable of oxidizing another molecule of PGAL
(fig. 9.7), and the glycolytic pathway can keep operating.

Energy Yield of Fermentation

Notice that fermentation consists of the glycolytic pathway plus
one additional reaction in which the end product of the glycolytic pathway, pyruvate, accepts hydrogen atoms and becomes
reduced. For this reason, only two ATP molecules are produced
and only 2.1% (14.6/686) of the energy of glucose is converted
to the energy of ATP during fermentation. This is much less
than the 36 ATP—representing 39% of the energy of glucose—
that are formed during aerobic cellular respiration.

Usefulness of Fermentation

Despite its low yield of ATP, fermentation does have its place,
because it can provide a *rapid burst* of ATP energy. In our own
bodies it is the muscle cells that are most likely to carry on fermentation. When the muscles are working vigorously over a
short period of time, as when we run, fermentation provides a
way to produce ATP even though oxygen is temporarily in limited supply. At first, the blood carries away all the lactate (lactic
acid) formed in the muscles, but eventually lactate begins to
build up in the muscles, changing the pH and causing muscle
fatigue and discomfort. When we stop running, the body has a
chance to catch up; while the lactate is converted back to pyruvate, by the removal of hydrogen atoms, which can now be
transported to the electron transport system. In the meantime,
we are said to be in **oxygen debt,** as signified by the fact that
we keep on breathing very heavily for a time.

Fermentation allows yeast cells to grow and divide anaerobically for a time. Eventually, though, they are killed by the
very alcohol they produce, if the initial glucose level is high.
Presumably, human beings were delighted to discover this form
of fermentation, as the ethyl alcohol produced has been consumed in great quantity for thousands of years. In addition, we
use the CO$_2$ to make bread rise.

Fermentation, which involves the reduction of pyruvate to lactate,
or to carbon dioxide and alcohol, gives a net yield of only two
ATP. Although this is a very low yield, it provides a way for a
cell to produce ATP energy even when oxygen is not available.

Alcoholic Beverages

Wine, beer, and whiskey production all require yeast fermentation. To produce wine, grape juice is allowed to ferment. After the grapes are picked, they are crushed and the juice is collected. In the old days, wine makers simply relied on spontaneous fermentation by wild yeasts that were on the grape skins, but now most add specially selected cultures of yeast. It is common practice to maintain the temperature at about 20° C for white wines and 28° C for red wines. Fermentation ceases after most of the sugar has been converted to alcohol. Various methods are used to clarify the wine, that is, to remove any suspended materials. Also, many fine wines improve when they are allowed to "age" during barrel or bottle storage.

Brewing beer is more complicated than producing wine. Usually grains of barley are first *malted;* that is, allowed to germinate for a short time so that amylase enzymes are produced that will break down the starch content of the grain. After the germinated grains have been crushed and mixed with water, the *malt wort* is separated from the spent grains and traditionally boiled with hops (an herb derived from the hop plant) to give flavor to the beer. Then the *hop wort* is seeded with a strain of yeast that converts the sugars in the wort to alcohol and carbon dioxide. At the end of fermentation, the yeast is separated from the beer, which is then allowed to mature for an appropriate period. After filtration and pasteurization, the beer is packaged.

Fermentation of grapes (above right), barley (above left), and corn produces wine, beer, and whiskey respectively.

The production of whiskey (from grains), brandy (from grapes), and rum (from molasses) differs from wine and beer production chiefly in that the alcohol is removed from the fermented substance by distillation. Most often, in the United States, corn or rye is used in the production of whiskey. These grains are ground up and mashed to release their starch content. Amylase enzymes are added to convert the starch to fermentable sugars. Then yeast is added so that fermentation can occur. Following fermentation, the alcohol is concentrated by distillation. Heating causes the alcohol to become gaseous and rise in a column where

it condenses to a liquid before entering a collecting vessel. The alcohol content of the liquid in the collecting vessel is much higher following this distillation process. The distillate is usually stored, or aged, quite often in an oak barrel, to improve the aroma and taste of the final product.

Figure 9.7

Fermentation consists of the glycolytic pathway followed by a reduction reaction during which pyruvate accepts hydrogen atoms from NADH and becomes reduced. This "frees" NAD$^+$ so that it can return to the glycolytic pathway to pick up more electrons as it becomes NADH.

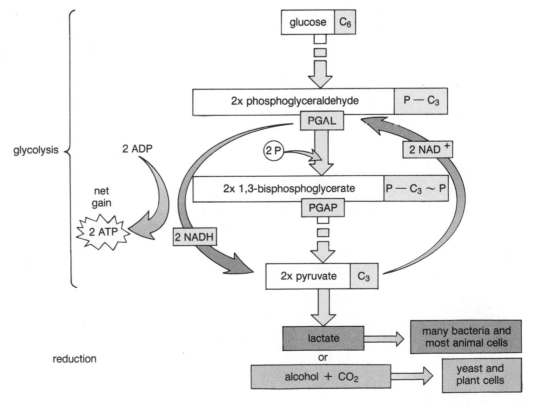

Figure 9.8

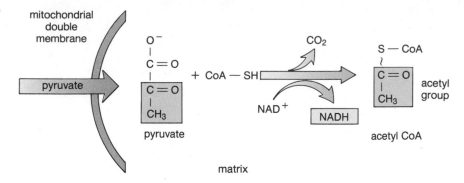

Transition reaction. Inside mitochondria, the oxidation of pyruvate by NAD^+ is accompanied by the release of CO_2. The remaining two-carbon (acetyl) group binds to coenzyme A (CoA) forming acetyl CoA. This reaction occurs twice per molecule of glucose broken down.

Evolutionary Perspective

Fermentation has probably been present just about as long as there have been living things on earth. At first, there was no oxygen in the atmosphere, but there were organic molecules in the ocean where the first cell(s) arose. Any cell that could utilize the energy of these molecules anaerobically would be at a competitive advantage, even if the method were not the most efficient. The presence of some form of fermentation in nearly all organisms shows how important this ancient competitive edge must have been to the continuance of life. Apparently, in the ancient past, only cells capable of fermentation survived, passing on the trait to later organisms.

Since fermentation is very inefficient compared to aerobic respiration, you might think it should have simply disappeared sometime during the history of life on earth. However, as with the two photosystems involved in photosynthesis, it is apparently customary for evolution to simply build on what has already been accomplished. In this instance, we see that pyruvate, the end product of glycolysis, can be metabolized either anaerobically or aerobically. Glycolysis, then, may have initially evolved as a part of fermentation, but today it is usually a preliminary stage to aerobic respiration.

Fermentation is an ancient pathway that first allowed organisms to capture the energy stored in organic molecules. The fact that most living organisms have inherited an ability to carry out the process shows that it must have given ancestral cells a competitive advantage over cells that could not ferment.

Pyruvate Metabolism inside the Mitochondria

If oxygen is available, fermentation does not occur and instead the pyruvate from glycolysis enters a mitochondrion. Here the transition reaction and the Krebs cycle occur. The electrons made available are passed to the electron transport system by either NAD^+ or FAD.

Transition Reaction

The **transition reaction** (fig. 9.8) takes its name from the fact that it connects glycolysis to the Krebs cycle. In this reaction pyruvate is oxidized and broken down to one molecule of CO_2

and a two-carbon acetyl group. The two-carbon acetyl group is transferred to coenzyme A (CoA)—a large molecule that contains a nucleotide and a portion of one of the B vitamins. This combination of acetyl group and CoA is called **acetyl CoA.** The inputs and outputs of the transition reaction are as follows:

inputs	outputs
pyruvate	CO_2
CoA	acetyl CoA
NAD^+	NADH

Since the transition reaction occurs once for each molecule of pyruvate, it occurs twice for each glucose molecule.

Krebs Cycle (Citric Acid Cycle)

The acetyl CoA produced during the transition reaction then enters the **Krebs cycle.** The **Krebs cycle** (fig. 9.9) is named for Hans Krebs, a British scientist who received the Nobel Prize for his part in determining these reactions in the 1930s. This cyclical series of reactions begins with a six-carbon organic compound called citric acid (citrate); for this reason the series is also known as the **citric acid cycle.** The enzymes that carry out the reactions of the Krebs cycle are located in the matrix of mitochondria (fig. 9.4)

The two-carbon acetyl group carried by coenzyme A enters the Krebs cycle and later leaves it as two molecules of CO_2. During the cycle, oxidation occurs, most of the hydrogen atoms are donated to NAD^+, but in one instance they are donated to **FAD,** another coenzyme of oxidation and reduction in cells. Some of the energy of oxidation is used immediately to form ATP by substrate-level phosphorylation, as in glycolysis (fig. 9.6). The inputs and outputs of the Krebs cycle are as follows:

inputs	outputs
acetyl group	$2CO_2$
ADP + P	ATP
3 NAD^+	3 NADH
FAD	$FADH_2$

Since the Krebs cycle occurs once for acetyl CoA, it occurs twice for each glucose molecule.

Figure 9.9

Krebs cycle. The net result of this cycle is the oxidation of an acetyl group to two molecules of CO_2. The energy of oxidation allows the immediate formation of one molecule of ATP. Most of the energy, however, resides in high-energy electrons, which are accepted by three molecules of NAD^+, as they become NADH, and one molecule of FAD, as it becomes $FADH_2$. These coenzymes then take the hydrogen atoms ($e^- + H^+$) to the electron transport system. (It should be kept in mind that the Krebs cycle turns twice per glucose molecule.)

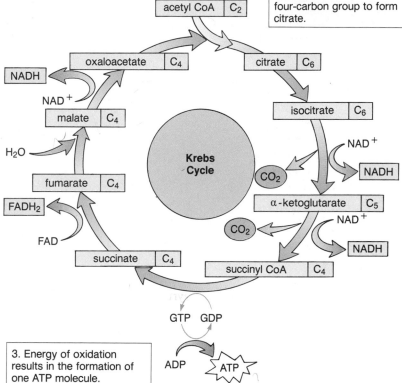

1. The cycle begins when an acetyl group carried by CoA combines with a four-carbon group to form citrate.

5. As the cycle returns to its starting point, another NADH is formed, which will take hydrogen atoms to the electron transport system.

2. Oxidation by NAD^+ accompanied by the production of CO_2. NADH will carry the hydrogen atoms to the electron transport system.

4. Oxidation by FAD, produces $FADH_2$. This coenzyme also carries hydrogen atoms to the electron transport system.

3. Energy of oxidation results in the formation of one ATP molecule.

Electron Transport System

As a result of the transition reaction and the Krebs cycle, the carbons of glucose have been completely oxidized to six molecules of CO_2. The hydrogen atoms ($e^- + H^+$) that were removed from the substrates of these metabolic pathways are donated by NADH and $FADH_2$ to the **electron transport system,** a series of carriers and enzymes located on the cristae of the mitochondria (fig. 9.4). Some of the carriers are **cytochromes,** and therefore the system is also called the cytochrome system.

A diagram of the electron transport system (fig. 9.10) illustrates that high-energy electrons are delivered to the top of chain and low-energy electrons leave at the bottom. As electrons pass from carrier to carrier, oxidation occurs, and the energy released is used to form ATP molecules. This process is sometimes called **oxidative phosphorylation** because oxygen receives the energy-spent electrons from the chain. An enzyme called cytochrome oxidase splits and reduces molecular oxygen (O_2) to water:

$$\frac{1}{2}O_2 + 2e^- + 2H^+ \longrightarrow H_2O$$

Oxygen is, then, the final acceptor for electrons during cellular respiration.

Chemiosmotic Phosphorylation

Today, we understand the process of ATP formation in greater detail. Certain carriers of the electron transport system accept not only electrons but also hydrogen ions. They deposit these hydrogen ions into the intermembrane space located between the inner and outer membrane of the mitochondrion. Both NADH and $FADH_2$ deliver hydrogen atoms ($e^- + H^+$) to the chain (figs. 9.10 and 9.11). These coenzymes serve, then, as a source for some of the H^+ that are pumped into the intermembrane space; other hydrogen ions are taken directly from the matrix and are pumped into the intermembrane space so that an H^+ reservoir is created. As H^+ enter the intermembrane space, an electrochemical gradient builds up across the inner membrane, located between the inner membrane space and the matrix. There are **F_1 particles** located in the inner membrane and they contain a channel through which hydrogen ions flow from the intermembrane space into the matrix down this gradient. The kinetic energy of the hydrogen ion flow is used by an ATP synthetase enzyme to carry out phosphorylation of ADP to ATP (fig. 9.11). Many today, therefore, use the term **chemiosmotic phosphorylation** for ATP formation associated with the electron transport system.

Figure 9.10

Electron transport system. NADH + H⁺ delivers two hydrogen atoms to a carrier at the top of the chain. Each pair of electrons delivered by NADH + H⁺ is ultimately responsible for the formation of three ATP molecules. Those delivered by FADH₂ are responsible for the formation of two ATP molecules. ATP formation is the result of chemiosmotic phosphorylation (*see* fig. 9.11*b* and *c*): the carriers in the chain deposit H⁺ ions in the intermembrane space of mitochondria, creating an H⁺ reservoir. The energy released as these hydrogen ions flow down their electrochemical gradient is used to form ATP from ADP and Ⓟ. Oxygen is the final acceptor for the electrons and it is reduced to water.

1. NADH delivers hydrogen atoms (e⁻ + H⁺) to the electron transport system and NAD⁺ leaves. The H⁺ are deposited in the intermembrane space of a mitochondrion but the electrons are passed to the next carrier.

2. FADH₂ also delivers hydrogen atoms to the electron transport system. Every time this carrier receives two electrons from either FAD or from higher up in the chain, two H⁺ are deposited in the intermembrane space and the electrons are passed to the next carrier.

3. Cytochrome oxidase complex receives electrons from the previous carrier and passes them on to oxygen. Four H⁺ are taken from the matrix—only two are taken by oxygen and the other two are deposited in the intermembrane space.

NADH
H⁺ → 2H
NAD⁺

ox.
NADH dehydrogenase
red.
2e⁻

2H⁺
ATP

FADH₂
coenzyme Q
ox.
red.
FAD
2e⁻

2H⁺
ATP

ox.
cytochrome b-c₁ complex
red.
2e⁻

H⁺ reservoir (intermembrane space)

ox.
cytochrome c
red.
2e⁻

ATP

4H⁺
½O₂
ox.
cytochrome oxidase complex
red.
2H⁺
H₂O

Hydrogen atoms (e⁻ + H⁺) are made available to the electron transport chain primarily by NADH. As electrons are passed from one carrier to the next, some carriers pump H⁺ into the intermembrane space. The resulting electrochemical gradient is used to carry out chemiosmotic phosphorylation of ADP to ATP.

Most of the ATP produced during cellular respiration is produced by the electron transport system. As mentioned previously, the chemiosmotic theory of ATP production (p. 105) was proposed in the 1960s by the British biochemist Peter Mitchell, who later received a Nobel Prize for his work.

Recycling of NAD⁺ and FAD

The coenzymes NAD⁺ and FAD are constantly being reduced and oxidized inside the cell:

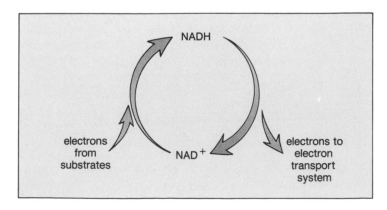

NADH

electrons from substrates

NAD⁺

electrons to electron transport system

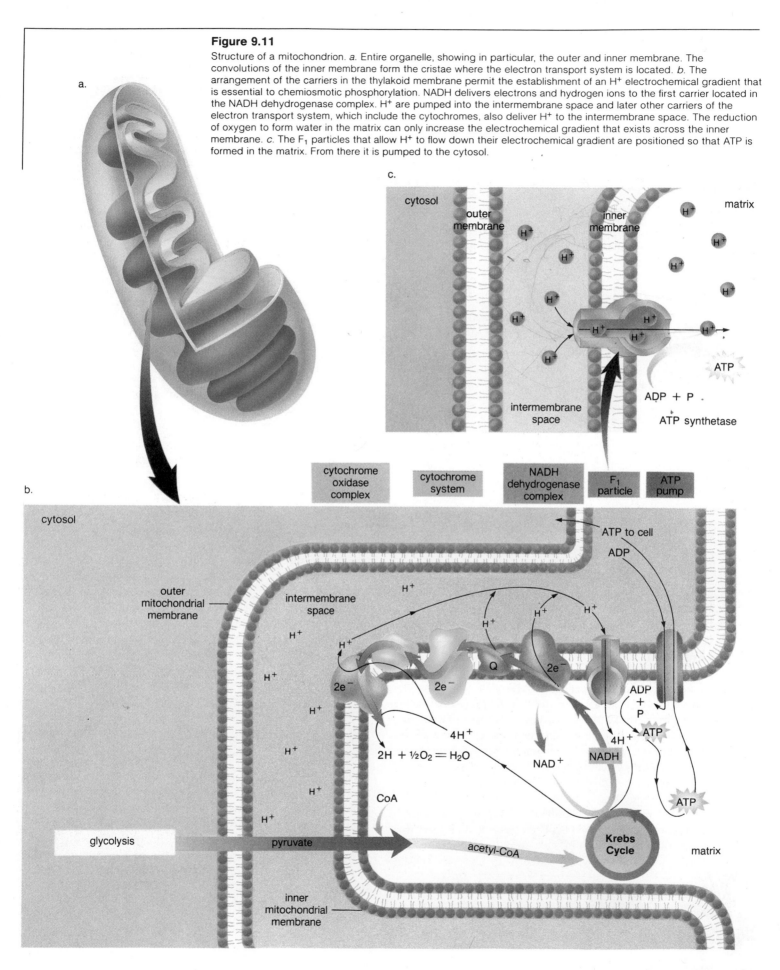

Figure 9.11

Structure of a mitochondrion. *a.* Entire organelle, showing in particular, the outer and inner membrane. The convolutions of the inner membrane form the cristae where the electron transport system is located. *b.* The arrangement of the carriers in the thylakoid membrane permit the establishment of an H⁺ electrochemical gradient that is essential to chemiosmotic phosphorylation. NADH delivers electrons and hydrogen ions to the first carrier located in the NADH dehydrogenase complex. H⁺ are pumped into the intermembrane space and later other carriers of the electron transport system, which include the cytochromes, also deliver H⁺ to the intermembrane space. The reduction of oxygen to form water in the matrix can only increase the electrochemical gradient that exists across the inner membrane. *c.* The F_1 particles that allow H⁺ to flow down their electrochemical gradient are positioned so that ATP is formed in the matrix. From there it is pumped to the cytosol.

a.

c.

cytosol

outer
membrane

H⁺

H⁺

H⁺

inner
membrane

matrix

H⁺

H⁺

H⁺

H⁺

H⁺

H⁺

H⁺

H⁺

H⁺

H⁺

ATP

ADP + P

intermembrane
space

ATP synthetase

| cytochrome oxidase complex | cytochrome system | NADH dehydrogenase complex | F_1 particle | ATP pump |

b.

cytosol

ATP to cell

ADP

outer
mitochondrial
membrane

intermembrane
space

H⁺

H⁺

H⁺

H⁺

H⁺

H⁺

H⁺

Q

$2e^-$

$2e^-$

$2e^-$

$2e^-$

ADP
+
P

$4H^+$

$2H + \frac{1}{2}O_2 = H_2O$

NAD⁺

NADH

ATP

ATP

CoA

glycolysis

pyruvate

acetyl-CoA

Krebs
Cycle

matrix

inner
mitochondrial
membrane

This means that the cell needs only a limited supply of these coenzymes because they are used over and over again.

Energy Yield of Glycolysis and Cellular Respiration

As we have been suggesting all along, it is possible to calculate the ATP yield for the complete breakdown of glucose to carbon dioxide and water. Table 9.1 shows the number of molecules of ATP produced by each part of the process.

Substrate-Level Phosphorylation

Per glucose molecule, four ATP are formed directly by substrate-level phosphorylation: two during glycolysis and two during two turns of the Krebs cycle.

Oxidative Phosphorylation

Per glucose molecule ten molecules of NADH and two molecules of $FADH_2$ take hydrogen atoms ($e^- + H^+$) to the electron transport system:

	NADH	$FADH_2$
from glycolysis	2 NADH	—
from transition reaction (two turns)	2 NADH	—
from Krebs cycle (two turns)	6 NADH	2 $FADH_2$
Total	10 NADH	2 $FADH_2$

For each molecule of NADH produced *inside* the mitochondria by the Krebs cycle, three ATP are produced by the electron transport system but for each $FADH_2$ there are only two ATP. Figure 9.10 explains the reason for this difference: $FADH_2$ delivers its hydrogens to the transport system at a lower level, and therefore only four H^+, instead of six H^+, are pumped into the intermembrane space per molecule of $FADH_2$.

What about the ATP yield of the NADH generated *outside* the mitochondria by the glycolytic pathway? Although NADH cannot cross mitochondrial membranes, there is a "shuttle" mechanism, which allows electrons to be delivered to the electron transport system inside the mitochondria. In most types of cells, the shuttle consists of an organic molecule, which can cross the outer membrane, accept the electrons, and deliver them to an FAD molecule in the inner membrane. However, this FAD can produce only two ATP molecules; therefore, in most cells the NADH produced in the cytosol results in the production of only two ATP instead of three ATP.

Table 9.1 Energy Yield of Glycolysis and Cellular Respiration per Glucose Molecule

Pathway	ATP Yield per Glucose Molecule Substrate-Level Phosphorylation	Oxidative Phosphorylation	Total ATP
Glycolysis	2 ATP (net)*	2 NADH = 4 ATP†	6 ATP
Transition reaction (per glucose)		2 NADH = 6 ATP	6 ATP
Krebs cycle (per glucose)	2 ATP	6 NADH = 18 ATP 2 $FADH_2$ = 4 ATP	24 ATP
Total	4 ATP	32 ATP =	36 ATP

*The only amount that is produced under anaerobic conditions.
†In most cells each NADH produced in the cytosol results in only two rather than three ATP because the electrons are shuttled to an FAD coenzyme. In the heart and the liver, a shuttle system delivers the electrons to an NAD$^+$ coenzyme, and in these cells each NADH does result in the production of three ATP.

Efficiency of Energy Transformation

It's interesting to consider how much of the energy in a glucose molecule eventually becomes available to the cell. The difference in energy content between the reactants (glucose and oxygen) and the products (carbon dioxide and water) is 686 Kcal. In other words, this is the total amount of energy available for the production of high-energy phosphate bonds. A high-energy phosphate bond has an energy content of 7.3 Kcal, and 36 of these are produced during glucose breakdown, so 36 phosphates are equivalent to a total of 263 Kcal. Therefore, 263/686 or 39% of the available energy is transferred from glucose to ATP. The rest of the energy is lost in the form of heat. In birds and mammals, in particular, some of this heat is used to maintain a body temperature above that of the environment. In other organisms, including plants, there are few if any measures to conserve heat, and the body temperature generally fluctuates according to that of the external environment.

The energy yield of 36 ATP molecules represents 39% of the total energy that is available in a glucose molecule.

The Metabolic Pool and Biosynthesis

Figure 9.12 shows how the primary organic compounds of cells make up a metabolic pool. Cells can oxidize, or degrade, other molecules besides glucose to release energy. For example, a fat molecule is oxidized to glycerol and three fatty acids when it is used as an energy source. Inside the cell, glycerol is easily broken down to PGAL and thereby can enter the glycolytic pathway. The fatty acids are degraded to acetyl groups, which can enter the Krebs cycle. If a fatty acid contains 18 carbons, it breaks down to give nine acetyl groups. Calculation shows that these nine groups would produce 108 ATP molecules. For this reason, fats are a very efficient form of stored energy, because there are three long-chain fatty acids per fat molecule.

The energy yield of protein is equivalent to carbohydrates. Every protein is broken down to amino acids, which undergo deamination, the removal of the amino group. The

number of carbons remaining now determines just where the molecule will enter the Krebs cycle (fig. 9.12). The amino group is converted to ammonia (NH_3) and excreted.

The substrates making up the pathways in figure 9.12 can also be used as starting materials for synthetic reactions. In other words, compounds that enter the pathways are oxidized to substrates that can be used for biosynthesis. This is the cell's **metabolic pool,** in which one type of molecule can be converted to another. In this way, carbohydrate intake can result in the formation of fat. PGAL molecules may be converted to glycerol molecules and acetyl groups may be joined together to form fatty acids. Fat synthesis will follow. This explains why it is that one can get fat from eating too much candy, ice cream, and cake.

Some metabolites of the Krebs cycle can be converted to amino acids. Plants are able to synthesize all of the amino acids they need. Animals, however, lack some of the enzymes necessary for synthesis of all amino acids. Humans (and white rats), for example, can synthesize ten of the common amino acids, but cannot synthesize the other ten. The amino acids that cannot be synthesized must be supplied by the diet; they are called the essential amino acids. The nonessential amino acids are the ones that can be synthesized. It is quite possible for animals to eat large quantities of protein and still suffer from protein deficiency, if their diets do not contain adequate quantities of the essential amino acids.

All the reactions involved in cellular respiration are part of a metabolic pool; the metabolites from the pool can be used either as energy sources or as substrates for various synthetic reactions.

There are other degradative pathways aside from those discussed in this chapter or depicted in figure 9.12, which are needed to supply substrates for biosynthesis. For example, a pathway called the pentose phosphate shunt converts glucose to pentose, rather than to pyruvate. NADPH is a byproduct of the pentose phosphate shunt. The pentose sugars are needed for nucleotide formation, and the NADPH is used in synthetic reactions. Therefore this pathway is also a very important one.

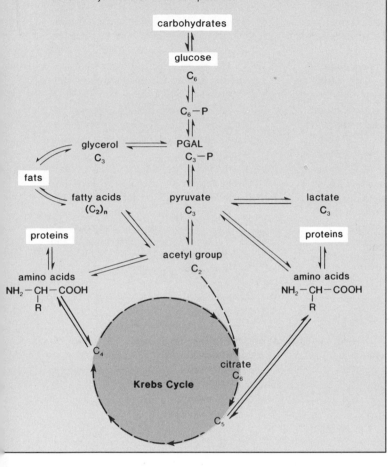

Figure 9.12

The metabolic pool concept. Carbohydrates, fat, and proteins can be used as energy sources, and enter degradative pathways at specific points. Degradation produces metabolites that can also be used for synthesis of other compounds.

Summary

1. In cells, the energy of carbohydrate molecules, in particular, is converted to ATP energy.

2. Glucose metabolism often provides energy. The glycolytic pathway anaerobically produces pyruvate, which either accepts electrons within the cytosol (fermentation), or becomes fully oxidized in mitochondria. Many bacteria and most animal cells carry on lactate fermentation; yeast and plant cells carry on alcoholic fermentation. Both kinds of fermentation only give a net yield of two ATP molecules.

3. When glucose is completely oxidized, at least 36 ATP molecules are produced. Three subpathways are required: glycolysis, the Krebs cycle, and the electron transport system. The transition reaction connects glycolysis to the Krebs cycle. Oxidation involves the removal of hydrogen atoms (e^- + H^+) from substrate molecules usually by the coenzyme NAD^+.

4. Glycolysis is the breakdown of glucose to 2 molecules of pyruvate. This pathway is a series of enzymatic reactions that occurs in the cytosol. Oxidation by

NAD^+ releases enough energy to immediately give a net gain of two ATP by substrate-level phosphorylation. NADH either gives its hydrogen atoms to pyruvate or takes them to a mitochondrion.

5. Fermentation involves substrate-level phosphorylation during the glycolytic pathway, followed by the reduction of pyruvate by NADH to either lactate, or alcohol and CO_2. The reduction process "frees" NAD^+, so that it can accept more hydrogen atoms from the glycolytic pathway.

6. Although fermentation results in only two ATP, it still serves a purpose: In vertebrates, it provides a quick burst of ATP energy for short-term, strenuous muscular activity. The accumulation of lactate puts the individual in oxygen debt, because oxygen is needed when lactate is completely metabolized to CO_2 and H_2O. Fermentation evolved before aerobic respiration, and its continued presence exemplifies how evolution usually works—the newly evolved process is added onto those that have already evolved.

7. Pyruvate from glycolysis can enter the mitochondrion where the transition reaction takes place. During this reaction, oxidation occurs as CO_2 is removed. NAD^+ is reduced, and CoA receives the two-carbon acetyl group that remains. Since the reaction must take place twice per glucose molecule, two NADH result.

8. The acetyl group enters the Krebs cycle, a cyclical series of reactions located in the mitochondrial matrix. Complete oxidation follows, as two CO_2 molecules, three NADH molecules, and one $FADH_2$ molecule are formed. The cycle also produces one ATP molecule by substrate-level phosphorylation. The entire cycle must turn twice per glucose molecule.

9. The final stage of pyruvate breakdown involves the electron transport system located in the inner membrane of mitochondria. The electrons received from NADH and $FADH_2$ are passed down a chain of carriers until they are finally received by oxygen, which combines with hydrogen ions to produce water. As the electrons pass down the chain, ATP is produced. The term oxidative phosphorylation is sometimes used for ATP production by the electron transport system.

10. Some of the electron carriers pump H^+ into the intermembrane space setting up an electrochemical gradient. H^+ flow down this gradient at F_1 particles and the energy released is used to form ATP molecules from ADP and $\textcircled{P}$. Therefore, the term chemiosmotic phosphorylation is preferred by some for ATP production by the electron transport system.

11. In the mitochondria, for each NADH molecule donating electrons to the electron transport system, three ATP molecules are produced. However, in the cytosol each NADH formed results in only two ATP molecules. This is because its hydrogen atoms must be shuttled across the mitochondrial outer membrane by a molecule that can cross it. In most cells, these hydrogen atoms are then taken up by FAD. Each molecule of $FADH_2$ results in the formation of only two ATP because the electrons enter the electron transport system at a lower level.

12. Once NAD^+ and FAD have delivered electrons to the chain they are "free" to turn around and accept more hydrogens. Therefore, the cell only needs a limited supply of these coenzymes because they are used over and over again.

13. Of the 36 ATP formed by aerobic cellular respiration, four are the result of substrate-level phosphorylation, and the rest are produced by oxidative phosphorylation. The energy for the latter comes from the electron transport system.

14. Carbohydrates, proteins, and fats can be broken down by entering the degradative pathways at different locations. These pathways also provide metabolites needed for the synthesis of various important substances. Degradation and synthesis, therefore, both utilize the same metabolic pool of reactants.

Objective Questions

For questions 1 to 8, identify the pathway involved and match to the terms in the key:

Key:
   a. Glycolysis
   b. Krebs cycle
   c. Electron transport system

1. carbon dioxide given off B
2. water formed C
3. PGAL A
4. NADH becomes NAD^+ C
5. oxidative phosphorylation C
6. cytochrome carriers C
7. pyruvate A
8. FAD becomes $FADH_2$ B

9. Which of these is *not* true of C fermentation?
   a. Net gain of only two ATP.
   b. Occurs in cytosol.
   c. NADH donates electrons to electron transport system.
   d. Begins with glucose.

10. The transition reaction D
   a. connects glycolysis to the Krebs cycle.
   b. gives off CO_2.
   c. utilizes NAD^+.
   d. All of these.

11. The greatest contributor of electrons to C the electron transport system is
   a. oxygen.
   b. glycolysis.
   c. the Krebs cycle.
   d. the transition reaction.

12. Substrate-level phosphorylation takes A place in the
   a. glycolytic pathway and the Krebs cycle.
   b. electron transport system and the transition reaction.
   c. glycolytic pathway and the electron transport system.
   d. Krebs cycle and the transition reaction.

13. Fatty acids are broken down to B
   a. pyruvate molecules, which take electrons to the electron transport system.
   b. acetyl groups, which enter the Krebs cycle.
   c. amino acids, which excrete ammonia.
   d. All of these.

14. Of the 36 ATP molecules that are D produced during the complete breakdown of glucose, most are formed by
   a. chemiosmotic phosphorylation.
   b. the electron transport system.
   c. substrate-level phosphorylation.
   d. Both (a) and (b).

For questions 15 to 20, match the items below to one of the locations in the key:

Key:
   a. matrix of the mitochondria
   b. cristae of the mitochondria
   c. between the inner and outer membrane of the mitochondria
   d. in the cytosol

15. electron transport system B
16. Krebs cycle A
17. glycolysis D
18. transition reaction A
19. H^+ reservoir C
20. F_1 particles B

Study Questions

1. What types of organisms carry on cellular respiration? What is the function of this process?
2. Glycolysis results in pyruvate molecules. Contrast the fate of pyruvate under anaerobic conditions and aerobic conditions.
3. What is the overall chemical equation for the complete breakdown of glucose to carbon dioxide and water? How does a human cell acquire the needed substrates, and what happens to the products?
4. What are the three pathways involved in the complete breakdown of glucose to carbon dioxide and water? What reaction is needed to join two of these pathways?

GLYCOLOSIS - TRANS, REACTION. KREBS - ELEC. TRANS, SYST.

5. Outline the main reactions of glycolysis, emphasizing those that produce energy.
6. Contrast glycolysis to fermentation. What is the benefit of pyruvate reduction during fermentation? What type of organisms carry out lactate fermentation and what types carry out alcoholic fermentation?
7. Give the substrates and products of the transition reaction. Where does it take place?
8. What happens to the acetyl group that enters the Kreb's cycle? What are the other steps in this cycle?
9. What is the electron transport system, and what are its functions? Relate these functions to the location of this system.
10. Explain how, when, and where chemiosmotic phosphorylation takes place in mitochondria.
11. Calculate the energy yield of glycolysis and cellular respiration per glucose molecule. Distinguish between substrate-level phosphorylation and oxidative phosphorylation.
12. Give examples to support the concept of the metabolic pool.

Thought Questions

1. When one form of energy is converted into another, there is always some loss of useful energy. Use this principle to explain why the process of cellular respiration causes energy to flow through a community of organisms.
2. The endosymbiotic theory says that mitochondria were once independent prokaryotes. Which cell must have evolved first, the host cell or the prokaryote? Why?
3. List similarities and differences between the processes of photosynthesis and of cellular respiration.

Selected Key Terms

cellular respiration (sel'u-lar res''pī-ra'shun) 125
pyruvate (pi'roo-vāt) 125
glycolysis (gli-kol i-sis) 125
fermentation (fer''men-ta'shun) 126
substrate-level phosphorylation (sub'strāt lĕv-al fos''fōr-i-la'shun) 128

oxygen debt (ok'si-jen det) 130
transition reaction (tran-zish'un re-ak'shun) 132
acetyl CoA (as''ĕ-til-ko-en'zīm A) 132
Krebs cycle (citric acid cycle) (Krebz si'k'l) 132

cytochrome (si'to-krōme) 133
oxidative (chemiosmotic) phosphorylation (ok''si-da'tiv fos''for-i-la'shun) 133
F₁ particle (par'tĕ-k'l) 133
metabolic pool (mĕt'a-bol-ic pōol) 137

Suggested Readings for Part 1

Alberts, B., et al. 1983. *Molecular biology of the cell.* New York: Garland Publishing.

Allen, R. D. February 1987. The microtubule as an intracellular engine. *Scientific American.*

Avers, C. J. 1981. *Cell biology.* 2d ed. New York: D. Van Nostrand.

Baker, J.J.W., and Allen, G. E. 1981. *Matter, energy, and life.* 4th ed. Reading, MA: Addison-Wesley.

Berns, M. W. 1983. *Cells.* 3d ed. New York: Holt, Rinehart & Winston.

Cronquist, A. 1982. *Basic botany.* 2d ed. New York: Harper & Row, Publishers, Inc.

Dautry-Varsat, A., and Lodish, H. F. May 1984. How receptors bring proteins and particles into cells. *Scientific American.*

de Duve, C. May 1983. Microbodies in the living cell. *Scientific American.*

Dickerson, E. March 1980. Cytochrome and the evolution of energy metabolism. *Scientific American.*

Dustin, P. August 1980. Microtubules. *Scientific American.*

Grivell, L. A. March 1983. Mitochondrial DNA. *Scientific American.*

Karplus, M., and McCammon, J. A. April 1986. The dynamics of proteins. *Scientific American.*

Lake, J. A. July 1981. The ribosome. *Scientific American.*

Miller, K. R. October 1979. The photosynthetic membrane. *Scientific American.*

Ostro, M. J. January 1987. Lipsosomes. *Scientific American.*

Porter, K. R., and Bonneville, M. A. 1973. *Fine structure of cells and tissues.* 4th ed. Philadelphia: Lea and Febiger.

Porter, K. R., and Tucker, J. B. March 1981. The ground substance of the living cell. *Scientific American* (offprint 1494).

Raven, H., et al. 1986. *Biology of plants.* 4th ed. New York: Worth.

Rothman, J. September 1985. The Compartmental organization of the Golgi apparatus. *Scientific American.*

Scientific American. October 1985. The molecules of life.

Sharon, N. November 1980. Carbohydrates. *Scientific American.*

Sheeler, P., and Bianchi, D. E. 1980. *Cell biology: Structure, biochemistry, and function.* New York: John Wiley and Sons.

Stryer, L. 1981. *Biochemistry.* 2d ed. San Francisco: W. H. Freeman.

Unwin, N. February 1984. The structure of proteins in biological membranes. *Scientific American.*

Wolfe, S. L. 1980. *Biology of the cell.* Belmont, CA: Wadsworth.

Yougan, D. C., and Mars, B. L. June 1987. Molecular mechanisms of photosynthesis. *Scientific American.*

CRITICAL THINKING

CASE STUDY

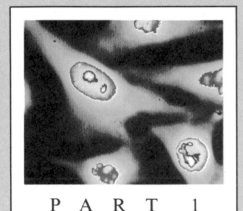

PART 1
The Cell

The Golgi complex discussed on page 66 in chapter 5 is an organelle that processes and packages proteins that will be used for specialized functions.

Experiments indicate that proteins synthesized at the endoplasmic reticulum enter the Golgi saccules (*see* figure) at the forming face and exit at the maturing face. If proteins are modified as they pass through the Golgi, what would you expect to find in the Golgi that could bring about this modification?

Prediction 1 Enzymes that can modify the proteins should be found in the Golgi complex.

Result 1 When the Golgi complex is isolated from cells and purified, a wide variety of enzymes are found that would be able to modify the proteins. Some can add sugar molecules and others can add phosphate groups.

Why is finding an enzyme in association with a particular organelle a strong argument that a specific biochemical reaction occurs in the organelle? If a specific enzyme is not *found, does this mean that a certain reaction does not occur?*

In another series of experiments, the Golgi saccules were separated from each other. This is accomplished by centrifuging the Golgi saccules in a sucrose density gradient in which cell fragments separate if these parts vary in density. Three groups (fractions) of the Golgi saccules were found. If each of these fractions contains saccules that carry out specific modifications, what would you expect to find?

Prediction 2 Specific enzymes will be found in the different fractions.

Result 2 Subsequent analysis showed that each of the three fractions contained specific enzymes.

In what ways could these data be misleading about Golgi structures? Could the Golgi have more than three types of saccules? Fewer?

While these results suggest that specific reactions occur in different parts of the Golgi, they do not indicate exactly *where* the reactions are occurring. It is possible to determine where a reaction is located by doing this (see fig. 6.5*a*): (1) A purified Golgi enzyme is injected into a mouse so that the mouse's immune system manufactures antibodies to that enzyme. (2) A substance, which is visible under the electron microscope, is joined to the antibody molecules. (3) This modified antibody preparation, which will act as a stain, is then added to cells, the cells are sectioned and viewed under an electron microscope.

For each of the antibody preparations, would you expect to find only portions of the Golgi complex stained?

Prediction 3 The stain should be localized within specific saccules.

Result 3 Electron micrographs showed that each antibody preparation stained specific saccules. It became apparent that the Golgi complex has three compartments: an inner portion, a middle portion, and an outer portion.

Summarize the data so far. Why do data show that "specific reactions occur in different parts of the Golgi?" What are the different parts?

You will recall that the Golgi complex has two faces, a forming face and a maturing face. It could be that proteins move through the Golgi in this manner: a saccule containing the protein gets pushed forward by new ones at the forming face. (Old saccules would constantly break down at the maturing face.) Or it could be that proteins move via the vesicles from one saccule to the next. This is called the "hopping" hypothesis because the vesicles would hop from the outer edge of one saccule to the next. (Notice that the figure shows vesicles at the edges of the saccules.)

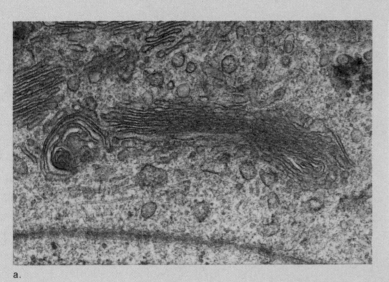

a.

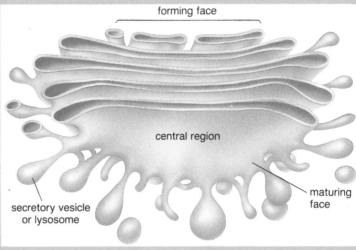

forming face

central region

secretory vesicle
or lysosome

maturing
face

b.

*Which of these hypotheses is best
supported by the data already presented?
Which, therefore, would you expect to be
supported by further data?*

To test which of these two models accounts for the movement of substances through the Golgi, two cells were fused (i.e., plasma membranes were joined to produce one large cell containing contents of both original cells). One cell was a mutant cell and the other was normal. The mutant cell could produce protein G, but could *not* add the sugar galactose to this protein. Prior to fusion, this cell was exposed to radioactive precursors so that all of the protein G in the Golgi was radioactively labeled. The normal cell could add galactose to protein G as it passes through the Golgi complex. The fused cell contained Golgi complexes from both cells. Those from the mutant cell contained radioactive protein G (but could not add galactose) while Golgi from the normal cell could add galactose to the protein (but did not contain radioactive protein G). Knowing that vesicles would be able to move from one Golgi complex to the other, what would you expect to happen if the hopping model is correct?

Prediction 4 If hopping can occur (even between Golgi complexes), then radioactive protein G with galactose added should be found in the fused cell.

Result 4 Careful analysis indicated that radioactive protein G with galactose added was found. Therefore, the hopping model was supported.

*In this test of the "hopping" hypothesis,
why is it necessary to use cells that have a
mutation preventing the addition of
galactose? What would happen if this
mutation were not present?*

These diverse experiments provide a more detailed description of the Golgi complex and indicate that there are several distinct biochemical compartments containing specific enzymes. Proteins appear to be transported from saccule to saccule via "hopping" of vesicles.

Rothman, J. E. September 1985. The compartmental organization of the Golgi apparatus. *Scientific American.*

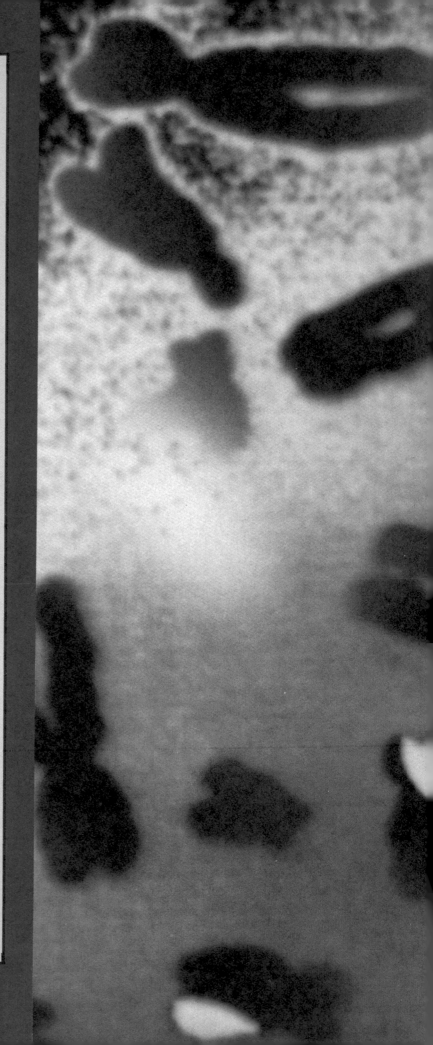

PART 2

Genetic Basis of Life

Chromosomes, at the time of cell division, are highly compacted and appear as rodlike structures that can be easily counted. Half of the total chromosome number came from the male parent and half from the female parent. The chromosomes contain the genes, which control the cell and the characteristics of the individual. Therefore, it is important that each new cell receive a copy of each and every chromosome.

In the nondividing cell, the chromosomes decompact and become extended. Not all the genes are active in every cell and this causes cells to vary in structure and function.

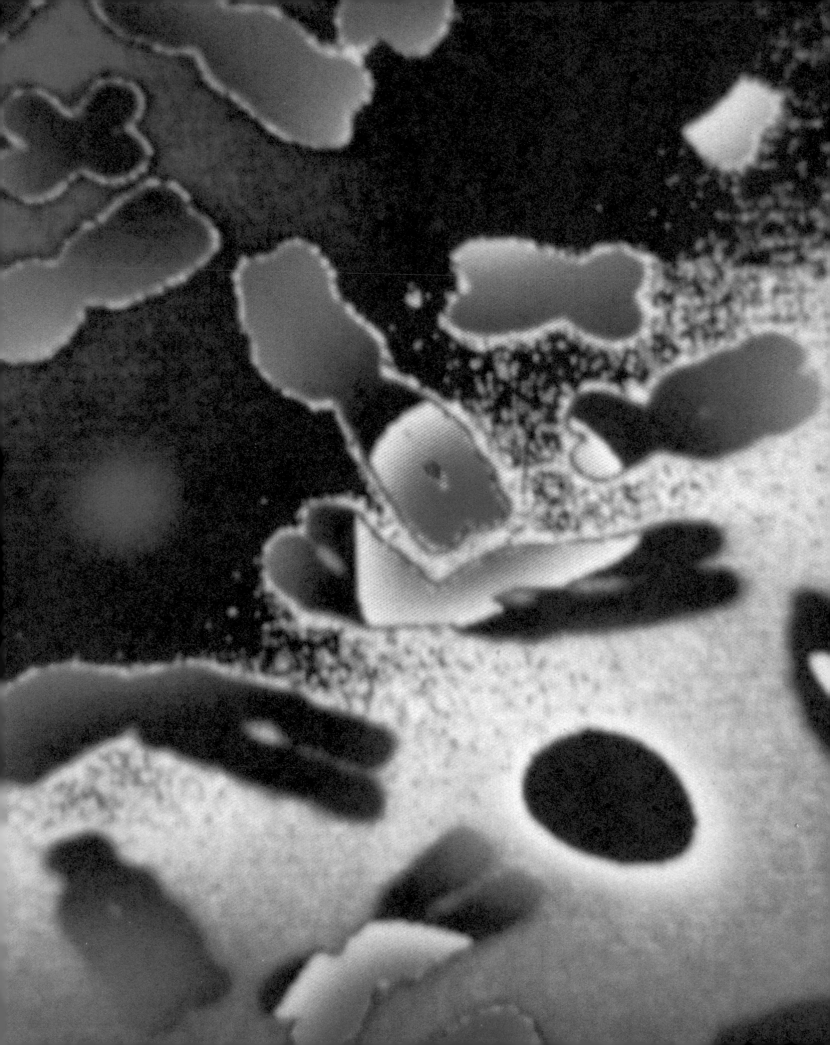

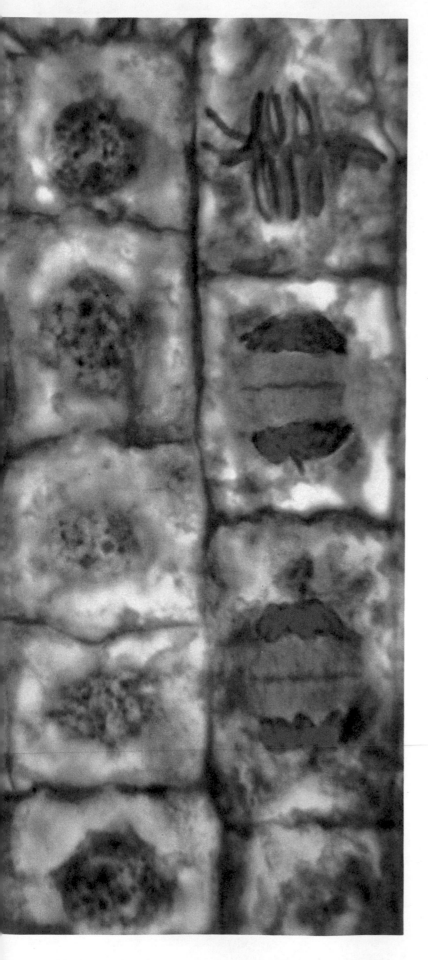

CHAPTER 10

Cell Reproduction: Binary Fission and Mitosis

Your study of this chapter will be complete when you can:

1. *List two benefits of and two requirements for the types of cell division discussed in this chapter.*

2. *Describe the prokaryotic chromosome and the process of binary fission.*

3. *Describe a duplicated eukaryotic chromosome, and distinguish between the diploid and haploid number of chromosomes.*

4. *State the difference between chromatin and chromosome, and explain how they relate to each other.*

5. *State the phases of the cell cycle, and describe what happens during each phase.*

6. *Discuss how the cell cycle might be controlled, and contrast the behavior of normal cells with those of cancer cells.*

7. *Draw a series of diagrams illustrating the stages of mitosis in animal cells, and tell what happens during each stage. Describe cytokinesis in animal cells.*

8. *State at least one difference between animal and plant cell mitosis. Describe cytokinesis in plant cells.*

9. *Relate binary fission and mitosis to the process of asexual reproduction among single-celled organisms and to growth and repair among multicellular forms.*

During mitotic cell division, the daughter cells receive a full complement of chromosomes. The process is very orderly; it has to be to ensure that the daughter cells receive one of each chromosome and not two of one kind and none of another. Cell division within this plant root tip will produce new cells so that the root can grow.

As we mentioned in the introductory chapter, the life cycle of all organisms includes reproduction. We have a tendency to think of reproduction in terms of multicellular organisms even though reproduction always involves single cells. For example, all organisms begin life as a single cell—when that cell divides in single-celled organisms it produces two new individuals. In other types of organisms, cell division is a part of the growth process that produces the multicellular form we recognize as the organism (fig. 10.1). Cell division is also important in multicellular forms for renewal and repair. For example, our own bodies are always producing new blood cells and skin cells to replace those that are worn out or damaged. In summary, then, simple cell division has these purposes:

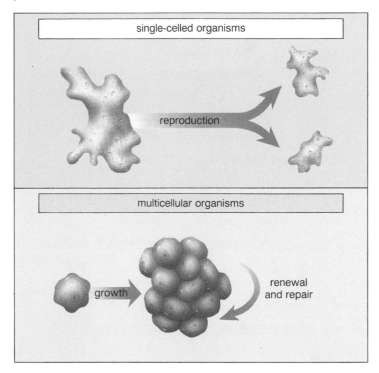

Cell division occurs in single-celled organisms when they reproduce and simple cell division is also involved in the growth and repair of multicellular forms.

You may recall that we previously touched on at least one advantage of cell division (p. 57) but let's consider it again. A small cell has a more adequate plasma membrane/volume ratio than does a large cell. Everything that enters and exits a cell must cross the plasma membrane, and as the cell increases in size, there comes a point when the plasma membrane surface area is not adequate to meet the needs of the cell. Cell division keeps the surface area/volume ratio favorable for the efficient survival of the cell. Another advantage that we have not considered previously concerns only multicellular organisms. Cell division allows specialization of parts to occur. In our bodies, there are specialized cells for various functions like digestion, respiration, and excretion. The single cell of a unicellular organism cannot become nearly so specialized, because the single cell alone must carry out all these functions.

Figure 10.1
Cell reproduction occurs in both single-celled organisms and in multicellular organisms. a. When a single-celled organism divides, reproduction occurs. This is a photograph of a paramecium dividing. Magnification, ×200. b. Life begins as a single cell in multicellular organisms. When this cell divides, a multicellular organism results. This is a photograph of a slipper-shell mollusk embryo after only a single division has occurred. Magnification, ×140.

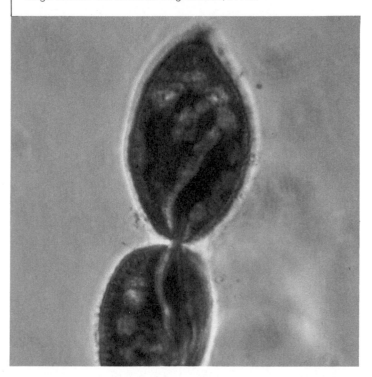

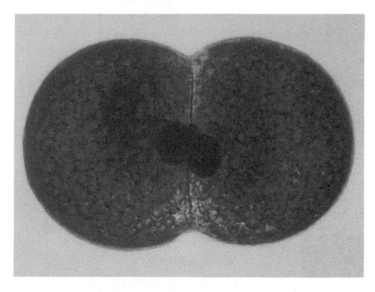

Cell division maintains a favorable plasma membrane/volume ratio for the cell and allows specialization of cells to occur in multicellular organisms.

All cells, both prokaryotic and eukaryotic, contain genetic material in the form of DNA, and a cell cannot function properly unless it receives a full copy of its species' DNA. Cell division must produce a mechanism by which copies of DNA are passed from the so-called **mother cell** to the **daughter cells.** It is also important that the cytoplasmic contents be divided. Only when the two cells are separate, has reproduction taken place.

Cell division must provide a mechanism for distribution of DNA to the daughter cells, and for a division of the cytoplasm.

Prokaryotic Chromosomes and Cell Division

Prokaryotic Chromosomes

Prokaryotes (i.e., bacteria) are single-celled organisms (p. 60) that lack a nucleus. They have a single chromosome that is compacted into an electron-dense irregularly shaped area called the **nucleoid region,** which is not enclosed by a membrane. When stretched out, the chromosome is seen to be a circular loop attached to the inside of the plasma membrane. Its length may be about 500 times the length of the cell and that's why it needs to be compacted inside the cell.

The prokaryotic chromosome is a single long loop of DNA that is compacted into an area called a nucleoid region.

Cell Division

Cell division in prokaryotes is termed **binary fission** because division (fission) produces two (binary) daughter cells that are identical in content and size to the mother cell. Before division takes place, the DNA replicates so that there are two chromosomes attached to the inside of the plasma membrane (fig. 10.2a and b). The replicated chromosomes are separated by an elongation of the cell that pushes the chromosomes apart. When the cell is approximately twice its original length, the plasma membrane grows inward and a cell wall forms, dividing the cell into two approximately equal portions (fig. 10.2c–e).

Cell division in prokaryotes is by binary fission. Replication of DNA is followed by elongation of the cell so that the two chromosomes are separated before an ingrowth of the plasma membrane and formation of a cell wall separates the cytoplasm.

Eukaryotic Chromosomes and Cell Division

Eukaryotic Chromosomes

Eukaryotes have a true nucleus (fig. 5.4) that is bounded by a double membrane called the nuclear envelope. The nuclear envelope contains pores that allow substances to pass between the nucleoplasm and the cytoplasm (fig. 5.7). The DNA (and associated proteins) within a nucleus that is not undergoing division is a tangled mass of thin threads called **chromatin.** At the time of division, the DNA becomes highly coiled and condensed, and it is easy to see the individual **chromosomes.** The reading for this chapter describes the manner in which the

Figure 10.2
Chromosome anatomy and reproduction in a prokaryote. The chromosome is a single loop of DNA that may be 500 times longer than the length of the cell. It is compacted within the nucleoid area shown in the micrograph in (f). a. The chromosome is attached to the inside of the plasma membrane and replication of DNA is beginning. b. Replication of DNA is complete. c. As the cell begins to elongate, the replicated chromosomes are pushed apart by the elongation. d. Ingrowth of the plasma membrane and growth of new wall have begun. e. There are now two daughter cells. f. Electron micrograph of the stage drawn in (e). Sometimes a prokaryotic cell will have several nucleoid areas, because several replications can be occurring at once. The third chromosome may begin at the site of the second before replication has even ended, with the production of a second chromosome. Magnification, ×87,500.

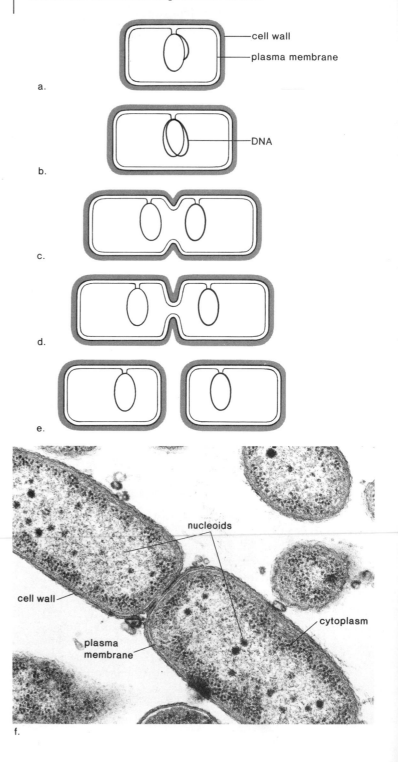

coiling and condensation take place. In addition to chromatin, the nucleus contains at least one nucleolus that is attached to, and formed by, special regions of particular chromosomes. You'll recall that the nucleolus is involved in the production of ribosomes, which are found in the cytoplasm.

When the chromosomes are highly coiled and condensed at the time of cell division, it is possible to photograph and count them. Each species has a characteristic number (table 10.1) and you'll note that the chromosome number for humans is 46. This is called the full or **diploid** (**2N**) number of chromosomes that occurs in all cells of the body. The diploid number includes two chromosomes of each kind. Half the diploid number, called the **haploid** (**N**) number of chromosomes, contains only one of each kind of chromosome. Sperm and egg cells have the haploid number of chromosomes as discussed later in this chapter.

Cell Division

Cell division in eukaryotes involves nuclear division and **cytokinesis**, division of the cytoplasm. Before nuclear division takes place, DNA replicates and therefore each chromosome is duplicated and has two identical parts called **sister chromatids** (fig. 10.3). Sister chromatids are genetically identical; that is, they contain exactly the same genes. Sister chromatids are constricted and are attached to each other at a region called the **centromere.**

During mitosis, the type of nuclear division considered in this chapter, these sister chromatids separate and each daughter cell receives one. In this way, each daughter cell gets a complete copy of the organism's genetic material. Once chromatids have separated, they are called chromosomes; therefore, each daughter cell following mitosis also has a complete set (2N number) of chromosomes.

In many eukaryotes, mitosis is associated with growth and repair of tissues. In these organisms there is another type of cell division called **meiosis,** which reduces the diploid number of chromosomes to the haploid (N) number. In humans, meiosis causes the sperm and egg to be haploid. When these cells, called the **gametes,** join to form a zygote at the time of fertilization, the diploid number is restored. Then the zygote repeatedly undergoes mitosis to give the multicellular form. Altogether, then, the human life cycle looks like this:

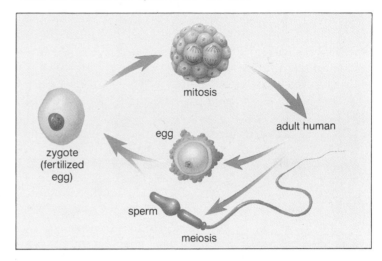

Table 10.1 Diploid Chromosome Numbers in Some Fungi, Plants, and Animals		
Type of Organism	**Name of Organism**	**Chromosome Number**
Fungi	*Aspergillus nidulans* (mold)	8
	Neurospora crassa (mold)	14
	Saccharomyces cerevisiae (yeast)	34
Plants	*Vicia faba* (broad bean)	12
	Zea mays (corn)	20
	Solanum tuberosum (potato)	48
	Nicotiana tabacum (tobacco)	48
Animals	*Felis domesticus* (cat)	38
	Equus caballus (horse)	64
	Gallus gallus (chicken)	78
	Pan troglodytes (chimp)	48
	Canis familiaris (dog)	78
	Rana pipiens (frog)	26
	Musca domestica (housefly)	12
	Homo sapiens (humans)	46

Eukaryotic chromosomes are located within the nucleus of the cell.

Figure 10.3

Duplicated chromosomes. DNA replication has occurred and each chromosome is held together at a region called the centromere.
a. Photomicrograph of highly coiled and compacted chromosome, typical of a nucleus about to divide. Magnification, ×8,400.
b. Diagram of a chromosome that is not as condensed. One chromatid is indicated by a box. A duplicated chromosome contains sister chromatids, each containing copies of the same genes.

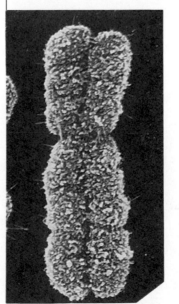

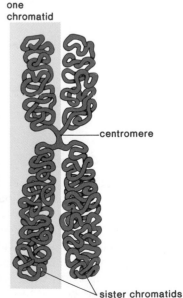

one chromatid

centromere

sister chromatids

a.

b.

What's in a Chromosome?

When early investigators decided that the genes were on the chromosomes, they had no idea of chromosome composition. By the mid-1900s, it was known that chromosomes are made up of both DNA and protein. Only in recent years, however, have investigators been able to produce models suggesting how chromosomes are organized.

A eukaryotic chromosome is more than 50% protein of two types: histone and nonhistone proteins. The nonhistone proteins are acidic (have negatively charged R groups) and are only loosely associated with DNA. The functions of the nonhistone proteins are being studied and some are known to be regulators of gene activity. Much more is known about the histone proteins, which are basic (have positively charged R groups) and bind strongly to the negatively charged phosphate groups of DNA. There are five different primary types of histone molecules designated H1, H2A, H2B, H3, and H4, and all five types are found in all tissues of an organism. Remarkably, the amino acid sequences of H3 and H4 vary little between organisms. For example, H4 of peas is only two amino acids different from H4 of cattle. This similarity suggests that mutations in the histone proteins have been selected during the evolutionary process and that the histones therefore have very important functions.

A human cell contains 46 chromosomes, and the length of the DNA in each is about 5 cm on the average. Therefore a human cell contains at least 2 m of DNA. Yet all of this DNA is packed into a nucleus that is about 5 μm in diameter. The histones are responsible for packaging the DNA so that it can fit into such a small space. First (fig. 1b) the DNA double helix is wound at intervals around a core of eight histone molecules (two copies each of H2A, H2B, H3, and H4) giving the appearance of a string of beads. Each bead is called a nucleosome and the nucleosomes are said to be joined by "linker" DNA. This string is coiled tightly into a fiber that has six nucleosomes per turn (fig. 1c)—the H1 histone appears to mediate this coiling process. The fiber loops back and forth (fig. 1d and e) and can condense to give a highly compacted form (fig. 1f) characteristic of metaphase chromosomes.

An interphase nucleus (fig. 2) shows darkly stained chromatin regions called heterochromatin and lightly stained chromatin regions called euchromatin. Heterochromatin (fig. 1e) is genetically inactive. The mechanism of inactivation is unknown but may involve methylation (adding methyl groups—CH_3—to the bases of DNA). Euchromatin (fig. 1c) is genetically active. It's believed that in this form the nucleosomes may allow access to the DNA, so that a gene may be turned on and become active in dictating protein synthesis in the cell.

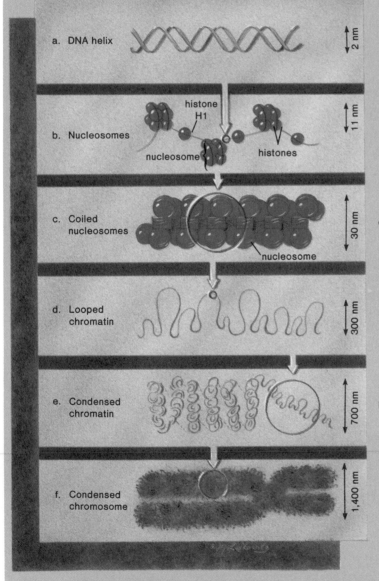

a. DNA helix — 2 nm

b. Nucleosomes — 11 nm
histone H1
nucleosome
histones

c. Coiled nucleosomes — 30 nm
nucleosome

d. Looped chromatin — 300 nm

e. Condensed chromatin — 700 nm

f. Condensed chromosome — 1,400 nm

Figure 1
Levels of chromosome structure.

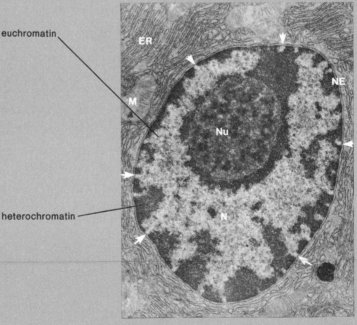

euchromatin
ER
M
Nu
NE
N
heterochromatin

Figure 2
Eukaryotic nucleus. The nucleus contains chromatin, DNA at two different levels of coiling and condensation. Euchromatin is at the level of coiled nucleosomes (30 nm) and heterochromatin is at the level of condensed chromatin (700 nm).

There are two types of nuclear division among eukaryotes. Following mitosis, the daughter cells have the diploid number of chromosomes. Following meiosis, the daughter cells have the haploid number of chromosomes; in humans these cells become the gametes.

Cell Cycle

By the 1870s, microscopy could give detailed and accurate descriptions of the movement of chromosomes during mitosis but there were no adequate microscopic techniques for studying the detailed activities of the cell during the period between divisions. Because there was little visible activity between divisions, this period of time was dismissed as a resting stage and later termed **interphase.** When it was discovered in the 1950s that DNA synthesis occurs during interphase, the **cell cycle** concept was formulated.

Cells grow and divide during a cycle that has four phases (fig. 10.4). The entire cell division phase, including both mitosis and cytokinesis, is termed the M phase (M = mitotic). The period of DNA replication is termed the S phase (S = synthesis) of the cycle; as mentioned previously, replication of DNA produces duplicated chromosomes. The proteins associated with DNA are also synthesized during this phase. There are two other phases of the cycle. The period of time following the *M phase* before the *S phase* begins is termed the *G_1 phase,* and the period of time following the S phase before the M phase begins is termed the *G_2 phase.* At first, not much was known about these phases and they were thought of as G = gap phases. Now we know that during the G_1 phase, the cell grows in size, and the cellular organelles increase in number. During the G_2 phase, synthesis of various enzymes and other types of proteins occurs. Some biologists today prefer the designation G = growth for these two G phases. In any case, it is now known that interphase consists of the G_1, S, and G_2 phases.

Normally, newly formed cells differentiate as cell specialization takes place (fig. 10.4). After differentiation has taken place, many cells, such as muscle and nerve cells, seem to lose the ability to divide. Others, such as liver cells, will divide when repair of the organ is needed and still others, such as epithelial cells, divide continuously throughout the life of the organism.

Cells undergo a cycle that includes interphase, consisting of G_1 (organelle duplication begins), S (DNA replication), and G_2 (synthesis of various proteins); and mitosis followed by cytokinesis.

Control of the Cell Cycle

Investigations are underway to determine what controls the cell cycle, in particular, what makes a cell go through the different phases. Typically, cells are grown in a growth medium that contains nutrients such as glucose, amino acids, vitamins, and blood serum—in short, everything they need to keep on dividing. Even so, normal cells in culture exhibit contact inhibition and stop dividing once they have made contact with other cells. If some of these cells are removed and placed in a new medium, they will start dividing again, but regardless, the total number of divisions for any particular human cell is usually a maximum of 50 to 100.

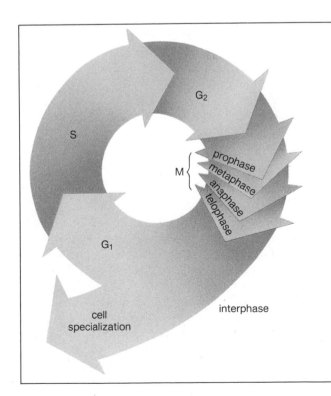

Phase	Main Events	Vicia faba	Homo sapiens (cultured fibroblasts)
G_1	Organelles begin to double in number	4.9 hours	6.3 hours
S	Replication of DNA	7.5	7.0
G_2	Synthesis of proteins	4.9	2.0
M	Mitosis	2.0	0.7
	Total:	19.3 hours	16.0 hours

Figure 10.4

The cell cycle consists of four phases (G_1, S, G_2, and M for mitosis), which are indicated in the figure and described in the accompanying table. The length of the different phases varies both between species and between different cell types in the same individual. The approximate length of the phases for the broad bean (*Vicia faba*) and humans (*Homo sapiens*) are given.

Many cells will not divide at all unless minute amounts of growth factors, such as certain hormones or proteins, are present in the medium. Hormones are regulatory factors that affect cell metabolism, and there is evidence that cytoplasmic factors affect cell division. Lester Goldstein and David Prescott have transplanted nuclei from one amoeba to another. When S phase nuclei are placed in G_2 cells, synthesis slows down and when G_1 nuclei are placed in S phase cells, DNA replication begins. Perhaps both inhibitory and stimulatory substances are involved, and if so, their presence or absence during G_1 seems to be critical. If a cell passes through G_1, it is certain to divide. But if the cell is denied nutrients and growth factors, it becomes arrested in G_1 and enters a phase called G_0, during which it can survive for long periods of time. Perhaps lack of nutrients prevents synthesis of a particular protein that triggers the rest of the cell cycle. On the other hand, aging cells contain inhibitory proteins that directly stop the cell cycle.

It is of interest to note that cancer cells do not exhibit the same cell cycle control that normal cells do. As discussed in chapter 17 this can be traced to an altered metabolism that allows them to repeatedly enter the cell cycle.

Normally cells exhibit cell cycle control, and cytoplasmic proteins are apparently involved in this control.

Mitosis

Mitosis is *division of the eukaryotic nucleus so that the daughter cell nuclei acquire the same number and kinds of chromosomes present in the mother cell nucleus.* Before mitosis begins, DNA has replicated and the chromosomes have duplicated:

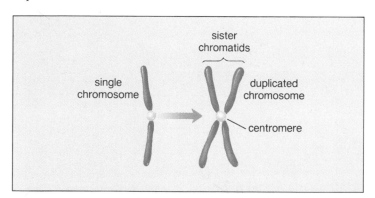

single chromosome → duplicated chromosome

sister chromatids

centromere

During mitosis a microtubular structure, the **spindle apparatus,** forms to bring about an orderly distribution of chromosomes to the daughter cell nuclei. (There is no spindle during the process of binary fission in prokaryotic cells.) Mitosis, also called karyokinesis, is a continuous process that is arbitrarily divided into four stages for convenience of description. These stages are prophase, metaphase, anaphase, and telophase.

Mitosis in Animal Cells

One of the distinctions between animal and plant cell mitosis is the presence of **centrioles** in animal cells but not in plant cells. Duplication of a single pair of centrioles begins prior to mitosis and results in two pairs of centrioles during prophase.

Prophase (fig. 10.5)

It is apparent during **prophase** that nuclear division is about to occur, because the chromatin has condensed and the chromosomes become visible structures (see reading, p. 148). As the chromosomes continue to compact, the nucleolus and nuclear envelope disappear. *The chromosomes are duplicated and composed of two sister chromatids held together at a centromere* (fig. 10.3). Counting the number of centromeres in diagrammatic drawings gives the diploid number of chromosomes for the cell depicted.

Polar spindle fibers, composed of bundles of microtubules, begin to appear between the pairs of centrioles that are moving away from one another. The centrioles, which mark the poles of the spindle apparatus, give off short radiating fibers called **asters.** At one time it was believed that the centrioles were the organizing centers for spindle assembly, but this no longer appears to be the case. Plant cells lack centrioles, but they still produce spindles during mitosis. Also, when centrioles are destroyed by a laser beam, the spindle continues to form normally.

Prophase is recognized by the presence of condensed chromosomes that have no particular orientation. The nucleolus has disappeared and the spindle has appeared.

Metaphase (fig. 10.6)

As **metaphase** begins, the fragmentation of the nuclear envelope that began in late prophase continues, and the polar spindle fibers now enter the nuclear area. Specialized structures called kinetochores develop on either side of each centromere. There are microtubules, called *kinetochore fibers* that extend from the kinetochores. They interact with the polar fibers in such a way that the chromosomes are located at the equator (middle) of the spindle apparatus. The chromatids of a chromosome dangle from the centromere, but the kinetochores are orientated so that only one chromatid will eventually go to each pole.

Metaphase is the easiest phase to recognize because it is marked by a fully formed spindle apparatus with the chromosomes located at the equator of the apparatus.

Anaphase (fig. 10.7)

During **anaphase,** the centromeres divide and the chromatids separate. The mechanism by which the centromeres divide is unknown but it's been suggested that the centromeres are merely a region where the DNA did not replicate. Replication in this area would then allow the chromatids to separate. *Once the chromatids have separated, they are called daughter chromosomes.*

The daughter chromosomes now move toward opposite poles of the spindle. There seem to be two processes that account for this movement. A lengthening of the polar fibers acts to push the chromosomes away from the equator, and a shortening of the kinetochore fibers acts to pull the chromosomes toward the poles (fig. 10.8).

Figure 10.5

Prophase. *a.* The chromosomes are now distinct and each one is composed of two chromatids. The nucleolus has disappeared. As the pairs of centrioles move apart, the polar spindle fibers appear. The nuclear envelope is beginning to fragment. *b.* Prophase in a whitefish (animal) embryonic cell. Magnification, ×280.

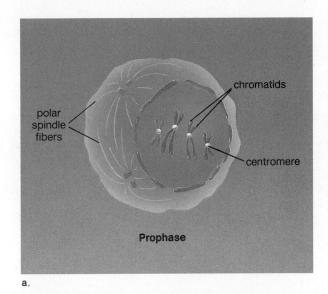

Prophase

a.

b.

Figure 10.6

Metaphase. *a.* Duplicated chromosomes are located at the equator of the cell. The centrioles are at the poles surrounded by the asters and the spindle apparatus is fully formed. The nuclear envelope is in fragments. *b.* Metaphase in a whitefish (animal) embryonic cell. Magnification, ×280.

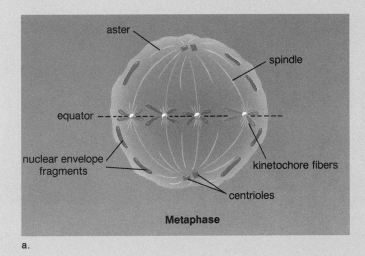

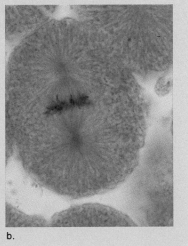

Metaphase

a.

b.

Figure 10.7

Anaphase. *a.* The centromeres have divided and the sister chromatids are now separating; each is now called a chromosome. Therefore, the diploid number of chromosomes will be at each pole of the spindle apparatus. *b.* Anaphase in whitefish (animal) embryonic cell. Magnification, ×500.

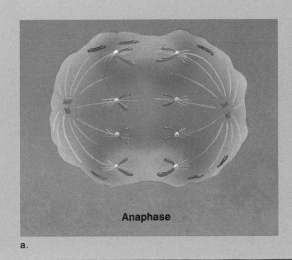

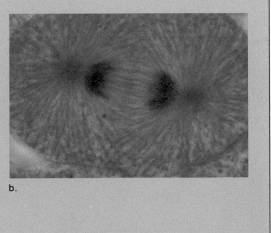

Anaphase

a.

b.

Figure 10.8

Spindle apparatus anatomy and function. *a.* Artist's representation of the spindle apparatus from an animal cell. The poles (yellow) contain the centrioles with radiating fibers called asters. Polar fibers reach from the poles to the equator, where they overlap. The chromosomes (blue) are moving toward the poles. *b.* Chromosome movement. Spindle fibers are composed of microtubules that can lengthen by the addition of tubulin dimers and shorten by the subtraction of tubulin dimers. It's believed that the polar fibers lengthen at the equator due to the addition of tubulin subunits and that they also slide past one another in the region of overlap. This pushes the chromosomes apart. Also the kinetochore fibers shorten by disassembling at the poles; this pulls the chromosomes toward the poles.

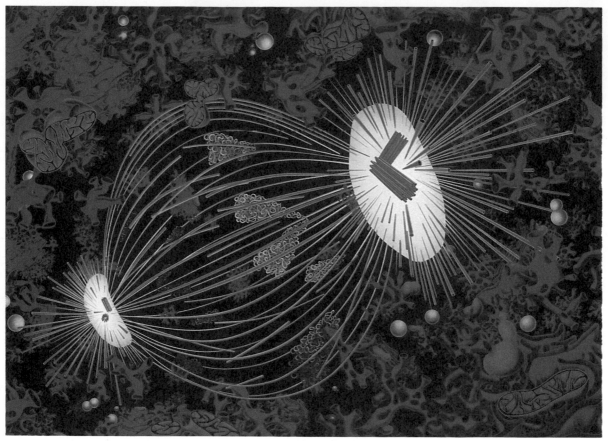

a.

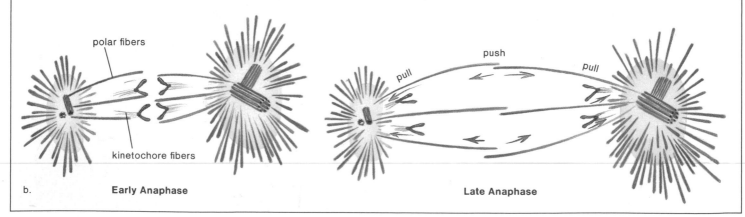

b.   **Early Anaphase**   **Late Anaphase**

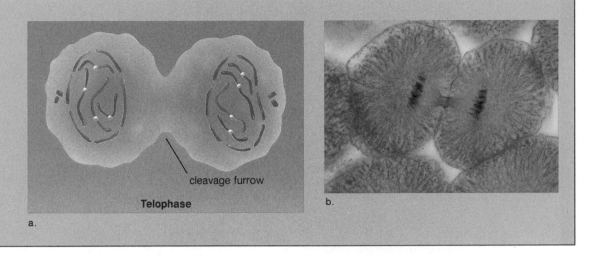

cleavage furrow

Telophase

a.

b.

Anaphase is recognized by the movement of chromosomes toward each pole of the spindle apparatus.

Telophase (fig. 10.9)

As the polar spindle fibers elongate during **telophase,** new nuclear envelopes form around the daughter chromosomes. Each daughter nucleus contains the same number and kinds of chromosomes as in the original mother cell. Remnants of the spindle apparatus are still visible between the two nuclei.

Soon, distinct chromosomes will become more diffuse chromatin, and a nucleolus will reappear in each daughter nucleus. Cytokinesis now begins and eventually there will be two individual daughter cells.

Telophase is recognized by the presence of daughter chromosomes within daughter nuclei. Remnants of the spindle apparatus can still be seen but the nucleus is taking on the appearance of an interphase nucleus.

Cytokinesis in Animal Cells

Cytokinesis, or cytoplasmic cleavage, usually accompanies mitosis. In animal cells, as telophase draws to a close, a **cleavage furrow,** an indentation of the membrane between the two daughter nuclei, forms and divides the cytoplasm. The equator of the spindle apparatus marks where the furrow will form. A bundle of microfilaments slowly constricts the cell, forming first a narrow bridge, and finally completely separating the cell into the two daughter cells (fig. 10.10).

Cytokinesis in animal cells is accomplished by a furrowing process.

Figure 10.10
Cytokinesis in an animal cell. The single cell becomes two cells by a furrowing process that gradually separates the two daughter cells.

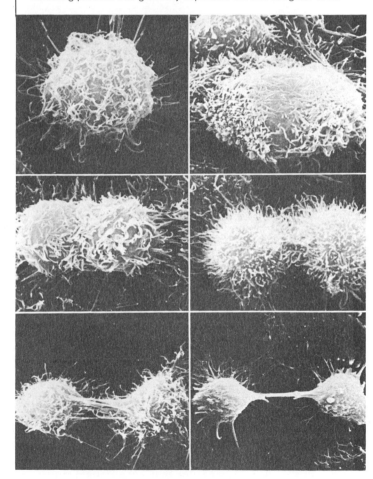

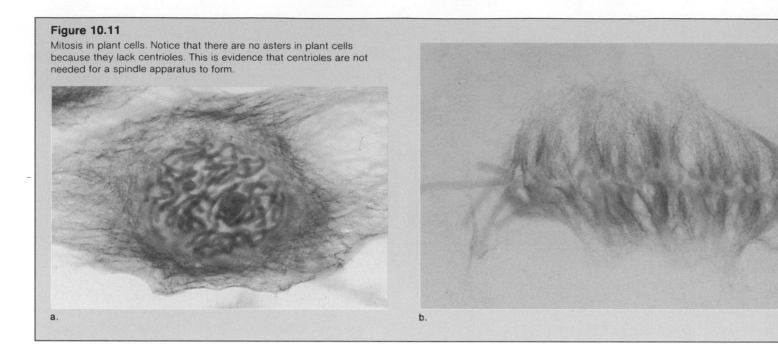

Figure 10.11

Mitosis in plant cells. Notice that there are no asters in plant cells because they lack centrioles. This is evidence that centrioles are not needed for a spindle apparatus to form.

a.

b.

Mitosis in Plant Cells

Figure 10.11 illustrates mitosis in plants and you will note that there are exactly the same stages in their cells as in animal cells. As mentioned previously, there are no centrioles or asters in plant cells.

Certain plant tissue, called meristem tissue, retains the ability to divide throughout the lifetime of a plant. Meristem tissue is found in root and shoot tips and in stems, accounting for the ability of trees to grow larger each growing season.

Cytokinesis in Plant Cells

Cytokinesis in plant cells occurs by a different process than that seen in animal cells. The rigid cell wall that surrounds plant cells does not permit cytokinesis by furrowing. Instead, vesicles, largely derived from the Golgi complex, travel down the polar spindle fibers to the region of the equator. These vesicles fuse forming a **cell plate.** Their membrane completes the plasma membrane for both cells (fig. 10.12) and they also release substances that signal the formation of plant cell walls. These walls are later strengthened by the formation of cellulose fibrils.

A spindle forms during mitosis in plant cells but there are no centrioles or asters. Cytokinesis in plant cells involves the formation of a cell plate.

Importance of Binary Fission and Mitosis

Binary fission and mitosis ensure that each daughter cell receives the same number and kinds of chromosomes as were present in the mother cell; therefore, the daughter cells are genetically identical to the mother cell. Fission and mitosis differ chiefly in that binary fission does not utilize a spindle apparatus whereas mitosis does utilize such a spindle.

Evolution of the Spindle

Observation of nuclear division in primitive eukaryotes, such as dinoflagellates and fungi, makes it possible to suggest the steps by which mitosis may have evolved. At first, division may have taken place in a manner similar to binary fission, except that the separating chromosomes were attached to the nuclear envelope instead of the cell membrane. The purpose of the first involvement of microtubules may have been to support the nuclear envelope. Later, the microtubules became attached to the chromosomes themselves. As microtubules began to take a more active role in separating the chromosomes, the nuclear envelope became less important. Indeed, we have seen that the nuclear envelope plays no role in the most advanced eukaryotes.

Asexual versus Sexual Reproduction

We have observed that single-celled prokaryotes reproduce by binary fission. Single-celled eukaryotic protists (e.g., protozoans and algae) reproduce by cell division that involves mitosis. Reproduction by binary fission and mitosis is termed **asexual reproduction** (table 10.2).

Table 10.2 Functions of Cell Division

Type of Organism	Cell Division	Function
Prokaryotes (bacteria)	Binary fission	Asexual reproduction
Protists Some fungi	Mitosis and cytokinesis	Asexual reproduction
Some fungi Plants Animals	Mitosis and cytokinesis	Primarily growth and repair. Asexual reproduction in some.
	Meiosis and fertilization	Sexual reproduction

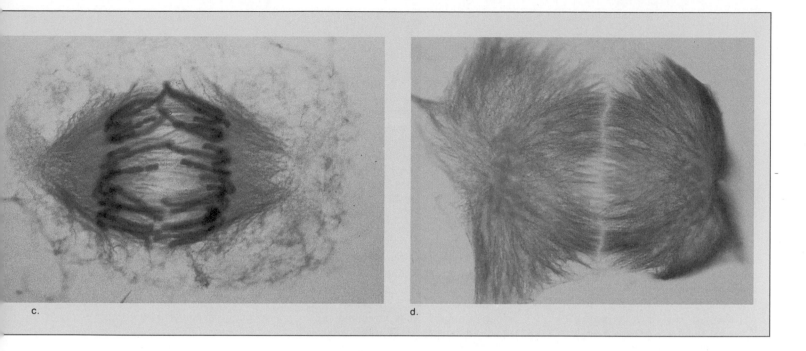

c.

d.

Figure 10.12

Cytokinesis in a plant cell involves the formation of a cell plate. *a.* Electron micrograph showing the stage that corresponds to (*c*). The nuclear envelopes of the daughter cells have already reformed. CP = cell plate; CW = cell wall; Pl = plasma membrane; N = nucleus; SpF = spindle fiber. *b.* Vesicles containing polysaccharides are at the center of the cell. *c.* Cell plate forms at the cell equator and extends to the plasma membrane. *d.* Daughter cell plasma membranes are complete and cell wall has formed.

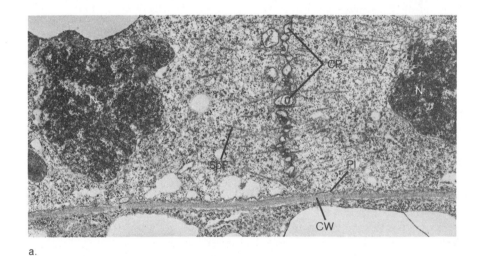

a.

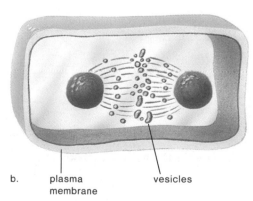

b.   plasma   vesicles
     membrane

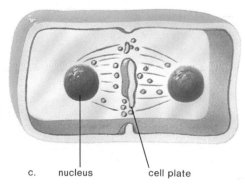

c.   nucleus   cell plate

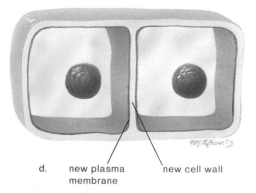

d.   new plasma   new cell wall
     membrane

Although some multicellular forms do reproduce asexually, generally speaking, mitosis occurs during the growth and repair of tissues. Multicellular forms usually carry on **sexual reproduction,** which requires sex cell (gamete) formation and fertilization. Meiosis is a form of cell division that reduces the chromosome number from diploid to haploid and is involved in sex cell formation and sexual reproduction. Meiosis is discussed in the next chapter.

Prokaryotes reproduce asexually by binary fission and single-celled eukaryotes reproduce asexually by mitosis. Mitosis in multicellular eukaryotes is usually involved in growth and repair, and meiosis occurs during sexual reproduction.

Summary

1. In single-celled organisms, simple cell division allows the organism to reproduce. In multicellular organisms, this type of cell division is primarily for growth and repair.
2. Cell division keeps the cell surface area/ cell volume ratio at a level appropriate to the life of the cell and its occurrence leads to specialization of cells.
3. The prokaryotic chromosome is a single long loop of DNA attached to the inside of the plasma membrane. Replication of DNA is followed by an elongation of the cell that pushes the chromosomes apart. Ingrowth of the plasma membrane and formation of new cell wall material divides the cell in two.
4. Each eukaryote species has a characteristic number of chromosomes. The total number is called the diploid number and half this number is the haploid number. Between nuclear divisions, the chromosomes are not distinct and are collectively called chromatin. Chromatin and chromosomes have levels of coiling and condensation that are explained in the reading on page 148.
5. Replication of DNA precedes cell division and the duplicated chromosomes are each composed of two sister chromatids held together at a centromere.
6. Among eukaryotes, cell division involves nuclear division (karyokinesis), and division of the cytoplasm (cytokinesis). There are two types of nuclear division. Mitosis is simple nuclear division in which the chromosome number stays constant, and meiosis is nuclear division in which the chromosome number is reduced by one-half. Only mitosis is considered in this chapter.
7. The cell cycle includes interphase and mitosis. Interphase has three phases. During G_1 the organelles increase in number, during S, DNA replication occurs, and during G_2 various proteins are synthesized. Specialized cells leave the cycle.
8. Control of the cell cycle is under investigation. In cells grown in tissue culture it appears that various proteins in the cytoplasm may control the cycle.
9. Mitosis, nuclear division by which the daughter nuclei receive the same number and kinds of chromosomes as the mother cell, has four stages, which are described here for animal cells.
   Prophase—Duplicated chromosomes are distinct, the nucleolus is disappearing, the nuclear envelope is fragmenting, and spindle fibers are forming as centriole pairs separate. Asters develop around the centrioles.
   Metaphase—Spindle apparatus is fully formed. Duplicated chromosomes move to the equator, and are attached to the polar fibers of the spindle by the kinetochore fibers.
   Anaphase—Centromeres separate and daughter chromosomes move toward the poles, perhaps by a push-pull mechanism. Push = polar fibers elongate and slide past one another at the equator. Pull = kinetochore fibers shorten at the poles.
   Telophase—Nuclear envelopes reform, chromosomes begin changing back to chromatin, the nucleoli reappear, and spindle fibers disappear. Cytokinesis quickly follows telophase and completes the furrowing process that divides the cytoplasm.
10. Mitosis in plant cells is somewhat different from mitosis in animal cells because plant cells lack centrioles and therefore asters. Even so, the spindle apparatus forms and the same four stages are observed. Cytokinesis in plant cells involves the formation of a cell plate from which the plasma membrane and cell wall are completed.
11. Binary fission in single-celled prokaryotes, and mitosis in single-celled eukaryotic protists and fungi, allow these organisms to reproduce asexually. Mitosis in multicellular eukaryotes is for the purpose of growth and repair. Multicellular forms generally carry on sexual reproduction, which requires meiosis, the type of cell division involved in gamete formation.

Objective Questions

1. What feature in prokaryotes substitutes for the fibers of the spindle apparatus in eukaryotes?
   a. centrioles with asters
   b. fission instead of cytokinesis
   c. elongation of plasma membrane
   d. looped DNA
2. How does a prokaryotic chromosome differ from a eukaryotic chromosome? A prokaryotic chromosome is
   a. shorter and fatter.
   b. looped and one piece.
   c. never replicates.
   d. All of these.
3. The diploid number of chromosomes
   a. is the 2N number.
   b. was in the mother cell and is in the two daughter cells, following mitosis.
   c. varies according to the particular organism.
   d. All of these.

For questions 4 to 6, match the descriptions below to the terms in the key:

Key:
   a. centriole
   b. chromatid
   c. chromosome
   d. centromere

4. point of attachment for sister chromatids
5. found at poles; centers of asters
6. condensed chromatin
7. If a mother cell has 14 chromosomes prior to mitosis, how many chromosomes will the daughter cells have?
   a. 28
   b. 14
   c. 7
   d. Any number between 7 and 28.

8. In which stage of mitosis are the chromosomes moving toward the poles?
   a. prophase
   b. metaphase
   c. anaphase
   d. telophase
9. Interphase
   a. is the same as prophase, metaphase, anaphase, and telophase.
   b. includes G_1, S, and G_2.
   c. requires the use of polar fibers and kinetochore fibers.
   d. rarely occurs.
10. Cytokinesis
   a. is mitosis in plants.
   b. requires the formation of a cell plate in plant cells.
   c. is the longest part of the cell cycle.
   d. is half a chromosome.

Study Questions

1. What are the benefits and problems associated with cell division?
2. Describe the prokaryotic chromosome and the process of binary fission.
3. A human cell contains at least 2 m of DNA yet all of this DNA is packed into a nucleus that is about 5 μm in diameter. Explain.
4. Describe the cell cycle, being sure to include a reference to interphase.
5. How do cancer cells differ from normal cells in relation to the cell cycle?
6. Define the following words: chromosome, chromatin, chromatid, centriole, cytokinesis, and centromere.
7. Why is duplication of a chromosome required before mitosis occurs?
8. Describe the events that occur during the stages of mitosis.
9. Contrast cytokinesis in animal cells with that in plant cells.
10. Why is binary fission in prokaryotes and mitosis in single-celled eukaryotes a form of asexual reproduction?

Thought Questions

1. Correlate the function of DNA (DNA dictates protein synthesis) with the necessity that each daughter cell receive a copy of each and every gene.
2. Considering the function of centrioles discussed in chapter 5, why would you expect these organelles to be present in animal cells and not in plant cells.
3. Show that cell plate formation, leading to the origination of new plasma membrane and new cell wall, can be compared to the process of exocytosis discussed on page 86.

Selected Key Terms

binary fission (bi′na-re fish′un) 146
chromatin (kro′mah-tin) 146
chromosome (kro′mo-sōm) 146
diploid (2N) (dip′loid) 147
haploid (N) (hap′loid) 147
cytokinesis (si″to-ki-ne′sis) 147
sister chromatid (sis′ter kro′mah-tid) 147

meiosis (mi-o′sis) 147
interphase (in′ter-fāz) 149
cell cycle (sel si′k′l) 149
mitosis (mi-to′sis) 150
spindle apparatus (spin′d′l ap″ah-ra′tus) 150
centriole (sen′tri-ol) 150

aster (aṣ′ter) 150
cleavage furrow (clē-vaj fur′o) 153
cell plate (sel plāt) 154
asexual reproduction (a-seks′u-al re″pro-duk′shun) 154
sexual reproduction (seks′u-al re″pro-duk′shun) 155

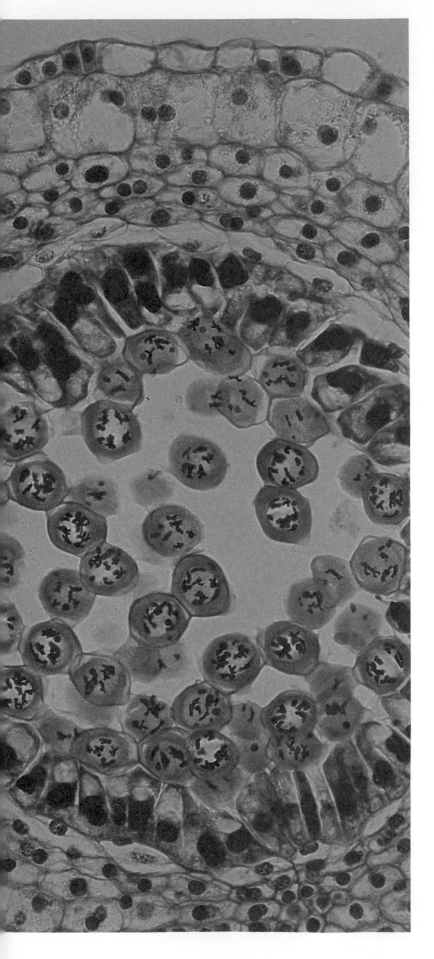

CHAPTER 11

Cell Reproduction: Meiosis

Your study of this chapter will be complete when you can:

1. *State in general the role of meiosis in the life cycle of plants and animals.*

2. *Give an overview of the process of meiosis and its stages, placing an emphasis on the main events.*

3. *Describe the stages of meiosis I in detail.*

4. *Describe synapsis (bivalent formation), and tell how crossing-over occurs.*

5. *Describe the stages of meiosis II in detail.*

6. *Describe the mammalian life cycle, and compare spermatogenesis to oogenesis.*

7. *Compare the process of meiosis to that of mitosis.*

8. *Describe the manner in which sexual reproduction brings about variation and contributes to the evolutionary process.*

When mature, the pollen grains of plants contain a sperm nucleus. It's not surprising, then, that meiotic cell division, which reduces the chromosome number, occurs during the production of pollen grains. This is a cross section of a flower part and each circular cell will be a pollen grain. The dark threads are the chromosomes.

amete formation and fusion are integral parts of **sexual reproduction.** Obviously, if the gametes such as the sperm and egg pictured in figure 11.1 contained the same number of chromosomes as the body cells, the number of chromosomes would double with each new generation. Within a few generations, the cells of the individual would be nothing but chromosomes! The early cytologists (biologists who study cells) realized this and Pierre-Joseph van Beneden, a Belgian, was gratified to find in 1883 that the sperm and egg of the worm *Ascaris* each contained only two chromosomes while the cells of the embryo and individual always had four chromosomes.

There is a special type of nuclear division, called **meiosis,** that reduces the chromosome number; in human males meiosis occurs only in the testes where sperm are produced, and in females it occurs only in the ovaries during the formation of the egg. Meiosis is very orderly so that each gamete receives one of each kind of chromosome.

Meiosis is a special type of cell division that occurs during the process of gamete production in organisms.

Meiosis

Chromosomes come in pairs, which can be recognized because the members of a pair of chromosomes look alike:

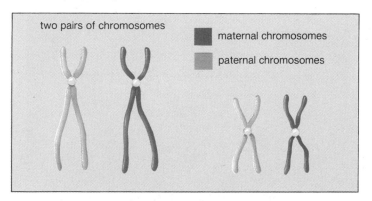

two pairs of chromosomes

maternal chromosomes

paternal chromosomes

The members of a pair are called **homologous chromosomes** not only because they look alike, but also because they carry genes for the same traits. One member of each pair is contributed by the male parent and the other is contributed by the female parent. Each **gamete** (sex cell, that is, the sperm or the egg) contains one of each kind of chromosome. During meiosis, the chromosome number is reduced from the diploid (2N) number to the haploid (N) number, but in such a way that the daughter nuclei receive one of each kind of chromosome. It is important for a gamete to have one member from each pair of homologous chromosomes because only in that way can a gamete pass on a copy of each kind of gene to the offspring.

The life cycle of all sexually reproducing organisms has two stages: the diploid stage in which there are two chromosomes of each kind and the haploid stage in which there is only one chromosome of each kind. In animals the haploid stage consists only of the gametes, but in plants the haploid stage is a multicellular one that produces the gametes (fig. 11.2). In both types of life cycles, it is **fertilization** of the egg by the sperm that restores the chromosome number to the full diploid number once again.

Meiosis is a form of nuclear division that reduces the diploid number of chromosomes to the haploid number found in gametes.

An Overview of Meiosis

Meiosis requires two nuclear divisions and produces four daughter cells, each having one of each kind of chromosome and therefore half the total number of chromosomes present in the mother cell nucleus. Figure 11.3 presents an overview of meiosis indicating its two nuclear divisions, meiosis I and meiosis II.

Prior to meiosis, DNA replication has occurred; therefore each chromosome has two sister chromatids. During meiosis I, something new happens that does not occur in mitosis. The homologous chromosomes, or **homologues,** actually come together and pair, forming a so-called **bivalent (tetrad)** that contains four chromatids in close association. Now **crossing-over** of genetic material may occur between the nonsister chromatids of the bivalent. Due to crossing-over, the sister chromatids of a chromosome are no longer identical (*see* fig. 11.5).

The occurrence of crossing-over temporarily holds the homologues together so that they interact with the spindle fibers as if they were one chromosome! Eventually, the chromosomes of each bivalent separate and this assures that one chromosome of each kind reaches the daughter nuclei. The separation process has no restrictions; either chromosome of a homologous pair may occur in a daughter nucleus with either chromosome of any other pair. In other words, each homologous pair separates independently of all the other pairs.

During meiosis I, homologues pair and crossing-over occurs before the paired chromosomes independently separate into the daughter nuclei. The chromosome number is reduced by one-half.

Figure 11.1
During sexual reproduction, the sperm fertilizes the egg. If these gametes carried the diploid number of chromosomes then the zygote would have double the chromosome number of the parents. However, since each gamete carries the haploid number of chromosomes, the zygote has the same number of chromosomes as the parents.

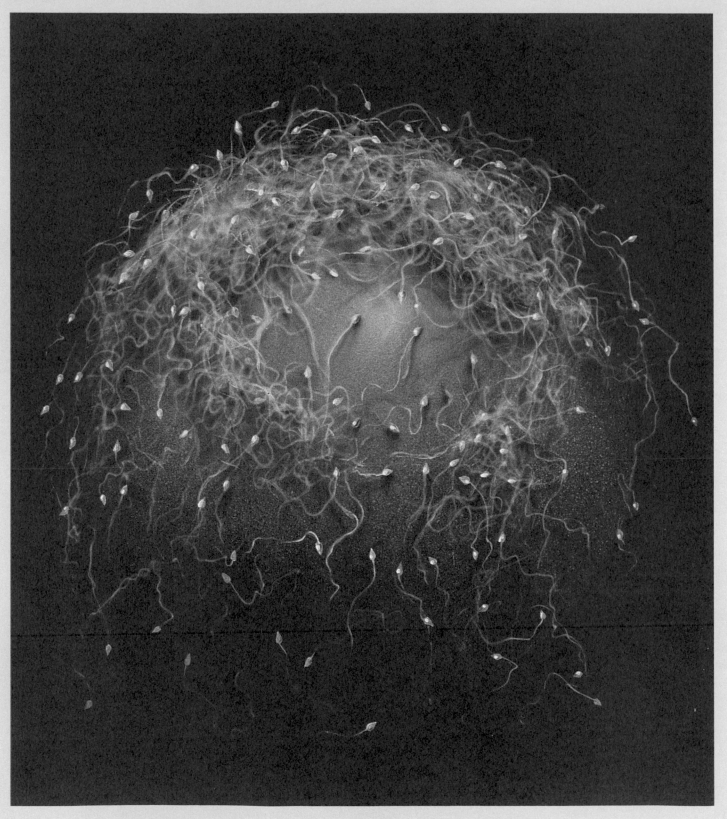

At the completion of the first cell division, (fig. 11.3), the chromosomes in the nucleus of each daughter cell still consist of duplicate chromatids. Therefore, no replication need occur between divisions. During meiosis II, the sister chromatids separate, so that the nucleus in each of four daughter cells has the haploid number of chromosomes. The events of meiosis II are, then, similar to those of mitosis.

No DNA replication occurs between meiosis I and meiosis II. During meiosis II, the sister chromatids separate and each of the four daughter cells contains the haploid number of chromosomes, that is, one of each kind of chromosome.

Figure 11.2

All sexually reproducing organisms have both a diploid and haploid stage in their life cycles. In animals the haploid stage only consists of the gametes, but in plants the haploid stage is multicellular and is called the haploid generation. This generation produces the gametes.

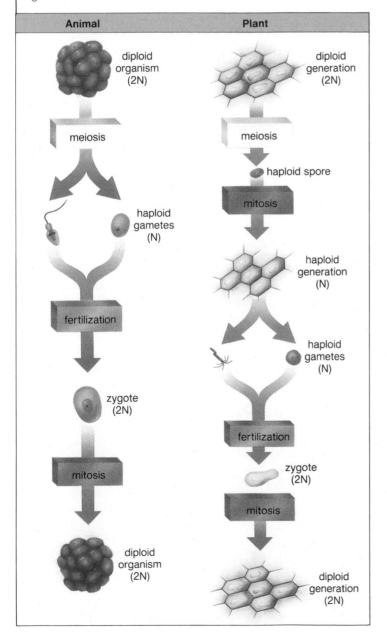

Figure 11.3

Overview of meiosis, indicating meiosis I and meiosis II. The number of chromosomes in each nucleus can be determined by counting the centromeres and the homologous chromosomes can be determined by noting the placement of the centromeres. During meiosis I the homologues separate and this means that the daughter cells have one of each kind of chromosome. During meiosis II, the chromatids separate and there are four haploid daughter cells. The daughter cells are not genetically similar to the mother cell for two reasons—they have only one of each kind of chromosome and crossing-over (represented in the figure by exchange of colors) between nonsister chromatids has occurred.

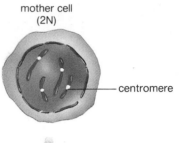

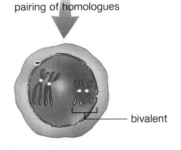

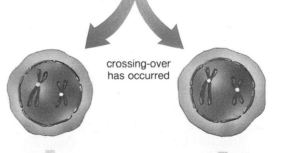

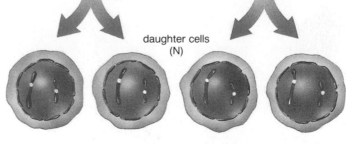

Figure 11.4

Meiosis consists of two divisions, meiosis I and meiosis II, and each of these has four stages. Members of homologous pairs have different colors to indicate that one member of each pair is a maternal chromosome and the other is a paternal chromosome. During meiosis I, the homologous chromosomes separate, and during meiosis II, sister chromatids separate.

Meiosis I

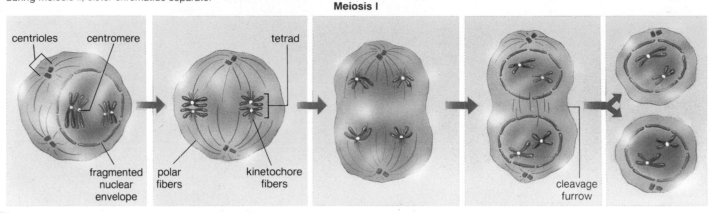

Prophase I

Replication of DNA has occurred and the chromosomes are duplicated. The homologous chromosomes pair and while they are in synapsis, crossing-over between nonsister chromatids occurs. The chromosomes are condensing as prophase continues. Spindle fibers appear as the centrioles move apart and the nuclear envelope begins to fragment. Prophase I is the longest stage of meiosis and takes up most of the time required for the process.

Metaphase I

The spindle apparatus is now fully formed and occupying the center of the cell. Homologue pairs (bivalents) are at the equator of the spindle. The kinetochores of the homologues point to opposite poles. Alignment of the homologues at the equator is completely random.

Anaphase I

The centrioles of sister chromatids remain intact and the homologous chromosomes separate. Like mitosis, movement is due to a push provided by spindle elongation and a pull due to kinetochore disassembly.

Telophase I

The haploid number of chromosomes is at each pole. Assortment is independent and all possible combinations of chromosomes may occur at the poles. This figure shows only two possible combinations—what are the other possible combinations? Usually cytokinesis occurs simultaneously with telophase.

Daughter Cells

Now let's look at the stages of meiosis in detail.

Meiosis I

Meiosis I (fig. 11.4) is divided into the same four stages as mitosis but the prophase stage is much more complicated.

Prophase I

Because replication has occurred, the chromosomes consist of sister chromatids held together at a centromere. Each one of these duplicated but still extended chromosomes begins to pair with its homologue. During this process, called **synapsis,** the homologous chromosomes line up side by side and a nucleoprotein lattice (called the synaptonemal complex) appears between them (fig. 11.5). This lattice holds the members of a bivalent (tetrad) together in such a way that the DNA of the nonsister chromatids is aligned in sequence. Now **crossing-over,** an exchange of genetic material between nonsister chromatids, can occur. Then, the lattice begins to break down and the homologous chromosomes begin to move apart but not too far because they are held together by **chiasmata** (sing. chiasma), regions where the nonsister chromatids are still attached due to crossing-over (fig. 11.6).

During prophase I, synapsis occurs and a nucleoprotein lattice develops between homologues. This close association assists crossing-over between nonsister chromosomes of the bivalent.

We should mention that throughout all these events of prophase I, the chromosomes have been condensing so that by now they have the characteristic appearance of metaphase chromosomes.

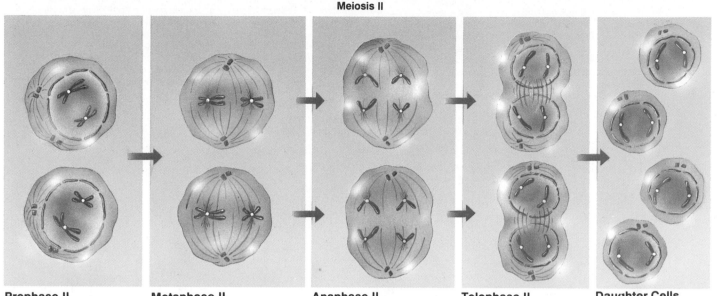

Meiosis II

Prophase II

No replication of the DNA occurs between meiosis I and meiosis II—the chromosomes were duplicated earlier. Condensation of chromosomes begins once again. The spindle forms and the nuclear envelope fragments once more.

Metaphase II

The duplicated and fully condensed chromosomes are aligned individually at the equator. The kinetochore fibers of the sister chromatids point in opposite directions. This is in preparation for their separation.

Anaphase II

Sister chromatids separate just as they do during mitosis and that's why it is said that the events of meiosis II are like those of a mitotic division. The chromatids, now called chromosomes, move toward opposite poles.

Telophase II

Each daughter nucleus contains the haploid number of chromosomes and these chromosomes are single. The daughter nuclei are not genetically similar to the mother cell. They have the haploid number of chromosomes, and due to crossing-over during prophase I, these chromosomes may carry a new rearrangement of the genetic material.

Daughter Cells

Metaphase I

The spindle apparatus forms as it did during mitosis but at the equator during metaphase I we find chromosome bivalents with this appearance:

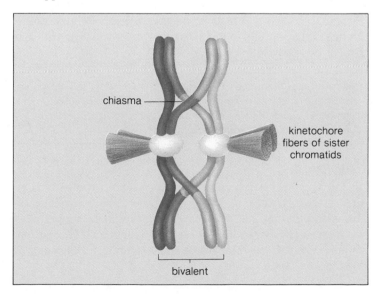

chiasma

kinetochore fibers of sister chromatids

bivalent

Notice that the kinetochore fibers extend from only one side of the centromere that holds the sister chromatids together and that the bivalent is held together by chiasmata.

There is no set way for the chromosomes to align themselves at the equator of the spindle. The maternal homologue of one bivalent may be oriented toward one pole and the maternal or paternal homologue of another bivalent may be aligned in like manner. Therefore, in metaphase I the chromosomes will assort independently of one another and any possible combinations of chromosomes can be found in the daughter cells.

Bivalents are present at the equator during metaphase I. The alignment is random and therefore any possible combination of chromosomes can occur in the daughter cells.

Anaphase I

During anaphase I, the association between the members of a bivalent is disrupted when the chiasmata, which mark points of exchange between the nonsister chromatids, cease to exist because crossing-over is completed (fig. 11.5*d*). Now the homologous chromosomes separate as the centromeres of each bivalent

Figure 11.5

Crossing-over occurs during prophase I. *a.* First, the homologous chromosomes pair up and a nucleoprotein lattice, called the synaptonemal complex, develops between them. This is an electron micrograph of the complex that "zippers" the members of the bivalent together so that corresponding genes are in alignment. *b.* This diagrammatic representation of crossing-over shows only two places where nonsister chromatids 1 and 3 have come into contact. Actually, the other two nonsister chromatids would most likely also be crossing-over. *c.* The chromatids break at the region of crossing-over and when they rejoin there has been an exchange of genetic material represented here by an exchange of color. The chiasmata indicate where crossing-over has occurred and due to the chiasmata the members of a bivalent remain together during metaphase I even though the synaptonemal complex has broken down. *d.* Chromosomes in the final daughter cells may have a new arrangement of genetic material due to crossing-over, which occurred between nonsister chromatids in prophase I.

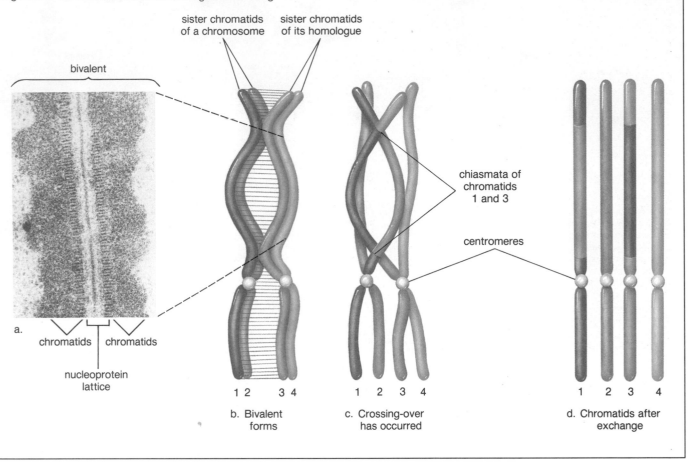

sister chromatids of a chromosome

sister chromatids of its homologue

bivalent

chiasmata of chromatids 1 and 3

centromeres

a.

chromatids | chromatids

nucleoprotein lattice

1 2 3 4

b. Bivalent forms

1 2 3 4

c. Crossing-over has occurred

1 2 3 4

d. Chromatids after exchange

Figure 11.6

Chiasmata (arrows) of a chromosome bivalent, from a testis cell of a grasshopper. The chiasmata mark the places where crossing-over (exchange of genetic material) between nonsister chromosomes of the bivalent has occurred. The chiasmata hold the members of the bivalent together until separation of homologues occurs in anaphase I of meiosis.

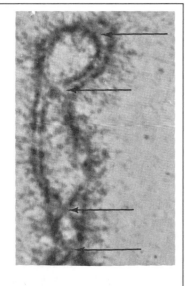

move to opposite poles. Due to this separation, the chromosome number is reduced from the diploid to the haploid number and the daughter cells will receive one of each kind of chromosome. Notice that at the end of anaphase I each chromosome still consists of two chromatids.

Homologous chromosomes separate during anaphase I. This is the means by which the diploid number is reduced to the haploid number and also the reason why the daughter cells will receive one of each kind of chromosome.

Telophase I

In some species there is a telophase I stage at the end of the first division. If there is a telophase, the nuclear envelope reforms and the nucleoli reappear. Also, this stage may or may not be accompanied by cytokinesis and daughter cell formation. In other species, the telophase I stage is omitted at the end of the first division.

Figure 11.7

Life cycle of an animal. *a.* In animals such as mammals, meiosis is involved in gamete production, the only haploid stage in the cycle. Fertilization restores the full number so that the adult has the diploid number of chromosomes. *b.* Meiosis occurs during spermatogenesis (production of sperm) and during oogenesis (production of eggs). In spermatogenesis each meiotic event produces four viable sperm, whereas in oogenesis each meiosis produces one egg and at least two nonfunctional polar bodies.

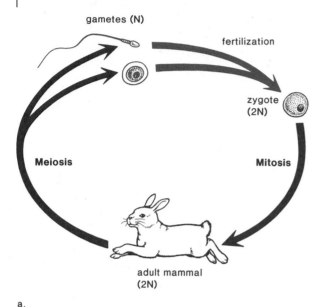

a.

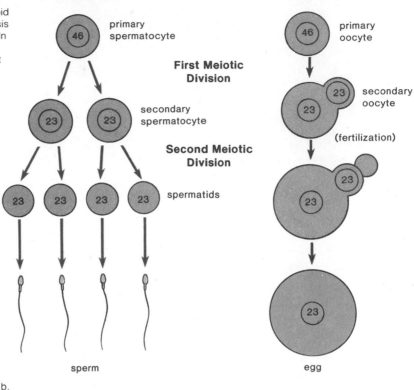

b.

Interkinesis

Following telophase I there is a short period called interkinesis before meiosis II begins. Interkinesis is similar to interphase between mitotic divisions except that no replication of DNA occurs. Replication is unnecessary because the chromosomes are already duplicated.

Cytokinesis does not necessarily follow telophase I. However, there is a short period known as interkinesis before meiosis II begins. DNA does not replicate because the chromosomes are already duplicated.

Meiosis II

The phases of **meiosis II** are referred to as prophase II, metaphase II, anaphase II, and telophase II. This second division of meiosis varies considerably from species to species but it is essentially a mitoticlike division in which the chromatids separate and become independent chromosomes during anaphase II. At the end of telophase II, following cytokinesis, there are four cells. These cells are not genetically identical to the mother cell because (1) they are haploid, and (2) crossing-over has resulted in different combinations of the genes on the daughter chromosomes. In other words, **genetic recombination** has occurred.

The haploid cells, which are the product of meiosis I and II, go on to mature and become gametes in animals. In plants they become spores that divide to give the haploid generation.

Meiosis II is similar to a mitotic division in that sister chromatids separate during anaphase.

Meiosis in Humans

Humans have a life cycle that is typical of all other mammals (fig. 11.7). Meiosis is confined to the germ line, those cells that produce the gametes and which remain separate and distinct from the somatic (body) cells. In females the germ cells are located in the ovaries where eggs are produced in a process called **oogenesis.** In males, the germ cells are located in the testes where sperm are produced in a process called **spermatogenesis.** It is these gametes that carry genes from one generation to the next. As described in the reading on page 166, in 1868 Charles Darwin attempted unsuccessfully to explain how traits are passed from generation to generation by way of the gametes. Now, of course, it is known that the chromosomes within the gametes carry these traits to the new individual.

Oogenesis and Spermatogenesis

Figure 11.7*b* contrasts oogenesis with spermatogenesis. In females, the first meiotic cell division produces two cells, one of which is much smaller than the other. The smaller cell is called a **polar body,** and remains attached to the larger cell. The larger cell, called the **oocyte,** goes on to form the egg. The second meiotic division of the oocyte actually does not occur unless fertilization takes place. Regardless, complete oogenesis in females

Pangenesis

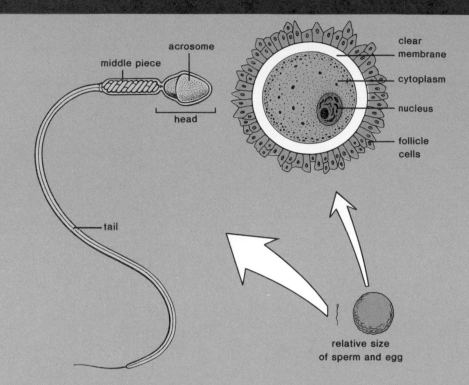

The egg and sperm cells are often referred to as *germ cells* because they are the beginnings, or germs, of new individuals. The germ cells represent the only connecting thread between successive generations. Accordingly, the mechanism of hereditary transmission must operate across this slender connecting bridge. In 1868, Charles Darwin proposed an ingenious theory to account for the transmission of traits by the germ cells. Darwin thought that all organs of the body sent contributions to the germ cells in the form of particles, or gemmules. These gemmules, or minute delegates of bodily parts, were supposedly discharged into the bloodstream and became ultimately concentrated in the gametes. In the next generation, each gemmule reproduced its particular bodily component. The theory was called *pangenesis*, or "origin from all," since all parental bodily cells were supposed to take part in the formation of the new individual.

Darwin's fanciful explanation of inheritance, however, has no foundation of factual evidence. Indeed, many observations argue against it. In 1909 W. E. Castle of Harvard University provided one of the earliest convincing observations refuting the theory of pangenesis. Castle successfully removed the ovaries of an albino guinea pig and grafted in their place the ovaries of a black guinea pig. The albino foster mother was then mated to an albino male. This pair produced several litters of young, and all the offspring were black. Ordinarily when albinos are crossed only albino offspring are produced. It is obvious, then, that the eggs produced by the transplanted ovaries (from the black guinea pig) were not changed, or instructed otherwise, by their new residence in the albino foster mother guinea pig.

Darwin recognized that the egg and sperm must have a means of carrying information to the next generation, but did not realize such information resides in the genes located on the chromosomes.

Notwithstanding Darwin's simplistic view, the mechanism of inheritance does reside in the gametes. Although the relatively large egg cell and the comparatively small sperm cell are as strikingly unalike in appearance as any two cells can be, they are equivalent in certain important components—their nuclei and, more particularly, their *chromosomes*. The chromosomes are, in fact, the vehicles for transmitting the blueprints of traits from parents to offspring. Deoxyribonucleic acid (DNA) [is] . . . the *chemical carrier* of the hereditary information.

From Volpe, E. Peter, *Biology and Human Concerns*, 3d ed. © 1975, 1979, 1983 Wm. C. Brown Publishers, Dubuque, Iowa. All Rights Reserved. Reprinted by permission.

---

produces from each original cell only one egg and at least two nonfunctional polar bodies that disintegrate. In males, on the other hand, spermatogenesis results in four viable sperm from each original cell.

From the drawing above, it is apparent that the sperm and egg are adapted to their functions. The sperm is a tiny, flagellated cell that is adapted for swimming to the egg. The egg is a large cell that awaits the arrival of the sperm and contributes most of the cytoplasm and nutrients to the zygote. The sperm and egg represent the haploid stage in the human life cycle. When fertilization occurs, the resulting zygote has the diploid number of chromosomes. The zygote develops into the organism.

Meiosis in humans, like other mammals, occurs during spermatogenesis in males and oogenesis in females. The sperm and egg are the only haploid stages and fertilization provides the zygote with the diploid number of chromosomes.

Comparison of Meiosis and Mitosis

Figure 11.8 compares mitosis to meiosis. The differences between these nuclear divisions can be categorized in these ways.

Occurrence

Meiosis occurs only in cells that will eventually give rise to the gametes, but mitosis occurs in all types of cells in the body.

Process

The following are distinctive differences between these two processes:

1. Chromosomes replicate their DNA once before both mitosis and meiosis. But chromosomes undergo only one division in mitosis and two divisions in meiosis.
2. Homologous chromosomes pair and undergo crossing-over during meiosis but not in mitosis.
3. Paired homologous chromosomes (bivalents) align at the equator of metaphase I in meiosis; individual

Figure 11.8

Comparison of mitosis and meiosis. For simplicity's sake crossing-over has been omitted from meiosis.

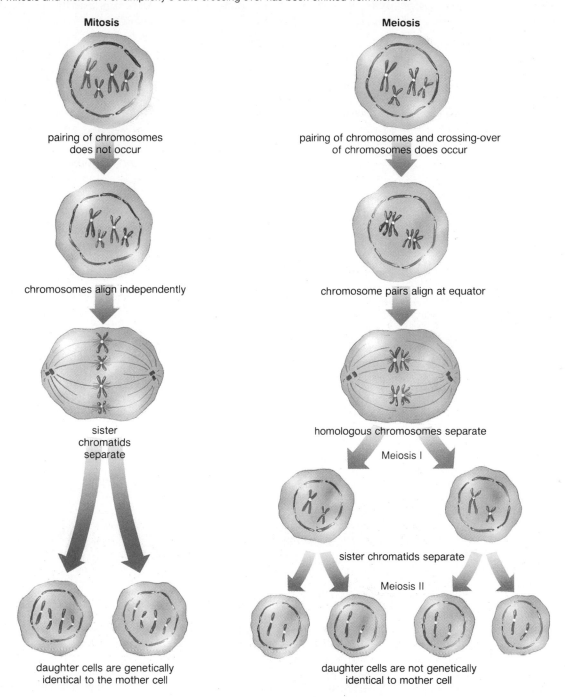

Mitosis	Meiosis
pairing of chromosomes does not occur	pairing of chromosomes and crossing-over of chromosomes does occur
chromosomes align independently	chromosome pairs align at equator
sister chromatids separate	homologous chromosomes separate
	Meiosis I
	sister chromatids separate
	Meiosis II
daughter cells are genetically identical to the mother cell	daughter cells are not genetically identical to mother cell

chromosomes composed of two sister chromatids align at the equator in metaphase of mitosis.

4. Homologous chromosomes separate from one another and move to opposite poles at anaphase I in meiosis; sister chromatids of individual chromosomes separate and move to opposite poles of mitotic anaphase.

5. At anaphase II, chromatids separate and move to opposite poles. Thereafter they are called chromosomes.

Daughter Nuclei and Cells

The genetic consequence of mitosis and meiosis are quite different.

1. Four daughter cells are produced from a single cell that undergoes meiosis; mitosis results in two daughter cells.

2. The four daughter cells formed by meiosis are haploid; the daughter cells produced by mitosis have a chromosome number identical to their mother cell.

3. The mitotic daughter cells are genetically identical to each other and to their parental cell. Meiosis produces haploid daughter cells that differ genetically from each other and from the diploid mother cell due to crossing-over and independent alignment of bivalents at anaphase I.

Mitosis, which occurs in all body parts, produces two daughter cells that are genetically identical to the mother cell. Meiosis, which occurs only in structures involved in sexual reproduction, produces four daughter cells that are not genetically identical to the mother cell.

Importance of Meiosis

Meiosis provides a mechanism for the reduction in chromosome number from the diploid to the haploid number found in gametes. Not only is the chromosome number halved precisely, each gamete receives one and only one copy of each homologous pair of chromosomes.

Meiosis also helps assure that the zygote will have a different combination of alleles than has either parent. As a result of *independent assortment of chromosomes,* the chromosomes are distributed to the gametes in various combinations. The total number of possible combinations is 2^N where N is the haploid number of chromosomes. In humans, where N = 23, the number of possible chromosome combinations produced by meiosis is a staggering 2^{23} or 8,388,608. And this does not even consider the variations that are introduced due to *crossing-over.* For example, if we assume only one crossover occurs within each bivalent (actually many crossovers usually occur), then in humans crossing-over and independent assortment will generate 4^{23} different kinds of gametes or 70,368,744,000,000.

This is just the variation introduced by meiosis but if the entire sexual reproductive process is considered, there is another element that introduces variation. Due to *fertilization,* the chromosomes donated by the parents are combined, and in humans this means that $(2^{23})^2$ or 70,368,744,000,000 chromosomally different offspring are possible, even assuming no crossing-over. If crossing-over occurs once, then $(4^{23})^2$ or 4,951,760,200,000,000,000,000,000,000 genetically different offspring are possible for every couple.

There's no doubt about it. A sexually reproducing population has a tremendous storehouse of variations, some of which may be advantageous for evolution, particularly when the environment is changing. Asexually reproducing organisms like prokaryotes depend primarily on mutations to generate variation and this is sufficient because they produce great numbers of offspring within a limited amount of time. Mutations are still the raw material for variation among sexually reproducing organisms but the shuffling of genetic material during meiosis and fertilization produces new and different combinations, some of which may allow adaptation to a changing environment.

Summary

1. Meiosis is involved in the production of gametes during sexual reproduction. It ensures that the chromosome number in offspring stays constant generation after generation.
2. The nucleus contains pairs of chromosomes, called homologous pairs (homologues). Due to the process of meiosis, the haploid daughter cells (gametes) receive one of each kind of chromosome.
3. In the life cycle of animals, only the gametes are haploid but in the life cycle of plants, there is a multicellular haploid generation that produces the gametes, in addition to a diploid generation.
4. Meiosis requires two cell divisions and results in four daughter cells. Replication of DNA takes place before meiosis begins. During meiosis I, the chromosome homologues pair (called a bivalent) and crossing-over between nonsister chromatids occurs before they independently separate. The daughter cells receive one member of each pair of homologous chromosomes. There is no replication of DNA during interkinesis. During meiosis II, the duplicated chromosomes separate in a manner similar to mitosis. The four resulting daughter cells are not genetically identical to the mother cell; they are haploid and their chromosomes carry a different arrangement of the genetic material due to crossing-over.
5. Meiosis I is divided into four stages: Prophase I—Bivalents form and crossing-over occurs as chromosomes condense; the nuclear envelope fragments. Metaphase I—Bivalents independently align at the equator. Anaphase I—Homologous chromosomes separate. Telophase I—Nuclei become haploid, having received one duplicated chromosome from each homologous pair.
6. Meiosis II is divided into four stages: Prophase II—Chromosomes condense and the nuclear envelope fragments. Metaphase II—Duplicated chromosomes align individually at the equator. Anaphase II—Chromatids separate. Telophase II—Four haploid daughter cells are genetically different from the mother cell.
7. During the life cycle of humans and many other animals, only the gametes are haploid and fertilization restores the diploid number. Meiosis is involved in spermatogenesis and oogenesis. Whereas spermatogenesis produces four sperm per meiosis, oogenesis produces one egg and two to three nonfunctional polar bodies.

8. Meiosis I can be compared to mitosis in this manner:

Mitosis	Meiosis I
Prophase	
No pairing of chromosomes	Pairing of homologous chromosomes
Metaphase	
Duplicated chromosomes at equator	Bivalents at equator
Anaphase	
Chromatids separate	Homologous chromosomes separate
Telophase	
Daughter nuclei are diploid	Daughter nuclei are haploid

9. Meiosis II can be compared to mitosis in this manner:

Mitosis	Meiosis II
Metaphase	
Diploid number of duplicated chromosomes at equator	Haploid number of duplicated chromosomes at equator
Anaphase	
Chromatids separate	Chromatids separate
Telophase	
Two daughter cells	Four daughter cells

10. Mutations are the primary source of genetic variation among asexually reproducing organisms. But among sexually reproducing organisms, meiosis produces gamete variation by independent assortment of homologous chromosomes and crossing-over. Fertilization also contributes to zygote variation. Variation is important to the process of evolution; it promotes the possibility of adaptation to a changing environment.

Objective Questions

1. A bivalent (tetrad) is
   a. a homologous chromosome.
   b. the paired homologous chromosomes.
   c. a duplicated chromosome composed of sister chromatids.
   d. the two daughter cells after meiosis I.
2. If a mother cell has 12 chromosomes, then the daughter cells following meiosis will have
   a. 12 chromosomes.
   b. 24 chromosomes.
   c. 6 chromosomes.
   d. Any one of these.
3. At the equator during metaphase of *mitosis,* there are the
   a. diploid number of duplicated chromosomes.
   b. haploid number of duplicated chromosomes.
   c. diploid number of bivalents.
   d. haploid number of bivalents.
4. At the equator during metaphase I of meiosis, there are
   a. singled chromosomes.
   b. unpaired duplicated chromosomes.
   c. bivalents (tetrads).
   d. 23 chromosomes.
5. At the equator during metaphase II of meiosis, there are
   a. singled chromosomes.
   b. unpaired duplicated chromosomes.
   c. bivalents (tetrads).
   d. always 23 chromosomes.
6. Gametes contain one of each kind of chromosome because
   a. the homologous chromosomes separate during meiosis.
   b. the chromatids never separate during meiosis.
   c. two replications of DNA occur during meiosis.
   d. crossing-over occurs during prophase I.
7. Crossing-over occurs between
   a. sister chromatids of the same chromosomes.
   b. two different bivalents.
   c. nonsister chromatids of a bivalent.
   d. two daughter nuclei.
8. During which stage of meiosis do homologous chromosomes separate?
   a. prophase II
   b. telophase II
   c. metaphase I
   d. anaphase I
9. Fertilization
   a. is also a source of variation during sexual reproduction.
   b. is fusion of the gametes.
   c. occurs in both animal and plant life cycles.
   d. All of these.
10. Which of these is not a difference between spermatogenesis and oogenesis in humans?

	Spermatogenesis	Oogenesis
a.	Occurs in males	Occurs in females
b.	Produces four sperm per meiosis	Produces one egg per meiosis
c.	Produces haploid cells	Produces diploid cells
d.	Always goes to completion	Does not always go to completion

Study Questions

1. Why did early investigators predict that there must be a reduction division in the sexual reproduction process?
2. What are the events of meiosis that account for the production of four daughter cells with the haploid number of chromosomes having a different genetic makeup compared to those in the mother cell?
3. Draw and explain a series of diagrams that illustrate the happenings of meiosis I.
4. Draw and explain a series of diagrams that illustrate synapsis and crossing-over. In your drawings indicate what holds the sister chromatids together. Indicate what holds the homologues together.
5. Draw and explain a series of diagrams that illustrate the happenings of meiosis II.
6. Draw a diagram to illustrate the life cycle of humans and compare spermatogenesis to oogenesis.
7. Construct a chart to describe the many differences between meiosis and mitosis.
8. What accounts for the genetic similarity between daughter cells and the mother cell following mitosis and the genetic dissimilarity between daughter cells and the mother cell following meiosis?
9. List the ways in which sexual reproduction contributes to variation among members of a population. What is the evolutionary significance of this variation?

Thought Questions

1. You'll recall that the evolutionary process is conservative and tends to add new processes onto old processes. Use this concept to explain the existence of meiosis I and meiosis II.

2. There are protists that can carry on either asexual or sexual reproduction. Asexual reproduction occurs when environmental conditions are favorable and sexual reproduction occurs when they are unfavorable. Why is this consistent with your knowledge of these two processes?

Selected Key Terms

homologous chromosome (ho-mol′o-gus kro′mo-sōm) 159
gamete (gam′ēt) 159
fertilization (fer″ti-li-za′shun) 159
homologue (hom′o-log) 159
bivalent (tetrad) (bi-va′lent) (tet′rad) 159

crossing-over (kros′ing o′ver) 162
meiosis I (mi-o′sis wun) 162
synapsis (si-nap′sis) 162
chiasmata (sing. **chiasma**) (ki-as′mah-tah) 162
meiosis II (mi-o′sis too) 165

genetic recombination (jĕ-net′ik re″kom-bĭ-na′shun) 165
oogenesis (o″o-jen′ĕ-sis) 165
spermatogenesis (sper″mah-to-jen′ĕ-sis) 165
polar body (po′lar bod′e) 165
oocyte (o′o-sīt) 165

CHAPTER 12

Mendelian Patterns
of Inheritance

Your study of this chapter will be complete when you can:

1. *Discuss the ideas about heredity when Mendel began his experiments.*
2. *Outline the general methodology that Mendel used while doing his experiments.*
3. *Using words to indicate the phenotype and letters to indicate the genotype, diagram the type of monohybrid and dihybrid cross Mendel did.*
4. *State Mendel's Law of Segregation and follow the reasoning he used to arrive at this law.*
5. *Explain the use of a monohybrid and dihybrid testcross to determine the genotype of an individual.*
6. *State Mendel's Law of Independent Assortment and follow the reasoning Mendel used to arrive at this law.*
7. *Do one-trait and two-trait genetics problems either by employing the use of the Punnett square or the laws of probability.*

What causes a giraffe to have such a long neck? The genes inherited from its parents, of course. This may seem obvious to us but it took many years of research to determine that the sperm and egg carry inheritory factors from parent to offspring. This is universally true, whether one is talking about giraffes, peas, or people.

It is very difficult to come up with a theory of heredity when you can't see what you are talking about. It is possible to see the sperm and the egg and maybe even the chromosomes, if you have a good microscope, but you can't see the **genes,** the units of heredity within the chromosomes. Considering this will help you fully appreciate the work of Gregor Mendel (fig. 12.1) who discovered certain laws of heredity in the early 1860s.

When Mendel began his work, most plant and animal breeders believed in a *blending theory of inheritance* because it supported the idea that both sexes contribute equally to a new individual. They felt that parents of contrasting appearance always produce offspring of intermediate appearance. Therefore, according to this theory a cross between plants with red flowers and plants with white flowers would yield only plants with pink flowers. When red and white flowers reappeared in future generations they mistakenly attributed this to an instability of the genetic material.

The blending theory of inheritance presented a real problem to Charles Darwin (chapter 19) who wanted to give his theory of evolution a genetic basis. If populations contained only intermediate individuals and normally lacked variations, how could diverse forms evolve?

At the time Mendel began his study of heredity, a blending theory of inheritance was popular.

Figure 12.1

Mendel working in his garden. Mendel selected 22 varieties of the edible pea *Pisum* for his genetic experiments. He grew and tended the plants himself. For each experiment, he observed as many offspring as possible. For example, for a cross that required him to count the number of round seeds to wrinkled seeds, he observed and counted a total of 7,324 peas.

Figure 12.2

In the garden pea, each flower produces both male and female gametes. The pollen grains, which upon maturity contain sperm, are produced within the anther of the stamens. The ovule, within the ovary of the pistil eventually contains an egg. Following pollination, when pollen is deposited on the stigma (the top portion of the pistil), the pollen grain develops a tube through which a sperm reaches the egg. Self-pollination is the rule because the sexual structures of the plant are entirely enclosed by petals. Sometimes Mendel allowed the plant to self-pollinate so that the male and female gametes of the same flower produced offspring. At other times, he performed cross-pollination, so that male and female gametes from different flowers produced offspring. He removed the pollen-producing anthers of one plant and brushed the stigmas with the anther from another plant.

Once the ovules had developed into seeds (peas), they could be observed or planted, in order to observe the results of a cross. The open pod shows the results of a cross involving smooth or wrinkled seed coat.

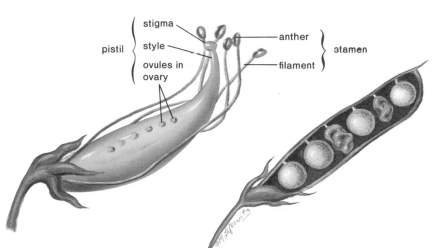

pistil { stigma
style
ovules in ovary

stigma — anther
filament

stamen

Gregor Mendel

Gregor Mendel was an Austrian monk who taught natural science at a local technical high school and had previously gone to the University of Vienna to study science and mathematics. Most likely, then, he had previously decided that his breeding experiments would have a statistical basis. He prepared for his experiments carefully.

He did preliminary studies with various animals and plants and then chose to work with the garden pea *Pisum sativum* (fig. 12.2). Mendel's use of the garden pea was a good choice. Peas normally self-pollinate but can be cross-pollinated for the sake of the experiment. Many different varieties of the peas were available, and Mendel chose 22 for his experiments. When these varieties self-pollinated, they were **true-breeding**—the offspring were like the parent plants, and like each other. We may also note that pea plants are easy to cultivate and have a short generation time. In contrast to his predecessors, Mendel studied the inheritance of relatively simple and easily distinguishable traits, such as seed shape, seed color, and flower color (table 12.1). Initially, Mendel followed the inheritance of individual traits and he kept careful records of the numbers of offspring that expressed each parental characteristic.

He used his understanding of the mathematical laws of probability to interpret the results. In other words, Mendel simply wanted to observe objective facts, and if he did have personal beliefs, he set them aside for the sake of the experiment. This is one of the qualities that makes his experiments as applicable today as they were in 1860.

Mendel carefully designed his experiments and previously decided that he wanted to gather statistical results.

Monohybrid Inheritance

Having made sure that his pea plants were normally true-breeding, Mendel was now ready to perform a cross-pollination experiment between two strains. For these initial experiments, Mendel chose varieties that differed in only one trait. Crosses that involve only one trait are called **monohybrid crosses.** If a blending theory of inheritance was correct, then the cross should yield offspring with an intermediate appearance compared to the parents. For example, the offspring of a cross between a tall plant and short plant would be intermediate in height.

Mendel called the original parents the **P generation** and the first generation offspring the **F₁** (for filial) **generation** (fig. 12.3). He performed reciprocal crosses. First, he dusted the pollen of tall plants on the stigmas of short plants, and then he dusted the pollen of short plants on the stigmas of tall plants. In both cases, all F₁ offspring resembled the tall parent.

Certainly these results were contrary to those predicted by the blending theory of inheritance. Rather than being intermediate, the offspring were tall, and resembled only one parent. Did these results mean that the other characteristic (i.e., shortness) had disappeared permanently? Apparently not, because when Mendel allowed the F₁ plants to self-pollinate, ¾ of the **F₂ generation** were tall and ¼ were short, a 3:1 ratio (fig. 12.3).

Table 12.1 Traits and Characteristics Studied by Mendel

Trait	Characteristics		F₂ Results*	
			Dominant	Recessive
stem length	tall	short	787	277
pod shape	inflated	constricted	882	299
seed shape	round	wrinkled	5,474	1,850
seed color	yellow	green	6,022	2,001
flower position	axial	terminal	651	207
flower color	purple	white	705	224
pod color	green	yellow	428	152

*All of these give approximately a 3:1 ratio. For example

$$\frac{787}{277} \cong \frac{3}{1}$$

Figure 12.3

The monohybrid cross allowed Mendel to deduce that individuals have two discrete and separate genetic factors for each trait. In true-breeding plants that self-pollinate, the two factors are the same (see P plants). If these plants are crossed, all the offspring show the same characteristic, indicating that it is dominant to the other, recessive characteristic (see F_1 plants). If these hybrid plants are allowed to self-pollinate, the recessive characteristic reappears. On the average, three out of four F_2 plants show the dominant characteristic and one shows the recessive (see F_2 plants), giving a 3:1 ratio.

Mendel's Experimental Design and Results

P (Parental Generation) Mendel crossed two true-breeding varieties that differed in one characteristic, such as length of stem.

TT X tt

F_1 (First Filial Generation) All offspring were tall, and therefore resembled only one of the parents. Mendel allowed these plants to self-pollinate.

Tt

F_2 (Second Filial Generation) Among the offspring 3/4 were tall but 1/4 were short. Therefore, Mendel said there are two factors (represented here by T and t) for each trait in the individual. These segregate into the gametes.

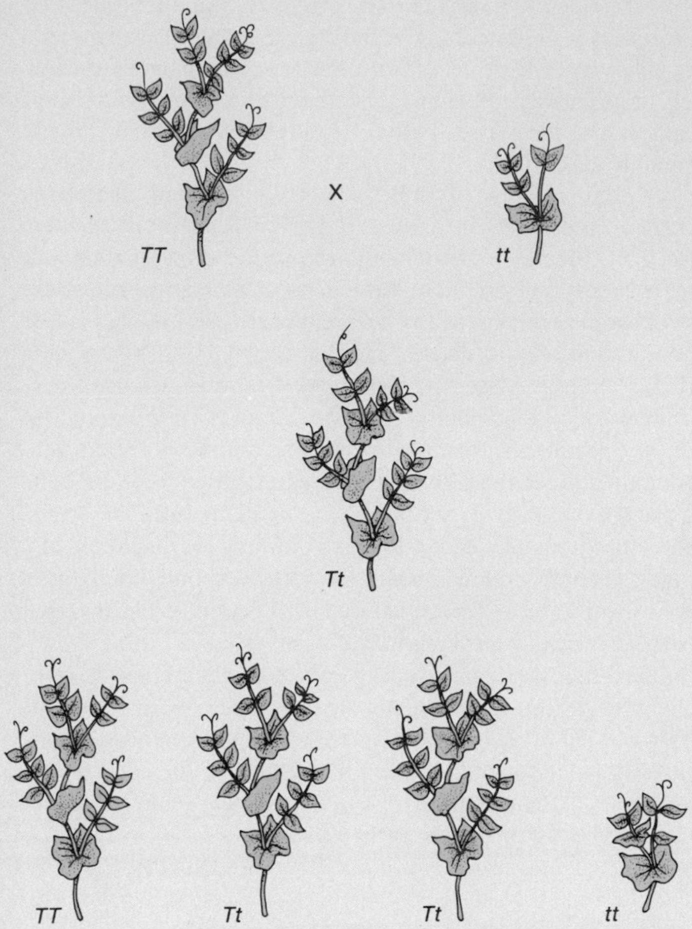

TT Tt Tt tt

key:

T = tall

t = short

Genotypes and Phenotypes

TT = tall

Tt = tall

tt = short

Mendel counted a lot of plants. For this particular cross, he counted a total of 1,064 plants, of which 787 were tall and 277 were short. In all crosses that he performed, he found a 3:1 ratio in the F_2 generation. The characteristic that had disappeared in the F_1 generation reappeared in ¼ of the F_2 offspring (table 12.1).

His mathematical approach led Mendel to interpret these results differently from previous breeders. He knew that the same ratio was obtained among the F_2 time and time again for each trait, and he sought an explanation for these results. A 3:1 ratio among the F_2 offspring was possible if:

1. the F_1 contained two separate copies of each hereditary factor, one of these being dominant and one being recessive,
2. the factors separated when the gametes were formed and each gamete carried only one copy of each factor, and
3. random fusion of gametes occurred upon fertilization.

In this way, Mendel arrived at the first of his laws of inheritance:

Mendel's Law of Segregation. Each organism contains two factors for each trait, and the factors segregate during the formation of gametes so that each gamete contains only one factor from each pair of factors. When fertilization occurs, the new organism will have two factors for each trait, one from each parent.

Mendel drew these conclusions even though he had almost no knowledge of cytology including meiosis, which is responsible for the segregation of chromosomes (and Mendel's factors) prior to the formation of gametes. As we discussed on page 154, each gamete eventually receives only one of each pair of chromosomes because the homologous chromosome pairs separate during meiosis I.

Modern Terminology

Figure 12.3 also shows how we now interpret the results of Mendel's experiments on inheritance of stem length in peas. Each pea plant is said to have two **alleles,** alternative forms of a gene, that controls the length of the stem. In genetic notation, the alleles are identified by letters, the **dominant allele** (so named because of its ability to mask the expression of another allele) with an uppercase (capital) letter and the **recessive allele** with the same letter, but lowercase (small). For example, there is an allele for tallness (T) and an allele for shortness (t). One of these alleles occurs on each chromosome of a homologous pair at a particular location that is called the **gene locus** (fig. 12.4). During meiosis, first the homologous chromosomes separate and then the chromatids separate. Therefore there is only one allele for each trait in the gametes (*see* fig. 13.8).

In Mendel's cross, the original P parents were true breeding; therefore, the tall plants had two alleles for tallness (TT) and the short plants had two alleles for shortness (tt). When an organism has two identical alleles, as these had, we say that the organism is **homozygous.** Because the first parents were homozygous, all gametes produced by the tall plant contained the allele for tallness (T), and all gametes produced by the short plant contained an allele for shortness (t).

After cross-pollination, all the individuals of the resulting F_1 generation had one allele for tallness and one for shortness (Tt). When an organism has two different alleles at a gene locus, we say that it is **heterozygous.** Although the plants of the F_1 generation had one of each type of allele, they were all tall. When only one allele in a heterozygous individual is expressed, we say that this allele is the dominant allele. The allele that is not expressed in a heterozygote is a recessive allele.

Genotype and Phenotype

It is obvious from our discussion that two organisms with different allele combinations for a trait may have the same outward appearance (TT and Tt peas are both tall). For this reason, it is necessary to distinguish between the alleles present in an organism and the appearance of that organism.

The word **genotype** refers to the alleles an individual receives at fertilization. Genotype may be indicated by means of letters or by means of short descriptive phrases, as in table 12.2. Genotype TT is called homozygous dominant and tt is called homozygous recessive. Genotype Tt is called heterozygous.

The word **phenotype** refers to the physical appearance of the individual. Homozygous dominant and heterozygous individuals both show the dominant phenotype and are tall, while the homozygous recessive individual shows the recessive phenotype and is short.

When one is doing genetics problems, it is always necessary to know whether the genotype (genes of the individual) or phenotype (appearance of the individual) is being considered.

We will now digress from Mendel's work with peas to show that these results are applicable to other organisms.

Figure 12.4
Diagrammatic representation of a homologous pair of chromosomes before and after replication. *a.* The letters represent alleles, i.e., alternate forms of a gene. Each allelic pair, such as *Gg* or *Zz,* is located on homologous chromosomes at a particular gene locus. *b.* Following replication, each sister chromatid carries the same alleles in the same order.

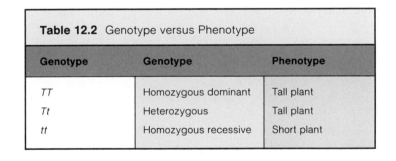

Table 12.2 Genotype versus Phenotype

Genotype	Genotype	Phenotype
TT	Homozygous dominant	Tall plant
Tt	Heterozygous	Tall plant
tt	Homozygous recessive	Short plant

Genetics Problems

When solving a genetics problem, it is first necessary to know which characteristic is dominant. For example, the following key indicates that unattached earlobes are dominant to attached earlobes:

key: E = unattached earlobes
e = attached earlobes

Suppose a homozygous man with unattached earlobes marries a woman who has attached earlobes. What types of earlobes will the children have? In row 1 below, P represents the parental generation, and the letters in this row are the genotypes of the parents. Row 2 shows that the gametes of each parent have only one type of allele for earlobes; therefore, all the children (F₁ offspring) will have a heterozygous genotype that will result in unattached earlobes:

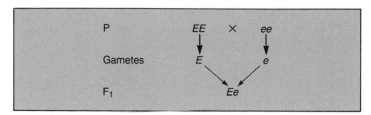

If these children marry someone with the same genotype as themselves, will their children have unattached earlobes? Pre-

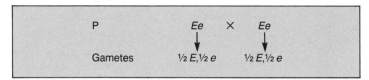

viously there was only one possible type of gamete for each parent because they were homozygous. The F_1 are heterozygous, so two possible types of gametes can occur for each; ½ will contain an E and ½ will contain an e. When designating the gametes, it is necessary to keep in mind that although the individual may have two alleles for each trait, *each gamete carries only one allele for each trait*. You will also want to keep in mind that no two letters in a gamete may be the same. This is true of single-trait crosses as well as multiple-trait crosses. Practice Problems 1 will help you learn to designate gametes.

Practice Problems 1*

1. For each of the following genotypes, give all possible gametes, noting the proportion of each for the individual.
   a. *WW*
   b. *Ww*
   c. *Tt*
   d. *TT*
2. For each of the following, state whether it represents a genotype or a gamete.
   a. *D*
   b. *GG*
   c. *P*

Punnett Square

Once the gametes have been designated, there are two possible ways to calculate the results of a cross. The first is to use a **Punnett square.** This simple method was introduced by a prominent poultry geneticist, R. C. Punnett, in the early 1900s. Figure 12.5

*Answers to Practice Problems appear on page 805 of appendix D.

Figure 12.5
Mendel's Law of Segregation illustrated in a cross between humans. Heterozygous individuals produce two types of gametes in equal amounts. Therefore, an offspring of two heterozygotes has one chance in four of showing the recessive phenotype.

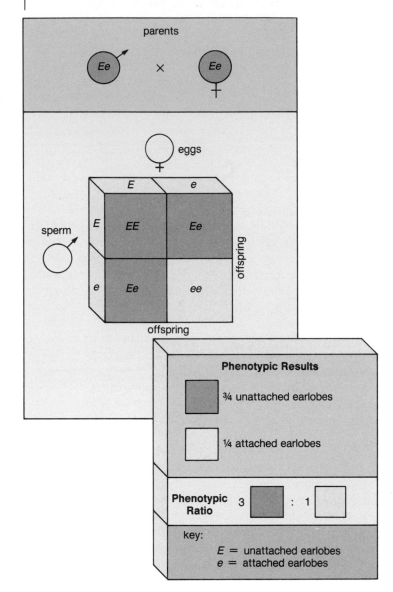

shows how the results of the cross under consideration (*Ee* × *Ee*) may be determined by using a Punnett square. The results show that the expected proportions of offspring are ¼ *EE* to ½ *Ee* to ¼ *ee* giving a 1:2:1 ratio of genotypes. However, in genetics problems you are usually asked for the ratio of phenotypes. In this example, ¾ have unattached earlobes and ¼ have attached earlobes giving a 3:1 phenotype ratio.

These are only expected proportions, not absolute numbers. For example, if a large number of heterozygous individuals produced 200 offspring, approximately 150 would have unattached earlobes and approximately 50 would have attached earlobes. In terms of genotypes, approximately 50 would be *EE,* about 100 would be *Ee,* and the remaining 50 would be *ee.*

Laws of Probability

Another method of calculating the expected ratios uses probability and is based on: *the chance or probability that two or more independent events will occur together is the product (multiplication) of their chances of occurring separately.*

In the cross just considered ($Ee \times Ee$), what is the chance of obtaining either an E or an e from a parent?

The chance of $E = ½$

The chance of $e = ½$

Therefore the probability of receiving these genotypes is as follows:

1. The chance of $EE = ½ \times ½ = ¼$
2. The chance of $Ee = ½ \times ½ = ¼$
3. The chance of $eE = ½ \times ½ = ¼$
4. The chance of $ee = ½ \times ½ = ¼$

Now we have to realize that: *the chance of an event that can occur in two or more independent ways is the sum (addition) of the individual chances.*
Therefore,

The chance of offspring with unattached earlobes
($EE, Ee,$ or eE) is ¾
(Add [1], [2], and [3] from above)
The chance of offspring with attached earlobes (ee) is ¼
(Simply [4] from above)

Because the gametes combine at random, it is necessary to observe a very large number of offspring before this 3:1 ratio can be obtained. Only in that way can all possible types of sperm have a chance to fertilize all possible types of eggs. This being the case, the most useful interpretation in humans is to consider the chances that a child will have a particular trait. Both methods tell us that each child has three chances in four (75%) of having unattached earlobes, and one chance in four (25%) of having attached earlobes. One should also realize that *"chance has no memory."* For example, if two heterozygous parents already have a child with attached earlobes, the next child still has a one-in-four chance of having attached earlobes. Although this may not seem important as far as earlobes are concerned, it is important when calculating the chances of a child inheriting a genetic disease. (Chapter 14 concerns this topic.)

When doing a genetics problem, it is assumed that all possible types of sperm fertilize all possible types of eggs. The results may be expressed as a probable phenotypic ratio; it is also possible to state the chances of an offspring showing a particular phenotype.

The Monohybrid Testcross

In order to test his ideas about the segregation of alleles, Mendel crossed his F_1 tall (dominant phenotype) plants with true-breeding, short (recessive phenotype) plants. He reasoned that half the offspring should be tall, as you can see by using the following method to represent the cross:

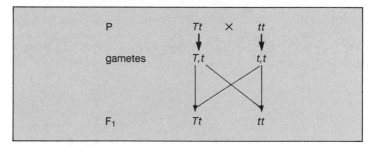

Mendel achieved the results he predicted when he did the cross. Therefore, his hypothesis that alleles segregate when gametes are formed was supported. It was Mendel's experimental use of simple dominant and recessive traits that allowed him to formulate and test the law of segregation.

In a **testcross,** an individual with a dominant phenotype is crossed with an individual having the recessive phenotype. Today, the testcross allows one to determine whether an individual is homozygous dominant or heterozygous. Both of these individuals show the dominant phenotype and it is not possible to determine the genotype by inspection. For example, in a certain strain of mice, black coat (B) is dominant over white coat (b). Therefore, the genotype of a black-coated mouse may not be known; it can be either BB or Bb and is represented as B____ . A testcross of a black-coated mouse (B____) with a white-coated mouse (bb) will indicate whether the black mouse is BB or Bb. Figure 12.6 shows the expected results when a black-coated homozygous dominant (BB) and a black-coated heterozygote (Bb) are each crossed with a white mouse (bb). In the first case, there are only black offspring, but in the second, there is a 50% chance that each offspring will have a white coat. If white-coated offspring appear, it indicates that the genotype of the black-coated parent is Bb rather than BB.

The results of a testcross allow one to determine whether an individual who expresses the dominant phenotype is heterozygous or homozygous dominant. If any of the offspring of a testcross expresses the recessive phenotype, the parent with the dominant phenotype must be heterozygous.

Practice Problems 2*

1. In rabbits, if B = dominant black allele and b = recessive white allele, which of these genotypes could a white rabbit have? Bb, BB, bb
2. In horses, trotter (T) is dominant over pacer (t). A trotter is mated to a pacer and the offspring is a pacer. Give the genotype of all horses.
3. In humans, freckles is dominant over no freckles. A man with freckles is married to a woman with freckles, but the children have no freckles. What chance did each child have for freckles?
4. In pea plants, yellow peas is dominant over green peas. Give the genotype of all plants that could possibly produce green peas when crossed with a heterozygote. (Y = yellow, y = green)

---

*Answers to Practice Problems appear on page 805 of appendix D.

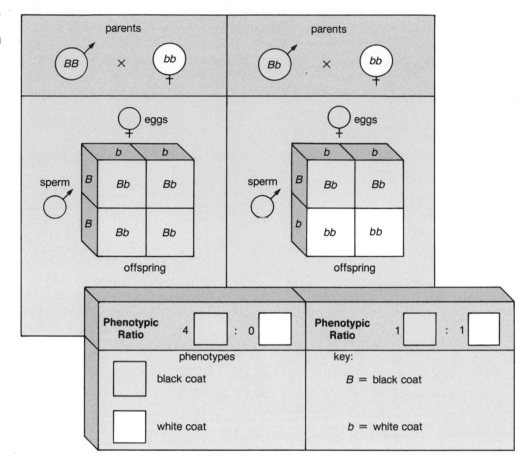

Figure 12.6

Representation of a testcross to determine if the individual showing the dominant trait is homozygous or heterozygous. *a.* Because all offspring show the dominant characteristic, the individual is most likely homozygous, as shown. *b.* Because the offspring show a 1:1 ratio of phenotypes, the individual is heterozygous, as shown.

Dihybrid Inheritance

Mendel performed a second series of experiments in which he crossed true-breeding plants that differed in two traits. Such crosses are called **dihybrid crosses** because they involve two traits. For example, he crossed tall plants having inflated pods with short plants having constricted pods (fig. 12.7). The F₁ plants showed both dominant characteristics. As before, Mendel then allowed the F₁ plants to self-pollinate. There were two possible results in the F₂ generation:

1. If the dominant factors (*TI*) always segregated into the gametes together, and the recessive factors (*ti*) always stayed together, then there should be two phenotypes among the F₂ plants: tall plants with inflated pods and short plants with constricted pods.
2. If the four factors segregated into the gametes independently, then there should be four phenotypes among the F₂ plants: tall plants with inflated pods, tall plants with constricted pods, short plants with inflated pods, and short plants with constricted pods.

Figure 12.7 shows that Mendel observed four phenotypes among the F₂ plants, supporting the second hypothesis. Therefore, Mendel formulated his second law of heredity as follows:

Mendel's Law of Independent Assortment: Members of one pair of factors segregate (assort) independently of members of another pair of factors. Therefore, all possible combinations of factors can occur in the gametes.

Today we know that all possible combinations of alleles located *on different chromosomes* occur in the gametes because homologous pairs of chromosomes separate independently during meiosis (*see* fig. 13.8). Let us use Mendel's Law of Independent Assortment to designate the gametes for the F₁ genotype: *TtIi.* Only one letter of each kind may appear in a gamete, for example, there may be either a *T* or a *t*. The letter *T* or *t* may be present with either one of the other letters. Therefore, the possible types of gametes for this individual are *TI, Ti, tI, ti.* Depending on the genotype, there may be fewer possible types of gametes. For example, the genotype *TtII* has only two possible types of gametes: *TI* or *tI.*

Figure 12.7

The dihybrid cross allowed Mendel to deduce that members of one pair of factors segregate into the gametes independently of members of another pair of factors. In true-breeding plants that self-pollinate, the two factors for each trait are the same (see P). If these plants are crossed, the offspring show both dominant traits (see F_1). If these F_1 plants are allowed to self-pollinate, all possible phenotypes appear among the offspring (see F_2). On the average, nine out of sixteen offspring show both dominant characteristics, three show a dominant trait and a recessive trait, three show the other recessive trait and the other dominant trait, and one shows both recessive traits. This gives a 9:3:3:1 ratio of phenotypes.

Mendel's Experimental Design and Results

P (Parental Generation) Mendel crossed two true-breeding varieties that differed in two traits, such as length of stem and type of pod.

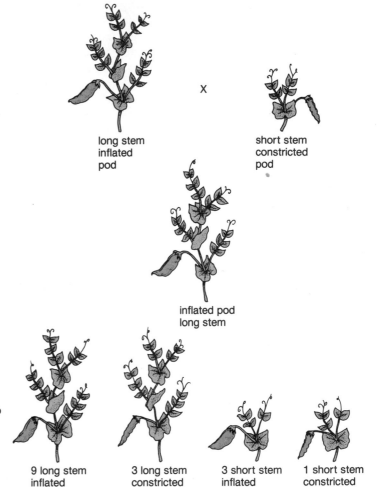

long stem
inflated
pod

X

short stem
constricted
pod

F_1 (First Filial Generation) All offspring were tall with inflated pods. Mendel allowed these plants to self-pollinate: $F_1 \times F_1$.

inflated pod
long stem

F_2 (Second Filial Generation) Four phenotypes were observed in the ratio noted. This showed that the factors had segregated into the gametes independently of one another.

9 long stem
inflated
pod

3 long stem
constricted
pod

3 short stem
inflated
pod

1 short stem
constricted
pod

Practice Problems 3*

1. For each of the following genotypes, give all possible gametes, noting the proportion of each for the individual.
   a. *ttGG*
   b. *TtGG*
   c. *TtGg*
   d. *TTGg*
2. For each of the following, state whether a genotype or a gamete is represented.
   a. *TT*
   b. *Tg*
   c. *IiCC*
   d. *TW*

Probability Revisited

Mendel always observed a 9:3:3:1 ratio among the F_2 generation of a dihybrid cross (fig. 12.7). He realized that this could only occur if each characteristic was being inherited separately from any other. Then, it is possible to apply again the laws of

probability mentioned on page 177. For example, we know the F_2 results for two separate monohybrid crosses would be:

1. The chance of tallness = ¾
   The chance of shortness = ¼
2. The chance of inflated pods = ¾
   The chance of constricted pods = ¼

The probabilities for the two-trait cross are, therefore,

The chance of tallness and inflated pods = ¾ × ¾ = ⁹⁄₁₆
The chance of tallness and constricted pods = ¾ × ¼ = ³⁄₁₆
The chance of shortness and inflated pods = ¾ × ¼ = ³⁄₁₆
The chance of shortness and constricted pods
   = ¼ × ¼ = ¹⁄₁₆

And the phenotypic ratio is 9:3:3:1.

Again, since all possible male gametes must have an equal opportunity to fertilize all possible female gametes to even approximately achieve these results, a large number of offspring must be counted.

*Answers to Practice Problems appear on page 805 of appendix D.

Genetics Problems Again

The fruit fly, *Drosophila melanogaster* (fig. 13.10*a*), less than one-half the size of a housefly, is a favorite subject for genetic research because it has easily recognizable mutant characteristics. "Wild-type" flies have long wings and gray bodies. Mutant flies exist that have vestigial (short) wings and ebony (black) bodies. The key for a cross involving these traits is L = long wing, l = short wing, G = gray body, and g = black (ebony) body.

In the P generation flies (fig. 12.8), only one type of gamete is possible for each fly, because they are homozygous. Therefore, all the F_1 will have the same genotype ($LlGg$) and the same phenotype (long wings and gray bodies).

If a heterozygote ($LlGg$) reproduces with another heterozygote, each fly has four possible types of gametes: $LG, Lg, lG,$ and lg. The Punnett square (fig. 12.8) shows the expected results of such a cross assuming that all possible types of sperm have an opportunity to fertilize all possible types of eggs. Notice that $9/16$ of the offspring have a gray body and long wings, $3/16$ have a gray body and short wings, $3/16$ have a black body and long wings, and $1/16$ have a black body and short wings. The 9:3:3:1 phenotype ratio is always expected in a dihybrid cross when a heterozygote is crossed with another heterozygote and simple dominance is present in both genes.

Dihybrid Testcross

A plant or animal that shows two dominant traits can be either heterozygous or homozygous for each one. Its genotype can be determined by a testcross with a homozygous recessive individual. A heterozygous fly with long wings and gray body can form four different type gametes and a fly with short wings and black body can form one gamete:

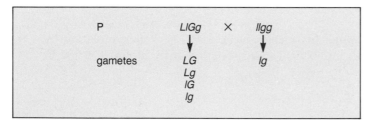

Figure 12.9 shows that the expected phenotypic ratio for this cross is ¼ gray body with long wings, ¼ gray body with short wings, ¼ black body with long wings, and ¼ black body with short wings; or 1:1:1:1.

The purpose of a testcross is to determine the genotype of the F_1 fly. What would be the results if the F_1 fly was not heterozygous for one or both traits? If the fly had an $LLGg$ genotype, it would form only two type gametes, LG and Lg. Or suppose the fly being tested is not heterozygous at all—it has the genotype $LLGG$. In this case, the fly would form only one type of gamete, LG. What would be the results if these flies were crossed with a fly that was recessive in both traits ($llgg$)?

Figure 12.8

Mendel's Law of Independent Assortment illustrated in a fruit fly cross. Each F_1 fly ($LlGg$) produces four types of gametes, because all possible combinations of alleles can occur in the gametes. Therefore, all possible phenotypes appear among the F_2 offspring.

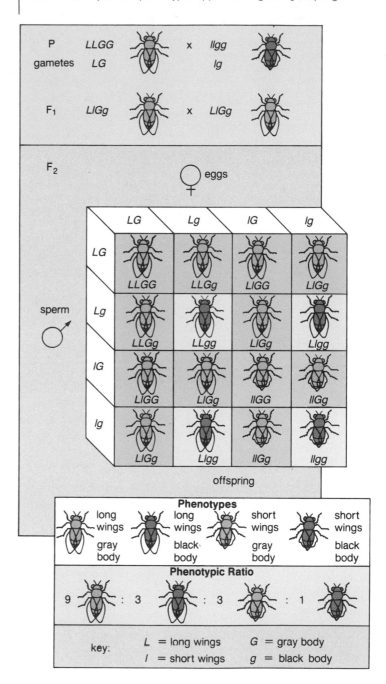

Figure 12.9

Representation of a dihybrid testcross to determine the genotype of a fly showing two dominant characteristics. If the fly is heterozygous dominant for both traits, and is crossed with another that is recessive for both traits as in the case illustrated here, the expected ratio of phenotypes would be a 1:1:1:1. What would be the results if the individual crossed with the recessive homozygote were homozygous dominant for both traits? Homozygous dominant for one trait, but heterozygous for the other?

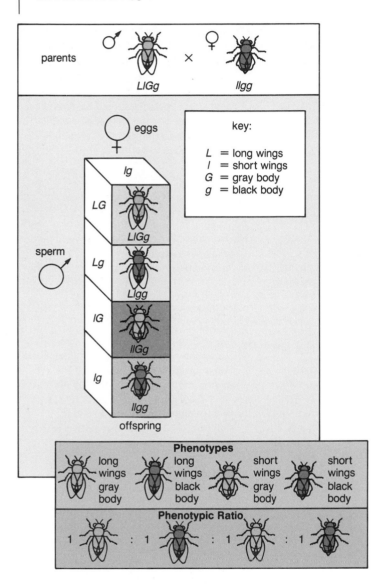

Table 12.3 gives examples of crosses we have studied thus far, along with the results. When these types of crosses are done, these phenotypic ratios are observed.

Table 12.3 Common Crosses Involving Simple Dominance	
Examples	**Phenotypic Ratios**
$Tt \times Tt$	3:1 (dominant to recessive)
$Tt \times tt$	1:1 (dominant to recessive)
$TtYy \times TtYy$	9:3:3:1 (9 dominant; 3 mixed; 3 mixed; 1 recessive)
$TtYy \times ttyy$	1:1:1:1 (all possible combinations in equal number)

Practice Problems 4*

1. In horses, B = black coat, b = brown coat; T = trotter, t = pacer. A black pacer mated to a brown trotter produces a black trotter offspring. Give all possible genotypes for this offspring.
2. In fruit flies, long wings (L) is dominant over short wings (l), and gray body (G) is dominant over black body (g). In each instance, what are the most likely genotypes of the previous generation, if a student gets the following phenotypic results?
   a. 1:1:1:1
   b. 9:3:3:1
3. In humans, short fingers and widow's peak are dominant over long fingers and continuous hairline. A heterozygote is married to a heterozygote. What is the chance that any child will have the same phenotype as the parents?

*Answers to Practice Problems appear on page 805 of appendix D.

Summary

1. At the time Mendel began his hybridization experiments, the blending theory of inheritance was popular. This theory stated that whenever the parents were distinctly different, the offspring would be intermediate between them.
2. Mendel, unlike preceding plant breeders, decided to do a statistical study, most likely because he had a mathematical background.
3. Mendel chose the garden pea for several reasons: many true-breeding varieties were available, they were easy to cultivate, had a short generation time, and could be cross-pollinated even though they normally self-pollinated. He studied only distinctly different traits, kept careful mathematical records, and interpreted his results mathematically.
4. When Mendel performed his monohybrid crosses, he found that the F_1 plants resembled only one of the parents. However, the characteristic of the other parent reappeared in about ¼ of the F_2. Mendel saw that these 3:1 results were possible if the F_1 hybrid contained two factors for each trait, one being dominant and the other recessive, and the factors segregated into the gametes.

5. Mendel's Law of Segregation states that there are two factors for each trait in the individual, and these segregate into gametes that then have only one factor for each trait.

6. Both the Punnett square and the laws of probability can be used to calculate the expected phenotypic ratio of a cross. In practice, a large number of offspring must be counted in order to observe the expected results, because only in that way can it be ensured that all possible types of sperm have fertilized all possible types of eggs.

7. Because humans do not produce a large number of offspring, it is best to use a predicted ratio as a means of determining the chances of an individual inheriting a particular characteristic.

8. Mendel also crossed F_1 plants with homozygous recessive plants. The results indicated that the recessive factor was present in the F_1 (i.e., that it was heterozygous). Today, we call this a testcross, because it is used to test whether an individual showing the dominant characteristic is homozygous dominant or heterozygous.

9. When Mendel performed his dihybrid crosses, the F_1 individuals were dominant in both traits, but there were four phenotypes among the F_2 offspring. This allowed Mendel to deduce the Law of Independent Assortment, which states that the members of one pair of factors segregate independently of those from another pair. Therefore, all possible combinations of factors can occur in the gametes.

10. The Punnett square and laws of probability for a dihybrid cross similar to Mendel's show that $9/16$ of the F_2 offspring have the two dominant traits, $3/16$ have one dominant trait with one recessive trait, $3/16$ have the other dominant trait with the other recessive trait, and $1/16$ have both recessive traits, giving a 9:3:3:1 ratio.

11. The dihybrid testcross allows one to test whether an individual showing two dominant characteristics is homozygous dominant for both traits or for one trait only, or is heterozygous for both traits.

12. Table 12.3 lists the most typical crosses and the expected results. Learning the table saves one the trouble of working out the results repeatedly for each of these crosses.

Objective Questions

For questions 1 to 4, match the cross with the results in the key:

Key:
  a. 3:1
  b. 9:3:3:1
  c. 1:1
  d. 1:1:1:1

1. *TtYy* $\times$ *Tt Yy*
2. *Tt* $\times$ *Tt*
3. *Tt* $\times$ *tt*
4. *TtYy* $\times$ *ttyy*
5. Which of these could be a gamete?
  a. *GgLlB*
  b. *GLlB*
  c. *GlB*
  d. None of these.
6. Which of these properly describes a cross between an individual that is homozygous dominant for hairline but heterozygous for finger length and an individual that is recessive for hairline and fingers? (*Key: W* = widow's peak; *w* = continuous hairline; *S* = short fingers; *s* = long fingers)
  a. *WwSs* $\times$ *WwSs*
  b. *WWS* $\times$ *wwSs*
  c. *Ws* $\times$ *ws*
  d. *WWSs* $\times$ *wwss*

7. In peas, yellow seeds (*Y*) is dominant to green seeds (*y*). In the F_2 of a monohybrid cross in which a dominant homozygote is crossed with a recessive homozygote, you would expect
  a. plants that produce three yellow seeds to every green seed.
  b. plants with one yellow seed for every green seed.
  c. only plants with either the genotype *YY* or *yy*.
  d. Both (a) and (c).
8. In humans, pointed eyebrows are dominant to smooth eyebrows. Mary's father has pointed eyebrows, but she and her mother have smooth. What is the genotype of the father?
  a. *BB*
  b. *Bb*
  c. *bb*
  d. Any one of these.
9. In guinea pigs, smooth coat (*S*) is dominant over rough coat (*s*) and black coat (*B*) is dominant over white coat (*b*). In the cross *SsBb* $\times$ *SsBb*, how many of the offspring will have a smooth black coat?
  a. 9 only
  b. about 9 of every 16
  c. only 1 out of 16
  d. 6 of every 16

10. In horses, *B* = black coat, *b* = brown coat; *T* = trotter, *t* = pacer. A black trotter that has a brown pacer offspring is
  a. *BT*.
  b. *BbTt*.
  c. *bbtt*.
  d. *BBtt*.
11. In tomatoes, red fruit (*R*) is dominant to yellow fruit (*r*) and tallness (*T*) is dominant to shortness (*t*). A plant that is *RrTT* is crossed with a plant that is *rrTt*. What are the chances of an offspring being heterozygous for both traits?
  a. none
  b. ½
  c. ¼
  d. $9/16$
12. If a plant with an *RrTt* genotype is crossed with a plant that is *rrtt*,
  a. all the offspring will be tall with red fruit.
  b. 75% will be tall with red fruit.
  c. 50% will be tall with red fruit.
  d. 25% will be tall with red fruit.

Additional Genetics Problems

1. If a man homozygous for widow's peak (dominant) marries a woman homozygous for continuous hairline (recessive), what are the chances the children will have widow's peak? Will have a continuous hairline?

2. John has unattached earlobes like his father, but his mother has attached earlobes. What is John's genotype?

3. In humans, the allele for short fingers is dominant over that for long fingers. If a person with short fingers who had one parent with long fingers marries a person with long fingers, what are the chances for each child to have short fingers?

4. In a fruit fly experiment (see key on page 180), two gray-bodied fruit flies produce mostly gray-bodied offspring but some offspring have black bodies. If there are 280 offspring, how many do you predict will have gray bodies and how many will have black bodies? How many of the 280 offspring do you predict will be heterozygous? If you wanted to test whether a particular gray-bodied fly was homozygous dominant or heterozygous, what cross would you do?

5. In rabbits, black color is due to a dominant allele B and brown color to its recessive allele b. Short hair is due to the dominant allele S and long hair is due to its recessive allele s. In a cross between a homozygous black, long-haired rabbit and a brown, homozygous short-haired one, what would the F_1 generation look like? The F_2 generation? If one of the F_1 rabbits reproduced with a brown, long-haired rabbit, what phenotypes, and in what ratio, would you expect?

6. In horses, black coat (B) is dominant to brown coat (b), and being a trotter (T) is dominant to being a pacer (t). A black horse who is a pacer is crossed with a brown horse who is a trotter. The offspring is a brown pacer. Give the genotypes of all these horses.

7. The complete genotype of a gray-bodied, long-winged fruit fly is unknown. When this fly is crossed with a black-bodied, short-winged fruit fly, the offspring all have a gray body but about half of them have short wings. What is the genotype of the gray-bodied, long-winged fly? (See key on page 180.)

8. In humans, widow's peak (pointed) hairline is dominant to continuous (smooth) hairline, and short fingers are dominant to long fingers. If an individual who is heterozygous for both traits is married to an individual who is recessive for both traits, what are the chances that a child will also be recessive for both traits?

Study Questions

1. How did Mendel overcome the difficulty of not being able to see the factors? In what way did he set aside any personal beliefs he may have had?

2. How did the monohybrid crosses performed by Mendel refute the blending theory of inheritance?

3. Using the monohybrid cross Mendel did as an example, trace his reasoning to arrive at the Law of Segregation.

4. Define these terms: allele, dominant allele, recessive allele, genotype, homozygous, heterozygous, and phenotype.

5. Use a Punnett square and the laws of probability to show the results of a cross between two individuals who are heterozygous for one trait. What are the chances of an offspring having the dominant phenotype? The recessive phenotype?

6. In what way did a monohybrid testcross support Mendel's Law of Segregation?

7. How is a monohybrid testcross used today?

8. Using the dihybrid cross Mendel did as an example, trace his reasoning to arrive at the Law of Independent Assortment.

9. Use a Punnett square and the laws of probability to show the results of a cross between two individuals who are heterozygous for two traits. What are the chances of an offspring being homozygous dominant for both traits? Recessive for both traits?

10. If an individual is a heterozygote for two traits, what would be the results of a testcross? If the individual is heterozygous in only one trait? If the individual is homozygous dominant in both traits?

Thought Questions

1. The Mendelian pattern of inheritance is found among all sexually reproducing organisms. Why?

2. If some of the traits studied by Mendel had been controlled by alleles located on the same chromosome, such as G and Z in figure 12.4, what might have been Mendel's second law?

3. It's been said that Mendel's results are almost too close to the expected ratio. Why is it doubtful that Mendel could have gotten a perfect 3:1 ratio for the F_2 in table 12.1?

Selected Key Terms

gene (jēn) 172
P generation (jen''ĕ-ra'shun) 173
F₁ generation (jen''ĕ-ra'shun) 173
F₂ generation (jen''ĕ-ra'shun) 173
segregation (seg''re-ga'shun) 174
allele (ah-lēl') 175

dominant allele (dom'i-nant ah-lēl') 175
recessive allele (re-ses'iv ah-lēl') 175
gene locus (jēn lo'kus) 175
homozygous (ho''mo-zi'gus) 175
heterozygous (het''er-o-zi'gus) 175
genotype (jen'o-tīp) 175

phenotype (fe'no-tīp) 175
Punnett square (pun'et skwār) 176
testcross (test'kros) 177
independent assortment (in''de-pend'ent ah-sort'ment) 178

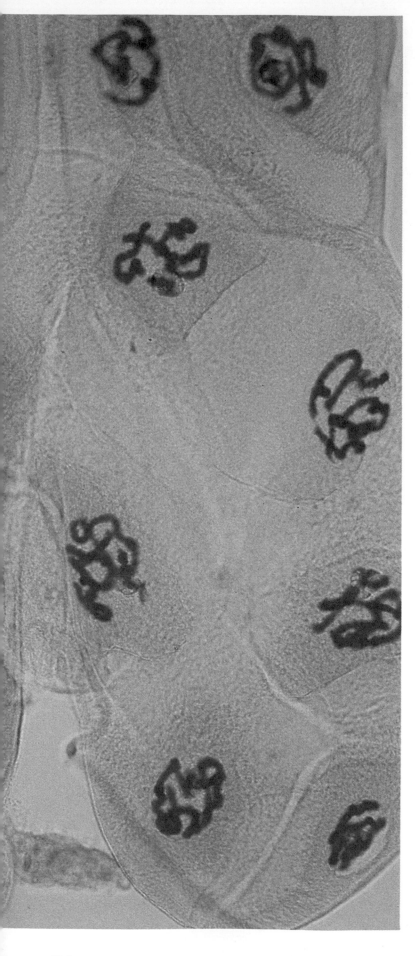

CHAPTER 13

Chromosomes and Genes

Your study of this chapter will be complete when you can:

1. *Recognize and do genetics problems involving multiple alleles, incompletely dominant or codominant alleles, epistatic genes, polygenes, and pleiotropic genes.*

2. *Discuss with examples the relative influence of the genotype and the environment on the phenotype.*

3. *Support the chromosomal theory of inheritance by listing similarities in the behavior of chromosomes and genes.*

4. *Explain the normal sex chromosome makeup of human males and females and the manner in which sex is determined.*

5. *Describe the experimental results that allowed investigators to determine the existence of X-linked genes.*

6. *Do genetics problems involving sex-linked genes.*

7. *Explain why linked genes do not obey Mendel's laws, and do genetic linkage problems.*

8. *Use the results of crosses involving linked genes to determine the order of genes on a chromosome.*

9. *Distinguish between chromosome mutations caused by a change in chromosome number and those caused by a change in chromosome structure.*

10. *Distinguish between polyploidy and aneuploidy, and give examples to show the importance of each.*

11. *Define and give examples of how inversions, translocations, deletions, and duplications may occur.*

The genes are on the chromosomes. Gregor Mendel did his work with garden peas raised in a garden plot. Later researchers found it convenient to use the fruit fly as their experimental material because the flies could be cultured in the laboratory. These cells taken from the salivary gland of a fruit fly have been stained so that the chromosomes are highly visible.

Beyond Mendel's Experiments

Inheritance that follows the patterns explained by Mendel's experiments is sometimes called "simple Mendelian inheritance" and for good reason. There are actually many more complicated patterns, some of which we will now examine.

Multiple Alleles

All traits discussed thus far have been controlled by two alleles; for example the alleles T and t control stem length in the garden pea. It is possible, however, for a gene to have three or more alleles, although a diploid individual can have a maximum of two alleles per trait. One of the best known examples of a **multiple allele** system is the one that controls A-B-O blood types in humans. The phenotypes listed in table 13.1 are produced by combinations of three different alleles: I^A, I^B, and i. Both I^A and I^B are fully expressed in the presence of the other. On the other hand, both I^A and I^B are dominant to i and only the genotype ii produces a person with type O blood.

Table 13.1 Blood Groups

Blood Types	Possible Genotypes
A	$I^A I^A$, $I^A i$
B	$I^B I^B$, $I^B i$
AB	$I^A I^B$
O	$i i$

$I, i =$ immunogen gene

An examination of possible matings between individuals with different blood types sometimes produces surprising results. For example:

If parents have the blood types

$$I^A i \qquad \times \qquad I^B i$$

children can be

$$I^A I^B, \qquad i i, \qquad I^A i, \qquad I^B i$$

Therefore, from this particular mating every possible phenotype (AB, O, A, B blood type) is possible.

Degrees of Dominance

In his experiments with the garden pea, Mendel only observed instances in which one allele was completely dominant over the other allele. However, **incomplete dominance** can be illustrated by a cross between a true-breeding, red-flowered snapdragon strain and a true-breeding, white-flowered strain (fig. 13.1). The offspring have pink flowers and, offhand, you might suppose that this result would support the blending theory of inheritance popular when Mendel did his work. However, if you let these F_1 plants self-pollinate, the F_2 generation has a phenotype ratio

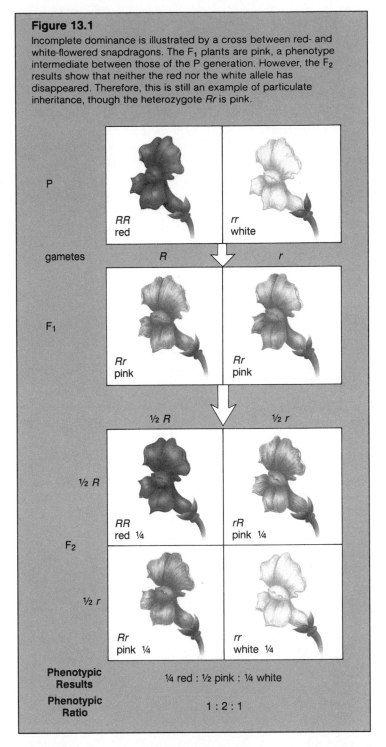

Figure 13.1

Incomplete dominance is illustrated by a cross between red- and white-flowered snapdragons. The F_1 plants are pink, a phenotype intermediate between those of the P generation. However, the F_2 results show that neither the red nor the white allele has disappeared. Therefore, this is still an example of particulate inheritance, though the heterozygote Rr is pink.

P

RR red   rr white

gametes   R   r

F_1

Rr pink   Rr pink

½ R   ½ r

½ R

F_2

RR red ¼   rR pink ¼

½ r

Rr pink ¼   rr white ¼

Phenotypic Results   ¼ red : ½ pink : ¼ white

Phenotypic Ratio   1 : 2 : 1

of 1 red: 2 pink: 1 white-flowered plant. Because the parental phenotypes reappear in the F_2, it is obvious that we are still dealing with an example of particulate inheritance of the type described by Mendel. However, red is incompletely dominant to white, and therefore the genotype Rr is pink instead of red.

In **codominance** the heterozygote doesn't show a phenotype intermediate between the two homozygous parents, but shows both characteristics, because both alleles are fully expressed. For example, we have just seen that I^A and I^B are expressed equally with regard to blood type.

In some instances, one allele is incompletely dominant over the other, and in other instances, members of an allelic pair are codominant.

Epistatic Genes

Epistatic genes affect the phenotypic expression of alleles that are at a different gene locus. One of the first observations of **epistasis** ("covering up") occurred when William Bateson and R. C. Punnett crossed two varieties of white-flowered peas. The F_1 flowers were unexpectedly purple in color! When these were allowed to self-pollinate, a 9:7 ratio (purple to white) in the F_2 resulted. Further investigation confirmed that a plant needed a dominant allele for two genes before the color would develop and that therefore $9/16$ of the plants were purple with the genotype $C\_\_P\_\_$ and $1/16$ were white with either the genotype $C\_\_pp$ or $ccP\_\_$.

Today it is possible to give these results a biochemical explanation. Production of plant pigments involves a metabolic pathway in which each step is catalyzed by an enzyme that is the product of a gene. Inheritance of at least one dominant allele results in a functional enzyme for a particular step, but inheritance of recessive alleles results in a nonfunctional enzyme for another step. If any one step is blocked, the plant lacks pigment. This is the usual manner in which an epistatic gene can interfere with the expression of another gene.

A somewhat similar situation occurs in humans and other animals. If an individual inherits the allelic pair (*aa*) that causes albinism, which is the inability to produce the pigment melanin, it does not matter what genes for eye color and hair color are inherited. These genes will not be expressed and the individual will be an albino, lacking pigment in all parts of the body (fig. 13.2).

Figure 13.2
Albino skunks. The inheritance of two recessive alleles prevented the production of melanin and resulted in these animals that have no color in the skin, hair, or eyes.

Polygenic Inheritance

It wasn't long after the rediscovery of Mendel's work that researchers became interested in another case where the F_1 was intermediate in appearance between the P parents. But when these F_1 individuals were allowed to interbreed, the F_2 generation showed continuous variation in phenotypes. Their distribution is illustrated by a bell-shaped curve indicating that there are many intermediate forms that differ slightly and quantitatively from each other.

For example, it's possible that skin color in humans is controlled by three pairs of alleles such that each dominant allele adds a unit of darkness to the skin of the individual:

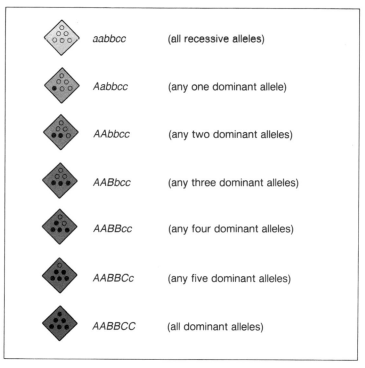

aabbcc	(all recessive alleles)
Aabbcc	(any one dominant allele)
AAbbcc	(any two dominant alleles)
AABbcc	(any three dominant alleles)
AABBcc	(any four dominant alleles)
AABBCc	(any five dominant alleles)
AABBCC	(all dominant alleles)

This would mean that there would be 7 possible shades of skin color and that the distribution of the colors in a population would follow a bell-shaped curve (fig. 13.3).

This type of inheritance pattern is called **polygenic inheritance** because one trait, for example, skin color, is being controlled by several genes that segregate independently of each other. Many other human traits (including eye color, height, weight, and blood pressure) depend on polygenic inheritance.

To take another example, H. Nilsson-Ehle studied the inheritance of seed color in wheat. When he crossed white and dark red varieties, the F_1 were intermediate between the parents. The F_2 consisted of seven phenotypes, ranging from white to dark red, with kernels of intermediate color being the most common. Only $1/64$ were red. He concluded that three allelic pairs control seed color in wheat, and that each allele adds quantitatively to the final phenotypic color. These results are contrary to any of Mendel's studies. Mendel had deliberately chosen discontinuous variations—short or tall plants, yellow or green seeds, round or wrinkled seeds—phenotypes that are immediately apparent.

Polygenic traits are typically influenced by the environment; for example, the height and weight of plants and animals is in part due to the type of nurture they receive.

In polygenic inheritance, a trait is controlled by several genes, and each allele adds quantitatively to the phenotype. The variations are continuous—there are many intermediate phenotypes—and the phenotypic distribution follows a bell-shaped curve.

Pleiotropy

Thus far we have considered examples of one or more allelic pairs that affect a single trait. **Pleiotropy** is the opposite of this situation; a pleiotropic gene affects several different phenotypic characteristics. This can occur because the same biochemical pathway is present in various developing organs and because biochemical pathways often interact.

Mendel himself observed that one of his factors simultaneously affected the color of the flowers, seeds, and axils of the leaves (fig. 13.4). In *Drosophila* too, investigators have noted a single gene locus that affects eye color, body proportions, wing size, wing vein arrangement, body hairs, size and arrangement of bristles, shape of testes and ovaries, viability, rate of growth, and fertility.

In humans, some of the most significant medical syndromes are due to pleiotropic genes. For example, in Marfan syndrome, individuals tend to be tall and thin with long legs, arms, and fingers. They are nearsighted and the wall of the aorta is weak, causing it to enlarge and eventually split. All of these effects can be traced to the inability of the individual to produce normal connective tissue.

In pleiotropy, the alleles of a single gene locus affect many phenotypic traits. This can be contrasted to polygeny, in which a single trait is affected by several genes (fig. 13.5).

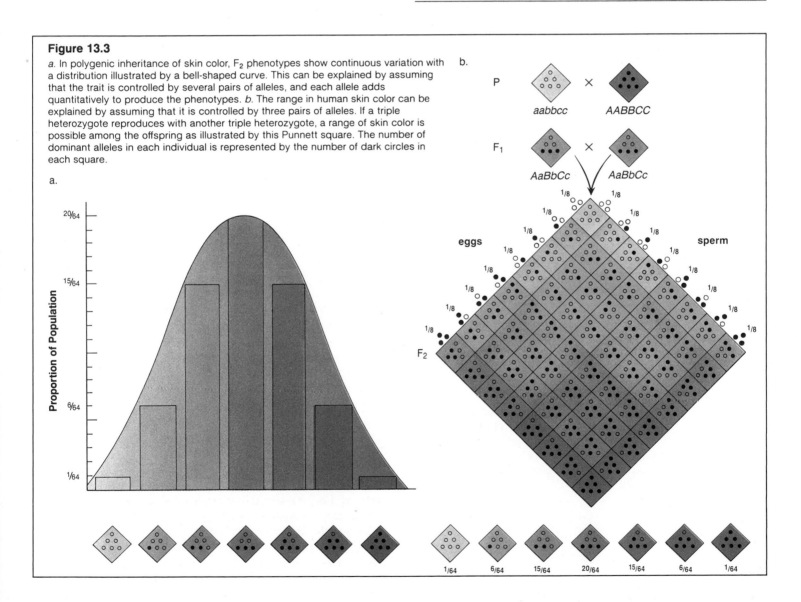

Figure 13.3

a. In polygenic inheritance of skin color, F_2 phenotypes show continuous variation with a distribution illustrated by a bell-shaped curve. This can be explained by assuming that the trait is controlled by several pairs of alleles, and each allele adds quantitatively to produce the phenotypes. *b.* The range in human skin color can be explained by assuming that it is controlled by three pairs of alleles. If a triple heterozygote reproduces with another triple heterozygote, a range of skin color is possible among the offspring as illustrated by this Punnett square. The number of dominant alleles in each individual is represented by the number of dark circles in each square.

Figure 13.4
Mendel observed an instance of pleiotropy in the garden pea. *a.* A
plant sometimes had red flowers, red color on the leaves where
noted, and brown seeds, all effects of one gene. *b.* The plant
without this gene lacked these characteristics.

a.

b.

Practice Problems 1*

1. A woman with blood type B gives birth to a child with blood type B. Could a man with blood type A be the father of this child?

2. What would be the results of a testcross involving incomplete dominance when snapdragons with pink flowers are crossed with white-flowered ones?

3. Breeders of dogs note various colors among offspring that range from white to black. If the coat color were controlled by three pairs of alleles, how many different shades would be possible?

4. Investigators note that albino tigers usually have crossed eyes. What two inheritance patterns can account for a syndrome such as this?

*Answers to Practice Problems appear on page 805 of appendix D.

Genes and the Environment

There is no doubt that both the genotype and the environment affect the phenotype. The relative importance of genetic and environmental influences on the phenotype can vary but in some instances the environment seems to have an extreme effect.

In the water buttercup, *Ranunculus peltatus* (fig. 13.6), the submerged part of the plant has a different appearance from that part above water. Apparently, the presence or absence of a water environment dramatically influences the phenotype.

In humans, if the drug thalidomide is taken during the second month of pregnancy, a seriously malformed baby lacking arms or legs, or having stublike appendages can result (fig. 13.7). This phenotype can be due to a rare recessive mutation in humans, but in this instance it is caused by the environmental effect of thalidomide—the genotype is normal!

The relative effect of genes and the environment on the phenotype varies; there are examples of extreme environmental influence.

Figure 13.5

Diagram illustrating that one gene locus can affect many characteristics of the individual (pleiotropy), and one characteristic can be controlled by several gene loci (polygeny).

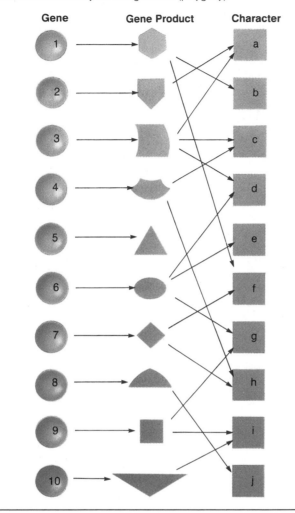

Figure 13.5

Diagram illustrating that one gene locus can affect many characteristics of the individual (pleiotropy), and one characteristic can be controlled by several gene loci (polygeny).

Gene   **Gene Product**   **Character**

Chromosomes and Genes

Mendel was rediscovered in 1900, at a time when the scientific community was ready to appreciate the work he had done so many years before. Three botanists, Karl Correns (German), Hugo de Vries (Dutch), and Erich Tschermak von Seysenegg (Austrian), working independently, discovered the laws of segregation and independent assortment. Upon surveying the scientific literature, they found that Gregor Mendel had already worked out these laws before them.

We could say that Mendel was "ahead of his time" to explain why his work lay unknown in dusty scientific libraries and archives for nearly 35 years. The scientific community caught up to Mendel because the cellular basis of inheritance had been worked out in the intervening years. The behavior of chromosomes in mitosis had been described by 1875 and meiosis had been described in the 1890s. By 1902, both Theodor Boveri (German) and Walter S. Sutton (U.S.) had independently noted

Figure 13.6

Environmental influence on the phenotype of the water buttercup is obvious. The submerged leaves are thin and finely divided, whereas those that lie above water are broad and flat.

Figure 13.7

Environmental influence can be so severe as to cause a phenotype that is contrary to the genotype. Children with shortened, deformed limbs were born to women who took the drug thalidomide in the early months of pregnancy. Because this condition is not due to genes, it cannot be passed on.

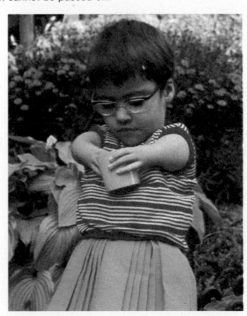

Figure 13.8

Mendel's Laws of Segregation and Independent Assortment have a chromosomal basis. The Law of Segregation holds because homologous chromosomes, and therefore allelic pairs, segregate during meiosis I. Consequently, there is only one chromosome and one allele of each kind in the gametes. The Law of Independent Assortment holds because homologous pairs of chromosomes, and therefore allelic pairs, segregate independently of other pairs during meiosis I.

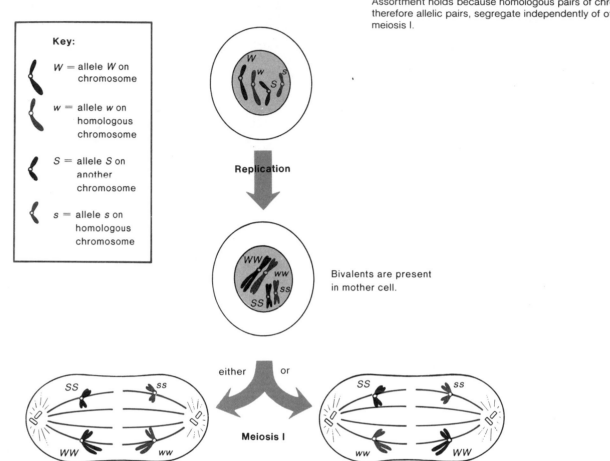

Key:

W = allele W on chromosome

w = allele w on homologous chromosome

S = allele S on another chromosome

s = allele s on homologous chromosome

Replication

Bivalents are present in mother cell.

either or

Meiosis I

Alternative types of separation are equally likely when bivalents segregate independently of each other.

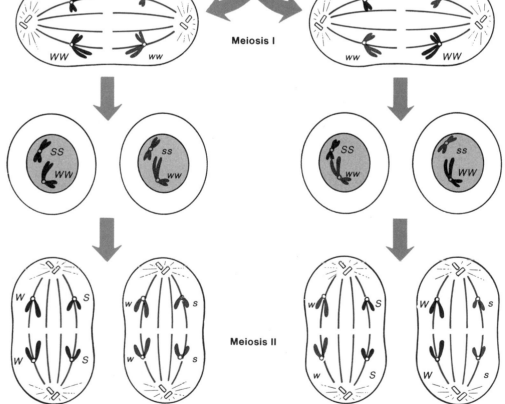

Meiosis II

Each cell contains one of each kind of chromosome. All possible combinations of chromosomes and alleles will be found in the gametes.

WS   ws   wS   Ws

the parallel behavior of genes and chromosomes, and had proposed that Mendelian factors are located on the chromosomes (fig. 13.8). The parallel behavior of chromosomes and alleles suggests that:

1. Both chromosomes and factors (now called alleles) exist in pairs in diploid cells.
2. Both homologous chromosome pairs and allelic pairs segregate during meiosis so that the gametes have one-half the total number.
3. Both homologous chromosome pairs and allelic pairs segregate independently so that the gametes contain all possible combinations.
4. Fertilization restores both the diploid chromosome number and the paired condition for alleles in the zygote.

The hypothesis that genes are on the chromosomes led to many experiments. Leading the way was a group of *Drosophila* geneticists at Columbia University, headed by Thomas Hunt Morgan. Their discovery of a mutant *Drosophila* with white, rather than red eyes played an important role in testing the **chromosome theory of inheritance.** These flies are even better subjects for genetic studies than are garden peas. They can be easily and inexpensively raised in simple laboratory glassware; females mate only once and then lay hundreds of eggs during their lifetime; the generation time is short, taking only about 10 days when conditions are favorable.

The chromosome theory of inheritance states that the genes are on the chromosomes, and therefore, they behave similarly during meiosis and at fertilization.

Chromosomal Sex Determination

By the time Morgan began his studies, it had already been determined that in many animal species chromosomes differ between the sexes. All but one of the pairs of chromosomes in males and females is the same; these are called the **autosomes.** The pair that is different is called the sex chromosomes. *Drosophila* and humans have the same sex chromosomal pattern: the **sex chromosomes** in females are XX and those in males are XY. Because males can produce two different types of gametes—those that contain an X and those that contain a Y, normally they determine the sex of the new individual (fig. 13.9). Although the chances of any human couple having a boy or a girl are 50:50, there are actually more boys born than girls. For reasons that are not clear, the ratio of boys to girls is 106:100 among newborns.

As you would suspect, the sex chromosomes carry genes that determine sex. Progress in identifying these genes has been slow, but recent findings are discussed in the reading in chapter 14, "Active Y Chromosomes and Inactive X Chromosomes."

The cells of males and females contain both autosomes and sex chromosomes. In many animals, males have XY sex chromosomes and females have XX sex chromosomes.

Figure 13.9
In this Punnett square, the sperm and eggs are shown as carrying only a sex chromosome. (Actually, they also carry twenty-two autosomes.) The offspring are either male or female, depending on whether they receive an X or Y chromosome from the male parent.

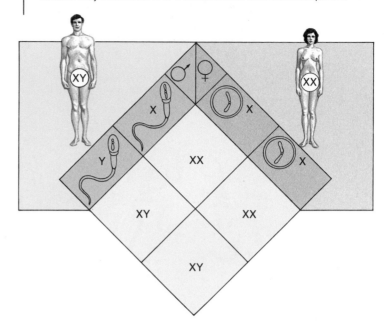

Sex-Linkage and the Chromosomal Theory of Inheritance

In addition to sex-determination genes, sex chromosomes carry genes that control traits of the individual that have nothing to do with sex characteristics. These genes are called *sex-linked genes.* Most of the known sex-linked genes are on the X chromosome and so are called **X-linked genes.**

Discovery of X-Linkage

Genetic crosses with *Drosophila* provided good evidence that the genes are on the chromosomes. Morgan took the newly discovered mutant male with white eyes and crossed it to a red-eyed female:

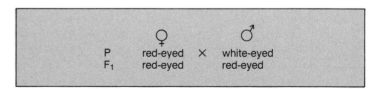

From these results he knew that red eyes are dominant to white eyes. He then crossed the F_1 flies. In the F_2 generation there was the expected 3:1 ratio but it struck him as odd that all of the white-eyed flies were males!

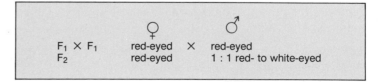

Figure 13.10

Representation of the cross that showed that the white-eye allele in *Drosophila* was linked to the X chromosome. *a.* Both males and females have four chromosomes, three pairs of autosomes, and one pair of sex chromosomes. Males are XY and females are XX.
b. When doing genetic problems, males do not have a superscript attached to the Y chromosome because they lack the gene for eye color on that chromosome. Among the offspring in this cross, 50% of the male offspring have white eyes. Because the male offspring receive a Y chromosome from the male parent, they express whichever X chromosome is inherited from the female parent. In contrast, 50% of the female offspring are red-eyed heterozygotes. When an X^r was received from the female parent, it was masked by the X^R from the male parent.

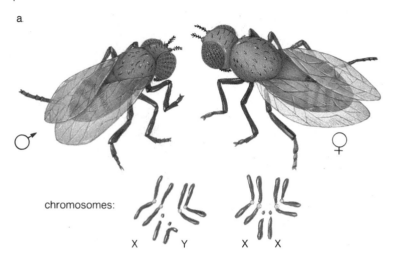

a.

chromosomes:

X Y   X X

Obviously, a major difference between the male flies and the female flies was their sex chromosomes. Could it be possible that an allele for eye color was on the Y chromosome but not on the X? This idea could be quickly discarded because normal females have red eyes and they have no Y chromosome. Perhaps an allele for eye color was on the X, but not the Y chromosome. Figure 13.10 indicates that this explanation would match the results obtained in the experiment. These results supported the chromosomal theory of inheritance by showing that the behavior of a specific pair of alleles corresponds exactly with that of specific chromosomes, the X chromosomes in *Drosophila*.

Some X-linked genes can control traits that have nothing to do with sex characteristics. Males have only one copy of these genes, and if they inherit a recessive allele, it will be expressed.

Sex-Linked versus Autosomal Genetics Problems

You'll recall that when doing autosomal genetics problems, we represent the genotypes of males and females similarly. For example, in humans,

Key:
W = widow's peak
w = continuous hairline

Genotypes:
WW, Ww, or ww

parents

♂ X^RY × ♀ X^RX^r

red-eyed   red-eyed

eggs ♀

	X^R	X^r
sperm ♂ X^R	X^RX^R	X^RX^r
Y	X^RY	X^rY

offspring

key:
X^R = red eyes
X^r = white eyes

Phenotypic Ratio

females: red-eyed

males: 1 red-eyed
      1 white-eyed

Genotypes and Phenotypes

X^RX^R	female red-eyed	X^RY	male red-eyed
X^RX^r	female red-eyed	X^rY	male white-eyed
X^rX^r	female white-eyed		

b.

But when we do sex-linked problems, males and females must be indicated by sex. For example, in humans,

Key:
X^B = normal vision
X^b = color blindness

Genotypes:
Males: X^BY and X^bY
Females: X^BX^B, X^BX^b, and X^bX^b

Also, when indicating the results of a cross involving a sex-linked gene you should give the phenotypic ratios for males and females (fig. 13.10*b*).

Color blindness and other genetic disorders in humans are discussed in more detail in the following chapter but in the meantime, we want to note that individuals with a heterozygous genotype, such as the X^BX^b female above, are called carriers. **Carriers** do not show a genetic disorder, yet they are capable of passing on an allele for the disorder.

Figure 13.11

Each homologous pair of chromosomes carries a number of genes. The alleles on a chromosome form a linkage group because they tend to go together into the gametes. This simplified chromosome map shows the relative positions of some of the genes on *Drosophila* chromosome 2. The distances between the genes (the numbers = map units) are equivalent to the percentage of crossing-over events that occurs between the various alleles. For example, the crossing-over frequency between gray body and long wings would be 48.5 − 31.0 = 17.5%. This means that 17.5% of all gametes would carry recombinant gametes. Notice that the chromosome theory of inheritance includes the concept that alleles are arranged linearly along a chromosome at specific loci.

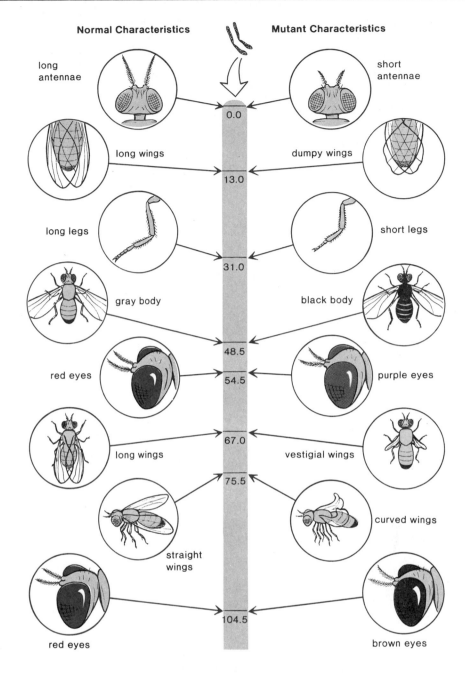

Normal Characteristics — Mutant Characteristics

long antennae — 0.0 — short antennae
long wings — 13.0 — dumpy wings
long legs — 31.0 — short legs
gray body — 48.5 — black body
red eyes — 54.5 — purple eyes
long wings — 67.0 — vestigial wings
— 75.5 —
straight wings — curved wings
red eyes — 104.5 — brown eyes

Practice Problems 2*

1. Using the key for *Drosophila* eye color shown in figure 13.10, give the three possible genotypes for females and the two possible genotypes for males. What are the possible gametes for these males?
2. Which *Drosophila* cross below will produce white-eyed males? In what ratio?
   a. $X^R X^R \times X^r Y$
   b. $X^R X^r \times X^R Y$
3. A woman is color blind. What are the chances that her sons will be color blind? If she is married to a man with normal vision, what are the chances that her daughters will be color blind? Will be carriers?
4. Both parents are right-handed, (R = right-handed, r = left-handed) and have normal vision. Their son is left-handed and color blind. Give the genotype of each person.

Genetic Linkage and Chromosome Mapping

Drosophila probably has thousands of different genes controlling all aspects of its structure, biochemistry, and behavior. Yet it has only four chromosomes. This paradox led investigators like Sutton to believe that each chromosome must carry a large number of genes. For example, it is now known that genes controlling eye color, wing type, body color, leg length, and antennae type are all located on chromosome 2 (fig. 13.11). These genes are said to form a **linkage group** because they are found on the same chromosome.

It is easy to predict that crosses involving linked genes will not give the same results as those involving unlinked genes. For example, suppose you were doing a cross between a gray-bodied,

*Answers to Practice Problems appear on page 806 of appendix D.

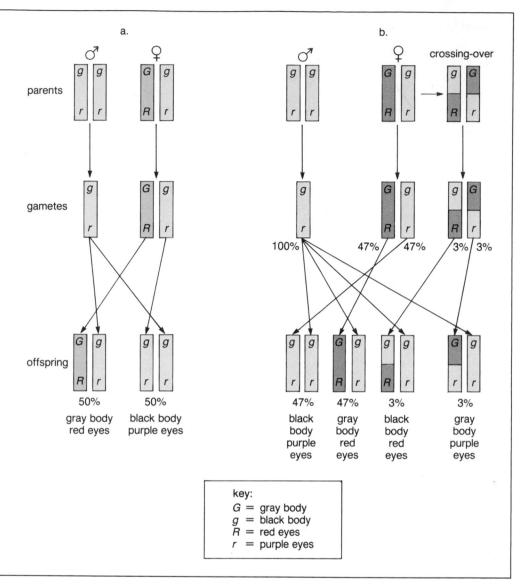

red-eyed heterozygote and a black-bodied fly with purple eyes. These genes are on chromosome 2, therefore they are linked (fig. 13.11). Using the following

Key:
G = gray body   R = red eyes[1]
g = black (ebony) body   r = purple eyes

you predict that the results will be 1:1 instead of 1:1:1:1 as was the case for unlinked genes. The reason of course, is that linked alleles tend to stay together and do not segregate independently as predicted by Mendel's laws because they are the same chromosome. Under these circumstances, the heterozygote will form only two types of gametes and produce offspring with only two phenotypes (fig. 13.12*a*).

When you do the cross, however, you find that a very small number of offspring show recombinant phenotypes (i.e., those that are different from the original parents). Specifically, you find that 47% of the offspring have gray bodies and red eyes, 47% have black bodies and purple eyes, 3% have gray bodies and purple eyes, and 3% have black bodies and red eyes (fig. 13.12*b*). What could have happened?

You'll recall that crossing-over can occur between homologous chromosomes when they are paired during meiosis, and that crossing-over produces recombinant gametes (fig. 13.13). In this instance, crossing-over produced a very small number of *recombinant gametes* (fig. 13.12*b*) and *recombinant phenotypes*. Take a look at figure 13.11 again and you'll notice that our two sets of alleles are very close together. Doesn't it stand to reason that the closer together two genes are, the less likely they are to cross over? This is exactly what various crosses have repeatedly shown.

---

[1]*Drosophila* has several genes for eye color. Red/white is on the X chromosome; red/purple is on chromosome 2.

Figure 13.13

Crossing-over causes an exchange of alleles between nonsister chromatids of a bivalent during meiosis I. (Notice that this illustration is a continuation of figure 12.4 in the previous chapter.) Whenever crossing-over occurs, the gametes will contain recombinant chromosomes.

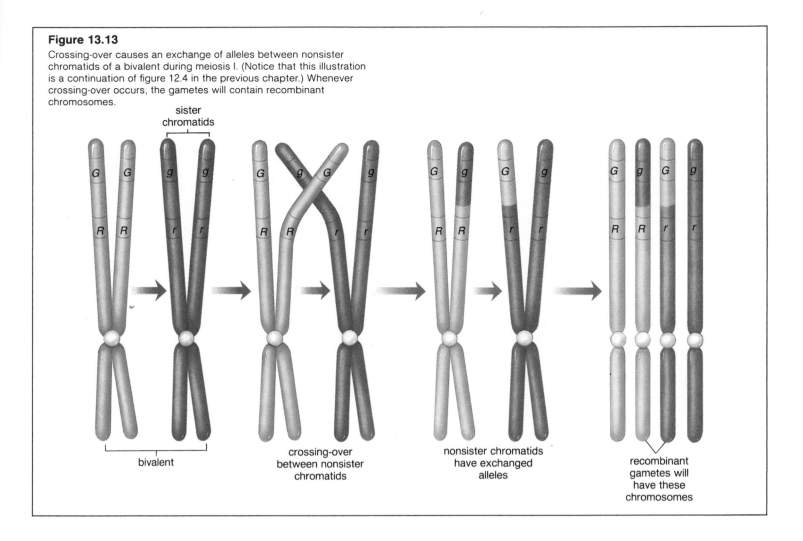

sister chromatids

bivalent

crossing-over between nonsister chromatids

nonsister chromatids have exchanged alleles

recombinant gametes will have these chromosomes

All the genes on one chromosome form a linkage group that tends to stay together except when crossing-over occurs.

Chromosome Mapping

You can use the percentage of recombinants to map the chromosomes, because there is a direct relationship between the frequency of crossing-over and the percentage of recombinant phenotypes. In our example, a total of 6% of the offspring were recombinants, and for the sake of mapping the chromosomes, it is assumed that 1% of crossing-over = 1 map unit. Therefore, the allele for black body and the allele for purple eyes are 6 map units apart.

Suppose you want to determine the order of any three genes on the chromosomes. To do so, you can perform crosses that tell you the map distance between all three pairs of alleles. Considering just the mutant alleles below, if you know that:

1. The distance between the black-body and purple-eye alleles = 6 map units,
2. The distance between the purple-eye and vestigial-wing alleles = 12.5 units, and
3. The distance between the black-body and vestigial-wing alleles = 18.5 units, then the order of the alleles must be:

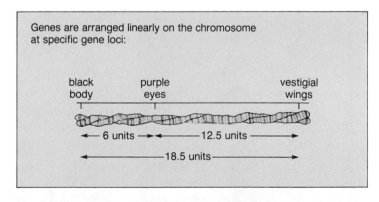

Genes are arranged linearly on the chromosome at specific gene loci:

black body   purple eyes   vestigial wings

← 6 units → ← 12.5 units →
← 18.5 units →

Of the three possible orders (any one of the three genes can be in the middle), only this order gives the proper distance between all three allele pairs. Because it is possible to map the chromosomes, we know that genes are indeed located on chromosomes, they have a definite location, and they occur in a definite order.

Various methods are used to map human chromosomes and these will be discussed in the next chapter.

The crossing-over frequency (equivalent to the recombinant phenotype frequency) in crosses involving linked genes indicates the distance between genes on the chromosomes.

Practice Problems 3*

1. When *AaBb* individuals were allowed to self-breed, the phenotypic ratio was just about 3:1. What ratio was expected? What may have caused the observed ratio?
2. In a cross involving a heterozygote in which long wings and red eyes are on one member of the homologous pair of chromosomes and dumpy wings and purple eyes are on the other, what total percentage of recombinant phenotypes are expected according to figure 13.11?
3. In two sweet pea strains, *B* = blue flowers, *b* = red flowers, *L* = long pollen grains, and *l* = round pollen grains. In a cross between a heterozygous plant with blue flowers and long pollen grains and a plant with red flowers and round pollen grains, 44% of offspring are blue, long; 44% are red, round; 6% are blue, round; and 6% are red, long. How many map units separate these two sets of alleles?
4. Investigators performed crosses that indicated bar-eye and garnet-eye alleles are 13 map units apart, scalloped-wing and bar-eye alleles are 6 units apart, and garnet-eye and scalloped-wing alleles are 7 units apart. What is the order of these alleles on the chromosome?

*Answers to Practice Problems appear on page 806 of appendix D.

Chromosome Mutations

Recombinations that occur during meiosis and fertilization increase the amount of variation among the members of a species, as do mutations. Mutations are permanent gene or chromosome changes that can be passed on to offspring if they occur in cells that will become gametes. Chromosome mutations are often associated with reduced fertility, because they may prevent normal synapsis and segregation of homologous chromosomes.

There are two categories of chromosome mutations: those that affect the number of chromosomes and those that affect the structure of the chromosomes.

Changes in Chromosome Number

Polyploidy

Eukaryotes are generally diploid with two sets of chromosomes, but there are some mutants that have more than two complete sets. These are called **polyploids** (poly, many; ploid, sets). Polyploid organisms are named according to the number of sets of chromosomes they have. Triploids (3N) have three of each kind of chromosome, tetraploids (4N) have four sets, pentaploids (5N) have five sets, and so on.

Polyploidy is a major evolutionary mechanism in plants, and it is estimated that 47% of all flowering plants are polyploids. Among these are many of our most important crops, such as wheat, corn, cotton, sugarcane, bananas, and potatoes (fig. 13.14). Also, many attractive flowers, such as chrysanthemums and daylilies are polyploids.

Figure 13.14

Many food sources are specially developed, polyploid plants. Among these are seedless (*a*) watermelons and (*b*) bananas. They are infertile (seeds poorly developed) triploids, but can be propagated by asexual means. Polyploidy makes plants and their fruits larger, such as these (*c*) jumbo McIntosh apples.

a.
b.
c.

It would appear from its common occurrence that polyploidy must in some way have been advantageous in plant evolution. Certainly polyploids are often larger and more vigorous than are their diploid counterparts but there may also be another explanation. Doubling the chromosome number following hybridization (cross between two species) increases the fertility of the plant:

---

allopolyploidy scenario

A + B ⟶ 27 chromosomes ⟶ 54 chromosomes

(12) (15)   (infertile)   (fertile)

Hybridization increases variation because it brings about new combinations of genes, but unfortunately hybrids are often sterile, because the chromosomes from the two species cannot pair during meiosis. Doubling of the chromosome number gives two of each kind of chromosome from each species, which allows meiosis to occur. This type of polyploidy—doubling the chromosome number of a hybrid—is termed **allopolyploidy.** Allopolyploidy in plants increases the chances of hybridization leading to a new species, and therefore contributes to plant evolution. This mode of speciation generally will not work for animals because they have sex chromosomes. Not only do abnormal sex chromosome numbers cause anatomical and physiological difficulties but XXYY males would always have sons.

Polyploids (having more than one set of chromosomes) are common in plants. Allopolyploidy (hybridization + doubling chromosome number) is believed to have contributed to speciation in plants.

Aneuploidy

Aneuploidy is an excess or deficiency of a particular chromosome: **trisomy** ($2N + 1$) occurs when an individual has an extra copy of a chromosome and **monosomy** ($2N - 1$) occurs when

an individual is missing one chromosome. Trisomics (individuals that have a trisomy) are known in both plants and animals. In animals, autosomal monosomies and trisomies are generally lethal, although a trisomic individual is more likely to survive than a monosomic one. The survivors are characterized by a distinctive set of physical and mental abnormalities called a syndrome. The most common autosomal trisomy among humans is trisomy 21 (Down syndrome), which is discussed in the next chapter.

Usually, the cause of aneuploidy is nondisjunction during meiosis (fig. 13.15). **Nondisjunction** can occur when either the members of a homologous pair or the sister chromatids fail to separate and instead go into the same daughter cell. When these gametes are fertilized by normal gametes, they either have an extra chromosome (trisomy) or lack a chromosome (monosomy). Nondisjunction is more common during meiosis I than meiosis II. Nondisjunction can also occur during mitosis.

Aneuploidy occurs when one or more individual chromosomes is absent or present in excess. Aneuploidies generally cause abnormalities, particularly in animals.

Changes in Chromosome Structure

There are various agents in the environment, such as radiation, certain organic chemicals, or even viruses, that can cause chromosomes to break apart. Ordinarily when breaks occur in chromosomes, the two broken ends reunite just as they were before. Sometimes, however, the broken ends of one or more chromosomes do not rejoin in the same pattern as before, and this results in a change in chromosome structure. The existence of such mutations can be readily detected in chromosomes taken from larval salivary gland cells of *Drosophila* (fig. 13.16). In these cells, the chromosomes duplicate repeatedly, but instead of separating, they remain side by side to produce what is called giant

---

Figure 13.15

Nondisjunction can occur during meiosis I if the members of a homologous chromosome pair fail to separate, or during meiosis II if the sister chromatids fail to separate completely. In either case, the resulting zygote following fertilization will be a monosomic (a lethal condition) or trisomic (sometimes lethal).

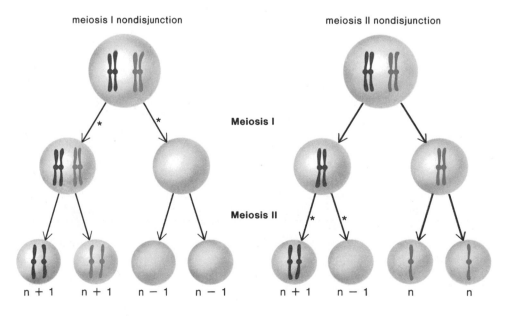

Figure 13.16

Giant polytene chromosomes from salivary gland cells of *Drosophila* are composed of many hundreds of chromatids that lie side by side. Each of these chromosomes has a characteristic banding pattern; the darker bands are places where DNA is more condensed. Sometimes it is possible to make out chromosomal mutations in these chromosomes; for example, a duplication leads to the increased size of a particular band.

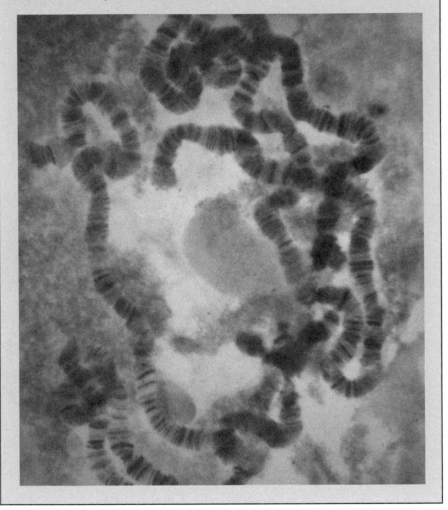

Figure 13.17

Types of chromosome mutations. *a.* Inversion is when a piece of chromosome breaks loose and then rejoins in the reversed direction. *b.* Translocation is the exchange of chromosome pieces between nonhomologous pairs. *c.* Deletion is the loss of a chromosome piece. *d.* Duplication occurs when the same piece is repeated within the chromosome.

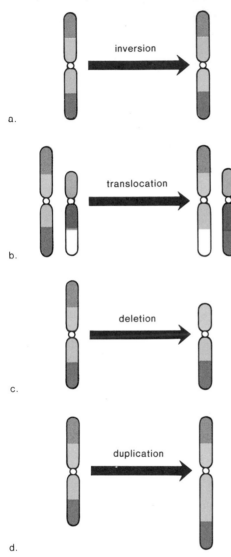

polytene chromosomes. The large size of these chromosomes makes it possible to see some of the types of chromosome mutations we will be discussing.

An **inversion** (fig. 13.17) occurs when a segment of a chromosome is turned around 180°. You might think this is not a problem because the same genes are present, but the new position might lead to altered gene activity. Also, crossing-over might be prevented because the chromosome pair cannot align properly during synapsis; if crossing-over does occur, duplications and deletions might result.

A **translocation** is the movement of a chromosomal segment from one chromosome to another, nonhomologous chromosome. Translocation heterozygotes usually have reduced fertility, again due to production of abnormal gametes. A translocation can also cause a Down syndrome child, as we will discuss in chapter 14.

A **deletion** occurs when an end of a chromosome breaks off or when two simultaneous breaks lead to the loss of a segment. Deletions are usually lethal when homozygous (both members of a pair of chromosomes affected), although exceptions have been detected in corn and other organisms. Even when heterozygous (one member of a pair of chromosomes affected), a deletion often causes abnormalities. An example of a heterozygous deletion in humans is the cri du chat (cat's cry) syndrome. The affected individual has a small head, is mentally retarded, and has facial abnormalities. Abnormal development of the glottis and larynx results in the most characteristic symptom—the infant's cry resembles that of a cat.

A **duplication** is the presence of a chromosome segment more than once in the same chromosome. There are two ways a duplication can occur. A broken segment from one chromosome can simply attach to its homologue, but the presence of a duplication in the middle of the chromosome is more likely due to unequal crossing-over. This can occur when homologues are slightly mispaired and simultaneously produce a gene duplication and a deletion:

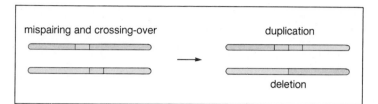

Multiple copies of genes that occur because of duplication can now mutate differently and thereby provide additional genetic variation for the species. For example, there are several genes for human globin (globin is a part of hemoglobin, which is present in red blood cells and carries oxygen), and these may have arisen by a process of duplication followed by different mutations. The evidence for this is the great similarity in the amino acid sequences of the different globins that still remain closely linked on the same chromosome.

Chromosome mutations can be the result of structural changes. Although most of these lead to abnormal gametes and offspring, some no doubt have played a role in evolutionary history.

Summary

1. Some genes have more than two alleles, although each person has only two. In humans, there are three alleles for blood type (I^A, I^B, i), and because each person can have any two, there are four blood types.
2. With incomplete dominance, the F_1 individuals are intermediate between the parental types; this is not a blending because the parental phenotypes reappear in F_2. With codominance, the F_1 individuals show both characteristics.
3. Epistatic genes affect the phenotypic expression of alleles at a different gene locus and can cause an allele to not be expressed at all.
4. Some traits are controlled by many genes and are therefore called polygenic. Each allele adds quantitatively to the phenotype. Therefore, there are many intermediate F_2 phenotypes. The distribution of these continuous variations follows a bell-shaped curve.
5. Pleiotropic genes affect more than one trait. Many syndromes are caused by pleiotropic genes.

6. The relative influence of the genotype and environment on the phenotype can vary, but there are examples of extreme environmental influence.
7. The chromosome theory of inheritance says that the genes are located on the chromosomes, which accounts for the similarity of their behavior during meiosis and fertilization.
8. Sex determination in animals is dependent upon the chromosomes. Usually females are XX and males are XY.
9. Solid experimental support for the chromosome theory of inheritance came when Morgan and his group were able to determine that the white-eye allele in *Drosophila* is on the X chromosome.
10. Genes on the X chromosome are called X-linked genes. Such genes are not on the Y chromosomes, therefore when doing X-linked genetics problems, it is the custom to indicate the sexes by using sex chromosomes and to indicate the alleles by superscripts attached to the X. The Y is blank because it does not carry these genes.

11. All the genes on a chromosome form a linkage group. Linked genes do not obey Mendel's laws because they tend to go into the gametes together. Crossing-over can cause recombinant gametes and recombinant phenotypes to occur. The percentage of recombinants is used to map the chromosomes.
12. Chromosome mutations fall into two categories: changes in chromosome number and changes in chromosome structure.
13. Changes in chromosome number are of two kinds: polyploidy and aneuploidy. Polyploidy occurs when the eukaryotic individual has more than two complete sets of chromosomes. Aneuploidy is an excess or deficiency in an individual chromosome.
14. Changes in chromosome structure include inversions, translocations, deletions, and duplications. Although these generally disrupt synapsis and segregation, there are instances in which they seem to have played a positive role in the evolutionary history of organisms.

Objective Questions

For questions 1 to 4, match the phrases below to the items in the key:

Key:
   a. multiple alleles
   b. incomplete dominance
   c. polygenes
   d. epistatic gene
   e. pleiotropic gene

1. A cross between oblong and round squash produced oval squash.

2. Although most people have an IQ of about 100, IQ generally ranges from about 50 to 150.
3. Investigators noted that whenever a particular species of plant had narrow instead of broad leaves, it was also short and yellow, instead of tall and green.
4. In rabbits, the allele C is dominant to the allele c^{ch}, which is dominant to the allele c^h, which is dominant to the allele c. Each individual has only two of these alleles.

5. When a white-eyed *Drosophila* female occurs
   a. both parental flies could have red eyes.
   b. the female parent could have red eyes but the male parent has to have white eyes.
   c. both the male and female parent must have white eyes.
   d. Both (a) and (b).

6. Investigators found that a cross involving the mutant genes *a* and *b* produced 30% recombinants, involving *a* and *c* produced 5% recombinants, involving *c* and *b* produced 25% recombinants. Which is the correct order of the genes?
   a. *a, b, c*
   b. *a, c, b*
   c. *b, a, c*
7. A boy is color blind (sex-linked recessive) and has a continuous hairline (autosomal recessive). Which could be the genotype of his mother?
   a. *bbww*
   b. X^bYWw
   c. bbX^wX^w
   d. X^BX^hWw
8. Which two of these chromosome mutations are most likely to occur when homologous chromosomes are undergoing synapsis?
   a. inversion and translocation
   b. deletion and duplication
   c. deletion and inversion
   d. duplication and translocation
9. Investigators do a dihybrid cross between two heterozygotes and get about a 3:1 ratio among the offspring. The reason must be due to
   a. polygenes.
   b. pleiotropic genes.
   c. linked genes.
   d. epistatic genes.
10. In snakes, the recessive genotype *cc* causes the animal to be albino despite the inheritance of the dominant allele (*B* = black). What would be the results of a cross between two snakes with the genotype *CcBB*?
    a. all black snakes
    b. all white snakes
    c. 9:3:3:1
    d. 3 black:1 albino

Additional Genetics Problems

1. Chickens that are homozygous for the frizzled trait have feathers that are weak and stringy. When raised at low temperatures, the birds have circulatory, digestive, and hormonal problems. What inheritance pattern explains these results?
2. Chickens having black feathers are crossed with chickens having white feathers and the result is chickens that appear to have blue feathers. What inheritance pattern explains these results and what key do you suggest for this cross?
3. In radish plants, the shape of the radish may be long (*LL*), round (*ll*), or oval (*Ll*). If oval is crossed with oval, what proportion of offspring would also be oval?
4. A man of blood type A and a woman of blood type B produce a child of type O. What are the genotypes of the man, woman, and child? If the couple were to have more children, what possible blood types could be produced?
5. In *Drosophila*, *S* = normal, *s* = sable body, *W* = normal, *w* = miniature wing. In a cross between a heterozygous normal fly and a sable-bodied, miniature-winged fly the results were 99 normal flies, 99 with a sable body and miniature wings, 11 with a normal body and miniature wings, and 11 with a sable body and normal wings. What inheritance pattern explains these results? How many map units separate the genes for sable body and miniature wing?
6. In *Drosophila*, the gene that controls red- (dominant) versus white-eye color is on the X chromosome. What are the expected phenotypic results if a heterozygous female is crossed with a white-eyed male?
7. Bar-eye in *Drosophila* is dominant and sex-linked. What phenotypic ratio is expected for reciprocal crosses between pure-breeding flies?
8. In humans, color blindness is a sex-linked, recessive trait, and widow's peak (autosomal) is dominant over continuous hairline. A man with normal vision and a widow's peak married a woman who is color blind and has a continuous hairline. The woman has a girl who has a widow's peak and is color blind. What are the genotypes of all persons involved? Is this man the girl's father? How do you know?
9. In cats, *S* = short hair, *s* = long hair, X^c = yellow coat, X^C = black coat, X^CX^c = tortoiseshell (calico) coat. If a long-haired, yellow male is crossed with a tortoiseshell female, homozygous for short hair, what will be the expected phenotypic results?

Study Questions

1. Explain inheritance by multiple alleles. List the human blood types and give the possible genotypes for each.
2. Compare the F_2 phenotypic ratio for a cross involving a completely dominant allele, an incompletely dominant allele, and codominant alleles.
3. Show that if a dihybrid cross between two heterozygotes involves an epistatic gene, you will not get a ratio of 9:3:3:1.
4. Explain why traits controlled by polygenes show continuous variation that can be measured quantitatively, have many intermediate forms and a distribution in the F_2 generation that follows a bell-shaped curve.
5. If you were studying a particular population of organisms, how might you recognize that several traits are being affected by a pleiotropic gene?
6. Give examples to show that the environment can have an extreme effect on the phenotype despite the genotype.
7. What is the chromosome theory of inheritance? List the ways in which genes and chromosomes behave similarly during meiosis and fertilization.
8. How is sex determined in humans? Which sex determines the sex of the offspring?

9. How did a *Drosophila* cross involving an X-linked gene help investigators show that certain genes are carried on certain chromosomes?

10. Show that a dihybrid cross between a heterozygote and recessive homozygote involving linked genes will not give the expected 1:1:1:1 ratio. What is the significance of the small percentage of recombinants that occurs among the offspring?

11. What are the two types of chromosome mutations? What two types of changes in chromosome number were discussed? What four types of changes in chromosome structure were discussed?

12. In what way may polyploidy have assisted plant evolution? Why don't you generally find polyploidy in animals? Why would you expect a monosomy to be more lethal than a trisomy?

Thought Questions

1. Morgan deduced that the white-eye allele is on the X chromosome of *Drosophila*. In what way did his procedure differ from the "standard" scientific method described in chapter 2?

2. Do you suppose it is possible that repeated chromosomal duplications could have brought about polygenic inheritance? Explain.

3. It's said that a pleiotropic gene affects more than one trait. Would it be more appropriate to say that a pleiotropic gene affects a single trait that has multiple effects? Why might the first description have arisen?

Selected Key Terms

multiple allele (mul'tı-p'l ah-lēl) 185
incomplete dominance (in''kom-plēt' dom'ı-nans) 185
codominance (ko-dom'ı-nans) 185
epistatic gene (epistasis) (ĕp''ĭ-stat'ik jēn) 186

polygenic inheritance (pol''ĕ-jēn'ik in-her'ĭ-tans) 186
pleiotropy (pli-ot'ro-pe) 187
chromosome theory of inheritance (kro'mo-sōm the'o-re uv in-her'ĭ-tans) 191
autosome (aw'to-sōm) 191

sex chromosome (seks kro'mo-sōm) 191
X-linked gene (eks linkt jēn) 191
carrier (kar'ē-er) 192
linkage group (lingk'ij grōōp) 193
polyploid (polyploidy) (pol'e-ploid) 196
nondisjunction (non''dis-junk'shun) 197

CHAPTER 14

Human Genetic Disorders

Your study of this chapter will be complete when you can:

1. Describe how a karyotype is prepared, what it consists of, and how it is used.
2. Explain what amniocentesis and chorionic villi testing are, and why they are sometimes recommended prior to the birth of a child.
3. Describe Down syndrome, its symptoms, its causes, and its relation to the age of the mother.
4. List the different types of sex chromosome aneuploidies seen in adult humans.
5. Name and describe three methods that assist researchers in mapping the human chromosomes.
6. Tell how to recognize three patterns of inheritance (autosomal dominant, autosomal recessive, X-linked recessive) when viewing a human pedigree chart.
7. Predict the chances that a couple will have a child with an autosomal recessive, autosomal dominant, and X-linked recessive disorder when supplied with their genotypes.

This child, who is participating in the Special Olympics, has Down syndrome. Investigators are only now discovering which genes on chromosome 21 bring about the characteristics of this inherited condition. The hope is that one day it will be possible to control the effects of the genes so that the characteristics will not appear.

The aim of those working in the field of human genetics is to predict, diagnose, treat, and perhaps even cure human genetic disorders. It's a lot easier to predict and diagnose an illness if you can connect a symptom with a cause. For example, measuring blood pressure and noting that it is high allows a physician to predict circulatory difficulties, and taking an X ray allows physicians to see that the source of great pain is kidney stones. In recent years, researchers have come closer and closer to this level of objectivity in the field of medical genetics.

The aim of medical genetics is to predict, diagnose, treat, and perhaps even cure human genetic disorders.

Before we begin, let's remember that the genes are composed of DNA (a sequence of nucleotides) and that DNA controls protein synthesis:

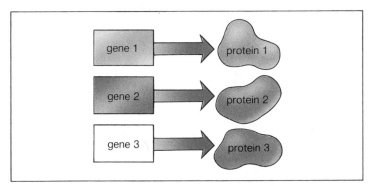

The diagram suggests that if protein 1 is abnormal, so is gene 1, and so forth. Therefore, removing proteins from the body and analyzing them for unusual characteristics is the second best thing to analyzing the genes themselves.

The genes are on the chromosomes, and therefore you would expect that some inherited disorders could be predicted by simply looking at the chromosomes. Let's start, then, with the chromosomes.

Human Chromosomes

To clearly view and count the chromosomes, a cell can be treated and photographed just prior to dividing, as described in figure 14.1. The chromosomes can then be cut out of the photograph and arranged in pairs. The members of a pair have the same size, shape (influenced by the position of the centromere) and banding pattern. In the resulting display of paired chromosomes, called a **karyotype,** autosomal chromosomes are usually ordered by size and numbered from largest to smallest; the sex chromosomes are identified separately. Karyotyping can even be done prior to birth to determine if a fetus has an abnormal chromosome number or structure (fig. 14.2).

Chromosome Abnormalities

Chromosomal abnormalities in humans tend to be either aneuploidies (changes in the number of individual chromosomes), or changes in chromosome structure.

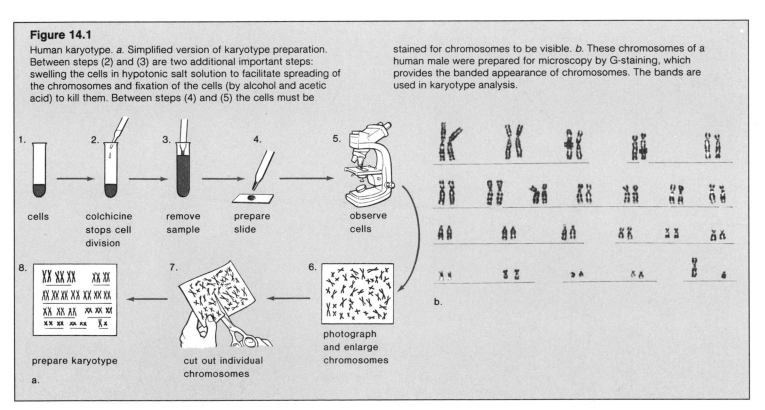

Figure 14.1

Human karyotype. *a.* Simplified version of karyotype preparation. Between steps (2) and (3) are two additional important steps: swelling the cells in hypotonic salt solution to facilitate spreading of the chromosomes and fixation of the cells (by alcohol and acetic acid) to kill them. Between steps (4) and (5) the cells must be stained for chromosomes to be visible. *b.* These chromosomes of a human male were prepared for microscopy by G-staining, which provides the banded appearance of chromosomes. The bands are used in karyotype analysis.

1. cells
2. colchicine stops cell division
3. remove sample
4. prepare slide
5. observe cells
6. photograph and enlarge chromosomes
7. cut out individual chromosomes
8. prepare karyotype

a.

b.

Figure 14.2

Detecting Birth Defects. A new method call chorionic villi sampling (*a*) allows physicians to collect human embryonic cells as early as the fifth week of pregnancy. The doctor inserts a long thin tube through the vagina into the uterus. With the help of ultrasound, which gives a picture of the uterine contents, the tube is placed between the lining of the uterus and the chorion. Then suction is used to remove a sampling of the chorionic villi cells. Chromosomal analysis and biochemical tests to detect some genetic defects can be done immediately on these cells. Before the development of chorionic villi sampling, physicians had to wait until about the sixteenth week of pregnancy to perform amniocentesis. In amniocentesis (*b*), a long needle is passed through the abdominal wall to withdraw a small amount of amniotic fluid along with fetal cells. Since there are only a few cells in the amniotic fluid, testing must be delayed until cell culture has caused them to grow and multiply. Therefore it may be another two to four weeks until the prospective parents are told whether their child has a genetic defect.

In fetoscopy, another possible procedure, the physician uses an endoscope to view the fetus and can withdraw chorionic villi cells or blood for prenatal diagnosis.

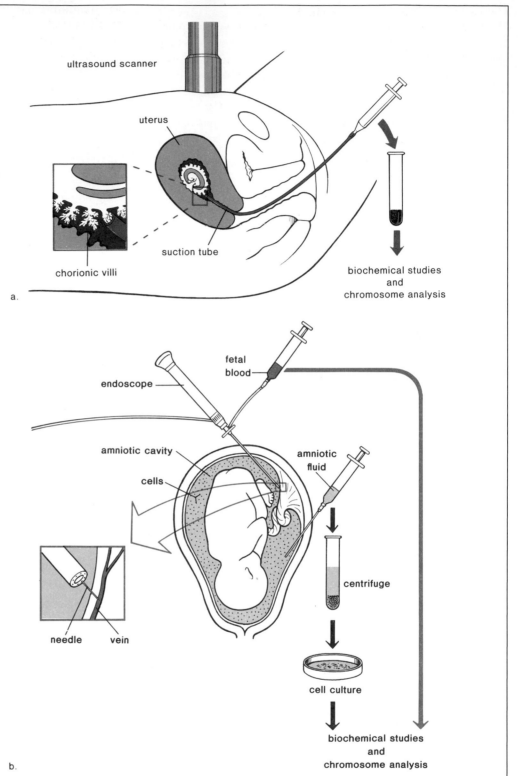

Down Syndrome

The most common autosomal aneuploidy (table 14.1) in humans is trisomy 21, also called **Down syndrome,** because it was first described by John L. Down in 1866. This syndrome (fig. 14.3) is generally characterized by mental retardation, a distinctive palm print, and a common facial appearance that includes an extra fold of the eyelids. Although most persons with Down syndrome live only until the middle teens, some live much longer and are able to become partially independent. There is no cure possible for Down syndrome but it is hoped that eventually there may be a treatment for some of its symptoms.

Persons with Down syndrome usually have three copies of chromosome 21, because nondisjunction (fig. 13.15) resulted in a gamete with two copies instead of one. Chance of nondisjunction increases rapidly with age starting at about age 35 in women (fig. 14.4) and age 55 in men. The likelihood of nondisjunction beginning at a younger age in women can be explained

Table 14.1 Frequency and Effects of the Most Common Aneuploidies in Humans

Syndrome	Sex	Chromosomes	Frequency		Expected Viability and Fertility
			Abortuses	Births	
Down	M or F	Trisomy 21	1/40	1/700	15 years, rarely reproductive
Patau	M or F	Trisomy 13	1/33	1/15,000	< 6 months
Edward	M or F	Trisomy 18	1/200	1/5,000	< 1 year
Turner	F	XO	1/18	1/5,000	Sterile
Metafemale	F	XXX (or XXXX)	0	1/700	Limited
Klinefelter	M	XXY (or XXXY)	0	1/2,000	Sterile
XYY	M	XYY	?	1/2,000	—

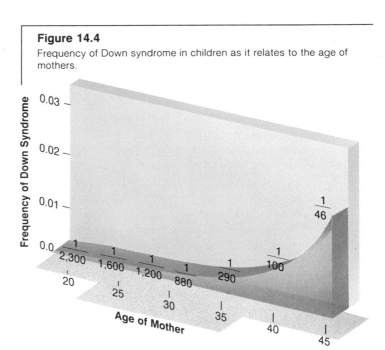

Figure 14.4

Frequency of Down syndrome in children as it relates to the age of mothers.

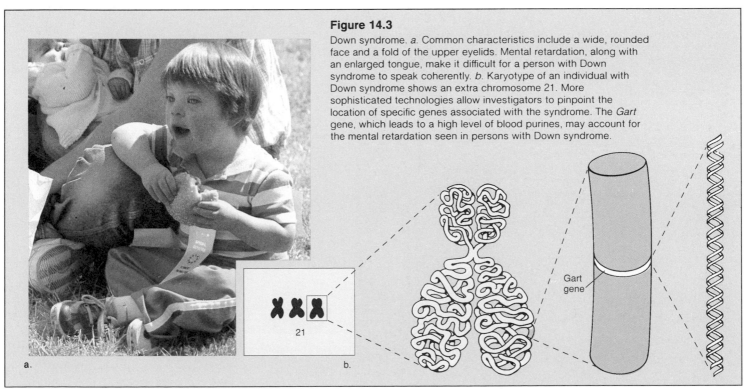

Figure 14.3

Down syndrome. *a.* Common characteristics include a wide, rounded face and a fold of the upper eyelids. Mental retardation, along with an enlarged tongue, make it difficult for a person with Down syndrome to speak coherently. *b.* Karyotype of an individual with Down syndrome shows an extra chromosome 21. More sophisticated technologies allow investigators to pinpoint the location of specific genes associated with the syndrome. The *Gart* gene, which leads to a high level of blood purines, may account for the mental retardation seen in persons with Down syndrome.

Hagelston/Leggitt

in this manner. Even at birth, the female ovary contains all the eggs that will ever become mature. These eggs are kept in a state of suspended animation and do not undergo complete meiosis unless the egg is fertilized by a sperm. Because the eggs age as the female ages, the probability that the chromosomes will fail to divide normally apparently increases with time. Amniocentesis followed by chromosome analysis can indicate whether an unborn child has Down syndrome, and therefore this procedure should be done whenever the expectant mother is older than 35 and/or her spouse is older than 55.

Down syndrome is always due to an extra chromosome 21 but in about 5% of cases the extra chromosome is attached to another chromosome, often chromosome 14. This abnormal chromosome arose because a translocation occurred between chromosome 14 and 21 in one of the parents, or even generations earlier. Therefore, when Down syndrome is due to a translocation, it is not age related, and instead it tends to "run in the family" of either the father or the mother.

It is known that the genes causing Down syndrome are located on the lower third of chromosome 21 (fig. 14.3b), and there has been a lot of research to discover which specific genes are responsible for the characteristics of the syndrome. Thus far, researchers have discovered several genes that may account for various conditions seen in persons with Down syndrome. For example, they have located the genes most likely responsible for an increased tendency toward leukemia, cataracts, Alzheimer's disease, and mental retardation. The latter gene, dubbed the *Gart* gene, causes an increased level of purines in the blood, a finding that is associated with mental retardation. It is hoped to someday find a way to control the expression of the *Gart* gene before birth, so that at least this particular symptom of Down syndrome will not appear.

Investigators are beginning to identify the genes whose presence in triplicate cause the symptoms of Down syndrome (trisomy 21).

Other Autosomal Abnormalities

You'll recall that a chromosome deletion is responsible for a syndrome known as cri du chat (p. 198), in which the infant has a cry that sounds like a cat and shows other abnormalities, including mental retardation. Chromosome analysis shows that a portion of chromosome 5 is missing (deleted), although the other chromosome 5 is normal.

Two other trisomies, besides that of chromosome 21, are seen with some regularity in humans. Individuals with trisomy 13 (Patau syndrome) and trisomy 18 (Edward syndrome) have an average lifespan of less than six months. Heart and nervous system defects prevent normal development.

Sex Chromosome Aneuploidies   Sex chromosome aneuploidies (table 14.1) are due to nondisjunction in the sex chromosomes. Nondisjunction of sex chromosomes during oogenesis can lead to an egg with either two X chromosomes or no X chromosome. Nondisjunction of the sex chromosomes during spermatogenesis can result in a sperm that has no sex chromosome, both an X and a Y chromosome, two X chromosomes, or two Y chromosomes. Assuming that the other gamete is normal, the zygote could develop into an individual with one of the conditions noted in figure 14.5.

Turner syndrome is a monosomy of the X chromosome; these individuals are designated as XO—the O signifies the absence of a second sex chromosome. Turner females are short, have a broad chest, and may have congenital heart defects. Because the ovaries never become functional, Turner females do not undergo puberty or menstruate, and there is a lack of breast

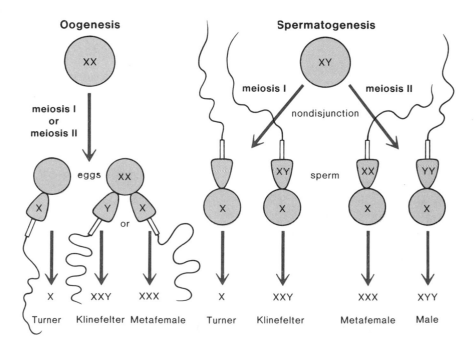

Figure 14.5

(*Left*) nondisjunction of sex chromosomes during oogenesis followed by fertilization with normal sperm results in the conditions noted. (*Right*) nondisjunction of sex chromosomes during spermatogenesis followed by fertilization of normal eggs results in the conditions noted.

development (fig. 14.6). Although no overt mental retardation is reported, Turner females show reduced skills in interpreting spatial relationships.

When an egg having two X chromosomes is fertilized by an X-bearing sperm, a **metafemale** having three X chromosomes results. It might be supposed that the XXX female would be especially feminine, but this is not the case. Although there is in some cases a tendency toward learning disabilities, most metafemales have no apparent physical abnormalities except they may have menstrual irregularities including early onset of menopause.

When an egg having two X chromosomes is fertilized by a Y-bearing sperm, a male with **Klinefelter syndrome** (XXY) results. Affected individuals are males, but the testes are underdeveloped and there may be some breast development (fig. 14.6). These phenotypic abnormalities are not apparent until puberty, although some evidence of subnormal intelligence may be apparent before this time.

XYY males may also result from nondisjunction during spermatogenesis. Affected males are usually taller than average, suffer from persistent acne, and tend to have barely normal intelligence. At one time, it was suggested that these men were likely to be criminally aggressive, but it has since been shown that the incidence of such behavior among them is no greater than among XY males.

One sex chromosome monosomy (Turner syndrome) and three trisomies (Klinefelter, metafemale, and XYY) are known in human beings.

Examination of these abnormal sex chromosome patterns indicates how sex is determined in humans. Turner syndrome (XO) produces a person who is more female than male. Klinefelter syndrome (XXY) produces a person who is more male than female. Clearly, the Y chromosome is not neutral in humans. Rather, it seems to bear genes that cause maleness, as discussed in the reading on page 210.

Mapping the Human Chromosomes

It is not too difficult to connect a disorder to an X chromosome or to a chromosome present in excess (e.g., realizing that Down syndrome is caused by trisomy 21), but how do you determine which particular genes are at fault? Similarly, recombinant offspring will allow you to determine linkage groups, but you cannot ask humans to reproduce just to study the incidence of crossing-over. So, it is not surprising that in the early 1970s only about 70 genes had been assigned to the human X chromosome and very few autosomal genes had been assigned to a specific human chromosome.

Human-Mouse Hybrid Cell Data

The groundwork for change began when techniques were developed for growing mammalian cells in culture. In the 1960s, it was discovered that human and mouse cells, when cultured together, may occasionally fuse to form a "cell hybrid" that contains the chromosomes from both species (fig. 14.7). For reasons still not understood, as the cells grow and divide, some of the human chromosomes are lost and eventually there are clones of daughter cells that contain only a few human chromosomes, each of which can be recognized by their distinctive banding pattern (fig. 14.1b). The specific human chromosomes that are retained is random. When the cells with only a few human chromosomes in them are collected, you can analyze for human proteins to determine which genes are associated with which chromosome. Sample data that show this matching process are given in figure 14.7.

Sometimes it is possible to obtain a hybrid cell that contains only one human chromosome or even a portion of a chromosome. This technique has been very helpful to those researchers who have been studying the genes that are located on chromosome 21 (fig. 14.3).

Genetic Marker Data

A new mapping technique works directly with DNA and uses **genetic markers** to tell if an individual has a defective allele. The exact location of the defective allele is not known, but the

Figure 14.6
a. A male with Klinefelter syndrome (XXY) has immature sex organs and shows breast development. b. A female with Turner syndrome (XO) is distinguished by a thick neck, short stature, and immature sexual features.

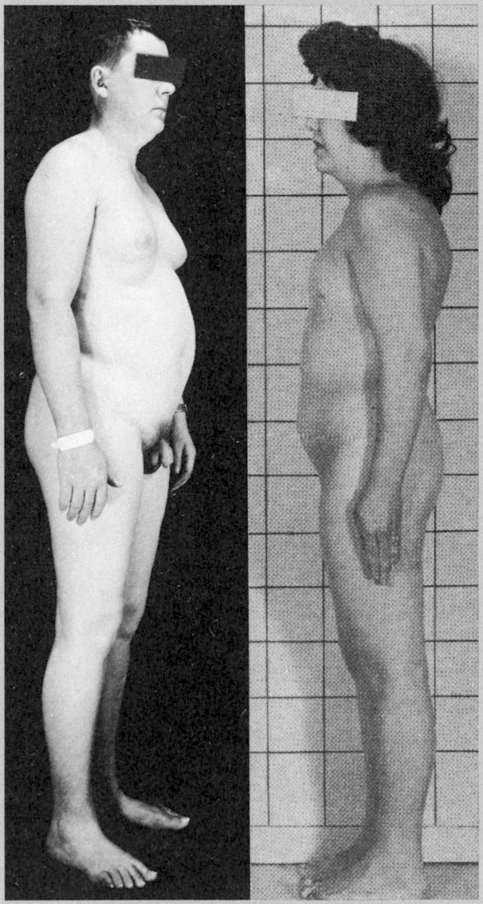

Figure 14.7

a. Mapping the human chromosomes with the help of numan-mouse hybrid cells. In the presence of a fusing agent, human fibroblast cells sometimes join with mouse tumor cells to give hybrid cells, whose nucleus contains both types of chromosomes. Cell division of the hybrid cell produces cells that have lost most of their human chromosomes. After a time this process stops and cell division then produces a clone of identical cells, which are examined. Each clone has a different complement of human chromosomes, allowing the investigator to study these chromosomes separately from all other human chromosomes. *b.* Sample data that are collected when cloned cells are analyzed for their protein content. It is important that these human chromosomes be distinguishable from mouse proteins. You will note that protein a is found in cloned cell 4 that contains chromosome 17 but does not contain the X chromosome. Therefore the gene coding for this protein must be on chromosome 17. Because proteins b and c are always present when the X

chromosome is present, the genes that code for these proteins must be on the X chromosome. Protein d is absent, even when both chromosomes are present (clones 2,3,5), indicating that the gene coding for this protein is not on either of these chromosomes.

Human Protein or Chromosome	Hybrid Cells				
	Clone 1	Clone 2	Clone 3	Clone 4	Clone 5
protein a	+	+	+	+	+
protein b	−	+	+	+	+
protein c	−	+	+	−	+
protein d	+	−	−	+	−
chromosome 17	+	+	+	+	+
X chromosome	−	+	+	−	+

b.

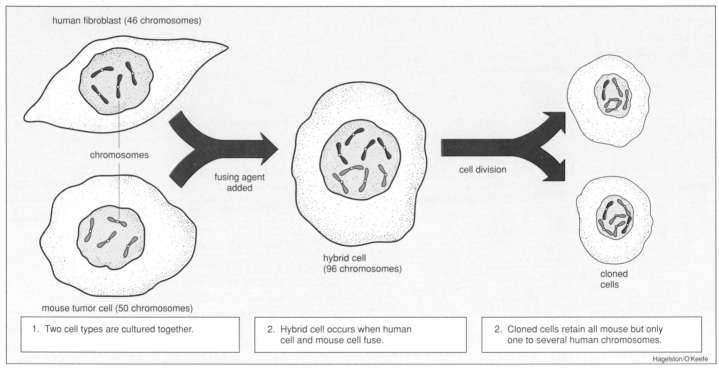

human fibroblast (46 chromosomes)

chromosomes

mouse tumor cell (50 chromosomes)

fusing agent added

hybrid cell (96 chromosomes)

cell division

cloned cells

1. Two cell types are cultured together.

2. Hybrid cell occurs when human cell and mouse cell fuse.

2. Cloned cells retain all mouse but only one to several human chromosomes.

Hagelston/O'Keefe

a.

close proximity of the allele and marker can be assumed because they are almost always inherited together. For a marker to be dependable, it should be inherited with the defective allele at least 98% of the time.

Detection of genetic markers makes use of special bacterial enzymes, called restriction enzymes,[1] each of which cuts the DNA strand at a specific nucleotide sequence, producing a particular pattern of DNA fragments. A marker, which is really a DNA mutation, alters the normal pattern of DNA fragments resulting from restriction enzyme use. Notice in figure 14.8 the different sizes of the fragments—the polymorphism that exists—in the normal and the affected individual. Scientists refer to the differences in the observed fragment lengths as "restriction

fragment length polymorphisms (RFLPs)." They usually discover these polymorphisms by comparing the DNA fragment pattern in a large number of closely related normal and affected individuals. The current tests for sickle-cell anemia, Huntington disease, and Duchenne muscular dystrophy are all based on the presence of a marker.

DNA Probe Data

On occasion, it has been possible to determine the unique sequence of nucleotides in a gene. Then a known radioactive sequence of nucleotides can be prepared in the laboratory and used as a probe. A **DNA probe** seeks the chromosome, among any others, that shares the same sequence of nucleotides and binds to it. The chromosomes can then be exposed to a photographic film that shows this pair bound together.

[1]These enzymes are called restriction enzymes because bacteria use them to restrict the growth of viruses whenever viral DNA enters bacteria. Scientists extract the enzymes from bacterial cells and use them whenever they wish to cleave DNA in the laboratory. See also chapter 18.

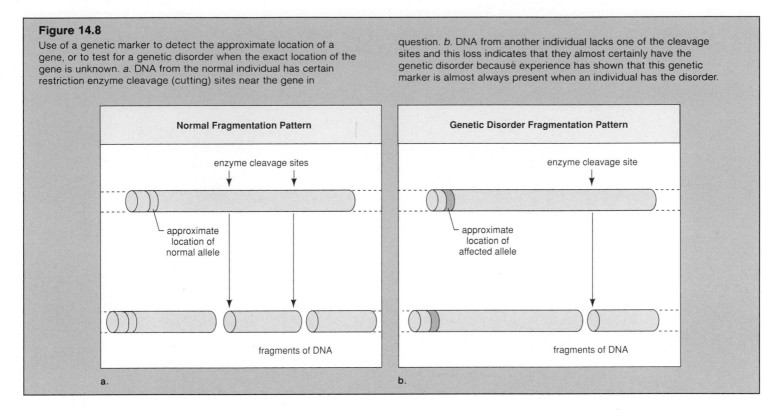

Figure 14.8

Use of a genetic marker to detect the approximate location of a gene, or to test for a genetic disorder when the exact location of the gene is unknown. *a.* DNA from the normal individual has certain restriction enzyme cleavage (cutting) sites near the gene in question. *b.* DNA from another individual lacks one of the cleavage sites and this loss indicates that they almost certainly have the genetic disorder because experience has shown that this genetic marker is almost always present when an individual has the disorder.

DNA probes, when available, more efficiently map chromosomes, and even test whether an individual is a carrier for, or is affected by a genetic disorder. This type of test is expected one day to replace those based on genetic markers previously discussed.

Several new techniques—human-mouse hybrid cell preparations and use of genetic markers and probes—are now making it possible to map the human chromosomes at a faster rate than was formerly possible.

Genetic Disorders

Most human birth defects are at least partially genetic in origin. Some of these disorders are controlled by a single set of alleles and therefore they are inherited in a simple Mendelian manner.

When studying human genetic disorders, biologists often construct pedigree charts that show the pattern of inheritance of a characteristic within a family. Let's contrast two possible patterns of inheritance in order to show how it is possible to determine whether the characteristic is an autosomal dominant or recessive characteristic:

pattern I   pattern II

In both patterns, males are designated by squares and females by circles. Shaded circles and squares are affected individuals. A line between a square and a circle represents a marriage. A vertical line going downward leads, in these patterns, to a single child. (If there is more than one child they are placed off a horizontal line.) Which pattern of inheritance do you suppose represents an autosomal dominant and which represents a recessive characteristic?

In Pattern I, the child is affected but neither parent is; this can happen if the characteristic is recessively inherited. What are the chances that any offspring from this union will be affected? Because the parents are heterozygous, the chances are 1 in 4, or 25% (see table 12.3). Notice that the parents could also be called **carriers** because they have a normal phenotype, but are capable of having a child with a genetic disorder. (See figure 14.9 for other ways to recognize a recessive pattern of inheritance.)

In Pattern II, the child is affected, as is one of the parents. When a characteristic is dominant, an affected child usually has at least one affected parent. Of the two patterns, this one shows a dominant pattern of inheritance. What are the chances that any offspring from this union will be affected? Because this is a heterozygote and homozygous recessive cross, the chances are 50% (see table 12.3). (See figure 14.10 for other ways to recognize a dominant pattern of inheritance.)

Autosomal Recessive Disorders

There are many autosomal recessive disorders and only three of the better known will be discussed. Others that are well known are albinism (p. 186); galactosemia (accumulation of galactose in the liver and mental retardation); thalassemia (production of

Active Y Chromosomes and Inactive X Chromosomes: The Functions of Sex Chromosomes in Body Cells

Although the Y chromosome in mammals has few gene loci, it has a powerful effect on sex determination. For years, it has been proposed that there must be a gene located on the Y chromosome that brings about maleness. Embryos begin life with no evidence of their sex, but about the third month of development, males can be distinguished from females. Researchers have recently reported finding a gene called testis determining factor gene (TDF) on the Y chromosome. In other words, the embryo is basically female and will automatically develop into a female unless TDF promotes the development of testes instead of ovaries. The importance of TDF is also exemplified by the fact that when the gene is lacking from the Y chromosome, the individual is a female even though the chromosomal inheritance is XY. On the other hand, if the gene is present in an XX individual due to translocation, this person is a male.

Another allele that has been found on the Y chromosome is called H-Y+. This allele directs the synthesis of a specific kind of protein molecule, called the H-Y antigen, that is present in the membranes of virtually all cells of a male, but not in those of a female. It is called an antigen because females produce antibodies against it. To test for maleness, it is possible to suspend a sample of white blood cells in a solution that contains some of these antibodies. If the cells carry H-Y antigen, indicating the person is a male, the antibodies bind with them.

Aside from identifying gene loci on the sex chromosomes that cause a person to be male or female, biologists have been interested in how males manage with only one X chromosome while females have two X chromosomes. That question turned out to have a rather unexpected answer.

Years ago, M. L. Barr observed a consistent difference between nondividing cells taken from female and male mammals, including humans. Females, but not males, have a small, darkly staining mass of condensed chromatin adhering to the inner edge of the nuclear envelope (part *a* in the figure). This dark-staining spot is called a Barr body after its discoverer.

Barr body

a.

a. Electron micrograph illustrates that a Barr body is a dark-staining spot that can be viewed microscopically. *b.* The coat of a calico cat is orange and black. The alleles for coat color occur at a gene locus on the X chromosome. Presumably, the black patches occur where cells have Barr bodies carrying the allele for

b.

orange, and the orange patches occurs where cells have Barr bodies carrying the allele for black coat. The white areas are due to a separate gene.

In 1961, Mary Lyon, a British geneticist proposed that the Barr body is a condensed, inactive X chromosome. Her hypothesis has been tested and found to be valid. One of the X chromosomes is inactivated in the cells of female embryos, but which one of the two is determined by chance. About 50% of the cells have one X chromosome active and 50% have the other X chromosome active. The female body, therefore, is a mosaic, with "patches" of genetically different cells. For example, human females who are heterozygous for an X-linked recessive form of ocular albinism, have patches of pigmented and nonpigmented cells at the back of the eye. Women heterozygous for Duchenne muscular dystrophy have patches of normal muscle tissue and degenerative muscle tissue (the normal tissue increases in size and strength to make up for the defective tissue), and women who are heterozygous for

hereditary absence of sweat glands have patches of skin lacking sweat glands. Inactivation of an X chromosome does not occur in female oocytes, most likely because both are needed for normal development of the immature egg.

The female calico cat (part *b*) also provides phenotypic proof of the Lyon hypothesis. In these cats, an allele for black coat color is on one X chromosome and a corresponding allele for orange coat color is carried on the other X chromosome. The patches of black and orange in the coat can be related to which X chromosome is in the Barr bodies of the cells found in the patches.

The existence of Barr bodies also explains why metafemales and persons with Klinefelter syndrome don't show a greater degree of abnormality than they do. The extra X chromosomes form Barr bodies.

abnormal amounts of hemoglobin); and xeroderma pigmentosum (lack of ability to repair ultraviolet-induced damage). The homozygous recessive phenotype is more likely to occur among a group of people that tend to marry each other, which may explain why these disorders are sometimes more prevalent among members of a particular ethnic group.

Cystic Fibrosis

Cystic fibrosis is the most common lethal genetic disease among Caucasians in the United States. About 1 in 20 Caucasians is

a carrier and about 1 in 2,000 children born to this group has the disorder. In these children the mucus in the lungs and digestive tract is particularly thick and viscous. In the digestive tract, the thick mucus impedes the secretion of pancreatic juices, and food cannot be properly digested; large, frequent, and foul-smelling stools occur. A few individuals have been known to survive childhood, but most die from recurrent lung infections.

In the past few years, much progress has been made in our understanding of cystic fibrosis. First of all, it was discovered that chloride ions fail to pass through plasma membrane

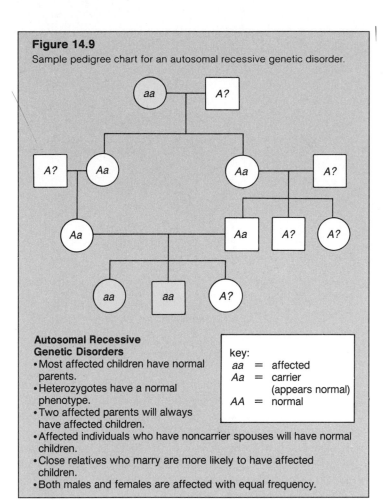

Figure 14.9
Sample pedigree chart for an autosomal recessive genetic disorder.

Autosomal Recessive Genetic Disorders
- Most affected children have normal parents.
- Heterozygotes have a normal phenotype.
- Two affected parents will always have affected children.
- Affected individuals who have noncarrier spouses will have normal children.
- Close relatives who marry are more likely to have affected children.
- Both males and females are affected with equal frequency.

key:
aa = affected
Aa = carrier (appears normal)
AA = normal

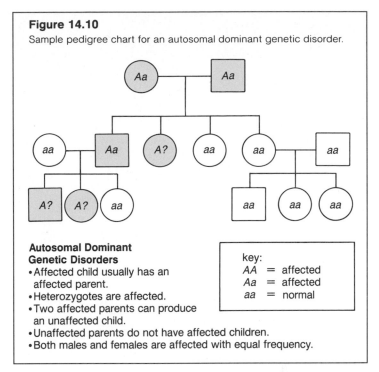

Figure 14.10
Sample pedigree chart for an autosomal dominant genetic disorder.

Autosomal Dominant Genetic Disorders
- Affected child usually has an affected parent.
- Heterozygotes are affected.
- Two affected parents can produce an unaffected child.
- Unaffected parents do not have affected children.
- Both males and females are affected with equal frequency.

key:
AA = affected
Aa = affected
aa = normal

channels in these patients. Ordinarily, after chloride has passed through the membrane, water follows. It is believed that lack of water in the lungs causes the mucus to be so thick. Secondly, the gene for cystic fibrosis which is located on chromosome 7 has been isolated and soon we should know how its product functions.

Tay-Sachs

Tay-Sachs is the best known genetic disease among U.S. Jewish people, most of whom are of central and eastern European descent. At first, it is not apparent that a baby has Tay-Sachs. However, development begins to slow down between four and eight months of age, as neurological impairment and psychomotor difficulties become apparent. Ophthalmologic examination reveals a characteristic red spot and yellowish accumulation in the region of the retina called the fovea. The child gradually becomes blind and helpless, develops uncontrollable seizures, and eventually becomes paralyzed. There is no treatment or cure for Tay-Sachs disease and most affected individuals die by the age of three or four.

Tay-Sachs disease results from a lack of the enzyme hexosaminidase A (Hex A) and the subsequent storage of its substrate, a glycosphingolipid, in lysosomes. Although more and more lysosomes build up in many body cells (fig. 14.11), the primary sites of storage are the cells of the nervous system, which accounts for the onset and progressive deterioration of psychomotor functions.

There is a test to detect carriers of Tay-Sachs. The test uses a sample of serum, white blood cells, or tears to determine whether Hex A activity is present. Affected individuals have no detectable Hex A activity. Carriers have about half the level of Hex A activity found in normal individuals. Prenatal diagnosis is also possible following either amniocentesis or chorionic villi sampling.

Phenylketonuria (PKU)

Phenylketonuria (PKU) occurs in 1 in 20,000 births and so is not as frequent as the disorders previously discussed. When it does occur, the parents are very often close relatives. Affected individuals lack an enzyme that is needed for the normal metabolism of the amino acid phenylalanine and an abnormal breakdown product, a phenylketone, accumulates in the urine. Newborns are routinely tested, and if they lack the necessary enzyme, they are placed on a diet low in phenylalanine. This diet must be continued until the brain is fully developed or else severe mental retardation develops. If a woman who is homozygous recessive for PKU wishes to have a normal child, she should resume her limited diet several months before getting pregnant; otherwise, she runs a high risk of having a microcephalic child.

Autosomal Dominant Disorders

There are many autosomal dominant disorders and only two of the better known will be discussed below. Others that are well known are Marfan syndrome (p. 187); achondroplasia (dwarfism); brachydactyly (abnormally short fingers); porphyria (inability to metabolize porphyrins from hemoglobin breakdown); and hypercholesterolemia (elevated levels of cholesterol in blood).

Figure 14.11

Tay-Sachs disease. *a.* When Hex A is present, glycosphingolipids are broken down. *b.* When Hex A is absent, these lipids accumulate in lysosomes and lysosomes accumulate in the cell. *c.* Electron micrograph of cell crowded with lysosomes.

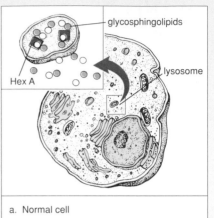

a. Normal cell

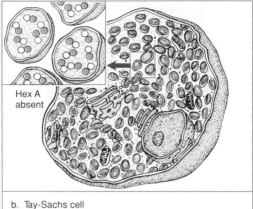

b. Tay-Sachs cell

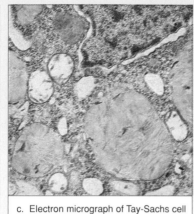

c. Electron micrograph of Tay-Sachs cell

Neurofibromatosis (NF)

Neurofibromatosis (NF), sometimes called von Recklinghausen disease, is one of the most common genetic disorders, affecting roughly 1 in 3,000 people, including an estimated 100,000 in the United States. It is seen equally in every racial and ethnic group throughout the world.

At birth, or later, the affected individual may have six or more large tan spots on the skin. Such spots may increase in size and number and get darker. Small benign tumors (lumps) called neurofibromas may occur under the skin, or deeper. Neurofibromas are made up of nerve cells and other cell types.

This genetic disorder shows *variable expressivity;* in most cases, symptoms are mild, and patients live a normal life. In some cases, however, the effects are severe. Skeletal deformities, including a large head are seen, and eye and ear tumors can lead to blindness and hearing loss. Many children with NF have learning disabilities and may be overactive.

Only recently, researchers have been able to determine that the gene responsible for the disorder is located on chromosome 17. They believe the gene is rather large, because of its varying effects and because about half of all NF cases are the result of new mutations in one of the parents.

Huntington Disease

As many as 1 in 10,000 persons in the United States have **Huntington disease (HD),** a neurological disorder that affects specific regions of the brain. Most individuals who inherit the allele appear normal until middle age. Then, minor disturbances in balance and coordination lead to progressively worse neurological disturbances until the victim becomes insane before death occurs.

Much has been learned about Huntington disease. The gene is located on chromosome 4, and there is a test, of the type described in figure 14.8, to determine if the dominant gene has been inherited. Because treatment is not available, however, few may want to have this information.

Research is being conducted, though, to determine the underlying cause of the disorder. It is known that the brain of a Huntington victim produces more than the usual amount of quinolinic acid, an excitotoxin that can overstimulate certain nerve cells. It is believed to lead to the death of these cells and the subsequent symptoms of Huntington disease. Researchers are looking for chemicals that can block quinolinic acid's action or inhibit quinolinic acid synthesis.

Some of the best known genetic disorders in humans are inherited in a simple Mendelian manner. Pedigree charts show the pattern of inheritance in a particular family.

Degrees of Dominance

The field of human genetics also has examples of incomplete dominance and codominance. For example, when a curly-haired Caucasian person reproduces with a straight-haired Caucasian person, their children will have wavy hair. We have already mentioned that the multiple alleles controlling blood type are codominant. An individual with the genotype $I^A I^B$ has the blood type AB. Skin color, you'll recall, is controlled by polygenes, and therefore it is possible to observe a range of skin colors among the members of a population in which Caucasians and blacks reproduce with one another.

Sickle-Cell Anemia

Sickle-cell anemia is an example of a human disorder that is controlled by incompletely dominant alleles. Individuals with the genotype $Hb^A Hb^A$ are normal, those with the $Hb^S Hb^S$ genotype have sickle-cell anemia, and those with the $Hb^A Hb^S$ genotype have **sickle-cell trait,** a condition in which the cells are sometimes sickle shaped. Two individuals with sickle-cell trait can produce children with all three phenotypes, as indicated in figure 14.12.

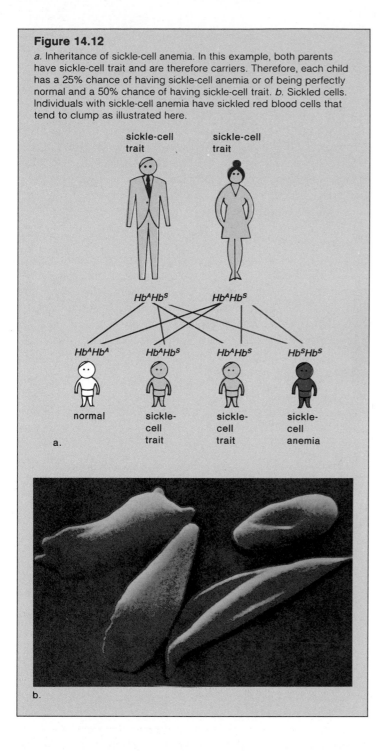

Figure 14.12

a. Inheritance of sickle-cell anemia. In this example, both parents have sickle-cell trait and are therefore carriers. Therefore, each child has a 25% chance of having sickle-cell anemia or of being perfectly normal and a 50% chance of having sickle-cell trait. *b.* Sickled cells. Individuals with sickle-cell anemia have sickled red blood cells that tend to clump as illustrated here.

sickle-cell trait

sickle-cell trait

Hb^AHb^S Hb^AHb^S

Hb^AHb^A Hb^AHb^S Hb^AHb^S Hb^SHb^S

normal   sickle-cell trait   sickle-cell trait   sickle-cell anemia

a.

b.

Among black Africans, the sickled cells seem to give protection against the malaria parasite, which uses red blood cells during its life cycle. Although infants with sickle-cell anemia often die, those with sickle-cell trait are protected from malaria, especially from ages two to four. This means that in Africa these children survive and grow to reproduce and pass on the allele to their offspring. As many as 60% of blacks in malaria-infected regions of Africa have the allele. In the United States, about 10% of the black population carries the allele. There is a test that can be done to detect its presence and prenatal testing is also possible.

The red blood cells in persons with sickle-cell anemia cannot easily pass through small blood vessels. The sickle-shaped cells either break down or they clog blood vessels, and the individual suffers from poor circulation, anemia, and sometimes internal hemorrhaging. Jaundice, episodic pain in the abdomen and joints, poor resistance to infection, and damage to internal organs are all symptoms of sickle-cell anemia.

Persons with sickle-cell trait do not usually have any difficulties, unless they undergo dehydration or mild oxygen deprivation. At such time, the cells become sickle shaped, clogging their blood vessels and leading to pain and even death. A study of the occurrence of sudden deaths during army basic training showed that a person with sickle-cell trait was 40 times more likely to die, compared to normal recruits. This has caused the Army Medical Corps to advise drill instructors to train recruits more gradually, to give them enough to drink, and to make allowances for heat and humidity when planning their workouts.

Sickle-cell anemia is an example of a human genetic disorder that is controlled by incompletely dominant alleles.

X-Linked Genetic Disorders

A number of genetic disorders are controlled by X-linked genes and three of the better known will be discussed below. Others that are well known are Lesch-Nyhan syndrome (compulsive self-mutilation and mental retardation); one type of manic depressive psychosis (mood swings from extreme elation to severe depression); and ichthyosis (skin disorder that causes scaling).

Color Blindness

In humans there are three genes involved in distinguishing color, one for each of the three different types of cones, which are the receptors for color vision. Two of these genes are X-linked, one affects the green-sensitive cones and the other affects the red-sensitive cones. About 6% of men in the United States are color blind due to a recessive mutation involving green perception and about 2% are color blind due to a mutation involving red perception. Figure 14.13 pertains to the more prevalent form of color blindness.

Hemophilia

There are about 100,000 hemophiliacs in the United States. Most have hemophilia A, caused by the absence or minimal presence of a particular clotting factor called factor VIII. **Hemophilia** is called the bleeder's disease because the affected person's blood does not clot. Not only do hemophiliacs bleed externally after an injury, but they also suffer from internal bleeding, particularly around joints. Hemorrhages can be stopped by transfusions of fresh whole blood (or plasma) or concentrates of the clotting protein. Unfortunately, some hemophiliacs have contracted AIDS from the use of whole blood instead of a purified form of the concentrate.

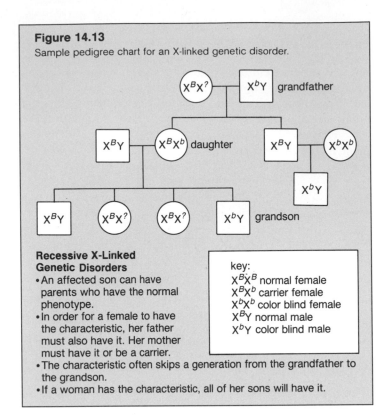

Figure 14.13

Sample pedigree chart for an X-linked genetic disorder.

Recessive X-Linked Genetic Disorders

- An affected son can have parents who have the normal phenotype.
- In order for a female to have the characteristic, her father must also have it. Her mother must have it or be a carrier.
- The characteristic often skips a generation from the grandfather to the grandson.
- If a woman has the characteristic, all of her sons will have it.

key:
$X^B X^B$ normal female
$X^B X^b$ carrier female
$X^b X^b$ color blind female
$X^B Y$ normal male
$X^b Y$ color blind male

At the turn of the century, hemophilia was common in the royal families of Europe, and all of the affected males could trace their ancestry to Queen Victoria of England (fig. 14.14). Because none of the queen's ancestors or relatives was affected, it seems that the recessive allele she carried arose by mutation either in Victoria or one of her parents. Her daughters Alice and Beatrice were carriers and introduced the allele into the ruling houses of Russia and Spain. Alexis, the last heir to the Russian throne before the Russian Revolution, was a hemophiliac. The British royal family has no hemophiliacs, because Victoria's eldest son, King Edward VII, did not receive the allele, and therefore could not pass it on to any of his descendants.

Muscular Dystrophy

This disorder, as the name implies, is characterized by a wasting away of the muscles. The most common form, **Duchenne muscular dystrophy,** is an X-linked recessive disorder that occurs in about 1 in 25,000 male births. Symptoms, such as waddling gait, toe walking, frequent falls, and difficulty in rising may appear as soon as the child starts to walk. Muscle weakness intensifies, until the individual is confined to a wheelchair. Death usually occurs during the teenage years.

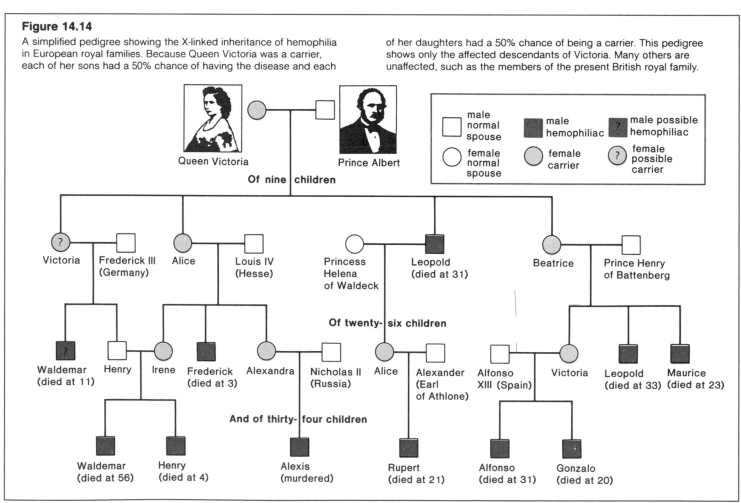

Figure 14.14

A simplified pedigree showing the X-linked inheritance of hemophilia in European royal families. Because Queen Victoria was a carrier, each of her sons had a 50% chance of having the disease and each of her daughters had a 50% chance of being a carrier. This pedigree shows only the affected descendants of Victoria. Many others are unaffected, such as the members of the present British royal family.

Recently, the gene for muscular dystrophy was isolated and it was discovered that the absence of a protein, now called dystrophin, is the cause of the disorder. Much research determined that dystrophin is involved in the release of calcium from the calcium storage sacs in muscle fibers. The lack of dystrophin causes a leakage of calcium into the cell, which promotes the action of an enzyme that dissolves muscle fibers (*see* chapter 40). When the body attempts to repair the tissue, fibrosis occurs, and cuts off the blood supply so that more and more cells die.

There is now a test to detect carriers for Duchenne muscular dystrophy.

Sex-Influenced Traits

Some genes not located on the X or Y chromosomes are expressed differently in the two sexes, and therefore they are referred to as **sex-influenced traits.** Pattern baldness (fig. 14.15) is caused by an autosomal allele that is dominant in males due to the presence of testosterone, the male sex hormone. Heterozygous women with adrenal tumors develop pattern baldness, but hair returns when the tumor is removed. This hair loss can be explained by the fact that an abnormally large amount of testosterone is produced by the malfunctioning adrenal gland. The adrenal glands normally produce some testosterone, even in women.

Another sex-influenced trait is index finger length. An index finger equal to or longer than the fourth finger is dominant in females but recessive in males.

Sex-linked traits are observed in humans. Both color blindness and hemophilia are X-linked recessive traits. Sex-influenced traits such as pattern baldness are also observed.

The Future

In this chapter we have indicated types of research that have increasingly enabled us to predict, diagnose, and treat human genetic disorders. As mentioned previously, it is hoped that one day it will be possible to cure such disorders. Steps toward this end have been taken and will be outlined in chapter 18.

Figure 14.15

Pattern baldness is a sex-influenced characteristic. The presence of only one *allele* for baldness causes the condition in the male, whereas the condition does not occur in the female unless she possesses both alleles for baldness.

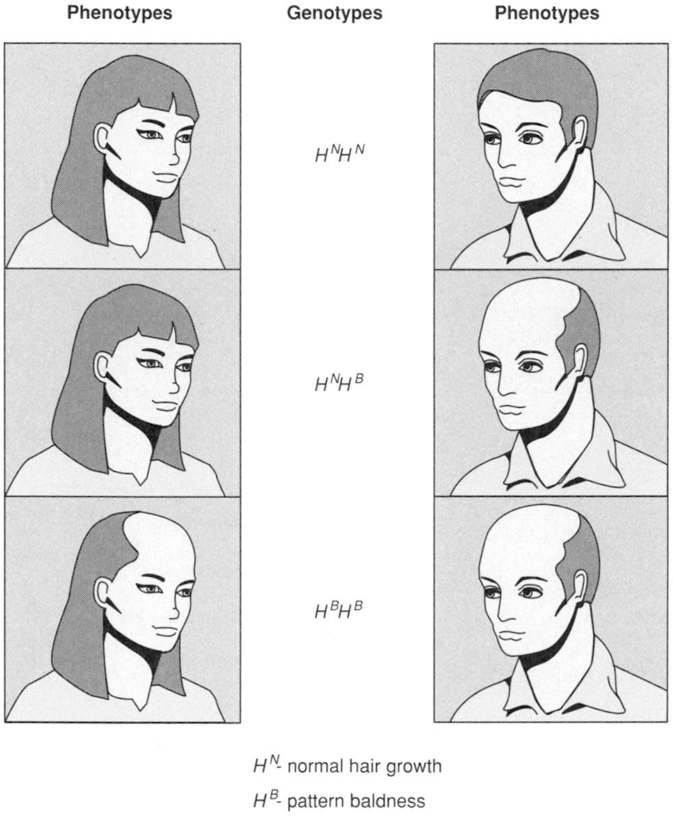

| Phenotypes | Genotypes | Phenotypes |

$H^N H^N$

$H^N H^B$

$H^B H^B$

H^N normal hair growth
H^B pattern baldness

Summary

1. It is possible to treat and photograph the chromosomes of a cell so that they can be cut out and arranged in pairs. The resulting karyotype can be used to diagnose chromosomal disorders.

2. Amniocentesis and chorionic villi testing allow physicians to recover fetal cells, whose chromosome content can be analyzed for any possible abnormalities.

3. Down syndrome (trisomy 21) is the most common autosomal aneuploidy. The occurrence of this syndrome is related to the mother's age, and women greater than 35 years of age may be encouraged to have amniocentesis done. Most often Down syndrome is due to nondisjunction during gamete formation, but in a small percentage of cases, there has been a translocation between chromosome 14 and 21, in which chromosome 21 becomes attached to chromosome 14.

4. Turner syndrome (XO) is a monosomy for the X chromosome. There are several trisomies: metafemales are XXX, Klinefelter syndrome is XXY, and there are also XYY males.

5. Mapping the human chromosomes was almost impossible until the advent of human-mouse hybrid cell data. Because clones of these cells contain a few, variable human chromosomes, it is possible to test for certain proteins. If present they can be connected with a particular chromosome. Genetic markers and DNA probes are other techniques for mapping human chromosomes.

6. When studying human genetic disorders, biologists often construct pedigree charts to show the pattern of inheritance of a characteristic within a family. The particular pattern indicates the manner in which the disorder is inherited. Sample charts are given for autosomal recessive, autosomal dominant, and X-linked recessive disorders.

7. Cystic fibrosis, lysosomal storage diseases, and PKU are autosomal recessive disorders that have been studied in detail.

8. Neurofibromatosis (NF), and Huntington disease are autosomal dominant disorders that have been well studied.

9. Sickle-cell anemia is a human disorder that is controlled by incompletely dominant alleles.

10. Color blindness, hemophilia, and muscular dystrophy are X-linked recessive disorders. Some traits, like pattern baldness, are sex-influenced traits, controlled by autosomal genes.

Objective Questions

For questions 1 to 3, match the conditions in the key with the descriptions below:

Key:
- a. Down syndrome
- b. Turner syndrome
- c. Klinefelter syndrome
- d. XYY

1. Male with underdeveloped testes and some development of the breasts.
2. Trisomy 21
3. XO female
4. Down syndrome
   - a. is always caused by nondisjunction of chromosome 21.
   - b. shows no overt abnormalities.
   - c. is more often seen in children of mothers past the age of 35.
   - d. Both (a) and (c).

5. Which of these would be of no help in mapping the human chromosomes?
   - a. chorionic villi testing
   - b. genetic marker
   - c. man-mouse hybrids
   - d. DNA probes

For questions 6 to 9, match the conditions in the key with the descriptions below:

Key:
- a. cystic fibrosis
- b. Huntington disease
- c. hemophilia
- d. Tay-Sachs

6. autosomal dominant
7. most often seen among Jewish people
8. X-linked recessive
9. thick mucus in lungs and digestive system
10. Determine if the characteristic possessed by the darkened males and females is an autosomal dominant, autosomal recessive, or X-linked recessive condition.

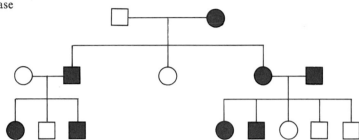

Additional Genetics Problems

1. A hemophiliac man is married to a homozygous normal woman. What are the chances that the sons will be hemophiliacs? That the daughters will be hemophiliacs? That the daughters will be carriers?
2. A son with cystic fibrosis is born to a couple who appear to be normal. What are the chances that any child born to this couple will have cystic fibrosis?
3. A man who is heterozygous for Huntington disease is married to a normal woman. What are the chances that a child will inherit the allele for Huntington disease? If the child does inherit the allele, what are the chances that she/he will pass it on?

4. Determine if the characteristic possessed by the darkened squares (males) and circles (females) is dominant, recessive, or sex-linked recessive. Write in the genotypes for the starred individuals.

Key:

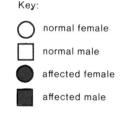

normal female

normal male

affected female

affected male

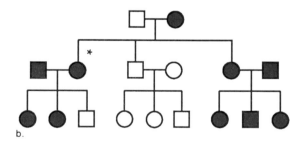

b.

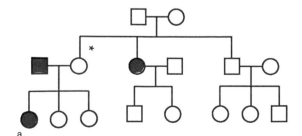

a.

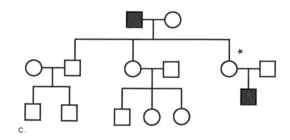

c.

*Answers to Additional Genetics Problems appear on page 806 of appendix D.

Study Questions

1. What does the normal human karyotype look like?
2. What are the characteristics of Down syndrome? What is the most frequent cause of this condition and how might it arise due to a translocation?
3. What is the only known sex chromosome monosomy in humans? Name and describe three sex chromosome trisomies.
4. Describe three means available to researchers to map the human chromosomes.
5. How might you distinguish an autosomal dominant from an autosomal recessive trait when viewing a pedigree chart?
6. Describe the symptoms of cystic fibrosis, Tay-Sachs, and PKU. For any one of these, what are the chances of two carriers having an affected child?
7. Describe the symptoms of neurofibromatosis (NF) and Huntington disease. How do they illustrate variable expressivity? For either of them, what are the chances of a heterozygote and a normal individual having an affected child?
8. Explain how sickle-cell anemia is inherited. What are the symptoms of sickle-cell trait and sickle-cell anemia?

If two persons with sickle-cell trait are married, what are the chances of having a child with sickle-cell trait, with sickle-cell anemia? What race of people is more likely to have this condition? Why?

9. Explain how color blindness, hemophilia and Duchenne-type muscular dystrophy are inherited. What are the symptoms of these conditions? What are the chances of having an affected male offspring if the mother is a carrier and the father is normal?
10. What is a sex-influenced trait? Give two examples of such traits.

Thought Questions

1. How could a woman who has a hemophiliac brother be absolutely certain that her unborn child will be free of this condition?
2. Sometimes genetic disorders are called genetic diseases. Why is the latter terminology confusing?
3. If a genetic disorder was controlled by polygenes, might the resulting continuous variation look like variable expressivity? Why or why not?

Selected Key Terms

karyotype (kar'e-o-tīp) 203
Down syndrome (down sin'drōm) 205
Turner syndrome (tur'ner sin'drōm) 206
metafemale (met''ah-fe'māl) 207
Klinefelter syndrome (klīn'fel-ter sin'drōm) 207
XYY male (eks wi wi māl) 207

carrier (ka'ē-er) 209
cystic fibrosis (sis'tik fi-bro'sis) 210
Tay-Sachs disease (ta saks' di-zēz) 211
phenylketonuria (PKU) (fen''il-ke''to-nu're-ah) 211
Huntington disease (HD) (hunt'ing-tun di-zēz') 212

sickle-cell anemia (sik'l sel ah-ne'me-ah) 212
hemophilia (he''mo-fil'e-ah) 213
Duchenne muscular dystrophy (du-shen' mus'ku-lar dis'tro-fe) 214
sex-influenced trait (seks in-flu'enst trāt) 215

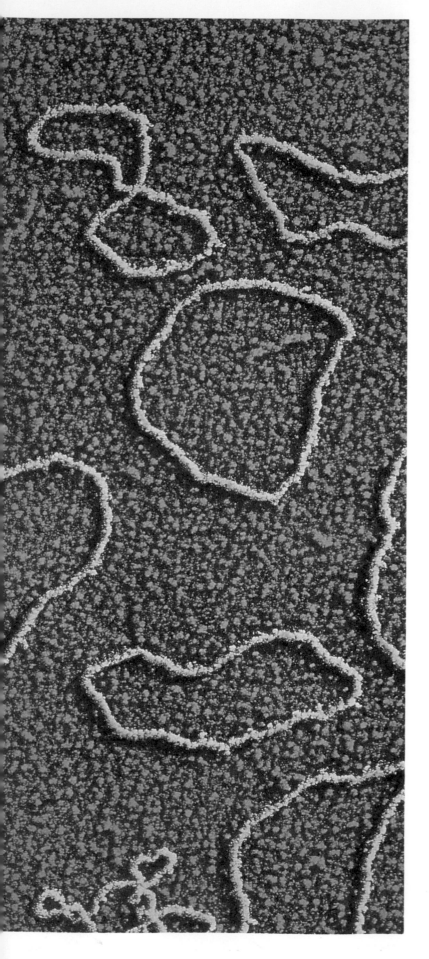

CHAPTER 15

DNA: The Genetic Material

Your study of this chapter will be complete when you can:

1. List the requirements for a substance to serve as genetic material.

2. Describe the transformation experiment of F. Griffith and his surprising results and conclusion.

3. Tell how O. Avery and colleagues showed that DNA is the transforming substance.

4. Describe the experiment of Hershey and Chase with T2 viruses and how it showed that DNA is the genetic material.

5. Explain Chargaff's rules for DNA and tell why they are significant.

6. Describe the Watson and Crick model of DNA and how it fits the Chargaff and the Franklin data.

7. Describe the semiconservative manner in which Watson and Crick suggested that DNA replicates.

8. Tell how Meselson and Stahl demonstrated that DNA replication is semiconservative.

9. Contrast the process of DNA replication in prokaryotes and eukaryotes.

10. Discuss new findings regarding DNA biochemistry.

DNA is the genetic material. Here we see plasmids, extrachromosomal rings of DNA found in bacteria. Work with plasmids has greatly advanced our knowledge of how DNA functions, including biochemical differences between prokaryotes and eukaryotes.

Even though Morgan and his colleagues were able to confirm that the genes were on the chromosomes and were even able to map the *Drosophila* chromosomes, they still didn't know just what genes consisted of. You can well imagine, then, that the search for the genetic material was of utmost importance to biologists at the beginning of the twentieth century. They knew that this material must

1. Be able to *store information* that is used to control both the development and the metabolic activities of the cell or organism.
2. Be stable so that it can *be replicated* with high fidelity during cell division and be transmitted from generation to generation.
3. Be able to *undergo rare changes* called **mutations** in order to generate genetic variability that is acted upon during evolution.

> The genetic material must be able to store information, be replicated, and undergo mutations.

Search for the Genetic Material

Knowledge about the chemistry of DNA was absolutely essential in order to come to the conclusion that this was the genetic material. In 1869, the Swiss chemist Friedrich Miescher removed nuclei from pus cells (these cells have little cytoplasm) and found that they contained a chemical he called nuclein. *Nuclein,* he said, was rich in phosphorus and had no sulfur, properties that distinguished it from protein. Later, other chemists did further work with nuclein and said that it contained an acidic substance they called **nucleic acid.** Soon it was realized that there were two types of nucleic acids: **DNA** and **RNA.**

When it was discovered, early in the twentieth century, that nucleic acids contain four types of nucleotides an unfortunate idea called the *tetranucleotide* (four nucleotide) *hypothesis* arose. It said that DNA is composed of repeating units and each unit always had just one of each of the four different nucleotides. In other words DNA could have no variability between species and therefore could not be the genetic material!

Transformation of Bacteria

In 1928, the bacteriologist Frederick Griffith did an experiment with a bacterium (*Streptococcus pneumoniae* or pneumococcus for short) that causes pneumonia in mammals. He noticed that when the bacteria are grown on culture plates, some, called S strains, produce shiny, smooth colonies and others, called R strains, produce colonies that have a rough appearance. Under the microscope, S strain bacteria have a mucous (polysaccharide) coat and R strain do not. When Griffith injected mice with an S strain of bacteria they died and when he injected mice with an R strain, they did not die (fig. 15.1). In an effort to determine if the smooth coat alone was responsible for the virulence (ability to kill) of the S strains, he injected mice with heat-killed S strain bacteria. The mice did not die.

Finally, Griffith injected the mice with a mixture of heat-killed S strain and live R strain bacteria. Most unexpectedly, the mice died and living S strain bacteria were recovered from their bodies! Griffith concluded that some substance necessary to the synthesis of a mucous coat must have passed from the dead S strain bacteria to the living R strain so that R strain became *transformed:*

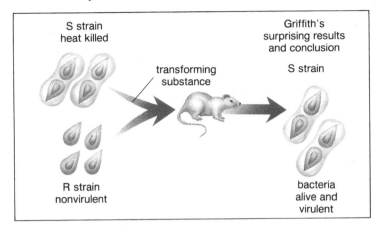

This change in the phenotype of the R strain bacteria must be due to a change in their genotype. Indeed, couldn't the transforming substance that passed from S strain to R strain be genetic material? Reasoning such as this was done by investigators at the time and prompted them to begin to look for the transforming substance in order to determine the chemical nature of the hereditary material.

The Transforming Substance

Obviously, it is not convenient to look for the transforming substance in mice, so it is not surprising that the next group of investigators, led by Oswald Avery, worked in vitro (in laboratory glassware). After sixteen years of research, this group published a paper that demonstrated that the transforming substance was DNA. This was their evidence:

1. DNA alone from S strain bacteria caused R strain bacteria to be transformed. And the DNA they used was pure—99.98% pure!
2. Enzymes that degrade proteins had no effect upon transformation, nor did RNase, an enzyme that digests RNA.
3. Enzymatic digestion of the transforming substance with DNase, an enzyme that digests DNA, did prevent transformation.
4. The molecular weight of the transforming substance was so large that it must contain about 1,600 nucleotides! Certainly enough for some genetic variability.

These experiments showed not only that DNA is the hereditary material, but also that DNA controls the biosynthetic properties of a cell. While these experiments seem quite convincing, it must be remembered that at the time biologists knew proteins were very complex, but they still were not sure about DNA. Some thought that perhaps these results pertained only to this experiment.

Figure 15.1

Griffith's transformation experiment. *a.* S strain of pneumococcus is virulent and kills its host, a mouse. *b.* R strain is not virulent and does not kill a mouse. *c.* Heat-killed S strain is not virulent and does not kill a mouse. *d.* If heat-killed S strain and R strain are both injected into a mouse, it dies because the R strain has been transformed into a virulent S strain.

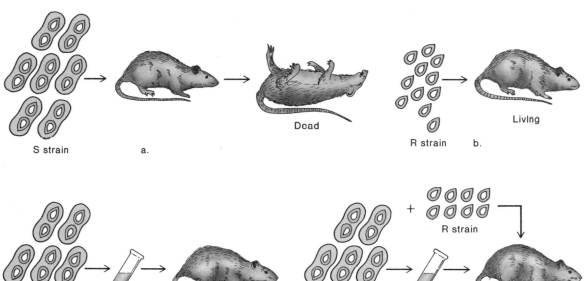

S strain   a.   Dead   R strain   b.   Living

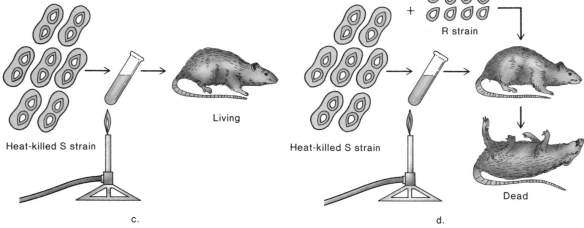

Heat-killed S strain   Living   R strain   Heat-killed S strain   Dead

c.   d.

Figure 15.2

Bacteria and bacteriophages. Bacteria and bacteriophages were the experimental material of choice for determining the physical and chemical characteristics of the genetic material. You'll recall that bacteria are prokaryotes (fig. 5.3), one-celled organisms whose genetic material is found within nucleoids. The most intensely studied species of bacterium is *Escherichia coli,* which normally lives within our own gut. A few thousand of these bacteria can be placed in a liquid medium containing only a few salts and an energy source such as the sugar glucose and within a few hours there will be as many as 3×10^9 bacteria per ml. Bacteriophages (or phages), viruses that attack bacteria, consist only of a protein coat surrounding a nucleic acid core. Enormous populations of a phage can be easily obtained—up to 10^{11} per ml or more in an infected bacterial culture is not unusual. The T virus depicted here in close-up and attacking an *E. coli* cell, is designated T2 (the T simply means "type"). There are also other types of bacteriophages that attack *E. coli* and they too have served as good experimental material. Magnification, ×50,000.

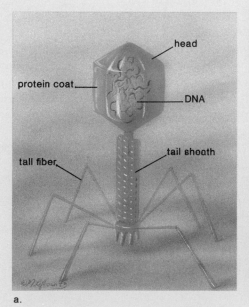

head

protein coat

DNA

tail fiber

tail sheath

a.

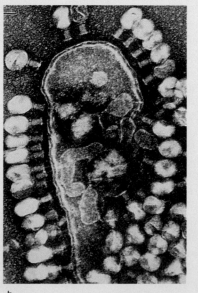

b.

Bacterial cells can be transformed when DNA passes from one to the other. Transformation changes the biosynthetic properties of the transformed cell.

Genetic Material of Viruses

In the twentieth century, many geneticists who were interested in determining the chemical nature of the genetic material began to work with **bacteriophages,** viruses that attack bacteria (fig. 15.2). In 1952 two experimenters, Alfred D. Hershey and Martha Chase, chose a bacteriophage known as T2 as their experimental material. They decided to see which of the bacteriophage components—protein or DNA—entered bacterial cells and directed reproduction of the virus (fig. 23.3*b*). Which ever did this, they reasoned, was the genetic material.

Figure 15.3
Hershey and Chase Experiment. Phage T2 is composed of DNA and coat protein only. It was reasoned that whichever of these enters a bacterium and controls phage reproduction was the genetic material. *a.* In this experiment, ^{32}P was used to label phage DNA. Even though the cells were agitated in a blender, the radioactively labeled material (i.e., DNA) was located inside the cell and reproduction proceeded normally. DNA was the genetic material. *b.* In this experiment, ^{35}S was used to label phage coat. When the cells were agitated in a blender, the radioactively labeled material (i.e., protein coats) was removed from the bacterial cells. Since reproduction proceeded normally, protein was not the genetic material.

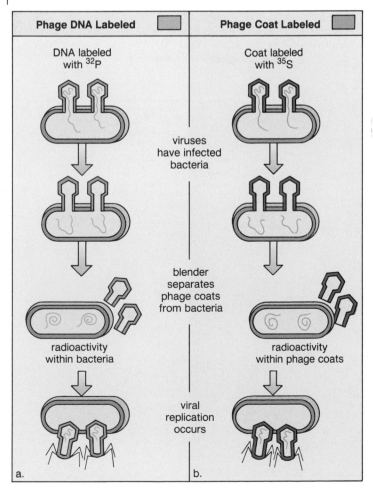

There is a chemical distinction between DNA and protein. Phosphorus (P) atoms are generally absent but sulfur *(S) is present in amino acids* of protein. On the other hand, sulfur (S) is absent but phosphorus *(P) is present in DNA* in high amounts. Therefore, it was possible for Hershey and Chase to prepare two batches of phages; one that had their DNA core radioactively labeled with ^{32}P and another that had their protein coats radioactively labeled with ^{35}S. Radioactive phages were allowed to briefly attach to *Escherichia coli* bacterial cells; then once the infection process had started, most of the adhering phage coats were sheared from the bacterial cells by agitation in a kitchen blender. Centrifugation then caused the bacterial cells to collect as a pellet at the bottom of the tube. In the one experiment (fig. 15.3*a*), they found that most of the ^{32}P-labeled DNA remained in the pellet:

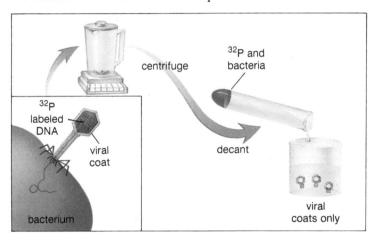

In the other experiment (fig. 15.3*b*), they found that most of the ^{35}S-labeled protein remained in the phage coats. These results indicated that the DNA of the virus (and not the protein) enters the host where viral replication takes place. Therefore, DNA is the genetic material of bacteriophage T2, directing protein coat synthesis and allowing replication to occur.

Biologists, in general, were persuaded that DNA is the genetic material by this experiment, possibly because recent chemical studies of the composition of DNA had finally shown that the tetranucleotide hypothesis was not true.

The Hershey and Chase experiment showed that DNA and not protein is the genetic material of the T2 bacteriophage.

Structure of DNA

During the same period of time that biologists were using viruses to show that DNA is the genetic material, biochemists were busy trying to decipher its structure.

Chargaff's Rules

With the development of new chemical techniques in the 1940s, it was possible for Erwin Chargaff to analyze in detail the base content of DNA. For at least twenty years previously, the tetranucleotide hypothesis was believed by many. This hypothesis stated that DNA was composed of repeating units, each unit having four nucleotides—one nucleotide for each of the four bases: **adenine (A)**, **thymine (T)**, **cytosine (C)**, and **guanine (G)**. According to this hypothesis, DNA would contain approximately 25% of each kind of base in every species.

A sample of Chargaff's data is shown in table 15.1 and you can see that while some species, for example, *E. coli* and *Zea mays* (corn) do have approximately 25% of each base in their DNA, most do not. He did find that the amount of A was always equal to the amount of T and the amount of G was always equal to the amount of C, regardless of the species.

Chargaff's data showed that

1. The base composition of DNA differs from species to species. For example, *Bacillus subtilis,* a type of bacterium, does not have the same amount of A, T, C, or G, as does *E. coli* (table 15.1).
2. In each species, however, the percent of A equals the percent of T and also the percent of G equals the percent of C. Further, 50% of the bases were **purines** (A + G) and 50% were **pyrimidines** (T + C). These relationships are called **Chargaff's rules.**

The first finding means that when comparisons between species are made, DNA does have the *variability* required of the genetic material. The second finding means that within the species, DNA has a *constancy* required of the genetic material.

X-Ray Diffraction Data

M.H.F. Wilkins and Rosalind Franklin at King's College in London studied the structure of DNA by use of X rays. They found that if a concentrated, viscous solution of DNA was made, it could be separated into fibers. Under the right conditions, the

Table 15.1 Chargaff's Data: % Base Composition of DNA in Various Species

Species	A	T	G	C
Homo sapiens	31.0	31.5	19.1	18.4
Drosophila melanogaster	27.3	27.6	22.5	22.5
Zea mays	25.6	25.3	24.5	24.6
Neurospora crassa	23.0	23.3	27.1	26.6
Escherichia coli	24.6	24.3	25.5	25.6
Bacillus subtilis	28.4	29.0	21.0	21.6

In any particular species listed, the amount of A = T and the amount of G = C. Between species, however, the A,T and G,C percentages differ. For example, in humans (*Homo sapiens*) the A,T percentage is about 31% but in fruit flies (*Drosophila melanogaster*) the percentage is about 27%.

fibers were enough like a crystal (a solid substance whose atoms are arranged in a definite manner) so that an X-ray pattern could be formed on a photographic film. Franklin's picture of DNA showed that DNA was a helix (fig. 15.4). The helical shape was indicated by the crossed (X) pattern in the center of the photograph. The dark portions at the top and bottom of the photograph indicated that some portion of the helix was being repeated over and over again. It was possible for Franklin to estimate certain dimensions of the molecule as indicated in figure 15.5.

X-ray diffraction data showed that DNA is most likely a helix with certain definite molecular measurements.

Figure 15.4

a. When a crystal is X rayed, the way in which the beam is diffracted reflects the pattern of the molecules in the crystal. The closer together two repeating structures are in the crystal, the farther from the center the beam will be diffracted. *b.* The diffraction pattern of DNA produced by Rosalind Franklin. The crossed (X) pattern in the center told investigators that DNA was a helix and the dark portions at the top and bottom told them that some feature was being repeated over and over again. Watson and Crick figured that this feature was the hydrogen-bonded bases.

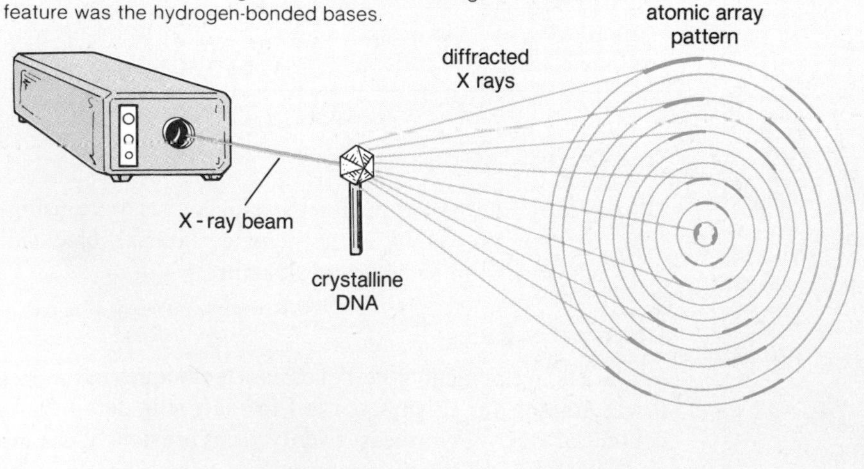

atomic array pattern

diffracted X rays

X-ray beam

crystalline DNA

a.

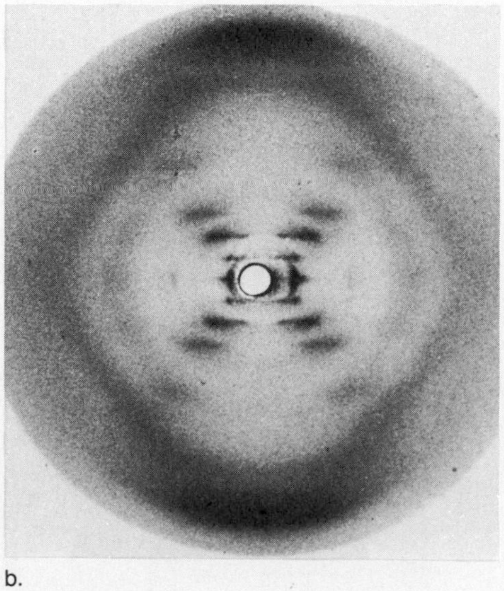

b.

Watson and Crick Model

In the early 1950s, James Watson, an American, was on a post-doctoral fellowship at Cavendish Laboratories in Cambridge, England. There he met the biophysicist Francis H. C. Crick. Using the data we have just presented, they constructed a model of DNA that fit the data (fig. 15.5).

Watson and Crick knew, of course, that DNA is a polymer of nucleotides (figs. 4.20 and 4.21), but they did not know how the nucleotides were arranged within the molecule. This is what they decided:

1. The Watson and Crick model shows that DNA is a double helix with sugar-phosphate backbones on the outside and paired bases on the inside. This arrangement fit the mathematical measurements provided by the X-ray diffraction data: the spacing between base pairs is 0.34 nm and the overall helix repeats in about 3.4 nm, i.e., there are about ten layers of stacked bases for every complete turn of the DNA double helix.

2. Chargaff's rules said that A = T and G = C. The model shows that A is hydrogen bonded to T, and G is hydrogen bonded to C (fig. 15.6). This so-called **complementary base pairing** means that a purine is always bonded to a pyrimidine. Only in this way will the molecule have the width (2 nm) dictated by its X-ray diffraction pattern, since two pyrimidines together are too narrow and two purines together are too wide.

The Watson and Crick double helix model of DNA is like a twisted ladder. The sugar-phosphate backbone of the two DNA strands make up the sides of the ladder; the hydrogen-bonded bases make up the rungs or steps of the ladder.

Figure 15.5

a. Watson and Crick model of DNA. The DNA double helix resembles a twisted ladder. Sugar phosphate backbones, represented here by S—P (Sugar–Phosphate), make up the sides of the ladder and hydrogen-bonded bases make up the rungs of the ladder. The correctness of the model is substantiated in that the noted dimensions match those that were determined by the X-ray diffraction pattern. b. A space-filling model of DNA is a more realistic representation of the molecule. Notice the close stacking of the paired bases, as determined by the X-ray diffraction pattern of DNA. Color code for atoms: yellow = P, dark blue = C, red = O, turquoise = N, and white = H.

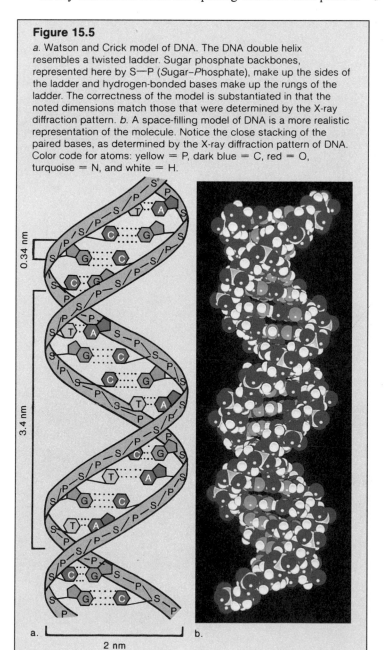

Figure 15.6

Complementary base pairing. In the Watson and Crick model of DNA, the bases are bonded as shown. This is in keeping with Chargaff's rules and it means that a purine (A and G) is always bonded to a pyrimidine (T and C). Only in this way does the molecule maintain a constancy of diameter as determined by its X-ray diffraction pattern. Watson and Crick also said that the DNA strands ran antiparallel to one another (notice that the sugar-phosphate groups are oriented in different directions) and this introduces complications, which are described in the reading on page 226.

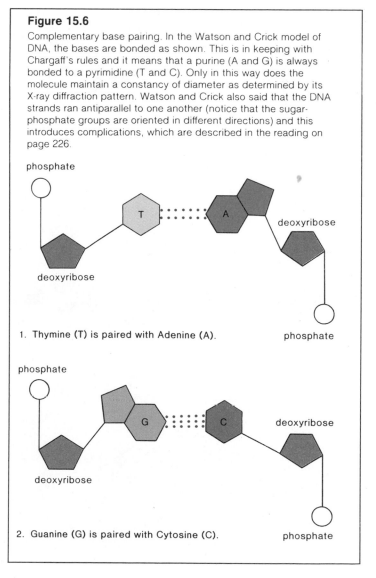

1. Thymine (T) is paired with Adenine (A).

2. Guanine (G) is paired with Cytosine (C).

Biological Significance

The Watson and Crick model also allows for variability, as dictated by Chargaff's data. The paired bases occur in any order:

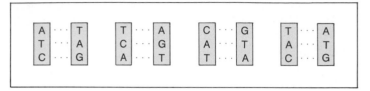

The variability that can be obtained is overwhelming. For example, it's been calculated that the human chromosome con-tains on the average about 140 million base pairs. Since any of the four possible nucleotides can be present at each nucleotide position, the total number of possible nucleotide sequences is $4 \times 140 \times 10^6$ or over 500 million possible arrangements. No wonder each species has its own base percentages!

DNA Replication

The Watson and Crick model suggested that DNA can be replicated by means of complementary base pairing. During **replication,** each parental DNA strand serves as a template for a new strand (fig. 15.7). A **template** is most often a mold used to

Figure 15.7

Replication of DNA as proposed by Watson and Crick. In this simplified view, the two strands of a DNA molecule separate and each serves as template for the formation of a new complementary strand. While this is basically correct, the actual process has complications, which are described in the reading on page 226.

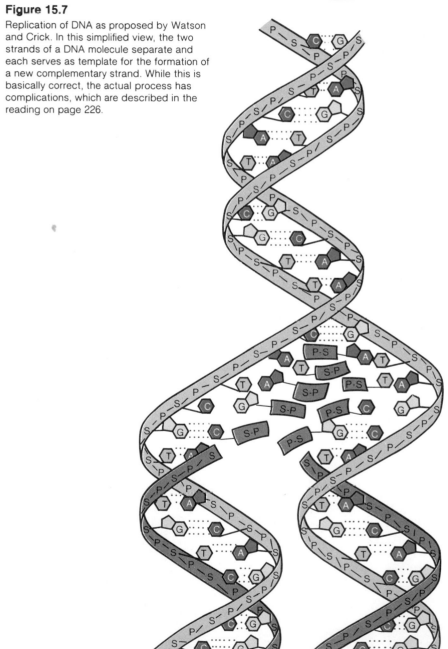

Region of parental DNA helix. (Both backbones are light.)

Region of replication (simplified). Parental DNA is unwound and unzipped. New nucleotides are pairing with those in parental strands.

Region of completed replication. Each double helix is composed of an old parental strand (light) and a new daughter strand (dark). Notice that each double helix is exactly like the other and also like the original parental double helix.

produce a shape complementary to itself. Replication requires the following steps:

1. The two strands that make up DNA must be unwound and "unzipped" (i.e., the weak hydrogen bonds between the paired bases break). There is a special enzyme called helicase that unwinds the molecule.
2. New complementary nucleotides, always present in the nucleus, move into place by the process of complementary base pairing.
3. The complementary nucleotides become joined together so that again DNA is double stranded. Steps (2) and (3) are carried out by the enzyme **DNA polymerase.**
4. When the process is finished, two complete double-stranded DNA molecules are present, identical to each other and to the original molecule.

Although DNA replication can be easily explained in this manner, it is actually an extremely complicated process involving many steps and enzymes. Some of the more precise molecular events are explained in the reading on page 226.

Replication Is Semiconservative

The replication process is termed **semiconservative** because each new double helix has one parental strand and one new strand. In other words, one of the parental strands is conserved or present in each daughter double helix. This was experimentally confirmed by Matthew Meselson and Franklin Stahl in 1958.

These investigators grew bacteria in a medium enriched by ^{15}N so that it became incorporated into their DNA. Ordinary nitrogen (^{14}N) has an atomic weight of 14 and ^{15}N has an atomic weight of 15 and is therefore heavier. It is possible to distinguish "light" DNA containing only ^{14}N from "heavy" DNA containing ^{15}N by centrifuging them in a tube containing a gradient that can separate them on the basis of density (fig. 15.8a).

The experimental design (fig. 15.8b) was to grow the bacteria in the ^{15}N so that in the beginning, the DNA was heavy/heavy (both strands contain ^{15}N) and then switch the bacteria to a medium of ^{14}N. If DNA replication is semiconservative,

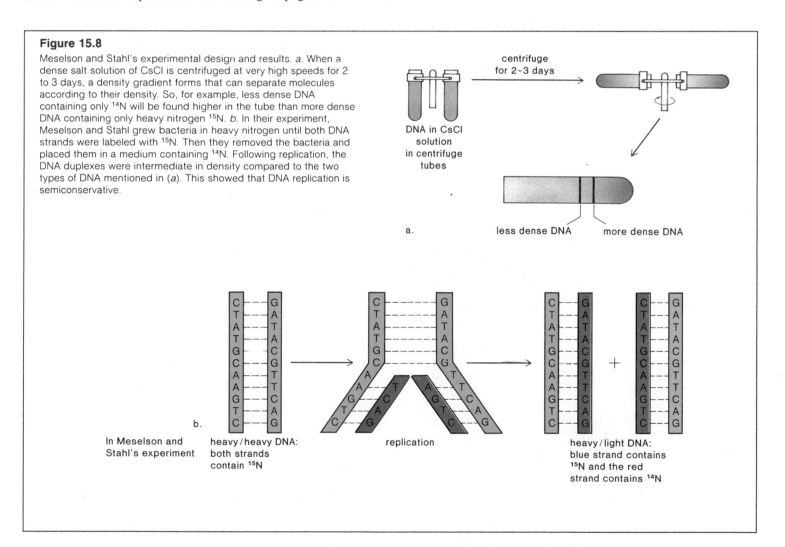

Figure 15.8

Meselson and Stahl's experimental design and results. a. When a dense salt solution of CsCl is centrifuged at very high speeds for 2 to 3 days, a density gradient forms that can separate molecules according to their density. So, for example, less dense DNA containing only ^{14}N will be found higher in the tube than more dense DNA containing only heavy nitrogen ^{15}N. b. In their experiment, Meselson and Stahl grew bacteria in heavy nitrogen until both DNA strands were labeled with ^{15}N. Then they removed the bacteria and placed them in a medium containing ^{14}N. Following replication, the DNA duplexes were intermediate in density compared to the two types of DNA mentioned in (a). This showed that DNA replication is semiconservative.

DNA in CsCl solution in centrifuge tubes

centrifuge for 2-3 days

a.

less dense DNA   more dense DNA

b.

In Meselson and Stahl's experiment

heavy/heavy DNA: both strands contain ^{15}N

replication

heavy/light DNA: blue strand contains ^{15}N and the red strand contains ^{14}N

Aspects of DNA Replication

Watson and Crick realized that the strands in DNA had to be antiparallel in order to allow for complementary pairing of the bases. This opposite polarity of the strands introduces complications for DNA replication, as we shall now see. First, it is important to take a look at a deoxyribose molecule in which the carbon atoms are numbered (fig. *a*). Use the structure to see that one of the strands of DNA (fig. *b*) runs in the 5′ → 3′ direction and the other runs in the 3′ → 5′ direction. You can tell the 5′ end because that's where the last phosphate is attached.

During replication (fig. *b*), one nucleotide is joined to another. Each new nucleotide already has a phosphate group attached to the 5′ carbon atom and it is joined to the 3′ carbon atom of a sugar already in place. Therefore it is said that *DNA polymerase synthesizes the replicated strand in the 5′ → 3′ direction.* This presents a problem at the replication fork (fig. *c*) where DNA is unwound and unzipped and where replication is occurring.

At a replication fork, only one of the new (daughter) strands will run in the 5′ → 3′ direction (the template for this strand, of course, runs in the 3′ → 5′ direction). The new strand running in the 5′ → 3′ direction can be synthesized continuously and is called the *leading strand.* But what about the situation when the 5′ → 3′ parental strand is serving as the template? Synthesis of the new strand must also be in the 5′ → 3′ direction, and therefore synthesis has to begin at the fork. (Although synthesis is in the 5′ → 3′ direction, this new daughter strand will, in the end, run from 3′ → 5′, opposite to its template—do you see why?) Because replication of the 5′ → 3′ parental strand must begin again repeatedly as the DNA molecule unwinds and unzips, replication is discontinuous; indeed, replication of this strand results in segments called Okazaki fragments, after the Japanese scientist who discovered them. Discontinuous replication takes more time than continuous replication and therefore the new strand in this case is called the *lagging strand.*

The fact that DNA polymerase can only add a new nucleotide to the free 3′ end of a nucleotide presents another problem: *DNA polymerase cannot start the synthesis of a new DNA chain* at the origin of replication. (In the lagging strand there are many origins of replication.) Here, RNA polymerases lay down a short amount of RNA, called an RNA primer, that is complementary to the DNA strand being replicated. Now DNA polymerase can add DNA nucleotides in the 5′ → 3′ direction. Later, while proofreading, DNA polymerase removes the RNA primer and replaces it with complementary DNA nucleotides. Another enzyme, called DNA ligase, can join the 3′ end of each fragment to the 5′ end of another.

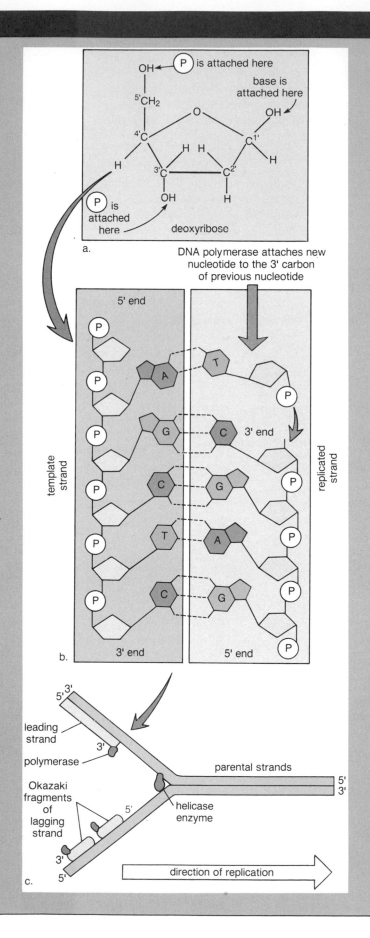

then the next generation of DNA would be heavy/light (one strand contains ^{15}N and the other contains ^{14}N). Of course, in the next generation you will have some strands that are completely light/light. This is exactly what happened, showing that DNA replication is semiconservative.

During DNA replication, DNA becomes unwound and unzipped, and a new strand forms complementary to each original strand. This is called semiconservative replication.

Accuracy of Replication

It doesn't always happen that the bases pair correctly with one another as replication proceeds. For bacteria, it's been estimated that a mistake occurs as often as one in 10^4 times the bases pair. This is far greater than the final error rate observed, because DNA polymerase itself has a "proofreading" function—it checks each pairing as soon as it occurs. If it finds that a mistake has been made, it removes the incorrect nucleotide and replaces it with a correct one. After proofreading, the average probability of an error in the insertion of a new nucleotide is about 10^{-8} to 10^{-12}. If an error does occur and is not corrected, then a gene mutation has occurred. Gene mutations are covered in more detail in chapter 16.

Prokaryotic versus Eukaryotic Replication

You'll recall that bacteria have a single circular loop of DNA (fig. 10.2) that must be replicated before the cell divides. In these organisms, replication begins at an origin of replication and then spreads out in both directions. Therefore replication is bidirectional:

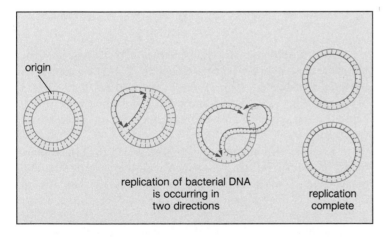

origin

replication of bacterial DNA is occurring in two directions

replication complete

Bacterial cells are able to replicate their DNA at a rate of about 10^6 base pairs per minute and about 40 minutes is required to replicate the complete chromosome. Because bacterial cells are able to divide as often as once every 20 minutes, it is possible for a new round of DNA replication to begin even before the previous round has been completed!

In eukaryotes, DNA replication begins at numerous origins of replication along the length of the chromosome and the so-called replication "bubbles" spread bidirectionally until they

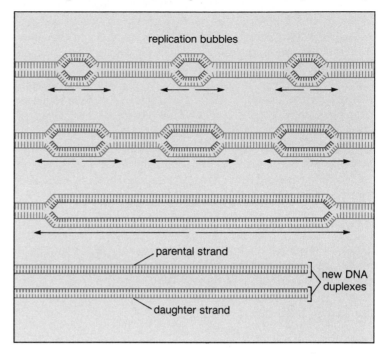

replication bubbles

parental strand

new DNA duplexes

daughter strand

meet. Notice that there is a V shape wherever DNA is being replicated, this is called a **replication fork.**

Although eukaryotes replicate their DNA at a slower rate—500 to 5,000 base pairs per minute—there are many individual origins of replication and eukaryotic cells complete the replication of the entire DNA (in humans over six billion base pairs) in a matter of hours!

Generally speaking, DNA replication is rapid and accurate. If an error in base pairing does take place, then a mutation has occurred.

Other Aspects of DNA Biochemistry

There have been several unexpected discoveries regarding DNA biochemistry and we will discuss some of these.

B-DNA and Z-DNA

The Watson and Crick model said that DNA is a right-handed smooth double helix with ten stacked bases for each complete turn of the molecule. This form of DNA is called B-DNA. But other forms of DNA have been found.

In 1979, Alexander Rich, Andrew Wang, and their colleagues discovered a new form of DNA—a left-handed double helix that is not smooth. The backbones zig-zag as they make their turns and therefore this form of DNA is called Z-DNA.

Z-DNA is narrower (1.8 nm) than B-DNA, and it contains twelve nucleotide base pairs per turn. It is also grooved differently from B-DNA, as you can tell by examining figure 15.9.

The discovery of Z-DNA has excited biologists because it suggests that DNA is a far more dynamic molecule than previously expected. Apparently, a B-DNA segment can change to the Z-DNA configuration and vice versa. Could it be that the Z configuration is actually the active form and the more familiar B configuration is the inactive form of DNA? Some evidence suggests that this is the case. Z-DNA has been found in genetically active cells and proteins have been discovered in *Drosophila* that bind specifically to Z-DNA or induce a transition of B-DNA to the Z-DNA configuration. Perhaps these proteins are regulatory proteins that function to turn genes on and off. Turned on genes are directing protein synthesis and turned off genes are not doing so.

DNA Sequences

Classical geneticists who make chromosomal maps would have predicted that almost all of DNA would be genes of some sort. However, it turns out that at most 70% of DNA seems to function as genes. The rest either has no function or has a function we haven't been able to figure out yet.

Repetitive DNA

Repetitive DNA contains the same nucleotide sequences over and over again. Regions of eukaryotic chromosomes adjacent to the centromere are made up of very long stretches of highly repetitive DNA in which 5 to 15 base pairs are repeated 100,000 to 1 million times. Other, moderately repetitive DNA consists of 1,000 to 1,500 base pairs repeated only 10 to 3,000 times.

The highly repetitive DNA has base sequences (5 to 15 base pairs) that are much too short to be multiple copies of a single gene. Since it is found in the region of the centromere, it is possible that it has a structural role connected with this location. Some moderately repetitive DNA (1,000 to 1,500 base pairs) however, does seem to be multiple copies of genes for ribosomal RNA, ribosomal proteins, and histones. It can be speculated that the many copies of these genes allow a faster production rate of these substances than would otherwise be possible.

Moveable Genetic Elements

Classical geneticists always assumed that the genes are fixed in a definite order on the chromosomes. It came as a surprise, then, when it was proven that organisms have moveable genetic elements called **transposons.** Such elements are often called "jumping genes" because they are able to move from one location to another in the genome (all the genes that the organism possesses). The movement of DNA from one place to another is called *transposition.*

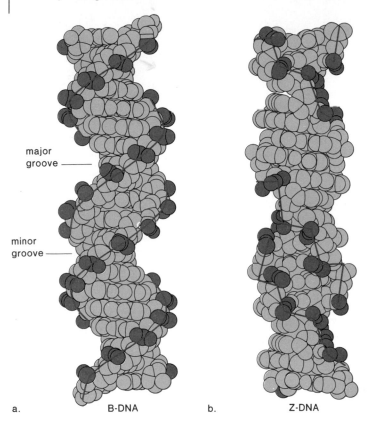

Figure 15.9
Two forms of the double helix. *a.* B-DNA is the form that corresponds to the structure proposed by Watson and Crick. B-DNA is said to be a right-handed spiral because the backbone (red line connecting red phosphate groups) goes up from left to right. B-DNA has both major and minor grooves. *b.* Z-DNA is a left-handed spiral with a backbone that zig-zags as it goes from right to left. Z-DNA has only minor grooves.

major groove

minor groove

a.   B-DNA   b.   Z-DNA

All transposons contain a gene that allows them to move, but some contain other types of genes as well. For example, in bacteria, transposons contain one or more genes that make a bacterium resistant to antibiotics. This is of some concern because transposons have even been known to move between bacterial cells.

The first jumping genes were discovered in corn by Barbara McClintock in the 1940s. She called them controlling elements because when inserted into a new location they altered the activity of nearby genes, including those, for instance, that affected the color of kernels (fig. 15.10). For this discovery she received the Nobel Prize in 1983. Transposons have now been found in fruit flies and humans in addition to bacteria and corn.

New findings regarding DNA biochemistry have shown that it is more variable in its form and function than originally thought.

Figure 15.10

a. Barbara McClintock reported forty years ago that her corn experiment results could only be explained by the presence of moveable genetic elements, now called "jumping genes." She was so ahead of her time that the scientific community did not recognize her achievement for many years. She received a Nobel Prize in 1983. *b*. An example of Indian corn showing kernels that have unusual colors. This was McClintock's experimental material that prompted her to suggest that there were moveable genetic elements.

a.

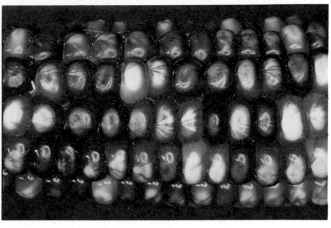
b.

Summary

1. Early work on the biochemistry of DNA wrongly suggested that it was composed of units each having only one A, T, C, and G. This so-called tetranucleotide hypothesis meant that DNA lacked the variability necessary for the genetic material.

2. Frederick Griffith injected strains of pneumococcus into mice and observed that smooth (S) was virulent but rough (R) was not. However, when a heat-killed S strain was injected with a live R strain, virulent S strain bacteria were recovered from the dead mice. He said that the R strain had been transformed by some transforming substance passing from the dead S strain to the live R strain.

3. Twenty years later, Oswald Avery and colleagues reported that the transforming substance was DNA. They showed that purified DNA (and not protein, for example) was capable of bringing about the transformation.

4. Hershey and Chase turned to bacteriophage T2 as their experimental material. In two separate experiments they labeled the protein coat with ^{35}S and the DNA with ^{32}P and then showed that the radioactive P alone was largely taken up by the bacterial host and that reproduction of viruses proceeded normally. This convinced most people that DNA was the genetic material.

5. In the meantime, work on the biochemistry of DNA was progressing. Erwin Chargaff did a chemical analysis of the molecule and showed that it was variable enough to be the genetic material. Further, he found that A = T and G = C and that the amount of purine equals the amount of pyrimidine. These relationships are known as Chargaff's rules.

6. Rosalind Franklin prepared an X-ray photograph of DNA that showed it was helical, had repeating structural features, and had certain dimensions.

7. James Watson and Francis Crick built a model of DNA in which the sugar-phosphate molecules make up the sides of a twisted ladder and the complementary paired bases were the rungs of the ladder. This model was in keeping with Chargaff's rules and the dimensions provided by the Franklin X-ray diffraction pattern.

8. The Watson and Crick model immediately suggested a method by which DNA could be replicated. The two strands unwind and unzip, and each parental strand acts as a template for a new (daughter) strand. In the end, each new duplex is like each other and like the parental duplex. The actual process is more complicated and involves many enzymes. The enzyme DNA polymerase joins the nucleotides together and is also capable of proofreading to make sure the bases have been paired correctly.

9. Replication is semiconservative since each new duplex contains an old (parental) strand and a new (daughter) strand. Meselson and Stahl demonstrated this by the following experiment: bacteria were grown in heavy nitrogen (^{15}N) and then switched to light nitrogen (^{14}N). The density of the DNA following replication was intermediate between these two, as measured by centrifugation of the molecules through a salt gradient.

10. Replication in prokaryotes is bidirectional from one point of origin and proceeds until there are two copies of the circular chromosome. Replication in eukaryotes is also bidirectional but there are many points of origin and many bubbles (places where the DNA strands are separate and replication is occurring). Replication occurs at the ends of the bubbles—at replication forks.

11. Since the time of Watson and Crick, it has been discovered that DNA is more variable than realized and that not all of it functions as genes in the classical sense. As examples of the former, we now know of the existence of Z-DNA and transposons, and as examples of the latter, we now know there is highly repetitive DNA with a function yet to be determined.

Objective Questions

For questions 1 to 4, match the names in the key to the statements below:

Key:

  a. Griffith
  b. Chargaff
  c. Meselson and Stahl
  d. Hershey and Chase

1. A = T and G = C.
2. Only the DNA from T2 enters the bacteria.
3. R strain bacteria became an S strain through transformation.
4. DNA replication is semiconservative.
5. If 30% of an organism's DNA is thymine then
   a. 70% will be purines.
   b. 20% will be guanine.
   c. 30% will be adenine.
   d. Both (b) and (c).

6. If you grew bacteria in heavy nitrogen and then switched them to light nitrogen, how many generations after switching would you have some light/light DNA?
   a. Never, because replication is semiconservative.
   b. first generation
   c. second generation
   d. only the third generation
7. If one strand of DNA has the base sequence ATCGTA, what will the complementary strand have?
   a. TAGCAT
   b. ATCGTA
   c. CAGTCT
   d. All of these.
8. The double helix model of DNA resembles a twisted ladder in which the rungs of the ladder are

  a. a purine paired with a pyrimidine.
  b. A paired with G and C paired with T.
  c. sugar-phosphate paired with sugar-phosphate.
  d. Both (a) and (b).
9. Cell division requires that the genetic material be able to
   a. store information.
   b. be replicated.
   c. undergo rare mutations.
   d. All of these.
10. In a DNA molecule,
    a. the bases are covalently bonded to the sugars.
    b. the sugars are covalently bonded to the phosphates.
    c. the bases are hydrogen bonded to one another.
    d. All of these.

Study Questions

1. List and discuss the requirements for genetic material.
2. What is the tetranucleotide hypothesis and why did this hypothesis hinder the acceptance of DNA as the genetic material?
3. Describe Griffith's experiments with pneumococcus, his surprising results, and conclusion.
4. How did Avery and his colleagues demonstrate that the transforming substance is DNA?

5. Describe the experiment of Hershey and Chase and explain how it shows that DNA is the genetic material.
6. What is the difference between a base, a nucleotide, and a nucleic acid? Which of these pertains to Chargaff's rules and what are the rules?
7. What is an X-ray diffraction pattern? What information was provided by DNA's diffraction pattern?
8. Describe the Watson and Crick model of DNA structure. How did it fit the data provided by Chargaff and the X-ray diffraction pattern?

9. Explain how DNA replicates semiconservatively. What role is played by helicase? By DNA polymerase?
10. How did Meselson and Stahl demonstrate semiconservative replication?
11. List and discuss differences between prokaryotic and eukaryotic replication of DNA.
12. What are Z-DNA, repetitive DNA, and transposons, and why are they significant?

Thought Questions

1. What does it mean to say that DNA stores information? How did the Hershey and Chase experiment show that DNA stores information?

2. How many strands of DNA are in a chromatid, in a duplicated chromosome, in a bivalent?
3. Imagine that DNA replication is completely conservative, that is, two parental strands stay together and

somehow produce a daughter helix composed of two completely new strands. In how many generations would light/light DNA have appeared in the Meselson and Stahl experiment?

Selected Key Terms

DNA (deoxyribonucleic acid) (de-ok″se-ri″bo-nu-kle′ik as′id) 219
RNA (ribonucleic acid) (ri″bo-nu-kle′ik as′id) 219
bacteriophage (bak-te′re-o-fāj″) 221
adenine (ad′ĕ-nīn) 222
thymine (thi′min) 222

cytosine (si′to-sin) 222
guanine (gwan′in) 222
purine (pu′rin) 222
pyrimidine (pi-rim′ĭ-din) 222
complementary base pairing (kom″plĕ-men′tă-re bās pār′ing) 223
replication (rĕ″pli-ka′shun) 224

template (tem′plāt) 224
DNA polymerase (de′en-a pol-im′er-ās) 225
semiconservative replication (sem″e-kon-ser′vah-tiv rĕ″pli-ka′shun) 225
transposon (trans-po′zun) 228

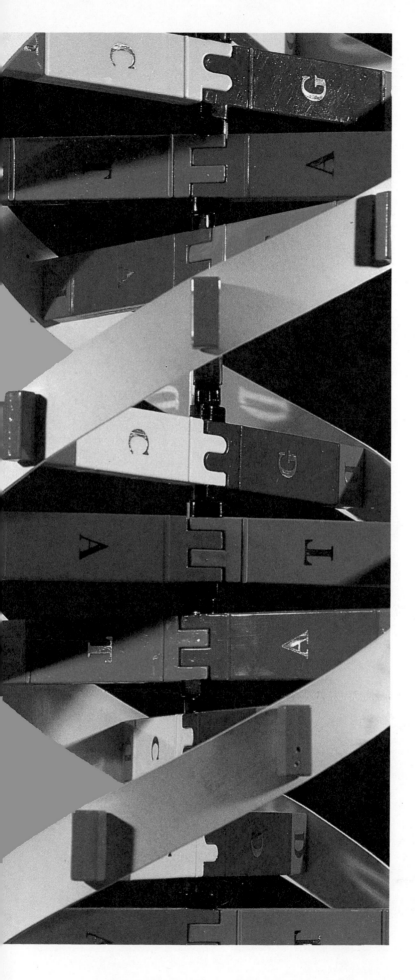

CHAPTER 16

Gene Activity

Your study of this chapter will be complete when you can:

1. *List and discuss the early studies that led to the recognition of gene activity.*
2. *Draw and explain a diagram that outlines the central dogma of biology.*
3. *List the biochemical differences between RNA and DNA.*
4. *Show that the DNA code is triplet, degenerate, and universal.*
5. *Describe the process by which RNA is made complementary to DNA.*
6. *List and discuss two ways that mRNA is processed before it leaves the eukaryotic nucleus.*
7. *Describe the role of ribosomes, mRNA, tRNA, and amino acids during protein synthesis.*
8. *When given a DNA coding strand and a table of codons, determine the mRNA codons, possible tRNA anticodons, and the sequence of amino acids in the resulting protein.*
9. *Discuss the different types of gene mutations, the rate of mutation, and the manner in which DNA is protected from mutation.*

DNA stores genetic information in what part of the molecule? This model can be used to demonstrate that it must be in the sequence of bases. The bases, A, T, G, and C can occur in any order, depending on the organism. They contain a code that functions to determine the sequence of amino acids in a protein. This is the genetic information stored within DNA.

The genes are composed of DNA, a molecule that lies within the nucleus of a eukaryotic cell (fig. 16.1). From its headquarters, DNA controls the metabolism of the cell and ultimately brings about the phenotype of the individual. How does DNA manage to store the necessary information and how does it exert its control? Just as it took some time for investigators to determine that DNA is the genetic material so it also took some time to discover exactly what a gene is and does. Knowledge about the action of genes preceded that of their makeup.

The Activity of Genes

In the early 1900s, the English physician, Sir Archibald Garrod, suggested that there is a relationship between inheritance and metabolic diseases. He introduced the phrase, **"inborn error of metabolism"** to dramatize this relationship. Garrod observed that family members often had the same disorder and he said this inherited defect could be caused by the lack of a particular enzyme in a metabolic pathway. Since it was known at the time that enzymes are proteins, Garrod was among the first to hypothesize a link between genes and proteins.

One Gene–One Enzyme Hypothesis

Many years later in 1941, G. W. Beadle and E. L. Tatum performed a series of experiments on *Neurospora crassa,* the red bread mold, which reproduces by means of spores. Normally, the spores become mold capable of growing on minimal medium (containing only a sugar, mineral salts, and the vitamin biotin) because it can produce all the enzymes it needs. In their experiments, Beadle and Tatum used X rays to induce mutations in asexually produced haploid spores. Some of the X-rayed spores could no longer become mold capable of growing on minimal medium; however, growth was possible on medium enriched by certain metabolites. In the example given in figure 16.2, the mold can grow only when supplied with enriched medium that includes all metabolites or C and D alone. Since C and D are a part of this hypothetical pathway

$$A \xrightarrow{\ 1\ } B \xrightarrow{\ 2\ } C \xrightarrow{\ 3\ } D$$

in which the numbers are enzymes and the letters are metabolites, it is concluded that the mold lacks enzyme 2. Beadle and Tatum further found that each of the mutant strains had only

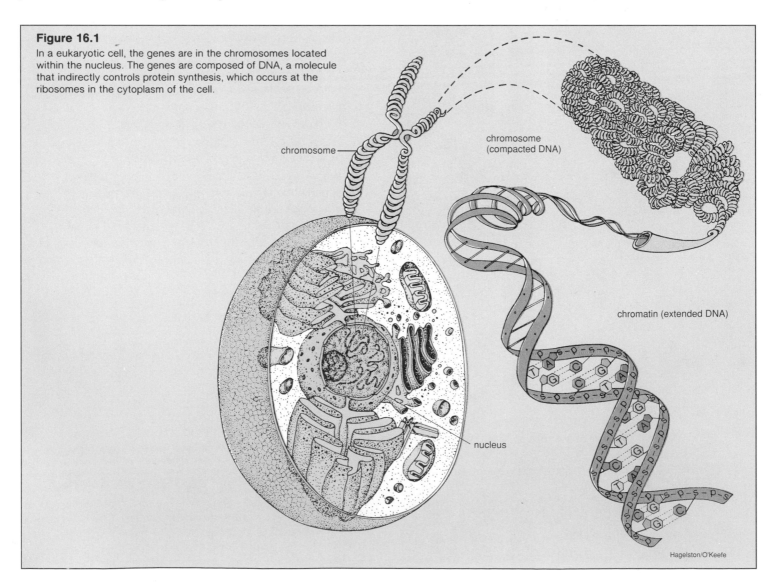

Figure 16.1

In a eukaryotic cell, the genes are in the chromosomes located within the nucleus. The genes are composed of DNA, a molecule that indirectly controls protein synthesis, which occurs at the ribosomes in the cytoplasm of the cell.

chromosome

chromosome (compacted DNA)

chromatin (extended DNA)

nucleus

Hagelston/O'Keefe

one defective gene leading to one defective enzyme and one additional growth requirement. Therefore, they proposed that each gene specifies the synthesis of one enzyme. This is called the **one gene–one enzyme hypothesis.**

One Gene–One Polypeptide Hypothesis

The one gene–one enzyme hypothesis suggests that a genetic mutation causes a change in the structure of a protein. To test this idea, Linus Pauling and Harvey Itano decided to see if the hemoglobin in the red blood cells of persons with sickle-cell anemia has a different structure from that in the red blood cells of normal individuals (fig. 16.3). You'll recall that proteins are polymers of amino acids, some of which carry a charge. These investigators decided to see if there was a charge difference between normal hemoglobin, Hb^A, and sickle-cell hemoglobin, Hb^S. To determine this they subjected hemoglobin collected from

normal individuals, sickle-cell trait individuals, and sickle-cell anemia individuals to electrophoresis, a procedure that separates similar-size molecules according to their charge. Here is what they found:

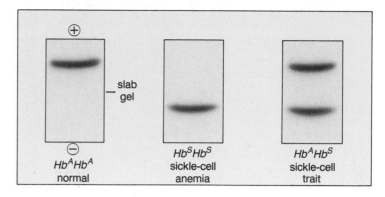

Figure 16.2
Beadle and Tatum experiment with *Neurospora crassa. a.* When haploid spores are X-rayed, some are no longer able to germinate on minimal medium; however, these can germinate on enriched medium. In this example, the mycelia produced will not grow on minimal medium and metabolite A, or on minimal medium and metabolite B, but will grow on minimal medium and metabolite C, and on minimal medium and metabolite D. This shows that enzyme 2 is missing from the hypothetical pathway.

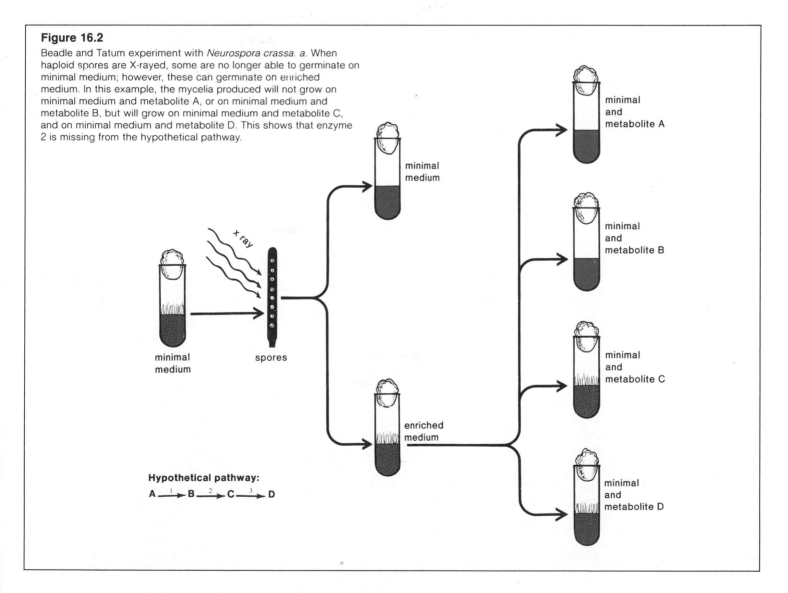

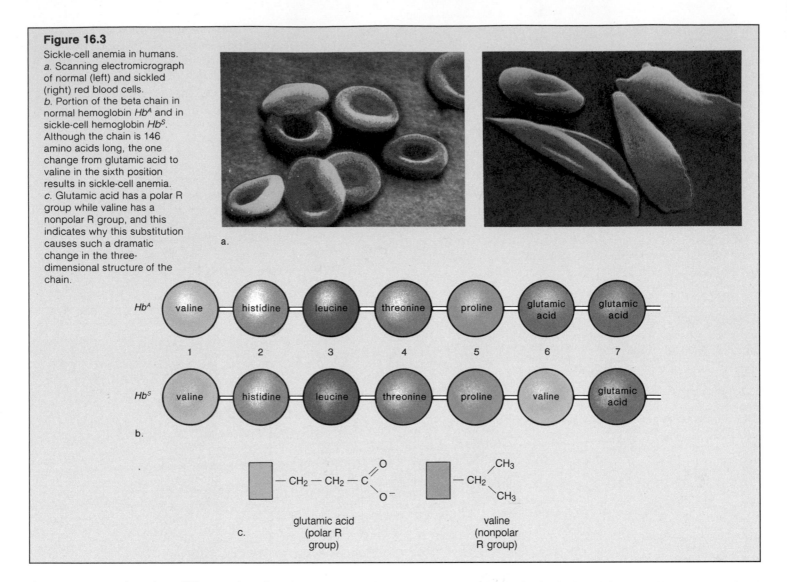

Figure 16.3

Sickle-cell anemia in humans.
a. Scanning electromicrograph of normal (left) and sickled (right) red blood cells.
b. Portion of the beta chain in normal hemoglobin Hb^A and in sickle-cell hemoglobin Hb^S. Although the chain is 146 amino acids long, the one change from glutamic acid to valine in the sixth position results in sickle-cell anemia.
c. Glutamic acid has a polar R group while valine has a nonpolar R group, and this indicates why this substitution causes such a dramatic change in the three-dimensional structure of the chain.

As you can see there is a difference in migration rate toward the positive pole between normal hemoglobin and sickle-cell hemoglobin. Further, hemoglobin from those with sickle-cell trait separates into two distinct bands, one corresponding to that for Hb^A and the other corresponding to that for Hb^S hemoglobin. Pauling and Itano therefore demonstrated that a mutation leads to a change in the structure of a protein.

Several years later, V. M. Ingram was able to determine the structural difference between the two types of hemoglobin, a molecule that always contains two alpha (α) polypeptide chains and two beta (β) chains (fig. 4.18*b*). In position 6 of the beta chain, normal hemoglobin contains negatively charged glutamic acid and sickle-cell hemoglobin contains nonpolar valine. This explains the slower migration of Hb^S toward the positive pole during electrophoresis. It also causes Hb^S to be less soluble and to precipitate out of solution, especially when environmental oxygen is low. At these times, the Hb^S molecules stack up into long, semirigid rods that push against the plasma membrane and distort the red blood cell into the sickle shape.

The alpha chains in sickle-cell hemoglobin are normal and only the beta chains are affected; therefore, there must be a separate gene for each of these two types of chains. A refinement of the one gene–one enzyme hypothesis was needed and it was replaced by the **one gene–one polypeptide hypothesis.**

Each gene contains the information for making one polypeptide of a protein, a molecule that may contain one or more different polypeptides. A mutated gene can lead to a defective polypeptide chain.

The Central Dogma of Biology

DNA is a linear polymer of nucleotides and proteins are linear polymers of amino acids. It seems plausible, then, that the nucleotide sequence of DNA somehow determines the order of amino acids in proteins. However, the DNA double helix cannot be the direct template for protein synthesis, because DNA is found in the nucleus (or the nucleoid in prokaryotes) and protein synthesis occurs in the cytoplasm. Therefore another molecule must carry out this task and the most likely candidate is RNA, a molecule found in both the nucleus and the cytoplasm.

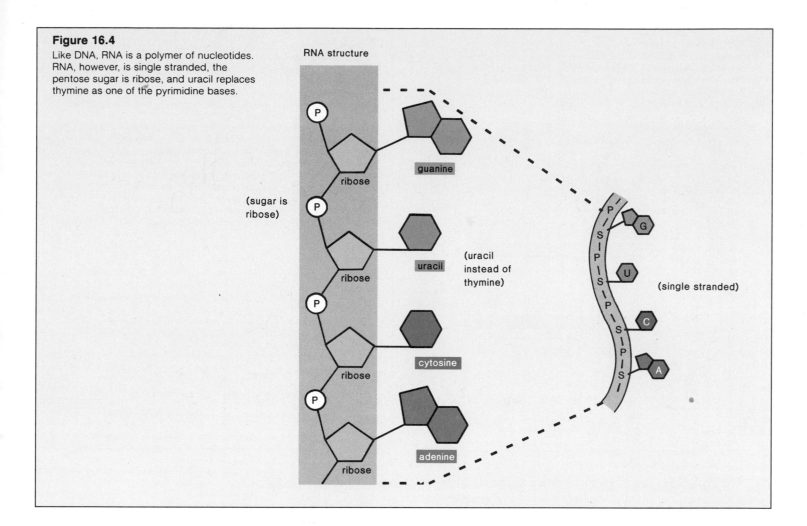

Figure 16.4
Like DNA, RNA is a polymer of nucleotides. RNA, however, is single stranded, the pentose sugar is ribose, and uracil replaces thymine as one of the pyrimidine bases.

RNA structure

(sugar is ribose)

guanine
ribose

uracil
ribose

(uracil instead of thymine)

cytosine
ribose

adenine
ribose

(single stranded)

RNA

Like DNA, RNA (fig. 16.4) is a polymer of nucleotides; however, in RNA, the sugar component is *ribose* (table 16.1). Also, RNA contains the bases adenine (A), cytosine (C), and guanine (G) as does DNA, but instead of thymine (T) it contains the base uracil (U). *Uracil* is a pyrimidine that can hydrogen bond to adenine like thymine does. Finally, RNA is single stranded and does not form a double helix in the same manner as DNA.

There are three different classes of RNA, each with specific functions in protein synthesis: messenger RNA (mRNA), ribosomal RNA (rRNA), and transfer RNA (tRNA).

RNA and DNA

By 1958 the **central dogma** of molecular biology had been formulated in this manner:

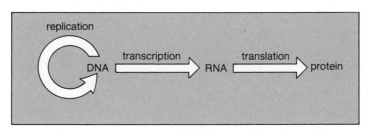

Table 16.1 DNA Compared to RNA

	DNA	RNA
Function	Genes; controls protein synthesis	Helper to DNA; involved in protein synthesis
Sugar	Deoxyribose	Ribose
Bases	Adenine, guanine, thymine, cytosine	Adenine, guanine, uracil, cytosine
Strands	Double stranded with base pairing	Single stranded
Helix	Yes	No

This diagram indicates that DNA not only serves as a template for its own replication, it is also a template for RNA formation. The arrow between DNA and RNA indicates that RNA is transcribed off the DNA template. **Transcription** means making an RNA that is complementary to a portion of DNA. Following transcription, RNA moves into the cytoplasm. Photographs are available that show that radioactively labeled RNA moves from the nucleus to the cytoplasm where protein synthesis occurs.

Figure 16.5

Transcription and translation in protein synthesis. Transcription occurs when DNA acts as a template for RNA synthesis. Translation occurs when the sequence of codons (groups of three bases) carried by a type of RNA called messenger RNA (mRNA) dictates the sequence of amino acids in a polypeptide.

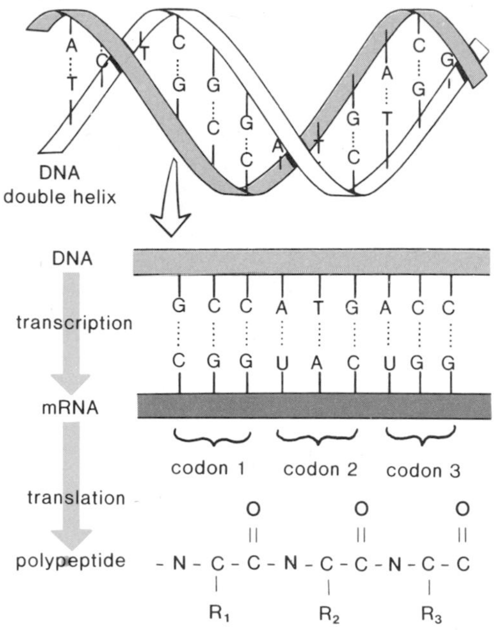

Figure 16.6

The first line of this figure assumes that the reading frame of DNA is three bases at a time and the bases CAT are repeated over and over again. If the base G is added (line 2) we change the reading frame so that all the bases thereafter read TCA. Or if a base A is deleted at the pointer (line 3) we change the reading frame so that all the bases thereafter read ATC.

To prove that DNA contains a triplet code, Francis Crick and colleagues treated the bacteriophage T_4 with a chemical agent (acridine dye) that usually causes a single-base insertion or deletion. In keeping with the assumption displayed in the figure, Crick found that either one of these frameshift mutations is so severe that the mutant gene cannot code for a functional protein. Also in keeping with the existence of a triplet code, Crick discovered that recombinants between these two mutations sometimes gave a "pseudo-wild-type" gene like the one on line 4. The first two codons, CAG and TCT, are wrong, but thereafter the insertion and deletion cancel each other out and the original reading frame is restored. Therefore, a functional protein sometimes results. Finally, Crick discovered that a set of three adjacent insertions or three adjacent deletions could produce a pseudo-wild-type gene (line 5) and a functional protein. This can only be if the code is indeed a triplet code.

1. Wild-type:	CAT	CAT	CAT	CAT	CAT
2. Add a base:	CAG	TCA	TCA	TCA	TCA
3. Delete a base:	CAT	C△TC	ATC	ATC	ATC
4. Cross #2 and #3:	CAG	TCT	CAT	CAT	CAT
5. Add 3 bases:	CAG	GGT	CAT	CAT	CAT

The arrow between RNA and protein in the diagram indicates all proteins are synthesized according to RNA's instructions. **Translation** means that the information carried by RNA (in nucleotides) is used to make a protein (of amino acids). You can see that these terms are apt because transcribing a document means making a close copy of it, while translating a document means putting it in an entirely different language (fig. 16.5).

> The central dogma of biology states that DNA serves as a template for its own replication and for the transcription of complementary RNA. Then the information carried by RNA is translated into protein during protein synthesis.

The Genetic Code

The central dogma suggests that the sequence of nucleotides in DNA and RNA must, in some way, direct the ordering of amino acids in a protein. It would seem that DNA must have a code for each of the 20 amino acids found in proteins. But can 4 nucleotides provide enough combinations to code for 20 amino acids? If each code word, called a **codon,** were 2 bases, such as AG, there could be only 16 codons (4^2)—not enough to code for 20 amino acids. But if each codon were 3 bases, such as AGC, there would be 64 codons (4^3)—more than enough to code for 20 different amino acids. It should come as no surprise, then, to learn that the genetic code is a **triplet code** and that each codon consists of 3 nucleotide bases (fig. 16.6).

Cracking the Code

In 1961, Marshall Nirenberg and J. Heinrich Matthei performed an experiment that laid the groundwork for cracking the DNA code. First, they found a cellular enzyme could be used to construct a synthetic RNA (one that does not occur in cells), and then they found that the synthetic polymer could be translated in a cell-free system (a test tube that contains "freed" cytoplasmic contents of a cell). Their first synthetic RNA was composed only of uracil and the protein that resulted was composed only of the amino acid phenylalanine. Therefore, the triplet code for phenylalanine was known to be UUU. Later, a cell-free system was developed by Nirenberg and Philip Leder in which only three nucleotides at a time were translated; in that way it was possible to assign an amino acid to each of the RNA codons (fig. 16.7).

A number of important properties of the genetic code can be seen by careful inspection of figure 16.7:

1. The genetic code is degenerate. This means that most amino acids have more than one codon; leucine, serine, and arginine have six different codons. The degeneracy of the code is probably a protective device that reduces the potentially harmful effects of mutations.
2. The genetic code is definite. Each triplet codon has only one meaning.
3. The code has "start" and "stop" signals. There is only one start signal but three stop signals.

Figure 16.7

These messenger RNA (mRNA) codons are complementary to those in DNA. The DNA code is degenerate; there is more than one codon for most of the twenty amino acids. Notice that in this chart each of the codons is comprised of a letter from the first, second, and third position. For example, find the square where C (from the first position) and A (from the second position) come together and then look across to the right, you will notice the letters for the third position of the codons for histidine and glutamine.

First Base	Second Base				Third Base
	U	C	A	G	
U	UUU phenylalanine	UCU serine	UAU tyrosine	UGU cysteine	U
	UUC phenylalanine	UCC serine	UAC tyrosine	UGC cysteine	C
	UUA leucine	UCA serine	UAA *stop*	UGA *stop*	A
	UUG leucine	UCG serine	UAG *stop*	UGG tryptophan	G
C	CUU leucine	CCU proline	CAU histidine	CGU arginine	U
	CUC leucine	CCC proline	CAC histidine	CGC arginine	C
	CUA leucine	CCA proline	CAA glutamine	CGA arginine	A
	CUG leucine	CCG proline	CAG glutamine	CGG arginine	G
A	AUU isoleucine	ACU threonine	AAU asparagine	AGU serine	U
	AUC isoleucine	ACC threonine	AAC asparagine	AGC serine	C
	AUA isoleucine	ACA threonine	AAA lysine	AGA arginine	A
	AUG (*start*) methionine	ACG threonine	AAG lysine	AGG arginine	G
G	GUU valine	GCU alanine	GAU aspartic acid	GGU glycine	U
	GUC valine	GCC alanine	GAC aspartic acid	GGC glycine	C
	GUA valine	GCA alanine	GAA glutamic acid	GGA glycine	A
	GUG valine	GCG alanine	GAG glutamic acid	GGG glycine	G

The triplet genetic code is made of three-base code words called codons. There are 64 codons: one is a start signal and three are stop signals. All the rest code for amino acids.

The genetic code given in figure 16.7 is just about universally used by living things. This suggests that the code dates back to the very first organisms on earth and that all living things are related. Once the code became established, changes in it would have been very disruptive, and therefore it has remained unchanged over eons of time.

Figure 16.7 applies to the genes located in the nucleus. What about the mitochondria? They have genes and carry on protein synthesis also. As it turns out, the code in mitochondria is slightly different from that utilized in the nucleus. This is consistent with the endosymbiotic theory that says mitochondria were once independent organisms. It also means that eukaryotes have a type of cytoplasmic inheritance, inheritance dependent on genes outside the nucleus. It would seem that these genes are inherited only through the maternal line, because the egg and not the sperm contributes organelles to the new individual. This is an area of intense investigation at the moment.

Transcription

Transcription means making an RNA copy of a portion of DNA. In many instances it is **messenger RNA (mRNA)** that is being produced. Messenger RNA is so named because it carries a message from DNA to the ribosomes in the cytoplasm, where protein synthesis occurs. The message is, of course, a sequence of codons that indicates the proper sequence of amino acids in a protein. Transcription is the first step required for **gene expression,** the process by which gene products are made. Translation is also required before the protein product is realized (fig. 16.5).

Steps in Transcription

Following transcription (fig. 16.8), mRNA has a sequence of bases complementary to DNA; wherever A, T, G, or C is present in the DNA template, U, A, C, or G is incorporated into the mRNA molecule. A segment of the DNA helix unwinds and unzips, and complementary RNA nucleotides pair with DNA nucleotides of one strand. When these RNA nucleotides are joined together by an **RNA polymerase,** an mRNA molecule results. This molecule now carries a series of codons that will be used to order the sequence of amino acids in a protein.

These specific steps are required for transcription:

1. Template binding: The enzyme RNA polymerase begins transcription at a specific location called a **promoter** region. The strand that is to be transcribed is called the *sense strand* and the other DNA strand is called the *antisense strand*.
2. RNA chain initiation: RNA synthesis begins when the first two ribonucleotides are joined. The promoter region is not transcribed.
3. RNA chain elongation: the chain lengthens (synthesis is in the 5′ → 3′ direction, p. 226). Only a portion of the RNA is bound (through complementary base pairing) to DNA at a time and the rest dangles off to the side (fig. 16.9).
4. RNA chain termination: Eventually the RNA polymerase encounters a stop signal, a specific nucleotide sequence, and the completed transcribed molecule of RNA, called an **RNA transcript,** is released from the DNA template.

RNA Transcript Processing in Eukaryotes

Most genes in eukaryotes are interrupted by segments of DNA that are not part of the gene. These portions are called **introns** because they are *intra*gene segments. The other portions of the gene are called **exons** because they are ultimately *ex*pressed.

Figure 16.8

During transcription, complementary RNA is made off of a DNA template. The DNA helix unwinds and unzips, and complementary RNA nucleotides are joined together by RNA polymerase. This is called transcription because the codons present in DNA are transcribed into RNA nucleotides. This is the message that mRNA takes to the cytoplasm.

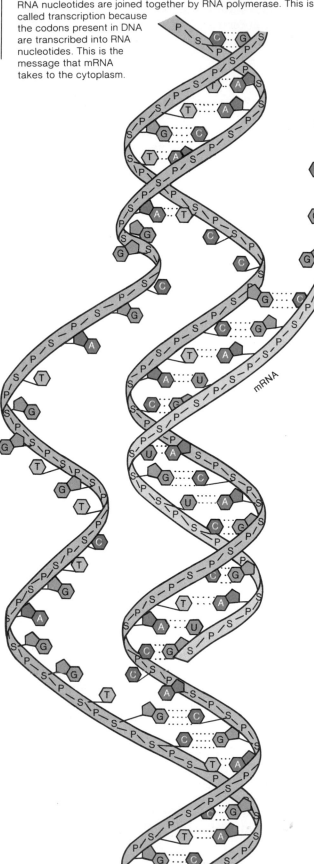

Figure 16.9

a. Numerous RNA transcripts extend from a horizontal gene in an amphibian egg cell. *b.* The strands get progressively longer because transcription begins to the left. The dark dots along the DNA are RNA polymerase molecules that are joining RNA nucleotides together. The dots at the end of the strands are spliceosomes involved in RNA processing (*see* fig. 16.10).

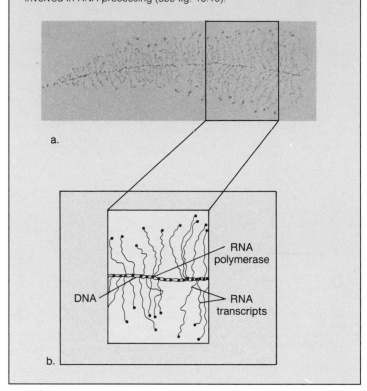

That eukaryotes had **split genes** was discovered by comparing the length of specific mRNA molecules in the cytoplasm with those in the nucleus. For example, the initial mRNA transcript for beta globin contains about 1,500 nucleotides, while the cytoplasmic mRNA contains about 700 nucleotides. It is clear that some nucleotides (the introns) were excised before the transcript left the nucleus. It is now believed that splicing of RNA may be done by *spliceosomes,* a complex that contains several kinds of ribonucleoproteins (fig. 16.10). Splicing of RNA is a delicate operation that must be extremely precise.

RNA processing in eukaryotes not only includes removing introns, but also includes an earlier addition of nucleotides at each end. A series of adenine nucleotides is added to one end of the RNA transcript, forming what is called a poly A tail, and a modified GTP (guanosine triphosphate), called a cap, is added to the other end. The significance of these two steps is unknown, although the tail and the cap appear to be necessary for efficient translation of mRNA.

There has been much speculation about the possible role of introns in the eukaryotic genome. It's possible that introns allow crossing-over within a gene during meiosis. It's also possible that introns divide a gene up into domains that can be joined in different combinations to give novel genes and protein products, a process that perhaps facilitates evolution.

Figure 16.10

mRNA processing. *a.* DNA contains both exons (coding sequences) and introns (noncoding sequences). *b.* Both of these are transcribed. *c.* The first event in processing is the addition of a cap (modified GTP molecule) to the 5' end of the mRNA and a poly A tail (many adenine molecules) to the 3' end of mRNA. *d.* The next event in processing is the excision of the introns and the splicing together of the exons. This is accomplished by complexes called spliceosomes. *e.* Now the mRNA molecule is ready to leave the nucleus.

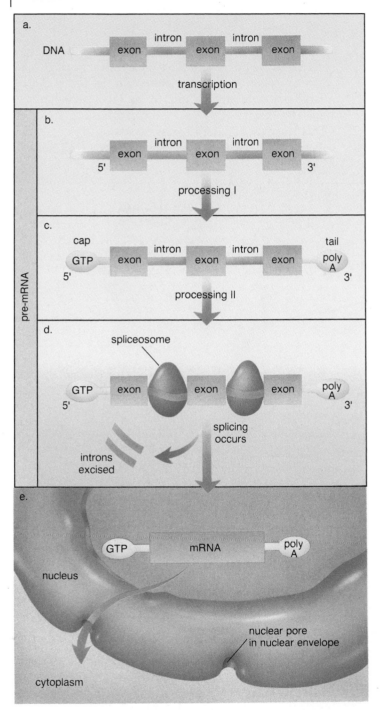

Genetic information, stored in DNA, is transferred to RNA during transcription. Messenger RNA carries a sequence of codons that direct the order of amino acids during protein synthesis. In eukaryotes, RNA is processed before it leaves the nucleus.

Translation

Protein synthesis occurs at the ribosomes, where a type of RNA called **ribosomal RNA (rRNA)** is found. There is another type of RNA called **transfer RNA (tRNA)** that is also involved in translation.

Ribosomal RNA and Ribosomal Proteins

Cells contain a very large number of ribosomes; bacterial cells have about 10,000 and eukaryotic cells contain many times more. Each ribosome is composed of two subunits that join together when protein synthesis begins (fig. 16.14). In prokaryotes, the large subunit contains two rRNA molecules and 34 different proteins; the smaller one contains one rRNA molecule and 20 different proteins. Eukaryotic cells have larger ribosomes and each subunit contains longer rRNA molecules and more proteins. In these cells, rRNA is transcribed off of rRNA genes found in nucleoli. Here there are multiple copies of rRNA genes so that many copies can be made in a short period of time. The newly synthesized rRNA is immediately packaged with ribosomal proteins that are manufactured in the cytoplasm. The subsequent subunits enter the cytoplasm separately and remain separate until a ribosome attaches to an mRNA molecule.

Transfer RNA

At the ribosomes, the message carried by mRNA is translated into a sequence of amino acids. How can the one language (nucleic acids) be translated into the other language (protein)? Naturally, a translator that knows both languages is needed. Transfer RNA molecules, so named because they transfer amino acids from the cytoplasm to the ribosomes, are such molecules. The structure of a tRNA is generally drawn as a flat cloverleaf but the space-filling model shows the molecule's actual shape (fig. 16.11). One end of each tRNA molecule contains an **anticodon,** a group of three bases that is complementary to a specific codon. The other end binds to one of the 20 amino acids, the one, of course, that is coded for by that codon.

How does a tRNA molecule get joined to the right amino acid? This all important operation is done by a number of *amino acid activating enzymes,* each one specific for one kind of amino acid and its tRNA. This is an energy-requiring process that utilizes ATP. Once the amino acid ~ tRNA complex is formed, it travels through the cytoplasm to a ribosome where protein synthesis is occurring.

Although there are 61 codons, there are only about 30 tRNA molecules. All 61 codons are read, however, because many tRNA molecules can bind to more than one codon. The pairing of the first two bases is set but there is some leeway on the third

Figure 16.11

Structure of a transfer RNA (tRNA) molecule. *a.* Complementary base pairing indicated by dashes occurs between nucleotides that make up the molecule and this causes the molecule to form its characteristic loops. The anticodon that pairs with a particular mRNA codon in a complementary manner occurs at one end of the folded molecule and the appropriate amino acid is attached at the other end of the molecule. *b.* Space-filling model of tRNA molecule.

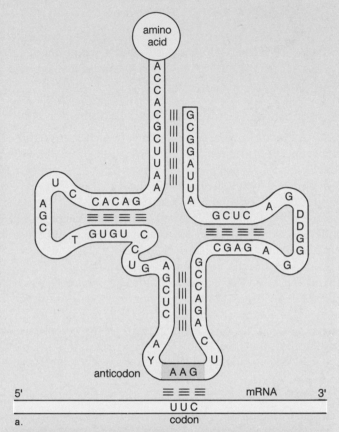

a.

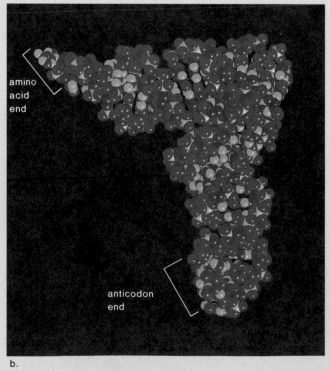

b.

base—this is known as the wobble effect. For example, there is one tRNA molecule that will pair with the codons GGU, GGC, or GGA, but it doesn't introduce an error because all of these code for the amino acid glycine.

Translation Process

The process of translation must be extremely orderly so that the amino acids of a polypeptide are sequenced correctly. Protein synthesis involves three steps—initiation, elongation, and termination.

1. Initiation of translation: A small ribosome subunit attaches to the mRNA in the vicinity of the **start codon.** The first, or initiator, tRNA pairs with this codon. Then a large ribosome subunit joins to the small subunit and translation begins.

2. Chain elongation: Each ribosome contains two sites, called the P (for polypeptide) site and the A (for amino acid) site:

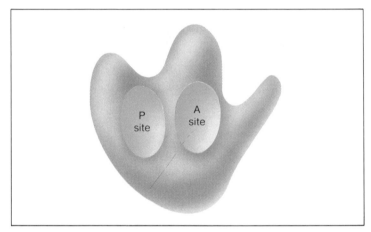

During elongation (fig. 16.12), a tRNA with attached polypeptide chain is at the P site and an amino acid ~ tRNA complex has just arrived at the A site. The polypeptide chain is transferred and attached by a new peptide bond to the newly arrived amino acid. A cellular enzyme (peptidyl transferase) and the use of energy is needed to bring about this transfer. Then, **translocation** occurs: the mRNA along with the polypeptide-bearing tRNA moves from the A site to the now empty P site.

This cycle is repeated at a rapid rate (about 15 amino acids each second in *E. coli*).

3. Chain termination: Termination of polypeptide chain synthesis occurs at a **stop codon,** codons that do not code for an amino acid. The polypeptide chain is enzymatically cleaved from the last tRNA and it leaves the ribosome, which dissociates into its two subunits.

Polysomes

As soon as the initial portion of mRNA has been translated by one ribosome, it is free to have another ribosome attach to it. Therefore, several ribosomes are often attached to and translating the same mRNA. The entire complex is called a **polysome** (fig. 16.13).

Figure 16.12

Chain elongation during protein synthesis. *a.* A tRNA bearing a peptide chain is already positioned on a ribosome (at the P site) *b.* A tRNA with the anticodon UUU recognizes its codon and binds to it (at the A site). *c.* The peptide chain is transferred and a peptide bond is formed with the new amino acid. As translocation occurs, a tRNA leaves. *d.* The tRNA bearing the peptide chain is now at the P site and another tRNA amino acid complex is approaching the ribosome. The same sequence of events (*a–d*) will now reoccur.

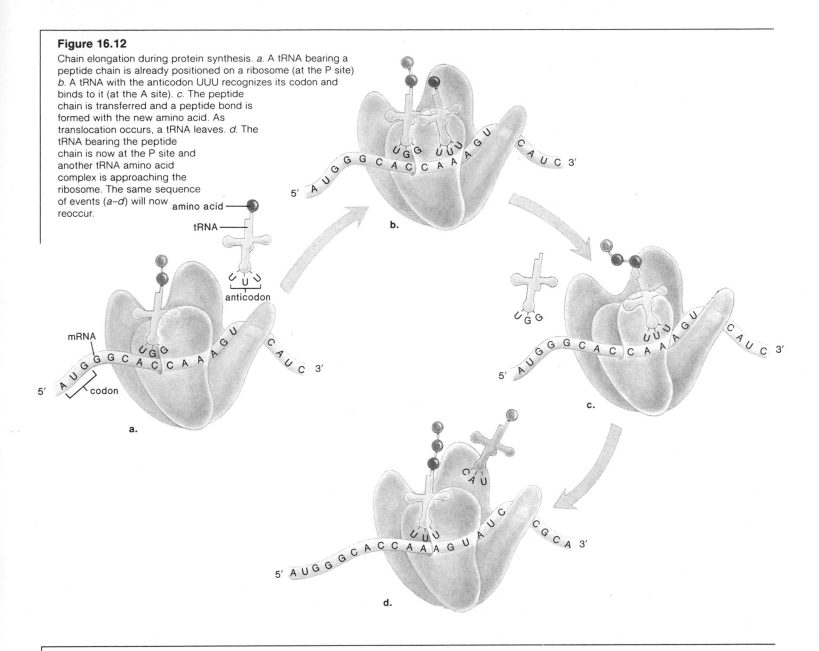

Figure 16.13

Polysome structure. *a.* Several ribosomes move along an mRNA at a time. They function independently of each other so that several polypeptides can be made at the same time. *b.* Electron micrograph of a polysome. Magnification, ×400,000.

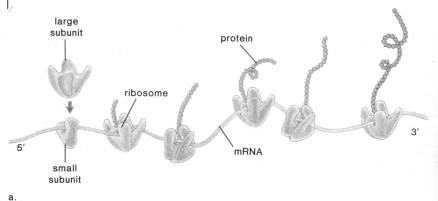

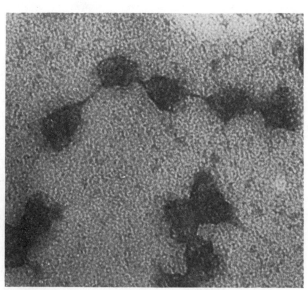

Figure 16.14

Summary of protein synthesis. Transcription occurs in the nucleus and translation occurs in the cytoplasm (blue). During translation, the codons borne by mRNA dictate the order of the amino acids in the polypeptide.

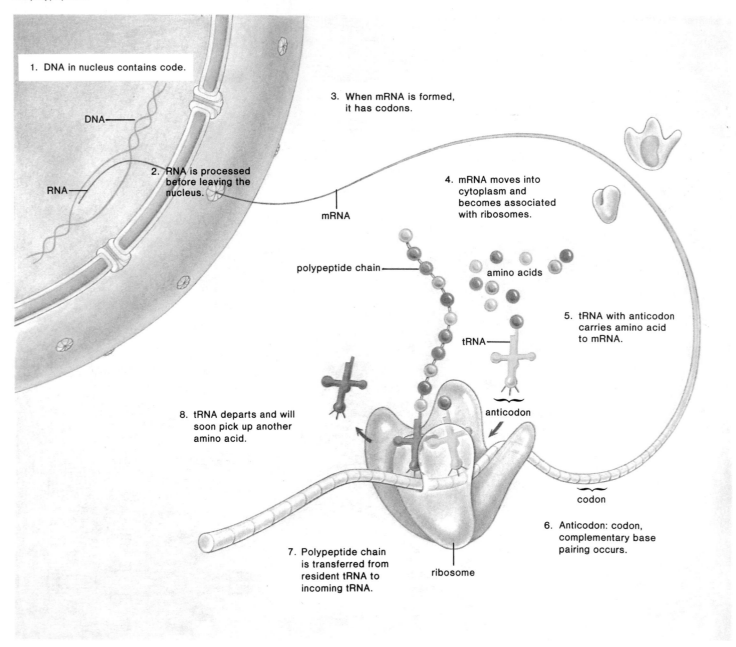

1. DNA in nucleus contains code.

DNA

RNA

3. When mRNA is formed, it has codons.

2. RNA is processed before leaving the nucleus.

mRNA

4. mRNA moves into cytoplasm and becomes associated with ribosomes.

polypeptide chain

amino acids

5. tRNA with anticodon carries amino acid to mRNA.

tRNA

anticodon

8. tRNA departs and will soon pick up another amino acid.

codon

6. Anticodon: codon, complementary base pairing occurs.

7. Polypeptide chain is transferred from resident tRNA to incoming tRNA.

ribosome

Polysomes are found both free within the cytoplasm and also seemingly bound to the endoplasmic reticulum. Actually they are held there because the growing polypeptide is in the process of entering the endoplasmic reticulum. It may be modified here or in the Golgi complex for incorporation into a cellular structure or for export from the cell.

During translation, the codons of mRNA base pair with the anticodons of tRNA molecules carrying specific amino acids. The order of the codons determines the order of the tRNA molecules and determines the sequence of amino acids in a polypeptide.

Protein Synthesis in Review

The following list, along with table 16.2 and figure 16.14 provides a brief summary of the events involved in protein synthesis.

1. DNA contains a sequence of bases that code for a sequence of amino acids in a protein. The code is a triplet code in which the code words, codons, are groups of three bases.

2. During transcription, one strand of DNA (the sense strand) serves as a template for the formation of messenger RNA (mRNA). Messenger RNA, therefore, has a series of bases that are complementary to those in DNA.
3. Messenger RNA is processed before it leaves the nucleus during which time the introns are removed.
4. Messenger RNA carries a sequence of codons to the ribosomes, which are composed of ribosomal RNA (rRNA) and proteins.
5. Transfer RNA (tRNA) molecules, each of which is bonded to a particular amino acid, have anticodons that pair complementarily to the codons in mRNA.
6. During translation, the linear sequence of codons of the mRNA determines the order in which the tRNA molecules and their attached amino acids arrive at the ribosomes, and this determines the primary structure (linear sequence of amino acids) of a protein.

Table 16.2 Participants in Protein Synthesis

Name of Molecule	Special Significance	Definition
DNA	Code (cell genotype)	Sequence of DNA bases in threes
mRNA	Codon	Complementary sequence of RNA bases in threes
tRNA	Anticodon	Sequence of three bases complementary to codon
rRNA	Ribosomes	Site of protein synthesis
Amino acids	Building blocks for protein	Transported to ribosomes by tRNAs
Protein	Enzyme (cell phenotype)	Amino acids joined in a predetermined order

Practice Problem

This is a segment of a DNA molecule. What are (1) the messenger RNA codons, (2) the possible tRNA anticodons, and (3) the sequence of amino acids in the polypeptide?

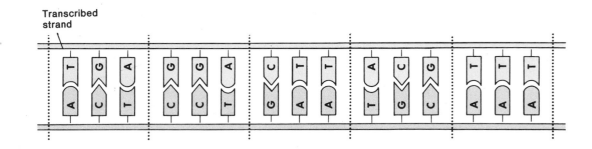

Transcribed strand

*Answers to Practice Problems are given on page 806 of appendix D.

Definition of a Gene and Gene Mutations

Classical (Mendelian) geneticists thought of a gene as a particle on a chromosome. To molecular geneticists, however, a gene is a sequence of DNA nucleotide bases that code for a product. Most often the gene product is a polypeptide chain. However, there are exceptions because rRNA and tRNA, for example, are gene products themselves. They are involved in protein synthesis, but they do not code for protein.

Gene Mutations

A gene is a sequence of DNA nucleotide bases and a **gene mutation** is a change in that sequence. Table 16.3 lists the different types of mutations so you can become familiar with the terminology that pertains to mutations.

Types of Gene Mutations

Frameshift Mutations **Frameshift mutations** can occur when a single base is inserted into or deleted from a gene. Both of

Table 16.3 Mutations

Type of Mutation	Definition
Chromosome mutation*	An alteration in the number or arrangement of genes on chromosomes (fig. 13.17) often has a drastic effect on the phenotype.
Gene mutation	An alteration in the nucleotide sequence of a gene or in the expression of the gene; may have an effect on the phenotype.
Germinal mutation	A mutation that occurs in the gametes and can be transmitted to the offspring.
Somatic mutation	A mutation that occurs in the body cells, can cause cancer, but cannot be transmitted to the offspring.

*See page 197.

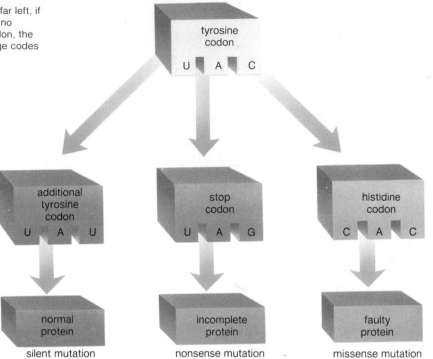

Figure 16.15
The effect of a substitution mutation can vary. Starting at far left, if the base change codes for the same amino acid, there is no noticeable effect; if the base change codes for a stop codon, the resulting protein will be incomplete; and if the base change codes for a different amino acid, a faulty protein is possible.

these change the reading frame of the bases (fig. 16.6 lines 2 and 3) and therefore either of these is expected to ultimately result in a nonfunctioning protein.

Substitutions   When one base is substituted for another, the results can be variable. For example, in figure 16.15, if UAC is changed to UAU there will be no noticeable effect, because both of these codons code for tyrosine. Therefore, this is called a silent mutation. However, if UAC is changed to UAG, the result could very well be a drastic one because UAG is a stop codon. If this substitution occurs early in the gene, the resulting protein may be much too short and unable to function. Such an effect is called a nonsense mutation. Finally, if UAC is changed to CAC then histidine will be incorporated into the protein instead of tyrosine. A change in one amino acid may not have an effect if the change occurs in a noncritical area or if the two amino acids have the same chemical properties. However, in this instance the polarity of tyrosine and histidine differ, and therefore, this substitution may very well have a deleterious effect on the functioning of the protein. You will recall that the occurrence of valine instead of glutamic acid in the beta (β) chain of hemoglobin results in sickle-cell anemia (fig. 16.3), for example.

Rate of Gene Mutations

Normally, gene mutations don't occur very often; for example, a frequency of 10^{-8} to 10^{-5} per generation is often quoted. There are several protective mechanisms against the occurrence of mutations. First, the bases are on the interior of the DNA molecule. Then, the supercoiling of the molecule in eukaryotes also lends stability. Finally, during replication DNA polymerases proofread the new strand against the old strand and detect any mismatched pairs, which are then replaced with the correct nu-

cleotides. In the end there is usually only one mistake for every 1 billion nucleotide pairs replicated.

Organisms, including humans, are continually exposed to **mutagens,** environmental substances that cause mutations, such as radiation (e.g., radioactive elements and X rays) and organic chemicals (e.g., certain pesticides and cigarette smoke). If these mutagens bring about a mutation in the gametes, then the offspring of the individual may be affected. On the other hand, if the mutation occurs in the body cells, then cancer may be the result. Ultraviolet (UV) radiation is a mutagen that everyone is exposed to; it easily penetrates the skin and breaks the DNA of underlying tissues. Wherever there are two thymine molecules next to one another, ultraviolet radiation may cause them to bond together forming thymine dimers:

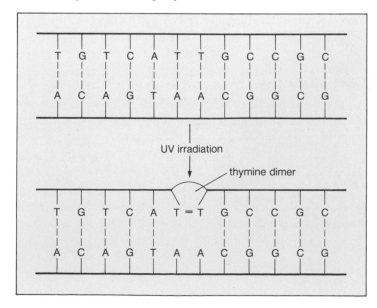

Usually, these dimers are removed from damaged DNA by special enzymes called **repair enzymes.** Repair enzymes are constantly monitoring DNA and repairing any irregularities such as this. The importance of this action is exemplified in that individuals with the condition known as xeroderma pigmentosum lack some of these repair enzymes and as a consequence have a high incidence of skin cancer.

Significance of Mutations

Although most mutations we observe are harmful, they still provide the raw material for evolutionary change. Gene mutations, along with recombinations of genes, produce a new genotype whose phenotype might be better adapted to the environment. Better adapted individuals are expected to have more surviving offspring than those less adapted and in that way the new genotype will become prevalent; in this way evolution can occur.

Gene mutations are an alteration in the nucleotide sequence of a gene. The spontaneous rate of mutation is low because DNA polymerases proofread the new strand against the old strand during replication and because DNA repair enzymes constantly monitor and repair any irregularities.

Summary

1. Several investigators contributed to recognition of gene activity. A. Garrod is associated with the phrase "inborn errors of metabolism" because he suggested that some of his patients had inherited their inability to carry out certain enzymatic reactions.

2. G. W. Beadle and E. L. Tatum X-rayed spores of *Neurospora crassa* and found that certain of the subsequent cultures lacked a particular enzyme needed for growth on minimal medium. Since they found that the mutation of one gene resulted in the lack of a single enzyme, they suggested the one gene–one enzyme hypothesis.

3. L. Pauling and H. Itano found that the chemical properties of the beta chain of sickle-cell hemoglobin differed from that of normal hemoglobin, and therefore the one gene–one polypeptide chain hypothesis was formulated. Later V. M. Ingram showed that the biochemical change was due to the substitution of the amino acid valine (nonpolar) for glutamic acid (polar).

4. The central dogma of biology says that: (1) DNA serves as a template for its own replication, and (2) as a template for RNA formation during transcription; and (3) the complementary sequence of nucleotides in mRNA orders the correct sequence of amino acids of a polypeptide during translation.

5. RNA differs from DNA in these ways: (1) the pentose sugar is ribose, not deoxyribose; (2) the base uracil replaces thymine; (3) RNA is single stranded.

6. DNA has a code to indicate the proper sequence of amino acids in a protein. It is a triplet code and each codon (code word) consists of three bases. The code is degenerate, that is, there is more than one codon for most amino acids. There are also one start and three stop codons.

7. The genetic code is just about universal; one exception is mitochondrial DNA has been found to have different codons than nuclear DNA for certain amino acids.

8. Transcription requires: template binding (RNA polymerase begins transcription at a promoter region); RNA chain initiation (first two nucleotides join); chain elongation (chain lengthens); and chain termination (RNA is released because of stop signal).

9. In eukaryotes, mRNA is processed following transcription: (1) a cap composed of a modified GTP molecule is placed at the 5′ end and a poly A tail is placed at the 3′ end, and (2) introns are excised by spliceosomes.

10. Translation requires the use of mRNA, rRNA, and tRNA. Ribosomal RNA (rRNA) is found in the ribosomes; transfer RNA (tRNA) is found in the cytosol. Both of these are transcribed off DNA. Each tRNA has an anticodon at one end and an amino acid at the other; there are amino acid activating enzymes that see to it that the correct amino acid is attached to the correct tRNA. When a tRNA binds with its codon at the ribosome, the correct amino acid is positioned for inclusion in a polypeptide.

11. Translation requires these steps. During initiation, mRNA, the first (initiator) tRNA, and the ribosome all come together in the proper orientation at a start codon. During elongation, as the tRNAs recognize their codons, the growing peptide chain is transferred by peptide bonding to the next amino acid in the chain. During termination at a stop codon, the polypeptide is cleaved from the last tRNA; the ribosome dissociates, and the mRNA leaves.

12. Many ribosomes move along an mRNA at a time. Collectively this is called a polysome. Some proteins enter the endoplasmic reticulum and later the Golgi complex for modification.

13. In molecular terms, a gene is a sequence of DNA nucleotide bases and a gene mutation is a change in this sequence. Frameshift mutations result when a base is added or deleted and the result is a nonfunctioning protein. Substitution mutations can range in their effects depending on the particular codon change. Gene mutation rates are rather low because DNA polymerase proofreads the new strand during replication and because there are repair enzymes that constantly monitor the DNA.

14. Gene mutations are the raw material for the evolutionary process and along with recombination they bring about greater adaptation to an environment and a change in the most prevalent phenotype of the species.

Objective Questions

For questions 1 to 4, match the investigator to the phrase in the key:

Key:
- a. one gene–one enzyme hypothesis
- b. inborn errors of metabolism
- c. one gene—one polypeptide hypothesis
- d. triplet code

1. Sir Archibald Garrod 𝐵
2. Francis Crick 𝐷
3. G. W. Beadle and E. L. Tatum 𝐴
4. Linus Pauling and Harvey Itano 𝐶
5. Considering the following pathway, if Beadle and Tatum found that *Neurospora* could not grow if metabolite A is provided but could grow if B, C, or D is provided, what enzyme would be missing?

$$A \xrightarrow{1} B \xrightarrow{2} C \xrightarrow{3} D$$

   - a. enzyme 1
   - b. enzyme 2
   - c. enzyme 3
   - d. All of these.

6. The central dogma of biology
   - a. states that DNA is a template for all RNA production.
   - b. states that DNA is a template only for DNA replication.
   - c. states that translation precedes transcription.
   - d. pertains only to prokaryotes because humans are unique.
7. If the sequence of bases in DNA is TAGC, then the sequence of bases in RNA will be
   - a. ATCG.
   - b. TAGC.
   - c. AUCG.
   - d. Both (a) and (b).
8. RNA processing is
   - a. the same as transcription.
   - b. an event that occurs after RNA is transcribed.
   - c. the rejection of old worn out RNA.
   - d. Both (b) and (c).
9. During protein synthesis, an anticodon on tRNA pairs with
   - a. DNA nucleotide bases.
   - b. rRNA nucleotide bases.
   - c. mRNA nucleotide bases.
   - d. other tRNA nucleotide bases.
10. If the DNA codons are CAT CAT CAT, and a guanine base is added at the beginning, then which would result?
    - a. G CAT CAT CAT
    - b. GCA TCA TCA T
    - c. frameshift mutation
    - d. Both (b) and (c).

Study Questions

1. What made Garrod think there were inborn errors of metabolism?
2. Explain Beadle and Tatum's experimental procedure.
3. How did Linus Pauling and Harvey Itano know that the chemical properties of Hb^S differed from those of Hb^A? What change would a substitution of valine for glutamic acid cause in the chemical properties of these molecules?
4. Draw a diagram for the central dogma of biology and tell where the various events occur in a eukaryotic cell.
5. What are the chemical differences between RNA and DNA?
6. How did investigators reason that the code must be a triplet code and in what manner was the code cracked? Why is it said that the code is degenerate, definite, and universal?
7. What are the specific steps that occur during transcription of RNA off a DNA template?
8. Name two ways in which mRNA is processed before leaving the eukaryotic nucleus.
9. Compare the functions of mRNA, rRNA, and tRNA during protein synthesis. What are the specific events of translation?
10. Why is a frameshift mutation always expected to result in faulty protein, while a substitution of one base for another may not always result in a faulty protein?

Thought Questions

1. From an examination of the genetic code, why might you have predicted the wobble effect?
2. Why is it reasonable that new and different genetic codes did not evolve during the history of life?
3. What is the minimum number of different kinds of amino acid activating enzymes that need to be present in the cytoplasm?

Selected Key Terms

central dogma (sen'tral dog'mah) 235
transcription (trans-krip'shun) 235
translation (trans''-lā'shun) 236
codon (ko'don) 236
triplet code (trip'let kōd) 236
messenger RNA (mRNA) (mes'en-jer ar'en-a) 237

RNA polymerase (ar'en-a pol-im'er-ās) 237
intron (in'tron) 237
exon (eks'on) 237
ribosomal RNA (rRNA) (ri'bo-sōm''al ar'en-a) 239
transfer RNA (tRNA) (trans'fer ar'en-a) 239

anticodon (an''ti-ko'don) 239
polysome (pol'e-sōm) 240
gene mutation (jēn mu-ta'shun) 243

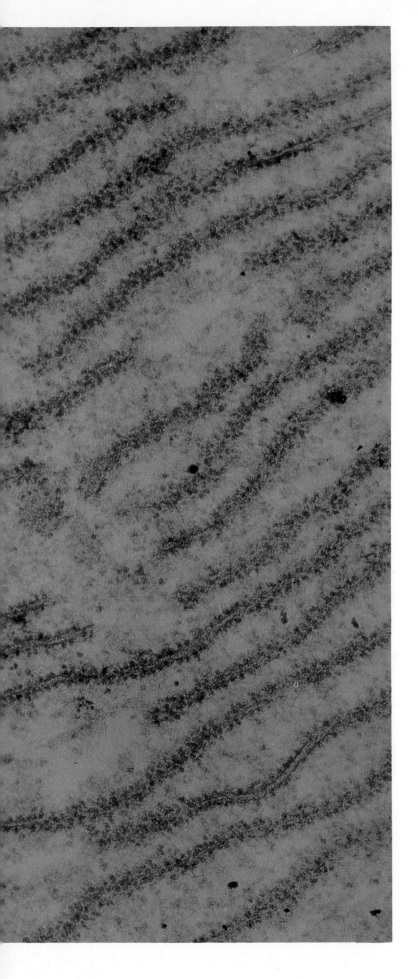

C H A P T E R 17

Regulation of Gene Activity

Your study of this chapter will be complete when you can:

1. *Explain the necessity for control of gene expression in prokaryotes and eukaryotes.*

2. *List the levels of control of gene expression in eukaryotic cells.*

3. *List and define the components of an operon and explain why the lac operon is an inducible operon and the tryp operon is a repressible operon.*

4. *Give evidence that genes are regulated at all levels of control in eukaryotes.*

5. *Use Barr bodies as an example that heterochromatin is genetically inactive.*

6. *Use giant (polytene) chromosome puffs as an example that euchromatin is genetically active.*

7. *Describe cancer as a failure in genetic control and the development of cancer as a two-step process of initiation and promotion.*

8. *Discuss how proto-oncogenes become oncogenes and describe the function of oncogenes in cells.*

9. *Relate the presence and activity of oncogenes to the two-step process of cancer development.*

A low-power electron micrograph illustrates how prolific rough endoplasmic reticulum (ER), a site for protein synthesis, can be within the cytoplasm. Regulation of the expression of genes is not confined to the nucleus, it also involves how many copies of a protein are made at the rough ER.

Cells don't need the same enzymes (and possibly other proteins) all the time. Suppose, for example, the environment is supplying the amino acid tryptophan—wouldn't it be advantageous to not produce the enzymes needed for tryptophan synthesis? Indeed, a medium containing tryptophan can be inoculated with two strains of *E. coli*. The first strain is capable of turning off its genes that code for tryptophan-producing enzymes, and the second, a mutated strain, is unable to do so. After a few days, the culture will contain only cells of the first type—they have overgrown the cells of the second type. It takes energy to carry out transcription and translation, and using energy unnecessarily puts the mutated cells at a disadvantage. It's to be expected, then, that prokaryotic cells normally do have a way to regulate gene expression.

Or, let's consider eukaryotic cells in a multicellular organism. Each mature cell is specialized for a particular task: for example, in humans, cells lining the gut produce a lot of digestive enzymes while muscle cells need the contractile proteins actin and myosin. Numerous experiments of the type discussed in the reading on page 249 indicate that each type of specialized cell contains a full complement of genes. In other words, all human cells have genes for digestive enzymes and muscle proteins but they are only expressed in certain cells of the body. This tells us that eukaryotic cells also must have a way to regulate the action of genes.

Cells are able to regulate the expression of genes, so that at any particular time only certain ones are fully turned on and producing a protein product.

Regulation of Gene Action

Regulation of gene activity pertains to all the steps involved in protein synthesis and protein function. In eukaryotic cells, there are four primary levels of control of gene activity (fig. 17.1):

1. **Transcriptional control:** There are a number of mechanisms that serve to control which genes are transcribed and/or the rate at which transcription occurs.
2. **Post-transcriptional control:** Differential processing of mRNA (fig. 16.10) and also the speed with which mRNA leaves the nucleus can affect the amount of gene expression finally realized.
3. **Translational control:** The life expectancy of mRNAs (how long they exist in the cytoplasm) can vary, as can their ability to bind ribosomes. It is also possible that some mRNAs may need additional changes before they are translated at all.
4. **Post-translational control:** Even after translation has occurred, the polypeptide product may have to undergo additional changes before it is biologically functional. Also, a functional protein is subject to feedback control in the manner described in figure 7.7.

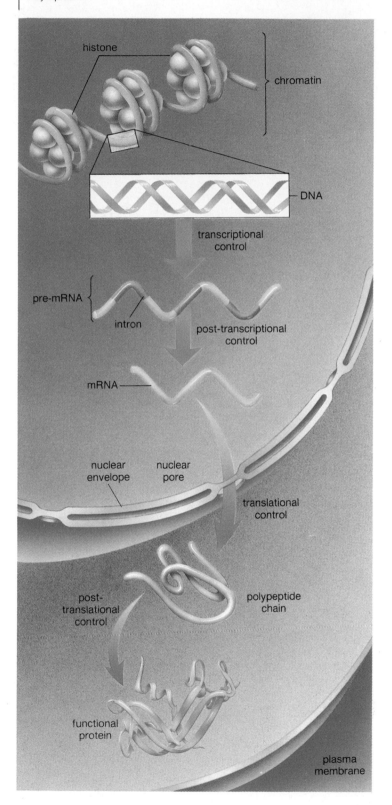

Figure 17.1

Levels at which control of gene expression occurs in eukaryotic cells. Transcriptional and post-transcriptional control occur in the nucleus; translation and post-translational control occur in the cytoplasm.

Totipotency of Differentiated Cells

A multicellular organism begins life as a zygote that undergoes cell division to produce many cells. These cells then become differentiated and assume a particular structure and function. You might think at first that a parceling out of genes brings about differentiation, i.e., different cells receive different genes. But, after you have studied this chapter, you will most likely hypothesize that the same genes are *present* in all cells and differentiation comes about because different genes are *active* in different cells.

Investigators have been testing these two hypotheses for many years. First, they showed that the amount of DNA remains constant in most cells (an exception is the sex cells [gametes], which have half the amount of DNA). This fact suggests that the hypothesis that genes are parceled out in different cells is incorrect. Then, investigators progressed to more challenging experiments to show that the nuclei in multicellular organisms are totipotent. *Totipotent* means "all powerful" and this expression is used to indicate that any cell in a multicellular organism is capable of acting like a zygote—it contains all the genetic information required to bring about complete development of the organism.

Adult plant tissue can be dissociated to give single cells, and if these cells are cultured in nutrient medium, they will multiply and form an undifferentiated mass called a *callus*. If the right mix of hormones is added, the callus will go on to develop into a mature plant. This experiment, which has been carried out with carrot, tobacco, and several other plants supports the hypothesis that adult plant cells are totipotent. Therefore, differentiation must be the result of regulation of gene expression rather than the parceling out of genes.

Totipotency has been demonstrated in animals, too. For example, when a nucleus from an intestinal cell of a tadpole is transplanted into a frog's egg whose own nucleus has been destroyed, development of a tadpole occurs, sometimes followed by further development into an adult frog (*see* figure). Of course, you might argue that support for totipotency would be stronger if the nucleus had been taken from the intestinal cell of a fully grown frog. This is true and, as yet, such experiments have never resulted in adult frogs. Apparently all the genes are present in fully differentiated cells but we just don't know how to turn on the genes that are normally inactive.

A more direct way to determine that different types of cells are expressing different types of genes is to look at the mRNAs. For example, the mRNAs found in liver cells are expected to be different from those found in kidney cells if these cells vary in the genes that have been turned on. It's possible to use nucleic acid hybridization techniques to see if this is the case. During nucleic acid hybridization, a hybrid molecule is formed between mRNA and DNA. The technique is as follows: DNA from an organism is heated so that the hydrogen bonds between the two strands break and the strands separate. This is called denaturation. Now, a large excess of RNA from a particular cell type is added and the amount of hybridization between DNA and RNA is noted (*see* figure). In one such experiment it was found that 4.5% of the total DNA forms hybrids with nuclear RNA from mouse liver cells and 4.0% of the total DNA forms hybrids with nuclear RNA from mouse kidney cells. If these RNA molecules were different in the two cells then when both types of RNA are added to denatured DNA, a total of 8.5% should hybridize. The actual results were less than this: 7.5% hybridize, which means that some of the RNA molecules were the same but most were different. Similar results were obtained when liver cells were compared to muscle and brain cells. These experiments support the hypothesis that different genes are active in different cells of a multicellular organism and that certain genes are turned on and certain others are turned off in specialized cells.

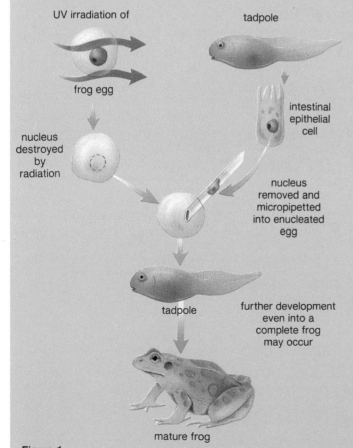

Figure 1
Totipotency experiment. (*Left*) the haploid nucleus of a frog's egg is destroyed by ultraviolet light. Now it can receive a diploid nucleus taken from an intestinal cell of a tadpole (*right*). In some cases, the reconstituted cell can develop into an adult frog.

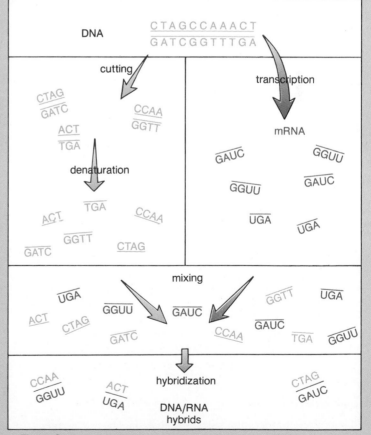

Figure 2
DNA/RNA hybridization. (*Above*) an organism's DNA is cut into fragments and denatured. Transcribed mRNA is obtained from a particular cell type. (*Middle*) the two are mixed. (*Below*) RNA and DNA hybridize whenever they find a complementary strand. Many of the DNA strands do not hybridize because they were not transcribed in this cell.

Genetic Control in Prokaryotes

Usually, regulation of gene expression in prokaryotes occurs at the level of gene transcription. As mentioned on page 237, when a gene is to be transcribed, an enzyme called RNA polymerase attaches to a special DNA sequence called a **promoter** and thereafter this enzyme moves along the DNA, joining mRNA nucleotides together. If RNA polymerase can be prevented from attaching to the promoter, then transcription will not occur.

Obviously a barrier molecule would prevent RNA polymerase from attaching to the promoter. Such barrier molecules exist; they are rather bulky protein molecules called repressors. When a **repressor** binds to a short sequence of DNA called an **operator,** it prevents RNA polymerase from attaching to the promoter, which is located adjacent to it. A repressor is encoded by a gene (called a **regulator**) that is separate from the genes being regulated:

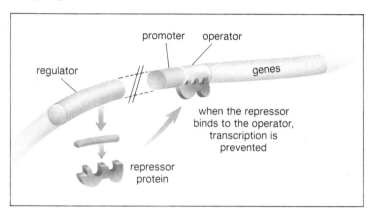

The elements that have just been described make up a unit of genetic operation called an operon. The **operon** model explains regulation, especially in prokaryotes. It suggests that several genes, located in sequence on a chromosome, are controlled by the same regulatory factors. Altogether, an operon includes:

regulator: a gene that codes for a repressor molecule, which binds to the operator and prevents RNA polymerase from binding to the promoter.
promoter: a short sequence of DNA where RNA polymerase first attaches when a gene is to be transcribed.
operator: a short sequence of DNA where the repressor binds, preventing RNA polymerase from attaching to the promoter. Often called the on/off switch of transcription.
structural genes: one to several genes of a metabolic pathway that are transcribed as a unit. (These genes are called **structural genes** because they determine the structure of proteins in a metabolic pathway.)

In prokaryotes, the operon model explains some kinds of regulation of transcription of structural genes.

The *Lac* Operon

In 1961, the French microbiologists, Francois Jacob and Jacques Monod, published the results of a long series of experiments showing that *E. coli* was capable of regulating the expression of genes necessary for lactose metabolism. They coined the word operon and later received a Nobel Prize for their investigations.

Ordinarily, *E. coli* uses glucose as its energy source; however, if it is denied glucose and given the milk sugar lactose instead, it will immediately begin to make the three enzymes needed for the metabolism of lactose. These three enzymes are encoded by genes designated by *lac z, lac y,* and *lac a: lac z* codes for an enzyme called β-galactosidase that breaks down the disaccharide lactose to glucose and galactose; *lac y* codes for a permease that facilitates the entry of lactose into the cell; and *lac a* codes for an enzyme called transacetylase that has an accessory function in lactose metabolism.

The three genes are adjacent to one another on the chromosome and are under the control of a single promoter and a single operator (fig. 17.2). The regulator, located some distance ahead of the promoter, codes for a *lac* operon repressor that ordinarily binds to the operator and prevents transcription of the z, y, and a genes. But when *E. coli* is switched to a medium containing lactose, it binds to the repressor and the repressor undergoes a change in shape that *prevents* it from binding to the operator. Because the repressor is unable to bind to the operator, RNA polymerase can carry out transcription and the three enzymes will be produced.

Because lactose brings about production of enzymes, it is called an **inducer** of the *lac* operon: the enzymes are said to be inducible enzymes; and the entire unit is called an **inducible operon.**

The *Tryp* Operon

Jacob and Monod found that other operons in *E. coli* were usually turned on rather than off. For example, the prokaryotic cell ordinarily produces five enzymes that are needed for the synthesis of the amino acid tryptophan. However, if tryptophan is present in the medium, these enzymes are no longer produced. In this operon model, the regulator codes for a repressor that ordinarily is unable to attach to the operator. The repressor has a binding site for tryptophan, and if tryptophan is present it binds to the repressor. Now a change in shape allows the repressor to bind to the operator. The enzymes are said to be repressible and the entire unit is called a **repressible operon.** Tryptophan is called the **corepressor.**

In the *lac* operon, the repressor ordinarily binds to the operator but is unable to do so when lactose is present. In the *tryp* operon, the repressor ordinarily does not bind to the operator but is able to do so when tryptophan is present. Therefore, some operons are inducible and some are repressible.

Genetic Control in Eukaryotes

As mentioned previously and described in figure 17.1, there are four levels of genetic control in eukaryotic cells.

Transcriptional Control

We might expect that transcriptional control in eukaryotes would (1) involve the organization of chromatin and (2) also include the use of regulatory proteins such as those we have just observed in prokaryotes.

Figure 17.2

The *lac* operon, a model for an inducible operon. *a.* The regulator gene codes for a repressor that is normally active. When active, the repressor binds to the operator and prevents RNA polymerase from attaching to the promoter. Therefore, transcription of the three structural genes does not occur. *b.* When lactose is present, it binds to the repressor and this changes its shape, preventing it from binding to the operator. Now RNA polymerase binds to the promoter; transcription and translation of the three structural genes follows.

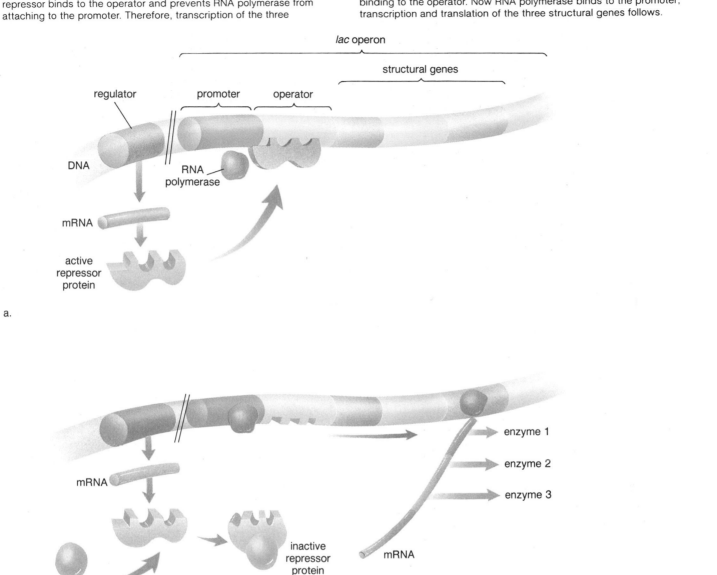

Chromatin and the Regulation of Transcription

In the reading, *What's in a Chromosome* on page 148, it is described how DNA in eukaryotes first winds around histone molecules and then how this beaded string coils, giving a fiber that is then compacted to varying degrees. Chromatin that is highly compacted and visible with the light microscope during interphase is called **heterochromatin.** More diffuse chromatin is called **euchromatin** (fig. 17.3).

In order to demonstrate that heterochromatin is genetically inactive, we can refer to the **Barr body,** a highly condensed structure. In mammalian females, one of the X chromosomes is found in the **Barr body** (p. 210), and chance alone determines which of the X chromosomes it will be. For a given gene, if a female is heterozygous, 50% of her cells will have Barr bodies containing one allele and the other 50% will have the other allele. There is now evidence that this causes the body of heterozygous females to be mosaic, with patches of genetically different cells. The mosaic is exhibited in various ways: some females show columns of both normal and abnormal dental enamel; some have patches of pigmented and nonpigmented cells at the back of the eye; some have patches of normal muscle tissue and degenerative muscle tissue. It is significant that Barr bodies are not found in the cells of the female gonads, where the genes of both X chromosomes may be needed for development of the immature

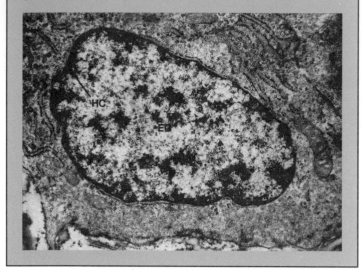

Figure 17.3
Electron micrograph of a nucleus that contains some heterochromatin (HC), highly compacted and genetically inactive DNA. Euchromatin (EC) is more diffuse and more likely to be genetically active. Magnification, ×25,400.

egg. Likewise, it is clear that the inactivation of one X in the female zygote lowers the dosage of gene products to that seen in males. Most likely, development has become adjusted to this lower dosage, and a higher dosage would bring about abnormalities. We certainly know that in regard to other chromosomes, any imbalance causes a greatly altered phenotype.

In contrast to heterochromatin, euchromatin is genetically active. In the salivary glands and other tissues of *Drosophila* larvae, the chromosomes duplicate and reduplicate many times without dividing mitotically. The homologues, each consisting of about a thousand sister chromatids, synapse together to form giant chromosomes (fig. 13.16) called **polytene chromosomes** (poly = many; tene = thread). Polytene chromosomes are visible, genetically active, interphase chromosomes, whereas the chromosomes that appear during mitosis are highly compacted and genetically inactive.

A useful feature of polytene chromosomes is their banding patterns, because it is possible to associate a particular gene with a particular band. It is observed that as a *Drosophila* larva develops, first one and then another of the chromosomes' bands bulge out, forming chromosome puffs (fig. 17.4). The use of radioactive uridine, a label specific for RNA, indicates that DNA is being actively transcribed at **chromosome puffs.** It would appear that the chromosome is decondensing at the puffs so that transcription enzymes can have access to a section of DNA. Perhaps it is the histones that regulate this change in structure.

It is of interest, too, that the hormone ecdysone, which plays an important role in the life cycle of insects, has been shown to induce the occurrence of chromosome puffs, which always appear at specific regions of specific chromosomes. This hormone may bind to DNA and in that way bring about transcription (fig. 41.2*b*).

Varying the amounts of chromatin is another form of transcriptional control. When the gene product is either tRNA or rRNA, only an increased amount of transcription will permit an increase in the amount of gene product. In *Xenopus* (a frog) germ cells that are producing eggs, the number of nucleoli actually increases to nearly 1,000. This means that there is a total of over one million copies of rRNA genes, much more than normal. An increase in the number of copies of a gene is called **gene amplification,** and in this case gene amplification ensures an immense number of ribosomes in the egg cytoplasm to support protein synthesis during the early stages of development.

Transcriptional control in eukaryotes may involve changes in chromatin structure and amount.

Regulatory Proteins and Transcription

There is evidence of regulatory proteins in eukaryotes that attach to promoter regions before RNA polymerase can attach. These proteins are believed to help RNA polymerase bind to the promoter, and therefore it suggests that eukaryotic genes are normally turned off and regulatory proteins such as these are needed to turn them on.

A unique feature in eukaryotes is the presence of DNA sequences called enhancers. An **enhancer** increases the frequency of structural gene transcription up to several hundredfold. Enhancers need not be located near the gene(s) they are affecting. Most likely they are attachment sites for regulatory proteins that help loosen chromatin structure so that transcription can begin.

There is evidence of regulatory proteins in eukaryotic cells, but there are no definite operon models as for prokaryotic cells.

Post-Transcriptional Control

In eukaryotes, mRNA molecules (and other RNAs) are processed before they leave the nucleus and pass into the cytoplasm (fig. 16.10). Differential excision of introns and splicing of mRNA can vary the type of mRNA that leaves the nucleus. For example, both the hypothalamus and thyroid gland produce a hormone called calcitonin. However, the mRNA that leaves the nucleus is not the same in both types of cells; radioactive labeling studies show that they vary because of a difference in mRNA splicing (fig. 17.5). Evidence of different patterns of mRNA splicing have been found in other cells, such as those that produce neurotransmitters, muscle regulatory proteins, and antibodies.

The speed of transport of mRNA from the nucleus into the cytoplasm can ultimately affect the amount of gene product realized per unit time following transcription. There is evidence that there is a difference in the length of time it takes various mRNA molecules to pass through a nuclear pore.

Post-transcriptional control in eukaryotes involves differential mRNA processing and factors that affect the length of time it takes mRNA to travel to the cytoplasm.

Figure 17.4

Structure of giant polytene chromosome. *a.* Electron micrograph of a midge (*chironomus*) chromosome that has a puff in one region. *b.* Artist's interpretation of a puffed region. *c.* Puffs contain loops of DNA where enzymes can transcribe mRNA off the DNA template.

Supporting this interpretation is the observation that varying regions of the chromosomes exhibit puffs as development of *Drosophila* and other insects proceeds.

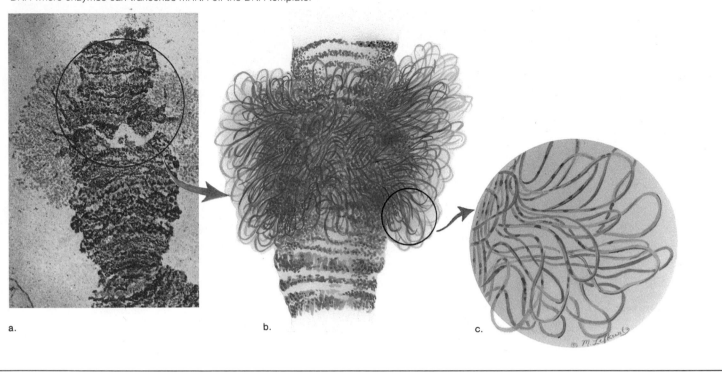

a.

b.

c.

Figure 17.5

Variable RNA splicing can bring about different gene products. Notice that the introns that have been excised from the mRNA on the left are different from those that were excised on the right.

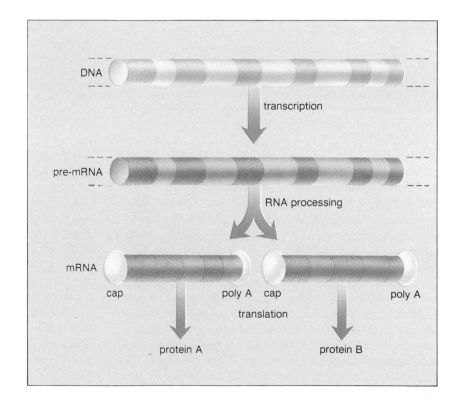

Translational Control

The eggs of frogs and certain other animals contain mRNA that is not translated at all until fertilization occurs. These mRNAs are called *masked messengers* because apparently the cell is unable to recognize them as available for translation. As soon as fertilization occurs, unmasking occurs, and there is a rapid burst of specific gene product synthesis.

Obviously, the longer a particular mRNA remains in the cytoplasm, the more product there will eventually be. During maturation, mammalian red blood cells eject their nucleus, and yet they continue to synthesize hemoglobin for several months thereafter. This means that globin mRNA is able to persist all this time.

The hormone prolactin promotes milk production in mammary glands, but it does so primarily by affecting the length of time the mRNA for casein, a major protein in milk, is translated. The presence of prolactin causes a 25-fold increase in the length of time casein mRNA is translated.

Translational controls directly affect whether an mRNA is translated and the amount of gene product eventually produced.

Post-Translational Control

Some proteins are not at all active immediately after translation. For example, bovine proinsulin is at first biologically inactive. After the single long polypeptide folds into a three-dimensional structure, a sequence of about 30 amino acids is enzymatically removed from the middle of the molecule. This leaves two polypeptide chains that are bonded together by disulfide (S — S) bonds and results in an active protein.

Finally, the metabolic activity of genes is often under feedback control, as we described in figure 7.7.

Post-translational controls affect the activity of a protein product, whether it is functional, and the length of time it is functional.

Cancer, a Failure in Genetic Control

Cancer cells exhibit characteristics that indicate they have experienced a severe failure in the control of genetic expression.

Cancer cells exhibit uncontrolled and disorganized growth. Normal cells only divide about fifty times, but cancer cells enter the cell cycle (fig. 10.4) over and over again and never differentiate.

In tissue culture, normal cells grow in only one layer because they adhere to the glass, and they stop dividing once they make contact with their neighbors, a phenomenon called **contact inhibition.** Cancer cells have lost contact inhibition and grow in multiple layers, most likely because of cell surface changes (fig. 17.6). In the body, a cancer cell divides to form a growth, or **tumor,** that invades and destroys neighboring tissue (fig. 17.7). The cells are disorganized because they don't differentiate into the tissues of the organ and therefore can never help fulfill the function of the organ. To support their growth, cancer cells release a growth factor that causes neighboring blood vessels to

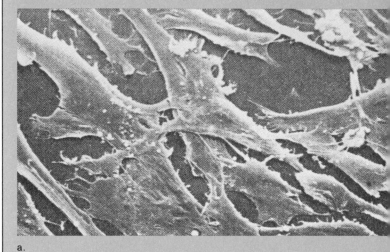
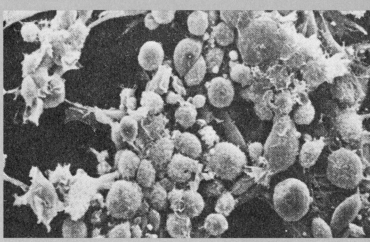

Figure 17.6
Transformation from normal to cancerous cells. *a.* Normal fibroblasts are flat and extended in cell culture. *b.* After being infected with a cancer-causing virus, the cells become round and cluster together in piles.

a.

b.

branch into the cancerous tissue. This phenomenon has been termed **vascularization,** and some modes of treatment are aimed at preventing the occurrence of vascularization.

Cancer cells detach from the tumor and spread around the body. Cancer cells produce proteolytic enzymes that allow them to invade underlying tissues. They are also motile, most likely because they lack bundles of intact microfilaments and have a disorganized internal cytoskeleton. After traveling through the blood or lymph, cancer cells start new tumors elsewhere in the body. This process is called metastasis. If the original tumor is found before metastasis has occurred, the chances of a cure are greatly increased. This is the rationale for the early detection of cancer. A tumor is considered benign (noncancerous) until metastasis has occurred. *Benign tumors* are those that remain in one place; *malignant tumors* are those that metastasize.

Cancer cells grow and divide uncontrollably and they metastasize, forming new tumors wherever they locate.

Figure 17.7

In the body, cancer cells form a tumor, a disorganized mass of cells undergoing uncontrolled growth. As a tumor grows, it invades underlying tissues, and some of the cells leave this primary tumor and move through layers of tissue into blood or lymphatic vessels. After traveling through these vessels, the cells start new tumors elsewhere in the body. A carcinoma is a cancer that begins in epithelial tissue; a sarcoma is one that begins in connective tissue.

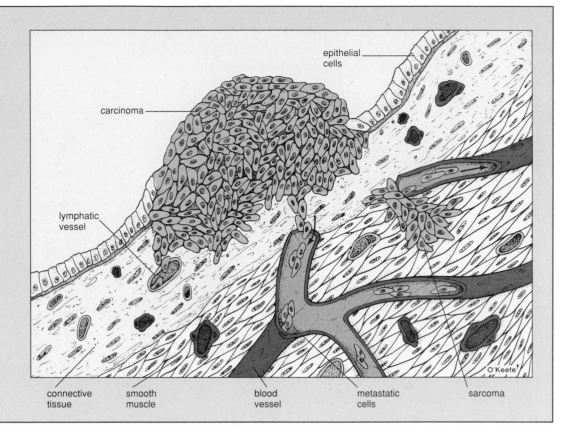

Causes of Cancer

Peyton Rous, whose investigations are discussed below, first described the development of cancer as a two-step process involving (1) initiation and (2) promotion.

During initiation, a mutation occurs that makes cancerous growth a possibility at a later date. **Carcinogens,** agents able to contribute to the development of cancer, are *initiators.* For example, the papilloma virus that causes genital warts is now associated with the development of cervical cancer. Radiation, including ultraviolet (UV) radiation, radon (a radioactive gas made by the natural decay of radium), and X rays damage the normal bonding patterns between DNA nucleotides. Chemical carcinogens such as pesticides are known to bring about base sequence changes in DNA. Cigarette smoke, which contains the carcinogens listed in table 17.1, plays a significant role in the development of lung cancer.

A *promoter* of cancer is any influence that triggers a cell to start to grow in an uncontrolled manner. It's possible that promotion can simply involve a second change in the DNA brought about by another one of the same factors just discussed. In other words, cumulative DNA changes can finally result in uncontrolled growth. Or a promoter might provide the environment that causes transformed cells to form a tumor. For ex-

Table 17.1 Carcinogens in Cigarette Smoke

Aminostilbene	N-Dibutylnitrosamine
Arsenic	2, 3-Dimethylchrysene
Benz (a) anthracene	Indenol (1,2,3-cd) pyrene
Benz (a) pyrene	5-Methylchrysene
Benzene	Methylfluoranthene
Benzo (b) fluoranthene	B-Napthylamine
Benzo (c) phenanthrene	Nickel compounds
Cadmium	N-Nitrosodiethylamine
Chrysene	N-Nitrosodimethylamine
Dibenz (a,c) anthracene	N-Nitrosomethylethylamine
Dibenz (a,e) fluoranthene	N-Nitrosonanabasine
Dibenz (a,h) acridine	Nitrosonornicotine
Dibenz (a,j) acridine	N-Nitrosopiperidine
Dibenz (c,g) carbazole	N-Nitrosopyrrolidine
	Polonium-210

From Steven B. Oppenheimer "Advances in Cancer Biology" *American Biology Teacher*, 49(1):13, January 1987. Copyright © 1987 National Association of Biology Teachers, Reston, VA. Reprinted by permission.

ample, there is some evidence to suggest that a diet rich in saturated fats and cholesterol is a possible promoter for cancer. Considerable time may elapse between initiation and promotion and this is why cancer is more apt to be seen in older rather than younger individuals.

One model suggests that carcinogenesis involves a genetic or chromosomal mutation (initiation) and a second influence that triggers cancerous growth (promotion).

Figure 17.8
Summary of the development of cancer. A virus can bring an oncogene into a cell. A normal gene, called a proto-oncogene, can become an oncogene due to a mutation caused by a chemical or radiation. The oncogene either expresses itself to a greater degree than normal or else expresses itself inappropriately. Thereafter, the cell can become cancerous. Cancer cells are usually destroyed by the immune system, and the individual develops cancer only when the immune system fails to perform this function.

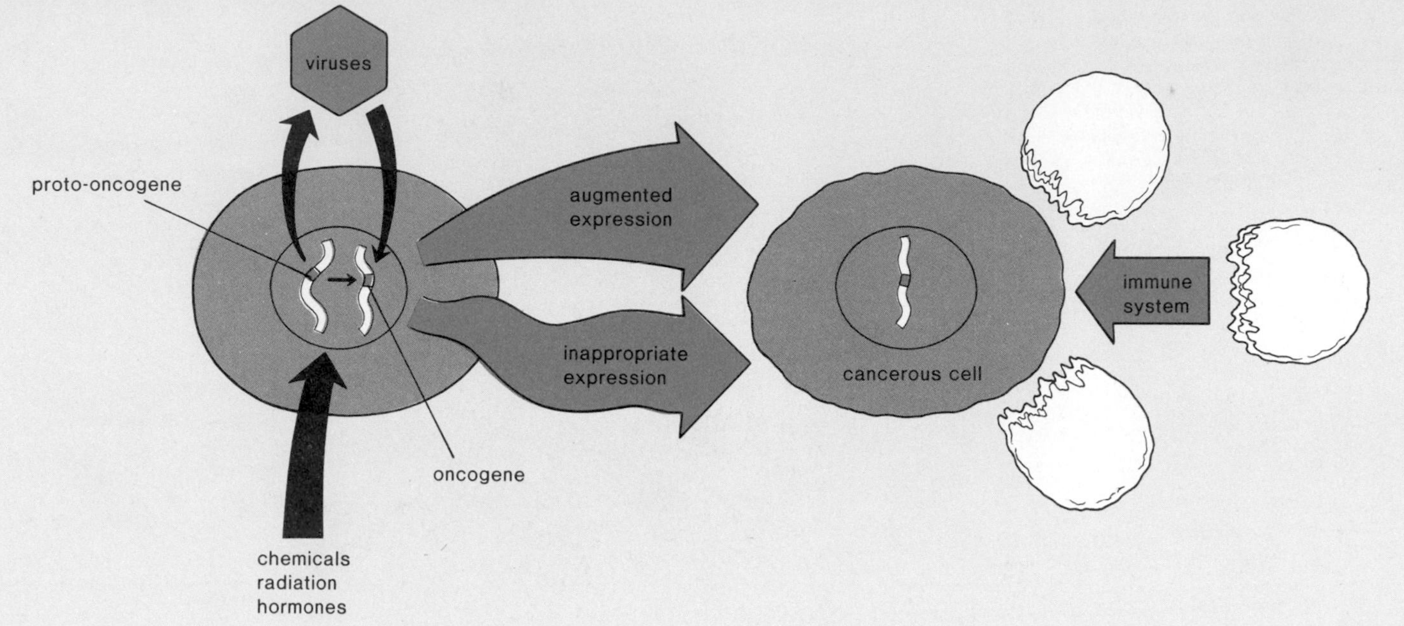

Oncogenes

Knowledge of oncogenes (*oncos* is Greek for cancer), has increased during this century. In 1920, Peyton Rous reported that an RNA virus, later named Rous sarcoma virus (RSV), was capable of causing sarcoma, a type of cancer, in chickens. It wasn't until 1966 that the scientific community realized the value of his contribution and awarded this investigator a Nobel Prize. In the meantime, it was clear that RSV inserts a DNA transcript of its RNA genome, which includes an oncogene (named *src* for sarcoma gene), into the host chromosome. Investigators prepared radioactive copies of the oncogene, knowing that these copies would bind to the host chromosome wherever the *src* gene was located. To their surprise, they found that binding occurred in two locations—one of them where viral DNA had inserted and another in the normal part of the chromosome! This showed that the virus was carrying a gene that was normal in chickens, but it had become an oncogene due to a mutation. Not only that, it was also discovered that the *src* gene codes for an enzyme that promotes growth of the cell and that the gene is normally inactive in mature cells! Therefore, the mutation may only have involved putting the *src* gene under the control of a viral promoter that turned it on.

Ongoing investigations with Rous sarcoma virus have shown that this virus carries a normal host gene that has undergone a mutation. The mutation has turned the normally inactive gene into an active gene.

It is now clear that cells contain proto-oncogenes, genes that can be transformed into oncogenes (fig. 17.8). These genes are not alien to the cell; they are normal, essential genes that have undergone a mutation that leads to augmented or inappropriate expression. R. A. Weinberg and colleagues have shown that an oncogene that causes both lung cancer and bladder cancer differs from a normal gene by a change in only one nucleotide. It is now believed that almost any type of mutation can convert a proto-oncogene into an oncogene. For instance, in addition to a gene mutation, a chromosomal rearrangement may place a normally genetically inactive structural gene next to an active promoter. If this structural gene is a proto-oncogene, it may now become an oncogene. Alternatively, an oncogene can be introduced into a cell by a virus. All cancer-causing viruses discovered so far are retroviruses like the RSV discussed earlier.

Having one oncogene may not cause a cell to become cancerous. Rather, the introduction of another oncogene or an enhancer may be needed to activate the cell to form a tumor (fig. 17.9). This is in keeping with the two-step process involving an initiator and promoter mentioned previously.

Function of Oncogenes   Several oncogenes have been found and studied and all of these are mutated forms of normal genes that regulate growth and cell division, but they don't all have the same type of function. There are numerous elements that affect the growth and development of cells. These include growth factors and growth factor inhibitors that act on the surface of the cell, receptors for these substances, proteins that carry signals

Figure 17.9
Possible scenarios for development of cancer due to viral infections. *a.* The normal cell contains one oncogene but is not cancerous. The virus also contains an oncogene, which it passes to the cell. Now the cell may begin to produce a protein that causes the cell to take on the characteristics of a cancerous cell. *b.* The normal cell contains an oncogene but is not cancerous. The virus contains an enhancer (p. 252) that it brings into the cell. Now the cell may begin to produce a protein that causes the cell to take on the characteristics of a cancerous cell.

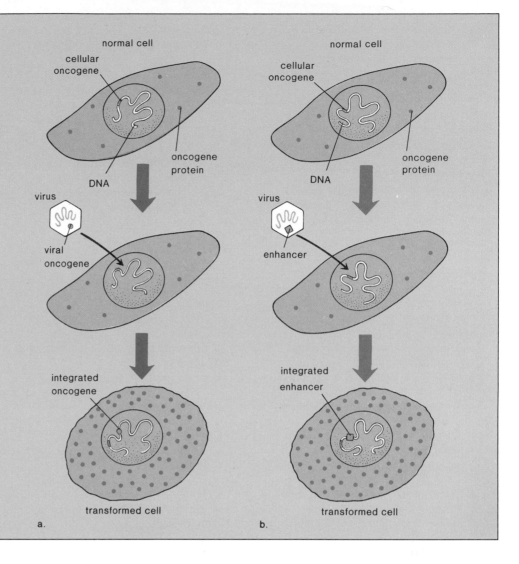

from the receptors, and nuclear functions that regulate the response to these influences. Oncogenes are genes that cause one of these elements to malfunction in such a way that the cell divides repeatedly to produce a tumor. Exactly how this comes about is not yet known.

Genes that encode elements necessary to growth and division of cells sometimes become oncogenes due to a gene or chromosome mutation.

Summary

1. In prokaryotes, genes are turned on and off as appropriate for environmental conditions. In eukaryotes, specialized cells have different genes turned on. In any case, both types of cells are able to regulate the expression of genes.
2. The following levels of control of gene expression are possible in eukaryotes: transcriptional control, post-transcriptional control, translational control, and post-translational control.
3. Regulation in prokaryotes usually occurs at the level of transcription, and the operon model developed by Jacob and Monod is a useful one. In this model, a regulator gene codes for a repressor, which sometimes binds to the operator. When it does so, RNA polymerase is unable to bind to the promoter and transcription of structural genes cannot take place.
4. The *lac* operon is an example of an inducible operon because when lactose, the inducer, is present it binds to the repressor; the repressor is unable to bind to the operator; and transcription of structural genes takes place.
5. The *tryp* operon is an example of a repressible operon because when tryptophan, the co-repressor, is present it binds to the repressor; the repressor is able to bind to the operator; and transcription of structural genes does not take place.
6. All four levels of control have been observed in eukaryotes. First we note that DNA is coiled and compacted to various degrees; an aspect of transcriptional control is therefore the degree to which chromatin is diffuse enough to allow transcription to take place.

7. Highly compacted heterochromatin is genetically inactive, as exemplified by Barr bodies. Less compacted euchromatin is genetically active, as exemplified by polytene chromosome puffs.
8. Gene amplification is the replication of just one or a few genes such that there are a greater number of copies of these genes (i.e., rRNA genes in the oocyte) than were originally present in the zygotic nucleus.
9. Regulatory proteins as well as DNA sequences, called enhancers, play a role in controlling transcription in eukaryotes.

10. Post-transcriptional control refers to variations in RNA processing and the speed with which a particular mRNA leaves the nucleus.
11. Translational control affects mRNA translation and the length of time it is translated. Post-translational control affects whether a protein is active and how long it is active.
12. Development of cancer is a failure in genetic control. Cancer cells have two main characteristics: they exhibit uncontrolled and disorganized growth, and they metastasize.

13. The development of cancer is often described as a two-step process involving initiation and promotion. Carcinogens, such as radiation and certain chemicals, are initiators and may also be promoters. Promoters can also be agents that affect the environment of the cell.
14. Oncogenes are simply proto-oncogenes that have undergone a mutation that causes them to be inappropriately expressed. Proto-oncogenes code for cellular elements that affect growth and cell division.

Objective Questions

1. Which type of prokaryotic cell would be more successful as judged by its growth potential?
   a. One that is able to express all its genes all the time.
   b. One that is unable to express any of its genes any of the time.
   c. One that expresses some of its genes some of the time.
   d. One that only divides when all types of amino acids and sugars are present in the medium.
2. Which of these is mismatched?
   a. post-translational control—nucleus
   b. transcriptional control—nucleus
   c. translational control—cytoplasm
   d. post-transcriptional control—nucleus
3. When lactose is present
   a. the repressor is able to bind to the operator.
   b. the repressor is unable to bind to the operator.
   c. transcription of *lac y, lac z,* and *lac a* genes occurs.
   d. Both (b) and (c).

4. When tryptophan is present
   a. the repressor is able to bind to the operator.
   b. the repressor is unable to bind to the operator.
   c. transcription of structural genes occurs.
   d. Both (b) and (c).
5. RNA processing varies in different cells. This is an example of
   a. transcriptional control of gene expression.
   b. post-transcriptional control of gene expression.
   c. translational control of gene expression.
   d. post-translational control of gene expression.
6. A scientist adds radioactive uridine (label for RNA) to a culture of *E. coli* and examines an autoradiograph. Which type of chromatin will be labeled?
   a. heterochromatin
   b. euchromatin
   c. Both (a) and (b).
   d. Neither (a) nor (b).

7. If Barr bodies were genetically active, females heterozygous for an X-linked gene would
   a. be mosaics.
   b. not be mosaics.
   c. die.
   d. Both (a) and (c).
8. Which of these might cause a proto-oncogene to become an oncogene?
   a. exposure of cell to radiation
   b. exposure of cell to certain chemicals
   c. viral infection of the cell
   d. All of these.
9. A cell is cancerous. Where might you find an abnormality?
   a. only in the nucleus
   b. only in the plasma membrane receptors
   c. only in cytoplasmic reactions
   d. in any part of the cell concerned with growth and cell division
10. Viruses are apt to contribute to the development of cancer when they
    a. cause an infectious disease.
    b. place an oncogene in the genome of the host.
    c. are retroviruses.
    d. Both (b) and (c).

Study Questions

1. How do prokaryotes and eukaryotes differ regarding their need for control of genetic expression?
2. What are the four levels of control of genetic expression in eukaryotes? In what part of the cell do you find each level?
3. Name and state the function of the four components found in operons.
4. Explain the operation of the *lac* operon in a way that stresses it is an inducible operon.

5. Explain the operation of the *tryp* operon in a way that stresses it is a repressible operon.
6. Explain why Barr bodies show that heterochromatin is genetically inactive. What is DNA methylation?
7. Explain why giant (polytene) chromosome puffs show that euchromatin is genetically active. What is gene amplification?

8. What do regulatory proteins do in eukaryotic cells? What are enhancers?
9. Give examples of post-transcriptional, translational, and post-translational control in eukaryotes.
10. What two characteristics are shared by all cancer cells? Describe the development of cancer as a two-step process.
11. What is an oncogene and what is the function of oncogenes in cells?

Thought Questions

1. Why is it proper to describe the binding of the repressor to the operator as reversible? Of what benefit is this reversibility?

2. Why couldn't you use an antiviral drug to treat cancer caused by a virus?

3. Using the information given in this chapter, develop a scenario to describe the development of lung cancer and death of an individual caused by smoking cigarettes.

Selected Key Terms

promoter (pro-mo′ter) 250
repressor (re-pres′or) 250
operator (op′er-a-tor) 250
regulator gene (reg′u-la″tor jēn) 250
operon (op′er-on) 250

structural gene (struk′tūr-al jēn) 250
inducer (in-dūs′er) 250
inducible operon (in-dūs′i-b′l op′er-on) 250
repressible operon (re-pres′i-b′l op′er-on) 250

corepressor (ko″re-pres′or) 250
heterochromatin (het″er-o-kro′mah-tin) 251
euchromatin (u-kro′mah-tin) 251
Barr body (bahr bod′e) 251
carcinogen (kar-sin′o-jen) 255
oncogene (ong′ko-jēn) 256

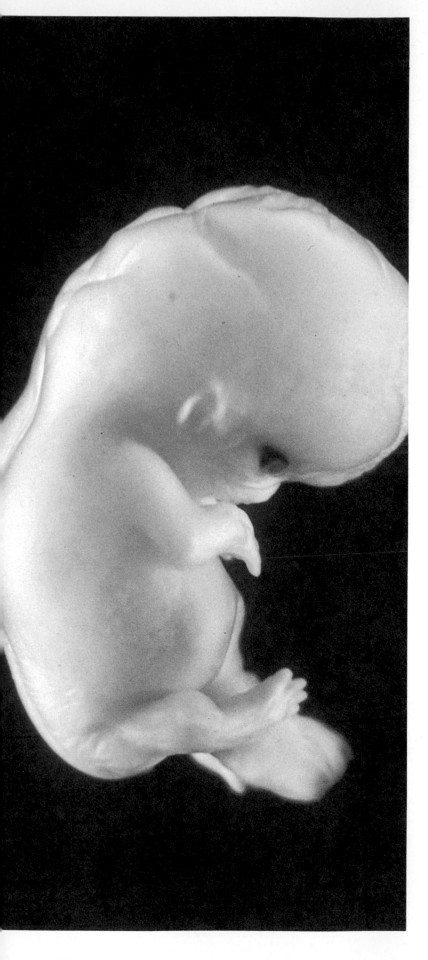

CHAPTER 18

Recombinant DNA and Biotechnology

Your study of this chapter will be complete when you can:

1. *Describe how the use of a vector permits cloning of a gene.*
2. *Describe the two-step process by which recombinant DNA is prepared.*
3. *Be able to state and recognize laboratory procedures involving DNA that are presently possible.*
4. *Categorize and list some of the biotechnology products currently on the market.*
5. *Explain why plants lend themselves to genetic manipulation and also discuss any problems involved.*
6. *List some of ways plants may be modified as a result of recombinant DNA technology.*
7. *Describe a method for genetically engineering animals and state any problems involved.*
8. *Outline a procedure by which gene therapy might be conducted in humans.*
9. *Describe a procedure by which it is possible to sequence the bases of DNA.*
10. *Describe some of the advances that are expected from the applications of DNA technology.*

Will it be possible one day to cure human genetic diseases before birth? Biotechnology not only involves the mass production of gene products such as insulin by bacterial cells, it also involves the transplantation of genes into organisms. If an egg is injected with a foreign gene it may possibly be copied and distributed to all the cells of an organism. In this way a human could conceivably be given a functional gene to make up for a defective gene.

M endel's original research was rediscovered in 1900, and in the intervening years, geneticists have made startling advances in their understanding of the molecular mechanisms of heredity. This basic research has even led to a new era in biotechnology.

Biotechnology, the use of a natural biological system to produce a product or achieve an end desired by human beings, is not new. Plants and animals have been bred to give a particular phenotype since the dawn of civilization. Also, the biochemical capabilities of microorganisms have been exploited for a very long time. For example, the baking of bread and the production of wine are dependent on the presence of yeast cells to carry out fermentation reactions.

Today, however, biotechnology is first and foremost an industrial process (fig. 18.1) that provides products due to our ability to genetically engineer bacteria. The product can be an organic chemical of interest, a protein that is useful as a vaccine, or a drug to promote human health. Engineered bacteria need not be confined to a chemical plant or laboratory; they may soon be released into the environment to clean up pollutants, increase the fertility of the soil, or kill insect pests. Biotechnology even extends beyond unicellular organisms; it is now possible to alter the genotype and subsequently the phenotype of plants and animals. An agricultural revolution of unequaled magnitude is not inconceivable even in the near future. And there is no need to stop here; gene therapy in humans may be just over the horizon. Let's begin our discussion, however, with the basic procedures and applications of biotechnology.

Genetic Engineering of Cells

Genetic engineering can produce cells that are **transformed;** they contain a foreign gene and are capable of producing a new and different protein. Often, **vectors** are used to carry the foreign gene into a *host* cell such as a bacterium, yeast cell, plant cell, or animal cell, including human cells.

Vectors

One common type of vector is a plasmid. **Plasmids** are small accessory rings of DNA, found in some bacteria, which carry genes not present in the bacterial chromosome. The reading on page 264 tells how plasmids were originally discovered by investigators studying the sex life of the intestinal bacterium *E. coli.*

Plasmids that are used as vectors have been removed from bacteria and have had a foreign gene inserted into them (fig. 18.2). Treated cells will take up a plasmid, and after it enters, the host cell continues to reproduce as usual. Whenever the host reproduces, the plasmid, including the foreign gene, is copied. Eventually, there are many copies of the plasmid and therefore many copies of the foreign gene. This gene is now said to have been **cloned.**

Plasmids are not taken up by animal cells, but viruses can be used as vectors for these cells. After a virus has attacked a cell, the DNA is released from the virus and enters the cell proper. Here it may direct the reproduction of many more viruses. Each virus derived from a viral vector will contain a copy of the foreign gene. Therefore viral vectors also allow cloning of a particular gene.

Figure 18.1
Biotechnology is an industrial endeavor. *a.* Laboratory procedures must be adapted to mass produce the product. *b.* Microbes are grown in huge tanks, called fermenters because they were first used for yeast fermentation to produce wine. *c.* The product is purified. *d.* The product is packaged.

a.

b.

c.

d.

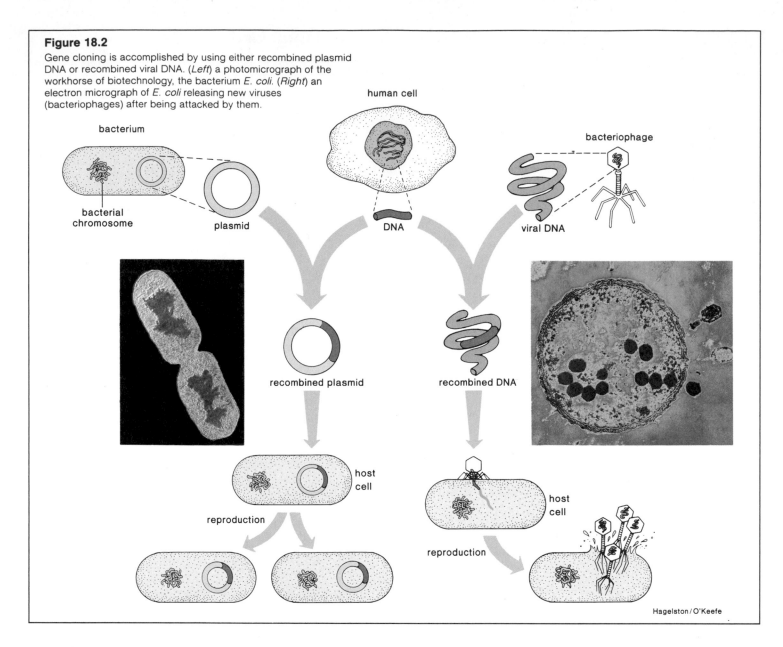

Figure 18.2

Gene cloning is accomplished by using either recombined plasmid DNA or recombined viral DNA. (*Left*) a photomicrograph of the workhorse of biotechnology, the bacterium *E. coli*. (*Right*) an electron micrograph of *E. coli* releasing new viruses (bacteriophages) after being attacked by them.

human cell

bacterium

bacteriophage

bacterial chromosome

plasmid

DNA

viral DNA

recombined plasmid

recombined DNA

host cell

host cell

reproduction

reproduction

Hagelston / O'Keefe

Viruses are also commonly used as vectors to introduce a foreign gene into bacteria (fig. 18.2). In fact bacteriophages have been used to create **genomic libraries.** A genomic library is a collection of engineered bacteriophages that together carry all the genes of a species. Since each bacteriophage carries only a short sequence, it would take about ten million bacteriophages to carry all the genes of a mouse.

Recombinant DNA

The introduction of foreign DNA into plasmid or viral DNA to give a cloning vector is a two-step process (fig. 18.3). First the plasmid (or viral) DNA is cut open, and then the foreign DNA is inserted into this opening. Both of these steps require a specific type of enzyme.

The first type of enzyme has the ability to cut a DNA molecule into discrete pieces. Such enzymes occur naturally in some bacteria, where they stop viral reproduction by cutting up viral DNA. They are called **restriction enzymes** because they *restrict* the growth of viruses. In 1970, Hamilton Smith at Johns Hopkins University isolated the first restriction enzyme; now more than 175 different restriction enzymes have been isolated and purified. Each one cuts the DNA at a specific cleavage site. For example, the restriction enzyme called Eco RI always cuts double-stranded DNA when it has this sequence of bases:

$$. . . . .\text{G A A T T C}. . . . .$$
$$. . . . .\text{C T T A A G}. . . . .$$

Furthermore, this enzyme always cleaves each strand between a G and an A in this manner:

$$. . . .\text{G} \qquad\qquad \text{A A T T C}. . . . .$$
$$. . . .\text{C T T A A} \qquad\qquad \text{G}. . . . .$$

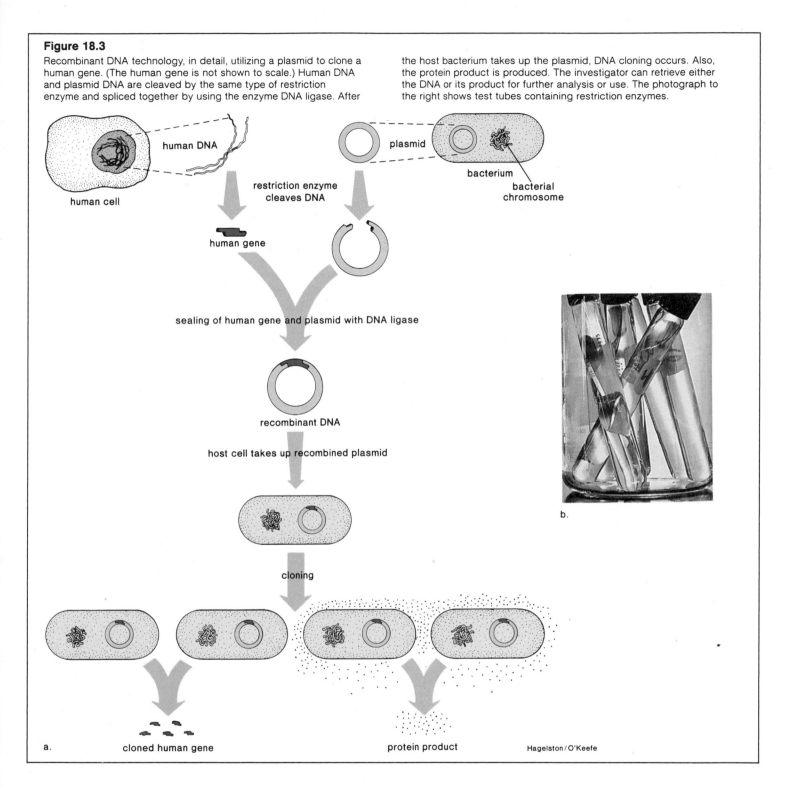

Figure 18.3

Recombinant DNA technology, in detail, utilizing a plasmid to clone a human gene. (The human gene is not shown to scale.) Human DNA and plasmid DNA are cleaved by the same type of restriction enzyme and spliced together by using the enzyme DNA ligase. After the host bacterium takes up the plasmid, DNA cloning occurs. Also, the protein product is produced. The investigator can retrieve either the DNA or its product for further analysis or use. The photograph to the right shows test tubes containing restriction enzymes.

human cell

human DNA

plasmid

bacterium

bacterial chromosome

restriction enzyme cleaves DNA

human gene

sealing of human gene and plasmid with DNA ligase

recombinant DNA

host cell takes up recombined plasmid

cloning

cloned human gene

protein product

a.

b.

Hagelston/O'Keefe

Notice that there is now a gap into which a piece of foreign DNA can be placed if it ends in bases complementary to those exposed by the restriction enzyme. To assure this, it is only necessary to cleave the foreign DNA with the same type restriction enzyme. The single-stranded but complementary ends of the two DNA molecules are called "sticky ends" because they adhere by complementary base pairing. They thus facilitate insertion of foreign DNA into vector DNA.

The second enzyme needed for preparation of a vector is **DNA ligase,** a cellular enzyme that seals any breaks in a DNA molecule. Genetic engineers use this enzyme to seal the foreign piece of DNA into the vector. Gene splicing is now complete and a **recombinant DNA** molecule has been prepared. A recombinant DNA molecule is one that contains DNA from two different sources.

Bacterial cells will take up a recombined plasmid, especially if they are treated with calcium chloride to make them more permeable. Thereafter if the inserted foreign gene is replicated and actively expressed, the investigator can later recover either the cloned gene or a protein product.

Aside from these procedures just discussed, others are common in biotechnology, including:

1. Preparation of a gene in vitro (within laboratory glassware) by using cellular components. Bacteria lack the cellular machinery to excise introns; therefore, this method can be used to prepare mammalian genes: the appropriate mRNA is isolated and single cDNA copies are made of it by using reverse transcriptase. Then double-stranded cDNA is made by using DNA polymerase.

2. Alternatively, a gene can be manufactured; that is, the nucleotides can be joined together in the correct sequence or, if desired, a mutated gene can be prepared by altering the sequence. This method can also be used to prepare mammalian genes that lack introns.

3. Joining the regulatory regions of a viral or bacterial gene to an isolated or machine-made structural gene so that transcription is ensured. Genes spliced into vectors very often are accompanied by regulatory regions, and mammalian genes can be put under the control of bacterial promoters.

4. Insertion of a prepared gene directly into a host cell rather than into a vector.

5. Lysing the bacterial cell and extracting the product. Often bacterial cells cannot carry out post-translational changes and therefore do not excrete the product. Such changes can be done in vitro after the product is extracted.

The Discovery of Plasmid and Viral Vectors

It wasn't until 1946 that investigators began to study bacterial genetics in earnest. This is surprising because bacteria make wonderful laboratory subjects. They are simply organized and easily broken apart, making it easy to get at and analyze their biochemical machinery. They also have relatively few genes, and because one cell produces so many descendants, it's easy to detect very rare mutants. But investigators did not believe bacteria could be prime candidates for genetic analysis because no routine sexual method had been observed by which recombinants were produced. It was the study of recombinant offspring that had allowed Mendel to deduce his laws of genetics, and Morgan to map the *Drosophila* chromosome.

In 1946 Joshua Lederberg and Edward Tatum went looking for recombinants among bacteria. They mixed two strains of *E. coli* together; one strain required the growth factors biotin and methionine and the other required threonine and leucine. The mixture was plated on a minimal medium that lacked all four growth factors because they reasoned that only those bacteria that had experienced genetic recombination would grow. A small but significant number of colonies were obtained when cell-to-cell contact had occurred. When each strain was exposed to only extracts from the other strain, no colonies resulted; therefore, this was not simply another example of *transformation,* a process first observed by Griffith in 1928 (p. 219).

Subsequently it was discovered that these bacteria were undergoing a form of *conjugation,* which allowed genes to be transferred from one to the other. In the conjugation process, one bacterium, the ''male,'' develops one or more sex pili (pilus, sing.), surface structures that make contact with a ''female.'' (It is now believed that as the sex pili retract, a conjugation bridge of some sort forms through or on which DNA passes from one cell to another.) In 1953, William Hayes found that after conjugation, the ''females'' had become ''males'' and could also form sex pili. Later, Lederberg decided to designate bacteria capable of forming pili as F+ (fertility positive) and those that were incapable as F−, and it was said that the F+ bacteria had passed a fertility factor (F factor) into the F− bacteria. This is what caused them to become F+ and capable of forming sex pili (see figure).

It turned out that the F factor was a plasmid (a small extrachromosomal ring of DNA)—the very first plasmid to be discovered in bacteria. The F plasmid includes genes responsible for allowing bacteria to carry out conjugation. Occasionally, however, it was noted that an F plasmid can transfer genes from the main bacterial chromosome, also. Such F plasmids are no longer independent of the host chromosome; instead they are integrated into the host chromosome. The bacterial genes that are transferred bring about recombinant genotypes and this explains the results observed by Lederberg and Tatum reported above. Strains that have the F plasmid integrated are called Hfr strains because they bring about a *high frequency* of recombination.

Francois Jacob and Elie Wollman thought of a way to use Hfr strains to map the *E. coli* chromosome. During conjugation, the integrated F plasmid breaks open and a single strand begins to move through the conjugation bridge, dragging with it the bacterial genes that are attached. It takes about 90 minutes for the entire bacterial chromosome to be transferred. Jacob and Wollman could interrupt this mating process (they used a kitchen blender to force the mating pairs apart) after different time intervals, and now they could determine which genes per unit time had been transferred and therefore the order of the genes in the *E. coli* chromosome! This is called the interrupted mating technique for mapping the *E. coli* chromosome.

Many other types of plasmids are now known. Among the most prevalent and well studied are the R plasmids that confer resistance to antibiotics. In most cases, antibiotic resistance is possible because the R plasmid codes for enzymes that inactivate the drug. These plasmids can be transferred from one bacterium to another and this allows more and more bacteria to become resistant. The indiscriminate use of antibiotics selects for the survival of those bacteria that have the R plasmid and ensures that most bacteria will be resistant. Therefore, antibiotics should only be prescribed and taken when absolutely necessary.

Later, still another way for genetic recombination to occur among bacteria was discovered. Joshua Lederberg and Norton Zinder found in 1952, when working with *Salmonella typhimurium,* that recombination was occurring without cell-to-cell contact and even in the presence of an enzyme that could destroy free DNA. This ruled out the possibility of transformation as the mechanism of transfer and it was discovered that transfer was being accomplished by a virus. The transfer of genetic information between bacteria via viruses is called *transduction.* It occurs because the viral genome contains host genes that were accidently picked up during the reproductive cycle of the virus within the host cell.

Cells can be genetically engineered (transformed) to produce a new protein product. Often vectors are used to carry the new gene into the cell.

Today's Biotechnology Products

Table 18.1 shows at a glance the types of biotechnology products that are now available. There are three categories in the table: hormones and similar types of proteins, DNA probes, and vaccines. Organic chemicals are also mentioned (p. 267) and monoclonal antibodies are discussed in chapter 34.

Hormones and Similar Types of Proteins

One impressive advantage of biotechnology is that it allows mass production of proteins that are very difficult to obtain otherwise. For example, human growth hormone was previously extracted from the pituitary gland of cadavers and it took 50 glands to obtain enough for one dose. Also insulin was previously extracted from the pancreas glands of slaughtered cattle and pigs; it was expensive and sometimes caused allergic reactions in recipients. And few of us knew of tPA (tissue plasminogen activator), a protein present in the body in minute quantities that activates an enzyme to dissolve blood clots. Now tPA is a product that is used to treat heart attack victims.

A study of the list of prospective protein products indicates that some of the most troublesome and serious afflictions in humans may soon be treatable: clotting factor VIII will be available for hemophilia; human lung surfactant will be available for premature infants with respiratory distress syndrome; atrial natriuretic factor may be helpful to many with hypertension. And the list will grow because bacteria (or other cells) can be engineered to produce virtually any protein.

Hormones are also being produced for use in animals. No longer is it necessary to feed steroids to farm animals; they can be given growth hormone and it produces a leaner meat that is

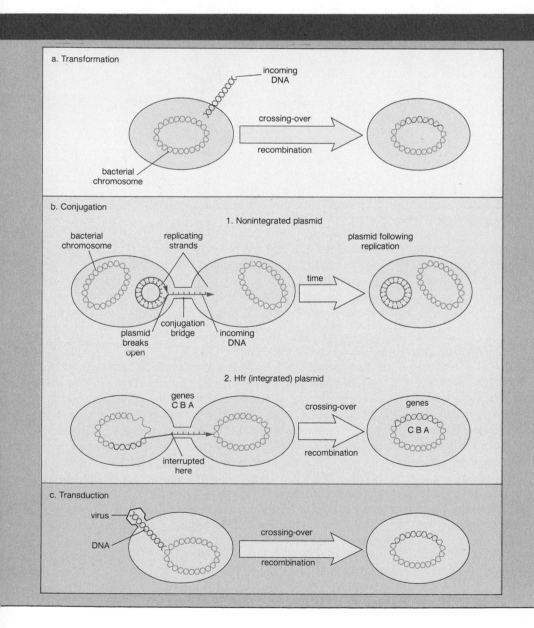

Recombinant genotypes in bacteria can be achieved in three ways. *a.* Transformation. The cell simply takes up DNA that is added to the medium and following crossing-over it becomes incorporated into the chromosome. *b.* Conjugation. (1) An independent nonintegrated plasmid such as an F plasmid breaks up, and one strand enters a F⁻ bacterium by way of a conjugation bridge. (2) If the plasmid has been integrated into the host chromosome as in Hfr plasmids, it can bring host genes with it into the Hfr⁻ bacterium. *c.* Transduction. A virus can introduce bacterial genes that were picked up previously from a former host. When these genes become incorporated into the bacterial chromosome, a recombinant genotype has occurred.

Table 18.1 Representative Biotechnological Products

Hormones and Similar Types of Proteins	DNA Probes	Vaccines
For Treatment of Humans	*For Diagnostics in Humans*	*For Human Use*

Insulin
Growth hormone
tPA (tissue plasminogen activator)
Interferon (alpha, beta*, gamma*)
Erythropoietin
Interleukin-2
Clotting factor VIII*
Human lung surfactant*
Atrial natriuretic factor*
Superoxide dismutase*
Alpha-l-antitrypsin*
	Legionnaires' disease
	Walking pneumonia
	Sickle-cell anemia
	Cystic fibrosis
	Tendency for emphysema, thalassemia,
	  retinoblastoma
	Hemophilia B
	Sexually transmitted diseases
	Polycystic kidney disease
	Paternity testing
	Detection of criminals
	Monitoring bone marrow transplants
	Susceptibility to atherosclerosis, diabetes, and
	  hypertension
	Periodontal disease
	Huntington disease*
	Duchenne muscular dystrophy*
	Cancers (chronic myelogenous leukemia, T- and
	  B-cell cancer, others*
		Hepatitis B
		Herpes* (oral and genital)
		AIDS*
		Hepatitis A*
		Group B meningococci*
		Hepatitis non-A and non-B*
		Malaria*

For Treatment of Animals		*For Animal Use*
Growth hormone (cows, pigs, chickens*) Lymphokines*		Hoof-and-mouth disease Scours Brucellosis Rift Valley fever Herpes Influenza Pseudorabies Feline leukemia virus* Rabies*

* Expected in the near future. These may be available now, but they are not presently made by recombinant DNA technology.

more healthful to humans. Also, when cows are given bovine growth hormone (bGH) they produce 25% more milk than usual, which should make it possible for the dairy farmer to maintain fewer cows and cut down on his overhead expenses.

DNA Probes

Biotechnology is contributing greatly to diagnostics by making DNA probes (table 18.1) and monoclonal antibodies.[1] To construct a DNA probe you need only: (1) determine the sequence of amino acids in a polypeptide, (2) deduce the sequence of nucleotides in DNA, and (3) prepare this sequence and clone it. If the cloned DNA is now denatured, the double helix will unzip, giving single strands. Each of these can be used as a **DNA probe** because it will seek out and bind to any complementary DNA strand present in body fluids or removed from the body's cells. DNA probes can be used presently in paternity suits (consider that one chromosome of each pair is inherited from the father) and sometimes in court cases to identify an individual who was

[1]Even though monoclonal antibodies are not produced by recombinant DNA technology, they are still considered to be a biotechnological product. Monoclonal antibodies are discussed in chapter 34.

at the scene of a crime if body cells were left behind. They can also be used to diagnose an infection by indicating if the gene of an infectious organism is present. When available, DNA probes can tell us whether a gene coding for a hereditary defect or causing a cell to be cancerous is present.

Vaccines

Recombinant DNA technology can produce pure and therefore safe vaccines. Bacteria and viruses have surface proteins and a gene for just one of these proteins can be placed in a plasmid. The host cell for the plasmid will produce many copies of the surface protein, which then can be used as a vaccine. Right now the only recombinant vaccine on the market is for hepatitis B, but you can see that the ones soon expected are of great importance: a vaccine for malaria and another for AIDS. Malaria has been a scourge on humankind for many thousands of years, and AIDS is a deadly disease known to modern humans.

Vaccines have also been produced for the inoculation of farm animals. Each of the illnesses listed—hoof-and-mouth disease, scours, etc.—causes an untold number of illnesses and deaths each year. These were formerly a severe drain on the time, energy, and resources of farmers.

Figure 18.4

Possible biotechnology scenario. *a*. A protein is removed from a cell and the amino acid sequence is determined. (This protein is gonadotrophic releasing hormone.) From this the sequence of nucleotides in DNA can be deduced (p. 237). *b*. The DNA synthesizer can be used to string nucleotides together in the correct order. *c*. A small section of the gene can be used as a DNA probe to test fetal cells for an inborn error of metabolism. (The manufactured gene could be placed in a bacterium to produce more of the protein, which then could possibly be used in treatment of the defect.)

c.

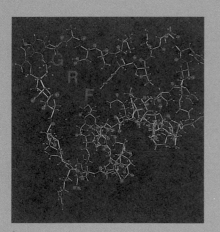
a.

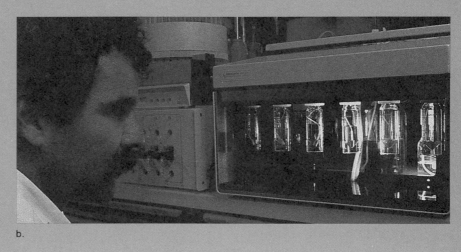
b.

Organic Chemicals

Organic chemicals are often synthesized by having catalysts act on precursor molecules or by using bacteria to carry out the synthesis.

Today, it is possible to go one step further and manipulate the genes that code for synthetic enzymes. For example, biochemists discovered a strain of bacteria that is especially good at producing phenylalanine, an organic chemical needed to make aspartame, the dipeptide sweetener better known as Nutra-Sweet™. They isolated, altered, and formed a vector for the appropriate genes so that various bacteria could be geared up to produce phenylalanine.

Biotechnology products include hormones and similar types of proteins, DNA probes, and vaccines. These products are of enormous importance to the field of medicine.

Transgenic Organisms

Thus far we have considered only the insertion of foreign genes into bacteria. If a foreign gene were inserted into an immature cell, would it be expressed in the cells of a multicellular organism? Indeed, this has been achieved. These organisms are called **transgenic organisms** because a gene has been transplanted into them. Similarly, suppose you inserted a gene into the cells of an individual who has an inborn error of metabolism, would these cells produce the deficient protein? We call this procedure **gene therapy** and it is the subject of extensive research.

Plants

Plants in particular lend themselves to genetic manipulation because it is possible to grow plant cells in tissue culture, and each one can be stimulated to produce an entire plant (fig. 18.5). Previously, tissue culture has been used to provide genetic carbon copies of strawberry, asparagus, and oil palm plants for commercial use. It has also been used to isolate any variant cells to give different strains of carrots, celery, and wheat. Indeed, plant tissue culture can be used to screen for cells that have a particular property, such as resistance to a herbicide. It is obviously less time-consuming and less costly to screen millions of cells in a flask than to screen adult plants.

Besides these procedures, it is also possible to apply recombinant DNA technology to plant cells. The presence of the plant cell wall hampers the possibility of plasmid uptake, and so it is necessary to work with "naked" cells, called **protoplasts,** that have had their walls removed (fig. 18.6).

In one experiment, protoplasts from two different species were fused and it was possible to achieve mature plants. A pomato plant (potato × tomato fusion) produced both types of vegetables, but they were small and the seed quality was poor.

Figure 18.5
Cloning of entire plants from tissue cells.
a. Sections of carrot root are removed, and
thin slices are placed in a nutrient medium.
b. After a few days, the cells form a callus,
a lump of undifferentiated cells. *c.* After
several weeks, the callus begins sprouting
carrot plants. *d.* Eventually the cloned carrot
plants can be moved from culture medium
to potting soil.

a.

b.   c.   d.

Genetic engineers have introduced genes that convey resistance to herbicides and insects into a variety of protoplasts. The vector of choice is the Ti (tumor-inducing) plasmid from the bacterium *Agrobacterium tumefaciens.* Ordinarily the Ti plasmid invades plant cells and causes a cancerlike growth called crown gall disease, but when used as a vector, the plasmid is engineered to lack virulence. Plants that develop from genetically engineered protoplasts have the desired characteristic and breed true. Thus far these plants are experimental and not yet field tested. In one instance, the gene for the enzyme luciferase was transplanted from a firefly into a tobacco plant. Whenever the plant was sprayed with luciferin, it glowed, proving that the inserted gene was indeed present and active (fig. 18.7).

Several problems remain before there will be transformed cereal plants for commercial use. Tobacco plants are dicots whereas the cereals are monocots (fig. 29.6). Only recently has it been possible to get engineered corn, a monocot, to regenerate from a protoplast. Monocot protoplasts do not ordinarily take up the Ti plasmid in tissue culture. Techniques are being developed to have these cells take up genetic material directly. For example, lasers can be used to make tiny self-sealing holes in plasma membranes through which genetic material can enter. Individual micro-injection has been used to insert foreign genes into protoplasts, but obviously it would be more efficient to simply irradiate cells with a laser beam while they are suspended in a liquid containing foreign DNA.

Since the necessary techniques are, or should soon be perfected, it is believed that within a relatively short period of time agriculturally significant plants will be produced that have the characteristics listed in table 18.2. In the meantime, bacteria have been engineered to promote the health of plants when they are applied to them. For example, field tests conducted to test the ability of genetically engineered bacteria, called ice-minus bacteria, to live on plants and protect the vegetative parts of the

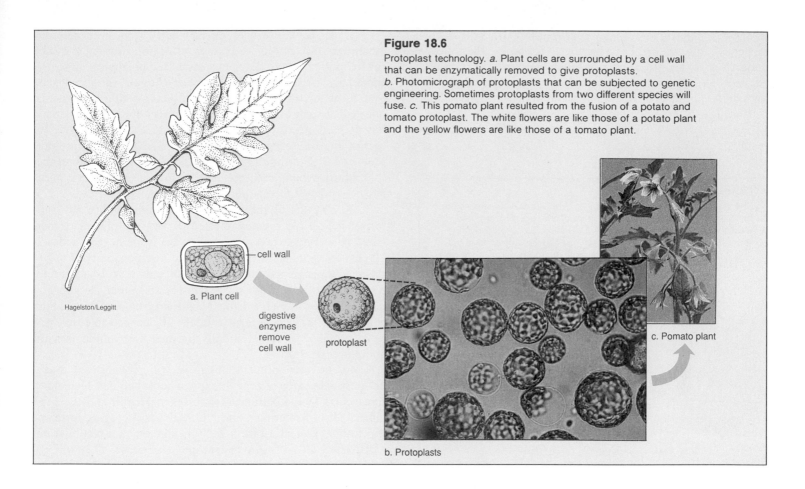

Figure 18.6

Protoplast technology. *a.* Plant cells are surrounded by a cell wall that can be enzymatically removed to give protoplasts. *b.* Photomicrograph of protoplasts that can be subjected to genetic engineering. Sometimes protoplasts from two different species will fuse. *c.* This pomato plant resulted from the fusion of a potato and tomato protoplast. The white flowers are like those of a potato plant and the yellow flowers are like those of a tomato plant.

Hagelston/Leggitt

cell wall

a. Plant cell

digestive enzymes remove cell wall

protoplast

c. Pomato plant

b. Protoplasts

Figure 18.7

The transgenic plant glows when sprayed with luciferin because its cells contain the protein luciferase, a firefly enzyme that acts on the chemical luciferin, which then emits light.

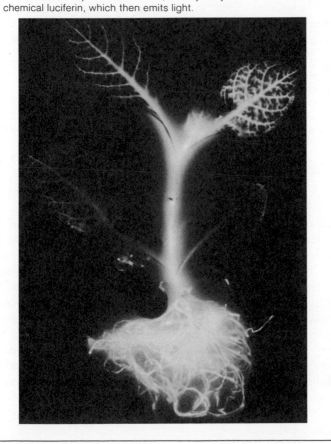

Table 18.2 Biotechnology in Plants

Due to biotechnology, plants of the future may
- have a natural resistance to various diseases and pests so that fewer pesticides will be needed.
- possess the ability to fix nitrogen so that less fertilizer will need to be applied to crops.
- have increased ability to grow under unfavorable environmental conditions such as lack of water or in salty soil. Less irrigation of crops would then be needed.
- have seed proteins that contain all the essential amino acids so that a complete protein source will be available for vegetarians.

plants from frost damage have been successful. Also, a bacterium that normally colonizes the roots of corn plants has now been endowed with genes (from another bacterium) that codes for an insect-killing toxin. Field tests are scheduled.

There are those who are very much concerned about the deliberate release of genetically engineered microbes (GEMs) into the environment. Ecologists point out that these bacteria might displace those that normally reside in an ecosystem and the effects could be deleterious. Others rely on past experience with GEMs, primarily in the laboratory, to suggest that these fears are unfounded.

Animals Including Humans

Genetic engineering of animals is becoming quite frequent. For example, researchers at the University of Minnesota are trying to engineer a superwalleye by inserting an extra growth hormone gene into the egg. The Minnesota legislature, the project's major funder, hopes the superfish will someday cause fishing enthusiasts to vacation in the state. Again, however, the researchers want to be as certain as possible that the fish will not be harmful to the environment. To this end they are investigating the possibility of having the gene turn off once the fish reach a certain size or age.

Figure 18.8

The virus selected for gene therapy is a retrovirus. Retroviruses have an RNA chromosome. When this chromosome enters a cell, reverse transcription, during which RNA is transcribed to cDNA, precedes reproduction of the virus (fig. 23.4b). You can see this adds to the steps needed for the virus to serve as a vector for the purpose of gene therapy.

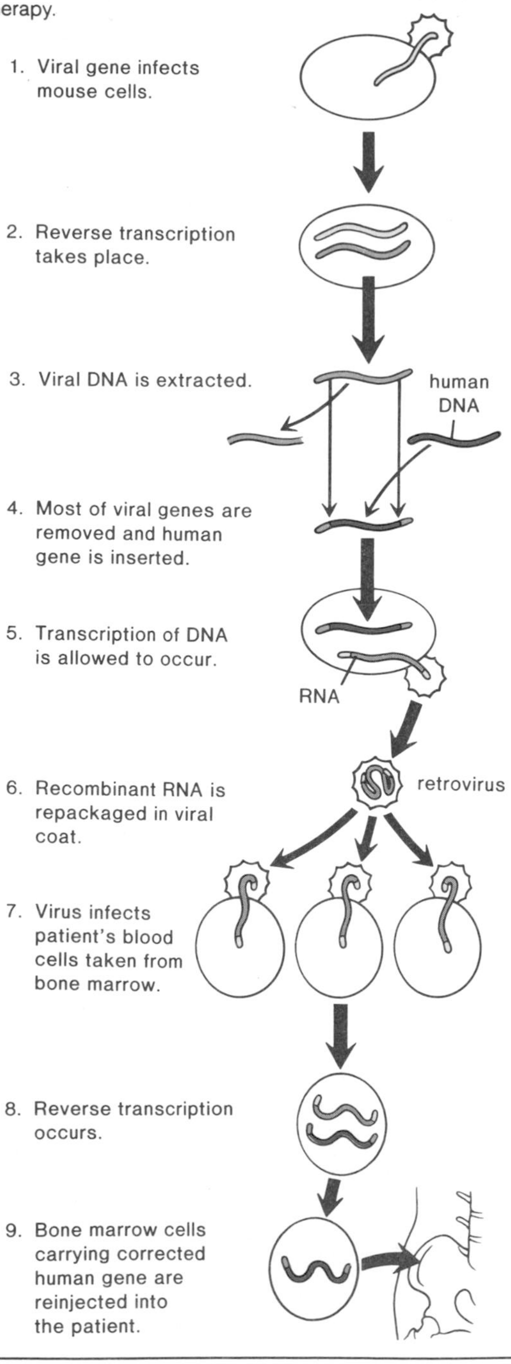

1. Viral gene infects mouse cells.

2. Reverse transcription takes place.

3. Viral DNA is extracted.

human DNA

4. Most of viral genes are removed and human gene is inserted.

5. Transcription of DNA is allowed to occur.

RNA

6. Recombinant RNA is repackaged in viral coat.

retrovirus

7. Virus infects patient's blood cells taken from bone marrow.

8. Reverse transcription occurs.

9. Bone marrow cells carrying corrected human gene are reinjected into the patient.

Genes have also been microinjected into the zygotes of laboratory mice, and these zygotes, after implantation into surrogate mothers, have developed normally. Further, they sometimes have the phenotype desired. For example, the sheep gene that codes for the milk protein beta lactoglobulin (BLG) was injected into mouse zygotes. (Mice normally produce a milk that has no BLG at all.) Of 46 offspring successfully weaned, 16 carried the BLG sequence and the females among them later produced BLG-rich milk. One produced BLG at five times the concentration found in sheep's milk. Some of the transgenic mice passed on the BLG gene to their offspring.

The results of these types of experiments are not uniform—not all the zygotes injected survive the procedure; not all of those born express the gene to the same degree; and not all pass the gene on to their offspring. The possibility of using this genetic engineering method in humans is still believed to be quite distant because successful results cannot be guaranteed. However, there is a distinct possibility of using this genetic engineering method in farm animals. It's conceivable that genetic engineers could eventually produce farm animals with many improved traits.

In the near future, though, we will probably see much progress in the area of gene therapy in adult humans. Even now, researchers at the National Institutes of Health and the Memorial Sloan-Kettering Cancer Center are preparing for clinical tests to correct a rare hereditary disease that results from a deficiency of the enzyme adenosine deaminase. They will inoculate bone marrow stem cells as described in figure 18.8. Other investigators are looking at the possibility of using liver cells rather than bone marrow stem cells to correct some inborn errors of metabolism. In the opinion of many researchers, it is unethical to delay human gene therapy experiments any longer. However, some lay people in particular have the exact opposite opinion. Most people believe that the issue should be publicly discussed until there is a general consensus to guide us all.

Transgenic organisms have been achieved. This genetic engineering feat is expected to revolutionize agriculture and animal husbandry. Work has also begun in the area of human gene therapy.

The Future

Laboratory Research

The techniques, enzymes, vectors, and hosts that are needed for biotechnology were discovered by investigators doing basic research in the laboratory. Basic research will continue and it would appear that eventually biotechnology experiments will be carried out in cell-free systems. In that case, it will be possible to bypass the need for vectors and hosts completely, because procedures like cloning will simply take place in vitro. This possibility is being actively pursued.

The United States government has recently become committed to providing the funds necessary to have the entire human genome sequenced (fig. 18.9). The human genome is estimated to contain up to 100,000 genes. Until this time, only about 100

Figure 18.9

Gel electrophoresis and DNA sequencing. In order to perform gel electrophoresis, a semisolid material similar to gelatin or agar has previously been molded into a slab. The slab has a small well in it, and a mixture of proteins or DNA fragments has been placed in the well. Electrodes are now attached to both ends of the slab and the current is turned on. Each macromolecule then migrates toward the electrode of opposite charge at a rate determined by its charge and size. The greater the charge and/or the smaller the size, the further the molecule migrates per unit time. Proteins vary as to their charge but all DNA fragments have a negative charge. *a.* This photograph gives an example of DNA gel electrophoresis in which the investigator sees orange bands because the gel has been stained with a DNA-binding dye that fluoresces orange in ultraviolet light. The length (size) of the DNA at each band can be determined by comparing the bands observed to those made previously by DNA molecules whose lengths were already known. *b.* DNA can be sequenced in a manner that was developed by Frederick Sanger who shared a Nobel Prize with Walter Gilbert for their work in this area. First, the DNA is divided up into fragments by using restriction enzymes. Then, each of these fragments is individually sequenced

after being separated into single strands. A copy of the strand to be sequenced is placed in each of four test tubes. The test tube contains everything necessary to DNA synthesis (DNA polymerase and all four types of nucleotides) and a radioactive primer that labels the 5′ end of the newly replicated strand of DNA. You need four test tubes because a different nucleotide analog (a dideoxyribonucleoside triphosphate) is added to each. These stop DNA synthesis after they are in position because the next phosphate group cannot attach to the analog. For example, in the first test tube, replication will produce synthesized DNA strands of different lengths, each length representing a place where an A analog was attached. In this instance, gel electrophoresis is followed by autoradiography (production of a photograph) that tells the investigator what these newly synthesized different lengths are. Comparing the results of electrophoresis for all four test tubes (corresponding to the bases adenine, cytosine, thymine, and guanine) allows the investigator to determine the sequence of the bases (simply read up from the bottom crossing over whenever necessary to determine the next nucleotide.)

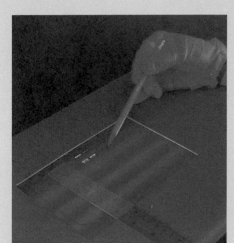

a.

b.

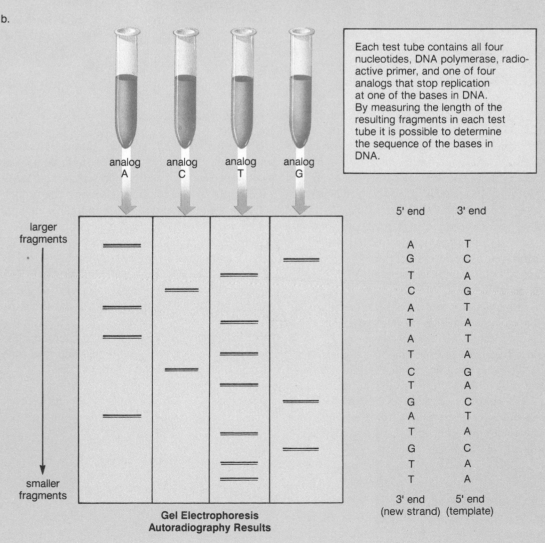

Each test tube contains all four nucleotides, DNA polymerase, radioactive primer, and one of four analogs that stop replication at one of the bases in DNA. By measuring the length of the resulting fragments in each test tube it is possible to determine the sequence of the bases in DNA.

analog A   analog C   analog T   analog G

larger fragments

smaller fragments

Gel Electrophoresis Autoradiography Results

5′ end	3′ end
A	T
G	C
T	A
C	G
A	T
T	A
A	T
T	A
C	G
T	A
G	C
A	T
T	A
G	C
T	A
T	A

3′ end (new strand)   5′ end (template)

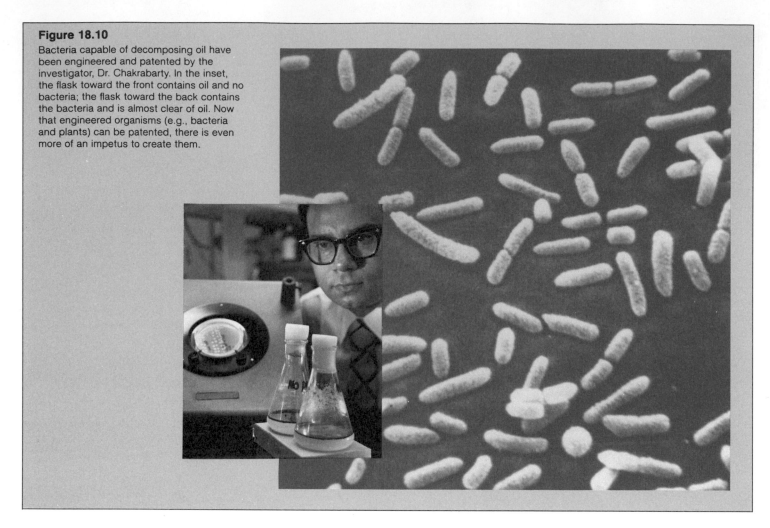

Figure 18.10
Bacteria capable of decomposing oil have been engineered and patented by the investigator, Dr. Chakrabarty. In the inset, the flask toward the front contains oil and no bacteria; the flask toward the back contains the bacteria and is almost clear of oil. Now that engineered organisms (e.g., bacteria and plants) can be patented, there is even more of an impetus to create them.

human genes have ever been sequenced completely because of the time needed to accomplish this task. Now there are automated DNA sequencers that make use of computers to accomplish **gene sequencing** at a fairly rapid rate. Even so, it's still estimated that the job will take about thirty years unless even newer and better technology is developed. In any case, it is difficult to imagine the size of the computer data base that will be needed to store this information. The amount of paper to record the results will be staggering, requiring on the order of 200 books the size of an encyclopedia volume.

Some researchers believe that it will be of little benefit to simply know the sequence of the bases in the human genome. Others believe that this knowledge will take us one step closer to being able to manufacture genes (and their control elements) without having to slice them out in a hit or miss fashion from their neighbors. Then it will be possible to study the protein, coded by the gene, to determine how it functions normally in the cell. This knowledge will contribute greatly, for example, to our understanding of how the immune system functions and how defective genes wreak havoc on the body.

Applied Research

It's easy to predict there will be at least three major biotechnology revolutions, one in the area of medical diagnostics and treatment, another in agriculture, and a third in improvement of the environment.

In medicine, human gene therapy could possibly become routine. Diagnostic DNA probes may be available to indicate whether a genetic disease is present and, if feasible, the defect will simply be corrected by inserting the proper gene into the body. In some instances, if the individual has inherited only a tendency toward a particular condition, proper steps may be taken to prevent its development. For example, if you know you have a tendency toward emphysema, you could be careful to avoid polluted air. If a medical condition does develop and there is no genetic cure, then new and novel medicines such as the ones mentioned previously might be available to keep the condition under control.

In agriculture, plants may be given the ability to grow and produce under adverse conditions, and also plants may thrive with a limited amount of fertilizer, pesticides, herbicides, and water. Farm animals may be healthier and more productive, and therefore fewer animals could possibly be needed to produce the same amount of food. As a result, agriculture could make a decreased contribution to the degradation of the environment.

Bacteria, even today, have been fashioned to decompose oil and various contaminants (fig. 18.10) but thus far they have not been used. In the future it may be possible to simply order the correct strains to clean up any contaminated area. Environmental pollution may then cease to be a great threat to continuance of the human species.

The picture painted here is one-sided and very optimistic. Could it be that we would use our new found knowledge to fashion and change the human genome to suit an idealized model, and that we will be even less tolerant than we are today of individual differences? Could the modified plants, animals, and microbes cause environmental problems that we cannot even begin to imagine? Could irresponsible countries develop deadly engineered bacterial and viral strains for use as military weapons? These are the types of questions we have to try to answer as the new era of biotechnology descends upon us.

Summary

1. Biotechnology, the use of natural biological systems to produce products or achieve ends desired by human beings, is being expanded greatly by DNA technology.
2. To achieve a genetically engineered cell, a vector is first prepared. Both plasmids and viruses are used as vectors to carry a foreign gene into a cell. When the plasmid replicates or the virus reproduces, the foreign gene is cloned.
3. The preparation of recombinant DNA is a two-step process. First, restriction enzymes are used to cleave plasmid DNA and to cleave foreign DNA. The "sticky ends" produced facilitate the insertion of foreign DNA into vector DNA. Second, the foreign gene is sealed into the vector DNA by DNA ligase.
4. Many other laboratory procedures are available to facilitate recombinant DNA technology. For example, it is possible to produce cDNA using reverse transcriptase; to manufacture a gene using a DNA-synthesizing machine; and to attach regulatory genes to structural genes before they are inserted into plasmid or viral DNA.
5. Transformed cells produce products of interest to humans such as hormones, DNA probes, and vaccines.
6. Transgenic organisms have also been made. Plants lend themselves to genetic manipulation because it is possible to grow an adult plant from a cell that originally was maintained in tissue culture.
7. It is hoped that plants having the following characteristics may be developed using recombinant DNA technology: natural resistance to various diseases and pests, ability to fix nitrogen, increased ability to grow in arid and salty soils, production of seeds possessing all the essential amino acids.
8. Genetic engineering of animals is being intensely investigated. Genetic engineering of animal zygotes has not as yet led to uniform results and the genetic engineering of the human zygote has not yet begun.
9. Human gene therapy is expected in the near future. A retrovirus may serve as a vector that carries a normal gene into bone marrow stem cells removed from a patient with a genetic disorder. Once the normal gene has been incorporated into the DNA of the stem cells, they can be injected back into the patient.
10. Laboratory techniques such as gel electrophoresis plus newly developed techniques are expected to be used to sequence the entire human genome.
11. Recombinant DNA technology may very well revolutionize almost every area of human concern—medicine, agriculture, animal husbandry, and cleanup of the environment.

Objective Questions

1. Which of these is a true statement?
   a. Both plasmids and viruses can serve as vectors.
   b. Plasmids can carry recombinant DNA but viruses cannot.
   c. Vectors carry only the foreign gene into the host cell.
   d. All of these statements are true.
2. Using this key, put the steps in the correct order to achieve a plasmid carrying recombinant DNA.

Key:

Step 1 use restriction enzymes
Step 2 use DNA ligase
Step 3 remove plasmid from parent bacterium
Step 4 introduce plasmid into new host bacterium
   a. 1,2,3,4
   b. 4,3,2,1
   c. 3,1,2,4
   d. 2,3,1,4
3. Which of these is a benefit to having insulin produced by biotechnology?
   a. It can be mass produced.
   b. It is nonallergenic.
   c. It is less expensive.
   d. All of these.
4. Why is it easier to genetically engineer plants as compared to animals?
   a. Plants can be grown from a single cell in tissue culture.
   b. *E. coli* plasmids can serve as vectors in plants.
   c. Agriculturally significant plants have been engineered.
   d. All of these.
5. Which of these is a true statement?
   a. Numerous transgenic monocot plants have been produced.
   b. Bacteria have been engineered to give plants with desirable characteristics.
   c. Since animal zygotes are so easily manipulated, human zygotes will soon be manipulated also.
   d. All of these are true statements.
6. Which of these would you not expect to be a biotechnology product?
   a. modified enzyme
   b. DNA probe
   c. protein hormone
   d. steroid hormone
7. What is the benefit of using a retrovirus as a vector in gene therapy?
   a. It is not able to enter cells.
   b. It incorporates the foreign gene into the host chromosome.
   c. It eliminates a lot of unnecessary steps.
   d. All of these.
8. Gel electrophoresis
   a. Measures the size of plasmids.
   b. Tells whether viruses are infectious.
   c. Measures the charge and size of proteins and DNA fragments.
   d. All of these.
9. Which of these is incorrectly matched?
   a. protoplast—plant cell engineering
   b. base analogs—sequencing DNA
   c. viruses—genomic libraries
   d. DNA ligase—a biotechnology product
10. After a plasmid has been cloned
   a. you can remove the cloned foreign gene and sequence it.
   b. it does not express itself and no protein product results.
   c. you can use it to carry out human gene therapy.
   d. All of these.

Study Questions

1. What are the similarities and differences between using a plasmid and using a virus as a vector for gene cloning?
2. What is the methodology for producing recombinant DNA to be used in gene cloning?
3. List and explain other types of laboratory procedures that are used in recombinant DNA experiments.
4. Categorize and give examples of types of biotechnology products available today.
5. Why are plants good candidates for genetic engineering and what are the problems involved in using this method to achieve agriculturally significant plants?
6. In what ways is it hoped that plants will soon be modified?
7. What types of bacteria have been produced by genetic engineering to provide plants with desirable characteristics? Give a pro and a con for releasing these into the environment.
8. Describe an experiment to produce transgenic animals. Why is this procedure not expected any time soon in humans?
9. Outline a method of gene therapy in humans.
10. Describe a method of sequencing DNA and state a pro and a con for sequencing the entire human genome.
11. Discuss the possible future benefits and possible pitfalls of biotechnology.

Thought Questions

1. The size of DNA fragments that result when an individual's entire genomic DNA is completely digested by restriction enzymes is unique to the individual. What does this mean about each individual's DNA?
2. Why might it be helpful to know a normal sequence of bases for any particular human gene?
3. The firefly gene for the enzyme luciferase was placed in the cells of a tobacco plant (fig. 18.7) where it would function if provided with luciferin. Why is it expected that animal genes will also function in plants?

Selected Key Terms

biotechnology (bi″o-tek-nol′o-je) 261
genetic engineering (je-net′ik en″ji-nēr′ing) 261
transformed (trans-formd′) 261
vector (vek′tor) 261
plasmid (plaz′mid) 261

cloned (klōnd) 261
genomic library (je-nom′ik li′brer-e) 262
restriction enzyme (re-strik′shun en′zīm) 262
DNA ligase (de′en-a li′gās) 263
recombinant DNA (re-kom′bi-nant de′en-a) 263

DNA probe (de′en-a prōb) 266
transgenic organism (trans-jen′ik orgah-nizm) 267
gene therapy (jēn thēr-ah-pe) 267
protoplast (pro′to-plast) 267
gene sequencing (jēn se′kwens-ing) 272

Suggested Readings for Part 2

Antebi, E., and Fishlock, D. 1986. *Biotechnology: Strategies for life.* Cambridge, MA: The MIT Press.

Bishop, J. M. March 1982. Oncogenes. *Scientific American.*

Chambon, P. May 1981. Split genes. *Scientific American.*

Chilton, M. June 1983. A vector for introducing new genes into plants. *Scientific American.*

Cohen, S. N., and Shapiro, J. A. February 1980. Transposable genetic elements. *Scientific American.*

Darnell, J. E. October 1983. The processing of RNA. *Scientific American.*

Dickerson, R. E. December 1983. The DNA helix and how it is read. *Scientific American.*

Doolittle, R. F. October 1985. Proteins. *Scientific American.*

Drlica, K. 1984. *Understanding DNA and gene cloning.* New York: John Wiley and Sons.

Elkington, J. 1985. *The gene factory: Inside the science and business of biotechnology.* New York: Carroll and Graf Publishers.

Feldman, M., and Eisenback, L. November 1988. What makes a tumor cell metastatic? *Scientific American.*

Glover, D. M. 1984. *Gene cloning: The mechanics of DNA manipulation.* New York: Chapman and Hall.

Grivell, L. A. March 1983. Mitochondrial DNA. *Scientific American.*

Kieffer, G. H. 1987. *Biotechnology, genetic engineering, and society.* Reston, VA: National Association of Biology Teachers.

Klug, W. S., and Cummings, M. R. 1986. *Concepts of genetics.* 2d ed. Westerville, Ohio: Charles E. Merrill Publishing Co.

Kornbert, R. D., and Klug, A. February 1981. The nucleosome. *Scientific American.*

Mendel, G. 1965. Experiments in plant hybridization. In *The origins of genetics,* by C. Stern and E. Sherwood. San Francisco: W. H. Freeman Co.

Murray, A. W., and Szostak, J. W. November 1987. Artificial chromosomes. *Scientific American.*

Nomura, M. January 1984. The control of ribosome synthesis. *Scientific American.*

Patterson, D. August 1987. The causes of Down syndrome. *Scientific American.*

Prescott, D. 1988. *Cells.* Boston: Jones and Bartlett.

Ptashne, M. January 1989. How gene activators work. *Scientific American.*

Radman, M., and Wagner, R. August 1989. The high fidelity of DNA duplication. *Scientific American.*

Ross, J. April 1989. The turnover of messenger RNA. *Scientific American.*

Shepard, J. F. May 1982. The regeneration of potato plants from leaf-cell protoplasts. *Scientific American.*

Smith-Klein, C., and Kish, V. 1988. *Principles of cell biology.* New York: Harper & Row, Publishers, Inc.

Stahl, F. W. February 1987. Genetic recombination. *Scientific American.*

Steitz, J. A. June 1988. "Snurps." *Scientific American.*

Tompkins, J. S., and Rieser, C. June 1986. Special report: Biotechnology. *Science Digest.*

Torrey, J. G. July–August 1985. The development of plant biotechnology. *American Scientist.*

Weinberg, R. A. November 1983. A molecular basis of cancer. *Scientific American.*

———. September 1988. Finding the anti-oncogene. *Scientific American.*

White, R., and Lalouel, J. February 1988. Chromosome mapping with DNA markers. *Scientific American.*

CRITICAL THINKING

CASE STUDY

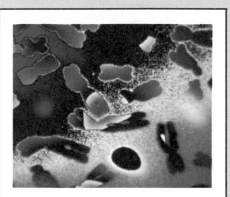

PART 2
Genetic Basis of Life

A study of chromosome puffs (described on page 252 in chapter 17) supports the central dogma of biology and suggests that specific genes are active only at certain times. It was proposed, sometime ago, that the puffs are sites where DNA has unwound and where mRNA transcription is taking place. There are two possible ways to test this hypothesis:

1. Look for a protein product that can be associated with a specific chromosome puff.
2. Determine if RNA transcription is occurring where there is a chromosome puff.

Investigators taking the first approach were fortunate in finding two species of midge flies whose phenotypes differ. Cells within the salivary glands of *Chironomus pallidivittatus* produce a granular secretion (contains protein granules) while these same cells in *Chironomus tentans* produce an agranular secretion (does not contain protein granules). In which of these species would you expect to find (a) chromosome puff(s) in salivary gland cells, which are producing a secretion?

Prediction 1 When the secretion is being produced, a puff or puffs should be apparent in *C. pallidivittatus,* whereas a corresponding type puff will not be present in *C. tentans.*

Result 1 Observation reveals that this is the case and supports the hypothesis that puffs are sites of gene activation.

> *Why wouldn't it suffice just to witness puffing in* C. pallidivittatus? *What role does* C. tentans *play in the experiment?*

In another experiment, hybrid offspring were produced by interbreeding flies of the two midge species. In hybrid flies, one member of each homologous pair of chromosomes is inherited from *C. pallidivittatus* and the other is inherited from *C. tentans*. Which homologous pair member do you predict would show a chromosome puff in salivary gland cells producing a secretion?

Prediction 2 Puffs would be apparent on the chromosome inherited from *C. pallidivittatus* but not on the chromosome inherited from *C. tentans.*

Result 2 Careful observation supports the prediction. On the homologue inherited from *C. pallidivittatus*, puffs were apparent when secretion was being produced. On the other homologue inherited from *C. tentans*, puffs were not observed at the corresponding chromosome site.

> *Why is it helpful to follow up Prediction 1 and Result 1 with Prediction 2 and Result 2? Also, what would you predict about the concentration of granules in secretions of the hybrid midge fly?*

So far, we have supported the hypothesis by showing that there is a relationship between chromosome puffs and a protein product. A much stronger argument for gene activation can be made if it is determined that RNA is being synthesized at chromosome puffs. A procedure, called autoradiography, can be used and, in this experiment, it requires these steps: (1) Radioactive uridine (a precursor of the base uracil, which is incorporated into RNA but not DNA) is injected into a midge fly. (2) Salivary glands are removed from the flies and sectioned for microscopic examination. (3) Each section is washed free of radioactive uridine so that any radioactivity is now within RNA molecules. (4) A photographic emulsion placed over the section is exposed by the radiation so that (5) following photographic development, dark spots indicate the presence of RNA molecules. Where do you predict that dark spots will appear in microscopic slides of active salivary gland tissue from *C. pallidivittatus?*

Prediction 3 Many dark spots should appear over chromosome puffs. None should appear on unpuffed regions of the chromosome.

Result 3 As shown in figure 1*a*, the concentration of dark spots is much greater on the puffed region of the chromosome. These results can be used to suggest that RNA synthesis is occurring predominantly at the puffs.

Actinomycin D is an inhibitor of RNA synthesis. What do you predict would be the results of autoradiography if a midge fly is exposed to radioactive uridine and actinomycin D at the same time?

Prediction 4 When treated with both actinomycin D and radioactive uridine, no RNA synthesis should be apparent at the puffs.

Result 4 As shown in figure 1*b*, an autoradiograph now shows much fewer dark spots than previously. The low concentration of the dark spots indicates little RNA synthesis. As predicted, actinomycin D has apparently inhibited RNA synthesis at puffs.

> *Why was this last experiment performed? Does comparing figures 1a and 1b seem more convincing than 1a alone? Why?*

The results of these experiments provide strong support for the hypothesis that chromosome puffs represent sites of gene activation and that gene activation involves unwinding DNA so that RNA synthesis can occur.

> *Could these experiments also be used to support the hypothesis that specific genes are active only at certain times? What other experiments could be done to support this hypothesis? (See also page 252 of the text.)*

Beerman, W., and Clever, U. April 1964. Chromosome puffs. *Scientific American*.

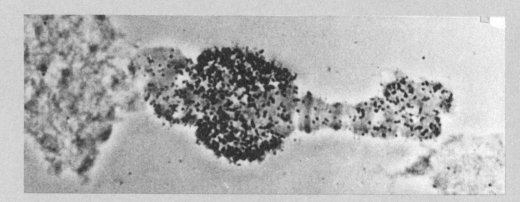

a.

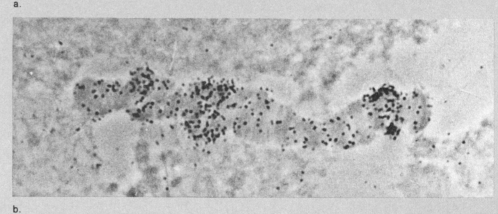

b.

Figure 1
A comparison of chromosome puffs. *a*. Chromosome puffs treated only with radioactive uridine. *b*. Chromosome puffs treated with actinomycin D and radioactive uridine.

P A R T 3

Evolution and Diversity

Blue-footed boobies live on the Galápagos Islands where Charles Darwin visited before working out his theory of natural selection to account for the diversity of life's many forms.

Boobies are large seabirds with a tapered body and long wings and tail, all appropriate for plunge-diving into the water to catch fish. This is made easy by their long bill, which has serrated edges and a sharp point. They also have inflatable air sacs just beneath the skin, which absorb the shock of impact and make them more buoyant. An asset because they have to come up quickly before their air runs out.

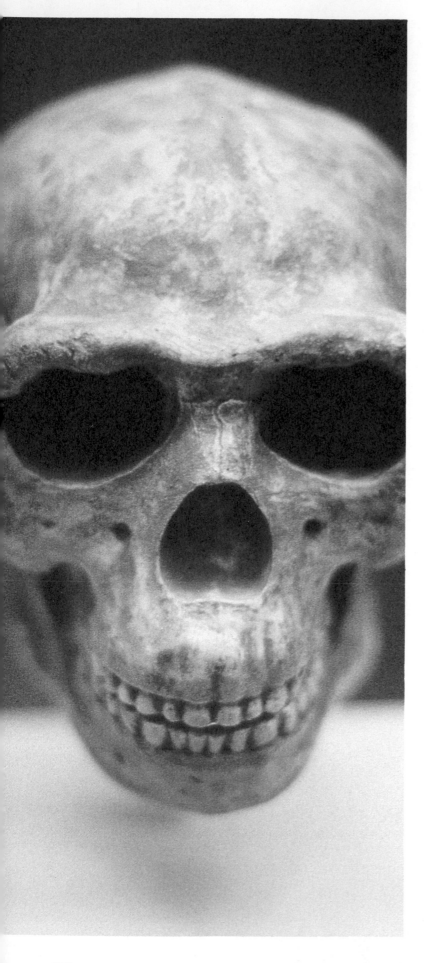

CHAPTER 19

Darwin and Evolution

Your study of this chapter will be complete when you can:

1. *Contrast pre-Darwinian ideas about evolution with post-Darwinian ideas about evolution.*

2. *State the contribution of Linnaeus, Buffon, and E. Darwin, naturalists of the mid-eighteenth century, to evolutionary theory.*

3. *Describe the catastrophe hypothesis of Cuvier and outline Lamarck's ideas on evolution.*

4. *Describe C. Darwin's background and the path of the voyage of the* Beagle.

5. *Divide Darwin's observations into two categories: (1) geology and fossils, and (2) biogeography. Tell how each category influenced his thinking.*

6. *Describe the process of natural selection as proposed by Darwin.*

Modern-day organisms don't always match their fossils exactly because evolution, in the meantime, has brought about structural changes. This is a human skull remarkably different from that of modern humans because of the heavy eyebrow ridges and the protruding muzzle. The large cranial capacity, however, convinces us though that we are indeed related to this species of humans, called Homo erectus.

harles Darwin (fig. 2.9) was only twenty-two in 1831 when he accepted the position of naturalist aboard the British naval ship *H.M.S. Beagle* that was going to sail around the world (fig. 19.1). The captain was hopeful that Darwin would find evidence of the biblical account of creation. However, the results of Darwin's observations were just the opposite, as you can tell by examining table 19.1.

Although it is often believed that Darwin forged this revolution all by himself, biologists during the preceding century were slowly beginning to accept the idea of **evolution,** that is that organisms change through time.

Figure 19.1

The voyage of the *H.M.S. Beagle* and what Darwin observed. Map (*a*) shows the journey of the *Beagle* around the world. As Darwin traveled along the east coast of South America he saw (*b*) an animal called a rhea, which he later noted looked like the African ostrich, and (*c*) the sparse vegetation of the Patagonian desert. Along the west coast he saw (*d*) the Andes Mountains with strata containing fossil animals, and (*e*) the lush vegetation of a rain forest. In the Galápagos Islands he saw (*f*) marine iguanas that have large claws to help them cling to rocks and blunt snouts for eating seaweed and (*g*) a finch that has the ecological niche of a woodpecker. Lacking a long beak to ferret out insects from trees, it uses a long stick instead.

b.

c.

d.

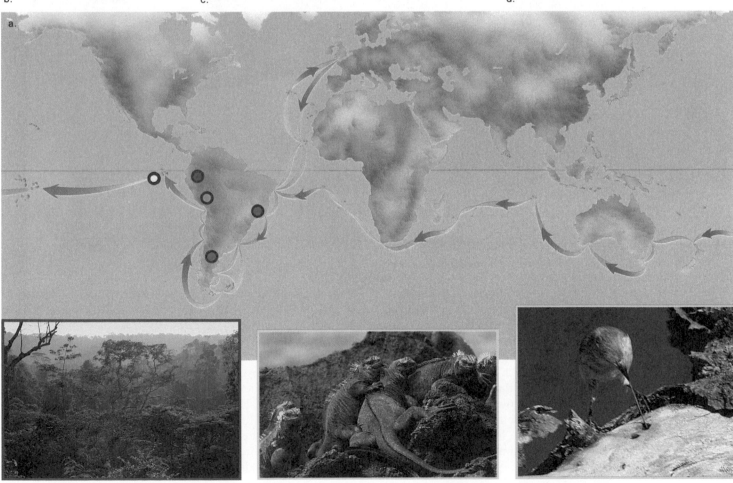

a.

e.

f.

g.

Table 19.1 Concepts Contrasted

Pre-Darwinian Concepts	Post-Darwinian Concepts
1. Earth is relatively young—age was measured in thousands of years	Earth is relatively old—age is now measured in billions of years
2. Fixity of species—organisms do not change and the number of species has remained the same	Organic evolution—organisms change and several new species can arise from a single ancestor
3. A creator previously decided the structure and function of each type of organism	Adaptation to environment explains structure and function of each type of organism
4. Observation and experimentation are unnecessary to substantiate self-evident truth	Observation and experimentation are used to substantiate theories of evolution

Evolution before Darwin

Mid-Eighteenth Century

Taxonomy, the science of classifying organisms, had played a very large role in shaping biology for many years prior to Darwin's trip. Chief among the taxonomists was Carolus Linnaeus (1707–1778), who gave us the binomial system of nomenclature (p. 2) and developed a system of classification that was fixed and rigid. It would appear that Linnaeus was influenced by the religious belief that every type of organism had been separately created independent of the others. The ideas of *separate creation and fixity of species* went hand in hand in the minds of most people at this time and lent support to the possibility of classifying living things (fig. 19.2).

Linnaeus was a botanist and his chief aim in life was to give a name to every plant known to him. Because he was a Swede, you might think that he was aware of plants only in Sweden, but actually many of his students brought back specimens that they had encountered on their travels to distant places. Botanists in America even sent him plants to classify.

For most of his life Linnaeus did not even consider the possibility of evolution, but there is evidence he did eventually perform hybridization experiments, which made him think that at least species, if not higher categories of classification, might change over time.

George Louis Leclerc, better known by his title Comte de Buffon (1707–1788), was a French naturalist who devoted many years of his life to writing a 44-volume natural history that described all known plants and animals. In order to enliven the description, he also included his thoughts on other matters such as a theory for the formation of the earth and his ideas about evolution. Here and there he mentioned that the following factors could influence evolutionary change: direct influences of the environment, migration, geographical isolation, overcrowding, and the struggle for existence. However, he presented no supportive data nor did he suggest a general mechanism by which

Figure 19.2
Does classifying organisms into genera and species make them seem intransmutable and fixed? a. *Calandrinia ciliata*. b. *Sedum spathifolium*.

a.

b.

evolution might occur. In fact, Buffon seemed to vacillate on the matter and in public, he often professed to believe in separate creation and the fixity of species.

Erasmus Darwin (1731–1802), Charles Darwin's grandfather, was a physician and naturalist. His writings on both botany and zoology contained many comments, although they were mostly in footnotes and asides, that mentioned the possibility of evolution. He based his conclusions on changes undergone by animals during development, artificial selection by humans, and the presence of vestigial organs (organs that were functional in an ancestor but are reduced and nonfunctional in a descendant). Again he offered no definite theory of evolution.

Figure 19.3

Figure 19.3
One of George Cuvier's first contributions was to compare the structure of a *Mastodon* to a living elephant. Given only a single fossil bone, he was able to suggest the anatomy of an entire *Mastodon*. He said these animals were shorter than modern elephants but with massive, pillarlike legs. The skull was lower and flatter and of generally simpler construction than in modern elephants.

In the mid-eighteenth century there was intense interest in classifying known plants and animals. Still, some of the naturalists (e.g., Buffon and E. Darwin) at this time thought it was plausible that evolution had occurred and though they proposed no general theory, they did put forth tentative suggestions as to possible mechanisms.

Late Eighteenth Century

George Cuvier and Catastrophism

In addition to taxonomy, biologists prior to Darwin were quite interested in comparative anatomy. Explorers and collectors were traveling about the world and bringing back not only currently existing species to be classified, but they were also bringing back **fossils,** remains of once living organisms, to be studied. George Cuvier (1769–1832) was a distinguished vertebrate (animals with backbones) zoologist who was the first to use comparative anatomy to develop a system of classifying animals. He recognized that structure and function are intimately related and went on to suggest that a single fossil bone was all he needed to deduce the entire anatomy of an animal (fig. 19.3). Because Cuvier was a staunch advocate of separate creation and fixity of species, he faced a real problem when geological evidence showed a progression of life forms during the history of the earth. Without much evidence to support it, he said that during the history of the world there had been a series of world-wide catastrophes. After each catastrophe, which caused mass extinctions to occur, the world had been repopulated by surviving species and this gave the appearance of change over time. This explanation of the history of life came to be known as **catastrophism.**

Figure 19.4
Lamarck is most famous for having suggested that the long neck of the giraffe is due to continual stretching to reach a food source. Each generation of giraffes stretched the neck a little further and passed on this characteristic to the next generation. With the advent of modern genetics, it became possible to explain why this so-called inheritance of acquired characteristics would not be possible.

Lamarck's Theory of Evolution

The first coherent theory of evolution was proposed in 1809 by the French naturalist Jean Baptiste de Lamarck (1744–1829). Although Lamarck was a contemporary of Cuvier, his ideas were entirely different, perhaps because he was an invertebrate (animals without backbones) zoologist who was quite impressed by the progression of life forms observed in geological rock strata. He stated clearly that more complex organisms were descended from less complex organisms and went on to suggest that this demonstrated a built-in drive toward perfection. In those days it was believed that humans represented perfection, and so Lamarck felt that all organisms were on their way to achieving this final perfect state. In order to account for the constant presence of simple and less complex organisms, he suggested that spontaneous generation was always producing a fresh supply of less complex organisms.

Lamarck also believed that plants and animals became adapted to their environment. To demonstrate how this might be possible, he proposed an **inheritance of acquired characteristics.** One example that he gave, and for which he is most famous, is that the long neck of a giraffe developed over time because generations of these animals kept stretching their necks to reach food high up in trees (fig. 19.4). Today, we know that in higher animals, at least, the **inheritance of acquired characteristics** is not possible because the germ cells that produce the sex cells are entirely separate from the somatic cells that are

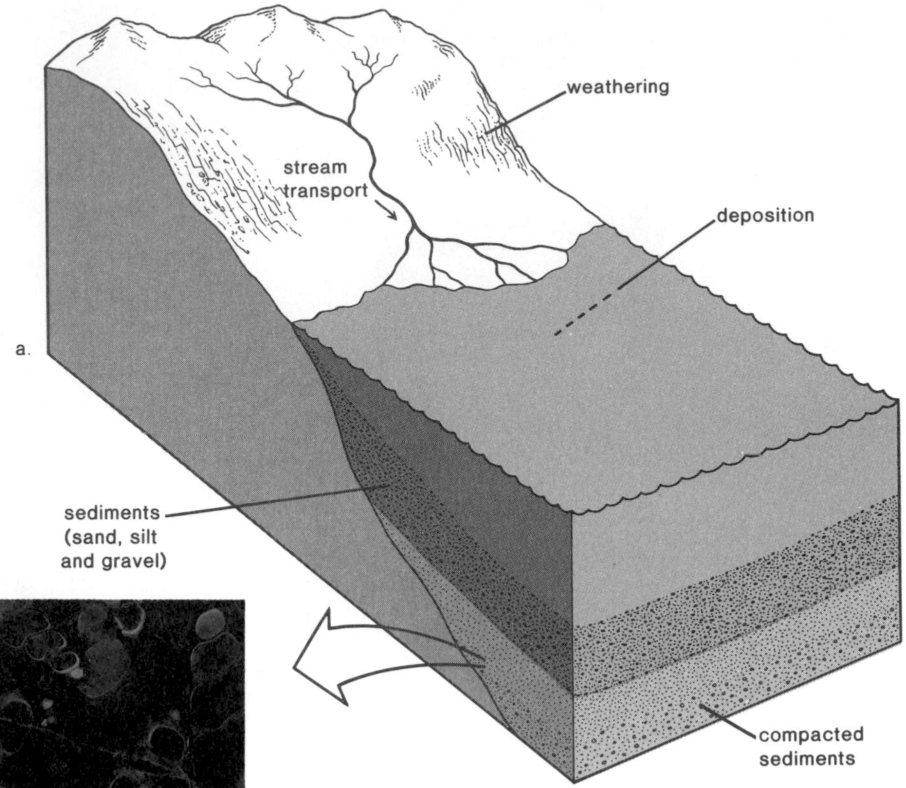

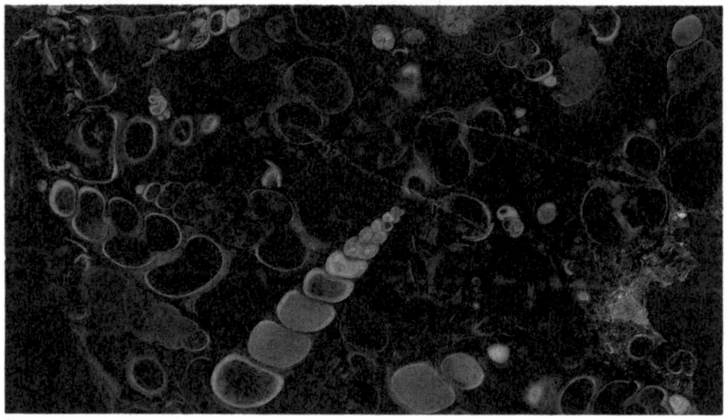

Figure 19.5

How is it possible for fossils of marine animals to be found on land? *a.* This diagram shows how water brings sediments into the sea; the sediments become compacted to form sedimentary rock. Fossils are often trapped in this rock and as a result of a later geological upheaval, the rock may be located on what is now land. *b.* Fossil freshwater snails, *Turitella.*

a.

weathering

stream transport

deposition

sediments (sand, silt and gravel)

compacted sediments

b.

present in the body in general. There is no means by which somatic cell changes can pass to the germ cells. This, of course, is the reason why Darwin's theory of pangenesis discussed earlier, page 166, is not feasible.

In the late eighteenth century, Cuvier and Lamarck differed sharply in regard to evolution. To explain the history of life as revealed by the fossil record, Cuvier proposed a series of catastrophes. Lamarck stated that more complex organisms evolved from less complex organisms and that over time, organisms became adapted to their environment. But his ideas regarding the mechanisms of evolution are largely unsupported.

Darwin's Theory of Evolution

When Darwin signed on as naturalist aboard the *H.M.S. Beagle,* he did have a suitable background. He was an ardent student of nature and had been a collector of insects since his early years. His sensitive nature prevented him from studying medicine and he went to divinity school at Cambridge. Even so, he attended

many lectures in both biology and geology and he was also tutored in these subjects by a friend and teacher, the Reverend John Henslow. As a result of arrangements made by Henslow, Darwin had spent the summer of 1831 doing field work with Adam Sedgwick, a geologist at Cambridge, and it was Henslow who recommended Darwin for the post aboard the *Beagle.* The trip was to take five years and traverse the southern hemisphere (fig. 19.1) where life is most abundant and varied. Along the way, Darwin encountered forms of life that were very different from those found in his native England.

Evolutionary Change

Even though it was not his original intent, Darwin began to realize and gather evidence that life forms do change over time and from place to place.

Geology and Fossils

Darwin had taken with him Charles Lyell's *Principles of Geology,* which presented arguments to support a theory of geological change proposed by James Hutton. Hutton believed the earth was not static and unchanging; rather, it was subject to

Figure 19.6

a. A giant armadillo, known only by the study of fossil remains. Darwin found such fossils and came to the conclusion that this extinct animal must be related to the living form. The giant armadillo weighed 2 tons. *b.* A modern armadillo weighs about 10 pounds.

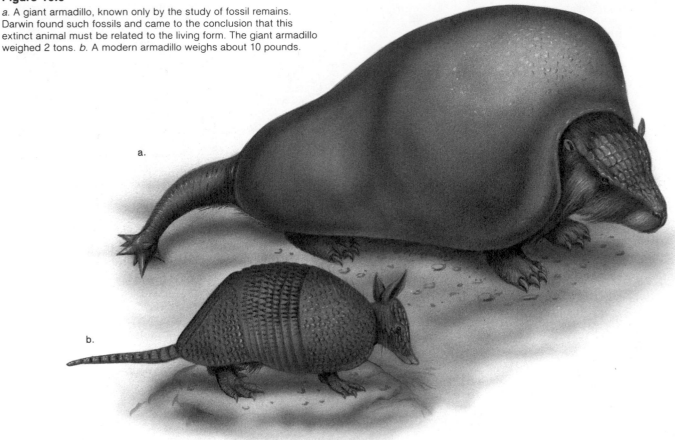

continuous cycles of erosion and uplift. Weathering causes erosion, dirt and rock debris are washed into the rivers and transported to oceans. There the loose sediments are deposited in thick layers and are converted eventually into sedimentary rocks (fig. 19.5). The sedimentary rocks, which often contain fossils, are uplifted from below sea level to form land. Hutton's ideas were called the **uniformitarian theory** of geology because he believed that the forces mentioned always acted at a uniform rate. Hutton's general ideas about continual geological change are still accepted today, although modern geologists realize that rates of change have not always been uniform. Lyell, unlike many of his peers, believed that cumulative effects of all these changes took many years and that the earth was very old.

Darwin found evidence to support Lyell's statements. When he explored what is now Argentina, he saw raised beaches that occurred along the coast for great distances. When he got to the Andes, he was impressed by their great height. In Chile he found marine shells well inland and witnessed the effects of an earthquake that caused the land to rise several feet. At the same time that Darwin was making geological observations, he was collecting specimens of fossils. For example, on the east coast of South America, he found the fossil remains of a giant ground sloth and an armadillo (fig. 19.6). Relying on Lyell's suggestion that geological changes take place slowly, Darwin began to think that there would have been enough time for slow biological evolution to occur and that living species could therefore be descended from extinct species.

Darwin's study of geology, influenced by Lyell, caused him to conclude that present day species are related to extinct species known only by their fossil remains.

Biogeography

Biogeography is the study of the geographic distribution of life forms on earth. Darwin could not help but compare the animals of Africa to those of South America. These tropical continents were occupied by similar animals that were apparently unrelated. He noted, for example, that the African ostrich and the South American rhea (fig. 19.7) although similar in appearance, were actually different animals. Could it be that these two animals had separate lines of descent, one in Africa and the other in South America?

On a finer scale, Darwin saw how closely allied animals replaced each other as he proceeded southward over the continent of South America. For example, the greater rhea found in the north was replaced by the lesser rhea in the south. Could these two forms be related in some way? When he got to the Galápagos Islands he saw island-by-island variation in organisms. The Galápagos Islands are a small group of volcanic islands that lie off the western coast of South America. The few types of plants and animals found there seemed to be slightly different from species Darwin had observed on the mainland, and they also varied from island to island!

Figure 19.7
The African ostrich (a) and the South
American rhea (b) are not closely related.
They look and behave similarly because
they have been exposed to the same type
of environment.

a.
b.

Figure 19.8
Each of the Galápagos Islands had its own form of tortoise. *a.* The
tortoises with short necks fed on low-lying vegetation. *b.* Those with
long necks were able to reach up and feed on cacti.

a.
b.

Tortoises   Each of the Galápagos Islands seemed to have its own type of tortoise (fig. 19.8) and Darwin began to wonder if this could be correlated with a difference in vegetation between islands. Long-necked tortoises seemed to inhabit only dry areas where food was scarce, most likely because the longer neck was helpful in reaching cacti. In moist regions with relatively abundant ground foliage, short-necked tortoises were found. Had the tortoises become adapted (suited) to the different environments of the islands?

Finches   Although the finches (fig. 19.9) on the Galápagos Islands reminded Darwin of a mainland finch, there were many more types. Even today, we can observe ground-dwelling finches that have different sized beaks, depending on the size of the seeds they feed on, and a cactus-eating finch that has a more pointed beak. The beak size of the tree-dwelling finches also varies according to the size of their insect prey. But the most unusual of the finches is a woodpecker-type finch. This bird has a sharp beak to chisel through tree bark, but lacks a woodpecker's long tongue to probe for insects. To make up for this, the bird carries a twig or cactus thorn in its beak and uses it to poke into crevices. Once an insect emerges, the finch drops this tool and seizes the insect with its beak.

Darwin speculated that all these different species of finches could have descended from one type of mainland finch. In other words, a mainland finch was the common ancestor to all the other types. Had speciation occurred because the separate islands allowed populations of birds to evolve independently? Could each type of finch have come into existence on a different island?

Biogeography had a powerful influence on Darwin and allowed him to see that it is possible for species to be related and for one species to even give rise to many others.

Once Darwin had decided that species do change, he turned his attention to the development of a theory by which evolution might occur.

Natural Selection

The theory that Darwin developed to provide a mechanism by which evolution might occur was nonteleological. Lamarck's theory had been **teleological,** that is, organisms were striving toward perfection and on the way they tried to acquire certain characteristics. In contrast, a nonteleological mechanism is one

Figure 19.9

The ancestral form from which the Galápagos finches are descended most likely had a strong beak capable of crushing seeds. Each of the present-day thirteen species of finches has a bill adapted to a particular way of life. For example, (a) the large tree finch grinds fruit and insects with a parrotlike bill. The small ground finch (b) has a pointed bill and eats tiny seeds and ticks picked from iguanas. The woodpecker finch (c) has a stout, straight bill that chisels through tree bark to uncover insects, but because it lacks a woodpecker's long tongue it uses a tool—usually a cactus spine or a small twig—to ferret insects out.

a.

b.

c.

in which the end result is not predetermined. For example, do certain tortoises have long necks because they wanted to have them (teleological), or do they have them because circumstances brought about this characteristic over time (nonteleological)?

Darwin's proposed mechanism of evolution, called **natural selection,** occurs because some organisms leave more fertile offspring than others. The characteristics of these organisms will therefore be more common in the next generation. The following important elements are necessary for natural selection to occur.

Struggle for Existence

In Darwin's time, a socioeconomist, Thomas Malthus, had stressed the reproductive potential of human beings. He proposed that death and famine were inevitable because the human population tended to increase faster than the supply of food. Darwin applied this concept to all organisms and saw that the available resources were not sufficient for all to survive. He calculated the reproductive potential of elephants. Assuming a life span of about 100 years and a breeding span of from 30 to 90 years, a single female will probably bear no fewer than six young. If all these young survived and continued to reproduce at the same rate, after only 750 years, the descendants of a single pair of elephants would number about 19 million! Such reproductive potential, Darwin said, necessitates a struggle for existence and only certain organisms will survive and reproduce.

Variations

Individual members of a population vary in their functional, physical, and behavioral characteristics (fig. 19.10). Some of these variations can be passed on from generation to generation. Darwin was never able to determine the cause of variations, nor how they are passed on. Today, we realize that genes determine the appearance of an organism and that mutations can cause new variations to arise.

Figure 19.10

Variations among individuals of a population. It is easy for us to note that humans vary one from the other, but the same is true for populations of any type of organism.

Survival of the Fittest

Whereas Darwin emphasized that only certain organisms survive to reproduce, modern evolutionists emphasize *differential reproduction.* Certain organisms can acquire a greater share of available resources; if they have the ability to reproduce, then their chances of successful reproduction are greater than those that are not as well equipped to capture resources.

Figure 19.11
Artificial selection of dogs. All dogs are descended from the wolf (*Canis lupus*), which began to be domesticated about 14,000 years ago. In evolutionary terms, the process of diversification has been exceptionally rapid. Several factors may have contributed: (1) the wolves under domestication were separated one from the other because human settlements were separate; (2) humans in each tribe selected for whatever traits appealed to them and artificial selection of dogs even continues today.

timber wolf

Boston terrier

beagle

dalmation

French bulldog

Chihuahua

Shetland sheepdog

bloodhound

red chow

Darwin noted that in **artificial selection,** humans choose which plants or animals will reproduce. This selection process brings out certain traits. For instance, there are today many varieties of dogs, each descended from the wolf (fig. 19.11). In a similar way, several varieties of vegetables can be traced to a single type (fig. 19.12).

In nature, interactions with the environment determine which members of a population that will reproduce to a greater degree than other members. In contrast to artificial selection, natural selection is not teleological. Selection occurs only because certain members of a population happen to have a variation that makes them more suited to the environment. For example, any variation that increases the speed of a hoofed animal will help it escape predators and live longer, a variation that reduces water loss will help a desert plant survive, and one that increases the sense of smell will help a wild dog find its prey. Therefore, we would expect organisms with these traits to eventually reproduce to a greater extent.

Adaptation

Natural selection causes in a population of organisms, and ultimately a species, to become adapted to its environment. The process is slow but each subsequent generation will include more individuals that are adapted than the previous generation. It is interesting to observe that as the species is changing, so is the environment. Environmental changes may result in different characteristics being adaptive; evolution is an ongoing process of organisms adapting to changing environments.

We can recognize **adaptation** particularly when unrelated organisms living in a particular environment display similar characteristics. For example, manatees, penguins, and sea turtles have flippers while most fishes, including sea skates, have flattened bodies, which allow them to move efficiently through the water (fig. 19.13).

Origin of Species

After the *Beagle* returned home in 1836, Darwin waited over twenty years to publish his ideas. During these intervening years, he collected data, formed hypotheses, and used deductive reasoning to substantiate his theory that evolution does occur. He was prompted to publish his findings after he received a letter from another naturalist, Alfred Russel Wallace, who had also come to a theory of natural selection.

Although supporting papers by Darwin and another by Wallace were read to a meeting of the Linnaean Society of London on the same day in 1858, only Darwin later presented detailed evidence in support of his theory of evolution. He described his experiments and reasonings at great length in his book, *On the Origin of Species*. Darwin's theory may be summarized as follows:

1. In a population, there are many more individuals produced each generation than can survive and reproduce.
2. There are heritable variations among the members of a population.
3. Individuals with adaptive characteristics are more likely to successfully reproduce than are those with other characteristics. This is called differential reproduction.
4. An even greater proportion of individuals in succeeding generations have common adaptive characteristics due to the occurrence of differential reproduction.
5. The end result of organic evolution is many different species each adapted to specific environments.

The occurrence of biological evolution is supported by many different types of evidence presented in the next chapter. Darwin's original proposed mechanism of evolution, termed natural selection, has undergone and is still undergoing modifications. Current views on what causes evolutionary change are discussed in chapter 21.

Figure 19.12
Humans have been able to develop several types of vegetables from a single species of *Brassica oleracea*: (*a*) Chinese cabbage (*b*) brussels sprouts, and (*c*) kohlrabi.

a.

b.

c.

Figure 19.13
The same type of environment produces similar adaptations in unrelated groups of organisms. Manatees, penguins, and sea turtles all have flippers, while most fishes, including sea skates, have flattened bodies.

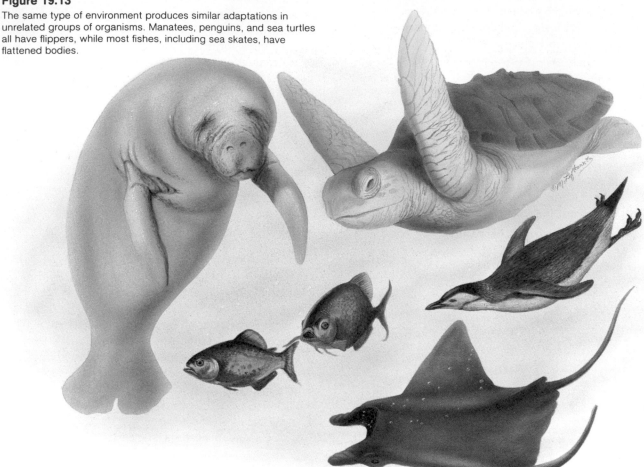

Summary

1. Charles Darwin came to the realization that evolution does occur after taking a trip around the world as naturalist aboard a British naval ship, *H.M.S. Beagle* from 1831 to 1836.

2. It is possible to contrast pre-Darwinian concepts about evolution with post-Darwinian concepts as is done in table 19.1.

3. A century before Darwin's trip, classification of organisms had been a main concern of biology. Linnaeus, like others, believed most of his life in separate creation and the fixity of species. Gradually, some naturalists began to put forth tentative suggestions that evolution does occur. For example, Comte de Buffon and Erasmus Darwin, who lived in the eighteenth century, had various disconnected ideas about the process of evolution.

4. George Cuvier and Jean Baptiste de Lamarck were contemporaries in the late eighteenth century who differed sharply on evolution. To explain the history of life as witnessed by the fossil record, Cuvier proposed a series of catastrophes that caused some organisms to become extinct and others (the survivors) to seemingly take their place. Lamarck, however, had a two-prong theory for evolution: (1) complex organisms were descended from less complex ones due to a built-in drive toward perfection, and (2) organisms became adapted to their environment by the inheritance of acquired characteristics. Lamarck's hypotheses regarding the mechanism of evolution are largely unsupported today.

5. Darwin's trip involved two primary types of observations: (1) geology and fossils, and (2) biogeography. Relying on Lyell's evidence that the earth was very old and changes in the earth's crust occur slowly, Darwin began to see that fossils represented a history of life, and that today's living species are descended from extinct species.

6. In the same manner, he saw that the various tortoises and finches on the Galápagos Islands were descended from a common ancestor, and that evolution can cause new species to come into existence. This meant that separate creation and fixity of species was not a plausible explanation for the diversity of life.

7. Natural selection is the mechanism that Darwin proposed for how evolution comes about. He was influenced by Malthus who stressed the consequences of the great reproductive potential of human beings. Due to natural selection, those members of a population that have a favorable variation are selected to reproduce and in this way a population gradually becomes adapted to an environment.

8. The occurrence of evolution is supported by all sorts of evidence, which is discussed in chapter 20. The mechanism of evolution proposed by Darwin has been modified and is still being modified, as discussed in chapter 21.

Objective Questions

1. Which of these is a correct statement?
   a. Linnaeus was a strong supporter of the occurrence of evolution but Comte de Buffon was rigidly against his proposal.
   b. Cuvier, who was a contemporary of Linnaeus, developed a mechanism to explain how life arose as a single cell and evolved to give the complex plants and animals of today.
   c. Both Cuvier and Lamarck said that catastrophes had occurred and that's why so many organisms have become extinct.
   d. Lamarck said that (1) evolution had occurred, and (2) organisms became adapted to their environment.
2. Which of these is a correct statement?
   a. No one prior to Darwin even suggested that evolution might occur.
   b. Darwin, unlike other investigators, provided evidence for the occurrence of evolution and proposed a theory to explain how it might occur.
   c. Lamarck's theory of inheritance of acquired characteristics is fully substantiated today.
   d. Linnaeus stated that classification schemes are impossible if evolution occurs.
3. According to Lamarck's theory of acquired inheritance,
   a. if a man lost his hand, then his children would also be missing a hand.
   b. if a man lost his hand, then his children would not be missing a hand.
   c. changes in phenotype are passed on by way of the genotype to the next generation.
   d. Both (a) and (c).
4. Why was it helpful to Darwin to learn that Lyell thought the earth was very old?
   a. An old earth has more fossils than a new earth.
   b. It gave time for evolution to occur slowly.
   c. Upheavals are necessary to cause evolution to occur.
   d. Natural selection doesn't occur in one generation, it takes several generations.
5. When he arrived at the Galápagos Islands, Darwin studied the
   a. giant armadillos and sloths.
   b. monkeys and tigers.
   c. tortoises and finches.
   d. All of these.
6. The tortoises and finches were adapted to
   a. feeding on different food sources.
   b. swimming in the ocean.
   c. gathering food on the mainland.
   d. All of these.
7. All the finches on the various islands are
   a. completely unrelated.
   b. descended from a common ancestor.
   c. in competition with one another because they feed on similar foods.
   d. Both (a) and (c).
8. Which of these are required for natural selection to work?
   a. variation
   b. struggle for existence
   c. survival of the fittest
   d. All of these.
9. The end result of natural selection is
   a. overpopulation.
   b. adaptation to the enviroment.
   c. origin of new species.
   d. Both (b) and (c).
10. Malthus in his writings stressed
    a. occurrence of variations.
    b. struggle for existence.
    c. survival of the fittest.
    d. adaptation to the environment.

Study Questions

1. In general, contrast pre-Darwinian ideas about evolution with post-Darwinian ideas about evolution.
2. Cite naturalists who made a contribution to biology in the mid-eighteenth century and state their beliefs about evolution.
3. In general, contrast the ideas of Cuvier with those of Lamarck (both of the late eighteenth century).
4. What reading did Darwin do and what observations did he make regarding geology? How did these influence his conclusions about fossils?
5. What observations did Darwin make regarding biogeography? How did these influence his conclusions about the origin of species?
6. What are the essential features of the process of natural selection as proposed by Darwin?

Thought Questions

1. Relying on your knowledge of the scientific method as discussed in chapter 2, why would you expect scientists to be modifying Darwin's original proposal of how natural selection functions?
2. What contribution does modern genetics make to Darwin's theory of natural selection?

Selected Key Terms

evolution (ev″o-lu′shun) 281
taxonomy (tak-son′o-me) 282
fossil (fos′il) 283
catastrophism (kah-tas′tro-fizm) 283

inheritance of acquired characteristics (in-her′i-tans uv ah-kwīrd′ kar″ak-ter-is′tiks) 283
uniformitarian theory (u″ni-for″mi-tār′e-an the′o-re) 285
biogeography (bi″o-je-og′rah-fe) 285

teleological (te″le-o-loj′i-kal) 286
natural selection (nat′u-ral sĕ-lek′shun) 287
artificial selection (ar″ti-fish′al sĕ-lek′shun) 289
adaptation (ad″ap-ta′shun) 289

CHAPTER 20

Evidences for Evolution

Your study of this chapter will be complete when you can:

1. *Explain in general what the fossil record is and what it tells us about the history of life.*

2. *Explain what is meant by mass extinction and discuss possible causes for the Permian and Cretaceous extinctions.*

3. *Explain why it is possible to divide the world into six biogeographical regions, and relate the distribution of life forms to the process of continental drift.*

4. *Contrast homologous structures with analogous structures.*

5. *Discuss the relationship between comparative anatomy, taxonomic groups, and the construction of evolutionary trees.*

6. *Relate the concept of common ancestor to evolutionary trees and indicate how ancestry is traced by way of common ancestors.*

7. *Tell what is meant by a molecular clock and relate this concept to the construction of evolutionary trees.*

8. *Relate evolutionary trees to overall patterns of macroevolution.*

9. *Define divergent, parallel, convergent, and coevolution.*

10. *Explain why the fossil record, biogeography, comparative anatomy, and biochemical studies are evidences for evolution.*

The best-known primates are monkeys, apes, and humans. The fossil record shows that these organisms evolved in the order stated. The evidence that humans are closely related to monkeys and apes is based on structure, function, and behavior. The behavior of the titi monkey shown here is most remarkable. Such monkeys are monogamous and live in a family group consisting of a father, mother, and offspring. The father grooms (cleans) his offspring and carries it on his back through the trees. To signify that their territory is occupied, the couple sits side-by-side at dawn and sings a duet consisting of calls that can be heard by their neighbors. At night, the family sleeps on the same tree limb with tails entwined.

any of these evidences for the occurrence of evolution were known to Darwin. Since that time, new evidences, such as those relating to the biochemistry of organisms, have been added to those formerly known. Further, even those that were known to Darwin have now been studied in greater depth.

Fossil Record

The fossil record tells us that *species are not immutable*. In other words, they change. The species we see about us today are not the ones that have always existed. This is to be expected if life forms are continually evolving.

Fossils (fig. 20.1) are the remains of organisms that lived long ago. Usually when an organism dies it is either consumed by scavengers or it undergoes bacterial decomposition. Occasionally, it is buried quickly and in such a way that decomposition is never completed or is completed so slowly that the organism may leave an imprint, or a cast, which clearly indicates its external and perhaps its internal structure.

Preserved fossils are most often found in sedimentary rock (fig. 19.5). Weathering produces sediments that are carried by streams and rivers into the oceans and other large bodies of water. There, they slowly settle and are converted into sedimentary rock. Later, sedimentary rocks are uplifted from below sea level to form new land and researchers are able to search for fossil remains trapped in the rocks.

Figure 20.1

Fossils can be used to trace the history of life. Those featured here are placed in the order in which they would appear in the fossil record. *a.* Prokaryotic cell fossils are about 3.5 billion years old. This is a filamentous form. *b.* Eukaryotic cell fossils are about 1.5 billion years old. Magnification, ×250. *c.* Invertebrate brachiopod (lamp shell) fossils are about 395 million years old. *d.* Fern fossils date from the Carboniferous Period, which lasted from 350 to 286 million years ago. *e.* Vertebrate dinosaur fossils are about 190 million years old. *f.* Placental mammals date as early as 65 million years ago, but this fossil rabbit is from the Pleistocene Epoch, which began about 2 million years ago.

Table 20.1 The Geological Time Scale—Major Divisions of Geological Time with Some of the Major Evolutionary Events of Each Geological Period

Era	Period	Millions of Years Ago	Plant Life	Animal Life
Cenozoic	Quaternary	— 2.5 —	Increase in number of herbaceous plants	Age of Human Civilization
	Tertiary		Dominance of land by flowering plants	First hominids appear Dominance of land by mammals and insects
Mesozoic	Cretaceous	— 65 — —130—	Flowering plants spread Cone-bearing trees decline	Dinosaurs become extinct
	Jurassic	—180—	First flowering plants appear	First mammals and birds appear Age of Dinosaurs
	Triassic	—230—	Dominance of land by ferns and cone-bearing trees	First dinosaurs appear
Paleozoic	Permian	—280—	First seed plants	Expansion of reptiles Decline of amphibians
	Carboniferous	—350—	Age of great coal-forming forests including club mosses, horsetails, and ferns	First reptiles appear Age of Amphibians
	Devonian	400	Low-lying primitive, vascular plants appear on land	First amphibians move onto land First insects appear Age of Fishes
	Silurian	—435—		
	Ordovician	500	Unicellular marine algae abundant	First fishes (jawless) appear Age of Invertebrates
	Cambrian	—600—		
Precambrian (Proterozoic)		700 1,000 —2,500—	Origin of nonshelled invertebrates Origin of complex (eukaryotic) cells	
Archeozoic		3,500 —4,600—	Oldest fossil (prokaryotic cells)	
			Formation of earth	

History of Life

During the last century, geologists and **paleontologists** (those who study fossils) noticed that the fossil deposits form recognizable layers or **strata** (fig. 19.1e), and each layer has its own mix of fossils. Boundaries between the strata, where one mix of fossils gives way to another, provide the basis for dividing geologic time as shown in table 20.1. We now know that many of these abrupt transitions are due to mass extinctions as explained in the reading on the next page. Notice that the names of the largest geological time divisions, the eras, pertain to life forms:

Cenozoic	Modern life
Mesozoic	Middle life
Paleozoic	Ancient life
Proterozoic	First life
Archeozoic	Beginning life

The strata alone cannot provide dates for fossil deposits; they only tell us the sequence of the organisms over time. Dates can be calculated, however, from a study of the radioactive isotopes in the rocks of each stratum. Radioactive isotopes decay at a constant rate; therefore, by determining how much decay has taken place since the rocks first formed, a date can be assigned to the rocks. This tells us that the solar system is 4.6 billion years old and that the history of life began about 3.5 billion years ago. A large portion of this time was devoted to the evolution of prokaryotes because eukaryotic cells didn't evolve until about 1.5 billion years ago. It was during this time that most of the metabolic pathways evolved. Single-celled organisms remained in the oceans and increased in complexity, until multicellular organisms evolved about 650 million years ago. Fossils of multicellular organisms are much easier to detect than those of simpler ones; therefore, the history of life is better documented from this time on (table 20.1).

Plants invaded the land environment about 400 million years ago, and they were followed by fungi, invertebrates, and finally vertebrates. The Age of Reptiles lasted from 300 million to 65 million years ago, when the dinosaurs died out. Mammals

Mass Extinctions

Extinctions are commonplace events in the fossil record; even the extinction of two complete families (there may be many genera in a family) per million years would be considered usual. The reasons for these extinctions might very well be an inability of individual species to adapt to a changing environment. However, the fossil record also gives evidence of the periodic occurrence of major or mass extinctions. In fact, the divisions of the geological time scale (table 20.1) are based on abrupt changes in the fossil record, some of which are now considered to have been due to mass extinctions. During the time of a mass extinction, as many as 10–20 families might become extinct.

There were four mass extinctions during the Paleozoic Era. Those at the end of the Cambrian, Ordovician, and Devonian Periods were serious, but not nearly as impressive as the one at the end of the Permian Period. It is estimated that perhaps more than 90% of all marine species living at the time may have become extinct. An explanation for this occurrence is con-tinental drift, specifically the formation of the supercontinent Pangaea (p. 297 and see fig. 20.6a). Certainly one large continent has less coastline than several continents and so loss of habitat could have killed off much marine life.

One well-studied mass extinction occurred at the end of the Cretaceous Period. This extinction affected land organisms more than marine organisms, and it is familiar to most of us as the time when the dinosaurs became extinct. The cause of this extinction may not have been due only to continental drift. Walter and Luis Alvarez and their colleagues found that Cretaceous clay contains an abnormally high level of iridium (see figure a), an element that is rare on earth but more common in asteroids. They hypothesize that it is possible a comet or asteroid collided with the earth. The result of such a bombardment would have been similar to a worldwide atomic bomb explosion. A cloud of dust would have mushroomed into the atmosphere, shading out the sun and causing plants to freeze and die. Mass extinction of an-imals and plants would have followed (see figure b).

In 1984, paleontologists David Raup and John Sepkoski of the University of Chicago discovered that the fossil record of marine animals shows that mass extinctions have occurred every 26 million years and surprisingly, astronomers can offer an explanation. Our solar system is in the Milky Way, a starry galaxy that is 100,000 light years[1] in diameter and 1,500 or so light years thick. Our sun moves up and down as it orbits in the Milky Way. Astronomers predict that this vertical movement will cause our solar system to approach certain other members of the Milky Way every 26–33 million years, producing an unstable situation that could lead to comet bombardment of the earth.

It appears that there may be multiple reasons for mass extinctions of organisms during the history of the earth.

[1]One light year, the distance light travels in a year, is about six trillion miles.

a.

b.

a. Separating the Mesozoic and Cenozoic strata is a layer of iridium (see placement of coin), an element that is more common in asteroids. It is hypothesized that if an asteroid struck the earth, there would have been a worldwide cloud of dust, which would have deposited the iridium on the planet when it settled.
b. Because the dust would have obliterated the sun, vegetation would have died off and this would have eventually led to the extinction of many animals, including the dinosaurs.

had their origins some 150 million years ago, but it was not until the dinosaurs had vanished that they became abundant. Direct ancestors to humans (hominids) do not appear until about 3 million years ago, which is very recent, compared to how long life has been evolving. If the history of the earth is measured using a 24-hour time scale that starts at midnight, humans do not appear until one half minute before the next midnight (fig. 20.2).

We will be discussing all these events in greater detail when we study the diversity of life in chapters 23–28.

The fossil record in broad terms traces the history of life, and more specifically, allows us to study the lineage of individual organisms.

Biogeography

Biogeography, the study of the distribution of plants and animals about the world, has shown us that the many forms of life are distributed in particular ways.

Biogeographical Regions

As early as 1857, Philip Sclater, a British bird specialist, had divided the world into six *biogeographical regions.* Alfred Wallace (p. 289) was also studying the distribution of life forms when he independently developed the concept of natural selection. Later, he supported Sclater's division of the world into the biogeographical regions depicted in figure 20.3. The land areas in each region have characteristic plants and animals, and the fos-

Figure 20.2

The outer ring of this diagram shows the history of the earth as it would be measured on a 24-hour time scale starting at midnight. (The inner ring shows the actual years starting at 4½ billion years ago.) If the history of the earth is measured as if it had all happened in 24 hours, the Cambrian Period wouldn't start until 8 P.M.! This means that a very large portion of life's history is devoted to the evolution of single-celled organisms. The first multicellular organisms do not appear until just after 8 P.M. and humans are not on the scene until less than a minute before midnight.

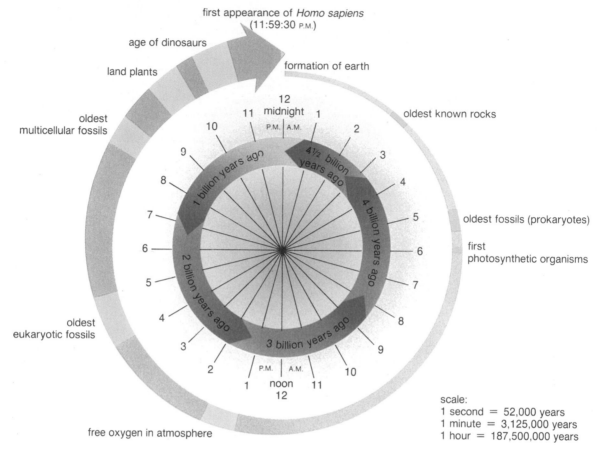

first appearance of *Homo sapiens*
(11:59:30 P.M.)

age of dinosaurs

land plants

oldest
multicellular fossils

formation of earth

oldest known rocks

oldest fossils (prokaryotes)

first
photosynthetic organisms

oldest
eukaryotic fossils

free oxygen in atmosphere

scale:
1 second = 52,000 years
1 minute = 3,125,000 years
1 hour = 187,500,000 years

...sils of that region sometimes resemble the living organisms only of that region. This may be explained by the fact that the biogeographical regions are separated by impassable barriers. Therefore, it is not surprising that each has a unique mix of plants and animals.

There may be similar environmental conditions in two different regions, but even then, the same plants and animals are not necessarily found. Instead, these organisms only resemble each other because they are similarly adapted. For example, plants of the cactus family are found in the deserts of southwestern North America, while members of the spurge family are found in the deserts of Africa. Both types of plants have made similar adaptations to arid habitats. And there are North American mammals and African mammals that are similarly adapted to living in a grassland environment (fig. 20.4).

Continental Drift

The presence of biogeographical regions cannot explain the distribution of life on earth completely. For example, reptilian ancestors have a wider distribution than do mammalian ones,

rhododendron plants are found in both eastern Asia and eastern North America, and marsupials are diverse in Australia, and were diverse in South America. These distributions are explainable on the basis of continental drift.

Alfred Wegener, a German earth scientist, proposed the theory of **continental drift** about 1915. He used geological evidence to suggest that the continents have undergone large movements over the past 300 million years.

We now know that the continents are on plates that move, or drift, relative to one another. There are ridges in the oceans where volcanic activity forms new crust (fig. 20.5). This crust spreads slowly away from the ridges until it is consumed, often at oceanic trenches. Plates are mammoth pieces of crust that act like conveyor belts, carrying along the drifting continents. The study of the movement of plates is called plate tectonics.

Geological upheavals are common wherever the plates meet one another; this is where one finds mountain building, volcanos, and earthquakes. For example, the San Andreas fault in California is at the boundary of two plates, and an earthquake is expected there soon. Mt. St. Helens, which erupted in recent years, is also near the edge of a plate.

Figure 20.3

Major biogeographical regions of the world are, or have been, separated either by water or impassable land barriers. This causes each region to have distinctive plants and animals. The photographs depict distinctive animals that are found in the various regions.

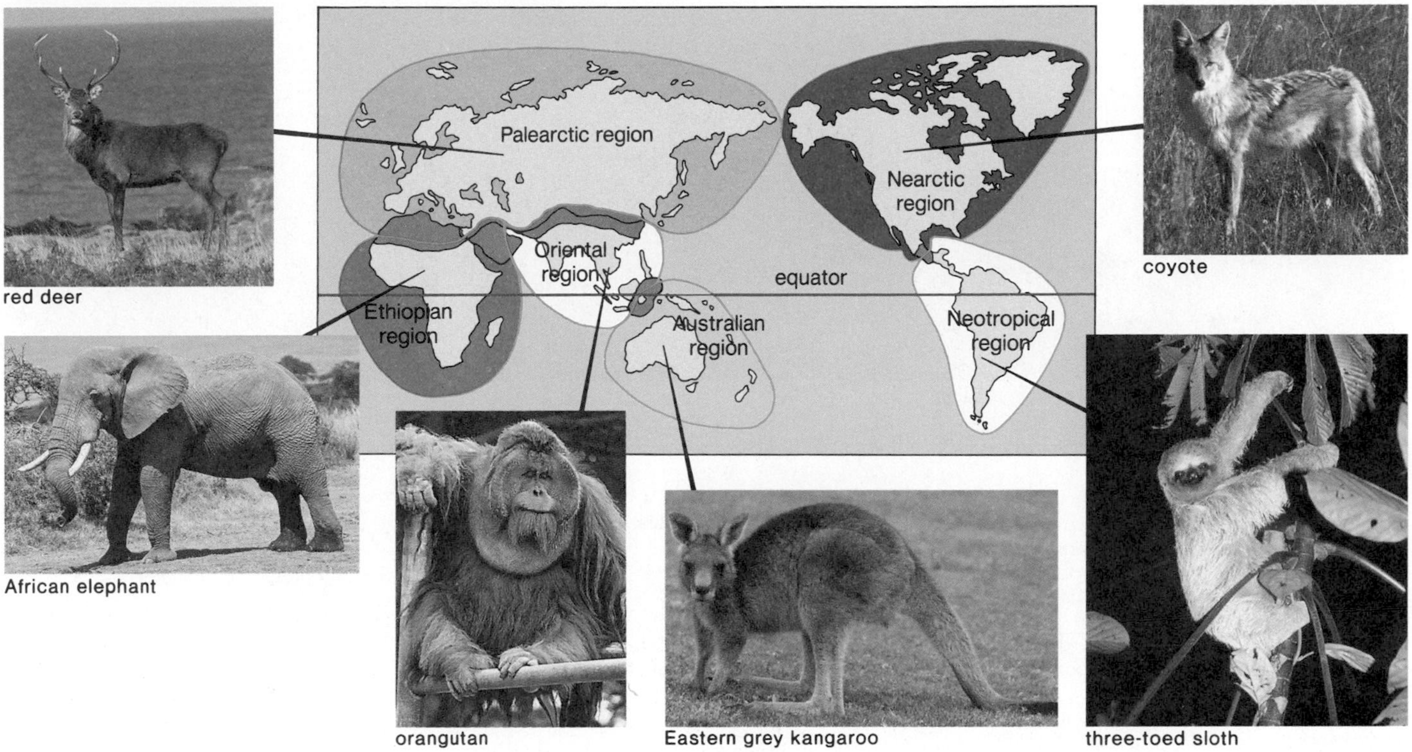

Palearctic region

Nearctic region

Oriental region

equator

Ethiopian region

Australian region

Neotropical region

red deer

coyote

African elephant

orangutan

Eastern grey kangaroo

three-toed sloth

Figure 20.4

Unrelated plants and animals are similarly adapted when the environments are similar. Cacti in North American deserts (a) and spurges in African deserts (b) are both adapted to arid environments. Coyotes in North American grasslands (c) and jackals in African grasslands (d) are both carnivorous predators that prey on small animals and also eat carrion.

a.   b.   c.   d.

Continental drift dramatically affected the evolution of life on earth. About 230 million years ago, near the end of the Permian Period, all the land masses came together and formed a supercontinent called Pangaea (pangaea means all land) (fig. 20.6a). There was now less shoreline and also the ocean basins increased in depth so that altogether the amount of shallow water habitat close to the continents was greatly reduced. This is where you find most marine life; therefore, many marine species most likely became extinct. Due to the joining of the continents, land species that had been isolated from one another were now competing for food and habitat. Another factor affecting both aquatic and land organisms at this time was a climatic change

Figure 20.5

Diagram illustrating plate tectonics. The continents drift because they are on plates, or mammoth pieces of crust that are formed at midoceanic ridges and disappear at oceanic trenches.

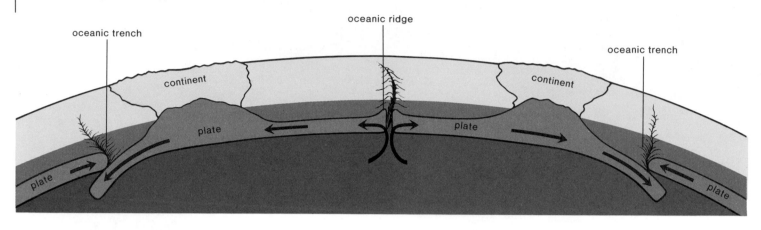

Figure 20.6

History of the placement of today's continents. *a.* About 200–250 million years ago all the continents were joined into a supercontinent call Pangaea. *b.* When the joined continents of Pangaea first began moving apart, there were two large continents called Laurasia and Gondwana. *c.* By 65 million years ago, all the continents began to separate and this process is continuing today. *d.* North America and Europe are presently drifting apart at a rate of about 2 cm per year.

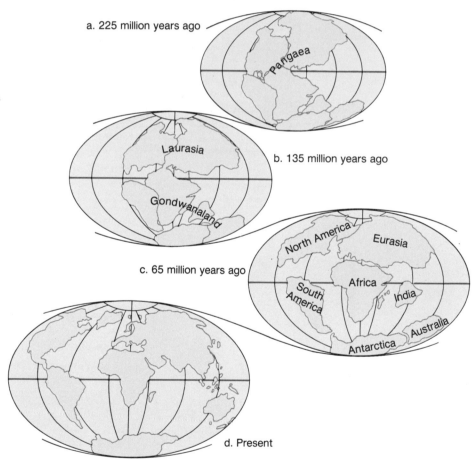

due to altered oceanic currents. Predictably, the fossil record indicates a **mass extinction** occurred at the close of the Permian Period as discussed in the reading on page 295.

During the early Mesozoic Era, Pangaea began to break up and geographical isolation between continents occurred once more. As the continents drifted apart (fig. 20.6*b-d*), each became separate evolutionary areas once again, and the plants and animals of the different regions diverged.

It is possible to explain the distributions we spoke of earlier on the basis of continental drift. Reptilian ancestors have a wider distribution than mammalian ones because reptiles were evolving when Pangaea was forming, whereas mammals didn't start diversifying until after the continents started moving apart. Rhododendron plants could have existed at one time throughout North America and Asia because these two continents were once

Figure 20.7
A chick (a) and pig (b) embryo at comparable early developmental stages have many features in common because they share many general characteristics, the details of which are refined at later developmental stages.

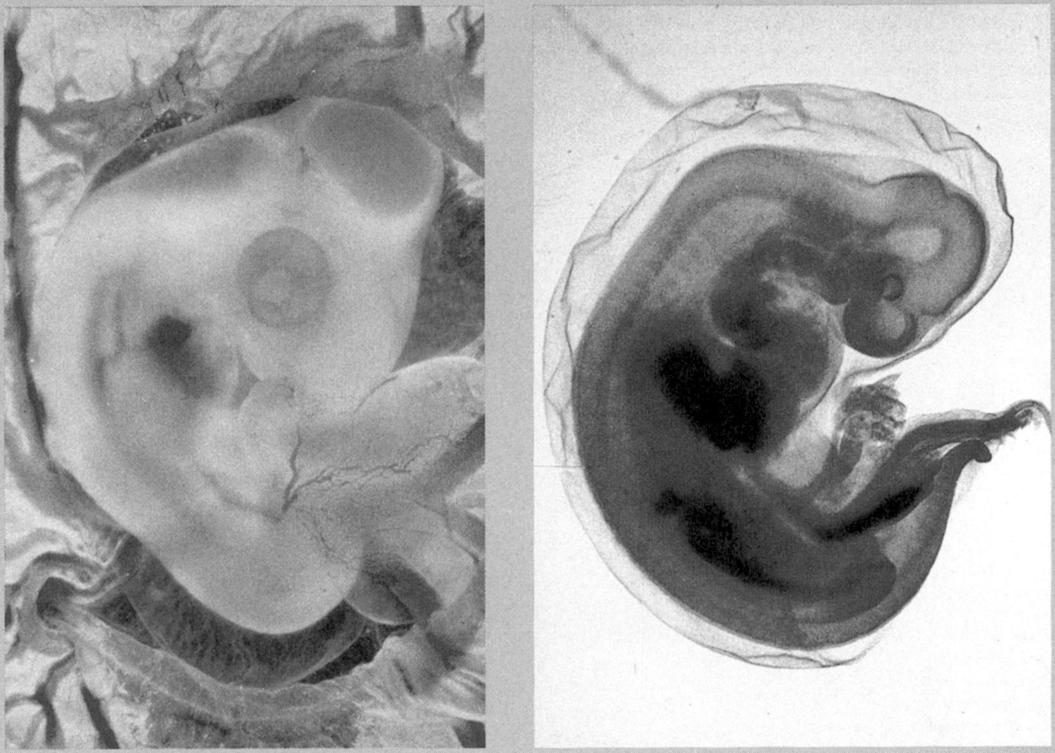

a.

b.

joined into Laurasia. The presence of marsupials in Australia and South America can also be explained by continental drift. Until about 53 million years ago, it would have been possible for animals to migrate from Australia to South America by way of Antarctica. This migration path was lost when the two continents moved away from Antarctica.

The distribution of life forms on earth (biogeography) is explainable on the basis of independent evolution during more recent geological periods and the movement of continents for over 300 million years.

Anatomy and Embryology

Related species share a *unity of plan*. For example, reproductive organs of all flowering plants are basically similar even though they vary in some details. All vertebrates, at some time during development, have a supporting dorsal rod, called a notochord, and exhibit gill pouches (fig. 20.7).

Homologous and Analogous Structures

Structures that are derived from a common ancestor are called **homologous structures.** Among adult terrestrial vertebrates, we can note that their forelimbs are organized similarly and contain the same bones (fig. 20.8), even though the animals themselves are adapted to different ways of life. The explanation for this remarkable unity of plan is that these vertebrates are descended from a common ancestor. The basic forelimb plan originated with this ancestor, and was modified in the descendant groups as each one continued along its own evolutionary pathway.

Sometimes organisms have structures that are constructed differently but appear to be similar because they serve the same function. An insect's wing is a stiffened membrane supported by veins with hardened walls. A bird's wing has an internal bony skeleton covered by muscle, skin, and feathers. Even so, both types of wings have the same broad flattened shape because they enable the animal to fly through the air. Structures such as these that serve a similar function but are not derived from a common ancestor are called **analogous structures.** While analogous structures do not indicate close genetic relationships, they are still evidence for evolution because they show the degree to which structures can be similarly modified to suit a particular environment.

Vestigial Structures

An organism may have structures that are underdeveloped and seemingly useless, but may be fully developed and functional in related organisms. These are called **vestigial structures.** Some vestigial structures in humans are remnants of a nictitating membrane, which in birds can be drawn over the eye, and caudal vertebrae, which are more fully developed and functional in mammals that have tails.

Figure 20.8

The unity of plan among vertebrates is exemplified by the bones in the forelimbs. (The bones are color coded.) Although the same bones are present, the specific design details of the limbs are different. All these vertebrates are descended from a common ancestor, but each is evolving separately.

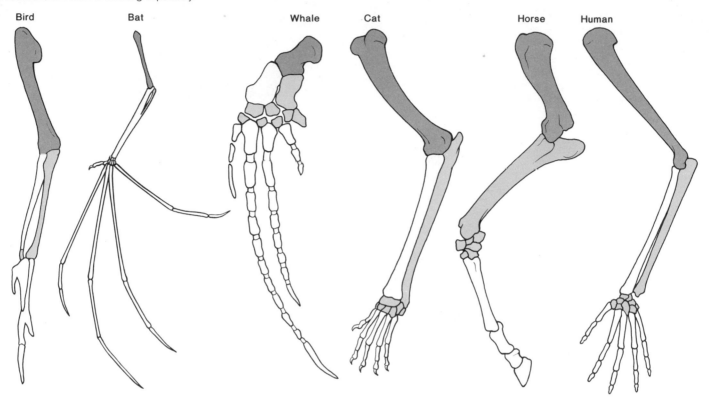

Bird   Bat   Whale   Cat   Horse   Human

Evolutionary Trees

Comparative anatomy is the study of the structures of different organisms, both living and extinct. This information helps in classifying organisms into **taxonomic categories** (table 20.2) and makes it possible to construct **evolutionary trees** (fig. 20.9). These trees can tell us how the members of a group are believed to be related and how their history might be traced by way of **common ancestors.** Common ancestors can be found at the points of divergence in an evolutionary tree. In the diversity chapters that follow we will have occasion to construct evolutionary trees showing the presumed relationships of major groups of organisms.

Organisms that are descended from a common ancestor share a unity of plan that indicates they are closely related. Comparative anatomy helps investigators classify organisms and construct evolutionary trees.

Most biologists believe that evolution takes place gradually and that divergence of two groups occurs slowly. This belief is reflected in the overall design of an evolutionary tree. The concept of gradualism is presently being challenged, however, as discussed in the reading on the next page.

Table 20.2 Hierarchy of Classification

Category	Description	
Species	A type of organism distinguishable from all other types	
Genus	Contains related species	
Family	Contains related genera	
Order	Contains related families	more characteristics in common / characteristics in common
Class	Contains related orders	are more distinctive
Phylum (animals) Division (plants)	Contains related classes	
Kingdom	Contains related phyla	

Comparative Biochemistry

Almost all living organisms use the same basic biochemical molecules, including DNA, ATP, and many identical or nearly identical enzymes. It would seem that these molecules evolved very early in the evolution of life and have been passed on ever since.

Figure 20.9

Evolutionary trees can show the relationship between different taxonomic categories. The organisms in family A and genera 1 and 2 are closely related as are those in family B and genera 3 and 4. A common ancestor can be found wherever a branch point occurs in the tree.

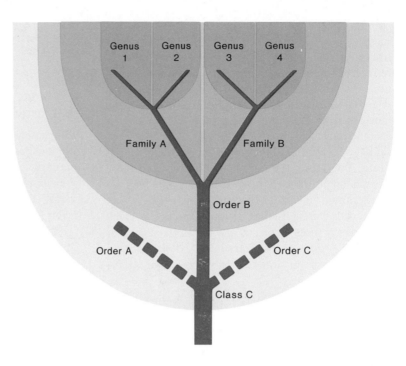

What Kind of Evolutionary Tree?

Most evolutionists have assumed that evolution occurs gradually and that a single species changes slowly—over millions of years—giving rise to new species. This so-called *gradualistic model* of evolutionary change implies that the fossil record contains a plentiful supply of *intermediate forms,* which are fossil remains having characteristics of two different groups. Some intermediate forms, such as *Archeopteryx* (fig. 28.11), have been found but not an overwhelming number.

In contrast to the gradualistic model, Niles Eldredge of the American Museum of Natural History and Stephen Jay Gould of Harvard University, upon close examination of the fossil record, have proposed a *punctuated equilib-*

rium model of evolutionary change. The record shows instances in which species tend to remain the same for a long time and then suddenly, new species appear. In other words, an *equilibrium* phase (long periods without change) is *punctuated* by a rapid burst of change during which some species become extinct and new species arise. For example, a rapid change of species may have followed major environmental upsets caused by continental drift. According to the punctuated equilibrium model, there would be few intermediate forms in the fossil record because such forms exist only during the period of rapid change before a quiet period returns.

On a smaller scale, there is evidence among living species that the punctuated equilibrium

model might be working. Peter Grant tagged 1,500 medium ground finches on Daphne Major, one of the Galápagos Islands. In 1977 there was a severe drought and the majority of these birds died. The birds that survived were of a larger size than those that died. This suggests that an ecological crisis is quite capable of causing a drastic change in species anatomy. Perhaps on a larger scale, speciation itself is apt to occur after periods of crises.

These two different hypotheses concerning the rate of evolution affect the manner in which an evolutionary tree is depicted. Figure *a* shows the type of evolutionary tree that is consistent with gradualism and figure *b* shows the type of evolutionary tree consistent with the punctuated equilibrium model.

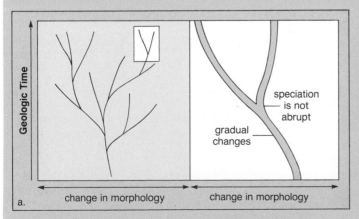

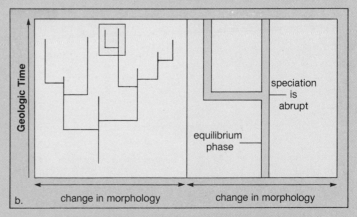

Evolutionary trees. *a.* This is the usual diagram for an evolutionary tree when it is assumed that evolution takes place gradually.
b. This is the usual diagram for an evolutionary tree when it is

assumed that evolution takes place according to the punctuated equilibrium model, with periods of relatively rapid speciation, followed by long periods with little change.

Molecular Clocks

Analyses of amino acid sequences in certain proteins like hemoglobin and cytochrome c have been determined for various animals, as have DNA nucleotide differences between animals. The rationale is that *background mutations,* sometimes called neutral mutations because they have no effect on the phenotype, take place at a constant rate. When two animals first diverge from a common ancestor, the genes and proteins of the organisms involved would have been nearly identical. But as time went by, each would have accumulated their own changes in gene and protein structure. Therefore, the degree of difference in these molecules provides some indication of how long ago they diverged from a common ancestor. In other words, DNA and amino acid differences can be used as a kind of **molecular clock** to indicate evolutionary time. Figure 20.10 shows the results of one such study of DNA differences. Trees based on biochemical data only tell you the length of time two groups of animals have been diverging. The actual dates for divergence can be acquired from the fossil record when the evolutionary trees based on biochemical data agree with those based on the fossil record. Whenever the same conclusions are drawn from independent data, they substantiate scientific theory, in this case evolution, even more.

Evolutionary Relationships

Biochemical methods are now sometimes used to help classify organisms. For example, molecular taxonomy shows that the flamingo is more closely related to storks than to geese. Whereas biochemists formerly restricted their studies to living organisms, they have recently begun to study extinct organisms and even fossils. For example, investigators extracted proteins and DNA from a scrap of muscle on the pelt of a 140-year-old museum specimen of a quagga (fig. 20.11), an African animal that became extinct a century ago. Cloning provided enough

Figure 20.10

Evolutionary tree of certain primate species based on a biochemical study of their genomes. The length of the branches indicates the relative number of nucleotide pair differences that were found between groups. With the help of the fossil record it is possible to suggest a date at which each group diverged from the other.

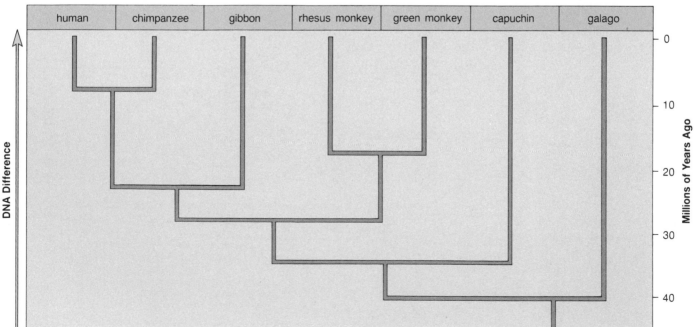

DNA (*see* fig. 18.2) to establish that the quagga was a zebra, not a true horse. Such studies are possible only because all living things share the same types of chemical molecules.

Biochemistry provides evidence for the unity of life forms and also can be used to construct evolutionary trees. These trees agree, in general, with those based on the fossil record and anatomical data.

Figure 20.11
The quagga was striped on the front end like a zebra, but had a solid, chestnut-colored coat like a horse on the hind end. Although it became extinct a century ago, scientists have been able to clone a portion of its DNA chemically extracted from tissue removed from museum hides. Comparative biochemical studies indicate that the quagga was a zebra.

Patterns of Evolution

The manner in which organisms evolve results in evolutionary patterns, which we will now discuss.

Macroevolution

Macroevolution is a term used to describe evolution of a species or even higher taxonomic categories. If you take a look at any type evolutionary tree you will notice a certain pattern. For a long time it seems as if nothing much is happening and then you come to a point where it branches. It's not surprising, then, that we will mention two patterns of macroevolution—adaptive radiation and phyletic evolution (fig. 20.12). **Adaptive radiation** is called evolution in space. It occurs when a single ancestor gives rise to a number of new species. **Phyletic evolution,** in contrast, is called evolution in time because it refers to gradual changes in a single lineage. While phyletic evolution may account for certain genotypic and phenotypic changes, it probably never actually results in speciation, the origination of new species.

Adaptive Radiation

There are 13 species of finches that live on the Galápagos Islands. All these birds are believed to be descended from a type of mainland finch that happened to invade the islands (p. 286). As the parent population increased in size, daughter populations were established on the various islands. Each population was then subjected to the process of natural selection as it became adapted to the particular conditions on its island. In time, the various populations became so genotypically different

---

Figure 20.12
Two patterns of evolutionary change. *a.* The population splits into different groups that will eventually achieve the status of recognizable species. Darwin refers to this form of change as "the origin of species;" today it is generally called adaptive radiation.

b. The population of organisms undergoes change from one state to another. Darwin refers to this type of change as "descent with modifications;" today it is called phyletic evolution. Phyletic evolution is not believed to result in speciation.

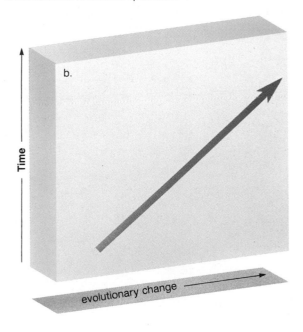

Figure 20.13

Just like Darwin's finches, all these different species of honeycreepers are believed to be descended from a common finchlike ancestor that underwent adaptive radiation on the Hawaiian Islands.

that where they are now found together on the same island, they do not interbreed and may become even more different (see fig. 21.19).

The wide variety of Hawaiian honeycreepers resulted from an adaptive radiation similar to that of Darwin's finches (fig. 20.13). They too are descended from a common finchlike ancestor and they vary in beak size and shape.

Adaptive radiation has also been observed in plants. On the island of New Caledonia, various species of *Dacrydium* have adapted to various microclimates. For example, *D. araucariodes,* with narrow, incurved leaves, is apparently adapted to a dry environment, whereas *D. taxoides,* with broad leaves, seems adapted to a moist environment (fig. 20.14).

Higher Taxonomic Categories   Adaptive radiation is observed not only among species but among higher taxonomic categories also. Both reptiles and mammals underwent large-scale adaptive radiations as they became adapted to different ways of life. Mammals began their radiation after many reptiles, including the dinosaurs, had become extinct and a large number of habitats were relatively empty. Therefore, mammals were free to evolve in the same directions as the reptiles had evolved (fig. 20.15).

Phyletic Evolution

The fossil record of the modern horse covers some 60 million years (fig. 20.16) and has been used to illustrate phyletic evolution. It begins with *Eohippus,* a dog-sized mammal with a small head and small, low-crowned molars. The feet, with four toes on each front foot and three toes on each hind foot, were

Figure 20.14

Adaptive radiation in *Dacrydium* plants on the island of New Caledonia.
a. *D. araucariodes* is adapted to a dry envrionment.
b. *D. taxoides* is adapted to a moist environment.

a.

b.

Figure 20.15

Adaptive radiation among reptiles and mammals. Both reptiles and mammals evolved into the types of forms shown. However, the mammals did not begin their adaptive radiation until the reptiles had declined. Therefore, each group shows similar adaptations because they were subjected to the same selective pressures.

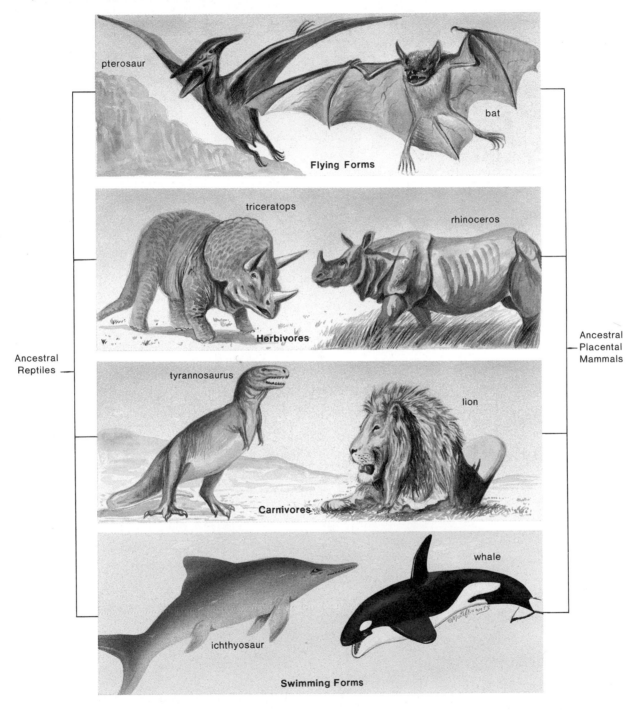

padded. *Equus,* the modern horse, is much larger than *Eohippus*. It has an increased skull size, bigger molars with flattened high crowns, and a single-hoofed toe on each foot.

These anatomical changes are attributed to a change in environment. *Eohippus* was adapted to the forestlike environment of the Eocene, an epoch of the Tertiary Period. For ex-

ample, the low-crowned teeth, with a relatively simple pattern of enamel, were appropriate for browsing on leaves. This small animal could have hidden among the trees for protection. In the Miocene and Pliocene Epochs, grasslands began to replace the forests. Now the ancestors of the modern horse, *Equus,* were subjected to selective pressure for the development of strength,

Figure 20.16

Evolution of the horse during the Tertiary and Quaternary Periods (table 20.1). The modern horse evolved from the dog-sized *Eohippus* but there were a number of adaptive radiations along the way. The other lineages became extinct and only *Equus* is alive today. Whereas *Eohippus* was adapted to a forest environment, *Equus* is adapted to a grassland environment.

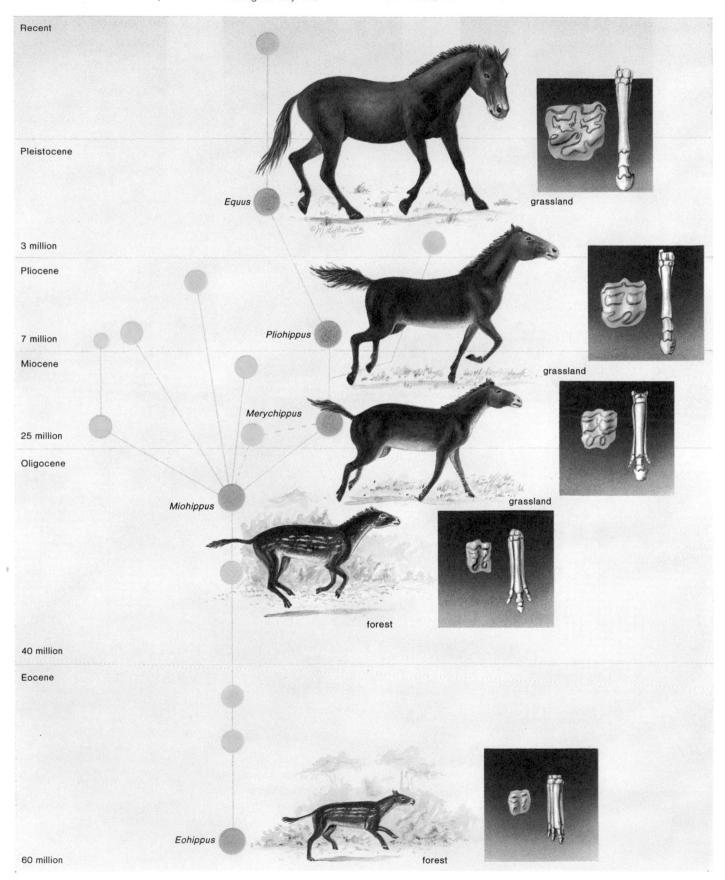

Recent

Pleistocene

Equus

grassland

3 million

Pliocene

7 million

Pliohippus

grassland

Miocene

Merychippus

grassland

25 million

Oligocene

Miohippus

forest

40 million

Eocene

Eohippus

forest

60 million

intelligence, speed, and durable grinding teeth. A large size provided the strength needed for combat, a large skull made room for a larger brain, elongated legs ending in hooves gave speed to escape enemies, and the durable grinding teeth enabled the animals to feed efficiently on grasses.

By picking and choosing only certain fossils, it is possible to trace the lineage of the modern horse as if all these changes occurred step-by-step. However, it would be necessary to forego mentioning a number of other fossils. As *Eohippus* evolved into *Equus,* there were several radiations. These horses became extinct except for the lineage that led to *Equus.*

As mentioned previously, perhaps we should use the term phyletic evolution for small changes in genotype and phenotype that indicate evolution is occurring but do not actually result in speciation. The modern view is that only divergence, discussed next, results in speciation.

Other Patterns of Evolution

When we view the world of living things, we can pick out certain other patterns of evolution that are the end result of the evolutionary process. **Divergent evolution** occurs when a single lineage splits to give new forms. For example, the adaptive radiation of Darwin's finches (fig. 19.9) and the Hawaiian honeycreepers (fig. 20.13) are examples of divergent evolution.

Parallel evolution is an evolutionary pattern that is seen in groups of organisms that are closely related. After they diverge from a common ancestor these organisms continue to resemble one another because they have been subjected to the same adaptive pressures. For example, the African ostrich and the South American rhea (fig. 19.7) are birds that are quite alike because they are adapted to similar types of environments.

Convergent evolution is an evolutionary pattern that is seen in groups of organisms that are distantly related. In this case, the organisms only resemble one another because they are secondarily adapted to the same type of environment. For example, figure 19.13 shows that mammals, reptiles, and birds all have flippers because they are all adapted to an aquatic environment.

In **coevolution,** two unrelated groups act as selective agents for each other. As the bat becomes more skillful in catching the moth, the moth becomes more skillful in escaping the bat (fig. 44.8). One of the most frequently used examples of coevolution is the adaptation that developed between flowers and pollinators. Bee-pollinated flowers produce nectar, have an ultraviolet coloration (fig. 39.8), and a sweet smell that attracts the bee. To make the bee's food-gathering work easier, it has a sucking tongue for collecting nectar, and pollen baskets on the upper segment of the third pair of legs. In this instance, coevolution benefits both the flower and the pollinator. The flower uses the bee to achieve cross-fertilization, and the bee uses the flower as a source of food.

During macroevolution we find examples of divergent evolution, parallel evolution, convergent evolution, and coevolution.

Summary

1. The fossil record provides evidence of the history of life in general, and the lineages of organisms in particular. The earth is about 4.6 billion years old, and the history of life began about 3.5 billion years ago. Because multicellular forms didn't appear until about 700 million years ago, most of life's history was devoted to the evolution of single-celled organisms. Humans don't appear in the record until quite recently (2–3 million years ago).
2. The earth is divided into six biogeographical regions that are separated by either water or impassable land barriers. Each region has its own mix of organisms because they evolved separately in each region. Between the regions you can find unrelated organisms that look similar because they have adapted to the same types of environments.
3. Some of the details of the geographic distribution of organisms can be accounted for by considering the consequences of continental drift. The continents are on massive plates that move, carrying the continents with them. Mammals didn't appear until the supercontinent Pangaea began to break up. Therefore each continent has its own particular types of mammals.
4. The fossil record gives evidence of mass extinctions; two of these have been studied in detail. The extinctions at the end of the Permian Period were catastrophic, and this is the time when Pangaea was forming. Loss of habitat caused the extinction of many forms of marine life. The extinctions at the end of the Cretaceous Period, when the dinosaurs died out, may have been due to asteroid or comet bombardment.
5. The study of anatomy and embryology shows that various groups of organisms are related. Organisms in the same group share a unity of plan and go through similar embryological stages. Homologous structures that have a similar construction indicate an evolutionary relationship, but analogous structures that just have the same function do not.
6. Comparative anatomy helps in classifying organisms and in constructing evolutionary trees. Evolutionary trees indicate how the members of a group are believed to be related and how their ancestry might be traced by way of common ancestors.
7. Almost all living organisms make use of the same sorts of biochemical molecules. An analysis of protein or DNA structure can show how closely related two types of organisms are. The more differences in animo acid sequences or DNA nucleotide sequences there are between them, the longer ago they shared a common ancestor. When this information is used to construct evolutionary trees, it is found that they agree in general with those based on the fossil record and comparative anatomy. This lends support to the occurrence of evolution.

8. Two patterns of macroevolution are adaptive radiation and phyletic evolution. Today, it is generally recognized that adaptive radiation occurs when a number of new species arise from a common ancestor. Darwin's finches underwent adaptive radiation on the Galápagos Islands and another finch gave rise to the honeycreepers on the Hawaiian Islands.

9. Phyletic evolution may include small phenotypic changes but is no longer believed to be involved in speciation. The evolution of the horse is now known to contain at least three periods of adaptive radiation.

10. Adaptive radiation involves divergent evolution. Other patterns of evolution are parallel, convergent, and coevolution.

Objective Questions

1. Which of these would be a true statement in regard to the fossil record?
   a. Eukaryotic cells evolved before prokaryotic cells.
   b. Most of the history of life was devoted to the evolution of single-celled organisms.
   c. It's surprising how early in the history of life humans appear.
   d. Land organisms evolved before marine organisms evolved.
2. Which of these best describes strata?
   a. sedimentary rock layers that contain fossils
   b. sedimentary rock layers that all date from the same historical time
   c. molten rock that contains radioactive material and is dangerous to the health of humans
   d. All of these.
3. What is believed to have caused the Permian extinctions?
   a. formation of the oceans
   b. formation of Pangaea
   c. formation of fossils
   d. formation of asteroids

4. Different biogeographical regions
   a. contain the same mix of plants and animals.
   b. each have their own mix of plants and animals.
   c. may have plants, and also animals, that are similarly adapted.
   d. Both (b) and (c).
5. The presence of marsupials in both Australia and South America has been explained on the basis of
   a. independent evolution on both continents.
   b. land connections by way of Antarctica.
   c. continental drift.
   d. Both (b) and (c).
6. Which of these would be used by investigators constructing evolutionary trees?
   a. comparative anatomy
   b. fossil record
   c. biochemical evidence
   d. All of these.
7. Wherever there is a branch point in an evolutionary tree there is a
   a. common ancestor.
   b. new phylum.
   c. new biogeographical region.
   d. All of these.

8. The same bones are found in the human arm and the bird's wing. These are
   a. both homologous and analogous structures.
   b. only homologous structures.
   c. only analogous structures.
   d. None of these.
9. The fact that there are DNA nucleotide differences between organisms shows that
   a. evolution occurs.
   b. mutations occur.
   c. genotype and phenotype differences are related.
   d. All of these.
10. Which of these best explains the concept of a molecular clock?
    a. It's a time-keeping device that tells organisms when to mutate.
    b. It's all the changes that have occurred in metabolic pathways since organisms first evolved.
    c. It's an increasing number of amino acid or DNA differences since two groups of organisms shared a common ancestor.
    d. All of these.

Study Questions

1. What is the fossil record, and what does it tell us about the history of life?
2. How do paleontologists explain the abrupt transitions between one rock stratum and another?
3. What is the significance of being able to divide the earth into six biogeographical regions?
4. How can the various distributions of plants and/or animals be related to continental drift?
5. What are mass extinctions, and in what way is the Permian extinction related to continental drift?

6. Contrast homologous structures with analogous structures by giving examples.
7. Draw a hypothetical evolutionary tree and point to the location of common ancestors. Assign taxonomic categories to the branches of your tree.
8. What is a molecular clock and how is it used to construct evolutionary trees? What is the significance of agreement between evolutionary trees based on the fossil record and those based on biochemical evidence?

9. Explain why the fossil record, biogeography, comparative anatomy, and biochemical studies are evidence of the occurrence of evolution.
10. What are two patterns of macroevolution? Which one is believed to involve more rapid speciation? Give examples of this type.
11. Define divergent, parallel, convergent, and coevolution.

Thought Questions

1. From your knowledge of the scientific method, show that the controversy between gradualists and those that support the punctuated equilibrium method is to be expected.

2. How is it possible that there could be "neutral mutations" that have no affect on the phenotype?

3. On the basis of religious faith, some people believe in separate creation of organisms. How are religious beliefs different from scientific beliefs?

Selected Key Terms

biogeography (bi''o-je-og'rah-fe) 295
continental drift (kon''ti-nen'tal drift) 296
mass extinction (mas eks-ting'shun) 298
homologous structure (ho-mol'o-gus struk'tūr) 299
analogous structure (ah-nal'o-gus struk'tūr) 299
vestigial structure (ves-tij'e-al struk'tūr) 299

taxonomic categories (tak''so-nom'ik kat'ĕ-go''rez) 300
evolutionary tree (ev''o-lu'shun-ar''e tre) 300
common ancestor (kom'un an'ses''tor) 300
macroevolution (mak''ro-ev''o-lu'shun) 303
adaptive radiation (ah-dap'tiv ra-de-a'shun) 303
phyletic evolution (fi-let'ik ev''o-lu'shun) 303

divergent evolution (di-ver'jent ev''o-lu'shun) 307
parallel evolution (par'ah-lel ev''o-lu'shun) 307
convergent evolution (kon-ver'jent ev''o-lu'shun) 307
coevolution (ko-ev''o-lu'shun) 307

CHAPTER 21

The Evolutionary Process

Your study of this chapter will be complete when you can:

1. State the Hardy-Weinberg law along with its conditions.
2. Give a definition of evolution based on population genetics.
3. Tell the two steps necessary for evolution according to the synthetic theory.
4. Tell how variations are produced and maintained in a population.
5. Define balanced polymorphism and give examples pertaining to sickle-cell anemia and Cepaea snails.
6. Discuss genetic drift and give examples of the founder effect and bottlenecks.
7. State the steps required for adaptation by natural selection in modern terms.
8. Distinguish between disruptive, stabilizing, and directional selection by giving examples.
9. Explain the biological concept of a species, and give examples of premating and postmating reproductive isolating mechanisms.
10. Explain the steps required for allopatric and sympatric speciation to occur.

Bleached lizard from White Sands region, New Mexico. Organisms are adapted to their environment. The white appearance of this lizard causes it to blend into the background, protecting it from would-be predators.

The modern theory of organic evolution emphasizes that individuals are members of populations. **A population** is all the members of the same species that live in one locale (fig. 21.1). A population could be all the green frogs in a frog pond, all the field mice in a barn, or all the English daisies on a hill. Each member of a population is assumed to be free to reproduce with any other member, and when reproduction occurs, the genes of one generation are passed on in the manner described by Mendel's laws. Therefore, in this so-called Mendelian population of sexually reproducing individuals, the various alleles at all the gene loci in the individuals make up the **gene pool** of the population. It is customary to describe the gene pool of a population in terms of **gene frequencies.**

For example, suppose it is known that one-fourth of all flies in a *Drosophila* population are homozygous dominant for long wings, one-half are heterozygous, and one-fourth are homozygous recessive for short wings. Using the key given in figure 12.8, we can describe the population in this manner:

$$\tfrac{1}{4}LL + \tfrac{1}{2}Ll + \tfrac{1}{4}ll$$

By inspection, it is obvious that the frequency of either allele in the gene pool is 50%, or 0.5.

It is now possible to calculate the expected allele frequencies for this gene locus in the next generation. The homozygous dominant individuals will produce one-fourth of all the gametes of the population, and these gametes will all carry the dominant allele, *L*. The heterozygotes will produce one-half of all the gametes; one-fourth will carry the dominant allele, *L,* and one-fourth will carry the recessive allele, *l*. The homozygous recessives will produce one-fourth of all the gametes and they will carry the recessive allele, *l*. Therefore, in summary, one-half of the gametes will be *L* and one-half will be *l*, as is expected if the frequency for each allele in the gene pool is 50%.

Assuming that all possible gametes have an equal chance to combine with one another, then as the Punnett square shows, the next generation will have exactly the same ratio of genotypes as the previous generation and the frequency of each allele will still be 50% or 0.5:

	½ L	½ l
½ L	¼ LL	¼ Ll
½ l	¼ Ll	¼ ll

Results: ¼ *LL* + ½ *Ll* + ¼ *ll*

This example indicates that dominant alleles do not necessarily take the place of recessive alleles, recessive alleles do not disappear, and sexual reproduction, in and of itself, cannot bring about a change in the allele frequencies of the population.

Hardy-Weinberg Law

The constancy of the gene pool was independently recognized in 1908 by G. H. Hardy, an Englishman, and W. Weinberg, a German physician. They presented a mathematical interpretation, discussed in the reading on the next page. In nonmathematical terms, the **Hardy-Weinberg law** *states that the frequencies of alleles in a population will remain the same in each succeeding generation as long as certain conditions are met.*

Figure 21.1
a. Thomson gazelles in the Kalahari Desert of Africa. *b.* A population of mule ears in bloom in the Santa Rosa Mountains, Nevada. In each population all individuals contribute alleles to a common gene pool. While they may appear to be very similar in appearance, genetic variations are present as they are in all populations.

Population Genetics

B ecause individuals live and reproduce within populations, it is at the population level where evolutionary change is most obvious. When biologists began to recognize this fundamental premise, they became interested in population genetics. G. H. Hardy and W. Weinberg recognized that the binomial equation, expressed here as

$$p^2 + 2pq + q^2$$

could be used to directly calculate the genotype and allele frequencies of a population.

Consider that any present generation of a large population is the result of random mating between members of the previous generation; each type of sperm had an equal chance to fertilize each type of egg. Assign the letter p to the dominant allele and the letter q to the recessive allele. Random mating will result in three genotypes:

	p	q
p	p^2	pq
q	pq	q^2

Results: $p^2 + 2pq + q^2$

p^2 = homozygous dominant individuals

q^2 = homozygous recessive individuals

$2pq$ = heterozygous individuals

To use the above binomial equation to calculate the genotype frequencies in a population, it is necessary to know the frequency of each allele in the population. (Usually, these frequencies are given in decimal rather than fractional form because the frequencies of the two alleles are not always easily converted to a fraction.) For example, suppose the dominant allele (p) has a frequency of 0.7 and the recessive allele (q) has a frequency of 0.3. Then the genotype frequencies of the population would be

p^2 (homozygous dominant)	= 0.49
q^2 (homozygous recessive)	= 0.09
$2pq$ (heterozygous)	= 0.42
	1.00

Notice that the sum, $p + q$ (frequencies of the two alleles) must equal 1.00 and the sum, $p^2 + q^2 + 2pq$ (frequencies of the genotypes) must also equal 1.00.

When the frequency of a recessive phenotype is known, it is possible to calculate the genotype frequencies and the allelic frequencies.[1] For example, suppose by inspection we determine that 16% of a human population has a continuous hairline caused by a recessive allele. This automatically means that 84% of the population has the dominant phenotype, a widow's peak. Of these, how many have the homozygous dominant genotype? How many have the heterozygous genotype?

To answer these questions, first convert 16% to a decimal. Then we know that $q^2 = 0.16$ and, therefore, $q = 0.4$. Since $p + q = 1.0$, we know that $p = 0.6$ and, therefore, p^2 (frequency of the population that is homozygous dominant) = 0.36. To determine the frequency of the heterozygote, we simply realize that thus far we have accounted for only 0.52 of the population ($p^2 + q^2 = 0.52$). The remainder, $2pq$ (1−0.52), or 0.48 will have the heterozygous genotype. Or, if you prefer, calculate that $2pq = 0.48$. In summary, we have found that

homozygous recessive	= 0.16	= 16% have a continuous hairline
homozygous dominant	= 0.36	
heterozygous	= 0.48	= 84% have a widow's peak

[1]Additional problems are given on page 327.

Knowing the proportion of the recessive phenotype, exemplified by the albino deer shown here, allows one to calculate the frequency of both the recessive and dominant alleles in a population.

We have considered only examples involving two alleles, but it is possible to use more complicated equations to calculate expected frequencies when there are more than two possible alleles at a gene locus.

Conditions

These are the conditions that must be met in order for the Hardy-Weinberg law to hold:

1. The population is large enough to be unaffected by random gene changes. If the population is large, small changes in allele frequencies are not expected to affect significantly the genotype frequencies of the next generation.
2. Mating is random. Every individual must have an equal opportunity to reproduce with any other individual in the population.
3. There is no gene flow. Immigration and emigration of individuals must not produce any change in the gene pool of the population.

4. No mutations occur, or else there must be mutational equilibrium. Genetic changes in one direction must be balanced by an equal number of mutations in the opposite direction.
5. There is no natural selection. No one phenotype has a greater chance of reproductive success than another.

Evolution

Because all these conditions are rarely, if ever, met, *we do expect phenotype and allele frequency changes in each subsequent generation.* That is, we expect evolution to occur. For example, suppose that 75% of a bear population has the dominant phenotype for a heavy coat of hair and 25% has the recessive phenotype for a light coat. We predict that the second generation will have the exact same phenotype frequencies as the first generation. When we sample the next generation, we find a change in the phenotype distribution (fig. 21.2). This indicates there

Figure 21.2

Without evolution, (*a*) the gene pool is constant generation after generation, but (*b*) if certain gametes are selected for reproduction at the expense of others, gene pool frequencies do not remain constant and evolution does occur.

a.　25% = light coat　**First Generation**　75% = heavy coat

b.　37.5% = light coat　**Second Generation**　62.5% = heavy coat

has been a change in the allele frequencies at a particular gene locus and that evolution is occurring in this population. *Evolution is occurring when there is a change in allele frequencies in a population from generation to generation.*

The concept of gene pool offers a way in which to recognize when evolution has occurred. The frequency of genes in the gene pool of a population stays constant unless evolution has occurred.

Synthetic Theory of Evolution

The explanation of evolution in terms of modern genetic principles is a synthesis; it takes data and hypotheses from many sources and blends them into a whole. According to the synthetic theory, the evolutionary process requires two steps (table 21.1):

1. Production of genotype and therefore phenotype variations among members of a population.
2. Sorting out and reduction of the variations passed on to next generation.

Table 21.1 Mechanism of Evolution*

Produce Variations	Reduce Variations
Mutations	Genetic drift
Gene flow	Natural selection
Recombination	

*Charles Darwin suggested that variations are reduced by natural selection. Later evolutionists have suggested that genetic drift also reduces variation.

Variations

Previously we assumed phenotype and genotype differences between members of a species. Now we are concerned with the production and maintenance of these variations.

Figure 21.3

A mutation in sheep produced the Ancon breed, represented by the ewe. This breed, which has shorter legs than most breeds, can be maintained only by inbreeding because the unique characteristic is transmitted as a recessive gene.

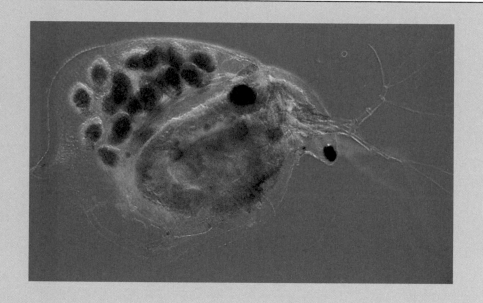

Figure 21.4

The water flea *Daphnia* normally requires a water temperature of about 20°C to survive, but one mutant requires a temperature between 25°C and 30°C. The head of this multicellular animal is to the right and a brood chamber is to the left. Usually, eggs develop into the young we see here without benefit of fertilization. When environmental conditions are poor, sexual reproduction provides the variations that may be required for the species to adjust to a changing environment.

Production of Variations

Genetic variations within the gene pool of sexually reproducing diploid organisms have three sources: mutation, gene flow, and recombination.

Mutation   There are chromosome mutations (p. 243) and there are gene mutations (p. 196) as discussed previously. Mutations are the only original source of allele changes. Even so, mutations do not contribute greatly to the immediate occurrence of evolution. First of all, mutation rates are quite low. A rate of $10^{-8} - 10^{-5}$ per generation is common in sexually reproducing organisms. Second, a new mutation is likely to be adaptively neutral or harmful (fig. 21.3).

Nevertheless, a mutation might eventually contribute to evolution. For example, the water flea *Daphnia* (fig. 21.4) normally thrives at temperatures around 20° C and cannot survive at 27° C or more. There is, however, a mutant strain of *Daphnia* that requires temperatures between 25° C and 30° C and cannot survive at 20° C. If bodies of water should happen to be only at the higher temperature, only the mutant would survive.

Gene Flow   Gene flow occurs when migrating individuals reproduce with members of a different population. As a result, new genes may be introduced into the gene pool of this population. Gene flow usually occurs between populations belonging to the same species. These populations may be different enough to be called **subspecies** or races. For example, six subspecies of the rat snake *Elaphe obsoleta* are recognized (fig. 21.5). Notice that each has been given a third name, the common way to designate subspecies. Because these snakes interbreed wherever their ranges meet, they all share the same gene pool.

On occasion, gene flow occurs between populations that are separate species. The resulting genotype is called a **hybrid.** If each species is adapted to entirely different environments, the hybrids may be at a disadvantage. On the other hand, if each species is inbred and carries undesirable or even harmful genes, the hybrid may be more viable and is said to have **hybrid vigor (heterosis).** Human beings have used heterosis to produce high-yield crops. Hybrid seeds are used in planting wheat, rice, and rye, and they have helped grow more food for the ever-increasing human population.

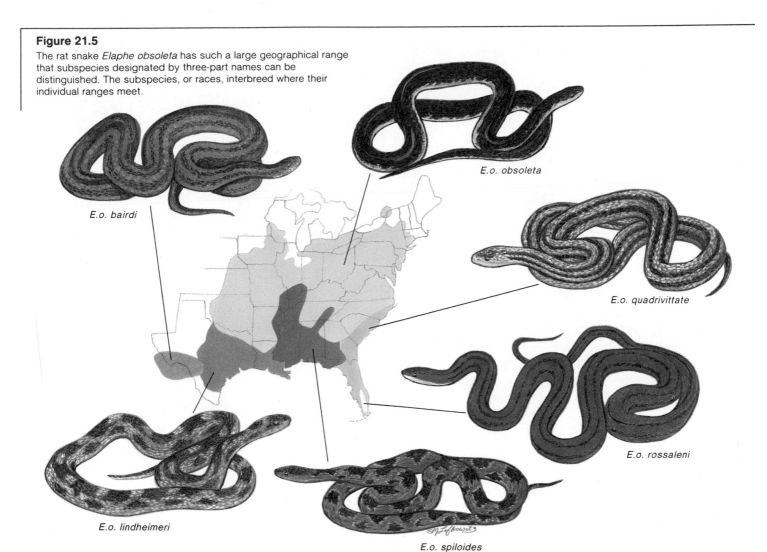

Figure 21.5
The rat snake *Elaphe obsoleta* has such a large geographical range that subspecies designated by three-part names can be distinguished. The subspecies, or races, interbreed where their individual ranges meet.

E.o. bairdi

E.o. obsoleta

E.o. quadrivittate

E.o. rossaleni

E.o. lindheimeri

E.o. spiloides

Recombination   Recombination of alleles occurs during meiosis due to crossing-over between homologous chromosomes and independent assortment of chromosomes. It also occurs at fertilization when two different gametes join. Recombination gives the offspring a genotype and possibly a phenotype that differs from either parent (fig. 21.6). Mutation is the ultimate source of variation, but the recombined genotype supplied by sexual reproduction is usually of much greater importance. This is because the entire phenotype is subject to natural selection, and individual alleles, particularly if recessive, may not change the phenotype very much.

Among prokaryotes, mutations are more important, because recombination does not regularly occur. Also, because these organisms are haploid, with only one copy of each gene, mutations can be immediately tested by natural selection. Haploidy is not disadvantageous to prokaryotes, because they produce a very large number of offspring within a short period of time. Harmful mutations may temporarily reduce the population size, but are not likely to eliminate the population.

The source of the variations in members of a population with different phenotypes is mutation, gene flow, and recombination.

Figure 21.6
Recombination of alleles occurs during sexual reproduction. Recombination causes these litter mates to be genetically distinct as reflected in their varying phenotypes.

Figure 21.7

There is a correlation between the distribution of sickle-cell anemia and the distribution of malaria. *a.* Frequency of the sickle-cell anemia allele *Hb*^S in Africa and adjoining countries. In the homozygous condition, this allele causes sickle-cell anemia. Sickle-shaped cells inhibit parasite reproduction. *b.* Geographic distribution of malaria in Africa and adjoining countries is about the same as the distribution of *Hb*^S suggesting that sickle-shaped cells must offer some adaptive advantage in these regions.

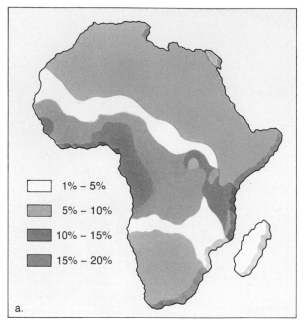

1% – 5%
5% – 10%
10% – 15%
15% – 20%

a.

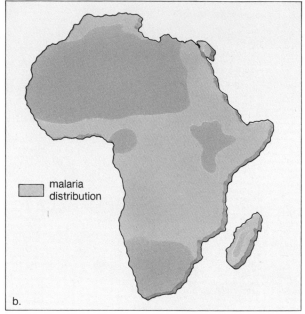

malaria distribution

b.

Maintenance of Variations

Even after evolution has occurred and the amount of variation has been reduced, some variations are always maintained. The more variation in a population, the greater its potential to adapt to new changes in its environment. Each population seems to strike a balance between loss of variations, so as to become better adapted to a particular environment, and retention of variations, so as to avoid extinction. Populations very closely adapted to certain environments may not be able to adapt to new conditions and may become extinct.

There are both genetic and ecological reasons for the maintenance of variations and we will discuss some of these here.

Diploidy and Pleiotropy Diploidy and pleiotropy (p. 187) are purely genetic reasons for the maintenance of variations. Diploidy permits the maintenance of a hidden reserve of recessive genes that are not often exposed to selection. The dominant phenotype may have a selective advantage and yet many of these individuals may be heterozygotes that carry the recessive allele. In pleiotropy, one gene locus controls more than one trait (fig. 13.5). It's possible that one of the characteristics brought about by a pleiotropic gene may actually be maladaptive. However, this characteristic is maintained because another of the traits controlled by the pleiotropic gene is adaptive.

Heterozygote Superiority In this example of variation maintenance, we can find both genetic and ecological reasons for variation. Individuals with sickle-cell anemia have the genotype *Hb*^S *Hb*^S and tend to die of this condition at an early age. Individuals who are heterozygous for the sickle-cell trait, *Hb*^A *Hb*^S,

are better off, but they are still expected to be at a disadvantage when compared to individuals who are homozygous normal *Hb*^A *Hb*^A. However, when geneticists studied populations of black Africans, in whom the disease is most common, they found that the recessive allele, *Hb*^S, has a higher frequency (0.2 to as high as 0.4 in a few areas) than would have been predicted. Further research showed that the heterozygote has an adaptive advantage over the normal homozygote in regions with malaria (fig. 21.7). The malarial parasite enters normal red blood cells to complete its life cycle, but it is unable to make use of sickle cells. For this reason, more heterozygotes survive than do homozygotes, and this **heterozygote superiority**[1] maintains the frequency of the sickle-cell allele in the population.

The maintenance of all three phenotypes (normal, sickle-cell trait, and sickle-cell anemia) in the population is an example of balanced polymorphism. **Polymorphism** is the coexistence of two or more forms in a population and **balanced polymorphism** occurs when the frequency of the varied forms (and the alleles involved) remains the same in each generation.

British Land Snail British land snails (*Cepaea nemoralis*) offer an example of variation maintenance that seems to have primarily an ecological basis because it is due to the snail having a wide geographic range. The shell of these snails can have various colors, decorated with as many as five bands that are usually dark in color, but may be pink or white. Song thrushes feed on the snails that are more easily seen. They break open the snails on a rock, eat the soft parts, and leave the shells behind.

[1]This terminology is customarily used but in this text the *Hb*^A and *Hb*^S alleles are considered to be codominant.

Figure 21.8

a. Graph showing the relationship between predation and phenotype in the snail *Cepaea*. Thrushes (birds) are more apt to feed on dark, unbanded snails in low-vegetation areas (light circles) and light-banded snails in forest areas (dark circles). As a result of these environmental effects, both phenotypes are maintained throughout the range of the snail. *b.* Photograph of two *Cepaea* snails.

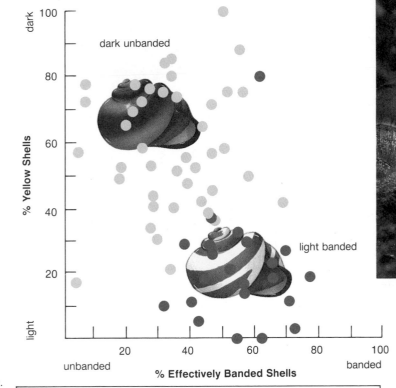

b.

a.

key:

● forest areas—oakwoods, mixed deciduous woods

● low vegetation areas—hedgerows, rough herbage

These rocks are called "thrush anvils." In low-vegetation areas (grass fields and hedgerows), dark, unbanded shells tend to be found, and in forest areas, light-colored, banded shells tend to be found around these anvils (fig. 21.8). Because the snails have a geographic range including both low-vegetation and forest areas, particular shell phenotypes are better at camouflaging the snails in one habitat than another. Selection will favor the phenotype that is best in each habitat and the polymorphism will continue.

There are both genetic and ecological reasons why populations maintain a storehouse of variations. Balanced polymorphism can be maintained due to heterozygote superiority (sickle-cell anemia) or wide geographic range (British land snail).

Genetic Drift

Genetic drift and natural selection both act in such a way that variation in a population is sorted out and reduced (fig. 21.9). **Genetic drift** is a change in the gene pool that occurs purely as a result of chance events. In very large populations, these random changes are insignificant. For example, Mendel reported slight deviations from his predicted 3:1 and 9:3:3:1 ratios.

However, in its most extreme form, genetic drift can lead to the loss of certain alleles, making the homozygous condition the norm (fig. 21.10). This is most apt to happen when only a few offspring from an originally large population mate to produce the next generation. Therefore, as in figure 21.9, drift occurs following a chance sampling of the original population. "Chance sampling" means only a few individuals, among all the phenotypes available, produce offspring. Genetic drift is not expected to result in adaptation to the environment because it is chance events rather than natural selection that determines which individuals will reproduce.

Founder Effect

It has been observed that genetic drift is particularly evident after a new population is started by only a few members of an originally large population. The **founder effect** is genetic drift that occurs when a small number of individuals, representing a

Figure 21.9

The genes in the gene pool recombine to give various genotypes that develop into various phenotypes. *a.* A chance sampling of these phenotypes can lead to genetic drift. *b.* Selection of adapted phenotypes can lead to adaptation to the environment. Mutation, gene flow, and recombination are sources of genetic variation in each generation.

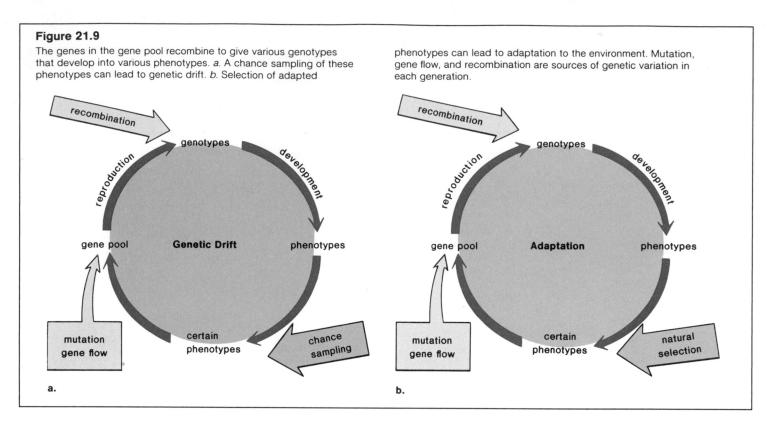

a.

b.

Figure 21.10

Possible random changes in the frequency of an allele over time. In this example, allele *A* is completely eliminated from the population. Therefore, allele *a* is now "fixed" in the population. By "fixing" certain alleles, genetic drift reduces the amount of variation present in the gene pool.

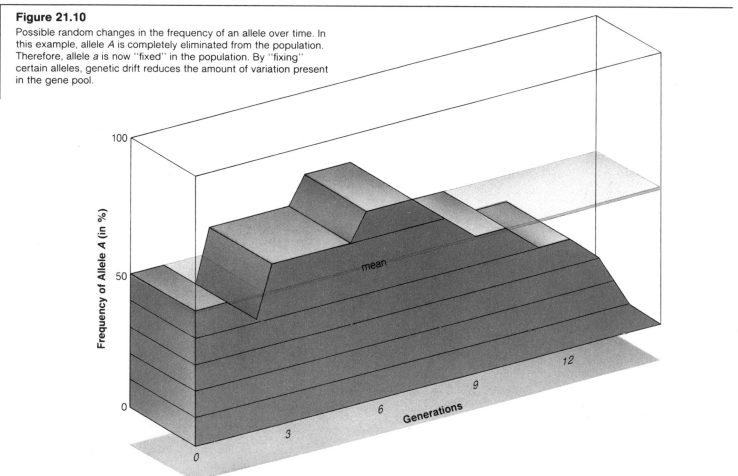

Figure 21.11
Various colonies of the California cypress (*a* and *b*) differ from one
another in shape of tree, type of bark, color foliage, and size of
cones. These variations in features are believed to be due at least in
part to the founder effect.

a.

b.

Figure 21.11
Various colonies of the California cypress (*a* and *b*) differ from one
another in shape of tree, type of bark, color foliage, and size of
cones. These variations in features are believed to be due at least in
part to the founder effect.

fraction of the gene pool, found a colony and only certain of the
original population's alleles are passed on to the next generation. As a result, genetic variation will be severely reduced in
the new population. For example, it has been observed that the
trees in California cypress groves (colonies) (fig. 21.11) tend to
have similar phenotypes. In some colonies, there are longitudinally shaped trees and in others there are pyramidally shaped
trees. The bark is rough in some colonies and smooth in others.
The leaves are gray to bright green or bluish, and the cones may
be small or large. Each colony is believed to have been started
by a single or a few trees. These differences between the colonies do not seem to be due to adaptations to the environment.

Human populations that were begun by a few individuals
and have remained relatively isolated are expected to show some
evidence of the founder effect. The Amish of Lancaster, Pennsylvania are such a group, and it is believed that as many as one
in fourteen individuals carry a recessive allele that causes an
unusual form of dwarfism (it affects only lower arms and legs)
and polydactylism (extra fingers). In the population at large,
only one in one thousand individuals has this allele; most likely
a founder of the Amish colony carried the gene.

Bottlenecks

Sometimes a population is subjected to near extinction because
of a natural disaster, such as an earthquake, fire, or slaughter
by humans. Chance alone determines which individuals survive
these unfavorable times that act like a **bottleneck,** preventing
the majority of genotypes from participating in the production
of the next generation.

As discussed in the reading on page 320, the large genetic
similarity found in cheetahs may be the result of a severe bottleneck in the past. It's also speculated that such an effect may
account for the high rate of Tay-Sachs disease among Ashke-
nazic Jews. Sometime during the Middle Ages, these people may
have experienced a severe reduction in their numbers.

Genetic drift is exaggerated when founder effects and bottlenecks
occur. In these instances, only a few individuals contribute to the
gene pool of the new population or one recovering from a disaster.

Copycat Cheetahs

If you have trouble telling one cheetah from another, you're not alone. When researchers at the National Institutes of Health checked blood samples of 55 cheetahs from two separate areas of South Africa, they found them to be almost genetically identical.

The scientists took blood from cheetahs born wild in Namibia and the Transvaal Province of South Africa and from offspring bred in a research center in Pretoria. They analyzed 47 different enzymes, each of which can come in several different forms. But all the cheetahs carried exactly the same form of every one of the 47 enzymes. By contrast, in sample populations of household cats, only 78% of the enzymes are identical; in human populations only 68% of them match exactly. In another test of more than 150 proteins, 97% of them matched in the cheetahs.

Such remarkably high levels of genetic uniformity are usually found only in specially bred laboratory mice. "You need at least twenty generations of inbreeding—brother-sister

A South African cheetah at rest with look-alike cubs.

mating—before you lose all genetic variability, the way the cheetah has," says geneticist Stephen O'Brien.

O'Brien's group theorizes the cheetah population was nearly wiped out generations ago. Perhaps, says O'Brien, they were slaughtered by nineteenth-century cattle farmers protecting their herds, captured by Egyptians as pets four thousand years ago, or decimated by the same mysterious cataclysm that caused the great mammalian extinction tens of thousands of years ago. During the crisis, "for every 1,000 animals, there were maybe three or four left," says Mitchel Bush, a vet at the National Zoo in Washington, D.C. "Then as the population expanded there was a lot of forced inbreeding and a limited source of new genes."

The inbreeding has now taken its toll by reducing the cheetahs' reproductive capacity. According to David Wildt, a reproductive physiologist at the zoo, the cheetahs show sperm counts averaging less than 10% that of lions and tigers, and 70% of the sperm they do have is defective. That sort of abnormality, seen previously only in inbred livestock and laboratory mice, would explain why zookeepers have had trouble getting the endangered animal to breed.

Reprinted from "Copycat Cheetahs" in *Science 83*, p. 6, October 1983. Copyright 1983 by The American Association for the Advancement of Science.

Natural Selection

Natural selection (fig. 21.9*b*) is the process by which populations become adapted to their environments. In chapter 19 we outlined how Darwin explained evolution by natural selection. Here we restate these steps in the context of modern evolutionary theory. In evolution by natural selection, the fitness of an individual is measured by how reproductively successful its offspring are in the next generation.

Evolution by natural selection requires:

1. Variation. The members of a population differ from one another.

2. Inheritance. Many of these differences are heritable, genetic differences.

3. Differential adaptedness. Some of these differences affect how well an organism is adapted to its environment.

4. Differential reproduction. Individuals that are better adapted to their environment are more likely to reproduce, and their fertile offspring will make up a greater proportion of the next generation.

Types of Natural Selection

The individuals in populations are subjected to the process of natural selection, or selective pressures. As an outcome, three types of selection (fig 21.12) are possible for any particular trait.

Disruptive Selection   In **disruptive selection,** two extreme phenotypes for a given trait are favored over the average phenotype in a population, resulting in polymorphism. This can happen, as discussed previously, when a population is exposed to very different conditions in various parts of its range. The population tends to become divided into subpopulations, as illustrated in figure 21.5.

We have also discussed the British land snail, *Cepaea nemoralis* that has different phenotypes because its range includes both forest areas and low-vegetation areas (fig. 21.8).

Stabilizing Selection   **Stabilizing selection** helps conserve the typical phenotype and tends to eliminate atypical phenotypes. It improves adaptation of the population to those aspects of the environment that remain constant.

With stabilizing selection, extreme phenotypes are selected against, so that the intermediate phenotypes are more prevalent than otherwise. As an example, consider that the birth weight of most infants is about 3 kg to 3.5 kg (7 lbs. to 8 lbs.). The death rate is higher for infants considerably smaller or larger than this.

Directional Selection   **Directional selection,** on the other hand, is seen when the environment is changing. One extreme phenotype may be better adapted to the changed environment and so will be selected. Stabilizing and directional selection can and usually do occur in the same population at the same time—certain traits are affected by one type, while other traits are affected by the other type of selection. In order to determine which type of selection is affecting which traits, one must be aware of the population's previous composition. For example, if we note that an earlier population of giraffes had medium-length necks, and the more recent population has long necks, we know that directional selection has taken place.

Industrial Melanism   **Industrial melanism** is an example of directional selection. Before the industrial revolution in England, collectors of the peppered moth, *Biston betularia,* noted that

Figure 21.12

One of three types of natural selection can affect the distribution curve for a particular trait among individuals of a population. The curves show that the characteristics for a trait vary continuously throughout the population. The arrow(s) point to the characteristic that is favored by selection. The shading indicates those characteristics that are not favored and will disappear from the population. In disruptive selection (*a*), two widely different phenotypes are favored and a separate curve for each is eventually seen. In stabilizing selection (*b*), the curve eventually constricts because the most common phenotype is favored. In directional selection (*c*), an extreme phenotype is favored and the curve eventually moves to the right.

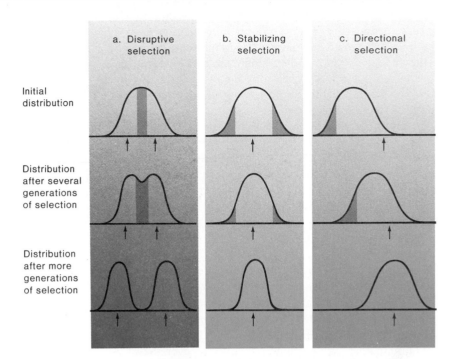

a. Disruptive selection   b. Stabilizing selection   c. Directional selection

Initial distribution

Distribution after several generations of selection

Distribution after more generations of selection

most moths were light-colored, although occasionally a dark-colored (melanistic) moth was captured. Several decades after the industrial revolution, however, black moths made up 90% of the moth population in air-polluted areas.

Moths rest on the trunks of trees during the day (fig. 21.13); if they are seen by predatory birds, they are eaten. As long as the trees in the environment were light in color, the light-colored moths lived to reproduce. But when the trees turned black from industrial pollution, the dark-colored moths survived and reproduced to a greater extent than the light-colored moths. The dark-colored phenotype then became the more frequent one in the population. If pollution is reduced and the trunks of the trees regain their normal color, the light-colored moths should increase in number. To provide experimental evidence that body color influences which moths are captured by birds, H.B.D. Kettlewell of Oxford University released equal numbers of the light- and dark-colored moths into two areas: (1) a nonindustrial, unpolluted area and (2) a highly industrial, polluted area. The released moths were marked so Kettlewell could identify them when he later recaptured his moths. Kettlewell set up blinds in each area from which he observed birds preying on moths. Birds in the unpolluted area preyed on dark-colored moths more than light-colored moths, and birds in polluted areas preyed on light-colored moths more than dark-colored moths. From the unpolluted area, he recaptured 14.6% of the light-colored moths and only 4.7% of the melanistic form. From the polluted area, he recaptured 27.5% of the melanistic form and only 13% of the light-colored moths. This showed that the better adapted moths in each area were more likely to avoid being eaten by predatory birds, which were acting as a selective agent.

Figure 21.13

Dark and light forms of the peppered moth (*Biston betularia*) *a*. The light form is visible on a soot-blackened oak tree in Birmingham, England. *b*. The dark form is visible on a light, lichen-coated oak tree in an unpolluted region.

a.

b.

Bacteria and Insects Indiscriminate use of antibiotics and pesticides has caused some bacteria and insects to become resistant to these chemicals. The mutation enabling them to survive the unfavorable environment is already present before exposure; the chemicals are merely acting as selective agents. For example, when bacteria are grown on medium containing streptomycin, a few survive. These few can grow on medium both with and without streptomycin. In any case, they are resistant to the antibiotic and the trait will now increase in the population as the resistant organisms survive and reproduce.

Another example is the human struggle against malaria, a disease caused by an infection of the liver and red blood cells (fig. 21.14). The dread *Anopheles* mosquito transfers the disease-causing protozoan *Plasmodium* from person to person. In the early 1960s, international health authorities thought that malaria would soon be eradicated. A new drug, chloroquine, was more effective than the quinine that had been used previously against *Plasmodium,* and DDT spraying had reduced the mosquito population. But in the mid-1960s, *Plasmodium* showed signs of chloroquine resistance and, worse yet, mosquitoes were becoming resistant to DDT. A few drug-resistant parasites and a few DDT-resistant mosquitoes had survived and multiplied, making the fight against malaria more difficult than ever. New tactics have to be devised. Recombinant DNA procedures have enabled researchers to identify proteins that will hopefully be effective as vaccines.

There are three major types of natural selection:

1. Disruptive selection favors extreme phenotypes at the expense of the intermediate. This type of selection may be common when geographically diverse populations are subjected to different environmental circumstances.
2. Stabilizing selection, common in relatively unchanging environments, favors the intermediate phenotype so that it increases in frequency at the expense of the extreme phenotypes.
3. Directional selection favors one of the extreme phenotypes so that this phenotype becomes increasingly more common in the population. Directional selection would be likely in a changing environment.

Speciation

A change in gene frequency within a gene pool, either by means of genetic drift or natural selection, can be considered evidence of evolution, but it does not constitute **speciation,** the origin of a species. Speciation requires additional steps and this section will consider these steps.

First, we have to be able to recognize speciation once it has occurred. There are two primary criteria by which a species can be recognized: (1) the members of each species resemble each other and are structurally different in some respects from the members of any other species, and (2) the members of a species are able to interbreed and produce fertile offspring only among themselves (fig. 21.15). It is said that each species is *reproductively isolated.* For example, the red maple and the sugar maple are both found in the eastern half of the United

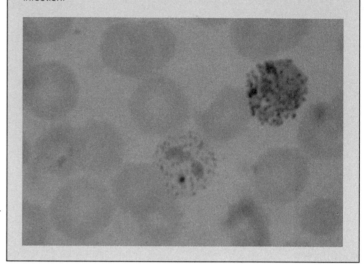

Figure 21.14
The parasite that causes malaria is shown bursting from red blood cells, which it infects at one point in its life cycle. This parasite is transmitted from person to person by *Anopheles* mosquitos, which are becoming resistant to DDT. The parasite itself is becoming resistant to chloroquine, the prescribed medicine to treat an infection.

Figure 21.15
A geep, a crossbreed between a goat and a sheep. Does such a curiosity contradict the biological definition of a species? No, because this geep was produced not by sexual reproduction, but by the *in vitro* mingling of cells from a goat embryo and a sheep embryo.

States (fig. 21.16). They remain distinct from one another because they do not interbreed. This criterion for recognizing a species is known as the *biological concept of a species.*

Reproductive Isolating Mechanisms

A **reproductive isolating mechanism** is any structural, functional, or behavioral characteristic that prevents successful reproduction from occurring. Table 21.2 lists the mechanisms by which reproductive isolation is maintained.

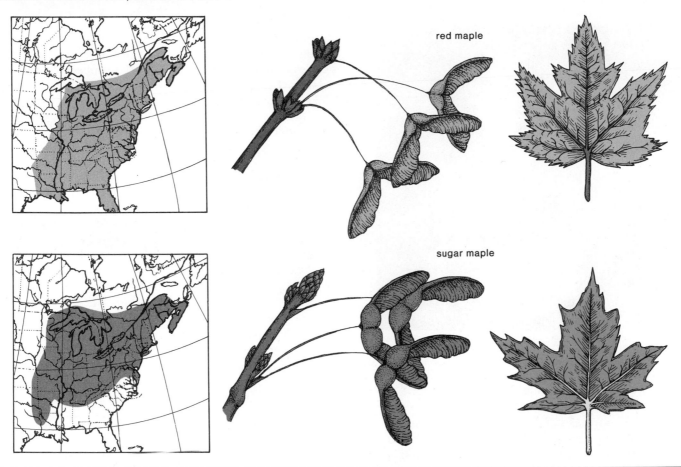

Figure 21.16
The red maple and sugar maple do not interbreed; therefore, these two species remain separate in the same locales, The upper illustration shows U.S. distribution of the red maple, and the lower shows U.S. distribution of the sugar maple. The anatomical differences between the two plants are also shown.

red maple

sugar maple

Premating Isolating Mechanisms

Premating isolating mechanisms are those that prevent reproduction from being attempted.

Habitat Isolation. When two species occupy different habitats, even within the same geographic range, they will be less likely to meet and attempt to reproduce. For example, the scarlet oak (*Quercus coccinea*) and the black oak (*Q. velutina*) can be found growing over the same very wide area. Hybrids are rare, however, because the scarlet oak is found in relatively moist, low areas with acidic soil, whereas the black oak is found in drier, well-drained habitats.

Temporal Isolation. Two species can occur in the same locale, but if each reproduces at a different time of year, they will not attempt to mate. For example, *Reticulitermes hageni* and *R. virginicus* are two species of termites. The former has its mating flights in March through May, whereas the latter mates in the fall and winter months.

Behavioral Isolation. Many animal species have courtship patterns that allow males and females to recognize one another. Male fireflies are recognized by females of their species by the pattern of their flashing and similarly male crickets are recognized by females of their species by their chirp.

Table 21.2 Reproductive Isolating Mechanisms

Isolating Mechanisms	Example
Premating Habitat	Species at same locale occupy different habitats
Temporal	Species reproduce at different seasons or different times of day
Behavioral	In animals, courtship behavior differs or they respond to different songs, calls, pheromones, or other signals
Mechanical	Genitalia unsuitable for one another
Postmating Gametic mortality	Sperm cannot reach or fertilize egg
Zygote mortality	Hybrid dies before maturity
Hybrid sterility	Hybrid survives but is sterile and cannot reproduce
F_2 fitness	Hybrid is fertile but F_2 hybrid has lower fitness

Figure 21.17

Orchids of the genus *Ophrys* have a peculiarly-shaped lower petal that is brown and black and hairy in appearance. The plant produces no nectar, but male bees, under the impression they have found a female of their species, visit it regularly and attempt to mate with the flower. In this way these bees only pollinate one particular species of orchid.

Mechanical Isolation. When animal genitalia or plant floral structures are incompatible, reproduction cannot occur. For example, male dragonflies have claspers that are suitable for holding only the females of their own species. Flowers of the orchid *Ophrys* have a strange and well-known mechanical isolation strategy. Certain species resemble female bees and attract only male bees of that species whose female it resembles (fig. 21.17). This assures that the bee will be pollinating only that particular species of orchid!

Postmating Isolating Mechanisms

Postmating isolating mechanisms prevent hybrid offspring from developing or breeding, even if reproduction has been successful.

Gametic Isolation. Even if the gametes meet, they may not fuse to give a zygote. In animals, the sperm of one species may not be able to survive in the reproductive tract of another species, or the egg may have receptors suitable to only sperm of its species. In plants, the stigma controls which pollen grains will successfully complete pollination and which will not.

Zygote Mortality, Hybrid Sterility, and F_2 Fitness. There are several frog species of the genus *Rana*, and when interspecies zygotes form, they fail to complete development, or else the offspring are frail. As is well known, a cross between a horse and donkey gives a mule, which is sterile and cannot reproduce. In some cases, the hybrid mules are fertile, but the F_2 generation is not fertile. This has also been observed in both evening primrose and cotton plants.

The biological definition of a species emphasizes that its members are able to interbreed and produce fertile offspring only among themselves. There are several mechanisms that keep species reproductively isolated from one another.

Formation of a Species

Speciation occurs when populations diverge from one another to the extent that members of one population are no longer capable of successful reproduction with members of another population. **Divergence** means that the gene pool and phenotypes for the two populations are now quite different.

Allopatric Speciation

Allopatric speciation (fig. 21.18) occurs when populations become geographically isolated.

Stage I Geographic isolation takes place. Imagine, for example, a grassland that contains several subpopulations of a particular grass (fig. 21.18*a*). Animals are allowed to overgraze a central portion of the grassland until it becomes a desert (fig. 21.18*b*). The resultant gap prevents the grasses in one area from reproducing with the grasses in the other area.

Stage II Reproductive isolation occurs. Imagine that surface mining has caused the soil in one area only to become contaminated with heavy metals like lead and copper. The grasses living in this area exhibit an increase in the frequency of genes that make them tolerant of these metals. Now the government initiates a reclamation effort that does away with the desert that separated the two groups of grasses. Even though the grasses are no longer separated, the two groups do not interbreed—the grasses tolerant of heavy metals do not reproduce with those that are not tolerant. At first, only *postmating isolating mechanisms,* represented by broken arrows in figure 21.18*c* may be evident. For example, cross-fertilization may occur but the seeds are not viable. Later, *premating isolating mechanisms,* shown in figure 21.18*d* by the absence of arrows, are evident. For example, one type of grass might flower in the spring, and the other might flower in the fall. Natural selection favors this difference. Plants exhibiting the premating isolating mechanism have more offspring than those not exhibiting the mechanism. Eventually, there are two separate species of grasses where before there was only one species.

Our example assumes that a large population was divided in half, but actually it has been found that speciation occurs more rapidly in small populations than in very large ones. Consider, for example, that speciation was favored in the Galápagos Islands, because only a few finches from the mainland arrived at these islands. In larger areas, speciation is promoted in **peripheral isolates,** which are small populations that occur at the fringes of a larger population and become geographically isolated from them.

Figure 21.18

Allopatric speciation usually occurs in two stages. In (a), subpopulations of a single species are represented by circles; the arrows indicate that crossbreeding may occur when individuals migrate from one subpopulation to another. Stage I (b), begins when two groups of subpopulations become geographically isolated, which makes any further exchange of genes between them impossible. The isolated groups adapt to local conditions and gradually diverge genetically. In Stage II (c), the populations again make contact, but reproduction, whenever attempted (broken arrows), is unsuccessful. Natural selection then favors the development of premating isolating mechanisms. In (d), isolation is complete and two separate species coexist.

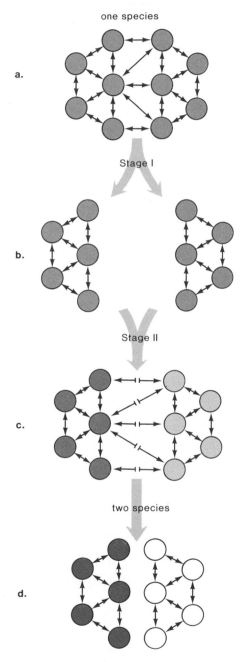

Figure 21.19

Beak depth differences among three species of ground finches. a. On the large islands of Isabela and Santa Cruz, all three species are present and each has a distinctive beak size. b. Only the species *G. fortis* is on Daphne Major. c. Only the species *G. fuliginosa* is on Los Hermanos. In these instances, both species tend to have the same intermediate beak depth.

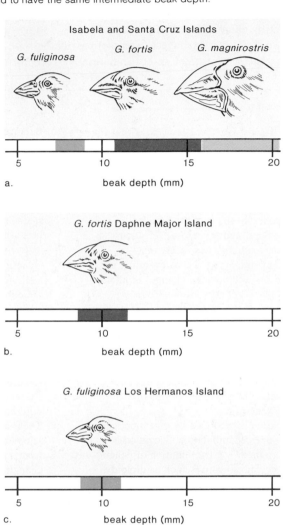

fortis, and *G. magnivostris,* with bills adapted to feeding on small-, medium-, and large-sized seeds, respectively (fig. 21.19). When all three species occur on the same island, they do not mate with each other and their bill sizes (beak depth) are quite distinctive. However, when *G. fortis* and *G. fuliginosa* are on separate islands, their bills tend to be the same intermediate size because there is no selection pressure to have larger or smaller beak size, perhaps because of lack of competition from other species.

Sympatric Speciation

Sympatric speciation is reproductive isolation without prior geographic isolation. The best evidence for this type of speciation is found among plants, where speciation can occur by means of polyploidy. Hybridization, followed by a doubling of the

Character Displacement When newly developed species come into contact again, competition can cause them to become more phenotypically different from one another. This process, called **character displacement,** is again exemplified by Darwin's finches. There are three species of ground finches, *G. fuliginosa, G.*

chromosome number, has been hypothesized in plants (p. 197). This creates a new species because the plant can only self-pollinate and can no longer breed with the parental strains.

Sympatric speciation has also been hypothesized in animals. It is noted, for example, that two types of corn borer moth do not mate with one another because each responds to a different pheromone mixture. (Some insects locate the opposite sex by detecting chemicals called pheromones.) It is suggested that these two species may have arisen sympatrically, with simultaneous genetic mutations in both pheromone-receptor cells in males and pheromone-producing cells in females. Sympatric speciation is believed to be much less common than allopatric speciation because of the low probability of different mutations in the two sexes at the same time.

Speciation is believed to occur in one of two ways. Allopatric speciation requires geographic isolation before reproductive isolation can be achieved, but sympatric speciation does not require prior geographic separation.

Summary

1. The Hardy-Weinberg law states that the allele and genotype frequencies in a population will remain the same in each succeeding generation as long as certain conditions are met.
2. The conditions are: large population size, random mating, no gene flow, no mutations, and no natural selection. These conditions are rarely met.
3. This allows us to see that evolution is expected and involves changes in allele frequencies in populations from generation to generation.
4. The synthetic theory of evolution says that the process of evolution requires: (1) production of genotype and phenotype variations among members of a population and (2) sorting out and reduction of the variations passed on to next generation.
5. Variation is produced by mutation, gene flow, and recombination. Variation is maintained by a combination of genetic and ecological causes such as diploidy, pleiotropy, heterozygote superiority, and geographical differences.
6. Balanced polymorphism is an example of variation maintained in populations. For example, in some parts of Africa, the allele for sickle-cell anemia and the normal allele are in balance, because of heterozygote superiority. Also, *Cepaea* snails have two distinct phenotypes because they occur in both forest and low-vegetation areas.
7. Genetic drift occurs when changes in allele frequencies within a population are due to chance occurrences. Founder effects and bottlenecks are particularly influential types of genetic drift.
8. Natural selection is expected to result in adaptation to the environment. In the modern context, natural selection requires: heritable variation, differential adaptedness of certain variations, and differential reproduction.
9. There are three types of natural selection: disruptive selection, which results in polymorphism due to a changing or varied environment; stabilizing selection, which favors the intermediate phenotype in a nonchanging environment; and directional selection, which favors an extreme phenotype in a changing environment.
10. Two examples of directional selection are industrial melanism in the peppered moth and resistance to chemicals in some bacteria and insects.
11. The biological concept of a species emphasizes that a species is reproductively isolated from other species. There are various premating and postmating reproductive isolating mechanisms.
12. Allopatric speciation requires geographic isolation before reproductive isolation occurs. Geographic isolation includes the possibility of peripheral isolates. Following speciation, character displacement can cause species to become even more dissimilar.
13. Sympatric speciation does not require geographic isolation for reproductive isolation to occur. The occurrence of polyploidy in plants is an example of this type of speciation.

Objective Questions

1. Which of these would contribute to evolution, defined as a change in allele frequencies, to occur?
   a. mutations
   b. genetic drift
   c. gene flow
   d. All of these.
2. If 16% of a population has the recessive phenotype and 84% has the dominant phenotype, what percentage of the next generation is expected to be homozygous recessive, assuming no evolution?
   a. 4%
   b. 16%
   c. 36%
   d. 84%
3. What do mutations and recombination have in common?
   a. They increase variation.
   b. They cause genetic drift to occur.
   c. They cause natural selection to occur.
   d. Both (b) and (c).
4. Sickle-cell anemia occurs in an African population that exhibits balanced polymorphism for the allele. What causes this?
   a. a varied geographical range
   b. heterozygote superiority
   c. maintenance of variation by sympatric speciation
   d. allopatric speciation
5. A human population has a higher than usual percentage of individuals with a genetic disease. The most likely explanation is
   a. natural selection.
   b. allopatric speciation.
   c. reproductive isolation.
   d. genetic drift.
6. In modern terms, the offspring of better adapted individuals are expected to make up a larger proportion of the next generation. This statement is to be associated with
   a. genetic drift.
   b. differential reproduction.
   c. natural selection.
   d. Both (b) and (c).

7. During the evolution of humans, brain size gradually increased. This is an example of
   a. disruptive selection.
   b. stabilizing selection.
   c. directional selection.
   d. All of these.
8. In polluted areas, the dark form of the peppered moth increased in numbers due to
   a. a selective agent.
   b. natural selection.

   c. disruptive selection.
   d. Both (a) and (b).
9. Which comes first during allopatric speciation?
   a. geographic isolation
   b. premating isolating mechanism
   c. postmating isolating mechanism
   d. character displacement

10. In both allopatric and sympatric speciation
    a. geographic isolation is required.
    b. reproductive isolation is required.
    c. parallel evolution is required.
    d. All of these.

Population Genetics Problems

1. Twenty-one percent of a population is homozygous dominant, 50% is heterozygous, and 29% is homozygous recessive. What percentage of the next generation is predicted to be homozygous recessive?

2. If 1% of a human population has the recessive phenotype, what percentage has the dominant phenotype?
3. Four percent of the members of a population of pea plants are short

(recessive characteristic). What is the frequency of both the recessive allele and the dominant allele? What are the genotype frequencies in this population?

Study Questions

1. Define species, population, variation, and allele frequency.
2. State the Hardy-Weinberg law along with its conditions. Define evolution, using this law as a basis.
3. According to the synthetic theory of evolution, what are the two steps required for evolution to take place?
4. What are three ways that variation is produced? How is it maintained?
5. What is balanced polymorphism and what causes it to occur in certain African populations? Among *Cepaea* snails?

6. What is meant by genetic drift and how is it exaggerated by founder effects and bottlenecks?
7. State the steps necessary for adaptation by natural selection according to modern thinking?
8. What are three observed ways by which natural selection operates? Give examples of each of these.
9. What is the biological definition of a species?

10. Name and give examples for pre- and postmating reproductive isolating mechanisms.
11. Describe the two stages of allopatric speciation and give a hypothetical example of this type of speciation.
12. Describe sympatric speciation and give a hypothetical example of this type of speciation.

Thought Questions

1. Why is it more appropriate to relate speciation to adaptive radiation rather than phyletic evolution?

2. What are the major differences between Darwin's version of evolution by natural selection and the modern restatement of this?

3. Why is genetic drift not expected to result in adaptation to the environment? Could it result in adaptation?

Selected Key Terms

gene pool (jēn pōol) 311
gene frequencies (jēn fre′kwen-sēz) 311
Hardy-Weinberg law (har′de vīn′berg law) 311
subspecies (sub′spe-sēz) 314
hybrid (hi′brid) 314
hybrid vigor (hi′brid vig′or) 314
polymorphism (pol″e-mor′fizm) 316

heterozygote superiority (het″er-o-zi′gōt su-pe″re-or′i-te) 316
genetic drift (jĕ-net′ik drift) 317
founder effect (fown′der ĕ-fekt′) 317
disruptive selection (dis-rup′tiv sĕ-lek′shun) 320
stabilizing selection (sta′bil-i-zing sĕ-lek′shun) 320

directional selection (di-rek′shun-al sĕ-lek′shun) 320
speciation (spe″se-a′shun) 322
allopatric speciation (al″o-pat′rik spe″se-a′shun) 324
sympatric speciation (sim-pat′rik spe″se-a′shun) 325

CHAPTER 22

Origins of Life

Your study of this chapter will be complete when you can:

1. Trace the steps by which chemical evolution produced the protocell.
2. Describe any supporting experiments or data that support the occurrence of chemical evolution.
3. Explain why the protocell would have been a heterotrophic fermenter.
4. Describe the difference between the protocell and a true cell.
5. Describe how genes may have first formed.
6. Give dates for the oldest prokaryotic fossils, evolution of cyanobacteria, and the eukaryotic cell.
7. State the significance of the fact that prokaryotic evolution occurred over such a long period.
8. Explain how the presence of oxygen changed the world so that now life comes only from life.

It's hard to imagine a time when the earth was devoid of all living things. The first organisms came into existence in the ocean but nearby rocks are thought to have played a role also.

Today we do not believe that life arises spontaneously from nonlife and we say that "life comes only from life." But if this is so, how did the first form of life come about? We can assume that the first form of life was very simple—a single cell (or cells) that could grow, reproduce, and mutate. Since it was the very first living thing, it had to come from nonliving chemicals. Could there have been an increase in the complexity of the chemicals—could a **chemical evolution** have produced the first cell(s)? This chapter reviews the evidence for such a chemical evolution; it is based on our knowledge of the primitive earth (fig. 22.1) and on experiments that have been performed in the laboratory.

Chemical Evolution

The sun and the planets probably formed from aggregates of dust particles and debris about 4.6 billion years ago. Intense heat produced by gravitational energy and radioactivity of some atoms caused the earth to become stratified into a core, mantle, and crust. Heavier atoms of iron and nickel became the molten liquid core, and dense silicate minerals became the semiliquid mantle. The unstable mantle caused the thin crust to move continually, so there were no stable land masses during the time life was evolving.

Primitive Atmosphere

The **primitive atmosphere** was not the same as today's atmosphere. Harold Urey proposed in the 1950s that the earliest atmosphere must have contained a lot of hydrogen (H_2) because it is the most abundant element in our solar system. Later, it was suggested that lightweight atoms, including hydrogen, would have been lost as the earth formed because its gravitational field was not strong enough to hold them. It is now thought that the primitive atmosphere was produced after the earth formed by outgassing from the interior, particularly by volcanic action. In that case, the atmosphere would have consisted mostly of water vapor (H_2O), nitrogen gas (N_2), and carbon dioxide (CO_2), with only small amounts of hydrogen (H_2) and carbon monoxide (CO). The primitive atmosphere with little if any free oxygen was a **reducing atmosphere** as opposed to the **oxidizing atmosphere** of today. This was fortuitous because oxygen (O_2) attaches to organic molecules, preventing them from joining together to form larger molecules.

The primitive atmosphere was a reducing atmosphere that contained little if any oxygen gas, O_2.

At first the earth was so hot that water was present only as a vapor that formed dense, thick clouds. Then as the earth cooled, water vapor condensed to liquid water, and rain began to fall (fig. 22.1a). This rain was in such quantity that it produced the oceans of the world. Of all the planets in the solar system, only earth has liquid water; this may be related to the distance of the earth from the sun. The reading on page 334 tells what would happen if we moved the earth closer or farther away from the sun.

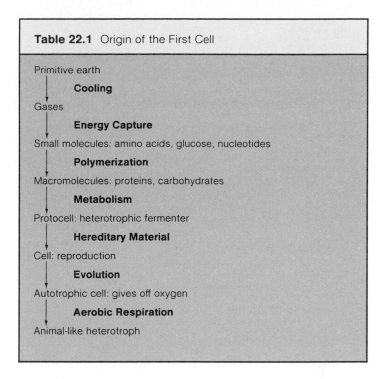

Table 22.1 Origin of the First Cell

Primitive earth
↓   **Cooling**
Gases
↓   **Energy Capture**
Small molecules: amino acids, glucose, nucleotides
↓   **Polymerization**
Macromolecules: proteins, carbohydrates
↓   **Metabolism**
Protocell: heterotrophic fermenter
↓   **Hereditary Material**
Cell: reproduction
↓   **Evolution**
Autotrophic cell: gives off oxygen
↓   **Aerobic Respiration**
Animal-like heterotroph

Simple Organic Molecules

The atmospheric gases, dissolved in rain, were carried down into newly forming oceans. The remaining steps shown in table 22.1 took place in and about the sea, where life is believed to have originated.

As early as the 1920s, Soviet biochemist A. I. Oparin proposed that organic molecules could be produced from the gases of the primitive atmosphere in the presence of strong outside **energy sources** present on the primitive earth. These energy sources included heat from volcanoes and meteorites, radioactivity from the earth's crust, powerful electric discharges in lightning, and solar radiation, especially ultraviolet radiation (fig. 22.1b).

In 1953 Stanley Miller provided support for Oparin's ideas through an ingenious experiment (fig. 22.2). Miller placed a mixture resembling a strongly reducing primitive atmosphere (methane, ammonia, hydrogen, and water) in a closed system, heated the mixture, and circulated the gases past an electric spark (simulating lightning). After a week's run, Miller discovered that a variety of amino acids and organic acids had been produced. Since that time, other investigators have achieved similar results by utilizing other less reducing combinations of gases dissolved in water.

These experiments indicate that the primitive gases not only could have but probably did react with one another to produce simple organic compounds that accumulated in the ancient seas. Neither oxidation (there was no free oxygen) nor decay (there were no bacteria) would have destroyed these molecules and they would have accumulated in the oceans for

Figure 22.1

A model for the origin of life.

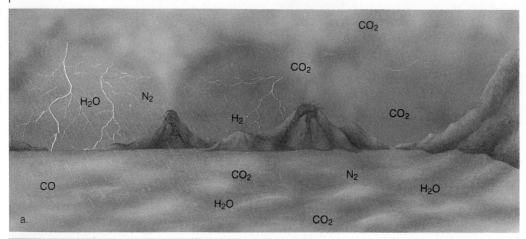

a. The primitive atmosphere contained gases, including water vapor that escaped from volcanoes; as the latter cooled, some gases were washed into the ocean by rain.

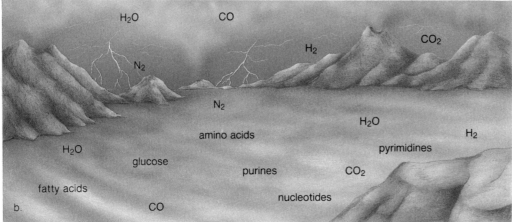

b. The availability of energy from volcanic eruption (shown here) and lightning allowed gases to form simple organic molecules.

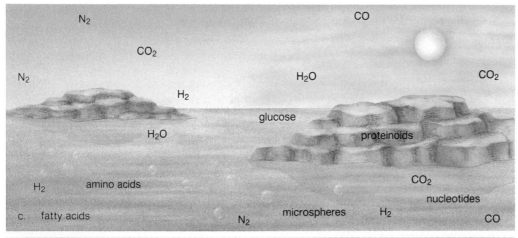

c. Amino acids splashed up onto rocky coasts could have polymerized into polypeptides (proteinoids) that would have become microspheres when they reentered the water.

d. Eventually various types of prokaryotes and then eukaryotes evolved. Some of the prokaryotes were oxygen-producing photosynthesizers. The presence of oxygen in the atmosphere was needed for aerobic respiration to evolve.

Figure 22.2

In Miller's experiment, gases were (a) admitted to the apparatus, (b) circulated past an energy source (electric spark), (c) and cooled to produce (d) a liquid that could be withdrawn. Upon chemical analysis the liquid was found to contain various simple organic molecules.

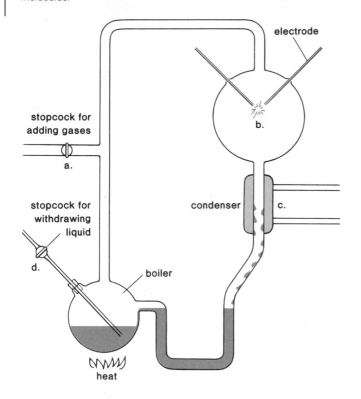

contains iron and zinc that may have served as inorganic catalysts for polypeptide formation. In addition, clay has some tendency to collect energy from radioactive decay and to discharge it when the temperature and/or humidity changes. In this way, energy may have been provided for polypeptide synthesis. Cairns-Smith believes that nucleic acid polymerization would also have been facilitated within clay. We will return to this idea shortly.

Small organic molecules polymerized to produce macromolecules. This could have occurred on heated rocks or in clay.

Biological Evolution

At first there was a **protocell,** which contained organic macromolecules bounded by a membrane. Several hypotheses have been suggested as a means by which a protocell could have developed.

Protocells

Fox has shown that when proteinoids are exposed to water they form **microspheres** (fig. 22.3*a*), which have properties similar to today's cells (table 22.2), including a membrane that is selectively permeable. Therefore he supports a **cell-first hypothesis,** meaning that the protocell came first—before true proteins and nucleic acids.

If lipids are made available to microspheres, they tend to become associated with them so that the outer boundary is a lipid-protein membrane. Even so, it is also possible that a lipid membrane formed first and that the protocell developed afterward. You'll recall that phospholipid molecules will automatically form droplets called **liposomes** in a liquid environment (fig. 22.3*b*). Perhaps the first membrane formed in this manner rather than by way of microspheres.

Some researchers support the work of Oparin, who was one of the original researchers in this area. As early as 1938, Oparin showed that under appropriate conditions of temperature, ionic composition, and pH, concentrated mixtures of macromolecules tend to give rise to complex units called **coacervate droplets.** Coacervate droplets have a tendency to absorb and incorporate various substances from the surrounding solution. Eventually, a semipermeable-type boundary may form about the droplet.

Heterotroph Hypothesis

We might now ask, how did the protocell carry on nutrition so that it could grow? Nutrition would have been no problem because the protocell existed in the ocean, which at that time was an organic soup containing simple organic molecules that could have served as sources of building blocks and energy. Therefore, the protocell would have been a **heterotroph,** an organism that takes in preformed food. Notice that by this theory heterotrophs are believed to have preceded **autotrophs,** organisms that make their own organic food.

hundreds of millions of years. With the accumulation of these simple organic compounds, the oceans became a thick, hot **organic soup,** containing a variety of organic molecules.

Cooling caused water vapor to turn to rain that formed the oceans. Here, as Miller and others have shown, atmospheric gases could have reacted with one another under the influence of an outside energy source to produce simple organic molecules.

Macromolecules

The newly formed organic molecules would have combined to form still larger molecules and macromolecules (fig. 22.1*c*). We know that **polymerization** in modern cells requires the presence of enzymes, but of course some other mechanism could have promoted the formation of the first polymers. Sidney Fox of the University of Miami has shown that amino acids will polymerize abiotically when exposed to dry heat. Further, these so-called **proteinoids** contain amino acids that join in a preferred manner. He suggests that amino acids collected in shallow puddles along the rocky shore and the heat of the sun caused proteinoids to form as drying took place.

Graham Cairns-Smith of Glasgow University believes instead that clay may have been especially helpful in causing polymerization to occur. Clay attracts small organic molecules and

Figure 22.3

Protocells may have resembled (a) microspheres, which have a variety of cellular characteristics (table 22.2) or (b) liposomes, which automatically form when phospholipids are placed in water.

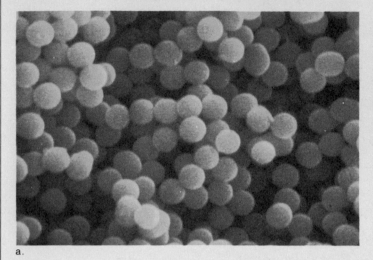
a.

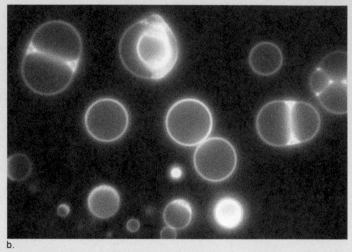
b.

Table 22.2 Microspheres Possess Many Properties Similar to Contemporary Cells

Stability (to standing, centrifugation, sectioning)
Microscopic size
Variability in shape but uniformity in size
Numerousness
Stainability
Ultrastructure (electron microscope)
Double-layered boundary
Selective passage of molecules through boundary
Catalytic activities
Patterns of association
Propagation by "budding" and fission
Growth by accretion
Motility
Propensity to form junctions and to communicate

From Sidney W. Fox, "Chemical Origins of Cells" in *Chemical & Engineering News*, 49, 50, Dec. 6, 1971. Copyright © 1971 American Chemical Society. Reprinted by permission.

At first the protocell may very well have made use of preformed ATP, but as this supply dwindled, natural selection would have favored any cells that could extract energy from carbohydrates in order to transform ADP to ATP. Glycolysis is a common metabolic pathway in living things and this testifies to its early evolution in the history of life. Since there was no free oxygen, we can assume that the protocell must have carried on a form of fermentation.

It would seem logical that the protocell at first contained limited ability to break down organic molecules and that it would have taken millions of years for glycolysis to have evolved completely. It is of interest that Fox has shown that microspheres have some catalytic activities and Oparin found that coacervates do incorporate enzymes if they are available in the medium. It's possible then that the protocell did contain polypeptides with some enzymatic activity.

The protocell could have been a simple sphere that contained polypeptides having some degree of enzymatic ability. It would most likely have been a heterotrophic fermenter.

True Cells

The protocell would have been capable of growing, but without genes, it could not have passed on its characteristics nor could it have mutated and evolved. Once it did contain genes, it would have been a true cell.

All cells today have genes composed of DNA. The central dogma of genetics states that the flow of information is always from DNA → RNA → protein. Is it possible that this sequence developed step by step? Graham Cairns-Smith suggests that in clay, RNA nucleotides and polypeptides became associated in such a way that the polypeptides were ordered by and helped synthesize RNA. This would mean that at least the RNA → protein step occurred before enclosure by a membrane and that the first genes were composed of RNA, not DNA. Once these genes were incorporated by the protocell it would have become a true cell.

Even today there are viruses that have RNA genes and the enzyme reverse transcriptase that uses RNA as a template to form DNA. Perhaps with time, reverse transcription occurred within the protocell and this is how DNA genes arose.

Once the protocells acquired genes they became true cells, capable of evolving and producing other types of cells.

Fossils of Primitive Cells

There is no recorded history for evolution prior to the true cell; the earliest fossils dated about 3.5 billion years ago are presumed to be prokaryotic cells. They are found in fossilized

Figure 22.4
The oldest prokaryotic fossils. *a.* This Precambrian filamentous
prokaryotic fossil was found in (*b*) layers of a stromatolite. *c.* Living
stromatolites located in shallow waters off the shores of western
Australia.

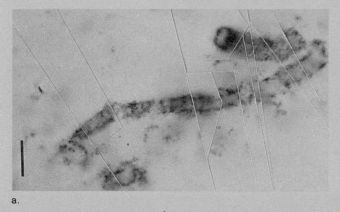

a.

b.

c.

stromatolites (*stroma* = bed and *lithos* = stone), pillarlike
structures composed of sedimentary layers containing communities of photosynthetic microorganisms (fig. 22.4). Living
stromatolites exist even today in shallow waters off the west coast
of Australia.

Autotrophs

The autotrophs found in ancient stromatolites are believed to
have been anaerobic photosynthesizers that did not give off
oxygen. They most likely arose because once the supply of organic molecules in the organic soup began to decline, natural
selection would have favored cells that could use solar energy
to produce carbohydrates. A billion years later, oxygen-
producing photosynthesizers (cyanobacteria) evolved. These
autotrophs would have possessed chlorophyll *a* and would have
photosynthesized, making their own food by using carbon dioxide
and water, and releasing O_2 as a by-product.

With the evolution of oxygen-producing photosynthesizers, the atmosphere was changed forever (fig. 22.1*d*). The
importance of this event to maintaining the mild temperature
of the earth is discussed in the reading on the next page.

Aerobic Respiration

The presence of oxygen in the atmosphere meant that most environments were no longer suitable for anaerobic organisms and
they began to decline in importance. The photosynthetic cyanobacteria would have then proliferated, and eventually the
atmosphere would have contained a stable amount of oxygen.
Still, it was another billion years before aerobic respiration and
the eukaryotic cell evolved. As table 20.1 and fig. 22.5 show, a
large portion of the history of life was devoted to prokaryote
evolution, during which time various metabolic pathways came
into existence.

The presence of oxygen in the atmosphere permitted life
forms to invade the land. Oxygen in the upper atmosphere forms
ozone (O_3), which filters out the ultraviolet rays of the sun.
Before the formation of this so-called **ozone shield,** the amount
of radiation would have destroyed land-dwelling organisms.

It was the evolution of autotrophs that caused oxygen to enter the
atmosphere. Oxygen allows aerobic respiration to occur and
forms an ozone shield that protects the earth from ultraviolet
radiation.

Figure 22.5

Some of the major events that occurred during Precambrian history (table 20.1). The entire Precambrian Period takes up seven-eighths of the history of life on earth! (BYA = billions of years ago)

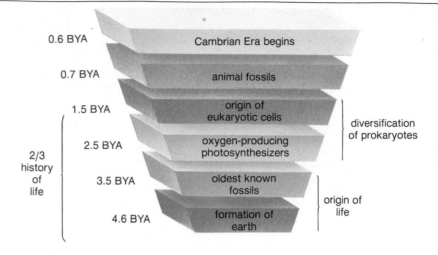

0.6 BYA — Cambrian Era begins
0.7 BYA — animal fossils
1.5 BYA — origin of eukaryotic cells
2.5 BYA — oxygen-producing photosynthesizers
3.5 BYA — oldest known fossils
4.6 BYA — formation of earth

2/3 history of life

diversification of prokaryotes

origin of life

Why Is the Earth neither Too Hot nor Too Cold?

The Earth is quite literally poised between fire and ice. Consider, for example, what would happen if we somehow moved the Earth slightly closer to the sun.

As the oceans grew warmer, more and more water vapor would begin to steam into the atmosphere. Once there, the vapor would begin to act like glass in a greenhouse, preventing heat from radiating back into space. So the Earth would grow warmer still, until oceans began to dry and the carbonate rocks—the limestones and dolomites—began to bake in the heat and release carbon dioxide.

The greenhouse effect caused by this gas is famous: By burning fossil fuels we are already releasing enough carbon dioxide to warm the climate measurably. The carbonate rocks, however, contain billions upon billions of tons of this gas, enough to trigger a "runaway greenhouse." In the end, our planet would become a twin of unfortunate Venus, the next planet inward to the sun: a gaseous, dry, searing hell, its surface covered with clouds, oppressed by a massive atmosphere of carbon dioxide, and hot enough to melt lead.

Suppose, on the other hand, we moved the Earth further out from the sun. As the planet grew colder, glaciers would grind southward over Canada, Europe, and Siberia, while sea ice crept northward from Antarctica. The ice would reflect more sunlight back into space, cooling the planet even more. Step by step, the ice would extend toward the equator. In the end, the Earth would gleam brilliantly—but its oceans would be frozen solid.

Thus, the climate is balanced precariously indeed—so precariously that many geologists now believe that tiny, cyclic variations in the Earth's orbit, known as the Milankovitch cycles, were enough to have triggered the ice ages.

But geologists also have found fossilized marine microbes in rocks just about 3.5 billion years old and they assure us that the oceans of the Earth have remained warm and liquid throughout its 4.6-billion-year history.

Perhaps this is a lucky accident—after all, if the Earth had not formed at just the right distance from the sun to have liquid oceans, we would not be here to worry about it. But the astrophysicists point out that things aren't quite that simple.

The sun, they say, is a quiet and stable star. But like others of its type, it is inexorably getting hotter with age. In fact, it is about 40% brighter now than when the Earth was born. So how could the climate possibly stay constant? If the Earth is comfortable now, then billions of years ago, under a colder sun, the oceans should have been frozen solid. But they were not. On the other hand, if the oceans were liquid then, why has the sun not broiled us into a second Venus by now?

One theory, advocated by a number of biologists, is that the early Earth started out with a good deal more carbon dioxide in the air than it has now, which gave enough of a greenhouse effect to keep the planet warm even when the sun was cool. If nothing had changed, the greenhouse eventually would have "run away" as it did on Venus. (There is some evidence that Venus started out with oceans much like ours.) But fortunately for us, about three billion years ago, certain prokaryotes [cyanobacteria] devised a way of taking carbon dioxide out of the air and turning it into organic carbon compounds. We call that process photosynthesis. In the eons since, as the algae and their descendants, the plants, have evolved and multiplied, the decreasing levels of atmospheric carbon dioxide have just about kept pace with the warming sun. Thus, say the biologists, it was life that saved the world for life.

M. Mitchell Waldrop. First appeared in *Science 83*. Reprinted by permission of the author.

We have shown that life could have come into existence by means of chemical evolution. However, we do not believe that this same chemical evolution is occurring today for the following reasons:

1. Certain energy sources, particularly ultraviolet radiation, are unavailable. The ozone layer now acts as a shield to prevent these rays from reaching the earth in great quantity.

2. While the first atmosphere was believed to be a reducing one and thus promoted the buildup of organic molecules, today's atmosphere is an oxidizing atmosphere, that tends to break down organic molecules.

3. Living organisms, present today, would use any newly formed organic molecules for food.

Today, life is believed to come only from life because if any organic molecules should form, they would be oxidized by oxygen or utilized by preexisting life.

Summary

1. The unique conditions of the primitive earth allowed chemical evolution to occur and produce the first cell(s).
2. The gases of the primitive atmosphere probably escaped from volcanoes and included water vapor, nitrogen gas, and carbon dioxide, with small amounts of hydrogen and carbon monoxide.
3. The primitive atmosphere was a reducing atmosphere with no free oxygen gas.
4. At first the earth was very hot but when it cooled, the water vapor condensed and the rains began to fall. They carried the gases into the newly formed oceans.
5. In the presence of an outside energy source such as ultraviolet radiation, lightning, and radioactivity, the gases reacted with one another to produce small organic molecules such as amino acids and glucose. This scenario is supported by experiments performed by Stanley Miller and others.
6. All of these organic molecules made the oceans an organic soup.
7. The next step would have been the formation of macromolecules by polymerization. Two hypotheses are offered; Sidney Fox has produced proteinoids by heating amino acids with dry heat and suggests that this could have originally occurred by solar heating on rocky shores. Graham Cairns-Smith believes that amino acid polymerization could have taken place in clay where radioactivity would have supplied the necessary energy.
8. Biological evolution began with the formation of the protocell, a membrane-bound sphere that could grow.
9. The proteinoids produced by Fox become microspheres when they are placed in water. Microspheres have many properties similar to cells and they will become associated with lipids in the medium, accounting perhaps for today's lipid-protein membrane.
10. Phospholipids readily form spherical liposomes and this may have been the origin of the first plasma membrane.
11. Oparin many years ago showed that coacervate droplets automatically form when concentrated mixtures of macromolecules are held at the right temperature, ionic composition, and pH. They, too, eventually form a membrane.
12. The protocell must have been a heterotrophic fermenter living on the preformed organic molecules in the organic soup. Microspheres and coacervates have limited enzymatic activity.
13. The protocell would not have been a true cell until it contained genes. Cairns-Smith believes that polypeptides and RNA became associated in clay in such a way that the polypeptides acted as an enzyme to facilitate RNA formation.
14. The earliest fossil evidence for cells is found in fossilized stromatolites. These fossils are believed to be anaerobic photosynthesizing bacteria.
15. The first prokaryote photosynthesizers did not give off oxygen but eventually cyanobacteria, which do give off oxygen, evolved. At least a billion years was required for this to occur.
16. After another billion years, aerobic respiratory pathways had evolved and eukaryote fossils date back to this time (1.5 billion years ago). Prokaryotes existed alone on the surface of the earth for at least two billion years. During this time all the metabolic pathways found in modern cells developed.
17. The presence of oxygen in the atmosphere also allowed the formation of the ozone shield, which filters out ultraviolet rays. This along with the fact that oxygen attacks small organic molecules helps account for the inability of cells to originate by chemical evolution today. Also, since living things already exist, any organic molecules that might arise are promptly utilized as food.

Objective Questions

1. Which of these gives a possible sequence of organic chemicals leading up to the protocell?
   a. inorganic gases, amino acids, polypeptide, microsphere
   b. inorganic gases, glucose, liposomes, genes
   c. water, salts, protein, oxygen
   d. Both (a) and (b).
2. Which of these did Stanley Miller place in his experimental system to show that organic molecules could have arisen from inorganic molecules on the primitive earth?
   a. microspheres
   b. purines and pyrimidines
   c. the primitive gases
   d. All of these.
3. Microspheres
   a. look like a eukaryotic cell.
   b. are simple sacs.
   c. have some cellular characteristics.
   d. Both (b) and (c).
4. Which of these is the chief reason the protocell was probably a fermenter?
   a. It didn't have any enzymes.
   b. The atmosphere didn't have any oxygen.
   c. Fermentation provides the most amount of energy.
   d. All of these.
5. Which of these could the protocell have possibly been able to do?
   a. mutate
   b. evolve
   c. grow
   d. photosynthesize
6. The first genes were most likely composed of
   a. protein.
   b. RNA.
   c. DNA.
   d. phospholipids.
7. Which of these is a true statement?
   a. Eukaryotes evolved before prokaryotes.
   b. Cyanobacteria evolved before eukaryotes.
   c. The true cell evolved before the protocell.
   d. Prokaryotes didn't evolve until 1.5 billion years ago.
8. Most of the history of life concerns the evolution of
   a. prokaryotes.
   b. eukaryotes.
   c. photosynthesizers.
   d. plants and animals.

9. What was happening during prokaryotic evolution?
   a. development of organelles
   b. development of specialization
   c. development of metabolic pathways
   d. All of these.

10. Which of these is a true statement?
    a. The primitive atmosphere was an oxidizing one and today's is a reducing one, making photosynthesis possible.
    b. The primitive atmosphere was 20% oxygen just like it is today.
    c. The reducing primitive atmosphere contributed to the origin of life and the oxidizing one of today would hinder it.
    d. It took so long for prokaryote evolution because the primitive atmosphere screened out the ultraviolet radiation from the sun.

Study Questions

1. What were the gases of the primitive atmosphere and how did its composition contribute to the origin of life?
2. Describe Stanley Miller's experiment. What hypothesis did it substantiate?
3. Why is the primitive ocean called an organic soup?
4. How does Sidney Fox believe that polymerization first occurred? What does Graham Cairns-Smith believe?
5. How might the plasma membrane have come into existence?
6. Why is it believed that the protocell was a heterotrophic fermenter?
7. On what theoretical grounds is it believed that the first genes were RNA genes?
8. The oldest prokaryotic fossils are found in what type of structures? How old are they?
9. When did oxygen-producing photosynthesizers evolve? How did the presence of oxygen in the atmosphere influence the history of life on earth?

Thought Questions

1. What would be the definitive experiment to show that an abiotic origin of life is possible?
2. Based on the information provided in this chapter, what types of bacterial nutrition do you expect to be discussed in the next chapter?
3. Based on the information provided in the reading, what argument might you put forward for saving the tropical rain forests?

Selected Key Terms

chemical evolution (kem'ĭ-kal ev''o-lu'shun) 329
primitive atmosphere (prim'ĭ-tiv at'mos-fēr) 329
reducing atmosphere (re-dūs'ing at'mos-fēr) 329

oxidizing atmosphere (ok''si-dī'zing at'mos-fēr) 329
energy source (en'er-je sor's) 329
organic soup (or-gan'ik sōōp) 331
polymerization (pol''ĭ-mer''ĭ-za'shun) 331
proteinoid (pro'te-in-oid'') 331
protocell (pro'to-sel) 331

microsphere (mi-kro-sfēr) 331
cell-first hypothesis (sel ferst hi-poth'ĕ-sis) 331
liposome (lip'o-som) 331
heterotroph (het'er-o-trōf'') 331
autotroph (aw'to-trōf) 331
ozone shield (o'zōn shēld) 333

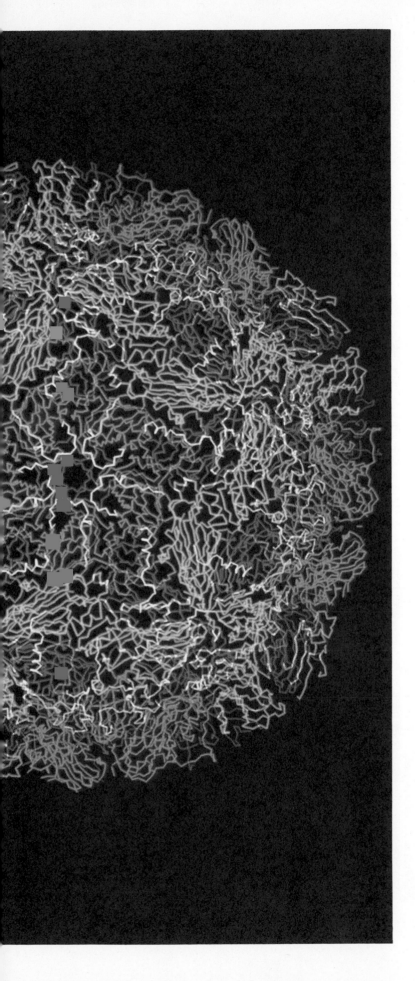

CHAPTER 23

Viruses and Monera

Your study of this chapter will be complete when you can:

1. *Describe in general the structure of viruses, including bacteriophages and animal viruses.*

2. *Describe the bacteriophage lytic and lysogenic life cycles.*

3. *Compare the life cycle of animal viruses to bacteriophage life cycles.*

4. *State how RNA retroviruses differ from other viruses and discuss their significance to human beings.*

5. *Compare viroids and prions to viruses.*

6. *Describe the anatomical characteristics of prokaryotic cells.*

7. *Tell how bacteria reproduce and relate this to their means of achieving genetic variation.*

8. *Explain how bacteria differ in their tolerance and need for oxygen.*

9. *List the different types of autotrophic bacteria, and specifically contrast the primitive and advanced types of photosynthesis.*

10. *Describe the nutrition of most heterotrophic bacteria, and give examples of different types of symbiotic bacteria.*

11. *Classify the different types of monerans.*

12. *Describe the anatomy and physiology of the cyanobacteria, and discuss their historical significance.*

Computer graphics representation of the three-dimensional structure of a foot-and-mouth disease virus. All viruses, regardless of their shape, have two portions—an outer coat of protein and inner core of nucleic acid. This particular virus uses RNA as its genetic material and typically infects the feet and mouths of hoofed animals, although it can also infect humans.

In this part, it is our aim to discuss living organisms from the simple to the complex and from the primitive (earliest evolved) to the most advanced (most recently evolved). We can discuss how living organisms might be related only in the broadest of terms, since detailed information is often lacking. It is also important to remember that no living group of organisms is the direct ancestor of another living group of organisms, although it is possible for two living groups to have shared a common ancestor.

It is curious that we begin our discussion with viruses when they are not even included in the classification table found in appendix A. We begin with viruses only because they are on the borderline between living and nonliving things.

Viruses

In 1892 Dimitri Ivanowsky, a Russian biologist, was performing experiments with tobacco plants infected with tobacco mosaic disease, a condition that takes its name from the wrinkled and mottled appearance of the infected leaves. After transmitting the disease to healthy plants by rubbing them with juice extracted from diseased plants, Ivanowsky passed the infective extract through a fine-meshed porcelain filter. To his surprise, the filtrate was still infective. This meant that the disease-causing agent was smaller than any known bacteria. Disease-causing agents that could pass through filters came to be known as filterable viruses and later simply as **viruses.**

In 1935 W. M. Stanley discovered that viruses are remarkably different from bacteria and that they have a noncellular organization. Today, we know that they are tiny particles (5 nm to 200 nm) composed of at least two parts: *an outer coat of protein and an inner core of nucleic acid*. The coat, called a **capsid,** is sometimes surrounded by a membranous outer envelope. The core of nucleic acid may be either DNA or RNA. Although viruses cannot be seen with the light microscope, the electron microscope has permitted a study of their structure (fig. 23.1). Notice how the capsid is made up of repeating protein subunits.

Viruses are capable of reproduction, but only within living cells; therefore, they are called *obligate parasites*. In the laboratory, some animal viruses are maintained by injecting them into live chick embryos (fig. 23.2). Outside living cells, viruses are nonliving and can be stored in much the same way as chemicals are stored. Therefore, it is proper to ask if viruses should be considered alive.

Figure 23.1
Viruses: micrographs below and drawings above. *a.* The tobacco mosaic virus is an RNA virus that attacks tobacco. The drawing shows how the nucleic-acid core circles within a capsid composed of individual proteinaceous units. *b.* The adenovirus is a DNA virus that causes respiratory and intestinal infections in humans. Magnification, ×250,000. *c.* T₄ bacteriophage, a virus that attacks bacteria, has a complex structure. Magnification, ×120,000. *d.* The influenza virus is an RNA virus whose protein capsid is enclosed by a membranous envelope.

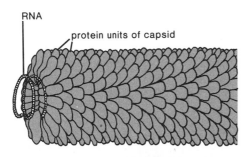

RNA

protein units of capsid

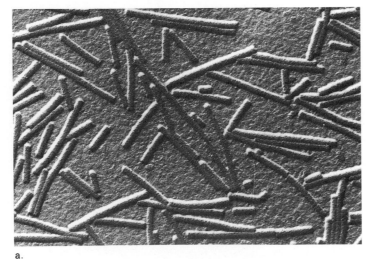

a.

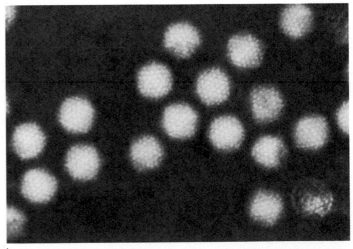

b.

Viruses typically have a specific host range. Certain ones attack only plants; others attack only animals—either birds or mammals; and the viruses called bacteriophages attack only bacteria. The human disease-causing viruses attack only certain types of cells. It is now known that there must be a match between a protein in the outer coat or envelope of the virus and the cell's outer surface before a virus can gain entry.

Viruses are noncellular obligate parasites that always have a coat of protein and a nucleic acid core.

Life Cycles

Bacteriophages

Two types of bacteriophage life cycles, termed the **lytic** and the **lysogenic** cycle (fig. 23.3) have been carefully studied.

Lytic Cycle   When a bacteriophage collides with an *E. coli* cell, it may attach to specific receptors by means of its protein tail fibers. An enzyme digests away part of the bacterial cell wall and the viral DNA enters the bacterial cell by way of the tail. Once inside, the viral DNA brings about disintegration of host DNA and takes over the operation of the cell. Viral DNA replication, utilizing the nucleotides within the host cell, produces many copies of viral DNA. Transcription occurs and

Figure 23.2

Inoculation of live chick eggs with virus particles. A virus reproduces only inside a living cell, not because it uses the cell as nutrients but rather because it takes over the machinery of the cell.

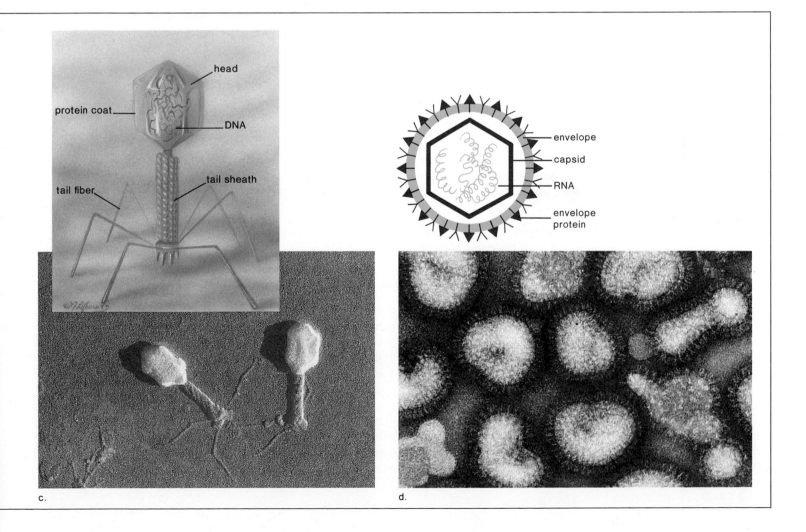

c.

d.

Figure 23.3

Lytic versus lysogenic life cycle. In the lytic cycle, a bacteriophage reproduces within the host bacterial cell and then the cell is lysed (broken open), allowing the viral particles to escape. In the lysogenic cycle, the DNA of a bacteriophage integrates into the host DNA and becomes a prophage. Thereafter, the prophage is replicated along with the bacterial chromosome and is passed to all the daughter cells. When and if the viral DNA leaves the chromosomes, the lysogenic cycle can be followed by the lytic cycle.

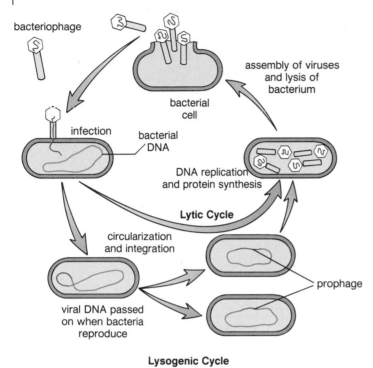

bacteriophage

assembly of viruses and lysis of bacterium

bacterial cell

infection

bacterial DNA

DNA replication and protein synthesis

Lytic Cycle

circularization and integration

prophage

viral DNA passed on when bacteria reproduce

Lysogenic Cycle

mRNA molecules utilize host ribosomes to bring about the production of multiple copies of coat proteins. Viral DNA and capsids are *assembled* to produce about 100 viral particles. In the meantime, viral DNA has directed the synthesis of lysozyme, an enzyme that disrupts the cell wall, and the particles are released.

Lysogenic Cycle   Some bacteriophages do not immediately undergo a lytic life cycle. Instead, the viral DNA becomes integrated into the bacterial DNA. In this stage it is called a *prophage*. The prophage is replicated along with the host DNA and all subsequent cells, called *lysogenic cells,* carry a copy of the prophage. Certain environmental factors such as ultraviolet radiation can induce the lytic cycle. The prophage leaves the bacterial chromosome; replication of viral DNA, production of capsids, assembly, and cell lysis follow.

During the lytic cycle of a bacteriophage, the bacterial cell dies when the viral particles burst from the cell. During the lysogenic cycle, viral DNA integrates itself into bacterial DNA for an indefinite period of time.

Animal Viruses

Some animal viruses have a membranous outer *envelope* as well as a mode of entry and exit from the host cell that are different from bacterial viruses (fig 23.4a). The entire virus enters by endocytosis and *uncoating,* which releases viral nucleic acid from the envelope and capsid, occurs inside the cell. Following multiplication and assembly, the virus exits from the cell by exocytosis, also called *budding.* The process of budding, which does not necessarily kill the host cell, provides the virus with its membranous envelope. This envelope often contains glycoproteins that are coded for by viral DNA; these glycoproteins interact with the host cell's plasma membrane and permit entry of the virus into the cell.

RNA Animal Viruses   Some animal viruses have RNA genes, that is, only RNA is enclosed within the capsid. In most of these viruses, the single strand of RNA serves as a template for the production of double-stranded RNA. This unique molecule then serves as a template for the replication of multiple copies of the original genetic material and for the transcription of mRNA molecules. Special enzymes, termed *RNA replicase and transcriptase,* are coded for by the RNA genes.

Other RNA viruses called **retroviruses** (fig. 23.4b) have an enzyme called *reverse transcriptase* that carries out RNA → DNA transcription. Following replication, the resulting double-stranded DNA (called cDNA because it contains a copy of the viral RNA) becomes integrated into the host chromosome for an indefinite period of time before reproduction of the virus occurs. Because of the process of integration, newly formed viruses sometimes include RNA copies of host genes, which they then carry into a new host cell.

Retroviruses are of extreme interest because these are the viruses that are apt to bring cancer-causing oncogenes into a host cell (fig. 17.9). The AIDS virus is also a retrovirus.

Animal viruses often have a membranous envelope that they acquire by budding from the host cell. Retroviruses are RNA viruses that carry out RNA → DNA transcription and integrate this cDNA into the host cell.

Viroids and Prions

Viroids and prions are very unusual infectious particles that can be compared to viruses. **Viroids** differ from viruses in that they consist only of a short chain of naked RNA. The number of nucleotides present is not sufficient to code for a capsid and there is no evidence that the viroid RNA is ever transcribed into a protein. We know of their presence only because they cause disease in plants; the first viroid to be recognized is the cause of potato spindle tuber disease. Viroids are also known to have attacked coconut trees in the Philippines and chrysanthemums in the United States. The manner in which they cause disease is not known, but it is hypothesized that they are reproduced by the host cell and that they may interfere with gene regulation in these cells.

Figure 23.4

Life cycles of animal viruses. *a.* DNA virus. After entering by endocytosis, the virus becomes uncoated. The DNA then codes for proteins, some of which are capsid (coat) proteins and some of which are envelope proteins. Assembly follows replication of the DNA. When the virus exits by budding, it becomes enclosed by an envelope made up of host-cell plasma membrane lipids and viral envelope proteins. *b.* RNA retrovirus. The life cycle includes steps not seen in (*a*). The RNA genes are transcribed to cDNA (DNA copied off of RNA) that becomes integrated into the host DNA. Transcription produces many copies of the RNA genes, which also serve to direct the synthesis of three types of proteins: the enzyme reverse transcriptase, capsid (coat) protein, and envelope protein. Again the virus buds from the host cell.

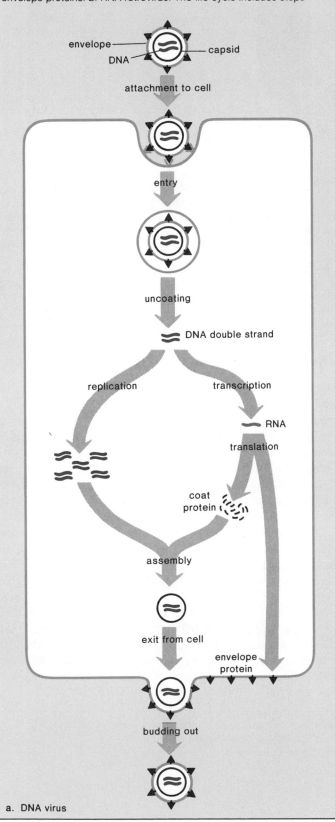

a. DNA virus

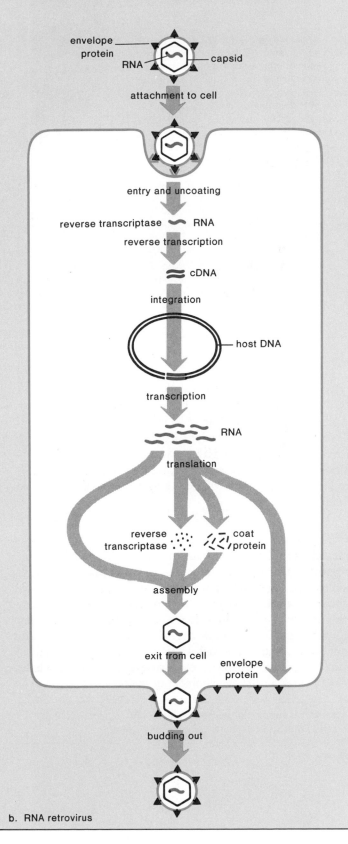

b. RNA retrovirus

Antibiotics and Antiviral Drugs

An antibiotic is a chemical that selectively kills bacteria when it is taken into the body as a medicine. Since the introduction of the first antibiotics in the 1940s, there has been a dramatic decline in deaths due to pneumonia, tuberculosis, and other infections.

Most antibiotics are produced naturally by soil microorganisms. Penicillin is made by the fungus *Penicillium;* streptomycin, tetracycline, and erythromycin are all produced by a bacterium, *Streptomyces.* Sulfa, a chemotherapeutic agent rather than an antibiotic, an analog of a bacterial growth factor, can be produced in the laboratory.

A few antibiotics are metabolic inhibitors specific for bacterial enzymes. This means that they poison bacterial enzymes without harming host enzymes. Penicillin blocks the synthesis of the bacterial cell wall; streptomycin, tetracycline, and erythromycin block protein synthesis; and sulfa prevents the production of a coenzyme.

There are problems associated with antibiotic therapy. Some patients are allergic to antibiotics and their reaction to them may even be fatal. Antibiotics not only kill off disease-causing bacteria, they also reduce the number of beneficial bacteria in the intestinal tract. The latter

Penicillium chrysogenum, from which the antibiotic penicillin is prepared.

may have held in check a pathogen that now is free to multiply and invade the body. The use of antibiotics sometimes prevents natural immunity from occurring, leading to the necessity for recurring antibiotic therapy. Most important, perhaps, is the growing resistance of certain strains of bacteria. While penicillin used to be 100% effective against hospital strains of

Staphylococcus aureus, today it is far less effective. Tetracycline and penicillin, long used to cure gonorrhea now have a failure rate of more than 20% against certain strains of gonococcus. Most physicians believe that antibiotics should only be administered when absolutely necessary. Some believe that if antibiotic use is not strictly limited then resistant strains of bacteria will completely replace present strains and antibiotic therapy will no longer be effective at all. They are very much opposed to the current practice of adding antibiotics to livestock feed in order to make animals grow fatter because resistant bacteria are easily transferred from animals to humans.

The development of antiviral drugs has lagged far behind the development of antibiotics. Viruses lack most enzymes and instead utilize the metabolic machinery of the host cell. Rarely has it been possible to find a drug that successfully interferes with viral reproduction without also interfering with host metabolism. One such drug, however, called vidarabine, was approved in 1978 for treatment of viral encephalitis, an infection of the nervous system. Acyclovir (ACV) seems to be helpful in treating genital herpes, and the drug AZT (azidothymidine) is being used in AIDS patients.

Prions differ markedly from viruses. They consist solely of a glycoprotein having only one polypeptide of about 250 amino acids. The DNA that codes for the polypeptide is not found in the host cell, so it is not known how prions manage to be reproduced. Like viroids, these particles are known only because they have been associated with a disease. It's possible that they are the cause of nervous system diseases previously termed *slow-virus diseases* because it takes a long time for symptoms to gradually worsen.

Viral Infections

Viruses are best known for causing infectious diseases in animals and plants. In plants, infectious diseases can only be controlled by burning those plants that show symptoms of disease. In animals, especially humans, viral diseases are controlled by administering vaccines and, only recently, by the administration of antiviral drugs. Some well-studied viral diseases in humans are influenza, mumps, measles, polio, rabies, and infectious hepatitis. As mentioned above, retroviruses cause AIDS and have been implicated in the development of cancer.

Viruses cause diseases of bacteria, plants, and animals. They have also been implicated in the development of cancer.

Kingdom Monera

Kingdom Monera is one of the five kingdoms recognized in this text (see table on page 3). A kingdom is the largest category of classification. There is a classification of organisms in the appendix that you may wish to refer to from time to time. Kingdom Monera contains the **prokaryotes,** the first types of cells to evolve. These organisms provide the first fossils—dated 3.5 billion years ago—and the fossil record indicates that the prokaryotes were the only type of living organisms for at least two billion years (fig. 22.5).

In our classification system, the prokaryotes are the **bacteria** and are divided into two groups, the archaebacteria and the eubacteria. It is important to realize that the eubacteria include the cyanobacteria, which were formerly called blue-green algae (fig. 23.5). Monerans are small single cells (*moneres* = single) that generally range in size from 1μm to 10 μm in length (fig. 5.2) and from 0.2 μm to 0.3 μm in width. Since they are microscopic, it is not always obvious that they are abundant in the air, water, soil, and on most objects. It has even been suggested that the combined number of all bacteria would exceed that of any other type of organism on earth.

Kingdom Monera includes the various types of bacteria, which are microscopic single-celled prokaryotes.

Figure 23.5

Electron micrograph following freeze-fracture of a cyanobacterium cell *Anabaena cylindrica*. Typical of prokaryotic cells, there is no membrane-bound nucleus or other types of organelles found in eukaryotic cells. Cyanobacteria carry out photosynthesis in the same manner as plants but there are no chloroplasts. The chlorophyll is found in an extensive array of thylakoids (photosynthetic membranes).

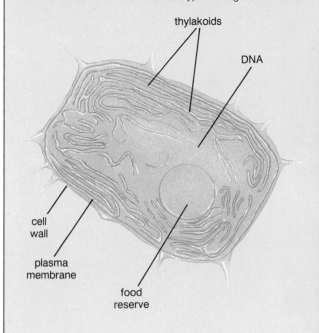

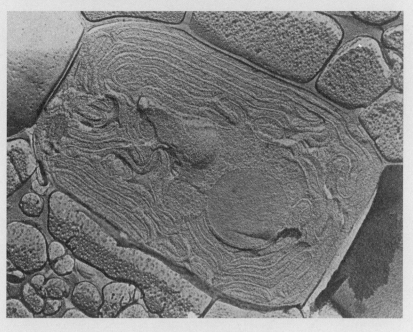

Structure

Prokaryotic cells lack a nucleus and do not have the cytoplasmic organelles found in eukaryotic cells, other than ribosomes (fig. 5.3). The eubacteria have a cell wall composed of the unique molecule peptidoglycan, which may be surrounded by a gelatinous capsule. Some bacteria can move by means of flagella, and some can adhere to surfaces by means of pili.

Bacteria occur in three basic shapes (fig. 23.6): *rod* (bacillus), *round* or spherical (coccus), and *spiral* (e.g., helical shape called a spirillum). The rods and bacilli may form chains of a length typical of the particular bacterium.

Prokaryotic cells lack a nucleus and most of the other organelles found in eukaryotes. They occur in three basic shapes.

Reproduction

Prokaryotes reproduce asexually by means of **binary fission,** a process depicted in figure 10.2. The single circular chromosome replicates, then the two copies separate as the cell enlarges. Newly formed plasma membrane and cell wall separate it into two cells. Mitosis, which requires the formation of a spindle apparatus, does not occur in prokaryotes.

As discussed in the reading on page 264, genetic recombination can occur among bacteria in three ways. *Conjugation* has been observed between bacteria when the so-called male cell passes DNA to the female by way of a sex pilus. *Transformation* occurs when a bacterium picks up from the medium DNA released by dead bacteria; during *transduction,* bacteriophages carry portions of bacterial DNA from one cell to another.

Since prokaryotes have no simple method for achieving genetic recombination as do plants and animals, *mutation* becomes a most important avenue for evolutionary change. With a generation time as short as 20 minutes under favorable conditions, mutations are generated and distributed throughout a population more quickly than in eukaryotes. Also prokaryotes are haploid and so mutations are immediately subjected to natural selection for any possible beneficial effect.

Prokaryotes reproduce asexually by binary fission. Mutations are the chief means of achieving genetic variation.

When faced with unfavorable environmental conditions, some bacteria can form **endospores.** During spore formation, the cell shrinks, rounds up within the former plasma membrane, and secretes a new thicker wall inside the old one (fig. 23.7). Endospores are amazingly resistant to extremes in temperature, drying, and harsh chemicals, including acids and bases. When conditions are again suitable for growth, the spore absorbs water, breaks out of the inner shell, and becomes a typical bacterial cell, reproducing once again by binary fission.

Figure 23.6

Scanning electron micrographs of bacteria. *a.* Spherical-shaped bacteria. *b.* Rod-shaped bacteria. *c.* Spiral-shaped bacteria with flagella used for locomotion. See figure 5.3 for generalized drawings of bacteria.

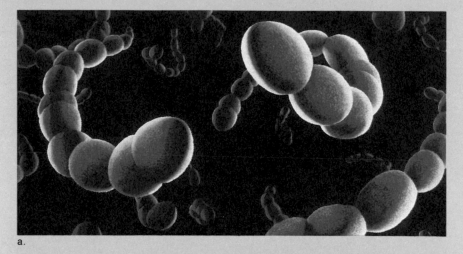

a.

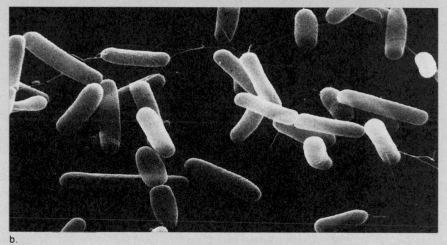

b.

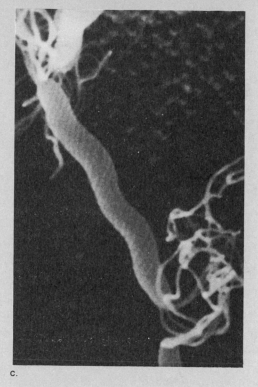

c.

Figure 23.7

Clostridium botulinum containing an endospore. This organism causes the type of food poisoning known as botulism. Sterilization, a process that kills all living things, even endospores, is used whenever all bacteria must be killed. Sterilization can be achieved by use of an autoclave, a container that maintains steam under pressure. Magnification, ×55,000.

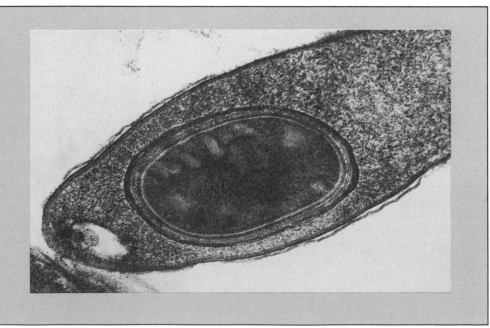

Figure 23.8

Giant tubeworms are a part of the community of organisms found near the deep ocean ridges and vents. These communities have been given such names as Dandelions, the Rose Garden, and the Garden of Eden.

Metabolism

Some bacteria are *obligate anaerobes* and are unable to grow in the presence of oxygen. A few serious illnesses, such as botulism, gas gangrene, and tetanus, are caused by anaerobic bacteria. Some other bacteria, called *facultative anaerobes* are able to grow in either the presence or the absence of oxygen. Most bacteria, however, are *aerobic* and, like animals, require a constant supply of oxygen to carry out cellular respiration.

Every type of nutrition is found among bacteria except for holozoism (eating whole food). There are both autotrophic and heterotrophic bacteria.

Autotrophic Bacteria

Some autotrophic bacteria are photosynthetic; they use light as a source of energy to produce their own food. These photosynthetic bacteria can be divided into two types—those that are primitive (evolved first) and those that are more advanced (evolved later):

Primitive	Advanced
Photosystem I only	Photosystems I and II
Do not give off O_2	Do give off O_2
Unique type of chlorophyll	Type of chlorophyll found in plants

The green sulfur and purple sulfur bacteria carry on the primitive type of photosynthesis. They do not give off oxygen because they do not use water as an electron donor; instead, they use hydrogen gas (H_2) and hydrogen sulfide (H_2S). These bacteria usually live in anaerobic conditions such as the muddy bottom of marshes and cannot photosynthesize in the presence of oxygen. The cyanobacteria (see figures 23.5 and 23.11) carry on the advanced type of photosynthesis in the same manner as plants (chapter 8).

Aside from the photosynthetic bacteria, the **chemosynthetic** bacteria are also autotrophs. Chemosynthetic bacteria oxidize inorganic compounds such as ammonia, nitrites, and sulfides, and trap the small amount of energy released from these oxidations to use in the reactions that synthesize carbohydrates. Just recently it's been discovered that chemosynthetic bacteria support entire communities a mile and a half below sea level where light never penetrates. At the mid-ocean ridge system, hot minerals spew out of the inner earth providing hydrogen sulfide to chemosynthetic bacteria. The organic molecules produced by the bacteria support the growth of giant tubeworms (fig. 23.8), clams, and crabs. There is evidence that such bacteria also reside within the bodies of the worms.

Nitrifying bacteria are also chemosynthesizers that oxidize ammonia (NH_3) to nitrites (NO_2^-) and nitrites to nitrates (NO_3^-). We will mention these important bacteria again when we discuss the nitrogen cycle.

Autotrophic bacteria include those that are photosynthetic and those that are chemosynthetic. Among the photosynthetic bacteria there are those that evolved first and do not give off oxygen and those (cyanobacteria) that evolved later and do give off oxygen.

Heterotrophic Bacteria

The majority of bacteria are aerobic heterotrophs and feed on dead organic matter by secreting digestive enzymes and absorbing the products of digestion. These so-called **saprophytic decomposers** can break down such a large variety of molecules that there probably is no natural organic molecule that cannot be digested by at least one bacterial species. The decomposing bacteria play a critical role in recycling matter in ecosystems

Table 23.1 Symbiotic Relationships

Name	Definition	Examples
Mutualism	Both organisms benefit	Bacteria in plant nodules Dinoflagellates in coral animals
Commensalism	One organism gains and the other is unharmed	Bacteria on human skin
Parasitism	One organism gains and the other is harmed	Bacterial, fungal, and protozoan infections

and making inorganic molecules available to photosynthesizers (fig. 1.5). The metabolic capabilities of various heterotrophic bacteria are exploited by human beings who use them to perform services ranging from the digestion of sewage and oil to the production of such products as alcohol, vitamins, and even antibiotics. Also, by means of gene splicing, bacteria can now be used to produce substances such as human insulin and growth hormone.

Heterotrophs may be free-living or **symbiotic** (table 23.1), forming mutualistic, commensalistic, or parasitic relationships. Certain of the **nitrogen-fixing** bacteria reduce atmospheric nitrogen (N_2) to ammonia (NH_3) and are *mutualistic;* they live in the root nodules of soybeans, clover, and alfalfa and provide organic nitrogen to their hosts (fig. 23.9). Plants are unable to fix atmospheric nitrogen so these bacteria perform a useful function for plant nutrition. The intestinal bacteria in humans release vitamins K and B_{12}, which we can use, and in the stomachs of cows and goats other bacteria digest cellulose, enabling these animals to feed on grass. *Commensalistic* bacteria live on organisms and cause them no harm, but *parasitic* bacteria are responsible for a wide variety of plant and animal diseases. Common human infections caused by bacteria include strep throat, diphtheria, typhoid fever, and gonorrhea.

The majority of bacteria are free-living heterotrophs (saprophytic decomposers) that contribute significantly to recycling matter through ecosystems. Many are also symbiotic heterotrophs, including those that cause disease.

Classification of Bacteria

Bacteria are a diverse group of organisms that vary widely in structure and metabolism and this diversity has made it more difficult to classify them according to evolutionary lines. Recently, however, comparisons of amino acid sequence in proteins and of base sequences in DNA and RNA have begun to give an indication of possible relationships. These metabolic studies indicate that prokaryotes split very early into two types: the archaebacteria and the eubacteria.

Figure 23.9
While there are some free-living bacteria that carry on nitrogen-fixation, those of the genus *Rhizobium* invade the roots of legumes with the resultant formation of nodules. Here the bacteria convert atmospheric nitrogen to a source that the plant can use. These are nodules on the roots of a soybean plant.

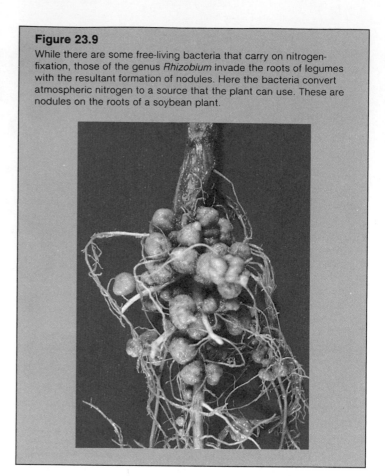

Archaebacteria

Most likely these organisms were the earliest of the prokaryotes. Their cell wall and plasma membrane do not have the same composition as do the same structures in the eubacteria. Some believe that they should even be placed in a different kingdom.

Many **archaebacteria** are able to live in the most extreme of environments, perhaps representing the kinds of habitats that were available when the earth first formed. The methanogens are anaerobic and live in swamps and marshes, producing methane, known as marsh gas. They also live in the guts of organisms, including humans. The halophiles live where it is salty, such as the Great Salt Lake in Utah. Curiously, a type of rhodopsin pigment (related to the one found in our own eyes) allows halophiles to carry on a primitive type of photophosphorylation for ATP production. The thermoacidophiles live where it is both hot and acidic. Those that live in the hot sulfur springs of Yellowstone National Park obtain energy by oxidizing sulfur.

Eubacteria

Most of our previous discussions pertain to the most common types of **eubacteria,** which can be divided in many different ways. One way to classify eubacteria is to consider their reaction to the **Gram stain.** After applying this stain, the gram-positive bacteria appear purple and the gram-negative bacteria appear red (fig. 23.10). This difference is dependent on the construction of the cell wall. The gram-positive cell has a single thick-wall

Figure 23.10
Photomicrograph of bacteria that have been subjected to Gram staining. The gram-positive bacteria are purple and the gram-negative are red. These bacteria are *Staphylococcus aureus* and *Escherichia coli*, respectively.

Figure 23.11
Diversity among the cyanobacteria. *a.* In *Chroacoccus* single cells are grouped together in a common gelatinous sheath. Magnification, ×30,115. *b.* Filaments of cells occur in *Oscillatoria*. Magnification, ×45,480.

a.

b.

layer while the gram-negative wall has several thinner layers. Some gram-negative bacteria such as the rickettsias, chlamydias, and mycoplasma are much smaller than the average size of bacteria. The rickettsias, which cause Rocky Mountain spotted fever and typhus, and the chlamydias, which cause a sexually transmitted urethritis, actually grow inside other cells. The mycoplasmas live in the body fluids of plants and animals but are unusual in that they don't have cell walls.

Among the photosynthetic bacteria, the cyanobacteria are of special importance; therefore, we will discuss them further here.

Cyanobacteria

Cyanobacteria (fig. 23.11), formerly called blue-green algae, are the most prevalent of the photosynthetic bacteria. The cyanobacteria carry on photosynthesis in a manner similar to that of plants; they possess chlorophyll *a* and evolve oxygen. They also have other pigments that can mask the color of chlorophyll, giving them, for example, not only blue-green but also red, yellow, brown, or black colors. We mentioned in chapter 22 that the cyanobacteria are believed to be responsible for first introducing oxygen into the primitive atmosphere.

Cyanobacteria may be unicellular, filamentous, or colonial. The filaments and colonies are not considered multicellular because each cell is independent of the others. Cyanobacteria lack any visible means of locomotion, although some glide when in contact with a solid surface and others oscillate (sway back and forth). Some cyanobacteria have a special advantage because they possess heterocysts, thick-walled cells without nuclei where nitrogen fixation occurs. The ability to photosynthesize and also fix atmospheric nitrogen (N_2) means that their nutritional requirements are minimal.

Cyanobacteria are common in fresh water, soil, and on moist surfaces but are also found in habitats devoid of higher forms of life, such as hot springs. They are symbiotic with a number of organisms such as liverworts, ferns, and even at times invertebrates like corals. In association with fungi, they form **lichens** that can grow on rocks. A lichen (fig. 23.12) is a symbiotic relationship in which the cyanobacterium provides organic nutrients to the fungus while the latter possibly protects and furnishes inorganic nutrients to its partner. It is also possible that the fungus is parasitic on the algal component. Lichens help transform rocks to soil; other forms of life then may follow. It is presumed that cyanobacteria were the first colonizers of land during the course of evolution. Once established, their mass could have provided a physical as well as chemical substrate for the ultimate growth of plants.

Cyanobacteria are ecologically important in still another way. If care is not taken in the disposal of industrial, agricultural, and human wastes, phosphates and nitrates drain into lakes and ponds, resulting in a "bloom" of these organisms. The surface of the water becomes turbid and light cannot penetrate to lower levels. When a portion of the cyanobacteria die off, the decomposing bacteria use up the available oxygen, causing fish to die from lack of oxygen.

Cyanobacteria are photosynthesizers that sometimes can also fix atmospheric nitrogen. They first introduced oxygen into the atmosphere and probably first colonized land. Even today, in association with fungi, they form lichens, which are important soil formers.

Figure 23.12

Lichens contain fungal cells and photosynthetic cells (either cyanobacterial cells or green algal cells) in close association. The photosynthesizer provides food but whether the fungus contributes to the relationship is debatable. This lichen, *Peltigera apathosa*, contains cyanobacterial cells and illustrates that lichens can superficially resemble low-lying plants. The cyanobacteria lie just beneath the surface surrounded by fungal hyphae.

Summary

1. All viruses have at least two parts: an outer coat (capsid) of protein and inner core of nucleic acid.

2. Viruses reproduce inside living cells and therefore they are obligate parasites. Since they are parasites they cause diseases in plants and animals.

3. Bacteriophages exhibit two primary life cycles—lytic and lysogenic.

4. During the lytic cycle the bacteriophage takes over the operation of the cell. The cycle consists of (1) attachment and release of DNA into the cell, (2) disintegration of host DNA, (3) viral DNA replication, (4) coat protein formation, (5) assemblage of virus, (6) production of lysozyme, and (7) release of new viruses from lysed host.

5. In the lysogenic cycle, viral DNA integrates into bacterial DNA for an indefinite period of time but can sometimes undergo the lytic cycle when stimulated.

6. Animal viruses have an outer membranous coat and enter cells by endocytosis. Uncoating occurs, releasing the viral nucleic acid; after replication, coat protein formation, and assembly, the viruses bud from the cell.

7. There are RNA animal viruses that never have a DNA stage. However, the RNA retroviruses possess an enzyme, reverse transcriptase, that causes the production of cDNA that becomes integrated into host DNA. These viruses sometimes carry cancer-causing oncogenes and the AIDS virus is a retrovirus.

8. Viroids, which have been known to cause disease in certain plants, consist of only a short chain of naked RNA. Prions consist only of a glycoprotein and yet they are thought to cause slow-virus nervous system diseases.

9. Kingdom Monera contains single-celled prokaryotes, i.e., the bacteria. Prokaryotic cells lack a nucleus and other cytoplasmic organelles found in eukaryotic cells. They do have a cell wall composed of peptidoglycan.

10. Bacteria have three basic shapes: rod (bacillus), round (coccus), and spiral (spirillum).

11. Bacteria reproduce asexually by binary fission and since they do not have a regular means of genetic recombination, their chief method for genetic variation is mutation.

12. Some bacteria form endospores, which are extremely resistant to destruction; their genetic material can thereby survive unfavorable conditions.

13. Bacteria differ in their tolerance and need for oxygen. There are obligate anaerobes, facultative anaerobes, and aerobic bacteria.

14. Some bacteria are autotrophic, being either photosynthetic or chemosynthetic. There are the primitive photosynthesizers that do not give off oxygen and the advanced photosynthesizers (cyanobacteria) that do give off oxygen. Chemosynthetic bacteria oxidize inorganic compounds like sulfides to acquire energy to make their own food. Surprisingly, these bacteria support communities at mid-oceanic ridges.

15. Most bacteria are aerobic heterotrophs, being saprophytic decomposers that are absolutely essential to the cycling of nutrients in ecosystems. Their metabolic capabilities are so vast that they are used by humans both to dispose of substances and to produce substances.
16. Many heterotrophic bacteria are symbiotic. For example, the mutualistic nitrogen-fixing bacteria live in nodules on the roots of legumes. Some symbiotes, however, are parasitic and cause plant and animal diseases.
17. Bacteria are divided into the archaebacteria and the eubacteria. The archaebacteria have such a different type of cell wall and plasma membrane from the other bacteria that they should perhaps be in a different kingdom. They live under harsh conditions—anaerobic marshes, salty lakes, hot sulfur springs.
18. Most bacteria are eubacteria and can be divided up in various ways. One way is to consider which are gram positive and which are gram negative.
19. Of special interest are the cyanobacteria, which were the first organisms to photosynthesize in the same manner as plants. They are sometimes symbiotic with fungi in lichens.

Objective Questions

1. Which of these are found in all viruses?
   a. membranous coat and capsule
   b. capsid and nucleic acid
   c. tail fibers and head
   d. All of these.
2. Which step in the lytic cycle follows attachment of virus and release of DNA into the cell?
   a. production of lysozyme
   b. disintegration of host DNA
   c. assemblage
   d. DNA replication
3. Which of these is a true statement?
   a. Viruses carry with them their own ribosomes for protein formation.
   b. New viral ribosomes form after viral DNA enters the cell.
   c. Viruses use the host ribosomes for their own ends.
   d. Viruses do not need ribosomes for protein formation.
4. During the lysogenic cycle, viral DNA
   a. is lysed and disappears.
   b. becomes incorporated into host DNA.
   c. replicates to give many copies.
   d. bursts from the cell.

5. RNA retroviruses have a special enzyme that
   a. disintegrates host DNA.
   b. polymerizes host DNA.
   c. transcribes viral RNA to cDNA.
   d. translates host DNA.
6. Viroids and prions
   a. have a more complicated structure than viruses.
   b. unlike viruses do not cause disease.
   c. are simpler than viruses and do cause disease.
   d. Both (a) and (b).
7. The organisms in kingdom Monera are
   a. prokaryotes and bacteria.
   b. either archaebacteria or eubacteria.
   c. cyanobacteria and viruses.
   d. Both (a) and (b).
8. Facultative anaerobes
   a. require a constant supply of oxygen.
   b. are killed in an oxygenated environment.
   c. do not always need oxygen.
   d. are photosynthetic.

9. Cyanobacteria, unlike other types of bacteria that photosynthesize,
   a. do not give off oxygen.
   b. do give off oxygen.
   c. do not have chlorophyll.
   d. do not have a cell wall.
10. Chemosynthetic bacteria
    a. are autotrophic.
    b. use the rays of the sun to acquire energy.
    c. oxidize inorganic compounds to acquire energy.
    d. Both (a) and (c).
11. Heterotrophic bacteria
    a. always cause disease.
    b. serve many useful functions.
    c. are usually saprophytic decomposers.
    d. Both (b) and (c).
12. Prokaryotic cells
    a. do not have a nucleus.
    b. evolved first in the history of the earth.
    c. reproduce by binary fission.
    d. All of these.

Study Questions

1. Describe the general anatomy of viruses and tell why they are obligate parasites?
2. Describe both the lytic and lysogenic cycles of bacteriophages.
3. How do animal viruses differ in structure and life cycle from bacteriophages?
4. What are the unique features of retroviruses?
5. How do viroids and prions differ from viruses?

6. What is the general anatomy of bacteria and how do they reproduce? What are the main two types?
7. How do bacteria achieve genetic recombination and why is the occurrence of mutations the more common source of variations?
8. How do some bacteria manage to survive the severest of unfavorable environments?
9. How do bacteria differ in their tolerance and need for oxygen?

10. What are the differences between the primitive type of photosynthesis and the advanced type of photosynthesis found in bacteria?
11. What are chemosynthetic bacteria and where have they been found to support whole communities?
12. Discuss the importance of heterotrophic bacteria and give examples of their symbiotic relationships with other types of organisms.
13. Discuss the importance of cyanobacteria in the history of the earth.

Thought Questions

1. Support the contention that viruses are "nothing but pieces of DNA that escaped from a cell."

2. If you were discussing the diversity of bacteria, which aspects of their biology would you stress and why?

3. What argument would you use to suggest that the archaebacteria evolved before the eubacteria?

Selected Key Terms

virus (vi′rus) 338
lytic cycle (lit′ik si′kl) 339
lysogenic cycle (li-so-jen′ik si′kl) 340
retrovirus (ret″ro-vi′rus) 340
viroid (vi′roid) 340

bacterium (bak-te′re-um) 342
endospore (en′do-spōr) 343
chemosynthetic (ke″mo-sin-thet′ik) 345
saprophytic decomposer (sap″ro-fit′ik de″kom-pōz′er) 345
symbiotic (sim″bi-ot′ik) 346

nitrogen-fixing (ni′tro-jen fiks-ing) 346
archaebacteria (ar″ke-bak-te′re-ah) 346
eubacteria (u″bak-te′re-ah) 346
cyanobacteria (si″ah-no-bak-te′re-ah) 347
lichen (li′ken) 348

CHAPTER 24

Protista and Fungi

Your study of this chapter will be complete when you can:

1. Describe the eukaryotic cell, and tell how and when it is believed to have evolved.
2. List and describe the types of organisms that are placed in the kingdom Protista.
3. Tell how the protozoans are grouped and describe a member of each group.
4. Describe the life cycle for Plasmodium vivax, *the cause of malaria.*
5. List the different types of algae in the kingdom Protista.
6. Describe the three types of life cycle found among living organisms. For each, name an alga that has this life cycle and describe specifically.
7. Tell how the green algae are grouped and give an example for each group. Contrast the manner in which these examples reproduce.
8. Describe the three types of seaweeds and contrast their characteristics. Tell how Fucus *reproduces.*
9. List, describe, and state the importance of euglenoids, dinoflagellates, and diatoms.
10. Describe the funguslike protista and tell why they are called funguslike.
11. Describe the anatomical features of organisms in kingdom Fungi and show the basis on which they are classified.
12. Describe the life cycle of a black bread mold in detail and the life cycle of a sac fungus and club fungus in general.

Fungi are quite a diverse group. This is the reproductive structure of a morel, which is a prized delicacy especially in Europe. Morels have to be grown out-of-doors, and no one has succeeded in developing a method for growing them indoors so that they can be harvested more than once a year.

The protists (fig. 24.1) and fungi are *eukaryotes*. The eukaryotic cell contains a nucleus and all the other organelles studied in chapter 5 and listed in table 5.1. The fossil record indicates that the eukaryotic cell evolved about 1.5 billion years ago. Evidence suggests that the eukaryotic cell acquired its organelles gradually. It may be that the nucleus and at least a primitive form of mitosis arose once there were several chromosomes, the DNA being too extensive to be contained within a single chromosome. As discussed on page 70, the mitochondria of the eukaryotic cell probably were once free-living aerobic prokaryotes and the chloroplasts probably were free-living photosynthetic prokaryotes. The theory of *endosymbiosis* says that the nucleated cell engulfed these prokaryotes, which then became organelles.

Only eukaryotes are ever multicellular organisms with specialized tissues. It is this feature that allows them to become the complex organisms that we will be studying in future chapters. We can also note that eukaryotes are generally aerobic organisms. It's interesting that just about the time that eukaryotes evolved and became complex, the amount of oxygen in the atmosphere was increasing to its present level and the ozone shield was being formed. Perhaps the origin of the eukaryotic cell and the rise of multicellularity occurred as a response to the environmental circumstance of increased oxygen in the atmosphere.

All the other kingdoms to be studied contain eukaryotes. The first eukaryotes were single aerobic cells that only later developed multicellularity.

Kingdom Protista

Unicellular organisms are predominant in **kingdom Protista** and even the multicellular forms lack the tissue differentiation that is seen in more complex organisms. This makes it reasonable to assume that the organisms placed in this kingdom evolved before those that are found in kingdom Fungi, Plants, and Animals. We will rest content with this basic observation; it is not our intent to explore the possible evolutionary relationships among the protists themselves nor to say for certain which of them may have given rise to fungi, plants, and animals.

The protists (fig. 24.1) have been grouped primarily according to their mode of nutrition into the following categories: (1) protozoans that are heterotrophic and ingest their food like animals, (2) algae that are photosynthetic like plants, and (3) slime molds and water molds that have some characteristics similar to the fungi.

Protists, even if only one cell, should not be considered simple organisms. Each single cell must be organized to cope with the environment so that it alone can carry out all the functions assumed by various tissues and organs in more complex organisms.

Figure 24.1

Dark-field micrographs (*see* page 59) illustrating protist diversity. *a.* Several paramecia. They are representative of the protista that are animal-like in that they ingest their food. Magnification, ×63. *b.* Volvox colonies with young escaping. These organisms are representative of the protista that are plantlike in that they carry on photosynthesis. This particular protist is colonial, being a loose association of cells that show some specialization. Magnification, ×120. *c.* A slime mold plasmodium moves slowly along the forest floor using dead organic matter as its food. In this way, slime molds are like fungi, which also feed on dead organic matter.

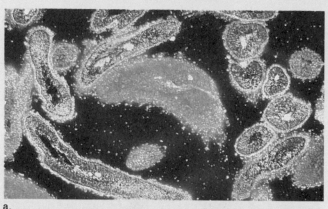

a.

b.

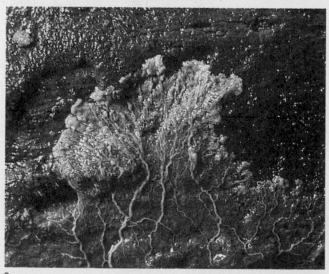

c.

The protista contain usually single-celled organisms that are grouped according to their mode of nutrition. There are the heterotrophic protozoans, the photosynthetic algae, and slime molds and water molds, which somewhat resemble fungi.

Protozoans

Protozoans are animal-like heterotrophic unicellular organisms that are characteristically motile. Some are colonial; a **colony** is a loose association of cells, each of which must still take care of its own physiological needs. On occasion, some cells of a colony specialize for reproduction.

In addition to the usual organelles, protozoans have specialized vacuoles. For example, they take in food by simple diffusion or transport or by means of phagocytosis, then digest the food in food vacuoles. Also, freshwater protozoans continually gain water by osmosis and eliminate it by means of "contractile" vacuoles. These vacuoles expand with water drawn from the cytoplasm and then appear to "contract," releasing the water through a temporary opening in the cell membrane.

Although asexual reproduction involving binary fission and mitosis is the rule, many protozoa also reproduce sexually during some part of their life cycle. Sexual exchange is preceded by meiosis.

The protozoans to be studied are placed in four groups according to their type of locomotor organelle (table 24.1).

Amoeboids

Typically, the **amoeboids** (phylum Sarcodina) move about and feed by means of cytoplasmic extensions called **pseudopodia.** In *Amoeba proteus* (fig. 24.2), a commonly studied member of this group, the pseudopodia form when the cytoplasm streams forward in particular directions. When *Amoeba* feed, they **phagocytize;** the pseudopodia surround and engulf the prey, which may be algae, bacteria, or other protozoans.

Amoeba proteus is found in fresh water, but most amoebas are marine, being either radiolarians or foraminiferans. The radiolarians float near the ocean surface despite an internal skeleton composed of silica or strontium sulfate. Their skeletons are intricate, exquisite, and of almost infinite variety. The foraminiferans have an internal skeleton usually made up of calcium carbonate (fig. 24.3). When they feed, pseudopodia extend

Table 24.1 Types of Protozoans

Protozoans	Locomotor Organelles	Example
Amoeboid	Pseudopodia	*Amoeba*
Ciliate	Cilia	*Paramecium*
Zooflagellate	Flagella	*Trypanosoma*
Sporozoa	No locomotor organelles	*Plasmodium*

Figure 24.2

Amoeba proteus is a protozoan that moves by formation of pseudopodia. Note also the unique organelles, including the food vacuoles and contractile vacuole. This artist's representation is based on electron micrographs of this organism. Arrows indicate the flow of cytoplasm as a newly formed pseudopodium takes shape.

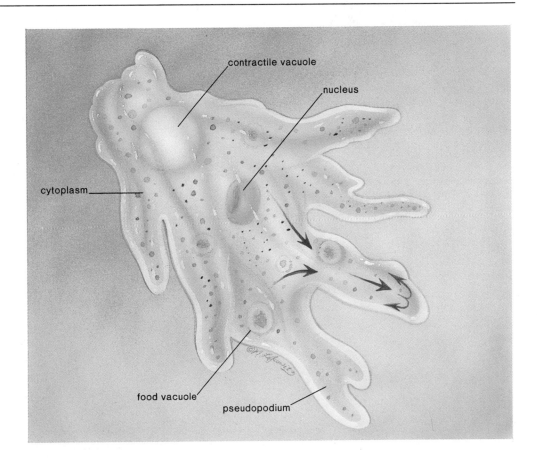

through holes called foramina. These protozoa are more often found in the ooze of the ocean floor. They have been so populous through the ages that following a land upheaval, their accumulated shells formed the White Cliffs of Dover along the southern coast of England.

Ciliates

The **ciliates** (phylum Ciliophora), such as those in the genus *Paramecium* (fig. 24.4), are the most complex of the proto-

zoans. Hundreds of cilia project through tiny holes in the outer covering, or pellicle. There is an array of organelles that perform special functions. For example, when a *Paramecium* feeds, food is swept down a gullet to the cell mouth below which food vacuoles form. Following digestion, the soluble nutrients are absorbed by the cytoplasm, and the indigestible residue is eliminated at the anal pore.

The diversity of ciliates is quite remarkable. The barrel-shaped didinia expand in order to consume paramecia much larger than themselves. Suctorians have an even more dramatic way of getting food. They rest quietly on stalks until a hapless victim blunders into them. Then they promptly paralyze it and use their tentacles like straws to suck it dry. *Stentor* might be the prettiest of the ciliates. It resembles a giant blue vase decorated with stripes when viewed under the microscope.

Ciliates have two types of nuclei; a large macronucleus and one or more small micronuclei. The macronucleus controls the normal metabolism of the cell, while the micronuclei are concerned with reproduction. Following meiotic division of the micronuclei, two ciliates may exchange these in a sexual process called conjugation (fig. 24.5).

Zooflagellates

Protozoans that move by means of flagella (phylum Zoomastigina) are called **zooflagellates** to distinguish them from unicellular algae that also have flagella. The flagella of these eukaryotic cells have the characteristic 9 + 2 microtubular structure discussed on page 71.

Many zooflagellates enter into symbiotic relationships (table 23.1). *Trichonympha collaris* lives in the gut of termites; it contains a bacterium that enzymatically converts the cellulose of wood to soluble carbohydrates easily digested by the

Figure 24.3

Shell of a foraminiferan, an amoeboid that has existed for more than 500 million years, time enough for hundreds of lineages to have evolved, diversified, and become extinct. In large shells like this one, each story of the shell is multichambered. Within each chamber lies not only cytoplasm but also symbiotic algae. The algae use the foram's waste products to make food that they share with their host. Without the algae, a foram could not grow so large, live as long, or produce as many offspring.

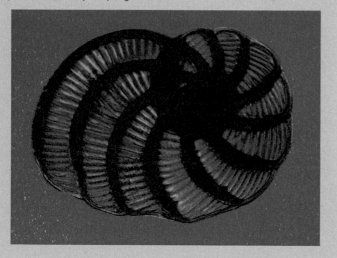

Figure 24.4

Diagram illustrating the principal structures found in *Paramecium caudatum*. Despite its complexity, a paramecium is a single-celled organism. Trichocysts are poisonous, threadlike darts that are used for defense and capturing prey. Arrows show path of food vacuoles from formation to discharge of nondigestible particles.

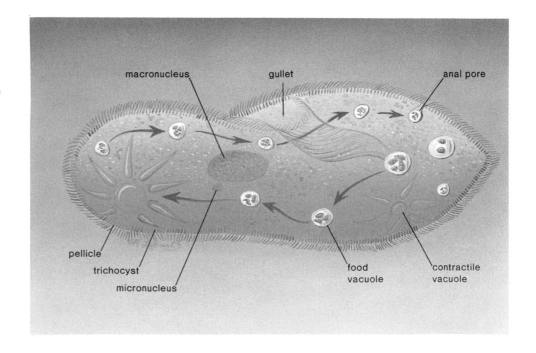

insect. The trypanosomes (fig. 24.6) cause African sleeping sickness and are transmitted to the blood of vertebrates by the tsetse fly. The tsetse fly, which becomes infected when it takes a blood meal from a diseased animal, passes on the disease when it feeds on another victim. The white blood cells in an infected animal accumulate around the blood vessels leading to the brain and cut off the circulation. The lethargy characteristic of the disease is caused by an inadequate supply of oxygen for normal brain alertness.

Sporozoa

The **sporozoans** (phylum Sporozoa) are also parasitic protozoans whose complicated life cycle almost always involves the formation of infective spores. The most widespread of the human parasites is *Plasmodium vivax,* the causative agent of one type of malaria. When a human is bitten by an infected *Anopheles* mosquito, the parasite eventually invades the blood cells. The chills and fever of malaria occur when the infected cells burst and release toxic substances into the blood (fig. 24.7).

Malaria is still a major killer of humans despite extensive efforts to control it. As discussed on page 322, resurgence of this disease is caused primarily by the development of insecticide-resistant strains of mosquitoes, and parasite resistance to anti-malarial drugs. Biotechnicians are presently trying to develop a vaccine utilizing a specific protein from the protozoan.

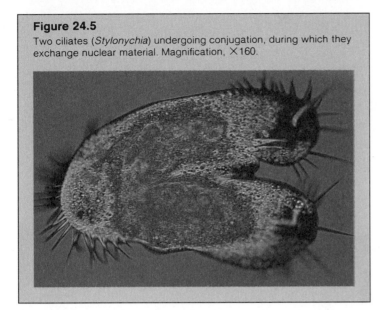

Figure 24.5
Two ciliates (*Stylonychia*) undergoing conjugation, during which they exchange nuclear material. Magnification, ×160.

Figure 24.6
Trypanosome infection. *a.* A stained blood smear from a patient suffering with African sleeping sickness, showing trypanosomes among the blood cells. Magnification, ×360. *b.* Structure of trypanosome as revealed by the electron microscope. The flexible pellicle allows undulating movement of the cell.

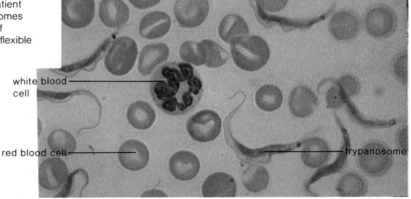

white blood cell

red blood cell — trypanosome

a.

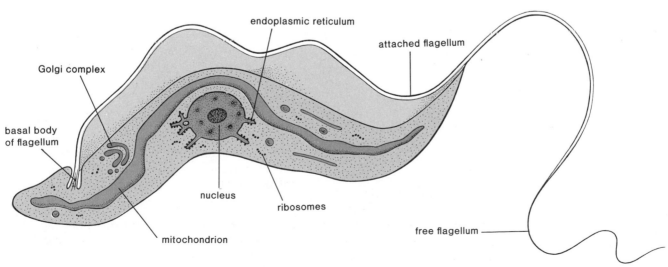

endoplasmic reticulum

attached flagellum

Golgi complex

basal body of flagellum

nucleus

ribosomes

mitochondrion

free flagellum

b.

Figure 24.7

Life cycle of *Plasmodium vivax,* a protozoan that causes one type of malaria. *Asexual reproduction in humans:* The bite of an infected *Anopheles* mosquito releases minute, spindle-shaped sporozoites that pass from the blood into the liver where asexual spores are produced. These invade and reproduce inside the red blood cells. Periodic chills and fever are caused by toxins that pour into the bloodstream when the spores are released. After a time, some of these spores become cells capable of developing into gametes if taken up by another mosquito. *Sexual reproduction in mosquito:* In the host stomach, the gametes mate and a wormlike zygote works its way through the stomach wall to encyst in the outer layer. Many divisions produce sporozoites that find their way to the salivary glands of the mosquito, and the cycle repeats.

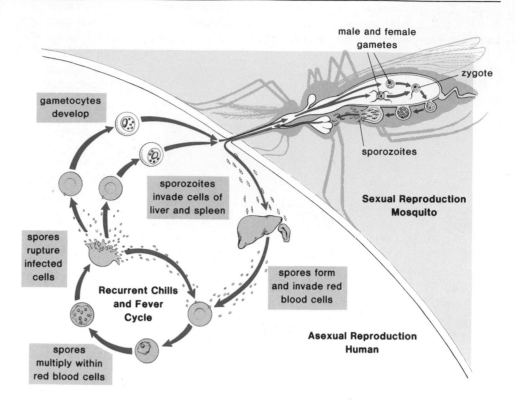

Protozoans are generally single cells that are heterotrophs and ingest food as animals do. They are grouped according to their mode of locomotion.

Algae

Algae is a term that is used for aquatic organisms that photosynthesize as do terrestrial plants. Algae produce the food that maintains communities of organisms both in the oceans and in bodies of fresh water. They are commonly named for the type of pigment they contain; therefore there are green, golden-brown, brown, and red algae. All algae contain green chlorophyll, but they may also contain other pigments that mask the color of the chlorophyll. Algae are grouped according to their pigmentation, and biochemical differences such as the chemistry of the cell wall and the chemical compound used to store excess food.

Life Cycles

There are three types of life cycles exhibited by living organisms and we find all these types among the algae: in the **haplontic cycle,** the adult is haploid; in **alternation of generations,** a haploid generation alternates with a diploid generation; in the **diplontic cycle** the adult is always diploid. These cycles are diagrammed in figure 24.8 and are further contrasted in table 24.2.

In the first two cycles, meiosis produces **spores,** structures that mature to a haploid generation. Plants, which are studied in the next chapter, use the alternation of generation life cycle.

Table 24.2 Life Cycles

Name	Chromosome Number in Adult(s)	Spores
Haplontic	Haploid only	Yes
Alternation of Generations	Haploid ⇄ Diploid	Yes
Diplontic	Diploid only	No

In the diplontic cycle, meiosis produces gametes, the only haploid structures found within the cycle. Animals, including human beings, follow the diplontic cycle.

In general, an organism exhibits one of three life cycles (haplontic, alternation of generations, or diplontic). All three of these are seen among the algae.

Green Algae

There are single-celled, colonial, filamentous, and multicellular green algae (phylum Chlorophyta). It is sometimes suggested that the green algae are ancestral to the first plants because both of these groups possess chlorophylls *a* and *b,* both store reserve food as starch, and both have cell walls that contain cellulose.

Flagellated Green Algae *Chlamydomonas* (fig. 24.9) is a single-celled green alga, which has been studied in detail by means of the electron microscope. It has a definite cell wall and a single, large, cup-shaped chloroplast that contains a pyrenoid

Figure 24.8

Organisms have one of three types of life cycles. *a. Haplontic* is typical of algae and fungi. Notice that the adult is haploid, the gametes are sometimes isogametes (look alike), and the only diploid part of the cycle is the zygote, which undergoes meiosis to produce spores. *b. Alternation of generations* is typical of plants. Notice that there are two generations, the sporophyte generation (2N) produces the spores by meiosis and the gametophyte generation (N) produces gametes. *c. Diplontic* is typical of animals. Notice that the adult is always diploid and meiosis produces heterogametes (egg and sperm). The earliest cycle may have been the haplontic, which could have led to both alternation of generations and the diplontic cycle.

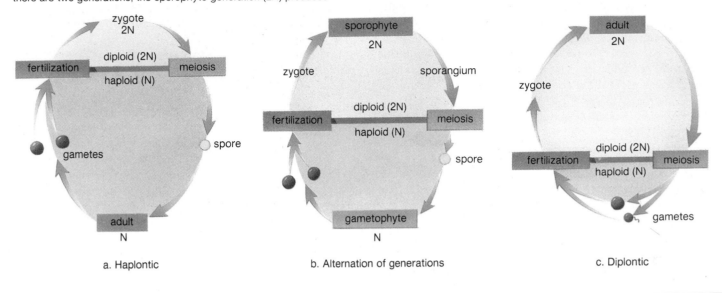

a. Haplontic

b. Alternation of generations

c. Diplontic

Figure 24.9

a. The structure of *Chlamydomonas*, a motile green alga.
b. Chlamydomonas has the haplontic life cycle. During sexual reproduction, mature adults produce gametes that fuse to give a zygote. The zygote becomes a thick walled zygospore that enters a period of dormancy. Upon germination, meiosis produces zoospores that grow to become haploid adults. These individuals usually reproduce asexually, with each of the original zoospores giving rise to many individuals.

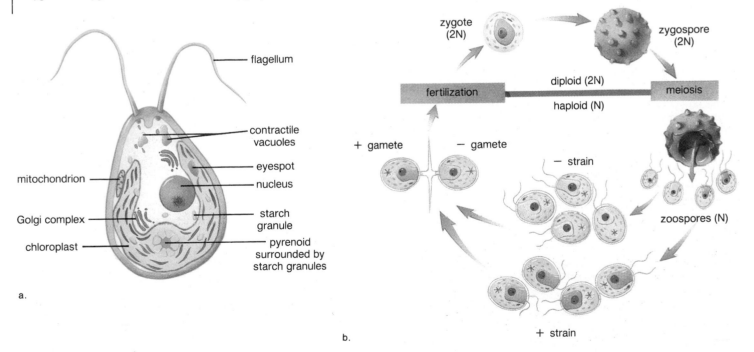

where starch is stored. A red-pigmented eye spot is sensitive to light and helps bring the organism into the light where photosynthesis can occur. The two flagella project from the anterior end and move the cell freely toward the light. *Chlamydomonas* has both animal- and plantlike characteristics in that it is motile and yet makes its own food; it is believed that such an organism may be a good example of what the most primitive protist was like.

Chlamydomonas is a good example of an organism that follows the haplontic life cycle (fig. 24.8). Usually this protist practices asexual reproduction and the adult divides to give *zoospores* (flagellated spores) that resemble the parent cell. During sexual reproduction, gametes of two different strains come into contact and join to form a zygote. A heavy wall forms around the zygote, and it becomes a resistant zygospore able to survive until conditions are favorable, at which time it germinates and produces four zoospores by meiosis. The gametes shown in figure 24.9 are **isogametes:** that is, they look exactly alike.

The life cycle of *Chlamydomonas* illustrates the primary differences between asexual and sexual reproduction (table 24.3). Sexual reproduction is simply reproduction that involves the use of gametes. Distinct and separate sexes are not required and heterogametes (dissimilar gametes, such as egg and sperm) are not required. Sexual reproduction aids the process of evolution because it offers an additional means to produce variations in addition to mutations.

Colonial Green Algae A number of **colonial** (loose association of cells) forms occur among the flagellated green algae. A *Volvox* colony is a hollow sphere with thousands of cells arranged in a single layer surrounding a watery interior. The cells of a *Volvox* colony, each one of which resembles a *Chlamydomonas* cell, cooperate in that the flagella beat in a coordinated fashion. Some cells are specialized for reproduction, and each of these can divide asexually to form a new daughter colony (fig. 24.10*a*). This daughter colony resides for a time within the parental colony. A daughter colony leaves the parental colony by releasing an enzyme that dissolves away a portion of the matrix of the parental colony, allowing it to escape. During sexual reproduction there are **heterogametes**—large nonmotile eggs and small flagellated sperm (fig. 24.10*b*).

Filamentous Green Algae Filaments are end-to-end chains of cells that form after cell division occurs in only one plane.

Spirogyra (fig. 24.11), found in green masses on the surfaces of ponds and streams, has chloroplasts that are ribbonlike and are arranged in a spiral within the cell. These species practice a form of sexual reproduction called **conjugation,** during which the contents of the cells act as isogametes. The two filaments line up next to one another and the contents of the cells of one filament move into the cells of the other filament, forming 2N zygotes. These zygotes survive the winter and in the spring undergo meiosis to produce new haploid filaments.

Among the filamentous algae, there are forms that use heterogametes during sexual reproduction. For example, the genus *Oedogonium* contains filamentous algae in which the cells are cylindrical with netlike chloroplasts. Sexual reproduction (fig. 24.12) involves the production of heterogametes (sperm and egg cells).

Table 24.3 Asexual versus Sexual Reproduction

	Asexual	Sexual
Number of parents	One	Two
Gametes	None	Gametes
Recombination of genes	Does not occur	Does occur

Figure 24.10

The genus *Volvox* contains colonial green algae. *a.* The adult often contains daughter colonies, which are asexually produced by special cells. Magnification, ×30. *b.* During sexual reproduction, colonies produce a definite sperm and eggs. Some produce either sperm or eggs, and others produce both sperm and eggs.

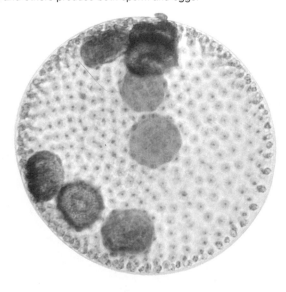

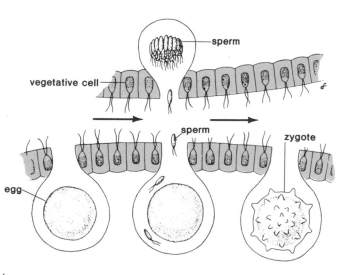

a.

b.

Multicellular Sheets Multicellular *Ulva* is commonly called sea lettuce because of a leafy appearance (fig. 24.13). *Ulva* shows alternation of generation in the same way as terrestrial plants except that (1) the spores are flagellated, (2) there are isogametes, and (3) both generations look exactly alike. In plants, one generation is typically dominant over (longer lasting than) the other.

Figure 24.11

Members of the genus *Spirogyra* are filaments of cells. *a.* Anatomy of one cell. *b.* Micrograph depicting conjugation between cells of two filaments. Magnification, ×50.

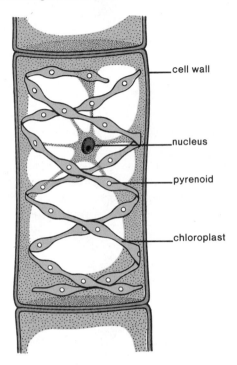

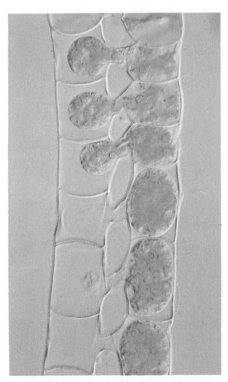

Green algae are a diverse group that have some of the same characteristics as plants. All three types of life cycles (fig. 24.8) are seen but *Ulva* has a life cycle that has two distinct generations like that of plants.

Seaweeds

Multicellular green algae (such as *Ulva*), red algae, and brown algae are all seaweeds. Although all seaweeds have chlorophyll, this green pigment is sometimes masked by red or brown pigments. While at one time it was believed that these accessory pigments determined the depth at which these algae are found, systematic studies have shown that there is no difference in the overall depth distribution of red, brown, and green seaweeds.

Although the seaweeds are multicellular, as are plants, this does not mean that they are closely related to plants. Multicellularity most likely arose independently in several different lines.

Figure 24.12

Members of the genus *Oedogonium* are filaments that reproduce both asexually and sexually. During sexual reproduction depicted here, the motile sperm swim to the stationary eggs. The zygote is the only diploid portion of the cycle; all the other structures are haploid.

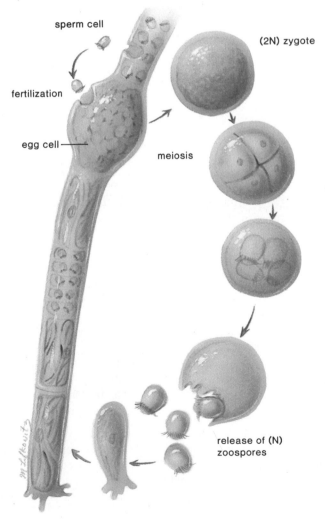

Figure 24.13

a. Ulva, a green alga. *b.* Members of the genus *Ulva* undergo alternation of generations in which the sporophyte generation and the gametophyte generation have the same appearance. The sporophyte generation produces spores by meiosis and the gametophyte generation produces gametes that upon fertilization give a zygote.

a.

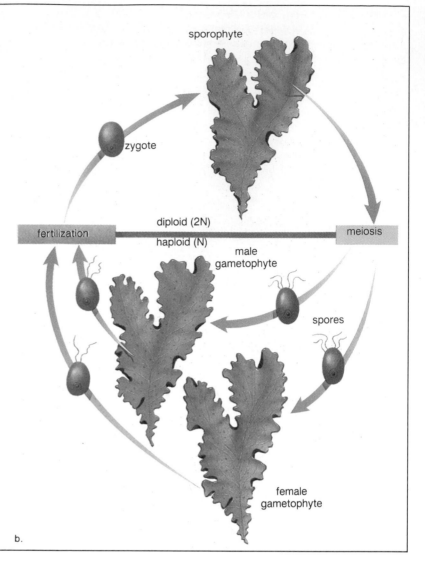

b.

Brown Algae   Brown algae (phylum Phaeophyta) range from small plants with simple filaments to large plants between 50 and 100 meters long (fig. 24.14). Large brown algae are often observed along the rocky shoreline in the north temperate zone where they are pounded by waves as the tide comes in and are exposed to drying at low tide. These plants are firmly anchored by holdfasts, and when the tide is in, their broad flattened blades are buoyed by air bladders. When the tide is out, they do not dry out because their cell walls contain a mucilaginous, water-retaining material.

Most brown algae have the alternation of generation life cycle but some species of *Fucus* are unique in that they have the diplontic cycle (fig. 24.8) in which meiosis produces gametes and the adult is always diploid, as in animals.

Red Algae   Like the brown algae, the red algae (phylum Rhodophyta) are multicellular, but they occur chiefly in warmer seawaters, growing both in shallow waters and as deep as light penetrates. Some forms of red algae are filamentous, but more often they are complexly branched (fig. 24.15) with the branches

having a feathery, flat, and expanded or ribbonlike appearance. Coralline algae are red algae that have cell walls impregnated with calcium carbonate. In some instances, they contribute as much to the growth of coral reefs as coral animals do.

The seaweeds are multicellular protista that are grouped according to the type of accessory pigment they contain.

Other Types of Algae

There are several groups of exclusively unicellular algae in kingdom Protista. They also contain chloroplasts and carry out photosynthesis in a manner similar to plants.

Euglenoids   Euglenoids (phylum Euglenophyta) are freshwater organisms having both animal-like and plantlike characteristics (fig. 24.16). Their animal-like characteristics include motility and a flexible body wall. Euglenoids move by means of flagellae, typically having one much longer than another, that project to the anterior. Because they are bounded by a flexible pellicle instead of a rigid cell wall, they can assume different shapes as the underlying cytoplasm undulates or contracts.

Figure 24.14

Diversification among the brown algae. Both *Laminaria,* a type of kelp, and *Fucus,* known as "rockweed" are examples of brown algae that grow along the shoreline. In deeper waters, giant kelps such as *Nereocystis* and *Macrocystis* often form spectacular underwater forests. Individuals of *Sargassum natans* sometimes break off from their holdfasts and form floating masses where life forms congregate in the ocean. Brown algae provide food and habitat for marine organisms and have even been harvested for human food and sources of fertilizer in several parts of the world. They are also a source of algin, a pectinlike material that is added to ice cream, sherbet, cream cheese, and other products to give them a stable, smooth consistency.

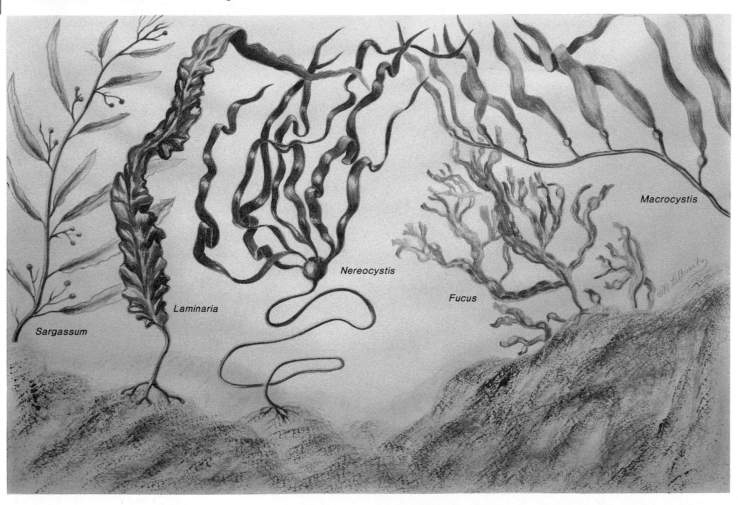

Figure 24.15

Representative red alga (*Sebdenia polydacrla*). Red algae are smaller and more delicate than brown algae. They are economically important because colloidal substances extracted from the outer layer of their cell walls are used in foods to produce a smooth consistency and help retain moisture. One colloid, agar, is used in culture media for bacteria and fungi.

Figure 24.16

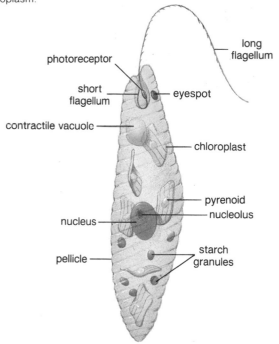

Euglena anatomy. *Euglena* is typical of those protozoans that have both animal-like and plantlike characteristics. A very long flagellum propels the body, which is enveloped by a flexible pellicle made of protein. A photoreceptor allows *Euglena* to find light after which photosynthesis can occur in the numerous chloroplasts. In addition to the pyrenoids, which store starch, there are starch granules in the cytoplasm.

photoreceptor
long flagellum
short flagellum
eyespot
contractile vacuole
chloroplast
pyrenoid
nucleus
nucleolus
pellicle
starch granules

Figure 24.17

Scanning electron micrograph of a dinoflagellate envelope. The cell can be either naked or covered by a cellulose envelope, sometimes divided into plates like this. There are two flagella, one circles the cell within a transverse groove, while the other is free and extends posteriorly.

Their plantlike characteristics include the possession of chloroplasts. A photoreceptor at the base of one flagellum, shaded by an eyespot, allows euglenoids to judge the direction of light. After they move toward light, photosynthesis takes place. Carbohydrates are stored as starch.

Dinoflagellates　Dinoflagellates (phylum Pyrrophyta) (fig. 24.17) have two flagella, one is free but the other is located in a transverse groove that encircles the animal. The beating of these flagella causes the organism to spin like a top. The cell wall, when present, is frequently divided into polygonal plates of cellulose closely joined together. At times there are so many of these organisms in the ocean that they cause a condition called "red tide." The toxins given off in these red tides cause widespread fish kills and can cause a paralysis in humans who eat shellfish that have fed on the dinoflagellates.

Usually the dinoflagellates are an important source of food for small animals in the ocean. They also live as symbiotes within the bodies of some invertebrates. For example, corals usually contain large numbers of these organisms and this allows them to grow much faster than otherwise.

Diatoms　Diatoms (phylum Chrysophyta) (fig. 24.18) have a golden-brown accessory pigment in their chloroplasts that can mask the color of chlorophyll. The structure of a diatom is often compared to a box because the cell wall has two halves, or valves, with the larger valve acting as a "lid" for the smaller valve. When diatoms reproduce, each receives only one old valve. The new valve fits inside the old one.

The cell wall has an outer layer of silica, a common ingredient of glass. The valves are covered with a great variety of striations and markings that form beautiful patterns when observed under the microscope. Diatoms are the most numerous of all unicellular algae in the oceans. As such, they serve as an important source of food for other organisms. Their remains, called diatomaceous earth, accumulate on the ocean floor and are mined for use as filtering agents, soundproofing materials, and scouring powders.

Slime Molds and Water Molds

These two types of protista have funguslike characteristics. They live on dead organic matter and produce windblown spores during one phase of their life cycle.

Slime Molds

There are two types of **slime molds** (phylum Myxomycota): cellular and acellular (fig. 24.1c). They both have an amoeboid stage that lives on rotting logs or dead agricultural crops. In cellular slime molds the total mass is composed of individual amoeboid cells, but in acellular ones it is multinucleated and called a **plasmodium.** In another part of their life cycle, slime molds form fruiting bodies, stalks that produce windblown spores. The spores germinate to give cells that join to form a zygote that will eventually begin the cycle again.

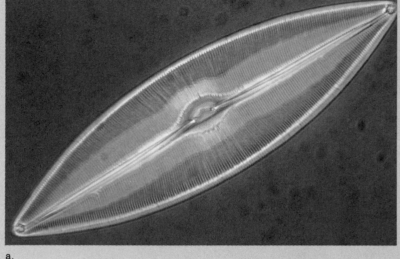

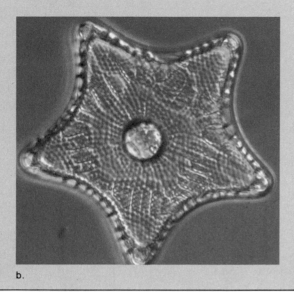

a.

b.

Water Molds

Some **water molds** (phylum Oomycota) live in the water where they parasitize fish, forming furry growths on their gills. In spite of their common name, others live on land and parasitize insects and plants. A water mold was responsible for the potato famine in the 1840s that caused many Irish to come to the United States. Most water molds are saprophytic and live off dead organic matter, however.

Water molds have a threadlike body like that of fungi but the cells are diploid and not haploid and their sexual life cycle is diplontic (fig. 24.8). Also, the cell walls are largely composed of cellulose quite unlike the fungi.

Kingdom Fungi

Fungi, like bacteria, are most commonly saprophytic decomposers that assist in the recycling of nutrients in ecosystems. Some fungi are parasitic, causing serious diseases in plants and animals. The bodies of all fungi, except unicellular yeast, are made up of filaments called **hyphae.** A hypha is an elongated cylinder, containing a mass of cytoplasm and many haploid nuclei, which may or may not be separated by cross walls. A collection of hyphae is called a **mycelium.**

Fungi reproduce in accordance with the haplontic life cycle (fig. 24.8) and most are adapted to life on land in that they produce windblown spores. Also most fungi reproduce both asexually and sexually; classification is largely based on the mode of sexual reproduction.

Fungi are saprophytic heterotrophic eukaryotes composed of hyphal filaments (a mycelium). Fungi produce spores during both sexual and asexual reproduction and the major groups of fungi are distinguishable on the basis of sexual reproduction.

Black Bread Molds

Black bread molds (division Zygomycota) belonging to the genus *Rhizopus* are often used as an example of this group. These molds exist as a whitish or grayish haploid mycelium on bread or fruit (fig. 24.19). During asexual reproduction, some hyphae grow upright and bear a spherical *sporangium* within which thousands of spores are formed.

During sexual reproduction, hyphae of two different strains (usually called plus and minus) reach out to one another and form a diploid zygote that darkens as it enlarges into a *zygospore*. After remaining dormant for several months, the zygospore germinates. Now the nucleus of the zygospore undergoes meiosis to produce a short haploid hypha, which immediately forms a sporangium with the subsequent release of windblown spores.

Sac Fungi

There are many different types of sac fungi (division Ascomycota) (fig. 24.20), most of which produce asexual spores, called **conidia.** During sexual reproduction, sac fungi form spores called

Figure 24.19

Life cycle of the black bread mold *Rhizopus*. (*Lower left*) asexually, a mycelium gives rise to spore-producing sporangia. (*Above*) sexually, the tip ends of hyphae from opposite mating strains can fuse, giving a zygote, the only diploid portion of the cycle. After a period of dormancy, meiosis is followed by germination and production of a sporangium. Windblown spores, an adaptation to land, produce mycelia.

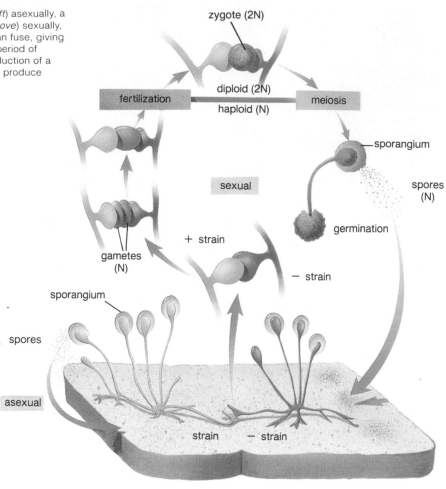

Figure 24.20

Sac and club fungi. *a*. Colorful cup fungi (ascomycetes). *b*. Bracket fungi (basidiomycetes) growing on a tree limb. *c*. Mushrooms (basidiomycetes) of the genus *Mycena* have bell-shaped caps.

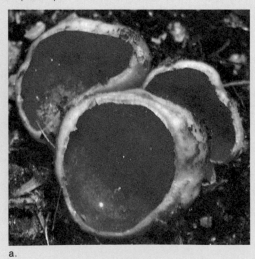

a.

b.

c.

ascospores within saclike cells called asci. In most species, the asci are supported within **fruiting bodies,** a collection of specialized hyphae. In cup fungi, the fruiting body takes the shape of a cup.

Yeasts are sac fungi that do not form fruiting bodies. In fact, yeasts are different from all other fungi in that they are unicellular and most often reproduce asexually by budding. Yeasts, as you know, carry out fermentation as follows:

$$glucose \longrightarrow carbon\ dioxide + alcohol$$

In baking, the carbon dioxide from this reaction makes bread rise. In the production of wines and beers, it is the alcohol that is desired, and the carbon dioxide escapes into the air.

Blue-green molds, notably *Penicillium,* are also sac fungi. This mold grows on many different organic substances, such as bread, fabrics, leather, and wood. It is used by humans to provide the characteristic flavor of Camembert and Roquefort cheeses; more important, it produces the antibiotic penicillin. Another mold, the red bread mold *Neurospora,* was used in the experiments that helped decipher the function of genes.

Unfortunately, sac fungi are also the cause of chestnut tree blight and Dutch elm disease, resulting in the death of most of these trees in the United States. Also, powdery mildew, apple scab, and ergot, a disease of cultivated cereals are caused by sac fungi.

The fungus portion of lichens (fig. 24.21) is usually a sac fungus, while the algal part of this symbiosis is usually a green alga. Lichens can live on bare rock or in poor soil and are able to survive great extremes of heat, cold, and dryness in all regions of the world. Reindeer moss is a lichen that is an important food source for arctic animals.

Club Fungi

Among the club fungi (division Basidiomycota), (fig. 24.20*b* and *c*), asexual reproduction is accomplished by formation of conidia. As a result of sexual reproduction, members of this group form club-shaped structures called *basidia,* often within fruiting bodies. The mushroom, puffball, and bracket (shelf) fungi are club fungi; the visible portions of these are actually fruiting bodies, and the mycelia lie beneath the surface. On the underside of a mushroom cap, the basidia project from the gills. Within each basidium, a diploid nucleus undergoes meiosis to produce basidiospores, which are windblown (fig. 24.22).

Club fungi are economically important. Mushrooms are commercially raised and sold as a delicacy. Rusts and smuts are parasitic club fungi that attack grains, resulting in great economic loss and necessitating expensive control measures. They do not have conspicuous fruiting bodies, and generally occur as vegetative hyphae that produce spores of various kinds. On the other hand, the mycelia of club fungi that lie beneath the soil

Figure 24.21
Lichens. Lichens can be either low lying or leafy or branchlike. All types are a symbiotic relationship between algal cells and fungal filaments.

often form beneficial symbiotic relationships with plants, notably pine trees. These *mycorrhizae* (fungus roots) help the trees garner nutrients from the soil and grow at a faster rate. Therefore, foresters make sure the tree roots are exposed to the fungi before they are planted (fig. 24.23).

Other Types of Fungi

Some fungi (division Deuteromycota) cannot be assigned to a definite group because the sexual portion of the life cycle has not been observed. For this reason these fungi are sometimes called "imperfect fungi." The fungi that cause ringworm and athletes foot belong to this group, as does the yeast *Candida albicans,* which causes thrush, a mouth infection, and moniliasis, a fairly common vaginal infection in females who take the birth control pill.

During sexual reproduction, the black bread molds produce spores in sporangia; the sac fungi produce spores in saclike cells; and the club fungi produce spores in club-shaped structures. The sac and club fungi typically have fruiting bodies. The sexual life cycles for certain fungi that parasitize humans are unknown.

Figure 24.22

Life cycle of a mushroom. Beginning at lower right, hyphae from two different strains fuse and produce a mycelium in which there are haploid nuclei from both strains. The mycelium gives rise to a mushroom consisting of a stalk and cap. The gills on the underside of the cap are lined with basidia (club-shaped structures). After nuclei fuse to give a diploid nucleus, meiosis in each basidium produces basidiospores, which are windblown. If conditions are favorable, each basidiospore germinates into hyphae. (*Upper right*) scanning electron micrograph of a basidium and basidiospores.

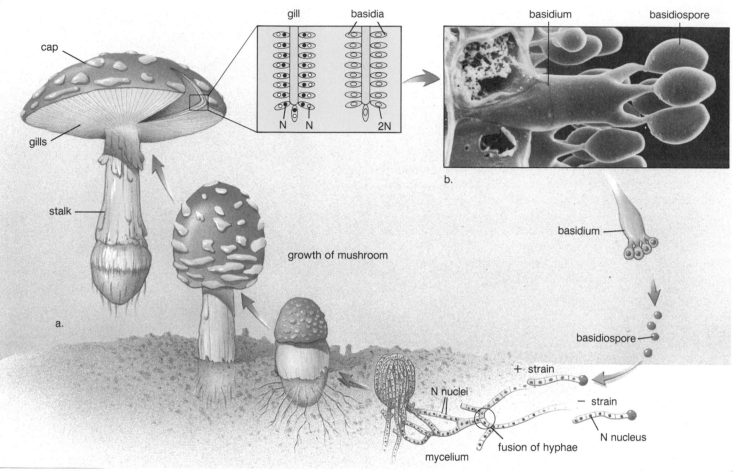

Figure 24.23

Mycorrhizae (fungus roots). *a.* One-year-old loblolly pine seedlings with the normally low levels of naturally occurring mycorrhizae. *b.* One-year-old loblolly seedlings after being inoculated with mycorrhizae. Note the denser root systems in (*b*).

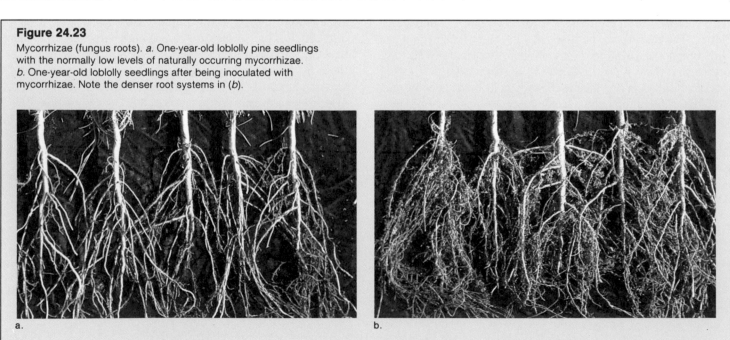

Summary

1. Protists and fungi are eukaryotes. The eukaryotic cell has a nucleus and other organelles, divides mitotically, and apparently led to multicellularity.

2. Kingdom Protista contains (1) protozoans which, like multicellular animals, are heterotrophic and ingest food, (2) algae, which carry on photosynthesis in the same manner as plants, and (3) slime molds and water molds that have some of the characteristics of fungi.

3. The protozoan cell contains vacuoles that are specialized to carry on functions like digestion of food and storage of excess water (contractile vacuole).

4. The protozoans are grouped according to the type of locomotor organelle: the amoeboids move and feed by forming pseudopodia; the ciliates move by means of cilia; the zooflagellates have flagella; the sporozoans have no locomotor organ and are parasitic (e.g., malaria).

5. Algae are aquatic organisms that photosynthesize as do higher plants. They are often named for the types of pigment they contain. The three different types of life cycles (haplontic, alternation of generations, and diplontic, fig. 24.8) are found among the algae.

6. Green algae possess chlorophyll *a* and *b*, store reserve food as starch, and have cell walls of cellulose as do higher plants. Green algae are divided into those that are flagellated (e.g., *Chlamydomonas*, which is haplontic with isogametes and *Volvox*, which is colonial); filamentous (e.g., *Spirogyra*, which undergoes conjugation and *Oedogonium*, which is haplontic with heterogametes); and multicellular sheets (e.g., *Ulva*, which exhibits alternation of generations).

7. Some algae are seaweeds. These include the green alga *Ulva*, the brown algae such as rockweed and kelps, and red algae, which are more delicate than the brown algae. It is of interest that the brown alga *Fucus* undergoes the diplontic cycle.

8. There are three types of algae that are exclusively unicellular: euglenoids, which have both plant- and animal-like characteristics; dinoflagellates; and diatoms, which are golden-brown algae and are important to food chains in the ocean.

9. The slime molds and water molds are protista with funguslike characteristics. Slime molds, both cellular and acellular, have an amoeboid stage and then form fruiting bodies that produce spores. Although most water molds with thread-like bodies are saprophytic, they are better known for parasitizing fish in aquatic habitats and parasitizing plants, most notably causing the Irish potato famine of the 1840s.

10. Kingdom Fungi contains organisms that are usually saprophytic decomposers but also cause diseases in plants and animals. The body is made up of hyphae, collectively called a mycelium. Fungi are classified according to specific features of the sexual life cycle.

11. The black bread molds exemplify the haplontic cycle. The spores are windblown, an adaptation to land. As a part of their sexual life cycle, sac and club fungi form fruiting bodies, conspicuous structures where windblown spores are produced by meiosis. There are other fungi, sometimes termed imperfect fungi because the sexual life cycle is not known, that cause some well-known diseases in humans.

Objective Questions

1. Which of these organisms is not classified as a protist?
   a. protozoans
   b. algae
   c. slime molds
   d. fungi

2. Protozoans are principally grouped according to mode of
   a. food ingestion.
   b. sexual reproduction.
   c. locomotion.
   d. type of chloroplast.

3. Sexual reproduction of the malarial parasite occurs
   a. in the mosquito.
   b. in humans.
   c. in red blood cells.
   d. Both (b) and (c).

4. Which of these is not a green alga?
   a. *Volvox*
   b. *Fucus*
   c. *Spirogyra*
   d. *Oedogonium*

5. *Chlamydomonas* forms zoospores
   a. during the asexual life cycle.
   b. during the sexual life cycle.
   c. in a fruiting body.
   d. Both (a) and (b).

6. Which of these is the most sturdy of the seaweeds?
   a. green
   b. red
   c. brown
   d. golden-brown

7. Euglenoids
   a. have both plant- and animal-like characteristics.
   b. carry on photosynthesis.
   c. have a flexible body wall.
   d. All of these.

8. Fungi have a body that is
   a. composed of hyphae.
   b. colonial.
   c. composed of tissues.
   d. flagellated.

9. Which of these is mismatched?
   a. black bread mold—zoospores
   b. sac fungi—fruiting body
   c. club fungi—conidia
   d. slime mold—sporangia

10. Yeasts are
    a. tiny protozoans.
    b. colorless algae.
    c. a type of water mold.
    d. a type of fungus.

Study Questions

1. Describe the eukaryotic cell and tell how it may have evolved.
2. What are the three types of protista and what are their distinguishing characteristics?
3. Describe and diagram the protozoan cell.
4. In what manner are the protozoans grouped? Give an example of each group.
5. Describe the life cycle of *Plasmodium vivax*, the causative agent of malaria.
6. What are the three types of life cycles found among living things? How do they differ? Show that *Chlamydomonas* and black bread mold have the haplontic cycle; that *Ulva* undergoes alternation of generations. What type of cycle does *Fucus* have?
7. Contrast how *Volvox, Spirogyra,* and *Oedogonium* reproduce, mentioning anything that is particularly distinctive.
8. List the distinguishing features of the euglenoids, dinoflagellates, and diatoms, mentioning any anatomical and ecological aspects that are of importance.
9. Why are slime molds and water molds sometimes called the funguslike protista?
10. Describe the anatomical features of the fungi and tell how fungi are classified.
11. Describe the life cycle of the black bread mold.
12. Tell how sac and club fungi reproduce sexually.

Thought Questions

1. Speculate on the benefits of following the diplontic life cycle as compared to the other two life cycles.
2. Prokaryotes and eukaryotes are both diverse groups of organisms. Contrast the type of diversity seen in each group.
3. Speculate on the benefits of being multicellular as opposed to being unicellular.

Selected Key Terms

protozoan (pro-ti-zo'en) 353
amoeboid (ah-me'boid) 353
haplontic cycle (hap-lon'tik si'kl) 356
alternation of generations (awl''ter-na'shun uv jen''ĕ-ra'shunz) 356

diplontic cycle (dip-lon'tik si'kl) 356
spore (spōr) 356
colonial (ko-lo'ne-al) 358
heterogamete (het''er-o-gam'ēt) 358
conjugation (kon''ju-ga'shun) 358
euglenoid (u-gle'noid) 360

dinoflagellate (di-no-flaj'ĕ-lāt) 362
diatom (di'ah-tom) 362
fungus (fun'gus) 363
hypha (hi'fa) 363
mycelium (mi-se'le-um) 363
fruiting body (frōōt'ing bod'e) 365

Plant Diversity

Your study of this chapter will be complete when you can:

1. List the characteristics shared by all plants.

2. Describe the bryophytes, give the most significant stages of the moss life cycle, and tell why you would expect to find bryophytes in moist locations.

3. Describe the structure and the most likely life cycle for the rhyniophytes, the first vascular plants.

4. List the nonseed vascular plants alive today, give the life cycle of the fern, and tell why you would expect to find these plants in moist locations.

5. Describe the life cycle of seed plants, and tell why seed plants can live in relatively dry locations.

6. List and describe the four divisions of gymnosperms and give the life cycle of the pine tree.

7. Contrast dicots to monocots, and give the life cycle of flowering plants.

8. Discuss the human and ecological relevance of the bryophytes, nonseed vascular plants, conifers, and flowering plants.

9. Compare the adaptations of these same groups of plants in such a way as to explain why angiosperms are more diverse and widespread than the other plant groups.

Many exotic plants grow in the tropics. This shows the growth pattern for Heliconia rostrata, a plant that has brightly colored bracts (leaf-shaped structures). If the tropical rain forests are destroyed we stand to lose many unique plants, some of which have not even been discovered yet!

Plants can be distinguished from algae even by a most cursory examination (fig. 25.1). However, it is possible to more specifically list the characteristics of plants (table 25.1).

Characteristics of Plants

Plants are photosynthetic organisms adapted to living on land. (There are aquatic plants, today, but they are secondarily adapted to living in the water. The same can be said for a number of reptiles such as sea turtles, or mammals such as whales.) Green algae contain many different types of organisms, some of which resemble plants. However, in the classification system adopted by this text, plants have suitable adaptations for a land existence; for example, they all protect the embryo from desiccation, or drying out. The green alga, *Ulva,* which is macroscopic with specialized parts, lacks these adaptations and, therefore, is not a plant. Table 25.2 lists the plants we will be considering.

A land existence offers some advantages to plants. One advantage is the great availability of light for photosynthesis, because even clear water filters light. Another advantage is that carbon dioxide is present in higher concentrations and diffuses more readily in air than in water. The land environment, however, requires some adaptations to deal with the constant threat of desiccation. For example, in flowering plants, the body, gametes, zygote, and embryo are all protected from drying out. The leaves are covered by a waxy cuticle interrupted on the un-

Figure 25.1
Representatives of the dominant types of land plants on earth today. *a.* Most plants today are flowering plants. This is a tiger lily. *b.* Mosses are low lying and usually found in moist locations. This is hairy-cap moss. *c.* Ferns were much more prominent in ancient times. This is bracken fern. *d.* The gymnosperms are seed plants as are flowering plants. This is a conifer in which the seeds are borne on cones.

a.

b.

c.

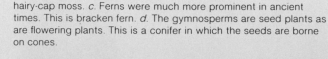

d.

derside by stomata, which are small openings that can be opened and closed to regulate gas exchange and water loss. Flowering plants have an internal skeleton to oppose the force of gravity and to lift the leaves toward the sun, and a vascular system that transports water up to and nutrients down from the leaves.

Table 25.1 Characteristics of Plants

1. Contain chlorophyll *a* and *b,* and carotenoids. Store reserved food as starch, and have cellulose cell walls.

2. Lack the power of motion or locomotion by means of contracting fibers.

3. Are multicellular and have cells specialized to form tissues and organs.

4. Have sex organs with an outer layer of nonreproductive cells to prevent desiccation of developing gametes.

5. Protect the developing embryo from drying out, by providing it with water and nutrients within the female reproductive structure.

6. Have a life cycle that is described as alternation of generations.

Table 25.2 A Classification of Plants

	Common Name	Approximate Number of Species
Nonvascular Plants		
Division Bryophyta	Liverworts and mosses	24,000
Vascular Plants		
Seedless Plants		
Division Psilophyta	Whisk ferns	10 to 13
Division Lycophyta	Club mosses	900
Division Sphenophyta	Horsetails	15
Division Pterophyta	Ferns	12,000
Seed Plants		
Gymnosperms		
Division Coniferophyta	Conifers	550
Division Cycadophyta	Cycads	100
Division Ginkgophyta	Ginkgos	1
Division Gnetophyta	Gnetae	70
Angiosperms		
Division Anthophyta	Flowering plants	235,000

From Campbell, *Biology.* Copyright © 1987 Benjamin/Cummings Publishing Company, Menlo Park, California. Reprinted by permission.

Plants are multicellular photosynthesizers that are adapted to living on land. They all protect the embryo from drying out.

Alternation of Generations

All plants have a two-generation life cycle known as alternation of generations:

1. The diploid sporophyte generation produces haploid spores by meiosis.
2. The haploid gametophyte generation produces gametes and protects the embryo.

In plants, one of the generations is dominant; in other words, it is larger and exists for a longer period of time.

All plants have a life cycle that shows an alternation of generations; some have a dominant gametophyte generation and some have a dominant sporophyte generation.

Nonvascular Plants

The nonvascular plants are restricted to the bryophytes; all other plants have vascular tissue.

Bryophytes

The **bryophytes** (division Bryophyta) include liverworts and mosses. Although liverworts (fig. 25.2), with flattened, lobed bodies, are widespread and well known, they represent but a

Figure 25.2

A liverwort, *Marchantia.* Male and female gametophyte generations are flat, lobed structures that fork as they grow. The umbrellalike structures rising from the body of the plants are gametophores that in the male contain antheridia, where sperm are produced, and in the female contain archegonia, where eggs are produced.

Figure 25.3

Life cycles of plants. All plants have a life cycle that shows an alternation of generations, but they differ as to which generation is dominant. *a*. In bryophytes, the gametophyte generation is dominant; therefore, more space is allotted to this generation in the diagram. *b*. In vascular plants, the sporophyte generation is dominant; therefore, more space is allotted to this generation in the diagram. *c*. Heterospory is seen in vascular plants that produce seeds. Note there is a separate male and female gametophyte generation.

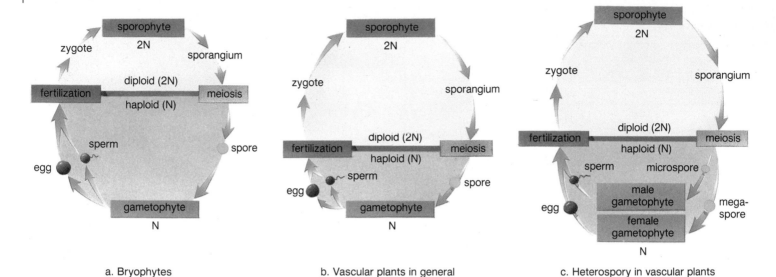

a. Bryophytes   b. Vascular plants in general   c. Heterospory in vascular plants

small fraction of the total number of bryophyte species. Most species of liverworts are "leafy" and look somewhat like mosses, but examination shows that the body of a liverwort has a distinct top and bottom surface, with numerous **rhizoids** (rootlike hairs) projecting into the soil. In contrast, a moss has a stemlike structure with radially arranged, leaflike structures (fig. 25.1*b*). Rhizoids anchor the plant and absorb minerals and water from the soil. Because bryophytes do not have vascular tissue, *they lack true roots, stems, and leaves.* Instead, they have rhizoids, stemlike structures, and leaflike structures.

Moss Life Cycle

In mosses, the gametophyte generation is dominant—it is the longer lasting generation (fig. 25.3*a*). In some mosses, there are separate male and female gametophytes (fig. 25.4). At the tip of a male gametophyte are **antheridia,** in which swimming sperm are produced. After a rain or heavy dew, the sperm swim to the tip of a female gametophyte, where eggs have been produced within the **archegonia.** Antheridia and archegonia are both multicellular structures, and each has an outer layer of jacket cells that protects the enclosed gametes from desiccation. After an egg is fertilized, it is retained within the archegonium and begins development as the sporophyte generation. The *sporophyte generation,* which is parasitic on the gametophyte, consists of a *foot* that grows down into the gametophyte tissue, a *stalk,* and an upper capsule or *sporangium,* where meiosis occurs and haploid spores are produced. In some species of mosses, a

hoodlike covering is carried along upward by the growing sporophyte. When this covering and the capsule lid fall off, the spores are mature and ready to escape. The release of spores is controlled by one or two rings of "teeth" that project inward from the margin. The teeth close the opening when the weather is wet but curl up and free the opening when the weather is dry. This appears to be the mechanism that allows spores to be released at times when they are more likely to be dispersed by the wind.

When a spore lands on an appropriate site, it germinates. The single row of cells that first appears branches, giving an algalike structure called a *protonema.* After about three days of favorable growing conditions, new moss plants can be seen at intervals along the protonema. Each of these consists of the rootlike rhizoids and the upright shoots of a moss gametophyte generation. The gametophytes produce gametes and the moss life cycle begins again.

Adaptation of Bryophytes

Bryophytes are generally found in locations where there is some moisture. Even so, the embryo is protected from drying out by remaining within the archegonium. Also, the organism is dispersed to new locations as windblown spores. The bryophytes, however, do not have vascular tissue for water transport. Also, the sperm must swim in external moisture to reach the egg. It is for these reasons that bryophytes are generally found in moist locations.

Figure 25.4

Moss life cycle. The male gametophyte generation (a leafy shoot) produces swimming sperm in antheridia. They need external water to reach egg-producing archegonia within the female shoot. Following fertilization, the 2N zygote develops into the sporophyte generation that produces N spores by meiosis. These spores develop into the separate male and female gametophyte generations.

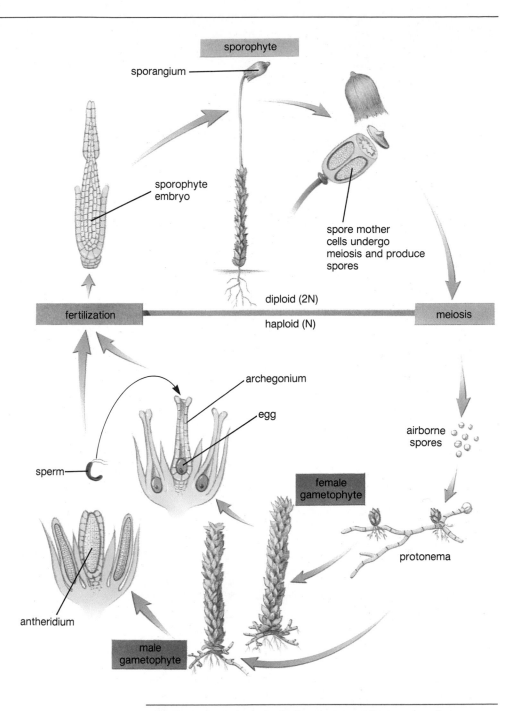

Importance of Bryophytes

Certain bryophytes colonize rocks and slowly convert them to soil that can be used for the growth of other organisms.

Sphagnum, also called bog or peat moss, has commercial importance. This moss has special nonliving cells that can absorb moisture, which is why peat moss is often used in gardening to improve the water-holding capacity of the soil. In some areas, like bogs, where the ground is wet and acidic, dead mosses, especially sphagnum, accumulate and do not decay. This accumulated moss, called peat, can be used as a fuel.

The bryophytes include the inconspicuous liverworts and mosses, plants that have a dominant gametophyte generation. Bryophytes lack vascular tissue, and fertilization requires an outside source of moisture. Windblown spores disperse the species.

Vascular Plants

All the other plants to be studied are vascular plants (also called **tracheophytes**), which are believed to have evolved sometime during the late Silurian Period (table 25.3).

Table 25.3 The Geological Time Scale: Some Major Evolutionary Events of Plants

Era	Period	M.Y.A.	Major Biological Events—Plants
Cenozoic	Quaternary		Increase in number of herbaceous plants
Cenozoic	Tertiary	2	Dominance of land by angiosperms — Age of Angiosperms
		65	
Mesozoic	Cretaceous	130	Angiosperms spread
Mesozoic	Jurassic		First angiosperms appear — Age of Gymnosperms
Mesozoic		180	
Mesozoic	Triassic		Dominance of land by gymnosperms and ferns
		230	
Paleozoic	Permian	280	Land covered by forests of primitive vascular plants
Paleozoic	Carboniferous	350	Age of great coal-forming forests, including clubmosses, horsetails, and ferns — swamp forests
Paleozoic	Devonian	400	Expansion of primitive vascular plants over land
Paleozoic	Silurian	435	Primitive vascular plants appear on land
Paleozoic	Ordovician		First plant fossils
Paleozoic	Cambrian	500	
Paleozoic		600	Unicellular marine algae abundant — Age of Algae

Figure 25.5

Rhynia major, the simplest and earliest-known vascular plant, thrived about 350–400 million years ago. The leafless stem was green and carried on photosynthesis. The terminal sporangia apparently released their spores by splitting longitudinally.

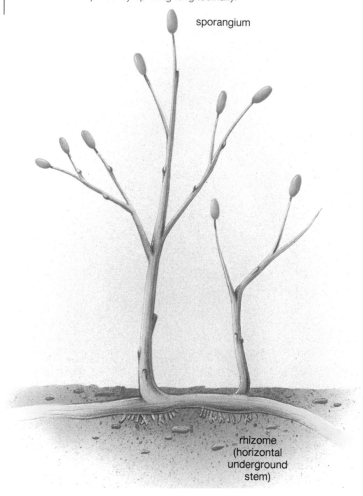

sporangium

rhizome (horizontal underground stem)

You might suppose that the bryophytes evolved before the vascular plants but actually they appear later in the fossil record than do the tracheophytes. Two suggestions have been proposed: (1) both the bryophytes and tracheophytes evolved separately from a green algal ancestor, or (2) the bryophytes, which are nonvascular, evolved from a vascular ancestor. The later suggestion is not held in high regard because it would mean that the bryophytes lost a feature (vascular tissue) that they formerly had.

There is an extinct group of plants, **rhyniophytes** (fig. 25.5), known from the fossil record, that may have been the ancestral tracheophytes because they provide the first evidence of vascular tissue. The tracheophytes have two types of vascular tissue. **Xylem** conducts water and minerals up from the soil, and **phloem** transports organic nutrients from one part of the body to another. Because they have vascular tissue, the specialized body parts of tracheophytes can properly be called roots, stems, and leaves.

Among the tracheophytes, the sporophyte, or diploid generation is dominant (fig. 25.3b). This is the generation that has vascular tissue. Xylem, with its strong-walled cells, supports the body of the plant against the pull of gravity. The tallest organisms in the world are vascular plants—the redwood trees of California.

The tracheophytes include those plants that are most diverse and widely distributed on land. One advantage of having a dominant sporophyte generation relates to it being diploid. If a faulty gene is present, it can be masked by a functional gene.

Figure 25.6

The Paleozoic swamp forests (*see* fig. 25.9) contained the primitive tracheophytes: whisk ferns, lycopod trees, and treelike giant horsetails. Today the whisk ferns (*a*) are represented by *Psilotum;* the club mosses (*b*) are represented by *Lycopodium;* and the horsetails (*c*) are represented by *Equisetum*. Their giant relatives died out during the Permian Period.

a.

b.

c.

In addition, possession of vascular tissue is an important adaptation to the land environment. As we shall see, the advanced tracheophytes also possess a means of reproduction that is suitable to a dry environment.

The tracheophytes evolved during the Silurian Period and the presence of vascular tissue helps explain why these plants are the most diverse and widely distributed of the plants.

Vascular Plants without Seeds

There are both nonseed plants and seed plants among the tracheophytes. The earliest vascular plants did not produce seeds.

Primitive Vascular Plants

The primitive vascular plants (fig. 25.6) include the whisk ferns (division Psilophyta), club mosses (division Lycophyta), horse-tails (division Sphenophyta). The psilopsids are of particular interest because they may be the most primitive. *Psilotum* (fig. 25.6*a*) has been classified as a whisk fern, though currently there is some question as to whether it is instead a true fern. Whatever its correct taxonomic designation, it bears considerable resemblance to the extinct rhyniophytes. The sporophyte consists of stems with scalelike structures, but no leaves. There is a horizontal stem (lacking roots), from which rhizoids grow, and green, photosynthetic, upright branches with tiny, scalelike structures that grow upward. Sporangia are located on the branches. The gametophyte generation is separate, smaller than the sporophyte, and water dependent. In fact, the life cycle of *Psilotum* is very close to that of the fern, which is discussed next.

Figure 25.7

Fern life cycle. The heart-shaped prothallus (gametophyte generation much enlarged here) produces swimming sperm in antheridia. They need external water in order to reach egg-producing archegonia. Following fertilization, the 2N zygote develops into a sporophyte generation—the familiar fern plant with large fronds. Spores produced by meiosis are released from sori located on the underside of the fronds. Each can germinate to give a gametophyte generation.

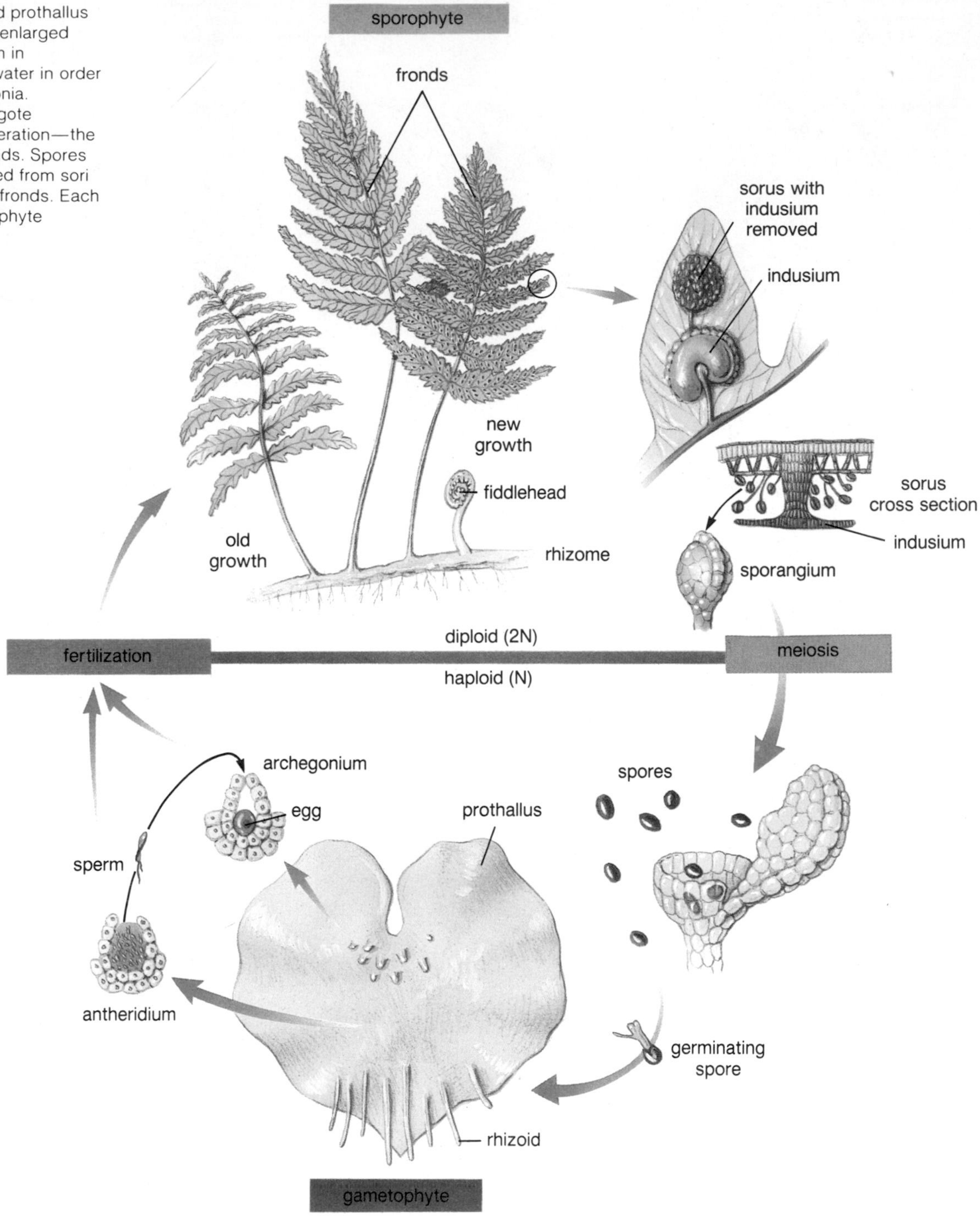

Ferns

Ferns (division Pterophyta) vary in appearance. Many are low-lying, but there are also tall tree ferns in the tropics. Figure 25.7 shows the life cycle of a common fern of the temperate zone, with a horizontal stem (rhizome) from which hairlike roots project downward and large leaves (fronds) project upward. Young fronds grow in a curled-up form called fiddleheads that unroll as they grow. The fronds are often subdivided into a large number of leaflets.

The fern plant represents the dominant sporophyte generation. Sporangia develop in clusters called **sori** (sing., sorus) (fig. 25.8a) on the underside of the leaflets. Within the sporangia, meiosis occurs and spores are produced. A band of thickened cells on the rim, or *annulus,* (fig. 25.8b) snaps in response to moisture changes and flings the windblown spores out. The gametophyte generation is a tiny (1–2 cm), heart-shaped structure called a *prothallus*. Archegonia develop at the notch, and antheridia are at the tip. Spiral-shaped sperm swim from the antheridia to the archegonia, where fertilization occurs. A

Figure 25.8

Fern sporophyte anatomy. *a.* A photomicrograph of the underside of a leaflet showing sori. *b.* A scanning electron micrograph of sporangia within a sorus. When the rim (annulus) contracts, a sporangium breaks open and the spores are released (*see* fig. 25.7). Magnification, ×300.

a.

b.

Figure 25.9

Carboniferous swamps are believed to have contained plants with fernlike foliage (*left*), treelike club mosses (*left*), and treelike horsetails (*right*).

zygote begins its development inside an archegonium, but the embryo soon outgrows the available space. As a distinctive first leaf appears above the prothallus and the roots develop below it, the sporophyte generation becomes visible. Often the sporophyte tissues and the gametophyte tissues are distinctly different shades of green. The young sporophyte grows and develops into a mature sporophyte, the familiar fern plant.

Adaptation of Ferns   Ferns are apt to be found in moist habitats because they have a water-dependent gametophyte generation. This generation lacks vascular tissue and is separate from the sporophyte generation. Swimming sperm require an outside source of water in which to swim to the eggs in the archegonia (fig. 25.7). Once established, some ferns, like the bracken fern *Pteridium aquilinum,* can spread by means of vegetative reproduction into drier areas. As the rhizomes grow horizontally in the soil, the fiddleheads grow up as new fronds.

Importance of Nonseed Tracheophytes

During the Carboniferous Period (table 25.3), the horsetails, club mosses, and ferns were abundant, very large, and treelike (fig. 25.9). For some unknown reason, a large quantity of these plants died and did not decompose completely. Instead they were compressed and compacted to form the coal that we still mine and burn today. (Oil was formed similarly, but most likely formed in marine sedimentary rocks that included animal remains.)

In the nonseed tracheophyte plants such as ferns, there is a dominant vascular sporophyte generation. These plants are usually found in a moist environment, because of an independent, nonvascular, gametophyte generation that produces swimming sperm.

Seed Plants

As Pangaea (see fig. 20.6) formed during the late Paleozoic Era, mountain ranges arose, and deserts appeared on their leeward sides. Land that had been swampy became much drier (table 25.3). A mass extinction occurred, and among those that all but vanished were the nonseed tracheophytes that had made up the great swamp forests of the Paleozoic Era. Now the seed plants, with their many adaptations to life on land, came into their own. During the Mesozoic Era, the more primitive **gymnosperms** dominated. By the Cenozoic Era, the **angiosperms** had become the dominant plants on earth. Some of these plants are trees with extensive root systems and enlarged stems—a result of the ability of vascular tissue to produce *secondary growth.*

The seed plants, to a greater extent than any of their predecessors, produce **heterospores** (fig. 25.3*c*), called microspores and megaspores, instead of identical homospores. Each microspore develops into a **pollen grain,** which is the immature male gametophyte, while it is still retained within a microsporangium. After they are released, pollen grains develop into mature,

Figure 25.10

Representatives of the lesser-known gymnosperm divisions. *a.* Cycads resemble palm trees but are gymnosperms, that produce cones. Male plants have pollen cones and female plants have seed cones. *b.* Ginkgos exist only as a single species, the maidenhair tree. This photograph of a female plant features the seeds. Male plants have pollen cones. *c.* There are three genera of gnetophytes—this is *Welwitschia mirabilis,* a very unusual plant of Africa. In this photograph, we see seed cones at the edges of the two enormous frayed leaves.

sperm-bearing, male gametophytes. Therefore, the *pollen grain* in seed plants *replaces external swimming sperm* in nonseed plants. **Pollination,** the transfer of the male gametophyte to the vicinity of the female gametophyte and the resulting fertilization, are dependent on wind or animals, rather than an external source of water.

A megasporangium is located within an ovule and produces only one functional megaspore. Megaspores develop into egg-bearing female gametophytes that are still retained within ovules. After fertilization, the zygote becomes an embryonic plant enclosed within a seed. While nonseed plants are dispersed by means of spores, in seed plants the seeds serve as the dispersal package. A **seed** contains an embryonic sporophyte generation, plus stored food, enclosed within a protective seed coat. Seeds are resistant to adverse conditions, such as dryness or temperature extremes.

There are two groups of seed plants, the **gymnosperms** ("naked seeds") and the **angiosperms** ("enclosed seeds"). Angiosperm seeds are enclosed in fruits, but gymnosperm seeds are not.

The seed plants didn't begin to flourish until the Mesozoic Era, when geological upheavals brought about a change in the climate and fostered their diversification.

Gymnosperms

There are four divisions of **gymnosperm** plants:

Gymnosperms:

    cycads (division Cycadophyta)
    ginkgos (division Ginkgophyta)
    gnetophytes (division Gnetophyta)
    conifers (division Coniferophyta)

Cycads are cone-bearing, palmlike plants found today mainly in tropical and subtropical regions (fig. 25.10*a*). During the Mesozoic Era, cycads were so numerous that this period is sometimes referred to as the Age of Cycads and Dinosaurs, and it is believed that dinosaurs fed on cycad seeds. Today there are a limited number of cycad species, and only one occurs in the United States.

Only one species of **ginkgo,** the maidenhair tree (fig. 25.10*b*) survives today. The maidenhair tree was largely restricted to ornamental gardens in China, until it was discovered that it does quite well in polluted areas. Because female trees produce rather smelly seeds, it is the custom to use only male

Figure 25.11
Conifer trees are usually evergreen and produce pollen and seed
cones. *a.* Hemlock tree. *b.* Seed cones.

a.

b.

trees, propagated vegetatively, in city parks. The maidenhair tree is recognized by its fan-shaped leaves with a forked vein pattern.

The **gnetophytes** contain only three genera of plants. They look quite different from one another, though they share certain anatomical features (related to their gametophyte generation and vascular tissue) that make them seem more related to angiosperms than other gymnosperms. Only one genus is found in the United States; *Ephedra* is a many-branched shrub with small, scalelike leaves and is found largely in desert regions. *Welwitschia* belongs to a genus of very unusual plants found in Africa. Most of the plant exists underground. Only a few centimeters of trunk occurs above ground, but it can be over a meter wide! There are only two enormous, straplike leaves that grow for hundreds of years and may cover most of the exposed portion of the trunk (fig. 25.10*c*).

The first seed-producing plants were the gymnosperms, which produce naked seeds. The four divisions of these plants are probably not closely related.

Conifers

The largest group of gymnosperms is the cone-bearing **conifers,** (fig. 25.11) which include the pine, cedar, spruce, fir, and redwood trees. These trees have needlelike leaves that are well adapted to withstand not only hot summers but also cold winters and high winds, and most are evergreen.

Pine Life Cycle The life cycle illustrated in figure 25.12 is a good example of a conifer's life history. The sporophyte generation is dominant, and its sporangia are located on the scales of the cones. There are two types of cones—male and female.

Typically, the male cones are quite small and develop near the tips of lower branches. Each scale of the male cone has two or more microsporangia on the underside. Within these sporangia, each meiotic cell division produces four **microspores,** and each microspore develops into a pollen grain, which is the *male gametophyte generation.* The pollen grain has two lobular wings and is carried by the wind. Pine trees release so many pollen grains during pollen season that everything in the area may be covered with a dusting of yellow, powdery, pine pollen.

The mature female cones are large and located near the top of the tree. Each scale of the female cone has two ovules that lie on the upper surface. Each ovule is surrounded by a thick, layered coat, with an opening at one end. Within the ovule, meiosis produces four megaspores. Only one of these spores develops into a *female gametophyte,* with two to six archegonia, each containing a single large egg lying near the ovule opening.

During pollination, pollen grains are transferred from the male to the female cones. Once enclosed within the female cone, the pollen grain develops a pollen tube that slowly grows toward the ovule. The pollen tube discharges two nonflagellated sperm.

Figure 25.12

Life cycle of a pine tree. The mature sporophyte (pine tree) has female pine cones, which produce megaspores that develop into female gametophyte generations, and male pine cones, which produce microspores that develop into male gametophyte generations (mature pollen grains). Following fertilization, the immature sporophyte generation is present in seeds located on the female cones.

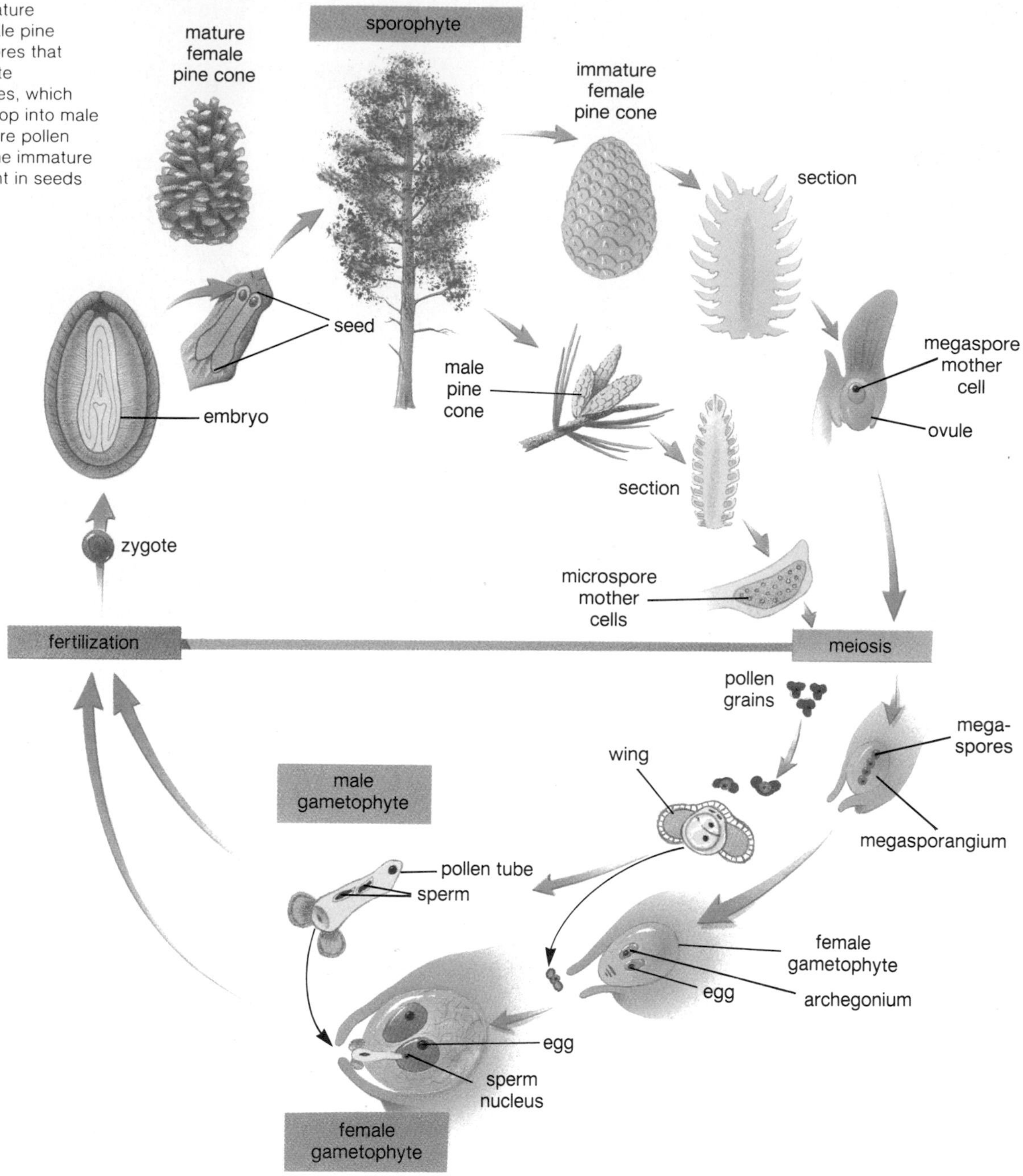

One of these fertilizes an egg and the other degenerates. Fertilization takes place fifteen months after pollination, and is an entirely separate event from pollination, which is simply the transfer of pollen.

After fertilization, the ovule matures and becomes the seed, composed of the embryo, its stored food, and a seed coat. Finally, in the third season, the female cone, by now woody and hard, opens to release its seeds, whose wings are formed from a thin, membranous layer of the cone scale. When a seed germinates, the sporophyte embryo develops into a new pine tree and the cycle is complete.

Adaptation of Gymnosperms The reproductive pattern of conifers has several important innovations not found in the plants that have been considered so far. These differences make the conifers better adapted to a dry environment. Transfer of pollen grains by wind and growth of the pollen tube eliminate the requirement of surface water for swimming sperm. Enclosure of the dependent female gametophyte in a cone protects it during its development, and shelters the developing zygote as well. Finally, the embryo is protected by the seed and provided with a store of nutrients that support development for the first period of its growth following germination. All of these factors increase the chance for reproductive success on land.

Importance of Gymnosperms Conifers grow on large areas of the earth's surface and are economically important. They supply much of the wood used for construction of buildings and production of paper. They also produce many valuable chemicals, such as those extracted from resin, a waxy substance that protects the conifers from attack by fungi and insects.

Perhaps the oldest and the largest trees in the world are conifers. Bristlecone pines in the Nevada mountains are known to be more than 4,500 years old, and a number of redwood trees in California are 2,000 years old and more than 90 meters tall.

Conifers are the most typical example of a gymnosperm. In their life cycle, windblown pollen grains replace swimming sperm. Following fertilization, the seed develops from the ovule, a structure that has been protected within the body of the sporophyte plant. The seeds are uncovered and dispersed by the wind.

Angiosperms

Angiosperms (division Anthophyta) or flowering plants, are an exceptionally large and successful group of plants. They range in size from tiny, pond-surface plants only 0.5 mm in diameter, to very large trees. The oldest angiosperm fossils come from the Cretaceous Period (table 25.3) of the Mesozoic Era, but flowering plants didn't diversify until the Cenozoic Era. The continents drifted to their present locations during the Cenozoic Era and the climate became colder than that of the Mesozoic Era. Perhaps the angiosperms were better able to cope with these changes and this led to their present dominance. It can also be pointed out, however, that the angiosperms have two innovations, both related to the evolution of the **flower,** where the reproductive structures are located. The flower attracts insects and birds that aid in pollination (pollination by wind also occurs), and it produces seeds enclosed by fruit. There are many different types of **fruits,** some of which are fleshy (e.g., apple), and some of which are dry (e.g., peas in a pod). Fruits are sometimes specialized to aid in dispersal of seeds. Fleshy fruits may be eaten by animals, which transport the seeds to a new location and then deposit them when they defecate.

Angiosperms have well-developed vascular and supporting tissues that make them well adapted for terrestrial life. Their xylem tissue contains xylem vessel cells as well as tracheids. Other vascular plants, including virtually all gymnosperms, have only tracheids in their xylem. Whereas the gymnosperms are softwood trees, **woody** angiosperms are hardwood trees. Nonwoody angiosperms are called **herbaceous** plants.

The angiosperms became the dominant land plants in the Cenozoic Era. They have flowers, which attract pollinators and produce seeds enclosed by fruits. Also, their vascular tissue is more complex than that of the gymnosperms.

Angiosperms are divided into two classes: **dicotyledons** (dicots) and **monocotyledons** (monocots) (fig. 29.6). The dicots are either woody or herbaceous, have flower parts usually in fours and fives, net leaf veins, vascular bundles arranged in a circle within the stem, and two cotyledons, or seed leaves. Dicot

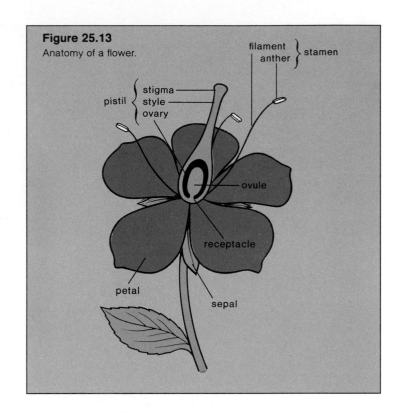

Figure 25.13
Anatomy of a flower.

families include many familiar plant groups, such as the buttercup, mustard, maple, cactus, pink, pea, and rose families. The rose family includes roses, apples, plums, pears, cherries, peaches, strawberries, raspberries, and a number of other shrubs. The monocots are almost always herbaceous, with flower parts in threes, parallel leaf veins, scattered vascular bundles in the stem, and one cotyledon, or seed leaf. Monocot families include lilies, palms, orchids, irises, and grasses. The grass family includes wheat, rice, corn (maize), and other agriculturally important plants.

Flowering Plant Life Cycle

A flower (fig. 25.13) consists of several kinds of highly modified leaves that are arranged in concentric rings and attached to a modified stem tip, the receptacle. The sepals, which form the outermost ring, are frequently green and quite similar to ordinary foliage leaves. They enclose the flower before it opens. Next are the petals, which are often large and colorful. Their color often helps attract pollinators. Within the petals, the stamens form a whorl around the pistil. **Stamens** are the pollen-producing portion of the flower. Each stamen has a slender filament with an **anther** at the tip. In the anther, meiosis within microsporangia produces microspores. Each microspore divides and becomes the male gametophyte, or pollen grain with two haploid nuclei. One nucleus is the tube nucleus, and the other is the generative nucleus. The pollen grains are either blown by the wind or carried by pollinators to the pistil. The **pistil** of most flowers has three parts—the **stigma, style,** and **ovary.** The ovary contains from one to many **ovules,** depending on the species of plant. Each ovule contains a megasporangium, where meiosis results in four megaspores, only one of which survives. The latter develops into the embryo sac, with eight haploid nuclei. At this point, the female gametophyte is ready to be fertilized (fig. 31.3).

The Living Library of Plants

T he California condor, the Maryland darter, the Florida panther, and other animals struggling to survive are not the only endangered species. Largely because of man's encroachment, many, perhaps dozens of American plant species are disappearing each year. Indeed, botanists estimate that some 3,000 of the 22,000 species of higher plants native to the United States may be facing extinction. Around the world, as many as 40,000 plant species are in trouble.

Now help is on the way, at least for America's vegetation, in the form of the Center for Plant Conservation, which has its headquarters at Harvard University's Arnold Arboretum. Started with a grant of $500,000, the center has an unprecedented program, by far the most comprehensive to date, that aims to preserve every kind of threatened plant in the United States.

Through a network of 18 affiliated botanical gardens and horticultural research facilities in 14 states, the center this summer coordinated the collection of 92 threatened species. By growing these rare plants, the center expects eventually to reintroduce some into their natural habitats and to satisfy the needs of both researchers and collectors. Collectors have

Texas bladderpod, *Lesquerella pallida*, is a potential source of seed oil but may soon become extinct.

contributed to the near extinction of several species. One victim is the Knowlton cactus, the first endangered species cataloged by the center. Says Donald Falk, the center's administrative director: "Collectors will go out and decimate populations, uprooting the cactus to send it back to live on windowsills."

Why spend money and energy to save, say, the frostweed of the small whorled pogonia?

Medical benefits alone, says Frank Thibodeau, the center's scientific director, could justify the center's efforts: "Well over a quarter of all prescription medicines in the United States are based on plant products." He points, for example, to antitumor alkaloids found in the Madagascar periwinkle that are now used in the treatment of childhood leukemia and Hodgkin's disease. "The question," says Thibodeau, "is whether you're willing to bet that there isn't another important drug out there among those 3,000 plants or whether you're willing to hold the plants long enough to study them."

Then, too, some of the plants may have as yet undiscovered characteristics important to agriculture: for example, resistance to disease or drought. Using new recombinant DNA techniques, scientists look forward to identifying the genes that confer these traits and transferring them from wild plants to crop plants. By preserving the endangered species, says Falk, "we're building a genetic library." Thibodeau considers the library essential "even if it turned out that these plants have no other identifiable value."

When pollen grains are transferred to the stigma, each develops a pollen tube, which penetrates the style and ovary. The generative nucleus enters the pollen tube and divides to give two sperm nuclei that travel to an opening in the ovule. One sperm nucleus fertilizes the egg nucleus and the other unites with two other nuclei (the polar nuclei) of the female gametophyte to form endosperm, which is food for the embryo. This so-called double fertilization is unique to angiosperms. The mature ovule or seed contains an embryo and food enclosed within a protective seed coat. The wall of the ovary and sometimes adjacent parts develop into a fruit that surrounds the seeds. Therefore, angiosperms are said to have *covered seeds*.

In angiosperms, the reproductive structures are located in the flower, which consists of highly modified leaves.

Adaptation

Like the gymnosperms, angiosperms are well adapted to a land environment. They have unique innovations based on the production of flowers. Their flowers attract appropriate animal (e.g., insects) pollinators that increase the efficiency of pollination by their habit of visiting particular types of plants (p. 488). The fruits produced by flowers are often specialized to help with dispersal of the seeds.

It should also be mentioned that their vascular tissue contains vessel cells in addition to tracheids, and the improvement in water flow allows them to thrive in drier habitats. These are all selective advantages for the angiosperms.

Importance of Angiosperms

The angiosperms are the major producers in most terrestrial ecosystems. They provide food that sustains most of the animals on land, including humans. Still, it has been observed, that humans use relatively few types of plants for food. Grasses, alfalfa, and clover are also used as food, or forage, for livestock.

Humans use angiosperms for many other functions. Their wood becomes the timber used for construction and the making of furniture. It is also used for fuel (firewood), particularly in poorer countries. Flax and cotton are sources of natural fiber for making cloth, and cellulose can be treated to yield rayon. Plant oils are not only used in cooking, but also in making perfumes and medicines. Spices are from various parts of plants, peppercorns are small berrylike structures from a vine; cinnamon comes from the bark of a tree; cloves are dried flower buds. Various drugs come from angiosperm plants, including morphine and heroin from the juice of the poppy, and marijuana from the leaves and flowers of the plant *Cannabis*.

Extinctions

Many flowering plants are now on the verge of extinction. This is caused not by a change in climate or bombardment by comets, but by the activities of humans. There are many arguments for attempting to preserve all plants, some of which are discussed in the reading.

Table 25.4 Adaptation Summary

Plant	Generations	Reproduction	Representative
Nonseed Plants: Windblown spores disperse the species			
Bryophytes	Both generations lack vascular tissue.	Swimming sperm require a source of outside moisture.	Moss
Primitive tracheophytes	Sporophyte generation has vascular tissue; gametophyte generation lacks vascular tissue and is separate and independent of the sporophyte.	Swimming sperm require a source of outside moisture.	Fern
Seed Plants: Seeds disperse the species			
Gymnosperms (naked seeds)	Gametophyte generation is retained and protected from desiccation by sporophyte, which has vascular tissue.	Pollen grains replace swimming sperm. Windblown seeds.	Pine
Angiosperms (enclosed seeds)	Adapted in the same manner as gymnosperms.	Insect pollination. Fruits aid dispersal.	Flower

Comparisons between Plants

We have seen that plants are adapted to a land existence. Some prefer a moist location and others can tolerate rather dry conditions. Table 25.4 compares the adaptations of the nonseed and seed plants in these regards.

The role of the haploid and diploid stages in the life cycle of various plants may be correlated with an adaptation to moist versus dry environments (fig. 25.14). Mosses, with a dominant gametophyte and small, dependent sporophyte, are adapted to moist locations and have a limited distribution elsewhere. Ferns have well-developed sporophyte bodies and vascular tissue, but still require very wet conditions for growth of a small, independent gametophyte and fertilization by swimming sperm.

The gymnosperms and angiosperms are widely distributed on land because the large, dominant sporophyte is well adapted to relatively dry environments. Furthermore, the delicate spores, gametophytes, gametes, zygotes, and embryos are enclosed within protective coverings produced by the sporophyte plant.

Figure 25.14

The relative importance of the haploid (N) and diploid generation (2N) among plants. In mosses, the gametophyte generation is dominant and larger than the sporophyte; in ferns, the sporophyte generation is dominant and separate; in conifers and flowering plants, the sporophyte generation is dominant and the gametophyte generation is dependent on the sporophyte.

Key

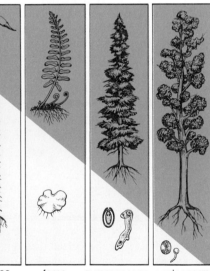

mosses   ferns   gymnosperms   angiosperms

Summary

1. Plants are photosynthetic organisms adapted to a land existence. Among the various adaptations, all plants protect the developing embryo from desiccation.

2. All plants have the alternation of generations life cycle. Some types of plants have a dominant gametophyte generation, others have a dominant sporophyte generation.

3. The bryophytes, which include the liverworts and mosses, are nonvascular plants, and therefore lack true roots, stems, and leaves.

4. In the moss life cycle, the gametophyte generation is dominant. Antheridia on the male shoot produce swimming sperm that need external water in order to reach the eggs in the archegonia of the female shoots. Following fertilization, the dependent sporophyte generation consists of a foot, stalk, and capsule where windblown spores are produced by meiosis. Each spore germinates to give either a male or female gametophyte shoot.

5. Mosses colonize rocks, helping to create soil. Humans use sphagnum in gardens to hold moisture and it becomes a fuel called peat in areas with wet and acid bogs.

6. Vascular plants (tracheophytes) arose during the Silurian Period of the Paleozoic Era. The extinct rhyniophytes may be the ancestral tracheophytes. These plants had photosynthetic stems (no leaves) with sporangia at their tips. Most likely, the life cycle was similar to today's ferns. The sporophyte generation, which is diploid and has vascular tissue, is the dominant generation in tracheophytes.

7. The primitive vascular plants include the whisk ferns, club mosses, and horsetails. While rather small and limited in diversity today, they were more diverse and quite large during the Carboniferous Period. Their life cycle resembles that of the ferns.

8. In ferns, there is a separate and water-dependent gametophyte generation (the heart-shaped prothallus) that produces swimming sperm in antheridia and eggs in archegonia. Following fertilization, the zygote develops into the sporophyte generation, which has large fronds. On the underside of the fronds are sori, each containing several sporangia. Here meiosis produces windblown spores, each of which develops into a prothallus.

9. The nonseed tracheophytes made up the great, coal-producing forests of the Paleozoic Era.

10. The Mesozoic Era saw many geological changes as Pangaea formed and then broke apart. A mass extinction occurred that paved the way for the diversification of the seed plants. Seed plants that are trees have especially well-developed roots and stems—due to secondary growth of vascular tissue. Seed plants produce heterospores and separate male and female gametophyte generations. The pollen grain replaces external swimming sperm and the female gametophyte is retained within the ovule that develops into the seed.

11. Three divisions of gymnosperms (meaning naked seeds) contain little known plants: cycads, ginkgos, and gnetophytes. Conifers, a fourth division, are the best known of the gymnosperms.

12. Male and female cones are produced by the sporophyte plant. On the underside of a male cone scale there are two sporangia that produce microspores; each becomes a pollen grain, the male gametophyte generation. On the upper surface of a female cone scale there are two ovules where meiosis produces one megaspore that develops into the female gametophyte generation. After windblown pollination, the pollen grain develops a tube through which sperm reach the egg-containing ovule. After fertilization, the ovule matures to be the seed.

13. Conifers are adapted to a dry environment because they produce pollen grains instead of swimming sperm, and the female gametophyte generation is protected by the sporophyte generation. The seed protects the developing embryo.

14. Conifers are of great economic importance. They are a source of softwood and valuable chemicals.

15. Angiosperms (meaning covered seeds) are more diverse than the other types of plants. Perhaps their success is due to climatic changes in the Cenozoic Era.

16. In a flower, the microsporangia develop within the anther portion of a stamen, and the megasporangia develop within ovules located in the ovary of the pistil. Pollination brings the mature male gametophyte (pollen grain) to the female gametophyte (within the ovule). Angiosperms exhibit double fertilization: one sperm nucleus fertilizes the egg and the other joins with the polar nuclei to form the endosperm. The ovule develops into the seed, and the ovary becomes the fruit.

17. Angiosperms have two unique features dependent upon the presence of flowers: (1) animal pollination, which increases the chance of appropriate fertilization and (2) fruit production, which can help with the dispersal of seeds. Also, their more complex vascular tissue aids water flow and permits these plants to better withstand dryness.

18. Angiosperms provide most of the food that sustains terrestrial life and they are the source of many products used by humans.

19. When we compare the plants, we see that they differ in their adaptations to a moist versus a dry environment. The more adaptations they have to withstand a moderate amount of dryness, the greater is their diversity and dispersal on land. Also, we can correlate this indication of success with the dominance of the sporophyte generation.

Objective Questions

1. Which of these are characteristics of plants?
   a. multicellular with specialized tissues and organs
   b. photosynthetic and contain chlorophyll *a* and *b*
   c. protect the developing embryo from drying out
   d. All of these.

2. In the alternation of generations life cycle
   a. the sporophyte generation is always dominant.
   b. the gametophyte generation produces the spores.
   c. the gametophyte generation is haploid.
   d. spores develop into the sporophyte generation.

3. In the moss life cycle, the sporophyte generation is
   a. leafy green shoots.
   b. the heart-shaped prothallus.
   c. a foot, stalk, and capsule.
   d. the dominant generation.

4. The rhyniophytes
   a. are a flourishing group of plants today.
   b. had large leaves like today's ferns.
   c. had sporangia at the tips of their branches.
   d. All of these.

5. You are apt to find ferns in a moist location because they have
   a. a water-dependent sporophyte generation.
   b. flagellated spores.
   c. swimming sperm.
   d. All of these.

6. Which of these is mismatched?
   a. pollen grain—male gametophyte
   b. ovule—female gametophyte
   c. seed—immature sporophyte
   d. pollen tube—spores

7. In the life cycle of the pine tree, the ovules are found on the
   a. needlelike leaves.
   b. female pine cone.
   c. male pine cone.
   d. All of these.

8. Which of these is mismatched?
   a. anther—produces microsporangia
   b. pistil—produces pollen
   c. ovule—becomes seed
   d. ovary—becomes fruit

9. Which of these plants contributed the most to our present-day supply of coal?
   a. bryophytes
   b. primitive tracheophytes
   c. conifers
   d. angiosperms

10. In seed plants, which of these is an adaptation to a dry environment?
   a. complex vascular tissue
   b. pollen grain to replace swimming sperm
   c. retention of female gametophyte generation within the ovule
   d. All of these.

Study Questions

1. What are the characteristics that define plants?
2. List all the ways that angiosperm plants are adapted to a land existence.
3. Draw a diagram to describe the life cycle of the moss, and point out significant features of this cycle.
4. What plants comprise the primitive tracheophytes, and at what time in the history of the earth were they larger and more abundant than today?
5. Draw a diagram to describe the life cycle of the fern and point out significant features of this cycle.
6. List three innovations that are observed in the life cycle of seed plants but are not seen in the life cycle of ferns.
7. List and describe the four divisions of gymnosperms.
8. Describe the life cycle of the pine, being careful to point out the three innovations you mentioned in question 6.
9. Describe innovations found in flowering plants. Which are directly related to the life cycle of these plants?
10. For each group of plants studied, list one feature with ecological relevance and one with human relevance.
11. Draw a diagram that shows the increasing dominance of the sporophyte generation among the four groups of plants studied. How is this feature related to adaptation to a relatively dry environment?

Thought Questions

1. By what steps could the alternation of generations life cycle have evolved from the haplontic cycle?
2. If the bryophytes evolved from vascular plants, what features would they have "lost"?
3. What criteria could be used to determine whether one group of organisms is more successful than another group?

Selected Key Terms

bryophyte (bri′o-fīt) 371
rhizoid (ri′zoid) 372
antheridia (an″ther-id′ī-ah) 372
archegonia (ar″kĕ-go′ne-ah) 372
tracheophyte (trā′kē-a-fīt) 373
xylem (zi′lem) 374

phloem (flo′em) 374
pollen grain (pol′en grān) 377
pollination (pol″ī-na′shun) 378
seed (sēd) 378
gymnosperm (jim′no-sperm) 378
conifer (ko′ni-fer) 379

angiosperm (an′je-o-sperm) 381
fruit (frōot) 381
woody (wood′e) 381
herbaceous (her-ba′shus) 381
dicotyledon (di-kot″ī-le′don) 381
monocotyledon (mon″o-kot″ī-le′don) 381

CHAPTER 26

Animals: Sponges to Roundworms

Your study of this chapter will be complete when you can:

1. *List the characteristics of animals.*
2. *Draw an evolutionary tree that shows the nine major animal phyla.*
3. *Compare these phyla in terms of body plan, type of symmetry, number of germ layers, level of organization, and presence of coelom.*
4. *Describe the way of life of sponges and their anatomical features. Indicate those features that are not shared by other animals.*
5. *Describe the cnidarian way of life, their two body forms, and other anatomical features using a hydra as an example.*
6. *Describe the way of life and the anatomical features of a free-living flatworm such as a planarian.*
7. *Describe the life cycle of tapeworms, emphasizing anatomical changes that accompany the parasitic way of life.*
8. *Describe the way of life and anatomical features of roundworms.*

Sea anemones were at one time thought to be plants and even today are sometimes called the "flowers of the sea" because of their attractive appearance. But they are animals. The projections you see are tentacles, which capture small prey and stuff it down the centrally located mouth. Several of these sac-shaped animals appear in this photograph.

hereas plants are multicellular, photosynthetic organisms, animals are *multicellular, heterotrophic organisms* that must take in preformed food (fig. 26.1). Unlike the fungi, which rely on external digestion, animals ingest their food and have a central cavity where it is digested. Animals follow the diplontic life cycle (fig. 24.8), in which the adult is always diploid. In this cycle, meiosis is necessary to produce haploid gametes, which join to give a zygote that develops into the adult. Table 26.1 lists the characteristics of animals.

Figure 26.1
The animal kingdom is diverse. A hydra (*a*) and a bobcat (*b*) are both multicellular heterotrophic organisms that must take in preformed food. Note the radial symmetry of the aquatic hydra that has tentacles for capturing small prey. Cats stalk their prey and it is important not to be seen. The black-spotted brown coat of a bobcat blends in well with dense vegetation. Locating prey by sound may be required and the bobcat has ear tufts that are thought to help hearing.

a.

b.

Animals are traditionally divided into **vertebrates**—animals with backbones—and **invertebrates**—animals without backbones. The great majority of animal species—97%—are invertebrates, and many of these live in the sea, where early animal evolution occurred. The evolutionary history of animals (see table 27.1) indicates that there was a literal explosion of invertebrate diversification at the start of the Cambrian Period. All the major invertebrate phyla are represented in the fossil record of this time. There is earlier evidence of animal fossils, but their relationship to the Cambrian fossils is uncertain. Apparently, the Cambrian diversification was so rapid, it cannot be traced in the fossil record.

Diversification of invertebrates is apparent in the fossil record of the Cambrian Period.

Without an adequate fossil record, biologists have developed possible evolutionary trees, such as the one in figure 26.2, based on a study of present-day animal groups. Although classification systems differ,[1] most authorities recognize approximately thirty animal phyla. We will consider only nine, those that most authorities believe are the major ones. The animals we will be studying are the product of the evolutionary process that has been continuing since all the major phyla first evolved.

Sponges

Most often, sponges (phylum Porifera, 8,000 species[2]) are found attached to a substratum in shallow, coastal waters. Sponges are **sessile filter feeders;** they stay in one spot and filter their food from the water that enters the central cavity through *pores* in the body wall. The body wall has two layers of cells; *epidermal cells* make up the outer layer, and *collar cells* make up the inner layer (fig. 26.3). Water enters the sponge through pores and moves to a central cavity. The flagella of the collar cells beat to keep the water moving and it exits from the sponge by way of the *osculum,* which is a single, larger opening. Although it might seem as if sponges can't do much, even a simple one, only 10 cm tall, is estimated to filter as much as 100 liters of water each day. The food particles carried in the water are

[1] A classification of organisms is given in appendix A.

[2] The approximate number of known species will be given beside each phylum name.

Table 26.1 Characteristics of Animals

1. Are heterotrophic and usually acquire food by ingestion followed by digestion.
2. Typically have the power of motion or locomotion by means of contracting fibers.
3. Are multicellular, and most have cells specialized to form tissues and organs.
4. Have a life cycle that is described as diplontic, in which the adult is always diploid.
5. Practice sexual reproduction, and produce an embryo that undergoes stages of development.

Figure 26.2

Evolutionary tree of the animal kingdom. All animals are believed to be descended from protists; however, the sponges may have evolved separately from the rest of the animals.

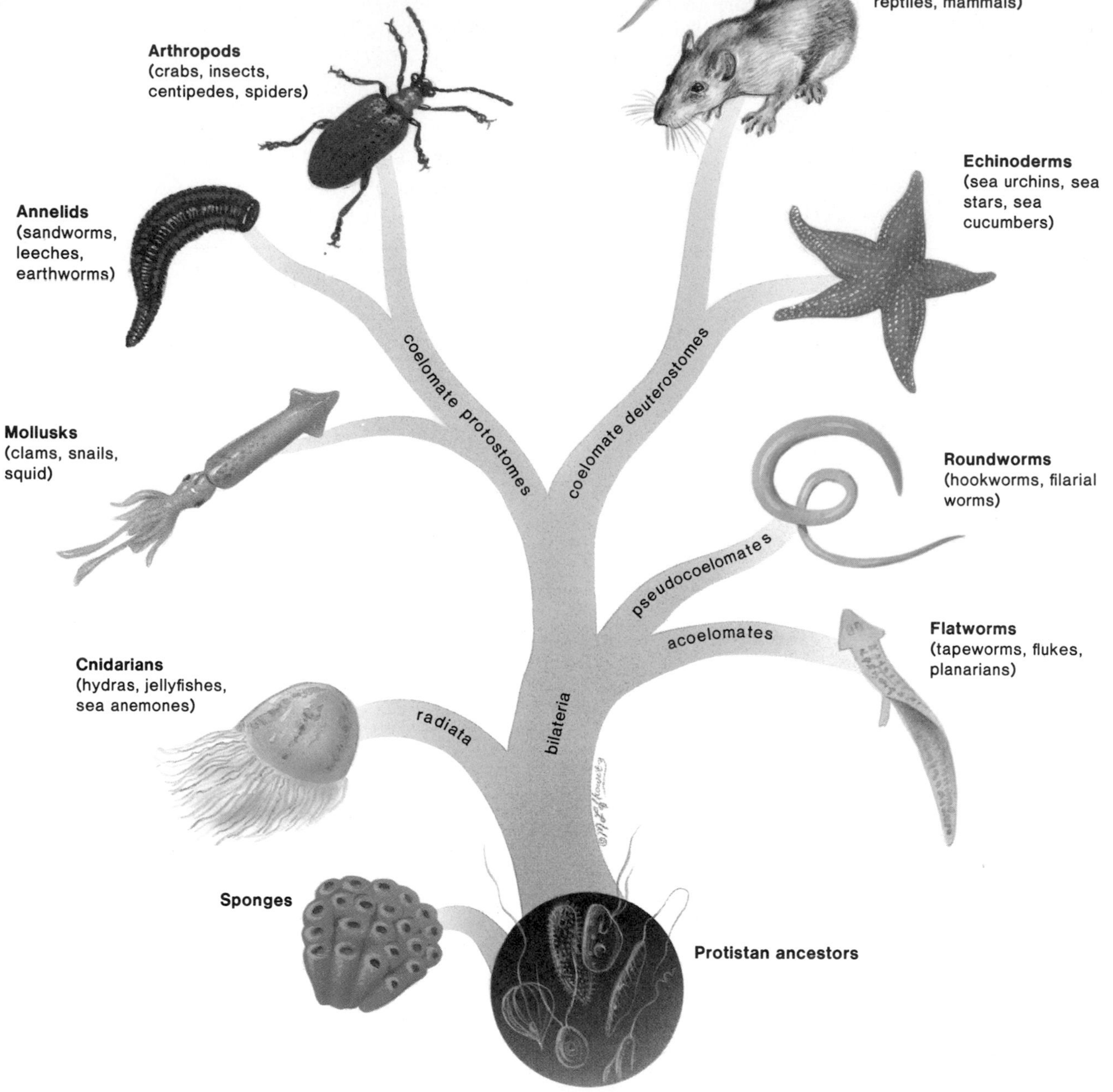

Arthropods
(crabs, insects, centipedes, spiders)

Chordates
(fishes, amphibians, reptiles, mammals)

Annelids
(sandworms, leeches, earthworms)

Echinoderms
(sea urchins, sea stars, sea cucumbers)

Mollusks
(clams, snails, squid)

Roundworms
(hookworms, filarial worms)

Cnidarians
(hydras, jellyfishes, sea anemones)

Flatworms
(tapeworms, flukes, planarians)

Sponges

Protistan ancestors

coelomate protostomes

coelomate deuterostomes

pseudocoelomates

acoelomates

radiata

bilateria

trapped by the collar cells, which pass them on to the *amoeboid cells* for digestion. The amoeboid cells also act as a circulatory device to transport nutrients from cell to cell, and they produce the skeletal fibers and gametes. Fertilization results in a zygote that develops into a ciliated larva (independent embryonic stage), capable of swimming to a new location. Sponges also reproduce asexually by budding, a process that can produce large colonies of sponges. It is not surprising, then, that sponges can regenerate, or regrow, an entire organism from a small portion.

Many sponges are simple and small, with pores leading directly from the outside water into the central cavity. Other sponges are quite large and more complex, with canals leading to internal pores. Sponges are classified according to the type of skeletal material they contain. **Spicules,** which act as an *internal skeleton* in some sponges, are made either of calcium carbonate or silicate. Other sponges have skeletal fibers made of spongin, a fibrous protein. Natural sponges are prepared by beating spongin-containing sponges until all of the living cells are removed and only the skeleton remains. Commercial sponges today, however, are largely synthetic.

Figure 26.3

Generalized sponge anatomy. The wall contains two layers of cells: the outer epidermal cells and the inner collar cells. The collar cells (enlarged) have flagella and they beat, moving the water through pore cells as indicated by the arrows. Food particles in the water are trapped by the collar cells and digested within their food vacuoles. Amoeboid cells transport nutrients from cell to cell; spicules comprise an internal skeleton of some sponges.

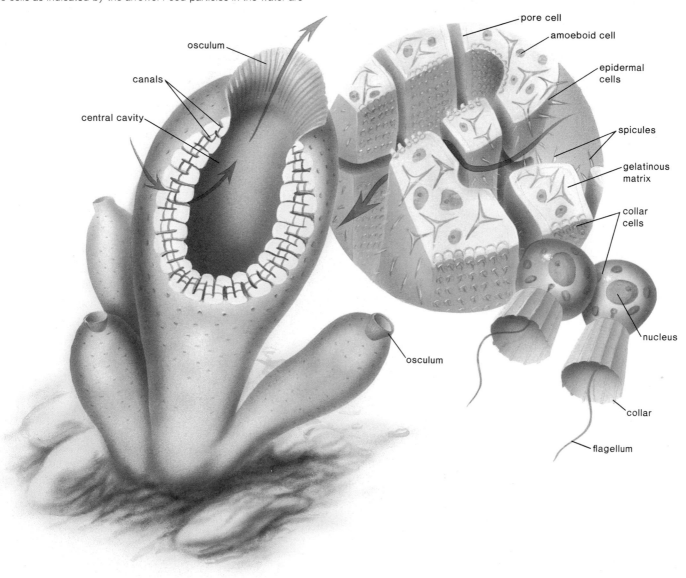

Comments

In many ways, sponges resemble a colony of protozoan cells more than a multicellular animal. Like a colony, sponges have a flexible body organization. In 1907, H. V. Wilson cut sponges into small pieces and squeezed the pieces inside bags so that individual cells were separated from each other. These cells, if left undisturbed, moved about, found other cells of their type, and reaggregated to give small, but normally functioning sponges. Sponge cells of one kind can even change their structure and function to be like another cell in the organism. Such plasticity has not been demonstrated in any other type of animal.

Sponges have a cellular level of organization and don't have true tissues. Also, they don't have fully developed muscle and nerve fibers, and movement is limited to the beating of the flagella, constriction of the osculum, and larval-stage swimming. For these reasons, sponges are believed to be an early branch in animal evolution, and are not believed to have contributed to the evolution of more complex animals. It is possible that they and the other truly multicellular animals evolved independently from different ancestral forms.

Sponges have a cellular level of organization. Movement is limited to the beating of the flagella, constriction of the osculum, and larval-stage swimming. Sponges are classified according to the type of skeletal material they contain.

Cnidarians

Cnidarians (phylum Cnidaria, 11,000 species) are tubular or bell-shaped animals that, like sponges, reside mainly in shallow coastal waters, except for the ocean-going jellyfishes.

Cnidarians have two tissue layers. The outer tissue layer is a protective *epidermis*. The inner layer, the *gastrodermis* secretes digestive juices into the internal cavity called the **gastrovascular cavity.** This cavity serves as a location for digestion (gastro) and also as a means to transport (vascular) nutrients to cells that line the cavity. Some of the cells that make up the tissue layers have muscle fibers, and between them is a jellylike packing material, the **mesoglea.** Within the mesoglea, nerve cells form a *nerve net,* a connecting network throughout the body. Having both muscle fibers and nerve fibers, these animals are capable of movement, for example the movement of their tentacles. Unlike sponges, cnidarians capture their prey. Most cnidarians have special stinging cells that are fluid-filled capsules containing a long, spiraled, threadlike fiber called a **nematocyst.** When the trigger of a stinging cell is touched, the nematocyst is discharged. Some of these trap the prey, and others have spines that penetrate and inject a paralyzing substance. Once the prey has been subdued, the tentacles about the cnidarian's mouth maneuver it into an internal cavity, where it is digested by secreted enzymes. Cnidarians also use their nematocysts for defense.

Cnidarians are **radially symmetrical,** that is splitting them lengthwise through the center always results in two equal halves because their bodies are arranged around a central axis. Most other animals are **bilaterally symmetrical;** only one lengthwise cut through the center gives two equal but opposite halves. This gives them a definite right and left half, which the cnidarians do not have.

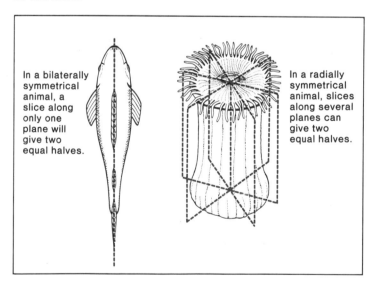

In a bilaterally symmetrical animal, a slice along only one plane will give two equal halves.

In a radially symmetrical animal, slices along several planes can give two equal halves.

Two basic body forms are seen among the cnidarians (fig. 26.4). The medusa is a bell-shaped, jellyfish form that swims about with the mouth directed downward. In the more slender and usually sessile polyp, the mouth is usually directed upward. Some cnidarians pass through both the medusa and the polyp

Figure 26.4

The two body forms of cnidarians. The drawings indicate the manner in which the two tissue layers (epidermis and gastrodermis), separated by a packing material, the mesoglea, surround the central gastrovascular cavity. *a.* The polyp form, with the mouth side upward, is usually attached to surfaces. *b.* The medusa form, with the mouth side downward, is free-swimming.

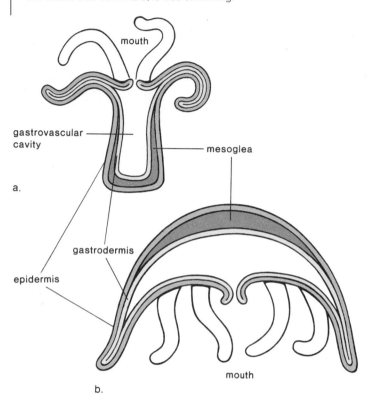

forms during their lives. (In such cases, the polyp is an asexual stage while the medusa is a sexual stage in the life cycle.) In other cnidarians, one stage is dominant and the other is reduced or, in yet other species, may be absent altogether.

Hydrozoans

In most hydrozoans, the polyp stage is dominant as is certainly the case in **hydras,** which are freshwater cnidarians. Even though most hydrozoans are marine, freshwater hydras are usually studied to show the typical cnidarian features (fig. 26.5).

Freshwater hydras can reproduce asexually and sexually. They reproduce asexually by forming buds, which are small outgrowths from the side of the animal. When hydras reproduce sexually, sperm from the testes swim to an egg within an ovary. Fertilization and early development occur within the ovary, after which, the embryo is encased within a hard, protective shell that allows it to survive until conditions are right for it to emerge and become a new polyp.

One of the most unusual hydrozoans is the Portuguese man-of-war (*Physalia*) (fig. 26.6), which is a colony of polyps although it looks as if it might be an odd-shaped medusa. Some polyps are specialized as a gas-filled float that provides buoyancy to keep the colony afloat. Others are specialized for feeding

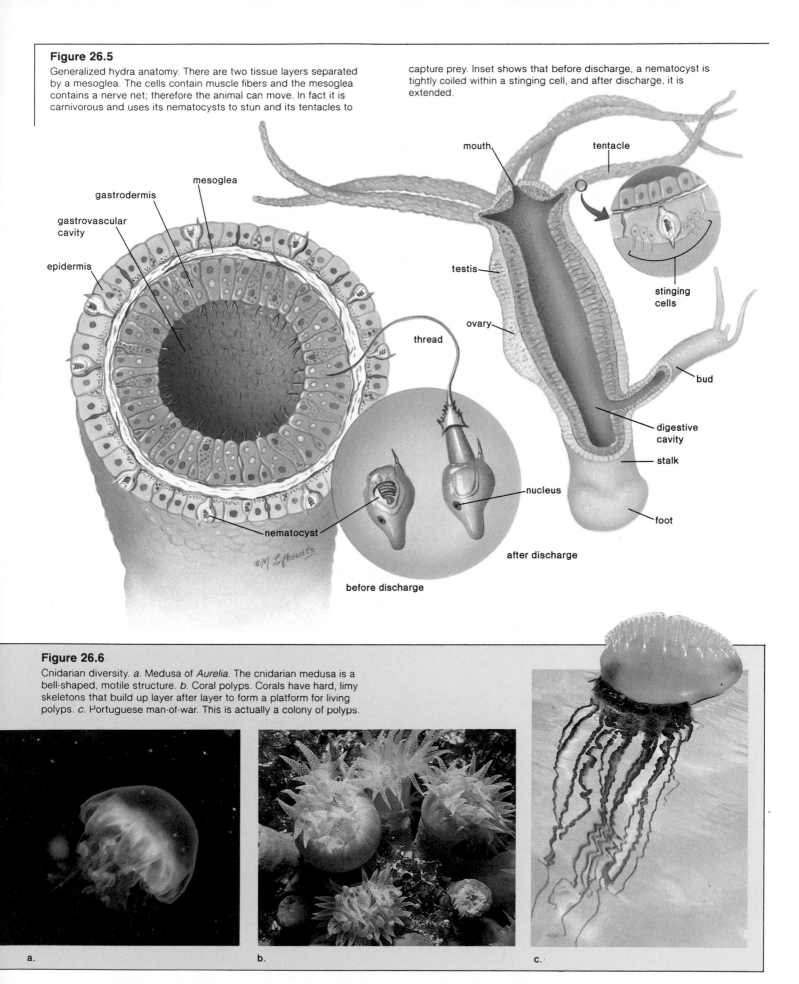

Figure 26.5

Generalized hydra anatomy. There are two tissue layers separated by a mesoglea. The cells contain muscle fibers and the mesoglea contains a nerve net; therefore the animal can move. In fact it is carnivorous and uses its nematocysts to stun and its tentacles to capture prey. Inset shows that before discharge, a nematocyst is tightly coiled within a stinging cell, and after discharge, it is extended.

mouth

tentacle

mesoglea

gastrodermis

gastrovascular cavity

epidermis

testis

stinging cells

ovary

thread

bud

nematocyst

nucleus

digestive cavity

stalk

foot

before discharge

after discharge

Figure 26.6

Cnidarian diversity. *a.* Medusa of *Aurelia*. The cnidarian medusa is a bell-shaped, motile structure. *b.* Coral polyps. Corals have hard, limy skeletons that build up layer after layer to form a platform for living polyps. *c.* Portuguese man-of-war. This is actually a colony of polyps.

a.

b.

c.

or reproduction. The Portuguese man-of-war also has stinging polyps armed with numerous nematocysts. Swimmers who accidentally come upon a Portuguese man-of-war can receive painful, even at times serious injuries from these stinging polyps.

Other Cnidarians

There are two other classes of cnidarians. One of these includes the "true *jellyfishes*," in which the medusa stage is the dominant phase, and the other contains the *sea anemones* and *corals,* which are polyps having no medusa stage (fig. 26.6). Sea anemones and corals are more often found in warm waters, where the anemones are the "flowers of the sea," and the corals are famous for building reefs. Coral polyps are rather small and inconspicuous but they secrete hard and limy skeletons that remain after the animals die and degenerate. New generations of live corals grow on the skeletal remains of past generations. Live reef-building corals are found in sunlit waters because their walls contain symbiotic dinoflagellates that carry on photosynthesis. Coral formations provide sheltered habitats, within which a great variety of organisms thrive.

Comments

Cnidarians have the tissue level of organization. They have two tissue layers, the epidermis and the gastrodermis. These are derived from two embryonic **germ layers,** the **ectoderm** and **endoderm,** respectively. (One of the first events during animal development is the establishment of layers of cells called germ layers.) Some animals, like the cnidarians, have only two germ layers, whereas more complex animals have three layers: ectoderm, endoderm, and **mesoderm,** which lies between the first two.

Some cells of the epidermis and gastrodermis contain muscle fibers and the mesoglea has nerve cells arranged to give a nerve net. Cnidarians are capable of motion.

The *radially symmetrical* cnidarians have a **sac body plan;** the mouth must serve as both an entrance for food and an exit for wastes. More complex animals are usually bilaterally symmetrical and have a **tube-within-a-tube body plan,** with both a mouth and an anus.

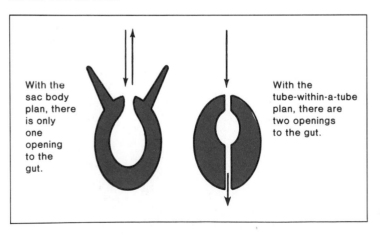

With the sac body plan, there is only one opening to the gut.

With the tube-within-a-tube plan, there are two openings to the gut.

The cnidarians are radially symmetrical and have a sac body plan. There are two tissue layers, the epidermis and gastrodermis, derived from the embryonic germ layers, the ectoderm and endoderm. The presence of nematocysts is a unique feature of cnidarians.

Flatworms

The flatworms (phylum Platyhelminthes, 12,700 species) take their name from the ribbonlike appearance of their bodies. There are free-living flatworms as well as parasitic groups, and we will look at both.

Whereas the adult cnidarians are radially symmetrical, the flatworms are *bilaterally symmetrical.* Also, flatworms have *three tissue layers;* there is a solidly packed mesodermal layer between the ectoderm and endoderm, which has led to an increase in complexity. They have the organ level of organization to carry on the functions of the animal. However, they have a gastrovascular cavity with only one opening, the same as the cnidarians, and they have no specialized respiratory or circulatory structures. Carbon dioxide and oxygen diffuse through the body surface or through the lining of the gastrovascular cavity.

Planarians

Most **planarians** (class Turbellaria) are marine animals, but freshwater ones are commonly studied as representative of the group.

Planarians show **cephalization,** a definite head region with a brain and sense organs (fig. 26.7). The head often has lateral extensions that function as sense organs to detect potential food sources and/or enemies. Planarians also have two pigmented eyespots that are sensitive to light changes. Inside, a concentration of nerve cells functions as a primitive brain, and a lengthwise pair of nerve cords are joined by cross-connectives. Planarians are said to have a *ladder-type nervous system,* because the nerve cords plus the cross-connectives look like a ladder. When planarians move, the action of ventrally-located, ciliated cells seems to be aided by rhythmical muscular contractions.

A muscular pharynx is extended through the ventrally-placed mouth when a worm is feeding on small dead or living animals. A strong sucking action tears off chunks of food. These chunks are then drawn into the branching gastrovascular cavity, which is extensive enough to make a circulatory system unnecessary for food transport. A water-regulating organ, consisting of a series of interconnecting canals, runs the length of the body on each side. Water drawn into the canals by ciliary action within bulbous cells exits at pores located in the body wall. The beating of the cilia in each bulbous cell looks like the flickering of a flame, so this organ is called a *flame-cell system.*

When planarians are cut in half, regeneration occurs, and each half can develop into another worm. Planarians are **hermaphroditic,** that is, each individual has both male and female reproductive organs. Zygotes develop directly into small worms, without larval stages.

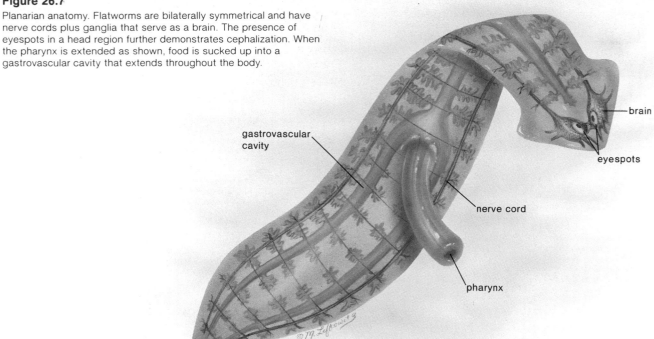

Figure 26.7

Planarian anatomy. Flatworms are bilaterally symmetrical and have nerve cords plus ganglia that serve as a brain. The presence of eyespots in a head region further demonstrates cephalization. When the pharynx is extended as shown, food is sucked up into a gastrovascular cavity that extends throughout the body.

brain

eyespots

gastrovascular cavity

nerve cord

pharynx

Flukes

Flukes (class Trematoda) are parasites inasmuch as they derive nourishment from a living host organism. Some flukes are external parasites, but the majority are endoparasites; that is, they live within the bodies of their hosts. Flukes often have two suckers by which they attach to and feed from host tissues.

Flukes that infect humans are commonly named for the organ they invade. For example, there are liver flukes and blood flukes. It is estimated that nearly half of all the people living in the tropics are infected by blood flukes of the genus *Schistosoma*. The accumulation of vast quantities of fluke eggs within the human body causes the symptoms of schistosomiasis—general weakness, anemia, diarrhea, and severely reduced resistance to other infections. Schistosome transmission depends on human feces reaching water, subsequent infection of the snail that is the intermediate host, and contact between the resulting larvae and human skin (fig. 26.8). Rice-paddy agriculture in tropical countries where human feces is used for fertilizer and people work in the water planting rice plants creates a nearly perfect environment for transmission of schistosome-causing flukes.

Tapeworms

Tapeworms (class Cestoda) are long (6–20 m), flattened worms that, as adults, live in the intestines of vertebrate animals, including humans. They cause problems because they excrete toxic wastes, absorb nutrients, and sometimes interfere with passage of food through the intestinal tract.

Instead of a head, a tapeworm has a structure called a **scolex.** The scolex has hooks and suckers that allow the worm to attach to the host's intestinal wall (fig. 26.9). The rest of the body is made of subunits called **proglottids.** When a proglottid is mature, it primarily contains male and female reproductive organs. Thousands (more than 100,000 in some cases) of eggs are stored in the uterus, which expands until it fills the entire proglottid. Such "ripe" proglottids detach and pass out with the host's feces, scattering fertilized eggs on the ground. If pigs or cattle should happen to ingest these, larvae develop and eventually become encysted in muscle that may be eaten as poorly cooked or raw meat by humans. The cyst wall is digested, and a bladder worm (cysticercus) develops into a new tapeworm that attaches to the intestinal wall. To prevent tapeworm infections, meat, especially pork, should be cooked thoroughly.

Comments

Although flatworms have a *sac body plan,* they are more complex than cnidarians. They have *three germ layers;* the presence of a mesoderm enables them to have the organ level of organization. The free-living forms exhibit *cephalization* and *bilateral symmetry,* both of which are typical of actively-moving, predaceous animals.

Although flatworms have a sac body plan, they are more complex than cnidarians because they have three germ layers and possess true organs. The free-living forms exhibit cephalization and bilateral symmetry.

The parasitic worms illustrate the modifications that occur in parasitic animals. Concomitant with the loss of predation, there is an absence of cephalization; the anterior end notably carries hooks and/or suckers for attachment to the host. There is an extensive development of the reproductive system at the

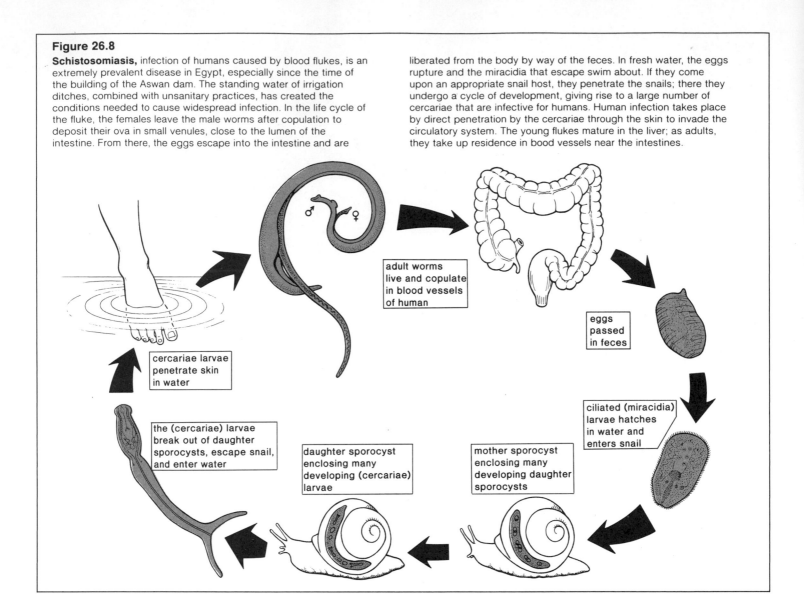

Figure 26.8
Schistosomiasis, infection of humans caused by blood flukes, is an extremely prevalent disease in Egypt, especially since the time of the building of the Aswan dam. The standing water of irrigation ditches, combined with unsanitary practices, has created the conditions needed to cause widespread infection. In the life cycle of the fluke, the females leave the male worms after copulation to deposit their ova in small venules, close to the lumen of the intestine. From there, the eggs escape into the intestine and are liberated from the body by way of the feces. In fresh water, the eggs rupture and the miracidia that escape swim about. If they come upon an appropriate snail host, they penetrate the snails; there they undergo a cycle of development, giving rise to a large number of cercariae that are infective for humans. Human infection takes place by direct penetration by the cercariae through the skin to invade the circulatory system. The young flukes mature in the liver; as adults, they take up residence in bood vessels near the intestines.

adult worms live and copulate in blood vessels of human

eggs passed in feces

cercariae larvae penetrate skin in water

ciliated (miracidia) larvae hatches in water and enters snail

the (cercariae) larvae break out of daughter sporocysts, escape snail, and enter water

daughter sporocyst enclosing many developing (cercariae) larvae

mother sporocyst enclosing many developing daughter sporocysts

expense of the other organs. Well-developed nerves and a gastrovascular cavity are not needed because the animal no longer seeks out and digests prey; instead it acquires nutrients from its host. Both flukes and tapeworms utilize a secondary, or intermediate host, to transport the species from primary host to primary host. The primary host contains the sexually mature adult; the secondary host(s) contain(s) the larval stage or stages.

Roundworms

Roundworms (phylum Nematoda, 12,000 species), as their name implies, have a smooth outside wall, indicating that they are not *segmented*. These worms, which are generally colorless and less than 5 cm long, occur almost anywhere—in the sea, in fresh water, and in the soil—in such numbers that thousands of them can be found in a very small area. They have various life-styles; some are carnivores with mouth parts to attack their prey, some are herbivores with stylets to pierce plant cell walls, and some feed on dead, organic matter that they predigest by secreting

digestive juices. Some are external parasites of plants and can cause extensive agricultural damage. Still others are internal parasites of animals; in humans they cause serious medical conditions.

The roundworms, like the animals to be studied in chapter 27, have the *tube-within-a-tube body plan*. The digestive tract is the inner tube within the rest of the animal, which is the outer tube. Roundworms also have a body cavity—a **pseudocoelom**—that is incompletely lined with mesoderm (fig. 26.10).

Ascaris

Ascaris, an intestinal parasite of vertebrates, including humans, is a commonly studied representative of roundworms. Females (20–35 cm) tend to be larger than males, but both sexes move by means of a *whiplike motion,* because all their muscles run lengthwise. The reproductive system is the most well developed of all the organ systems (fig. 26.11). Because mating produces embryos that mature in the soil, the parasite is limited to warmer environments. When the larvae in their protective covering are swallowed, they escape and burrow through the intestinal wall.

Figure 26.9

Tapeworm of a *Taenia* species. *a.* Scanning electron micrograph of tapeworm (*Taenia*) scolex. The scolex contains hooks and suckers that permit the animal to cling to the wall of the digestive tract. *b.* Life cycle, showing mature proglottid in detail. The ripe proglottid is little more than a sac of eggs.

a.

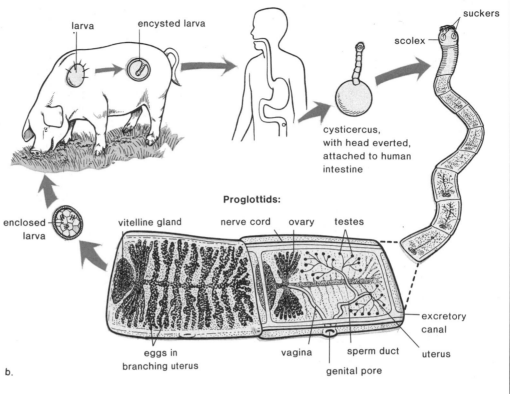

larva   encysted larva

suckers

scolex

cysticercus, with head everted, attached to human intestine

enclosed larva

Proglottids:

vitelline gland   nerve cord   ovary   testes

eggs in branching uterus

vagina   sperm duct

genital pore

excretory canal

uterus

b.

Figure 26.10

Presence of a coelom. A coelom is a body cavity surrounding the digestive organ or system. *a.* Flatworms are **acoelomate;** they have no body cavity and the mesoderm is packed solidly between the ectoderm and endoderm. *b.* The roundworms are **pseudocoelomate;** they have a body cavity in which the organs lie loose. Apparently this is disadvantageous to an increase in size and most of these worms are small. *c.* In coelomates the coelom develops as a cavity within the mesoderm; therefore it is completely lined with mesoderm. Such a coelom is often called a "true coelom." The organs in a **true coelom** are held in place by mesentery, assuring a more stable arrangement with less crowding. Further, the gut is muscular (muscles are derived from mesoderm) and shows specialization of parts not seen in the pseudocoelomates. In the true coelomates (all the other animals to be studied) the internal organs are more complex.

The coelom serves other functions as well. It allows the organs and the body wall to move independently of one another. This means an animal can stretch and bend without putting a strain on the internal organs. The coelom is fluid filled and this fluid protects and cushions the internal organs. In some animals, the fluid aids in the movement of materials such as metabolic wastes, in others this function is taken over by blood vessels. Not only metabolic wastes, but also sex cells may be deposited into the coelomic cavity before they are transported away by ducts. The gastrovascular cavity of acoelomates (cnidarians and flatworms) and the fluid-filled coelom of soft-bodied coelomates can act as a hydrostatic skeleton. It offers some resistance to the contraction of muscles and yet permits flexibility, so that the animal can change shape and perform a variety of movements.

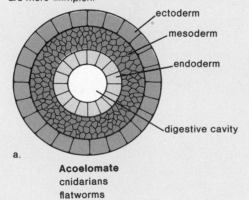

ectoderm
mesoderm
endoderm
digestive cavity

a.

Acoelomate
cnidarians
flatworms

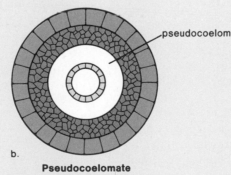

pseudocoelom

b.

Pseudocoelomate
roundworms

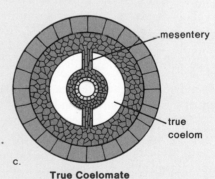

mesentery
true coelom

c.

True Coelomate
mollusks
annelids
arthropods
echinoderms
chordates

Figure 26.11

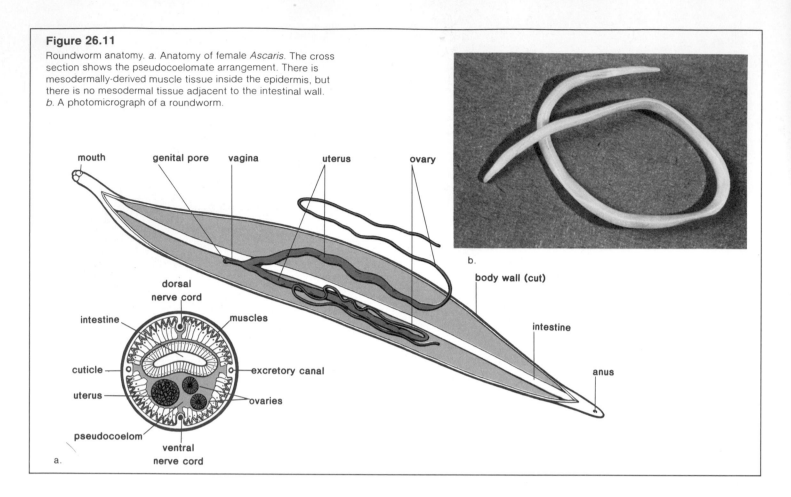

Roundworm anatomy. *a.* Anatomy of female *Ascaris.* The cross section shows the pseudocoelomate arrangement. There is mesodermally-derived muscle tissue inside the epidermis, but there is no mesodermal tissue adjacent to the intestinal wall. *b.* A photomicrograph of a roundworm.

mouth   genital pore   vagina   uterus   ovary

b.

body wall (cut)

dorsal nerve cord

intestine   muscles

cuticle   excretory canal

uterus   ovaries

pseudocoelom   ventral nerve cord

intestine

anus

a.

Then, they move to the liver, heart, and lungs. Within the lungs, further development takes place, and after about 10 days, the larvae migrate up the windpipe to the throat, where they are swallowed, allowing them to once again reach the intestine. After the mature worms mate, the female produces embryo-containing eggs that pass out with the feces. To complete this life cycle, as with other roundworms, feces must reach the mouth of the next host; therefore, proper sanitation is the best means to prevent infection with *Ascaris* and other parasitic roundworms.

Other Roundworm Infections

Trichinella is a roundworm that encysts in the muscles of pigs and humans (fig. 26.12). When humans eat poorly cooked pork, they may contract an infection called **trichinosis,** in which permanent cysts cause muscle damage.

Elephantiasis is caused by a type of roundworm called a filarial worm, which utilizes the mosquito as a secondary host. Because the adult worms reside in lymph vessels, draining of fluid is impeded, and the limbs of an infected human may swell to a monstrous size (fig. 26.13). When a mosquito bites an infected person, it passes larvae to new hosts.

Other roundworm infections are more common in the United States. Children frequently acquire a pinworm infection, and hookworm is seen in the southern states. Hookworm,

Figure 26.12

Trichinella larva embedded in a muscle. If meat like this were eaten raw or poorly cooked, these larvae would infect the consumer. Magnification, ×400.

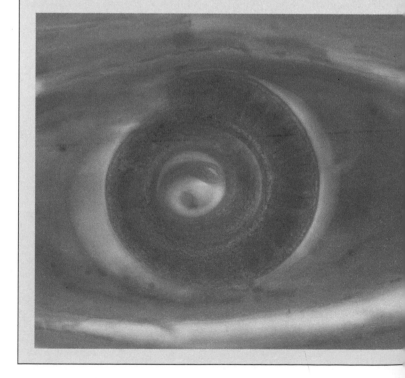

judged by some to be the most troublesome parasitic intestinal worm of humans, can be very debilitating because the worms, which attach themselves to the intestinal wall, feed on blood.

Roundworms possess two features not seen previously: a body cavity and a tube-within-a-tube body plan. The body cavity is a pseudocoelom rather than a true coelom. The tube-within-a-tube plan, in contrast to the sac plan, has both a mouth and an anus.

Table 26.2 presents a comparison of the phyla we have discussed in this chapter.

Table 26.2 Comparison of Phyla				
	Sponges	**Cnidarians**	**Flatworms**	**Roundworms**
Body plan	———	Sac	Sac	Tube-within-a-tube
Symmetry	Radial or none	Radial	Bilateral	Bilateral
Germ layers	———	2	3	3
Level of organization	———	Tissues	Organs	Organs
Body cavity	———	———	———	Pseudocoelom

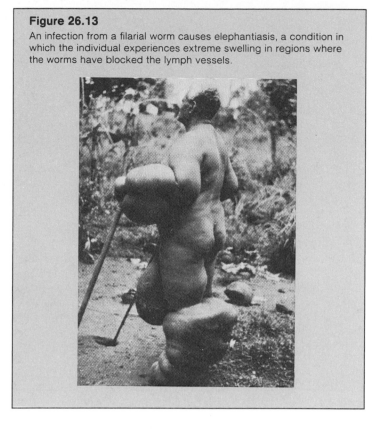

Figure 26.13
An infection from a filarial worm causes elephantiasis, a condition in which the individual experiences extreme swelling in regions where the worms have blocked the lymph vessels.

Summary

1. Animals are multicellular organisms that are heterotrophic and ingest their food. They follow the diplontic life cycle. Typically, they have the power of motion by means of contracting fibers.
2. It's possible to construct an evolutionary tree for animals, but this is largely based on a study of today's forms. All the major phyla studied today evolved during an explosive and short period of diversification at the start of the Cambrian Period.
3. Sponges are sessile filter feeders. They contain these types of cells: an outer layer of epidermal cells; pore cells that admit water, collar cells that keep it moving; and amoeboid cells that aid circulation of nutrients, produce skeletal material, and gametes.
4. Sponges may have evolved separately from other animals and they have features that set them apart. They lack tissues and don't have fully-developed muscle and nerve fibers. Movement is limited to the beating of flagella, constriction of the osculum, and larval-stage swimming.

5. Cnidarians are carnivores. An epidermis covers the animal, and the mesoglea lies between these two. Some cells contain muscle fibers and a nerve net lies in the mesoglea. They possess tentacles to capture prey and nematocysts to stun it. The prey is digested within the gastrovascular cavity, which is lined by a tissue called the gastrodermis.
6. Cnidarians can exist in the polyp form, medusa form, or alternate between the two. Hydras and sea anemones have a dominant polyp stage; jellyfishes have a dominant medusa stage.
7. Cnidarians possess the tissue level of organization, they are radially symmetrical, and have a sac body plan.
8. Flatworms may be free-living or parasitic. Freshwater planarians exemplify the features of flatworms in general, and free-living forms in particular. They have muscles, a ladder-type nervous system, and show cephalization. They take in food through an extended pharynx that leads to a gastrovascular cavity that extends throughout the body. There is a water-regulating organ that contains flame cells.

9. Flukes and tapeworms are parasitic. Tapeworms have a scolex with hooks and suckers for attaching to the host. The body is made up of proglottids, which when mature, contain thousands of eggs. If these eggs are taken up by pigs or cattle, the larvae become encysted in their muscles, after which they can possibly be eaten by humans.
10. In general, flatworms are bilaterally symmetrical, have three germ layers, and the sac body plan. It's the presence of mesoderm that accounts for their organ level of development.
11. Roundworms are mostly small and very diverse; they are present most everywhere in great numbers. The parasite, *Ascaris,* is studied as representative of the group. Infections can be caused by *Trichinella,* whose larva stage encysts in the muscle of humans. Elephantiasis is caused by a filarial worm blocking lymph vessels.
12. Roundworms have two features not seen previously: a body cavity and a tube-within-a-tube body plan. The body cavity is a pseudocoelom rather than a true coelom.

Objective Questions

1. Which of these is not a characteristic of animals?
   a. heterotrophic
   b. diplontic life cycle
   c. have contracting fibers
   d. single cells, colonial, or multicellular
2. The evolutionary tree of animals shows that
   a. cnidarians evolved from sponges.
   b. flatworms evolved from roundworms.
   c. both sponges and cnidarians evolved from protists.
   d. All of these.
3. Which of these animals has the organ level of organization?
   a. sponges
   b. cnidarians
   c. flatworms
   d. All of these.
4. Which of these sponge characteristics is not typical of animals?
   a. They practice sexual reproduction.
   b. They have the cellular level of organization.
   c. They do not have fully-developed muscle fibers.
   d. Both (b) and (c).
5. The two body forms seen in cnidarians are
   a. polyp and medusa.
   b. radial and bilateral symmetry.
   c. free-living and parasitic.
   d. asexual and sexual.
6. Which of these is mismatched?
   a. sponges—spicules
   b. tapeworms—proglottids
   c. cnidarians—nematocysts
   d. roundworms—cilia
7. Flukes and tapeworms
   a. show cephalization.
   b. have well-developed reproductive systems.
   c. have well-developed nervous systems.
   d. are plant parasites.
8. A gastrovascular cavity is seen in
   a. sponges and cnidarians.
   b. cnidarians and flatworms.
   c. flatworms and roundworms.
   d. None of these.
9. The presence of mesoderm
   a. restricts the development of a coelom.
   b. allows the development of a coelom.
   c. is seen only in acoelomates.
   d. Both (a) and (c).
10. *Ascaris* is a parasitic
   a. roundworm.
   b. flatworm.
   c. hydra.
   d. sponge.

Study Questions

1. What are the characteristics that separate animals from plants? from fungi?
2. What does the evolutionary tree in figure 26.2 tell you about the evolution of the animals studied in this chapter?
3. List the types of cells found in a sponge and describe their function.
4. What features make sponges different from the other organisms placed in the animal kingdom?
5. Explain how cnidarians carry on their carnivorous life-style.
6. What are the two body forms found in cnidarians? Explain how they function in the life cycle of various types of cnidarians.
7. What features suggest that free-living planarians are predaceous?
8. Describe the parasitic flatworms and give the life cycle of a tapeworm.
9. Describe the anatomy of a roundworm.
10. Compare the animals studied in this chapter in terms of body plan, symmetry, germ layers, level of organization, and presence of a coelom.

Thought Questions

1. The burst of Cambrian diversification has always puzzled biologists. Offer an explanation based on your knowledge of gradual versus punctuated equilibrium models of evolutionary progress.
2. Cnidarians are carnivorous but are they predaceous? Why would you ordinarily associate bilateral symmetry and not radial symmetry with being a predator?
3. What functions are served by a secondary host in a parasite's life cycle? Do your suggestions also apply to the life cycle of *Plasmodium vivax* (p. 356)?

Selected Key Terms

vertebrate (ver′tĕ-brāt) 387
invertebrate (in-ver′tĕ-brāt) 387
sessile filter feeder (ses′il fil′ter fēd′er) 387
spicule (spik′ūl) 388
cnidarian (ni-dah′re-an) 390
gastrovascular cavity (gas′′tro-vas′ku-lar kav′i-te) 390

mesoglea (mes′′o-gle′ah) 390
nematocyst (nem′ah-to-sist) 390
radially symmetrical (ra′de-al-e sĭ-met′re-kal) 390
bilaterally symmetrical (bi-lat′er-al-e sĭ-met′re-kal) 390

hydra (hi′drah) 390
sac body plan (sak bod′e plan) 392
tube-within-a-tube body plan (tūb with-in′ ah tūb bod′e plan) 392
cephalization (sef′′al-i-za′shun) 392
hermaphroditic (her-maf′′ro-dit′ik) 392
pseudocoelom (su′′do-se′lom) 394

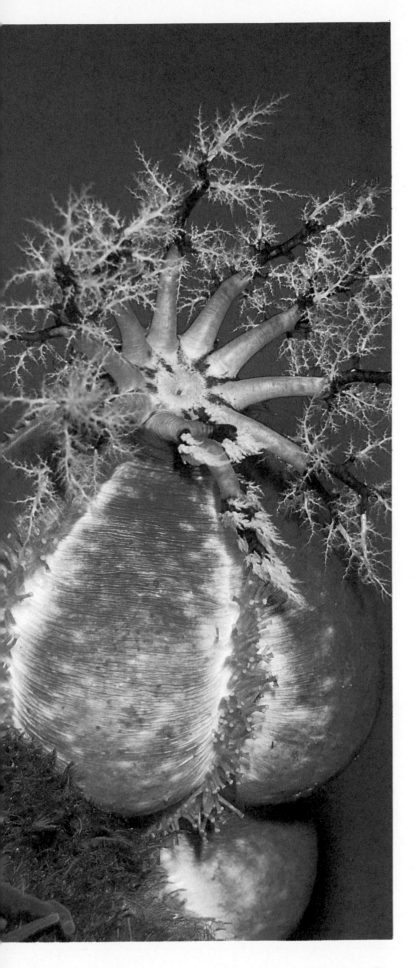

CHAPTER 27

Animals: Mollusks to Chordates

Your study of this chapter will be complete when you can:

1. *List the embryological differences between protostomes and deuterostomes and list the phyla in each group.*

2. *Describe the general characteristics of mollusks. Contrast the anatomy of the snail, clam, and squid, indicating how each is adapted to its way of life.*

3. *Describe the characteristics of annelids. Contrast the anatomy of predaceous polychaetes and earthworms.*

4. *Describe the characteristics of arthropods, and list the five major classes, giving examples of each.*

5. *Describe the anatomy of the grasshopper, indicating how it is adapted to a terrestrial existence.*

6. *Describe the characteristics of echinoderms, in particular, the sea stars.*

7. *List the three chordate characteristics and tell which animals are the primitive chordates.*

A blue and red sea cucumber, an echinoderm. Normally you would not suspect that this animal is at all related to vertebrates. What is the evidence? The early development of echinoderms is similar to vertebrates and their larval stage is bilaterally symmetrical. Later they take on the radial symmetry you see here.

All the major phyla of animals arose in the Cambrian seas (fig. 27.1). Many groups remained in the water, but some ventured onto land and became very successful there. In this chapter we will have an opportunity to contrast animal adaptations suitable to living in water with adaptations suitable to living on land.

Two Evolutionary Paths

The phyla we will study in this chapter are divided into the protostomes and deuterostomes (fig. 26.2). There are three major differences in the embryological development of these two groups of animals (fig. 27.2). First, cleavage of the very first cells occurs differently. In **protostomes,** spiral cleavage occurs and daughter cells sit in grooves formed by the previous cleavages. It may also be noted that the fate of these cells is fixed and determinate in protostomes; each can only contribute in one particular way to development. In **deuterostomes,** radial cleavage occurs and the daughter cells sit right on top of the previous cells. The fate of these cells is indeterminate; that is, if they are separated from each other, each cell can go on to become a complete organism. As development proceeds, a hollow sphere forms, and the indentation that follows produces an opening called the blastopore. In protostomes the blastopore becomes the mouth (*proto* = first, *stome* = mouth); in deuterostomes, this opening becomes the anus, and only later does a new opening form the mouth (*deutero* = second, *stome* = mouth). Finally, the coelom develops differently in the two groups. In protostomes, the mesoderm arises from cells located near the embryonic blastopore, and a splitting occurs that produces the coelom, called a schizocoelom. In deuterostomes, the coelom arises as a pair of mesodermal pouches from the wall of the primitive gut. The pouches enlarge until they meet and fuse, forming an enterocoelom.

Checking with the evolutionary tree (fig. 26.2), you can see which phyla are the protostomes and which are the deuterostomes:

Protostomes	Deuterostomes
annelids	echinoderms
arthropods	chordates
mollusks	

The **coelomate** animals are divided into two groups. In the protostomes, there is spiral cleavage, the blastopore becomes the mouth, and there is a schizocoelom. In the deuterostomes, there is radial cleavage, the blastopore becomes the anus, and there is an enterocoelom.

Figure 27.1

Artist's interpretation of the shallow seas of the Cambrian. During a short few tens of millions of years during this period, all the major phyla of animals apparently arose. The animals depicted here are found as fossils in the Burgess Shale, a formation of the Rocky Mountains of British Columbia that originated in the sea. Not all of the Cambrian animals survive today. They have become extinct, which is the fate of most species. The rest have been evolving since they first arose.

Figure 27.1

Figure 27.2

Coelomate animals are divided into two groups—the protostomes and the deuterostomes—based on the embryological evidence given here.

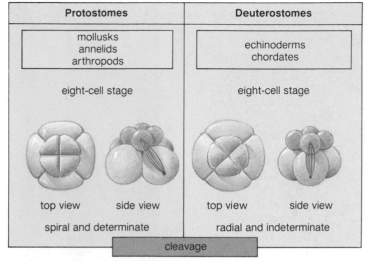

Protostomes	Deuterostomes
mollusks annelids arthropods	echinoderms chordates

eight-cell stage | eight-cell stage

top view side view | top view side view

spiral and determinate | radial and indeterminate

cleavage

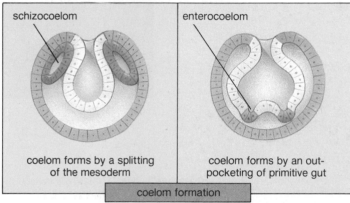

schizocoelom | enterocoelom

coelom forms by a splitting of the mesoderm | coelom forms by an out-pocketing of primitive gut

coelom formation

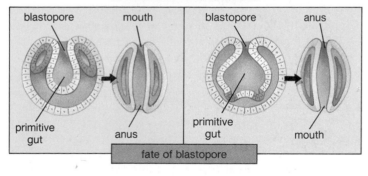

blastopore mouth | blastopore anus

primitive gut anus | primitive gut mouth

fate of blastopore

Mollusks

There are over 100,000 living species of **mollusks** (phylum Mollusca), more than twice the number of vertebrate species! You can well image, then, that there must be quite a number of life-styles among the mollusks, which include such animals as squids, octopuses, clams, scallops, oysters, snails, and slugs. All these animals share three common features: a visceral mass, a mantle, and a foot (fig. 27.3). The *visceral mass* contains the internal organs, including a highly specialized digestive tract, paired kidneys, and reproductive organs. The phylum name comes from the Latin word *mollis,* meaning soft, which refers to the visceral mass. The *mantle* is a covering that lies to either

Figure 27.3

a. Relationship of parts of a hypothetical mollusk. *b.* Photograph of a chiton (class Polyplacophora), a mollusk whose anatomy may have changed very little from its primitive state. On the upper surface is a row of eight overlapping plates. Below is a flat foot used for creeping along or clinging to rocks. The chiton scrapes algae and other plant food from rocks with its well-developed radula.

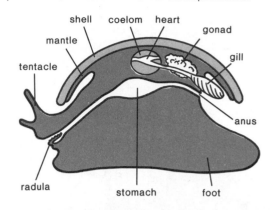

shell coelom heart gonad gill mantle tentacle radula stomach foot anus

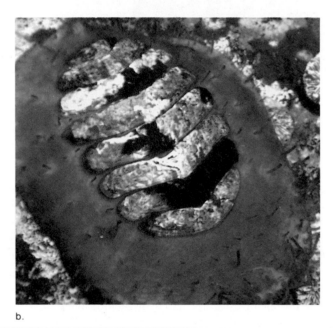

b.

side, but does not completely enclose, the visceral mass. It may secrete a shell and/or contribute to the development of gills or lungs. The space between the folds of the mantle is called the mantle cavity. The *foot* is a muscular organ that may be adapted for locomotion, attachment, food capture, or a combination of functions. Another feature often present is a *radula* (fig. 35.6), an organ that bears many rows of teeth and is used to obtain food.

The earliest mollusks (fig. 27.3) are presumed to have had a nervous system consisting of a *nerve ring and two longitudinal cords,* one serving the foot and the other the visceral mass. The *coelom is much reduced* and is largely limited to the region around the heart. In most mollusks, the heart pumps blood through vessels that empty into open spaces, called *sinuses,*

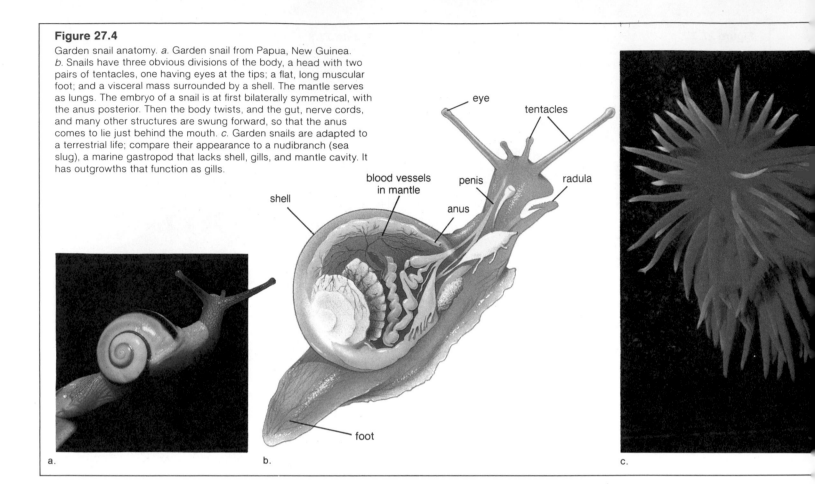

Figure 27.4

Garden snail anatomy. *a.* Garden snail from Papua, New Guinea. *b.* Snails have three obvious divisions of the body, a head with two pairs of tentacles, one having eyes at the tips; a flat, long muscular foot; and a visceral mass surrounded by a shell. The mantle serves as lungs. The embryo of a snail is at first bilaterally symmetrical, with the anus posterior. Then the body twists, and the gut, nerve cords, and many other structures are swung forward, so that the anus comes to lie just behind the mouth. *c.* Garden snails are adapted to a terrestrial life; compare their appearance to a nudibranch (sea slug), a marine gastropod that lacks shell, gills, and mantle cavity. It has outgrowths that function as gills.

shell
blood vessels in mantle
eye
tentacles
penis
anus
radula
foot

a.   b.   c.

before being collected in vessels and returned to the heart. This arrangement is called an *open circulatory system* because the blood is not always enclosed within blood vessels.

Snails and Relatives

Snails, whelks, conchs, periwinkles, and sea slugs are representative **gastropods** (class Gastropoda) that are found usually in marine but also freshwater habitats. In addition, garden snails (fig. 27.4) and slugs are gastropods adapted to terrestrial habitats. Many gastropods are herbivores and use their radula to scrape food from surfaces. Others are carnivorous, using their radula to bore through surfaces, such as bivalve shells, to obtain food.

Gastropods have an elongated, flattened foot for crawling. In most, there is a head region and a coiled shell to protect the visceral mass—sea slugs (nudibranchs) and terrestrial slugs, however, lack a shell. Although these animals begin life bilaterally symmetrical, the larva undergoes a *torsion,* or twisting, that brings the anus and mantle cavity downward, then forward, and around to a position above the head. Torsion positions the visceral mass squarely above the foot.

In aquatic gastropods, gills are found in the mantle cavity, but in those adapted to land, the mantle is richly supplied with blood vessels and functions as a lung when air is moved in and out through respiratory pores. As another adaptation to life out of water, terrestrial gastropod development does not include the swimming larval stage found in aquatic species.

Bivalves

Clams, oysters, scallops, and mussels are all **bivalves** (class Bivalvia). The term bivalve refers to their two shells, which are hinged and closed by powerful muscles. Clams use their hatchet-shaped foot for burrowing in sandy or muddy soil, and mussels use theirs for production of threads that attach them to nearby objects. Scallops both burrow and swim; rapid clapping of the valves releases water in spurts and causes the animal to move forward.

The bivalves have large gills that hang down in the mantle cavity and are used for both respiration and gathering food. They are **filter feeders** and glean food from the water that enters and exits the mantle by way of *siphons,* located at the posterior end. Debris trapped on the *gills* is swept toward the mouth by ciliary action.

Bivalves have no head. Their nervous system consists of *three pairs of ganglia* (nerve centers) located anteriorly, posteriorly, and in the foot. These nerve centers are connected by nerves (fig. 27.5).

Squids and Relatives

Cephalopods (class Cephalopoda), include the squids (fig. 27.6) octopuses, and nautiluses, all of which are marine, fast-moving, swimming organisms. Both squids and octopuses can squeeze their mantle cavity so that water is forced out, thus propelling them rapidly backward by a sort of *jet propulsion.* Cephalopods

Figure 27.5

a. Bivalve anatomy is exemplified by the anatomy of a clam. The shell and the mantle on one side have been removed. Notice the three-ganglion nervous system and the lack of cephalization, a characteristic that is consistent with filter-feeding animals. *b.* This diagram of gill structure shows the flow of water through the mantle cavity. Water (blue arrows) enters the mantle cavity via the incurrent siphon and passes through pores entering the gills before exiting via the excurrent siphon. In the meantime, food particles trapped in mucus on the gills passes to the mouth (brown arrows).

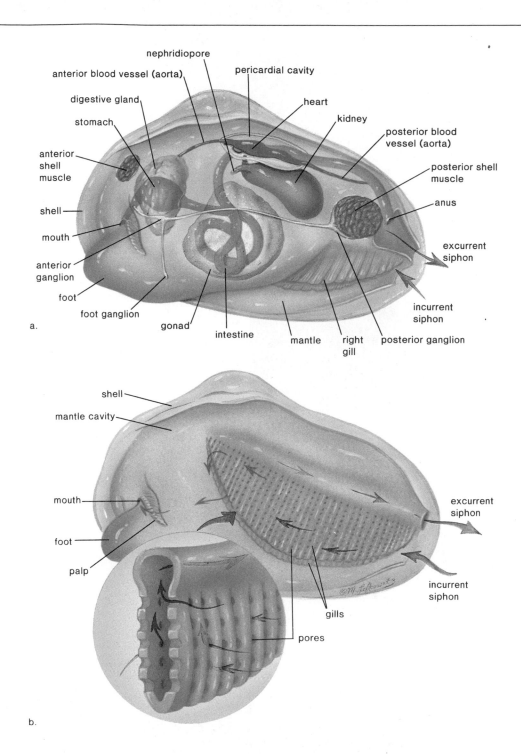

are adapted for a predatory way of life. The foot has become tentacles around the head (cephalopod means head-footed) and are used to seize prey. The powerful, parrotlike beak is used to tear apart the prey. They have well-developed sense organs, including focusing *camera-type eyes* that are very similar to those of vertebrates. Cephalopods, particularly octopuses, have well-developed brains and show a remarkable capacity for learning. For protection, they possess *ink sacs,* from which they can squirt a cloud of brown or black ink. This action often leaves a would-be predator completely confused.

Nautiluses are enclosed in shells, but the shell of squids is reduced and internal. Octopuses lack shells entirely.

Mollusks are a very diverse group, with a body plan composed of three parts: the foot, mantle, and visceral mass. The mostly herbivorous gastropods have a flattened foot, the filter-feeding bivalves have a two-part shell, and the predaceous cephalopods have tentacles about the head.

Figure 27.6

Squid anatomy. *a. Loligo pedlei.*
b. Squids live in the ocean and range
in size from 3.75 cm to 16.5 m. A
tough muscular mantle contains a
vestigial skeleton called the pen, and
surrounds the visceral mass, which is
elongated and lies in a horizontal
direction. Cephalization includes
vertebrate-type eyes that aid the
squid in recognizing its prey and in
escaping its enemies. The squid
moves quickly by jet propulsion when
large nerves signal the mantle to
contract and water to flow out
through the funnel which is a
modified siphon. *c.* Compare the
structure of the squid to the
chambered nautilus, a shelled
cephalopod with 94 small suckerless,
contractile tentacles for capturing
prey.

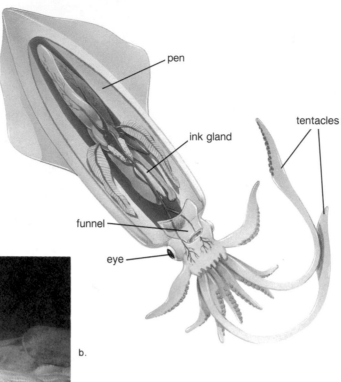

a.

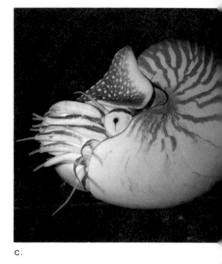

c.

b.

Figure 27.7

a. A sandworm such as *Neanthes* has fleshy lobes, the parapodia,
on each body segment. They are used for swimming and as
respiratory organs. Numerous chitinous bristles grow out from the
parapodia, and hence they are polychaetes, or "many bristled."
b. This worm is a predator; its small prey is captured by a pair of
strong chitinous jaws that evert with a part of the pharynx when the
worm is feeding. *c.* Compare (*b*) to the head region of a sedentary
tubeworm, which extends long ciliated tentacles into the water. The
cilia create currents, filter food particles, and move them toward the
mouth.

a.

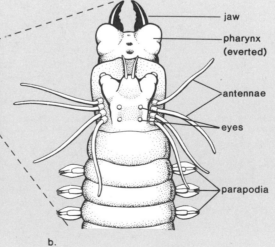

b.

c.

Annelids

Annelids (phylum Annelida, 10,000 species), represented by earthworms and many aquatic species, are **segmented** worms, as is externally evident in the rings that encircle the body. Internally, partitions divide the *well-developed, fluid-filled coelom* that acts as a **hydrostatic skeleton** (fig. 26.10). Annelids have both circular and lengthwise muscles. When lengthwise muscles contract, segments of the body shorten; when circular muscles contract, segments of the body elongate.

Annelids are *bilaterally symmetrical and have the tube-within-a-tube body plan*. The digestive tract shows specialization of parts along its length; for example, there may be a pharynx, stomach, and accessory glands. There is an extensive closed circulatory system, in which blood is always enclosed within blood vessels that run the length of the body, and branch to every segment. The nervous system consists of a *brain, connected to a ventral, solid, nerve cord, with a ganglion in each segment*. The excretory system consists of *paired* **nephridia,** which are coiled tubules in each segment that collect waste material from the coelom and excrete it through openings in the body wall.

Marine Annelids

Most annelids are marine and belong to the class Polychaeta, named for the presence of many setae (fig. 27.7). The **setae** are bristles that anchor the worm or help it move. In **polychaetes,** the setae occur in bundles on **parapodia,** paddlelike appendages found on most segments. Some polychaetes tend to be active, and, aided by their parapodia, wander about in search of prey. These worms have a definite head with sense organs. Other polychaetes are sedentary tube worms, with tentacles around their head. Water currents, created by the action of cilia, bring debris within reach of the tentacles, where it adheres. These worms are continuous filter feeders. Because the parapodia have a rich supply of blood vessels, they assist the process of gas exchange in polychaetes.

The polychaetes have breeding seasons, and only during these times do the worms have sex organs. Internal sex organs develop each year and produce either sperm or eggs. In some worms, the gametes can escape by way of the nephridia, whereas in others, the body literally breaks apart to release them. These events are synchronized, and this annual rhythm (behavior that occurs yearly) may be controlled by an internal biological clock. The zygote develops into a larval stage.

Earthworms

Earthworms (class Oligochaeta = few setae) (fig. 27.8), do not have well-developed heads or parapodia. Their setae protrude in clusters directly from the surface of their bodies. The placement of setae and other anatomical features indicate that the worms are segmented (table 27.1).

Figure 27.8

Earthworm anatomy. The drawing shows the internal anatomy of the anterior part of an earthworm's body. The closed circulatory system has five pairs of hearts. Earthworms are hermaphroditic; each worm has both male and female reproductive structures. The small sketch shows the location of the clitellum.

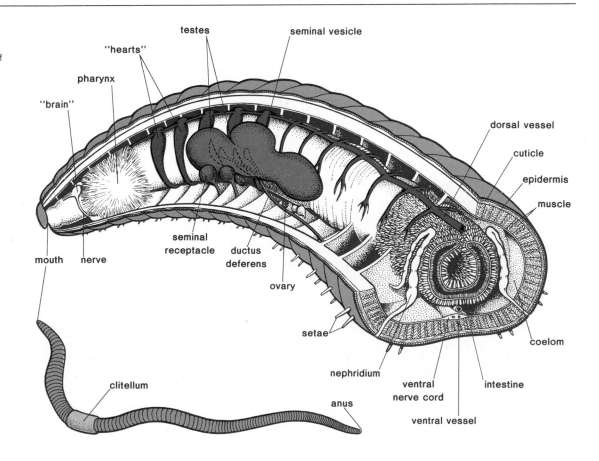

Table 27.1 Segmentation in the Earthworm

1. Body rings
2. Coelom divided by septa
3. Setae on each segment
4. Ganglia and lateral nerves in each segment
5. Nephridia in each segment
6. Branch blood vessels in each segment

Earthworms are restricted to soils containing adequate moisture because they use a *moist body wall for gas exchange.* When rainwater fills their underground burrows however, they must move up to the surface and remain there until the water percolates down and away from their burrows.

Earthworms feed by ingesting quantities of soil containing living and decaying organic matter. Much of what they eat passes through their bodies and is deposited as small piles of dirt (worm casts) on the soil surface. This activity of earthworms is important for soil turnover because, as Charles Darwin calculated, earthworms can carry and deposit as much as 7 to 16 English tons of soil per acre per year.

Earthworms and other **oligochaetes** are *hermaphroditic* (fig. 27.8), but self-fertilization does not occur. Two worms line up parallel to each other, facing opposite directions. The collarlike **clitellum** secretes mucus that protects the sperm as they pass between the worms. Several days later, the clitellum secretes a mucus sheath having a hard outer covering. As the worm backs out, eggs and the sperm received earlier are released into the sheath, where fertilization occurs. The sheath then closes, forming a cocoon in which the zygotes develop into young worms and eventually hatch from it.

Leeches

Leeches have the same body plan as other annelids, but among their modifications are two suckers, a small, oral one around the mouth and a large, posterior one. While some leeches are free-living, the ones most remembered are the blood-sucking species. Leeches are able to keep blood flowing and prevent clotting by means of a substance in their saliva known as hirudin, a powerful anticoagulant.

Annelids are segmented worms. Most of their organ systems show evidence of segmentation. These worms have a well-developed coelom divided by septa, a closed circulatory system, and a ventral, solid nerve cord. Marine annelids have adaptations for aquatic life and earthworms are adapted for terrestrial living.

Arthropods

Arthropods (phylum Arthropoda, 900,000 species) show such diversity and are adapted to so many different habitats that they

Figure 27.9
Fossil trilobite, an arthropod that has been extinct for about 225 million years. Note the regular segmentation of this animal named for the three lobes seen here. There are so many trilobites in the fossil record that 3,900 species have been recognized, and even the developmental stages of some species have been studied.

are often said to be the most successful of all animals. Arthropod means literally "jointed foot." Actually they have **jointed appendages** that are freely movable. Their bodies are segmented, and at one time, there was a pair of similar appendages on each body segment (fig. 27.9). In modern arthropods, some segments have fused into regions, such as a head, thorax, and abdomen (fig. 27.10). The appendages may be specialized for such functions as walking, swimming, reproducing, and eating.

Arthropods have a well-developed nervous system. There is a brain and a *ventral solid nerve cord.* The head bears various types of sense organs, including eyes of two types—compound and simple. The **compound eye** is composed of many complete visual units grouped in one structure. Each visual unit contains a separate lens and a light-sensitive cell.

The external skeleton of arthropods is composed primarily of **chitin,** a strong, flexible, nitrogenous polysaccharide. Because it is hard and nonexpandable, arthropods must **molt,** or shed, the exoskeleton as they grow larger. Before molting, the body secretes a new, larger exoskeleton, which is soft and wrinkled, underneath the old one. After enzymes partially dissolve and weaken the old exoskeleton, the animal breaks it open and wriggles out. The new exoskeleton then quickly expands and hardens.

Lobsters and Relatives

Members of the class Crustacea (40,000 species) are named for their hard shells. Although most are aquatic, they are very diverse in structure, habitat, and way of life. The small **crustaceans,** such as copepods and krill, live in the water where they feed on algae and serve as food for fish and whales. They are so numerous that despite their small size, some believe that we might be able to harvest them and use them for food. Larger crustaceans, such as shrimp, crabs, lobsters, and crayfish, are well known because they are frequently eaten by humans. These all have five pairs of walking legs, the first pair of which is often modified to form claws (fig. 27.11).

Figure 27.10

There are five major classes of Arthropods. *a.* Class Crustacea is represented by this crab with a large carapace and five pairs of legs. The first pair are pinching claws. *b.* Class Arachnida is represented by this land-dwelling scorpion, with four pairs of legs, poisonous claws, and stinging tail. *c.* Class Chilopoda is represented by this centipede, a carnivorous animal, with a pair of appendages on every segment. *d.* Class Diplopoda is represented by this millipede, a scavenger that seems to have two pairs of appendages on each segment because every two segments are fused. Class Insecta is represented by (*e*) a swallowtail butterfly with wings and (*f*) an ant with clearly defined head, thorax, and abdomen. Insects have three pairs of legs.

a.

b.

c.

e.

d.

f.

Figure 27.11
Decapods are the larger crustaceans and include lobsters, crayfish, shrimp, and crabs. *a.* Photograph of a shrimp. *b.* A shrimp has five pairs of walking legs. The head also bears five pairs of appendages. The first two are the sensory antennae, the next three are mouthparts for eating. Gills lie above the walking legs, along the thorax, and there are swimmerets attached to the abdomen. The last two segments bear the uropods and the telson, which make up a fan-shaped tail for swimming backwards. *c.* Notice how the first pair of walking legs are claws in the lobster.

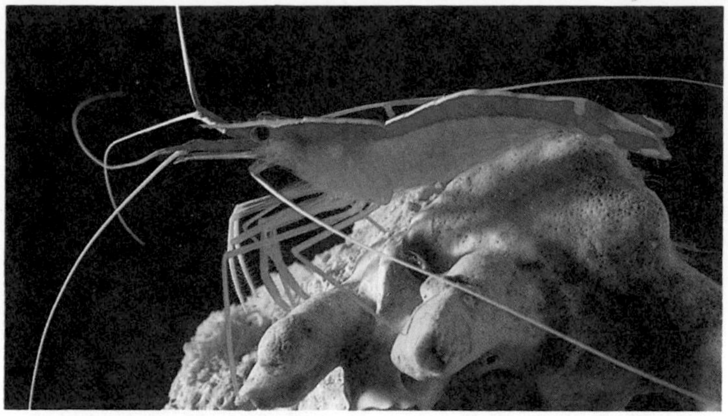

The sow bugs and pill bugs are crustaceans that live on land, even though they have few adaptations for terrestrial life and are restricted to moist places. Barnacles are aquatic sessile, filter-feeding crustaceans that live attached to a substrate. Goose barnacles have a stalk that looks like a long neck, while acorn barnacles resemble acorns attached to rocks. Barnacles begin life as free-swimming larvae, but they undergo a metamorphosis that transforms their swimming appendages to cirri, which are feathery structures that allow them to filter feed. They only extend these when they are submerged so most visitors to the seashore see only the calcareous plates that enclose the body of the barnacle.

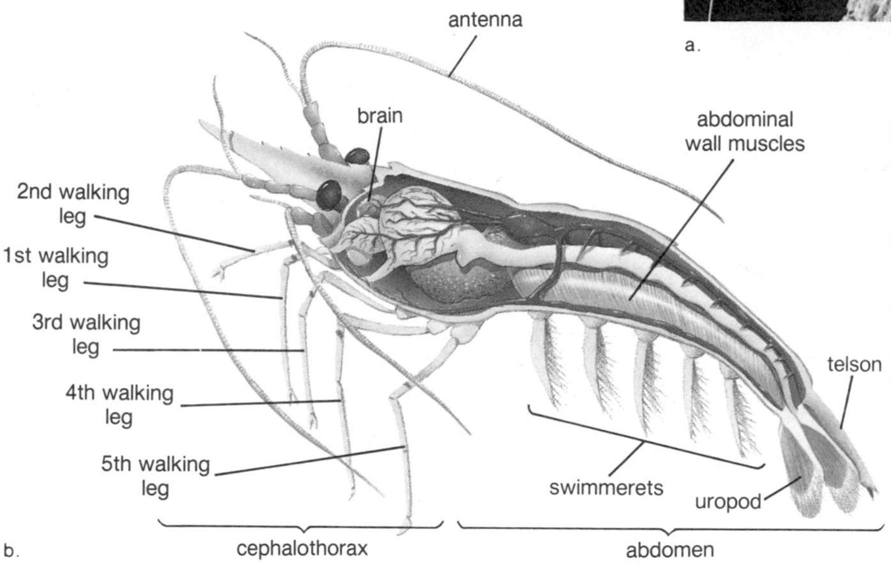

Insects

Insects (class Insecta, 1,000,000 species) include more known species than all other animal species combined. Nearly 90% of all arthropod species are insects. *Wings* enhance the insects' ability to survive by providing a new way of escaping enemies, finding food, facilitating mating, and dispersing the species.

The grasshopper (fig. 27.12) is often studied as a representative insect. There are three pairs of legs, one of which is suited to jumping. There are two pairs of wings. The forewings are tough and leathery, and when folded back at rest, protect the broad, thin hindwings. The first segment of the grasshopper bears on its lateral surface a large **tympanum** for the reception of sound waves.

The digestive system is suitable for a grass diet. The food is broken down by the grinding action of mouth parts and by the chemical action of saliva. The food is ground even more finely in the gizzard. Chemical digestion is completed by enzymes secreted by the gastric caeca. Excretion is carried out by **Malpighian tubules,** which extend into the hemocoel, a network of open spaces where blood flows around the organs. (The hemocoel represents the coelomic cavity in arthropods.) Malpighian tubules form a solid, nitrogenous waste that passes out of the animal by way of the digestive tract. Solid waste allows the animal to conserve water.

Respiration occurs when air enters small tubules called **tracheae** through openings in the exoskeleton called spiracles. The tracheae branch and rebranch, ending in moist areas adjacent to cells, where the actual exchange of gases takes place.

The grasshopper's reproductive system is adapted to life on land (fig. 27.12). The male passes sperm to the female by way of a penis, and fertilization is internal. In this way, both sperm and zygotes are protected from drying out. The female deposits the fertilized eggs in the ground with the aid of her ovipositor.

Figure 27.12

The anatomy of a female grasshopper illustrates many adaptations to life on land. *a.* Externally, the hard skeleton prevents loss of water. There are spiracles, openings in the skeleton, to admit air into tracheae (air tubes for respiration). The tympanum uses air waves for sound reception and the hopping legs and wings are for locomotion. The ovipositor deposits eggs in soil. *b.* Internally, the digestive system is adapted to digesting grass. The Malpighian tubules excrete a solid nitrogenous waster (uric acid). A seminal receptacle receives sperm from the male, which has a penis. Internal fertilization prevents the gametes from drying out.

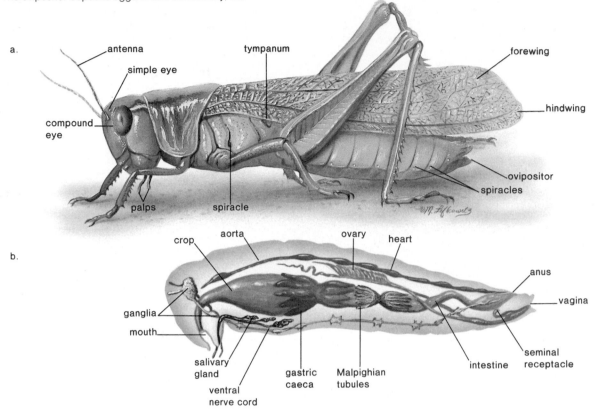

Grasshoppers undergo incomplete **metamorphosis,** a gradual change in form as the animal matures. The immature grasshopper, called a nymph, is recognizable as a grasshopper, even though it differs somewhat in shape and form from the adult. Other insects undergo complete metamorphosis involving drastic changes in form. At first, the animal is a wormlike larva with chewing mouthparts. It then forms a case or cocoon about itself and becomes a *pupa.* During this stage, the body parts are completely reorganized after which the *adult* emerges. This life cycle allows the larvae and adults to make use of different food sources.

Insects also show remarkable behavioral adaptations, exemplified by the social systems of bees, ants, termites, and other colonial insects. Insects are so numerous and so diverse that the study of this one group is a major specialty in biology called entomology.

Spiders and Relatives

Spiders, scorpions, ticks, and mites are **arachnids** (class Arachnida). Spiders are predaceous, usually spinning a web in which to trap insects. They have six pairs of appendages. The first pair are modified as fangs, with ducts from poison glands. The second pair have become basal parts for chewing. The other four are walking legs that end in claws.

The body of a spider doesn't show much obvious segmentation. The head and thorax are fused into a cephalothorax, and there is also an abdomen. Spiders don't have compound eyes; instead the eyes are simple. Spiders are adapted to living on land, as witnessed by the presence of **book lungs,** which are air pockets within a cavity surrounded by blood.

Scorpions, which are more common in warmer climates, are also predators. Ticks and mites are often parasites. Ticks suck the blood of vertebrates and are sometimes transmitters of diseases, such as Rocky Mountain spotted fever or Lyme disease. Chiggers have larvae that feed on the skin of vertebrates.

Centipedes and Millipedes

Centipedes (class Chilopoda) have a body composed of a head and trunk, the latter having many segments, with a pair of walking legs on each segment. They are carnivorous animals and the head bears antennae, and mouthparts with jaws.

Millipedes (class Diplopoda) have the same basic organization as centipedes but some segments are fused and therefore they appear to have two pairs of walking legs on each segment. Millipedes dwell in the soil, feeding on dead, organic matter.

Evolutionary Relationships among the Protostomes

The protostomes (annelids, arthropods, and mollusks) share embryological features that indicate they are related, but how they are related is not known. Mollusks are not segmented except for *Neopilina gelatheae* (*see* figure *a*). This animal was presumed to be extinct for the past 500 million years but ten living specimens were dredged up from a depth of more than 3,500 m in the Pacific Ocean near Costa Rico in 1952. There is a segmental arrangement of gills, nephridia, and muscles, which could indicate the animal is similar to a possible common ancestor of both mollusks and annelids. On the other hand, it is possible that the segmentation seen in *Neopilina* could have arisen independently in the molluskan line and that both mollusks and annelids are descended from nonsegmented flatworms.

A possible direct relationship between arthropods and annelids can also be found. *Peripatus* (phylum Onychophora) (*see* figure *b*) is called the walking worm, because it has 14–400 pairs of short, stumpy legs, and possesses organs with annelid or arthropod characteristics. Its excretory organs and musculature are more like those of annelids, and its respiratory and circulatory system are more like those of arthropods. It has a cuticle of chitin and modified appendages that serve as jaws. Although the appendages of *Peripatus* are not jointed, the fossil record does contain wormlike animals with jointed appendages. It's possible that arthropods and annelids share a common ancestor or perhaps arthropods are descended directly from annelids.

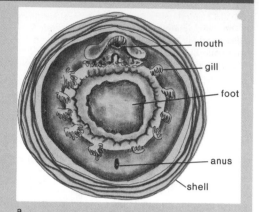

mouth
gill
foot
anus
shell

a.

b.

Figure 27.13

Echinoderm diversity. *a.* A sea star with one of its arms lifted, showing extended tube feet. The feet attach and contract, pulling the starfish along. When a starfish attacks a clam, it arches its body over the shell and by the concerted action of the tube feet, forces the clam to open. Then it everts a portion of its stomach to digest the contents of the clam. *b.* Sea urchins have fused skeletal plates and long, movable spines that protrude from their skin. Many organisms, ranging from sea stars to mammals, feed on sea urchins. For example, sea otters float on their backs and use stones to crack open urchins held on their chests. Sea urchin eggs are used for human food, especially in Japan where large quantities are imported each year from the west coast of the United States and Canada.

a.

b.

Figure 27.14

Sea star anatomy. Aside from the water vascular system shown here in color, there are gonads and digestive glands in each arm. The two-part stomach is in the central disc with the anus uppermost and the mouth hidden beneath, on the ventral surface.

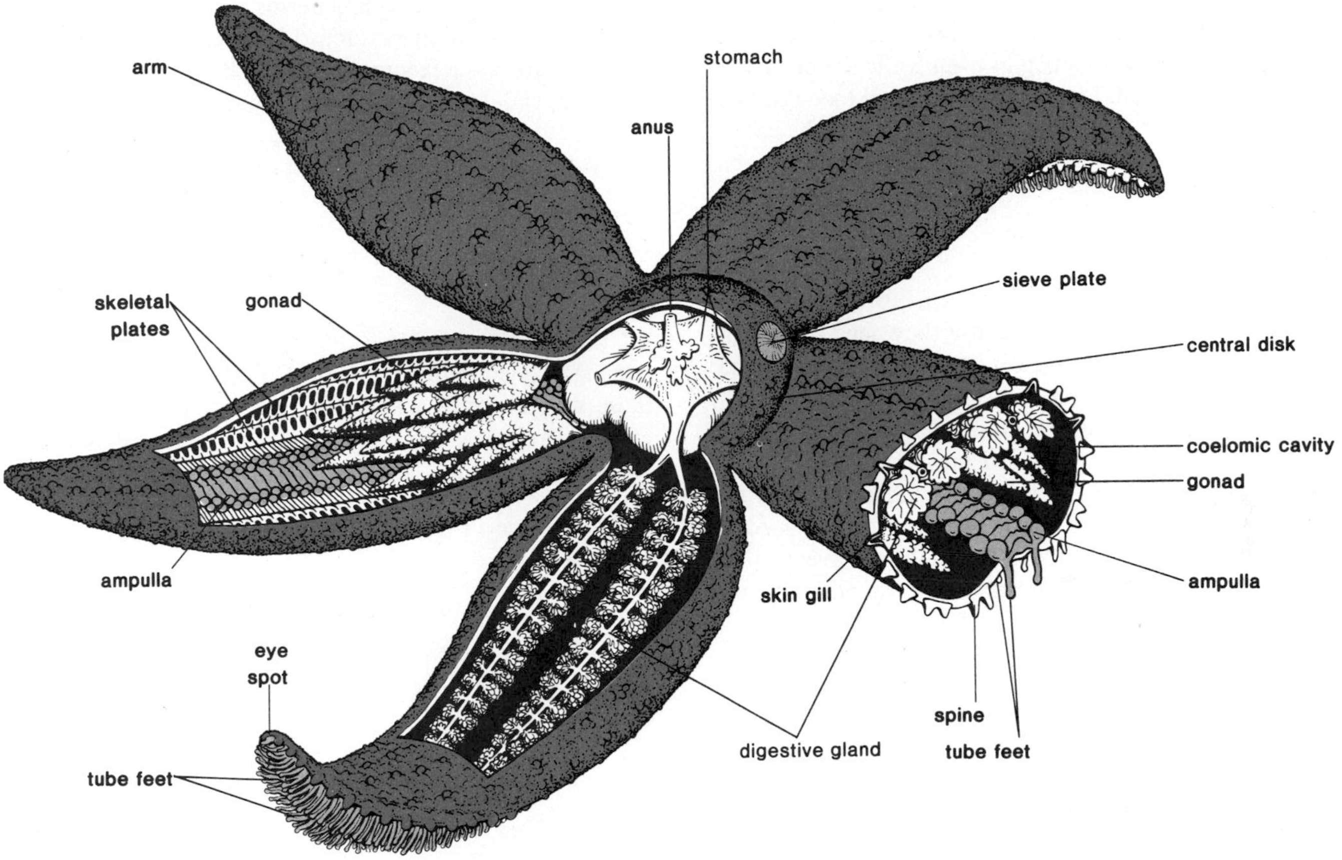

Arthropods are the most numerous and varied of all the animal phyla. They are segmented, with an external skeleton and jointed appendages. Several groups (insects, arachnids, centipedes, and millipedes) contain species that are adapted to terrestrial life.

Echinoderms

Echinoderms (phylum Echinodermata, 6,000 species) (fig. 27.13) are a marine group that includes sea stars (starfishes), sea urchins, sand dollars, sea cucumbers, and sea lilies. The *radial symmetry* of these animals is usually reflected in a five-part body organization. They have an *internal skeleton* (endoskeleton) consisting of spine-bearing plates. The spines stick out through the delicate skin (especially in sea urchins), which accounts for their name. Echinoderm comes from the Greek words meaning "spiny skin."

Echinoderms are deuterostomes, with a coelom that develops similarly to that of the chordates (fig. 27.2). Coelomic fluid circulates substances, and wandering amoeboid cells clean up particulate wastes. Gas exchange takes place through *gills* that are tiny, fingerlike extensions of the skin. Echinoderms have

a central *nerve ring and nerve branches* extending into the multiple body divisions. They are capable of coordinated but slow responses and body movements.

Movement is facilitated by **tube feet** (fig. 27.14), which are part of a *water vascular system*. Water enters by a *sieve plate* into an arrangement of major canals extending into each portion of the body. Small canals to the tube feet branch off the major canals. Each of the small canals ends in an expanded region called an *ampulla*. Contraction and expansion of the ampullae control the tube feet. By alternating the attachment and release of the tube feet, the animal moves.

Sexes are separate in echinoderms. They shed gametes into the water, where fertilization occurs. The zygotes become free-swimming, bilaterally symmetrical larvae that undergo metamorphosis to become radially symmetrical adults.

Sea Stars

Sea stars (starfishes) (class Asteroidea) are the most familiar of the echinoderms. The body of most sea star species is flattened and has a central disc with five, or a multiple of five, sturdy arms (rays) extending outward from it. Each arm has rows of tube feet in a groove that extends along its ventral surface.

Sea stars commonly feed on clams, oysters, and other bivalve mollusks. To feed, a sea star positions itself over a bivalve and attaches some of its tube feet to each side of the shell. By working its tube feet in alternation, it pulls the shell open. A very small crack is enough for the starfish to push the larger, lower part of its stomach out of its mouth and through the crack, so that it contacts the soft parts of the bivalve. The stomach secretes enzymes, and digestion begins even while the bivalve is attempting to close its shell. Later, partly digested food is taken into the sea star's body, where digestion continues in the upper part of the stomach and in the digestive glands found in each arm.

Other Echinoderms

There are several other classes of echinoderms. Brittle stars can move more rapidly than starfishes because they bend their long arms to push themselves along quickly, instead of relying on the slow action of the tube feet. When the long arms are injured, they are discarded, and new ones grow in their place. Sand dollars have skeletal plates that are fused into a single, flattened unit. The surface has a five-part, flowerlike pattern caused by pores in the plates. These pores permit the extension of special tube feet modified for respiration. Sand dollars have furlike spines, rather than the long, stiff movable kind found in sea urchins (fig. 27.13b). Surprisingly, sea urchin spines are used as molds for the production of artificial human blood vessels. To feed, sea urchins extrude teeth from a dental structure called Aristotle's lantern and scrape algae and other food from rocks.

Sea cucumbers, which have reduced skeletons and leathery bodies, differ greatly from other echinoderms. A sea cucumber lies on its side and traps food particles in mucus on the surface of tentacles set in a ring around its mouth. It puts one tentacle at a time into its mouth and scrapes off the food. If attacked by a predator, a sea cucumber can eject most of its internal organs. This behavior gives the sea cucumber an opportunity to move away and begin regeneration of the lost organs, while the startled predator is left with a writhing mass of body organs.

Unlike all other echinoderms, sea lilies have mouths directed upward and live attached by a stalk to a substrate. Most sea lilies are filter feeders, and they have branched arms with small appendages that sweep food out of the water.

Echinoderms are radially symmetrical with a spiny skin. They move by tube feet, which are part of their water vascular system. Their other body systems are rather primitive.

Chordates

Among the **chordates** (phylum Chordata, 39,000 species) are those animals with which we are most familiar (fishes, amphibians, reptiles, birds, and mammals), as well as the less familiar tunicates and lancelets. To be considered a chordate, an animal must have the following three basic characteristics at some time in its life history:

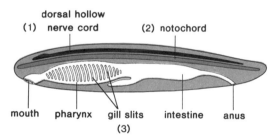

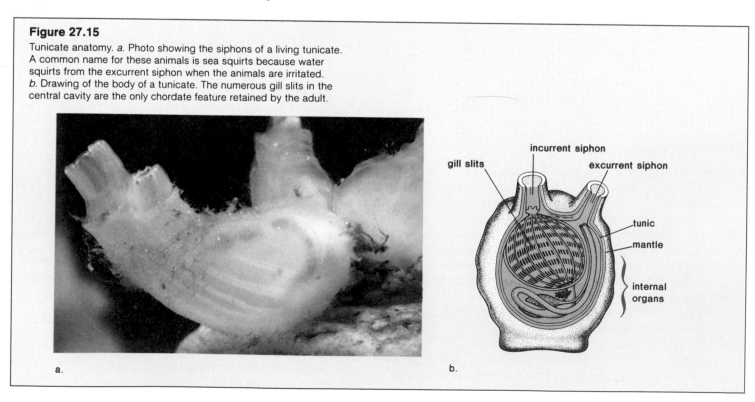

Figure 27.15

Tunicate anatomy. *a.* Photo showing the siphons of a living tunicate. A common name for these animals is sea squirts because water squirts from the excurrent siphon when the animals are irritated. *b.* Drawing of the body of a tunicate. The numerous gill slits in the central cavity are the only chordate feature retained by the adult.

1. A dorsal hollow nerve cord. By hollow, it is meant that the cord contains a canal that is filled with fluid.
2. A dorsal supporting rod called a **notochord.** This is replaced by the vertebral column in most adult vertebrates.
3. Gill pouches (slits). These are seen only during embryological development in most vertebrate groups.

Two groups of animals, **tunicates** (subphylum Urochordata) and **lancelets** (subphylum Cephalochordata) are invertebrate filter-feeding chordates. The tunicates have the three chordate characteristics only when they are larvae; as adults they are much modified (fig. 27.15). The lancelet is the only chordate to have the three characteristics as an adult (fig. 27.16). In vertebrates, the **notochord** is replaced by the vertebral column.

How the primitive chordates are related to the vertebrates has not been exactly determined. However, all three groups probably descended from an ancestor that was in existence sometime during the Cambrian Period, when the chordates appeared.

The invertebrate chordates include the tunicates and lancelets. Lancelets are the best example of a chordate that possesses the three chordate characteristics as an adult.

We are going to continue our study of the chordates in the next chapter, where we will trace the evolutionary history of humans, who are mammalian vertebrates.

Figure 27.16

The lancelet *Branchiostoma* (amphioxus). *a.* Diagrammatic illustration showing lancelets retain the three chordate characteristics as adults. Lancelets are filter feeders that take food from the water that passes out the numerous gill slits. *b.* Photomicrograph of several individuals in which segmentation of the musculature is obvious. Sometimes lancelets swim freely. Other times, they filter feed while partially buried in the bottom soil. Photograph by Carolina Biological Supply Company.

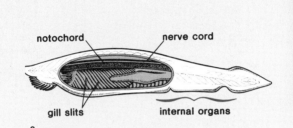

notochord   nerve cord

gill slits   internal organs

a.

b.

Summary

1. There are two groups of coelomate animals. In the protostomes, the blastopore becomes the mouth, and in the deuterostomes, it becomes the anus. Other differences between the two are spiral cleavage and a schizocoelom in protostomes, versus radial cleavage and an enterocoelom in deuterostomes.
2. Mollusks, annelids, and arthropods are protostomes. The body of a mollusk typically contains a visceral mass, a mantle, and a foot. Some have a head and a radula. There is a nerve ring and two cords, a reduced coelom, and an open circulatory system. Snails are adapted to life on land, squids to an active life, and clams to a sedentary life. Bivalves are filter feeders. Water enters and exits by siphons; food trapped on the gills is swept toward the mouth. The nervous system has three major ganglia.
3. Annelids are segmented worms. They have a well-developed coelom, closed circulatory system, ventral solid nerve cord, and paired nephridia. Polychaetes (many bristles) are marine worms that have parapodia. They may be predaceous, with a defined head region, or filter feeders, with ciliated tentacles to filter food from the water. Earthworms are oligochaetes (few bristles) that use the body wall for gas exchange. Leeches are also in this phylum.
4. Arthropods (the most varied and numerous of animals) have an external skeleton of chitin (necessitating molting) and jointed appendages. Their body segments are often fused. Cephalization, with compound and simple eyes and a ventral solid nerve cord, is typical. The five major classes are: crustaceans (e.g., lobster, shrimp, krill, sow and pill bugs, and barnacles); insects (e.g., butterflies, grasshoppers, bees, and beetles); arachnids (e.g., spiders, scorpions, ticks, and mites); centipedes; and millipedes.
5. Many insects are adapted to life on land. Grasshoppers have wings and hopping legs, a tympanum for sound reception, a specialized digestive system for a grass diet, excretion of solid, nitrogenous waste by Malpighian tubules, tracheae for respiration, internal fertilization, and incomplete metamorphosis.
6. Echinoderms and chordates are deuterostomes. Echinoderms (e.g., sea stars, sea urchins, sea cucumbers, and sea lilies) have radial symmetry and an internal skeleton with spines. Notable are their tiny skin gills, nerve ring with branches, and water vascular system for locomotion. Each arm of a sea star contains branches from the nervous, digestive, and reproductive systems.

7. Chordates (tunicates, lancelets, and vertebrates) have a notochord, dorsal hollow nerve cord, and gill pouches at one time in their life history. Lancelets are the only chordate to have the three characteristics in the adult stage. In vertebrates, the notochord is replaced by the vertebral column.

8. Table 27.2 compares anatomical features of the animal phyla studied in this chapter.

Table 27.2 Comparison of Animal Phyla

	Symmetry	Germ Layers	Body Plan	Coelom	Organs	Segmentation
Sponges				No	No	No
Cnidarians	Radial	2	Sac	No	No	No
Flatworms	Bilateral	3	Sac	No	Yes	No
Roundworms	Bilateral	3	Tube-within-a-tube	Pseudo-coelom	Yes	No
Mollusks	Bilateral	3	Tube-within-a-tube	Coelom (reduced)	Yes	No
Annelids	Bilateral	3	Tube-within-a-tube	Coelom	Yes	Yes
Arthropods	Bilateral	3	Tube-within-a-tube	Coelom (reduced)	Yes	Yes
Echinoderms	Radial	3	Tube-within-a-tube	Coelom	Yes	No
Chordates	Bilateral	3	Tube-within-a-tube	Coelom	Yes	Yes

Objective Questions

1. Which of these does not pertain to a protostome?
   a. blastopore becomes the anus
   b. spiral cleavage
   c. schizocoelom
   d. annelids, arthropods, and mollusks

2. Which of these is an adaptation to a terrestrial way of life in the snail?
   a. using the mantle instead of gills for respiration
   b. cephalization with antennae and eyes
   c. flat foot for crawling
   d. torsion so that anus is above the head

3. Which of these is mismatched?
   a. clam—gills
   b. lobster—gills
   c. grasshopper—book lungs
   d. polychaete—parapodia

4. Which of these is an incorrect statement?
   a. Spiders are carnivores in the phylum Arthropoda.
   b. Clams are filter feeders in the phylum Mollusca.
   c. Earthworms are scavengers in the phylum Arthropoda.
   d. Lancelets are filter feeders in the phylum Chordata.

5. A radula is a unique organ for feeding found in
   a. mollusks.
   b. annelids.
   c. arthropods.
   d. echinoderms.

6. Which of these is mismatched?
   a. crayfish—walking legs
   b. clam—hatchet foot
   c. grasshopper—wings
   d. earthworm—many tube feet

7. Which of these phyla contains animals that have both a well-developed coelom and a closed circulatory system?
   a. mollusk
   b. arthropod
   c. annelid
   d. echinoderm

8. Which of these is mismatched?
   a. mollusk—schizocoelom
   b. insects—hemocoel
   c. echinoderms—enterocoelom
   d. chordates—schizocoelom

9. The tube feet of echinoderms
   a. are secondary respiratory organs.
   b. are a part of the water vascular system.
   c. help pass sperm to females during reproduction.
   d. All of these.

10. Which of these is not a chordate characteristic?
    a. dorsal supporting rod, the notochord
    b. dorsal hollow nerve cord
    c. gill pouches
    d. vertebral column

Study Questions

1. Contrast the development of protostomes and deuterostomes in three ways. Tell which phyla belong to each group.
2. Contrast the phyla with respect to symmetry and segmentation.
3. Name at least three members of each phylum studied in this chapter.
4. What are the three characteristics of all mollusks? Contrast the way of life of the garden snail, clam, and squid, pointing out their different adaptations.
5. What are the general characteristics of annelids? How is the earthworm adapted to its way of life?
6. What are the general characteristics of arthropods? Name the five major classes and give an example of each.
7. List different ways in which the grasshopper is adapted to a terrestrial habitat.
8. What are the general characteristics of echinoderms? Explain how the water vascular system works in sea stars.
9. What three characteristics do all chordates have sometime in their life history? What happens to the notochord in vertebrates?

Thought Questions

1. Based on your knowledge of the various animals described in this chapter, what adaptations would you expect to see in a predaceous carnivore as compared to a filter feeder?
2. Why is it sometimes helpful to use embryological data to indicate a relationship, in addition to the appearance of the adults?
3. How is the manner in which human beings reproduce an adaptation to life on land?

Selected Key Terms

protostome (pro'to-stōm) 400
deuterostome (du'ter-o-stōm'') 400
coelomate (sēl'o-māt) 400
mollusk (mol'usk) 401
annelid (ah'nel-id) 405
segmented (seg-ment'ed) 405
hydrostatic skeleton (hi''dro-stat'ik skel'ĕ-ton) 405
nephridia (nĕ-frid'e-ah) 405
arthropod (ar'thro-pod) 406
jointed appendage (joint'ed ah-pen'dij) 406
chitin (ki'tin) 406
molt (mōlt) 406
crustacean (krus-ta'shun) 406
Malpighian tubule (mal-pig'ī-an tu'būl) 408
tracheae (tra'ke-a) 408
metamorphosis (met''ah-mor'fo-sis) 409
echinoderm (e-kin'o-derm) 411
chordate (kor'dāt) 412
notochord (no'to-kord) 413

C H A P T E R 28

Vertebrate Diversity and Human Evolution

Your study of this chapter will be complete when you can:

1. Tell why vertebrates are placed in the phylum Chordata and how they are modified as adults.

2. Draw an evolutionary tree for the vertebrates.

3. List the three living classes of fishes, and trace the evolution of amphibians from fishes.

4. Give examples of modern-day amphibians, and explain why they are not well adapted to life on land.

5. Give examples of modern-day reptiles, and explain why they are well adapted to life on land.

6. Give examples of modern-day birds, and explain how they are adapted for flight.

7. List and give examples of the three types of modern-day mammals.

8. Give the complete classification of humans. Distinguish (with examples) between prosimians, anthropoids, hominoids, hominids, and humans.

9. List anatomical features of *Dryopithecus*, and tell how most are retained by modern-day apes, but not humans.

10. Draw and discuss a hypothetical evolutionary tree for hominids.

11. Contrast the australopithecines, Homo habilis, Homo erectus, Homo sapiens neanderthalis, *and* Homo sapiens sapiens.

Over half of the vertebrates alive today are fishes. They come in all sizes and shapes. A coral reef is home for these small butterfly fish. Like other animals they have adaptations to escape predation and others to help capture their own food. The bands break up the body and make them less obvious to a predator. The small snout allows them to get into the crevices of the coral in order to collect food.

In this chapter, we are going to study the animals most closely related to us and trace our immediate ancestry. Even though we have a strong interest in other vertebrate animals, it is well to keep in mind that most animal species are invertebrates, and that it is our bias that makes us see a world mostly populated by vertebrates.

Vertebrates

Vertebrates are in the phylum Chordata and show, at some time in their life histories, the three chordate characteristics (notochord, dorsal hollow nerve cord, and gill slits). The embryonic notochord, however, is generally replaced by a *vertebral column* composed of individual vertebrae. The main axis of the internal skeleton consists not only of a vertebral column, but also a skull that encloses and protects the brain. There is an extreme amount of cephalization, with complex sense organs. The eyes develop as outgrowths of the brain. The ears are primarily equilibrium devices in aquatic vertebrates, and also function as sound wave receivers in land vertebrates.

Vertebrates belong to the phylum Chordata and have the three chordate characteristics at some time during their life. The notochord is replaced by the vertebral column.

Vertebrates have *bilateral symmetry,* the coelomic, *tube-within-a-tube body plan,* and are *segmented.* There is a *closed circulatory system,* in which blood is contained entirely within blood vessels. The kidneys are important excretory and water-regulating organs that conserve or rid the body of water as necessary. The sexes are generally separate, and reproduction is usually sexual.

Occasionally, similar adaptations are seen in both invertebrates and vertebrates; both insects and birds have wings, and both fishes and squids have fins. Such structures are analogous rather than homologous, even though they serve the same function, because their basic structural plans are completely different and they did not evolve from a common ancestor.

We have mentioned previously that the Cambrian rocks contain fossils of all the major animal phyla (fig. 27.1). But only invertebrate chordates are found; there is no record of vertebrate beginnings at this time. The evolutionary tree of vertebrates (fig. 28.1) begins with the fishes, which evolved later.

Fishes

There are three classes of living **fishes** (22,000 species): the jawless fishes (class Agnatha), the cartilaginous fishes (class Chondrichthyes), and the bony fishes (class Osteichthyes). The earliest vertebrate fossils are jawless fishes of the Ordovician Period. They had heavy, bony armor and were probably filter feeders. Living representatives of the **jawless fishes** are hagfishes and lampreys (fig. 28.2). They are cylindrical, up to a meter long, with smooth, scaleless skin. The hagfishes are scavengers, feeding mainly on dead fish. The lampreys may be parasitic, in which case the round, muscular mouth serves as a sucker. The lamprey uses this sucker to attach itself to another fish and sucks nutrients from the host's circulatory system. Marine parasitic lampreys gained entrance to the Great Lakes when a canal from the St. Lawrence River was deepened. The lamprey population grew quickly and caused extensive reduction in the trout population of the Great Lakes in the early 1950s.

The first jawed vertebrates in the fossil record are **placoderms** (fig. 28.3), a class of fishes of the Devonian Period (table 28.1) that has become extinct. Jaws are a distinct advantage because they allow the development of an active, herbivorous or carnivorous life-style. It is not surprising, then, that these fishes also had paired pectoral fins and pelvic fins. Paired fins allow a fish to balance and maneuver well in the water. Unknown early jawed fishes were ancestral to the placoderms and to today's jawed fishes.

Cartilaginous Fishes

Sharks, rays, and skates are **cartilaginous fishes** (fig. 28.4); they have a *skeleton of cartilage* instead of bone. Their bodies are covered with small, toothlike scales called denticles, which is why a shark's skin feels like sandpaper. The menacing teeth of sharks and their relatives are simply larger, specialized versions of these scales.

Rays and skates have horizontally flattened bodies. They swim slowly along the bottom and feed on animals that they dredge up. Sharks have beautifully streamlined bodies and are fast-swimming predators in the open sea.

Sharks are of extreme interest to some people because they believe sharks attack and eat most anything in the water. While this may be true of some sharks, there are other sharks that are not active predators of large prey. Instead they are filter feeders like the basking sharks and whale sharks that ingest tons of small prey, collectively called *krill.*

Bony Fishes

The earliest **bony fishes** (class Osteichthyes) are thought to have evolved in freshwater habitats because their earliest fossils are found in rocks from lakes and rivers. In addition to gills, they possessed simple, *saclike lungs* connected to the anterior end of their digestive tracts. These lungs probably functioned as supplementary gas exchange surfaces that allowed them to breathe air when the stagnant water was oxygen-deficient. These fishes gave rise to groups of bony fishes alive today.

Figure 28.1

Evolutionary tree of the vertebrates. Extinct jawless fishes are the oldest vertebrates in the fossil record. Although they had a well-developed, sometimes armored head, they were most likely bottom feeders. Fishes with jaws that appeared during the late Silurian Period evolved into the cartilaginous fishes and the bony fishes. The ray-finned fishes, which are most common today, are descended from bony fishes with lungs; amphibians came from fleshy-finned fishes and were dominant during the Carboniferous Period. Stem reptiles also appeared around this time, and from them arose the dinosaurs, who ruled during the Jurassic Period, and the first mammals, who did not come into their own until the Tertiary Period. Birds may very well be of dinosaurian origin.

Figure 28.2

The jawless fishes of today are not at all like their fossil ancestor (fig. 28.1). *a*. The hagfishes are scavengers with sensory tentacles around the mouth and a row of slime glands along each side. *b*. Parasitic lampreys attach themselves to fish with their suckerlike mouth, rasp away the flesh with their horny teeth, and suck blood.

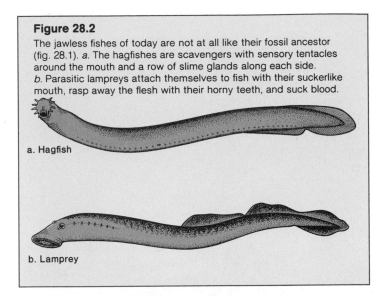

a. Hagfish

b. Lamprey

Figure 28.3

Painting of an extinct placoderm, the first type of jawed vertebrate in the fossil record. The appearance of jaws, which most likely evolved from the first anterior pair of gill arches, was a revolutionary event in the evolutionary history of vertebrates.

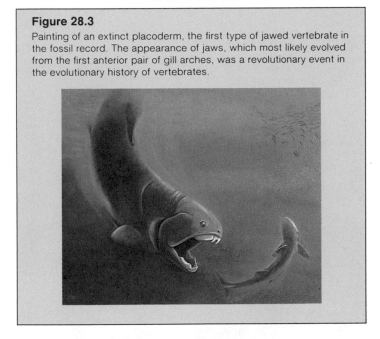

There are two subclasses of bony fishes alive today, the ray-finned fishes and the fleshy-finned fishes. The **ray-finned fishes** (subclass Actinopterygii), which have fins supported by spinelike rays, include most of today's bony fishes (fig. 28.5). In these fishes, the ancient lung eventually became a swim bladder. By secreting gases into or absorbing gases from this bladder, a fish can change its relative density and thus move up or down in the water. Modern bony fishes have broad, flattened scales covering some or all of their body surfaces. Their gills are located in an enclosed gill chamber that is covered and protected by a hard, bony flap, the operculum. Thus, they have only one gill opening on each side, instead of the separate openings for each gill slit that are found in sharks.

Table 28.1 The Geological Time Scale: Some Major Evolutionary Events of Animals

Era	Period	M.Y.A	Major Biological Events—Animals
Cenozoic	Quaternary	2	Age of Human Civilization
Cenozoic	Tertiary		First hominids appear
Cenozoic			Dominance of land by mammals and insects — Age of Mammals
Mesozoic	Cretaceous	65	Dinosaurs become extinct
Mesozoic		130	
Mesozoic	Jurassic		First mammals and birds appear
Mesozoic		180	
Mesozoic	Triassic		First dinosaurs appear — Age of Dinosaurs
Paleozoic	Permian	230	Reptiles appear and expand
Paleozoic		280	Decline of amphibians
Paleozoic	Carboniferous		
Paleozoic		350	Insects appear — Age of Amphibians
Paleozoic	Devonian		First amphibians move to land
Paleozoic		400	Age of Fishes
Paleozoic	Silurian	435	
Paleozoic	Ordovician		First fishes (jawless) appear
Paleozoic		500	Invertebrates dominate the seas
Paleozoic	Cambrian	600	Trilobites abundant — Age of Invertebrates

Figure 28.4
Cartilaginous fishes. *a.* The southern sting ray has a flattened shape and moves slowly across the bottom dredging up food. *b.* The bull shark is a streamlined predator that moves swiftly through the water to capture its prey.

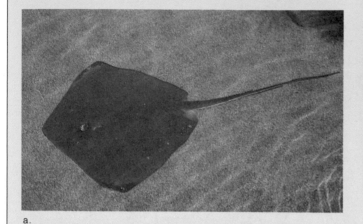
a.

b.

Figure 28.5
Ray-finned fishes. Most fishes today are bony, ray-finned fishes adapted to living in a great variety of habitats. For example, trout and stickleback are in freshwater; tuna and swordfish are in the open seas; parrot and angelfish live among the coral reefs. *a.* Queen angelfish. *b.* Rainbow trout.

a.

b.

Today, the **fleshy-finned fishes** (subclass Sarcopterygii) are represented by two groups, the lungfishes and only one species of "lobe-finned" fish. There are three genera of lungfishes—one each in Australia, Africa, and South America—that regularly breathe air to supplement gas exchange in their gills. The single species of lobe-finned fish is the coelocanth (fig. 28.6) long thought to be extinct but now known to exist off the southeast coast of Africa. The coelocanth does not have lungs.

The fleshy-finned fishes are of interest because they are thought to have been ancestral to the amphibians. Most likely they could breathe air while they crawled clumsily on stumpy fins from pond to pond during the Devonian Period, a time of alternating floods and droughts. Land plants and insects had evolved before land vertebrates, so there was an excellent food supply available on land for any animal that could remain ashore long enough to utilize it. Under these conditions, a selective pressure obviously favored those animals that developed land-dwelling abilities. Animals able to spend more time on land could avoid more vigorous competition in the water. Figure 28.6 compares the features of an ancestral fleshy-finned fish to those of the first amphibian and a coelocanth.

Figure 28.6

The fleshy-finned fishes of the Paleozoic Era underwent an adaptive radiation that included the lobe-finned fishes and the ancestors to the amphibians. *a.* The coelocanth is a living fossil, a lobe-finned fish that is alive today. Most likely it resembles the fleshy-finned fish that gave rise to the amphibians. Note the tuft at the end of its tail, no other fish has such a tail pattern. *b.* Comparison of the limbs of a fleshy-finned fish and the limbs of a primitive amphibian. They must have been awkward walkers, but legs were an improvement for locomotion, as is made clear by the side drawings.

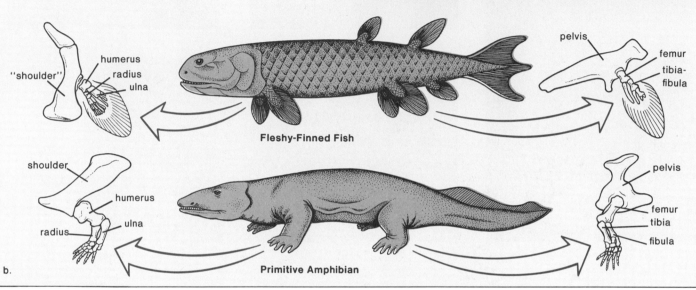

The most primitive fishes today, the lampreys and hagfishes, are jawless. Sharks, skates, and rays are cartilaginous fishes and all other types of fishes are bony fishes. Amphibians are descended from an ancient group of bony fishes that, unlike most today, had fleshy fins and lungs.

Amphibians

Improved locomotion on land and other adaptations for terrestrial life led to a diversification of **amphibians** (class Amphibia, 2,500 species) during the Carboniferous Period (table 28.1), now known as the Age of Amphibians. Many of these became extinct; amphibians are mainly represented today by newts and salamanders, frogs (fig. 28.7), and toads. These animals are not fully adapted to life on land. They have small, relatively *inefficient lungs,* and most of them also depend on gas exchange through the skin. For skin to function efficiently as a gas-exchanging surface, it must be quite thin and kept moist. In a dry environment, amphibians are in danger of desiccation and death. However, the most important factor binding amphibians to water is their method of reproduction. Amphibians, like most fishes, shed their eggs and sperm into the water where *external fertilization* takes place. The eggs are not protected against

dessication when exposed to air because they are only enclosed by a jelly coat. When young amphibians hatch from their eggs, they are tadpoles (aquatic larvae with gills) that feed and grow in the water. After they undergo *metamorphosis,* amphibians emerge from the water as adults that breathe air (fig. 28.7). The life history of these animals is divided between water and land as is accurately described by their class name, Amphibia, which means double life.

A few amphibians have special reproductive adaptations that enable them to reproduce on land. For example, some frogs carry their developing eggs in fluid-filled pouches. The young metamorphose very quickly, becoming tiny adults that are ready to live and grow on land.

Some time before the Permian Period, amphibians probably related to *Seymouria* (fig. 28.8) gave rise to the reptiles. Reptiles were to become the dominant organisms on earth during the next geological era.

Metamorphosis allows amphibians to switch from an aquatic to a land existence. There they are restricted to a moist environment because they use their skin for respiration and must return to the water for reproduction. Reptiles are descended from amphibians.

Figure 28.7

Frog metamorphosis, during which the animal changes from an aquatic to a terrestrial organism. *a.* Hatching of eggs. *b.* Tadpoles breathe by means of gills. *c.* Development of legs. *d.* Frogs breathe by means of lungs and their skin.

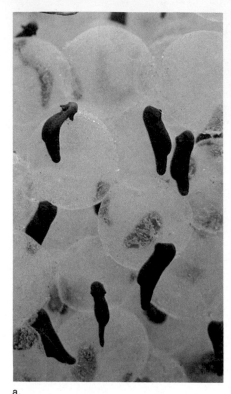

a.

b.

c.

d.

Figure 28.8

Drawing of *Seymouria,* a fossil animal related to a possible evolutionary link between amphibians and reptiles.

Reptiles

Reptiles (class Reptilia, 7,000 species) are largely represented today by turtles, crocodiles and alligators, lizards, and snakes (fig. 28.9). All of them have thick, scaly skin that contains keratin, which makes the skin impermeable to water. Because their lungs are efficient, there is no need to use the skin for gas exchange.

The reptiles' most outstanding adaptation to living on land is their means of reproduction. They have internal fertilization and lay eggs that are protected by leathery shells. This shelled egg eliminated the need for a swimming larval stage during development. The egg contains *extraembryonic membranes* that protect the embryo, remove nitrogenous wastes, and provide the embryo with oxygen, food, and water. These membranes are not part of the embryo itself and are disposed of after development is complete. One of the membranes, the amnion, is a sac that fills with fluid and provides a "private pond," within which the embryo develops (fig. 28.10).

The earliest reptiles, called stem reptiles, gave rise to several lines of descent during the Mesozoic Era (table 28.1), called the Age of Reptiles. One line produced the **dinosaurs** that underwent adaptive radiation and dominated the earth for more than 100 million years before their relatively sudden extinction,

Figure 28.9

Reptilian diversity. *a*. A snake has a tough, scaly skin that resists drying out. *b*. The slow-moving turtle has an external shell in addition to scaly skin.

a.

b.

Figure 28.10

Reptilian egg, a major advancement because it allowed these animals to reproduce on land. *a*. Baby crocodile hatching out of its shell. Note that the shell is leathery and flexible, not brittle like birds' eggs. *b*. Inside the egg, the embryo is surrounded by membranes. The chorion aids gas exchange; the yolk sac provides nutrients; the allantois stores waste; and the amnion encloses a fluid that prevents drying out and provides protection.

a.

chorion

amnion

albumen (egg white)

air space

embryo

egg shell

yolk sac

allantois

yolk

b.

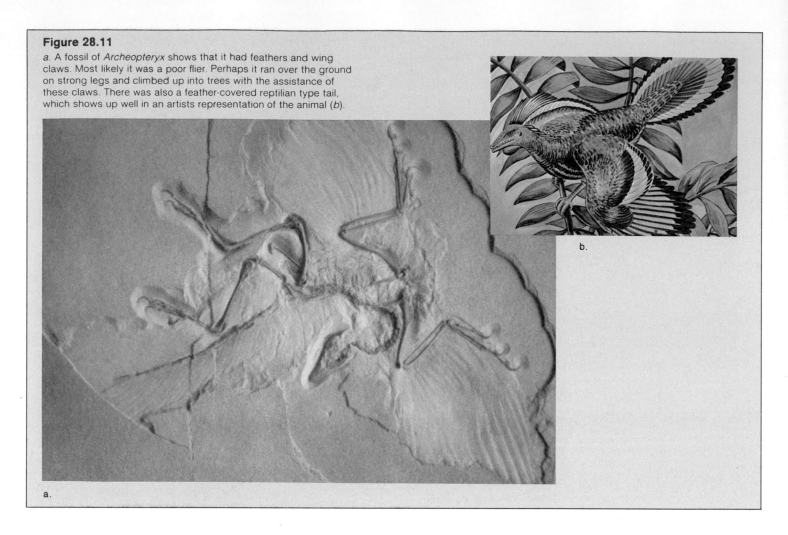

Figure 28.11

a. A fossil of *Archeopteryx* shows that it had feathers and wing claws. Most likely it was a poor flier. Perhaps it ran over the ground on strong legs and climbed up into trees with the assistance of these claws. There was also a feather-covered reptilian type tail, which shows up well in an artists representation of the animal (*b*).

possibly due to asteroid or comet bombardment of the earth, as discussed in the reading on page 295. **Birds** may be closely related to dinosaurs, although their poor fossil record makes it difficult to trace their history. *Archeopteryx* (fig. 28.11), the oldest known fossil bird, had several reptilian characteristics, including jaws with teeth and a long, jointed, reptilian tail. The bird was about the size of a crow and had rather feeble wings for its size, but it was well feathered.

There is another branch from the stem reptiles of interest. Mammal-like reptiles called *therapsids* (fig. 28.12) gave rise to the true mammals early in the Jurassic Period. The therapsids had mammal-like limbs, skulls, and teeth, and some of them resembled large dogs and were aggressive predators.

Reptiles are well adapted to a land environment. They have scaly skin that prevents loss of water and they can reproduce on land because they lay a shelled egg. Both mammals and birds are descended from reptilian ancestors.

Birds

Birds (class Aves, 8,900 species) are the only modern animals that have feathers, which are actually modified reptilian scales. Scales on the legs and feet of birds are other reminders of their reptilian ancestry, as are the claws at the ends of their toes.

Bird feathers are of two types. *Contour feathers* are attached to the wings in such a way that they overlap to produce a broad, flat surface beneficial to flight. *Down feathers* provide excellent insulation against body heat loss. This is important because birds maintain a constant, relatively high body temperature, which permits them to be continuously active even in cold weather.

There are many orders of birds, including birds that are *flightless* (ostriches), *web footed* (penguins), *divers* (loons), *fish eaters* (pelicans), *waders* (flamingos), *broad billed* (ducks), *birds of prey* (hawks), *vegetarians* (fowl), *shore dwellers* (sandpipers), *nocturnal* (owls), *small nectar feeders* (hummingbirds), and *songbirds,* the most familiar of the birds.

Nearly every anatomical feature of birds can be related to their ability to fly. They have hollow, very light bones, with internal air spaces that are a part of the respiratory system. A bird's head is relatively small and light because a light horny beak has replaced jaws equipped with teeth. A slender neck connects the head to a rounded, compact torso. The breastbone is enlarged and has a keel, to which the strong flight muscles are attached.

Flight requires well-developed sense organs. Birds have particularly acute vision and excellent muscular reflexes. They can land precisely on small tree branches and swoop from great heights to capture small animals that are spotted while cruising high above the unsuspecting prey.

Figure 28.12

Therapsid, a Jurassic reptile that had many mammalian features, including their characteristic means of locomotion and differentiated teeth. Most likely therapsids chewed their food before swallowing, which would have provided quick energy to maintain a warm temperature.

Figure 28.13

Unlike *Archaeopteryx*, modern birds have toothless beaks and an entirely feathered tail instead of a long bony tail. They retain the reptilian egg, except that the bird egg is hard-shelled, not leathery. Also, birds are usually devoted parents. Here, a song sparrow feeds its young.

Birds have internal fertilization and lay *hard-shelled eggs*. Many newly hatched birds require parental care (fig. 28.13) before they are able to fly away and seek food for themselves. Complex hormonal regulation and behavioral responses are involved in bird behavior. Some species migrate long distances twice a year, apparently using the sun and stars as compasses and even the earth's magnetic field to find their way on cloudy days.

Birds can maintain a constant internal temperature. All organ systems are adapted to allow birds to fly.

Mammals

The chief characteristics of **mammals** (class Mammalia, 4,500 species) are hair and milk-producing mammary glands (fig. 28.14). Hair provides insulation against heat loss and helps maintain a constant body temperature so mammals can be active in cold weather. Other characteristics include limb positioning that allows mammals to run faster than most other vertebrates, and well-developed sense organs and an enlarged brain, both important for an active life.

Mammary glands enable females to feed (nurse) their young without deserting them to find food. Nursing also creates a bond between mother and offspring that helps assure parental care while the young are helpless. In most mammals, the young are born alive after a period of development in the uterus, a part of the female reproductive tract. Internal development shelters the young and allows the female to move actively about while the young are maturing.

Figure 28.14

Mammals are animals that have hair and mammary glands. The mammary glands allow females to nurse their young. Nursing not only nourishes the young, it also builds up a strong mother-offspring relationship at the same time.

Mammals evolved during the Mesozoic Era from therapsid, mammal-like reptiles (fig. 28.12). Some of the earliest groups, represented today by the *monotremes* and *marsupials,* are not abundant today. Mammals remained relatively small until the reptiles declined. Then the marsupials, and later the *placental mammals* radiated into many of the habitats previously occupied by the large reptiles.

Monotremes

Monotremes (subclass Prototheria; order Monotremata) are represented by the duck-billed platypus and the spiny anteater, both of which are found in Australia. Mammals are classified according to means of reproduction. The **monotremes** lay their eggs, which resemble reptilian eggs, in burrows in the ground. The female incubates the eggs, as do birds, but after hatching, the young are dependent upon the milk that seeps from modified sweat glands on the abdomen of the female.

Marsupials

The young of **marsupial** (subclass Metatheria; order Marsupialia) mammals begin their development inside the female's body, but they are born in a very immature condition. Newborn young crawl up into a pouch on their mother's abdomen. Inside the pouch, they attach to nipples of mammary glands and continue their development.

Today, marsupial mammals are found mainly in Australia; only a few marsupials, such as the American opossum, are found outside that continent. In Australia, marsupials underwent adaptive radiation for several million years without competition from placental mammals, which arrived there only recently. Among the herbivorous marsupials, koala bears are tree-climbing browsers, and kangaroos are grazers. The Tasmanian "wolf" or "tiger," thought to be extinct, was a carnivorous marsupial about the size of a collie dog.

Placental Mammals

Developing **placental mammals** (subclass Eutheria) are dependent on the placenta, an organ of exchange between maternal blood and fetal blood. Nutrients are supplied to the growing offspring, and wastes are passed to the mother for excretion. While the fetus is clearly parasitic on the female, in exchange, she is free to move about as she chooses while the fetus develops.

Placental mammals (fig. 28.15) evolved during the Tertiary Period and entered the habitats left vacant when the dinosaurs became extinct. Table 28.2 lists the major orders of placental mammals, including those that are adapted to life in the water, on land, and in the air. In chapter 47 we will see that the various mammals are distributed in a specific manner about the world. For example, the hoofed mammals are likely to be found in grasslands, and primates are likely to be found in forests. We will now consider primate evolution, and in so doing, will discuss human evolution.

Figure 28.15
Most mammals are placental mammals that develop in the uterus of the female parent. Here they receive oxygen and nutrients by way of a special structure called the placenta, an area of exchange between maternal blood and fetal blood.

Table 28.2 Some Major Orders of Placental Mammals

Insectivora (moles, shrews)	Primitive; small, sharp-pointed teeth
Chiroptera (bats)	Digits support membranous wings
Carnivora (dogs, bears, cats, sea lions)	Canine teeth long; teeth pointed
Rodentia (mice, rats, squirrels, beavers, porcupines)	Incisor teeth grow continuously
Lagomorpha (rabbits, hares, pikas)	Chisel-like incisors; hindlegs longer than front; herbivorous
Perissodactyla (horses, zebras, tapirs, rhinoceroses)	Large, long-legged, one or three toes, each with hoof; grinding teeth
Artiodactyla (pigs, cattle, camels, buffalos, giraffes)	Medium to large; two or four toes, each with hoof; many with antlers or horns
Cetacea (whales, porpoises)	Medium to very large; forelimbs paddle-like; hind limbs absent
Primates (lemurs, monkeys, gibbons, chimpanzees, gorillas, humans)	Mostly tree-dwelling; head freely movable on neck; five digits, usually with nails; thumbs and/or large toes usually opposable

Adapted from table 18–5 *Essentials of Biology,* Second Edition by Willis H. Johnson, et al., copyright © 1974 by Holt, Rinehart and Winston, Inc., reprinted by permission of the publisher.

Placental mammals are far more numerous than the monotreme (egg laying) and marsupial (pouched) mammals. They are adapted to life in the air, water, and on land. Humans are classified as primates.

Primate Evolution

Primates are believed to have evolved (fig. 28.16) during the early Cenozoic Era from small insectivores (insect-eating mammals) that took up an **arboreal** life, thereby escaping predators on the ground and finding new sources of food in the angiosperm trees that were becoming prevalent.

Figure 28.16
Evolution of primates took place in the Cenozoic Era, which includes the Tertiary and Quaternary Periods. There are at least four lines of evolution among the primates who arose from mammalian insectivores. Prosimians include the lemurs and tarsiers; the monkeys include the New World and the Old World monkeys; apes include the gibbons and great apes (orangutan, gorilla, and chimpanzee); and humans.

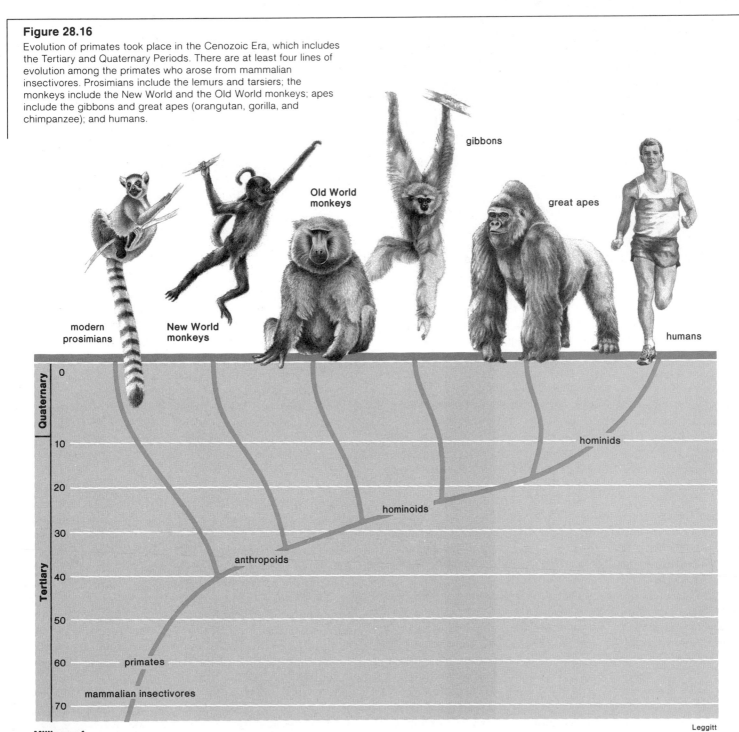

Figure 28.17

A tree shrew is a small animal with feet adapted for climbing in trees, where it feeds on fruit and insects. Notice that the ears are similar to the ears of the primates, being rounded with folded edges. A tree shrew's brain is relatively large, but the olfactory portion is surprisingly small. Tree shrews may resemble the ancient ancestors of modern primates.

The first primates may have been similar to the living tree shrews of Southeast Asia (fig. 28.17), which are rat-sized animals with a snout, claws, and sharp front teeth. By 50 million years ago, however, primates had evolved characteristics suitable to moving freely through the trees. Their limbs became adapted to swinging and leaping from branch to branch. While both hands and feet retained five functional digits (fingers and toes), the hands became especially dexterous and mobile. The thumbs were **opposable,** that is, they closed to meet the fingertips. Therefore, these animals could easily reach out and bring food, such as fruit, to the mouth. Claws were replaced by nails, which offer protection, but still allow a tree limb to be freely grasped and released.

A snout is common to those animals in which a sense of smell is of primary importance. In primates, the sense of sight is more important, so the snout has been considerably shortened, allowing the eyes to move to the front of the head. This has resulted in excellent binocular (three-dimensional) vision, permitting primates to make accurate judgments about the distance and position of adjoining tree limbs. The visual portion of the brain, located in the cerebral cortex, has also increased in size, as have those centers responsible for hearing and touch, which are also beneficial to survival when living in trees.

As primates increased in size, the care of multiple offspring from a single birth would have been a distinct liability. One live birth at a time became the norm and the period of postnatal maturation was prolonged, giving the immature young an adequate length of time to learn complex behavioral patterns.

Table 28.3 Classification of Primates

Category	Animals	Category	Animals
Phylum Chordata Subphylum Vertebrata Class Mammalia			
Order Primates	Prosimians Lemurs Tarsiers Anthropoids Monkeys Apes Humans	**Superfamily Hominoidea**	*Dryopithecus* Modern Apes Humans
		Family Hominidae	*Australopithecus Homo habilis Homo erectus Homo sapiens*
		Genus Homo (Humans)	*Homo habilis Homo erectus Homo sapiens*

Table 28.3 classifies the primates we will be discussing.

Prosimians

The first primates were **prosimians,** a term which means premonkeys. The living members of the group include lemurs and tarsiers (fig. 28.18). Lemurs, which have a squirrel-like appearance, are confined largely to the island of Madagascar. Their face is long and snoutlike; some digits have claws and some have nails. The big toe and thumb are mobile, but not opposable. The tarsiers, which are found in the Philippines and East Indies, are curious, mouse-sized creatures, with enormous eyes suitable for their nocturnal way of life. They have a flattened face, and their digits terminate in nails.

Anthropoids

Monkeys

Monkeys evolved from the prosimians about 38 million years ago during the Oligocene Epoch (table 28.4), when the weather was warm and vegetation was like that of a tropical rain forest. When Gondwana separated into South America and Africa two groups of monkeys evolved independently. There are *New World monkeys,* which have long, prehensile (grasping) tails and flat noses, and *Old World monkeys,* which lack such tails and have protruding noses. Two of the well-known New World monkeys are the spider monkey and the capuchin, the "organ grinder's" monkey. Some of the better-known Old World monkeys are now ground dwellers, such as the baboon, and the rhesus monkey, which has been used in medical research.

Apes and humans, the other **anthropoids,** evolved later.

Primates evolved from shrewlike insectivores and became adapted to living in trees, as exemplified by skeletal features, good vision, and aspects of their reproduction. Prosimians were the first primates to evolve and they were followed by monkeys, the first of the anthropoids.

Figure 28.18
The ring-tailed lemur, a prosimian. Many lemurs are endangered today, but not the ring-tailed (obviously named for the appearance of its tail). Lemurs have long, bushy tails and long hind limbs.

Table 28.4 Biological Events during Cenozoic Era

Million Years before Present	Epoch	Event	Duration of Period (Millions of Years)
3	Pleistocene	Late hominids and humans	2–3
6	Pliocene	Early hominids	3
27	Miocene	Early apes	21
38	Oligocene	Early monkeys	11
56	Eocene	Early primates	18
66	Paleocene	Placental mammals	10

Hominoids

Humans are more closely related to apes (fig. 28.19) than to monkeys. **Hominoids** include apes and humans. There are four types of apes: the *gibbon, orangutan, gorilla,* and *chimpanzee.* The gibbon is the smallest of the apes, with a body weight ranging from 5½ kg to 11 kg. Gibbons have extremely long arms that are specialized for swinging between tree limbs. The orangutan is large (75 kg), but nevertheless spends a great deal of time in trees. The gorilla, the largest of the apes (180 kg), spends most of its time on the ground. Chimpanzees, which are at home both in trees and on the ground, are the most human-like of the apes in appearance, and are frequently used in psychological experiments.

Hominoid Ancestor　During the Miocene Epoch, which began about 25 million years ago (table 28.4), the weather was changing. The continents were drifting into their present locations, and as a result, the weather was becoming cooler and drier. In places, the tropical forests were giving way to grasslands.

At this time, apes became abundant and widely distributed in Africa, Europe, and Asia. Among these Miocene apes, members of the genus *Dryopithecus* are of particular interest, because they are thought to be a possible *hominoid ancestor*—an ancestor that led to all other apes and eventually humans. **Dryopithecines** were forest dwellers that probably spent most of the time in the trees. The bones of their feet, however, indicate that when they did walk on the ground, they walked on all fours virtually all the time, using the knuckles of their hands to support part of their weight. Such "knuckle-walking" is retained by modern apes but not by humans, who stand erect (fig. 28.20). The skull of *Dryopithecus* had a sloping brow, heavy eyebrow ridges (called supraorbital ridges), and jaws that projected forward. These too are features that were retained by apes but not

humans. In contrast to apes, humans have a high brow and lack the eyebrow ridges of apes. The face is flat, and there is no muzzle, because the canine teeth are comparable in size to the other human teeth.

The first hominoid (ancestor to all apes and humans) may possibly have been *Dryopithecus,* whose characteristics are somewhat like those of today's apes.

Human Evolution

Hominids

Just when the human lineage split from that of apes is not known, but the common belief is that it occurred during the early Pliocene (table 28.4). The cooling trend that began in the Miocene continued during the Pliocene and grasslands became even more prevalent than before.

Figure 28.19

Ape diversity. *a.* Of the apes, gibbons are the most distantly related to humans. They dislike coming down from trees, even at watering holes. They will extend a long arm into the water and then drink collected moisture from the back of the hand. *b.* Orangutans are solitary except when they come together to reproduce. Their name means ''forest man''; early Malayans believed that they were intelligent and could speak but did not because they were afraid of being put to work. *c.* Gorillas are terrestrial and live in groups in which a silver-backed male, such as this one, is always dominant. *d.* Of the apes, chimpanzees sometimes seem the most humanlike.

b.

a.

c.

d.

Figure 28.20

a. As outlined in the table, the *Dryopithecines* had primitive characteristics that could have led to both modern-day apes and humans. Compare the modern human skeleton (*b*) to the modern ape (gorilla) skeleton (*c*). Humans are bipedal while modern-day apes are knuckle walkers. In the ape, the pelvis is very long and tilts forward, whereas in humans it is short and upright. The ape shoulder girdle is more massive than that of humans, and the head hangs forward because of the angle of attachment of the vertebral column to the skull. In apes, the foramen magnum (hole in skull through which the spinal cord passes) is well to the rear of the skull; in humans it is almost directly in the bottom center of the skull. Also, in humans the spine has an S-shaped curve, allowing a better weight distribution and improved balance when the body is upright. (*b*) and (*c*) From John Napier, ''The Antiquity of Human Walking.'' Copyright © 1967 Scientific American, Inc. All rights reserved.

Comparison of Dryopithecus, Modern Apes, and Modern Humans

Feature	Dryopithecus	Modern Apes	Modern Humans
Brain size	Small brain and skull	Slightly enlarged	Very much enlarged
Face	Sloping brow, heavy eyebrow ridges, and projection of face	Same as primitive	High brow, reduced eyebrow ridges, face flat
	Rectangular-shaped jaw with large molars and long canine teeth	Same as primitive	U-shaped jaw with small molars and shortened canine teeth
Locomotion	Quadrupedal locomotion Limbs of equal length Opposable thumb and toe	Same as primitive Forelimbs elongated Same as primitive	Bipedal locomotion Forelimbs shortened Opposable thumb retained

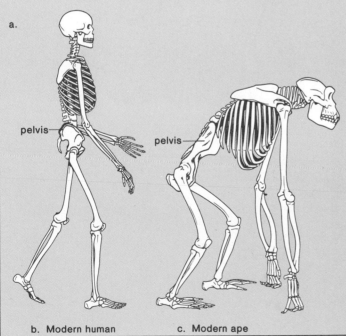

a.

pelvis

pelvis

b. Modern human   c. Modern ape

Some years ago, on the basis of only an upper jaw and a handful of molars, it was suggested that the fossil remains of *Ramapithecus* may be those of the first hominids. However, when fossilized skulls were found in the late 1970s and 1980s, it became apparent that the ramapithecines had close-set eye sockets, eyebrow ridges, protruding jaws, and flaring cheekbones. These features suggested that the ramapithecines were most likely ancestral orangutans.

Therefore, we do not know what the very first hominids were like, but they must have come down out of the trees and begun to assume an upright, **bipedal** posture (walking on two limbs). This would have enhanced survival on the savanna, because it would have allowed hominids to see over tall grass, to spot predators or prey, and would have left the hands free to perform other functions, such as throwing rocks. As discussed in the reading on page 434, such ideas are being tested in modern investigations.

> The first hominid (ancestor leading to humans) arose at a time when a change in weather reduced the size of the African forests making it advantageous to move on the ground.

New ideas regarding speciation have been used to interpret the hominid fossil record. Although formerly scientists attempted to place each hominid fossil in a straight line from the most primitive to the most advanced, it is now reasoned that several hominid species could have existed at the same time. Figure 28.21 indicates a possible revolutionary tree for the known **hominids.**

Australopithecines

The fossil remains of those classified in the genus *Australopithecus* (Southern Apeman) were found in southern Africa and have been dated from about 3.5 million years ago. The **australopithecines** (fig. 28.22) were 4 to 5 feet in height, with a brain that ranged in size from 300 to 500 cubic centimeters (cc). The pelvis definitely indicates that they were capable of bipedal locomotion; supporting this contention is the non-opposability of the big toe.

Three species of *Australopithecus* have been identified—*afarensis, africanus,* and *robustus.* The exact relationship between the three species is in dispute. Because of its small brain size, large canine teeth, and protruding face, it has been suggested that *afarensis* is the more primitive of the three. Moreover, because of certain skeletal limb features indicating that these people walked erect, some believe that *afarensis,* called Lucy (from the song "Lucy in the Sky with Diamonds") by her discoverers, is also ancestral to humans, as indicated in figure 28.21. It may also be noted that the combined characteristics of this fossil indicate that an enlarged brain was *not* needed for bipedal locomotion to evolve.

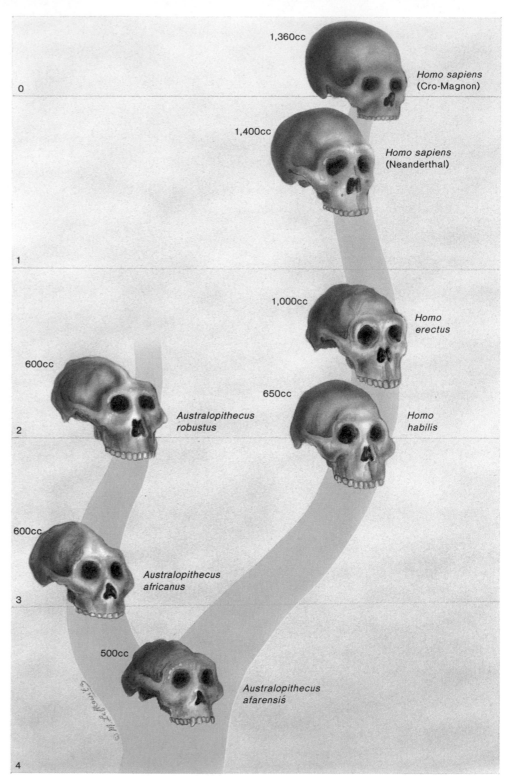

Figure 28.21

One possible evolutionary tree for hominids. Some authorities believe that *Australopithecus afarensis* is a common acestor for a line of australopithecines and for humans, as shown here. This means that the australopithecines and humans coexisted for a time. The numbers are the cranial capacities.

1,360cc

Homo sapiens
(Cro-Magnon)

1,400cc

Homo sapiens
(Neanderthal)

1,000cc

*Homo
erectus*

600cc

650cc

*Homo
habilis*

*Australopithecus
robustus*

600cc

*Australopithecus
africanus*

500cc

*Australopithecus
afarensis*

Millions of Years Ago

Figure 28.22

Australopithecus africanus. While at one time these hominids were thought to have been hunters, a study of cut marks left on animal bones suggests that they were most likely scavengers that sought and confiscated the kill of coexisting predators and scavengers.

Australopithecus robustus, as its name implies, was larger than *africanus.* Certain anatomical differences may be due to diet. The massive skull and face and the large cheek teeth of *robustus* indicate a vegetarian diet. The gracile facial features of *africanus* indicate a more varied diet, possibly including meat. Facial bones, including eyebrow ridges, facial muscles, and teeth, need not be as large in meat eaters. It is now believed that the australopithecines were not hunters but scavenged the meat that they ate.

One possible hominid evolutionary tree shows the bipedal *Australopithecus afarensis* is a common ancestor to two other members of this genus and humans. *Australopithecus africanus* was slight of build compared to *Australopithecus robustus.*

Humans

The rest of the fossils to be mentioned are in the genus *Homo,* which is the genus for all humans.

Homo Habilis

This newly discovered fossil is dated about 2 million years ago, making it a contemporary of *Australopithecus africanus.* **Homo habilis** is significant for several reasons. (1) Formerly it was believed that humans evolved less than 2 million years ago, but this fossil and other current evidence seem to suggest that they must have appeared *more* than 2 million years ago. (2) Almost certainly, humans evolved in Africa. (3) Formerly only hominid fossils with a brain capacity of 1,000 cc were designated humans. *Homo habilis* had a maximum brain capacity of 800 cc, and yet has been placed in the genus *Homo* because it not only used tools, but made them. *Homo habilis* means handyman; and was given this name because of the quality of the tools found with the bones.

At one time it was thought that only primitive people with a brain capacity of more than 1,000 cc could have made tools. Because *Homo habilis* had a smaller brain and yet made tools,

Modern Focus

One of the most striking features of paleoanthropology these days is the multidisciplinary nature of the pursuit. No longer is it a simple alliance between fossil hunters and archaeologists. Instead, a battery of sciences is being focused on the questions of human origins. In addition to geology, which provides the basic backdrop for the recovery of fossils, there is paleoecology, taphonomy (the study of the way bones become buried), primatology, molecular biology, neurophysiology, energetics, and many more.

Instead of pondering the imponderable, such as the causal link between bipedalism, tool-making and use, and expansion of brain size, there is now a growing tendency to ask specific, answerable questions. What are the energetic considerations of upright-walking as opposed to, say, knuckle-walking? Under what ecological circumstances might it be energetically possible to evolve a large brain? What can the surface of fossil teeth tell us about ancient diets? What does one need to know about an assemblage of fossil bones and stone artifacts to be sure that the association is not mere coincidence? What does the history of animals contemporary with our ancestors tell us about, say, important migrations of the past? And so on.

The question of 'why we stood upright' is of course an old one and has engendered responses such as, 'so as to gain more visibility over tall grasses' and, 'so as to free the hands for defensive purposes, because we lost our long, sharp canines.' Neither scenario is very testable. But to ask, 'What are the energetics of bipedalism as compared to quadrupedalism?' as Richard Taylor of Harvard University has is a more promising approach. Other researchers have used Taylor's data and have added observations of their own.

For example, Peter Rodman and Henry McHenry, of the University of California at Davis, and Richard Wrangham, at the University of Michigan, examined the feeding habits of orangutans, which forage mainly for fruit in trees, and chimpanzees, which do the same but usually from sources that are more meagre and more widely dispersed. McHenry and his colleagues wondered whether an ape, faced with food sources dispersed even more thinly than those exploited by chimpanzees, would forage more efficiently in energetic terms by knuckle-walking (see figure) or by bipedalism. It turns out, although bipedalism is not particularly efficient compared with fully developed quadrupedalism in animals moving at high speeds,

Modern researchers study the energetics of knuckle-walking in order to determine if the hominid ancestor would have walked this way or uprightly.

under the circumstances in which our ancestors evolved, that is, in the forest fringe with an ape-like ancestor, the evolution of upright-walking was a very suitable adaptation. At slow walking speeds, bipedalism compares very favourably with quadrupedalism.

It may be, therefore, that bipedalism arose as an energetic strategy for foraging for widespread food sources. The point here is that the approach adopted by Taylor and others allowed specific questions, predictions and answers, a truly scientific way of tackling the problem.

Another exploration in energetics comes from the work of Robert Martin, at University College, London. Addressing primates in general and humans in particular, he asked the questions, 'What dietary and ecological factors affect a species' energy budget and reproductive strategy?' and, 'How is this related to brain size?' Some strands of the argument are as follows. Animals that adopt the reproductive strategy of having just a few carefully nurtured offspring, as against many that are left to fend for themselves, generally live in environments blessed with a stable food supply. And species that are endowed with large brains are those

that feed on high-energy diets. How does the requirement for a stable environment and a high-energy diet match up to other conclusions about the environment of early hominids? Such questions are, thus, brought into scientific focus. . . .

There has always been much speculation on the putative appearance of the last common ancestor between human and apes, but only infrequently has it been inspired by rigorous scientific questioning. By applying the principles of allometry in the study of body and limb proportions in monkeys, apes, and humans, Leslie Aiello, again, of University College, London, has been able to dismiss some of the long-favoured 'models' for the human-ape ancestor because, it turned out, they were biologically incongruent. No, the last common ancestor was not like a chimpanzee; nor was it especially long in the forelimb, an adaptation to an arm-swinging life-style; and neither was it like a modern Old World monkey. The comparative study tells us, says Aiello, that the last common ancestor probably moved about in a slow suspensory climbing motion, reminiscent of that of the modern howler monkey, ironically a New World monkey. This is not to suggest an involvement of New World primates in human origins; however it gives some guidance in interpreting the fossils of any putative last common ancestor.

This new approach of asking specific questions often brings unexpected answers, as happened to Alan Walker of Johns Hopkins University, Baltimore. When he imaged under the electron microscope the tooth surface of fossil hominids and compared them with marks on modern teeth he found that the early hominids and *Homo habilis* could be classified with chimpanzees, that is they were fruit-eaters. With *Homo erectus* came an apparently marked change in diet, possibly with the inclusion of underground tubers, possibly with increased meat-eating. This approach, still being developed, brings modern researchers as close as they will ever come to the food actually eaten by the individual whose fossilized remains are in their study. Walker had asked, 'What do the surfaces of chimpanzees' teeth look like?' 'What do baboons' teeth look like?' 'What do pigs' teeth look like?' 'What do hyaenas' teeth look like?' And, 'How does this help us understand better the diet of our ancestors?' He got an answer.

are we to think that the making of tools preceded the evolution of the enlarged brain? This seems to be an unnecessary question. Increased brain capacity, no matter how slight, would have permitted better toolmaking; this combination would have been selected because toolmaking would have fostered survival in a grassland habitat. Therefore, as the brain became increasingly larger, tool use would have become more sophisticated.

Homo habilis, a fossil dated at least 2 million years ago, may not have been highly intelligent, but did make tools. Intelligence and the making of tools probably evolved together. If *Homo habilis* is indeed considered human, then humans evolved much earlier than previously thought.

Figure 28.23

Homo erectus. This wide-spread hominid was a successful hunter
who had a knowledge of fire.

Homo Erectus

Homo erectus (fig. 28.23) was prevalent throughout Eurasia and
Africa during the Pleistocene Epoch, also called the Ice Age,
because of the recurrent cold weather that produced the glaciers of this epoch. *Homo erectus* had an average brain size of
1,000 cc, but the shape of the skull indicates that the areas of
the brain necessary for memory, intellect, and language were
not well developed.

Whereas new fossil remains indicate that *Homo habilis*
had a build that more closely resembles that of apes, *Homo
erectus* had a posture and locomotion similar to that seen in
modern humans. The tools of *Homo erectus* were superior to
those of *Homo habilis,* in that they were made of flakes chipped
from a stone, rather than of the stone itself. Each flake was struck
off a stone "core" that had been previously shaped to give just
the right kind of flakes. Analysis of cut marks on the animal
bones near campsites indicates that these people were hunters.
They also possessed knowledge of fire and would have been able
to cook their meat in order to tenderize it.

Homo Sapiens

Neanderthal The **Neanderthals** (*Homo sapiens neanderthalis*) existed over 130,000 years ago during the last Ice Age,
and lived about 100,000 years. They are called Neanderthals
because they were first found in the Neanderthal Valley in Germany. Later, they were also found in most of Europe and in the
Middle East.

Surprisingly, the Neanderthal brain was, on the average,
slightly larger than that of modern humans. The Neanderthal
brain was 1,400 cc, whereas that of most humans today is
1,360 cc. The Neanderthal skeleton is robust, giving evidence

Figure 28.24

Homo sapiens neanderthalis. It is now believed that Neanderthals generally had the modern appearance depicted here.

that the skeletal muscles were very well developed. It's been hypothesized that a larger brain than ours was required for control of the extra musculature.

Early investigators believed that Neanderthals were apelike in appearance (fig. 28.24), but now it is believed that they had a more modernlike appearance except that they, like *Homo erectus,* had a low brow, wide nose, and receding chin. Like *Homo erectus,* too, the Neanderthal people were stone toolmakers, lived together in a kind of society, and hunted large game together.

The Neanderthals seem to have been more culturally advanced than *Homo erectus,* however. They buried their dead with flowers, as an expression of grief perhaps, and with tools,

as if they thought these would be needed in a future life. Some researchers have even suggested that Neanderthals had a religion and that because great piles of bear skulls have been found in their caves, the bear must have played a role in this religion.

It's possible that the Neanderthals evolved into modern humans, called **Cro-Magnon** because their remains appear first in Cro-Magnon, France, but it could also be that the Neanderthals were replaced by Cro-Magnon. Another theory is that modern humans evolved in one locale, perhaps Africa, and then interbred with other populations, including the Neanderthals, as they spread out into Europe and Asia.

Figure 28.25
Cro-Magnon people are the first to be designated *Homo sapiens sapiens*. Their tool-making ability and other cultural attributes, such as their artistic talents, are legendary.

Cro-Magnon Cro-Magnon (*Homo sapiens sapiens*) people lived about 40,000 years ago. Their brain capacity was similar to ours (about 1,360 cc). They were such accomplished hunters that some researchers believe they are responsible for the extinction, during the Upper Pleistocene Epoch, of many, large mammalian animals, such as the giant sloth, mammoth, saber-toothed tiger, and giant ox. Because language would have facilitated their ability to hunt such large animals, it's quite possible that meaningful speech began at this time. Cooperative hunting would also have led to socialization and the advancement of culture. Humans are believed to have lived in small groups, the men going out to hunt by day, while the women remained at home with the children. It's quite possible that a hunting way of life among prehistoric men has shaped behavior still seen today.

Cro-Magnon people had an advanced form of stone technology that included the making of compound tools; the stone was fitted to a wooden handle. They were the first to throw spears, enabling them to kill animals from a distance. And they were the first to make blades, knifelike objects that do not appear to have been chipped from a stone.

Cro-Magnon people lived during the Reindeer Age, a time when great reindeer herds spread across Europe. They used every part of the reindeer, including the bones and antlers, from which they sculpted many small figurines. They also painted beautiful drawings of animals on cave walls in Spain and France (fig. 28.25). Perhaps, they, too, had a religion, and these artistic endeavors were an important part of their form of worship.

If Cro-Magnon people did cause the extinction of many types of animals, it may account for the transition from a hunting economy to an agricultural economy about 12,000 to 15,000 years ago. This agricultural period extended from that time to about 200 years ago, when the Industrial Revolution began. At this time, many people began to live in cities, in large part divorced from nature, and endowed with the philosophy of exploitation and control of nature. Only recently have we begun to realize that the human population, like all other organisms with which we share an evolutionary history, should work with, rather than against, nature.

Homo erectus had a large brain, walked with a striding gait, and used fire. *Homo sapiens neanderthalis* was not as primitive as formerly thought. The Neanderthals may have evolved into or may have been replaced by Cro-Magnon, the first *Homo sapiens sapiens*. Cro-Magnon was an expert hunter. Hunting promoted language and socialization.

Summary

1. Vertebrates are in the phylum Chordata and have the three chordate characteristics as embryos. As adults, the notochord is replaced by the vertebral column.
2. Vertebrates include fishes, amphibians, reptiles, birds, and mammals.
3. The earliest, jawless fishes are represented today by hagfishes and lampreys. Sharks, skates, and rays are cartilaginous, and the bony fishes include all other well-known fishes.
4. The original bony fishes gave rise to two groups: today's ray-finned fishes and fleshy-finned fishes. The fleshy-finned fishes are represented today by three species of lungfishes and the coelacanth, a lobe-finned fish.
5. Fleshy-finned fishes gave rise to the amphibians during the Devonian Period (Mesozoic Era). Amphibians, represented today by frogs and salamanders, are not completely adapted to life on land. They must return to water to reproduce.
6. Reptiles (e.g., today's turtles, crocodiles, snakes, and lizards) evolved from amphibians. They lay a shelled egg, which allows them to reproduce on land.
7. One main group of ancient reptiles, the stem reptiles, produced a line of descent that evolved into both dinosaurs and birds. A different line of descent from stem reptiles evolved into the mammals.

8. Birds are feathered, which helps them maintain a constant body temperature. They are adapted for flight; their bones are hollow, their shape is compact, their breastbone is keeled, and they have well-developed sense organs.

9. Mammals are vertebrates with hair and mammary glands. The former helps them maintain a constant body temperature, and the latter allows them to nurse their young. Monotremes lay eggs; marsupials have a pouch in which the newborn matures; and the placentals, which are far more varied and numerous, retain offspring inside the uterus until birth.

10. Mammals remained small and insignificant while the dinosaurs existed, but when the latter became extinct during the Cretaceous Period (end of Mesozoic Era), mammals became the dominant land organisms.

11. Humans are primates, most species of which are adapted to an arboreal existence. Prosimians (e.g., today's lemurs and tarsiers) are more primitive than the anthropoids, which include monkeys, apes, and humans.

12. During the Miocene Epoch (Tertiary Period, Cenozoic Era), when the climate was becoming cooler and dryer, apes arose. *Dryopithecus* could be the hominoid ancestor that gave rise to all the apes and humans. The dryopithecines had a sloping brow, heavy eyebrow ridges, and distinct muzzle, because the molars were large and the canine teeth were large and pointed. Most likely, they were knuckle-walkers and had an opposable thumb. Most of these features are seen today in apes, but not in humans.

13. In the early Pliocene Epoch, a hominid ancestor (leading only to humans) began to walk erect when the savanna was replacing forests. We are certain that the australopithecines (*A. afarensis*, *A. africanus*, and *A. robustus*) walked erect, even though the brain was small.

14. It is possible that *A. afarensis* was a common ancestor to both the australopithecines and to those in the genus *Homo*. The first fossil to be placed in the genus *Homo* is *Homo habilis*, not because of a large brain, but because these individuals made tools. Perhaps tool making and brain development evolved together after this time.

15. The anatomy, posture, and locomotion of *Homo erectus* is humanlike. They had the knowledge of fire and probably were religious, a sign of the development of culture.

16. Two fossils are designated as *Homo sapiens*: Neanderthals and Cro-Magnon. Neanderthals are now thought to have been more modern than previously supposed. What happened to them is not definitely known. Cro-Magnon evolved into modern humans and definitely had culture.

Objective Questions

1. Which of these features is replaced by the vertebral column in vertebrates?
   a. notochord
   b. dorsal hollow nerve cord
   c. gill pouches
   d. All of these.

2. Birds and mammals evolved from
   a. fishes.
   b. amphibians.
   c. reptiles.
   d. a bird-mammal that existed in the Mesozoic Era.

3. Amphibians arose from
   a. cartilaginous fishes.
   b. jawless fishes.
   c. ray-finned fishes.
   d. bony fishes with lungs.

4. Which of these is not a feature of amphibians?
   a. dry skin that resists desiccation
   b. metamorphosis from a swimming form to a land form
   c. small lungs necessitating a supplemental way to achieve gas exchange
   d. reproduction in the water

5. Which of these are not reptiles?
   a. snakes
   b. penguins
   c. crocodiles
   d. lizards

6. Which of these animals are able to maintain a constant body temperature by internal processes?
   a. fishes and amphibians
   b. amphibians and reptiles
   c. reptiles and birds
   d. birds and mammals

7. Which of these is a true statement?
   a. In all mammals, offspring develop within the female.
   b. All mammals have hair and mammary glands.
   c. All mammals are land-dwelling forms.
   d. All of these are true.

8. Which of these gives the correct evolutionary order?
   a. anthropoids—prosimians—hominids—hominoids
   b. prosimians—hominoids—hominids—anthropoids
   c. hominids—anthropoids—prosimians—humans
   d. prosimians—anthropoids—hominoids—hominids

9. Which of these features of the dryopithecines is seen in apes today?
   a. face that projects forward
   b. eyebrow ridges
   c. knuckle-walking
   d. All of these.

10. *Homo habilis* most likely existed at the same time as the
    a. dinosaurs.
    b. australopithecines.
    c. Cro-Magnon.
    d. All of these.

11. Which of these is mismatched?
    a. australopithecines—walked erect
    b. *Homo habilis*—large brain
    c. *Homo erectus*—had fire
    d. *Homo sapiens*—flat face

12. The hominid ancestor most likely
    a. made tools.
    b. had a large brain.
    c. walked erect.
    d. All of these.

Study Questions

1. What are the three chordate characteristics and which is replaced by the vertebral column in vertebrates?
2. Draw a simplified evolutionary tree of vertebrates using figure 28.1 as a guide.
3. What are the three classes of living fishes and how are they related? The amphibians evolved from what type of fish?
4. In what ways are amphibians well adapted to a land existence and in what ways are they ill adapted?
5. In what ways are reptiles better adapted to a land existence than amphibians?
6. In what ways are birds adapted to flying?
7. What are the three subclasses of mammals and what are their primary characteristics?
8. List representative examples of all classes of vertebrates.
9. Give the complete classification of humans using table 28.3 as a guide. Distinguish between prosimians, anthropoids, hominoids, hominids, and humans.
10. Describe *Dryopithecus*, and tell in what ways the anatomy of modern apes and modern humans differ from that of this fossil.
11. Draw an evolutionary tree for hominids and describe each group in the tree.
12. Discuss the evolutionary relationship of erect walking, brain size, and tool making, and tell how these aspects relate to the hominids in your tree.

Thought Questions

1. Explain why there are so few types of amphibians and reptiles in existence today, compared to the very large number of species of birds and mammals.
2. Why would you have expected that angiosperms evolved before primates?
3. What features have humans retained that would make them good tree dwellers?

Selected Key Terms

amphibian (am-fib′e-an) 421
reptile (rep′tīl) 422
mammal (mam′al) 425
monotreme (mon′o-trēm) 426
marsupial (mar-su′pe-al) 426
placental mammal (plah-sen′tal mam′al) 426

primate (pri′māt) 427
hominoid (hom′ĭ-noid) 429
dryopithecine (dri′′o-pith′ē-sīn) 429
bipedal (bi′ped-al) 431
hominid (hom′ĭ-nid) 431
australopithecine (aw′′strah-lo-pith′ē-sīn) 431

Homo habilis (ho′mo hah′bĭ-lis) 433
Homo erectus (ho′mo ĕ-rek′tus) 435
Neanderthal (ne-an′der-thawl) 435
Cro-Magnon (kro-mag′non) 436

Suggested Readings for Part 3

Bonatti, E. March 1987. The rifting of continents. *Scientific American.*

Brock, T. D., and Madigan, M. T. 1988. *Biology of microorganisms.* 5th ed. Englewood Cliffs, New Jersey: Prentice Hall.

Cairns-Smith, A. G. June 1985. The first organisms. *Scientific American.*

Dickerson, R. E. September 1981. Chemical evolution and the origin of life. *Scientific American.*

Dodson, E. O., and Dodson, P. 1985. *Evolution: Process and product.* 3d ed. Boston: Prindle, Weber, and Schmidt.

Eckert, R., and Randall, D. 1983. *Animal physiology.* 2d ed. San Francisco: W. H. Freeman.

Feder, M. E., and Burggren, W. W. November 1985. Skin breathing in vertebrates. *Scientific American.*

Futuyma, E. J. 1986. *Evolutionary biology.* 2d ed. Sunderland, Ma: Sinauer.

Gallo, R. C. December 1986. The first human retrovirus. *Scientific American.*

Gosline, J. M., and De Mont, M. E. January 1985. Jet-propelled swimming in squids. *Scientific American.*

Grant, P. R. 1981. Speciation and the adaptive radiation of Darwin's finches. *American Scientist* 69:653.

Hadley, N. F. July 1986. The arthropod cuticle *Scientific American.*

Hay, R. L., and Leaky, M. D. February 1982. The fossil footprints of Laetoli. *Scientific American.*

Hickman, Z. P., and Roberts, L. S. 1988. *Integrated principles of zoology.* 8th ed. St. Louis: C. V. Mosby Co.

Hirsch, M. S., and Kaplan, J. C. April 1987. Antiviral therapy. *Scientific American.*

Hogle, J. M. March 1987. The structure of poliovirus. *Scientific American.*

Koehl, M. A. R. December 1982. The interaction of moving water and sessile organisms. *Scientific American.*

Margulis, L. 1982. Early life. Boston: Science Books International.

Mossman, D. J., and Sarjeant, W. A. S. January 1983. The footprints of extinct animals. *Scientific American.*

Raven, P. H. et al. 1986. *Biology of plants.* 4th ed. New York: Worth.

Rensberger, B. April 1982. Evolution since Darwin. *Science 82.*

Scientific American. 1978. *Evolution.* San Francisco: W. H. Freeman.

Simons, K. et al. February 1982. How an animal virus gets into and out of its host cell. *Scientific American.*

Stebbins, G. L., and Ayala, F. J. July 1985. The evolution of Darwinism. *Scientific American.*

Vidal, G. February 1983. The oldest eukaryotic cells. *Scientific American.*

Volpe, E. P. 1981. *Understanding evolution.* 4th ed. Dubuque, IA: Wm. C. Brown Publishers.

CRITICAL THINKING

CASE STUDY

PART 3
Evolution and Diversity

As described in chapter 21, balanced polymorphism is the maintenance of genetic variation due to environmental pressures. For example, in sickle-cell anemia, the allele frequency for *Hb^S* and *Hb^A* is maintained in African populations because of heterozygote superiority in the presence of the malaria parasite.

This case study considers an example of polymorphism in the mussel *Mytilus edulis* off the shores of Long Island. To the south of the island there is the Atlantic Ocean and to the north there is Long Island Sound, which separates Long Island from the mainland. The ocean waters are much more salty than those of the Sound because the Sound receives freshwater runoff from the mainland. We will be considering two populations of *M. edulis*—one that lives in the ocean and one that lives in the Sound.

Would you expect the allele frequencies to vary between the two populations? Why?

The cells of *Mytilus edulis* contain a type of enzyme called aminopeptidase I, which occurs in three variations distinguishable one from the other by certain physical properties. For each of these aminopeptidase I enzymes there is a different allele, designated as *lap* 94, *lap* 96, and *lap* 98. Considering our discussion so far, would you expect to find a different frequency, for instance, for *lap* 94 in the two populations?

Prediction 1 *Lap* 94 frequency should be different in the two populations because the environment varies for the two populations.

Result 1 Measurements indicate that the frequency of *lap* 94 allele is 55% in the population on the ocean side of the island but only 12% for the population in Long Island Sound.

What is the sum frequency of lap *96 and 98 in the ocean population? Which of the alleles (*lap *94 versus 96 and 98 taken together) seems to confer an advantage to mussels living in the ocean? What is the sum frequency of* lap *96 and 98 in the sound population? Which of the alleles (*lap *94 versus 96 and 98 taken together) seems to confer an advantage to mussels living in the sound?*

Mussel larvae (but not the adults, which are permanently attached to a substrate) are mobile and it has been noted that mussel larvae from the ocean migrate each year into the sound on prevailing ocean currents. Despite migration of these larvae, the frequency of *lap* 94 stays the same in the sound population.

Why might you have assumed that the frequency of lap *94 would continually rise in the Sound population? What explanation can you give for the frequency not rising? On the other hand, why wouldn't the allele be completely eliminated?*

The much higher frequency of *lap* 94 in the ocean population suggests that the *lap* 94 allele confers an advantage to mussels living in the ocean. What is the possible advantage? Additional research into the role of aminopeptidase I, an enzyme that breaks down polypeptides to amino acids in mussel cells, helps answer this question. In many marine organisms, intracellular free amino acids osmotically balance the salts in seawater. A cell with a low content of amino acids will tend to become dehydrated but a cell with a higher content of amino acids will not become dehydrated. Would you expect greater aminopeptidase I activity in mussels exposed to seawater of high salinity?

Prediction 2 Aminopeptidase I activity should be greater when mussels are in seawater of high salinity.

Result 2 Experimental results have shown that amino peptidase I activity is greater in mussels exposed to seawater of high salinity than when they are exposed to seawater of low salinity. Aminopeptidase I activity apparently helps a mussel respond to an increase in the osmolarity of the environment.

In regard to the *lap* alleles, *lap* 94 codes for an aminopeptidase I enzyme that is more active in breaking down polypeptides than the other two alleles. Mussels having which type of allele do you predict would be better able to respond to a high-salinity environment by producing amino acids?

Prediction 3 Mussels with the *lap* 94 allele should be able to quickly respond to a high-salinity environment by producing free amino acids because the aminopeptidase I enzyme coded for by this allele exhibits greater activity.

Result 3 Subsequent measurements reveal that this prediction is valid. Mussels with the *lap* 94 allele are able to control cellular osmolarity in waters of high salinity.

Summarize the data given in Result 2 and 3. Why does the presence of the lap *94 allele help mussels "control cellular osmolarity in waters of high salinity?"*

The presence of the *lap* 94 allele provides a distinct advantage in mussels living in high-salinity waters. However, if mussels adjusted to a high-salinity environment are placed in a low-salinity environment, they excrete waste products characteristic of amino acid breakdown. The loss of amino acids prevents their cells from taking in too much water. However, would you expect mussels with the *lap* 94 allele to be at a disadvantage when placed in a habitat that changes in salinity from high to low and back again?

Prediction 4 In some environmental conditions, such as one that has frequent changes in salinity, mussels with the *lap* 94 allele would be at a disadvantage.

Result 4 Investigations of nitrogen balance in *M. edulis* indicate that mussels with the *lap* 94 allele suffer more nitrogen loss than mussels with only alleles *lap* 96 and/or 98 in an environment like the entrance to Long Island Sound where salinity changes are great. When the salinity rises, they can quickly adjust but when it lowers, they lose more nitrogen than otherwise. When nitrogen loss is sufficiently great, mussels cannot retain sufficient amounts of this nutrient to survive.

Suppose the environment could provide nitrogen to make up for any lost by the mussels, would you expect the frequency of the lap *94 allele to vary significantly between the two populations? Why not? Since the frequency of the* lap *94 allele does vary between the two populations, what does this say about the availability of nitrogen in their diets?*

These experiments and observations strongly support the hypothesis that genetic variation is maintained by selection pressures acting on the *lap* alleles in different environments. A remaining question, however, is why should both the *lap* 96 and *lap* 98 alleles persist in the population? Both are apparently characterized by a lower rate of polypeptide breakdown. Is selection on some other property acting to maintain both alleles? Further studies will be necessary to investigate additional physiological roles of these two alleles.

P A R T 4

Plant Structure
and Function

The shoot system of a plant consists of stems, leaves, and reproductive structures. All are displayed here. The stem carries water to the leaves, which take in CO_2 and capture the energy of the sun to carry on photosynthesis. Then, the stem distributes organic food to the rest of the plant. The pea pod grew from a flower part that contained fertilized eggs, each one having developed into an embryo now located within a seed. When a seed is planted, a new pea plant arises.

The tendrils are also most amazing. They are leaflets modified to twine around solid objects and give support to the plant.

CHAPTER 29

Plant Structure

Your study of this chapter will be complete when you can:

1. Describe the overall organization of a flowering plant, distinguishing between the shoot system and the root system.

2. Describe the structure of epidermis (dermal tissue) and the modification of epidermis in various parts of the plant body.

3. Describe the structure and function of the parenchyma, collenchyma, and sclerenchyma cells in ground tissue.

4. Describe the structure and function of xylem and phloem in vascular tissue.

5. Contrast monocots and dicots in three ways.

6. Name the various types of meristem tissue and state their specific functions.

7. Give three functions of roots. Describe the structure and function of the three zones within a root tip. Describe the cross sections of herbaceous dicot and monocot roots.

8. Describe three patterns of root growth.

9. Give three functions of stems. Describe the cross sections of a herbaceous dicot and a monocot stem.

10. Explain how secondary growth of stems and roots occurs, and describe the structure and function of a woody stem.

11. Describe several ways in which stems can be modified.

12. Describe the structure and function of tissues found in a C_3 dicot leaf. Contrast C_3 leaves with C_4 leaves.

13. Describe several leaf modifications.

An onion plant with soil removed illustrates the three major organs of a plant: root, stem, and leaf. The roots take in water and minerals from the soil and these move through the stem to the leaves, which are green and carry on photosynthesis, a process that requires water, carbon dioxide, and solar energy.

In this chapter, we will study the organization and structure of flowering plants, which are the most complex, recently evolved, and widely distributed plants (see chapter 25).

Organization of a Plant Body

The body of a flowering plant is divided into two portions: the *root system* and the *shoot system* (fig. 29.1). The root system anchors the plant in the soil and the stem holds the leaves aloft to catch the rays of the sun. Both the roots and the stem transport water, inorganic nutrients, and organic nutrients. The roots absorb water and minerals and then these are transported in the stem to the leaves, which receive carbon dioxide from the air

and carry on photosynthesis. The sugar produced by photosynthesis is first transported in the stem to the rest of the plant and finally it reaches the root where it is stored as starch.

Tissue Types in Plants

Plants grow throughout their entire lives because there is **meristem** (embryonic) tissue located in the shoot and root apex. As apical meristem cells divide they give rise to three types of primary meristem tissue, one for each of these specialized tissues found in the body of a plant:

1. *dermal tissue:* forms the outer protective covering of a plant
2. *ground tissue:* fills the interior of a plant
3. *vascular (transport) tissue:* conducts water and nutrients in a plant

Figure 29.1

Organization of plant body. The body of a plant is divided into two main portions: the root system below ground and the shoot system containing the stems and leaves aboveground. A plant grows lengthwise by the production of new cells at the terminal bud (shoot tip) and root tip. The body of a plant contains three types of tissues. Dermal tissue covers the entire body of a nonwoody plant. Ground tissue makes up the bulk of the plant, and vascular tissue conducts water and nutrients to all body parts.

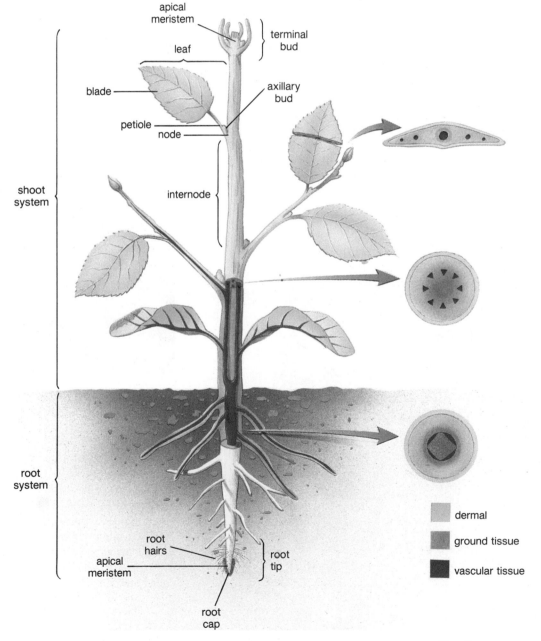

apical meristem

terminal bud

leaf

blade

axillary bud

petiole

node

shoot system

internode

root system

root hairs

apical meristem

root tip

root cap

dermal

ground tissue

vascular tissue

Figure 29.2

Epidermis is modified in different parts of a plant. *a.* Leaf epidermis contains stomata for gas exchange. Magnification, ×100. *b.* Root epidermis has root hairs to absorb water. *c.* Cork replaces epidermis in older woody stems.

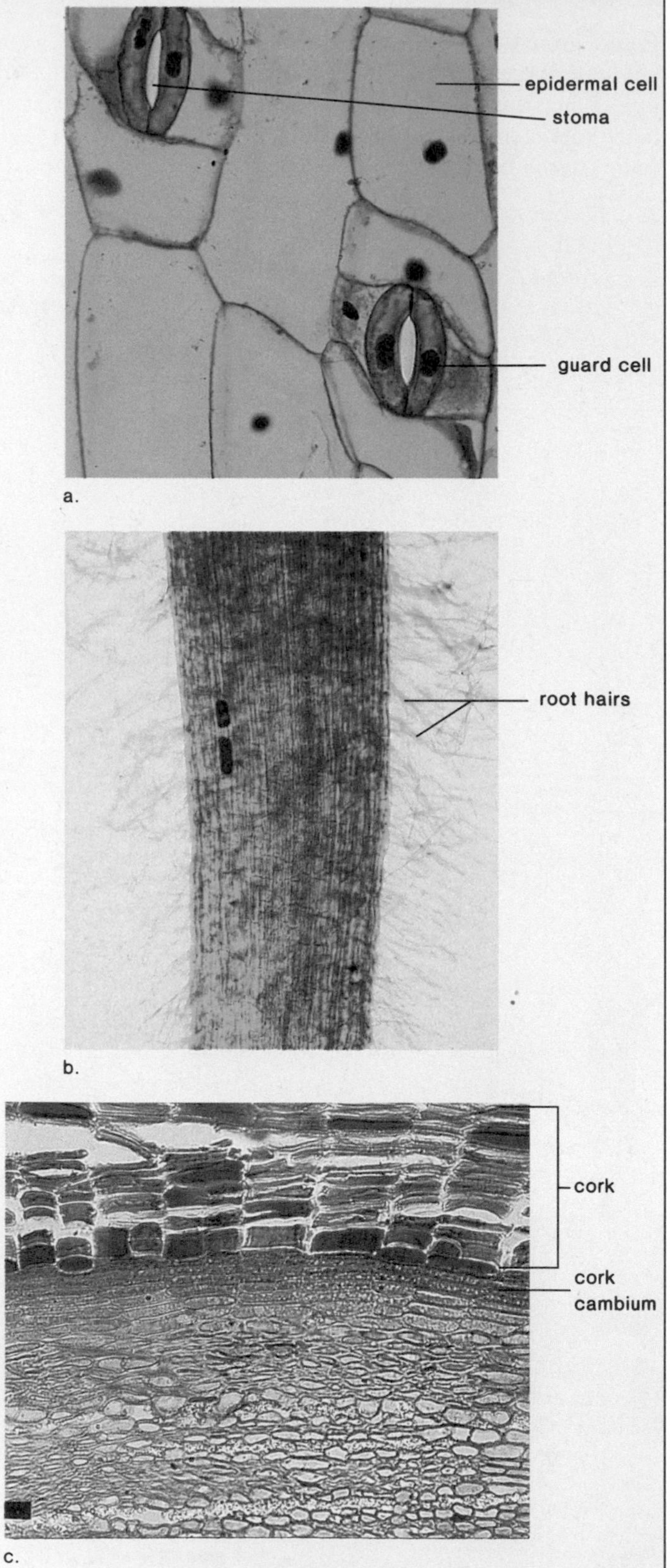

epidermal cell

stoma

guard cell

a.

root hairs

b.

cork

cork cambium

c.

Figure 29.3

Ground tissue cells. *a.* Parenchyma cells are the least specialized of the plant cells. This type of parenchyma cell is found in the stems and roots. Those in the leaf contain chloroplasts. *b.* Collenchyma cells. Notice how much thicker the walls are compared to those of parenchyma cells. Collenchyma cells give flexible support. Magnification, ×41,024. *c.* Sclerenchyma cells. Sclerenchyma cells have very thick walls and are nonliving—their only function is to give strong support.

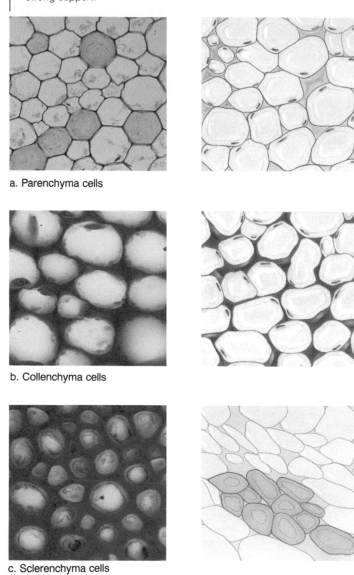

a. Parenchyma cells

b. Collenchyma cells

c. Sclerenchyma cells
   (fibers)

Dermal Tissue

The entire body of a nonwoody (herbaceous) and young woody plant is covered by a layer of **epidermis,** which contains epidermal cells. Epidermal cells are closely packed, and the epidermis protects the inner tissues of a plant. Also, the walls of epidermal cells that are exposed to air are covered with a waxy cuticle to minimize water loss. In addition to the epidermal cells, the epidermis may also contain specialized cells, such as guard cells (fig. 29.2*a*). The guard cells surround microscopic pores called **stomata** (sing., **stoma**). When the stomata are open, gas

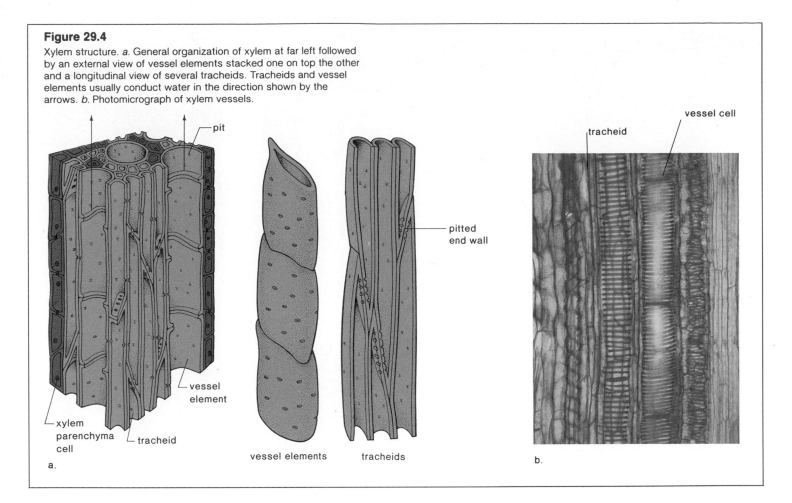

Figure 29.4

Xylem structure. *a*. General organization of xylem at far left followed by an external view of vessel elements stacked one on top the other and a longitudinal view of several tracheids. Tracheids and vessel elements usually conduct water in the direction shown by the arrows. *b*. Photomicrograph of xylem vessels.

pit

vessel element

xylem parenchyma cell

tracheid

a.

vessel elements

tracheids

pitted end wall

vessel cell

tracheid

b.

exchange can occur. The epidermal cells of roots have long, slender projections called root hairs (fig. 29.2*b*). The hairs increase the surface area of the root for absorption of water and minerals and also help to anchor the plant firmly in place.

In older woody plants, the epidermis of the stem is replaced by cork tissue. **Cork** (fig. 29.2*c*), the outer covering of the bark of trees, is made up of dead cork cells that may be sloughed off. New cork cells are made by meristematic tissue called cork cambium.

Ground Tissue

Ground tissue forms the bulk of a plant and contains parenchyma, collenchyma and sclerenchyma cells (fig. 29.3). **Parenchyma** cells correspond best to the typical plant cell depicted in figure 5.5. These are the least specialized of the cell types and are found in all the organs of a plant. They may contain chloroplasts and carry on photosynthesis or they may contain colorless plastids that store the products of photosynthesis. Parenchyma cells can divide and give rise to more specialized cells as when roots develop from stem cuttings placed in water.

Collenchyma cells are like parenchyma cells except they have thicker primary walls (fig. 6.15). The thickness is uneven; and usually involves the corners of the cell. Collenchyma cells often form bundles just beneath the epidermis, and give flexible support to immature regions of a plant body. The familiar strands of celery stalks are composed mostly of collenchyma cells.

Sclerenchyma cells have thick secondary cell walls (fig. 6.15) usually impregnated with lignin, which is an organic substance that makes the walls tough and hard. Most sclerenchyma cells are nonliving; their primary function is to support mature regions of a plant. Two types of sclerenchyma cells are fibers and sclereids. Although fibers do occasionally occur in ground tissue, most are found in vascular tissue to be discussed next. Fibers are long and slender and may occur in bundles that are sometimes commercially important. Hemp fibers can be used to make rope, and flax fibers can be woven into linen. However, flax fibers are not lignified and that is why linen is soft. Sclereids, which are shorter than fibers with more varied shapes, occur in seed coats and nut shells. They also give pears their characteristic gritty texture.

Vascular (Transport) Tissue

There are two types of vascular tissue. **Xylem** transports water and minerals from the roots to the leaves, and **phloem** transports organic nutrients, usually from the leaves to the roots. Xylem contains two types of conducting cells: *tracheids* and *vessel elements* (fig. 29.4). Both types of conducting cells are hollow and nonliving, but the vessel elements are larger, lack transverse end walls, and are arranged to form a continuous pipeline for water and mineral transport. The elongated tracheids with tapered

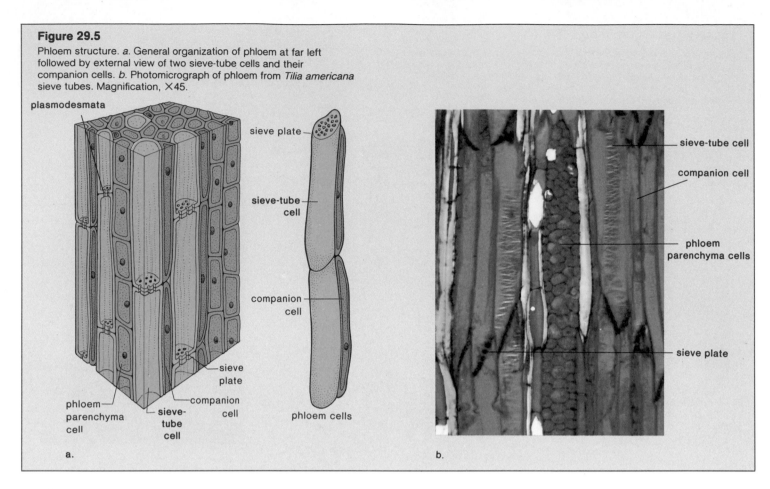

Figure 29.5
Phloem structure. *a.* General organization of phloem at far left followed by external view of two sieve-tube cells and their companion cells. *b.* Photomicrograph of phloem from *Tilia americana* sieve tubes. Magnification, ×45.

plasmodesmata

sieve plate

sieve-tube cell

companion cell

sieve plate

companion cell

sieve-tube cell

phloem parenchyma cell

phloem cells

a.

sieve-tube cell

companion cell

phloem parenchyma cells

sieve plate

b.

ends form a less obvious means of transport but water can move across the end and side walls because there are pits, or depressions where the secondary wall does not form.

The conducting cells of phloem are sieve-tube cells, each of which has a companion cell (fig. 29.5). *Sieve-tube cells* contain cytoplasm but no nuclei. These cells have channels in their end walls that in cross section makes them resemble a sieve. Plasmodesmata extend from one cell to the other through this so-called *sieve plate.* The smaller *companion cells* are more generalized cells and are closely connected to sieve-tube cells by numerous plasmodesmata. The nucleus of the companion cell controls and maintains the life of both cells.

It is important to realize that vascular tissue (xylem and phloem) extends from the root to the leaves, and vice versa. In the roots, the vascular tissue is located in the vascular cylinder. In the stem, it forms vascular bundles, and in the leaves, it is found in leaf veins.

Further Differentiation

In the organs of a plant, the three types of tissues are sometimes given special names (table 29.1). We will be using these terms to discuss the roots, stems, and leaves in the following sections.

Monocot and Dicot

Flowering plants are divided into the *monocots* and *dicots,* depending on the type of cotyledon(s) or seed leaf (leaves) that the embryonic plant has. Cotyledons provide nutrient molecules

for growing embryos before the true leaves begin photosynthesizing. Some embryos have one cotyledon, and plants with such embryos are known as monocotyledonous, or monocots. Other embryos have two cotyledons, and plants with these embryos are known as dicotyledonous, or dicots. Adult monocots and dicots have several other structural differences, as illustrated in figure 29.6.

Primary and Secondary Growth

All plants show primary growth, which increases the length of a plant. Primary growth occurs because shoot and root apical meristems are actively producing new cells. The following three primary meristems develop from apical meristem cells (fig. 29.7) and give rise to the tissues noted.

1. protoderm: produces dermal (epidermal) tissue
2. ground meristem: produces ground tissue
3. procambium: produces primary vascular tissue

Some dicots, notably woody trees, also have secondary growth, which increases the diameter of stems and roots and makes them thicker. Secondary growth is associated with these lateral meristems:

1. vascular cambium: produces secondary vascular tissue
2. cork cambium: produces new cork cells

As discussed on page 447, cork replaces epidermis in woody plants.

Table 29.1 Vegetative Organs and Major Tissues

	Roots	Stems	Leaves
Function	Absorb water and minerals Anchor plant Storage of materials	Transport water and nutrients Support leaves Helps store materials	Carry on photosynthesis
Tissue			
Epidermis*	Root hairs absorb water and minerals	Protect inner tissues	Stomata carry on gas exchange
Cortex†	Store products of photosynthesis and water	Carry on photosynthesis, if green	———
Vascular‡	Transport water and nutrients	Transport water and nutrients	Transport water and nutrients
Pith†	Store products of photosynthesis and water	Store products of photosynthesis	———
Mesophyll† Spongy layer Palisade layer	———	———	Gas exchange photosynthesis

Plant tissues belong to one of three tissue systems.
*Dermal tissue
†Ground tissue
‡Vascular tissue

Figure 29.6

Flowering plants are either monocots or dicots. Three features used to distinguish monocots from dicots are the number of cotyledons, arrangement of vascular bundles, and pattern of leaf veins. Not discussed in this chapter are the number of flower parts: monocots have flower parts in threes, dicots in fours or fives.

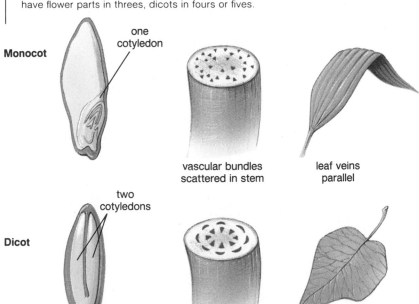

Monocot

one cotyledon

vascular bundles scattered in stem

leaf veins parallel

Dicot

two cotyledons

vascular bundles in a distinct ring

leaf veins form a net pattern

Figure 29.7

Diagram of shoot tip and root tip illustrating the arrangement of plant meristem tissues. Apical meristem gives rise to three primary meristems: protoderm, ground meristem, and procambium. Vascular cambium and cork cambium are present in those parts of the plant body where secondary growth occurs. Secondary growth increases the girth of the plant.

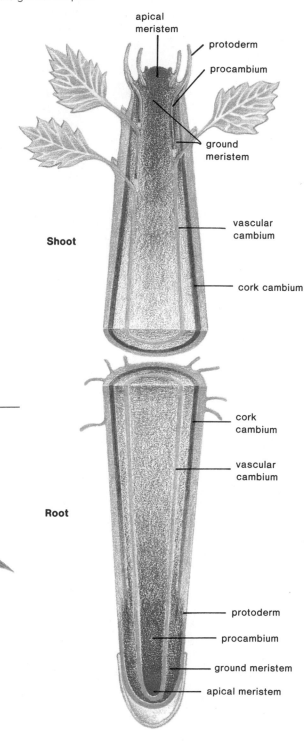

apical meristem

protoderm

procambium

ground meristem

Shoot

vascular cambium

cork cambium

cork cambium

vascular cambium

Root

protoderm

procambium

ground meristem

apical meristem

Figure 29.8
Dicot root tip. *a.* Root tip is divided into four zones, best seen in a longitudinal section such as this. *b.* Vascular cylinder of a dicot root contains the vascular tissue. Xylem is typically star shaped, and phloem lies between the points of the star. *c.* Casparian strip.

Because of the Casparian strip (waxy substance), water and solutes must pass through the cytoplasm of endodermal cells. In this way, endodermal cells regulate the passage of materials into the vascular cylinder.

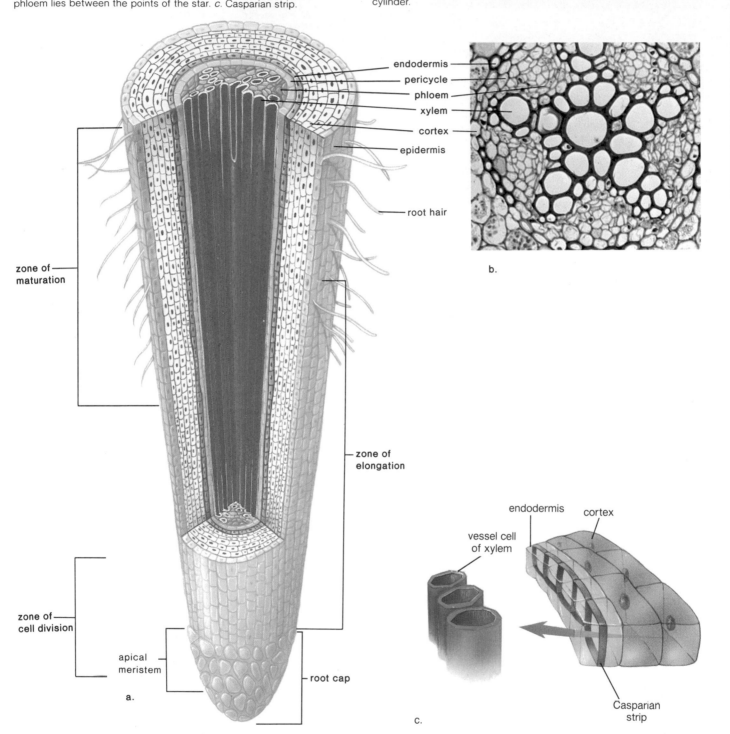

endodermis
pericycle
phloem
xylem
cortex
epidermis

root hair

b.

zone of
maturation

zone of
elongation

zone of
cell division

apical
meristem

a.

root cap

vessel cell
of xylem

endodermis   cortex

Casparian
strip

c.

Root System

The root system functions to anchor a plant in the soil, absorb water and minerals from the soil, and store the products of photosynthesis received from the leaves. The entire root system of a plant is quite extensive. In his study of rye roots in the 1930s, Dittmer estimated that a single four-month-old rye plant had roots totaling more than 600 km in length. There were literally millions of branched roots. He estimated that the plant had about 14×10^9 root hairs, and that the total surface area of roots and root hairs was almost 640 m².

Primary Growth of Dicot Roots

Figure 29.8, a longitudinal section of a dicot root, reveals zones where cells are in various stages of differentiation. Below the *zone of cell division* (apical meristem), cells are continuously being added to the *root cap,* which is a protective cover for the root tip. Cells in the root cap have to be replaced constantly, because they are ground off as the root pushes through rough soil particles. Above the zone of cell division, cells elongate as they become specialized. This region of a root is called the *zone of elongation.* Finally, there is a region in which cells are mature and specialized. This region is called the *zone of maturation.* The zone of maturation is recognizable even in a whole root, because root hairs are borne by many of the outer epidermal cells. Root hairs add tremendously to the total absorptive surface area of roots.

Tissues

Figure 29.8*b* shows a cross section of a root cut through the region of maturation to show the specialized tissues found there.

Epidermis   The epidermis is the outer layer of the root and consists of only a single layer of cells. The majority of epidermal cells are thin walled and rectangular, but in the region of maturation, many epidermal cells have root hairs. These project as far as 5–8 mm into the soil particles.

Cortex   Moving inward, next to the epidermis, are large, thin-walled, parenchyma cells that make up the *cortex* of the root. These irregularly shaped cells are loosely packed, and it is possible for water and minerals to move through the cortex without entering the cells. The cells contain starch granules, and the cortex functions in food storage.

Endodermis   The **endodermis** is a single layer of rectangular endodermal cells that forms a boundary between the cortex and the inner vascular cylinder. The endodermal cells fit snugly together and are bordered on four sides (fig. 29.8*c*) (but not the two sides that contact the cortex or the vascular cylinder) by a strip of waxy material known as the **Casparian strip.** This strip will not permit water and mineral ions to pass between adjacent cell walls. Therefore, the only access to the vascular cylinder is through the endodermal cells themselves, as shown by the arrow in figure 29.8*c*. It is said that the endodermis regulates the entrance of molecules into the vascular cylinder.

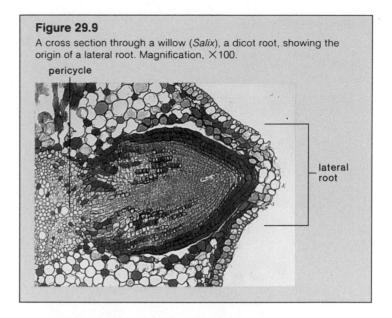

Figure 29.9

A cross section through a willow (*Salix*), a dicot root, showing the origin of a lateral root. Magnification, ×100.

pericycle

lateral root

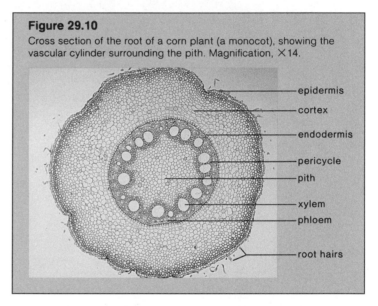

Figure 29.10

Cross section of the root of a corn plant (a monocot), showing the vascular cylinder surrounding the pith. Magnification, ×14.

epidermis

cortex

endodermis

pericycle

pith

xylem

phloem

root hairs

Vascular Cylinder   The *pericycle,* the first layer of cells within the *vascular cylinder,* have retained their capacity to divide and can start the development of branch or secondary roots (fig. 29.9). The main portion of the vascular cylinder, though, contains vascular tissue. The xylem appears star shaped in dicots (fig. 29.8*b*) because several arms of tissue radiate from a common center. The phloem is found in separate regions between the arms of the xylem.

Monocot Roots

Monocot roots often have pith, centrally located ground tissue. In a monocot root, pith is surrounded by a ring of alternating xylem and phloem bundles (fig. 29.10). They also have pericycle, endodermis, cortex, and epidermis. Typically, monocot roots do not undergo secondary growth.

Figure 29.11

Three types of roots. *a.* A taproot may have secondary roots in addition to a main root. *b.* A fibrous root has many secondary roots with no main root. *c.* Adventitious roots, such as prop roots, extend from the stem.

a. b. c.

The root system of a plant absorbs water and minerals that cross the epidermis and cortex before entering the endodermis, the tissue that regulates the entrance of molecules into the vascular cylinder.

Root Diversity

In some plants, notably dicots, the first or primary root grows straight down and remains the dominant root of the plant. This so-called *taproot* (fig. 29.11*a*) is often fleshy and stores food. Carrots, beets, turnips, and radishes have taproots that we consume as vegetables.

In other plants, notably monocots, a number of slender roots develop, with no single main root. These slender roots and their lateral branches make up a *fibrous root* system (fig. 29.11*b*). Most everyone has observed the fibrous root system of grasses and noted how these roots can hold the soil.

Sometimes a root system develops from an underground stem or from the base of an aboveground stem. These are *adventitious roots,* whose main function may be to help anchor the plant. If so, they are called *prop roots*. Mangrove plants have large prop roots that spread away from the plant to help anchor it in the marshy soil where mangroves are typically found. Corn plants also develop prop roots for support (fig. 29.11*c*).

Shoot System

The shoot system includes the stem and leaves. First, we will consider the growth of the stem, and in the next section, we will examine the leaf. A stem supports the leaves, flowers, and fruits; conducts material to and from the roots and leaves; and helps store water and the products of photosynthesis.

Even highly modified stems can be recognized by the presence of nodes, internodes, buds, and leaves (fig. 29.1). A *node* is the point on a stem at which leaves or buds are attached, and an *internode* is the segment of a stem between nodes.

Primary Growth of Stems

The terminal bud at the tip of each growing stem contains shoot meristem (fig. 29.12). Whereas root-tip meristem is protected by a root cap, shoot-tip meristem is protected by newly formed leaves within a bud. Shoot-tip meristem produces cells that will become the stem and the leaves. At first, the nodes are very close together, then internode growth occurs so that the nodes are distant from one another. This pattern of growth complicates the growth of stems, and it is not possible to divide the entire stem into zones of cell division, elongation, and maturation as with the root.

Axillary buds, which are usually dormant but may develop into branch shoots, are seen in the axes of mature leaves. Inactive buds are covered by protective bud scales. In the temperate zone, the terminal bud stops growing in the winter and

Figure 29.12

Terminal bud anatomy. In this micrograph, the apical meristem is surrounded by two sets of leaves, a younger pair and an older pair. Axillary buds are seen at the axes of the older pair of leaves. Axillary buds can give rise to side branches when they are active.

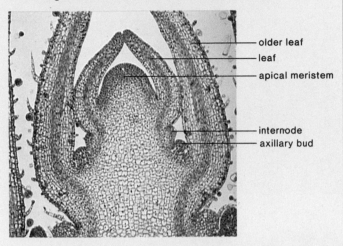

- older leaf
- leaf
- apical meristem
- internode
- axillary bud

is then protected by bud scales also. In the spring when growth resumes, these scales fall off and leave a scar. You can tell the age of a stem by counting these bud scale scars.

We will now discuss the tissues of mature nonwoody stems. Such stems, also called *herbaceous stems,* exhibit only primary growth.

Herbaceous Stems

The outermost tissue of herbaceous stems is the epidermis, which is covered by a waxy cuticle to prevent water loss. These stems have distinctive *vascular bundles* where xylem and phloem are found. In each bundle, xylem is typically found toward the inside of the stem and phloem is found toward the outside.

In the dicot herbaceous stem (fig. 29.13), the bundles are arranged in a distinct ring that separates the cortex from the central pith. The cortex is sometimes green and carries on photosynthesis, and the pith may function as a storage site. In the monocot stem (fig. 29.14), the vascular bundles are scattered throughout the stem, and there is no well-defined cortex or well-defined pith.

Figure 29.13

Herbaceous dicot stem anatomy. *a.* Cross section of alfalfa (*Medicago*) stem shows that the vascular bundles occur in a ring. *b.* Drawing of a section of the stem, with tissues in the vascular bundle and in the stem identified.

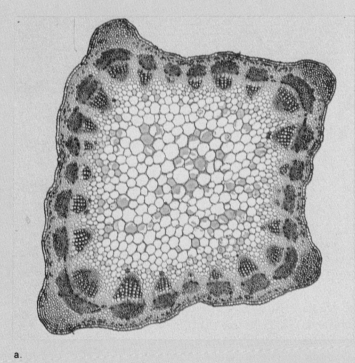

a.

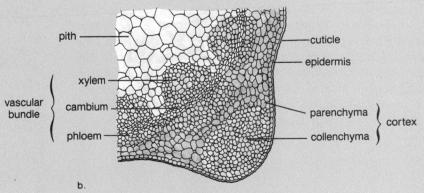

- pith
- cuticle
- epidermis
- xylem
- cambium
- parenchyma
- phloem
- collenchyma

vascular bundle

cortex

b.

Figure 29.14
Monocot stem anatomy. *a.* Cross section of corn (*Zea mays*), showing that the vascular bundles are scattered. *b.* Enlargement of stem showing vascular bundle in more detail. *c.* Enlargement of one vascular bundle shows the arrangement of tissues in a bundle. Photograph by Carolina Biological Supply Company.

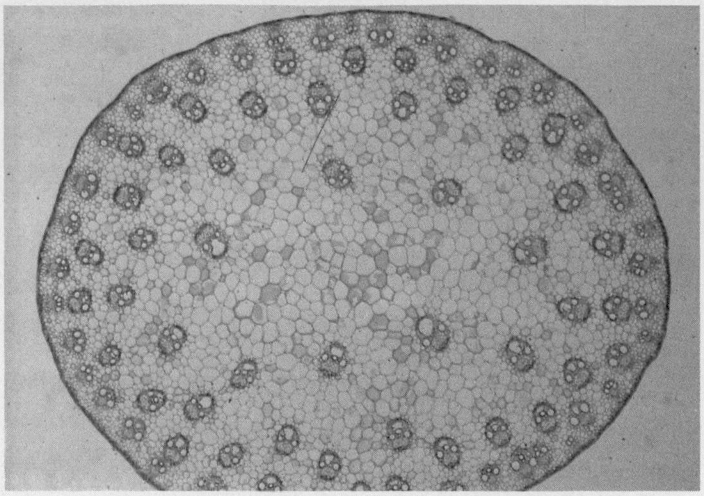

a.

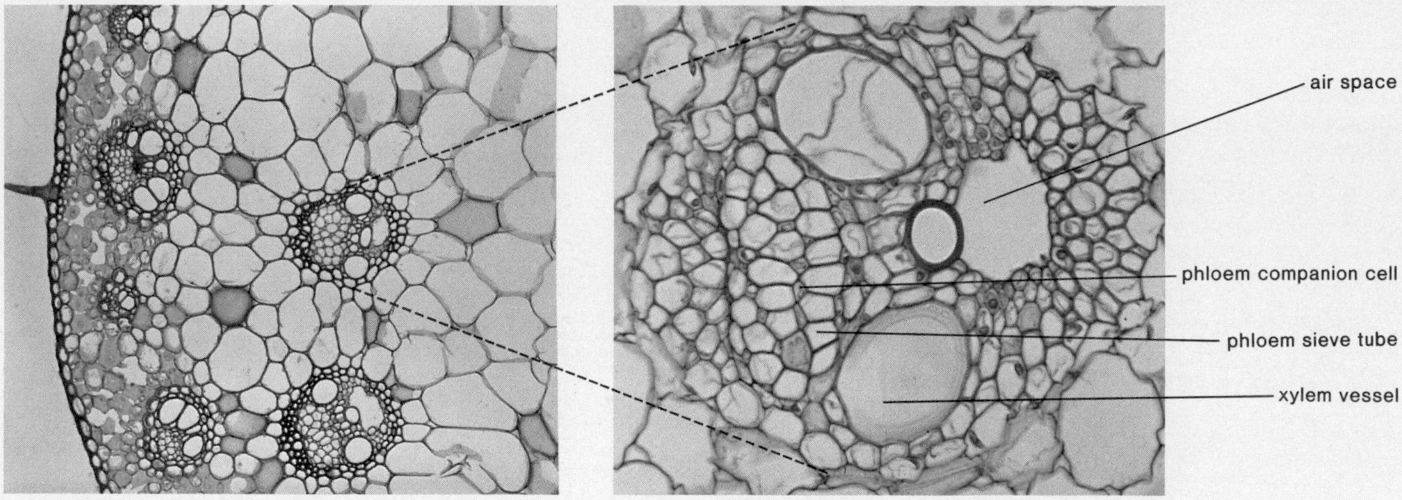

b.

c.

air space

phloem companion cell

phloem sieve tube

xylem vessel

Secondary Growth of Stems

Secondary growth of stems is seen primarily in woody plants, such as trees that live for many years. Almost all trees are dicots. Primary growth in woody plants occurs for a short distance behind the apical meristem, at which point secondary growth begins. You'll recall that secondary growth occurs because of the presence of meristematic tissue called vascular cambium and cork cambium.

Vascular cambium begins as meristematic cells between the xylem and phloem of each vascular bundle. Then these cells join to form a ring of meristematic tissue. The division of this meristematic tissue takes place mostly in a plane parallel to the surface of the plant. The secondary tissues produced by the vascular cambium, called secondary xylem and secondary phloem therefore, add to the girth of the stem instead of to its length (fig. 29.15).

Cork cambium is located beneath the epidermis. When it begins to divide, it produces tissue that disrupts the epidermis and replaces it with cork cells. Cork cells are impregnated with suberin, a fatty substance that makes them waterproof. Dead cork only allows gas exchange in pockets of loosely arranged cells, called **lenticels,** that do not become impregnated with suberin.

Cross Section of Woody Stem

As a result of secondary growth, a woody stem (fig. 29.16) has an entirely different type of organization than that of a dicot herbaceous stem. After secondary growth has continued for a time, it is no longer possible to make out individual vascular bundles. Instead, a woody stem has three distinct areas: the bark, the wood, and the pith. The bark contains cork, cork cambium, cortex, and phloem. Although secondary phloem is produced each year by vascular cambium, it does not build up. The outer

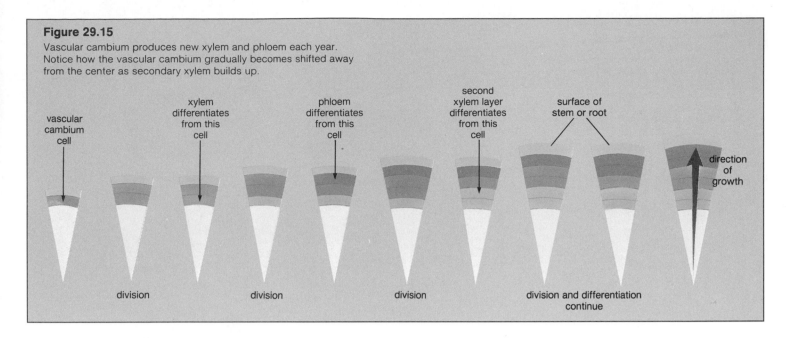

Figure 29.15

Vascular cambium produces new xylem and phloem each year. Notice how the vascular cambium gradually becomes shifted away from the center as secondary xylem builds up.

vascular cambium cell

xylem differentiates from this cell

phloem differentiates from this cell

second xylem layer differentiates from this cell

surface of stem or root

direction of growth

division   division   division   division and differentiation continue

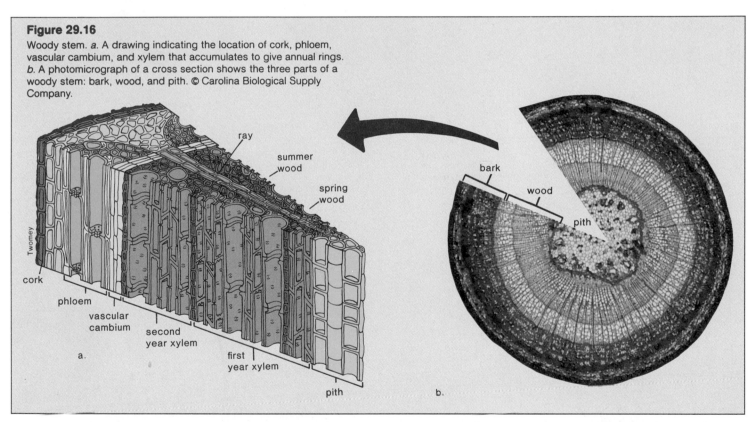

Figure 29.16

Woody stem. *a.* A drawing indicating the location of cork, phloem, vascular cambium, and xylem that accumulates to give annual rings. *b.* A photomicrograph of a cross section shows the three parts of a woody stem: bark, wood, and pith. © Carolina Biological Supply Company.

ray

summer wood

spring wood

Twomey

cork

phloem

vascular cambium

second year xylem

first year xylem

pith

a.

bark

wood

pith

b.

layers of phloem become crushed as they are pushed against the bark. Because the newly formed phloem is in the bark of a tree, even partial removal of the bark can seriously damage a tree.

The wood of trees that grow in seasonal climates, contains rings of secondary xylem that are called growth rings, because there is one for each season of growth. In temperate regions with one growing season per year these are called annual rings. It is easy to tell where one ring begins and another ends. In the spring, when moisture is plentiful, the xylem cells are much larger than later in the summer, when moisture is scarcer. In large trees,

only the more recently formed layer of xylem, the sapwood, functions in water transport. The older inner part, called the heartwood, becomes plugged with deposits, such as resins, gums, and other substances. Heartwood may help support a tree, although some trees stand erect and live for many years after the heartwood has rotted away.

Like stems, the roots of woody plants also show secondary growth. Figure 29.17 illustrates that the secondary growth of roots arises and occurs in the same manner as that of the stem.

Figure 29.17

Both roots and stems experience secondary growth. *a.* Young stem and root in which the vascular cambium is just forming a ring. *b.* One layer of secondary xylem and phloem produced by vascular cambium in both stem and root.

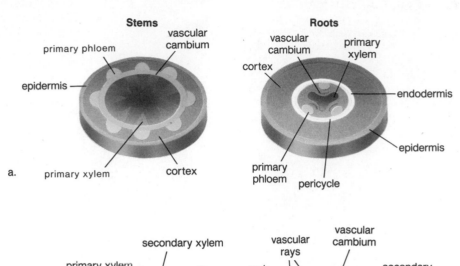

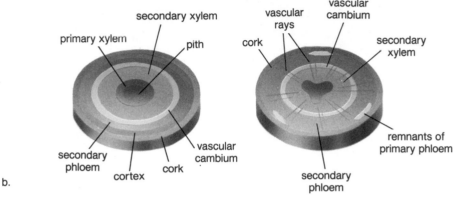

Figure 29.18

Stem modifications. *a.* A strawberry plant has aboveground, horizontal stems called stolons. Every other node produces a new shoot system. *b.* The underground horizontal stem of an iris is fleshy.

c. The underground stem of a potato plant has enlargements called tubers. We call the tubers potatoes. *d.* Corms are stem tissue covered by papery leaves.

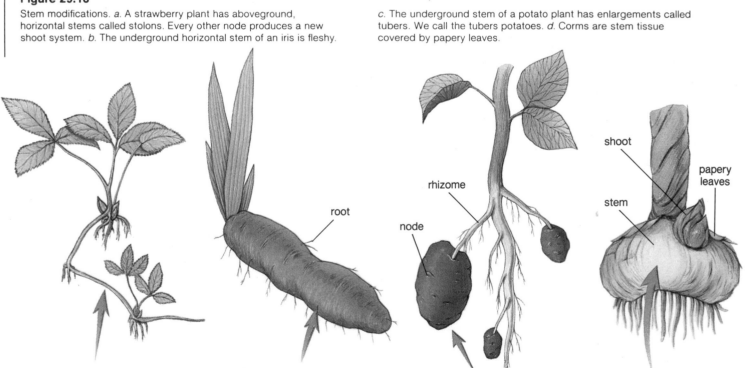

a. Stolon   b. Rhizome   c. Tuber   d. Corm

Types and Uses of Stems

Stem modifications are illustrated in figure 29.18. Aboveground horizontal stems, called **stolons** or **runners,** produce new plants where nodes touch the ground. The strawberry plant is a common example of this type of propagation.

Underground horizontal stems (**rhizomes**) may be long and thin, as in sod-forming grasses, or thick and fleshy as in *Iris*. Rhizomes survive the winter and contribute to asexual reproduction, because each node bears a bud. Some rhizomes have enlarged portions called tubers that function in food storage. For example, potatoes are tubers. The eyes of potatoes are buds that mark the nodes.

Corms are bulbous underground stems that lie dormant during the winter, just as rhizomes do. They also produce new plants the next growing season. Gladiolus corms are referred to as bulbs by lay persons, but the botanist reserves the term bulb for a structure composed of modified leaves.

Aboveground vertical stems can also be modified. For example, cacti have succulent stems modified for water storage. The tendrils of grape plants that twine around a support structure are modified stems.

Humans use stems for many purposes. The stem of the sugarcane plant is a primary source of table sugar. The spice, cinnamon and the drug, quinine are derived from the barks of two different plants. The softwood of gymnosperms is used for

Paper Is Secondary Xylem

Over 500 lb, or about 1.4 lb per day, is a close approximation of the paper used annually by each person in America. Figured on a per capita basis, we use paper products at nearly twice the rate of the next highest consumer, Canada, and at better than 25 times the rate of the Communist bloc nations taken as a whole.

Immediately, of course, we think of paper as a means of communication; historically, that was nearly its exclusive use for over a thousand years. Paper appears to have been invented in China around A.D. 105. Paper remained an exclusively Chinese product for about 500 years until it appeared in Japan. It wasn't until another 500 years had passed that it found its way westward into Egypt, and finally Europe in the twelfth century. Its use for other than communication purposes was virtually nil until the mid-1800s when paper bags and boxes were invented. Around the turn of this century, we find milk cartons, cups, plates, food wrappers, and all kinds of things being made of paper.

What is this wonderful stuff made of? Plants, of course! Actually, the fibers used in manufacturing paper come from rags, straw, grasses, old newspapers, and a number of other sources. But the ultimate source is always plant material, and most of it, better than 90%, is wood. Cellulose, the basic material for paper, is a material in the cell walls of secondary xylem, which is wood. You may have the feeling that paper manufacturing is very complicated. Actually the basic process is quite simple and hasn't changed much in its nearly 2,000-year history. There are four steps: (1) make the pulp, (2) make the stock (optional), (3) make wet paper, and (4) dry it out.

Pulp is simply a suspension of pulverized wood cells. In the trade these cells are called fibers, but as plant biologists we know that most of them are really tracheids and vessels. This suspension of fibers is made in either of two ways. The logs with the bark removed may simply be ground against a grinding stone in the presence of water. Paper made from this kind of pulp is relatively inexpensive but weak

Paper machine take-up spool at Potlatch Corporation, Lewiston, Idaho.

and generally of poor quality. Newsprint, for example, is primarily of this type. . . . Alternatively, the logs may be fragmented into chips and then pulverized, using hot sulfites, soda, or sulfates, depending on the various processes. These chemicals remove impurities (especially lignin) from the chips, leaving only the cellulose; consequently the yield is only about 40%–50% of the original wood, but the product is of much finer quality. The pulp is then bleached so that the paper will be white instead of wood colored. It can be used directly to make paper, but usually it passes through an intermediate step during which it is made into stock.

Pulp is refined into stock simply by further grinding. The purpose is to roughen and fray the extracted fiber fragments and to make them more uniform in size. As a result of this treatment, the fibers adhere to one another more readily, and the strength and quality of the paper are improved. Also dyes, strengtheners, and other additives may be introduced at this point.

The third step is to make wet paper out of the pulp or stock. This is accomplished by applying the pulp or stock to a wire-screen filter, draining off the liquid and thereby leaving only

the wood fibers, which form a thin, matted network on the surface of the screen. In times past this was done by simply immersing a screen in a vat that contains the stock, and then drawing the screen upward, causing the wood fibers to collect on it. This was done by hand, and one man could produce perhaps 750 sheets per day. Today you'll find manual operation replaced by a revolving wire-screen belt up to 30 feet wide which moves at rates up to 88 feet per second or about 60 miles per hour. This belt picks up the stock and drains off the water rapidly by using a vacuum on the underside. . . . At such rates one machine can turn out in less than 2 seconds the equivalent of one man working all day using the techniques of the early 1800s.

The final step is to press the paper and dry it. This is accomplished by squeezing the paper between heated rotating drums. Finally, the paper is collected in giant rolls.

From *Botany: A Human Concern* 2nd ed. by David L. Rayle and Hale L. Wedberg. Copyright © 1980 by Saunders College/Holt, Rinehart and Winston. Reprinted by permission of Holt, Rinehart and Winston, CBS Publishing.

Figure 29.19

a. Drawing of a leaf from a C₃ plant that is adapted to temperate climate. In a C₃ leaf the mesophyll is divided into a palisade and spongy layer. The bundle sheath cells lack chloroplasts. *b.* A scanning electron micrograph of the mesophyll. *c.* Drawing of a leaf vein from a C₄ plant that is adapted to a hot, dry climate. In a C₄ leaf, the mesophyll is arranged around the bundle sheath, and the bundle sheath cells have chloroplasts.

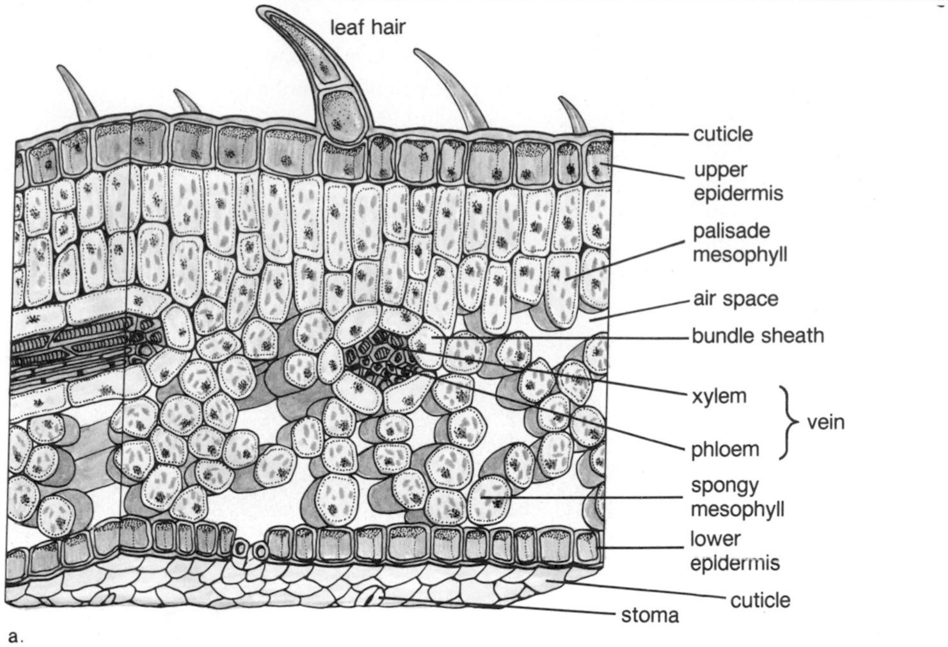

leaf hair

cuticle
upper epidermis
palisade mesophyll
air space
bundle sheath
xylem ⎱ vein
phloem ⎰
spongy mesophyll
lower epidermis
cuticle
stoma

a.

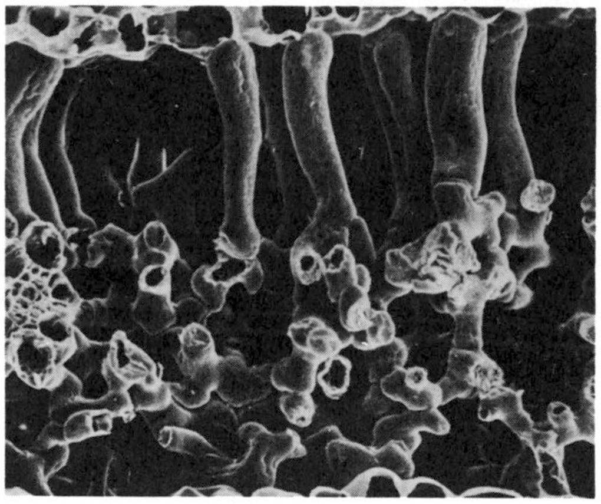

b.

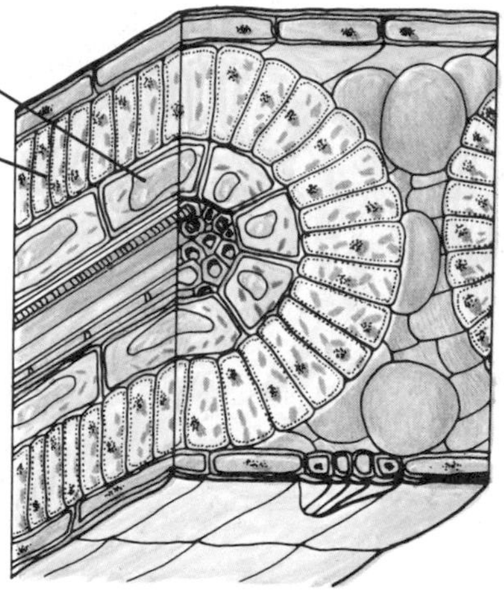

bundle sheath cell
mesophyll

c.

---

numerous purposes, including lumber for construction and the production of paper (see reading, page 457) and rayon. Hardwood from angiosperms is used for flooring, furniture, and interior finishes.

Within the shoot system, the stem transports water and nutrients between the leaves and the roots. Stem anatomy differs according to whether the plant is a monocot or dicot and whether it is herbaceous or woody. There are also various types of stem modifications.

Leaves

Leaves are the organs of photosynthesis in vascular plants. A leaf usually consists of a flattened **blade** and a **petiole,** which connects the blade to the stem. The blade may be single or composed of several leaflets. Externally, it is possible to see the pat-

tern of the **leaf veins** that contain vascular tissue. Leaf veins have a net pattern in dicot leaves and a parallel pattern in monocot leaves (fig. 29.6).

Figure 29.19*a* shows the cross section of a typical dicot leaf of a temperate zone plant. At the top and bottom is a layer of epidermal tissue that often bears protective hairs and/or glands that produce irritating substances. These features may prevent the leaf from being eaten by insects. Regardless of whether these additions are present, the epidermis always has an outer waxy cuticle that keeps the leaf from drying out. Unfortunately, it also prevents gas exchange, because the cuticle is not gas permeable. However, the epidermis, particularly the lower epidermis, contains openings called stomata that allow gases to move into and out of the leaf. Epidermal cells do not ordinarily contain chloroplasts; however, the guard cells surrounding the stomata do contain chloroplasts. These guard cells regulate the opening and closing of the stomata, as discussed on page 469.

Figure 29.20

Classification of leaves. *a.* Simple leaf. *b.* Pinnately compound leaf. *c.* Twice pinnately compound leaf. *d.* Palmately compound leaf.

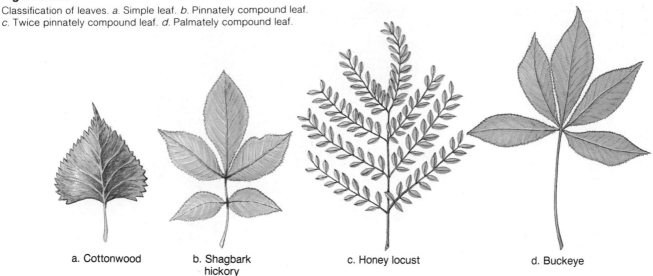

a. Cottonwood   b. Shagbark hickory   c. Honey locust   d. Buckeye

The body of a leaf is composed of **mesophyll tissue,** which has two layers of cells: the palisade layer contains elongated cells, and the spongy layer contains irregular cells bounded by air spaces. The parenchyma cells of the **palisade layer** and **spongy layer** have many chloroplasts and carry on most of the photosynthesis for the plant. The loosely packed arrangement of the cells in the spongy layer increases the amount of surface area for gas exchange.

The cells of the mesophyll are protected from drying out by a thin film of water that is constantly evaporating. Loss of water by evaporation at the leaves is called **transpiration.** At least 90% of water taken up by the roots is eventually lost by transpiration. This means that the total amount of water lost by a plant over a long period of time is surprisingly large. For example, a single *Zea mays* (corn) plant loses somewhere between 135 and 200 liters of water through transpiration during a growing season. If this water loss is multiplied by the number of corn plants in a heavily planted cornfield, one can understand why farming requires so much water. The amount of water required by different agricultural crops to produce a given quantity of food can also be compared. Millet requires about 225 kg of water for every kilogram of food produced, wheat requires about 500 kg of water per kilogram of food, and potatoes require about 800 kg of water per kilogram of food.

Leaf Veins

A cross section of a leaf shows that leaf veins consist of a strand of xylem and a strand of phloem surrounded by a **bundle sheath.** The bundle sheath differs in C_3 and C_4 plants. As discussed on page 119, these terms refer to the number of carbon atoms in the first detected molecule after the start of photosynthesis. The bundle sheath cells in the leaves of C_4 plants characteristically have chloroplasts, and some of the mesophyll cells are often closely packed around the bundle sheath in a radial fashion (fig. 29.19*c*). The bundle sheath cells in C_3 plants do not contain chloroplasts, and mesophyll tissue is divided into the palisade and spongy layer (fig. 29.19*a*).

Within the shoot system, the leaves carry on photosynthesis. Leaf anatomy differs according to whether the plant is a C_3 or C_4 plant. There are also various types of leaf modifications.

Types and Uses of Leaves

In a leaf, the blade can be simple and undivided or compound, with two or more separate leaflets making up the blade (fig. 29.20). The leaflets of a compound leaf can be arranged along the length of a central stalk (pinnately compound) or attached to the end of the petiole (palmately compound). Aside from these blade patterns, there are also various patterns of vascular arrangement in leaves, and innumerable combinations of overall leaf shape, margin, and base modifications. Leaves are adapted to environmental conditions. Plants that usually live in the shade tend to have broad, wide leaves, and those that live where it is dry tend to have reduced leaves with sunken stomata. The leaves of a cactus are the spines attached to the succulent stem (fig. 29.21). Other succulents, however, have leaves adapted to hold moisture.

Leaves can also be specialized for food storage. An onion is a **bulb,** with leaves surrounding a short stem. A head of cabbage has a similar construction, except the large leaves overlap one another. The petiole, too, can be thick and fleshy, as in celery and rhubarb.

Climbing leaves, such as those of peas and cucumbers, are modified into tendrils that can attach to nearby objects. The leaves of a few plants are specialized for catching insects. The leaves of a sundew have sticky epidermal hairs that trap insects and then secrete digestive enzymes. The Venus-flytrap has hinged leaves that snap shut and interlock when an insect triggers sensitive hairs. The leaves of a pitcher plant resemble a pitcher and have downward-pointing hairs that lead insects into a pool of digestive enzymes. Insectivorous plants commonly grow in marshy regions, where the supply of soil nitrogen is severely limited. Feeding on insects provides them with a source of organic nitrogen.

Figure 29.21

Leaf modifications. *a.* The spines of a cactus plant are leaves modified to protect the fleshy stem from animal consumption. *b.* The tendrils of a cucumber are leaves modified to attach the plant to a physical support. *c.* The Venus-flytrap has leaves that are modified to serve as a trap for insect prey. When triggered by an insect, the leaf snaps shut. Once shut, the leaf secretes digestive juices that break down the soft parts of the insect's body.

a.

b.

c.

Summary

1. The body of a flowering plant is divided into the root system and the shoot system, which contains stems and leaves. Both systems contain three types of tissues: dermal, ground, and vascular tissue.

2. Dermal tissue contains the epidermis, which is modified in different organs of the plant. In the roots, epidermal cells bear root hairs; in the leaves the epidermis contains guard cells. Cork replaces epidermis in woody plants.

3. Ground tissue contains parenchyma cells, which are thin walled and capable of photosynthesis when they contain chloroplasts. If they contain only colorless plastids, they serve as storage cells. Collenchyma cells have thicker walls for flexible support. Sclerenchyma cells are hollow, nonliving, support cells with secondary walls.

4. Vascular tissue consists of xylem that contains xylem vessel elements and tracheids, which are elongated and tapered with pitted end walls. Xylem transports water and minerals. Phloem contains sieve-tube cells, each of which has a companion cell. Phloem transports organic nutrients.

5. The primary growth of roots and shoots results in a lengthening of the plant and occurs when apical meristem produces three types of primary meristem, one for each of the types of tissues. Secondary growth, which causes a stem or root to increase in diameter, is due to the activity of vascular cambium and cork cambium.

6. A root anchors a plant, absorbs water and minerals, and stores the products of photosynthesis. A root tip shows three zones in the region of primary growth: the zone of cell division (apical meristem), the zone of elongation, and the zone of maturation.

7. A cross section of an herbaceous dicot root shows the epidermis (protects), cortex (stores), endodermis (regulates movement of molecules), and vascular cylinder (vascular tissue). In the vascular cylinder of a dicot, the xylem appears star shaped and the phloem is found in separate regions, between the arms of the xylem. In contrast, a monocot root has a ring of vascular tissue with alternating bundles of xylem and phloem surrounding a pith.

8. Three types of roots are taproots, fibrous roots, and adventitious roots.

9. Stems support leaves, etc., conduct materials to and from roots and leaves, and help store plant products. Primary growth of a stem is due to the activity of the apical meristem. In cross section, a nonwoody dicot has cortex tissue, vascular bundles in a ring, and an inner pith. Monocot stems have scattered vascular bundles, and the cortex and pith are not well defined.

10. Secondary growth of a temperate woody stem is due to vascular cambium, which produces new xylem and phloem every year, and cork cambium, which produces new cork cells when needed. Cork replaces epidermis in woody plants. The cross section of a woody stem shows bark, wood, and pith. The bark contains cork and phloem. Wood contains annual rings of xylem. Roots undergo secondary growth in the same manner as stems.
11. Stems are modified in various ways, such as horizontal aboveground and underground stems. Corms, cacti, and some tendrils are also modified stems.
12. A leaf, the organ of photosynthesis, has a blade and a petiole. The leaf veins form a net in dicot leaves and are parallel in monocot leaves.
13. A cross section of a leaf shows the epidermis, with stomata mostly below. Mesophyll tissue within the leaf has a palisade and spongy layer in C_3 leaves. In C_4 leaves, some mesophyll cells surround the bundle sheath cells, which contain chloroplasts.
14. One way to classify leaves is to note whether they are simple or compound. If compound, they may be either pinnately or palmately compound. Leaves are modified for various purposes. The spines of a cactus are leaves. Other succulents have fleshy leaves. An onion is a bulb with fleshy leaves, and the tendrils of peas are leaves. The Venus-flytrap has leaves that trap and digest insects.

Objective Questions

1. Which of these is an incorrect contrast between monocots and dicots?
   monocots   dicots
   a. one cotyledon—two cotyledons
   b. leaf veins parallel—net veined
   c. vascular bundles in a ring—vascular bundles scattered
   d. All are incorrect.
2. Which of these types of cells is most likely to divide?
   a. parenchyma
   b. meristem
   c. epidermis
   d. xylem
3. Which of these cells in a plant is apt to be nonliving?
   a. parenchyma
   b. collenchyma
   c. sclerenchyma
   d. epidermal
4. Root hairs are found in
   a. the zone of cell division.
   b. the zone of elongation.
   c. the zone of maturation.
   d. All of these.
5. Cortex is found
   a. in roots, stems, and leaves.
   b. in roots and stems.
   c. in roots only.
   d. in stems only.
6. Between the bark and the wood in a woody stem, there is a layer of meristem called
   a. cork cambium.
   b. vascular cambium.
   c. apical meristem.
   d. the zone of cell division.
7. Which part of a leaf carries on most of the photosynthesis of a plant?
   a. epidermis
   b. mesophyll
   c. epidermal layer
   d. guard cells
8. Annual rings are
   a. the number of internodes in a stem.
   b. the number of rings of vascular bundles in a monocot stem.
   c. the number of layers of secondary xylem in a stem.
   d. Both (b) and (c).
9. The Casparian strip is found
   a. between all epidermal cells.
   b. between xylem and phloem cells.
   c. on four sides of endodermal cells.
   d. within the secondary wall of parenchyma cells.
10. Which of these is a stem?
    a. taproot of carrots
    b. stolon of strawberry plant
    c. spines of cacti
    d. Both (b) and (c).

Study Questions

1. Contrast the root and shoot systems of a plant in as many ways as possible.
2. Contrast an epidermal cell with a cork cell. These cells occur in what type of plant tissue? Tell how epidermis is modified in various organs of a plant.
3. Contrast the structure and function of parenchyma, collenchyma, and sclerenchyma cells. These cells occur in what type of plant tissue?
4. Contrast the structure and function of xylem and phloem. Xylem and phloem occur in what type of plant tissue?
5. List three differences between monocots and dicots.
6. Contrast primary growth with secondary growth of stems and roots.
7. Name and state the function of the main tissues within each plant organ.
8. Name and discuss the zones of a root tip. Trace the path of water and minerals from the root hairs to xylem. Be sure to mention the Casparian strip.
9. Describe three basic types of roots and give examples of other root modifications.
10. Describe the cross section of a monocot, an herbaceous dicot, and a woody stem.
11. Discuss the adaptation of stems by giving several examples.
12. Describe the structure and organization of a C_3 dicot leaf, a C_4 plant.
13. Give several ways in which leaves are specialized.

Thought Questions

1. Develop a categorization system of your own for the tissues in a flowering plant, and place the various tissues studied in the different categories.

2. How are different organs of a flowering plant protected from drying out?

Selected Key Terms

meristem (mer'ĭ-stem) 445
epidermis (ep''ĭ-der'mis) 446
stoma (pl., **stomata**) (stō-ma) 446
cork (kork) 447
xylem (zi'lem) 447
phloem (flo'em) 447
endodermis (en''do-der'mis) 451

Casparian strip (kas-par'e-on strip) 451
vascular cambium (vas'ku-lar kam'be-um) 454
rhizome (rī'zōm) 457
blade (blād) 458

petiole (pet'ē-ōl) 458
palisade layer (pal-a-sād la'er) 459
spongy layer (spun'je la'-er) 459
bundle sheath (bun'd'l shēth) 459

CHAPTER 30

Nutrition and Transport in Plants

Your study of this chapter will be complete when you can:

1. *List the major nutrients for plants.*
2. *Trace two pathways of water from the soil to the vascular cylinder of a dicot root.*
3. *Relate water availability to soil type, and contrast field capacity with permanent wilting point.*
4. *Explain the mechanism by which water enters root cells, and relate root pressure to guttation.*
5. *Describe the cohesion-tension model of xylem transport.*
6. *Explain the mechanism by which stomata open and close.*
7. *Distinguish between macroelements and microelements needed by a plant, and describe how mineral requirements are determined.*
8. *Explain how acid deposition affects availability of minerals to a plant.*
9. *Describe the mechanism of mineral uptake, and trace the path of minerals from the soil to the xylem within the vascular cylinder.*
10. *Describe some adaptations of roots for nutrient uptake.*
11. *Discuss the pressure-flow model of phloem transport.*

The veins of this leaf are dramatically outlined in green due to a concentration of chloroplasts in certain cells only. The veins contain the vascular tissue where water is transported to a leaf and organic nutrients are transported away from a leaf. Leaves are the primary producers of organic food for a plant.

In order to grow, plants require only water, carbon dioxide, oxygen, and certain minerals. The minerals required by plants are absorbed by roots from the soil. After being dissolved in water, they are transported from the roots to the leaves in vascular tissue that is continuous throughout the body of the plant (fig. 30.1). Water carried to the leaves is needed for photosynthesis, although much is lost to evaporation through the stomata. The carbon dioxide required for photosynthesis enters a leaf through the stomata.

We don't usually think that plants need oxygen, because it is available to them as a by-product of photosynthesis. However, plants do occasionally need to take in oxygen through the stomata and by root hairs from small pockets of oxygen in the soil. Because roots don't photosynthesize, this is their only supply of oxygen. Oxygen is needed for plant cells to carry on cellular respiration, a process that makes ATP available to them (see chapter 9).

In this chapter, we will concentrate on how plants acquire water and minerals from the soil and how they transport these substances. We will also see how they transport the products of photosynthesis.

Water Uptake and Transport

Water Uptake

As figure 30.2 shows, water can enter the root of a flowering plant from the soil, simply by diffusing *between* the cells of the

epidermis and cortex via the porous cell walls. Eventually, however, the *Casparian strip* (fig. 29.8c) prevents further progress between the cells, and the water (and any substances dissolved in the water) must enter the endodermal cells if it is to reach the xylem. Alternatively, water can move directly into the root-hair cells by osmosis and then progress from cell to cell across the cortex and endodermis until it reaches the xylem. This pathway is facilitated by the plant cells being joined to one another by plasmodesmata.

Regardless of the route by which water crosses the epidermis and cortex, it must at some time enter root cells. The cytoplasm of root cells usually has a higher concentration of solutes than does soil water. Therefore, water enters root cells passively by osmosis.

Water Availability

After rain or irrigation, water in the soil drains away by gravity. The rapidity with which this happens can be correlated with the size of the *soil pores,* the spaces between the particles of the

---

Figure 30.2

Water and minerals can travel one of two pathways across the cortex to the xylem of the vascular cylinder. Pathway A shows that water and minerals can travel via porous cell walls between the cells of the cortex, but must eventually enter endodermal cells because of the Casparian strip (see also fig. 29.8). In this way, the endodermal cells control the passage of materials into the vascular cylinder and xylem. Pathway B shows that water and minerals can immediately enter a root hair and thereafter move from cell to cell to finally enter the xylem.

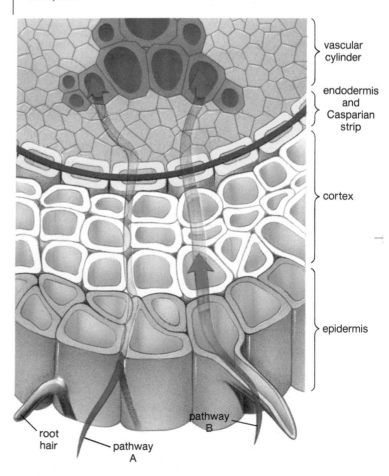

Figure 30.1

Leaves and roots are regions of a flowering plant body that are specialized to interact with the environment. The roots take up water and minerals. The leaves carry on gas exchange, but at the same time they lose water to the environment.

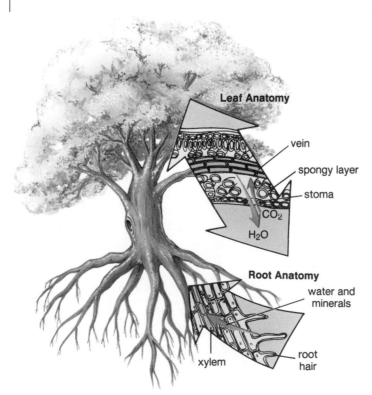

soil. Sandy soil, which drains rapidly, has the largest pores because its particles are rather large. Clay soil, which drains slowly, has the smallest pores because its particles are very tiny. The water remaining in the soil after draining is referred to as the *field capacity* of the soil. As you can see in figure 30.3, sand has a much smaller field capacity than does clay.

Another factor to consider is the *availability* of the water in the soil. Available water is water the roots are able to take up. The point at which water is no longer available to the roots is called the *permanent wilting point* of the soil. Available water lies somewhere between the field capacity of the soil and the permanent wilting point (fig. 30.3b). At this point, the soil, especially clay soil, still contains water, but it is held so tightly in the small pores and on the tiny particles that the roots are unable to remove it.

Water Transport

Once water enters xylem, it is transported to all parts of a plant. Transport need not be rapid, but the mechanism employed must be capable of transporting materials over long distances. You will recall from the previous chapter that xylem contains two types of conducting elements: vessels and tracheids. The vessels (fig. 30.4) offer the best route, as they form a continuous, completely hollow pipeline from the roots to the leaves.

When water enters root cells by osmosis, it creates root pressure. **Root pressure** tends to push xylem sap (water and minerals) upward. This can be shown by attaching a glass tube to the cut end of a short stem and watching the water rise. Water will rise as much as a meter above a cut tomato stem, and several times that height in some kinds of vines. Root pressure is also responsible for **guttation,** when drops of water are forced out of vein endings along the edges of leaves (fig. 30.5). Although root pressure may contribute to the upward movement of water in xylem vessels, it is not nearly as important as transpirational pull, which will be discussed next.

Cohesion-Tension Model of Xylem Transport

The **cohesion-tension model** of water transport explains how water is transported to great heights, against gravitational force. First, we must realize that water molecules have a great tendency to cling together because they are polar. The *cohesive property* of water means that a column of water can be pulled without breaking. In fact, it is easier to pull apart the molecules in fine wires made of some common metals than it is to pull apart a thin column of water in a small-diameter, airtight tube. Secondly, we must also remember that water is continuously lost by **transpiration** (evaporation) through the stomata in the leaves. When water evaporates from the mesophyll cells in a leaf, these

Figure 30.3

Availability of soil water. *a.* Root hairs absorb water from the pore spaces in soil. If the pores are large, as in sand, the water drains quickly. If the pores are small, as in clay, water does not drain because it adheres to the many small soil particles. *b.* Graph illustrating field capacity and permanent wilting point. Sand has a low field capacity because water drains from the soil. Clay has a high field capacity, but much of the water is unavailable to the plant. Loam has a perfect combination of qualities for plants. It is composed of 20% clay, 40% silt (particles of intermediate size), and 40% sand. Sandy, silty, and clay loam, respectively, have proportionately more sand, silt, or clay.

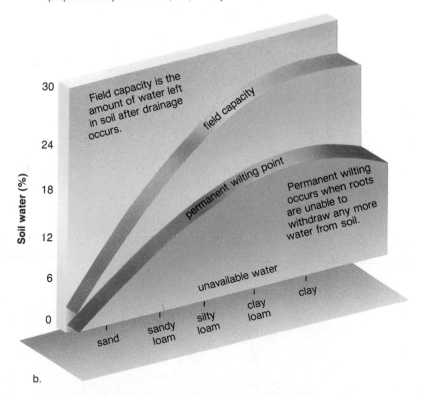

Figure 30.4

a. Cross section of xylem vessels from a cucumber root. Note the
sculptured appearance of the inside of the vessel elements. b. The
elements stacked one above the other form a continuous pipeline
that is hollow.

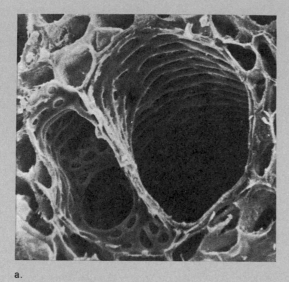

a.

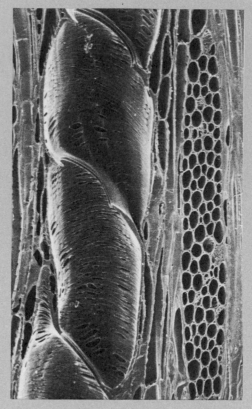

b.

Figure 30.5

Drops of guttation water on the edges of a strawberry leaf.
Guttation, which occurs at night, is thought to be due to root
pressure.

cells become less turgid. This loss of turgidity creates an osmotic gradient that causes water to move out of the xylem into these cells. Transpiration therefore creates a *tension* that can pull water upward in the xylem (fig. 30.6).

Stephen Hales, an English botanist working at the beginning of the eighteenth century, first suggested that transpiration pulls water up through plants. In 1914, Dixon and Joly coupled the idea of transpiration with their knowledge of the cohesion of water. Because water molecules tend to cling to one another, they postulated that transpiration could be the force causing water to rise in the hollow, continuous xylem vessels.

The following experiment shows that transpiration aids the movement of water in the xylem. A tube is filled with mercury and then quickly inverted and placed in a pan of mercury.

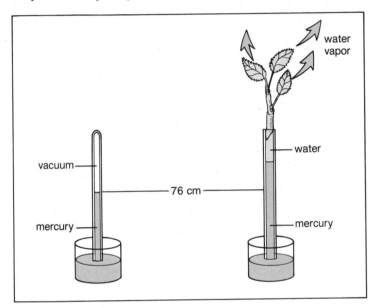

Atmospheric pressure (the weight of air) is sufficient to maintain the height of the mercury at 76 centimeters. (This height expressed in terms of water is equal to 10.4 meters.) This shows that atmospheric pressure even combined with root pressure could not raise water in xylem to the height of the tallest trees (120 meters). But if a twig with leaves is placed in the top of a tube that contains water above mercury, the mercury column rises higher than 76 centimeters. The explanation is that transpiration of water at the leaves exerts a pulling force on the contents of the tube and raises the height of the mercury above that expected. Consistent with the belief that water is pulled upward by transpiration rather than pushed upward by atmospheric pressure is the observation that water moves in the branches of a tree earlier in the day than does water in the trunk (fig. 30.7).

Although biologists, in general, accept the cohesion-tension model of water transport, there are many specific questions to be answered. For example, how do aquatic plants, or plants living in the desert or in cold climates manage to overcome varied difficulties? They must have particular adaptations for water transport, and these are areas of active research.

Figure 30.6

Cohesion-tension model of water transport. Water enters a plant at the root hairs and evaporates (transpires) at the leaves. Xylem vessels form a continuous pipeline from the roots to the leaves, and this pipeline is completely full of water. Therefore, transpiration exerts a pull on the water column that causes it to move upward.

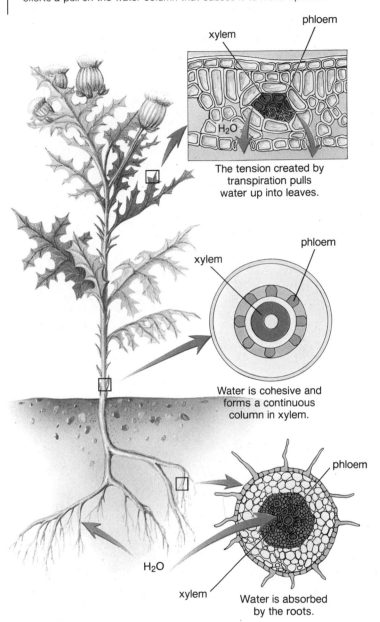

The tension created by transpiration pulls water up into leaves.

Water is cohesive and forms a continuous column in xylem.

Water is absorbed by the roots.

There is an important consequence of the way water is transported in plants. As mentioned in chapter 29, transpiration usually results in a loss of at least 90% of the water taken in at the roots. When the ground is dry and the plant is under water stress, the stomata (fig. 30.8) close. Now the plant will lose little water, because the leaves are protected against water loss by the waxy cuticle of the upper and lower epidermis (fig. 29.19).

However, when the stomata are closed, CO_2 cannot enter the leaves, and plants are unable to photosynthesize. Photosynthesis therefore requires an abundant supply of water so that the stomata can remain open and allow CO_2 to enter.

Movement of water in a plant is dependent upon the physical and chemical properties of water. Transpiration of water at the leaves causes water to flow upward in the xylem.

Figure 30.7

Velocity of water movement in a tree. The water movement in the upper branches begins earlier in the day than the water movement in the trunk. This is expected, if water is pulled upward, as is believed.

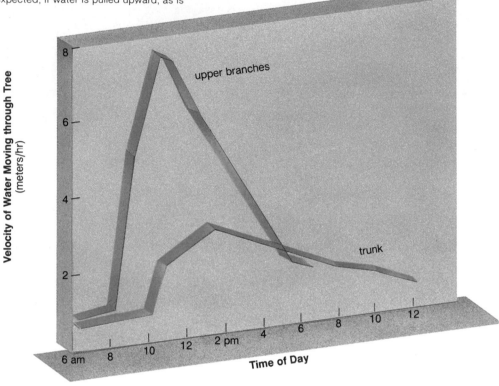

Figure 30.8

Opening of stomata. *a.* Scanning electron micrograph of an open stoma. *b.* When a stoma opens, first K⁺ and then water enters guard cells. This increases the turgor pressure of guard cells so that the stoma opens. *c.* When a stoma closes, first K⁺ and then water exits guard cells. This decreases the turgor pressure of guard cells so that the stoma closes.

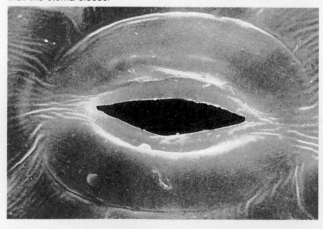

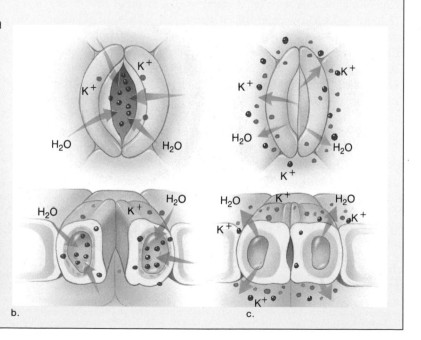

Regulation of Stomata Opening and Closing

Each stoma has two **guard cells** that regulate the opening and closing of the stomatal opening. Notice in figure 30.8 that the guard cells have the appearance of sausage-shaped balloons. The walls of the guard cells are thinner at the ends and sides away from the stomatal opening. When water enters guard cells and *turgor pressure increases,* the thicker inner walls do not move, but the thinner, outer side walls begin to stretch. The guard cells bend inward, causing the stoma to open. You can simulate this effect by placing your hands together and allowing your fingers to curve outward.

This explanation for the opening and closing of stomata has been known since 1856, but botanists are still gathering data to explain what causes the guard cells to take up or lose water. Because guard cells have chloroplasts, it was at first assumed that whenever photosynthesis was occurring, the accumulation of sugar in guard cells would cause water to enter these cells. In 1968, R. A. Fischer offered a different explanation. He floated small strips of lower epidermis tissue from a broad bean plant (*Vicia fava*) on various concentrations of potassium (K^+) chloride. He found that the guard cells took up K^+ *in the light, if the air lacked CO_2.* Once the guard cells take up K^+, water enters and the stomata open. Apparently the role of photosynthesis in stomata opening is to bring about a decreased concentration of CO_2 inside the leaf. Low CO_2 concentration prompts, in some unknown way, the uptake of K^+ by guard cells, leading to the opening of the stomata. Presumably, the movement of K^+ out of guard cells causes the guard cells to lose water and then the stomata close.

The connection between low CO_2 and uptake of K^+ is not known. It's possible that photosynthesis in guard cells leads to ATP production. ATP could then provide the energy needed for active transport of K^+ into the guard cells.

We have offered an explanation as to why stomata open when a plant is photosynthesizing. What might cause stomata to close even though sunlight is available for photosynthesis? Whenever leaf cells become water stressed, as when wilting occurs, the stomata close. This is caused by the hormone called abscisic acid (ABA), which is produced by cells in wilting leaves. ABA is believed to cause membrane leakage, leading to a loss of K^+ and then stomata closing.

If water is available stomata tend to open and the plant photosynthesizes. They close when a plant is water stressed even though light may be present for photosynthesis.

Mineral Requirements and Uptake

Mineral Requirements

Certain elements are considered essential to the health of plants and can be remembered by the mnemonic given in figure 30.9. These elements are divided into macroelements and microelements, as indicated in table 30.1. Studies show that those elements classified as macroelements are required in greater concentration than those classified as microelements.

Table 30.1 Essential Elements for Plants

Element	Chemical Symbol	Form Available to Plant	% of Plant's Dry Weight
Macroelements			
Hydrogen	H	H_2O	6
Carbon	C	CO_2	45
Oxygen	O	O_2, Co_2, H_2O	45
Nitrogen	N	NO_3^-,	1.5
Potassium	K	K^+	1.0
Calcium	Ca	Ca^{++}	0.5
Magnesium	Mg	Mg^{++}	0.2
Phosphorus	P	$H_2PO_4^-$, $HPO_4^=$	0.2
Sulfur	S	$SO_4^=$	0.1
Microelements			
Chlorine	Cl	Cl^-	0.01
Iron	Fe	Fe^{++}, Fe^{+++}	0.01
Boron	B	$BO_3^{\equiv}$, $B_4O_7^=$	0.002
Manganese	Mn	Mn^{++}	0.005
Zinc	Zn	Zn^{++}	0.002
Copper	Cu	Cu^+, Cu^{++}	0.006
Molybdenum	Mo	$MoO_4^=$	0.00001

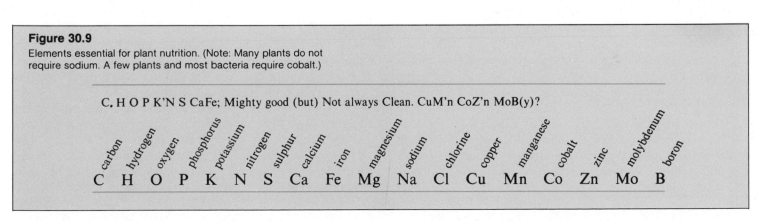

Figure 30.9

Elements essential for plant nutrition. (Note: Many plants do not require sodium. A few plants and most bacteria require cobalt.)

C. H O P K'N S CaFe; Mighty good (but) Not always Clean. CuM'n CoZ'n MoB(y)?

carbon   hydrogen   oxygen   phosphorus   potassium   nitrogen   sulphur   calcium   iron   magnesium   sodium   chlorine   copper   manganese   cobalt   zinc   molybdenum   boron

C   H   O   P   K   N   S   Ca   Fe   Mg   Na   Cl   Cu   Mn   Co   Zn   Mo   B

Carbon, hydrogen, and oxygen make up 96% of a plant's dry weight. Carbon dioxide is the source of carbon for a plant, and water is the source of hydrogen. Oxygen can come from oxygen gas, CO_2, or water.

The other essential elements listed in table 30.1 are minerals that come from the soil. Nitrogen is a common atmospheric gas, but plants are unable to make use of it. Instead most plants are dependent upon a supply of NO_3^- in the soil; chapter 48 discusses how the presence of NO_3^- in the soil is dependent upon the work of bacteria. However, as discussed on page 472, some plants have root nodules containing mutualistic bacteria that fix atmospheric nitrogen and make nitrogen compounds available to their host.

Table 30.2 outlines the functions of some of these elements in plant nutrition.

Determination of Mineral Requirements

If you burn a plant, the nitrogen is given off as ammonia and other gases, but most other mineral elements remain in the ash. However, the presence of a particular element in the ash does not necessarily mean that the plant normally requires it. For example, plants take up the man-made [239]plutonium and the naturally occurring [90]strontium, elements they do not require. These radioactive elements are subsequently found in their ashes. This uptake of radioactive elements by plants is unfortunate because it intensifies the concerns and worries of a nuclear power plant mishap. Complicating the situation still further, is the knowledge that if cows feed on plants containing [90]strontium, their milk will contain this radioactive element in even greater concentrations. Therefore, humans are likely to receive a concentrated dose of radioactivity when they drink the milk.

Because analysis of the ash does not allow an investigator to determine the mineral requirements of a plant, a method for making this determination was developed at the end of the nineteenth century. This method is called water culture, or **hydroponics** (fig. 30.10). Hydroponics allows plants to grow well if they are supplied with all the mineral nutrients they need. The investigator omits a particular mineral and observes the effect on plant growth. If growth suffers, it can be concluded that the omitted mineral is a required nutrient. This method has been more successful for macronutrients than for micronutrients. For studies involving the latter, the water and mineral salts used must be absolutely pure, and purity is difficult to attain, because even instruments and glassware might introduce micronutrients. Then, too, the element in question might already be present in the seedling used in the experiment. These factors complicate the determination of essential plant micronutrients by means of hydroponics.

Mineral Availability

As mentioned previously, the essential minerals that a plant needs are normally taken from the soil. Placement of minerals in the soil is critical, because if a root does not come within a few millimeters of an ion, no uptake can occur. The downward movement of minerals through soil is influenced by the quantity

Table 30.2 Functions of Some Essential Plant Nutrients

Element	Function
Nitrogen	Constituent of amino acids, proteins, coenzymes, nucleic acids, and chlorophyll
Phosphorus	Sugar phosphates, nucleotides, nucleic acids, phospholipids, coenzymes
Potassium	Not part of any known organic molecule; coenzyme and osmoregulator
Sulfur	Constituent of sulfur-containing amino acids, coenzyme A, and vitamins
Magnesium	Constituent of chlorophyll; coenzyme, especially in reactions involving ATP
Calcium	Constituent of cell wall, middle lamella, and membranes; cofactor for some enzymes; may detoxify large concentrations of heavy metals
Iron	Cofactor in enzymes of chlorophyll synthesis and other enzymes; constituent of cytochromes and ferredoxin, where it functions in electron transport

Modified by permission from D. K. Northington and J. R. Goodin, *The Botanical World*. St. Louis, 1984, Times Mirror/Mosby College Publishing.

and pH of water. Sometimes water makes mineral ions available to roots, but often it leaches, or removes, mineral ions from the soil zone in which roots grow. The lower the pH, the greater the leaching power of water.

Soil particles, particularly clay and any organic matter in soil, contain negative charges that attract positively charged ions, such as calcium (Ca^{++}), potassium (K^+), and magnesium

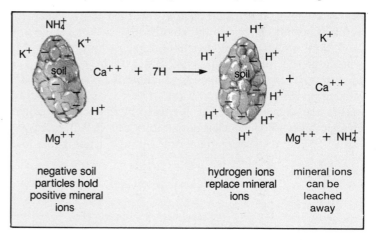

negative soil particles hold positive mineral ions	hydrogen ions replace mineral ions	mineral ions can be leached away

(Mg^{++}). This attraction keeps these ions at a soil level where they are available to plants. In acid soils, however, hydrogen ions (H^+) replace these positive ions so that the useful ions float free and are easily leached away by water.

On the other hand, in very acid soils, aluminum (Al^{+++}) and iron (Fe^{+++}) become available to plants. These mineral ions are insoluble above a pH of 4, but become soluble at more acid pH. They displace calcium, potassium, and magnesium on soil particles in the same way that hydrogen does. High levels of aluminum and iron in soils are toxic to plants.

Figure 30.10

Some plant nutrient deficiencies are easily diagnosed when plants are grown in a complete nutrient solution with elimination of one nutrient at a time. Sunflower plants respond rapidly to a deficiency of nitrogen (a), phosphorus (b), and calcium (c). Nutrient-deficient plants are shown on the left in each photograph; healthy plants with a complete nutrient solution are shown on the right.

a.

b.

c.

It is clear, then, that acid deposition has two effects that are deleterious to plants. It causes the removal of the mineral ions that plants need for proper nutrition, and it makes available other mineral ions that are toxic at high levels. The death of trees in certain portions of the United States is at least, in part, the result of acid deposition (fig. 49.10).

Mineral Uptake

Like water, mineral ions can move past the epidermis and through the cortex by way of porous cell walls (fig. 30.2). Eventually, because of the Casparian strip, they must enter the cytoplasm of the endodermal cells if they are to proceed further and enter the vascular cylinder. Plasma membranes are freely permeable to water but not to all mineral ions. Mineral ions that cannot pass through the plasma membranes of endodermal cells can go no further. In this way, the endodermis regulates the entrance of minerals into the vascular cylinder.

Mineral ions are often actively absorbed into the cytoplasm at the root-hair region of a young root. Only occasionally, when particular mineral ions are highly concentrated in soil water, such as after a fertilizer application, do mineral ions enter root cells by simple diffusion. Figure 30.11 presents data from one experiment showing that active transport is most likely involved in the uptake of minerals by a plant. The rate of respiration and the rate of bromine ion absorption by thin discs of

Figure 30.11

Bromine ion absorption by a potato tuber. As respiration increases, bromine ion absorption also increases. This shows that active transport is involved in nutrient uptake by plants.

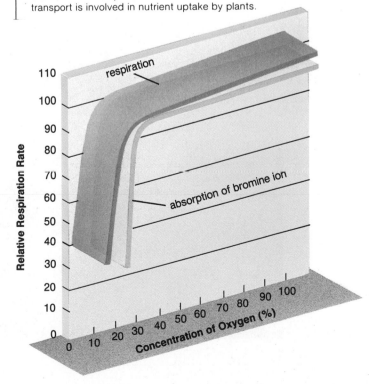

potato increased as more oxygen was made available by bubbling the gas through the culture solution. This result suggests that energy was being used to take up the bromine ions.

Plants possess an astonishing ability to concentrate ions; that is, to absorb them until they are many times more concentrated than in the surrounding medium. Experiments have shown that the concentration of certain minerals in roots is as much as 10,000 times greater than in the surrounding soil. Nutrition in animals is also dependent on this ability of plants to concentrate minerals.

Once mineral ions have entered root cells, they diffuse down a concentration gradient and enter the xylem within the vascular cylinder. From there they are distributed to the various organs of the plant.

Minerals follow the same path as water. They are taken up primarily by root hairs, move across the cortex, and then enter the xylem. From there, they are distributed to other plant organs.

Root Adaptations for Mineral Uptake

Two symbiotic relationships are known to assist roots in supplying mineral nutrients to the rest of a plant. Plants are unable to make use of nitrogen (N_2) in the air because they do not have the cellular enzymes to break the $N \equiv N$ bond. Therefore, plants are usually dependent on the nitrate (NO_3^-) that roots remove from the soil. But some roots, such as those of the legumes soybeans and alfalfa, are infected by bacteria in the genus *Rhizobium* that can break the $N \equiv N$ bond and reduce nitrogen to NH_4^+ for incorporation into organic compounds. (The reduction of N_2 to NH_4^+ process is called nitrogen fixation.) The bacteria live in **nodules** (fig. 30.12a) and are supplied with carbohydrates by the host plant. The bacteria, in turn, furnish their host with nitrogen compounds. This is such a useful service that research is now being directed toward finding ways to make more plants receptive to infection by nitrogen-fixing bacteria. Gene-splicing techniques are being developed to introduce the bacterial genes that control nitrogen fixation (called *nif* genes) into plant chromosomes. Less fertilizer would then be needed to grow crops.

The second symbiotic relationship involves fungi. The fungus grows around the root, often even penetrating to the level of the cortex. The fungus increases the surface area available for mineral and water uptake and breaks down organic matter, releasing nutrients that the plant can use. In return, the root furnishes the fungus with sugars and amino acids. These "fungus roots," or **mycorrhizae,** assist the growth and development of plants, but seem to be particularly essential to some plants. Orchid seeds, which are quite small and contain limited nutrients, will not germinate until a mycorrhizal fungus has invaded their cells. Nonphotosynthetic plants, such as Indian pipe, use their fungus roots to extract nutrients from nearby trees by way of mycorrhizal hyphae that connect their roots. In order to make tree seedlings grow better, they are deliberately infected with mycorrhizal fungi before they are transplanted (fig. 30.12b).

As a special adaptation, some plants have poorly developed roots or no roots at all, because minerals and water are supplied by other mechanisms. The carnivorous plants discussed on page 459 have poorly developed roots and rely on their leaves to absorb compounds containing minerals. **Epiphytes** are "air plants" that do not grow in soil at all but grow on larger plants that give them support (fig. 30.13a). They do not get nutrients from their host, however. Some epiphytes have roots that absorb moisture from the atmosphere, and many catch rain and minerals in special pockets at the base of their leaves. Other

---

Figure 30.12

Symbiosis and mineral nutrition. *a.* Nodules on soybean roots that contain nitrogen-fixing bacteria. Their red color is caused by a pigment produced while active nitrogen fixation is occurring.
b. Loblolly pine root systems. Roots at top were untreated and those at bottom were inoculated with a fungus. The fungus-covered mycorrhizal roots absorb minerals and water more efficiently.
Photograph by Carolina Biological Supply Company.

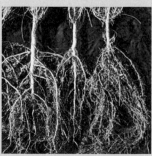
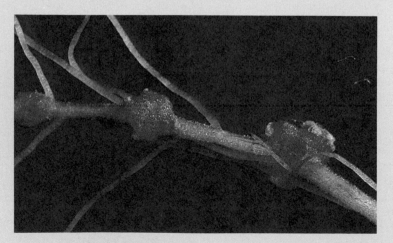

a.

b.

epiphytes are parasitic plants that attach to a host by way of rootlike projections called haustoria. The haustoria of mistletoe penetrate only the xylem of the host to remove water and minerals (fig. 30.13b). Other plants, such as dodders, broomrapes, and pinedrops send their haustoria into both the xylem and phloem of the host.

Transport of Organic Substances

As long ago as 1679, Marcello Malpighi suggested that bark is involved in translocating sugars from leaves to roots. He observed the results of removing a strip of bark from around a tree, called **girdling.** If a tree is girdled below the level of the majority of leaves, the bark swells just above the cut and sugar accumulates in the swollen tissue. We know today that when a tree is girdled, the phloem is removed but the xylem is left intact. Therefore, the results of girdling suggest that phloem is the tissue

that transports sugars. Radioactive tracer studies have confirmed this. When ^{14}carbon-labeled CO_2 is supplied to mature leaves, radioactively labeled sugar is soon found moving down the stem into the roots. This labeled sugar is found mainly in the phloem, not in the xylem. Radioactive tracer studies have also confirmed the role of phloem in transporting other substances, such as amino acids, hormones, and even mineral ions. Hormones are transported from their production sites to target areas where they exert their regulatory influences. In the autumn, before leaves fall, mineral ions are removed from the leaves and taken to other locations in the plant.

Chemical analysis of phloem sap shows that its main component is sucrose, and the concentration of nutrients is usually 10% to 13% by volume. It is difficult to take samples of sap from the phloem without injuring the phloem, but this problem was solved by using aphids (fig. 30.14), small insects that are phloem feeders. The aphid drives its stylet, which is a sharp mouthpart

Figure 30.13
a. Epiphytic bromeliad growing on a tree. An epiphyte is a plant that attaches itself to a larger plant but does not derive nourishment from it. Water and solutes fall into pockets formed by the bases of the leaves. b. Mistletoe derives water and minerals from a host plant by way of haustoria that penetrate the host tissues.

a.

haustoria

b.

Figure 30.14
Aphids are small insects that remove nutrients from phloem by means of a hypodermic-like mouthpart called a stylet. a. Aphid with stylet in place. b. When the aphid's body is removed, phloem sap is available to the experimenter.

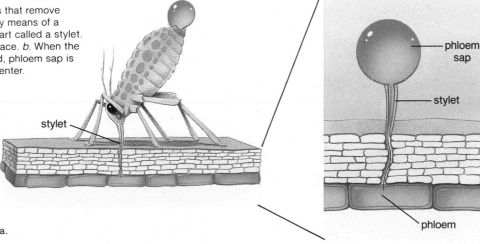

stylet

a.

phloem sap

stylet

phloem

b.

Figure 30.15

Phloem structure. *a.* Longitudinal section shows that sieve-tube cells in phloem lie end-to-end. *b.* Cross section of a sieve plate, across which plasmodesmata extend between sieve-tube cells. (The label cc = companion cell.)

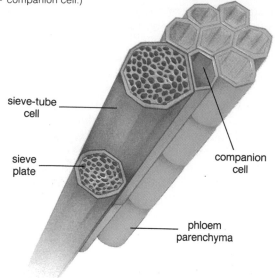

sieve-tube cell

sieve plate

companion cell

phloem parenchyma

a.

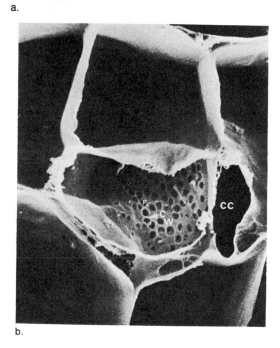

b.

Figure 30.16

Pressure-flow model of phloem transport. Sucrose and water enter sieve-tube cells at a source. Sieve-tube cells form a continuous pipeline from a source to a sink, where sucrose and water exit sieve-tube cells.

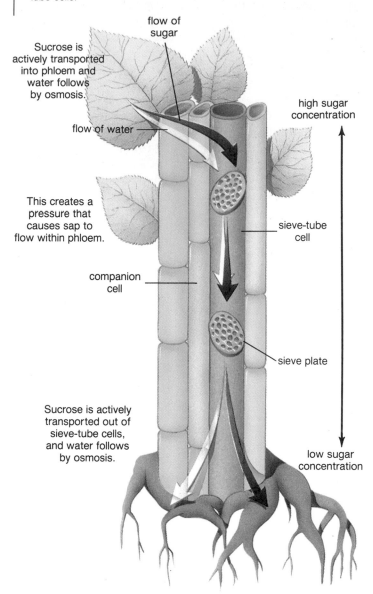

flow of sugar

Sucrose is actively transported into phloem and water follows by osmosis.

flow of water

high sugar concentration

This creates a pressure that causes sap to flow within phloem.

sieve-tube cell

companion cell

sieve plate

Sucrose is actively transported out of sieve-tube cells, and water follows by osmosis.

low sugar concentration

Pressure-Flow Model of Phloem Transport

An explanation of phloem transport must account for the movement of fairly large amounts of organic material for long distances, in a relatively short period of time. The movement rate of radioactively labeled ^{14}carbon-sugar has been determined by analysis of sap withdrawn through aphid stylets between two areas of the stem. Materials appear to move through the phloem at a rate of 60 cm to 100 cm per hour, and possibly up to 300 cm per hour.

A **pressure-flow model** currently offers an explanation for the movement of organic materials in phloem (fig. 30.16). During the growing season, the leaves are photosynthesizing and producing sugar. This sugar is actively transported into phloem, and water follows passively by osmosis. Transport and osmosis

that functions like a hypodermic needle, between the epidermal cells and withdraws sap from a sieve-tube cell. If the aphid is anesthetized by ether, the body may be carefully cut away, leaving the stylet. Phloem can then be collected and analyzed by the researcher.

The conducting cells of phloem are sieve-tube cells lined up end-to-end with their sieve plates abutting (fig. 30.15). Cytoplasm extends through the sieve plates of adjoining cells to form a continuous sieve-tube system that extends from the roots to the leaves, and vice versa.

are possible because sieve-tube cells have a living plasma membrane. Also, the energy needed for sucrose transport is provided by the companion cells. The buildup of water within the sieve-tube cells at the leaves creates a *pressure*.

At the roots (or other places in the plant) the sugar is transported out of the phloem and water follows passively by osmosis. This exit of sucrose and water at the roots means that the pressure created at the leaves will cause a *flow* of water from the leaves (higher pressure) to the roots (lower pressure). As the water flows, it brings the sucrose with it.

This model can be supported by the following experiment: Two bulbs are connected by a glass tube. The first bulb contains solute at a higher concentration than the second bulb. Each bulb is bounded by a selectively permeable membrane, and the entire apparatus is submerged in distilled water:

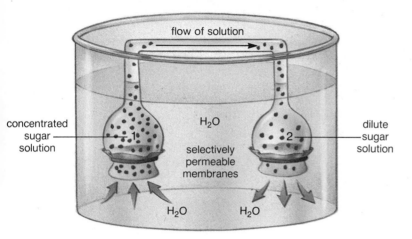

Distilled water flows into the first bulb because it has the higher concentration of solute. The entrance of water creates *pressure* that causes a *flow* toward the second bulb. The pressure flow will not only drive water toward the second bulb, but will provide enough drive to force water out through the membrane of the second bulb, even though the second bulb contains a higher concentration of solute than does the distilled water.

The pressure-flow model of phloem transport can account for any direction of flow in the sieve-tube cell system if we consider that the direction of flow is from *source to sink*. In young seedlings, the cotyledons containing reserve food are major sources of sucrose, and roots are a sink. Therefore, the flow will be from cotyledons to the root. In older plants, the most recently formed leaves that are actively photosynthesizing will be a source of sucrose for the shoot tip (a sink), and the lower leaves will be a source of sucrose for the root (a sink). When a plant is in fruit, it has been observed that phloem flow is monopolized by the fruits, little goes to the rest of the plant, and vegetative growth is slow.

The pressure-flow model of phloem transport suggests that phloem sap can move either up or down, as is appropriate for the plant at a particular time in its life cycle.

As with xylem transport, there are many questions to be answered regarding phloem transport. For example, what factors aside from source-to-sink relationships affect the direction and rate of phloem transport? Do plant hormones have an affect? These and many other questions are now being addressed by plant physiologists.

Summary

1. In order to grow, plants require water, carbon dioxide, oxygen, and certain minerals.
2. Water can enter the root by moving between the cells until it reaches the Casparian strip, after which it passes through an endodermal cell before entering xylem. Water can also enter root hairs, then passes through the cells of the cortex and endodermis to reach xylem.
3. Much of the water in soil is not available to plants. In sandy soils, it drains quickly, and in clay, it adheres to the tiny particles. Field capacity is the amount of water in a soil after drainage has occurred. At the permanent wilting point, no more water is available to the plant.
4. Water enters root cells by osmotic pressure, creating root pressure. Root pressure does not contribute greatly to upward movement of water in xylem, but it does cause guttation.
5. The cohesion-tension model of xylem transport states that transpiration creates a tension that pulls water upward in xylem. This transport works only because water molecules are cohesive. Most of the water taken in by a plant is lost through stomata by transpiration. Only when there is plenty of water will stomata remain open, allowing carbon dioxide to enter the leaf and photosynthesis to occur.
6. Stomata open when guard cells take up water stretching their thin side walls. Water enters the guard cells after K^+ has entered. Uptake of K^+ has been observed when there is a low level of CO_2 inside leaves. Perhaps photosynthesis in guard cells leads to ATP production and active transport of K^+.
7. Plant nutrition requires sixteen different elements. Carbon, hydrogen, and oxygen make up 96% of plant dry weight. The other necessary elements are taken up by the roots as mineral ions. Even nitrogen, which is present in the atmosphere, is taken as NO_3^- from the soil.
8. You can determine mineral requirements by growing plants in a solution that varies as to a missing mineral. If the plant grows poorly, then the missing mineral is a requirement for growth.
9. Roots can only take up the minerals that are available in soil. Acid deposition reduces the availability of some minerals and increases the availability of other minerals. Ions, such as calcium, potassium, and magnesium are displaced from soil particles by hydrogen ions and leached from the soil. Acid deposition increases the solubility and makes available toxic amounts of aluminum and iron.
10. Root hairs use active transport to take in mineral ions. After entering root hairs, minerals move across the cortex and endodermis and finally enter the xylem.

11. Plants have various adaptations to assist them in acquiring nutrients. Legumes have nodules infected with the bacterium *Rhizobium,* which make nitrogen compounds available to them. Many other plants have mycorrhizae, or fungus roots. The fungus gathers nutrients from the soil, and the root provides the fungus with sugars and amino acids. Some plants have poorly developed roots. Epiphytes live on, but do not parasitize trees, whereas mistletoe and some other plants parasitize their host.

12. The pressure-flow model of phloem transport states that sucrose is actively transported into phloem at a source and water follows by osmosis. The resulting increase in pressure creates a flow that moves water and sucrose to a sink.

Objective Questions

1. Which of these is not a nutrient for plants?
   a. water
   b. carbon dioxide gas
   c. mineral ions
   d. nitrogen gas
2. The Casparian strip affects
   a. how water and minerals move into the vascular cylinder.
   b. how water but not minerals move.
   c. how minerals but not water move.
   d. neither the flow of water nor of minerals into a plant.
3. Field capacity is
   a. the same as permanent wilting point.
   b. the amount of water in soil available to plants.
   c. the amount of water in soil after drainage takes place.
   d. higher for sandy soils than clay soils.

4. Water flows upward in xylem because
   a. root pressure pushes it all the way up to the leaves.
   b. sucrose enters xylem at the leaves and water follows.
   c. water naturally flows upward because of cohesion.
   d. transpiration pulls water from the roots to the leaves.
5. Stomata are usually open
   a. at night, when the plant requires a supply of oxygen.
   b. during the day, when the plant requires a supply of carbon dioxide.
   c. whenever there is excess water in the soil.
   d. All of these.
6. Which of these is not a mineral ion.
   a. NO^-
   b. Mg^+
   c. CO_2
   d. Al^{+++}

7. Acid deposition
   a. affects the availability of water to plants.
   b. affects the availability of oxygen.
   c. affects the availability of minerals.
   d. All of these.
8. The pressure flow model of phloem transport states that
   a. phloem sap always flows from the leaves to the root.
   b. phloem sap always flows from the root to the leaves.
   c. water flow brings sucrose from a source to a sink.
   d. Both (a) and (c).
9. Active transport is involved in
   a. stomatal opening.
   b. uptake of minerals by roots.
   c. uptake of water by roots.
   d. Both (a) and (b).
10. Root hairs do not play a role in
   a. oxygen uptake.
   b. mineral uptake.
   c. water uptake.
   d. carbon dioxide uptake.

Study Questions

1. Give two pathways by which water and minerals can cross the epidermis and cortex of a root. What feature allows endodermal cells to regulate the entrance of molecules into the vascular cylinder?
2. Tell how field capacity and permanent wilting point relate to water availability in soils. Why do soils differ as to water availability?
3. Describe and give evidence for the cohesion-tension theory of water transport.

4. What events precede the opening and closing of stomata by guard cells?
5. Name the elements that make up most of a plant's body. Why are some elements called microelements and some macroelements?
6. Give reasons why a plant needs nitrogen, phosphorus, and potassium.
7. Briefly describe two methods used to determine the mineral nutrients of a plant.

8. Explain how acid deposition affects availability of minerals to a plant, and why this can lead to plant death.
9. Contrast the manner in which water and minerals are absorbed by root cells.
10. Name two symbiotic relationships that assist plants in taking up minerals, and two types of plants that tend not to take up minerals by roots from soil.
11. What data are available to show that phloem transports organic compounds? Explain the pressure-flow model of phloem transport.

Thought Questions

1. In contrast to leaves, roots exclusively carry on cellular respiration. How do roots acquire the oxygen and sugar needed for ATP production?

2. Plants depend on the physical and chemical properties of water to move materials through their bodies. In what ways is this disadvantageous to them?

3. Animals also require the minerals listed in table 30.2. Why is that to be expected?

Selected Key Terms

root pressure (rŏŏt presh′ur) 465
guttation (gut-ta′shun) 465
cohesion-tension model (ko-he′zhun ten′shun mod′el) 465

transpiration (tran″spi-ra′shun) 465
hydroponics (hi″dro-pon′iks) 470
nodules (nod′ūlz) 472
mycorrhiza (mi″ko-ri′zah) 472

epiphyte (ep′ĭ-fīt) 472
girdling (ger′d′l-ing) 473
pressure-flow model (presh′ur flo mod′el) 474

CHAPTER 31

Reproduction in Plants

Your study of this chapter will be complete when you can:

1. Draw and explain a diagram that shows the life cycle of flowering plants.
2. State and describe the parts of a flower.
3. Describe the development of a female gametophyte in flowering plants from the megaspore mother cell to the production of an egg.
4. Describe the development of the male gametophyte in flowering plants from the microspore mother cell to the production of sperm.
5. Distinguish between pollination and fertilization. Explain the events of double fertilization in flowering plants.
6. Give a classification of fruits, with examples of different types of fruits.
7. Describe different means of dispersing seeds and fruits.
8. Describe the development of a dicot embryo during seed formation.
9. Contrast the structure and germination of a bean seed and that of a corn kernel.
10. Describe different ways of propagating plants asexually.

Male catkins of the Italian alder tree. Each catkin consists of numerous tiny flowers that produce windblown pollen. A cloud of pollen is seen here escaping from the catkins that hang down and are well exposed to the passing breezes.

The flowering plants, or angiosperms, are the most diverse and widespread of all the plants. Their life cycle is like that of the gymnosperms, except that the seeds are covered within fruits. Whereas the seeds of conifers develop on the scales of cones, those of the angiosperms develop within a flower. The structure of the flower allows them to produce seeds within fruits. The evolution of the flower has no doubt contributed heavily to the enormous success of angiosperms in another way. It permits pollination to take place not only by wind, but also by animals. Flowering plants that rely on animals have a symbiotic relationship with their specific pollinator. The flower provides nutrients for a pollinator, such as a bee, fly, beetle, bird, or even bat. When this animal visits flowers to gather food, it carries pollen from one flower to another, allowing pollination to occur.

Life Cycle of Flowering Plants

Flowering plants, like other seed plants, exhibit an *alternation of generations* life cycle in which the sporophyte generation produces heterospores, called microspores and megaspores (fig. 31.1). A microspore develops into an immature *male gameto-*

phyte called a pollen grain, which is either windblown or carried by an animal to the vicinity of the female gametophyte. When a pollen grain matures, it contains sperm that travel by way of a pollen tube to the female gametophyte, which is called an embryo sac. Therefore, flowering plants, like gymnosperms, are not dependent on an outside source of water for fertilization to take place.

A megaspore develops into a *female gametophyte,* the embryo sac, that is microscopic and retained within the body of the sporophyte. Each female gametophyte produces an egg that is fertilized by a sperm released by a mature pollen grain. The zygote becomes an embryonic plant enclosed within a seed. A seed contains the embryonic plant of the next sporophyte generation, plus stored food.

Figure 31.1 diagrams the life cycle of a flowering plant. In this life cycle:

1. The dominant generation is represented by the diploid sporophyte that bears flowers.
2. The flower produces microspores and megaspores that develop into the male and female gametophytes, respectively.

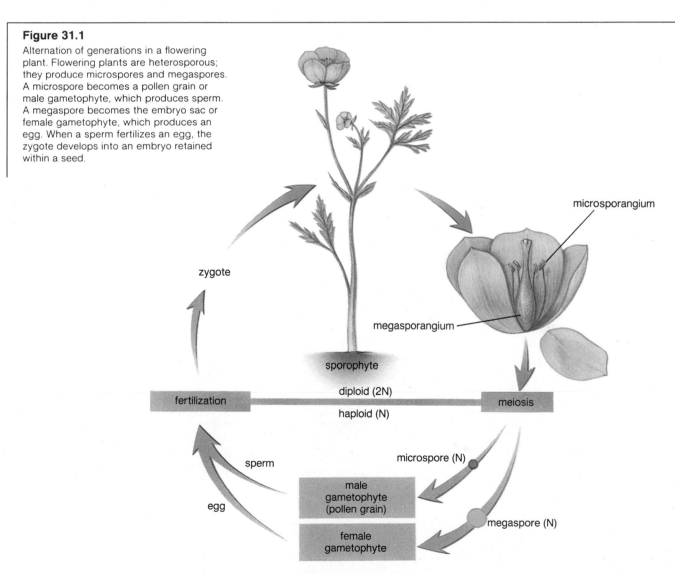

Figure 31.1

Alternation of generations in a flowering plant. Flowering plants are heterosporous; they produce microspores and megaspores. A microspore becomes a pollen grain or male gametophyte, which produces sperm. A megaspore becomes the embryo sac or female gametophyte, which produces an egg. When a sperm fertilizes an egg, the zygote develops into an embryo retained within a seed.

3. The mature male gametophyte is the pollen grain that releases sperm, one of which fertilizes the egg. The female gametophyte is retained within the flower and produces an egg.
4. Fertilization results in an embryo that eventually is enclosed within a seed covered by fruit.

In flowering plants, the flower is a reproductive structure that produces microspores and megaspores. These develop into separate gametophytes that produce sperm and egg. Fertilization results in seeds enclosed by a fruit.

The Flower

A flower develops within a bud. In many plants, the same shoot apex that previously formed leaves suddenly stops producing leaves and starts producing a flower. In other plants, axillary buds develop directly into flowers. Flower structures are modified leaves attached to a short stem tip called a **receptacle** (fig. 31.2). In monocots, flower parts occur in threes and multiples of three. In dicots, flower parts are in fours or fives and multiples of four or five. The **sepals,** which are the most leaflike of all the flower parts, are green and they protect the bud as the flower develops within. When the flower opens, there is an outer whorl of sepals and an inner whorl of **petals,** whose color accounts for the attractiveness of many flowers. The size, shape, and color of a flower is attractive to a specific pollinator. Wind-pollinated flowers often have no petals at all.

The reproductive structures of a flower lie within the petals. At the very center of a flower is the **pistil,** which is often vaselike in appearance. A pistil may be simple or compound. A simple pistil contains a single reproductive unit called a *carpel.* A carpel usually has three parts. The **stigma** is an enlarged sticky knob, the **style** is a slender stalk, and the **ovary** is an enlarged base. A compound pistil has multiple carpels that are often fused.

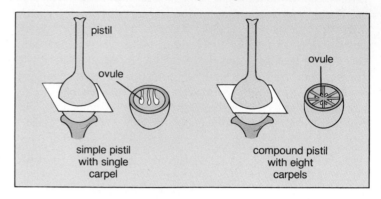

The ovary contains a number of **ovules** that play a significant role in the production of megaspores and female gametophytes. Grouped about the pistil are a number of **stamens,** each of which has two parts: the **anther,** a saclike container, and the **filament,** a slender stalk. Pollen grains develop in the anther.

Sometimes the pistil is called the female part of the flower and the stamens are called the male part of the flower, but this is not strictly correct. The pistil and stamens do not produce gametes; they produce megaspores and microspores, respectively. A microspore goes on to become a male gametophyte that produces sperm and a megaspore goes on to become a female gametophyte that produces an egg.

Development of the Gametophytes

Within a flower, there is a diploid megaspore mother cell in each ovule of the ovary (fig. 31.3). The megaspore mother cell undergoes meiosis, producing four haploid megaspores. Three of these

---

Figure 31.2

Parts of a typical flower, a sporophyte structure. Microspores are produced by the anthers and megaspores are produced by ovules.

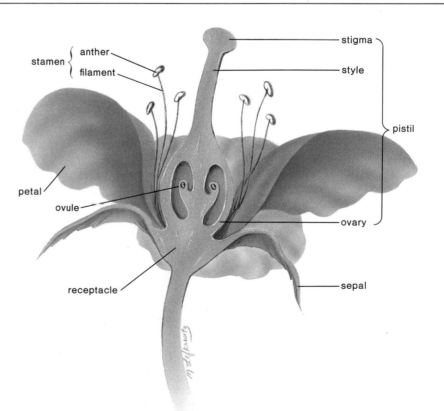

Figure 31.3

Life cycle of a flowering plant. Each ovule in the ovary of a pistil contains a megaspore mother cell that produces one functional haploid megaspore by meiosis. A megaspore divides mitotically three times and the resulting structure with eight nuclei and seven cells is the female gametophyte (embryo sac). One of these cells is an egg. An anther contains many microspore mother cells, each of which produces four haploid microspores by meiosis. Each microspore divides mitotically becoming a two-celled pollen grain.

When a pollen grain germinates, it contains a tube nucleus and two sperm nuclei. This is the mature male gametophyte. Following pollination, double fertilization occurs; one sperm nucleus unites with the egg nucleus, giving a zygote, and the other unites with the polar nuclei, giving a 3N endosperm nucleus. The zygote becomes an embryo, and the endosperm nucleus becomes stored food contained within a seed. Germination of the seed gives a sporophyte plant.

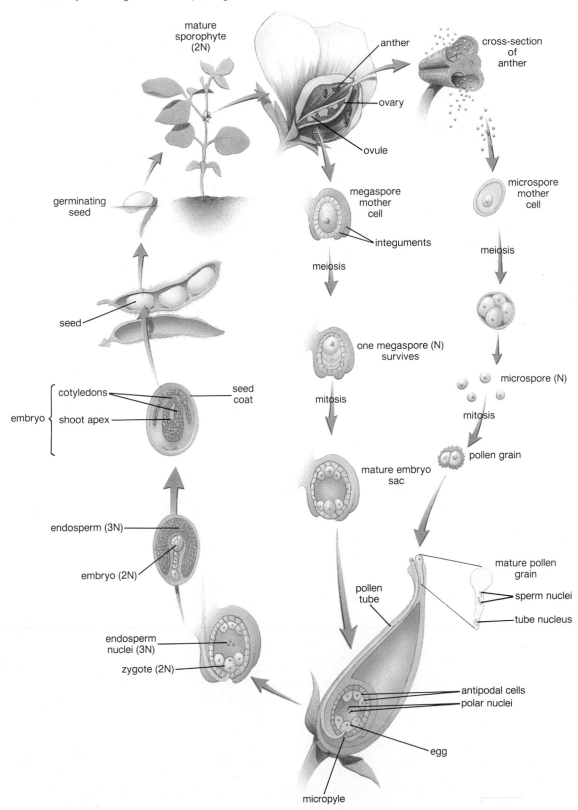

mature sporophyte (2N)

anther

cross-section of anther

ovary

ovule

germinating seed

megaspore mother cell

microspore mother cell

integuments

meiosis

meiosis

seed

one megaspore (N) survives

microspore (N)

cotyledons

seed coat

mitosis

mitosis

embryo

shoot apex

pollen grain

mature embryo sac

endosperm (3N)

mature pollen grain

embryo (2N)

pollen tube

sperm nuclei

tube nucleus

endosperm nuclei (3N)

zygote (2N)

antipodal cells

polar nuclei

egg

micropyle

disintegrate, leaving one functional megaspore, whose nucleus divides mitotically, until there are eight haploid nuclei. This is the female gametophyte, also called the *embryo sac,* and it frequently takes on the appearance shown in figure 31.4. At one end there are three cells called the antipodals; in the middle are two polar nuclei; and at the other end are three cells, the middle one of which is the egg cell. The entire embryo sac is enclosed by two layers called the integuments, which have a small opening, the *micropyle,* near the egg cell.

Male gametophytes are produced in the stamens. An anther contains four pollen sacs (fig. 31.5) with many microspore mother cells, each of which undergoes meiosis to produce four haploid microspores. A microspore divides mitotically, forming two cells enclosed within a finely sculptured wall. This structure is the pollen grain and contains a **tube cell** and a **generative cell.** The generative cell divides, either now or later, to give two sperm. The walls separating the pollen sacs in the anther break down when the pollen is ready to be released.

In flowering plants, an egg is found within the embryo sac, which is the female gametophyte and is located within an ovule. A sperm cell is found within the pollen grain, which is the male gametophyte.

Pollination

Pollination is simply the transfer of pollen from the anther to the stigma and should not be confused with fertilization. As mentioned previously and considered in the reading for this chapter, page 488, pollination is brought about by wind or with the assistance of a particular pollinator, an animal that carries the pollen from the anther to the stigma.

Fertilization

When a pollen grain lands on the stigma of the same species, it germinates, forming a pollen tube. The germinated pollen grain containing a tube nucleus and two sperm nuclei is the mature male gametophyte. The pollen tube grows as it passes between the cells of the stigma and style to reach the micropyle. Now

Figure 31.4

Embryo sac in a lily ovule. An embryo sac is the female gametophyte generation of flowering plants. It contains seven cells, one of which is the egg. The two polar nuclei have not yet migrated to the center of the sac.

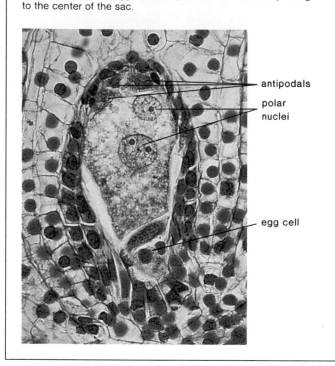

antipodals

polar nuclei

egg cell

Figure 31.5

Development of pollen grains. *a.* A mature anther showing that the walls between the pollen sacs have opened, allowing the pollen grains to be released. *b.* When released, a pollen grain has two nuclei, a tube nucleus and a generative nucleus. Magnification, ×255.

c. Photomicrograph of various pollen grains showing their diversity. The appearance of pollen grains varies with the species. Magnification, ×100. *d.* Electron micrograph of goldenrod pollen grains. Magnification, ×630.

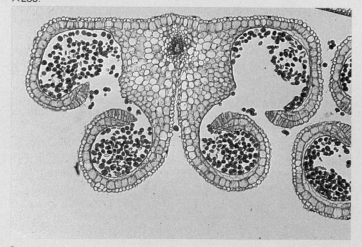

a.

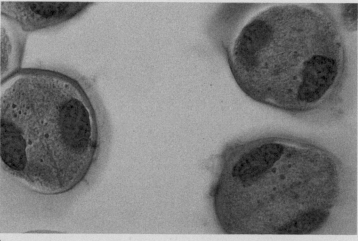

b.

double fertilization occurs. One sperm nucleus unites with the egg, forming a zygote, and the other sperm nucleus migrates to and unites with the polar nuclei, forming a 3N endosperm nucleus. The tube nucleus disintegrates.

Flowering plants practice double fertilization. One sperm nucleus unites with the egg nucleus giving a zygote, and the other unites with the polar nuclei, forming a 3N endosperm nucleus.

Seeds and Fruits

With double fertilization, a number of processes begin. (1) The endosperm nucleus divides, forming the endosperm. **Endosperm** is a nutrient material for the developing embryo and sometimes for the young seedling as well. (2) The **zygote** develops into an embryo, as we will be discussing in detail later. (3) The inte
guments of the ovule harden and become the seed coat. A **seed** is a structure formed by the maturation of the ovule; it contains a sporophyte embryo plus stored food. The ovary, and sometimes other floral parts, develop into a fruit. Although peas, beans, tomatoes, and cucumbers are commonly called vegetables, botanists categorize them as fruits. A **fruit** is a mature ovary that usually contains seeds.

Kinds of Fruits

As a fruit develops from an ovary, the ovary wall thickens to become the **pericarp.** Examples of the variety of fruits are shown in figure 31.6.

Most fruits are *simple fruits,* derived from a single pistil. In fleshy simple fruits, the pericarp is at least somewhat fleshy (table 31.1). Peaches and plums are good examples of fleshy fruits that are derived from a simple ovary. An apple develops from a compound ovary, but much of the flesh comes from the receptacle that grows around the ovary. It's more obvious that a tomato comes from a compound ovary, because in cross section you can see several seed-filled cavities.

Dry fruits have a pericarp that is dry. Legumes, such as peas and beans, produce a fruit that splits along two sides or seams. In the case of peas, we eat the seeds, but in the case of beans, we eat the fruit and seeds. Not all dry fruits split at maturity. The pericarp of a grain is tightly fused to the seed and cannot be separated from it. A corn kernel is actually a grain, as are the fruits of wheat, rice, and barley plants.

Some fruits are compound fruits and develop from several individual ovaries. In aggregate fruits, such as a blackberry, the berries are derived from separate ovaries of a single flower. The strawberry is also an aggregate fruit, but each ovary becomes an achene (table 31.1). The flesh is from the receptacle. In contrast, a pineapple comes from the fruit of many individual flowers attached to the same fleshy stalk. As the ovaries mature, they fuse to form a large, multiple fruit.

Fruit and Seed Dispersal

For plants to be successful, they have to disperse their seeds, preferably long distances from the parent plant. Wind, water, or various animals are all agents of dispersal for angiosperms.

Some seeds, such as the approximately one million produced by an orchid, are so small and light that they need no special adaptation to carry them far away. The somewhat heavier dandelion fruit uses a tiny "parachute" (fig. 31.7) for dispersal. The winged fruit of a maple tree, which contains two seeds, has been known to travel up to 10 kilometers (6 miles) from its parent. Some seeds are forcibly ejected from a fruit that remains attached to the parent. Touch-me-not has seed pods that swell as they mature. When the pods finally burst, the ripe seeds are hurled out.

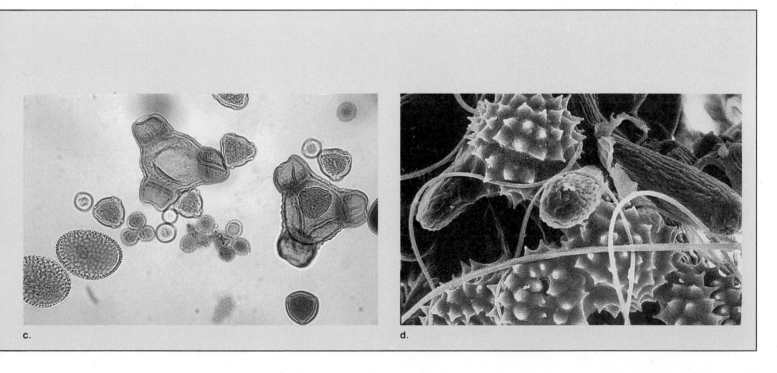

c.

d.

Figure 31.6

Fruit diversity. In general, fruits develop from the ovary and seeds develop from the ovules. Often the other parts disappear.

a.

Developing fruit of tomato plant and a tomato cut crosswise revealing ripened compound ovary and seeds.

b.

Flower of blackberry with several individual ovaries and aggregate fruit of blackberry. Each cluster is a berry from a single flower.

Table 31.1 Kinds of Fruits

Name	Description	Example
Simple Fruits	Develop from an individual ovary	
Fleshy	Pericarp is usually fleshy	
Drupe	From simple ovary with one pitted seed and soft "skin"	Peach, plum, olive
Berry	From compound ovary with many seeds	Grapes, tomato
Pome	From compound ovary; flesh is from accessory flower parts	Apple, pear
Dry	Pericarp is dry	
Follicle	From simple ovary that splits open down one side	Milkweed, peony
Legume	From simple ovary that splits open along both sides	Peas, beans, lentils
Capsule	From compound ovary with capsules that split in various ways	Poppy
Achene	From simple ovary with one-seeded small fruit; pericarp easily removed	Sunflower
Nuts	From simple ovary with one-seeded fruit; hard pericarp	Acorns, hickory nuts, chestnuts
Grain	From simple ovary with one-seeded small fruits; pericarp completely united with seed coat	Rice, oats, barley
Compound Fruits	Develop from a group of individual ovaries	
Aggregate fruits	Ovaries are from a single flower	Blackberries, raspberries, strawberries
Multiple fruits	Ovaries are from separate flowers clustered together	Pineapple

c.

Developing fruit of pea plant and dry fruit of pea plant developed from a simple ovary. Each pea developed from an ovule.

Figure 31.7
What you may think of as a dandelion seed is actually a one-seeded fruit. The dandelion inflorescence resembles a single flower, but actually it is a composite of many individual flowers called florets. Dandelion fruits are adapted to dispersal by the wind. Each fruit has a tiny "parachute" that keeps it afloat in the breezes.

The hooks and spines of clover bur, cocklebur, and stick-seed attach to the fur of animals and the clothing of humans. Birds and mammals sometimes eat fruits, including the seeds, which are then defecated (passed out of the digestive tract with the feces) some distance from the parent plant. Squirrels and other animals gather seeds and fruits, which they bury some distance away. Even those that are forgotten or abandoned aid the dispersal of the plant that produced them.

The fruit of the coconut palm can be dispersed by ocean currents and may land many hundreds of kilometers away from the parent plant.

Development of the Embryo

As the integuments of the ovule become the seed coat, the zygote is developing into an embryo. Early divisions of the zygote produce two parts; the upper part will be the embryo and the lower part is the suspensor, which anchors the embryo and transfers nutrients to it from the sporophyte plant. Soon the cotyledons, or seed leaves, can be seen. Figure 31.8 shows the development of a dicot embryo with its two cotyledons. At this point, the dicot embryo is "heart shaped." Later, when it becomes "torpedo shaped" it is possible to distinguish the shoot apex and the root apex. These contain apical meristem, the tissue that brings about primary growth in a plant; the shoot apical meristem is responsible for aboveground growth, and the root apical meristem is responsible for underground growth (fig. 29.1).

Monocots differ from dicots by having only one cotyledon. Another important difference between monocots and dicots is the manner in which nutrient molecules are stored in the seed. In a monocot, the cotyledon rarely stores food; rather, it absorbs food molecules from the endosperm and passes them to the embryo. During the development of a dicot embryo, the cotyledons usually store the nutrient molecules that the embryo uses. Therefore, in figure 31.8, we can see that the endosperm seemingly disappears. Actually, it has been taken up by the two cotyledons.

Regardless of whether or not the endosperm is taken up by the cotyledon(s), a mature seed contains an embryonic sporophyte plant and stored food. It is surrounded by a **seed coat** derived from the integuments of the ovule.

Some seeds will not germinate (begin to grow) until they have been dormant for a period of time. For seeds, dormancy is the time when growth does not occur, even though conditions may be favorable for growth. Seed dormancy may not occur in the moist tropics, but dormancy in the temperate zone may not be broken until seeds have been exposed to a period of cold weather. In deserts and dry tropic areas, germination will not occur until there is adequate moisture. This requirement helps ensure that seeds do not germinate until the most favorable growing season has arrived. The first sign of the onset of germination is the bursting of the seed coat due to the uptake of

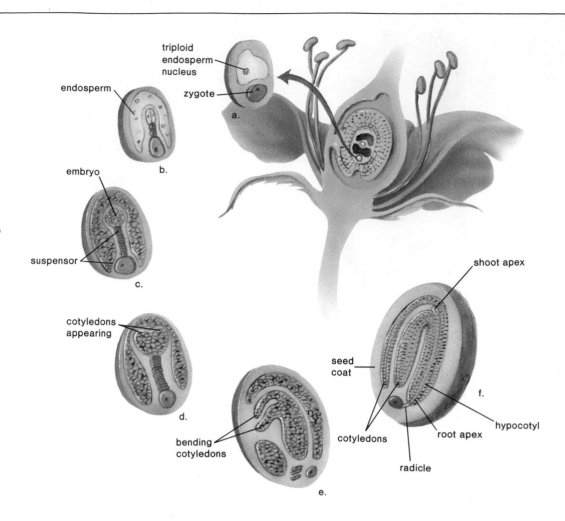

Figure 31.8

Stages in development of a dicot embryo and seed. *a.* The single-celled zygote lies beneath the endosperm nucleus. *b.* and *c.* The endosperm is a mass of tissue surrounding the embryo. The embryo itself is actually made up of the cells above the suspensor. *d.* The embryo becomes heart shaped as the cotyledons begin to appear. *e.* There is progressively less endosperm as the embryo differentiates and enlarges. As the cotyledons bend around, the embryo takes on a torpedo shape. *f.* The mature seed has a seed coat that protects the enclosed embryo. The embryo consists of the epicotyl, represented here by the shoot apex; the hypocotyl; and the radicle, which forms the first seedling root.

water and swelling of the cells. Water, bacterial action, and even fire act on the seed coat, allowing it to become permeable to water. Germination takes place if there is sufficient water, warmth, and oxygen to sustain growth.

Structure and Germination of Seeds

To illustrate the main features of seeds, we will contrast the bean seed (dicot) and a corn kernel (monocot). A bean seed (fig. 31.9*a*) has a small white scar called a *hilium,* where the ovule was attached to the ovary wall. Right next to the hilium is the *micropyle,* which is visible as a small pore.

If the halves (the cotyledons) of the seed are parted, you can see a rudimentary plant. The embryonic shoot bears young leaves and is called a **plumule.** The **epicotyl** is the portion of the stem above the attachment of the cotyledons, and the **hypocotyl** is the portion of stem below the attachment of the cotyledons. The **radicle** is the part of the embryo that contains the root apical meristem and becomes the first (primary) root of the seedling. All of these parts become more obvious when the seed germinates. As the dicot seedling emerges from the soil, the shoot is hook shaped to protect the delicate plumule (fig. 31.9*b*).

Corn kernels (fig. 31.10*a*) are actually fruits, and the outer covering is the pericarp. The bulk of the food-storage tissue is endosperm, and the cotyledon does not play a storage role. The

plumule and radicle are enclosed in protective sheaths called the coleoptile and coleorhiza, respectively. The plumule and the radicle burst through these coverings when germination occurs (fig. 31.10*b*).

Asexual Reproduction in Plants

Asexual reproduction, also known as **vegetative propagation,** is common in plants. In vegetative propagation, a portion of one plant gives rise to a completely new plant. Both plants now have identical genes. Some plants have aboveground horizontal stems, called stolons or runners, and others have underground stems, called *rhizomes,* that produce new plants. For example, strawberry plants grow from the nodes of stolons and violets grow from the nodes of rhizomes. White potatoes can be propagated in a similar manner. White potatoes are actually portions of underground stems, and each eye is a bud that will produce a new potato plant if it is planted with a portion of a swollen tuber. Sweet potatoes are modified roots and may be propagated by planting sections of the root. You may have noticed that the roots of some fruit trees, such as cherry and apple trees, produce "suckers," small plants that can be used to grow new trees.

Asexual reproduction has a great deal of commercial importance. Once a plant variety with desired characteristics has been developed through vegetative propagation, new plants can

Figure 31.9

A common garden bean. *a.* Seed structure. *b.* Germination and development of the seedling.

a.

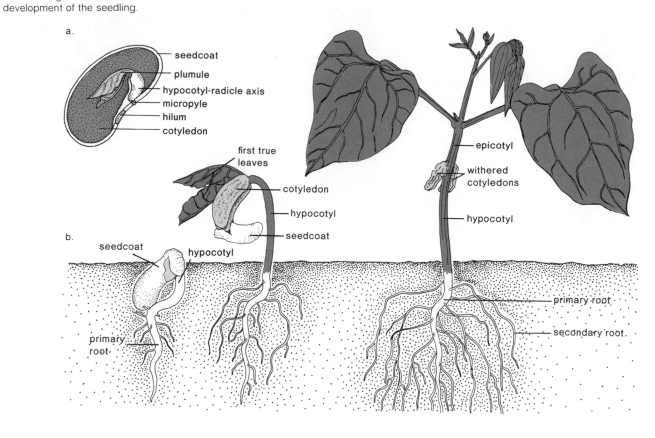

- seedcoat
- plumule
- hypocotyl-radicle axis
- micropyle
- hilum
- cotyledon

first true leaves

- cotyledon
- hypocotyl
- seedcoat

b.

seedcoat
hypocotyl

primary root

epicotyl

withered cotyledons

hypocotyl

primary root

secondary root

Figure 31.10

Corn. *a.* Grain structure. *b.* Germination and development of the seedling.

a.

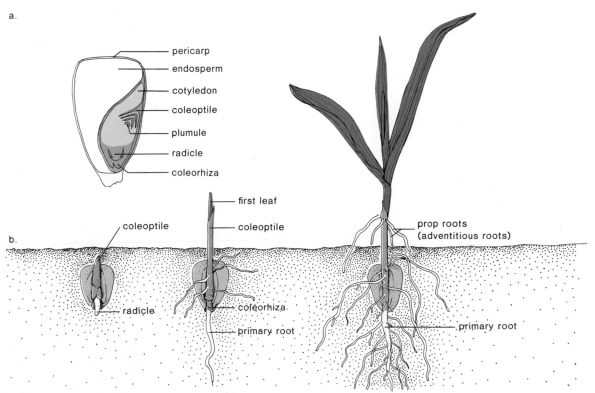

- pericarp
- endosperm
- cotyledon
- coleoptile
- plumule
- radicle
- coleorhiza

first leaf

coleoptile   coleoptile

b.

radicle

coleorhiza

primary root

prop roots (adventitious roots)

primary root

Coevolution between Plants and Their Pollinators

A plant and its pollinator(s) are adapted to one another. They have a mutualistic relationship in which each benefits—the plant uses its pollinator to ensure that cross-pollination takes place, and the pollinator uses the plant as a source of food. This mutualistic relationship came about through the process of coevolution; that is, the dependency of the plant and pollinator on each other is the result of suitable changes in structure and function in each of them. The evidence for coevolution is observational. For example, floral coloring and odor are suited to the sense perceptions of the pollinator, the mouth parts of the pollinator are suited to the structure of the flower, the type of food provided is suited to the nutritional needs of the pollinator, and the pollinator forages at the time of day that specific plants are open. The following are examples of this coevolution.

Bee-Pollinated Flowers

There are now 20,000 different species of bees that pollinate flowers. The best-known pollinators are the honeybees (fig. *a*). Bee eyes see a spectrum of light that is different from humans. The bee's visible spectrum is shifted so that they do not see red wavelengths but do see ultraviolet wavelengths. Bee-pollinated flowers are usually brightly colored, predominantly blue or yellow—not entirely red. They may also have ultraviolet shadings, called honey guides that highlight the portion of the flower that contains the reproductive structures (fig. 39.8). Bees have mouthparts that are fused into a long tube that contains a tongue. This tube is an adaptation for sucking up nectar provided by the plant, usually at the base of the flower. Bee flowers are delicately sweet and fragrant to advertise that nectar is present. The honey guides often point to a narrow floral tube large enough for the bee's feeding apparatus,

Flowers have adaptations that make them attractive to their pollinators.
a. A bee-pollinated flower is a color other than red (bees can't detect this color), and has a landing platform where the reproductive structures of the flower brush up against the bee's body.

but too small for other insects to reach the nectar. Bees also collect pollen as food for their larvae. Pollen clings on the hairy body of a bee, and the bees also gather it by means of bristles on their legs and store it in pollen baskets on the third pair of legs. Bee-pollinated flowers are sturdy and irregular in shape because they often have a "landing platform" where the bee can alight. The landing platform requires the bee to brush up against the anther and stigma as it moves toward the floral tube to feed. One type of orchid (genus *Ophrys*) has evolved a unique adaptation. The flower resembles a

b. A butterfly-pollinated flower is often a composite, containing many individual flowers. The broad expanse provides room for the butterfly to land, after which it lowers its proboscis into each flower in turn.

female bee and when the male of that species attempts to copulate with the flower, (fig. 21.17) the bee receives pollen.

Moth- and Butterfly-Pollinated Flowers

Contrasting these two types of flowers emphasizes the close adaptation between pollinator and flower. Both moths and butterflies have a long, thin, hollow proboscis, but they differ in other characteristics. Moths feed at night and have a well-developed sense of smell. The flowers they visit are visible at night because they are lightly shaded (white, pale yellow, or pink) and have a strong, sweet perfume, which helps attract moths. Moths hover when they feed, and their flowers have deep tubes with open margins that allow the hovering moths to

---

be supplied to gardeners and farmers. Cuttings can be taken from the plant, and the cut end can be treated to encourage it to grow roots, or a cutting can be grafted to the stem of a plant that has a root. Budding is the form of grafting most often used commercially. In this procedure, the axillary buds are grafted onto the stem of another plant. Today, entire plants can be produced by tissue culture (fig. 31.11), a technique that may eventually replace the older methods thus far discussed. Usually an embryonic *tissue* is removed from a plant and placed in a special *culture* medium. After the tissue has grown for a while, it is subdivided to produce many identical plants from a very small amount of starting cells.

Plants reproduce both asexually and sexually. Asexual reproduction occurs when a portion of one plant gives rise to an entirely new plant. Grafting, particularly budding, and tissue culture propagation have commercial importance today.

Figure 31.11

Each dish shows several shoots of Douglas Fir growing from a single cotyledon. Such cultures are part of a research project for cloning genetically improved trees. Subsequently, the shoots will be cut, rooted, and planted in the forest. Tissue culture propagation is expected to play an important part in bringing forest yields toward their theoretical maximum.

c. Hummingbird-pollinated flowers are curved back, allowing the bird to insert its beak to reach the rich supply of nectar. While doing this, the bird's forehead and other body parts touch the reproductive structures.

d. Bat-pollinated flowers are large, sturdy composites that can take rough treatment. Here the head of the bat is positioned so that its bristly tongue can lap up nectar.

reach the nectar with their long proboscis. Butterflies are active in the daytime and have good vision but a weak sense of smell. Their flowers have bright colors, even red, because butterflies can see the color red, but tend to be odorless. Unable to hover, butterflies need a place to land. Flowers that are visited by butterflies often have flat landing platforms on which butterflies can alight (fig. *b*). They also tend to be composites with many individual flowers clustered in a head. Each flower has a long, slender floral tube, accessible to the long, thin proboscis of butterflies.

Bird- and Bat-Pollinated Flowers

In North America, the most well-known bird pollinators are the hummingbirds. These tiny animals have good eyesight but do not have a well-developed sense of smell. Like moths, they hover when they feed. Typical flowers pollinated by hummingbirds are red, with a slender floral tube and margins that are curved back and out of the way. And although they produce copious amounts of nectar, they have little odor. As a hummingbird feeds on nectar with a long, thin beak, its head comes into contact with the stamens and pistil (fig. *c*).

Bats are adapted to gathering food in various ways, including feeding on the nectar and pollen of plants. Bats are nocturnal and have an acute sense of smell. Those that are pollinators also have keen vision and a long, extensible, bristly tongue. Typically, bat-pollinated flowers open only at night and are light colored or white. They have a strong, musty smell similar to the odor that bats produce to attract one another. The flowers are generally large, sturdy, and able to hold up when a bat inserts part of its head to reach the nectar. While the bat is at the flower, its head is dusted with pollen (fig. *d*).

Coevolution

These examples illustrate that coevolution has occurred, but how did it come about? About 200 million years ago, seed plants were just beginning to evolve, and insects were not as diverse as today, wind alone was used to carry the pollen, and wind pollination is a hit-or-miss affair. Beetles that visited plants to feed on vegetative leaves may have carried pollen from plant to plant by chance. Plants are immobile, and when insects transport their pollen from plant to plant, the plants are using the motility of animals for the purpose of achieving cross-pollination. Cross-pollination and fertilization no doubt resulted in a greater number of variations, in flowers, including features, such as production of nectar, that increased the likelihood of beetle visits. The habit that beetles have of feeding on reproductive parts may have led to other features of flowers. Protected ovules are less likely to be eaten than exposed ovules, and any plants that protected their ovules would have been selected. Today, all flowering plants have protected ovules within ovaries.

As cross-pollination and fertilization continued, more and more variations would have been seen. Eventually, increased specificity began to develop as angiosperm species and insect species became closely adapted to one another. Today there are some 235,000 species of flowering plants and over 700,000 species of insects. This diversity suggests that the success of angiosperms has meant the success of insects and vice versa.

Summary

1. Angiosperms produce flowers and often have a symbiotic relationship with an animal pollinator. They also produce seeds that are covered by fruits.
2. Flowering plants exhibit an alternation of generations life cycle that includes heterospores and separate male and female gametophytes. The pollen grain is the male gametophyte. The female gametophyte is called the embryo sac and remains within the body of the sporophyte plant.
3. Typical parts of a flower are: sepals that are green in color and form an outer whorl; petals, often colored, form an inner whorl; the pistil is in the center and contains the carpels, each consisting of stigma, style, and ovary; and the stamens, each having a filament and anther, are around the base of the carpels. The ovary contains ovules.
4. Each ovule contains a megaspore mother cell that divides meiotically to produce four haploid cells, only one of which survives and is the megaspore. The megaspore divides mitotically to produce the embryo sac (female gametophyte) that usually has eight nuclei. The two center nuclei are the polar nuclei, and one of the cells next to the micropyle is an egg cell.
5. The anthers contain microspore mother cells, each of which divides meiotically to produce four haploid microspores. Each of these divides mitotically to give a two-celled pollen grain. One cell is the tube cell, and the other is the generative cell. Later, the generative cell divides to give two sperm. A pollen grain is the male gametophyte. After pollination, the pollen grain germinates and, as the pollen tube grows, the sperm nuclei travel to the embryo sac. Pollination is simply the transfer of pollen from anther to stigma.

6. Angiosperms practice double fertilization. One sperm nucleus unites with the egg nucleus, forming a 2N zygote, and the other unites with the polar nuclei to form a 3N endosperm nucleus.

7. After fertilization, the endosperm nucleus forms the endosperm. The zygote becomes the sporophyte embryo. The ovule matures to become the seed (its integuments become the seed coat). The ovary becomes the fruit.

8. There are different types of fruits. Simple fruits are derived from a single ovary (which can be simple or compound). Some simple fruits are fleshy, such as a peach or apple. Others are dry, such as peas, nuts, and grains. Aggregate fruits develop from a number of ovaries of a single flower, and compound fruits develop from multiple ovaries of separate flowers.

9. Flowering plants have several ways to disperse seeds and fruits. They may be blown by the wind, attached to animals that carry them away, eaten by animals that defecate the seeds some distance away, or adapted to water transport.

10. As the ovule is becoming a seed, the zygote is becoming an embryo. After the first several divisions, it is possible to discern the embryo and the suspensor. The suspensor attaches the embryo to the ovule and supplies it with nutrients. The dicot embryo becomes heart shaped, then torpedo shaped. Once you can see the two cotyledons, it is possible to distinguish the shoot apex and the root apex, which contain the apical meristems. In dicot seeds, the cotyledons frequently take up the endosperm.

11. In a bean (dicot) seed prior to germination, you can distinguish the two cotyledons and the plumule, which is the shoot that bears leaves. Also present are the epicotyl, hypocotyl, and the radicle. In a corn kernel (monocot), the endosperm, cotyledon, plumule, and radicle are visible.

12. Many flowering plants reproduce asexually. Sometimes, the nodes of stems (either aboveground or underground) give rise to entire plants. Cuttings of plants will produce entire plants, assuming that the cutting can be encouraged to develop roots. Grafting and budding also allow the production of new plants. Tissue culture is the latest method for asexually producing plants.

Objective Questions

1. In flowering plants,
   a. the gametes become the gametophyte generation.
   b. the spores become the sporophyte generation.
   c. the sporophyte organism produces spores.
   d. Both (a) and (b).
2. The flower part that contains ovules is the
   a. carpel.
   b. stamen.
   c. sepal.
   d. petal.
3. The megaspore mother cell and the microspore mother cell
   a. both produce pollen grains.
   b. both divide meiotically.
   c. both divide mitotically.
   d. produce pollen grains and embryo sacs, respectively.

4. A pollen grain is a
   a. haploid structure.
   b. diploid structure.
   c. first diploid and then haploid structure.
   d. first haploid and then diploid structure.
5. Which of these is mismatched?
   a. polar nuclei—plumule
   b. egg and sperm—zygote
   c. ovule—seed
   d. ovary—fruit
6. Which of these is not a fruit?
   a. walnut
   b. pea
   c. green bean
   d. peach
7. Animals assist with
   a. both pollination and seed dispersal.
   b. only pollination.
   c. only seed dispersal.
   d. only asexual propagation of plants.

8. A seed contains
   a. a zygote.
   b. an embryo.
   c. stored food.
   d. Both (b) and (c).
9. Which of these is mismatched?
   a. plumule—leaves
   b. cotyledon—seed leaf
   c. epicotyl—root
   d. pericarp—corn kernel
10. Gametes are involved in
   a. both the asexual and sexual reproduction of plants.
   b. the asexual reproduction of plants.
   c. the sexual reproduction of plants.
   d. neither the asexual nor the sexual reproduction of plants.

Study Questions

1. Name two unique features of flowering plant reproduction.
2. Draw a diagram that illustrates the life cycle of seed plants, and tell how it applies to the flowering plants. Why don't flowering plants require a source of outside water for pollination?
3. Draw a diagram of a flower and name the parts.

4. Describe the development of a female gametophyte from the megaspore mother cell to the production of an egg.
5. Describe the development of the male gametophyte from the microspore to the production of sperm.
6. What is the difference between pollination and fertilization?
7. Distinguish between simple, fleshy and simple, dry fruits. Give an example of each type. What is an aggregate fruit, a multiple fruit?

8. Name several mechanisms of seed and fruit dispersal.
9. Describe the sequence of events as a dicot zygote becomes a seed. Describe the dormant and germinated seed of a dicot, of a monocot.
10. In what ways do plants reproduce asexually without human assistance? with human assistance?

Thought Questions

1. Defend the hypothesis that the type of fruit is suitable to the life strategy of the plant. For example, why do you think fruit trees have fleshy fruits and bean plants have dry fruits?

2. New plants can arise from the nodes of underground stems. Why is that?

3. Polyploidy can arise when gametes are diploid due to nondisjunction. When during the life cycle of a flowering plant could nondisjunction occur?

Selected Key Terms

pistil (pis'til) 480
ovary (o'vah-re) 480
ovule (o'vūl) 480
stamen (sta'men) 480
anther (an'ther) 480

pollination (pol''ĭ-na'shun) 482
endosperm (en'do-sperm) 483
zygote (zi'gōt) 483
seed (sēd) 483
fruit (frōōt) 483

plumule (plōō'mūl) 486
epicotyl (ep''ĭ-kot'il) 486
hypocotyl (hi''po-kot'il) 486
radicle (rad'ĭ-k'l) 486

Growth and Development in Plants

Your study of this chapter will be complete when you can:

1. Name the five major types of hormones, state which are stimulatory and which are inhibitory, and give a primary function for each type.
2. Describe two experiments that contributed to understanding the role that auxin plays in phototropism.
3. Indicate the manner in which plant hormones function according to gibberellin research.
4. Describe a tissue culture experiment demonstrating that plant hormones interact.
5. Describe three types of tropisms that occur in response to external stimuli. Indicate any involvement of plant hormones in these responses.
6. List some movements of plants that occur as a response to internal stimuli, and provide evidence that biological clocks function in plants.
7. Discuss the relationship of photoperiodism to flowering in certain plants.
8. Explain the phytochrome conversion cycle, and suggest possible functions of phytochrome in plants.

What causes buds to form and then later break their dormancy? It can be shown that plant hormones are involved in all aspects of plant growth and development. Responses to environmental stimuli is dependent upon changes in the proportions of hormones present in the body of the plant.

P lants often respond to external and internal stimuli with changes in their growth patterns. Some of the external factors that regulate growth are light, day length, gravity, and temperature. Among the principal internal factors that regulate growth are plant hormones, and we will begin our discussion with these. A **hormone** is a chemical messenger produced in small amounts by one part of the body that is active in a different part of the body. Generally, plant hormones are produced by the meristematic regions of a plant and are transported in vascular tissue. Responses may be observed in almost every part of the plant's body. Table 32.1 lists the types of hormones we will consider. These hormones sometimes interact to control particular physiological processes. Different combinations bring about different effects.

Each naturally occurring hormone has a specific chemical structure. Other chemicals, some of which differ only slightly from the natural hormones, also affect the growth of plants. These and the naturally occurring hormones are sometimes grouped together and called plant growth regulators. The reading on page 496 discusses the various uses of plant growth regulators.

Hormones that Promote Growth

Notice in table 32.1 that three groups of hormones (auxins, gibberellins, and cytokinins) promote activities associated with growth. We will consider these hormones first.

Auxins

The most common naturally occurring auxin is indolacetic acid (IAA), which is produced in shoot meristem, young leaves, and in flowers and fruits.

Early researchers observed that plants bend toward light, a phenomenon called *phototropism* (fig. 32.1). Around 1881, Charles Darwin and his son Francis reported on experiments they had performed with grass and oat seedlings. When these plants first appear, each seedling is covered by a sheath called the *coleoptile*. The leaves soon break through this covering of the tiny shoot (fig. 32.2). The Darwins found that if the coleoptile was kept intact, the seedlings would bend toward a unidirectional light source. But if the tip of the seedling was cut off or was covered by a black cap, the seedling would not respond to light. They concluded that some influence is transmitted from the coleoptile tip to the rest of the shoot to cause bending.

Building on the results of the Darwins and others, Frits W. Went performed still more experiments in 1926 (fig. 32.3). He cut off the tips of coleoptiles and placed them on agar (a gelatinlike material). After about an hour, he removed the tips and cut the agar into small blocks. When an agar block was placed to one side of a coleoptile without a tip, the shoot would bend away from that side. The bending occurred even though the seedlings were not exposed to light.

Table 32.1 Plant Hormones

Type	Primary Example	Notable Function
Promoters of Growth		
Auxins	Indolacetic acid (IAA)	Cell elongation
Gibberellins	Gibberellic acid (GA)	Stem elongation
Cytokinins	Zeatin	Cell division
Inhibitors of Growth		
Abscisic acid	Abscisic acid (ABA)	Dormancy
Ethylene	Ethylene	Abscission

Figure 32.1
Phototropism is easily observed in this field of sunflowers. All the flowers point toward the sun and even move as the sun moves—they track the sun from sunrise to sunset. This is obviously an adaptive trait of plants.

Went concluded that the agar blocks contained a chemical that had been produced by the coleoptile tips. It was this chemical, he decided, that had caused the shoots to bend. He named the chemical substance **auxin** after the Greek word *auxein,* which means "to increase." It can be shown that when a plant is exposed to unidirectional light, auxin is transported to the shady side. Elongation of the cells on this side brings about the characteristic bending toward the light (fig. 32.4). Auxin causes the affected cell to degrade some of the polysaccharides in the cell wall. The cell then elongates because it is less able to resist the expansion caused by osmotic movement of water into the cell. The direction of cell growth is dependent on where the wall is weakest.

Figure 32.2

a. An oat seedling is at first protected by a hollow sheath called a coleoptile. Later the leaves break through this sheath.
b. Experiments with oat seedlings show that the tip is necessary and must be exposed to light in order for oat seedlings to bend toward the light. If the tip of a seedling is cut off, the seedling will not bend toward unidirectional light. If the tip is covered by a black cap, the seedling will not bend toward the light.

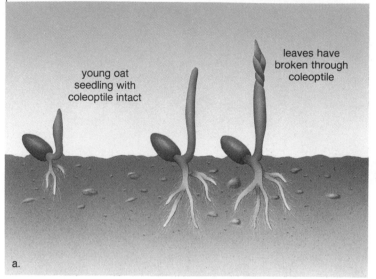

young oat seedling with coleoptile intact

leaves have broken through coleoptile

a.

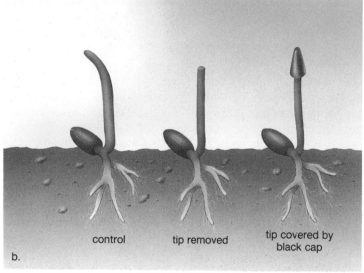

control

tip removed

tip covered by black cap

b.

Figure 32.3

This experiment shows that a chemical produced by the coleoptile tip causes the bending of a seedling toward the light.
a. Coleoptile tips are cut off and placed on agar (a gelatinlike material). *b.* After a time the tips are removed and the agar is cut into small blocks. A block is placed to one side of a coleoptile that lacks a tip. *c.* Bending occurs even in the absence of a light stimulus.

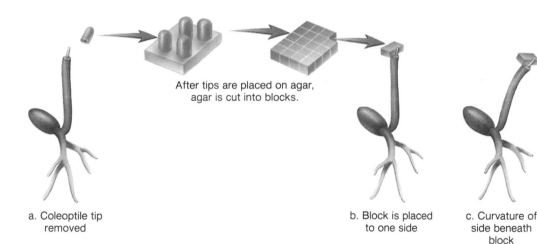

After tips are placed on agar, agar is cut into blocks.

a. Coleoptile tip removed

b. Block is placed to one side

c. Curvature of side beneath block

Other Effects of Auxin

Auxin has been found to affect many other aspects of plant growth. Auxin produced in a terminal bud at the apex of a plant prevents other bud development and growth for some distance from the apex. When a terminal bud is removed deliberately or by accident, the nearest axillary buds begin to grow, and the plant branches (fig. 32.5).

The application of a weak solution of auxin to a woody cutting will cause roots to develop. Auxin production by seeds also promotes the growth of fruit. As long as auxin is concentrated in leaves or fruits, rather than in the stem, leaves and fruits will not fall off. Therefore, trees can be sprayed with auxin to keep mature fruit from falling to the ground.

Auxin is also involved in the response of stems to gravity, as will be discussed later in more detail.

Researchers studying the bending of coleoptiles toward light discovered the presence of a hormone called auxin. Auxin is also involved in apical dominance and in the response of stems to gravity.

Gibberellins

We know of about seventy gibberellins that chemically differ only slightly. The most common is **GA$_3$** (the subscript designation distinguishes it from other gibberellins). **Gibberellins** are likely present in newly developing plant organs, because they are growth promoters that bring about cell division and enlargement of the resulting cells. When gibberellins are applied externally to plants, the most obvious effect is *stem elongation* (fig. 32.6).

Figure 32.4

Auxin is transported from the illuminated side to the shaded side of a coleoptile sheath as indicated by the arrows. This unequal concentration of auxin causes the cells on the shaded side to elongate. Elongation of the cells results in the bending of a seedling toward the light source.

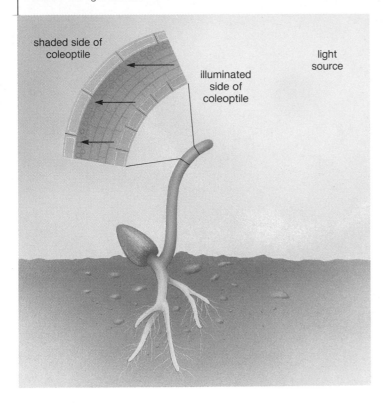

Figure 32.5

The trunk of this Jeffrey pine tree, which would normally be single, is forked because of the earlier removal of the terminal bud.

Figure 32.6

The plant on the right was treated with gibberellin; the plant on the left was untreated. Gibberellins are often used to promote stem elongation in economically important plants, but the exact mode of action still remains unclear.

Gibberellins were discovered in 1926, the same year that Went performed his classic experiments with auxin. Kurosawa, a Japanese scientist, was investigating a fungal disease of rice plants called "foolish seedling disease." The plants elongated too quickly, causing the stem to weaken and the plant to collapse. Kurosawa found that the fungus infecting the plants produced an excess of a chemical he called gibberellin named after the fungus (*Gibberella fujukuroi*).

It wasn't until 1956 that gibberellic acid was isolated from a flowering plant rather than from a fungus. Since minute amounts of externally applied gibberellic acid caused genetically dwarfed pea plants to grow to a normal size, it was thought that dwarfed plants would contain little gibberellin, but surprisingly they contained more gibberellin than normal plants. For some reason, they were unable to utilize the gibberellin present in their cells. As mentioned previously, it is often the balance of hormones in a plant cell that produces an effect, rather than the presence or absence of a particular hormone.

If applied at the appropriate concentration, gibberellins can cause most dicots and monocots to grow taller and faster. A cabbage plant will grow to 2 meters (6 feet tall) and bush beans become pole beans instead. Apparently, gibberellins, as well as the auxins, help regulate general plant development. The dormancy of seeds and buds can be broken by an application of gibberellin, and flowering can be induced in a mature plant.

Gibberellin research provides an example of how a plant hormone might act as a chemical messenger. Barley seeds have a large starchy endosperm that must be broken down into sugars to provide energy for growth. After the embryo produces gibberellin, cells inside the seed coat synthesize an enzyme that

Plant Growth Regulators

N ow that the formulas for many plant hormones are known, it is possible to synthesize them as well as related chemicals in the laboratory. Collectively, these substances are known as plant growth regulators. Many scientists hope plant growth regulators will bring about an increase in crop yield, just as fertilizers, irrigation, and pesticides have done in the past.

Since auxin was first discovered, researchers have found many agricultural and commercial uses for it. Auxins cause the base of stems to form new roots quickly, so that new plants are easily started from cuttings. When sprayed on apples, pears, and citrus fruits shortly before harvest, auxins prevent fruit from dropping too soon. Because auxins inhibit the growth of lateral buds, potatoes sprayed with an auxin will not sprout and thus have a longer storage life.

In high concentrations, some auxins are used widely in agriculture as herbicides to prevent the growth of broad-leaved plants. In addition to their use in weed control, the synthetic auxins known as 2,4D and 2,4,5T were used as defoliants during the Vietnam War. Even though 2,4D has been known for over 35 years, we do not yet know how it works. Apparently, it is structurally different enough from natural auxin that a plant's own enzymes cannot break it down. Its concentration rises until metabolism is disrupted, cellular order is lost, and the cells die.

The other plant hormones studied in this chapter also have agricultural and commercial uses. Gibberellins are used to stimulate seed germination and seedling growth of some grains, beans, and fruits. They will also increase the size of some mature plants. Treatment of sugarcane with as little as 2 ounces per acre increases the cane yield by more than 5 tons. The application of either auxins or gibberellins can cause an ovary and accessory flower parts to develop into fruit, even though pollination and fertilization have not taken place. In this way, it is sometimes possible to produce

Effect of gibberellic acid (GA₃) on Thompson seedless grapes (*Vitis vinifera*) (*Left*) control grapes. (*Right*) GA₃ was sprayed at bloom and at fruit set. Almost all grapes sold in stores are now treated with gibberellic acid.

seedless fruits and vegetables or bigger, more uniform bunches with larger fruit, as shown above.

Because cytokinins retard the aging of leaves and other organs, they are sprayed on vegetables to keep them fresh during shipping and storage. Such treatment of holly, for example, allows it to be harvested many weeks prior to its use as a holiday decoration.

Plant growth regulators are making it possible to grow plants from a few cells in laboratory glassware. It may even be possible to develop new varieties of food plants with particular characteristics, such as tolerance to heat, cold, toxins, or drought. Using gene-splicing techniques, it might eventually be possible to create plants capable of utilizing atmospheric nitrogen.

Several synthetic inhibitors are used to oppose the action of the auxins, gibberellins, and cytokinins normally present in plants. Some of these can cause leaf and fruit drop at a time convenient to the farmer. Removing leaves from cotton plants aids harvesting of cotton, and thinning the fruit of young fruit trees results in larger fruit as the trees mature. Retarding the growth of other plants sometimes increases their hardiness. For example, an inhibitor has been used to reduce stem length in wheat plants, so that the plants do not fall over in heavy winds and rain.

The commercial uses of ethylene were greatly increased with the development of ethylene-releasing compounds. Ethylene gas is injected into airtight storage rooms to ripen bananas, honeydew melons, and tomatoes. It will also degreen oranges, lemons, and grapefruit when the rind would otherwise remain green because of a high chlorophyll level. When sprayed on certain fruit and nut crops, ethylene increases the chances that the fruit will detach when the trees are shaken at harvest time.

Today, fields and orchards are often sprayed with synthetic growth regulators just as they are sprayed with pesticides.

---

breaks down the starch. It is hypothesized that gibberellin turns on the gene that codes for the necessary enzyme. In this particular instance, then, a plant hormone is functioning as illustrated in figure 41.14*b*.

Gibberellins cause stems to elongate and dwarf plants to grow to normal size. This hormone also causes barley seeds to produce an enzyme that breaks down endosperm.

Cytokinins

The **cytokinins** are a class of plant hormones that promote cell division; *cytokinesis means cell division.* These substances are derivatives of the purine adenine, one of the nitrogenous bases in DNA and RNA. A naturally occurring cytokinin, called *zeatin,* has been extracted from corn kernels. Kinetin also promotes cell division, but it has not been found to occur naturally.

The cytokinins were discovered as a result of attempts to grow plant tissue and organs in culture vessels (fig. 32.7) in the 1940s. It was found that cell division occurred when coconut milk (a liquid endosperm) and yeast extract were added to the culture medium. Although the effective agent or agents could not be isolated, they were collectively called cytokinins. Not until 1967 was the naturally occurring cytokinin, zeatin, isolated from coconut milk. Cytokinins have been isolated from various seed plants where they occur in the actively dividing tissues of roots and also in seeds and fruits.

Figure 32.7

Interaction of hormones. It is now clear that plant hormones rarely act alone; it is the relative concentrations of these hormones that produce an effect. The modern emphasis is to look for an interplay of hormones when a growth response is studied. For example, tissue culture experiments have revealed that auxin and cytokinin interact to affect differentiation during development. In a tissue culture that has no added amounts of these two hormones, tobacco strips develop into a callus of undifferentiated tissue. *a.* If the ratio of auxin to cytokinin is changed slightly, the callus produces roots.

b. Change the ratio again, and vegetative shoots and leaves may be produced. *c.* Yet another ratio results in floral shoots. *d.* This example illustrates that it is the *relative* proportions of hormones that probably bring about specific effects. To attribute a particular growth response to only one hormone is no doubt an oversimplification. Most likely, it will be many years before the interactions of hormones in determining the growth responses of plants will be well understood.

a.   b.   c.   d.

Plant tissue culturing is now common practice and researchers are well aware that the ratio of auxin to cytokinin and the acidity of the culture medium determine whether the plant tissue will form an undifferentiated mass, called a callus, or differentiate to form roots, vegetative shoots, leaves, or floral shoots (fig. 32.7). Recently, researchers reported that chemicals called oligosaccharins (chemical fragments released from the cell wall) are also effective in directing differentiation. They hypothesize that the function of auxin and cytokinin is actually to activate enzymes that release these more specific chemical messengers from the cell wall. They similarly believe that all plant hormones probably have such a function, which may explain the pleiotropy (many and various effects) of any specific plant hormone. If plant hormones do regulate the activity of enzymes, they would be functioning as shown in figure 41.14*a.*

Other Effects of Cytokinins

When a plant organ, such as a leaf, loses its natural color, it is most likely undergoing an aging process called *senescence.* During senescence, large molecules within the leaf are broken down and transported to other parts of the plant. Senescence need not affect the entire plant at once; for example, as some plants grow taller, they naturally lose their lower leaves. It has been found that senescence can be prevented by the application of cytokinins. Not only can cytokinins prevent death of plant organs, they can also initiate growth. Lateral buds will begin to grow despite apical dominance when cytokinin is applied to them.

Cytokinins promote cell division, prevent senescence, and initiate growth. The interaction of hormones is well exemplified by the effect of varying ratios of auxin and cytokinins on differentiation of plant tissues.

Plant Growth Inhibitors

As is evident in table 32.1, two types of plant hormones (ethylene and abscisic acid) inhibit growth. We would expect these hormones to be present and active whenever the plant faces conditions that are unfavorable for continued growth, such as lack of water or cold temperatures.

Ethylene

In the early 1900s, it was common practice to ready citrus fruits for market by placing them in a room with a kerosene stove. Because heat alone did not have the same effect, it was finally realized that an incomplete combustion product of kerosene, namely **ethylene,** was responsible for the ripening of fruit. Ethylene ripens fruit by increasing the activity of enzymes that soften fruit. For example, it stimulates the production of *cellulase,* an enzyme that hydrolyzes the cellulose of plant cell walls.

Ethylene, being a gas, moves freely through the air—a barrel of ripening apples can induce ripening of a bunch of bananas even some distance away. Its presence in the air can also

Figure 32.8

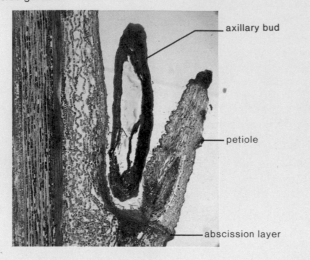

Abscission layer. Before a leaf falls, a special band of cells, called the abscission layer, develops at the base of the petiole, the leaf stem. Here the hormone ethylene promotes the breakdown of plant cell walls so that the leaf finally falls. Here also, a layer forms a leaf scar that protects the plant from possible invasion by microorganisms.

- axillary bud
- petiole
- abscission layer

Figure 32.9

Stomata are openings in the leaf epidermis that are opened and closed by guard cells. Because of transpiration, a plant ordinarily loses a great deal of water by way of the stomata; therefore, it is beneficial for the stomata to be closed when a plant is under water stress. The hormone ABA is believed to bring about the closing of the stomata when water stress occurs. As discussed in chapter 30, stomata close when K^+ and then H_2O leave the guard cell.

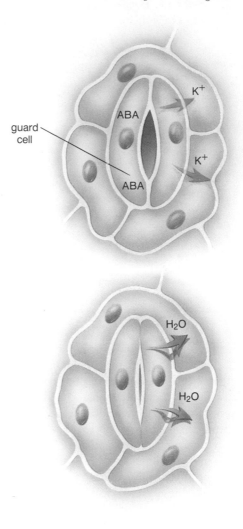

guard cell

ABA K^+

ABA K^+

H_2O

H_2O

retard the growth of plants in general. Homeowners who use natural gas to heat their homes sometimes report difficulties in growing house plants, for example. Ethylene is also present in automobile exhaust and it's just possible that plant growth, in general, is being affected by all the exhaust that enters the atmosphere. It only takes 1 part of ethylene per 10 million parts of air to bring about an inhibition of plant growth.

Other Functions

High concentrations of auxin stimulate the production of ethylene in shoot meristem. This means that certain plant responses attributed to auxin may really be due to the presence of ethylene. For example, ethylene is probably needed for apical dominance to occur.

Ethylene also inhibits stem elongation and prevents the maturation of young leaves. This effect can have survival value as when young seedlings are germinating (fig. 31.9). If the soil is compact, the dicot shoot produces more ethylene, leading to a thickening of the hook, which helps the young plant push up through the soil.

Ethylene is also involved in *abscission,* the dropping of leaves, fruits, and flowers from a plant. As mentioned previously, a relative decrease of auxin and perhaps gibberellin in these areas of the plant compared to the stem probably initiates abscission (fig. 32.8). But once the process of abscission has begun, ethylene is produced. This is believed to stimulate such enzymes as cellulase to cause leaf, fruit, or flower drop.

Abscisic Acid

Abscisic acid (**ABA**) is sometimes called the stress hormone, because it initiates and maintains seed and bud dormancy and brings about the closure of stomata. Although the external application of abscisic acid will promote abscission, this hormone is no longer believed to function naturally in that process.

Dormancy occurs when a plant organ readies itself for adverse conditions by stopping growth (even though conditions at the time are favorable for growth). For example, it is believed that abscisic acid moves from leaves to vegetative buds in the fall, and thereafter these buds are converted to winter buds. A winter bud is covered by thick and hardened scales. In the spring, dormancy is broken when seeds germinate, and buds begin to send forth new growth. A reduction in the level of abscisic acid and an increase in the level of gibberellins are believed to break seed and bud dormancy.

Abscisic acid is called the stress hormone not only because it promotes dormancy, but also because it brings about stomatal closing (fig. 32.9) when a plant is under water stress. In some unknown way, ABA causes K^+ to leave guard cells. Thereafter, the guard cells lose water, and the stomata close. Notice that this is the opposite sequence of events to those that cause stomata to open (fig. 30.8).

Figure 32.10
When an upright plant is turned on its side, in time the roots bend
down and the stem bends up. Roots therefore demonstrate positive
gravitropism and the stem demonstrates negative gravitropism.

Ethylene and abscisic acid are plant growth inhibitors. Ethylene
ripens fruits and is involved in abscission. Abscisic acid maintains
dormancy and causes stomata to close.

Plant Movements

As noted at the beginning of this chapter, plants respond to both
external and internal stimuli by changing their growth patterns.
These growth patterns often result in movement of some part
of the plant. Now that you have some knowledge of plant growth
hormones, it is possible to discuss this topic in more depth.

External Stimuli

Movements of plants toward or away from an external stimulus
is called a **tropism.** Three well-known tropisms are:

- phototropism, a movement in response to a light stimulus;
- gravitropism, a movement in response to gravity; and
- thigmotropism, a movement in response to touch.

Phototropism

As already discussed, plant stems bend toward the light (fig.
32.1). This is called a positive **phototropism.** A bending away
from light is called a negative phototropism. Roots, depending
on the species examined, are either insensitive to light or exhibit
negative phototropism.

Response to a stimulus requires a receptor, and in this case
the photoreceptor is believed to be a yellow pigment related to
the vitamin riboflavin. Following detection of light, the plant
hormone auxin migrates from the bright side to the shady side
of a stem. The cells on that side elongate faster than those on
the bright side (fig. 32.4), causing the stem to bend toward the
light.

Gravitropism

When a plant is placed on its side (fig. 32.10), the stem displays
negative **gravitropism** because it grows opposite to the direction
of gravity. Roots display positive gravitropism because they grow
in the same direction as gravity. This difference in growth response is explained in this manner:

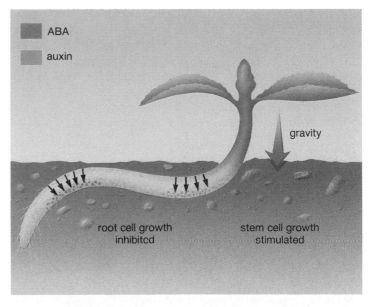

Gravity causes stem cells to grow more quickly (so they become
more elongated), and inhibits the growth of root cells leading
to the bending we observe.

A hormone could accumulate on the lower side of roots
and stems to cause these gravitropic bendings. In the case of
stems, it can be demonstrated that auxin accumulates on the

Figure 32.11
Coiling response of a pole green bean.

Figure 32.12
A prayer plant. *a.* The plant at noon. *b.* The same plant at 10 P.M., after "sleep" movements of its leaves have occurred.

lower side and in the case of roots, it is believed that the hormone ABA, which inhibits growth, may accumulate on the lower side.

How do plants detect a change in gravity? There is evidence that roots perceive gravity by the movement of starch grains in special root cap cells called statocytes. Due to gravity, the starch grains accumulate on the lower side of these cells. How and why this causes a redistribution of hormones is not known.

Thigmotropism

The response of a plant or plant part to contact is called a **thigmotropism** (*thigma* in Greek means touch). The presence of this response in plants is seen in the coiling of tendrils and the twining of a climbing plant stem. The response occurs because the cells in contact with an object such as a pole, grow less and those on the opposite side elongate (fig. 32.11).

The Venus-flytrap is a plant with specialized lobed leaves that snap shut when they are touched by an insect (fig. 29.21c). While the leaves are closed, glands on the leaf surface secrete enzymes that digest the insect's body. The leaf closes when the outer epidermal cells expand rapidly, due to an increase in turgor pressure. When the inner epidermal cells expand, the leaf opens again. There are other examples of touch responses due to turgor pressure changes. For example, leaflets of the sensitive plant (*Mimosa pudica*) fold when they are touched, because of turgor pressure changes in specialized cells at the base of the leaflets and the leaf itself.

Internal Stimuli

Some plant movements occur primarily in response to internal stimuli, as for example:

1. Helical movements—Although plants appear to be growing straight up, many actually spiral as they grow.
2. Nodding movements—Members of the legume family, such as garden beans, nod from side to side as the seedling pushes up through the soil.
3. Twining movements—In some plants, the visible twining of tendrils as they grow (even when they have not made contact with an object) is due to the presence of auxin. In other plants, similar movements are due to the presence of ethylene.
4. Contraction movements—Contractile roots of bulbs pull them deeper into the ground, to a point where there is no difference in daytime and nighttime temperatures.
5. Nastic "sleep" movements—Plant organs, such as leaves and flowers, sometimes bend up and then down. For example, prayer plants (fig. 32.12) have leaves that fold up and reopen daily, due to turgor pressure changes. These movements are controlled by a "biological clock," as we will discuss next.

Circadian Rhythms in Plants

Organisms exhibit periodic fluctuations that correspond to environmental changes. For example, your temperature and blood pressure tend to change with the time of day, and you become sleepy at a certain time of night. The prayer plant we mentioned earlier displays rhythmic "sleep" behavior (fig. 32.12). A biological rhythm with a 24-hour cycle is called a **circadian rhythm** (Latin *circa* = about and *dies* = day, i.e., about a day).

Circadian rhythms tend to persist, even if the appropriate environmental cues are no longer present. If by chance you take a transcontinental flight, you will likely suffer jet lag. It will take several days to become adjusted to the time change, because your body is still attuned to the day-night pattern of your previous environment. The internal mechanism by which a biological rhythm is maintained in the absence of appropriate environmental stimuli is termed a **biological clock.** Typically, if organisms are sheltered from environmental stimuli, their circadian rhythms continue, but the cycle is extended a bit. In prayer plants, the sleep cycle changes to 26 hours instead of 24. Therefore, it is believed that biological clocks are synchronized by external stimuli to give 24-hour rhythms. The length of daylight compared to the length of darkness, called the photoperiod, sets the clock. Temperature has little or no effect. This is adaptive because the **photoperiod** indicates seasonal changes better than do temperature changes. Spring and fall, in particular, can have both warm and cold days.

There are other examples of circadian rhythms in plants. For example, stomata usually open in the morning and close at night, and some plants secrete nectar at the same time of the day or night. But the primary usefulness of the biological clock appears to be to measure day-length changes, so that a plant is able to respond to the changing seasons.

Photoperiodism

Many physiological changes in plants are related to a seasonal change in day length. Such changes include seed germination, the breaking of bud dormancy, and the onset of senescence. A physiological response prompted by changes in the length of day or night is called **photoperiodism.** In some plants, photoperiodism influences flowering; violets and tulips flower in the spring, and asters and goldenrods flower in the fall.

In the 1920s, when U.S. Department of Agriculture scientists working in Beltsville, Maryland began to study photoperiodism in more detail, they decided to grow plants in the greenhouse, where they could artificially alter the photoperiod. This work led them to conclude that plants can be divided into three groups:

- **Short-day plants**—flower when the day length is shorter than a critical length. (Good examples are cocklebur, poinsettias, and chrysanthemums.)
- **Long-day plants**—flower when the day length is longer than a critical length. (Good examples are wheat, barley, clover, and spinach.)
- **Day-neutral plants**—flowering is not dependent on day length. (Good examples are tomatoes and cucumbers.)

Further, we should note that both a long-day and a short-day plant can have the same critical length (fig. 32.13). Spinach is a long-day plant having a critical length of fourteen hours; ragweed is a short-day plant with the same critical length. Spinach, however, flowers in the summer when the day length increases to fourteen hours or more, and ragweed flowers in the fall when the day length shortens to fourteen hours or less. We now know that some plants may require a specific sequence of different day lengths in order to flower.

In 1938, K. C. Hammer and J. Bonner began to experiment with artificial lengths of light and dark that did not necessarily correspond to a normal, 24-hour day. They discovered that the cocklebur, a short-day plant, will flower as long as the dark period is continuous for 8½ hours, regardless of the length of the light period. Further, if this dark period is interrupted by a brief flash of light, the cocklebur will not flower. (Interrupting the light period with darkness had no effect.) Similar results have also been found for long-day plants. They require a dark period that is shorter than a critical length regardless of the length of the light period. However, if a slightly longer-than-critical-length night is interrupted by a brief flash of light, long-day plants will flower. We must conclude, then, that it is the length of the dark period that controls flowering, not the length of the light period. Of course, in nature, shorter days always go with longer nights and vice versa.

Researchers have discovered that short-day plants require a period of darkness that is longer than a critical length and long-day plants require a period of darkness that is shorter than a critical length in order to flower.

Phytochrome

If flowering is dependent on day and night length, plants must have some way to detect these periods. Many years of research by U.S. Department of Agriculture scientists led to the discovery of a plant pigment called phytochrome. **Phytochrome** is a blue-green leaf pigment that alternately exists in two forms. Figure 32.14 indicates that

P_r (phytochrome red) absorbs red light (of 660 nm wavelength) and is converted to P_{fr}.
P_{fr} (phytochrome far-red) absorbs far-red light (of 730 nm wavelength) and is converted to P_r.

Direct sunlight contains more red light than far-red light; therefore, P_{fr} is apt to be present in plant leaves during the day. In the shade and at sunset, there is more far-red light than red light; therefore P_{fr} is converted to P_r as night approaches. There is also a slow metabolic replacement of P_{fr} by P_r during the night. It was thought for some time that this slow reversion might provide a means for the plant to measure the length of the night. However, it is now assumed that the active form of phytochrome, P_{fr}, signals a biological clock that brings about flowering.

Figure 32.13

Day length (night length) effect on two types of plants. *a.* Short-day (long-night) plant. *1.* When the day is shorter (the night longer) than a critical length, this type of plant flowers. *2.* It does not flower when the day is longer (night is shorter) than the critical length. *3.* It also does not flower if the longer-than-critical-length night is interrupted by a flash of light. *b.* Long-day (short-night) plant. *1.* When the day is shorter (the night longer) than a critical length, this type plant does not flower. *2.* It flowers when the day is longer (the night shorter) than a critical length. *3.* It flowers if the slightly longer-than-critical-length night is interrupted by a flash of light.

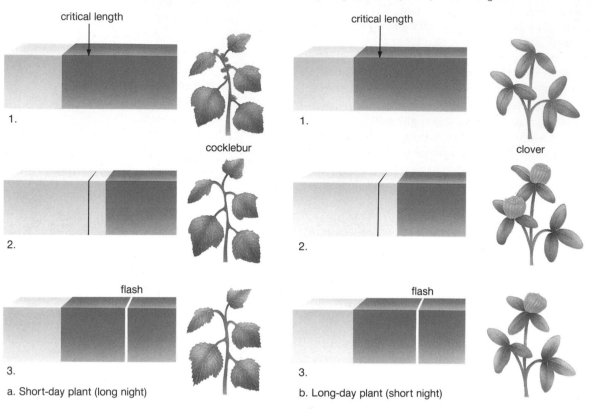

critical length

1.

cocklebur

2.

flash

3.

a. Short-day plant (long night)

critical length

1.

clover

2.

flash

3.

b. Long-day plant (short night)

Figure 32.14

The $P_r \rightleftarrows P_{fr}$ conversion cycle. The inactive form P_r is prevalent during the night. At sunset or in the shade when there is more far-red light, P_{fr} is converted to P_r. Also during the night, metabolic processes cause P_{fr} to be converted i..to P_r. P_{fr}, the active form of phytochrome, is prevalent during the day because at that time there is more red light than far-red light.

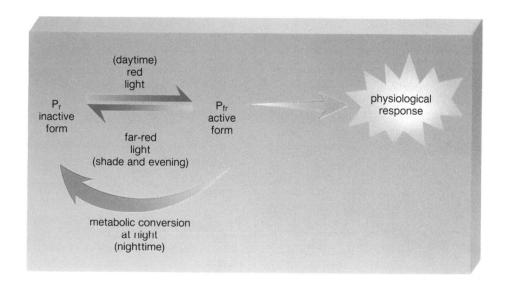

(daytime)
red
light

P_r
inactive
form

far-red
light
(shade and evening)

P_{fr}
active
form

physiological
response

metabolic conversion
at night
(nighttime)

Figure 32.15

Phytochrome has other functions besides regulating flowering. For example, if far-red light is prevalent, as it is in the shade, the stem of a seedling elongates and the leaves remain small (left-hand side of photo). However, if red light is prevalent, as it is in bright sunlight, the stem does not elongate and the leaves expand. These effects are due to phytochrome.

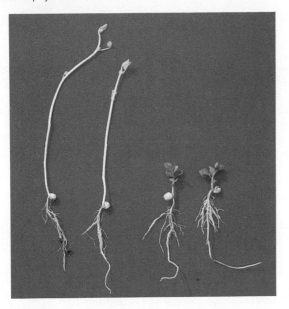

Just how this might happen is unknown. Hormones are probably involved, but this has not been proven. At one time researchers thought they might find a special flowering hormone called florigen, but such a hormone has never been discovered.

> Phytochrome is a leaf pigment that alternates between two forms (P_{fr} during day and P_r during night), apparently allowing a plant to detect photoperiod changes that may result in flowering.

Other Functions of Phytochrome

The $P_r \rightleftarrows P_{fr}$ conversion cycle is now known to control other growth functions in plants. It promotes seed germination and inhibits stem elongation, for example. The presence of P_{fr} indicates to some seeds that sunlight is present and conditions are favorable for germination. This is why some seeds must be partly covered with soil when planted. Germination of other seeds is inhibited by light so they must be planted deeper. Following germination, the presence of P_r indicates that stem elongation may be needed to reach sunlight. Seedlings that are grown in the dark *etiolate,* that is, the stem increases in length and the leaves remain small (fig. 32.15). Once the seedling is exposed to sunlight and P_r is converted to P_{fr}, the seedling begins to grow normally—the leaves expand and the stem branches.

Summary

1. Both stimulatory and inhibitory hormones help control certain growth patterns of plants. There are hormones that stimulate growth (auxins, gibberellins, cytokinins) and hormones that inhibit growth (ethylene and abscisic acid).

2. The auxins, which promote cell elongation, control phototropism. When a plant is exposed to light, auxin moves laterally from the bright to the shady side of a stem. Auxins also influence apical dominance, fruit maturation, prevention of leaf and fruit drop, and gravitropism.

3. Gibberellins promote stem elongation and external application of them can cause dwarf plants to reach normal size. They also help break seed and bud dormancy.

4. Cytokinins cause cell division, the effects of which are especially obvious when plant tissues are grown in culture. They can also prevent leaf senescence.

5. Ethylene causes fruit to ripen and promotes abscission. Abscisic acid is no longer believed to be involved in abscission, but it does promote bud dormancy and causes stomata to close when a plant is under water stress. ABA might also be involved in the negative gravitropism of roots.

6. It is typical for plant hormones to interact, and therefore it is probably an oversimplification to suggest that each has specific functions that do not involve other hormones.

7. Plants respond to external and internal stimuli by growth responses that bring about movement of plant parts.

8. Tropisms are responses to external stimuli. Positive phototropism occurs when auxin moves to the shady side of a stem. Negative gravitropism occurs when auxin moves to the lower side of a stem. Thigmotropism occurs when a plant part makes contact with an object. Tendrils and stems coil, and Venus-flytrap leaves snap shut.

9. Several plant movements such as helical, nodding, twining, contraction, and nastic movements are believed to be responses to internal stimuli. Also, plants exhibit circadian rhythms that are believed to be controlled by a biological clock. For example, bean seeds have rhythmic sleep patterns.

10. Photoperiodism is seen in some plants. For example, short-day plants flower only when the days are shorter than a critical length, and others, called long-day plants, flower only when the days are longer than a critical length. Actually, research has shown that it is the length of darkness that is critical. Interrupting the dark period with a flash of white light prevents flowering in a short-day plant and induces flowering in a long-day plant.

11. Phytochrome is a pigment that responds to both red and far-red light and is involved in flowering. Daylight causes phytochrome to exist as P_{fr}, but during the night, it is converted back to P_r by metabolic processes. Phytochrome probably communicates with a biological clock that in some unknown way brings about flowering.

12. In addition to being involved in the flowering process, P_{fr} promotes seed germination, leaf expansion, and stem branching. When P_r predominates, the stem elongates and grows toward sunlight.

Objective Questions

For questions 1 to 5, match the items below with the hormones in the key.

Key:
- a. auxin
- b. gibberellin
- c. cytokinin
- d. ethylene
- e. abscisic acid

1. One rotten apple can spoil the barrel.
2. Cabbage plants bolt (grow tall).
3. Stomata close when a plant is water stressed.
4. Sunflower plants all point toward the sun.
5. Coconut milk causes plant tissues to undergo cell division.

6. Which of these is a correct statement?
   - a. Both stems and roots show positive gravitropism.
   - b. Both stems and roots show negative gravitropism.
   - c. Only stems show positive gravitropism.
   - d. Only roots show positive gravitropism.
7. Short-day plants
   - a. are the same as long-day plants.
   - b. are apt to flower in the fall.
   - c. do not have a critical photoperiod.
   - d. All of these.
8. A plant requiring a dark period of at least 14 hours
   - a. will flower if a 14-hour night is interrupted by a flash of light.
   - b. will not flower if a 14-hour night is interrupted by a flash of light.
   - c. will not flower if the days are 14 hours long.
   - d. Both (b) and (c).
9. Phytochrome
   - a. is a plant pigment.
   - b. is present as P_{fr} during the day.
   - c. may communicate with a biological clock.
   - d. All of these.
10. Circadian rhythms
    - a. require a biological clock.
    - b. do not exist in plants.
    - c. are involved in the tropisms.
    - d. All of these.

Study Questions

1. Name three types of hormones that promote growth processes, and give several specific functions for each.
2. Name two types of hormones that inhibit growth processes, and give several specific functions for each.
3. Name three types of tropisms, and indicate the role of hormones in these plant movements.
4. Describe two experiments that led to the discovery of auxin.
5. Give experimental evidence to suggest that hormones interact when they bring about an effect.
6. List and describe some movements of plants that occur as a response to internal stimuli.
7. What is a biological clock, how does it function, and what is its primary usefulness in plants?
8. Define photoperiodism, and discuss its relationship to flowering in certain plants.
9. What is the phytochrome conversion cycle, and what are some possible functions of phytochrome in plants?

Thought Questions

1. Explain the adaptive value of using photoperiodic changes rather than seasonal temperature changes to time events in plants' lives.
2. Why do you think that some short-day plants near the northern edge of their range are able to produce mature seeds and fruits during some growing seasons but not during others?
3. Why is a biological clock assumed to be a necessary part of the flowering process?

Selected Key Terms

hormone (hor′mōn) 493
auxin (awk′sin) 493
gibberellin (gib-ber-el′in) 494
cytokinin (si″to-ki′nin) 496
ethylene (eth′i-lēn) 497
abscisic acid (ab-sis′ık as′id) 498

tropism (tro′pizm) 499
phototropism (fo-tot′ro-pizm) 499
gravitropism (grav″i-tro′pizm) 499
thigmotropism (thig-mot′ro-pizm) 500
circadian rhythm (ser″kah-de′an rith′m) 501

biological clock (bi-o-loj′ĕ-kal klok) 501
photoperiod (fo″to-pe′re-od) 501
photoperiodism (fo″to-pe′re-od-izm) 501
phytochrome (fi′to-krōm) 501

Suggested Readings for Part 4

Albersheim, P., and Darvill, A. G. September 1985. Oligosaccharins. *Scientific American.*

Barrett, S. C. H. September 1987. Mimicry in plants. *Scientific American.*

Bold, H. C. 1980. *Morphology of plants and fungi.* 4th ed. New York: Harper & Row, Publishers, Inc.

Brill, W. J. March 1977. Biological nitrogen fixation. *Scientific American.*

Epel, D. November 1977. The program of fertilization. *Scientific American.*

Heslop-Harrison, Y. February 1978. Carnivorous plants. *Scientific American.*

Jansen, W., and Salisbury, F. B. 1971. *Botany: An ecological approach.* Belmont, CA: Wadsworth.

Niklas, K. J. July 1987. Aerodynamics of wind pollination. *Scientific American.*

Raven, P. H., et al. 1986. *Biology of plants.* 4th ed. New York: Worth.

Rayle, D., and Wedberg, H. L. 1980. *Botany: A human concern.* Boxton: Houghton Mifflin.

Rost, R. et al. 1984. *Botany: A brief introduction to plant biology.* 2d ed. New York: Wiley.

Salisbury, F. B., and Ross, C. W. 1985. *Plant physiology.* 3rd ed. Belmont, CA: Wadsworth.

Shepard, J. F. May 1982. The regeneration of potato plants from leaf-cell protoplasts. *Scientific American.*

Tippo, O., and Sterm, W. L. 1977. *Humanistic botany.* New York: W. W. Norton and Co.

Zimmerman, M. H. March 1963. How sap moves in trees. *Scientific American.*

CRITICAL THINKING
CASE STUDY

PART 4
Plant Structure and Function

As described in chapter 30, the guard cells on either side of a stoma control whether a stoma is open or closed. When the guard cells swell with water and develop tugor pressure, a stoma is open (*see* fig. 30.8), and when the guard cells lose water, a stoma is closed. Usually stomata are open during the day and are closed during the night. When stomata are open, water is being transported in xylem to the leaves and the leaves are receiving CO_2 by way of the stomata. Water and CO_2 are the reactants needed for photosynthesis to occur.

Is it adaptive for plants to have their stomata open during the day? Why? Will the level of CO_2 inside the leaves be high or low when a plant is photosynthesizing? Why?

To test if both light and low level of CO_2 are needed to cause stomata to open, researchers exposed epidermal strips (the outer layer of leaf tissue containing epidermal cells and guard cells) floating on distilled water to CO_2-free and CO_2-laden air in both the light and in the dark. Should CO_2-free air in the presence of light result in a larger stomatal opening?

Prediction 1 CO_2-free air in the presence of light should result in a larger stomatal opening.

Result 1 As shown in the graph, exposure of epidermal strips to CO_2-free air in the light did result in larger stomatal opening.

Study the graph and consider first just the initial reactions of the guard cells. What happens to the size of the stomatal opening in CO_2-free air, regardless of whether it is light or dark? to the size in CO_2-laden air, regardless of whether it is light or dark? Which factor seems more pertinent to stomatal opening, concentration of CO_2 or presence of light?

Consider now the final reactions of the guard cells. What happens to a stoma in the light if CO_2-laden air is provided? Is it adaptive for the stomata of plants to close completely in the light when there is plenty of CO_2 and water available? Why? What happens to a stoma in the dark regardless of the amount of CO_2? In the dark a plant carries on cellular respiration and what gas is released? Now explain why a stoma provided with CO_2-free air in the dark initially starts to open and then retreats to partly closed.

Since a stoma never opens nor closes completely in the dark, is it possible that some additional factor (besides CO_2 concentration) not tested for, is involved in controlling stomatal movement?

To test whether inhibition of photosynthesis would have an effect on stomatal opening, inhibitors of photosynthesis were applied to epidermal strips exposed to light and CO_2-free air. In the light, a plant carries on both photosynthesis and cellular respiration; therefore, if photosynthesis is inhibited, the plant will only be respiring. Would you predict, then, that if photosynthesis is inhibited, a stoma will close?

Prediction 2 Inhibition of photosynthesis in the light should lead to stomatal closing.

Result 2 As predicted, inhibition of photosynthesis in the light did lead to stomatal closing.

Recall that the level of CO_2 inside the leaf is low when a plant is photosynthesizing. What is a possible role of photosynthesis in causing the stomata to open? Is this a direct role or an indirect role? Is it possible that photosynthesis has another effect not tested for?

This sequence of events has been suggested to explain stomatal opening: (1) low CO_2 concentration in the leaf in the presence of light leads to (2) high concentration of K^+ in guard cells, leads to (3) uptake of water by guard cells, leads to (4) opening of a stoma. If this sequence of events is to be expected, which of the experimental conditions noted in the graph should cause the greatest K^+ concentration in guard cells?

Prediction 3 CO_2-free air in the presence of light should result in the greatest K^+ concentration in guard cells.

Result 3 Of the experimental conditions noted in the graph CO_2-free air in the presence of light did result in the greatest K^+ concentration in guard cells. Indeed, for all conditions noted in the graph, measurements of the final K^+ concentrations did agreee with what would be predicted by the graph.

Explain why you would expect K^+ to initially increase and then decrease in guard cells of epidermal strips exposed to CO_2-free air in the dark. Give a sequence of events, corresponding to (1)–(4) above, that would explain why stomata, which are open during the day, close at night.

A logical question to ask is, "What is the source of the K^+ that enters the guard cells when stomata open?" The epidermal strips used in these experiments were floated on distilled water that lacked K^+; therefore, the only source is from the epidermal cells surrounding the guard cells.

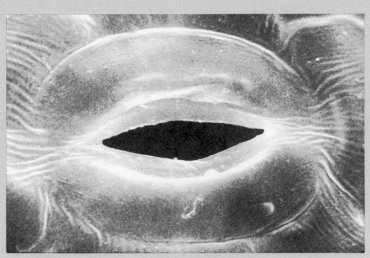

a.

a. This electron micrograph shows an open stoma and its guard cells. A stoma can vary as to the degree it is open. b. Epidermal strips (containing epidermal and guard cells only) floating in water were exposed to the four experimental conditions noted in the key. The experimenter recorded the degree of stomatal opening as related to time.

If these epidermal cells supply K^+ to the guard cells, what do you predict would happen to the level of K^+ in epidermal cells as the concentration in guard cells increases and as the stomata open?

Prediction 4 K^+ concentration in epidermal cells should decrease as the concentration in guard cells increases and as the stomata open.

Result 4 Experimental measurements show that there was a decrease in epidermal cells K^+ concentration as the concentration in guard cells increased and as the stomata opened.

If there was one epidermal cell per guard cell what might happen to this epidermal cell once it gives up its K^+? Why? Recognizing that there are many epidermal cells per guard cell explain why you would not expect epidermal cells to suffer dehydration due to the passage of K^+ from epidermal cells to the guard cells.

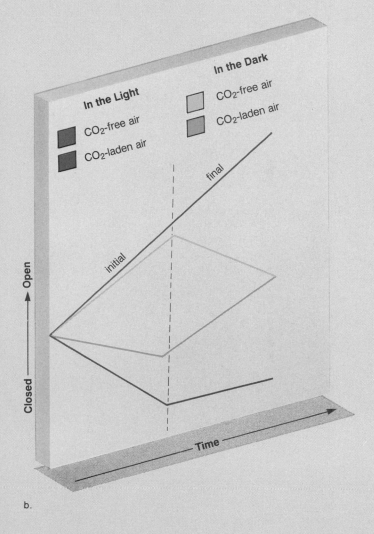

b.

The experiments described here support the hypothesis that in the light a low CO_2 concentration causes K^+ movement into guard cells leading to increased turgor pressure and opening of the stomata. Other questions are raised, however, by these investigations. For example, what is the mechanism whereby CO_2 controls K^+ movement in the light, and what are the factors that control K^+ movement in the dark. It is customary for experiments to generate new questions and lead to further hypotheses for investigation.

Pallaghy, C. K. 1971. Stomatal movement and potassium transport in epidermal strips of *Zea mays*: The effect of CO_2. *Planta* (Berl.) 101, 287–95, published by Springer-Verlag.

Rashke, K. 1975. Stomatal action. *Annual Revue Plant Physiology* 26: 309–40.

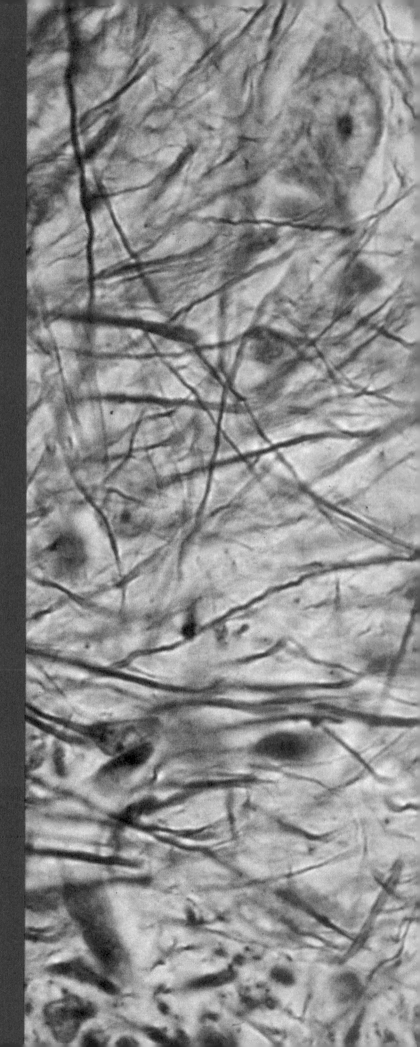

PART 5

Animal Structure and Function

Every system in an organism's body contributes to home-ostasis, the relative constancy of the internal environment even though external conditions may vary. The nervous system helps coordinate the activities of the other systems. The large cells in this photo are nerve cells whose fibers transmit nerve impulses. When nerve impulses are received muscles and glands react in a manner that helps maintain homeostasis.

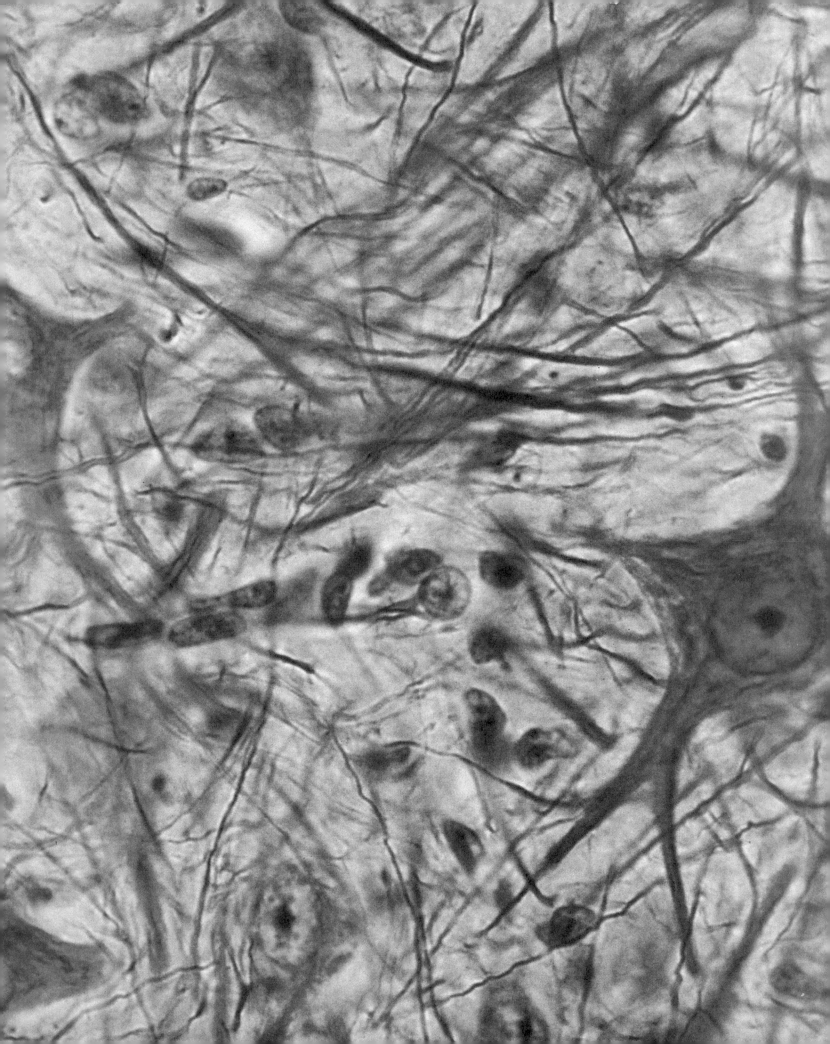

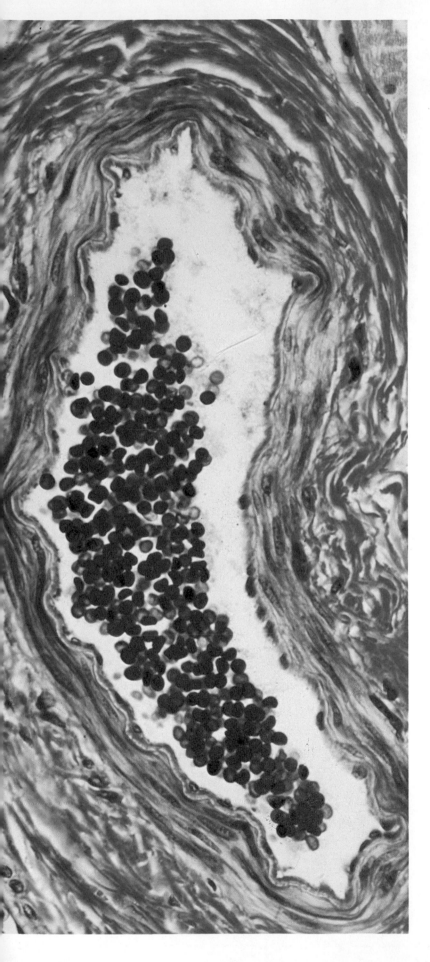

CHAPTER 33

Circulatory System

Your study of this chapter will be complete when you can:

1. Compare the circulatory systems of invertebrate and vertebrate animals.

2. Compare the structure and function of the three types of blood vessels.

3. Trace the path of blood through the mammalian heart and about the body.

4. Describe how the heart beats, and tell how the heartbeat is controlled.

5. List the factors that control the pressure and velocity of the blood in the vessels.

6. Describe the structure and function of the lymphatic system, including the lymph vessels.

7. List the components of human blood, and describe the function and source of each.

8. Draw and explain a diagram depicting capillary exchange within tissues.

9. List the three steps necessary for the clotting of blood.

10. List the ABO blood types, and describe a laboratory procedure for typing blood.

Human blood vessel with red blood cells. Red blood cells are red because they contain the respiratory pigment hemoglobin, which combines with oxygen. Red blood cells are packed full of hemoglobin and don't even contain a nucleus.

very cell requires a supply of oxygen and nutrient molecules and must rid itself of waste molecules. Single-celled organisms (fig. 33.1a) make these exchanges directly with the external environment that surrounds their bodies. Even some small, multicellular animals (fig. 33.1b and c) do not require an internal transport system in order to take care of the needs of their cells. Each cell is able to exchange materials directly with water or with nearby cells that are in contact with water.

The other larger and more complex multicellular animals, however, require a *blood circulatory system* to transport molecules, because the body cells are far from specialized regions of exchange with the external environment. Further, in these animals the body cells are surrounded by a fluid called **tissue fluid.** Blood and tissue fluid together create an internal environment for cells. Because the internal environment remains relatively constant, the cells are protected from extreme changes in the external environment. Claude Bernard, the great French physiologist, was the first to recognize the presence and relative constancy (*homeostasis*) of this *internal environment,* which he called *milieu intérieur.* Experimental studies on mammalian

circulatory systems during the mid-nineteenth century led him to conclude that these animals could be active in almost any type of external environment, because the internal environment stayed relatively constant.

Invertebrates

Gastrovascular Cavity

The sac body plan of invertebrates, such as cnidarians and flatworms, makes a circulatory system unnecessary. In cnidarians (fig. 26.5), cells are either a part of an external layer or they line the gastrovascular cavity. In either case, each cell is exposed to water and can independently exchange gases and get rid of wastes. The cells that line the gastrovascular cavity are specialized to carry out digestion. They pass nutrient molecules to other cells by diffusion. In flatworms (fig. 26.7), the trilobe gastrovascular cavity ramifies throughout the small and flattened body. No cell is very far from one of the three digestive branches, so nutrient molecules can diffuse from cell to cell. Similarly, diffusion meets the respiratory and excretory needs of the cells.

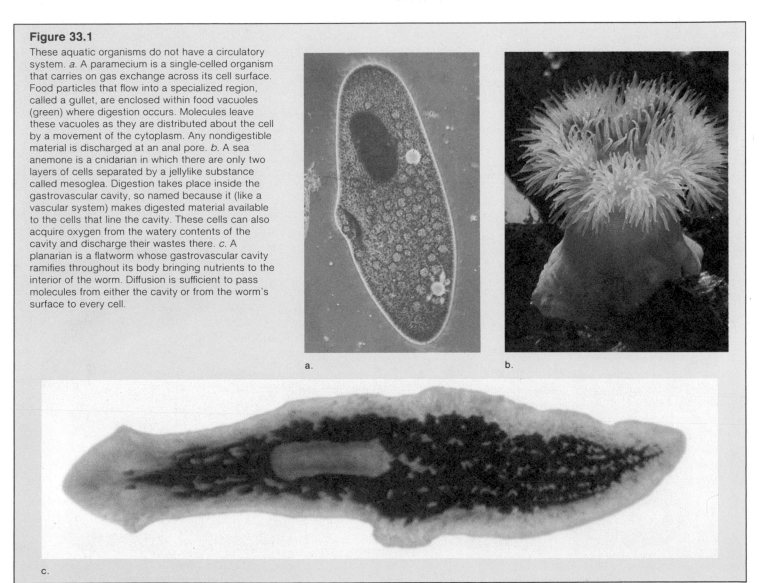

Figure 33.1
These aquatic organisms do not have a circulatory system. *a.* A paramecium is a single-celled organism that carries on gas exchange across its cell surface. Food particles that flow into a specialized region, called a gullet, are enclosed within food vacuoles (green) where digestion occurs. Molecules leave these vacuoles as they are distributed about the cell by a movement of the cytoplasm. Any nondigestible material is discharged at an anal pore. *b.* A sea anemone is a cnidarian in which there are only two layers of cells separated by a jellylike substance called mesoglea. Digestion takes place inside the gastrovascular cavity, so named because it (like a vascular system) makes digested material available to the cells that line the cavity. These cells can also acquire oxygen from the watery contents of the cavity and discharge their wastes there. *c.* A planarian is a flatworm whose gastrovascular cavity ramifies throughout its body bringing nutrients to the interior of the worm. Diffusion is sufficient to pass molecules from either the cavity or from the worm's surface to every cell.

a.

b.

c.

Open and Closed Circulatory Systems

Some of the larger invertebrates have an open circulatory system, while others have a closed system. In an *open circulatory system,* the blood is not always contained within blood vessels (fig. 33.2*a*). A heart pumps blood into vessels; then these vessels empty either into body cavities where blood bathes the internal organs or into sinuses located within the organs themselves. In an open circulatory system, the blood ebbs and flows in a sluggish manner.

In the grasshopper, for example, the dorsal heart pumps the blood into an anterior vessel, which empties into a body cavity termed a *hemocoel.* The hemocoel, appropriately named, is filled with blood. When the heart contracts, openings called *ostia* are closed; when the heart relaxes, the blood is sucked back into the heart by way of the ostia. The blood of a grasshopper is colorless because it does not contain hemoglobin or any other respiratory pigment. It does not carry oxygen, but only nutrients. Oxygen is taken to cells and carbon dioxide is removed from them by way of air tubes that are found throughout the body. Flight muscles require a very efficient means of receiving oxygen and the open circulatory system probably would not suffice.

In a *closed circulatory system,* the blood is always contained within blood vessels. The pumping of a heart keeps the blood moving in this system (fig. 33.2*b*). For example, in the segmented earthworm the five pairs of anterior hearts pump blood into the ventral blood vessel, which has a branch in every segment of the worm's body. Blood moves through these branches into capillaries where exchanges with tissue fluid take place. The blood then moves into branches of the dorsal blood vessel. This vessel returns the blood to the heart for repumping.

The earthworm has red blood, because it contains the respiratory pigment **hemoglobin.** Hemoglobin is dissolved in the blood and is not contained within cells. The earthworm has no specialized boundary for gas exchange with the external environment. Gas exchange takes place across the body wall, which must always remain moist for this purpose.

A gastrovascular cavity in small aquatic invertebrates serves as an internal circulatory system. Larger invertebrates do have a circulatory system. A closed circulatory system is more efficient than an open one.

Vertebrates

All vertebrate animals have a closed circulatory system (fig. 33.3). It consists of a strong, muscular heart, in which the **atria** (sing., atrium) primarily receive blood, and the muscular **ventricles** pump blood out through the blood vessels. There are three kinds of blood vessels (fig. 33.4): **arteries** carry blood away from the heart, **capillaries** exchange materials with tissue fluid, and **veins** return blood to the heart.

Arteries, the strongest of the blood vessels, are resilient: they are able to expand and constrict because their walls have substantial layers of both elastic and muscle fibers. *Arterioles* are small arteries whose constriction can be regulated by the nervous system. Arteriole constriction and dilation affect blood pressure. The more vessels are dilated, the lower the blood pressure.

Arterioles branch into *capillaries,* extremely narrow, microscopic tubes having a wall composed of only one layer of cells (fig. 33.4). Capillary beds (many capillaries interconnected) are

Figure 33.2

a. The grasshopper has an open circulatory system. A hemocoel is a body cavity filled with blood that freely bathes the internal organs. The heart keeps the blood moving, but this open system probably would not supply oxygen to wing muscles rapidly enough. They receive their oxygen directly from air tubes. *b.* The earthworm has a closed circulatory system. The dorsal and ventral vessels are joined by five pairs of hearts at the anterior and by branch vessels in the rest of the worm.

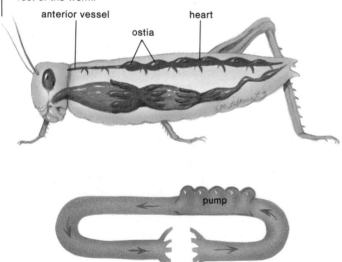

a. Open circulatory system

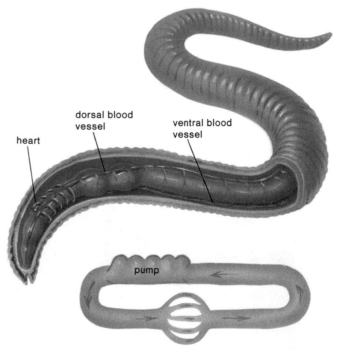

b. Closed circulatory system

Figure 33.3

All vertebrates have a closed circulatory system containing a heart (pumping device), arteries that take blood away from the heart, and veins that take blood to the heart. Capillaries are the smallest of the vessels and exchanges take place across their thin walls. Blood and the fluid that surrounds the cells (tissue fluid) make up an internal environment that remains relatively constant, because the circulatory system transports blood between the body's cells and specialized exchange boundaries with the external environment.

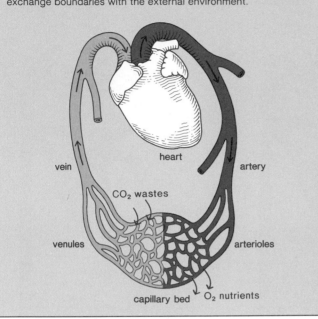

present in all regions of the body; consequently, a cut to any body tissue draws blood. Capillaries are the most important part of a closed circulatory system because exchange of nutrient and waste molecules takes place across their thin walls.

The distribution of blood in the various capillary beds is regulated by **sphincters,** circular muscles that open and close tubular structures. In this instance, when the sphincter relaxes, blood can enter a capillary bed. When the sphincter contracts, a capillary is closed and no blood can enter. Contraction and relaxation of sphincters is controlled by the nervous system. Not all capillary beds are open at the same time. After a person has eaten, the capillary beds around the digestive tract are usually open; during muscular exercise, the capillary beds of the skeletal muscles are open.

Venules and veins collect blood from the capillary beds and take it to the heart. First the **venules** drain the blood from the capillaries, and then they join together to form a *vein.* The wall of a vein is much thinner than that of an artery because the muscle and elastic fiber layers are poorly developed. Valves within the veins point, or open, toward the heart, preventing a backflow of blood (see fig. 33.12).

Comparison of Circulatory Pathways

Among vertebrate animals there are three different types of circulatory pathways. In fishes, blood follows a *one-circuit* (single-loop circulatory) pathway through the body. The heart has a single atrium and a single ventricle (fig. 33.5a). The pumping

Figure 33.4

The three types of vertebrate blood vessels differ in construction. *a.* The artery has a strong muscular wall. Arteries can constrict or dilate in order to help regulate blood pressure and the distribution of the blood. The veins have weak walls, but valves keep the blood moving toward the heart. A capillary has a thin wall of only one layer of cells, which facilitates the exchange of molecules between blood and tissue fluid, the fluid that surrounds the cells. *b.* A scanning electron micrograph of a major artery (MA) and a major vein (MV), illustrating the difference in wall thickness and vessel size. Magnification, ×305.

Kessel, R. G., and Kardon, R. H.: *Tissues and Organs: A Text-Atlas of Scanning Electron Microscopy.* © 1979 by W. H. Freeman and Co.

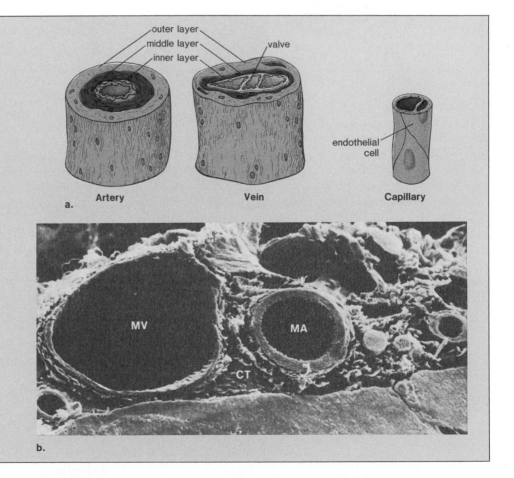

Figure 33.5

Comparison of circulatory systems in vertebrates. *a.* In a fish, the blood moves in a single loop. The heart has a single atrium and ventricle and pumps the blood into the gill region where gas exchange takes place. Blood pressure created by the pumping of the heart is dissipated after the blood passes through the gill capillaries. This is a disadvantage of this single-loop system. *b.* Amphibians have a double-loop system in which the heart pumps blood to both the lungs and the body itself. The system is not very efficient because oxygenated blood and deoxygenated blood mix in the single ventricle. *c.* The pulmonary and systemic systems are completely separate in birds and mammals since the heart is divided by a septum into a right and left half. The right side pumps blood to the lungs and the left side pumps blood to the body proper.

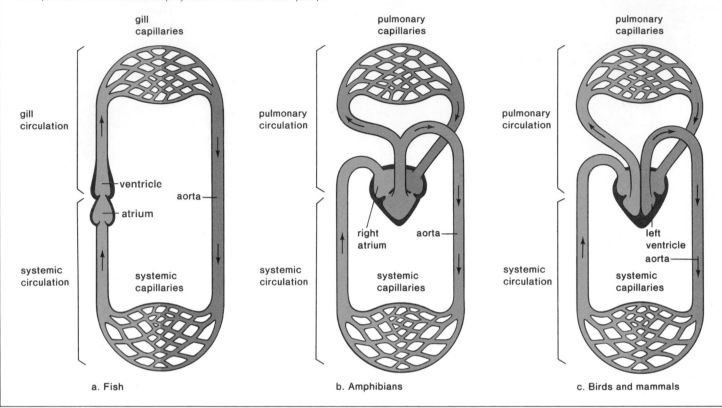

a. Fish b. Amphibians c. Birds and mammals

action of the ventricle sends blood under pressure only to the gills where it is oxygenated. After passing through gill capillaries, there is little blood pressure left to distribute the oxygenated blood to the tissues.

In contrast, the other vertebrates have a *two-circuit* (double-loop circulatory) pathway. The heart pumps blood to the lungs, called **pulmonary** circulation, and pumps blood to the tissues, called **systemic** circulation. This double pumping action assures adequate blood pressure and flow to both circulatory loops.

In amphibians the heart has two atria, but there is only a single ventricle (fig. 33.5b). Therefore, the pulmonary system is not separate from the systemic system and oxygenated blood mixes with deoxygenated blood in the single ventricle. The hearts of other vertebrates are partially (most reptiles) or completely (some reptiles, all birds, and mammals) divided into right and left halves (fig. 33.5c). The right ventricle pumps blood to the lungs and the left ventricle pumps blood to the rest of the body. This arrangement increases the likelihood of adequate blood pressure for both the pulmonary and systemic circulations.

In mammals, birds, and reptiles, the heart is divided into a right side (pumping deoxygenated blood) and a left side (pumping oxygenated blood); this ensures adequate blood pressure for both the pulmonary and systemic circulations.

Human Circulation

Humans, like all other vertebrates, have a closed circulatory system. Blood flows away from the heart through arteries and arterioles to capillary networks, where exchanges between blood and tissue fluid take place. Blood leaves the capillaries and returns to the heart through venules and veins. Like other mammals, humans have both *pulmonary* and *systemic* circulatory pathways (fig. 33.6). Humans also have a typically mammalian four-chambered heart, with right and left atria and right and left ventricles (fig. 33.7).

Blood Pathways

Pulmonary Circulation

All blood entering the right atrium of the heart is deoxygenated. Blood passes from the right atrium into the right ventricle, which then pumps it out through the pulmonary trunk. The pul-

Figure 33.6

Diagram of the human circulatory system indicating the path of blood. In order to trace blood from the right to the left side of the heart, you must begin at the lung capillaries. In order to trace blood from the gut capillaries to the right atrium, you must consider the hepatic portal system and hepatic vein.

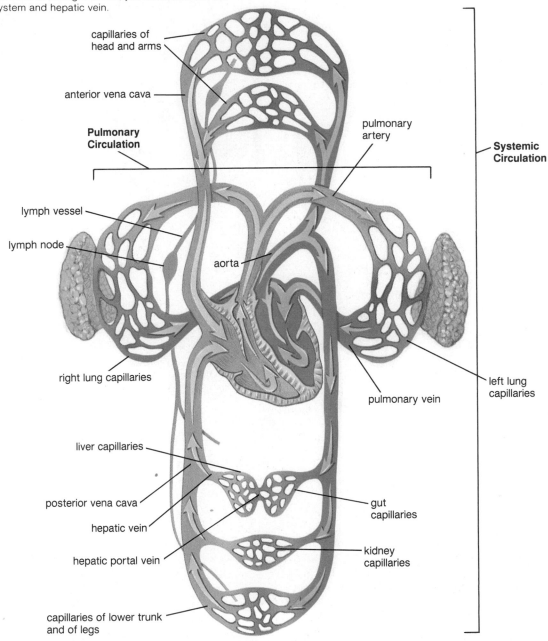

capillaries of head and arms

anterior vena cava

Pulmonary Circulation

pulmonary artery

Systemic Circulation

lymph vessel

lymph node

aorta

right lung capillaries

left lung capillaries

pulmonary vein

liver capillaries

posterior vena cava

hepatic vein

hepatic portal vein

gut capillaries

kidney capillaries

capillaries of lower trunk and of legs

monary trunk branches into the two pulmonary arteries that carry blood to arterioles and capillaries in the lungs. After passing through lung capillaries located around the alveoli (see fig. 36.8), blood returns to the left atrium of the heart through *pulmonary venules* and *veins*.

Systemic Circulation

Blood returning from the pulmonary circulation is oxygenated. This oxygenated blood passes from the left atrium into the left ventricle, which then pumps it out through the **aorta,** the large arterial trunk that supplies the entire systemic circulation (fig. 33.7).

The aorta sends branches to all parts of the body. The first branch is to the coronary arteries that are a part of a **coronary** circulation system for the heart. Heart cells do not exchange material with the blood being pumped through the chambers; therefore, blood flow through the coronary arteries is critical for normal functioning of the heart. Blockage of any coronary artery quickly results in a "heart attack" with resultant damaged heart muscle and impaired heart function.

Anterior to the heart, the aorta arches to the left and then passes posteriorly through the body. Branches off the arch of the aorta supply blood to the upper parts of the body. As the

Figure 33.7

Internal view of human heart. Vena cavae bring deoxygenated blood to right atrium and the right ventricle pumps it to lungs by way of pulmonary arteries. Pulmonary veins bring oxygenated blood to the left atrium and the left ventricle pumps it into the aorta. Note that the atrioventricular valves are supported by strings that prevent them from inverting into the atria when the ventricles are filled with blood.

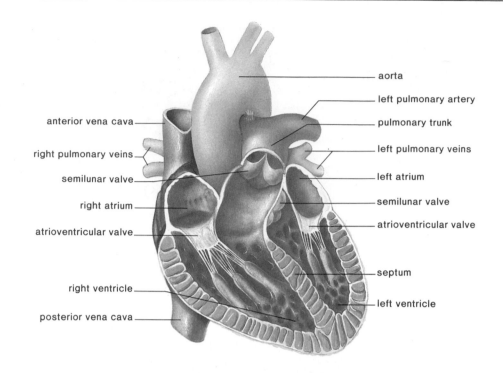

aorta

left pulmonary artery

pulmonary trunk

left pulmonary veins

left atrium

semilunar valve

atrioventricular valve

septum

left ventricle

anterior vena cava

right pulmonary veins

semilunar valve

right atrium

atrioventricular valve

right ventricle

posterior vena cava

William Harvey

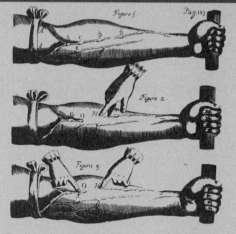

William Harvey (1578–1657) was the first to offer proof that the blood circulates in the body of humans and other animals. He was an English scientist of the seventeenth century, a time of renewed interest in the collection of facts, use of the hypothesis, experimentation, and respect for mathematics. The seventeenth century was the time of the scientific revolution.

After many years of research and study, Harvey hypothesized that the heart is a pump for the entire circulatory system and that blood flows in a circuit. In contrast to former anatomists, Harvey dissected not only dead but also live organisms and observed that when the heart beats, it contracts, forcing blood into the aorta. Had this blood come from the right side of the heart? To do away with the complication of the lungs (pulmonary circulation), Harvey turned to fishes and noticed that the heart first received and then pumped the blood forward. He observed that blood in the fetus passes directly from the right side of the heart through the septum to the left side. He felt confident that in mature, higher organisms, all blood moves from the right to the left side of the heart by way of the lungs.

Harvey then wanted to show an intimate connection between the arteries and veins in the tissues of the body. Again, using live lower

To show that the veins return blood to the heart, Harvey tied a ligature (*Figure 1*) above the elbow to observe the accumulation of blood in the veins; (*Figure 2*) blood can be forced past a valve from H to O, but (*Figure 3*) not in the opposite direction. Therefore it can be deduced that blood ordinarily moves toward the heart in the veins.

organisms, he demonstrated that if an artery is slit, the whole blood system empties, including arteries and veins. He measured the capacity of the left ventricle in humans and found it to be 2 ounces. Since the heart beats 72 times a minute, in one hour the left ventricle will force into the aorta no less than $72 \times 60 \times 2 = 8,640$ ounces = 540 pounds, or three times the weight of a heavy man! Could so much blood be created and consumed every hour? The same blood must return again and again to the heart.

Harvey also studied the valves in the veins and suggested their true purpose. By the use of ligatures (see the illustration), he demonstrated that a tight ligature on the arm causes the artery to swell on the side of the heart, a slack ligature causes the vein to swell on the opposite side. He said, "This is an obvious indication that the blood passes from the arteries into the veins . . . and there is an anastomosis of the two orders of vessels."

Harvey's methods showed how fruitful research might be done. He established that physical and mechanical evidence could provide data for a theory of circulation. However, he erred when he speculated on the function of the heart and lung. He thought the heart heated the blood, and the lung served to cool it or control the degree of heat. His basic method, however, contributed to the scientific revolution and set an example for others to follow.

Figure 33.8
Cardiac muscle tissue. *a.* Light micrograph showing its striated
appearance. Intercalated disks (see arrows) are present whenever
two cardiac muscle cells meet. *b.* Electron micrograph of disk.

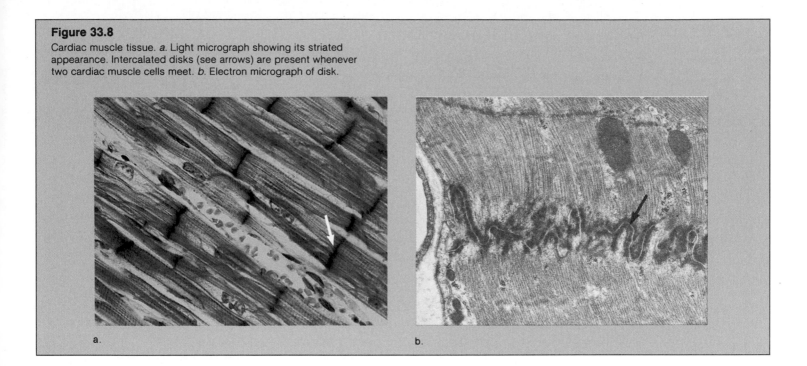

a.

b.

aorta descends through the abdominal cavity, branches are given off to the digestive organs, the kidneys, the body wall, the legs, and other posterior parts.

Blood returns from the systemic circulation to the right atrium of the heart by way of two large veins, the anterior (superior) **vena cava** and the posterior (inferior) vena cava. The anterior vena cava returns blood from the head, arms, and chest; the posterior vena cava returns blood from the remainder of the systemic circulation.

Portal System

A *portal* system is one that begins and ends in capillaries. The human body has only one such major system, the *hepatic portal system* (fig. 35.14). In this system, the first set of capillaries occurs at the digestive organs and the second occurs in the liver. Blood passes from the capillaries about the digestive organs into venules that join a major vein (the *hepatic portal vein*) that takes the products of digestion to the liver. Here, these products may be processed or stored until they are needed to maintain the constancy of blood composition within the *hepatic vein,* a vessel that leaves the liver to enter the vena cava.

The Human Heart

The human heart (fig. 33.7) is a pumping device that is remarkable for its efficiency and durability. Its regular and continual beating is essential to life; the most obvious indication of death is cessation of the heartbeat. Even though some other organs can survive brief pauses in circulatory supply, the heart can never stop its regular beating for even a short time because of the sensitivity of the brain to the lack of oxygen. If circulation to the brain stops for even five seconds, consciousness is lost. A four-minute stoppage of circulation to the brain results in the death of a large number of brain cells, and a nine- to ten-minute circulatory failure causes irreversible and massive brain damage leading usually to death.

During each minute of life, a normal, resting person's heart beats about seventy times and pumps out about five liters of blood, a volume approximately equal to the total blood volume in the body. All this work is done by a relatively small organ. The heart of a highly trained athlete weighs about 1% of total body weight but the average human heart weighs only 300 g, less than 0.5% of total body weight. The heart sustains its heavy work load and yet it never rests for more than a second. It keeps beating continuously from the time of embryonic life to the time of death.

Control of the Heartbeat

The contraction of the heart is intrinsic to the heart; it does not require outside nervous stimulation. After individual cardiac muscle cells are dissected from the heart, they still can beat spontaneously and rhythmically. These cells are **striated,** having both light and dark bands, and are branched (fig. 33.8*a*). When cardiac muscle cells mesh within the heart, they form the **myocardium,** the muscular wall of the heart. Where cardiac muscle cells meet, there is a band known as an *intercalated disk* (fig. 33.8*b*). Here portions of the two cell membranes fuse, and the excitation causing muscular contraction can be transmitted from one cell to another.

The heartbeat is normally regulated by the **pacemaker,** a small node of myocardial cells termed the **SA (sinoatrial) node.** The SA node, embedded in the right atrial wall (fig. 33.9), initiates a wave of excitation every 0.85 seconds. This wave of excitation spreads quickly throughout the atria, which contract while the ventricles are still resting. After the wave of excitation

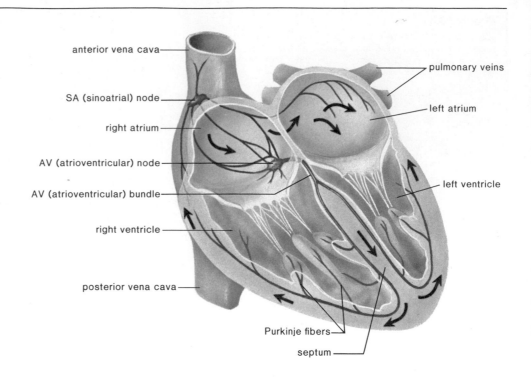

anterior vena cava

pulmonary veins

SA (sinoatrial) node

left atrium

right atrium

AV (atrioventricular) node

AV (atrioventricular) bundle

left ventricle

right ventricle

posterior vena cava

Purkinje fibers

septum

Table 33.1 The Heart Cycle		
Time	**Atria**	**Ventricles**
0.15 sec.	Systole	Diastole
0.30 sec.	Diastole	Systole
0.40 sec.	Diastole	Diastole

Figure 33.10
Close-up view of closed semilunar valves. Refer to figure 33.7 for their anatomical position in the heart.

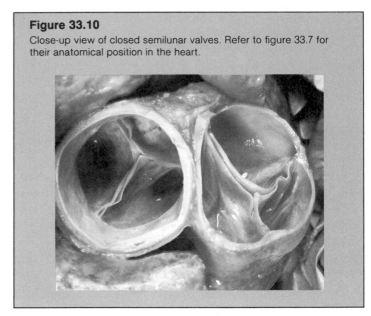

makes contact with the **AV** (atrioventricular) node, a second node located at the base of the right atrium, the ventricles contract. While the ventricles contract, the atria rest. Following ventricular systole, the entire heart rests. Notice in table 33.1, which gives the time sequences for these events, that contraction of a chamber is called **systole,** while relaxation is called **diastole.**

The heartbeat sounds, "lubb-dup," are caused by the closing of valves, first those that conduct blood from the atria to the ventricles (atrioventricular valves), then those that conduct blood from the ventricles to their respective arteries (semilunar valves) (fig. 33.10). These valves aid circulation of the blood through the heart because they permit only one-way flow and do not allow a backward flow of blood. The *pulse* is a wave of vibration that passes down the walls of the arterial blood vessels when the aorta expands following ventricle systole. Because there is one arterial pulse per ventricular systole, the arterial pulse rate can be used to determine the heart rate.

Although the beat of the heart is intrinsic, it may still be speeded up or slowed down by nervous stimulation. A heart rate center in the medulla oblongata portion of the brain alters the heartbeat by way of the autonomic nervous system. The latter is made up of two divisions: the parasympathetic system promotes normal activities, and the sympathetic system instigates responses associated with times of stress. The parasympathetic system causes the heartbeat to slow down, and the sympathetic system increases the heartbeat. Various factors, such as the relative need for oxygen or the level of blood pressure, determine which of these systems is activated.

Systemic Blood Flow

When the left ventricle contracts, blood is forced into the arteries under pressure. *Systolic pressure* results from blood being forced into the arteries during ventricular systole, and *diastolic pressure* is the pressure in the arteries during ventricular diastole. Human blood pressure is measured with a sphygmomanometer, which has a pressure cuff that permits measurement

Figure 33.11

Blood pressure in different parts of the human circulatory system. The pulse pressure (includes systolic and diastolic pressures) decreases as blood flows into smaller arteries and disappears as blood enters capillaries. The slow and even movement of the blood through the capillaries facilitates exchange of molecules there. The blood pressure in the veins is so low that it cannot account for the movement of blood in these vessels. Skeletal muscle contraction pushes the blood along from valve to valve in the veins.

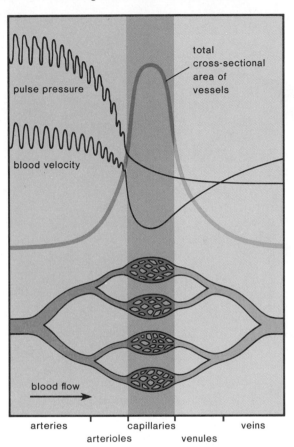

of the amount of pressure required to stop the flow of blood through an artery. Blood pressures normally are measured on the brachial artery, an artery of the upper arm, and are stated in millimeters of mercury (Hg). A blood pressure reading consists of two numbers, for example, 120/80, that represent systolic and diastolic pressures, respectively.

Blood flows from the left ventricle, the region of highest pressure, to the right atrium, the region of lowest pressure. Therefore, pressure is highest in the first part of the aorta, where it is usually about 140/120. Measurements of blood pressure at other various points along the arteries would show a progressive decrease in blood pressure (fig. 33.11). In other words, as blood flows through the aorta and then enters into smaller arteries and arterioles, pressure falls. Also, the difference between systolic and diastolic pressure at first decreases and then finally disappears in the capillaries, where there is but a even, steady flow of blood.

Blood pressure in the venules is very low (5 to 10 mm Hg) and finally falls to 2 mm Hg or less in the vena cava. The very low blood pressure in the veins is not sufficient to move blood back to the heart, especially from the limbs of the body. When skeletal muscles near veins contract, they put pressure on the collapsible walls of the veins and on the blood contained in these vessels. Veins, however, have *valves* (fig. 33.12) that prevent backward flow, and therefore pressure from muscle contraction is sufficient to help move blood through veins toward the heart. When we are inactive, blood moves very sluggishly through veins in certain parts of the body and can collect in the veins. For example, blood accumulates in the veins of the legs during long periods of standing.

In the human circulatory system, the beat of the heart supplies the pressure that keeps the blood moving in the arteries; pressure drops off in the capillaries; and skeletal muscle contraction pushes the blood in the veins back to the heart.

Figure 33.12

Cross section of a valve. *a.* When a valve is open, a result of pressure on the veins exerted by skeletal muscle contraction, the blood flows toward the heart in veins. *b.* Valves close when external pressure is no longer applied to them. Closure of the valves prevents the blood from flowing in the opposite direction.

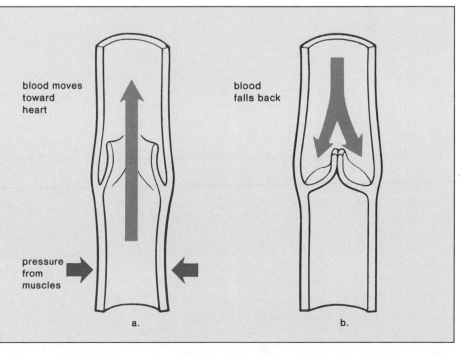

Circulatory System   **519**

Coronary Heart Disease

E ven though the United States has recently seen a decline in deaths due to cardiovascular disease, more than 800,000 people suffered heart attacks during 1987, and of these only 300,000 survived. It seems desirable, then, to identify those factors that predispose one to developing coronary heart disease.

A physician can make two routine measurements in the office that will indicate if a person is at greater risk of a heart attack than average. He or she can measure the patient's resting blood pressure, and if it is 160/95 or higher, there is reason for concern, and antihypertensive drugs can be prescribed. A physician can also measure the level of cholesterol in the blood. The Natonal Heart, Lung, and Blood Institute (NHLBI), which believes that all Americans over the age of 20 should be tested at least every five years, has recently set cholesterol-level standards for adults. If the cholesterol level is 200–239 mg/100 ml the person should take precautionary measures and be retested annually. If the level is greater than 240 mg/100 ml then more refined cholesterol testing is needed immediately.

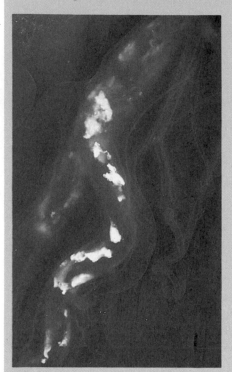

a.

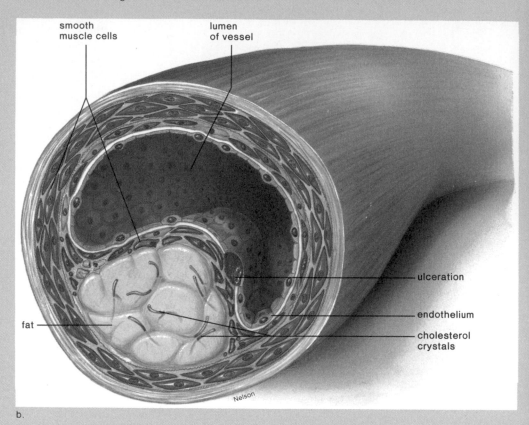

b.

a. Plaques (*yellow*) in the coronary artery of a heart patient. *b.* Cross section of an artery that is partially blocked by plaque. The plaque bulges out into the lumen of the artery, obstructing blood flow. *b.* From Kent M. Van De Graaff and Stuart I. Fox, *Concepts of Human Anatomy and Physiology*, 2d ed. Copyright © 1989 Wm. C. Brown Publishers, Dubuque, Iowa. All Rights Reserved. Reprinted by permission.

The Lymphatic System

The lymphatic system is a one-way system, rather than a circulatory system, and contains lymph veins and capillaries only. The system begins with lymph capillaries that lie near the blood capillaries. They take up any excess *tissue fluid*. Once tissue fluid enters lymph capillaries, it is called **lymph.**

Lymph vessels have a construction similar to that of cardiovascular veins, including valves. Lymph in humans moves through the lymph capillaries and vessels as a result of pressure applied by muscle contractions near the vessels. The valves prevent a backward flow of lymph. Lymph moves through smaller vessels that unite to form larger vessels, and finally, two major lymph ducts empty into large blood circulatory veins of the shoulders near the heart.

In addition to returning fluids from the tissues to the blood circulatory system, the lymphatic (fig. 33.13) system has several other important functions. Lymph vessels known as **lacteals** are present within the intestinal villi. The products of fat digestion enter the lacteals and are carried in lymph vessels to eventually enter the circulatory system.

Lymph nodes are small ovoid or round structures composed of lymphoid tissue. These nodes occur at certain strategic points along the medium-sized lymph vessels. **Lymphocytes,** a type of white blood cell, are packed into the spaces of a lymph node, which also filters and traps bacteria and other debris, helping keep the blood purified. When a local infection, such as a sore throat, is present, lymph nodes in that region may swell and become painful.

Cholesterol is a substance found within the cell membrane of all cells. Some of it is converted to hormones by certain glands. We now know that cholesterol is ferried in the bloodstream by either low-density lipoproteins (LDL) or high-density lipoproteins (HDL). LDL takes cholesterol to the tissues, and HDL transports cholesterol out of the tissues. The cholesterol-LDL molecular combination is atherogenic—it leads to atherosclerosis. When these molecules adhere to arterial walls a series of events occur that eventually result in an accumulated mass known as plaque (see the figure). This plaque can eventually grow so large that it hinders or even stops blood flow through the artery. A plaque also can break open and produce blood clots.

The NHLBI recommends that a person with a cholesterol level of over 240 mg/100 ml should be further tested to determine the level of LDL cholesterol. Those with an LDL level of 160 mg/100 ml and above are considered at high risk, and those between 130 and 159 mg/100 ml also require treatment. First and foremost, treatment consists of adopting a diet that is low in saturated fat and cholesterol (see table). Although the prescribed diet does not lower blood cholesterol in all persons, it is expected to do so for most individuals.

If diet alone does not bring down the cholesterol level, drugs can be prescribed. The drugs cholestyramine and colestipol act in the intestines to remove bile, which is derived in part from cholesterol, and a new drug, lavastatin, inhibits the production of cholesterol in the liver. These drugs do reduce the blood level of cholesterol, but the long-term side effects are not completely known and may be serious.

Aside from the presence of hypertension and an elevated blood cholesterol level, certain behaviors are also associated with the development of coronary heart disease. These are smoking cigarettes and excessive weight gain. Smoking cigarettes has been found to be a major contributor to the development of coronary heart disease, and filter cigarettes carry as great a risk as the nonfiltered brands. Excessive weight is accompanied by both increased blood pressure and atherosclerosis. Since it is very difficult for obese individuals to lose weight, it is recommended that weight control begin before it gets out of hand. No doubt, the prescribed low-cholesterol diet (see table) would be helpful.

Investigators have identified two behaviors that may help *reduce* the risk of heart attack and stroke. Exercise seems to be critical. Sedentary individuals have a risk of coronary heart disease that is about double those who are very active. One investigator, for example, recommends that his patients walk for one hour, three times a week. Reduction of stress is also desirable. The same investigator recommends meditating and doing yogalike stretching and breathing exercises every day to reduce stress.

Genetics also seem to play a role in the development of coronary heart disease—the disease tends to run in families. Although nothing can be done about this factor, the other factors can be controlled if the individual believes it is worth the effort.

Diet to Reduce Cholesterol Levels

EAT LESS	EAT MORE
Fatty meats	Fish, poultry, lean cuts of meat
Organ meats (liver, kidney, brain, pancreas), shrimp	
Sausage, bacon, processed meats	Fruit, vegetables, cereals, starches
Whole milk	Skim or low-fat milk
Butter, hard margarine	Soft margarine
High-fat cheese (bleu, cheddar)	Low-fat cheese
Ice cream and cream	Yogurt
Egg yolks	Egg whites
Animal fats and certain saturated oils	Vegetable fats
Commercial baked goods	Oat bran and other soluble fibers

Reprinted by permission from the 1986 *Medical and Health Annual*, copyright 1985, Encyclopaedia Britannica, Inc., Chicago, Illinois.

Along with lymphatic vessels lymph nodes are sometimes removed in cancer operations because lymph vessels may help spread any remaining cancer cells. However, this may result in *edema,* a swelling of a portion of the body caused by an accumulation of excess tissue fluid. An extreme example of edema is seen in elephantiasis when parasitic roundworms block lymphatic drainage.

The lymphatic system is a one-way system of vessels that returns excess tissue fluid to the blood circulatory system. The lymph nodes assist the infection-fighting capacity of the body.

Human Blood

Blood has two main portions: the liquid portion, called plasma, and cells. **Plasma** contains many types of molecules, including nutrients, wastes, salts, and proteins. The salts and proteins buffer the blood, effectively keeping the pH near 7.4. They also maintain blood's osmotic pressure so that water has an automatic tendency to enter blood capillaries.

Types of Blood Cells

The cells are of three types: red blood cells, or **erythrocytes;** white blood cells, or **leukocytes;** and **platelets,** or **thrombocytes.** Erythrocytes are so small that a cubic millimeter (mm^3) of blood contains five million of them (table 33.2). Each one is so packed

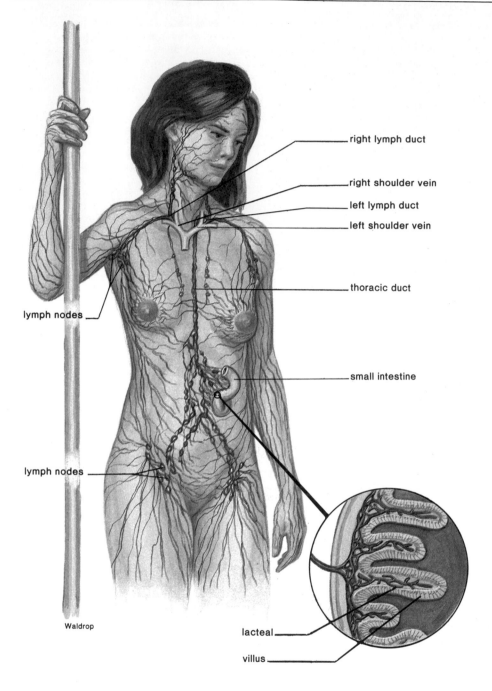

Figure 33.13

The lymphatic system. Lymph vessels flow into two main ducts; the left lymph duct drains lymph vessels from all parts of the body, except the upper right portion. Vessels from that area drain through the right lymph duct. These two large lymphatic ducts drain into the veins of the shoulder, thus returning fluid from the tissues to the circulatory system. Lymph vessels in the region of the small intestine are called lacteals. The lymph nodes filter debris and bacteria out of lymph.

right lymph duct

right shoulder vein

left lymph duct

left shoulder vein

thoracic duct

lymph nodes

small intestine

lymph nodes

Waldrop

lacteal

villus

with hemoglobin that there is no room for cellular organelles, not even a nucleus. Hemoglobin inside erythrocytes reversibly binds with oxygen to form *oxyhemoglobin.* Actually, oxygen combines with iron (Fe) found in the heme portion of hemoglobin (fig. 4.18*b*). When most of its hemoglobin is in the form of oxyhemoglobin, blood takes on a bright red color. Oxygenated blood, carrying oxyhemoglobin, is found in pulmonary veins and systemic arteries. Hemoglobin that is not carrying oxygen is called *reduced hemoglobin,* and deoxygenated blood, carrying reduced hemoglobin, is a darker, more purplish color. Deoxygenated blood is found in systemic veins and pulmonary

arteries. Thus, it is not correct to indicate *all* arteries by using a red color and *all* veins by using a bluish color since it is just the reverse in the pulmonary system.

Red blood cells are manufactured in the *red bone marrow* at the rate of two million cells per second. They live about four months, after which time they tend to become damaged from squeezing through small capillaries. Damaged erythrocytes are withdrawn from circulation as blood passes through the liver or spleen. The red blood cells are destroyed and hemoglobin is released. The iron is recovered and returned to the red bone marrow for reuse. The heme portions of the molecules undergo chemical degradation and are excreted by the liver as bile pigments in the bile. The bile pigments are primarily responsible for the color of feces.

Table 33.2 Numbers and Distribution of Cells in Normal Human Blood

Component	Number (per mm³ of blood)
Erythrocyte (red blood cell)	5,000,000
Leukocyte (white blood cell)	7,000
Thrombocytes (platelets)	200–400

White Blood Cell Types	% of Total White Blood Cells
Granular leukocytes	
Neutrophils	55–65
Eosinophils	2–3
Basophils	0.5
Agranular leukocytes	
Lymphocytes	25–33
Monocytes	4–7

Sometimes the mechanisms that normally balance erythrocyte production and destruction fail. This can result in *anemia,* a condition in which there are fewer than the normal number of erythrocytes (or, in some cases, lower hemoglobin content in the erythrocytes). Patients who have anemia often complain of being fatigued or chilled because cellular respiration is minimal due to a lack of oxygen. Fundamental physiological problems can cause anemia; therefore, it is important to identify and correct the specific underlying problem so that treatment is adequate.

White blood cells (fig. 33.14 and table 33.2) can be distinguished microscopically from red blood cells because they are usually larger, have a nucleus, and, without staining, appear white in color. With staining, white blood cells appear bluish and may have granules that may bind with certain stains. The latter type of white blood cell, called *granulocytes,* has a lobed nucleus and is of three types: **neutrophils** have granules that do not take up a dye; eosinophils have granules that take up the red dye eosin; and basophils have granules that take up a basic

Figure 33.14

Components of blood. *a.* Human blood cells as they appear when using the light microscope. *b.* Artist's representation of blood cells being released from a cut blood vessel. You can easily make out the biconcave red blood cells and the fuzzy looking white blood cells. *c.* Table outlining the function and source of components of blood.

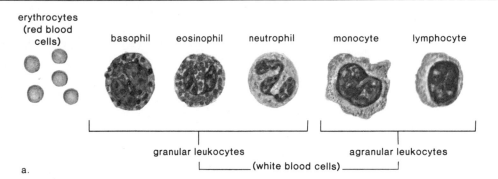

erythrocytes (red blood cells)   basophil   eosinophil   neutrophil   monocyte   lymphocyte   thrombocytes (platelets)

granular leukocytes   agranular leukocytes

(white blood cells)

a.

b.

c.

Blood	Function	Source
I. Cells		
Red blood cells	Transport oxygen	Bone marrow
White blood cells	Fight infection	Bone marrow and lymphoid tissue
Platelets	Clotting	Bone marrow
II. Plasma*		
Water	Maintains blood volume and transports molecules	Absorbed from intestine
Proteins	Maintains blood osmotic pressure and pH	
Albumin	Transport	Liver
Fibrinogen	Clotting	Liver
Globulins	Fight infection	Lymphocytes
Gases		
Oxygen	Cellular respiration	Lungs
Carbon dioxide	End product of metabolism	Tissues
Nutrients		
Fats, glucose, amino acids, etc.	Food for cells	Absorbed from intestinal villi
Salts	Maintain blood osmotic pressure and pH; aid metabolism	Absorbed from intestinal villi
Wastes		
Urea and ammonia	End products of metabolism	Tissues
Hormones, vitamins, etc.	Aid metabolism	Varied

Plasma is 90%-92% water, 7%-8% plasma proteins, and not quite 1% salts. All other components are present in even smaller amounts.

dye that stains them a deep blue color. *Agranulocytes*, or agranular leukocytes, white blood cells that have no granules, have a circular or indented nucleus; there are two types—the larger ones are called monocytes and the smaller ones **lymphocytes.**

When bacteria or viruses enter the body by way of a cut, a response called the inflammatory reaction occurs. Damaged tissue releases kinins that cause vasodilation and histamines that cause increased capillary permeability. Blood rushes to the area, and the neutrophils, which are amoeboid, squeeze through the capillary wall and enter the tissue fluid where they phagocytize foreign material. The thick yellowish fluid **pus** contains a large proportion of dead neutrophils. Later, monocytes also appear on the scene and are transformed into **macrophages** (fig. 34.1), large phagocytizing cells that often engulf material. The speed with which some macrophages consume bacteria has been timed at less than one-hundredth of a second.

The presence of bacteria and viruses in the body also causes lymphocytes to produce globulin proteins called antibodies. Each lymphocyte produces just one type of antibody, which is specific for one type of antigen. **Antigens,** which are most often proteins, but sometimes polysaccharides, are found in the outer covering of a parasite or are present in its toxin. **Antibodies** combine with their antigens in such a way that the antigens are rendered harmless. An individual is *actively immune* when the body has specific antibodies that can react to a disease-causing antigen. Chapter 34 deals with immunity and explores this topic in detail.

Capillary Exchange

No cell is far away from one or even several capillaries because capillaries are found throughout every tissue in the body. There is such a complex network of capillaries in various tissues that we can't begin to even imagine how many there are much less get an accurate count of the total number of capillaries. For example, an area of 1 mm² of guinea pig muscle during maximal exercise may contain more than 3,000 open capillaries. The lead in an ordinary pencil has a cross-sectional area of about 3 mm². Therefore a piece of muscle of that same size during maximal exercise would contain close to 10,000 open capillaries.

Because the cross-sectional area of capillaries is so large compared to the arterioles, blood flows very slowly through the capillaries. This slow rate allows adequate time for exchange of materials between blood and tissue fluid. Much of this exchange occurs by *diffusion* through one-cell thick capillary walls. While lipid-soluble substances can pass freely through the plasma membrane of the cells of the capillary walls, water-soluble substances usually diffuse through pores present in junctions between the cells of the capillary walls. Two forces control movement of fluid through the capillary wall: *osmotic pressure,* which tends to cause water to move from tissue fluid to blood; and *blood pressure,* which tends to cause water to move in the opposite direction. The osmotic pressure remains constant at about 25 mm Hg, but blood pressure is about 40 mm Hg at the arteriole end and only 10 mm Hg at the venule end of a capillary.

As figure 33.15 illustrates, blood pressure is higher than osmotic pressure at the arterial end of a capillary. This tends to force fluid out through the pores in the capillary wall. Midway along the capillary, where blood pressure is lower, the two forces essentially cancel one another, and there is no net movement of water. However, solutes now diffuse according to their concentration gradient; nutrients (glucose and oxygen) diffuse out of the capillary, and wastes (ammonia and carbon dioxide) diffuse into the capillary. Since proteins are too large to pass out of a capillary, *tissue fluid* tends to contain all components of plasma except proteins. At the venule end of a capillary, where the blood pressure has fallen to 10 mm Hg, osmotic pressure is greater and water tends to move into the capillary. Almost the same amount of fluid that left the capillary returns to it, but not quite. There is always some excess tissue fluid collected by the lymph capillaries.

Blood Clotting

Blood must be a free-flowing liquid if it is to circulate easily through blood vessels, but this liquidity can also cause serious problems. Any injury that breaks a large blood vessel can quickly lead to a serious loss of blood. This is countered by a complex *clotting* (coagulation) mechanism. Clots form temporary barriers to blood loss until a vessel's walls have healed.

When a blood vessel is injured, blood *platelets* begin to congregate near the cut or injury, forming a barrier known as the platelet plug. There are 200 to 400 platelets per mm³ of blood. Platelets are fragments of large bone marrow cells, called megakaryocytes, that disintegrate and discharge them into the bloodstream. When platelets come into contact with an injured vascular wall, they swell up, become sticky, and release certain chemicals. Some of these chemicals stimulate the blood vessel to constrict; some increase the tendency of platelets to form a plug; and some help initiate the process of blood clotting.

Blood clotting requires a complex and precise series of reactions that satisfy the requirement for a delicate balance between quick and efficient clot formation as a response to significant blood vessel damage and prevention of accidental formation of clots that interfere with normal circulation.

Since blood clotting is so complex, we will consider only the major events involved:

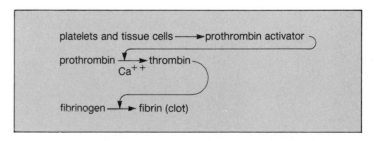

Prothrombin and fibrinogen are two proteins manufactured by the liver that are always present in the plasma. Injured tissues and platelets release *prothrombin activator* and this along with

Figure 33.15
Diagram of a capillary illustrating the exchanges that take place and the forces that aid the process. At the arterial end of a capillary, the blood pressure is higher than the osmotic pressure; therefore water, oxygen, and glucose tend to leave the bloodstream. At the venous end of a capillary, the osmotic pressure is greater than the blood pressure and, therefore, water, ammonia, and carbon dioxide tend to enter the bloodstream. Notice that the red blood cells and plasma proteins are too large to exit from a capillary.

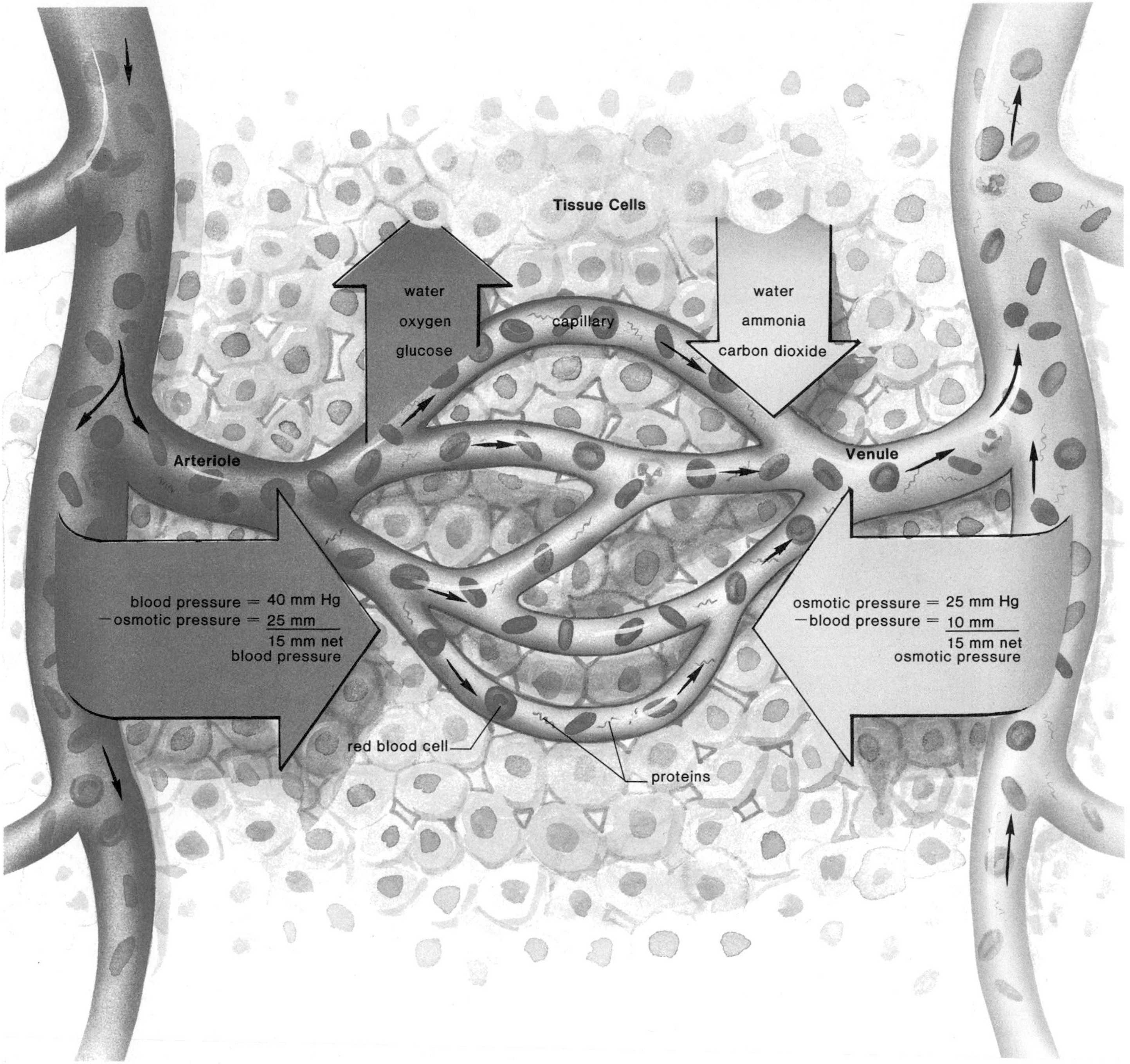

Tissue Cells

water
oxygen
glucose

capillary

water
ammonia
carbon dioxide

Arteriole

Venule

blood pressure = 40 mm Hg
−osmotic pressure = 25 mm
15 mm net
blood pressure

osmotic pressure = 25 mm Hg
−blood pressure = 10 mm
15 mm net
osmotic pressure

red blood cell

proteins

calcium ions (Ca^{++}) convert *prothrombin* into the enzyme thrombin. *Thrombin* in turn, acts as an enzyme that severs two short amino acid chains from each *fibrinogen* molecule. These activated fragments then join end to end, forming long threads of fibrin. *Fibrin* threads wind around the platelet plug in the damaged area of the blood vessel and provide the framework for the clot. Red blood cells are also trapped within the fibrin threads (fig. 33.16) and their presence makes a clot appear red. After the clot has formed, it shrinks, releasing a fluid. If clotting has occurred in a test tube (fig. 33.17), this fluid, called *serum,* is easily collected. Serum has the same composition as plasma, except that it lacks prothrombin and fibrinogen.

Figure 33.16

A scanning electron micrograph showing a red blood cell caught in the fibrin threads of a clot. Fibrin threads form a mesh that catches many blood cells. Magnification, ×3,000.

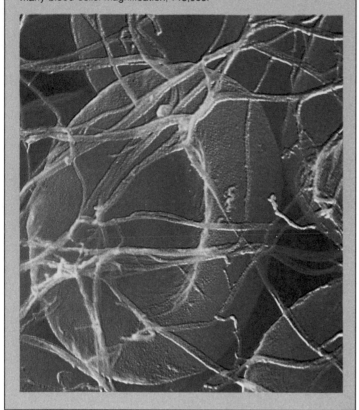

Blood Typing

ABO Grouping

Although there are at least twelve well-known blood-type identification systems, the *ABO grouping* and the *Rh* factor are most often used to determine blood type. Before the twentieth century, blood transfusions had been attempted, although sometimes the results were dire and even caused the death of a recipient. Concerned about these occurrences, a newly established physician of Vienna, Karl Lansteiner, began to examine the effect of mixing different samples of blood in the late 1800s. After months of laboratory research, he determined that there

Figure 33.17

When a blood clot contracts, serum is released. The clot is then a solid plug that prevents further blood loss.

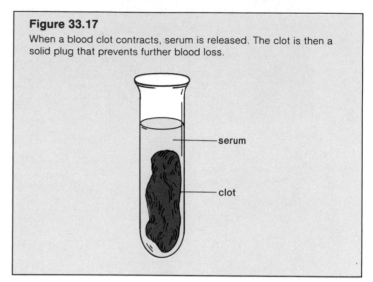

Figure 33.18

The standard test to determine ABO and Rh blood type consists of putting a drop of anti-A antibodies, anti-B antibodies, and anti-Rh antibodies on a slide. To each of these a drop of the person's blood is added. *a.* A possible reaction to any one of these is shown. Top, no reaction; therefore, no antigen to that antibody is present. Bottom, agglutination occurs; therefore, antigen is present. *b.* Several possible results. For example, in the top row there is no agglutination with either anti-A or anti-B, but there is agglutination with anti-Rh. Therefore, the person's blood type is O⁺.

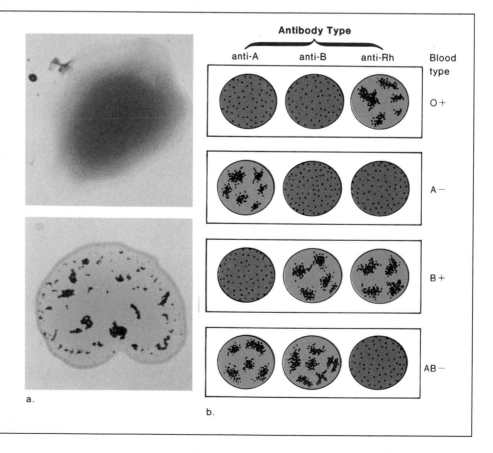

a.

b.

are four major blood groups among humans. He and his associates designated them as A, B, AB, and O. The ABO system, of course, still uses these same designations today. It is now known that blood type is based on a type of glycoprotein present on the red blood cells. Type A blood has A-type glycoproteins on the red blood cells; type B blood has B-type glycoproteins; type AB has both of these glycoproteins; and type O has neither of them. The A and B glycoproteins function as antigens; that is, they combine specifically with antibody molecules. When this particular antigen-antibody reaction occurs, the red blood cells clump or *agglutinate* (see fig. 33.18*a*).

When they are present, anti-A and anti-B antibodies are found in the plasma. Table 33.3 shows which type antibody is present in the plasma of which type of blood. For a donor to give blood to a recipient, the recipient must not have an antibody that would cause the donor's cells to clump. For this reason, it is important to determine each person's blood type. Figure 33.18 demonstrates a way to use the antibodies derived from plasma to determine the type of blood. If clumping occurs after a sample of blood is exposed to a particular antibody, the person has that type of blood.

Rh Factor

Another important antigen used in matching blood types is the *Rh factor*. Persons with this particular antigen on the red blood cells are Rh+ (Rh positive); those without it are Rh− (Rh negative). Rh-negative individuals do not normally make antibodies to the Rh factor, but they will make them when exposed to the Rh factor. It is possible to extract these antibodies and use them for blood-type testing.

The Rh factor is particularly important during pregnancy. If the mother is Rh negative and the father is Rh positive, the child may be Rh positive. The Rh-positive red blood cells begin leaking across into the mother's circulatory system as placental tissues normally break down before and at birth. This causes the mother to produce Rh antibodies. If the mother becomes pregnant with another Rh-positive baby, Rh antibodies (but not anti-A antibodies and anti-B antibodies) may cross the placenta and cause destruction of the child's red blood cells. This is called fetal erythroblastosis. This problem is solved by giving Rh-negative women an Rh-immune globulin injection just after the birth of an Rh-positive child. This injection contains Rh antibodies that attack the baby's red blood cells before these cells stimulate the mother's immune system to produce its own antibodies.

Human blood consists of liquid plasma, which transports nutrient and waste molecules, and cells. Red blood cells carry oxygen and white blood cells fight infection. Blood clotting involves the platelets, and blood typing is based on the types of antigens in the plasma membrane of the red blood cells.

Table 33.3 Blood Groups

Blood Type	Antigen or Red Blood Cells	Antibody in Plasma	% U.S.* Black	% U.S.* Caucasian
A	A	Anti-B	25	41
B	B	Anti-A	20	7
AB	A,B	None	4	2
O	None	Anti-A and anti-B	51	50

*Blood type frequency for other races is not available.

Summary

1. Some invertebrates do not have a transport system. The presence of a gastrovascular cavity allows diffusion alone to supply the needs of cells in cnidarians and flatworms.

2. Other invertebrates do have a transport system. Insects have an open system and earthworms have a closed one.

3. Vertebrates have a closed system in which arteries carry blood away from the heart to capillaries, where exchange takes place, and veins carry blood to the heart.

4. Fishes have a single circulatory loop because the heart with a single atrium and ventricle pumps blood only to the gills. The other vertebrates have both a pulmonary and a systemic circulation. Amphibians have two atria but a single ventricle. Birds and mammals, including humans, have a heart with two atria and two ventricles, in which oxygenated blood is always separate from deoxygenated blood.

5. The heartbeat in humans begins when the SA node (pacemaker) causes the two atria to contract, sending blood through the atrioventricular valves to the two ventricles. The AV node causes the two ventricles to contract, sending blood through the semilunar valves to the aorta. Now all chambers rest. The heart sounds, lubb-dup, are caused by the closing of the valves.

6. Blood pressure created by the beat of the heart accounts for the flow of blood in the arteries, but skeletal muscular contraction is largely responsible for the flow of blood in the veins, which have valves preventing a backward flow.

7. The lymphatic system is a one-way system. Lymphatic capillaries collect excess tissue fluid that moves in lymph veins to the blood circulatory veins. Lacteals absorb the products of fat digestion, and lymph nodes help combat infection.

8. Blood has two main parts: plasma and cells. Plasma contains mostly water (92%) and proteins (8%) but also nutrients and wastes.

9. The red blood cells contain hemoglobin and function in oxygen transport. The platelets and two plasma proteins, prothrombin and fibrinogen, function in blood clotting, an enzymatic process that results in fibrin threads.

10. Defense against disease depends on the various types of leukocytes. Neutrophils and monocytes are phagocytic and are especially responsible for the inflammatory reaction. Lymphocytes are involved in the development of immunity to disease.

11. When blood reaches a capillary, water moves out at the arterial end, due to blood pressure. At the venule end, water moves in, due to osmotic pressure. In between, nutrients diffuse out and wastes diffuse in.

12. Blood clotting is a complex process that includes three major events; platelets and injured tissue release prothrombin activator that enzymatically changes prothrombin to thrombin. Thrombin is an enzyme that causes fibrinogen to be converted to fibrin threads.

13. In the ABO blood system there are four types of blood: A, B, AB, and O, depending on the type of antigen present on the red blood cells. If the red blood cells have a particular antigen, they clump when exposed to the corresponding antibody. Table 33.3 tells which types of antibody are present in the plasma for the various ABO blood groups.

Objective Questions

1. Which one of these would you expect to be part of a closed, but not an open, circulatory system?
   a. ostia
   b. capillary beds
   c. hemocoel
   d. heart

2. In a one-circuit circulatory system, blood pressure
   a. is constant throughout the system.
   b. drops significantly after gas exchange has taken place.
   c. is higher at the intestinal capillaries than at the gill capillaries.
   d. cannot be determined.

3. In which of the following animals is the blood entering the aorta incompletely oxygenated?
   a. frog
   b. chicken
   c. monkey
   d. fish

4. Which of these factors has little effect on blood flow in arteries?
   a. the heartbeat
   b. blood pressure
   c. total cross-sectional area of vessels
   d. skeletal muscle contraction

5. In humans, blood returning to the heart from the lungs returns to the
   a. right ventricle.
   b. right atrium.
   c. left ventricle.
   d. left atrium.

6. Systole refers to the contraction of
   a. the major arteries.
   b. the SA node.
   c. the atria and ventricles.
   d. All of these.

7. A baby born to which of these couples is most likely to suffer from fetal erythroblastosis?
   a. Rh-positive mother and Rh-negative father
   b. Rh-negative mother and Rh-negative father
   c. Rh-positive mother and Rh-positive father
   d. Rh-negative mother and Rh-positive father

8. During blood typing, agglutination indicates that
   a. the plasma contains certain antibodies.
   b. the red blood cells carry certain antigens.
   c. the plasma contains certain antigens.
   d. the red blood cells carry certain antibodies.

9. Water enters capillaries on the venule side as a result of
   a. active transport from tissue fluid.
   b. an osmotic pressure gradient.
   c. increased blood pressure on this side.
   d. higher red blood cell concentration on this side.

10. Which of these associations is incorrect?
    a. white blood cells—infection fighting
    b. red blood cells—blood clotting
    c. plasma—water, nutrients, and wastes
    d. platelets—blood clotting

11. The last step in blood clotting
    a. requires calcium ions.
    b. occurs outside the bloodstream.
    c. converts prothrombin to thrombin.
    d. converts fibrinogen to fibrin.

Study Questions

1. Describe transport in those invertebrates that have no circulatory system; those that have an open circulatory system; and those that have a closed circulatory system.

2. Compare the circulatory systems of a fish, an amphibian, and a mammal.

3. Trace the path of blood in humans from the right ventricle to the left atrium; from the left ventricle to the kidneys and return to the right atrium; from the left ventricle to the small intestine and return to the right atrium.

4. Define these terms: pulmonary circulation, systemic circulation, portal system.

5. Describe the beat of the heart, mentioning all the factors that account for this repetitive process. Describe how the heartbeat affects blood flow; what other factors are involved in blood flow?

6. Describe the structure and function of the lymphatic system.

7. List the major components of blood and the function of each component.

8. Describe in detail the exchange of molecules between blood and tissue fluid at a capillary, including the forces that facilitate this exchange.

Thought Questions

1. Explain the advantages of a circulatory system with separate pulmonary and systemic loops over a system with a single loop, such as that of fishes.

2. Explain why a person with any ABO blood type can theoretically be given type O blood in an emergency.

3. Sometimes blood clots even when the blood vessel has not been cut. If the clot moves from the place of origin to another location, it is called an embolism. Explain why blood clots originating in the legs result in pulmonary embolisms.

Selected Key Terms

hemoglobin (he''mo-glo'bin) 512
artery (ar'ter-e) 512
capillary (kap'ĭ-lar''e) 512
vein (vān) 512
pulmonary (pul'mo-ner''e) 514
systemic (sis-tem'ik) 514
aorta (a-or'tah) 515

vena cava (ve'nah ka'vah) 517
myocardium (mi''o-kar'de-um) 517
SA node (es a nōd) 517
systole (sis'to-le) 518
diastole (di-as'to-le) 518
lymph (limf) 520
plasma (plaz'mah) 521

erythrocyte (ĕ-rith'ro-sīt) 521
leukocyte (lu'ko-sīt) 521
platelet (plāt'let) 521
neutrophil (nu'tro-fil) 523
macrophage (mak'ro-fāj) 524

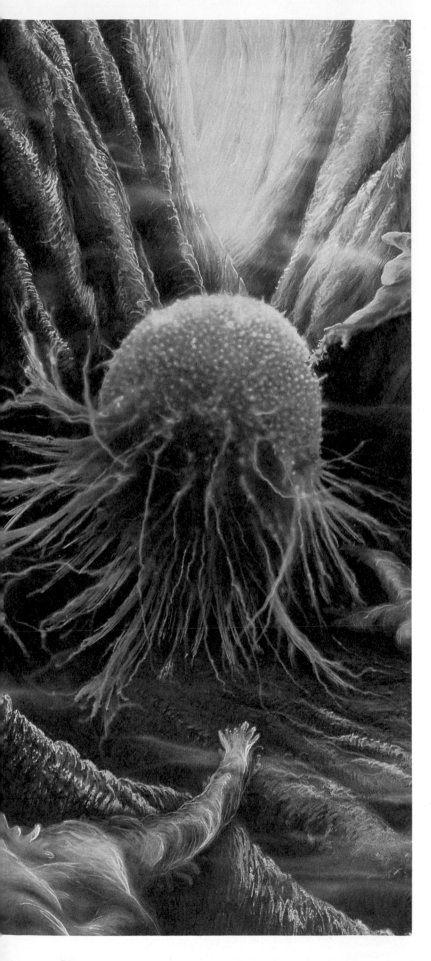

CHAPTER 34

Immunity

Your study of this chapter will be complete when you can:

1. *Name three general ways the body defends itself against infections, and give examples of each.*
2. *Contrast the maturation, structure, and function of B and T lymphocytes.*
3. *Explain the clonal selection theory as it pertains to B cells and T cells.*
4. *Describe the structure and function of an antibody.*
5. *Name the different types of T cells, and describe the function of each type.*
6. *Name and discuss three types of immunological side effects.*
7. *Contrast active immunity with passive immunity, and tell why the former lasts longer.*
8. *Tell, in general, how monoclonal antibodies are produced, and list ways in which they are used today.*
9. *Explain the rationale for the use of lymphokines and growth factors in cancer therapy.*

A disruptive cancer cell coursing through a lymphatic vessel is a threat to the body's integrity. The lymphatic system aids in the body's defense but may, on occasion, help spread cancer in the manner depicted here.

The mammalian body is prepared to protect itself from foreign substances and cells, including infectious microbes. The first line of defense is a *general* one available immediately after infection because it consists of mechanisms that are nonspecific. The second line of defense takes a little longer to come into play because it is highly *specific* and contains mechanisms that are tailored to resist a particular threat.

General Defense

The environment contains many organisms that are able to invade and infect the body. There are three general defense mechanisms that are useful against all of them: barriers to entry, phagocytes, and protective proteins.

Barriers to Entry

The skin and the mucous membrane lining of the respiratory and digestive tracts serve as mechanical barriers to possible entry by bacteria and viruses. Also, the secretions of the sebaceous glands in the skin contain chemicals that weaken or kill bacteria. The respiratory tract is lined by cilia that sweep mucus and any trapped particles up into the throat, where they may be swallowed. The stomach has an acid pH that inhibits the growth of many types of bacteria. The intestine and other organs, such as the vagina, contain a mix of bacteria normally residing there, which prevent potential pathogens from also taking up residence.

Phagocytic White Blood Cells

If microbes do gain entry into the body, as described in the section on the inflammatory reaction on page 524, other nonspecific forces come into play. For example, neutrophils (fig. 33.14) and *monocyte*-derived macrophages (fig. 34.1) are phagocytic white blood cells that can engulf bacteria and viruses upon contact. *Macrophages* engaged in fighting infection give off a chemical that stimulates general defense and also causes fever. Fever may also be protective because some microbes cannot tolerate a higher-than-normal body temperature.

Figure 34.1
Macrophages are the body's scavengers. This one (colored red) is phagocytizing bacteria (colored green).

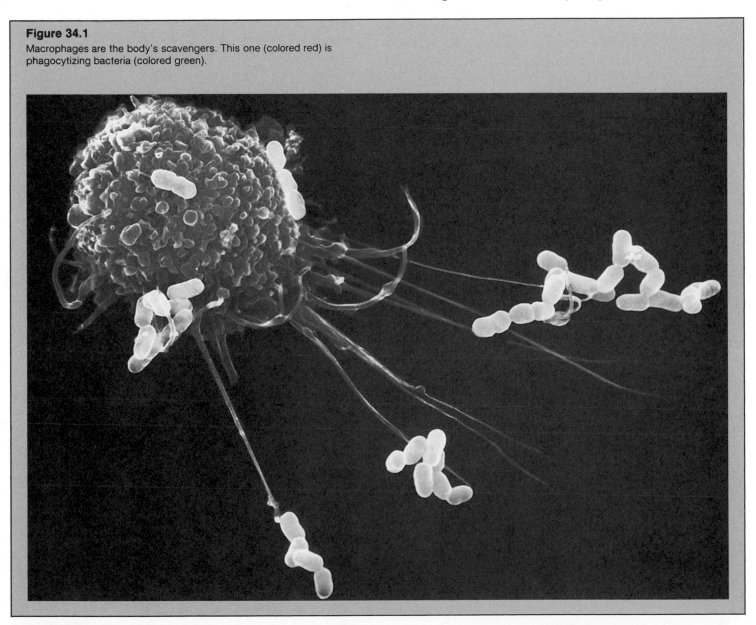

Figure 34.2

Complement is a series of proteins that is always present in the plasma. When activated, some of these form pores in bacterial cell walls and plasma membranes, allowing fluids and salts to enter until the bacteria eventually burst.

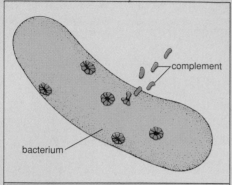

1. Complement proteins form pores in the bacterial cell wall and plasma membrane.

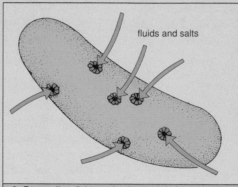

2. Pores allow fluids and salts to enter the bacterium.

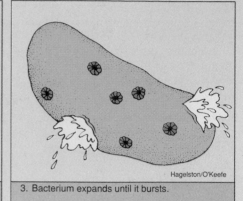

3. Bacterium expands until it bursts.

Hagelston/O'Keefe

Protective Proteins

Also present in the blood and tissues are several proteins that prevent the multiplication of bacteria and viruses. Complement works against bacteria, and interferon is specific against viruses.

Complement

Complement is a series or complex of proteins produced by the liver that is present in the plasma. When complement is *activated,* a cascade of reactions occur. Each protein molecule in the series activates the next one in a predetermined sequence. In the end, certain of the proteins form pores in bacterial plasma membranes. This allows fluids and salts to enter and causes the bacterial cell to burst (fig. 34.2).

If complement is unable to destroy the microbes directly it still can coat them and release chemicals that attract phagocytes to the scene. Although complement is a general defense mechanism, it also plays a role in specific defense as we shall see later.

Interferon

If a virus should escape phagocytosis and enter a tissue cell, then the infected cell will produce and secrete interferon. **Interferon** binds to receptors on noninfected cells, and this causes them to prepare for possible attack by producing substances that will interfere with viral replication.

A cell, which has bound interferon, will be protected against any type of virus; therefore, interferon should potentially be very useful in preventing viral infection. However, interferon is specific to the species; only human interferon can be used in humans. It used to be quite a problem to collect enough interferon for clinical and research purposes, but now interferon is made by recombinant DNA technology.

The first line of defense against disease is nonspecific. It consists of barriers to entry, phagocytic white blood cells, and protective proteins.

Specific Defense: The Immune System

Sometimes we are threatened by an illness that cannot be successfully counteracted by the general defense mechanisms. In such cases, the **immune system** becomes activated to provide a specific defense. The immune system (fig. 34.3) consists of about a trillion *agranular leukocytes* (lymphocytes and monocytes); the spleen, thymus, and lymph nodes, where lymphocytes and monocytes are produced; and the lymph vessels, where these cells are primarily found. Following maturation, agranular leukocytes are also found in the bloodstream.

The cells of the immune system respond to **antigens,** usually proteins (or polysaccharides) that they recognize as not normally found within the body. Since some antigens are part of a bacterial cell wall or viral coat, the immune system protects us from disease, but an antigen can also be part of a foreign cell or of cancerous cells. These, too, are destroyed by the immune system. Ordinarily, the immune system does not attack the body's own cells; therefore, it is said that the immune system is able to distinguish self from nonself.

Immunity usually lasts for some time. For example, once we have recovered from measles we usually cannot be infected a second time.

The immune system includes the lymphoid organs, as well as the agranular leukocytes, which respond to the presence of antigens.

Figure 34.3
The immune system includes all those organs in which lymphocytes
are produced and found, such as lymph nodes, thymus, bone
marrow, and spleen. The lymphocytes and monocytes within lymph
nodes help purify lymph, and those within the spleen help purify the
blood.

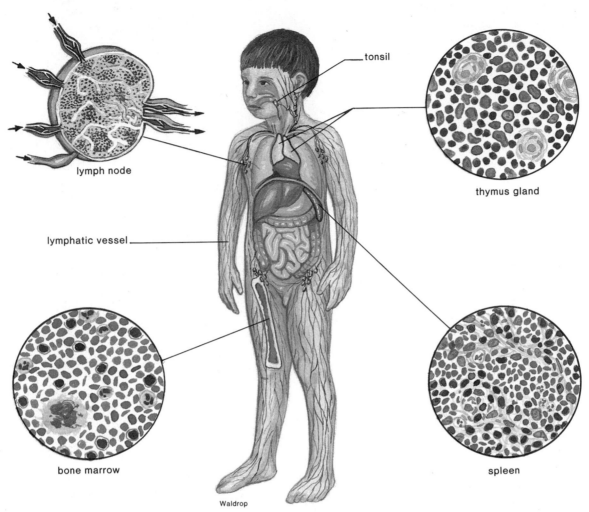

lymph node

tonsil

thymus gland

lymphatic vessel

bone marrow

spleen

Waldrop

The Lymphocytes

Immunity is primarily the result of the action of two types of
lymphocytes, the B lymphocytes and the T lymphocytes. Their
names are derived from their site of maturation.

All the different types of blood cells are derived from bone
marrow stem cells, which divide to give a specific precursor for
each cell type. The B cells finish their maturation in the bone
marrow; pre-T cells go by way of the blood to the thymus before
entering the lymph. Those cells that will become **T lymphocytes,** or T cells, pass through the *thymus* (fig. 34.3); those that
will become **B lymphocytes,** or B cells, do not pass through the
thymus and therefore are derived more directly from *bone
marrow* stem cells.

B cells and T cells have different functions. B cells produce **antibodies,** proteins that are capable of combining with and
inactivating antigens. These antibodies are secreted into the
blood and lymph. In contrast, T cells do not produce antibodies.
Instead, the T lymphocytes themselves directly attack cells that
bear antigens they recognize.

Lymphocytes are capable of recognizing an antigen because they have *receptor* molecules on their surfaces. The shape
of the receptors on any particular lymphocyte are complementary to a portion of one specific antigen. It is often said that the
receptor and antigen fit together as do a *lock and key*. It is estimated that during our lifetimes, we encounter a million different antigens, and we need the same number of different
lymphocytes to protect ourselves against those antigens. It is
remarkable that so much diversification occurs during the maturation process that, in the end, there is a different type of lymphocyte for each possible antigen. Despite this great diversity,
none of these cells is supposed to attack the body's own cells.
It's believed that if by chance a lymphocyte arises that could
respond to the body's own proteins, it normally is suppressed
and develops no further.

Figure 34.4

The clonal selection theory as it applies to B cells. The binding of antigen causes the appropriate B cell to divide, producing many mature plasma cells by the fifth day. Some of these mature cells actively secrete antibodies. Others form memory B cells, which retain the ability to secrete these antibodies at a future time.

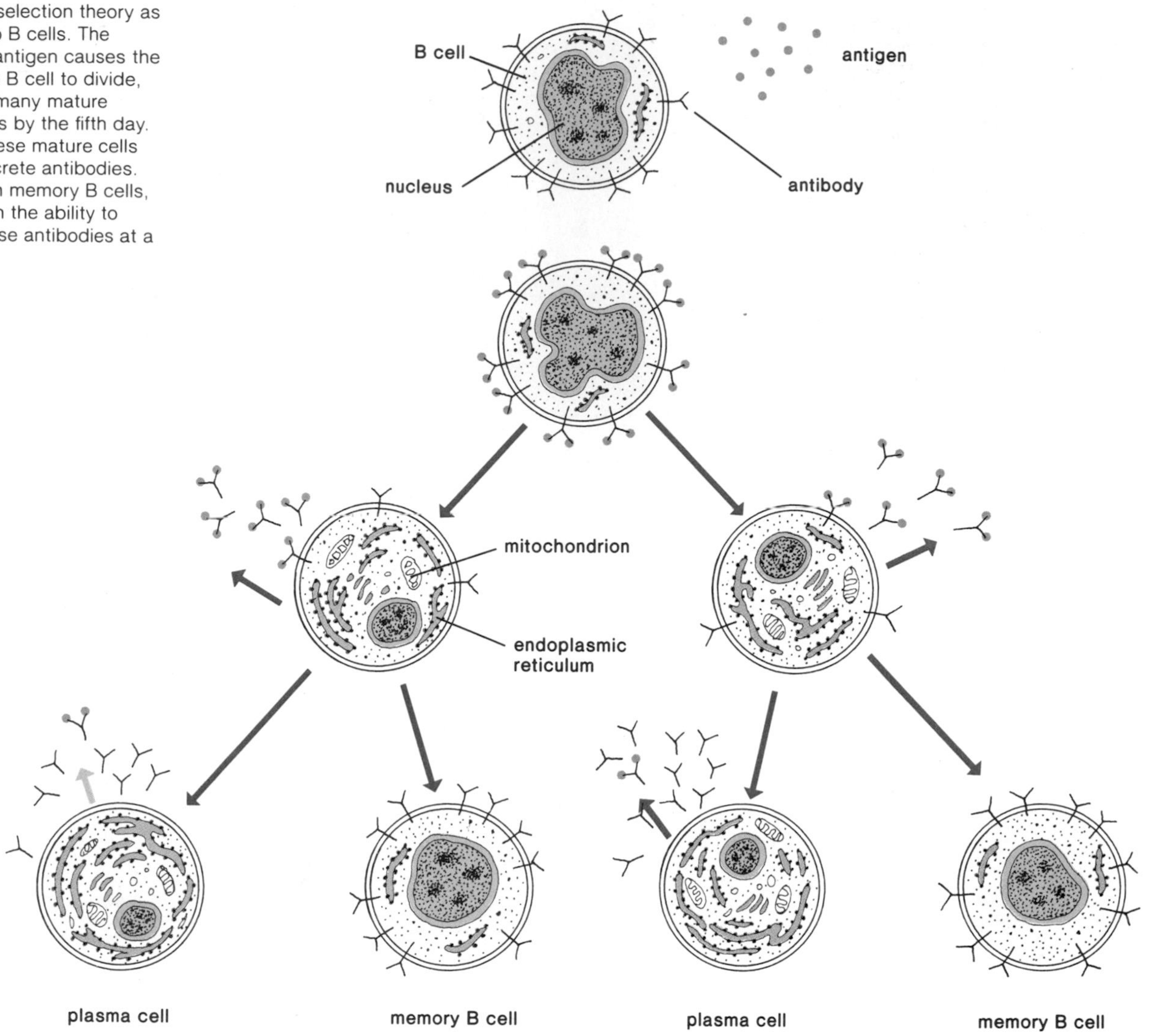

There are two types of lymphocytes. B cells produce and secrete antibodies that can combine with antigens. T cells directly attack antigen-bearing cells.

The Action of B Cells

The receptor present in a B cell is called a membrane-bound antibody because it is shaped exactly like an antibody. When a B cell encounters a bacterial cell or toxin bearing an antigen that combines with its receptor, the B cell is stimulated to divide and produce **plasma cells** that actively secrete antibodies (fig. 34.4). All of the plasma cells derived from one parent lymphocyte are called a clone, and they all produce the *same* type of antibody. Notice that the particular antigen determines exactly which lymphocyte will be stimulated to produce antibodies. This is called the *clonal selection theory* because the antigen has selected which B cells will produce a clone of plasma cells. (T cells also undergo clonal selection as will be discussed later.)

Once antibody production is high enough, the antigen disappears from the system, and the development of plasma cells ceases. However, some members of the clone do not participate in antibody production; instead they remain in the bloodstream as **memory B cells,** capable for some time of producing the antibody specific to a particular antigen. As long as these cells are present, the individual is said to be *actively immune* because a certain number of antibodies are always present and also because the memory cells can produce more plasma cells if the same antigen invades the system again.

Defense by B cells is called *antibody-mediated immunity* because B cells produce antibodies. Sometimes, it is also called humoral immunity because these antibodies are present in the bloodstream.

B cells are responsible for anitbody-mediated immunity. After they recognize an antigen, they divide to produce both antibody-secreting plasma cells and memory B cells.

Antibodies

An antibody is a Y-shaped protein molecule composed of two long, "heavy" chains and two short, "light" chains of amino acids. At the end of both upper arms of every antibody (fig. 34.5) is a short segment called the *variable region* because it varies in such a way that each antibody is specific for a particular antigen. Here the antigen binds to the antibody in a lock-and-key manner.

The other parts of each heavy and light chain are called the *constant regions*. However, the constant regions are not identical among all the antibodies. Instead, they are the same for different *classes* of antibodies. Most antibodies found in the blood belong to the class known as IgG (immunoglobulin G).

Antigen-Antibody Complexes   The antigen-antibody reaction can take several forms, but quite often the antigen-antibody reaction produces complexes of antigens combined with antibodies (fig. 34.5c). Such an antigen-antibody complex, sometimes called the *immune complex,* marks the antigen for destruction by other forces. For example, the complex may be engulfed by neutrophils or macrophages or it may activate a portion of blood serum called *complement.* As mentioned earlier, complement refers to a series of different proteins that stimulate phagocytic cells. This series of proteins also bring about lysis of bacterial cells—several of the proteins form pores in the cell wall that allows water and salts to enter the cell causing it to burst (fig. 34.2).

An antibody combines with its antigen in a lock-and-key manner. The antibody-antigen reaction can lead to complexes that contain several molecules of antibody and antigen.

Figure 34.5

Antibody structure. *a.* Computer model of an antibody. *b.* An antibody contains two heavy (long) amino acid chains and two light (short) amino acid chains. Part of each chain contains a variable region where a particular antigen is capable of binding with the antibody in a lock-and-key manner. The rest of each chain is a constant region. *c.* The antigen-antibody reaction often produces complexes of antigens combined with antibodies.

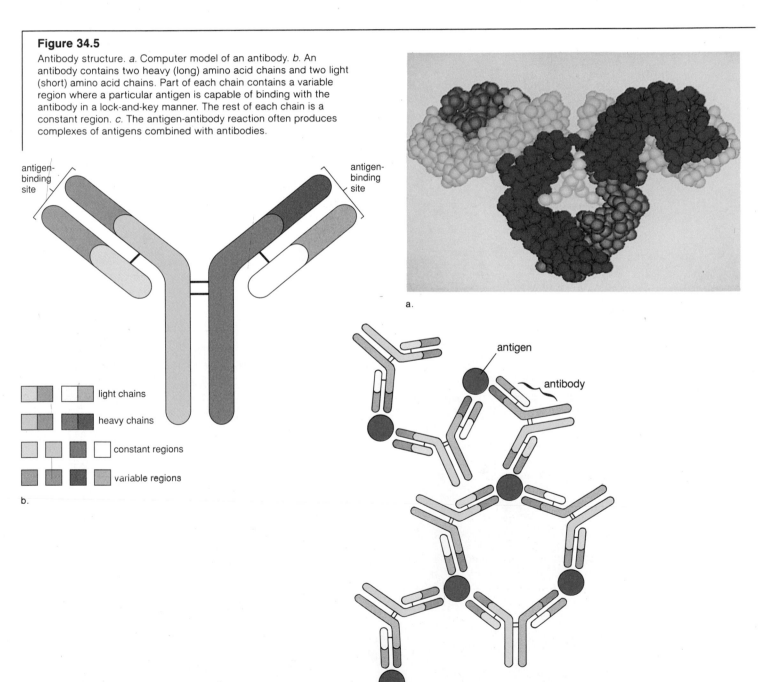

Figure 34.6

T cell activation and proliferation. *a.* Photomicrograph shows four T cells on the surface of a macrophage. *b.* Macrophage has ingested an antigen-bearing bacterium; later a protein fragment from the bacterium, along with the MHC complex, is presented to the T cell. This activates the T cell, which now divides to produce many cells capable of responding to this particular antigen. Activation and proliferation are enhanced by the presence of lymphokines produced by both macrophages and T cells.

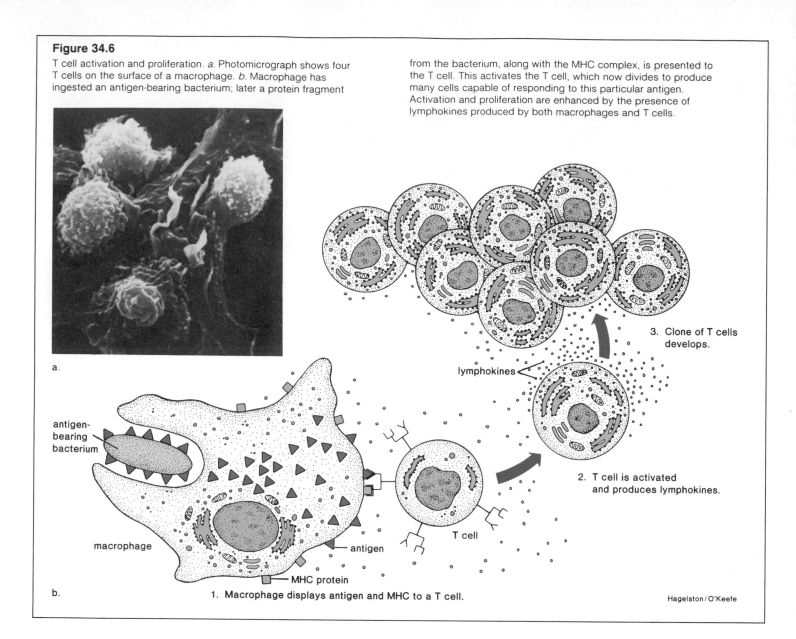

3. Clone of T cells develops.

lymphokines

2. T cell is activated and produces lymphokines.

T cell

antigen-bearing bacterium

macrophage

antigen

MHC protein

1. Macrophage displays antigen and MHC to a T cell.

a.

b.

Hagelston/O'Keefe

Table 34.1 Some Properties of B Cells and T Cells

Property	B Cells	T Cells
Antigen recognition	Direct recognition	Must be presented by macrophages
Early response	Produce plasma cells	Enlarge, multiply, and secrete lymphokines
Later response	Antibody production	Helper cells stimulate other cells; killer cells attack antigen-bearing cells
Final response	By memory cells	Memory cells; suppressor cells shut down immune response

The Action of T Cells

T cells have receptors just like B cells do. Unlike B cells, however, T cells are unable to recognize an antigen that is simply present in lymph or blood. Instead, the antigen must be presented to a T cell by a (monocyte-derived) macrophage. When a *macrophage* engulfs a bacterium or virus, it enzymatically breaks it down and displays only a protein fragment on its plasma membrane, together with an **MHC** (major histocompatibility complex) **protein** (fig. 34.6). The importance of these antigens was first recognized when it was discovered that they contribute to the specificity of tissues and make it difficult to transplant a tissue from one person to another. In other words, the donor and the recipient must be histo- (tissue) compatible (the same or nearly so) for a transplant to be successful without the administration of immunosuppressive drugs.

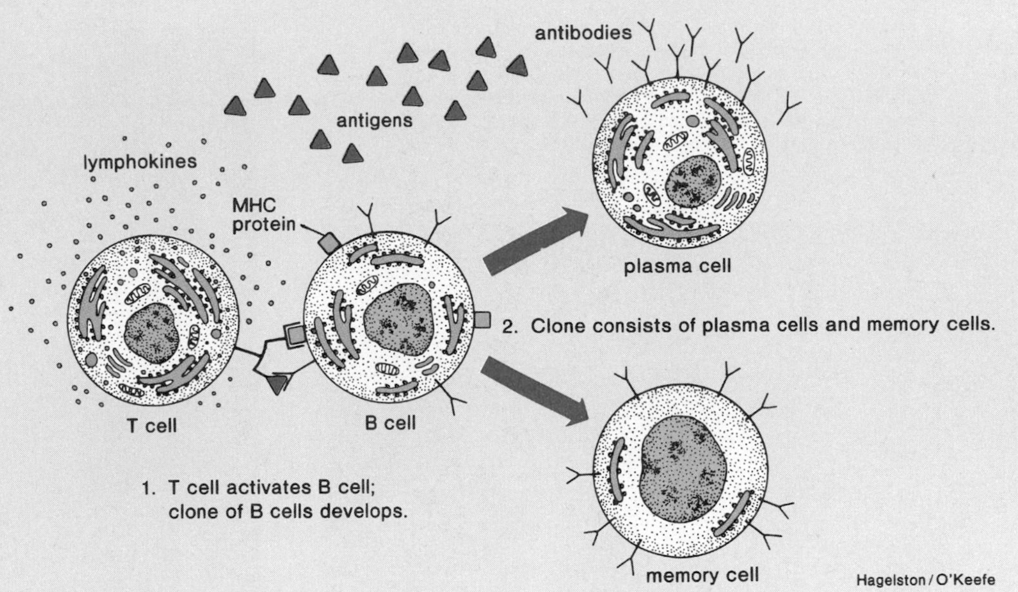

Figure 34.7
Helper T cells activate B lymphocytes. Among the four types of T cells, the helper T cells are the most important for stimulating the immune response. Here a T cell stimulates a B cell by direct contact, and the B cell goes on to proliferate—producing many plasma cells and memory cells that secrete antibodies specific to one antigen.

antibodies
antigens
lymphokines
MHC protein
plasma cell
2. Clone consists of plasma cells and memory cells.
T cell
B cell
1. T cell activates B cell; clone of B cells develops.
memory cell
Hagelston/O'Keefe

Once a T cell recognizes an antigen, it proliferates, giving rise to a clone of T cells that contains the types discussed next. Therefore, it can be seen that, like B cells, T cells also undergo clonal selection.

In order for a T cell to recognize an antigen, the antigen must be presented, along with an MHC protein, to the T cell by a macrophage.

Types of T Cells

When a T cell passes through the thymus, it becomes one of four possible types. The four different types of T cells that look alike but can be distinguished by their function are:

Helper T (T_H) cells are believed to orchestrate the immune response. Once they recognize an antigen, they enlarge and secrete **lymphokines,** including interferons and interleukins. Lymphokines are stimulatory chemicals for all types of immune cells, including B cells (fig. 34.7). Because the AIDS (acquired immune deficiency syndrome) viruses attack helper T cells, they inactivate the immune system. This is discussed in the reading on page 540.

Killer T (T_K) cells are the type of T cell that attack and destroy cells that bear a foreign antigen (fig. 34.8). *Cytotoxic T cells* mainly do away with virus-infected cells. *Natural killer cells* are thought to restrict their activities to tumor cells and perhaps also to cells infected by agents other than viruses. Regardless of this distinction, they both function similarly. They have storage granules that contain a chemical called *perforin,* which perforates plasma membranes. Perforin molecules form a pore in the membrane of cells bearing a foreign antigen. Water and salts can now enter, causing the cell to swell and eventually burst.

To provide a contrast with B cells, it is often said that T cells are responsible for *cell-mediated immunity*. But you can see that strictly speaking it is the T_K cells that destroy other cells and therefore are responsible for cell-mediated immunity.

Suppressor T (T_S) cells increase in number more slowly than the other two types of T cells just discussed. Once there is sufficient number, however, the immune response ceases. Following suppression, a population of **memory T (T_M) cells** also persists, perhaps for life.

T_K cells are responsible for cell-mediated immunity. T_H cells promote the immune response, T_S cells suppress the immune response, and T_M cells maintain immunity.

Immunological Side Effects and Illnesses

The immune system protects us from disease because it can tell self from nonself. Sometimes, however, the immune system is underprotective, as when an individual develops cancer, or is overprotective, as when an individual has allergies.

Allergies

Allergies are caused by an overactive immune system that forms antibodies to substances that are not usually recognized as being foreign substances. Unfortunately, allergies are usually accompanied by coldlike symptoms or, even at times, severe systemic reactions such as shock, a sudden drop in blood pressure.

Among the five varieties of antibodies—immunoglobulin A (IgA), IgD, IgE, IgG, and IgM—it is IgE that causes allergies. IgE antibodies are found in the bloodstream, but they, unlike the other types of antibodies, also reside in the membrane of certain types of cells found in the tissues, called *mast* cells. Some investigators contend that mast cells are basophils that have left the bloodstream and taken up residence in the tissues. In any case, when the *allergen,* an antigen that provokes an allergic reaction, attaches to the IgE antibodies on mast cells, these cells release histamine and other substances that cause secretion of mucus and constriction of the airways, resulting in

Figure 34.8

Cell-mediated immunity. *a.* Scanning electron micrograph showing killer T cells attacking cancer cell. *b.* Cancer cell is now destroyed. *c.* During the killing process, granules from a T cell fuse with the T cell plasma membrane and release the protein perforin. These combine to form pores in the target cell plasma membrane. Fluid and salts enter the target cell so that it eventually bursts.

a.

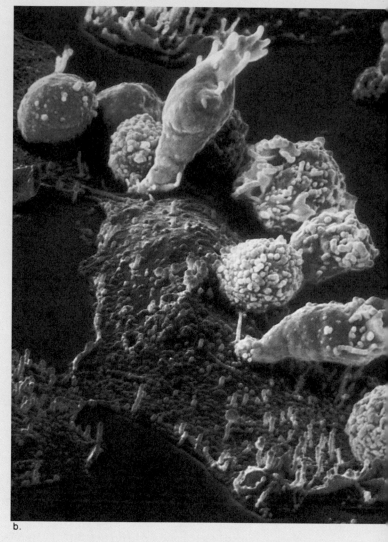

b.

the characteristic symptoms of an allergy. On occasion, basophils and other white blood cells release these chemicals into the bloodstream. The increased capillary permeability that results from this can lead to fluid loss and shock.

Allergy shots sometimes prevent the occurrence of allergic symptoms. Injections of the allergen cause the body to build up high quantities of IgG antibodies, and these combine with allergens received from the environment before they have a chance to reach the IgE antibodies located in the membrane of mast cells.

Allergic symptoms are caused by the release of histamine and other substances from mast cells.

Tissue Rejection

Certain organs, such as skin, the heart, and the kidneys, could easily be transplanted from one person to another if the body did not attempt to *reject* them. Rejection occurs when it is obvious that the transplanted organ is foreign to the individual, and for this reason the immune system reacts to it. At first T cells appear on the scene, and later antibodies bring about a disintegration of the foreign tissue.

Organ rejection can be controlled in two ways: careful selection of the organ to be transplanted and the administration of *immunosuppressive drugs*. In the first instance, it is best if the organ is made up of cells having the same type of MHC proteins as those on the cells of the prospective recipient. It is these that the T cell recognizes as foreign to the recipient. In regard to immunosuppression, there is now available a drug

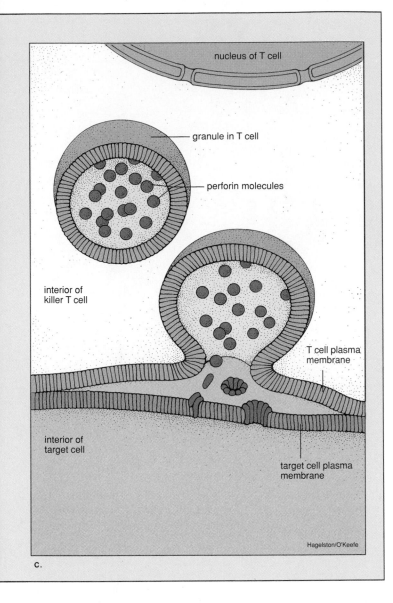

granule in T cell

perforin molecules

nucleus of T cell

interior of
killer T cell

T cell plasma
membrane

interior of
target cell

target cell plasma
membrane

Hagelston/O'Keefe

c.

called Cyclosporine, which suppresses cell-mediated immunity only as long as it is administered. This means that after the body has accepted the new organ, the drug can be withdrawn and immunity to disease will return. When immunity is suppressed the patient is constantly threatened by infectious diseases.

When an organ is rejected, the immune system is attacking cells that bear different MHC proteins from those of the individual.

Autoimmune Diseases

Certain human illnesses are believed to be due to the production of antibodies that act against an individual's own tissues. In the disease myasthenia gravis, such *autoantibodies* attack the neuromuscular junctions so that the muscles do not obey nervous stimuli. Muscular weakness results. In MS (multiple sclerosis), antibodies attack the myelin sheath of nerve fibers, causing various neuromuscular disorders. A person with SLE (systemic lupus erythematosus) forms various antibodies to different constituents of the body, including the DNA of the cell nucleus. The disease sometimes results in death, usually due to kidney damage. In rheumatoid arthritis it is the joints that are affected. When an autoimmune disease occurs, a viral infection of tissues has often set off an immune reaction to the body's own tissues in an attempt to attack the virus. There is evidence to suggest that Type I diabetes is also the result of this sequence of events.

Autoimmune diseases at times seem to be preceded by a viral infection that fools the immune system into attacking the body's own tissues.

Immunotherapy

The immune system can be manipulated to help people avoid or recover from diseases. Some of these techniques have been utilized for quite a long time, and some are relatively new.

Induced Immunity

Active immunity, which provides long-lasting protection against a disease-causing organism, develops after an individual becomes infected with a virus or bacterium. In many instances today, however, it is not necessary to suffer an illness to become immune because it is possible to be medically immunized against a disease. One recommended immunization schedule for children is given with figure 34.9. Immunization requires the use of **vaccines,** which are traditionally bacteria and viruses that have been treated so that they are no longer virulent (able to cause disease). New methods of producing vaccines are being developed. For example, it is possible to use the recombinant DNA technique to mass-produce a protein that can be used as a vaccine (table 18.1). Vaccination results in the production of antibodies (fig. 34.4).

Passive immunity occurs when an individual is given antibodies to combat a disease. Since these antibodies are not produced by the individual's B cells, passive immunity is short-lived. For example, newborn infants possess passive immunity because antibodies have crossed the placenta from their mother's blood. These antibodies soon disappear, however, so that within a few months infants become more susceptible to infections. Breast feeding prolongs the passive immunity an infant receives from its mother because there are antibodies in the mother's milk.

Vaccines can be used to make people actively immune. Passive immunity is short-lived because the antibodies are administered to, rather than made by the individual.

AIDS and Its Treatment

AIDS (acquired immune deficiency syndrome) is a sexually transmitted disease that is almost always fatal. It is characterized by the destruction of the immune system, which leads to various infectious diseases and a virulent type of skin cancer (Kaposi's sarcoma). Many victims also show early brain impairment with ever-increasing neuromuscular and psychological consequences. Since the number of deaths from AIDS reported thus far in the United States is at least 128,000 and victims rarely live longer than four years, there is great interest in developing techniques to treat those who are infected and to prevent the spread of infection to others.

Investigators have now identified several viruses that can cause AIDS. These are termed human immunodeficiency viruses and are numbered as HIV-1 and HIV-2, for example. The HIV viruses (see figure 1) primarily attack helper T cells because the viruses have an outer protein (gp-120) that combines with a molecule (called CD4 antigen) in the plasma membrane of these lymphocytes. Thereafter, the viruses enter helper T cells (fig. 23.4b). Monocytes have CD4 in fewer number, and are therefore less susceptible to viral infection. They act as a reservoir for the virus and distribute it about the body, including the lungs and brain. Neurons lack CD4 and therefore are rarely infected so research is focusing on the possibility that the viruses cause a chemical imbalance leading to the psychiatric disorders seen in AIDS patients.

The AIDS viruses are retroviruses, RNA viruses that have the capability of incorporating a DNA copy of their genome into the host cell DNA. When this occurs the virus becomes latent and does not multiply, and the individual does not display any symptoms of having AIDS. Eventually, about 1% of infected people come down with symptoms of AIDS. There is great interest in determining what triggers activation of the viruses so that they multiply and burst forth from the T cell, destroying it. It's possible that activation of the T lymphocyte by an antigen also activates the AIDS viruses. Certainly, AIDS is seen more often in individuals that have another type of sexually transmitted disease in addition to AIDS.

Since antibodies to the viruses are found in the bloodstream, it seems that at first the body is able to mount an offensive to the invader. Eventually, however, the body succumbs. Various explanations have been given:

1. Infected lymphocytes are known to fuse with others, some of which are not infected. If this is the way the viruses usually spread, then they need never return to the bloodstream where antibodies are located.

2. There is evidence that the AIDS viruses alter the MHC proteins that appear on the surfaces of infected cells. This would make T cells incapable of recognizing and destroying infected cells.

3. The AIDS viruses mutate so rapidly that antibody production lags behind and never catches up enough to be useful. Then, too, the viruses invade the very cells that are sent to destroy them, so that antibody production is never adequate.

Development of a treatment for AIDS is taking several avenues. There are drugs that will interfere with the viruses' normal reproductive cycle. The new drug AZT (azidothymidine) interferes with the ability of the viruses to incorporate their genomes into host DNA, and others are being developed that prevent viral replication. Most of these, like AZT, are abnormal nucleotides that take the place of normal nucleotides, preventing DNA replication. AZT has been found helpful even in infected persons who have no symptoms yet. One concern is that the drugs will also interfere with the production of lymphocytes themselves. In any case, it will be necessary to reconstitute the normal white blood cell count in AIDS victims, and it is hoped that the growth factors mentioned on page 542 will be helpful in this regard.

Two experimental vaccines have been developed against the AIDS viruses and are undergoing clinical trials. Both vaccines utilize gp-120, the outer viral coat protein labeled in the figure, as the antigen. One uses gp-120 directly. For the other, the gene coding for gp-120 is incorporated into the vaccinia virus. The hope is that after this virus enters the body, the gene will use the machinery of invaded cells to make gp-120 proteins, which will then appear on the cell surface and be recognized by the body's immune system.

A novel treatment based on the HIV viruses' strong attraction to the CD4 molecule has been proposed, but thus far has only been tried in laboratory experiments. First, CD4 proteins are made by recombinant DNA technology. Then copies of this protein are added to samples of the HIV viruses and helper T lymphocytes, the manufactured CD4 proteins attach to the viruses, preventing them from combining with the lymphocytes. Testing of this approach in animals and humans is planned.

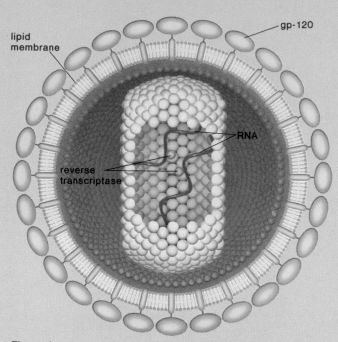

Figure 1

AIDS virus. The virus is covered by a membrane that is derived from the host cell plasma membrane and contains proteins coded for by virus (fig. 23.4b) one of these, gp-120, extends beyond the membrane.

While there is much to be learned about AIDS, we do at least have knowledge of its means of transmission. We know that AIDS can be transmitted by blood-to-blood contact. This puts intravenous drug abusers at great risk. Infected semen (and possibly saliva) also may pass the disease to a sexual partner when it comes in contact with the partner's bloodstream. This can happen if a woman is menstruating, if the partner has an open abrasion, or if the sexual act itself damages the mucosal lining.

There is little danger of contracting AIDS by casual encounters between individuals, such as shaking hands, touching the same utensils, or breathing the same air. However, the public is advised to avoid sexual promiscuity because it may cause them to unknowingly have sex with a carrier of AIDS. It is estimated that there may be as many as two million persons in the United States who are infected with AIDS, but who display no symptoms. It seems that the use of condoms is a protection against contracting AIDS. Even so, individuals can best protect themselves by practicing monogamy (always the same sexual partner) with someone who is free of the disease.

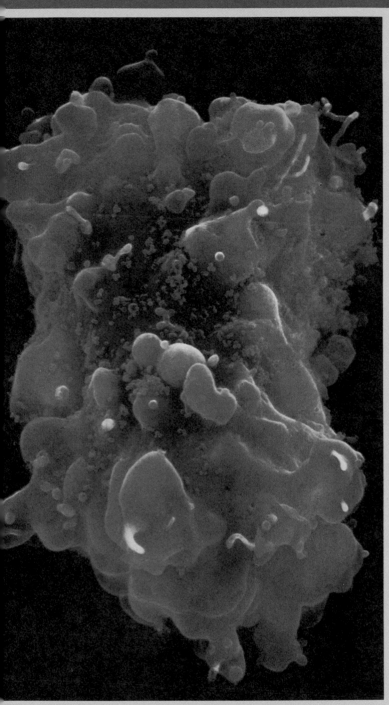

Figure 2
T cell (pink) attacked by AIDS virus (blue).

Figure 34.9
Suggested immunization schedule for infants and young children. Children who are not immunized are subject to childhood diseases that can cause serious health consequences.
Source: DHRS Immunization Program, State of Florida.

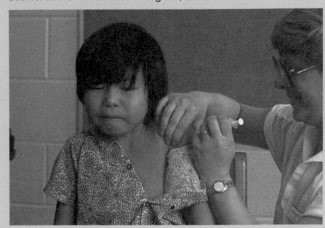

Age:	Immunization Schedule
2 months	1st DTP[a] 1st polio
4 months	2nd DTP 2nd polio
6 months	3rd DTP 2nd polio[b]
15 months	3rd DTP 2nd polio[b] 1st measles[c] 1st rubella[c,d] 1st mumps[c]
18 months and older	4th DTP 3rd polio[b]
4 to 6 years, before starting to school	DTP booster (the 5th immunization) polio booster (the 4th immunization)[b]
14 to 16 years and thereafter	Tetanus-diphtheria booster should be given every 10 years or following a dirty wound if a booster has not been given in the preceding 5 years.

[a]D = diphtheria, T = tetanus, P = pertussis (whooping cough)
[b]Some doctors give one additional dose when the child is 6 months old.
[c]Only one shot needed. Some doctors combine those vaccines in a single injection.
[d]rubella (German measles)

Monoclonal Antibodies

There is much hope that the technique of producing *mono-* (one) *clonal* (clone) antibodies will one day make plentiful amounts of pure and specific antibodies available. One method of producing monoclonal antibodies is depicted in figure 34.10. Lymphocytes are removed from the body and exposed in vitro (in laboratory glassware) to a particular antigen. Then they are fused with cancer cells because cancerous cells, unlike normal cells, will divide an unlimited number of times. The fused cells are called *hybridomas* (hybrid- refers to there being a fusion of two different cells, and -oma stands for the fact that one of the cells is a cancer [carcinoma] cell).

The antibodies produced by hybridoma cells are called **monoclonal antibodies** because all of them are the same type and all are produced by cells derived from the same parent cell. At present monoclonal antibodies are being used to allow quick and certain diagnosis of various conditions. For example, a specific hormone is present in the urine when a woman is pregnant. Monoclonal antibodies can be used to detect the hormone and indicate that the woman is pregnant. Some monoclonal antibodies may also be available to detect and fight certain types of cancer within the individual. It may even be possible to have them carry chemicals that will help destroy the cancerous cells.

Monoclonal antibodies are produced in pure batches—they are specific against just one antigen.

Lymphokines and Growth Factors

Lymphokines and blood cell growth factors are being investigated as possible adjunct therapy for cancer and AIDS because they potentiate the action of T cells (p. 537).

Lymphokines

Both interferon and various interleukins have been used as immunotherapeutical drugs, particularly to potentiate the ability of the individual's own T cells (and possibly B cells) to fight cancer.

Interferon is produced by leukocytes, fibroblasts, and probably most cells in response to a viral infection. When it is produced by T cells, it is called a *lymphokine.* Interferon is still being investigated as a possible cancer drug, but thus far it has proven to be effective only in certain patients, and the exact reasons cannot as yet be discerned. For example, interferon has been found to be effective in up to 90% of patients with a type of leukemia known as hairy-cell leukemia (because of the hairy appearance of the malignant cells).

When and if cancer cells carry an altered protein on their cell surface, we would expect that they should be attacked and destroyed by T_K (killer T) cells. Whenever cancer develops, it's possible that the T_K cells have not been activated. In that case, the use of lymphokines might awaken the immune system and lead to the destruction of the cancer. In one technique being investigated, researchers first withdraw T cells from the patient and activate the cells by culturing them in the presence of an interleukin. The cells are then reinjected into the patient, who is given doses of interleukin to maintain the killer activity of the T cells.

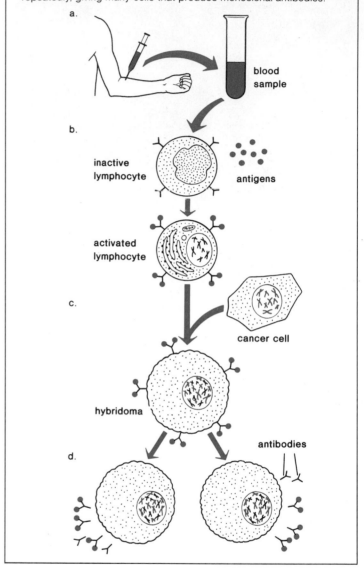

Figure 34.10

One possible method for producing human monoclonal antibodies. *a.* Blood sample is taken from a patient. *b.* Inactive lymphocytes from a sample are exposed to an antigen. *c.* Activated lymphocytes are fused with cancer cells. *d.* Resulting hybridomas divide repeatedly, giving many cells that produce monoclonal antibodies.

a.

b.

inactive lymphocyte   antigens

activated lymphocyte

c.   cancer cell

hybridoma

antibodies

d.

Growth Factors

Several growth factors that raise the blood cell count of particular white blood cells have been discovered. The best known of these is *GM-CSF* (granulocyte-macrophage colony-stimulating factor), which enhances production of macrophages, neutrophils, eosinophils, and possibly red blood cells. This growth factor has been given to patients with AIDS, and it did raise their white blood cell count. Presently, the growth factor is being produced by recombinant DNA technology, and more extensive clinical trials are expected soon.

Lymphokine therapy and white blood cell growth factors show some promise of potentiating the individual's own immune system.

Summary

1. The general defense of the body consists of barriers to entry, phagocytic white blood cells, and protective proteins.
2. Immunity is dependent upon the lymphocytes. Bone marrow stem cells produce precursor cells that seed the lymphnoid organs, such as the lymph nodes, where monocytes and lymphocytes are produced.
3. T but not B lymphocytes have passed through the thymus. B cells are responsible for antibody-mediated immunity, and T cells are responsible for cell-mediated immunity.
4. B lymphocytes can directly recognize an antigen, and thereafter they divide giving antibody-secreting plasma cells and memory cells.
5. An antibody is a Y-shaped molecule that has two binding sites. Each antibody is specific for a particular antigen.
6. Because specific lymphocytes respond to an antigen, the clonal selection theory states that the antigen selects which B or T lymphocytes will proliferate.
7. For T cells to be activated, macrophages must present an antigen, along with an MHC protein, to them. There are several types of T cells: T_K cells kill cells on contact; T_H cells stimulate B cells and other T cells; and T_S cells suppress the immune response. There are also memory T cells; these, along with the memory B cells, provide long-lasting immunity.
8. AIDS is an infection of T_H cells in which the monocytes probably serve as a reservoir for HIV viruses.
9. Immunity has certain unwanted side effects. Allergies are due to an overactive immune system that forms antibodies to substances not normally recognized as foreign.
10. T_K cells attack foreign tissues making immunosuppressive drugs necessary following transplant operations.
11. Autoimmune illnesses occur when antibodies form against the body's own tissues.
12. Immunity can be fostered by immunotherapy. Vaccines promote active immunity and monoclonal antibodies provide short-term passive immunity. Lymphokines, notably interferon, the interleukins, and white blood cell growth factors are used for therapy in cancer patients.

Objective Questions

1. Complement
   a. is a general defense mechanism.
   b. is a series of proteins present in the plasma.
   c. plays a role in destroying bacteria.
   d. All of these.
2. Which one of these does not pertain to T cells?
   a. have specific receptors
   b. have cell-mediated immunity
   c. stimulate antibody production by B cells
   d. have no effect on macrophages
3. Which one of these does not pertain to B cells?
   a. have passed through the thymus
   b. specific receptors
   c. antibody-mediated immunity
   d. synthesize and liberate antibodies
4. The clonal selection theory says that
   a. an antigen selects out certain B cells and suppresses them.
   b. an antigen stimulates the multiplication of B cells that produce antibodies against it.
   c. T cells select out those B cells that should produce antibodies regardless of antigens.
   d. T cells suppress all those B cells except the ones that should multiply and divide.
5. Plasma cells
   a. are the same as memory cells.
   b. are formed from blood plasma.
   c. are B cells that are actively secreting antibody.
   d. are inactive T cells carried in the plasma.
6. In order for a T cell to recognize an antigen, it must interact with
   a. complement.
   b. a macrophage.
   c. a B cell.
   d. All of these.
7. Antibodies combine with antigens
   a. at variable regions.
   b. at constant regions.
   c. only if macrophages are present.
   d. Both (a) and (c).
8. Which one of these is mismatched?
   a. Helper T cells—help complement react
   b. Killer T cells—active in tissue rejection
   c. Suppressor T cells—shut down the immune response
   d. Memory T cells—long-living line of T cells
9. Vaccines are
   a. the same as monoclonal antibodies.
   b. treated bacteria or viruses or one of their proteins.
   c. MHC proteins.
   d. All of these.
10. The theory behind the use of lymphokines in cancer therapy is
    a. if cancer develops, the immune system has been ineffective.
    b. lymphokines stimulate the immune system.
    c. cancer cells bear antigens that should be recognizable by T_K cells.
    d. All of these.

Study Questions

1. Distinguish between the general defense and the specific defense of the body against disease.
2. Contrast B and T cells in as many ways as possible.
3. What is the clonal selection theory?
4. Explain the process that allows a T cell to recognize an antigen.
5. List the four types of T cells, and state their functions.
6. What are lymphokines, and how are they used in immunotherapy?
7. Killer T cells are said to be responsible for what type of immunity?
8. Discuss allergies, tissue rejection, and autoimmune disease as they relate to the immune system.
9. Relate active immunity to the presence of plasma cells and memory cells.
10. How is active immunity achieved? How is passive immunity achieved?

Thought Questions

1. There are more specific cells in the immune system than any other system. What does this tell us about the environment of organisms?

2. The immune system has several different ways of attacking foreign antigens. Explain from an evolutionary point of view.

3. What would anti-antibodies be, and why would you expect B cells to produce these proteins? What could be the function of anti-antibodies?

Selected Key Terms

complement (kom′plĕ-ment) 532
interferon (in″ter-fer′on) 532
immune system (i-mūn sis′tem) 532
antigen (an′tĭ-jen) 532
T lymphocyte (te lim′fo-sīt) 533
B lymphocyte (be lim′fo-sīt) 533

antibody (an′tĭ-bod″e) 533
plasma cell (plaz′mah sel) 534
memory B cell (mem′o-re be sel) 534
MHC protein (em āch se pro′tein) 536
helper T cell (hel′per te sel) 537

lymphokine (lim′fo-kīn) 537
killer T cell (kil′er te sel) 537
suppressor T cell (sŭ-pres′or te sel) 537
memory T cell (mem′o-re te sel) 537
vaccine (vak′sēn) 539

CHAPTER 35

Digestive System and Nutrition

Your study of this chapter will be complete when you can:

1. *Contrast the characteristics of an incomplete and complete digestive tract.*

2. *Contrast the characteristics of a continuous and a discontinuous feeder.*

3. *Contrast the dendition and digestive tracts of mammalian herbivores and mammalian carnivores.*

4. *Name the parts of the human digestive system and describe, in general, their structure and function.*

5. *Outline the digestion of carbohydrates (e.g., starch), proteins, and fats in humans by listing the names and functions of the appropriate enzymes and by telling where in the human digestive tract these enzymes function.*

6. *List the contributions of the liver and pancreas to the digestive process in humans.*

7. *Name six functions of the liver.*

8. *Give the names and functions of hormones that affect the flow of digestive juices.*

9. *In general, tell how an animal's body makes use of the products of digestion.*

10. *Describe a balanced diet; describe how one achieves a balanced diet, and how one avoids too much fat, sugar, and salt in the diet.*

All animals require nutrients to serve as a source of building blocks and energy. Spiders capture and store prey within silken webs, which they spin. The silk is a fibrous protein with a strength almost equal to that of nylon. Most spiders renew their webs once a night but this is not a waste of materials because they roll up the old web and eat it.

A nimals are heterotrophic organisms that must take in preformed food. However, the various adaptations that have occurred to acquire and digest food are extremely varied. We will necessarily have to limit our discussion to a few examples. Some animals are **omnivores;** they eat both plants and animals. Others are **herbivores;** they feed only on plants. Still others are **carnivores;** they eat only other animals (fig. 35.1).

Animal Digestive Tracts

Generally, animals have some sort of gut (digestive tract). The gut may have specialized parts for storing, grinding, enzymatically digesting, absorbing nutrient molecules, and finally eliminating undigestible material.

Incomplete versus Complete Gut

The planarian is an example of an animal with an *incomplete* gut. It (fig. 35.2) is carnivorous and feeds largely on smaller aquatic animals. Its digestive system contains only a mouth, pharynx, and intestine. When the worm is feeding, the pharynx actually extends beyond the mouth. It wraps its body about the prey and uses its muscular pharynx to suck up minute quantities at a time. Digestive enzymes present in the *gastrovascular cavity* allow some extracellular (outside the cells) digestion to occur. Digestion is finished intracellularly (fig. 35.2) by the cells that line the cavity, which branches throughout the body. No cell in the body is far from the intestine and therefore diffusion alone is sufficient to distribute nutrient molecules.

Figure 35.1

A food chain having at least three links is represented here. The wildebeest (gnu) is a herbivore that feeds on photosynthetic grasses. The lion is a carnivore that feeds on herbivores. Both the herbivore and carnivore need to digest food. Digestion is necessary to break down food and to release the nutrient molecules that serve as building blocks and as a source of energy in these animals.

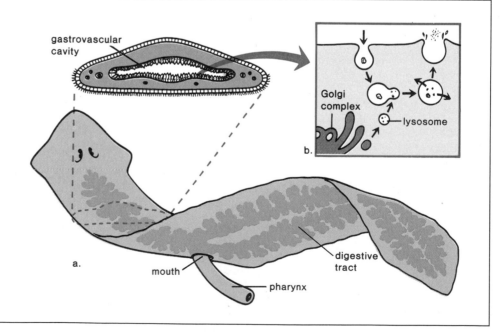

Figure 35.2

Planarians are rather simple aquatic organisms with (a) a very extensive nonspecialized digestive tract, the gastrovascular cavity. Incomplete digestive tracts have little specialization of parts because the single opening serves as both entrance and exit. The insert (b) emphasizes that the planarian relies on intracellular digestion to complete the digestive process. Phagocytosis produces a vacuole that joins with an enzyme-containing lysosome. Notice that the food particles are always surrounded by membrane; therefore, they are never part of the cytoplasm. The digested products pass from the vacuole before any nondigestible material is eliminated at the plasma membrane.

The digestive system of a planarian is notable for its lack of specialized parts. The digestive system is termed *incomplete* and saclike because the pharynx serves not only as an entrance for food, but also as an exit for nondigestible material. Specialization of parts does not occur under these circumstances.

Recall that the planarian has some parasitic relatives. The tapeworm (fig. 26.9), for example, has no digestive system at all—it simply absorbs nutrient molecules from the intestinal juices that surround its body. The body wall is highly modified for this purpose: it has millions of microscopic, fingerlike projections that increase the surface area for absorption.

In contrast to the planarian, an earthworm has a complete gut. Earthworms (fig. 35.3) feed mainly on decayed organic matter in dirt. The muscular *pharynx* draws in food with a sucking action. The *crop* is a storage area that has expansive thin walls. The *gizzard* has thick muscular walls for churning and grinding the food. Digestion is extracellular—in the *intestine*. The surface area for absorption of nutrient molecules is increased by an intestinal fold called the **typhlosole**. Undigested food passes out of the body at the anus.

Specialization of parts is obvious in the earthworm because the pharynx, crop, gizzard, and intestine each has a particular function in the digestive process. A complete digestive system leads to specialization of parts.

In contrast to the incomplete, saclike digestive system, the complete digestive system with both a mouth and an anus has many specialized parts.

Continuous versus Discontinuous Feeders

The clam is a sessile (fairly stationary) **filter feeder** that feeds *continuously*. Water is always moving into the mantle cavity by way of the incurrent siphon (slitlike opening) and depositing particles on the gills. The size of the incurrent siphon restricts the entrance of particles to a certain size, and only the smallest of these will adhere to the gills. Ciliary action moves suitably sized pieces to the labial palps, which force these pieces through the mouth into the stomach. The larger of these particles are immediately eliminated by way of the intestine and the anus. The smaller ones, however, are acted on by digestive enzymes secreted by a large digestive gland. Amoeboid cells are present throughout the tract and it is believed that these cells complete the digestive process by means of intracellular digestion.

There are filter feeders in many other phyla of the animal kingdom. For example, the marine fanworms (fig. 35.5) live in a tube and extend feathery tentacles when they are feeding. Ciliary action not only brings organic material to the tentacles, it also sends appropriately sized particles to the mouth.

Not all filter feeders are sessile. The baleen whale is an active filter feeder. The baleen, a curtainlike fringe that hangs from the roof of the mouth, filters krill, a small shrimp, from the water. This type of whale filters up to a ton of krill in a single feeding.

Another mollusk, the squid (fig. 35.4), is a *discontinuous* feeder with a feeding system that is entirely different from that of the clam. The body of a squid is streamlined and the animal moves rapidly through the water using jet propulsion (forceful expulsion of water from a tubular funnel). It is not surprising, then, that the squid is a carnivorous predator that feeds on such aquatic organisms as fish, shrimp, and worms. The head of a squid is surrounded by ten arms, two of which have developed into long slender *tentacles* whose suckers have toothed, horny rings. These tentacles seize the prey and bring it to the squid's *bearlike jaws*, which bite off pieces pulled into the mouth by the action of a *radula*, a toothy tongue (fig. 35.6). An esophagus leads to a *stomach* and *cecum* (an extension of the stomach), where digestion occurs. The squid also has a *digestive gland* that secretes digestive enzymes. Absorption takes place in the *intestine,* and undigested material passes out the *anus.*

Since the squid is a discontinuous feeder, it requires a storage area for its infrequent meals. The size of the stomach is supplemented by the cecum, which can also hold food. Discontinuous feeders, whether they are carnivores or herbivores, require such a storage area.

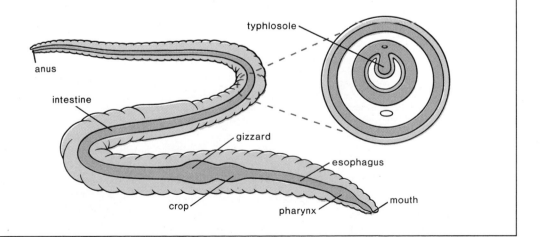

Figure 35.3
The earthworm has a complete digestive tract with both a mouth and an anus and many other specialized parts. The absorptive surface of the intestine is increased by an internal fold called the typhlosole (insert).

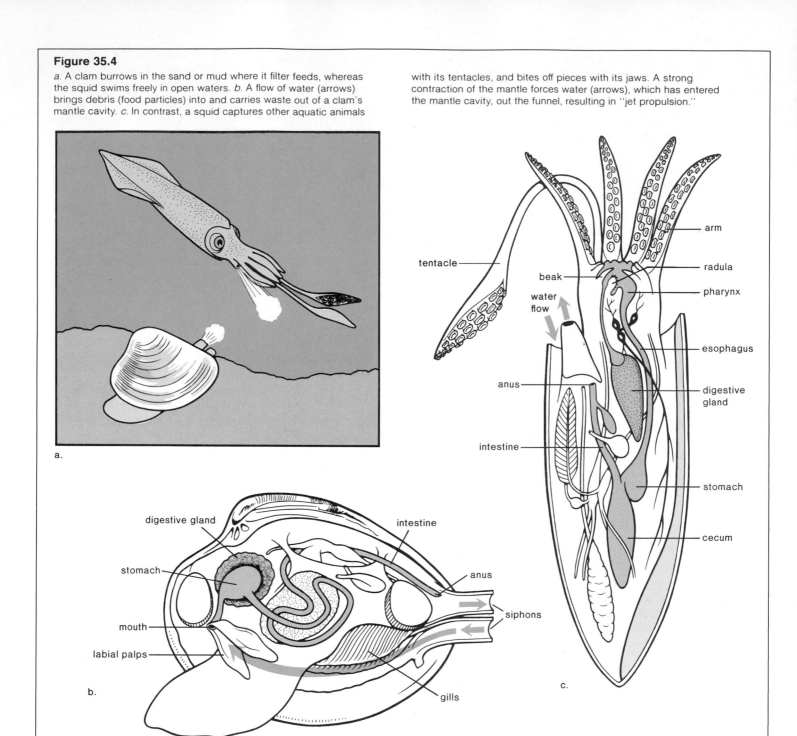

Figure 35.4

a. A clam burrows in the sand or mud where it filter feeds, whereas the squid swims freely in open waters. *b.* A flow of water (arrows) brings debris (food particles) into and carries waste out of a clam's mantle cavity. *c.* In contrast, a squid captures other aquatic animals with its tentacles, and bites off pieces with its jaws. A strong contraction of the mantle forces water (arrows), which has entered the mantle cavity, out the funnel, resulting in "jet propulsion."

Continuous feeders, such as sessile filter feeders, do not need a storage area for food; discontinuous feeders, such as herbivores and carnivores, do need a storage area for food.

Mammalian Herbivores, Carnivores, and Omnivores

Figure 35.7 contrasts the dentition (teeth) of three types of mammals. The herbivorous horse has large, strong, sharp front teeth for neatly clipping off blades of grass, its primary food source. The horse's other teeth are large and flat for grinding

up the grass so that it can be acted on by the enzymes of the digestive system. Plant material is difficult to digest because it contains a high proportion of cellulose. The stomach of a horse is supplemented by a cecum, and both of these contain a large bacterial population that can digest cellulose. However, a horse must chew its food thoroughly before swallowing. Other mammalian herbivores, like the cow and deer, chew their food at two different times. They graze quickly, sending only partially chewed plant material to a special part of the stomach, called a *rumen,* for bacterial action. This material, now called the *cud,* is regurgitated into the mouth and chewed once more when convenient for these *ruminants.*

Figure 35.5

A fanworm is a filter feeder that feeds by sweeping feathery appendages through the water and straining out food particles.

Figure 35.6

Longitudinal section of a radula from a snail. A radula contains rows of teeth that point posteriorly. When a snail is feeding, the radula extends and then scrapes against the surface of an object to bring minute particles of food into the mouth. The squid uses its beak to acquire larger-sized chunks of food and the radula propels them into the mouth.

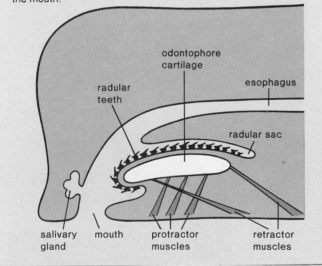

A carnivore bolts its food, swallowing it quickly in large-sized chunks. Its teeth are primarily adapted to tearing the food into pieces small enough to be swallowed. Meat is a richer and more accessible source of nutrients than is plant material. Therefore, the digestive system of carnivores lacks the digestive-tract specializations seen in herbivores.

The dentition of a mammal reflects its types of food. Herbivores need large, flat molars to grind their food; carnivores need incisors and canines that can tear off chunks; omnivores, such as humans, have less specialized teeth.

Figure 35.7

Comparison of the teeth of several mammals. *a.* The herbivorous horse is a grazing mammal that clips off plants with its incisors and grinds them with its flat molars. *b.* The lion is a carnivorous mammal that uses its teeth for killing and for tearing off chunks of flesh that are swallowed whole. *c.* The human is an omnivorous mammal that has nonspecialized teeth.

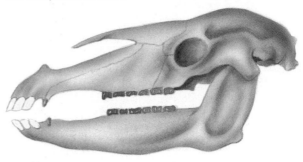

a. Horse

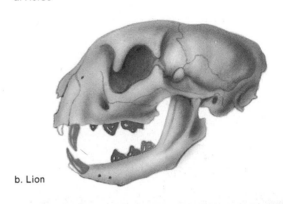

b. Lion

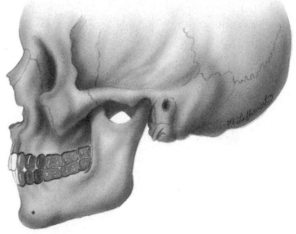

c. Man

Human Digestive Tract

The human digestive system is complex (fig. 35.8 and table 35.1) illustrating that the complete digestive system, in contrast to the incomplete one, permits specialization of parts. Each part has a specific function.

Mouth

Humans are omnivores; they feed on both plant and animal material. Therefore, their human dentition is nonspecialized, (fig. 35.7c) and able to deal adequately with both a vegetable and a

Figure 35.8
Human digestive system. The enlargement shows the four layers of
the small intestinal wall: the serosa, an outer membranous covering;
the muscularis with longitudinal and circular muscles; the
submucosa, a connective tissue layer and absorptive function.

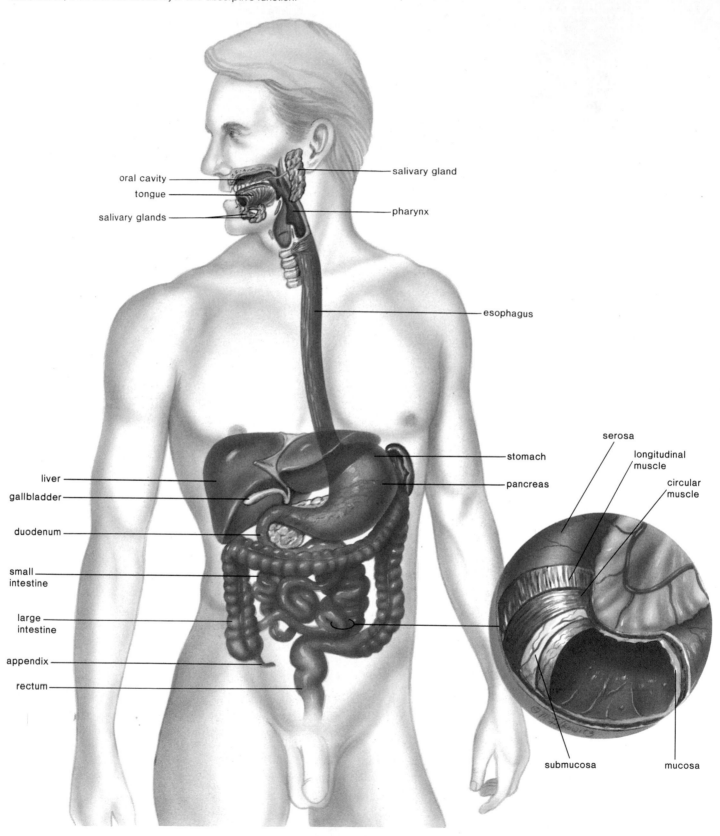

oral cavity

tongue

salivary glands

salivary gland

pharynx

esophagus

liver

gallbladder

duodenum

small intestine

large intestine

appendix

rectum

stomach

pancreas

serosa

longitudinal muscle

circular muscle

submucosa

mucosa

Table 35.1 Path of Food

Organ	Special Features	Functions
Mouth	Teeth	Chewing of food; digestion of starch to maltose
Esophagus		Passageway
Stomach	Gastric glands	Digestion of protein to peptides
Small intestine	Intestinal glands Villi	Digestion of all foods and absorption of unit molecules
Large intestine		Absorption of water
Anus		Defecation

meat diet. An adult human has 32 teeth. One-half of each jaw has teeth of four different types: two chisel-shaped *incisors* for biting; one pointed *canine* for tearing; two fairly flat *premolars* for grinding; and three *molars,* well flattened for crushing. The last molar, or wisdom tooth, may fail to erupt, or if it does, it is sometimes crooked and useless.

Three pairs of **salivary glands** (fig. 35.8) secrete saliva into the mouth where it mixes with and moistens the food, which is composed of carbohydrates, proteins, and fats. Saliva also contains an enzyme, **salivary amylase,** that begins the process of starch digestion by hydrolyzing some of the bonds between the glucose units that make up starch. The disaccharide maltose is a typical end product of salivary amylase digestion:

$$\text{starch} + H_2O \xrightarrow{\text{salivary amylase}} \text{maltose}$$

While in the mouth, food is manipulated by a muscular tongue that has touch and pressure receptors similar to those in the skin. The tongue also has *taste buds* on its surface, chemical receptors that are stimulated by the chemical composition of food. After the food has been thoroughly chewed and mixed with saliva, the tongue starts the process of swallowing by pushing the food back to the pharynx (fig. 35.9).

Pharynx and Esophagus

The digestive and respiratory passages diverge in the *pharynx* (fig. 35.10). Thus, swallowing poses a potential problem because, if food enters the *trachea* (windpipe), it could block the path of air to the lungs. Normally, however, swallowing involves a set of reflexes that closes the opening into the trachea. A flap of tissue, the **epiglottis,** covers the opening into the trachea as muscles move the food mass through the pharynx into the **esophagus,** a tubular structure that takes food to the stomach.

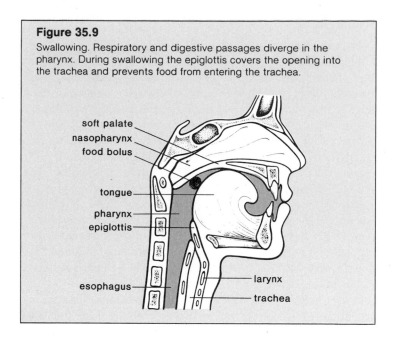

Figure 35.9
Swallowing. Respiratory and digestive passages diverge in the pharynx. During swallowing the epiglottis covers the opening into the trachea and prevents food from entering the trachea.

soft palate
nasopharynx
food bolus
tongue
pharynx
epiglottis
esophagus
larynx
trachea

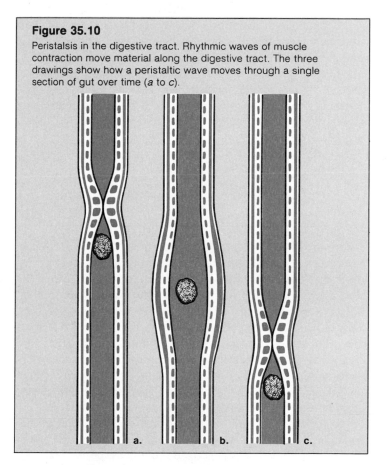

Figure 35.10
Peristalsis in the digestive tract. Rhythmic waves of muscle contraction move material along the digestive tract. The three drawings show how a peristaltic wave moves through a single section of gut over time (*a* to *c*).

a.   b.   c.

When food enters the esophagus, a rhythmical contraction of the esophagus wall termed **peristalsis** (fig. 35.10) begins; it pushes the food down the esophagus to the stomach (fig. 35.9).

Stomach

The stomach stores up to two liters of partially digested food, and even more in some cases. Because the stomach stores food, humans can periodically eat relatively large meals and spend the rest of their time at other activities. But the stomach is much more than a mere storage organ, as was discovered by William Beaumont in the mid-nineteenth century. Beaumont was an American doctor who had a French Canadian, Alexis St. Martin, as a patient. St. Martin had been shot in the stomach, and when the wound healed, he was left with a fistula, or opening, that allowed Beaumont to collect gastric (stomach) juices and to look inside the stomach to see what was going on there. Beaumont was able to determine that the muscular walls of the stomach (fig. 35.11) contract vigorously and mix food with juices that are secreted whenever food enters the stomach. He found that *gastric juice* contains *hydrochloric acid* (HC1) and a substance active in digestion. (This substance was later identified as pepsin.) He also found that the gastric juices are produced independently of the protective mucous secretions of the stomach. Beaumont's work, which was very carefully and painstakingly done, pioneered the study of the physiology of digestion.

So much hydrochloric acid is secreted by the stomach that it routinely has a pH of about 2.0. Such a high acidity usually is sufficient to kill bacteria and other microorganisms that might be in food. This low pH also stops the activity of salivary amylase, which functions optimally at the near-neutral pH of saliva, but it promotes the activity of pepsin. **Pepsin** is a hydrolytic enzyme that acts on protein to produce peptides.

$$\text{protein} + H_2O \xrightarrow{\text{pepsin}} \text{peptides}$$

By now the stomach contents have a thick, soupy consistency and are called *chyme.* At the base of the stomach is a narrow opening controlled by a *sphincter.* Whenever the sphincter relaxes, a small quantity of chyme squirts through the opening into the *duodenum,* the first part of the *small intestine* (fig. 35.8). When chyme enters the duodenum, it sets off a neural reflex that causes the muscles of the sphincter to contract vigorously and to temporarily close the opening. Then the sphincter relaxes again and allows more chyme to squirt through. The slow manner in which chyme enters the small intestine allows digestion to be more thorough than it otherwise would be.

Normally, the wall of the stomach and first part of the duodenum is protected by a thick layer of mucus, but if by chance gastric juice does penetrate the mucus, pepsin starts to digest the stomach or duodenal lining and an *ulcer* results. An ulcer is an open sore in the wall caused by the gradual disintegration of tissues. It is believed that the most frequent cause of an ulcer is oversecretion of gastric juice due to too much nervous stimulation. Persons under stress tend to have a greater incidence of ulcers.

Figure 35.11
This scanning electron micrograph gives you a greatly enlarged view of what Dr. Beaumont saw when he looked through the opening in St. Martin's stomach. The wall of the stomach has folds called rugae (Ru) that disappear when the stomach is full. The arrows indicate the openings to the gastric glands that produce gastric secretions, including hydrochloric acid and pepsinogen, a precursor that becomes pepsin in the stomach. The wall of the stomach has the same four layers as the small intestine, but with modifications. Notice how large the mucosa (MU) is because of the presence of a mucosal muscle (MM) layer; SU = submucosa; ML = a muscle layer; SE = serosa. Magnification, ×55.
Kessel, R. G., and Kardon, R. H.: *Tissues and Organs: A Text-Atlas of Scanning Electron Microscopy.* © 1979 by W. H. Freeman and Co.

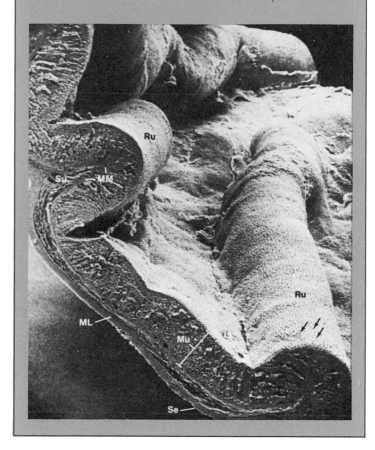

Before food enters the small intestine it passes through the mouth, esophagus, and stomach. In the mouth, salivary amylase digests starch to maltose; in the stomach pepsin digests protein to peptides.

Small Intestine

The human small intestine, a coiled tube about 3 meters long, is not a simple, smoothly lined organ; it has ridges and furrows on the inner surface that give it an almost corrugated appearance. On the surface of these ridges and furrows are small, fingerlike projections called **villi.** Cells on the surfaces of the villi have minute projections called microvilli. The villi and microvilli greatly increase the effective surface area of the small intestine. If the small intestine were simply a smooth tube, it would have to be 500 to 600 meters long to have a comparable surface area.

Figure 35.12

Anatomy of intestinal lining. *a.* The products of digestion are absorbed by villi, fingerlike projections of the intestinal wall, (*b*), each of which contains blood vessels and a lacteal.

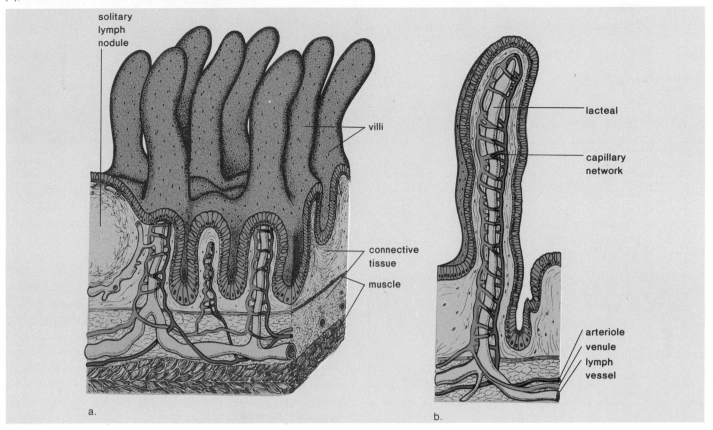

When the chyme enters the duodenum, the proteins and carbohydrates in it have been only partly digested, and fat digestion still needs to be carried out. Considerably more digestive activity is required before these nutrients can be absorbed through the intestinal wall. Two important accessory glands, the liver and the pancreas, send secretions to the duodenum (fig. 35.8). The liver produces **bile** that is stored in the **gallbladder** and sent to the duodenum by way of a duct. Bile looks green because it contains pigments that are products of hemoglobin breakdown. This green color is familiar to anyone who has observed the color changes of a bruise. Hemoglobin within the bruise is breaking down into the same types of pigments found in bile. Bile also contains bile salts, which are *emulsifying* agents that break up fat into fat droplets.

> bile salts
>
> fat ⟶ fat droplets

When fat is physically broken apart and caused to mix with water, it is emulsified. Emulsified fat is more easily acted on by enzymes.

The pancreas sends pancreatic juice into the duodenum, also by way of a duct. While the pancreas is an *endocrine* gland when it produces and secretes insulin into the bloodstream, it is an *exocrine* gland when it produces and secretes pancreatic juice into the duodenum. Besides digestive enzymes, pancreatic juice contains sodium bicarbonate ($NaHCO_3$), which neutralizes the chyme and makes the pH of the small intestine slightly basic. Pancreatic juice contains enzymes that act on every major component of food. **Pancreatic amylase** digests starch to maltose; **trypsin** digests protein to peptides; and **lipase** digests fat droplets to glycerol and fatty acids.

> pancreatic
> amylase
> starch + H_2O ⟶ maltose
>
> trypsin
> protein + H_2O ⟶ peptides
>
> lipase
> fat droplets + H_2O ⟶ glycerol + fatty acids

The cells that line the wall of the small intestine (fig. 35.12) produce the intestinal juice, which contains intestinal enzymes. These enzymes complete the digestion of protein and carbohydrates. Peptides, which result from the first step in protein

Table 35.2 Digestive Enzymes

Reaction	Enzyme	Gland	Site of Occurrence
starch + H_2O → maltose	a. Salivary amylase b. Pancreatic amylase	a. Salivary b. Pancreas	a. Mouth b. Small intestine
maltose + H_2O → glucose*	Maltase	Intestinal	Small intestine
protein + H_2O → peptides	a. Pepsin b. Trypsin	a. Gastric b. Pancreas	a. Stomach b. Small intestine
peptides + H_2O → amino acids*	Peptidases	Intestinal	Small intestine
fat + H_2O → glycerol + fatty acids*	Lipase	Pancreas	Small intestine

*Absorbed by villi.
Food is largely made up of carbohydrates (starch), protein, and fat. These very large macromolecules are broken down by digestive enzymes to small molecules that can be absorbed by intestinal villi. This table indicates the steps needed for carbohydrate digestion (starch and maltose), protein digestion (protein and peptides), and fat digestion (fat), and shows that they are all hydrolytic reactions.

digestion, are digested by peptidases to amino acids. Maltose, which results from the first step in starch digestion, is digested by maltase to glucose:

$$peptides + H_2O \xrightarrow{peptidases} amino\ acids$$

$$maltose + H_2O \xrightarrow{maltase} glucose$$

Other disaccharides, each of which is acted upon by a specific enzyme, are digested in the small intestine. Table 35.2 reviews reactions required for the digestion of food and the various digestive enzymes discussed in this chapter.

Absorption

The small intestine is specialized for absorption. Molecules are absorbed by the huge number of villi (fig. 35.12) that line the intestinal wall. Each villus contains blood vessels and a lymphatic vessel called a **lacteal.** Sugars and amino acids enter villi cells and then move absorbed into the bloodstream. Glycerol and fatty acids enter villi cells and are reassembled into fat molecules that move into the lacteals.

Absorption continues until almost all products of digestion have been absorbed. Absorption involves active transport and requires an expenditure of cellular energy.

The wall of the small intestine is lined by villi. The outer cells of the villi produce enzymes that finish the digestive process. Sugars and amino acids enter the blood vessel and reformed fats enter the lacteal of the villi.

Control of Secretions

Under normal circumstances, the digestive secretions are released *only* when food is present in the digestive tract. Saliva in the mouth and gastric juices in the stomach flow primarily in response to the taste, smell, or sometimes even the thought of food. Also, the presence of food in the stomach stimulates sense receptors that signal the brain to stimulate gastric glands in the stomach by way of nerves.

Perhaps the major factor in gastric secretory regulation, however, is the hormone *gastrin,* which is produced by the stomach itself. When proteins contact the stomach mucosa, gastrin-producing cells are stimulated to release gastrin into the bloodstream (fig. 35.13). As soon as gastrin circulates through blood vessels and reaches the acid- and enzyme-secreting cells of the stomach lining, those cells respond by secreting large quantities of HC1 and pepsin. As the stomach empties, both the neural reflexes and gastrin release subside, and less HC1 and pepsin are secreted.

Similarly, some duodenal cells produce the hormone *secretin* that stimulates the pancreas to release pancreatic juice, especially the sodium bicarbonate component. Other cells release a hormone called *CCK-PZ* (cholecystokinin-pancreozymin) that stimulates the release of bile. CCK-PZ stimulates the gallbladder to empty its contents through a duct into the duodenum. This same hormone also stimulates the pancreas, especially the enzyme secretion of the pancreas, as its full name suggests (fig. 35.13).

Liver

Blood vessels from the large and small intestine merge to form the hepatic portal vein that leads to the liver (fig. 35.14). The liver has numerous functions, including the following:

1. Detoxifies the blood by removing and metabolizing poisonous substances.
2. Makes the blood proteins.
3. Destroys old red blood cells and converts hemoglobin to the breakdown products in bile (bilirubin and biliverdin).
4. Produces bile that is stored in the gallbladder before entering the small intestine where it emulsifies fats.
5. Stores glucose as glycogen and breaks down glycogen to glucose between meals to maintain a constant glucose concentration in the blood.
6. Produces urea from amino groups and ammonia.

Figure 35.13

Hormonal control of digestive gland secretions. Especially after eating a protein-rich meal, gastrin produced by the lower part of the stomach enters the bloodstream and thereafter stimulates the upper part of the stomach to produce more digestive juices. Acid chyme from the stomach causes the duodenum to secrete secretin and CCK-PZ. Secretin and CCK-PZ stimulate the pancreas, and CCK-PZ alone stimulates the gallbladder to release bile.

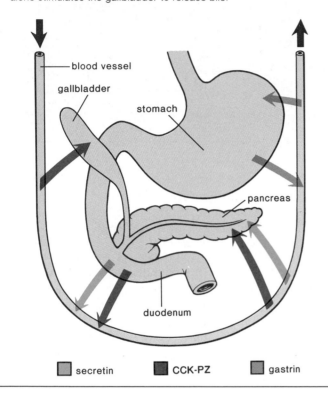

secretin   CCK-PZ   gastrin

Just now we are interested in the last two functions listed. The liver helps maintain the glucose content of blood at about 0.1% by removing excess glucose from the hepatic portal vein and storing it as glycogen. Between meals, glycogen is broken down and glucose enters the hepatic vein. Glycogen is sometimes called *animal starch* because both starch and glycogen are made up of glucose molecules joined together (fig. 4.8). If by chance the supply of glycogen and glucose runs short, the liver will convert amino acids to glucose molecules. Amino acids contain nitrogen in the form of amino groups, whereas glucose contains only carbon, oxygen, and hydrogen. Thus, before amino acids can be converted to glucose molecules, **deamination,** or the removal of amino groups from the amino acids, must take place. By an involved metabolic pathway, the liver converts these amino groups to urea, which has the following structure:

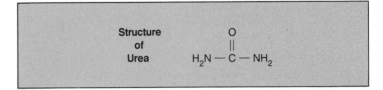

Structure of Urea

$$H_2N - \overset{\overset{\displaystyle O}{\|}}{C} - NH_2$$

Urea is the most common, nitrogenous waste product of humans, and after its formation in the liver, it is transported by the bloodstream to the kidneys where it is excreted.

Blood from the small intestine enters the hepatic portal vein, which goes to the liver, a vital organ that has numerous important functions as listed above.

Figure 35.14

Hepatic portal system, a system that begins and ends in capillaries. Blood leaving the digestive tract flows through the hepatic portal vein to the liver, where it enters spaces among the liver cells. Later, it enters the hepatic vein, which joins with the vena cava.

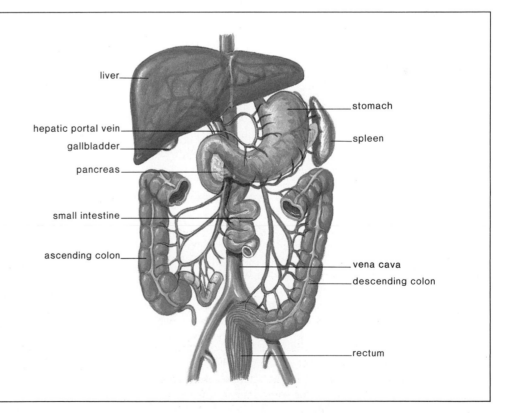

liver

hepatic portal vein

gallbladder

pancreas

small intestine

ascending colon

stomach

spleen

vena cava

descending colon

rectum

Digestive System and Nutrition   **555**

Figure 35.15

In order to have a balanced diet, it is recommended that you eat a variety of foods from each food group every day. *a.* Milk and milk products. *b.* Meat, fish, and poultry. *c.* Breads and cereals. *d.* Fruits and vegetables.

Liver Disorders

Jaundice When a person is jaundiced there is a yellowish tint to the skin due to an abnormally large amount of hemoglobin breakdown products in the blood. In *hemolytic jaundice,* red blood cells are being destroyed in increased amounts; in *obstructive jaundice,* there is an obstruction of the bile duct or damage to the liver cells, and this causes an increased amount of bilirubin to enter the bloodstream. Obstructive jaundice often occurs when crystals of cholesterol precipitate out of bile and form gallstones.

Jaundice is also frequently caused by liver damage due to viral hepatitis, a term that includes Type A, which is most often caused by eating shellfish from polluted waters, and Type B, which is commonly spread by means of blood transfusions, kidney dialysis, and injection with inadequately sterilized needles.

Cirrhosis This is a chronic disease of the liver in which the organ first becomes fatty, and then liver tissue is replaced by inactive fibrous scar tissue. This condition is common among alcoholics, where it is most likely caused by the excess demand placed upon the liver to detoxify alcohol and break it down. When liver cells and other types of cells break down alcohol it is converted to acetyl CoA and these molecules can be converted to fatty acids (fig. 9.12). Smooth endoplasmic reticulum increases dramatically in the liver of alcoholics—this may be the first step toward cirrhosis.

Large Intestine

About 1.5 liters of water enter the digestive tract daily as a result of eating and drinking. An additional 8.5 liters enter the digestive tract each day carrying the various substances secreted by the digestive glands. About 95% of this water is reabsorbed into cells of the large intestine, or *colon.* This water reabsorption is essential. Failure to reabsorb water can result in *diarrhea,* which can lead to serious dehydration and ion loss, especially in children.

In addition to water, the large intestine absorbs some sodium and other ions from the material passing through it. At the same time, colon cells excrete still other metallic ions into the wastes leaving the body. Thus, the colon functions both in water conservation and ion regulation. *Vitamin K,* produced by intestinal bacteria, is also absorbed in the colon.

The last 20 cm of the large intestine is the rectum, which terminates in the anus, an external opening. Digestive wastes (feces) eventually leave the body through the rectum and anus. Feces are about 75% water and 25% solid matter. Almost one-third of this solid matter is made up of intestinal bacteria. The remainder is undigested plant material, fats, waste products (such as bile pigments), inorganic material, mucus, and dead cells from the intestinal lining.

Appendicitis and Colostomy

Two serious medical conditions are associated with the large intestine. The small intestine joins the large intestine in such a way that there is a blind end on one side (fig. 35.9). This blind sac, or cecum, has a small projection about the size of the little finger, called the appendix. In the case of *appendicitis,* the appendix becomes infected and so filled with fluid that it may burst. The appendix should be removed before it bursts, because a burst appendix can lead to a serious, generalized infection of the abdominal lining.

The colon is also subject to the development of polyps, small growths that appear on epithelial tissue. Whether polyps are benign or cancerous, they can be individually removed along with a portion of the colon if necessary (called *colostomy*).

Nutrition

In order to be sure that the essential nutrients are included in the diet, it is necessary that we eat a balanced diet. A balanced diet is obtained by eating a variety of foods from the four food groups pictured in figure 35.15.

Food largely consists of proteins, carbohydrates, and fats. Therefore we will begin by considering these substances. Vitamins and minerals are also important nutrients and will be discussed later in this chapter.

Table 35.3 Analysis of Meal for 19-Year-Old Woman and Man

Food	Amount	Energy (Kcal)	Protein(g)	Calcium (mg)	Vitamin C (mg)	Vitamin A(IU)*
Roast beef (lean & fat)	3 oz	165	25	11	—	10
Baked potato	1 medium	145	4	14	31	Trace
Broccoli	½ cup	20	2.5	68	70	3,880
Roll	1	155	5	24	Trace	Trace
Margarine	1 pat	35	Trace	1	0	170
Milk, low-fat (1%)	1 cup	100	8	300	2	500
Apple	1 medium	80	Trace	10	6	120
Totals		700	44.5	428	109	4,680
RDA for 19-year-old woman			44	800	60	4,000
% RDA			101%	53%	182%	117%
RDA for 19-year-old man			56	800	600	5,000
% RDA			80%	52%	182%	94%

*From: *Decisions in Nutrition* by Vincent Hegarty. Copyright © 1988 Times Mirror/Mosby College Publishing, St. Louis, Missouri.

Proteins

Foods rich in protein include meat, fish, poultry, dairy products, legumes, nuts, and cereals. Following digestion of protein, amino acids enter the blood and are transported to the tissues. Most are incorporated into the structural proteins found in muscles, skin, hair, and nails. Others are used to synthesize such proteins as hemoglobin, plasma proteins, enzymes, and hormones.

Protein formation requires twenty different types of amino acids. Of these, nine are required in the diet because the body is unable to produce them. These are termed the **essential amino acids.** The body produces the other amino acids by simply transforming one type into another type. Some protein sources, such as meat, are *complete* in the sense that they provide all the different types of amino acids. Vegetables and grains do supply us with amino acids, but each one alone is an *incomplete* source because at least one of the essential amino acids is absent. However, it is possible to combine foods in order to acquire all of the essential amino acids. For example, the combination of cereal with milk or beans with rice will provide all the essential amino acids.

A complete source of protein is absolutely necessary to ensure a sufficient supply of the essential amino acids.

Even though we tend to put a lot of stress on protein intake, it is not difficult for most Americans to meet the daily requirement. For example, for the meal described in table 35.3, the woman has met her recommended dietary (or daily) allowance (RDA) for protein for the whole day with a *single* serving of meat. (In the United States, the RDAs are determined by the National Research Council, a part of the National Academy of Sciences.) Unfortunately, the manner in which we usually meet our daily requirement for protein is not always the most healthy way. This is because the foods that are richest in protein are apt to also be the ones that are richest in fat. In the sample meal

the roast beef serving provides 25 grams of protein and 165 Kcalories.[1] There are 4 Kcalories in each gram of protein; therefore, 65 of the Kcalories in the roast beef serving was from fat. Considering this, it is probably a good idea to depend on protein from plant origins (e.g., whole grain cereals, dark breads, legumes) because they are low in fat, to a greater extent than is the custom in this country.

[1]One kilocalorie (Kcal) is the amount of heat required to raise 1 gram of water 1° C.

Carbohydrates

The quickest, most readily available source of energy for the body is carbohydrates, which can be *complex* as in breads and cereals or *simple* as in candy, ice cream, and soft drinks. As mentioned previously, starches are digested to glucose, which is stored by the liver in the form of glycogen. Between eating, the blood glucose is maintained at about 0.1% by the breakdown of glycogen or by the conversion of amino acids to glucose. If necessary, these amino acids are taken from the muscles, even from heart muscle. To avoid this situation, it is suggested that the diet contain at least 100 grams of carbohydrate daily.

Actually, the dietary guidelines produced jointly by the U.S. Department of Agriculture and the Department of Health and Human Services recommend that we increase the proportion of carbohydrate Kcalories in the diet from the present level of 46% to 58%. Further, it is assumed that these carbohydrates will be of the complex type. The simple carbohydrates (e.g., sugars) are labeled "empty calories" by some because they only contribute to energy needs and weight gain and do not supply other nutritional requirements. Table 35.4 gives suggestions on how to cut down on your consumption of sugar as well as fat and sodium.

In contrast to simple sugars, complex carbohydrates are likely to be accompanied by a wide range of other nutrients and fiber. Fiber does not occur in meat or other animal products, and it is refined out of sugar and white flour. Some studies have suggested that "soluble fibers" found in fruit, oat bran, and beans lower serum cholesterol. It's reasoned that cholesterol binds to this type of fiber and is eliminated with the feces. Other studies have suggested that "insoluble fibers" found in wheat bran, whole wheat, beans, fruits, and vegetables reduce the risk of colon cancer. This type of fiber has a laxative effect, and it is reasoned that its presence reduces the amount of time that potential mutagens in the feces have to act on the colon.

While the diet should be *adequate* in the amount of fiber, a very high-fiber diet may be detrimental. Some evidence suggests that the absorption of iron, zinc, and calcium may be impaired by a high-fiber diet.

Carbohydrate is needed in the diet to maintain the blood glucose level. Complex carbohydrates, along with fiber, are considered to be beneficial to health.

Carbohydrates provide most of the dietary Kcalories. There are only 4 Kcalories per gram of carbohydrate as compared to 9 Kcalories per gram of fat, but since carbohydrates are the bulk of the diet, they provide the most calories.

Fats

Fats are present not only in butter, margarine, and oils, but also in foods high in protein. Fats of an animal origin tend to have saturated fatty acids, and those from plants tend to have un-

Table 35.4 Recommendations for Improved Diet

The Less-Fat Recommendations:

1. Choose lean meat, poultry, fish, dry beans, and peas. Trim fat off.
2. Eat eggs and such organ meats as liver in moderation. (Actually, these are high in cholesterol rather than fat.)
3. Broil, boil, or bake, rather than fry.
4. Limit your intake of butter, cream, hydrogenated oils, shortenings, coconut oil.

The Less-Salt Recommendations:

1. Learn to enjoy unsalted food flavors.
2. Add little or no salt to foods at the table and add only small amounts of salt when you cook.
3. Limit your intake of salty prepared foods such as pickles, pretzels, and potato chips.

The Less-Sugar Recommendations:

1. Eat less sweets such as candy, soft drinks, ice cream, and pastry.
2. Eat fresh fruit or canned fruit without heavy syrup.
3. Use less sugar—white, brown, raw—and less honey and syrups.

Source: American Dietetic Association based on *Dietary Guidelines for Americans 1980*, U.S. Department of Agriculture and Department of Health, Education, and Welfare.

saturated fatty acids. After being absorbed into lacteals, the products of fat digestion are transported by the blood to the tissues. The liver can alter ingested fats to suit the body's needs, except it is unable to produce the fatty acid *linoleic acid*. Since this is required for phospholipid production, it is considered to be an *essential fatty acid*.

While we need to ingest some fat in order to satisfy our need for linoleic acid, the dietary guidelines mentioned earlier suggest that we reduce the amount of fat from the typical 40% of total Kcalorie intake to 30%. Dietary fat has been implicated in cancer of the colon, pancreas, prostate, and breast (fig. 35.16). Many animal studies have shown that a high-fat diet stimulates the development of mammary tumors while a low-fat diet does not. Similarly it has been found that women who have a high-fat diet are more likely to develop breast cancer. Mammary tumors in animals and breast cancer in women produce abundant amounts of prostaglandins and fatty acids are precursors for these substances. Surprisingly, it's been discovered that reduced amounts of linoleic acid in the diet help prevent the occurrence of breast cancer. Linoleic acid is found in corn, safflower, sunflower, and other common plant oils, but it is not abundant in olive oil or in fatty fishes and marine animals.

As a nation we have in recent years increased our consumption of fat from plant sources and have decreased our consumption from animal sources, such as red meat and butter. Most likely this is due in part to new information concerning the link between saturated fatty acids and cholesterol with hypertension and heart attack.

Fats are the highest energy foods; they contain about 9 Kcalories per gram. Raw potatoes have about 0.9 Kcalories per gram, but when they are cooked in fat, the number of

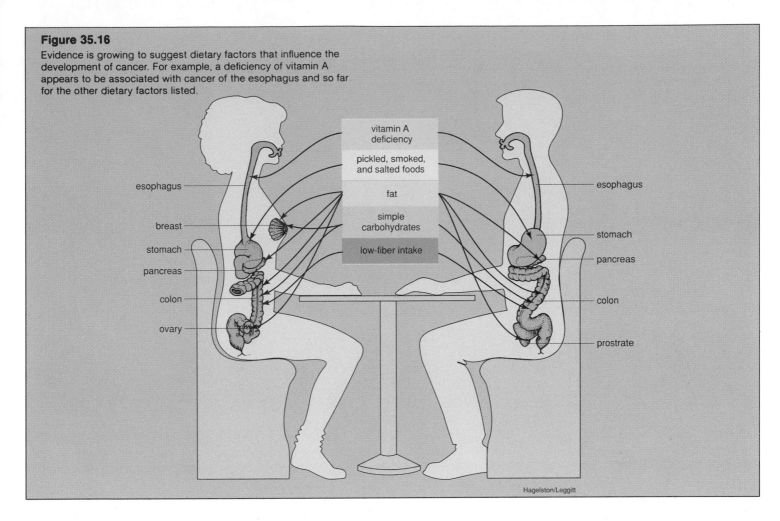

Figure 35.16

Evidence is growing to suggest dietary factors that influence the development of cancer. For example, a deficiency of vitamin A appears to be associated with cancer of the esophagus and so far for the other dietary factors listed.

vitamin A deficiency

pickled, smoked, and salted foods

fat

simple carbohydrates

low-fiber intake

esophagus

breast

stomach

pancreas

colon

ovary

esophagus

stomach

pancreas

colon

prostrate

Hagelston/Leggitt

Kcalories jumps up to 6 Kcalories per gram. Another problem for those trying to limit their Kcaloric intake is that fats are not always highly visible; butter melts on toast or potatoes, and salad dressings coat lettuce leaves. Table 35.4 gives suggestions about how to cut down on the amount of fat in the diet.

Fats have the highest caloric content, but they should not be avoided entirely because of the need for the essential fatty acid linoleic acid.

Vitamins and Minerals

Vitamins

Vitamins are organic compounds (other than carbohydrates, fats, and proteins) that the body is unable to produce but needs for metabolic purposes. Therefore, vitamins must be present in the diet; if they are lacking, various symptoms develop (fig. 35.17). There are many substances that are advertised as being vitamins, but in reality there are only thirteen vitamins (table 35.5). The table also gives the RDA vitamin values for a "typical" female and a "typical" male. In general, carrots, squash, turnip greens, and collards are good sources for vitamin A. Citrus fruits, other fresh fruits, and vegetables are natural sources of vitamin C. Fortified milk is the primary source of vitamin D, and whole grains are a good source of B vitamins.

By using the data given in tables 35.3 and 35.5, you can see that it is not difficult to acquire the RDAs for vitamins if your diet is balanced (fig. 35.15) because each vitamin is needed in small amounts only. Many vitamins are portions of coenzymes. For example, niacin is part of the coenzyme NAD[†], and riboflavin is a part of FAD. Coenzymes are needed in only small amounts because each one can be used over and over again.

The National Academy of Sciences suggests that we eat more fruits and vegetables in order to acquire a good supply of vitamins C and A because these two vitamins may help guard against the development of cancer. Nevertheless they discourage the intake of excess vitamins by way of pills because the practice can possibly lead to illness. For example, excess vitamin C can cause kidney stones, and can also be converted to oxalic acid, a molecule that is toxic to the body. Vitamin A taken in excess can cause peeling of the skin, bone and joint pains, and liver damage. Excess vitamin D can cause vomiting, diarrhea, loss of weight, and kidney damage. Megavitamin therapy should, therefore, always be supervised by a physician.

A properly balanced diet will include all the vitamins and minerals needed by most individuals to maintain health.

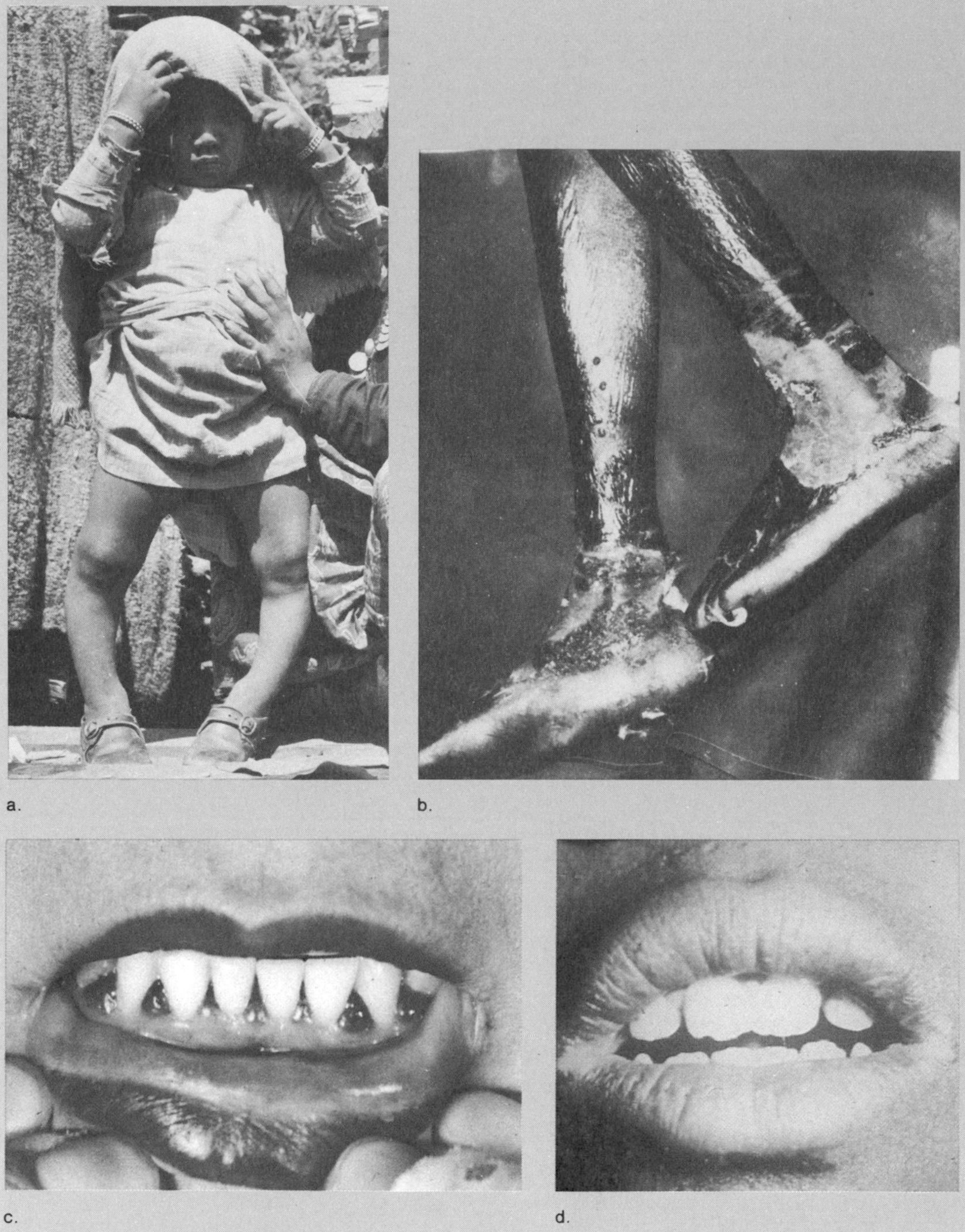

Figure 35.17
Illnesses due to vitamin deficiency. *a.* Bowing of bones (rickets) due to vitamin D deficiency. *b.* Dermatitis of areas exposed to light (pellagra) due to niacin deficiency. *c.* Bleeding of gums (scurvy) due to vitamin C deficiency. *d.* Fissures of lips (cheilosis) due to riboflavin deficiency.

Minerals

In addition to vitamins, various minerals are also required by the body. Some minerals (calcium, phosphorus, potassium, sulfur, sodium, chlorine, and magnesium) are recommended in amounts more than 100 mg per day. They serve as constituents of cells and body fluids and as structural components of tissues. For example, calcium is needed for the construction of bones and teeth and also for nerve conduction and muscle contraction. The microminerals (iron, manganese, copper, iodine, cobalt, and zinc) are recommended in amounts of less than 20 mg per day.

The microminerals are more likely to have very specific functions. For example, iron is needed for the production of hemoglobin, and iodine is used in the production of thyroxin, a hormone produced by the thyroid gland. As research continues, more and more elements have been added to the list of those considered essential. During the past three decades, molybdenum, selenium, chromium, nickel, vanadium, silicon, and even arsenic have been found to be essential to good health in very small amounts.

Table 35.5 Vitamins—Their Functions and Sources

Name	Body Functions‡	Major Food Source‡	RDA‡ Female*	RDA‡ Male	Deficiency†	Excess†
Fat-Soluble Vitamins						
Vitamin A	Vision; health of skin; growth of teeth, nails, hair, bones, and glands	Dairy products, liver, deep green and orange vegetables	800 mg	1,000 mg	Blindness, esp. night blindness	Headache, vomiting, peeling of skin, liver damage, anorexia, swelling of long bones
Vitamin D	Bones and teeth	Dairy products, egg yolk	7.5 mg	7.5 mg	Bone deformities, rickets in children	Vomiting, diarrhea, loss of weight, kidney damage
Vitamin E	Antioxidant—prevents cell damage	Oils, nuts, seeds	8 mg	10 mg	None known in humans	Relatively nontoxic
Vitamin K	Blood clotting	Green leafy vegetables, meats			Possibly severe bleeding	Synthetic forms at high doses may cause jaundice.
Water-Soluble Vitamins						
Vitamin C	Antioxidant; collagen formation; health of teeth and gums	Citrus fruits, green pepper, broccoli, cantaloupe	60 mg	60 mg	Scurvy (degeneration of skin, teeth, blood vessels)	Possibility of kidney stones
Thiamin (Vitamin B_1)	Nerve function; aids in energy metabolism	Whole grains and cereals, meats (especially pork)	1.1 mg	1.5 mg	Beriberi (nerve changes, edema, heart failure)	None reported
Riboflavin (Vitamin B_2)	Aids in energy metabolism; protects against skin and eye disorders	Whole or enriched grains, milk, eggs, cheese, meats, green vegetables	1.3 mg	1.7 mg	Cheilosis (cracks at corners of mouth), lesions of eye	None reported
Niacin	Aids in energy metabolism	Lean meats, fish, whole grains, cheese, peanuts, vegetables, eggs	14 mg	19 mg	Pellagra (skin and gut lesions; nervous and mental disorders)	Flushing, burning, and tingling around neck, face, and hands.
Vitamin B_6	Aids in amino acid—protein metabolism; nerve function	Liver, fish, nuts, meats, potatoes, some vegetables	2.0 mg	2.2 mg	Irritability, convulsions, kidney stones	None reported
Pantothenic acid	Aids in energy metabolism; nerve function	Most foods of plant and animal origin	—	—	Sleep disturbances, fatigue, impaired coordination, nausea (rare in humans)	None reported
Biotin	Aids in energy metabolism	Most foods of plant and animal origin	—	—	Fatigue, depression, nausea, dermatitis, muscular pains	None reported
Folacin	Synthesis of DNA, RNA (genetic material); prevents a type of anemia; red blood cell formation	Liver, green leafy vegetables, peanuts	400 mg	400 mg	Anemia, gastrointestinal disturbances, red tongue	None reported
Vitamin B_{12}	Red blood cell formation; synthesis of DNA and RNA; component of sheath around nerves	Foods of animal origin, fermented foods	3.0 mg	3.0 mg	Pernicious anemia, neurological disorders	None reported

*females 19–22, 120 lbs., 5′4″; males 19–22, 154 lbs., 5′10″
†From Nevin S. Scrimshaw and Vernon R. Young, "The Requirements of Human Nutrition." Copyright © 1976 Scientific American, Inc. All rights reserved.
‡From *Decisions in Nutrition* by Vincent Hegarty. Copyright © 1988 Times Mirror/Mosby College Publishing, St. Louis, Missouri.

Some individuals do not receive enough iron (in women), calcium, magnesium, or zinc in their diets. Adult females need more iron in the diet than males (RDA of 18 mg compared to 10 mg) because they lose hemoglobin each month during menstruation. Stress can bring on a deficiency of magnesium, and a vegetarian diet might lack zinc, which is usually obtained from meat. However, a varied and complete diet will usually supply the RDAs for minerals.

Calcium   There has been much interest of late in calcium supplements to counteract the possibility of developing *osteoporosis,* a degenerative bone disease that afflicts an estimated one-fourth of older men and one-half of older women in the United States. These individuals have porous bones that tend to break easily because their bones lack sufficient calcium. In 1984 a National Institutes of Health conference on osteoporosis advised postmenopausal women to increase their intake of calcium to 1,500 mg and all others to 1,000 mg (compared with the RDA of 800 mg).

However, recent studies have shown that calcium supplements cannot prevent osteoporosis after menopause even when the dosage is 3,000 mg a day. The reason is that the body becomes less able to take in calcium after about the age of thirty-five. The most effective defense against osteoporosis in older women is now believed to be *estrogen replacement* and *exercise.* In one survey, women ages thirty-five to sixty-five who took a 50-minute aerobics class three times a week lost only 2.5% of the density in their forearm bones, compared with 9.5% for women who did not exercise.

Young women can guard against the possibility of developing osteoporosis when they grow older by forming strong, dense bones before menopause occurs. Eighteen-year-old women are apt to get only 679 mg of calcium a day when the RDA is 800 mg. They should consume more calcium-rich foods such as milk and dairy products. Taking calcium supplements may not be as effective. A cup of milk supplies 270 mg of calcium while a 500 mg tablet of calcium carbonate provides only 200 mg. The rest is just not taken up by the body, it is not in a form that is *bioavailable.* An excess of bioavailable calcium can lead to kidney stones.

Dietary calcium and exercise, plus estrogen therapy if needed, are the best safeguards against osteoporosis.

Sodium   The recommended amount of sodium intake per day is 400–3,300 mg, and the average American takes in 4,000–4,700 mg, mostly as salt (sodium chloride). In recent years this imbalance has caused concern because high-sodium intake has been linked to hypertension in some people. About one-third of the sodium we consume occurs naturally in foods; another third is added during commercial processing; and we add the last third either during home cooking or at the table in the form of table salt.

Clearly it is possible for us to cut down on the amount of sodium in the diet by reducing the amount of salt we eat. Table 35.4 gives recommendations for doing so.

Summary

1. Some animals (e.g., planarians) have an incomplete digestive tract that shows little specialization of parts. Other animals (e.g., earthworms) have a complete digestive tract that does show specialization of parts.
2. Some animals are continuous feeders, (e.g., clam, which are sessile filter feeders); others are discontinuous feeders (e.g., squid). Discontinuous feeders need a storage area for food.
3. All mammals have teeth. Herbivores need teeth that can clip off plant material and grind it up. Also, the herbivore's stomach contains bacteria that can digest cellulose.
4. Carnivores need teeth that can tear and rip meat into pieces. Meat is easier to assimilate so the digestive system of carnivores has less specialization of parts and a shorter intestine than that of herbivores.
5. In the human digestive tract, food is chewed and manipulated in the mouth where salivary glands secrete saliva.

Saliva contains salivary amylase, which begins carbohydrate digestion.
6. Food then passes down the esophagus to the stomach. The stomach stores and mixes food with mucus and gastric juice to produce chyme. Pepsin begins protein digestion here.
7. Chyme gradually enters the duodenum where bile, pancreatic juice, and the intestinal secretions are found. Enzymes in the small intestine hydrolyze all of the organic nutrients. Table 35.2 summarizes the enzymes involved in digesting food.
8. Most nutrient absorption takes place in the small intestine, but some water and minerals are absorbed in the colon. Digestive wastes leave the colon by way of the anus.
9. The liver produces bile, which is stored in the gallbladder. The liver is also involved in the processing of absorbed nutrient molecules and in maintaining the blood concentration of nutrient

molecules, such as glucose. The liver converts ammonia to urea and breaks down toxins.
10. The pancreas produces digestive enzymes and hormones (insulin and glucagon) involved in the control of carbohydrate metabolism in the body.
11. There are three hormones that regulate digestive-tract secretions: gastrin, which stimulates acid- and enzyme-secreting cells in the stomach; secretin, which stimulates the pancreas to release sodium bicarbonate; and CCK-PZ, which stimulates the gallbladder to release bile and the pancreas to release digestive enzymes.
12. A balanced diet is required for good health. Food should provide us with all the necessary vitamins, minerals, amino acids, fatty acids, and an adequate amount of energy.

Objective Questions

1. Animals that feed discontinuously
   a. must have digestive tracts that permit storage.
   b. are able to avoid predators by limiting their feeding time.
   c. exhibit extremely rapid digestion.
   d. Both (a) and (b).
2. In which of the following types of animals would you expect the digestive system to be more complex?
   a. those with a single opening for the entrance of food and exit of wastes
   b. those with two openings, one serving as an entrance and the other as an exit
   c. only those complex animals that also have a respiratory system
   d. Both (b) and (c).
3. The typhlosole within the gut of an earthworm compares best to which of these organs in humans?
   a. teeth in the mouth
   b. esophagus in the thoracic cavity
   c. folds in the stomach
   d. villi in the small intestine
4. Which of these animals is a continuous feeder with a complete gut?
   a. planarian
   b. clam
   c. squid
   d. lion
5. The most common food digested in the human stomach is
   a. carbohydrate.
   b. fat.
   c. protein.
   d. nucleic acid.
6. Assuming normal body temperature, which of these combinations is most likely to result in complete digestion?
   a. fat, bile, sodium bicarbonate, lipase
   b. fat, bile, sodium bicarbonate, pepsin
   c. fat, HCl, maltase
   d. fat, bile, HCl, sodium bicarbonate, trypsin
7. Which association is incorrect?
   a. protein—trypsin
   b. fat—lipase
   c. maltose—pepsin
   d. starch—amylase
8. Most of the absorption of the products of digestion takes place in humans across the
   a. squamous epithelium of the esophagus.
   b. convoluted walls of the stomach.
   c. fingerlike villi of the small intestine.
   d. smooth wall of the large intestine.
9. The hepatic portal vein is located between the
   a. hepatic vein and the vena cava.
   b. mouth and the stomach.
   c. pancreas and the small intestine.
   d. small intestine and the liver.
10. Bile in humans
    a. is an important enzyme for the digestion of fats.
    b. is made by the gallbladder.
    c. emulsifies fat.
    d. All of these.
11. Which of these is not a function of the human liver?
    a. produce bile
    b. store glucose
    c. produce urea
    d. make red blood cells
12. The large intestine in humans
    a. digests all types of food.
    b. is the longest part of the intestinal tract.
    c. absorbs water.
    d. is connected to the stomach.

Study Questions

1. Contrast the incomplete with the complete gut using the planarian and earthworm as examples.
2. Contrast a continuous with a discontinuous feeder using the clam and squid as examples.
3. Contrast the dentition of the mammalian herbivore with the mammalian carnivore using the horse and lion as examples.
4. List the parts of the human digestive tract, anatomically describe them, and state the contribution of each to the digestive process.
5. State the location and describe the functions of both the liver and pancreas.
6. Assume that you have just eaten a ham sandwich. Discuss the digestion of the contents of the sandwich.
7. Discuss the absorption of the products of digestion into the circulatory system.
8. What are gastrin, secretin, and CCK-PZ? Where are they produced and what are their functions?
9. Give reasons why carbohydrates, fats, proteins, vitamins, and minerals are all necessary to good nutrition.

Thought Questions

1. Explain why the tube-within-a-tube body plan leads to specialization of parts.
2. Explain the difference between extracellular and intracellular digestion, and contrast the relative importance of each in planarian and human digestion.
3. Which would you expect to be longer: the digestive tract of a two-pound cat or that of a two-pound rabbit? Why?

Selected Key Terms

omnivore (om′nĭ-vōr) 546
herbivore (her′bĭ-vōr) 546
carnivore (kar′nĭ-vōr) 546
salivary amylase (sal′ĭ-ver-e am′ĭ-lās) 551
epiglottis (ep″ĭ-glot′is) 551
esophagus (ĕ-sof′ah-gus) 551
peristalsis (per″ĭ-stal′sis) 551

pepsin (pep′sin) 552
villi (vil′i) 552
bile (bīl) 553
gallbladder (gawl′blad-der) 553
pancreatic amylase (pan″kre-at′ik am′ĭ-lās) 553

trypsin (trip′sin) 553
lipase (lip′ās) 553
lacteal (lak′te-al) 554
essential amino acid (ĕ-sen′shal ah-me′no as′id) 557
vitamin (vi′tah-min) 559

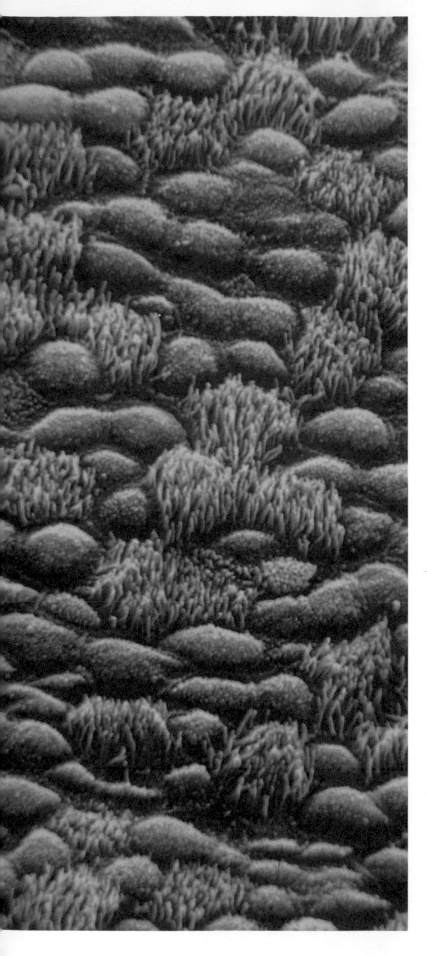

CHAPTER 36

Respiratory System

Your study of this chapter will be complete when you can:

1. *Describe the mechanism of gas exchange in a hydra and a planarian.*

2. *Compare the respiratory organs of aquatic and terrestrial animals.*

3. *Compare the various methods for ventilating the lungs among terrestrial vertebrates.*

4. *Compare the incomplete method of ventilation used by amphibians, reptiles, and mammals to the complete method used by birds.*

5. *Describe the path of air in humans, and describe in general the structure and function of all organs mentioned.*

6. *Describe how the breathing rate is controlled in humans.*

7. *Give the equations applicable to the exchange of gases in the lungs and tissues, and tell how the gases are transported in the blood.*

8. *Name three types of hemoglobin found in humans.*

Some of the cells that line the trachea (windpipe) are ciliated. The cilia beat and move foreign substances back up into the throat where it can be swallowed. These cilia sometimes disappear in smokers and the cigarette smoke causes their lungs to turn black.

E ach cell in an animal's body must acquire oxygen and rid itself of carbon dioxide. Some animals are small enough that each cell makes these exchanges directly with the external environment. Other animals are complex and have a respiratory system with a localized boundary across which gas exchange takes place. These animals also have a circulatory system that brings oxygen to and collects carbon dioxide from each cell.

Animal Respiratory Systems

A supply of oxygen must be present in the water or air in order for respiration to take place. Without respiration, an animal dies. This was first observed in 1774 by the English scientist Joseph Priestley, who collected oxygen by heating red mercuric oxide. His contemporary, the Frenchman Antoine Lavoisier, correctly deduced that both combustion by a lit candle and respiration by an animal remove oxygen from the air. This was one of the first times that a physiological process was explained with reference to a nonliving mechanism.

Gas exchange takes place by the physical process of *diffusion*. For diffusion to be effective, the gas-exchange region must be (1) moist, (2) thin, and (3) large in relation to the size of the body. We will now compare how these requirements are met in aquatic and terrestrial animals (fig. 36.1).

Aquatic Animals

Water fully saturated with air contains about twenty times less oxygen than air does. This means that aerobic aquatic animals need to have a more efficient means of acquiring oxygen than do terrestrial animals. Even so, some small aquatic organisms do *not* have a specialized region for gas exchange. Their organization is such that they have a large surface area in compar-

ison to their size. Therefore, in animals such as hydras and planarians, it is possible for each cell to carry out gas exchange with the water environment (fig. 36.2).

However, many aquatic organisms do have a specialized region for gas exchange. Most respire by means of vascularized (fig. 36.3) **gills,** which can be external or internal. Among invertebrates, gills are typically outgrowths of the body surface. They can be simple, such as those found in sea stars (fig. 27.14), or they can be finely divided, as in the clam (fig. 27.5). Among vertebrates, the gills of fishes are extensions of the digestive tract. Water enters the pharynx and moves through the gill slits past the gills, which may be protected by an outer flap called an *operculum* (fig. 36.3a). The gills are composed of *filaments* that are themselves folded into platelike *lamellae* (fig. 36.3b and c). In the capillaries of each lamella, the blood flows in a direction opposite to the movement of water across the gills. This *countercurrent flow* increases the amount of oxygen that can be taken up; as the blood in each lamella gains oxygen, it encounters water having a higher oxygen content (fig. 36.3d). Even though this is a highly effective means of acquiring oxygen, fishes expend up to 25% of their energy for breathing, compared to 1% to 2% in terrestrial mammals, because water is more dense and much less rich in oxygen than air.

Terrestrial Animals

The earthworm is an example of an invertebrate, terrestrial animal that uses its body surface (fig. 27.8) for respiration. Therefore, this surface must remain thin and moist and cannot serve as a protection against drying out. As long as the worm remains buried beneath the moist ground there is little danger, but an earthworm certainly cannot venture forth on a dry, hot day without dire consequences. Loss of too much water across the body's surface will lead to death.

Figure 36.1
Fishes live in water, but butterflies and birds live on land. *a.* The gills of a fish are kept moist by the water in which they function. When exposed to air, the gills collapse and stick together, making them useless. *b.* Insects have an internal system of rigid air tubes that take oxygen directly to the cells. *c.* Birds, like humans, inhale air into the lungs, vascularized cavities where oxygen is picked up by the blood, which then transports it to the cells.

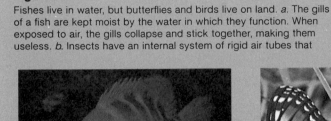

a.

b.

c.

Figure 36.2

Some aquatic animals do not have gills. Instead, each cell carries out its own gas exchange. Although they are multicellular, the organization of certain aquatic animals permits each cell to carry out its own gas exchange. Therefore, they do not need a circulatory system to transport gases to and away from the cells. *a.* In hydras, the outer layer of cells is in contact with its watery environment and the inner layer can exchange gases with the water in the gastrovascular cavity. *b.* In planarians, their flat bodies permit diffusion of gases to and away from the external environment.

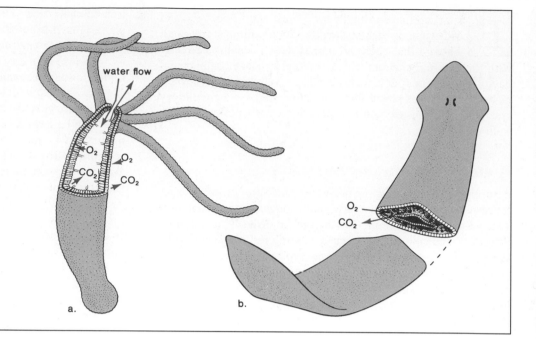

Figure 36.3

Anatomy of gills in detail. *a.* The operculum (folded back) covers and protects several layers of delicate gills. *b.* Each layer of gills has two rows of gill filaments. *c.* Each filament has many thin, platelike lamellae. Gases are exchanged between capillaries inside the lamellae and the water that flows between the lamellae. *d.* The blood in the capillaries flows in the direction opposite to that of the water. The blood takes up 90% of the oxygen in the water as a result of this countercurrent flow.

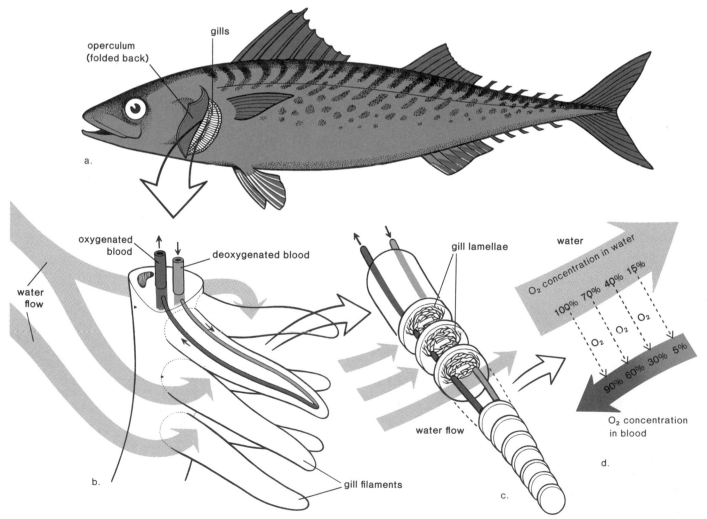

Figure 36.4

Respiratory system of insects. *a.* A system of air tubes, the tracheal system, extends throughout the body of an insect. The blood has no need to carry oxygen because the oxygen is brought to the cells by way of these tubes. *b.* Air enters the tracheae at openings called spiracles. From there it moves to the smaller tracheoles that take it to the cells where gas exchange takes place.

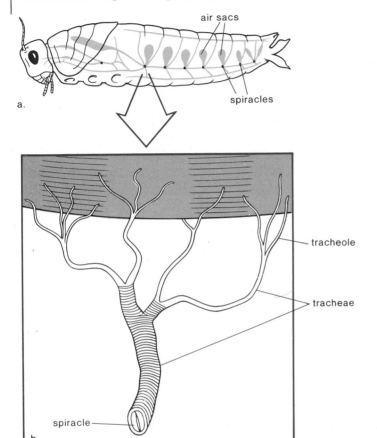

Insects and certain other terrestrial arthropods have a simple but specialized type of respiratory system known as a **tracheal system** (fig. 36.4). Oxygen enters the *spiracles,* valve-like openings on each side of the body, and moves directly to all cells of the body by way of tubes called tracheae that branch repeatedly into all parts of the body. The tracheae end in tiny channels (tracheoles) that are in direct contact with the body cells. Larger insects pump air in and out of the system by body movements, but even so, the efficiency of this system is limited. Notice that utilization of an independent gas-distribution system relieves the blood of this function. The blood of insects is colorless because oxygen is distributed to cells by the respiratory system, not the circulatory system.

Terrestrial vertebrates, in particular, have evolved vascularized internal cavities known as **lungs.** The lungs are not highly developed in amphibians (fig. 36.5) because these animals also make use of the skin for gas exchange. For example, during the courtship season, the male hairy frog (*Astylosternus robustus*) develops dorsal extensions that resemble hair. Extra oxygen is needed because courtship rituals go on for hours. The skin of amphibians, like that of earthworms, must stay thin and moist for gas exchange, and these animals are also restricted to a moist environment.

The inner lining of the lungs is more finely divided in reptiles than in amphibians (fig. 36.5). The lungs of birds and mammals are even more elaborately subdivided into small passageways and spaces. It has even been estimated that human lungs have a total surface area that is at least fifty times the skin's surface area. This extensive surface area of the lungs helps to assure adequate gas exchange.

Figure 36.5

The amphibian lung versus the reptilian lung. The amphibian lung (*left*) has fewer convolutions than does the reptilian lung (*right*).

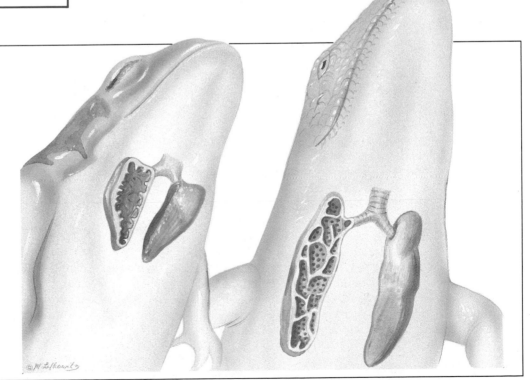

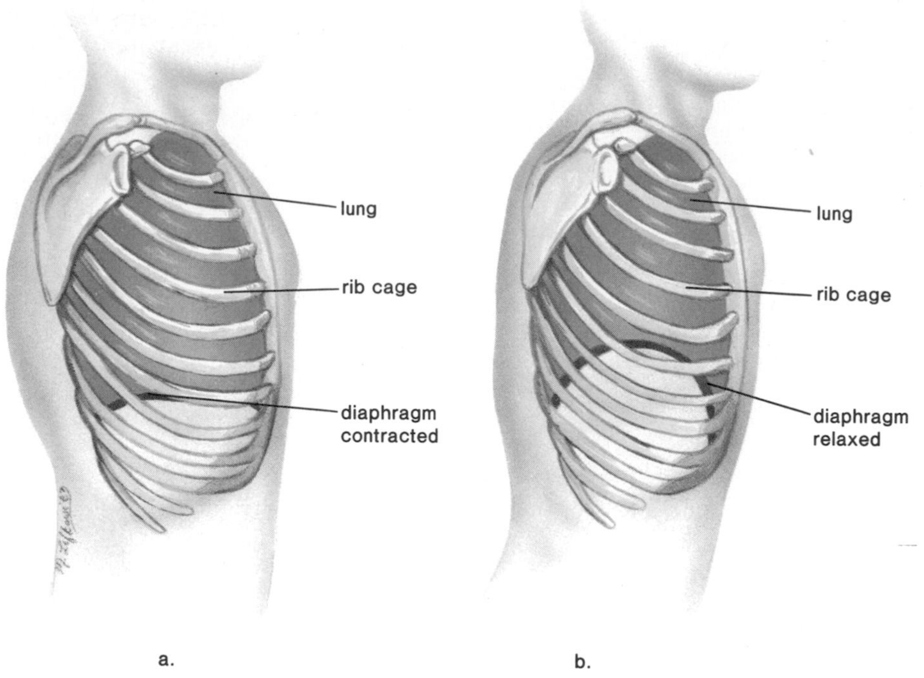

Figure 36.6
Breathing by negative pressure. *a*. Inhalation occurs as the rib cage moves up and out, and the diaphragm flattens. As the chest area expands, the lungs also expand and the air comes rushing in. *b*. Exhalation occurs as the rib cage moves down and in and the diaphragm rises. As the chest area is decreased, the lungs recoil, and air is pushed out.

lung

rib cage

diaphragm contracted

lung

rib cage

diaphragm relaxed

a.

b.

Some aquatic animals, like hydras and planarians, use their entire body surface, but many have a specialized region called gills for gas exchange. Among terrestrial animals, insects have a tracheal system, but most animals utilize lungs for gas exchange.

Terrestrial vertebrates ventilate the lungs by moving air in and out of the respiratory tract. The air becomes moistened as it moves along the tubes leading to the lungs, which protects the lungs from the drying effects of air. Frogs use *positive pressure* to force air into the respiratory tract. With the nostrils firmly shut, the floor of the mouth rises and pushes the air into the lungs. Reptiles, birds, and mammals use *negative pressure* (fig. 36.6). The lungs expand, and the air comes rushing in. Reptiles have jointed ribs that can be raised to expand the lungs. Mammals have a rib cage that is lifted up and out and a muscular **diaphragm** that is flattened. Both these actions increase the volume of the lungs, and air is thereby drawn in, during the process called **inhalation,** or inspiration. After inhalation, exhalation occurs. During **exhalation** (also called expiration), air is pushed out of the lungs. In reptiles, lowering the ribs exerts a pressure that forces air out. In mammals, the rib cage is lowered, and the diaphragm rises, forcing the air out of the lungs.

The lungs of amphibians, reptiles, and mammals are not completely emptied and refilled during each breathing cycle. Instead, the air coming in mixes with used air still in the lungs. While this helps conserve water, it also decreases gas-exchange efficiency. The high oxygen requirement of birds, whose body size must stay small, cannot be met by this system of *incomplete ventilation* i.e., incomplete replacement of spent air with fresh air. Therefore, they have developed a system of *complete ventilation* (fig. 36.7). Fresh, oxygen-rich air passes through the lungs in a one-way direction only and does not mix with used air. Incoming air is carried past the lungs by a bronchus that takes it to a set of posterior air sacs. The air then passes forward through the lungs into a set of anterior air sacs. From here, it is finally expelled.

In most animals, the exchange area is highly vascularized, being richly supplied with blood capillaries. Oxygen is carried from the exchange surface to the interior of the body by the blood circulatory system. The blood of all vertebrates contains the respiratory pigment **hemoglobin,** which helps carry the oxygen. It is hemoglobin that makes the blood red.

Terrestrial vertebrates have several different kinds of respiratory systems involving lungs. Amphibians use positive pressure, but reptiles, birds, and mammals use negative pressure. In contrast to the other vertebrates, birds have evolved a complete (one-way) ventilation system.

Human Respiratory System

Path of Air

The human gas-exchange, or respiratory, system (fig. 36.8) includes all those structures (table 36.1) that conduct air to and from the lungs. The lungs lie deep within the **thoracic** (chest) **cavity** where they are protected from drying out. As air moves through the **nose, trachea,** and **bronchi** to the lungs, it is filtered so it is free of debris, warmed, and humidified. By the time it

Figure 36.7

The lungs of amphibians, reptiles, and mammals are incompletely ventilated whereas those of birds are completely ventilated. *a.* Inhalation. There is residual "used" air (blue) in the lungs of amphibians, reptiles, and mammals. In birds, the posterior air sacs have been cleared of residual air. *b.* Exhalation. Used air is leaving the lungs of amphibians, reptiles, and mammals, although some will remain. Used air, leaving the lungs of birds, will be completely removed because fresh air (pink) enters by its own pathway as shown in (*a*).

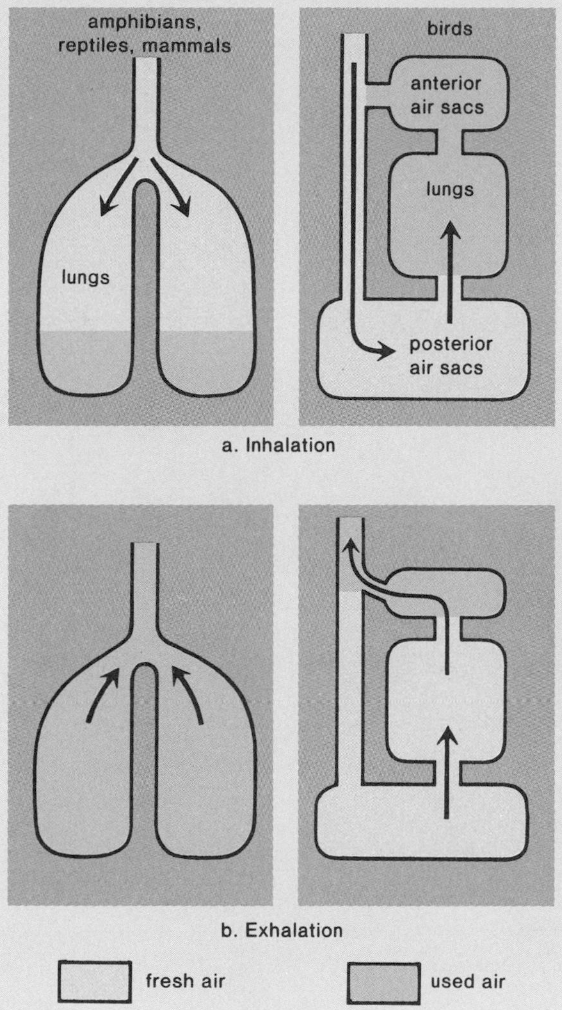

amphibians, reptiles, mammals

lungs

birds

anterior air sacs

lungs

posterior air sacs

a. Inhalation

b. Exhalation

fresh air   used air

reaches the lungs, it is at body temperature and is saturated with water. In the nose, hairs and cilia act as a screening device. In the trachea and the bronchi, cilia beat upward, carrying mucus, dust, and occasional bits of food that "went down the wrong way" into the throat or the **pharynx** where the accumulation may be swallowed or expectorated.

The hard and soft palates separate the nasal cavities from the mouth, but the air and food passages cross in the pharynx. This may seem inefficient, and there is danger of choking if food accidentally enters the trachea, but this arrangement does have the advantage of letting one breathe through the mouth in case

the nose is plugged up. In addition, it permits greater intake of air during heavy exercise when greater gas exchange is required.

Air passes from the pharynx through the **glottis,** an opening into the **larynx** or voice box. At the edges of the glottis, embedded in mucous membrane, are the *vocal cords*. These elastic ligaments vibrate and produce sound when air is expelled past them through the glottis from the larynx.

The larynx and the trachea are held permanently open to receive air. The larynx is held open by the complex of cartilages that form the Adam's apple. The trachea is held open by a series of C-shaped, cartilaginous rings that do not completely meet in the rear. When food is being swallowed, the larynx rises, and the glottis is closed by a flap of tissue called the **epiglottis.** A backward movement of the soft palate covers the entrance of the nasal passages into the pharynx. The food then enters the esophagus, which lies behind the larynx (fig. 35.9).

The trachea divides into two **bronchi** (sing., **bronchus**) that enter the right and left lungs; each branches into a great number of smaller passages called **bronchioles.** The two bronchi resemble the trachea in structure, but as the bronchial tubes divide and subdivide, their walls become thinner, and rings of cartilage are no longer present. Each bronchiole terminates in an elongated space enclosed by a multitude of air pockets or sacs called **alveoli** (sing., **alveolus**), which make up the lungs.

In humans, air moves through the nose, pharynx, and larynx to the trachea, which is held open by cartilaginous rings. Two bronchi divide into bronchioles which finally take air into the lungs. They contain a large number of air sacs, called alveoli, covered by an extensive network of capillaries.

Breathing

Humans breathe using the same mechanism as all other mammals. The volume of the thoracic cavity and lungs is increased by muscular contractions that lower the diaphragm and raise the ribs (fig. 36.6). These movements create a negative pressure in the thoracic cavity and lungs, and atmospheric pressure then forces air into the lungs. When rib and diaphragm muscles relax, air is exhaled as a result of increased pressure in the thoracic cavity and lungs.

Extensive experimentation shows that increased carbon dioxide (CO_2) and hydrogen ion (H^+) concentrations in the blood are the primary stimuli that increase breathing rate. The chemical content of the blood is monitored by *chemoreceptors* called the *aortic* and *carotid bodies,* specialized structures located in the walls of the aorta and the carotid arteries (fig. 36.9). These receptors are very sensitive to changes in CO_2 and H^+ concentrations, but are only minimally sensitive to a lower O_2 concentration. Information from the chemoreceptors goes to the respiratory center in the medulla oblongata portion of the brain, which then increases the breathing rate. This respiratory center, itself, is also sensitive to the chemical content of the blood reaching the brain.

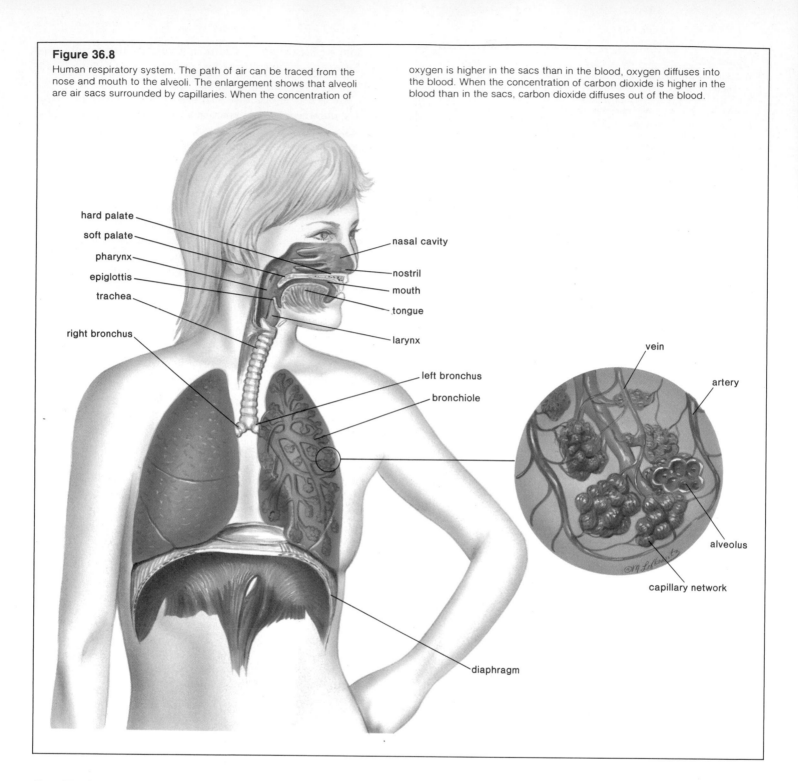

Figure 36.8

Human respiratory system. The path of air can be traced from the nose and mouth to the alveoli. The enlargement shows that alveoli are air sacs surrounded by capillaries. When the concentration of oxygen is higher in the sacs than in the blood, oxygen diffuses into the blood. When the concentration of carbon dioxide is higher in the blood than in the sacs, carbon dioxide diffuses out of the blood.

Gas Exchange and Transport

Diffusion primarily accounts for the exchange of gases between the air in the alveoli and the blood (fig. 36.10) in the pulmonary capillaries. Atmospheric air contains little CO_2, but blood flowing into the lung capillaries is almost saturated with the gas. Therefore, CO_2 diffuses out of the blood into the alveoli. The pattern is the reverse for oxygen. Blood coming into the pulmonary capillaries is oxygen poor and the alveolar air is oxygen rich; therefore, O_2 diffuses into the capillaries.

Transport of O_2 and CO_2

Most oxygen entering the blood combines with **hemoglobin** (Hb) in red blood cells to form oxyhemoglobin (HbO_2):

$$Hb + O_2 \longrightarrow HbO_2$$

Each hemoglobin molecule (fig. 4.18*b*) contains four polypeptide chains, and each chain is folded around an iron-containing structure called heme. It is actually the iron that forms a loose

Table 36.1 Path of Air

Structure	Function
Nasal cavities	Filters, warms, and moistens
Pharynx (throat)	Connection to larynx
Glottis	Permits passage of air
Larynx (voice box)	Sound production
Trachea (windpipe)	Passage of air to thoracic cavity
Bronchi	Passage of air to each lung
Bronchioles	Passage of air to each alveolus
Alveoli	Air sacs for gas exchange

Figure 36.9

When chemoreceptors within the aortic and carotid bodies are stimulated by a rise in the H^+ concentration of the blood, caused by an increased amount of CO_2, nerve impulses travel to the respiratory center of the brain and the breathing rate increases.

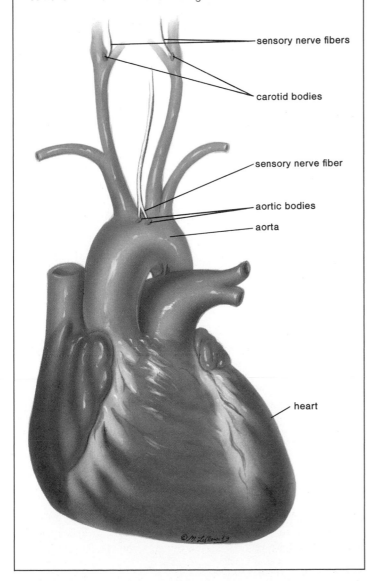

sensory nerve fibers

carotid bodies

sensory nerve fiber

aortic bodies

aorta

heart

association with oxygen. Since there are about 280 million hemoglobin molecules in each red blood cell, each cell is capable of carrying more than one thousand million molecules of oxygen.

Carbon monoxide, present in automobile exhaust, combines with hemoglobin more readily than does oxygen, and it stays combined for several hours regardless of the environmental conditions. Accidental death or suicide from carbon monoxide poisoning occurs because the hemoglobin of the blood is not available for oxygen transport.

Oxygen-binding characteristics of hemoglobin can be studied by examining oxyhemoglobin dissociation curves (fig. 36.11). These curves show the percentage of oxygen-binding sites of hemoglobin that are carrying oxygen at various oxygen partial pressures (PO_2). A partial pressure of a gas is simply the amount of pressure exerted by that gas among all the gases present. At the normal partial pressures of O_2 in the lungs, hemoglobin becomes practically saturated with O_2, but at the partial pressures in the tissues, hemoglobin quickly gives up much of its oxygen:

$$HbO_2 \longrightarrow Hb + O_2$$

The acid pH and warmer temperature of the tissues also promote this dissociation (breakdown). Hemoglobin that has given up its oxygen is called reduced hemoglobin.

In the tissues reduced hemoglobin combines with carbon dioxide to form *carbaminohemoglobin.* Most of the carbon dioxide, however, is transported in the form of the **bicarbonate ion,** HCO_3^-.

$$CO_2 + H_2O \longrightarrow H_2CO_3 \longrightarrow H^+ + HCO_3^-$$

Carbon dioxide combined with water forms *carbonic acid;* this dissociates to a hydrogen ion and the bicarbonate ion.

An enzyme in red blood cells, **carbonic anhydrase,** speeds up this reaction. The released hydrogen ions, which could drastically change the pH of the blood, are absorbed by the globin portions of hemoglobin, and the bicarbonate ions diffuse out of the red blood cells to be carried in the plasma. Reduced hemoglobin, which combines with a hydrogen ion, can be symbolized as HHb. The latter plays a vital role in maintaining the pH of the blood.

As blood enters the pulmonary capillaries, most of the carbon dioxide is present in plasma as the bicarbonate ion. The little free carbon dioxide remaining begins to diffuse out, and the following reaction is driven to the right.

$$H^+ + HCO_3^- \longrightarrow H_2CO_3 \longrightarrow H_2O + CO_2$$

Carbonic anhydrase also speeds up this reaction, during which hemoglobin gives up the hydrogen ions it has been carrying, HHb becoming Hb.

Figure 36.10

Respiration in humans and other mammals involves gas exchanges at the lungs and the tissues. In the lungs, CO_2 leaves the blood and O_2 enters the blood. At the tissues, O_2 leaves the blood and CO_2 enters the blood. Steps necessary for gas exchange are shown for the lungs (*insert above*) and for the tissues (*insert below*).

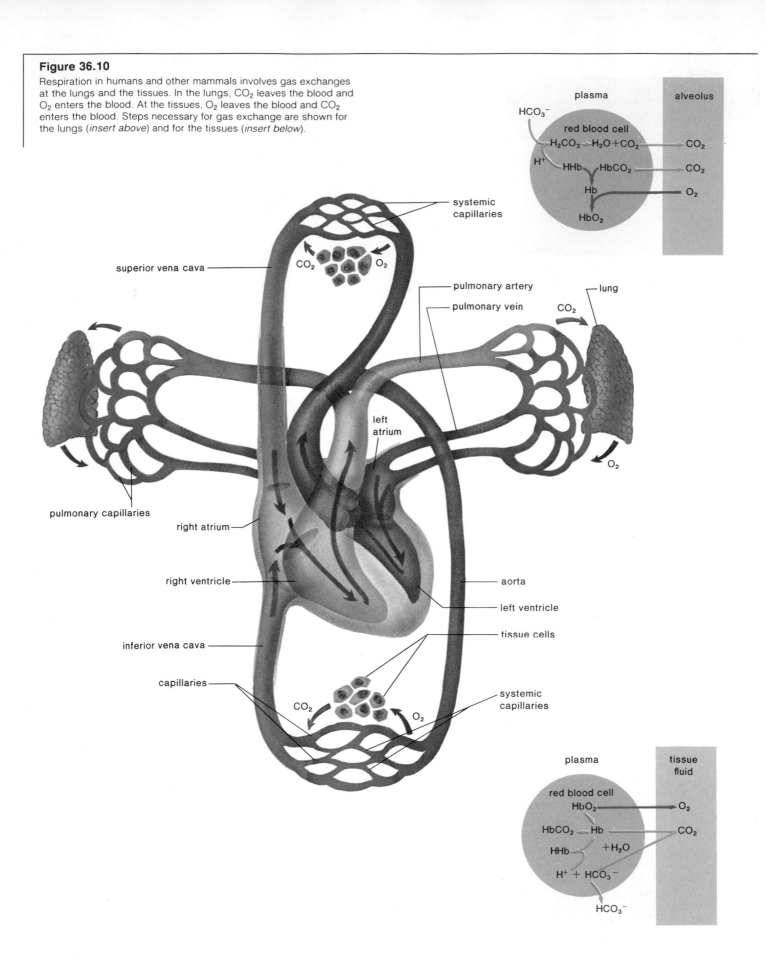

Figure 36.11

Properties of hemoglobin as revealed by hemoglobin dissociation curves. These curves show that as the partial pressure of oxygen (PO_2) decreases, hemoglobin gives up its oxygen more readily and this effect is promoted by both higher temperature (a), and higher acidity (b). These are the very conditions of the tissues compared to the lungs.

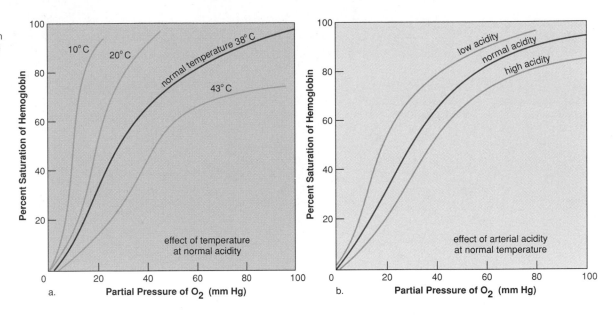

a.

b.

Figure 36.12

Fetal mammals, such as (a) a human fetus, have hemoglobin that has a higher affinity for oxygen than adult hemoglobin (b). This permits fetal hemoglobin to combine with oxygen at a lower partial pressure of O_2 than the mother's hemoglobin. Therefore, oxygen tends to leave the blood of the mother and moves into the blood of the fetus.

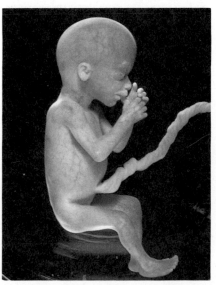

a.

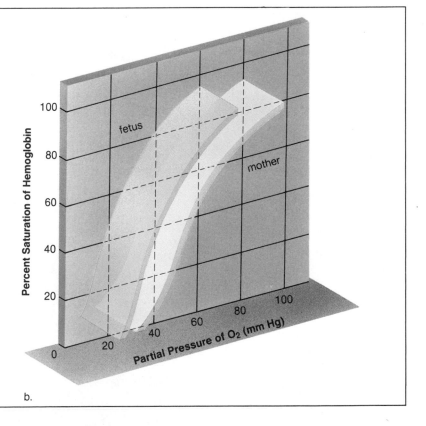

b.

Gas exchange in the lungs and the tissues is dependent upon the process of diffusion. Oxygen is transported in the blood by hemoglobin, and carbon dioxide is carried as the bicarbonate ion.

Types of Hemoglobin

Muscle cells contain an oxygen-binding pigment called **myoglobin,** which has a higher affinity for oxygen than does hemoglobin, and which tends to hold oxygen until the PO_2 falls

very low. Myoglobin provides an excellent reserve source of oxygen for muscle cells when they are contracting and metabolizing rapidly.

Fetal hemoglobin has a higher affinity for oxygen than adult hemoglobin at the PO_2 levels found in the placenta (fig. 36.12). This facilitates transfer of oxygen from mother's blood to fetal blood. After birth, fetal hemoglobin is gradually replaced with the adult type of hemoglobin.

Risks of Smoking versus Benefits of Quitting

Based on available statistics, the American Cancer Society informs us that smoking carries a high risk. Among the **risks of smoking** are the following:

Shortened life expectancy A twenty-five-year-old who smokes two packs of cigarettes a day has a life expectancy 8.3 years shorter than a nonsmoker. The greater the number of packs smoked, the shorter the life expectancy.

Lung cancer Smoking cigarettes is the major cause of lung cancer in both men and women. The frequency of lung cancer has risen in women of late because more women are smoking now. Autopsies of smokers have revealed the progressive steps by which cancer of the lung develops. The first event appears to be a thickening of the cells that line the bronchi. Then there is a loss of cilia so that dust and dirt settle in the lungs. Following this, cells with atypical nuclei appear in the thickened lining. A disordered collection of cells with atypical nuclei may be considered to be cancer in situ (at one location). The final step occurs when some cells break loose and penetrate the other tissues, a process called metastasis. This is true cancer (see figure *b*).

Cancer of the larynx, mouth, esophagus, bladder, and pancreas The chances of developing these cancers are from 2 to 17 times higher in cigarette smokers than in nonsmokers.

Emphysema Cigarette smokers have 4 to 25 times greater risk of developing emphysema. Damage is seen even in the lungs of young smokers. Smoking causes the lining of the bronchioles to thicken. If these become obstructed, the air within the alveoli is trapped. The trapped air very often causes the alveolar walls to rupture and the walls of the small blood vessels in the vicinity to thicken. If a large part of the lungs is involved, the lungs are permanently inflated and the chest balloons out due to this trapped air. The victim is breathless and has a cough. Since the surface area for gas exchange is reduced, not enough oxygen reaches the heart and brain. The heart works furiously to force more blood through the lungs, which may lead to a heart condition. Lack of oxygen for the brain may make the person feel depressed, sluggish, and irritable.

Coronary heart disease Cigarette smoking is the major factor in 120,000 additional U.S. deaths from coronary heart disease each year.

Reproductive effects Smoking mothers have more stillbirths, as well as more low-birthweight babies who are more vulnerable to disease and death. Children of smoking mothers are smaller, and underdeveloped physically and socially even seven years after birth.

The American Cancer Society also informs smokers of the **benefits of quitting.** These benefits include the following:

Risk of premature death is reduced If a smoker quits smoking and does not smoke for 10 to 15 years, the risk of death due to any one of the types of cancer mentioned approaches that of the nonsmoker.

Health of respiratory system improves The cough and excess sputum disappear during the first few weeks after quitting. As long as cancer has not yet developed, all the ill effects mentioned can reverse themselves and the lungs can become healthy again. In patients with emphysema, the rate of alveoli destruction is reduced and lung function may improve.

Coronary heart disease risk sharply decreases After only one year the risk factor is greatly reduced, and after 10 years an ex-smoker's risk is the same as that of nonsmokers.

The increased risk of having stillborn children and underdeveloped children disappears Even for women who do not stop smoking until the fourth month of pregnancy, such risks to infants is decreased.

People who smoke must ask themselves if the benefits of quitting outweigh the risks of smoking.

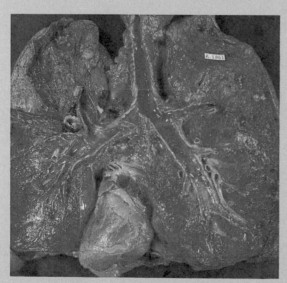

a.

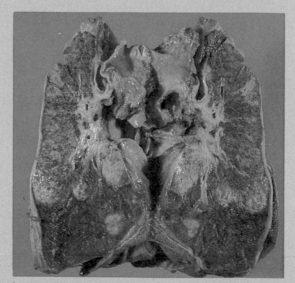

b.

a. Normal lungs with heart in place. Notice the healthy red color. *b.* Lungs of a heavy smoker. Notice how black the lungs are except where cancerous tumors are located.

American Cancer Society

Summary

1. Some aquatic animals like hydras and planarians use their entire body surface for gas exchange.
2. Most animals have a localized, special gas-exchange area. Most aquatic animals pass water over gills. On land, insects utilize tracheal systems and vertebrates have lungs.
3. Lungs are found inside the body, and this reduces water loss, but there is need for ventilation. Some vertebrates use positive pressure, but most inhale, using muscular contraction to produce a negative pressure that causes air to rush into the lungs. When the breathing muscles relax, air is exhaled.
4. Birds have a series of air sacs that allow a one-way flow of air over the gas-exchange area. This is called a complete ventilation system.
5. Table 36.1 lists the structures found in the human respiratory system.
6. Humans breathe by negative pressure, as do other mammals. During inhalation, the rib cage goes up and out, and the diaphragm lowers. During exhalation, the rib cage goes down and in, and the diaphragm rises.
7. The rate of breathing is dependent upon the amount of carbon dioxide in the blood, as detected by chemoreceptors such as the aortic and carotid bodies.
8. Gas exchange in the lungs and tissues is brought about by diffusion. Hemoglobin transports oxygen in the blood; carbon dioxide is mainly transported in plasma as the bicarbonate ion. The enzyme carbonic anhydrase found in red blood cells speeds up the formation of the bicarbonate ion.

Objective Questions

1. One problem faced by terrestrial animals with lungs, but not by freshwater aquatic animals with gills is
   a. gas exchange involves water loss.
   b. breathing requires considerable energy.
   c. oxygen diffuses very slowly in air.
   d. All of these.
2. Which of these exemplifies that not all active animals require that the circulatory system transport gases?
   a. mouse
   b. dragonfly
   c. trout
   d. sparrow
3. Birds have a more efficient lung than humans because the flow of air is
   a. the same during both inhalation and exhalation.
   b. in only one direction through the lungs.
   c. never backed up like in human lungs.
   d. not hindered by a larynx.
4. Which animal breathes by positive pressure?
   a. fish
   b. human
   c. bird
   d. frog
5. Which of these is a true statement?
   a. In lung capillaries, carbon dioxide combines with water to give carbonic acid.
   b. In tissue capillaries, carbonic acid breaks down to carbon dioxide and water.
   c. In lung capillaries, carbonic acid breaks down to carbon dioxide and water.
   d. In tissue capillaries, carbonic acid combines with hydrogen ions to form the carbonate ion.
6. Air comes into the human lungs because
   a. atmospheric pressure is less than the pressure inside the lungs.
   b. atmospheric pressure is greater than the pressure inside the lungs.
   c. although the pressures are the same inside and outside, the partial pressure of oxygen is lower within the lungs.
   d. the residual air in the lungs causes the partial pressure of oxygen to be less than outside.
7. If the digestive and respiratory tracts were completely separate in humans, there would be no need for
   a. swallowing.
   b. external nares.
   c. an epiglottis.
   d. a diaphragm.
8. To trace the path of air in humans you would place the trachea
   a. directly after the nose.
   b. directly before the bronchi.
   c. before the pharynx.
   d. Both (a) and (c).
9. In humans, the respiratory center
   a. is stimulated by carbon dioxide.
   b. is located in the medulla oblongata.
   c. controls the rate of breathing.
   d. All of these.
10. Carbon dioxide is carried in the plasma
    a. in combination with hemoglobin.
    b. as the bicarbonate ion.
    c. combined with carbonic anhydrase.
    d. All of these.

Study Questions

1. Compare the respiratory organs of aquatic animals to those of terrestrial animals.
2. How does the countercurrent flow of blood within gill capillaries and water passing across the gills assist respiration in fishes?
3. Why don't insects require a blood respiratory pigment, and why is it beneficial for the body wall of earthworms to be moist?
4. Explain the phrase, "breathing by using negative pressure."
5. Contrast incomplete ventilation in humans with complete ventilation in birds.
6. Name the parts of the human respiratory system and list a function for each part.
7. The concentration of what gas in the blood controls the breathing rate in humans? Explain.

8. How is most carbon dioxide carried in the blood? What role is played by carbonic anhydrase?

9. Describe the role that hemoglobin plays in transporting oxygen and carbon dioxide.

10. Describe gas exchange and in the process explain the equations given in figure 36.10.

Thought Questions

1. Considering the fact that humans practice incomplete ventilation, explain why there are limits to the length and/or diameter of tubes through which a person can breathe.

2. Inhalation is dependent on a difference between air pressure in the lungs and external air pressures. Make up a table to show the effect of higher and lower air pressures, for both the lungs and external air, on inhalation. Which of these applies to breathing at higher altitudes?

3. With reference to figure 36.11 explain why hemoglobin tends to give up oxygen when the tissues need it the most.

Selected Key Terms

diaphragm (di'ah-fram) 568
inhalation (in''hah-la'shun) 568
exhalation (eks''hah-la'shun) 568
hemoglobin (he''mo-glo-bin) 568
thoracic cavity (tho-ras'ik kav'ĭ-te) 568

trachea (tra'ke-ah) 568
bronchi (brong'ki) 568
pharynx (far'inks) 569
glottis (glot'is) 569
larynx (lar'inks) 569
epiglottis (ep''ĭ-glot'is) 569

bronchiole (brong'ke-ōl) 569
alveoli (al-ve'o-li) 569
bicarbonate ion (bi-kar'bo-nāt i'on) 571
carbonic anhydrase (kar-bon'ik an-hi'drās) 571
myoglobin (mi''o-glo'-bin) 573

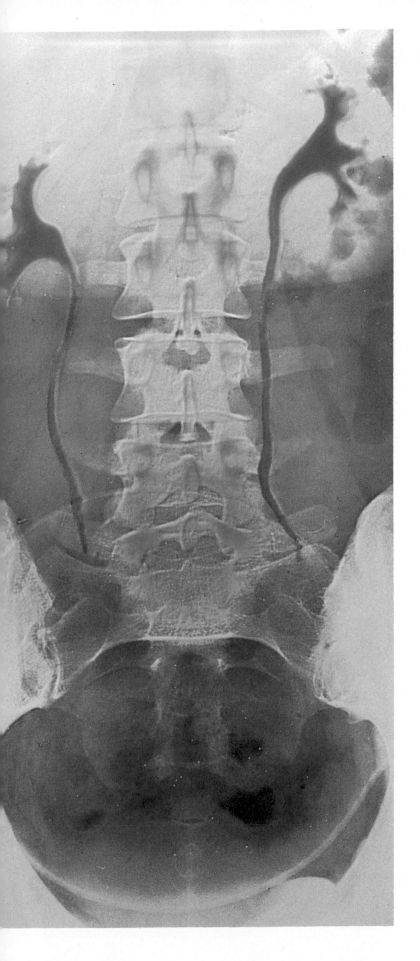

CHAPTER 37

Excretory System

Your study of this chapter will be complete when you can:

1. Relate the excretion of ammonia, urea, and uric acid to the animal's environment.

2. Give examples of how various animals regulate the water and salt balance of the body.

3. Contrast the manner in which marine bony fishes and freshwater bony fishes regulate the water and salt content of the blood.

4. Compare the operation of earthworm nephridia, insect Malpighian tubules, and human kidneys.

5. Trace the path of urine in humans, and describe in general the structure and function of each organ mentioned.

6. List the parts of the kidney nephron, and relate these to the macroscopic anatomy of the kidney.

7. Describe the three steps in urine formation, and relate these to the parts of the nephron.

8. Describe how water excretion is regulated and how the pH of the blood is adjusted by the kidneys.

9. Tell, in general, how the artificial kidney machine works.

X-ray enhanced picture of lower abdominal area of a human features the kidneys. Urine formed in the kidneys passes by way of ureters (here shown in red) to the bladder where it is stored before passing out of the body.

577

We have seen that the digestive and respiratory systems of animals allow molecules to enter and to exit from the internal environment. Now let us consider how certain other organs (fig. 37.1) remove nitrogenous wastes, water, and salts from the internal environment. This process is called **excretion**, the elimination of molecules that have taken part in metabolic reactions, and should not be confused with **defecation**, which is the elimination of nondigested materials, dead cells, and bacteria from the gut.

Excretion of Nitrogenous Wastes

The breakdown of various molecules, including nucleic acids and amino acids, results in nitrogenous wastes. For simplicity's sake, however, we will limit our discussion to amino acid metabolism.

Amino acids, derived from protein in food, can be used by cells for synthesis of new body protein or other nitrogen-containing molecules. The amino acids not used for synthesis are oxidized to generate energy or are converted to fats or carbohydrates that can be stored. In either case, the *amino groups* ($-NH_2$) must be removed (fig. 37.2) because they are not needed for any of these purposes. Once the amino groups have been removed from amino acids, they may be excreted from the body in the form of ammonia, urea, or uric acid, depending on the species (table 37.1). Removal of amino groups from amino acids requires a fairly constant amount of energy. However, the energy requirement for the conversion of amino groups to either ammonia, urea, or uric acid differs, as indicated in figure 37.2.

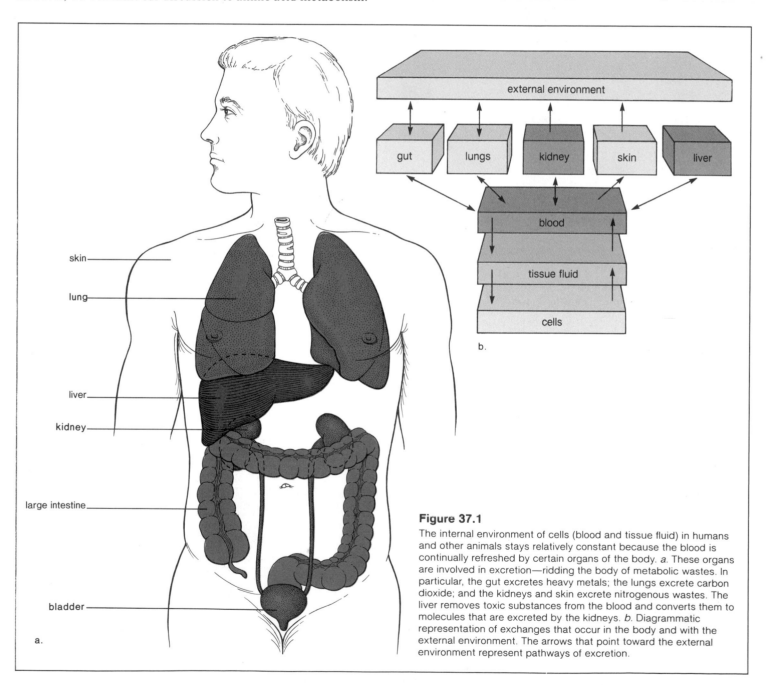

Figure 37.1

The internal environment of cells (blood and tissue fluid) in humans and other animals stays relatively constant because the blood is continually refreshed by certain organs of the body. *a.* These organs are involved in excretion—ridding the body of metabolic wastes. In particular, the gut excretes heavy metals; the lungs excrete carbon dioxide; and the kidneys and skin excrete nitrogenous wastes. The liver removes toxic substances from the blood and converts them to molecules that are excreted by the kidneys. *b.* Diagrammatic representation of exchanges that occur in the body and with the external environment. The arrows that point toward the external environment represent pathways of excretion.

Ammonia

Amino groups removed from amino acids immediately form **ammonia** (NH_3) by addition of a third hydrogen. Therefore, little or no energy is required to convert an amino group to ammonia. Ammonia is quite toxic and can only be used as a nitrogenous excretory product if a good deal of water is available to wash it from the body. The high solubility of ammonia permits this means of excretion in bony fishes, aquatic invertebrates, and amphibians whose gills and skin surfaces are in direct contact with the water of the environment.

Figure 37.2

Amino acid metabolism. Amino acids can join together to form a protein, or they can be broken down. The carbon skeleton can be used as an energy source, but the amino group must be excreted as either ammonia, urea, or uric acid. The energy and water requirements for these vary as indicated.

Urea

Terrestrial amphibians and mammals usually excrete **urea** as their main nitrogenous waste. Urea is much less toxic than ammonia and can be excreted in a moderately concentrated solution. This allows body water to be conserved, an important advantage for terrestrial animals with limited access to water.

However, production of urea requires the expenditure of energy. Urea is produced in the liver by a set of enzymatic reactions known as the *urea cycle*. In the cycle, some of whose reactions require ATP, carrier molecules take up carbon dioxide and two molecules of ammonia to finally release urea:

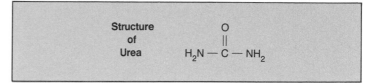

Uric Acid

Uric acid is excreted by insects, reptiles, birds (fig. 37.3) and some dogs (for example, dalmatians). Uric acid is not very toxic and is poorly soluble in water. Poor solubility is an advantage for water conservation because uric acid can be concentrated even more readily than can urea. In reptiles and birds, a dilute solution of uric acid passes from the kidneys to the *cloaca,* a common reservoir for the products of the digestive, urinary, and reproductive systems. After water is absorbed by the cloaca, the uric acid passes out with the feces.

Excretion of uric acid by reptiles and birds is correlated with the necessity to conserve water during development. Embryos of reptiles and birds develop inside shelled eggs that are completely enclosed (fig. 37.4). Before embryonic development begins, all nutrients and water for embryonic growth and metabolism must be inside the egg. The production of insoluble, relatively nontoxic uric acid is advantageous for shelled embryos because all nitrogenous wastes must also be stored inside the shell until hatching takes place. The uric acid is highly concentrated and stored during development inside a sac attached to the body of the embryo. The embryo leaves the uric acid inside the sac and eggshell upon hatching.

Table 37.1 Nitrogenous Waste Excretion in Relation to Habitat

Product	Habitat	Animals
Ammonia	Water	Aquatic invertebrates Bony fishes Amphibian larvae
Urea	Land	Adult amphibians Mammals
Uric acid	Land	Insects Birds Reptiles

Figure 37.3

Birds excrete uric acid, a solid material, as their nitrogenous waste. It is mixed with fecal material in a common repository for the urinary, digestive, and reproductive systems. Sea birds congregate in such numbers that their droppings build up to give a nitrogen-rich substance called guano. At one time, guano was harvested as a natural fertilizer.

Uric acid is synthesized by a long, complex series of enzymatic reactions that require expenditure of even more ATP than does urea synthesis. Here again, there seems to be a trade-off between the advantage of water conservation and the disadvantage of energy expenditure for synthesis of an excretory molecule.

Figure 37.4

Hard-shelled egg. All nutrients and water required during development must be enclosed by the eggshell because only oxygen, carbon dioxide, and water vapor diffuse freely through the shell. Nitrogenous wastes must also be collected and stored within a sac (allantois) until hatching time. At hatching time, the sac is detached from the body and remains inside the broken shell.

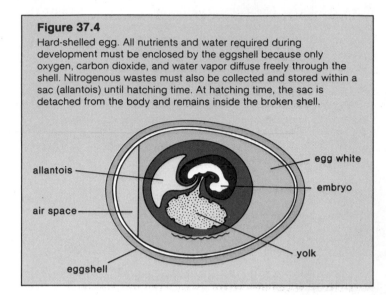

Figure 37.5

Comparison of relative salt concentration of animal body fluids and the ionic concentration in seawater (blue horizontal plane above) and fresh water (blue horizontal plane below). For example, marine invertebrates are the only animals to have fluids with the ionic concentration of seawater, and freshwater invertebrates are the only animals to approach the ionic concentration of fresh water.

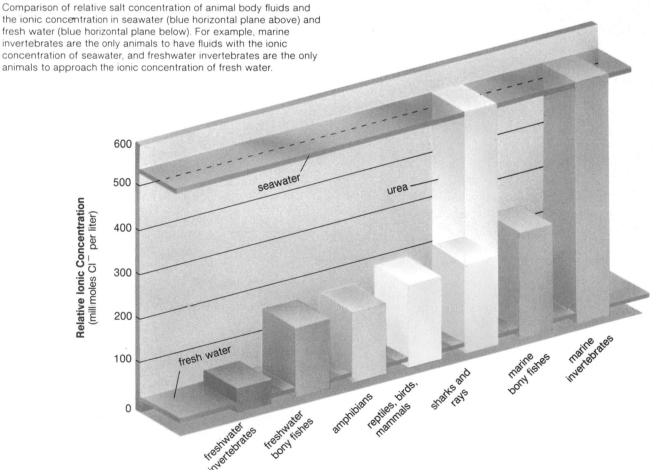

Animals excrete nitrogenous wastes derived from protein breakdown as either ammonia, urea, or uric acid. Ammonia requires the most and uric acid requires the least amount of water to excrete; uric acid requires the most and ammonia requires the least amount of energy to produce. Habitat plays a primary role in determining which type of molecule is excreted.

Osmotic Regulation

In addition to their role in excretion of nitrogenous wastes, excretory organs also have the important function of regulating the water and salt balance of the body. Figure 37.5 shows that among animals, only marine invertebrates and cartilaginous fishes, such as sharks and rays, have body fluids that are nearly isotonic to seawater. These organisms have little difficulty maintaining their normal salt and water balance. Surprising, though, is the observation that while they are isotonic, the body fluids of cartilaginous fishes do not contain the same amount of salt as does seawater. The answer to this paradox is that their blood contains a concentration of urea high enough to match the tonicity of the sea! For some unknown reason, this amount of urea is not toxic to them.

The body fluids of all bony fishes have only a moderate amount of salt. Apparently, their common ancestor evolved in fresh water, and only later did some groups invade the sea. Marine bony fishes (fig. 37.6a) are therefore prone to water loss and could become dehydrated. To counteract this, they drink seawater almost constantly. On the average, marine bony fishes swallow an amount of water estimated to be equal to 1% of their body weight every hour. This is equivalent to a human drinking about 700 ml of water every hour around the clock. While they get water by drinking, this habit also causes these fishes to acquire salt. Instead of forming a hypertonic urine, however, they actively transport sodium (Na^+) and chloride (Cl^-) ions into the surrounding seawater at the gills.

It is easy to see that the osmotic problems of freshwater bony fishes (fig. 37.6b) are exactly opposite to those of marine bony fishes. The body fluids of freshwater bony fishes are hypertonic to fresh water, and they are prone to gain water. These fishes never drink water but, instead, eliminate excess water through production of large quantities of dilute (hypotonic) urine. They discharge a quantity of urine equal to one-third their body weight each day. Because they tend to lose salts, they actively transport salts in across the membranes of their gills.

Figure 37.6
Water and salt balances in bony fishes.
a. Marine bony fish. b. Freshwater bony fish. The black arrows represent passive transport from the environment, and the colored arrows represent active transport by the fishes to counteract environmental pressures.

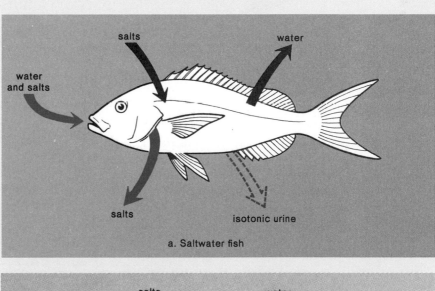

a. Saltwater fish

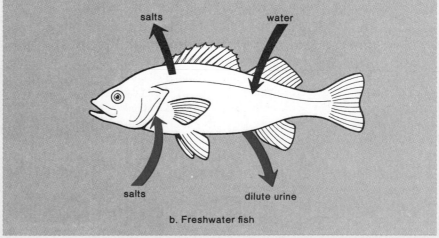

b. Freshwater fish

Figure 37.7

a. The flame-cell excretory system in planarians. Two or more branching tubules run the length of the body and open to the outside by pores. At the ends of side branches there are small flame cells, bulblike cells whose beating cilia causes fluid to enter the tubules that remove excess fluid from the body. *b.* The earthworm nephridium. The nephridium has a ciliated opening, the nephridiostome, that leads to a coiled tubule surrounded by a capillary network. Urine can be temporarily stored before being released to the outside via a pore termed a nephridiopore. Most segments contain a pair of nephridia, one on each side.

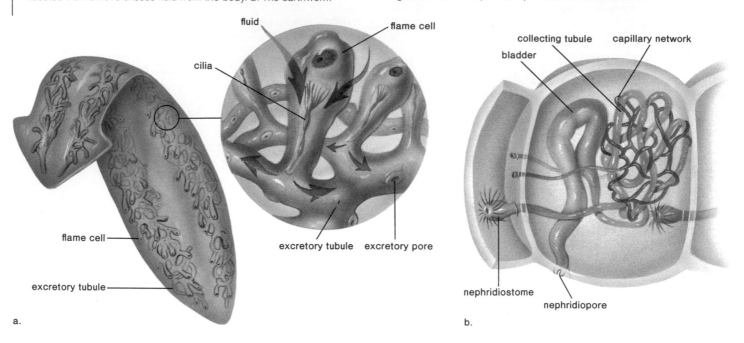

Excretory organs help regulate the water and salt balance of the body. For example, marine bony fishes drink water constantly and excrete salt from the gills. Freshwater bony fishes never drink water and excrete a dilute urine.

The difference in adaptation between marine and freshwater bony fishes makes it remarkable that some fishes actually can move between the two environments during the life cycle. Salmon, for example, begin their lives in freshwater streams and rivers, move to the ocean for a period of time, and finally return to fresh water to breed. These fish alter their behavior and their gill and kidney functions in response to the osmotic changes they encounter when moving from one environment to the other.

Freshwater invertebrates also have to rid the body of excess water. Planarians have a network of excretory tubules (fig. 37.7*a*) that open to the outside of the body through excretory pores. Located along the tubules are bulblike flame cells, each of which has a cluster of beating cilia that look like a flickering flame under the microscope. The beating of flame-cell cilia propels fluid through the excretory canals and out of the body.

Like marine bony fishes, some animals that evolved on land are also able to drink seawater despite its high toxicity. Birds and reptiles that live near the sea have a nasal salt gland that can excrete large volumes of concentrated salt solution. Mammals that live at sea, like whales, porpoises, and seals, most likely can concentrate their urine enough to drink salt water. Humans cannot do this, and they die if they drink only seawater.

Most terrestrial animals need to drink water occasionally, but the kangaroo rat manages to get along without drinking water at all. It forms a very concentrated urine, and its fecal material is almost completely dry. These abilities allow it to survive using metabolic water derived from the breakdown of nutrient molecules alone.

Excretory Organs of Earthworms and Insects

Most animals have a special organ that excretes nitrogenous wastes and is also involved in body fluid homeostasis. We will consider two of these, aside from the human kidney.

Earthworm Nephridia

The earthworm's body is divided into segments, and nearly every body segment has a pair of excretory structures called **nephridia** (sing., **nephridium**). Each nephridium (fig. 37.7*b*) is a tubule with a ciliated opening and an excretory pore. As fluid from the body cavity is propelled through the tubule by beating cilia, certain substances are reabsorbed and carried away by a network

Figure 37.8

Location of Malpighian tubules in a spider. The Malpighian tubules are attached to the gut and take up wastes from body fluids. The Malpighian tubules are named for the Italian microscopist Marcello Malpighi (1628–1694) who first described them.

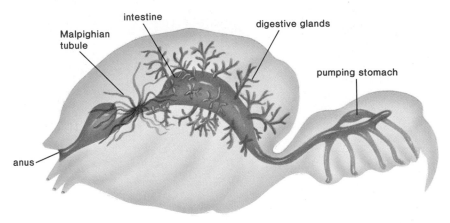

intestine

Malpighian tubule

digestive glands

pumping stomach

anus

of capillaries surrounding the tubule. This process results in the formation of a urine that contains only unwanted metabolic wastes, salts, and water.

Malpighian Tubules in Insects

Insects have a unique excretory system consisting of long, thin tubules, called **Malpighian tubules** (fig. 37.8), attached to the gut. Water and uric acid simply flow from the surrounding fluid into the tubules before moving into the gut. The water may be reabsorbed, but the uric acid eventually passes out of the gut. Insects living in water and insects eating large quantities of moist food reabsorb little water. But insects in dry environments reabsorb most of the water and excrete a dry, semisolid mass of precipitated uric acid. Mealworms, the larvae of a beetle that lives in dry flour, produce an excretory product that is so dry that it actually absorbs water from very humid air.

All animals have a primary excretory organ for excreting nitrogenous wastes: earthworms have nephridia; insects and spiders have Malpighian tubules; and humans have kidneys.

Human Kidney

The **kidneys** are bean-shaped, reddish-brown organs, each about the size of a fist. They are located one on either side of the vertebral column just below the diaphragm where they are partially protected by the lower rib cage. They are a part of the human urinary system (fig. 37.9), which is composed not only of the kidneys that make urine, but also of those organs that conduct urine out of the body. Each kidney is connected to a

ureter, a duct that carries urine from the kidney to the **urinary bladder,** where it is stored until it is voided from the body through the single **urethra.** In males, the urethra passes through the penis, and in females, it opens ventral to the opening of the vagina.

Structure

If a kidney is sectioned longitudinally (fig. 37.10), three major parts can be distinguished. The outer region is the **cortex,** which has a somewhat granular appearance. The **medulla** lies on the inner side of the cortex and is arranged in a group of pyramid-shaped regions, each of which has a striped appearance. The innermost part of the kidney is a hollow chamber called the **pelvis.** Urine formed in the kidney collects in the pelvis before entering the ureter.

Nephrons

Microscopically, each kidney is composed of about one million tiny tubules called **nephrons.** Some nephrons are located primarily in the cortex, but others dip down into the medulla. Each nephron (fig. 37.11) is made of several parts.

Urine is made by the nephrons in the kidneys. Thereafter, it passes through the ureters to the urinary bladder for storage, before leaving the body by way of the urethra.

The blind end of a nephron is pushed in on itself to form a cup-like structure called **Bowman's capsule.** Next, there is a region known as the **proximal** (near Bowman's capsule) **convoluted tubule** that leads to a narrow U-turn known as the **loop of Henle.** This is followed by the **distal** (far from Bowman's capsule) **convoluted tubule.** Several distal convoluted tubules enter one **collecting duct.** The collecting duct transports urine down through

Figure 37.9

Human urinary system. Urine is excreted by the kidneys and passes to the bladder by way of the ureters. After storage in the bladder, it exits when convenient by way of the urethra. These are the only organs in the body that ever contain urine.

Figure 37.10

Human kidney structure. A diagram of a longitudinal section of the human kidney, with an enlargement of one pyramid.

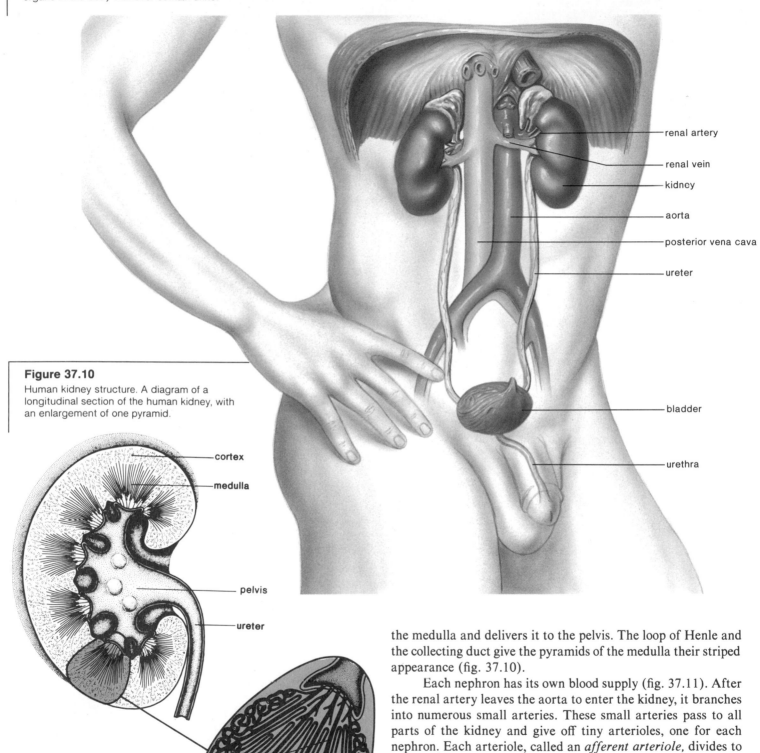

renal artery

renal vein

kidney

aorta

posterior vena cava

ureter

bladder

urethra

cortex

medulla

pelvis

ureter

collecting ducts

nephrons

the medulla and delivers it to the pelvis. The loop of Henle and the collecting duct give the pyramids of the medulla their striped appearance (fig. 37.10).

Each nephron has its own blood supply (fig. 37.11). After the renal artery leaves the aorta to enter the kidney, it branches into numerous small arteries. These small arteries pass to all parts of the kidney and give off tiny arterioles, one for each nephron. Each arteriole, called an *afferent arteriole,* divides to form a capillary tuft, the **glomerulus,** which is surrounded by Bowman's capsule. The glomerular capillaries drain into an *efferent arteriole,* which subsequently branches into a second

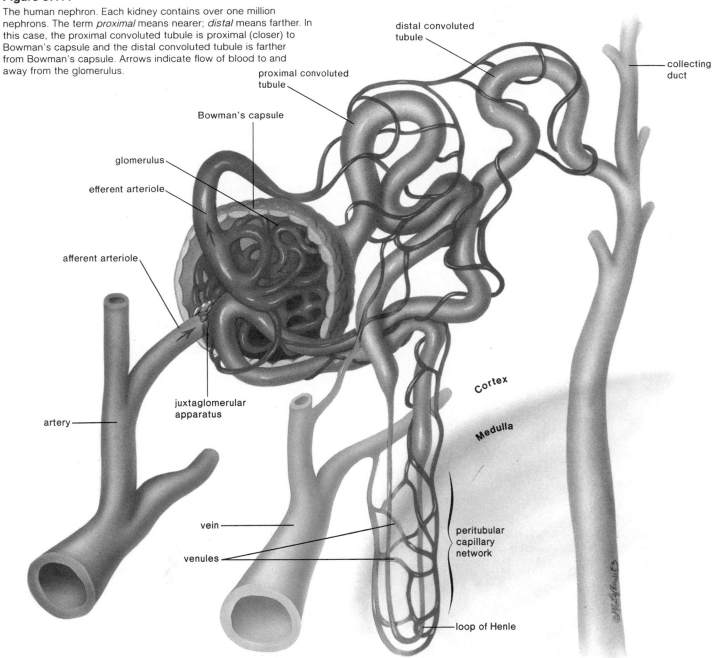

Figure 37.11

The human nephron. Each kidney contains over one million nephrons. The term *proximal* means nearer; *distal* means farther. In this case, the proximal convoluted tubule is proximal (closer) to Bowman's capsule and the distal convoluted tubule is farther from Bowman's capsule. Arrows indicate flow of blood to and away from the glomerulus.

distal convoluted tubule

collecting duct

proximal convoluted tubule

Bowman's capsule

glomerulus

efferent arteriole

afferent arteriole

Cortex

Medulla

juxtaglomerular apparatus

artery

peritubular capillary network

vein

venules

loop of Henle

capillary network around the tubular parts of the nephron. These capillaries, called *peritubular capillaries,* lead to venules that join the renal vein, a vessel that enters the vena cava.

Urine Formation

Human nephrons function somewhat like earthworm nephridia in that they exchange molecules with the blood. Urine production requires three distinct processes (fig. 37.12):

1. Pressure filtration at Bowman's capsule
2. Reabsorption, including selective reabsorption, at the proximal convoluted tubule in particular
3. Tubular secretion at the distal convoluted tubule in particular

Pressure Filtration

When blood enters the glomerulus, blood pressure is sufficient to cause small molecules, such as nutrients, water, salts, and wastes, to move from the glomerulus to the inside of Bowman's capsule, especially since the glomerular walls are 100 times more permeable than the walls of most capillaries elsewhere in the body. The molecules that leave the blood and enter Bowman's capsule are called the *glomerular filtrate.* Blood proteins and blood cells are too large to be part of this filtrate, and they remain in the blood as it flows into the efferent arteriole. Glomerular filtrate has the same composition as tissue fluid, and if this composition were unaltered in other parts of the nephron, death from loss of water (dehydration), loss of nutrients (starvation), and lowered blood pressure would quickly follow. However, *selective reabsorption* prevents this from happening.

Reabsorption

Reabsorption from the nephron to the blood takes place through the walls of the proximal convoluted tubule. Nutrients, water, and even some waste molecules diffuse passively back into the peritubular capillary network. Osmotic pressure further influences the movement of water. The nonfilterable proteins remain in the blood and exert an osmotic pressure. Also, after sodium (Na^+) is actively reabsorbed, chlorine (Cl^-) follows along, and these two ions together increase the osmotic pressure of the blood.

Selective reabsorption of nutrient molecules is brought about by active transport. The cells of the proximal convoluted tubule have numerous microvilli that increase the surface area and numerous mitochondria that supply the energy needed for reabsorption. Reabsorption is selective since only molecules recognized by carrier molecules are actively transported from the tubule to the blood.

The kidney has a threshold level for every substance that is reabsorbed. The threshold level of a substance is its normal level in the blood, and reabsorption occurs until this level is obtained. Thereafter, the substance will not be reabsorbed and will appear in the urine. For example, plasma normally is 100 mg% of glucose and 15 mg% of urea. After this amount is reabsorbed, any excess present in the filtrate will appear in the urine. In the end, nearly 100% of glucose is reabsorbed whereas only 50% of urea is reabsorbed.

Tubular Secretion

Tubular secretion is the means by which other nonfilterable wastes can be added to the fluid as it passes through the tubules. Toxic substances, such as foreign acids and bases that have been absorbed in the gut, are eliminated by tubular secretion. Peni-cillin and a number of other substances also are excreted in this way. But the process of tubular secretion is not so important to urine formation as are the first two steps studied.

Each step in urine formation is associated with a particular part of the nephron. Pressure filtration occurs at Bowman's capsule; reabsorption occurs at the proximal convoluted tubule; tubular secretion occurs at the distal convoluted tubule. The loop of Henle plays the primary role in the reabsorption of water so that the urine of humans is hypertonic.

Control of Water Excretion

Reptiles and birds rely primarily on the gut to reabsorb water, but mammals rely on the kidneys. The arrangement of the loop of Henle in relation to the collecting duct enables mammals to excrete a hypertonic urine. A concentration gradient is established in the medulla (fig. 37.11) that promotes the reabsorption of water from the descending limb of the loop of Henle and the collecting duct. It now appears that this gradient is due to the extrusion of salt, Na^+Cl^-, by the upper portion of the ascending limb (of the loop of Henle) and by the diffusion of urea from the collecting duct. Because the descending limb of the loop of Henle loses water the ascending limb has a high concentration of salt. You might think this would cause it to take up water, but the ascending limb is impermeable to water. Instead it loses salt first by diffusion and then by active extrusion.

The descending limb automatically loses water, but the amount of water that leaves the collecting duct is regulated by hormonal action. The hormone **ADH (antidiuretic hormone),** released by the posterior lobe of the pituitary, increases the permeability of the collecting duct so more water will leave it and be reabsorbed into the blood. If the osmotic pressure of the blood increases, the posterior lobe of the pituitary releases ADH, more water is reabsorbed, and consequently there is less urine. On the other hand, if the osmotic pressure of the blood decreases, the posterior lobe of the pituitary does not release ADH. The resulting impermeability of the collecting duct causes more water

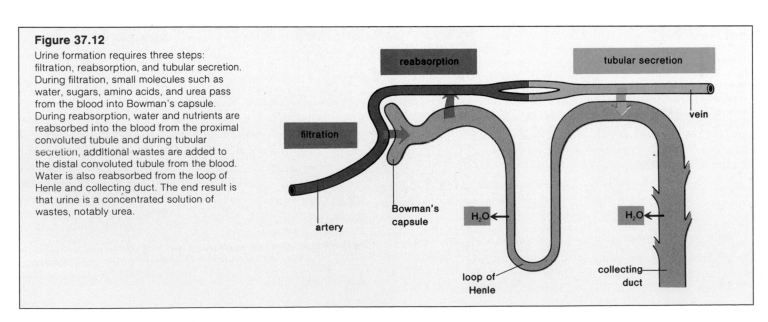

Figure 37.12
Urine formation requires three steps: filtration, reabsorption, and tubular secretion. During filtration, small molecules such as water, sugars, amino acids, and urea pass from the blood into Bowman's capsule. During reabsorption, water and nutrients are reabsorbed into the blood from the proximal convoluted tubule and during tubular secretion, additional wastes are added to the distal convoluted tubule from the blood. Water is also reabsorbed from the loop of Henle and collecting duct. The end result is that urine is a concentrated solution of wastes, notably urea.

Figure 37.13

Drawing of glomerulus and adjacent distal convoluted tubule. The afferent arteriole cells within juxtaglomerular apparatus (*circled*) are sensitive to blood pressure and release renin if this pressure falls below normal.

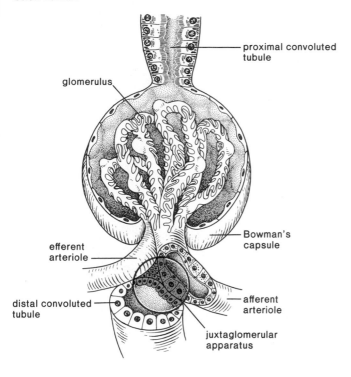

proximal convoluted tubule

glomerulus

Bowman's capsule

efferent arteriole

distal convoluted tubule

afferent arteriole

juxtaglomerular apparatus

to be excreted and more urine to be formed. Drinking alcohol causes diuresis (increased urine flow) because it inhibits the secretion of ADH. Beer drinking also causes diuresis mainly because of increased fluid intake. Drugs called diuretics are often prescribed for high blood pressure. These drugs cause increased urinary excretion and thus reduce blood volume and blood pressure.

The hormone **aldosterone,** which is secreted by the adrenal cortex, primarily maintains the sodium (Na^+) and potassium (K^+) balance of the blood. It causes the distal convoluted tubule to reabsorb Na^+ and to excrete K^+. The increase of Na^+ in the blood causes water to be reabsorbed, leading to an increase in blood volume and blood pressure.

Blood pressure is constantly monitored in the nephrons by the afferent arteriole cells within the *juxtaglomerular apparatus,* in order to maintain a fairly constant pressure. The juxtaglomerular apparatus (fig. 37.13) is located at the region of contact between the afferent arteriole and the distal convoluted tubule. It's believed that the afferent arteriole cells in the region secrete the enzyme renin, when the blood pressure is insufficient, to promote efficient filtration in the glomerulus (fig. 37.14). Renin is an enzyme that converts a large plasma protein called angiotensinogen into *angiotensin.* Angiotensin then stimulates the adrenal cortex to release aldosterone, which allows Na^+ to be reabsorbed and blood volume and pressure to rise.

A possible mechanism has also been suggested for suppressing renin secretion. Perhaps the distal tubule cells in the juxtaglomerular apparatus are sensitive to Na^+ concentration in urine and when there is a high Na^+ concentration in the tubule, they inhibit the afferent arteriole cells from secreting

Figure 37.14

A scanning electron micrograph of a section of kidney cortex, showing a glomerulus (the outer layer of Bowman's capsule has been removed). The holes surrounding the glomerulus are cross sections of tubules.

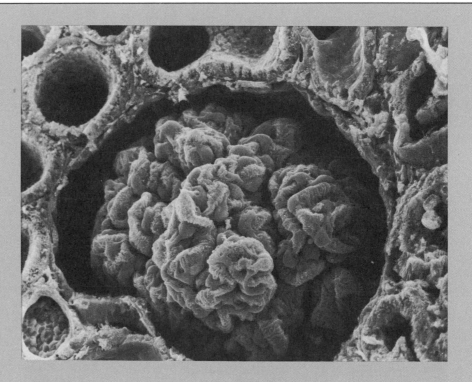

Figure 37.15

Kidney dialysis. *a.* As the patient's blood circulates through dialysis tubing, it is exposed to a solution. *b.* Wastes exit from the blood into the solution because of a preestablished concentration gradient. In this way, the blood is not only cleansed, but the pH can also be adjusted.

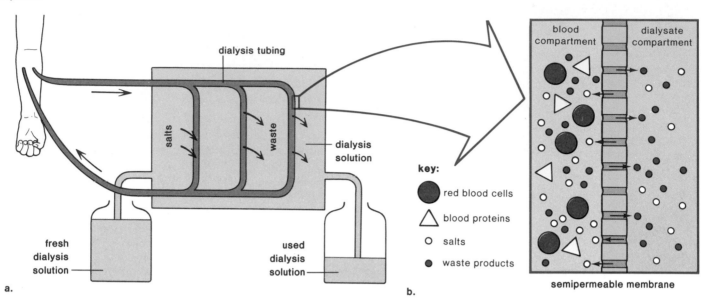

renin. If so, this mechanism may be faulty in certain individuals. For example, the renin-angiotensin-aldosterone system (fig. 41.11) seems to always be active in some patients with hypertension.

Adjustment of Blood pH

The kidneys help maintain the pH level of the blood within a narrow range, and the whole nephron takes part in this process. The excretion of hydrogen ions and ammonia, together with the reabsorption of sodium and bicarbonate ions, is adjusted to keep the pH within normal bounds. If the blood is acidic, hydrogen ions are excreted in combination with ammonia, while sodium and bicarbonate ions are reabsorbed. This will restore the pH because sodium ions promote the formation of hydroxyl ions, while bicarbonate takes up hydrogen ions when carbonic acid is formed.

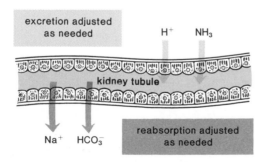

If the blood is basic, fewer hydrogen ions are excreted, and fewer sodium and bicarbonate ions are reabsorbed.

Reabsorption and/or excretion of ions (salts) by the kidneys illustrates their homeostatic ability: they maintain not only the pH of the blood but also its osmolarity.

Kidney Failure

The urinary tract is subject to attack by a number of different bacteria. If the infection is localized in the urethra, it is called *urethritis*. If it invades the bladder, it is called *cystitis*. And finally, if the kidneys are affected, it is called *nephritis*. Often nephritis occurs after a strep infection in some other part of the body. Antigen-antibody complexes reach the glomeruli from the arterioles and are trapped there, causing irritation that leads to glomerular inflammation and damage. The glomerular membrane may then become more permeable than usual. Therefore, albumin, white blood cells, or even red blood cells may appear in the urine. Also, glomerular damage sometimes leads to blockage of the glomeruli so no fluid moves into the tubules.

Kidney Dialysis

Persons suffering from either temporary or permanent renal failure can be treated with an artificial kidney machine, a process commonly called kidney dialysis (fig 37.15). Blood circulates from an artery in the patient's arm to the kidney machine and returns to a vein. In the process, the patient's blood is passed through a semipermeable, membranous tube that is in contact with a balanced salt solution, or *dialysis fluid*. Substances more concentrated in the blood diffuse into the dialysis fluid. Conversely, substances more concentrated in the dialysis fluid diffuse into the blood. Hence, the artificial kidney can be used either to extract substances from the blood, including waste products or toxic chemicals and drugs, or to add substances, such as ions, to the blood. In the course of a six-hour dialysis, from 50 to 250 grams of urea can be removed from a patient, which greatly exceeds the rate of urea clearance of normal kidneys. Therefore, a patient need undergo treatment only about twice a week.

Summary

1. Animals excrete nitrogenous wastes that differ as to the amount of water required to excrete them. Aquatic animals usually excrete ammonia, and land animals excrete either urea or uric acid.
2. Osmotic regulation is important to animals. Most have to balance their water and salt intake and excretion to maintain the normal concentration in the body fluids. Marine fishes constantly drink water, excrete salts at the gills, and pass an isotonic urine. Freshwater fishes never drink water, take in salts at the gills, and pass a hypotonic urine.
3. Animals often have an excretory organ. Earthworm nephridia exchange molecules with the blood in a manner similar to vertebrate kidneys. Malpighian tubules in insects take wastes and water from the hemocoel to the gut where only water is reabsorbed.
4. Kidneys are a part of the human urinary system. Microscopically, each kidney is made up of nephrons, each of which has several parts and its own blood supply.
5. Urine formation requires three steps: pressure filtration at Bowman's capsule where nutrients, water, and wastes enter the tubule; selective reabsorption when nutrients and some water are reabsorbed; and tubular secretion when additional wastes are added to the tubule.
6. Humans excrete a hypertonic urine. The ascending limb of the loop of Henle actively extrudes salt so that the medulla is hypertonic relative to the contents of the descending limb and the collecting duct. Therefore water has a tendency to diffuse out of these.
7. Two hormones are involved in maintaining the water content of the blood. The hormone ADH, which makes the collecting duct more permeable, is secreted by the posterior pituitary in response to an increase in the osmotic pressure of the blood.
8. The hormone aldosterone is secreted by the adrenal cortex, after the low Na^+ content of the blood and resultant low blood pressure of the blood has caused the kidneys to release renin. The presence of renin leads to the formation of angiotensin, which causes the adrenal cortex to release aldosterone. Aldosterone causes the kidneys to retain Na^+.
9. The kidneys adjust the pH of the blood by excreting or conserving H^+, NH_3, HCO_3^-, and Na^+, as appropriate.
10. During kidney dialysis, waste molecules diffuse out of the blood into the dialysis fluid. The composition of the dialysis fluid prevents nutrient molecules from leaving the blood.

Objective Questions

1. Which of these is mismatched?
   a. insects—excrete uric acid
   b. humans—excrete urea
   c. fishes—excrete ammonia
   d. birds—excrete ammonia
2. One advantage of urea excretion over uric acid excretion is that urea
   a. requires less energy to form.
   b. can be concentrated to a greater extent.
   c. is not a toxic substance.
   d. requires less water to excrete.
3. Freshwater bony fishes maintain water balance by
   a. excreting salt across their gills.
   b. periodically drinking small amounts of water.
   c. excreting a hypotonic urine.
   d. excreting wastes in the form of uric acid.
4. Animals with which of these is most likely to excrete a semisolid, nitrogenous waste?
   a. nephridia
   b. Malpighian tubules
   c. human kidneys
   d. All of these.
5. In which of these human structures are you least apt to find urine?
   a. large intestine
   b. urethra
   c. ureter
   d. bladder
6. Excretion of a hypertonic urine in humans is associated with
   a. Bowman's capsule.
   b. the proximal convoluted tubule.
   c. the loop of Henle.
   d. the distal convoluted tubule.
7. The presence of ADH causes an individual to
   a. excrete sugars.
   b. excrete less water.
   c. excrete more water.
   d. Both (a) and (c).
8. In humans, water is
   a. found in the glomerular filtrate.
   b. reabsorbed from the nephron.
   c. in the urine.
   d. All of these.
9. Pressure filtration is associated with
   a. Bowman's capsule.
   b. the distal convoluted tubule.
   c. the collecting duct.
   d. All of these.
10. In humans, glucose
   a. is in the filtrate and urine.
   b. is in the filtrate and not in urine.
   c. undergoes tubular secretion and is in urine.
   d. undergoes tubular secretion and is not in urine.

Study Questions

1. Relate the three primary nitrogenous wastes to the habitat of animals.
2. Contrast the osmotic regulation of a marine bony fish with that of a freshwater bony fish.
3. Give examples of how other types of animals regulate their water and salt balance.
4. Describe how the excretory organs of the earthworm and the insect function.
5. Give the path of urine, and give a function for each structure mentioned.
6. List the parts of a nephron, and give a function for each structure mentioned.
7. Describe how urine is made by telling what happens at each part of the nephron.
8. What role do ADH and aldosterone play in regulating the tonicity of urine? How does this effect blood pressure?
9. How does the nephron regulate the pH of the blood?
10. Explain how the artificial kidney machine works.

Thought Questions

1. Would an animal that lives in fresh water or on land be more likely to have a well-developed loop of Henle? Why?

2. Why would you predict that the tubules in the kidney of a freshwater fish would be better developed than the tubules of a marine fish?

3. Why is the term filtration used both in regard to a capillary in the tissues and to the glomerulus in the kidney tubule? What differences are there between filtration at these two locations?

Selected Key Terms

excretion (eks-kre′shun) 578
urea (u-re′ah) 579
uric acid (u′rik as′id) 579
nephridia (nĕ-frid′e-ah) 582
ureter (u-re′ter) 583
urinary bladder (u′rĭ-ner″e blad′der) 583
urethra (u-re′thrah) 583

nephron (nef′ron) 583
Bowman's capsule (bo′manz kap′sūl) 583
proximal convoluted tubule (prok′sĭ-mal kon′vo-lūt-ed tu′būl) 583
loop of Henle (lōōp uv hen′le) 583
distal convoluted tubule (dis′tal kon′vo-lūt-ed tu′būl) 583

collecting duct (kŏ-lekt′ing dukt) 583
glomerulus (glo-mer′u-lus) 584
antidiuretic hormone (ADH) (an″tĭ-di″u-ret′ik hor′mōn) 586
aldosterone (al″do-ster′ōn) 587

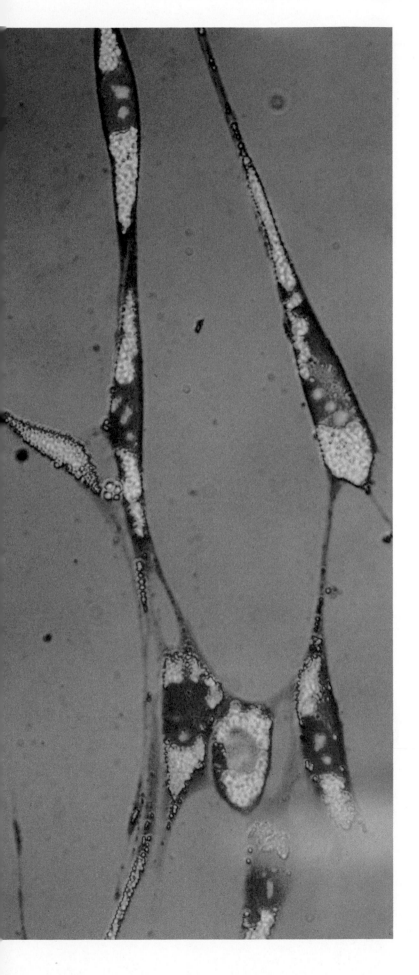

Nervous System

Your study of this chapter will be complete when you can:

1. *State, in general, the overall function of the nervous system.*

2. *Compare the organization and complexity of various invertebrate nervous systems to that of the human nervous system.*

3. *Describe, in general, the structure of a neuron; name three types of neurons and state their specific functions.*

4. *Describe the nerve impulse as an electrochemical change that is recorded as the action potential by the oscilloscope.*

5. *Describe the structure and function of a synapse, including transmission across a synapse.*

6. *Describe the structure and function of the peripheral nervous system.*

7. *Draw a diagram depicting a spinal reflex, and explain the function of all parts included.*

8. *State the parts of the autonomic nervous system; cite similarities and differences in structure and function of its two divisions.*

9. *Describe, in general, the structure and function of the central nervous system.*

10. *List the major parts of the brain, and give a function for each part.*

11. *Discuss, in general, current research in regard to learning and memory.*

12. *List several well-known excitatory and inhibitory neurotransmitters in the peripheral and central nervous systems.*

13. *Discuss and give examples of the effects of drugs at synapses and list five criteria for drug abuse.*

Neuron and glial cells from the brain as seen by using interference contrast microscopy. Neurons are cells specialized to conduct a nerve impulse and glial cells are supportive cells that aid the function of neurons. There are about five times as many glial cells as neurons in the nervous system.

The ability to respond to stimuli is a characteristic of all living things. In complex animals, such as vertebrates, this ability depends on the nervous, endocrine (hormonal), sensory, and musculoskeletal systems. Working together, the nervous and endocrine systems coordinate the actions of all the other systems of the body to produce effective behavior and to keep the internal environment within safe limits. Because hormones are transported in the blood, it may require seconds, minutes, hours, or even longer for these chemical messengers to produce their effects. The nervous system, on the other hand, communicates rapidly, requiring only thousandths of a second. The nervous system receives information and processes it before sending out signals to the muscles and glands so an appropriate response is possible. In this way the nervous system integrates and controls the other systems of the body.

Evolution of Nervous Systems

A comparative study of animal nervous systems (fig. 38.1) indicates what steps may have led to the complex nervous system of vertebrates. Hydras have a very simple nervous system that looks like a net of threads extending throughout the body. The net is actually composed of neurons (nerve cells), each one of which has processes that reach out to the other neurons. Impulses spread in all directions along this net that enables hydras to give a local response to a stimulus. External receptor cells that respond to stimuli, like chemicals or pressure, communicate with the *nerve net* that then causes cells to contract within the two layers making up the body of a hydra (fig. 26.5).

Planarians have a more complicated nervous system that resembles a *ladder*. However, there is a brain, a concentration of neurons that apparently acts as a relay station between sense organs and muscles. Two longitudinal nerve cords with crosslinks between them connect the brain to muscles and other parts of the body.

Segmented worms, such as the earthworm, have an even more elaborate nervous system. The brain and a large ventral nerve cord running along the midline of the body form a **central nervous system** that is responsible for integration and coordination. Nerves branch off from the nerve cord in each segment, and these nerves comprise a so-called **peripheral nervous system.**

This organization is typical of other advanced invertebrates and all of the vertebrates. However, when invertebrate nervous systems are compared to vertebrate systems (fig. 38.1*d*), it is apparent that there has been a vast increase in the number of neurons. For example, an insect's entire nervous system may contain a total of about one million neurons, while a vertebrate nervous system may contain many thousand to several billion times that number.

In complex animals the nervous system has two parts: the central nervous system and the peripheral nervous system.

Human Nervous System

The human nervous system (fig. 38.2), like that of the earthworm, is divided into the central and peripheral nervous systems. The *central nervous system* includes the brain and spinal cord (dorsal nerve cord), which lie in the midline of the body where the skull protects the brain and the vertebrae protect the

Figure 38.1

Evolution of the nervous system. *a.* The nerve net of a hydra, a cnidarian. *b.* The paired nerve cords with cross connections of a planarian, a flatworm, have a ladder appearance. *c.* The earthworm, a segmented worm, has a central nervous system consisting of the brain and a ventral solid nerve cord. It also has a peripheral nervous system consisting of nerves. *d.* A rabbit, like other vertebrates, has a dorsal, hollow nerve cord in its central nervous system.

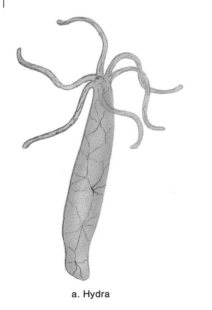

| a. Hydra | b. Planarian | c. Earthworm | d. Rabbit |

spinal cord. The *peripheral nervous system* contains both cranial nerves, which originate from the brain, and spinal nerves, which project from either side of the spinal cord. The peripheral nervous system is further divided into the somatic system and the autonomic (or visceral) system. The somatic division contains nerves that control skeletal muscles, skin, and joints. The autonomic division contains nerves that control internal organs. Before we continue our discussion of the nervous system, we must first examine the anatomy and functions of nerve cells, or **neurons.**

Neurons

Neurons are cells that vary as to size and shape (fig. 38.3), but they all have three parts: the dendrite(s), the cell body, and the axon. The **dendrites** receive information from other neurons and generally conduct nerve impulses *toward* the cell body. The **axon,** on the other hand, conducts nerve impulses *away* from the cell body. The cell body contains the nucleus and other organelles typically found in cells. One of the main functions of the cell body is to manufacture neurotransmitters which are chemicals stored in secretory vesicles at the ends of axons. When neurotransmitters are released they influence the excitability of nearby neurons.

Three types of neurons are shown in figure 38.3. **Sensory neurons,** each with one long dendrite and a short axon, take messages from sense organs to the central nervous system. **Motor neurons,** each with a long axon and short dendrites, take messages from the central nervous system to muscle fibers or glands. Because motor neurons cause muscle fibers or glands to react, they are said to innervate these structures. The third type of neuron, called an **interneuron,** is only found within the central nervous system. It conveys messages between various parts of the central nervous system, such as from one side of the brain or spinal cord to the other or from the brain to the cord, and vice versa. An interneuron has short dendrites and either a long or a short axon.

Figure 38.2
Overall organization of the nervous system in human beings.
a. Pictorial representation. The central nervous system (brain and spinal cord) and some of the nerves of the peripheral nervous system are shown. *b.* Diagrammatic representation: the central nervous system is at the top of the diagram, and the peripheral nervous system is below. The nerves of the peripheral nervous system belong either to the somatic nervous system or to the autonomic nervous system. The autonomic nervous system has two portions, the sympathetic and parasympathetic systems.

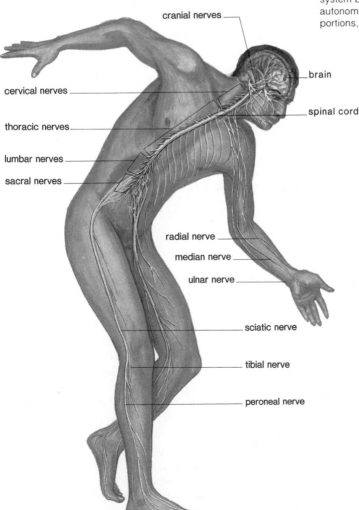

cranial nerves

cervical nerves

thoracic nerves

lumbar nerves

sacral nerves

brain

spinal cord

radial nerve

median nerve

ulnar nerve

sciatic nerve

tibial nerve

peroneal nerve

a.

Waldrop

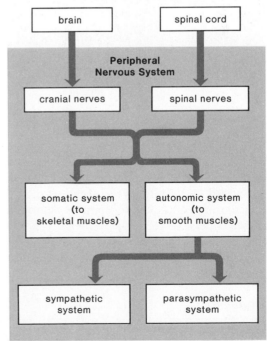

Central Nervous System

brain

spinal cord

Peripheral Nervous System

cranial nerves

spinal nerves

somatic system (to skeletal muscles)

autonomic system (to smooth muscles)

sympathetic system

parasympathetic system

b.

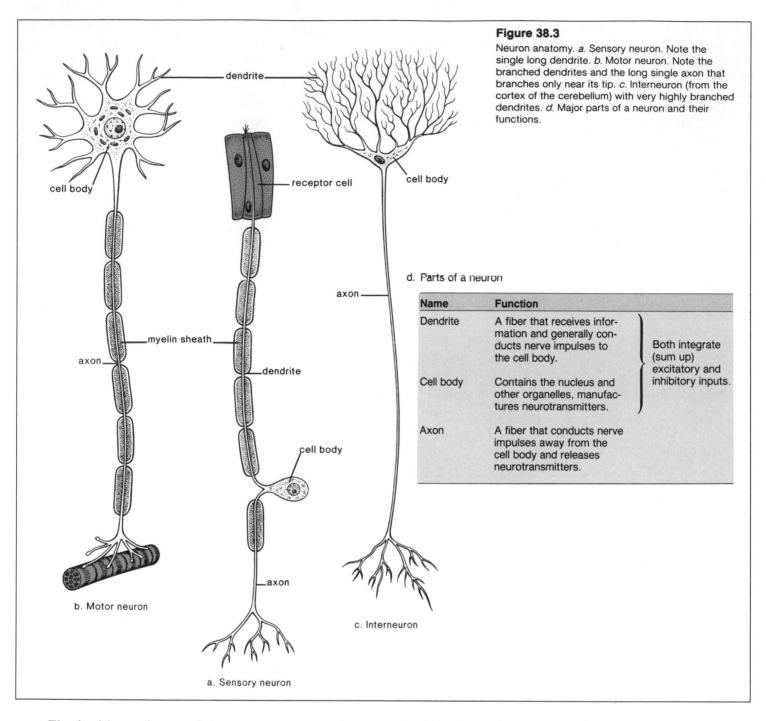

Figure 38.3

Neuron anatomy. *a*. Sensory neuron. Note the single long dendrite. *b*. Motor neuron. Note the branched dendrites and the long single axon that branches only near its tip. *c*. Interneuron (from the cortex of the cerebellum) with very highly branched dendrites. *d*. Major parts of a neuron and their functions.

d. Parts of a neuron

Name	Function	
Dendrite	A fiber that receives information and generally conducts nerve impulses to the cell body.	Both integrate (sum up) excitatory and inhibitory inputs.
Cell body	Contains the nucleus and other organelles, manufactures neurotransmitters.	
Axon	A fiber that conducts nerve impulses away from the cell body and releases neurotransmitters.	

b. Motor neuron

c. Interneuron

a. Sensory neuron

The dendrites and axons of these neurons are sometimes called fibers or processes. Most long fibers, whether dendrite or axon, are covered by a white **myelin sheath** (fig. 38.3) formed from the membranes of the tightly spiraled Schwann cells that surround these fibers.

Schwann cells are one of several types of glial cells in the nervous system. Glial cells service the neurons and have supportive and nutritive functions.

Nerve Impulse

An Italian investigator, Luigi Galvani, discovered in 1786 that a nerve could be stimulated by an electric current. But it was realized later that the speed of the nerve impulse is too slow to simply be due to the movement of electrons or current within a nerve fiber. In the early 1900s Julius Bernstein at the University of Halle, Germany, suggested that the nerve impulse is an electrochemical phenomenon involving the movement of unequally distributed ions on either side of a nerve cell membrane. It was not until 1939 that investigators developed a technique that enabled them to substantiate this hypothesis. A. L. Hodgkin and A. F. Huxley, English neurophysiologists, received the Nobel Prize in 1963 for their work in this field. They and a group of researchers headed by K. S. Cole and J. J. Curtis at Woods Hole, Massachusetts managed to insert a very tiny electrode into the giant axon of the squid *Loligo* (fig. 38.4). This internal electrode was then connected to a voltmeter and an *oscilloscope,* an instrument with a screen that shows a trace or pattern indicating a change in voltage with time (fig. 38.4*d*). *Voltage* is a measure of the electrical potential difference between two points, which in this case is the difference between two electrodes, one placed inside and another placed outside the axon. (An electrical potential difference across a membrane is called the membrane potential.) When a potential difference exists, we can say that a plus and minus pole exist; therefore, an oscilloscope indicates the existence of polarity and records polarity changes.

Resting Potential

When the axon is not conducting an impulse, the oscilloscope records a membrane potential equal to about −65 mV (millivolts), indicating that the inside of the neuron is more negative than the outside (fig. 38.4*b*). This is called the **resting potential** because the axon is not conducting an impulse.

The existence of this polarity can be correlated with a difference in ion distribution on either side of the axomembrane (plasma membrane of axon). As figure 38.5*a* shows, there is a higher concentration of sodium ions (Na⁺) outside the axon and a higher concentration of potassium ions (K⁺) inside the axon. The unequal distribution of these ions is due to the action of the sodium-potassium pump. This is an active transport system in the cell membrane that pumps sodium ions out of and potassium ions into the axon. The work of the pump maintains the unequal distribution of sodium and potassium ions across the membrane.

The pump is always working because the membrane is somewhat permeable to these ions and they tend to diffuse toward their lesser concentrations. Since the membrane is more permeable to potassium than to sodium there are always more positive ions outside the membrane than inside and this accounts for the polarity recorded by the oscilloscope. There are also large, negatively charged proteins in the axoplasm (cytoplasm of axon); altogether, then, the oscilloscope records that the axoplasm is −65 mV compared to tissue fluid. This is the resting potential.

Action Potential

If the axon is stimulated to conduct a nerve impulse by either an electric shock, a sudden change in pH, or a pinch, there is a rapid change in the polarity recorded as a trace on the oscilloscope screen. This change of polarity is called the **action potential** (fig. 38.5). First, the trace goes from −65 mV to +40 mV (called *depolarization*) indicating that the axoplasm is now more positive than tissue fluid. Then the trace returns to −65 mV again (called *repolarization*) indicating that the inside of the axon becomes negative again.

The action potential is due to the presence of special protein-lined channels in the axon membrane, which can open to allow either sodium or potassium ions to pass through (fig. 38.5*b* and *c*). These channels have gates called the "sodium gates" and the "potassium gates." During the depolarization phase of the action potential the sodium gates open and sodium rushes into the axon. Once this phase is complete, repolarization occurs. During repolarization, the potassium gates open and potassium rushes out of the axon.

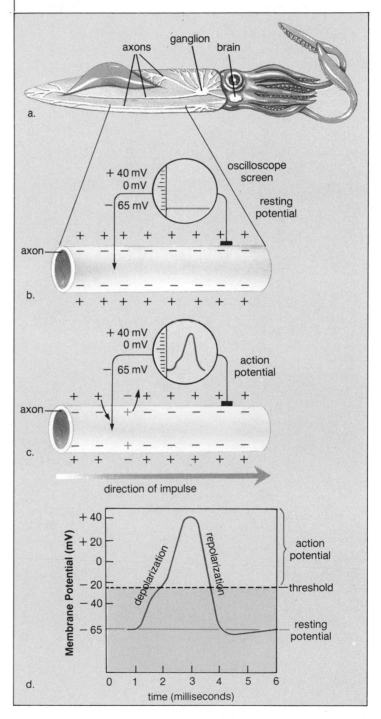

Figure 38.4

The original nerve impulse studies utilized giant squid axons and a voltage recording device known as an oscilloscope. *a.* The squid axons shown produce rapid muscular contraction so that the squid can move quickly. *b.* These axons are so large (about I mm in diameter) that a microelectrode can be inserted inside. When the axon is not conducting a nerve impulse, the electrode registers and the oscilloscope records a resting potential of −65 mV. *c.* When the axon is conducting a nerve impulse, the *threshold* for an action potential has been achieved and there is a rapid change in potential from −65 mV to +40 mV (called depolarization) followed by a return to −65 mV (called repolarization). *d.* Enlargement of action potential of nerve impulse.

Figure 38.5

The action potential is the result of an exchange of sodium (Na⁺) and potassium (K⁺) ions, and it is recorded as a change in polarity by an oscilloscope (as shown on the right). So few ions are exchanged for each action potential that it is possible for a nerve fiber to repeatedly conduct nerve impulses. Whenever the fiber rests, the sodium-potassium pump restores the original distribution of ions.

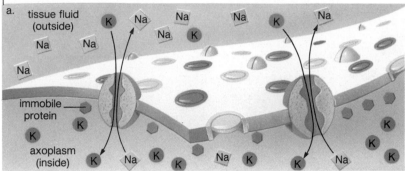

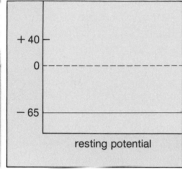

When a neuron is not conducting a nerve impulse, the sodium (Na⁺) and potassium (K⁺) gates are closed. The Na⁺/K⁺ pump maintains the uneven distribution of these ions across the axomembrane. The oscilloscope registers a resting potential of −65 mV inside compared to outside.

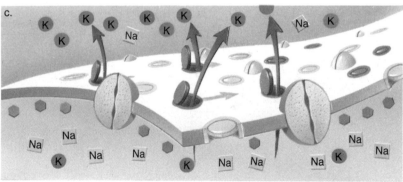

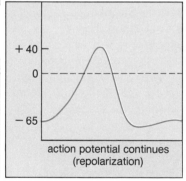

An action potential begins when the Na⁺ gates open and Na⁺ moves to the inside. The oscilloscope registers a depolarization as the axoplasm reaches +40 mV compared to tissue fluid.

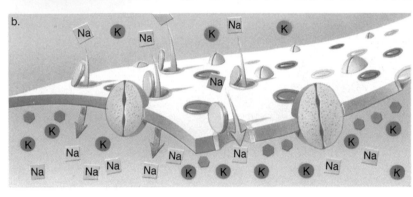

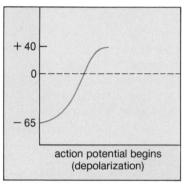

Recovery occurs as the Na⁺ gates close and the K⁺ gates open, allowing K⁺ to move to the outside. The oscilloscope registers repolarization as the axoplasm again becomes −65 mV compared to tissue fluid.

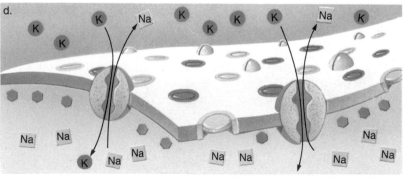

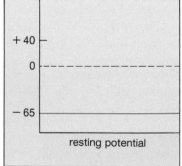

The oscilloscope registers −65 mV again but the Na⁺/K⁺ pump is working to restore the original sodium and potassium ion distribution illustrated in (a). The Na⁺ and K⁺ gates are now closed but will open again in response to another stimulus.

Figure 38.6

The high speed of the nerve impulse in myelinated fibers can be accounted for by the fact that the nerve impulse jumps from one node of Ranvier to the next.

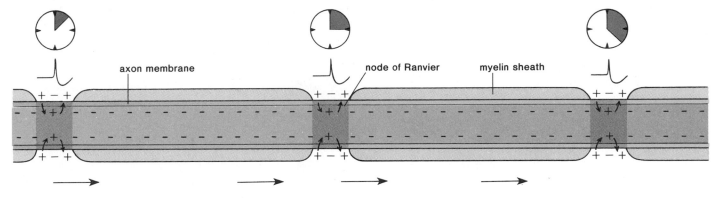

axon membrane   node of Ranvier   myelin sheath

Direction of Nerve Impulse

Notice that at the completion of an action potential (fig. 38.5d) the original ion distribution has been altered somewhat. There are now more sodium ions *inside* the membrane than before and more potassium ions *outside* the membrane than before. The sodium/potassium pump is able to restore the former distribution, however.

The oscilloscope records changes at only one location in a fiber, but actually the action potential travels along the length of a fiber. It is self-propagating because the ion channels are prompted to open whenever the membrane potential decreases in an adjacent area. In the laboratory, electrical stimulation is often used to cause a local change in the membrane potential that will trigger the nerve impulse. In organisms, the reception of neurotransmitters influences the membrane potential as we will discuss shortly.

In invertebrates, some fibers are larger than others; in vertebrates, most long fibers are myelinated. The nerve impulse travels faster in larger fibers and also in myelinated ones. In myelinated fibers, there are intervals, or gaps, between the Schwann cells called *nodes of Ranvier,* and the presence of these intervals greatly speeds up the conduction of nerve impulses. The nerve impulse may reach speeds of 200 meters per second compared to 20 meters per second for large, nonmyelinated ones, because the action potential jumps from one node of Ranvier to the next (fig. 38.6). This is called **saltatory** (*saltatory* means jumping) **conduction.**

All neurons transmit the same type of nerve impulse—an electrochemical change that is self-propagating along the fiber(s).

Transmission across a Synapse

In 1897 the English scientist Sir Charles Sherrington and others noted two important aspects of nerve impulse transmission between neurons. First, an impulse passing from one vertebrate nerve cell to another always moves in only one direction. Second, there is a very short delay in transmission of the nerve impulse from one neuron to another. This latter observation led to the hypothesis that there is a minute space between neurons. Sherrington called the region where the impulse moves from one neuron to another a **synapse,** meaning "to clasp." Synapses occur between the end of an axon and a dendrite or between the end of an axon and a cell body (fig. 38.7). A synapse has three components: a presynaptic membrane, a gap now called the synaptic cleft, and a postsynaptic membrane. As mentioned previously, there are vesicles at the ends of axons where **neurotransmitters,** also called transmitter substances, are stored. When nerve impulses reach a presynaptic membrane, the vesicles fuse with the membrane and discharge their contents into the synaptic cleft. The discharged neurotransmitter molecules diffuse across the synaptic cleft and bind in a lock-and-key manner to the postsynaptic membrane at *receptor sites.* This binding process alters the membrane potential of the postsynaptic membrane in the direction of either excitation or inhibition. If excitation occurs, the axoplasm becomes less negative, and if inhibition occurs, the axoplasm becomes more negative compared to tissue fluid. For example, if a neurotransmitter causes only potassium gates to open, potassium will exit and the axoplasm will become even more negative.

Summation and Integration

A dendrite or cell body is on the receiving end of many synapses. Whether the neuron fires (initiates a nerve impulse) or not depends on the summary, or net, effect of all the excitatory and inhibitory neurotransmitters it receives. If the amount of excitatory neurotransmitter received is sufficient to raise the membrane potential above the threshold level (fig. 38.4d), the neuron fires. If the amount of excitatory neurotransmitter received is insufficient, only local excitation occurs. For this reason, the dendrites and the cell body are the parts of a neuron where *integration* (a summing-up) occurs (fig. 38.3). This is the way that the nervous system fine-tunes its response to the environment.

Figure 38.7

Diagrammatic representation of a synapse at three different magnifications. *a.* Typically, a synapse is located wherever an axon is close to a dendrite or cell body. Drawing based on a photomicrograph shows that there are several synaptic endings per axon because of terminal branching of the axon. *b.* Drawing based on low-power electron micrographs shows that a branch ends in a terminal knob having numerous synaptic vesicles, each filled with a neurotransmitter substance. This drawing makes it clear that a synapse contains a cleft (space) between the axon and the dendrite.

c. Drawing based on high-power electron micrographs shows that a synapse consists of a presynaptic membrane, the synaptic cleft, and the postsynaptic membrane. When a nerve impulse reaches the synaptic vesicles, they move toward and fuse with the presynaptic membrane, discharging their contents. The neurotransmitter substance diffuses across the cleft and combines with a receptor. A nerve impulse may follow.

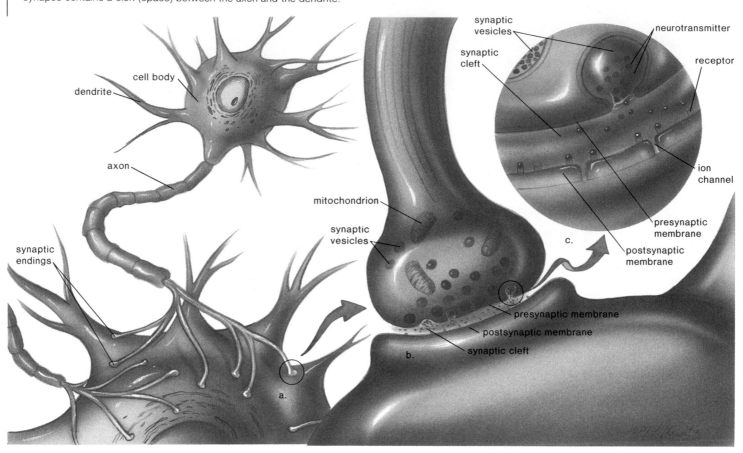

Neurotransmitter Substances

Acetylcholine (ACh) and **norepinephrine** (NE) are the primary excitatory neurotransmitters and are active in both the peripheral and central nervous systems. Examples of inhibitory substances, so far discovered only in the central nervous system, are given on page 607.

Once neurotransmitters have been released into a synaptic cleft, they remain active for only a short time. In some neurons the cleft contains enzymes that rapidly inactivate the neurotransmitter. For example, the enzyme acetylcholinesterase (AChE) or simply cholinesterase breaks down acetylcholine. A single molecule of AChE catalyzes the breakdown of 25,000 molecules of acetylcholine per second. The breakdown products are then taken up into the presynaptic neuron and used in the synthesis of new acetylcholine molecules. The enzyme monoamine oxidase breaks down noradrenalin after it is absorbed.

Transmission of a nerve impulse across a synapse is dependent on neurotransmitter substances that alter the potential difference across the postsynaptic membrane.

Peripheral Nervous System

The peripheral nervous system (PNS) contains **nerves** (fig. 38.2*b*), structures that contain only long dendrites and/or long axons. Each of these fibers is surrounded by myelin (fig. 38.3), and therefore these nerves have a white, shiny, glistening appearance. There are no cell bodies in nerves because cell bodies are found only in the central nervous system (CNS) or in the ganglia. **Ganglia** (sing., **ganglion**) are collections of cell bodies within the PNS.

Humans have 12 pairs of cranial nerves and 31 pairs of spinal nerves. **Cranial nerves** are either sensory nerves (having long dendrites of sensory neurons only), motor nerves (having long axons of motor neurons only), or mixed nerves (having both

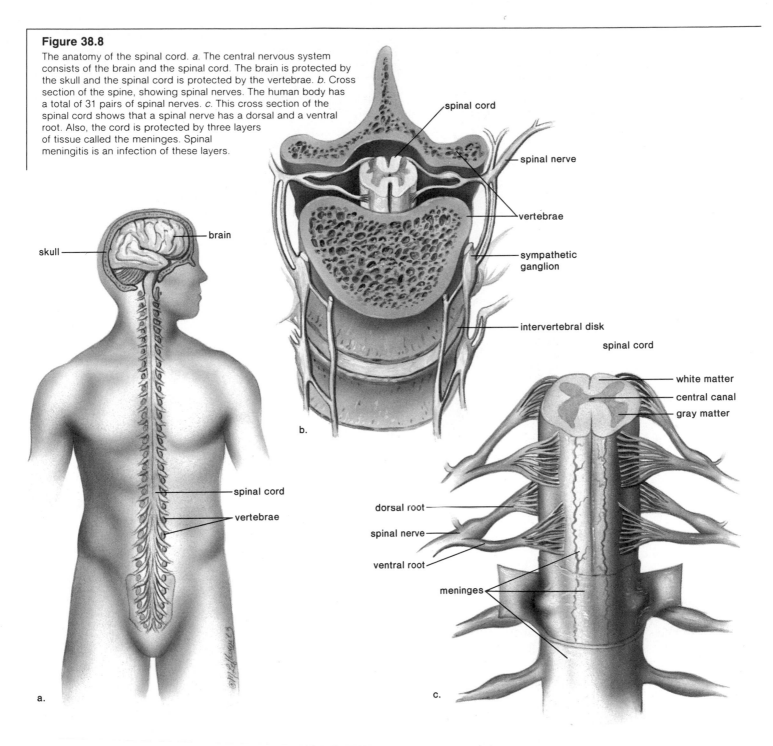

Figure 38.8

The anatomy of the spinal cord. *a.* The central nervous system consists of the brain and the spinal cord. The brain is protected by the skull and the spinal cord is protected by the vertebrae. *b.* Cross section of the spine, showing spinal nerves. The human body has a total of 31 pairs of spinal nerves. *c.* This cross section of the spinal cord shows that a spinal nerve has a dorsal and a ventral root. Also, the cord is protected by three layers of tissue called the meninges. Spinal meningitis is an infection of these layers.

long dendrites and long axons). All cranial nerves, except the vagus nerve, control the head, neck, and face. The *vagus nerve* controls the internal organs.

Spinal nerves (fig. 38.8) are all mixed nerves that take impulses to and from the spinal cord. Their arrangement shows that humans are segmented animals: there is a pair of spinal nerves for each segment. Spinal nerves project from the spinal cord, which is a part of the central nervous system. The spinal cord extends longitudinally down the back where it is protected by the vertebrae (sing., vertebra). The cord contains a tiny central canal filled with cerebrospinal fluid; gray matter consisting of cell bodies and short fibers; and white matter consisting of myelinated fibers.

In the PNS, cranial nerves take impulses to and/or from the brain and spinal nerves take impulses to and from the spinal cord.

Somatic Nervous System

The *somatic nervous system* includes all of those nerves that serve the musculoskeletal system and the exterior sense organs, including those in the skin. Exterior sense organs are **receptors** that receive environmental stimuli and then initiate nerve impulses. Muscle fibers are **effectors** that bring about a reaction to the stimulus. Muscle effectors are discussed in chapter 40, and receptors are discussed in chapter 39.

Figure 38.9

Diagram of a reflex arc shows the detailed composition of a spinal nerve. When the receptors in the skin are stimulated, nerve impulses (see arrows) move along a sensory neuron to the spinal cord. (Note that the cell body of a sensory neuron is in a ganglion outside the cord.) The nerve impulses are picked up by an interneuron, which lies completely within the cord, and pass to the dendrites and cell body of a motor neuron that lies ventrally within the cord. The nerve impulses then move along the axon of the motor neuron to an effector, such as a muscle fiber that contracts. The brain receives information concerning sensory stimuli by way of other interneurons, with long fibers in tracts that run up and down the cord within the white matter.

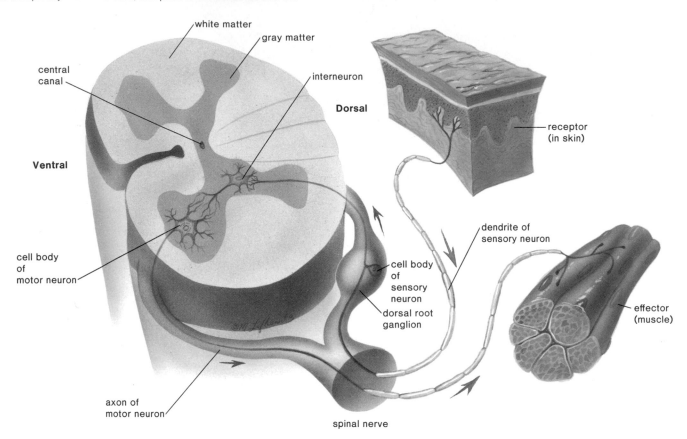

Reflex Arc

Reflexes are automatic, involuntary responses to changes occurring inside or outside the body. In the somatic nervous system, outside stimuli often initiate a reflex action. Some reflexes, such as blinking the eye, involve the brain, but others, such as withdrawing the hand from a hot object, do not necessarily involve the brain. Figure 38.9 illustrates the path of the second type of reflex action involving the spinal cord and a spinal nerve, called a *spinal reflex,* or *reflex arc.* Whenever a person touches a very hot object, a receptor in the skin generates nerve impulses that move along the dendrite of a sensory neuron toward the cell body and the CNS. The cell body of a sensory neuron is located in the dorsal root ganglion, just outside the cord. From the cell body, the impulses travel along the axon of the sensory neuron and enter the cord; there they may pass to many interneurons, one of which lies completely within the gray matter and connects with a motor neuron. The short dendrites and cell body of the motor neuron are in the ventral region (horn) of the gray matter, and the axon leaves the cord by way of the ventral root.

The nerve impulses travel along the axon to muscle fibers, which then contract so that the hand is withdrawn from the hot object. Various other reactions usually accompany a reflex response; the person may look in the direction of the object, jump back, and utter appropriate exclamations. This whole series of responses is explained by the fact that the sensory neuron stimulates several interneurons, which take impulses to all parts of the central nervous system, including the cerebrum which, in turn, makes the person conscious of the stimulus and his or her reaction to it.

The reflex arc is a major functional unit of the nervous system. It allows us to react to internal and external stimuli.

Autonomic Nervous System

The **autonomic nervous system** (fig. 38.10) is a part of the PNS. It is made up of motor neurons that control the internal organs automatically and usually without the need for conscious intervention. There are two divisions to the autonomic nervous system (table 38.1): the **sympathetic system** and the **parasympathetic**

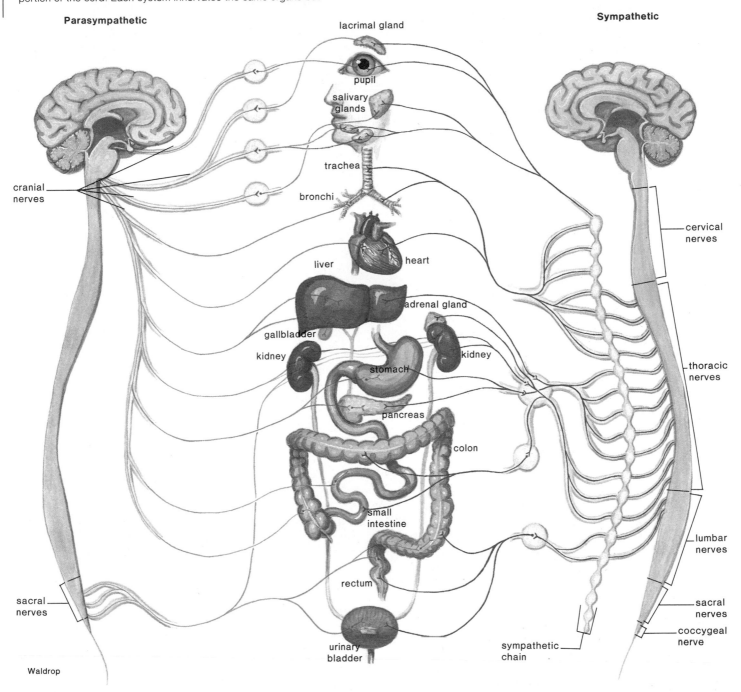

Parasympathetic

Sympathetic

lacrimal gland

pupil

salivary glands

cranial nerves

trachea

bronchi

liver   heart

adrenal gland

gallbladder

kidney   kidney

stomach

pancreas

colon

small intestine

rectum

sacral nerves

urinary bladder

cervical nerves

thoracic nerves

lumbar nerves

sacral nerves

coccygeal nerve

sympathetic chain

Waldrop

system. Both of these (1) function automatically and usually subconsciously in an involuntary manner; (2) innervate all internal organs; and (3) use two motor neurons for each impulse. The cell body of the first motor neuron is located in the CNS, and the cell body of the second motor neuron is located in a ganglion. The axon that occurs before the ganglion is called the preganglionic fiber, and the axon that occurs after the ganglion is called the postganglionic fiber.

Sympathetic Nervous System

Preganglionic fibers of the sympathetic nervous system arise from the middle, or thoracic-lumbar, portion of the cord and almost immediately terminate in ganglia that lie near the cord (fig. 38.10). The sympathetic nervous system is especially important during emergency situations and may be associated with

Table 38.1 Sympathetic versus Parasympathetic System

Sympathetic	Parasympathetic
Fight or flight	Normal activity
Norepinephrine is neurotransmitter	Acetylcholine is neurotransmitter
Postganglionic fiber is longer than preganglionic	Preganglionic fiber is longer than postganglionic
Preganglionic fiber arises from middle portion of cord	Preganglionic fiber arises from brain and lower portion of cord

the "fight or flight response." For example, it inhibits the digestive tract but dilates the pupil of the eye, accelerates the heartbeat, and increases the breathing rate. It is not surprising, then, that the neurotransmitter released by the postganglionic axon is *norepinephrine,* a chemical close in structure to epinephrine (adrenalin), a well-known heart stimulant.

Parasympathetic Nervous System

In the parasympathetic nervous system the preganglionic fibers arise from the brain and the bottom (sacral) portion of the cord system. This system is therefore often referred to as the craniosacral portion of the autonomic nervous system. The preganglionic fibers terminate in ganglia that lie near or within the organ (fig. 38.10). The parasympathetic system, sometimes called the "housekeeper system," promotes all the internal responses we associate with a relaxed state; for example, it causes the pupil of the eye to contract, promotes digestion of food, and retards the heartbeat. The neurotransmitter utilized by the parasympathetic system is acetylcholine.

The autonomic nervous system controls the functioning of internal organs without need for conscious control.
The sympathetic nervous system brings about those responses we associate with "fight or flight."
The parasympathetic nervous system brings about those responses we associate with normally restful activities.

Central Nervous System

The central nervous system (CNS) consists of the spinal cord and brain. As figures 38.8 and 38.11 illustrate, the CNS is protected by bone: the brain is enclosed within the skull, and the spinal cord is surrounded by vertebrae. Also, both the brain and the spinal cord are wrapped in three layers of protective membranes known as the **meninges;** spinal meningitis is a well-known infection of these coverings. The interior of the brain contains spaces known as *ventricles.* Like the central canal of the spinal cord, these are filled with cerebrospinal fluid.

Also, in the same manner as the spinal cord, the brain contains gray matter and white matter. The *gray matter* is made up of cell bodies and short fibers. The *white matter* consists of

the long, myelinated fibers of interneurons. Some of these are grouped together in bundles called **tracts** that run between the brain and the spinal cord. The tracts cross so that the left side of the brain controls skeletal muscles on the right side of the body and vice versa.

The CNS consists of the brain and the spinal cord and is the place where sensory information is received and motor control is initiated.

The Brain

The largest and most prominent portion of the human brain (fig. 38.11) is the *cerebrum.* Consciousness resides only in the cerebrum, which we will discuss later; the rest of the brain functions below the level of consciousness.

The Unconscious Brain

The unconscious brain is made up of a number of different organs. The **medulla oblongata** is the organ that lies closest to the spinal cord. It contains centers for regulating heartbeat, breathing, and vasoconstriction (blood pressure), and also reflex centers for vomiting, coughing, sneezing, hiccupping, and swallowing.

The **hypothalamus** is concerned with homeostasis, or the constancy of the internal environment, and contains centers for regulating hunger, sleep, thirst, body temperature, water balance, and blood pressure. The hypothalamus controls the pituitary gland and thereby serves as a link between the nervous and endocrine systems.

The medulla oblongata and the hypothalamus are both concerned with control of the internal organs.

The *midbrain* and *pons* contain tracts that connect the cerebrum with other parts of the brain. In addition, the pons functions together with the medulla to regulate breathing rate, and the midbrain has reflex centers concerned with head movements in response to visual and auditory stimuli.

The **thalamus** is the last portion of the brain for sensory input before the cerebrum. It serves as a central relay station for sensory impulses traveling upward from other parts of the cord and brain to the cerebrum. It receives all sensory impulses (except those associated with the sense of smell) and channels them to appropriate regions of the cortex for interpretation.

The thalamus has connections to various parts of the brain, and these projections are called the diffuse thalamic projection system. This system is an extension of the reticular formation (fig. 38.12), which is a complex network of cell bodies and fibers that extends from the medulla to the thalamus. Together they form the *ARAS* (ascending reticular activating system), which is believed to sort out incoming stimuli, passing on only those that require immediate attention. The thalamus is sometimes called the gatekeeper to the cerebrum because it monitors the sensory data to be sent on to the cerebrum. For this reason, it is possible to ignore extraneous sensory information unless it becomes important for us to pay attention to it.

Figure 38.11

The human brain. Note how large the cerebrum is, compared to the rest of the brain.

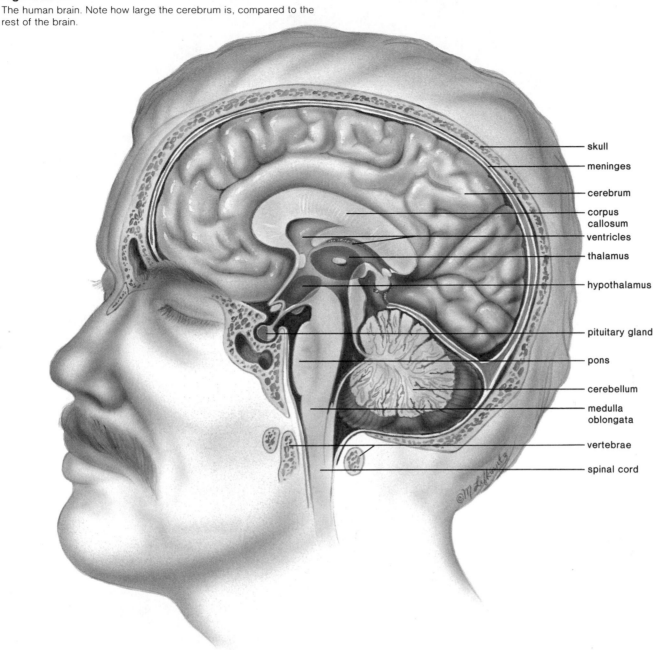

skull

meninges

cerebrum

corpus callosum

ventricles

thalamus

hypothalamus

pituitary gland

pons

cerebellum

medulla oblongata

vertebrae

spinal cord

The **cerebellum** (fig. 38.11), a bilobed structure that resembles a butterfly, is the second largest portion of the brain. It functions in muscle coordination by integrating impulses received from higher centers to ensure that all the skeletal muscles work together to produce smooth and graceful motions. The cerebellum is also responsible for maintaining normal muscle tone and transmitting impulses that maintain posture. It receives information from the inner ear indicating the position of the body, and sends impulses to those muscles whose contraction maintains or restores balance.

The thalamus is concerned with sensory reception, and the cerebellum is concerned with muscle control.

The Conscious Brain

In humans the cerebrum permits consciousness. A comparative study of vertebrates (fig. 38.13) indicates a progressive increase in the size of the cerebrum from fishes to humans. In fishes and amphibians, the cerebrum largely has an olfactory function, but in reptiles and other vertebrates, the cerebrum receives information from other parts of the brain and coordinates sensory data and motor functions. The outer portion of the cerebrum, the cerebral cortex, is most highly convoluted in mammals, especially humans. The cortex is gray in color and contains cell bodies and short fibers.

Figure 38.12

The reticular formation is a part of the ARAS, the ascending reticular activating system. The sensory information received by the reticular formation is sorted out by the thalamus before nerve impulses are sent to other parts of the brain.

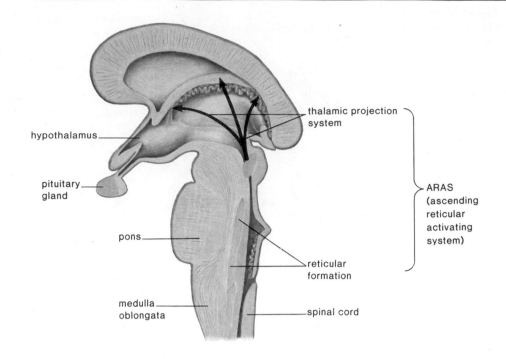

Figure 38.13

Comparison of brain sizes among vertebrates. Note the enlargement in the size of the cerebrum counterclockwise from bass to human.

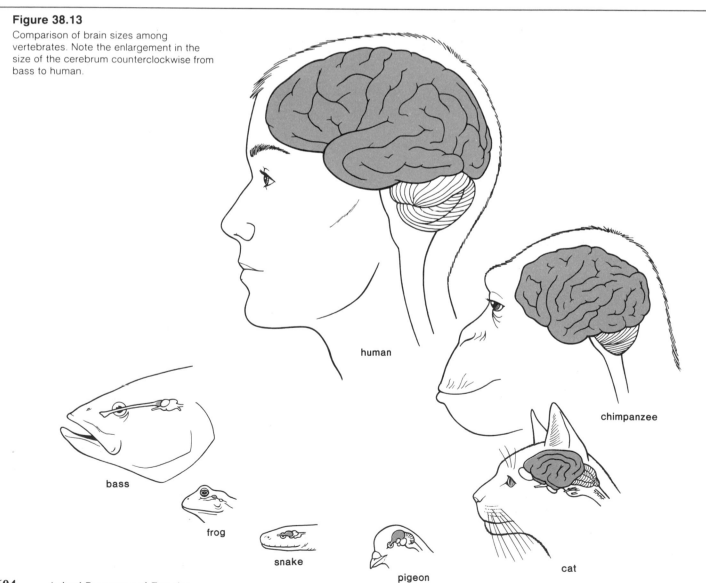

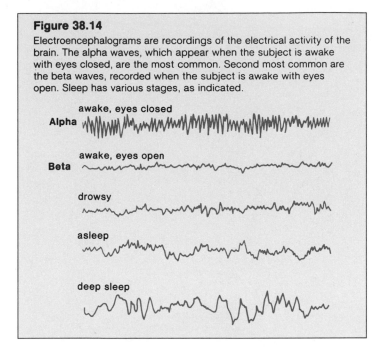

Figure 38.14
Electroencephalograms are recordings of the electrical activity of the brain. The alpha waves, which appear when the subject is awake with eyes closed, are the most common. Second most common are the beta waves, recorded when the subject is awake with eyes open. Sleep has various stages, as indicated.

awake, eyes closed
Alpha

awake, eyes open
Beta

drowsy

asleep

deep sleep

The cerebrum is divided into halves called the right and left cerebral hemispheres. The cerebral cortex of each half contains four types of lobes: frontal, parietal, temporal, and occipital. Different functions are associated with each lobe. For example, the *frontal lobe* controls motor functions and permits us to consciously control our muscles. The *parietal lobe* receives information from receptors located in the skin, such as those for touch, pressure, and pain. The *occipital lobe* interprets visual inputs. The *temporal lobe* has sensory areas for hearing and smelling.

Certain areas of the cerebral cortex have been "mapped" in great detail. For example, we know which portions of the frontal lobe control various parts of the body and which portions of the parietal lobe receive sensory information from these same parts. Each of the four lobes of the cerebral cortex contains an association area that receives information from the other lobes and integrates it into higher, more complex levels of consciousness. These areas are concerned with intellect, artistic and creative ability, learning, and memory.

Consciousness is the province of the cerebrum, the most highly developed portion of the brain. It is responsible for higher mental processes, including the interpretation of sensory input and the initiation of voluntary muscular movements.

There has been a great deal of testing to determine whether the right and left halves of the cerebrum serve different functions. These studies have tended to suggest that the left half of the brain is the verbal (word) half and the right half of the brain is the visual (spatial relation) and artistic half. However, other results indicate that such a strict dichotomy does not always exist between the two halves. In any case, the two cerebral hemispheres normally share information because they are connected by a horizontal tract called the *corpus callosum.*

Severing the corpus callosum can control severe epileptic seizures but results in the person having two brains, each with its own memories and thoughts. Today, the use of the laser during surgery permits more precise treatment without these side effects. *Epilepsy* is caused by a disturbance of the normal communication system between the ARAS and the cortex. In a *grand mal* seizure, the cerebrum becomes extremely excited. There is a reverberation of signals within the ARAS and cerebrum that continues until the neurons become fatigued and the signals cease. In the meantime, the individual loses consciousness, even while convulsions are occurring. Following an attack, the brain is so fatigued the person must sleep for a while.

EEG The electrical activity of the brain can be recorded in the form of an **electroencephalogram (EEG).** Electrodes are taped to different parts of the scalp, and an instrument called the electroencephalograph records the so-called brain waves (fig. 38.14).

When the subject is awake, two types of waves are usual: *alpha waves,* with a frequency of about 6–13 per second and a potential of about 45 microvolts, predominate when the eyes are closed, and *beta waves,* with higher frequencies but lower voltage, appear when the eyes are open.

During an eight-hour sleep there are usually five times when the brain waves become slower and larger than alpha waves. During each of these times, there are irregular flurries as the eyes move back and forth rapidly. When subjects are awakened during the latter, called *REM* (rapid eye movement) *sleep,* they always report that they were dreaming. The significance of REM sleep is still being debated, but some studies indicate that REM sleep is needed for memory to occur.

Arousal and sleep are controlled by specific structures in the brain. Also, there are specific centers in the medulla and the pons that bring about sleep and another in the midbrain that brings about arousal.

Limbic System

The **limbic system** involves portions of both the unconscious and conscious brain. It lies just beneath the cortex (fig. 38.15) and contains neural pathways that connect portions of the frontal lobes, temporal lobes, thalamus, and hypothalamus. Several masses of gray matter that lie deep within each hemisphere of the cerebrum, termed the *basal nuclei,* are also a part of the limbic system.

Stimulation of different areas of the limbic system causes the subject to experience rage, pain, pleasure, or sorrow. By causing pleasant or unpleasant feelings about experiences, the limbic system apparently guides the individual into behavior that is likely to increase the chance of survival.

Learning and Memory

The limbic system is also involved in the processes of learning and memory. Learning requires memory, but just what permits memory to occur is not definitely known. Investigators have been working with invertebrates, such as slugs and snails, because

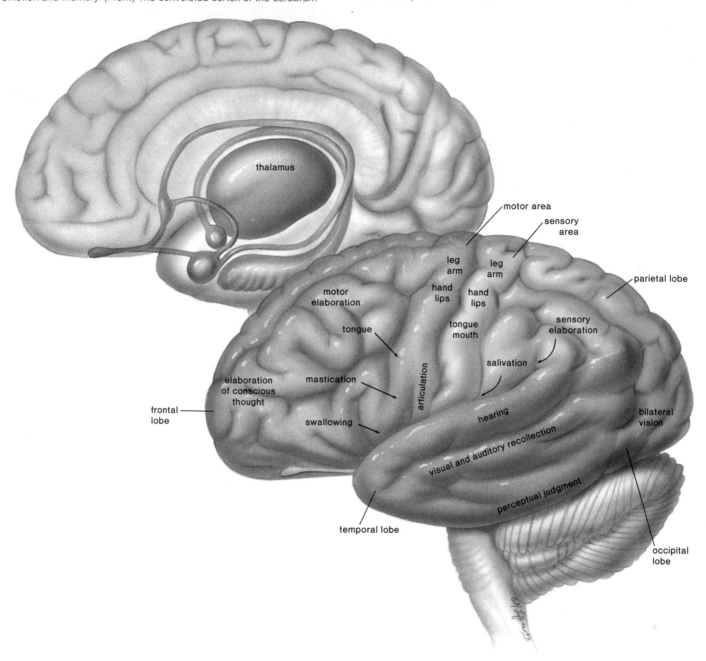

Figure 38.15

(*Rear*) The limbic system (in color), which includes portions of the cerebrum, thalamus, and hypothalamus, is concerned mainly with emotion and memory. (*Front*) The convoluted cortex of the cerebrum is divided into four lobes: frontal, temporal, parietal, and occipital. Further, it is possible to map the cerebrum since each particular area has a particular function.

their nervous systems are very simple and yet they can be conditioned to perform a particular behavior. In order to study this sample type of learning, it has been possible to insert electrodes into individual cells and to alter or record the electrochemical responses of these cells (fig. 38.16). This type of research has recently shown that learning is accompanied by an increase in the number of synapses, while forgetting involves a decrease in the number of synapses. In other words, the nerve-circuit patterns are constantly changing as learning, remembering, and forgetting occur. Within the individual neuron, learning involves a change in gene regulation, nerve protein synthesis, and an increased ability to secrete transmitter substances.

At the other end of the spectrum, some investigators have been studying learning and memory in monkeys. This work has led to the conclusion that the limbic system (fig. 38.15) is absolutely essential to both short-term and long-term memory. An example of short-term memory in humans is the ability to recall a telephone number long enough to dial the number; an example of long-term memory is the ability to recall the events of the day. It's believed that at first impulses move within the limbic circuit, but eventually the basal nuclei transmit the neurotransmitter ACh to the sensory areas where memories are stored. The involvement of the limbic system certainly explains why emotionally charged events result in our most vivid memories.

Figure 38.16

Individual nerve cells in a snail, *Hermissenda*, are being stimulated by microelectrodes, that produce the signals scientists previously recorded when a snail learns to avoid light. When this snail is freed, it automatically avoids the light and does not need to be taught like other snails. Normally, to teach snails to avoid light, they are placed on a table that rotates every time they venture toward light.

The fact that the limbic system communicates with the sensory areas for touch, smell, vision, etc., accounts for the ability of any particular sensory stimulus to awaken a complex memory.

The limbic system is particularly involved in the emotions and in memory and learning.

Neurotransmitters in the Brain

As discussed previously, neurotransmitters released at the end of axons affect the membrane potential of postsynaptic membranes. Some excitatory transmitters are amines, such as acetylcholine (ACh), norepinephrine (NE), serotonin, and dopamine. We have already had occasion to mention that ACh and NE are transmitters for the autonomic nervous system. *Serotonin* and *dopamine* are associated with behavioral states such as mood, sleep, attention, learning, and memory. The inhibitory transmitters include the amino acids gamma-aminobutyric acid (GABA) and glycine.

Both excitatory and inhibitory neurotransmitters are active in the brain.

In addition, a number of different types of peptides have been discovered in the CNS. The *endorphins,* molecules that are the body's natural opioids are of particular interest. They are called this because the psychoactive drugs morphine and heroin, which are both derived from opium, attach to the receptors for endorphin in the CNS. When endorphins are released they, like opium, produce a feeling of elation and reduce the sensation of pain. For example, if endorphins are present, neurons do not release substance P, a neurotransmitter that brings about the sensation of pain. Exercise has been associated with their presence, and this may account for the so-called "runner's high."

Neurotransmitter Disorders   It has been discovered that several neurological illnesses such as *Parkinson disease* and *Huntington disease* are due to an imbalance of neurotransmitters. Parkinson disease is a condition characterized by a wide-eyed, unblinking expression, an involuntary tremor of the fingers and thumbs, muscular rigidity, and a shuffling gait. All these symptoms are due to dopamine deficiencies. Huntington disease is characterized by a progressive deterioration of the individual's nervous system that eventually leads to constant thrashing and writhing movements and finally to insanity and death. The problem is believed to be malfunction of the inhibitory neurotransmitter GABA. Most recently it has been discovered that *Alzheimer disease,* a severe form of senility with marked memory loss found in 5% to 10% of all people over age sixty-five, is due to deterioration in cells of the basal nuclei that use ACh as a transmitter. It seems that a gene located on chromosome number 21, which is normally active only during development and which directs the production of a protein associated with neuron death, has been inexplicably turned on in these patients.

Treatment of individuals with brain disorders has formerly been directed toward restoring the proper balance of neurotransmitter substances. However, tissue that produces the missing neurotransmitter has been implanted in several patients, and researchers are hopeful that this technique will provide long-lasting cures.

Some neurological illnesses are associated with the deficiency of a particular neurotransmitter in the brain.

Action of Neurological Drugs   Drugs that alter the mood of a person either enhance or block the action of a particular neurotransmitter found in the ARAS or the limbic system. There are a number of different ways drugs can influence the action of neurotransmitters, some of which are shown in figure 38.18. It is clear, as outlined in table 38.2, that stimulants can either enhance the action of an excitatory transmitter or block the action of an inhibitory transmitter. On the other hand, depressants can either enhance the action of an inhibitory transmitter or block the action of an excitatory transmitter.

There are a wide variety of drugs that can be used to alter the mood and/or emotional state. The reading on page 608 discusses four of the most commonly abused drugs: alcohol, marijuana, cocaine, and heroin. Drug abuse occurs when a person

Four Drugs of Abuse

Alcohol

It is possible to drink alcoholic beverages in moderation, but they are often abused. Alcohol use becomes "abuse," or an illness, when alcohol ingestion impairs an individual's social relationships, health, job efficiency, or judgment. While it is general knowledge that alcoholics are prone to drink until they become intoxicated, there is much debate as to what causes alcoholism. Some believe that alcoholism is due to an underlying psychological disorder, while others maintain that the condition is due to an inherited physiological disorder.

Alcohol effects on the brain are biphasic; after consuming several drinks, blood alcohol concentration rises rapidly and the drinker reports feeling "high" and happy (euphoric). Later, after about 90 minutes, and lasting until about five to six hours after consumption, the drinker feels depressed and unhappy (dysphoric). On the other hand, if the drinker continues to drink in order to maintain a high blood level of alcohol, he or she will experience ever-increasing loss of control. Coma and death are even possible if a substantial amount of alcohol is consumed within a limited period.

Marijuana

The dried flowering tops, leaves, and stems of the Indian hemp plant *Cannabis sativa* contain and are covered by a resin that is rich in THC (tetrahydrocannabinol). The names *cannabis* and *marijuana* can apply either to the plant or to THC.

The effects of cannabis differ depending upon the strength and amount consumed, the expertise of the user, and the setting in which it is taken. Usually, the user reports experiencing a mild euphoria along with alterations in vision and judgment that result in distortions of space and time. Also involved may be the inability to concentrate, to speak coherently, and motor incoordination.

Intermittent use of low-potency cannabis is not generally associated with obvious symptoms of toxicity, but heavy use may produce chronic intoxication. Intoxication is recognized by the presence of hallucinations, anxiety, depression, rapid flow of ideas, body image distortions, paranoid reactions, and similar psychotic symptoms. The terms *cannabis psychosis* and *cannabis delirium* refer to such reactions.

The use of marijuana has ill effects (see table) but does not seem to produce physical dependence, but there may be a psychological dependence on the euphoric and sedative effects. Craving or difficulty in stopping may also occur as a part of regular heavy use.

Marijuana has been called a *gateway* drug because adolescents who have used marijuana tend to also try other drugs. For example, in a study of 100 cocaine abusers, 60% had also smoked marijuana for more than ten years.

Cocaine

Cocaine, currently the second most popular illegal drug after marijuana, is derived from the shrub *Erythroxylum cocoa*. *Cocaine* is sold in powder form and in the form of *"crack,"* a more potent extract. Users often use the word *rush* to describe the feeling of euphoria that follows intake of the drug. Snorting (inhaling through the nose) produces this effect in a few minutes; injection, within 30 seconds; and smoking in less than 10 seconds. Persons dependent upon the drug are, therefore, most likely to use the last method of intake. The rush only lasts a few seconds and is then replaced by a state of arousal that lasts from 5–30 minutes. Then the user begins to feel restless, irritable, and depressed. To do away with these symptoms the user is apt to take more of the drug, repeating the cycle over and over again until there is no more drug left. A binge of this sort can go on for days after which the individual suffers a *"crash."* During the binge period, the user is hyperactive and has little desire for food or sleep, but has an increased sex drive. During the crash period, the user is fatigued, depressed, irritable, has memory and concentration problems, and displays no interest in sex. Indeed, men are often impotent. Other drugs, such as marijuana, alcohol, or heroin, are often taken to ease the symptoms of the crash.

With continued cocaine use, the body begins to make less dopamine as a compensation for a seemingly excess supply. The user, therefore, now experiences *tolerance* (always needing more of the drug for the same effect), *withdrawal* (symptoms such as those described previously when a drug is not taken), and an intense *craving* for cocaine. These are indications that the person is highly dependent upon the drug or, in other words, that cocaine is extremely addictive.

Heroin

Heroin is derived from morphine, a derivative of *opium*. Heroin is usually injected. After intravenous injection, the onset of action is noticeable within 1 minute and reaches its peak in 3–6 minutes. There is a feeling of euphoria along with relief of pain. Side effects can include nausea, vomiting, dysphoria, and respiratory and circulatory depression leading to death.

Heroin binds to receptors meant for the body's own opioids, the endorphins. As mentioned previously, the opioids are believed to alleviate pain by preventing the release of a neurotransmitter termed substance P from certain sensory neurons in the region of the spinal cord. When substance P is released, pain is felt, and when substance P is not released, pain is not felt. Evidence also indicates that there are opioid receptors in neurons that travel from the spinal cord to the limbic system and that stimulation of these can cause a feeling of pleasure. This explains why opium and heroin not only kill pain but also produce a feeling of tranquility.

Individuals who inject heroin become physically dependent on the drug. With time, the body begins to produce less endorphins; now *tolerance* develops so that the user needs to take more of the drug just to prevent *withdrawal* symptoms. The euphoria originally experienced upon injection is no longer felt.

The withdrawal symptoms include perspiration, dilation of pupils, tremors, restlessness, abdominal cramps, goose flesh, defecation, vomiting, and increase in systolic pressure and respiratory rate. Those who are excessively dependent may experience convulsions, respiratory failure, and death. Infants born to women who are physically dependent also experience these withdrawal symptoms.

Name	Mode of Action	Associated Ill Effects	United States Statistics
Alcohol	Acts on the GABA receptor to increase membrane potential	Intoxication; hangovers; cirrhosis of the liver; brain damage	In males aged 25–44, intoxication plays a major role in these causes of death: accidents, homicide, suicide, and cirrhosis of the liver.
Marijuana	Classified as a hallucinogen; may, like LSD, have an effect on the action of serotonin	Chronic respiratory disease and lung cancer; brain impairment; reproductive dysfunctions	In 1985, 6,461 teenagers who were using illegal drugs attempted suicide.
Cocaine	Prevents the reuptake of dopamine at synapses	Death due to overdosing; damage to nasal tissues; possible brain damage	In 1986, cocaine killed at least 930 persons by cardiac arrest, respiratory failure, or brain hemorrhage.
Heroin	Binds to receptors for the body's own opioids, the endorphins, and enkephalins	Death due to overdosing or severe withdrawal symptoms	In 1986, heroin was responsible for at least 1,469 deaths.

Figure 38.17

Some drug actions at synapses. *a.* Drug stimulates the release of the neurotransmitter. *b.* Drug blocks the release of the neurotransmitter. *c.* Drug combines with the neurotransmitter, preventing its breakdown. *d.* Drug mimics the action of the neurotransmitter. *e.* Drug blocks the receptor so that the neurotransmitter cannot be received. Generally, only one of these actions occurs at a given synapse.

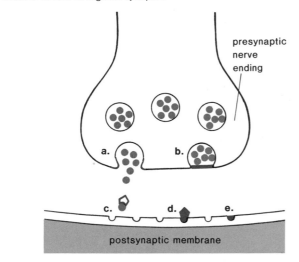

presynaptic nerve ending

postsynaptic membrane

Table 38.2 Results of Drug Action

Type of Drug	Type of Neurotransmitter	Action or Effect
Stimulant	Excitatory	Enhances
Stimulant	Inhibitory	Blocks
Depressant	Excitatory	Blocks
Depressant	Inhibitory	Enhances

takes excessive amounts of a drug under circumstances that increase the likelihood of a harmful effect. Individuals who are drug abusers are also apt to display a *physical dependence* on the drug (formerly called an addiction to the drug). Dependence is present when the person (1) spends much time thinking about the drug or arranging to get it; (2) often takes more of the drug than was intended; (3) is tolerant to the drug—that is, must increase the amount of the drug to get the same effect; (4) has withdrawal symptoms when he or she stops taking the drug; and (5) has a repeated desire to cut down on use.

Neurological drugs interfere with normal transmitter function in the brain, particularly in the ARAS and the limbic system. Drug abuse often results in physical dependence on the drug.

Summary

1. In humans, the nervous system, along with the endocrine system, regulates the other systems of the body and coordinates body functions.
2. A comparative study of the invertebrates shows a gradual increase in the complexity of nervous systems. The human nervous system, like that of the earthworm, is divided into the central and peripheral nervous systems.
3. Neurons, cells that conduct nerve impulses, have three parts: dendrite(s), cell body, and axon. The dendrites and the cell body receive information from the environment, and if it is excitatory enough, a nerve impulse travels down the axon to where neurotransmitters are stored in vesicles.
4. The nerve impulse, which is recognized when the resting potential becomes an action potential, is an electrochemical phenomenon involving the movement first of sodium and then of potassium ions across the cell membrane.

   Resting potential: Sodium-potassium pump at work, inside of the neuron is negative (-65 mV) compared to the outside of the neuron.

   Action potential: (1) Sodium ions move to the inside, making it positive compared to outside ($+40$ mV). (2) Potassium ions move to the outside, making the inside negative again compared to the outside (-65 mV).
5. The nerve impulse is self-propagating because the gated channels that allow sodium and potassium ions to flow down their concentration gradients are sensitive to a near-by decrease in the membrane potential.
6. Saltatory conduction occurs in myelinated fibers—the action potential jumps from node of Ranvier to node of Ranvier. This accounts for the great speed of the impulse in these fibers.
7. Transmission across a synapse usually requires neurotransmitters because there is a small space, the synaptic cleft, that separates neuron from neuron. The neurotransmitters released at the ends of axons may be either excitatory or inhibitory.
8. Acetylcholine and noradrenalin are well-known excitatory transmitters. After its release, acetylcholine is destroyed by acetylcholinesterase in the synaptic cleft.
9. Within the human peripheral nervous system, the somatic system includes cranial and spinal nerves. Spinal nerves extend from the spinal cord and contain both sensory and motor fibers. It is possible to use them to trace the path of a reflex arc from receptor to effector, as follows:

   Sensory neuron: receptor generates nerve impulses that travel in the dendrite to the cell body and then to the axon that enters the cord.

   Interneuron: transmits impulses from the dorsal to the ventral root of the cord.

   Motor neuron: impulses begin in the dendrites and cell body and then pass out of the cord by way of the axon that innervates an effector.
10. The autonomic nervous system controls internal organs and includes the sympathetic and parasympathetic nervous systems, which are contrasted in table 38.1.
11. The human central nervous system includes the spinal cord and the brain. In the brain, the medulla oblongata and hypothalamus regulate internal organs. The thalamus receives sensory input and passes it to the cerebrum by way of the ARAS. The cerebellum functions in muscle coordination.

12. A survey of vertebrates shows a continual evolutionary increase in the size of the cerebrum, with its highest development in humans. The highly convoluted cerebral cortex is divided into lobes, each of which has specific functions. In general, the cerebrum is responsible for consciousness, sensory perception, motor control, and all the higher forms of thought.

13. The EEG pattern varies according to our state of consciousness; whether we are fully awake, resting, or sleeping.

14. Research in invertebrates indicates that learning is accompanied by an increase in the number of synapses; research in monkeys indicates that the limbic system is involved. There are short-term and long-term memories; the involvement of the limbic system explains why emotionally charged events result in vivid long-term memories.

15. Various illnesses such as Parkinson, Huntington, and Alzheimer disease are due to an imbalance in a specific neurotransmitter in the brain. Drugs that people take to alter the mood and/or emotional state often interfere with normal neurotransmitter function in the brain, particularly the ARAS and limbic system.

Objective Questions

1. Which is the most complete list of animals that have both a central and a peripheral nervous system?
   a. hydra, planaria, earthworm, rabbit, human
   b. planaria, earthworm, rabbit, human
   c. earthworm, rabbit, human
   d. rabbit, human
2. Which of these are the first element and the last element in a spinal reflex?
   a. axon and dendrite
   b. sense organ and muscle effector
   c. ventral horn and dorsal horn
   d. motor neuron and sensory neuron
3. Which phrase does not belong with the others?
   a. cerebrum
   b. cerebral cortex
   c. cerebral hemispheres
   d. cerebellum
4. A mixed nerve takes nerve impulses
   a. to the CNS.
   b. away from the CNS.
   c. both to and away from the CNS.
   d. only inside the CNS.

5. Which of these correctly describes the distribution of ions on either side of an axon when it is not conducting a nerve impulse?
   a. Na^+ outside and K^+ inside
   b. K^+ outside and Na^+ inside
   c. charged protein outside; Na^+ and K^+ inside
   d. Na^+ and K^+ outside and water only inside
6. When the action potential begins, sodium gates open, allowing sodium ions to cross the axomembrane. Now the polarity changes to
   a. negative outside and positive inside.
   b. positive outside and negative inside.
   c. no difference in charge between outside and inside.
   d. Any one of these.
7. Transmission of the nerve impulse across a synapse is accomplished by
   a. the movement of sodium and potassium ions.
   b. the release of neurotransmitters.
   c. Both of these.
   d. Neither of these.

8. The autonomic nervous system has two divisions called the
   a. CNS and peripheral systems.
   b. somatic and skeletal systems.
   c. efferent and afferent systems.
   d. sympathetic and parasympathetic systems.
9. Synaptic vesicles
   a. are at the ends of dendrites and axons.
   b. are at the ends of axons only.
   c. are along the length of all long fibers.
   d. All of these.
10. Which of these is mismatched?
    a. cerebrum—consciousness
    b. thalamus—motor and sensory centers
    c. hypothalamus—internal environment regulator
    d. cerebellum—motor coordination

Study Questions

1. What is the overall function of the nervous and endocrine systems? How do they differ in regard to this function?
2. Trace the evolution of the nervous system by contrasting the organization of the nervous system in hydras, planarians, earthworms, and humans.
3. Describe the structure of a neuron and give a function for each part mentioned. Name three types of neurons and give a function for each.

4. What are the major events of an action potential and the ion changes that are associated with each event?
5. Describe the mode of action of a neurotransmitter at a synapse, including how it is stored and how it is destroyed.
6. Contrast the structure and function of the peripheral and the central nervous systems.
7. Trace the path of a spinal reflex.
8. Contrast the sympathetic and parasympathetic divisions of the autonomic nervous system.

9. Name the major parts of the human brain and give a principal function for each part. What is an EEG?
10. Define the limbic system and discuss its possible involvement in learning and memory.
11. Name several specific neurotransmitters and, in general, describe how various types of drugs can affect the action of neurotransmitters.

Thought Questions

1. Why would a neuron be able to respond to one type of neurotransmitter and not to another type?

2. Speculate about why there are two different motor neurons (one from the sympathetic and the other from the parasympathetic system) for internal organs and glands but not for skeletal muscles.

3. The limbic system is sometimes called the animal brain. Considering the function of the limbic system, why might this term be appropriate? What might explain the existence of the limbic system beneath the cerebral cortex?

Selected Key Terms

central nervous system (CNS) (sen'tral ner'vus sis'tem) 592

peripheral nervous system (PNS) (pĕ-rif'er-al ner'vus sis'tem) 592

neuron (nu'ron) 593

dendrite (den'drīt) 593

axon (ak'son) 593

myelin sheath (mi'ĕ-lin shēth) 594

resting potential (rest'ing po-ten'shal) 595

action potential (ak'shun po-ten'shal) 595

saltatory conduction (sal'tah-to''re kon-duk'shun) 597

synapse (sin'aps) 597

neurotransmitter (nu''ro-trans-mit'er) 597

acetylcholine (as''ĕ-til-ko'lēn) 598

norepinephrine (nor''ep-i-nef'ron) 598

nerve (nerv) 598

ganglia (gang'gle-ah) 598

reflex (re'fleks) 600

autonomic nervous system (aw''to-nom'ik ner'vus sis'tem) 600

sympathetic nervous system (sim''pah-thet'ik ner'vus sis'tem) 600

parasympathetic nervous system (par''ah-sim''pah-thet'ik ner'vus sis'tem) 600

medulla oblongata (mĕ-dul'ah ob''long-ga'tah) 602

hypothalamus (hi''po-thal'ah-mus) 602

thalamus (thal'ah-mus) 602

cerebellum (ser''ĕ-bel'um) 603

electroencephalogram (EEG) (e-lek''tro-en-sef'ah-lo-gram'') 605

limbic system (lim'bic sis'tem) 605

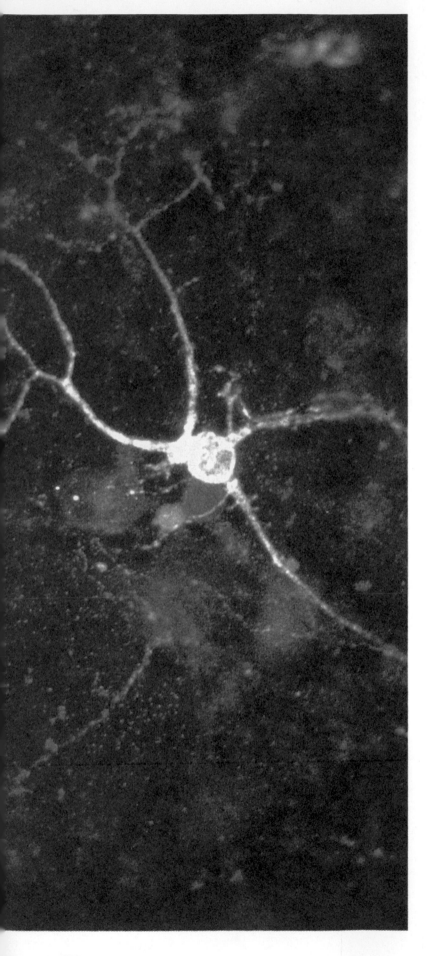

CHAPTER 39

Sense Organs

Your study of this chapter will be complete when you can:

1. State the function of sense organs.

2. Describe the receptors for smell and taste in humans.

3. Tell why the receptors for smell and taste are categorized as chemoreceptors.

4. In general, contrast the eye of arthropods with that of humans.

5. List the structures of the human eye, and give a function for each one.

6. Contrast the action of the sight receptors, the rods and cones.

7. Give examples of various types of mechanoreceptors in animals.

8. List the structures of the human ear, and give a function for each one.

9. Explain how the ear serves as an organ for static and dynamic equilibrium.

10. Describe the organ of Corti and how it functions to permit hearing.

Nerve impulses from the sense organs of the skin travel in neurons whose cell bodies are located in dorsal root ganglia lying just outside the spinal cord. This is a immunofluorescence micrograph of one such cell body, removed from a dorsal root ganglion and grown in culture.

Sense organs receive external and internal stimuli; therefore they are called **receptors.** Each type of receptor is designed to respond to a particular stimulus: for example, eyes respond only to light and ears only to sound waves. Receptors do not interpret stimuli—they act merely as transducers that transform the energy of stimuli into nerve impulses. Interpretation is the function of the brain, which has a specific region for receiving nerve impulses from each of the sense organs. Impulses arriving at a particular sensory area of the brain can be interpreted in only one way; those arriving at the visual area result in sight sensation, and those arriving at the olfactory area result in smell sensation, for example. Usually the brain is able to discriminate the type of stimulus, the intensity of the stimulus, and the origin of the stimulus. On occasion, the brain can be fooled, as when a blow to the eyes causes one to "see stars."

Interoceptors located within the body monitor such conditions as blood pressure, expansion of lungs and bladder, and movement of limbs. *Exteroceptors* are located near the surface of the animal and respond to outer stimuli. Both types of receptors send information to the brain and both are needed to maintain homeostasis and to promote appropriate behavioral actions. Interoceptors will be discussed later in connection with the internal organs; now we will limit our coverage to exteroceptors.

Chemoreceptors

The receptors responsible for taste and smell are termed **chemoreceptors** because they are sensitive to certain chemical substances in food, liquids, and air. Chemoreception is found universally in animals and is therefore believed to be the most primitive sense. Chemoreceptors are present all over the body of planaria and are concentrated on the auricles at the sides of the head. In arthropods, they are found on the antennae and mouthparts but in humans, unlike these animals, sense and smell have distinguishable receptors, as we will now discuss.

Smell

The receptors for the sense of smell are called the **olfactory cells** (*olfactus* means smell in Latin), located high in the roof of the nasal cavity in humans (fig. 39.1). The olfactory cells are actually modified neurons that synapse with the nerve fibers

Figure 39.1

Olfactory cell location and anatomy. (*Right*) the olfactory area in humans is located high in the nasal cavity. (*Left*) enlargement of the olfactory cells, which are modified neurons located between supporting cells. The axons of these cells synapse with sensory nerve fibers making up the olfactory nerve. The dendrites of these cells have special structures known as olfactory cilia that project into the nasal cavity where they are stimulated by chemicals in the air.

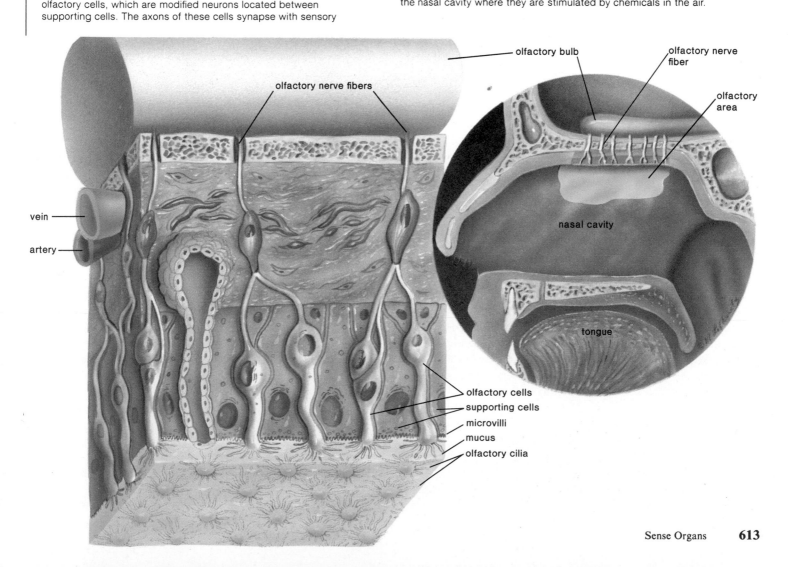

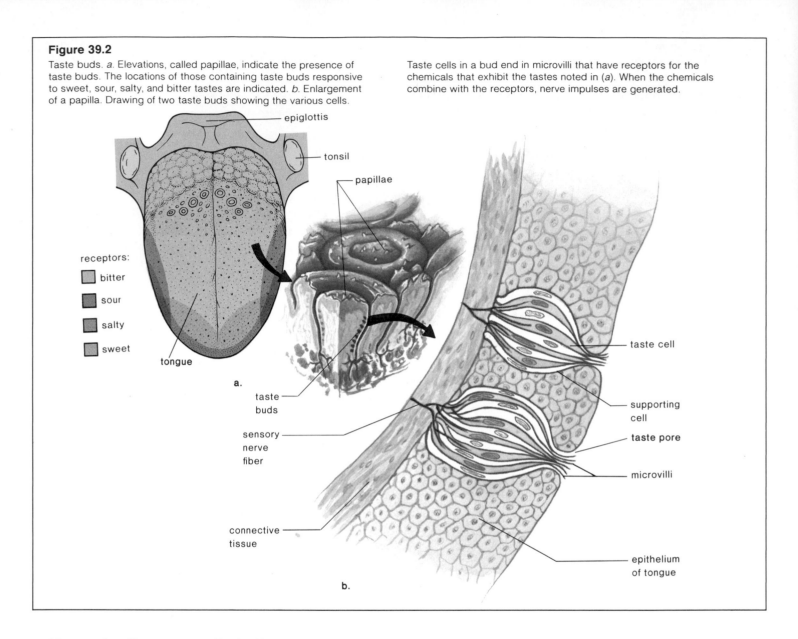

Figure 39.2
Taste buds. *a.* Elevations, called papillae, indicate the presence of taste buds. The locations of those containing taste buds responsive to sweet, sour, salty, and bitter tastes are indicated. *b.* Enlargement of a papilla. Drawing of two taste buds showing the various cells.

Taste cells in a bud end in microvilli that have receptors for the chemicals that exhibit the tastes noted in (*a*). When the chemicals combine with the receptors, nerve impulses are generated.

making up the olfactory nerve. Each olfactory cell has six to eight cilia that are stimulated by many chemicals in the air. Research, resulting in the stereochemical theory of smell, has shown that different smells may be related to the various shapes of the molecules rather than to the atoms that make up the molecules. These shapes fit specific olfactory sites on the olfactory cells' cilia.

The sense of smell is generally much more acute than the sense of taste. The human nose, for example, can detect one 25-millionth of 1 mg of mercaptan, the odoriferous chemical given off by a skunk. This averages out to approximately one molecule per sensory ending. Yet, humans have a weak sense of smell compared to other vertebrates, such as dogs.

Taste

The receptors of the sense of taste in mammals are the **taste buds.** Most of these are located on the tongue (fig. 39.2), but they are also present on the surface of the soft palate, pharynx, and epiglottis. Each taste bud contains a number of elongated cells. These cells have *microvilli* that project through the taste pore, an opening in the taste bud. It is the microvilli that are stimulated by various chemicals in the environment. Humans are believed to have four types of taste buds, each type stimulated by chemicals that result in a bitter, a sour, a salty, or a sweet sensation.

The sense of taste and the sense of smell supplement each other, creating a combined effect when interpreted by the cerebral cortex. For example, when one has a cold, food seems to lose its taste, but actually the ability to sense its smell is temporarily absent. This may work in reverse also. When we smell something, some of the molecules move from the nose down into the mouth and stimulate certain taste buds. Thus, part of what we refer to as smell is actually taste.

Photoreceptors

Many animals have light-sensitive receptors called **photoreceptors.** In its simplest form, a photoreceptor indicates only the presence of light and its intensity. The "eye spots" of planarians

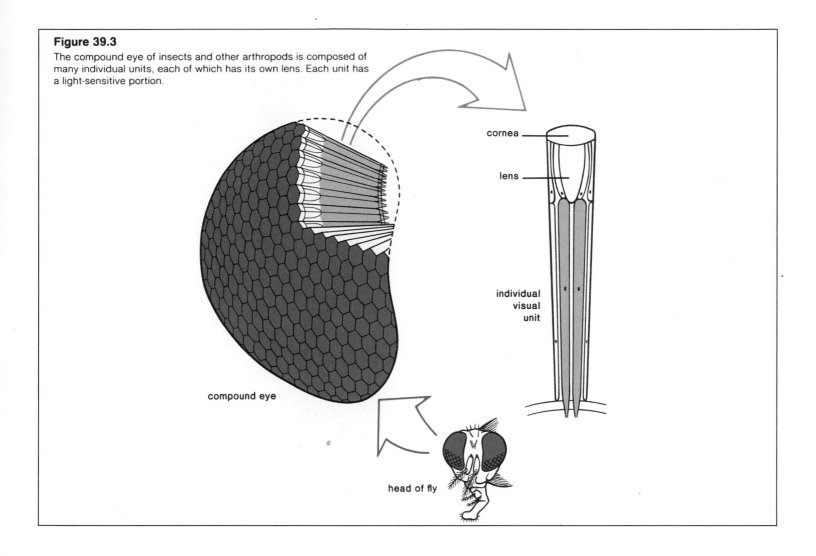

Figure 39.3
The compound eye of insects and other arthropods is composed of many individual units, each of which has its own lens. Each unit has a light-sensitive portion.

cornea

lens

individual visual unit

compound eye

head of fly

also allow the animal to determine the direction of light. More complex eyes have lenses that focus light on certain receptors. Arthropods have **compound eyes** composed of many independent visual units, each of which has its own lens (fig. 39.3) and views a separate portion of the object. How well the brain combines this information to see the entire object is not known, but compound eyes seem especially well suited to detecting motion, as anyone knows who has tried to catch an insect.

Vertebrates and certain mollusks, like the squid, have a *camera-type of eye.* There is a *single* lens that focuses an image of the visual field on the photoreceptors, which are closely packed together. All of the photoreceptors taken together can be compared to a piece of film in a camera. However, the eye is more complex than a camera, as we shall see.

Human Eye

The most important parts of the human eye and their functions are shown in table 39.1. The human eye (fig. 39.4), which is an elongated sphere about one inch in diameter, has three layers or coats. The outer layer, the **sclera,** is an opaque, white, fibrous layer that surrounds the transparent cornea, the window of the eye. The middle, thin, dark-brown layer, the **choroid,** contains many blood vessels and pigment that absorbs stray light rays.

Table 39.1 Function of the Parts of the Eye

Part	Function
Lens	Refraction and focusing
Iris	Regulates light entrance
Pupil	Opening in iris
Choroid	Absorbs stray light
Sclera	Protection
Cornea	Refraction of light
Humors	Refraction of light
Ciliary body	Holds lens in place
Retina	Contains receptors
Rods	Black-and-white vision
Cones	Color vision
Optic nerve	Transmits impulse
Fovea centralis	Region of cones in retina
Ciliary muscle	Accommodation

Figure 39.4

Anatomy of the human eye. Notice that the sclera becomes the cornea; the choroid becomes the ciliary body and iris. The ciliary body is composed of 70 to 80 radiating folds that contain the ciliary muscle and ligaments that hold and adjust the shape of the lens.

The retina contains the receptors for sight, and vision is most acute in the fovea centralis, where there are only cones. A blind spot exists where the optic nerve leaves the retina and there are no receptors.

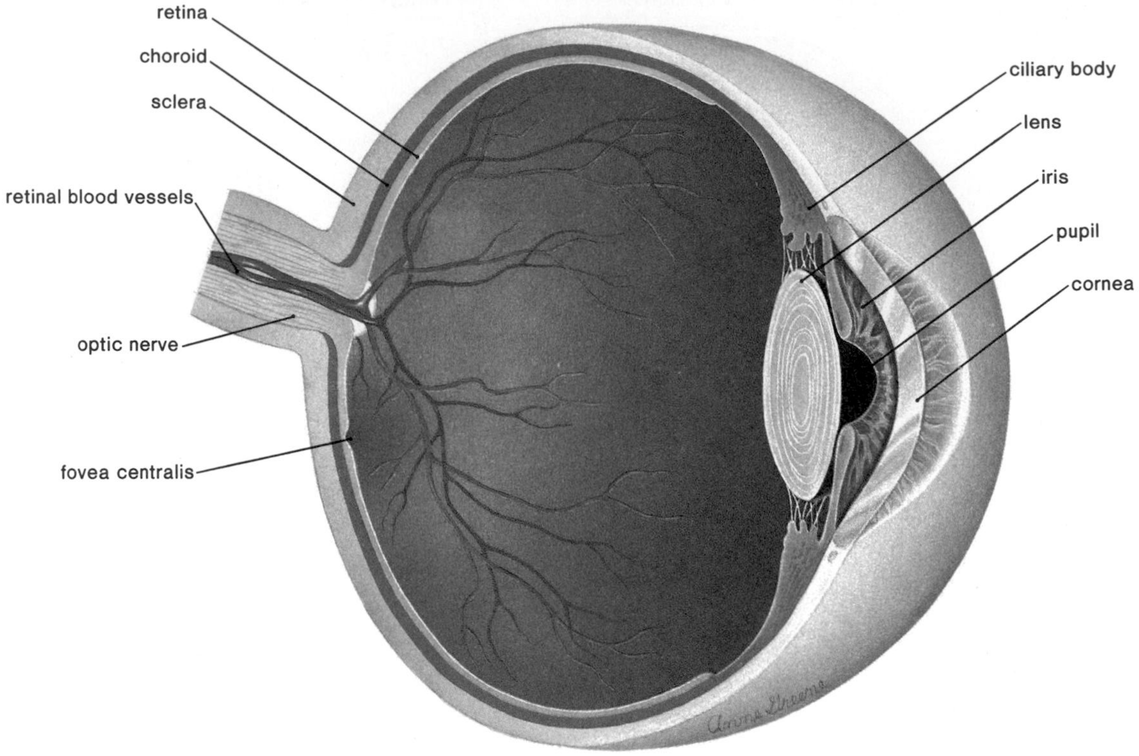

Toward the front of the eye, the choroid thickens and forms the ring-shaped ciliary body, and finally becomes a thin, circular, muscular diaphragm, the *iris,* that regulates the size of an opening called the *pupil.* The *lens,* which is attached to the ciliary body by ligaments, divides the cavity of the eye into two chambers. A basic, watery solution called *aqueous humor* fills the chamber between the cornea and the lens. A viscous, gelatinous material, the *vitreous humor,* fills the large cavity behind the lens.

The inner layer of the eye, the **retina,** contains the receptors for sight: the **rods** and **cones** (fig. 39.5). Nerve impulses initiated by the rods and cones are passed to the bipolar cells, which in turn pass them to the ganglionic cells. The fibers of these cells pass in front of the retina forming the **optic nerve,** which carries the nerve impulses to the brain. You'll notice in figure 39.5*b* that there are many more rods and cones than there are nerve fibers leaving ganglionic cells. This means that there will be considerable mixing of messages and a certain amount of integration before nerve impulses are sent to the brain. There are no rods and cones at the point where the optic nerve passes through the retina; therefore, this point is called the blind spot.

The center of the retina contains a special region called the fovea centralis (fig. 39.4), an oval, yellowish area with a depression where there are only cone cells. In the fovea centralis or fovea, color vision is most acute in daylight; at night, it is barely sensitive. At this time, the rods in the rest of the retina are active.

The human eye has three layers: the outer sclera, the middle choroid, and the inner retina. Only the retina contains receptors for sight.

Vision Process

Light rays entering the eye are bent (refracted) as they pass through the cornea, lens, and humors and brought to a focus on the retina. The lens is relatively flat when viewing distant objects, but rounds up for close objects because light rays must be bent to a greater degree when viewing a close object. These changes of the lens shape are called accommodation (fig. 39.6). With normal aging, the lens loses its ability to accommodate for close objects; therefore, persons frequently need reading glasses once they reach middle age.

The image on the retina is rotated 180° from the actual, but it is believed that this image is righted in the brain. In one experiment, scientists wore glasses that inverted and reversed the field. At first, they had difficulty adjusting to the placement

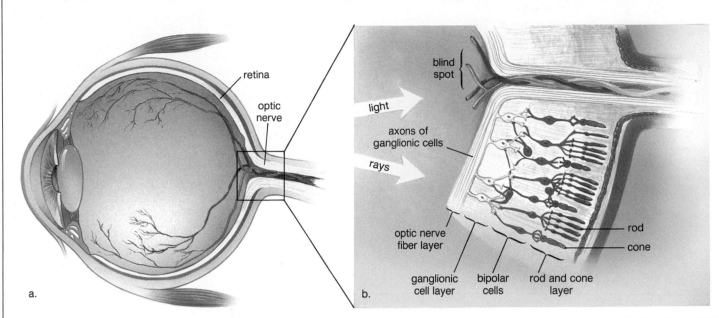

Figure 39.5

a. The retina is the inner layer of the eye. *b*. Rods and cones are located toward the back of the retina, followed by the bipolar cells and the ganglionic cells whose fibers become the optic nerve. Notice that rods share bipolar cells but cones do not. Cones, therefore, distinguish more detail. *c*. The photosensitive pigment is located in the disks of the outer segment of rods and cones. A disk is apparently a modified cilium. *d*. Scanning electron micrograph of rods and cones. The cones are responsible for color vision and the rods are responsible for night vision.

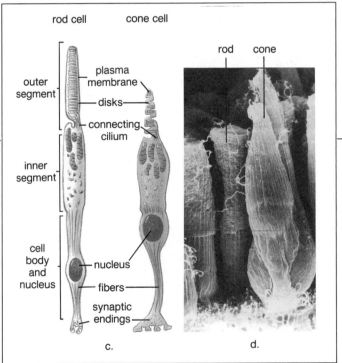

of objects, but they soon became accustomed to their inverted world. Experiments such as this suggest that if the retina sees the world "upside down," the brain has learned to see it right side up.

Rods   Only dim light is required to stimulate rods; therefore, they are responsible for *night vision* (fig. 39.7). The numerous rods are also better at detecting motion than cones are, but are unable to allow distinct and/or color vision. This causes objects to appear blurred and look gray in dim light. Many molecules of **rhodopsin** are located within the membrane of the disks (lamellae) found in the outer segment (fig. 39.5*c*) of the rods. Rhodopsin is a complex molecule that contains protein (opsin) and a pigment molecule called *retinal,* which is a derivative of vitamin A. When light strikes retinal it changes shape and opsin is activated. The reactions that follow eventually end after many molecules of GMP (guanosine monophosphate) have been converted to cyclic GMP.[1] Cyclic GMP, in turn, initiates nerve impulses in the rod that pass through the retina to the optic nerve. Each stimulus generated lasts about one-tenth of a second. This is why we continue to see an image if we close our eyes immediately after looking at an object. It also allows us to see motion if still frames are presented at a rapid rate, as in "movies."

[1]GMP is a nucleotide that contains the base guanine, the sugar ribose, and one phosphate group. In cyclic GMP, the single phosphate is attached to ribose in two places.

Cones and Color Vision   The cones, located primarily in the fovea and activated by bright light, detect the fine detail and the color of an object. *Color vision* depends on three different kinds of cones that contain either blue, green, or red receptors. Each pigment is made up of retinal and opsin, but there is a slight difference in the opsin structure of each, which accounts for their individual absorption patterns. Various combinations

Figure 39.6

Accommodation. *a.* When the eye focuses on a far object, the lens is flat because the eye muscles holding the lens are relaxed and the suspensory ligament is taut. *b.* When the eye focuses on a near object, the lens rounds up because the eye muscles contract, causing the suspensory ligament to relax. *c.* In nearsighted individuals, the eyeball is too long. They cannot see far because when viewing a distant object, the image is brought to focus in front of the retina. *d.* In farsighted individuals, the eyeball is too short. They cannot see near objects well because when viewing a near object the image is brought to focus behind the retina.

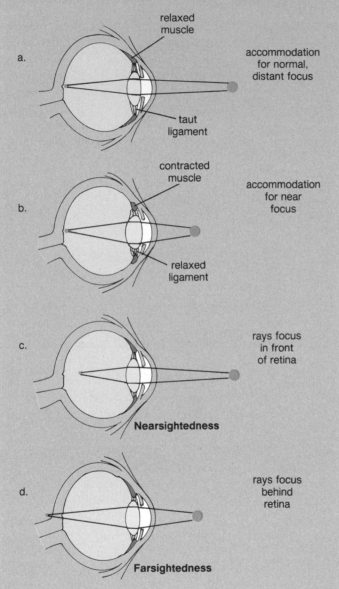

Figure 39.7

a. The high concentration of rhodopsin in the alligator's dark-adapted eye heightens sensitivity and gives the eye's interior a rose-colored appearance. *b.* After moments in the light, the concentration of rhodopsin drops and the inner eye appears pale gold. This cycle of bleaching in light and regenerating in darkness was the first clue to the chemistry of vision.

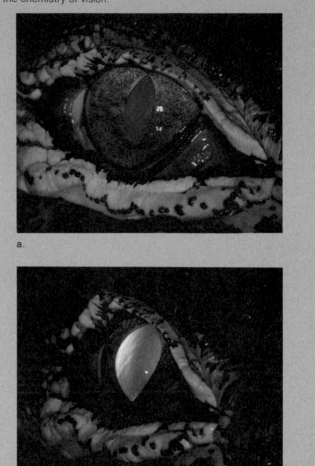

of receptors are believed to be stimulated by in-between shades of color, and the combined nerve impulses are interpreted in the brain as a particular color.

In the human eye, the receptors for sight are the rods and cones. The rods are responsible for vision in dim light, and the cones are responsible for vision in bright light and color vision. When a receptor is stimulated, nerve impulses are transmitted by the optic nerve to the brain.

Fishes, reptiles, and most birds are also believed to have color vision, but among mammals only humans and other primates have color vision. It would seem, then, that this trait must have been adaptive for life in trees, accounting for its retention in just these few mammals. Insects have color vision, but they make use of a slightly shorter range of the electromagnetic spectrum compared to humans. They can see the longest of the ultraviolet rays, and it is speculated that this may enable them to be especially sensitive to the reproductive parts of flowers (fig. 39.8) that reflect particular ultraviolet patterns.

Mechanoreceptors

Mechanoreceptors are sensitive to mechanical stimuli, such as pressure, sound waves, and gravity. Human skin (fig. 39.9) contains various types of mechanoreceptors, such as *touch receptors* and pressure receptors. A pressure receptor, called the Pacinian corpuscle, is shaped like an onion and consists of a series

Figure 39.8

Marsh marigold as seen by humans (*left*) and insects (*right*). Humans see no markings, but insects see distinct blotches because their eyes respond to ultraviolet rays whereas our eyes do not. These types of markings often highlight the reproductive parts of flowers where insects feed on nectar and pick up pollen at the same time.

Figure 39.9

a. Mechanoreceptors in skin include touch receptors (Meissner's corpuscles) and pressure receptors (Pacinian corpuscles). *b*. Human skin showing location of these receptors.

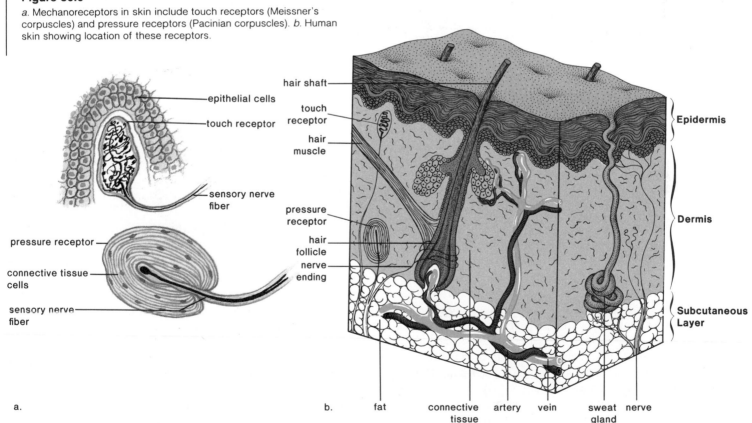

of concentric layers of connective tissue wrapped around the end (dendrite) of a sensory neuron. In contrast, *pain receptors* are only the unmyelinated ("naked") ends (dendrites) of the fibers of sensory neurons. Some pain receptors are especially sensitive to mechanical stimuli; others are most sensitive to temperature or chemicals.

Hair cells, which are named for the cilia they bear, are often mechanoreceptors in various specialized vertebrate sense organs. Fishes and amphibians have a series of such receptors, called the lateral line system, that detect water currents and pressure waves from nearby objects (fig. 39.10). Each receptor

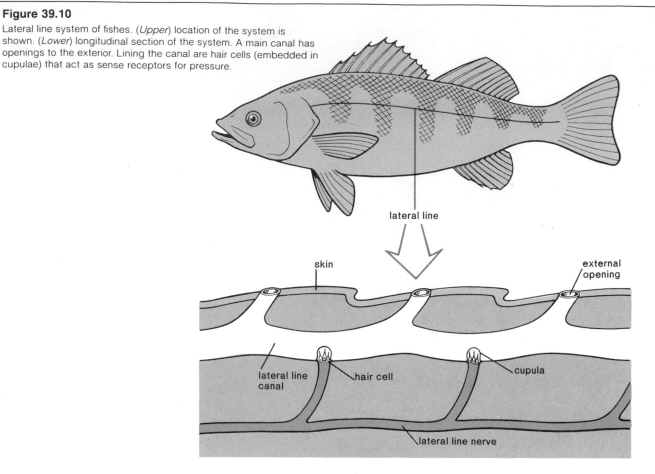

Figure 39.10

Lateral line system of fishes. (*Upper*) location of the system is shown. (*Lower*) longitudinal section of the system. A main canal has openings to the exterior. Lining the canal are hair cells (embedded in cupulae) that act as sense receptors for pressure.

lateral line

skin

external opening

lateral line canal

hair cell

cupula

lateral line nerve

is a collection of hair cells with cilia embedded in a gelatinous, wedge-shaped mass known as a *cupula* (pl., cupulae). This mass bends when disturbed, and the hair cells initiate nerve impulses.

Fishes also have ears, which are derived from a portion of the lateral line system and function mainly as equilibrium organs. In contrast, the inner ear of humans contains both equilibrium receptors and sound receptors.

Human Ear

Table 39.2 lists the most important parts of the human ear and their functions. The *inner* human ear contains the receptors for the senses of balance and hearing. The inner ear probably evolved from the lateral line system of fishes and amphibians. As figure 39.11 shows, the human ear has an outer and a middle portion, as well as an inner ear. The *outer ear* consists of the pinna (external flap) and *auditory canal*. The latter is lined with fine hairs that filter the air. Modified sweat glands are located in the upper wall of the canal; these secrete earwax that helps guard the ear against entrance of foreign materials.

The *middle ear* begins at the **tympanic membrane** (eardrum) and ends at a bony wall that has small openings covered by membranes, the *oval window* and the *round window*. Three small bones are located between the tympanic membrane and the oval window. Collectively called **ossicles,** individually they

Table 39.2 Function of the Parts of the Ear

Part	Function
Outer Ear	
Pinna	Collect sound waves
Middle Ear	
Tympanic membrane and ossicles	Amplify sound waves
Inner Ear	
Oval window	Initiate pressure waves
Cochlea	Transmit pressure waves
Organ of Corti	Receptor for hearing
Utricle and saccule	Static equilibrium
Semicircular canals	Dynamic equilibrium

are the hammer (malleus), anvil (incus), and stirrup (stapes), named for their structural resemblance to these objects. The Eustachian tube extends from the middle ear to the pharynx and permits the equalization of air pressure between the inside and outside of the ear. Chewing gum, yawning, and swallowing help move air through the Eustachian tubes during ascent and descent in airplanes or elevators.

Figure 39.11

Anatomy of the human ear. *a.* In the middle ear, the hammer, anvil, and stirrup amplify sound waves. Otosclerosis is a condition in which the stirrup becomes attached to the inner ear and unable to carry out its normal function. It can be replaced by a plastic piston, and thereafter the individual hears normally because sound waves are transmitted as usual to the cochlea, which contains the receptors for hearing. *b.* Inner ear. The sense organs for balance are in the inner ear; the vestibule contains the utricle and saccule, and the ampullae are at the bases of the semicircular canals. The receptors for hearing are also in the inner ear: the cochlea has been cut away to show the location of the organ of Corti.

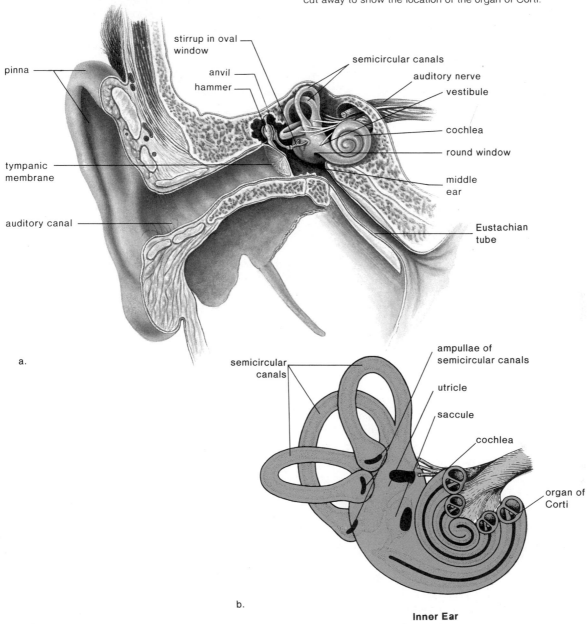

Inner Ear

Whereas the outer ear and middle ear contain air, the inner ear is filled with fluid. The inner ear has three anatomic areas: the first two, called the vestibule and the semicircular canals, are concerned with balance; the third, the cochlea, is concerned with hearing (fig. 39.11*b*).

Balance

The vestibule is a chamber between the semicircular canals and the cochlea. It contains two small sacs, the **utricle** and **saccule.** Within each of these are groups of little hair cells whose cilia are embedded in a capula with calcium carbonate granules known as *otoliths* above. When the head tilts, the otoliths press on the cilia in a certain direction, which initiates nerve impulses that inform the brain about the position of the head. Similar types of organs are also found in coelenterates, mollusks, and crustacea; these are known as statocysts (fig. 39.12). These organs only give information about position and do not result in a sensation of movement. They are therefore called *static equilibrium organs.*

Figure 39.12
Generalized statocysts as found in mollusks and crustaceans. A small particle, the statolith, moves in response to a change in the animal's position. When the statolith stops moving, it stimulates the closest cilia of hair cells. These cilia transmit impulses that indicate the position of the body.

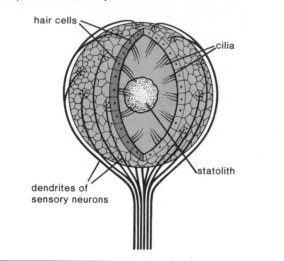

The **semicircular canals** (fig. 39.13) are *dynamic equilibrium organs* because they initiate a sensation of movement. At their bases are the ampullae, each one having hair cells with a cupula. As the body moves about, first there is a slight lag, and then the fluid within the canals moves; this causes the cupula to move and the cilia to bend. The three semicircular canals are each orientated in a different plane to detect movement in any direction. The brain integrates the impulses it receives from each of the three canals and thus can determine the direction and rate of movement. Very rapid or prolonged movements of the head may cause uncomfortable side effects, such as the dizziness and nausea of seasickness.

The outer ear, middle ear, and cochlea are necessary for hearing in humans. The vestibule, containing the utricle and the saccule, and the semicircular canals are concerned with the sense of balance.

Hearing

The **cochlea** within the inner ear resembles the shell of a snail because it is spiral-shaped (fig. 39.13). A cross section of the cochlea shows that it contains three canals: the vestibular, coch-

Figure 39.13
Location and anatomy of the organ of Corti. *a.* The organ of Corti is located in the cochlea. *b.* Cross section of the cochlea. The organ of Corti is located in the cochlear canal. *c.* The organ of Corti consists of hair cells with cilia that touch the tectorial membrane. Pressure waves cause the basilar membrane beneath the hair cells to vibrate.

When the cilia touch the tectorial membrane, nerve impulses are initiated and taken up by sensory nerve fibers within the auditory nerve. The intensity of the sound is determined by how many cells are stimulated; the quality of the sound is determined by which particular cells along the entire organ of Corti are stimulated.

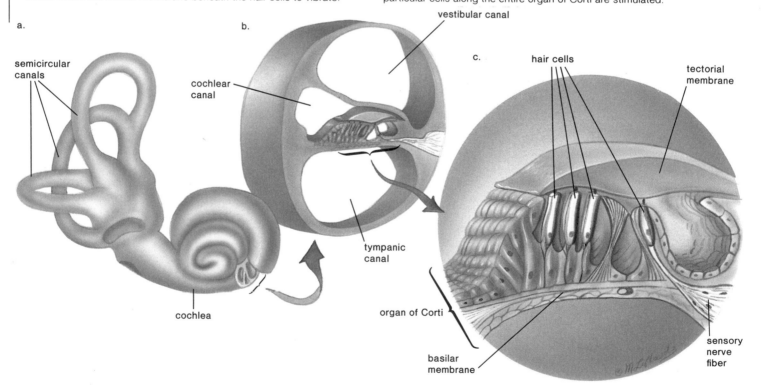

lear, and tympanic canals. Along the length of the *basilar membrane,* which forms the floor of the cochlear canal, there are at least 24,000 ciliated hair cells. Just above them is another membrane called the *tectorial membrane* (fig. 39.13c). The hair cells plus the tectorial membrane form the **organ of Corti,** the sense organ for hearing.

Sound waves reach the organ of Corti by way of the outer and middle ear. Ordinarily, sound waves do not carry much energy, but when a large number of waves strike the eardrum, it moves back and forth (vibrates) very slightly. The hammer transfers the pressure from the inner surface of the eardrum to the anvil, and thence to the stirrup (fig. 39.11a). The pressure is multiplied about twentyfold as it moves from the eardrum to the stirrup. The stirrup vibrates the oval window, which transmits pressure waves to the fluid in the inner ear. These waves cause the basilar membrane to move up and down and the cilia of the hair cells to rub against the tectorial membrane. Bending of the cilia initiates nerve impulses that pass by way of the auditory nerve to the brain, where the impulses are interpreted as a sound.

The organ of Corti is narrow at its base but widens as it approaches the tip of the cochlear canal. Various cells along it are sensitive to different wave frequencies, or pitches. Near its apex, the organ of Corti responds to low pitches, such as a bass drum, and near the base it responds to higher pitches, such as a bell or whistle. The neurons from each region along the organ lead to slightly different areas in the brain. The pitch sensation we experience depends on which brain area is stimulated. Volume is a function of the **amplitude** of sound waves. Loud noises (measured in decibels, db) cause the fluid of the cochlea to oscillate to a greater degree, and this, in turn, causes the basilar membrane to move up and down to a greater extent. The resulting increased stimulation is interpreted by the brain as loudness. It is believed that tone is interpreted by the brain according to the distribution of hair cells that are stimulated.

The sense receptors for sound are ciliated hair cells on the basilar membrane (the organ of Corti). When the basilar membrane vibrates, the delicate cilia touch the tectorial membrane, producing nerve impulses that are transmitted by the auditory nerve to the brain.

Picturing the Effects of Noise

We have an idea of what noise does to the ear," David Lipscomb [of the University of Tennessee Noise Laboratory] says. "There's a pretty clear cause-effect relationship." And these photomicrographs (*see* figure) of the cochlea's tiny structures graphically document noise trauma to the inner ear.

Hair cells transmit the mechanical energy of sound waves into those neural impulses that the brain interprets as sound. Loud noise can damage or destroy hair cells as these scanning electron micrographs illustrate.

Hair cells come in two varieties: a single row of inner cells and a triple row of outer ones. "Outer cells degenerate before inner cells," notes Clifton Springs, N.Y.-otolaryngologist Stephen Falk. The most subtle change wrought by noise is a development of vesicles, or blister-like protrusions along the walls of the hair cells' stereocilia. Continued assault by noise will lead to a rupturing of the vesicles and damage. In addition, the "cuticular plate"—base tissue supporting the stereocilia—may soften, followed by a swelling and ultimate degeneration of hair cells.

But sensory hair cells are not the only structures at risk. Adjacent inner-ear cells . . . may undergo vacuolation—development of degenerative empty spaces in cells. Even nerve fibers synapsing at the hair cells' roots may die.

a.

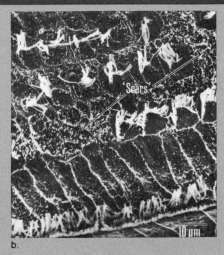

b.

Electron micrographs showing damage to the organ of Corti due to loud noise. *a.* Normal organ of Corti. The tectorial membrane has been removed to reveal the hair cells. *b.* Organ of Corti after 24-hour exposure to noise level typical of rock music. Note scars where cilia on hair cells have worn away. Magnification, ×1,400.

In the final phase of noise-induced cochlear damage, the organ of Corti—of which hair cells and supporting cells are a part—is completely denuded and covered by a layer of scar tissue.

Reprinted with permission from *Science News,* the weekly news magazine of science, copyright 1982 by Science Service, Inc.

Summary

1. Sense organs are transducers; they transform the energy of a stimulus to the energy of nerve impulses. It is the brain, not the sense organ, that interprets the stimulus.
2. Human olfactory cells and taste buds are chemoreceptors, sensitive to chemicals in water and air.
3. The human eye is a photoreceptor. The compound eye of arthropods is made up of many individual units, whereas the human eye is a camera-type eye with a single lens. Table 39.1 lists the parts of the eye and the function of each part.
4. The receptors for sight in humans are the rods and cones that are in the retina. The rods work in minimum light and detect motion, but do not detect color. The cones require bright light and do detect color.
5. When light strikes rhodopsin, a molecule composed of opsin and retinal, retinal changes shape and opsin is activated. Chemical reactions follow that produce nerve impulses, which are eventually picked up by the optic nerve.
6. There are three kinds of cones containing either blue, green, or red receptors. Each pigment is also made up of retinal and opsin but there is a difference in the opsin.
7. Mechanoreceptors include the touch and pressure receptors in the human skin.
8. Many mechanoreceptors are hair cells with cilia, such as those found in the lateral line of fishes as well as the inner ear of humans. Table 39.2 lists the parts of the ear and the function of each part.
9. The inner ear contains the sense organs for balance. Just like the statocysts of invertebrates, portions of the human inner ear contain calcium carbonate granules resting on hair cells. The movement of these granules gives us a sense of static equilibrium. Movement of fluid past hair cells in the semicircular canals gives us a sense of dynamic equilibrium.
10. Hair cells on the basilar membrane (the organ of Corti) are responsible for hearing. Pressure waves that begin at the oval window cause the basilar membrane to vibrate so that the cilia of the hair cells touch the tectorial membrane. This causes the hair cells to initiate nerve impulses that are carried by the auditory nerve to the brain.

Objective Questions

1. A receptor
   a. is the first portion of a reflex arc.
   b. initiates nerve impulses.
   c. responds to only one type of stimulus.
   d. All of these.
2. Which of these gives the correct path for light rays entering the human eye?
   a. sclera, retina, choroid, lens, cornea
   b. fovea centralis, pupil, aqueous humor, lens
   c. cornea, pupil, lens, vitreous humor, retina
   d. optic nerve, sclera, choroid, retina, humors
3. Which gives an incorrect function for the structure?
   a. lens—focusing
   b. iris—regulation of amount of light
   c. choroid—location of cones
   d. sclera—protection
4. Which of these contain mechanoreceptors?
   a. human skin
   b. lateral line of fishes
   c. statocysts of arthropods
   d. All of these.
5. Which association is incorrect?
   a. lateral line—fishes
   b. compound eye—arthropods
   c. camera-type eye—squid
   d. statocysts—sea stars
6. Which one of these wouldn't you mention if you were tracing the path of sound vibrations?
   a. auditory canal
   b. tympanic membrane
   c. semicircular canals
   d. cochlea
7. Which one of these correctly describes the location of the organ of Corti?
   a. between the tympanic membrane and the oval window in the inner ear
   b. in the utricle and saccule within the vestibule
   c. between the tectorial membrane and the basilar membrane in the cochlear canal
   d. between the outer and inner ear within the semicircular canals
8. Which of these is mismatched?
   a. semicircular canals—inner ear
   b. utricle and saccule—outer ear
   c. auditory canal—outer ear
   d. ossicles—middle ear
9. Retinal
   a. is sensitive to light energy.
   b. is a part of rhodopsin.
   c. is found in both rods and cones.
   d. All of these.
10. Both olfactory receptors and sound receptors have cilia, and they both
    a. are chemoreceptors.
    b. are mechanoreceptors.
    c. initiate nerve impulses.
    d. All of these.

Study Questions

1. In what ways are all receptors similar; in what ways are they different?
2. Discuss the structure and function of human chemoreceptors.
3. In general, how does the arthropod eye differ from that in humans? What types of animals have eyes that are constructed similarly to the human eye?
4. Name the parts of the eye, and give a function for each part.
5. Contrast the location and function of rods to that of cones.
6. What are the types of mechanoreceptors in human skin?
7. Describe how the lateral line system of fishes works and why it is considered to contain mechanoreceptors.
8. Describe the anatomy of the ear and how we hear.
9. Describe the role of the utricle, saccule, and semicircular canals in balance.

Thought Questions

1. If we connected a pain receptor in the foot to a nerve normally receiving impulses from a heat receptor in the hand, and then proceeded to stick a pin into the pain receptor in the foot, what sensation would the subject report?

2. Does it seem likely that our sound receptors evolved from lateral line receptors? Why or why not?

3. Explain why a hearing aid is ineffective in persons with noise-induced cochlear damage.

Selected Key Terms

receptor (re-sep′tor) 613
chemoreceptor (ke″mo-re-sep′tor) 613
olfactory cell (ol-fak′to-re sel) 613
taste bud (tāst bud) 614
photoreceptor (fo″to-re-sep′tor) 614
compound eye (kom′pownd i) 615
sclera (skle′rah) 615

choroid (ko′roid) 615
retina (ret′ĭ-nah) 616
rod (rod) 616
cone (kōn) 616
rhodopsin (ro-dop′sin) 617
mechanoreceptor
  (mek″ah-no-re-sep′tor) 618

tympanic membrane (tim-pan′ik mem′brān) 620
semicircular canal (sem″e-ser′ku-lar kah-nal′) 622
cochlea (kok′le-ah) 622
organ of Corti (or′gan uv kor′ti) 623

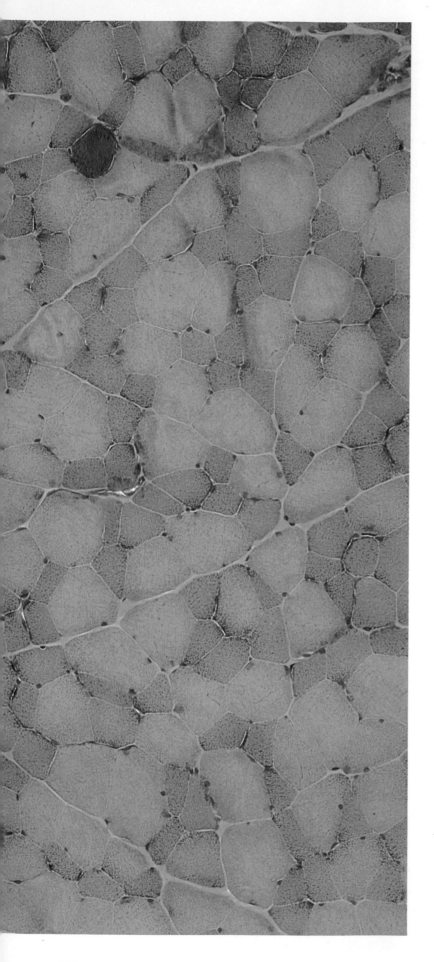

CHAPTER 40

Musculoskeletal System

Your study of this chapter will be complete when you can:

1. Distinguish between the three types of skeletons in the animal kingdom, and give examples of animals that have each type.
2. Give the functions of a skeletal system in animals.
3. Name the two parts of the human skeleton, and list the location of the bone for both.
4. Describe the anatomy of a long bone and the tissues found in a long bone.
5. Explain how bone is continually being broken down and rebuilt.
6. Name the different types of joints, and give an example of each.
7. Describe the anatomy and location of the three types of muscles found in the vertebrate body.
8. Describe how whole muscles work in antagonistic pairs.
9. Describe the anatomy of a muscle fiber and sarcomere.
10. Explain the sliding filament theory of muscle contraction.
11. List the sources of ATP energy for muscle contraction, and explain the occurrence of oxygen debt.
12. List the series of events that occur after a muscle fiber is innervated by nervous stimulation.

Light micrograph of cross section of human skeletal muscle showing that some fibers are darker than others. The darker ones are slow-twitching fibers, which provide steady pulling power for sports like swimming, running, and skiing. The lighter ones are fast-twitching fibers, which provide bursts of energy for sprinting, weight lifting, or swinging a golf club.

Skeleton Organization

The skeleton is the framework of the body. It helps protect the internal organs and assists in movement. To produce body movements, the force of muscular contractions must be specifically directed against other parts of the body. In some animals, such as flatworms and segmented worms, the pressure of muscular contractions is applied to fluid-filled body compartments (table 40.1). Then the animal is said to have a **hydrostatic skeleton.**

Animals with hydrostatic skeletons are not as capable of complex movements as are animals with a rigid, supporting exo- or endoskeleton. The protective external **exoskeleton** of arthropods is largely composed of chitin that must be periodically shed by molting as the animal grows (fig. 40.1). The internal **endoskeleton** of vertebrates is largely composed of cartilage and bone that grows with the animal. These tissues are storage areas for calcium and phosphorus; they are also the site of blood cell production in adult animals.

The endoskeleton of vertebrates is largely composed of cartilage and/or bone. It not only permits flexible movement, it also supports and protects the body, produces blood cells, and serves as a storehouse for certain inorganic salts.

Table 40.1 Classification of Skeletons

Type	Phylum	Example
Hydrostatic skeleton gastrovascular cavity	Cnidaria Platyhelminthes	Hydras, anemones Flatworms
coelom*	Annelida	Earthworms
Exoskeleton	Arthropoda	Lobsters, grasshoppers
Endoskeleton	Chordata	Vertebrates

*See figure 26.10.

Human Skeleton

The human skeleton (fig. 40.2) may be divided into two parts: the **axial skeleton** and the **appendicular skeleton.**

Axial Skeleton

The axial skeleton includes the skull, the vertebrae, the ribs, and the sternum. The **skull,** or cranium, is composed of many bones fitted tightly together in adults. In newborns, certain bones are not completely formed and instead are joined by membranous regions called **fontanels,** all of which usually close by the age of 16 months. The bones of the skull contain the **sinuses,** air spaces lined by mucous membrane. Two of these, called the mastoid sinuses, drain into the middle ear. Mastoiditis, a condition that can lead to deafness, is an inflammation of these sinuses. Whereas the skull protects the brain, the several bones of the face join together to support and protect the special sense organs and to form the jawbones.

The **vertebral column** extends from the skull to the pelvis and forms a dorsal backbone that protects the spinal cord (fig. 40.3). Normally, the vertebral column has four curvatures that provide more resiliency and strength than a straight column could. It is composed of many parts, called **vertebrae,** that are held together by bony facets, muscles, and strong ligaments. The vertebrae are named according to their location in the body.

There are *disks* between the vertebrae that act as padding (fig. 40.3). They prevent the vertebrae from grinding against one another and also absorb shock caused by movements such as running, jumping, and even walking. The presence of the disks allows motion between the vertebrae so that we can bend forward, backward, and from side-to-side. Unfortunately, these disks become weakened with age and can slip, or even rupture. This causes pain when the damaged disk presses against the spinal cord and/or spinal nerves. The body may heal itself, or else the disk can be removed surgically. If the latter occurs, the two adjacent vertebrae can be fused together but this will limit the flexibility of the body.

Figure 40.1

Arthropods have an external skeleton. In order to grow, they must periodically shed this exoskeleton, a process called molting. In this photo, a cicada is just emerging from its discarded skeleton. It will be more vulnerable than usual until its new exoskeleton has hardened.

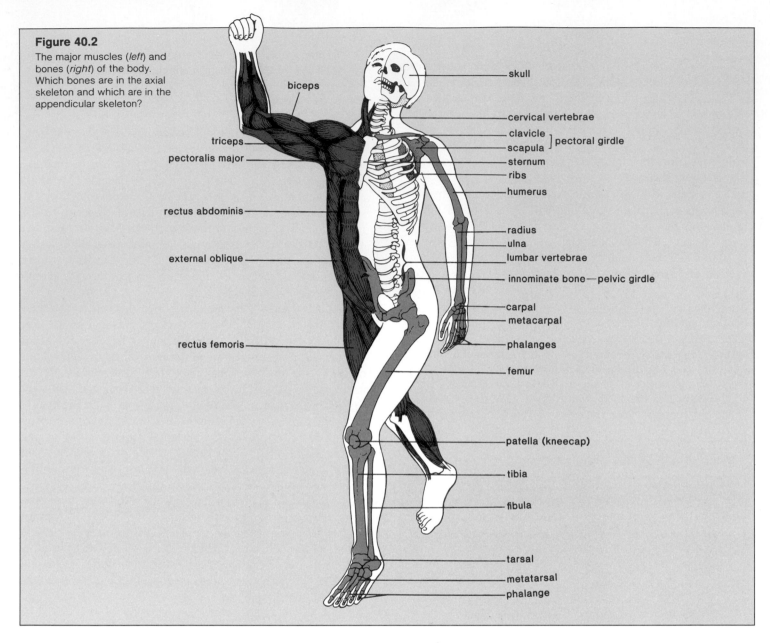

Figure 40.2
The major muscles (*left*) and bones (*right*) of the body. Which bones are in the axial skeleton and which are in the appendicular skeleton?

biceps

triceps

pectoralis major

rectus abdominis

external oblique

rectus femoris

skull

cervical vertebrae

clavicle ⎤ pectoral girdle
scapula ⎦

sternum

ribs

humerus

radius

ulna

lumbar vertebrae

innominate bone—pelvic girdle

carpal

metacarpal

phalanges

femur

patella (kneecap)

tibia

fibula

tarsal

metatarsal

phalange

The vertebral column, directly or indirectly, serves as an anchor for all the other bones of the skeleton. All the **ribs** connect directly to the thoracic vertebrae in the back, and all but two pairs connect either directly or indirectly via shafts of cartilage to the **sternum** (breastbone) in the front. The lower two pairs of ribs are called "floating ribs" because they do not attach to the sternum.

Appendicular Skeleton

The appendicular skeleton consists of the bones within the pectoral and pelvic girdles and the attached appendages. The pectoral (shoulder) girdle and appendages (arms and hands) are specialized for flexibility, while the pelvic girdle (hip bones) and appendages (legs and feet) are specialized for strength.

The components of the **pectoral girdle** (fig. 40.4) are loosely linked together by ligaments rather than firm joints. Each **clavicle** (collar bone) connects with the sternum in front and the **scapula** (shoulder blade) behind, but the scapula is freely mov-

able and held in place by muscles. This allows it to freely follow the movements of the arm. The single long bone in the upper arm (fig. 40.4), the **humerus,** has a smoothly rounded head that fits into a socket of the scapula. The socket, however, is very shallow and much smaller than the head of the humerus. Although this means that the arm can move in almost any direction, there is little stability. Therefore this is the joint that is most apt to dislocate. The opposite end of the humerus meets the two bones of the lower arm, the **ulna** and the **radius,** at the elbow. (The prominent bone in the elbow is the topmost part of the ulna.) When the arm is held so that the palm is turned frontward, the radius and ulna are about parallel to one another. When the arm is turned so that the palm is next to the body, the radius crosses in front of the ulna, a feature that contributes to the easy twisting motion of the lower arm.

The many bones of the hand increase its flexibility. The wrist has eight **carpal** bones that look like small pebbles. From these, five **metacarpal** bones fan out to form a framework for

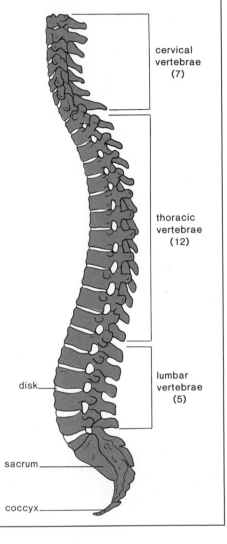

Figure 40.3
The vertebral column. The vertebrae are named according to their location in the column, which is flexible due to the presence of disks between the vertebrae. The numbers tell how many of each kind of vertebrae there are in the human body. Note the presence of the coccyx, the vestigial tailbone.

cervical vertebrae (7)

thoracic vertebrae (12)

lumbar vertebrae (5)

disk

sacrum

coccyx

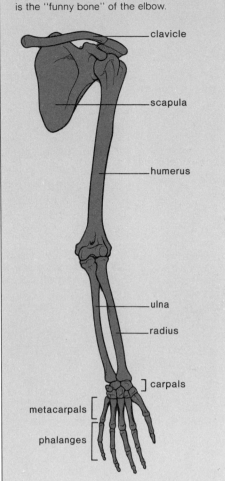

Figure 40.4
The bones of the pectoral girdle, arm, and hand. The lower end of the humerus is the "funny bone" of the elbow.

clavicle

scapula

humerus

ulna

radius

carpals

metacarpals

phalanges

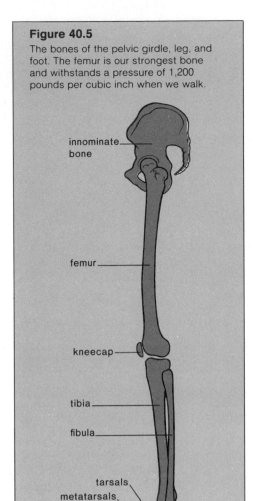

Figure 40.5
The bones of the pelvic girdle, leg, and foot. The femur is our strongest bone and withstands a pressure of 1,200 pounds per cubic inch when we walk.

innominate bone

femur

kneecap

tibia

fibula

tarsals
metatarsals
phalanges

the palm. The metacarpal bone that leads to the thumb is placed in such a way that the thumb can reach out and touch the other digits. (**Digits** is a term that refers to either fingers or toes.) Beyond the metacarpals are the **phalanges,** the bones of the fingers and the thumb. The phalanges of the hand are long, slender, and lightweight.

The **pelvic girdle** (fig. 40.5) consists of two heavy, large **innominate** bones, or hipbones. The innominate bones are anchored to the **sacrum** and together these bones form a hollow cavity, the pelvis. The weight of the body is transmitted through the pelvis to the legs and then onto the ground. The largest bone in the body is the **femur,** or thigh bone. Although the femur is a strong bone, it is doubtful that the femurs of a fairy-tale giant could support the increase in weight. If a giant were 10 times taller than an ordinary human being, he would also be about 10 times wider and thicker, making him weigh about 1,000 times as much. This amount of weight would break even giant-size femurs.

In the lower leg the larger of the two bones, the **tibia** (fig. 40.5), has a ridge we call the shin. Both of the bones of the lower leg have a prominence that contributes to the ankle—the tibia on the inside of the ankle and the **fibula** on the outside of the ankle. Although there are seven **tarsal** bones in the ankle, only one receives the weight and passes it on to the heel and the ball of the foot. If one wears high-heel shoes, the weight is thrown even further toward the front of the foot. The **metatarsal** bones form the arches of the foot. There is a longitudinal arch from the heel to the toes and a transverse arch across the foot. These provide a stable, springy base for the body. If the tissues that bind the metatarsals together become weakened, flatfeet are apt to result. The bones of the toes are called *phalanges,* just like those of the fingers, but in the foot the phalanges are stout and extremely sturdy.

The axial and appendicular skeletons contain the bones that are listed in table 40.2.

Table 40.2 The Major Bones of the Skeleton	
Part	**Bones**
Axial skeleton	Skull, vertebral column, sternum, ribs
Appendicular skeleton	
Pectoral girdle	Clavicle, scapula
Arm	Humerus, ulna, radius
Hand	Carpals, metacarpals, phalanges
Pelvic girdle	Innominate bone
Leg	Femur, tibia, fibula
Foot	Tarsals, metatarsals, phalanges

Long Bone Structure

A long bone, such as the femur, illustrates the principles of bone anatomy. When the bone is split open, as in figure 40.6a, the longitudinal section shows that it is not solid but has a cavity, called the medullary cavity, bounded at the sides by compact bone and at the ends by spongy bone. Beyond the spongy bone there is a thin shell of compact bone and finally a layer of cartilage.

Both cartilage and bone have cells separated by a matrix. The matrix of **cartilage** (fig. 40.6b), composed of protein and proteinaceous fibers, is more flexible than that of bone, which contains a protein framework in which salts of calcium and phosphorus are embedded. Bone occurs in two forms—compact bone and spongy bone. **Compact bone** (fig. 40.6c) consists of structural units called *Haversian systems*. Each system has bone cells within lacunae (cavities) arranged in concentric circles

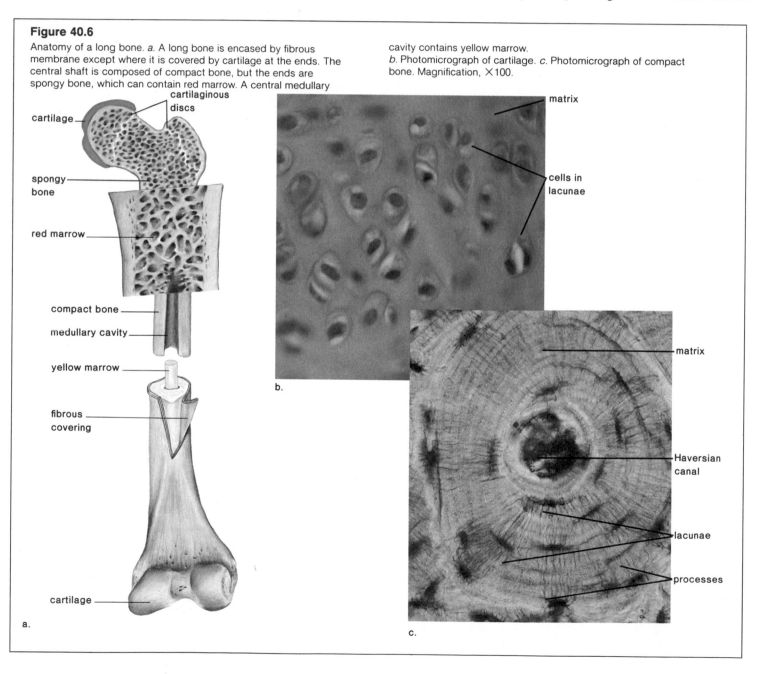

Figure 40.6

Anatomy of a long bone. *a.* A long bone is encased by fibrous membrane except where it is covered by cartilage at the ends. The central shaft is composed of compact bone, but the ends are spongy bone, which can contain red marrow. A central medullary cavity contains yellow marrow.
b. Photomicrograph of cartilage. *c.* Photomicrograph of compact bone. Magnification, ×100.

around tiny tubes called Haversian canals. These canals contain blood vessels and nerve fibers. Nutrients, gases, and wastes diffuse through tiny passages (processes) that connect the lacunae with the central canal. **Spongy bone** contains numerous bony bars and plates separated by irregular spaces. Although lighter than compact bone, spongy bone is still designed for strength. Just as braces are used for support in buildings, the solid portions of spongy bone follow lines of stress. The spaces in spongy bone are often filled with *red marrow,* a specialized tissue that produces blood cells. The medullary cavity of a long bone usually contains *yellow marrow,* which is a fat-storage tissue.

Growth and Development of Bones   Most of the bones of the skeleton are cartilaginous during prenatal development. Later, bone-forming cells known as **osteoblasts** replace cartilage with bone. At first, there is only a primary ossification center at the middle of a long bone but later, secondary centers form at the ends of the bones. There remains a *cartilaginous disk* between the primary ossification center and each secondary center. The length of a bone is dependent on how long the cartilage cells within the disk continue to divide. Eventually, though, the disks disappear and the bone stops increasing in size when the individual attains adult height.

In the adult, bone is *continually* being broken down and then built up again. Bone-absorbing cells, called **osteoclasts,** are derived from cells carried in the bloodstream. As they break down bone, they remove worn cells and deposit calcium in the blood. Apparently, after about three weeks, they disappear. The destruction caused by the work of osteoclasts is repaired by **osteoblasts.** As they form new bone, they take calcium from the blood. Eventually some of these cells get caught in the matrix they secrete and are converted to **osteocytes,** the cells found within Haversian systems (fig. 40.6c).

Because of continual renewal, the thickness of bones can change, depending on the amount of physical activity or a change in certain hormone balances (see chapter 41). In most adults, the bones become weaker due to a loss of mineral content. Strange as it may seem, adults seem to require more calcium in the diet than children in order to promote the work of osteoblasts. Many older women, due to a lack of estrogen, suffer from *osteoporosis,* a condition in which weak and thin bones cause aches and pains and fracture easily. A tendency toward osteoporosis may also be augmented by lack of exercise and too little calcium in the diet.

Bone is a living tissue and is always being rejuvenated.

Classification of the Joints

Bones are linked at the joints, which are often classified according to the amount of movement they allow. Some bones, such as those that make up the cranium, are sutured together and are *immovable.* Other joints are *slightly movable,* such as the joints between the vertebrae. The vertebrae are separated by disks, described earlier, that increase their flexibility. Similarly, the two hipbones are slightly movable where they are ventrally joined by cartilage. Owing to hormonal changes, this joint becomes more flexible during late pregnancy, which allows the pelvis to expand during childbirth.

Most joints are *freely movable* (**synovial**) **joints** in which the two bones are separated by a cavity. **Ligaments,** which are composed of fibrous connective tissue, form a capsule that binds the two bones to one another, holding them in place. In a "double-jointed" individual, the ligaments are unusually loose. The joint capsule is lined by synovial membrane, which produces *synovial fluid,* a lubricant for the joint. The knee is an example of a synovial joint (fig. 40.7). In the knee, as in other freely movable joints, the bones are capped by cartilage, but in the knee there are also crescent-shaped pieces of cartilage between the bones, called **menisci** (sing., **meniscus**). These give added stability, helping to support the weight placed on the knee joint. Unfortunately, athletes often suffer injury of the menisci, known as torn cartilage. The knee joint also contains 13 fluid-filled sacs called **bursae** (sing., **bursa**) which ease friction between tendons and ligaments and between tendons and bones. Inflammation of bursae is called bursitis. Tennis elbow is a form of bursitis.

There are other types of freely movable (synovial) joints. The knee and elbow joints are *hinge joints* because, like a hinged door, they largely permit movement in one direction only. More movable are the ball-and-socket joints; for example, the ball of the femur fits into a socket on the hipbone. *Ball-and-socket joints* allow movement in all planes and even a rotational movement.

Figure 40.7

The knee joint, an example of a freely movable synovial joint. Notice that there is a cavity between the bones that is encased by ligaments and lined by synovial membrane. The kneecap protects the joint. Synovial joints are subject to arthritis. In rheumatoid arthritis, the synovial membrane becomes inflamed and grows thicker. Degenerative changes take place that make the joint almost immovable and painful to use. There is evidence that these effects are brought on by an autoimmune reaction. In old-age arthritis, or osteoarthritis, the cartilage at the ends of the bones disintegrates so that the ends of the bones become rough and irregular. This type of arthritis is most likely to affect the joints that have received the greatest use over the years.

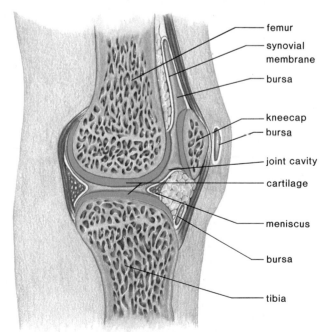

femur
synovial membrane
bursa
kneecap
bursa
joint cavity
cartilage
meniscus
bursa
tibia

Joints are classified according to degree of movement. Some joints are immovable, some are slightly movable, and some are freely movable.

Vertebrate Muscle Organization

Muscles perform three functions. Along with the skeleton they give support and structure to the body. When they contract, they account for movement of both the limbs and the internal organs. Contraction also releases heat that serves to warm the body. Vertebrate animals have three types of muscles: smooth, cardiac, and skeletal. The contractile cells of these tissues are called muscle fibers. Therefore, a muscle fiber is a cell that specializes in contraction.

Smooth muscle fibers are spindle-shaped cells, each having a single nucleus. Most often the cells are arranged in parallel to form flat sheets (fig. 40.8*a*). The *striations* (light and dark bands) seen in cardiac and skeletal muscle are not observed in this tissue, which is found in the walls of hollow internal organs like the stomach, intestine, and urinary bladder.

Cardiac muscle fibers also have only a single nucleus, but they are striated and branched (fig. 40.8*b*). This branching allows the fibers to interlock for greater strength. There are structures called intercalated disks where the individual fibers meet. These permit contractions to spread throughout the wall of the heart, the only place where cardiac muscle is found.

Skeletal muscle fibers are long and striated (fig. 40.8*c*). Each cell has many nuclei located just beneath the plasma membrane. Skeletal muscles are attached to and move the bones of the skeleton; therefore we will be studying their anatomy and physiology in more detail in this chapter.

Smooth muscle and cardiac muscle are said to be *involuntary* muscles because their contraction is most often controlled by the autonomic nervous system without conscious thought. In fact, heart muscle is capable of contracting without any outside nervous stimulation at all. Skeletal muscle is called *voluntary* muscle, and we often consciously control its contraction by way of the somatic nervous system. If the nerve to a skeletal muscle is damaged, it may very well degenerate and become paralyzed.

Vertebrate muscles are of three types: smooth, cardiac, and skeletal. These muscles differ as to their anatomy and mode of control.

Skeletal Muscle Contraction

Skeletal muscles, which make up over 40% of the body's weight, are attached to the skeleton by **tendons,** tough, fibrous bands of cords. When muscles contract, they shorten. Therefore, muscles can only pull; they cannot push. Because of this, skeletal muscles must work in *antagonistic pairs* (fig. 40.9). For example, one muscle of an antagonistic pair bends the joint and brings the limb toward the body. The other muscle then straightens

Figure 40.8

Vertebrate muscle fibers. *a.* Smooth muscle fibers have a single nucleus and lack striations. Magnification, ×250. *b.* Cardiac muscle fibers. Note the branching of the fibers, the central position of the nuclei, and the presence of intercalated disks, which are the complex junctions between adjacent individual cells. Magnification, ×250. *c.* Skeletal muscle fibers. Note striations and the peripheral location of nuclei in the multinucleate fibers. Magnification, ×100.

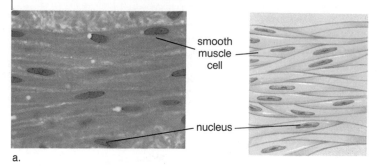

smooth muscle cell

nucleus

a.

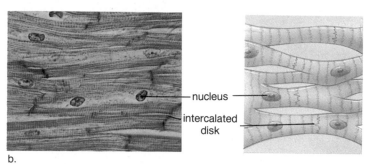

nucleus

intercalated disk

b.

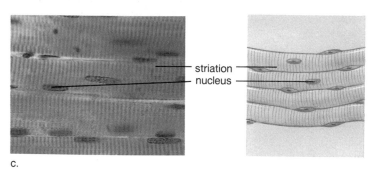

striation
nucleus

c.

the joint and extends the limb. Figure 40.9 illustrates this principle and compares the actions of muscles in the limb of an arthropod, where the muscles are attached to an exoskeleton, with those of a human, where the muscles are attached to an endoskeleton.

In the laboratory, it is possible to study the contraction of whole muscles and of their fibers. Customarily, the entire muscle or a portion thereof is mounted in an apparatus called a physiograph. The specimen is stimulated electrically, and when it contracts, its contraction pattern, which is called a myogram, is recorded (fig. 40.10).

Muscle Fiber Anatomy and Physiology

Present understanding of the microscopic anatomy and the physiology of muscle contraction has resulted in part from the research of two English scientists. In the late 1950s, A. F. Huxley

Figure 40.9

Antagonistic muscle pairs. Muscles can exert force only by shortening. Movable joints are supplied with double sets of muscles that work in opposite directions. As indicated by the arrows, flexor muscles move the lower limb toward the body and extensor muscles move the lower limb away from the body. *a*. Muscles in an insect's leg as an example of antagonistic muscles attached to the inside of an arthropod exoskeleton. *b*. Muscles in the human leg as an example of antagonistic muscles attached to a vertebrate endoskeleton.

rectus femoris

biceps femoris

extensor muscle

flexor muscle

a.

b.

Figure 40.10

Laboratory study of muscle fiber versus whole muscle contraction. *a*. When a muscle fiber is electrically stimulated in the laboratory it may contract. At first the stimulus may be so weak that no contraction occurs, but as soon as the strength of the stimulus reaches the threshold stimulus, the muscle fiber will contract and then relax. This action, which is called a muscle fiber twitch, is divided into 3 stages, as shown. Increasing the strength of the stimulus will not change the strength of the muscle fiber's contraction; therefore, it is said that a fiber obeys the all-or-none law. If a muscle fiber is given two stimuli in quick succession, it will not respond to the second because it is still recovering from the first stimulus. *b*. A whole muscle behaves quite differently. If the strength of the stimulus is increased, the strength of the response increases. This is because more and more fibers are being stimulated to contract. Also, quickly repeated stimuli to a whole muscle can result in a summation of twitches. This occurs because more fibers can begin to contract while others are relaxing. The muscle as a whole exhibits a greater and greater degree of contraction until tetanus, a sustained contraction, is reached. Eventually the muscle suffers fatigue and begins to relax. Fatigue in isolated muscles is caused by depletion of ATP.

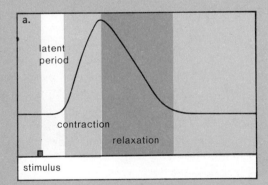

a.

latent period

contraction

relaxation

stimulus

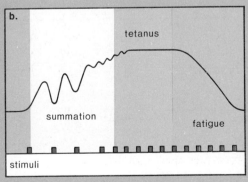

b.

tetanus

summation

fatigue

stimuli

Figure 40.11

Anatomy of a skeletal muscle. *a.* Whole muscle contains several bundles of muscle fibers bound together by connective tissue. *b.* A muscle fiber is packed full of myofibrils, the contractile elements in a fiber. A myofibril contains protein filaments arranged in such a way that striations are seen.

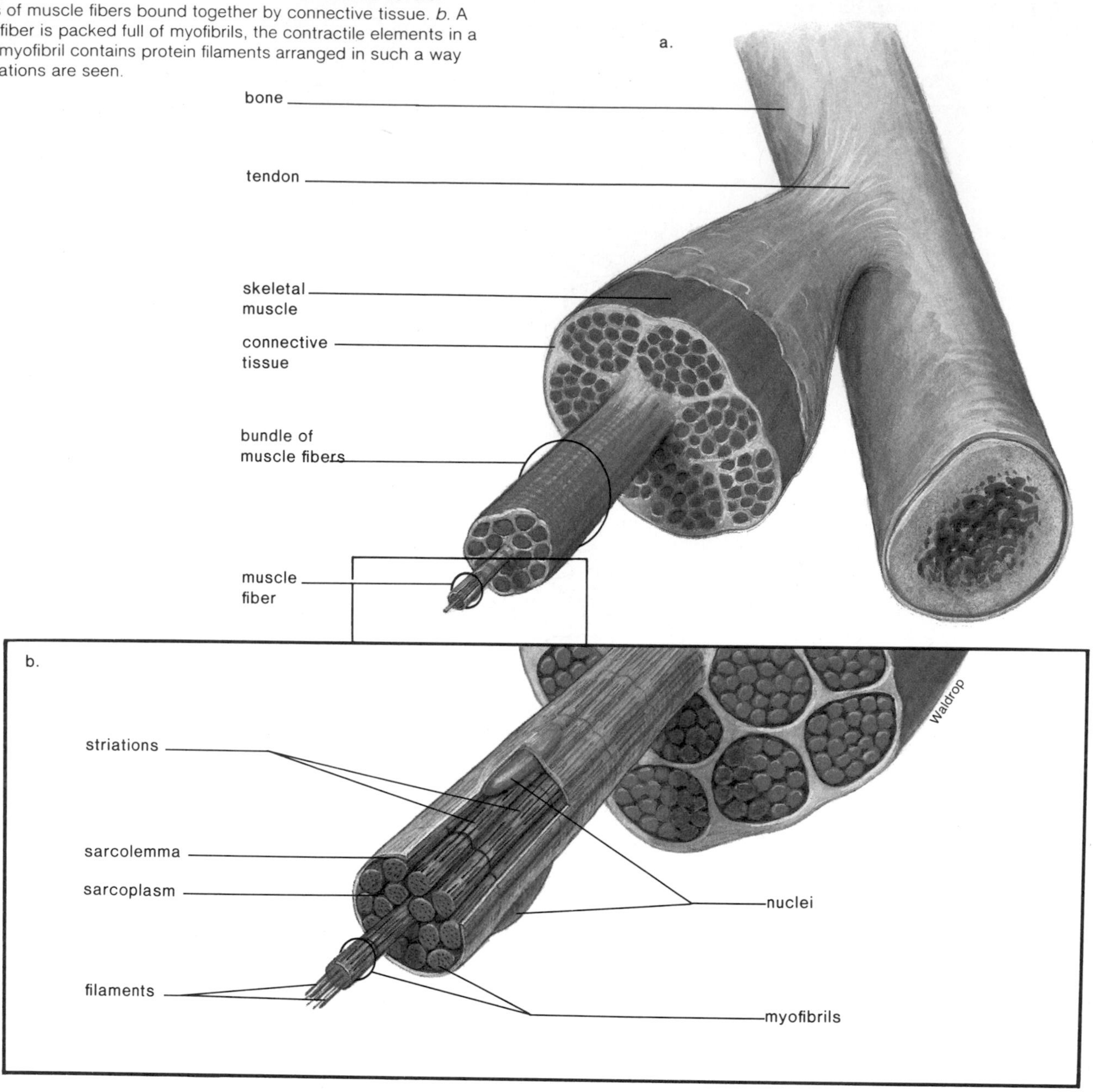

studied muscle contraction using mainly the light microscope. H. E. Huxley studied the fine structure of skeletal muscle with the electron microscope.

Muscle Fiber Structure

A whole skeletal muscle (fig. 40.11) is composed of a number of bundles of **muscle fibers.** Each fiber is a cell containing the usual cellular components, but special names have been assigned to some of these components, as indicated in table 40.3.

Table 40.3 Muscle Cell Terminology

Component	Term
Plasma membrane	Sarcolemma
Cytoplasm	Sarcoplasm
Endoplasmic reticulum	Sarcoplasmic reticulum

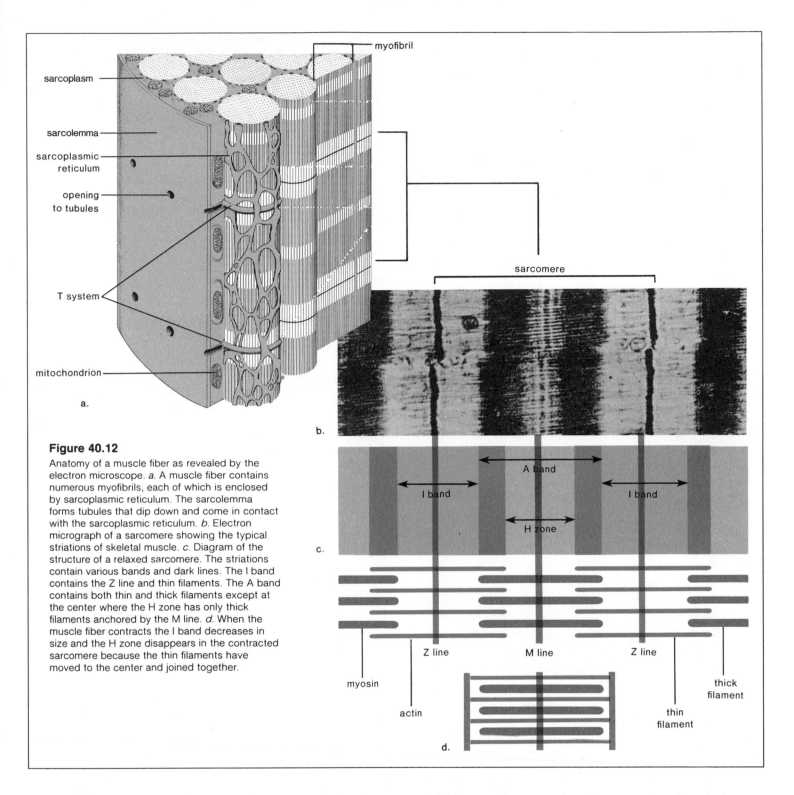

sarcoplasm

sarcolemma

sarcoplasmic reticulum

opening to tubules

T system

mitochondrion

a.

myofibril

sarcomere

b.

A band

I band

I band

H zone

c.

Z line   M line   Z line

myosin

actin

thick filament

thin filament

d.

Figure 40.12

Anatomy of a muscle fiber as revealed by the electron microscope. *a.* A muscle fiber contains numerous myofibrils, each of which is enclosed by sarcoplasmic reticulum. The sarcolemma forms tubules that dip down and come in contact with the sarcoplasmic reticulum. *b.* Electron micrograph of a sarcomere showing the typical striations of skeletal muscle. *c.* Diagram of the structure of a relaxed sarcomere. The striations contain various bands and dark lines. The I band contains the Z line and thin filaments. The A band contains both thin and thick filaments except at the center where the H zone has only thick filaments anchored by the M line. *d.* When the muscle fiber contracts the I band decreases in size and the H zone disappears in the contracted sarcomere because the thin filaments have moved to the center and joined together.

A muscle fiber has some unique anatomical characteristics. For one thing, it has a T (for tubule) system (fig. 40.12). The **sarcolemma,** or plasma membrane, forms these *tubules* that penetrate or dip down into the cell, so that they come into contact with, but do not fuse with, expanded portions of modified endoplasmic reticulum, termed the **sarcoplasmic reticulum.** The expanded portions of the sarcoplasmic reticulum, called calcium-storage sacs, contain calcium ions, Ca^{++}, which are es-

sential for muscle contraction. The sarcoplasmic reticulum encases hundreds, and sometimes even thousands of *myofibrils,* which are the contractile portions of the fibers.

Myofibrils

Myofibrils are cylindrical in shape and run the length of the muscle fiber. The light microscope shows that a myofibril has light and dark bands called striations; it is their striations that cause skeletal muscle to appear striated (fig. 40.8c). The electron microscope has shown that the striations of myofibrils are dependent on the placement of protein filaments that contain units called **sarcomeres** (fig. 40.12b).

Filaments

A sarcomere extends between two dark lines called the *Z lines* (fig. 40.12c). The *I band* is a light-colored region that takes in the sides of two sarcomeres; therefore, an I band includes a Z line. The center dark region of a sarcomere is called the *A band*. The A band is interrupted by a light center portion, the *H zone,* and a fine, dark stripe called the *M line* cuts through the H zone.

These bands and zones relate to the placement of filaments within each sarcomere. A sarcomere contains two types of filaments, *thin filaments* and *thick filaments* (fig. 40.12c). The thin filaments are attached to a Z line, and the thick filaments are anchored by an M line. The I band is light because it contains only thin filaments; the A band is dark because it contains both thin and thick filaments, except at the center in the lighter H zone, where only thick filaments are found.

Sarcomere contraction is dependent on the action of two proteins, **actin** and **myosin,** that make up the thin filaments and the thick filaments respectively (table 40.4). When a sarcomere contracts (fig. 40.12d), the actin filaments slide past the myosin filaments and approach one another. This causes the I band to become shorter and the H zone almost or completely to disappear. The movement of actin in relation to myosin is called the **sliding filament theory** of muscle contraction. During the sliding process, the sarcomere shortens, even though the filaments themselves remain the same length (fig. 40.13).

The overall formula for muscle contraction can be represented as follows:

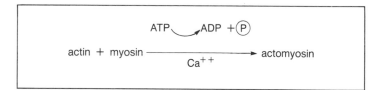

The participants in this reaction have the functions listed in table 40.5. Even though it is the actin filaments that slide past the myosin filaments, it is the latter that do the work. In the presence of Ca^{++}, portions of the myosin filaments, called *cross bridges* (fig. 40.13), reach out and attach to the actin filaments, pulling them along. (The term actomyosin in the above formula

stands for this joining of actin and myosin.) Following attachment, ATP is broken down and detachment occurs. Myosin brings about ATP breakdown, and therefore it is not only a structural protein, it is also an ATPase enzyme. The cross bridges attach and detach some 50 to 100 times as the thin filaments are pulled to the center of a sarcomere. If, by chance, more ATP molecules are not available, detachment cannot occur. This explains *rigor mortis,* permanent muscle contraction just after death.

The sliding filament theory states that actin filaments slide past myosin filaments because myosin has cross bridges that pull the actin filaments inward.

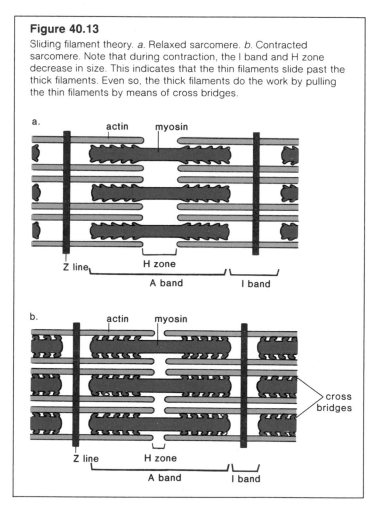

Figure 40.13

Sliding filament theory. *a.* Relaxed sarcomere. *b.* Contracted sarcomere. Note that during contraction, the I band and H zone decrease in size. This indicates that the thin filaments slide past the thick filaments. Even so, the thick filaments do the work by pulling the thin filaments by means of cross bridges.

Table 40.4 Contractile Elements	
Component	**Definition**
Myofibril	Muscle cell contractile subunit
Sarcomere	Functional unit of myofibril
Myosin	Thick filament
Actin	Thin filament

Table 40.5 Muscle Contraction	
Name	**Function**
Actin filaments	Slide past myosin
Myosin filaments	a. Enzyme that splits ATP b. Pulls actin by means of cross bridges
Ca^{++}	Needed for actin to bind to myosin
ATP	Supplies energy for bonding between actin and myosin = actomyosin

It is obvious from our discussion that ATP provides the energy for muscle contraction. In order to assure a ready supply, muscle fibers contain **creatine phosphate** (phosphocreatine), a storage form of high-energy phosphate. Creatine phosphate does not participate directly in muscle contraction. Instead, it is used to regenerate ATP by the following reaction:

$$\text{creatine} \sim P + ADP \longrightarrow ATP + \text{creatine}$$

Oxygen Debt

When all of the creatine phosphate has been depleted and there is no oxygen available for aerobic respiration, a muscle fiber can generate ATP by using fermentation. Fermentation, which is apt to occur during strenuous exercise, can supply ATP for only a short time because lactic acid buildup produces muscular aches and fatigue.

We have all had the experience of having to continue deep breathing for a few minutes following strenuous exercise. This continued intake of oxygen is required to complete the metabolism of lactic acid that has accumulated during exercise and represents an **oxygen debt** that the body must pay to rid itself of lactic acid. The lactic acid is transported to the liver, where one-fifth of it is completely broken down to carbon dioxide and water by means of the Krebs cycle and respiratory chain. The ATP gained by this respiration is then used to convert the remaining four-fifths of the lactic acid back to glucose.

Muscle contraction requires a ready supply of ATP. Creatine phosphate is used to generate ATP rapidly. If oxygen is in limited supply, fermentation produces ATP and results in oxygen debt.

Contraction and Relaxation Process

Nerves innervate muscles, and nerve impulses cause muscles to contract. The details of the process are as follows: a motor axon within a nerve has branches that go to several muscle fibers, which collectively are termed a *motor unit*. Each branch has terminal end knobs that contain synaptic vesicles filled with the neuromuscular transmitter ACh (acetylcholine). A terminal end knob lies in close proximity to the sarcolemma of the muscle fiber. This region, called a **neuromuscular junction** (fig. 40.14) has the same components as a synapse: a presynaptic membrane, a synaptic cleft, and a postsynaptic membrane. Only in

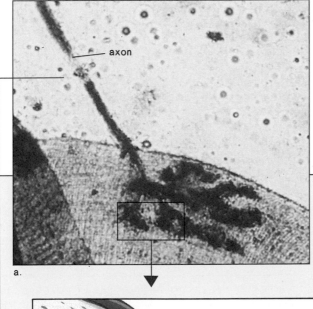

axon

a.

Figure 40.14

Muscle fiber innervation. *a.* Micrograph of a motor unit showing several terminal knobs. *b.* A neuromuscular junction occurs where a terminal end knob comes in close proximity to a muscle fiber. The knob contains synaptic vesicles (enlargement, *c*) filled with ACh. When these vesicles fuse with the presynaptic membrane, ACh diffuses across the synaptic cleft to the sarcolemma. If the stimulus is sufficient a muscle action potential begins.

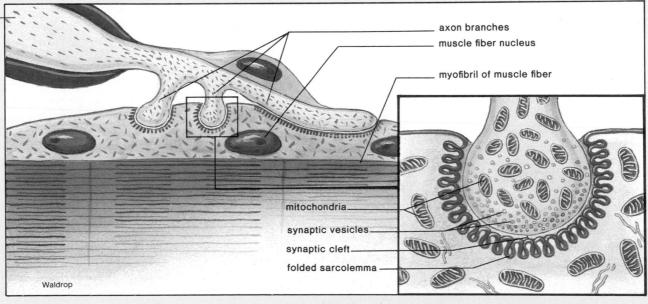

axon branches
muscle fiber nucleus
myofibril of muscle fiber
mitochondria
synaptic vesicles
synaptic cleft
folded sarcolemma

Waldrop

b.

c.

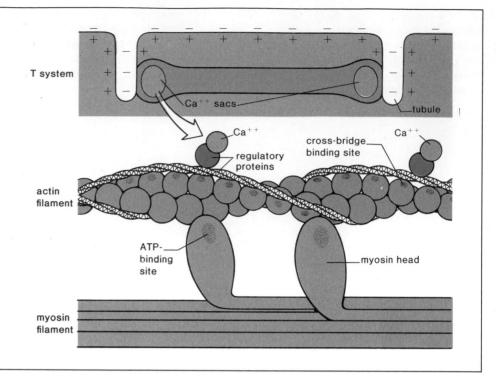

Figure 40.15

Detailed structure and function of muscle fiber contraction. After CA++ is released from its storage sacs, it combines with troponin, a protein that occurs periodically along tropomyosin threads on the actin filament. This causes the tropomyosin threads to shift their position so that cross-bridge binding sites are revealed along the actin filaments. The myosin filament extends its globular heads, forming cross bridges that bind to these sites. The breakdown of ATP by myosin causes the cross bridges to detach and reattach farther along the actin. In this way, the actin filaments are pulled along past the myosin filaments.

this case, the postsynaptic membrane is a portion of the sarcolemma of a muscle fiber. The sarcolemma, just like a neural membrane, is polarized; the inside is negatively charged and the outside is positively charged.

Nerve impulses cause synaptic vesicles to fuse with the presynaptic membrane and release ACh into the synaptic cleft. When ACh reaches the sarcolemma, the sarcolemma is depolarized. The result is a muscle action potential that spreads over the sarcolemma and down the T system (fig. 40.12a) to the calcium-storage sacs of the sarcoplasmic reticulum. When the action potential reaches a sac, Ca++ ions are released and diffuse into the sarcoplasm, after which they attach to the thin filaments. A thin filament is a twisted double strand of globular actin molecules. Associated with the actin filaments are two other proteins: *tropomyosin* forms threads that twist about the actin filaments, and *troponin* is located at intervals along the tropomyosin. The calcium combines with the troponin and this causes the tropomyosin threads to shift in position so that binding sites for myosin on the actin filament are exposed (fig. 40.15).

The thick myosin filament is a bundle of myosin molecules, each having a globular head. These heads form the cross bridges that attach to the binding sites on actin. Also, as the myosin molecules break down ATP, detachment and reattachment to a site farther along occurs. In this way actin filaments are pulled past the myosin filament.

Contraction ceases when nerve impulses no longer stimulate the muscle fiber. With the cessation of a muscle action potential, Ca++ ions are pumped back into their storage sacs by active transport. Relaxation now occurs.

A neuromuscular junction functions like a regular nerve synapse, except that a muscle action potential causes calcium to be released from calcium-storage sacs, and thereafter muscle contraction occurs.

Summary

1. The three types of skeletons that are found in the animal kingdom are the hydrostatic skeleton (flatworms and segmented worms), the exoskeleton (arthropods), and the endoskeleton (vertebrates).

2. A rigid skeleton gives support to the body, helps protect internal organs, and assists movement. In humans, the skeleton is also a storage area for calcium and phosphorus and the site of blood cell production.

3. The human skeleton is divided into two parts: (1) the axial skeleton, which is made up of the skull, ribs, and vertebrae; and (2) the appendicular skeleton, which is composed of the appendages and their girdles.

4. A longitudinal section of a long bone shows three types of tissue. Cartilage occurs at the ends of the bone; it is composed of cells within lacunae separated by a flexible matrix. Spongy bone containing red marrow occurs at the ends beneath the cartilage, and compact bone containing Haversian systems makes up the shaft of the bone. There is also a central cavity containing yellow marrow where fat is stored.

5. Bone is constantly being renewed; osteoclasts break down bone and osteoblasts build new bone. Osteocytes are in the lacunae of Haversian systems.

6. Joints are classified as: immovable, like those of the cranium; slightly movable, like those between the vertebrae; and freely movable synovial joints, like those in the knee and hip. In synovial joints, ligaments bind the two bones together forming a capsule in which there is synovial fluid.

7. There are three types of muscles in vertebrates: smooth muscle contains spindle-shaped cells, is involuntary, and is found in the walls of most hollow organs. Cardiac muscle, which makes up the heart, is striated and branched, and involuntary. Skeletal muscle is striated and voluntary.

8. Skeletal muscles can only pull bones; they cannot push. Therefore, they work in antagonistic pairs.

9. Muscle fibers contain a T system: the sarcolemma forms tubules that contact the sarcoplasmic reticulum. Here calcium ions are stored in storage sacs.

10. The sarcoplasmic reticulum encases myofibrils, long contractile units having longitudinal units called sarcomeres. The anatomy of the sarcomere shows the placement of the protein filaments actin and myosin. It is the placement of these that results in the striations of skeletal muscle. Actin (thin) filaments are attached to the Z line, and both actin and myosin (thick) filaments are in the A band. The H zone contains only myosin.

11. The sliding filament theory of muscle contraction says that the myosin filaments have cross bridges that attach to and detach from the actin filaments causing them to slide and the sarcomere to shorten. The H zone disappears as the actin filaments approach one another.

12. ATP is the only source of energy for muscle contraction. The globular heads of myosin break down ATP when they detach from the actin filament. Creatine phosphate is a storage form of high-energy phosphate in muscle tissue. When there is no more freely available ATP, the organism goes into oxygen debt if fermentation occurs in the muscle.

13. Nerve innervation of a muscle fiber begins at a neuromuscular junction. Here synaptic vesicles release ACh into the synaptic cleft. When the sarcolemma receives the ACh, a muscle action potential moves down the T system to the calcium-storage sacs.

14. After calcium is released, contraction occurs because calcium combines with troponin, a protein that occurs on tropomyosin threads on the myosin filament. This causes the threads to shift their position so that myosin binding sites on actin become available. When calcium is actively transported back into the storage sacs, muscle relaxation occurs.

Objective Questions

For questions 1 to 4, match the items below with the correct locations given in the key.

Key:
- a. upper arm
- b. lower arm
- c. pectoral girdle
- d. pelvic girdle
- e. upper leg
- f. lower leg

1. ulna
2. tibia
3. clavicle
4. femur
5. Spongy bone
   a. contains Haversian systems.
   b. contains red marrow where blood cells are formed.
   c. lends no strength to bones.
   d. All of these.

6. Haversian systems
   a. are found in compact bone.
   b. are subject to being broken down by osteoclasts and rebuilt by osteoblasts.
   c. contain calcium and phosphorus.
   d. All of these.
7. Which of these is mismatched?
   a. slightly movable joint—vertebrae
   b. hinge joint—hip joint
   c. synovial joint—elbow
   d. immovable joint—sutures in cranium
8. Skeletal muscle is
   a. involuntary.
   b. striated.
   c. branched.
   d. All of these.
9. In a muscle fiber
   a. the sarcolemma is connective tissue holding the myofibrils together.
   b. the T system contains calcium storage sacs.

   c. both filaments have cross bridges.
   d. All of these.
10. When muscles contract
    a. sarcomeres increase in size.
    b. myosin slides past actin.
    c. the H zone disappears.
    d. calcium is taken up by calcium storage sacs.
11. Which of these is a source of energy for muscle contraction?
    a. ATP
    b. creatine phosphate
    c. lactic acid
    d. Both (a) and (b).
12. Nervous stimulation of muscles
    a. occurs at a neuromuscular junction.
    b. results in an action potential that travels down the T system.
    c. causes calcium to be released from storage sacs.
    d. All of these.

Study Questions

1. What are the three types of skeletons found in the animal kingdom, how do they differ, and what are some animals having each type?
2. Give several functions of the skeletal system in humans.
3. Distinguish between the axial and appendicular skeletons.

4. List the bones that form the pectoral and pelvic girdles.
5. Describe the anatomy of a long bone, and tell how bone is continually being rejuvenated.
6. Name the different types of joints, and give an example of each type.

7. What are the three types of muscles in vertebrates, and how do they differ?
8. Why do muscles act in antagonistic pairs?
9. Describe the microscopic anatomy of a muscle fiber and the structure of a sarcomere. What is the sliding filament theory?

10. Give the function of each participant in the following reaction:

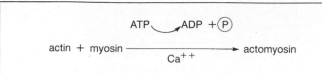

11. Discuss the availability and the specific role of ATP during muscle contraction. What is oxygen debt and how is it repaid?
12. What causes a muscle action potential? How does the muscle action potential bring about sarcomere and muscle fiber contraction?

Thought Questions

1. What is the value of having a jointed skeleton, and give examples by referring to different types of joints?

2. Muscles vary in their bulk. Why do the muscles that have the most bulk move the appendages?

3. Begin with a whole muscle and then list each smaller structure until you finally arrive at troponin. Compare your results with that of another student—are your lists the same?

Selected Key Terms

hydrostatic skeleton (hi″dro-stat′ik skel′ĕ-ton) 627
exoskeleton (ek″so-skel′ĕ-ton) 627
endoskeleton (en′do-skel′ĕ-ton) 627
axial skeleton (ak′se-al skel′ĕ-ton) 627
appendicular skeleton (ap″en-dik′u-lar skel′ĕ-ton) 627

cartilage (kar′tĭ-lij) 630
compact bone (kom′pakt bōn) 630
spongy bone (spun′je bōn) 631
synovial joint (sĭ-no′ve-al joint) 631
muscle fiber (mus′el fi′ber) 634
sarcolemma (sar″ko-lem′ah) 635
sarcoplasmic reticulum (sar″ko-plaz′mik rĕ-tik′u-lum) 635

myofibril (mi″o-fi′bril) 635
sarcomere (sar″ko-mēr) 636
actin (ak′tin) 636
myosin (mi′o-sin) 636
neuromuscular junction (nu″ro-mus′ku-lar junk′shun) 637

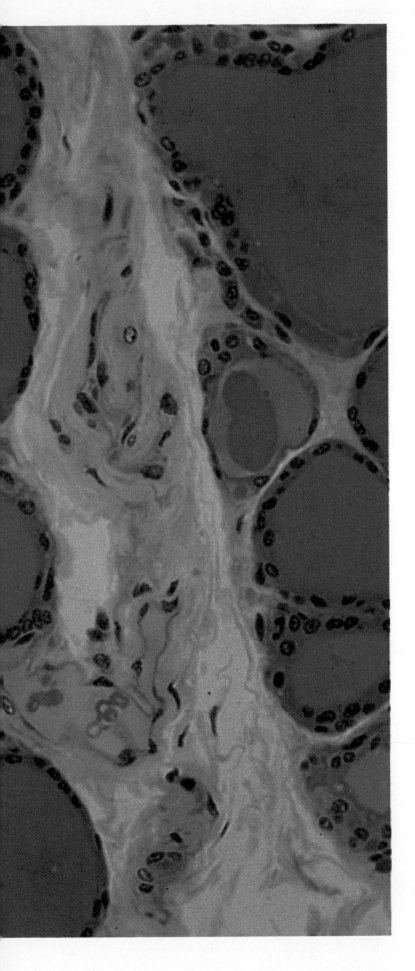

C H A P T E R   41

Endocrine System

Your study of this chapter will be complete when you can:

1. *Describe the location of each of the major endocrine glands of the human body.*
2. *Chemically distinguish between the two major types of hormones, and tell how each type brings about its effect on the cell.*
3. *Contrast the ways in which the hypothalamus controls the posterior and anterior pituitary.*
4. *List the most important hormones produced by each endocrine gland, and describe their most important effects.*
5. *Give examples of medical conditions associated with the overproduction or underproduction of specific hormones.*
6. *Explain the concept of control by negative feedback, and give an example that involves the hypothalamus, anterior pituitary, and a gland controlled by the anterior pituitary. Also give examples that do not involve this three-tiered system.*
7. *Discuss the relationship between the nervous system and the endocrine system. Give examples to show that there are regions of overlap between their functions.*
8. *Discuss the concept of chemical messengers in general, and group chemical messengers into three categories. Discuss why the term hormone could be broadened to include all these categories.*

The thyroid gland is composed of many hollow follicles filled with a protein rich fluid. The cells that line the follicles secrete thyroxin, the hormone produced by the thyroid gland.

Endocrine Glands

ertebrate animals have evolved two major systems, the nervous system (chapter 38) and the **endocrine system,** to coordinate the various activities of body parts. Both systems utilize chemical messengers to fulfill their functions, and in chapter 38, we discussed the neurotransmitters that are released by one neuron and influence the excitability of another. In this chapter, we will discuss the chemical messengers of the vertebrate endocrine system that are produced by the glands of internal secretion (fig. 41.1). These messengers are called **hormones** and are released directly into the bloodstream by these glands. Such glands, also called endocrine glands, can be contrasted to *exocrine* glands. Exocrine glands have ducts, and they secrete their products into these ducts for transport into body cavities; for example, the salivary glands send saliva into the mouth by way of the salivary ducts. Endocrine glands are ductless; they secrete their hormones directly into the bloodstream for distribution throughout the body.

The nervous system reacts quickly to external and internal stimuli; you will rapidly pull away your hand from a hot stove, for example. The endocrine system is slower because it takes time for a hormone to travel through the circulatory system to its target organ. You might think from the use of this terminology that the hormone is seeking out a particular organ but quite the contrary, the organ is awaiting the arrival of the hormone. Those cells that can react to a hormone have specific receptors that combine with the hormone in a lock-and-key manner. Therefore certain cells respond to one hormone and not to another depending on their receptors.

Figure 41.1

Anatomical location of the major endocrine glands in humans. The hypothalamus controls the anterior pituitary, which in turn controls the hormonal secretions of the thyroid, adrenal glands and gonads. The male gonads, the testis, are shown in the circle at the lower left.

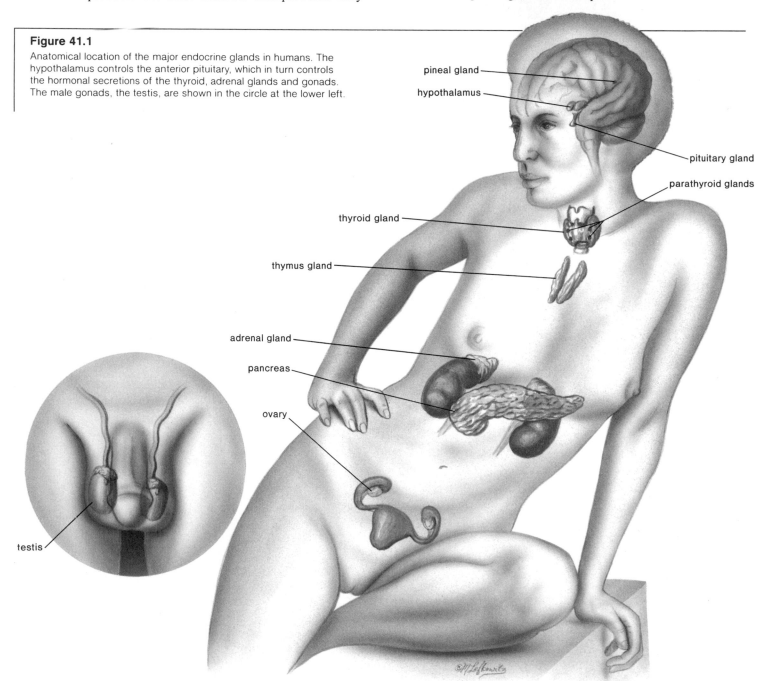

pineal gland

hypothalamus

pituitary gland

parathyroid glands

thyroid gland

thymus gland

adrenal gland

pancreas

ovary

testis

Sometimes a hormone is defined as a chemical produced by one set of cells that affects a different set. The problem with this, however, is that this particular definition would allow us to categorize all sorts of chemical messengers as hormones, even neurotransmitters! We will have more to say about this later but a certain amount of overlap between the nervous and endocrine systems is to be expected. The systems evolved together, and at the same time, no doubt making occasional use of the same chemical messengers and communicating not only with other systems but with each other as well. Later, we will give several examples of such associations between the two systems.

> Both the nervous and endocrine systems function to coordinate body parts and activities. Perhaps since they evolved at the same time, a certain amount of overlap between the chemical messengers utilized and the types of activities performed is to be expected.

Mechanism of Hormonal Action

Table 41.1 indicates that, among the vertebrates, most hormones are either *peptides, polypeptides,* or *proteins.* These molecules are coded for by genes and synthesized at the ribosomes (fig. 16.14). Eventually, they are packaged into vesicles at the Golgi complex and secreted at the plasma membrane (fig. 5.9). There are some other hormones, such as epinephrine, that are derived from the amino acid tyrosine; their production requires only a series of metabolic reactions within the cytoplasm.

From the viewpoint of hormonal action we can group together the hormones that are peptides, polypeptides, and proteins, and those derived from amino acids; we will call all of these peptide hormones.

The hormones produced by the adrenal cortex, the ovaries, and the testes (fig. 41.1) are *steroids.* Steroids are derived from cholesterol (fig. 4.13) by a series of metabolic reactions. These hormones are stored in fat droplets in the cell cytoplasm until their release at the plasma membrane.

There are some other types of molecules that also serve as chemical messengers but, in large part, the endocrine glands produce either peptides or steroids. These two types of molecules differ both in the way they are produced and also the way they act (fig. 41.2).

Peptide Hormones

The mode of action of these hormones was discovered in the 1950s by Earl W. Sutherland, who studied the effects of epinephrine on liver cells. Sutherland received a Nobel Prize for his hypothesis that when this hormone binds to a cell-surface receptor, the resulting complex activates an enzyme that produces cyclic AMP (fig. 41.2a). **Cyclic AMP** (cAMP) is a compound made from ATP but it contains only one phosphate group that is attached to adenosine at two locations. It promotes a whole series of chemical reactions within the cell.

We now know that most peptide hormones work in this manner. Since peptide hormones never enter the cell they are sometimes called the *first messengers,* while cAMP, which sets the metabolic machinery in motion, is called the *second mes-*

Figure 41.2

Cellular activity of hormones. *a.* Peptide hormones combine with receptors located in the plasma membrane. This promotes the production of cAMP, which in turn leads to activation of a particular enzyme. *b.* Steroid hormones pass through the plasma membrane to combine with receptors; the complex activates certain genes, leading to protein synthesis.

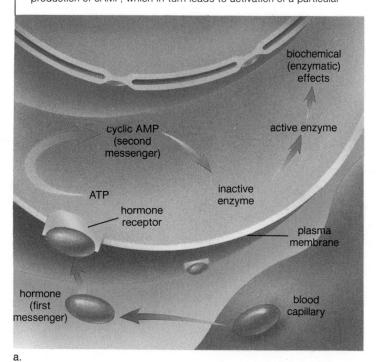

a.

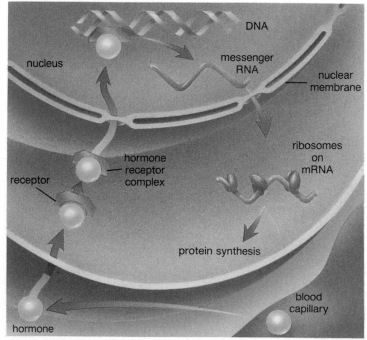

b.

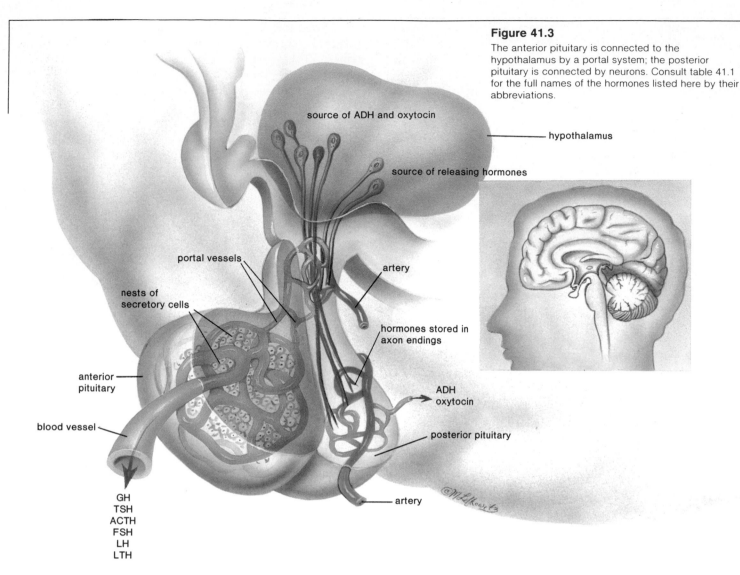

Figure 41.3
The anterior pituitary is connected to the hypothalamus by a portal system; the posterior pituitary is connected by neurons. Consult table 41.1 for the full names of the hormones listed here by their abbreviations.

source of ADH and oxytocin

hypothalamus

source of releasing hormones

portal vessels

artery

nests of
secretory cells

hormones stored in
axon endings

anterior
pituitary

ADH
oxytocin

blood vessel

posterior pituitary

GH
TSH
ACTH
FSH
LH
LTH

artery

senger (fig. 41.2*a*). Cyclic AMP activates only one particular enzyme in the cell but this enzyme in turn activates another, and so forth; in other words a series of enzymatic reactions, an *enzyme cascade,* is set in motion. The activated enzymes, of course, can be used over and over again. Therefore, at every step in the enzyme cascade more and more reactions take place—the binding of a single hormone molecule eventually results in a thousandfold response. We should also note that the same kind of peptide hormone can cause different responses in different cells because each type of cell has its own metabolic machinery.

Steroid Hormones

Steroid hormones do not bind to cell surface receptors; they can enter the cell freely. For example, scans of the reproductive organs clearly show that the hormones estrogen and progesterone enter the cells of the reproductive organs. Once inside, steroid hormones bind to receptors in the cytoplasm. The hormone-receptor complex then enters the nucleus, where it binds with chromatin at a location that promotes activation of particular genes. Transcription of DNA and translation of mRNA follow. In this manner, steroid hormones lead to protein synthesis.

Steroids act more slowly than peptides do because it takes more time to synthesize new proteins than to activate enzymes that are already present in the cell. But steroids have a more *sustained* effect on the metabolism of the cell than do peptide hormones.

Hormones are chemical messengers that influence the metabolism of the receiving cell. Peptide hormones activate existing enzymes in the cell, and steroid hormones bring about the synthesis of new proteins.

Hypothalamus and the Pituitary Gland

The hypothalamus is a portion of the brain (figs. 38.11 and 41.1) that regulates the internal environment. For example, it helps to control the heart rate, body temperature, and water balance, as well as the glandular secretions of the stomach and of the **pituitary gland.** The pituitary, a small gland (fig. 41.3) of about 1 cm in diameter, lies just below the hypothalamus and is divided into two portions called the posterior pituitary and the anterior pituitary.

Table 41.1 Vertebrate Endocrine Glands and Their Hormones

Endocrine Gland	Hormone Released	Chemical Structure of Hormone	Chief Function of Hormone
Hypothalamus	Hypothalamic-releasing hormones	Peptides	Regulate anterior pituitary hormones
Anterior pituitary	Thyroid-stimulating (TSH, thyrotropic)	Glycoprotein	Stimulates thyroid
	Adrenocorticotropic (ACTH)	Polypeptide	Stimulates adrenal cortex
	Gonadotropic follicle-stimulating (FSH)	Glycoprotein	Controls egg and sperm production
	luteinizing (LH)		Controls sex hormone production
	Lactogenic (LTH, prolactin)	Protein	Stimulates milk production and secretion
	Growth (GH, somatotropic)	Protein	Stimulates cell division, protein synthesis, and bone growth
	Melanocyte-stimulating (MSH)	Polypeptide	Regulates skin color in lower vertebrates
Posterior pituitary (storage of hypothalamic hormones)	Antidiuretic (ADH, vasopressin)	Peptide	Stimulates water reabsorption by kidneys
	Oxytocin	Peptide	Stimulates uterine muscle contraction and release of milk by mammary glands
Thyroid	Thyroxin and triiodothyronine	Iodinated amino acid	Increases metabolic rate, helps regulate growth and development
	Calcitonin	Peptide	Lowers blood calcium level
Parathyroids	Parathyroid (PTH)	Polypeptide	Raises blood calcium level
Adrenal cortex	Glucocorticoids (cortisol)	Steroids	Raise blood glucose level; stimulate breakdown of protein
	Mineralocorticoids (aldosterone)	Steroids	Stimulate kidneys to reabsorb sodium and excrete potassium
	Sex hormones	Steroids	Stimulate development of secondary sex characteristics (particularly in male)
Adrenal medulla	Epinephrine and norepinephrine	Catecholamines	Stimulate fight or flight reactions; raise blood glucose level
Pancreas	Insulin	Polypeptide	Lowers blood glucose level; promotes formation of glycogen, proteins, and fats
	Glucagon	Polypeptide	Raises blood glucose level; promotes breakdown of glycogen, proteins, and fats
Gonads			
Testes	Androgens (testosterone)	Steroids	Stimulate spermatogenesis, develop and maintain secondary male sex characteristics
Ovaries	Estrogen and progesterone	Steroids	Stimulate growth of uterine lining; develop and maintain secondary female sex characteristics
Thymus	Thymosins	Peptides	Stimulate production and maturation of T-lymphocytes
Pineal gland	Melatonin	Catecholamine	Involved in circadian and circannual rhythms; possibly involved in maturation of sex organs
Gut*	Gastrin	Peptide	Stimulates secretion of gastric juices
	Secretin	Peptide	Stimulates secretion of pancreatic juices
	Cholecystokinin-pancreazymin (CCK-PZ)	Peptide	Stimulates secretion of pancreatic juices and flow of bile from gallbladder

*The hormones produced by the gut are discussed in chapter 35

Insect Growth Hormones

This chapter mainly considers vertebrate hormones but all animals use chemical messengers to regulate the activities of their cells. These messengers are now being identified in a wide variety of invertebrate animals. The work on insect growth hormones began in the 1930s, and therefore they have been studied for some time.

Insects, like other arthropods, have an exoskeleton that hinders growth unless it is shed periodically. Typically, the larvae of insects molt (shed their skeleton) a number of times before they pupate and metamorphose into an adult (*see* figure). V. B. Wigglesworth, who performed experiments on a bloodsucking bug (*Rhodnius*) in the 1930s, showed that the brain was necessary for maturation to take place because it produced a hormone appropriately called *brain hormone*. Brain hormone stimulates the prothoracic gland, which lies in the region just behind the head, to secrete *ecdysone*, a steroid hormone. This hormone is also called the molt and maturation hormone because it promotes both molting and maturation. During larval stages, however, response to ecdysone is modified by the action of *juvenile hormone*. If juvenile hormone is present, the insect molts into another larval form; if it is minimally present, the insect pupates; and if it is absent, the insect undergoes metamorphosis into an adult.

Some plants produce compounds that are similar or identical to either ecdysone or juvenile hormone. These compounds apparently protect the plants by disrupting the development of the insects that feed upon them. Work is underway to extract these compounds and to use them as insecticides.

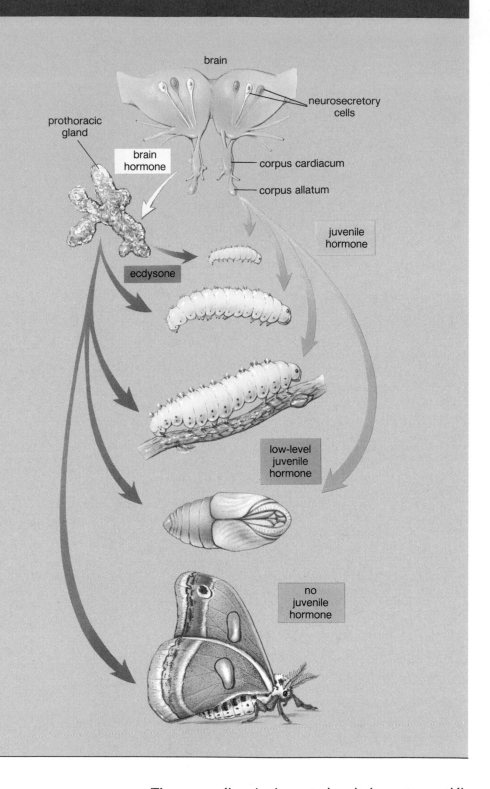

Three hormones are involved in insect development: brain hormone controls the secretion of ecdysone, the molt and maturation hormone, which causes the insect to pupate if the level of juvenile hormone is low, and to metamorphose if juvenile hormone is absent.

Posterior Pituitary

The two parts of the pituitary are connected to the hypothalamus by a means of a stalk. Neurons in the hypothalamus produce the hormones that are stored in and released from the posterior pituitary. The hormones pass from the hypothalamus through axons that terminate in the posterior pituitary (fig. 41.4).

The axon endings in the posterior pituitary store antidiuretic hormone (ADH), sometimes called vasopressin, and oxytocin. **ADH,** as discussed in chapter 37, promotes the reabsorption of water from the collecting duct, a portion of the kidney tubules. The hypothalamus contains other nerve cells that are sensitive to the osmolarity of the blood. When these cells determine that the blood is too concentrated, ADH is released into

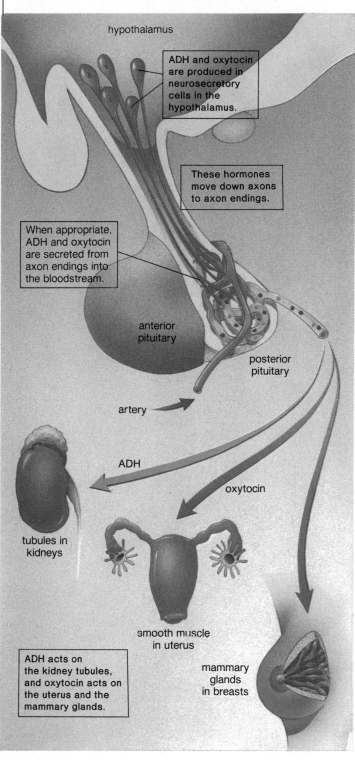

Figure 41.4
The hypothalamus produces the two hormones, ADH and oxytocin, that are stored in and secreted by the posterior pituitary.

hypothalamus

ADH and oxytocin are produced in neurosecretory cells in the hypothalamus.

These hormones move down axons to axon endings.

When appropriate, ADH and oxytocin are secreted from axon endings into the bloodstream.

anterior pituitary

posterior pituitary

artery

ADH

oxytocin

tubules in kidneys

smooth muscle in uterus

mammary glands in breasts

ADH acts on the kidney tubules, and oxytocin acts on the uterus and the mammary glands.

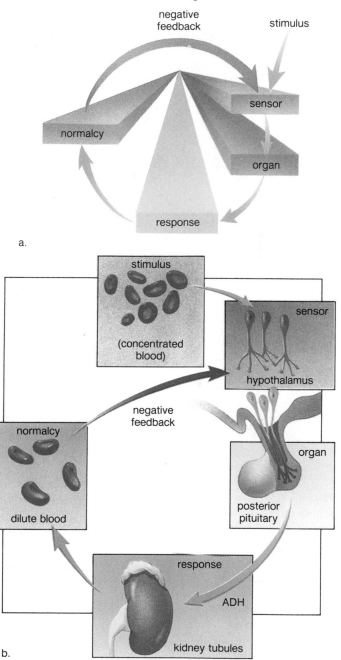

Figure 41.5
Control by negative feedback. *a.* The sensor detects a change in the environment and communicates with other cells to bring about a response. Once normalcy is achieved, the sensor is no longer stimulated to react. This is negative feedback because the activity of the organ ceases due to a product produced by that organ. *b.* As a specific example, there are neurons in the hypothalamus sensitive to osmolarity of the blood. When the blood is concentrated, they send signals to hypothalmic neurosecretory cells which release ADH from their axon endings. ADH increases the permeability of the collecting ducts in the kidneys so that more water is reabsorbed. Once the blood is diluted, ADH is no longer secreted.

negative feedback

stimulus

sensor

normalcy

organ

response

a.

stimulus

(concentrated blood)

sensor

hypothalamus

negative feedback

normalcy

organ

dilute blood

posterior pituitary

response

ADH

kidney tubules

b.

the bloodstream from the axon endings in the posterior pituitary. When the blood reaches the correct osmolarity and becomes dilute, the hormone is no longer released. This is an example of control by *negative feedback* (fig. 41.5) because the effect of the hormone (diluted blood) acts to shut down the release of the hormone. Negative feedback mechanisms regulate the activities of most endocrine glands.

Oxytocin is the other hormone that is made in the hypothalamus and stored in the posterior pituitary. Oxytocin causes the uterus to contract and is used to artificially induce labor. It also stimulates the release of milk from the mother's mammary glands when a baby is nursing.

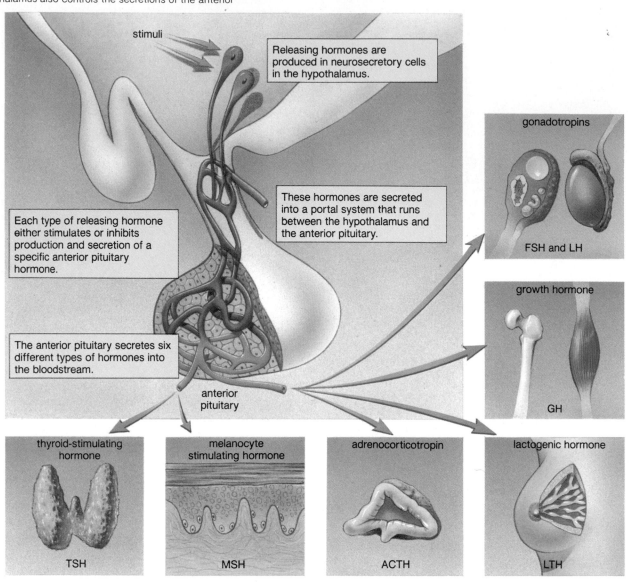

stimuli

Releasing hormones are produced in neurosecretory cells in the hypothalamus.

These hormones are secreted into a portal system that runs between the hypothalamus and the anterior pituitary.

Each type of releasing hormone either stimulates or inhibits production and secretion of a specific anterior pituitary hormone.

The anterior pituitary secretes six different types of hormones into the bloodstream.

anterior pituitary

gonadotropins

FSH and LH

growth hormone

GH

thyroid-stimulating hormone

TSH

melanocyte stimulating hormone

MSH

adrenocorticotropin

ACTH

lactogenic hormone

LTH

The neurons in the hypothalamus that produce and release ADH and oxytocin are unique in that they respond to nerve signals and yet they also produce a hormone. Therefore they are called *neurosecretory cells.* They are one example of the remarkable way the nervous system and the endocrine system are joined.

The posterior pituitary stores two hormones, ADH and oxytocin, both of which are produced by and released from neurons whose cell bodies are in the hypothalamus.

Anterior Pituitary

By the 1930s it had been determined that the hypothalamus also controlled the release of the anterior pituitary hormones. For example, electrical stimulation of the hypothalamus caused the release of anterior pituitary hormones, but *direct* stimulation of the anterior pituitary itself had no effect. Detailed anatomical studies showed that there was a *portal* system consisting of blood vessels connecting a capillary bed in the hypothalamus with one in the anterior pituitary (fig. 41.6). The hypothesis was put forward that a different set of neurosecretory cells in the hypothalamus synthesized a number of different releasing hormones that were sent to the anterior pituitary by way of the vascular portal system. There began a long and competitive struggle between two groups of investigators to identify one of these **"hypothalamic-releasing hormones"** that cause the anterior pituitary to release hormones. The group headed by R. Guillemin processed nearly 2 million sheep hypothalami, and the group headed by A. V. Schally processed more than 1 million pig hypothalami until each group announced, in November of 1969, that it had isolated and determined the structure of one of these hormones, a peptide

Figure 41.7

containing only three amino acids. To date, seven such releasing hormones have been identified; each of them causes the anterior pituitary either to secrete or to stop secreting a specific hormone that is produced in the anterior pituitary itself (table 41.1).

> The anterior pituitary is controlled by the hypothalamic-releasing hormones; they are produced in neurosecretory cells and pass to the anterior pituitary by way of a vascular portal system.

The anterior pituitary produces at least six different types of hormones (fig. 41.6), each by a distinct cell type. Three of these hormones have a direct effect on the body. **Growth hormone (GH),** or somatotropin, dramatically affects the physical appearance since it determines the height of the individual (fig. 41.7). If too little GH is produced during childhood, the individual becomes a pituitary dwarf and if too much is produced, the individual is a pituitary giant. In both instances, the individual has normal body proportions. However, on occasion, there is overproduction of growth hormone in the adult and a condition called *acromegaly* results. Since only the feet, hands, and face (particularly chin, nose, and eyebrow ridges) can respond, these portions of the body become overly large.

Growth hormone promotes cell division, protein synthesis, and bone growth. It stimulates the transport of amino acids into cells and increases the activity of ribosomes, both of which are essential to protein synthesis. In bones it promotes growth of the cartilaginous plates and causes osteoblasts to form bone. Evidence suggests that the effects on cartilage and bone may actually be due to hormones called somatomedins that are released by the liver. Growth hormone causes the liver to release somatomedins.

Lactogenic hormone (LTH), also called prolactin, is produced in quantity only after childbirth. It causes the mammary glands in the breasts to develop and produce milk. It also plays a role in carbohydrate and fat metabolism.

Melanocyte-stimulating hormone (MSH) causes skin color changes in lower vertebrates—it is present in humans but no one knows what its function might be. However, it is derived from a molecule that is also the precursor for both ACTH and the anterior pituitary endorphins. These endorphins are structurally and functionally similar to the endorphins produced in brain nerve cells.

Regulation of the Anterior Pituitary

The anterior pituitary is called the *master* gland because it controls the secretion of some other endocrine glands (fig. 41.6). As indicated in table 41.1, the anterior pituitary secretes the following hormones, which have an effect on other glands:

1. **TSH,** thyroid-stimulating hormone
2. **ACTH,** adrenocorticotropic hormone, a hormone that stimulates the adrenal cortex
3. **Gonadotropic hormones** (FSH and LH), which stimulate the gonads—the testes in males and the ovaries in females

TSH causes the thyroid to produce thyroxin; ACTH causes the adrenal cortex to produce cortisol; and gonadotropic hormones cause the gonads to secrete sex hormones. A three-tiered relationship exists between the hypothalamus, anterior pituitary, and other endocrine glands; the hypothalamus produces releasing hormones that control the anterior pituitary, and the anterior pituitary produces hormones that control the thyroid, adrenal cortex, and gonads. Figure 41.8 illustrates the negative feedback mechanism that controls the activity of these glands.

> The hypothalamus, anterior pituitary, and the other endocrine glands controlled by the anterior pituitary are all involved in a self-regulating negative feedback loop.

Thyroid Gland

The **thyroid gland** (fig. 41.9a) is a large gland that is located in the neck where it is attached to the trachea just below the larynx. The two hormones produced by the thyroid both contain iodine: **thyroxin,** or T_4, contains four atoms of iodine; it is secreted in greater amount but is less potent than triiodothyronine, or T_3,

Figure 41.8
TRH (thyroid releasing hormone) stimulates the anterior pituitary and TSH (thyroid stimulating hormone) stimulates the thyroid to secrete thyroxin. The level of thyroxin in the body is controlled by negative feedback: (a) the level of TSH exerts negative feedback control over the hypothalamus; (b) the level of thyroxin exerts feedback control over the anterior pituitary; and (c) the level of thyroxin exerts feedback control over the hypothalamus. In this way, thyroxin controls its own secretion. Cortisol and sex hormone levels are controlled in similar ways.

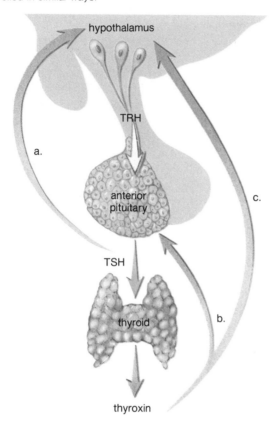

Figure 41.9
a. The thyroid gland is located in the neck in front of the trachea.
b. The four parathyroid glands are embedded in the posterior surface of the thyroid gland. Yet the parathyroids and thyroid glands have no anatomical or physiological connection with one another.
c. Regulation of parathyroid hormone secretion. A low blood level of calcium causes the parathyroids to secrete PTH, which causes the kidneys and gut to retain calcium and osteoclasts to break down bone. The end result is an increased level of calcium in the blood. A high blood level of calcium inhibits secretion of PTH.

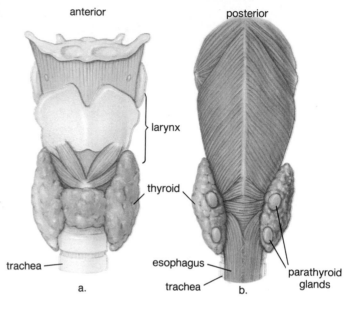

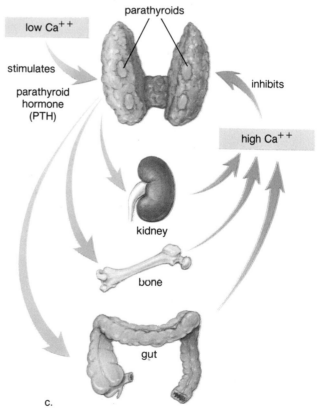

which has only three atoms of iodine. Iodine is actively transported into the thyroid gland and it may reach a concentration as much as 25 times greater than that of the blood (fig. 3.4).

Even before the structure of thyroxin was known, it was surmised that the hormone contained iodine because when iodine is lacking in the diet, the thyroid gland enlarges, producing a **goiter** (fig. 41.10). The cause of the enlargement becomes clear if we refer to figure 41.8. When there is a low level of thyroxin in the blood, the anterior pituitary is stimulated to produce TSH, thyroid-stimulating hormone. The thyroid responds by increasing in size but this increase in size is ineffective because active thyroxin cannot be produced without iodine. Later, it was found that the occurrence of goiter can be prevented by using iodized salt.

Thyroxin is necessary in vertebrates for proper growth and development. For example, without thyroxin, frogs do not metamorphose properly, and humans do not mature properly. Cretinism occurs in individuals who have suffered from *hypothyroidism* (low thyroid function) since birth. They show reduced skeletal growth, sexual immaturity, and abnormal protein metabolism. The latter leads to mental retardation.

Figure 41.10

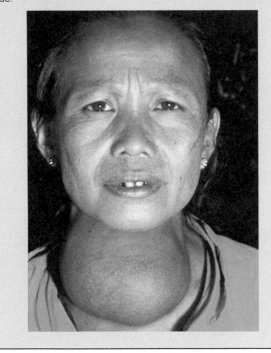

In general, thyroxin increases the metabolic rate in all cells; the number of respiratory enzymes increases, as does oxygen uptake. Hypothyroidism in adults produces the condition known as myxedema, which is characterized by lethargy, weight gain, loss of hair, slow pulse, and decreased body temperature. In *hyperthyroidism* (too much thyroxin), the individual becomes hyperactive, nervous, irritable, and suffers from insomnia. Usually, in this case the thyroid is enlarged, and the eyes protrude for some unknown reason. Therefore, this type of goiter is known as exophthalmic goiter. A common treatment is to remove a portion of the thyroid.

In addition to thyroxin, the thyroid gland also produces the hormone *calcitonin*. This hormone lowers the level of calcium in the blood and opposes the action of the parathyroid hormone, which we will discuss in the next section.

The anterior pituitary produces TSH, a hormone that promotes the production of thyroxin (and triiodothyronine) by the thyroid. Thyroxin helps regulate growth and development in immature animals and speeds up metabolism in all animals.

Parathyroid Glands

The four **parathyroid glands** are embedded in the posterior surface of the thyroid gland, as shown in figure 41.9b. They produce a hormone called **parathyroid hormone (PTH).** Under the influence of PTH, the calcium (Ca^{++}) level in the blood increases and the phosphate ($PO_4^=$) level decreases. PTH stimulates the *absorption* of calcium from the gut by activating

vitamin D, the *retention* of calcium (the excretion of phosphate) by the kidneys, and the *demineralization* of the bone by promoting the activity of the osteoclasts, the bone-reabsorbing cells. When the blood calcium level reaches the appropriate level, the parathyroid glands no longer produce PTH (fig. 41.9c).

If parathyroid hormone is not produced in response to low blood calcium, *tetany* results because calcium plays an important role in both nervous conduction and muscle contraction. In tetany, the body shakes from continuous muscle contraction. The effect is brought about by increased excitability of the nerves, which fire spontaneously and without rest.

Parathyroid hormone (PTH) maintains a high blood level of calcium. Its actions are opposed by calcitonin, which is produced by the thyroid.

Adrenal Glands

Each of the two **adrenal glands,** as the name implies (ad, near; renal, kidney) lie atop a kidney (fig. 41.1). Each consists of an inner portion, called the medulla, and an outer portion, called the cortex. These portions, like the anterior and posterior pituitary, have no connection with one another.

Adrenal Medulla

The adrenal medulla secretes **epinephrine** (adrenalin) and **norepinephrine** (noradrenalin) under conditions of stress. They bring about all those responses we associate with the "fight or flight" reaction: the blood glucose level and the metabolic rate increase, as do breathing and the heart rate. The blood vessels in the intestine constrict, and those in the muscles dilate. This increased circulation to the muscles causes them to have more stamina than usual. In times of emergency, the sympathetic nervous system *initiates* these responses but they are maintained by secretions from the adrenal medulla.

The adrenal medulla is innervated by just one neuron of the sympathetic nervous system. You'll recall that usually there are two neurons for each organ stimulated by the sympathetic nervous system. In this instance, what happened to the second one? It appears that the adrenal medulla may have evolved from a modification of the second neuron. Like the neurosecretory neurons in the hypothalamus, it also secretes hormones into the bloodstream.

The adrenal medulla releases epinephrine and norepinephrine into the bloodstream. These hormones help us and other animals cope with situations that threaten survival.

Adrenal Cortex

Although the adrenal medulla may be removed with no ill effects, the adrenal cortex is absolutely necessary to life. The two major classes of hormones made by the adrenal cortex are the *glucocorticoids* and the *mineralocorticoids*. The cortex also secretes a small amount of male and an even smaller amount of female sex hormones. All of these hormones are steroids.

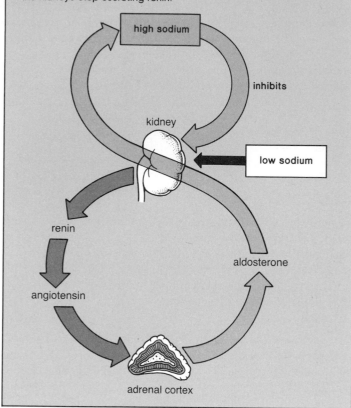

Figure 41.12
Insulin and glucagon have contrary effects. When blood glucose level is high, the pancreas secretes insulin. Insulin promotes the storage of glucose as glycogen, the use of glucose as an energy source, and the synthesis of proteins and fats as opposed to their use as energy sources. Therefore insulin lowers the blood glucose level. When the blood glucose level is low, the pancreas secretes glucagon. Glucagon acts in opposition to insulin in all respects; therefore glucagon raises the blood glucose level.

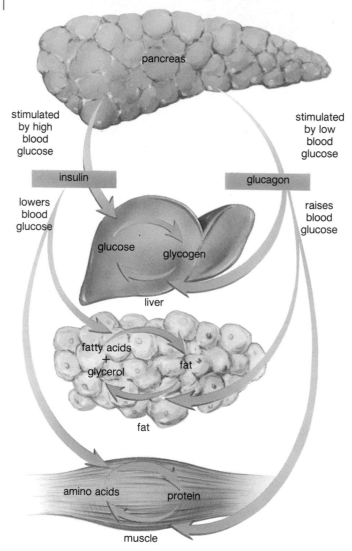

Of the various glucocorticoids, the hormone responsible for the greatest amount of activity is **cortisol.** Cortisol promotes the hydrolysis of muscle protein to amino acids which enter the blood. This leads to an increased level of glucose when the liver converts these amino acids to glucose. Cortisol also favors metabolism of fatty acids rather than carbohydrates. In opposition to insulin, therefore, cortisol raises the blood glucose level. Cortisol also counteracts the inflammatory response, which leads to the pain and swelling of joints in arthritis and bursitis. The administration of cortisol aids these conditions because it reduces inflammation.

The secretion of cortisol by the adrenal cortex is under the control of the anterior pituitary hormone ACTH. Using the same kind of system shown in figure 41.8, the hypothalamus produces a releasing hormone (CRH) that stimulates the anterior pituitary to release ACTH. ACTH in turn stimulates the adrenal cortex to secrete cortisol, which regulates its own synthesis by negative feedback of both CRH and ACTH synthesis.

The secretion of mineralocorticoids, the most significant of which is **aldosterone,** is not under the control of the anterior pituitary. Aldosterone regulates the level of sodium (Na$^+$) and potassium (K$^+$) ions in the blood, its primary target organ being the kidney where it promotes renal absorption of sodium and renal excretion of potassium. The level of sodium is particularly important to the maintenance of blood pressure, and the concentration of this ion indirectly regulates the secretion of aldosterone. When the blood level of sodium is low, the kidneys secrete renin. Renin is an enzyme that converts the plasma protein angiotensinogen to angiotensin, and this molecule stimulates the adrenal cortex to release aldosterone (fig. 41.11). This is called the renin-angiotensin-aldosterone system. The effect of this system raises the blood pressure in two ways. First, angiotensin constricts the arteries directly and secondly, when aldosterone causes the kidneys to reabsorb sodium, blood volume is raised as water is reabsorbed.

Cortisol, which raises the blood glucose level, and aldosterone, which raises the blood sodium level, are two hormones secreted by the adrenal cortex.

Figure 41.13

The pancreas is both an exocrine and an endocrine gland. It sends digestive juices to the duodenum by way of the pancreatic duct and the hormones insulin and glucagon into the bloodstream. In 1920, Frederick Banting, a physician who gave occasional physiology lectures decided to try to isolate insulin. Investigators before him had been unable to do this because the enzymes in the digestive juices destroyed insulin (a protein) during the isolation procedure. He hit upon the idea of tying off the pancreatic duct which he knew from previous research would lead to the degeneration only of the cells that produce digestive juices and not of the islets of Langerhans where insulin is made. J. J. Macleod made a laboratory available to him at the University of Toronto and also assigned a graduate student named Charles Best to assist him. Banting and

Best had limited funds and spent the summer working, sleeping, and eating in the lab. By the end of the summer they had obtained pancreatic extracts that did lower the blood glucose level in diabetic dogs. Macleod then brought in biochemists who purified the extract; and insulin therapy for the first human patient was begun in January 1922. Large-scale production of purified insulin from pigs and cattle followed; Banting and Macleod received a Nobel Prize in 1923. The amino acid sequence of insulin was determined in 1953. Insulin is presently synthesized using recombinant DNA technology (chapter 18). Banting and Best followed the required steps given below to identify a chemical messenger.

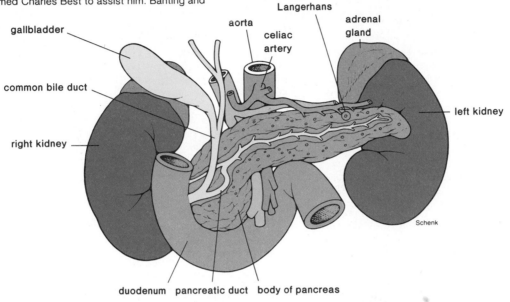

Steps	Example
Identify the source of the chemical	Pancreas is source
Identify the effect to be studied	Presence of pancreas in body lowers blood sugar
Isolate the chemical	Insulin isolation from pancreatic secretions
Show that the chemical alone has the effect	Insulin alone lowers blood sugar

Two medical conditions are associated with a malfunctioning of the adrenal cortex. In *Addison's disease,* the adrenal cortex is underactive. There is a reduction in blood pressure and blood glucose levels, and an unexplained, peculiar bronzing of the skin. In *Cushing's syndrome* the adrenal cortex is overactive. The muscles of the arms and legs waste away, and the blood glucose level and the blood pressure rise. Hypertension develops.

Pancreas

The **pancreas** is a long organ that lies transversely in the abdomen (fig. 41.13) between the kidneys and near the duodenum of the small intestine. It is composed of two types of tissues—one of these produces and secretes *digestive* juices that go by way of the pancreatic duct to the small intestine; the other type, called the *islets of Langerhans,* produces and secretes the hormones **insulin** and **glucagon** directly into the blood.

Insulin is secreted when there is a high level of glucose in the blood, which usually occurs just after eating. Insulin has three different actions: (1) It stimulates fat, liver, and muscle cells to take up and metabolize glucose; (2) It stimulates the liver and muscles to store glucose as glycogen; and (3) It promotes the build up of fats and proteins and inhibits their use as an energy source. Therefore, insulin is a hormone that promotes storage of nutrients so that they will be on hand during leaner times. It also helps to lower the blood glucose level. Glucagon is secreted from the pancreas in between eating, and its effects are opposite to those of insulin; glucagon stimulates the breakdown of stored nutrients and causes the blood sugar level to rise (fig. 41.12).

Diabetes Mellitus

Diabetes mellitus is a fairly common disease caused by a defect in insulin production or utilization. The most common laboratory test for **diabetes mellitus** (usually called diabetes) is to look

Anabolic Steroid Use

Anabolic steroids are essentially the male sex hormone testosterone and its synthetic derivatives. The latter were developed in the 1930s to prevent the atrophy of muscle tissue in patients with debilitating illnesses. In some cases, they are also given to burn victims and surgery patients to speed recovery. Steroids were first used for nonmedical reasons in World War II, when German doctors gave them to soldiers, perhaps in order to make them more aggressive. Following the war, the Soviet Union began dispensing steroids to athletes, and when the results were observed by a U.S. physician he recommended that the U.S. weightlifting team use them. In the 1960s, when their dangerous side effects became apparent, (*see* figure) steroid therapy fell into disfavor, and the Olympic Committee added anabolic steroids to its list of banned substances in 1973. Although the FDA (Federal Drug Administration) has also banned most steroids except for limited medical use, they are not *controlled* substances like cocaine

and heroin. Presently, they are smuggled into the United States from Mexico and to a lesser degree, Eastern Europe; this illicit trafficking has become a $100 million-a-year business. Recently Congress passed an anti-drug bill that makes the selling of steroids a felony offense with jail terms of three years for dispensing to adults and six years to children.

There is great concern because these drugs, originally taken primarily by professional athletes, are now being more widely used, even by children, to increase muscle size. The latest estimate is that one to three million Americans use steroids, a figure that has increased steadily since the early 1970s. No one has really conclusively determined the long-term harmful effects of steroid use because researchers are reluctant to do such studies on humans. The accompanying figure indicates the major side effects that have been observed in users. Often these side effects can be understood in view of the fact that testosterone is the

male sex hormone that maintains the secondary sex characteristics of males, such as the development of the testes and penis, and the deep voice and facial hair. Anabolic steroid use can also apparently cause death. Steroids have been implicated in the death of young athletes from liver cancer and a type of kidney tumor. Furthermore, because they cause the body to retain fluid, users often take diuretics; if they take excessive amounts that can rob the body of its proper electrolyte balance they may suffer a heart attack.

Recently, it has come to light that steroids may be addictive and might cause psychological problems. Psychotic side effects, sometimes called "roid mania," have been observed in abusers who experience wild aggression and delusions. For example, it has been reported that one user had a friend videotape him while he deliberately drove a car into a tree at 35 mph. Psychiatrists feel that the magnitude of the mental effects of steroids is only now being realized.

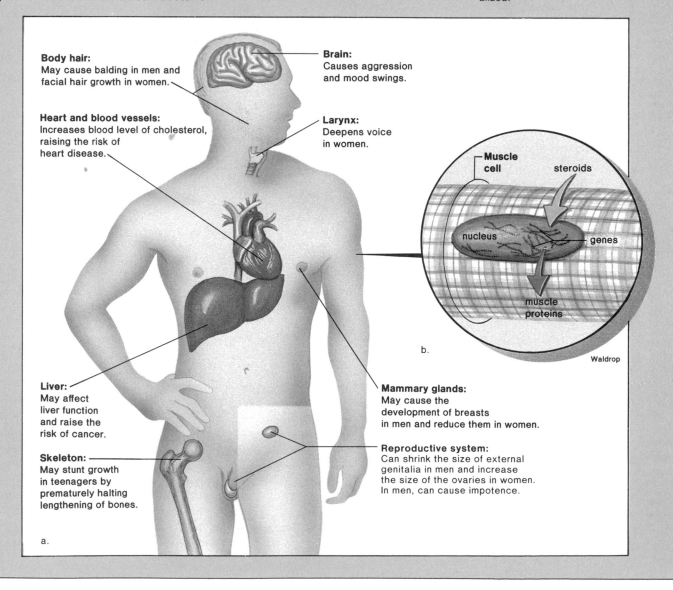

Body hair:
May cause balding in men and facial hair growth in women.

Heart and blood vessels:
Increases blood level of cholesterol, raising the risk of heart disease.

Liver:
May affect liver function and raise the risk of cancer.

Skeleton:
May stunt growth in teenagers by prematurely halting lengthening of bones.

Brain:
Causes aggression and mood swings.

Larynx:
Deepens voice in women.

Mammary glands:
May cause the development of breasts in men and reduce them in women.

Reproductive system:
Can shrink the size of external genitalia in men and increase the size of the ovaries in women. In men, can cause impotence.

Muscle cell   steroids

nucleus   genes

muscle proteins

Waldrop

a.

b.

for sugar, or glucose, in the urine. If there is glucose in the urine, it means that the blood glucose level is high enough to cause the kidneys to excrete glucose. This indicates that the liver is not storing glucose as glycogen and that cells are using proteins and fats as an energy source due to insufficient production or use of insulin. An individual with untreated diabetes will also be extremely thirsty because the solute content of the blood will be higher than normal. The breakdown of proteins and fat also leads to rapid weight loss, to the buildup of acids in the blood (acidosis), and to respiratory distress. It is acidosis that can eventually cause coma and death if the diabetic does not receive treatment.

There are two forms of diabetes. In type I diabetes, the pancreas does not produce insulin. This type of diabetes frequently follows a viral infection. Apparently, the infection sets off an autoimmune reaction that destroys the insulin-secreting cells in the islets of Langerhans. Patients with type I diabetes must either have daily insulin injections or receive a pancreas transplant. In type II diabetes, the most common type, the pancreas is producing insulin, but the cells of the body do not respond to it. At first, the cells lack the receptors necessary to detect the presence of insulin, and then later they become incapable of taking up glucose. The best treatment for this type of diabetes is a low-fat and low-sugar diet and regular exercise. If that fails, there are oral drugs that make the cells more sensitive to the effects of insulin or stimulate the pancreas to make more of it.

The pancreas secretes insulin, which acts to promote storage of nutrients, thereby lowering the blood glucose level. In diabetes mellitus, either insulin is not produced or it cannot be utilized; the excess glucose in the blood is excreted by the kidneys.

Other Endocrine Glands

Gonads

The gonads are the endocrine glands that produce the hormones that determine sexual characteristics. As we will discuss in detail in the following chapter, the *testes* produce the androgens—the most important one is called testosterone—which are the male sex hormones, and the *ovaries* produce estrogen and progesterone, the female sex hormones. The secretion of these hormones are under the control of the anterior pituitary.

The sex hormones bring about the secondary sex characteristics of males and females. Among other traits, males have greater muscle strength than do females. As discussed in the reading on page 654, athletes and others sometimes take so-called **anabolic steroids,** which are synthetic steroids that mimic the action of testosterone, in order to improve their strength and physique. Unfortunately, this practice is accompanied by harmful side effects.

Thymus

The *thymus* is a gland that extends from below the thyroid in the neck into the thoracic cavity. This organ grows during childhood but gradually decreases in size after puberty. Lymphocytes that have passed through the thymus are transformed into T cells. Colonies of T cells in the lymph nodes and other organs are apparently able to produce new T cells when stimulated by various thymus hormones called *thymosins*. There is hope that these hormones can be used in conjunction with lymphokine therapy to restore or stimulate T cell function in patients suffering from AIDS or cancer.

Pineal Gland

The *pineal gland* in amphibians and fish is located near the skin surface and is a "third eye" that transforms light directly into nerve impulses sent to the brain. In birds, the pineal gland, although located in the brain, still receives direct light stimulus; when light is absent the gland produces a hormone called *melatonin*. In mammals, the pineal gland is positioned in the brain in the third ventricle where it cannot receive direct light signals. However, the eyes send nerve impulses to a cervical ganglion that communicates with the pineal gland, which produces melatonin chiefly at night.

Most animals go through daily cycles called circadian rhythms and also yearly cycles called circannual rhythms. For example, the reproductive organs in hamsters shrink in size and weight as the amount of light decreases in fall and winter. This inhibits reproduction at a time of the year when food is not plentiful. The pineal gland is believed to bring about this regression of the gonads by secreting melatonin. Although a daily rhythm of melatonin level in the plasma and urine has also been detected in rats, they do not appear to have a yearly rhythm like that of hamsters that causes the gonads to atrophy.

At present, melatonin seems to have no function in humans, although its level in the blood does increase at night as with rats. Also, children with brain tumors that destroy the pineal gland experience early puberty. Therefore, the present emphasis of research is to see if melatonin is involved in human sexual development.

Still Other Hormones

Even organs that are not traditionally considered to be endocrine glands have been found to secrete hormones. For example, we have previously discussed the hormones produced by the stomach and the small intestine. And recently it's been discovered that the heart produces *atrial natriuretic factor*, which helps regulate the water and salt balance in the kidneys.

Chemical Messengers

In this chapter we have concentrated on describing the functions of the vertebrate endocrine glands and their hormonal secretions. We already know that hormones are only one type of chemical messenger. In fact, the concept of the chemical messenger has now been broadened to include at least the following three different categories of messengers (fig. 41.14):

Chemical messengers that act at *a distance between individuals*—many organisms release these chemical messengers called *pheromones* into the air or in externally deposited body fluids. These are intended as messages for other members of the species. For example, ants lay down a trail of pheromone to direct other ants to food; the female silkworm moth releases bombykol, a sex attractant that is received by male moth antennae

Figure 41.14

The three categories of chemical messengers. Pheromones are chemical messengers that act at a distance between individuals. Endocrine hormones and neurosecretions are typically carried in the bloodstream and act at a distance within the body of a single organism. Some chemical messengers have local effects only; they pass between cells that are adjacent to one another. This, of course includes neurotransmitter substances.

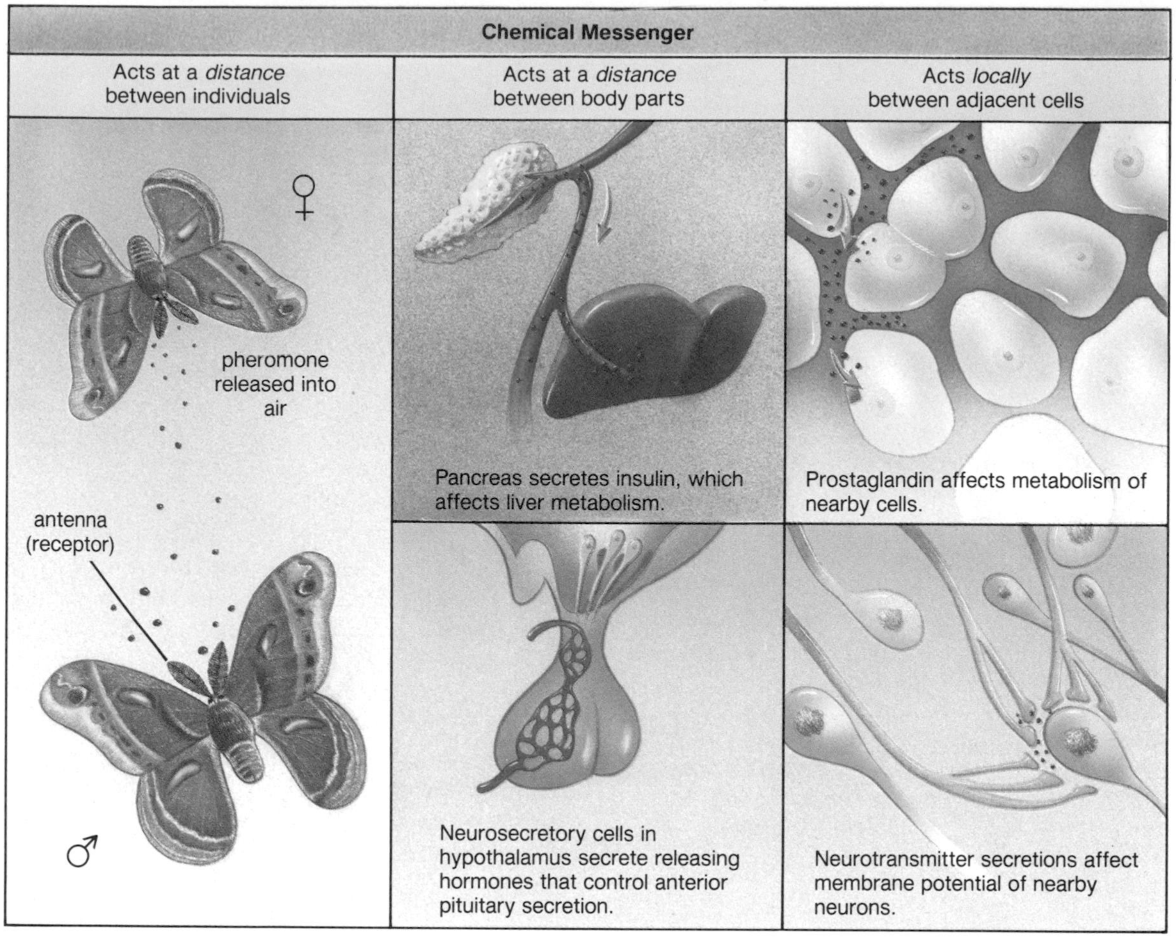

Chemical Messenger		
Acts at a *distance* between individuals	Acts at a *distance* between body parts	Acts *locally* between adjacent cells

pheromone released into air

antenna (receptor)

♀

♂

Pancreas secretes insulin, which affects liver metabolism.

Neurosecretory cells in hypothalamus secrete releasing hormones that control anterior pituitary secretion.

Prostaglandin affects metabolism of nearby cells.

Neurotransmitter secretions affect membrane potential of nearby neurons.

even several miles away. The chemical is so potent that it has been estimated that only 40 out of 40,000 receptors on the male antennae need to be activated in order for the male to respond. Mammals, too, release pheromones; the urine of dogs serves as a territorial marker, for example. Studies are being conducted (see reading on page 713) to determine if humans also have pheromones.

Chemical messengers that act at *a distance between body parts*—this category includes the endocrine secretions, which traditionally have been called hormones. It also includes the secretions of the neurosecretory cells in the hypothalamus—the production and action of ADH and oxytocin illustrate the close relationship between the nervous system and the endocrine system. Neurosecretory cells produce these hormones, which are released when these cells receive nerve impulses. As another example of the overlap between the nervous and endocrine system, consider that endorphins, on occasion, travel in the bloodstream but they act on nerve cells to alter their membrane potential.

Chemical messengers that act locally *between adjacent cells*—neurotransmitters belong in this category, as do substances that are sometimes called local hormones. For example, when the skin is cut, histamine is released by mast cells and promotes the inflammatory response. Nearby capillaries dilate allowing more blood to reach the site, and they become leaky so that cells of the immune system are able to leave the capillaries and attack incoming viruses and bacteria.

Prostaglandins are local messengers derived from cell membrane phospholipids. There are many different types of prostaglandins produced by many different tissues. In the uterus certain prostaglandins cause muscles to contract; therefore, they are implicated in the pain and discomfort of menstruation in some women. (Antiprostaglandin therapy is useful in these cases.) On the other hand, certain prostaglandins are being used to treat ulcers because they reduce gastric secretion, to treat hypertension because they lower blood pressure, and to prevent thrombosis because they inhibit platelet aggregation. Because the different prostaglandins may have contrary effects, however, it has been very difficult to standardize their use and, in most instances, prostaglandin therapy is still considered experimental.

Another kind of messenger is important in development; embryonic cells release *growth factors* that influence the development of adjacent cells. For example, nerve growth factor affects the development of the nervous system, and epidermal growth factor has effects on various types of cells.

Definition of a Hormone

Traditionally, a hormone was considered to be a secretion of an endocrine gland that was carried in the bloodstream to a target organ. In recent years, some scientists have broadened the definition of a hormone to include *all* types of chemical messengers. This change seemed necessary because those chemicals traditionally considered to be hormones have now been found in all sorts of tissues in the human body. For example, it is impossible for insulin produced by the pancreas to enter the brain because of the blood-brain barrier—a tight fusion of endothelial cells of the capillary walls that prevents passage of larger molecules like peptides. Yet, insulin has been found in the brain. It now appears that the brain cells themselves can produce insulin, which is used locally to influence the metabolism of adjacent cells. Also, some chemicals identical to the hormones of the endocrine system have been found in lower organisms, even in bacteria! A moment's thought about the evolutionary process helps explain this; these regulatory chemicals may have been present in the earliest cells and only became specialized as hormones as evolution proceeded.

Summary

1. The vertebrate endocrine glands whose locations are shown in figure 41.1 produce hormones that are secreted into the bloodstream and circulate about the body until they are received by their target organs.

2. There are two major types of compounds used as hormones; most hormones are either amino acids, peptides, or proteins; a few hormones are steroids.

3. The peptide protein hormones are usually received by a receptor located in the plasma membrane. Most often their reception activates an enzyme that changes ATP to cyclic AMP. Cyclic AMP then activates another enzyme that activates another and so forth. There is an amplification of effect at each step of this enzyme cascade.

4. The steroid hormones are lipid soluble and can pass through plasma membranes. Once inside the cytoplasm, they combine with a receptor molecule, and the complex moves into the nucleus where it combines with chromatin. Transcription and translation lead to protein synthesis; in this way steroid hormones alter the metabolism of the cell.

5. Neurosecretory cells in the hypothalamus produce ADH and oxytocin, which are stored in axon endings in the posterior pituitary until they are released. ADH secretion is controlled by a negative feedback mechanism in which concentrated blood stimulates ADH release and dilute blood inhibits its release. ADH causes the collecting ducts of the kidney to reabsorb water. Oxytocin causes the uterus to contract and milk to be released from the mammary glands.

6. The hypothalamus produces hypothalamic-releasing hormones that pass to the anterior pituitary by way of a portal system and either stimulate or inhibit the release of a particular anterior pituitary hormone.

7. The anterior pituitary produces six types of hormones (table 41.1). GH promotes body growth; LTH promotes breast development and production of milk following pregnancy; MSH causes skin color changes in lower vertebrates; TSH stimulates the thyroid; ACTH stimulates the adrenal cortex to produce corticoids; gonadotropic hormones (FSH and LH) stimulate the gonads. Because the anterior pituitary also stimulates certain other hormonal glands, it is sometimes called the master gland.

8. The thyroid gland produces thyroxin and triiodothyronine, hormones that contain iodine. These hormones play a role in growth and development of immature forms; in mature individuals they increase the metabolic rate. A goiter may develop in individuals who have inadequate iodine in the diet. The thyroid also produces calcitonin, which helps lower the calcium level of the blood.

9. The parathyroid glands raise the level of calcium in the blood by (1) activating vitamin D, which is involved in absorption of calcium from the gut; (2) stimulating retention of calcium by the kidneys and (3) stimulating its removal from the bones. PTH also decreases the blood phosphate level.

10. The adrenal medulla secretes epinephrine and norepinephrine, which bring about responses we associate with the "fight or flight" reaction.

11. The adrenal cortex produces the glucocorticoids (cortisol) and the mineralocorticoids (aldosterone). Cortisol stimulates hydrolysis of proteins to amino acids that are converted to glucose; in this way it raises the blood glucose level. It also counteracts the inflammatory reaction. Aldosterone causes the kidneys to reabsorb sodium: when the blood sodium level is low the kidneys secrete renin, which converts angiotensinogen to angiotensin, a molecule that stimulates the release of aldosterone. The main effect of this renin-angiotensin-aldosterone system is to raise the blood pressure.

12. The pancreas secretes insulin, which lowers the blood glucose level, and glucagon, which has opposite effects to insulin. In type I diabetes mellitus, insulin-secreting cells in the islets of Langerhans have been destroyed; in type II diabetes, the cells lack receptors for insulin or do not respond to it. Only type I requires daily injections of insulin, while type II requires a special diet.

13. The gonads produce the sex hormones, which are discussed in chapter 42; the thymus secretes thymosins, which stimulate T lymphocyte production and maturation; the pineal gland produces melatonin whose function in mammals is uncertain; it may affect development of the reproductory organs.

14. There are three categories of chemical messengers: those that act at a distance between individuals (pheromones); those that act at a distance within the individual (traditional endocrine hormones and secretions of neurosecretory cells); and local messengers (such as prostaglandins and neurotransmitters). Since there is great overlap between these categories perhaps the definition of a hormone should now be expanded to include all of them.

Objective Questions

Match the hormone in numbers 1 to 5 below to the correct gland in the key.

Key:
- a. pancreas
- b. anterior pituitary
- c. posterior pituitary
- d. thyroid
- e. adrenal medulla
- f. adrenal cortex

1. cortisol
2. GH
3. oxytocin storage
4. insulin
5. epinephrine
6. The anterior pituitary controls
   - a. the secretions of both the adrenal medulla and the adrenal cortex.
   - b. the secretion of both cortisol and aldosterone.
   - c. the secretion of thyroxin.
   - d. All of these.
7. Peptide hormones
   - a. are received by a receptor located in the plasma membrane.
   - b. are received by a receptor located in the cytoplasm.
   - c. bring about the transcription of DNA.
   - d. Both (b) and (c).
8. Aldosterone causes
   - a. the kidneys to release renin.
   - b. the kidneys to reabsorb sodium.
   - c. the blood volume to increase.
   - d. All of these.
9. Diabetes mellitus is associated with
   - a. too much insulin in the blood.
   - b. too high a level of glucose in the blood.
   - c. blood that is too dilute.
   - d. All of these.
10. The level of cortisol in the blood controls the secretion of
    - a. a releasing hormone from the hypothalamus.
    - b. ACTH from the anterior pituitary.
    - c. cortisol from the adrenal cortex.
    - d. All of these.
11. It is now clear that
    - a. there is a discrete distinction between the activities of the nervous and endocrine systems.
    - b. both the nervous and endocrine systems coordinate the activities of body parts.
    - c. the nervous system controls the secretions of all endocrine glands.
    - d. All of these.
12. One of the chief differences between pheromones and local hormones is
    - a. one is a chemical messenger and the other is not.
    - b. the distance over which they act.
    - c. one is made by invertebrates and the other is made by vertebrates.
    - d. All of these.

Study Questions

1. Give a definition of endocrine hormones that includes their most likely source, how they are transported in the body, and how they are received. What does "target" organ mean?
2. Categorize endocrine hormones according to their chemical makeup.
3. Tell how the two major types of hormones influence the metabolism of the cell.
4. Give the location in the human body of all the major endocrine glands. Name the hormones secreted by each gland, and describe their chief functions.
5. Explain the relationship of the hypothalamus to the posterior pituitary and to the anterior pituitary.
6. Explain the concept of negative feedback and give an example involving ADH.
7. Give an example of the three-tiered relationship between the hypothalamus, the anterior pituitary, and other endocrine glands. Explain why the anterior pituitary can be called the master gland.
8. Draw a diagram to explain the contrary actions of insulin and glucagon. Use your diagram to explain the symptoms of type I diabetes mellitus.
9. Categorize chemical messengers into three groups, and give examples of each group.
10. Give examples to show that there is an overlap between the mode of operation of the nervous system and that of the endocrine system. Explain why the traditional definition of a hormone may need to be expanded.

Thought Questions

1. With reference to figure 41.5a, explain why you would expect negative feedback to result in a fluctuation of hormonal blood levels about a mean value.
2. Name several hormones that control the concentration of blood glucose level, and show that this duplication is necessary.
3. Give pro and con arguments for combining the nervous and endocrine systems into one system and calling it the neuroendocrine system.

Selected Key Terms

endocrine system (en'do-krin sis'tem) 642
hormone (hor'mōn) 642
cyclic AMP (sik'lik AMP) 643
pituitary gland (pi-tu'i-tār''e gland) 644
thyroid gland (thi'roid gland) 649
goiter (goi'ter) 650
parathyroid gland (par''ah-thi'roid gland) 651
adrenal gland (ah-dre'nal gland) 651
pancreas (pan'kre-as) 653
diabetes mellitus (di''ah-be'tēz mē-li'tus) 653

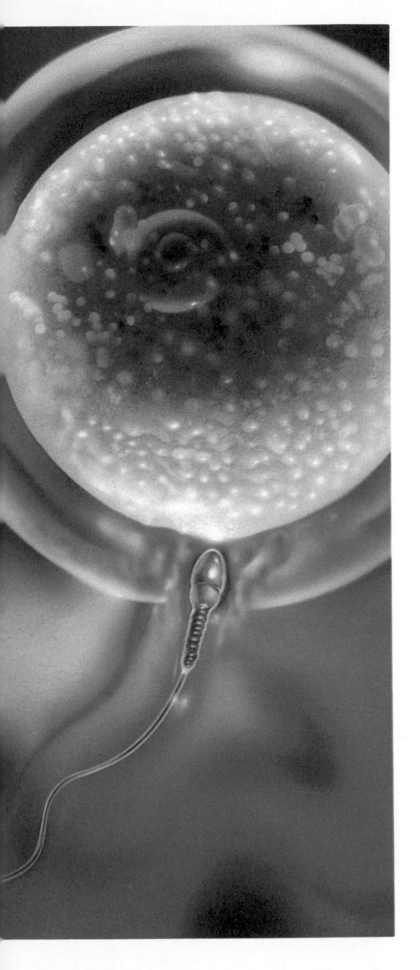

C H A P T E R   42

Reproduction in Animals

Your study of this chapter will be complete when you can:

1. *Contrast asexual reproduction with sexual reproduction and relate their advantages to environmental conditions.*

2. *Explain why aquatic animals are apt to practice external fertilization, whereas land animals are apt to practice internal fertilization.*

3. *State and give a function for the organs of the human male reproductive system. Trace the path of sperm from the testes to the penis.*

4. *Describe the microscopic anatomy of the testes, including the stages of spermatogenesis. Relate the action of the hormones FSH (ICSH) and LH to the functions of the testes.*

5. *Describe hormonal control in the male, mentioning the hypothalamus, anterior pituitary, and testosterone. Give several actions of testosterone.*

6. *State in sequence and give a function for the organs of the human female reproductive system.*

7. *Describe the microscopic anatomy of the ovary, including a description of the ovarian cycle.*

8. *Describe the ovarian and uterine cycle. Explain the relationship between the hormones GnRH, FSH, LH, and the female sex hormones. Give several functions of estrogen and progesterone.*

9. *Explain the cessation of the ovarian and uterine cycles in the pregnant female.*

10. *List several means of birth control, and compare their effectiveness in preventing pregnancy.*

11. *Describe the cause, symptoms, and mode of transmission of the following sexually transmitted diseases: AIDS, genital herpes, chlamydia, gonorrhea, and syphilis.*

Reproduction requires that a sperm fertilize an egg. The life strategy of every type animal includes the anatomy, physiology, and behavior, which makes fertilization possible.

A nimals expend a considerable amount of energy in reproduction. After all, reproduction ensures that the animals' genes are passed on to the next generation. The life cycle of any particular animal comes to an end, but its genes can be perpetuated as long as reproduction has taken place. In evolutionary terms, the most fit organisms are the ones that have produced the most offspring. They are the best adapted to the environment, and it is beneficial to the species that they reproduce more than other members of their cohort.

Patterns of Reproduction

There are two fundamental patterns of reproduction: asexual and sexual. We will examine the methodology and particular advantage of each pattern.

Asexual Reproduction

In asexual reproduction, there is only one parent (fig. 42.1), and the offspring tend to have the same genotype and phenotype as that parent. This lack of variation is not a disadvantage as long as the environment stays the same. Organisms that reproduce asexually often have the tremendous advantage of being able to produce a large number of offspring within a limited amount of time, when the organism has not differentiated to the point of forming sex cells.

Only certain methods of asexual reproduction are found among animals. Many flatworms can constrict into two halves, each of which becomes a new individual. This is a form of **regeneration** that is also seen among sponges and echinoderms. Chopping up a sea star does not kill it. Instead, each fragment grows into another animal, and the total number of sea stars increases in the process. Some coelenterates, such as hydras, reproduce by **budding** (fig. 26.5), during which the new indi-

vidual arises as an outgrowth (bud) of the parent. Insects have the ability to reproduce parthenogenetically (fig. 42.2). **Parthenogenesis** is a modification of sexual reproduction in which an unfertilized egg develops into a complete individual.

Sexual Reproduction

Sexual reproduction (fig. 42.1) involves gametes (sex cells). The gametes may be specialized into eggs or sperm, which are produced by the same or separate individuals. Earthworms (fig. 27.8) practice cross-fertilization, even though they are *hermaphroditic* and have both male and female sex organs. Among vertebrates, the sexes tend to be separate, and it is often easy to tell whether an animal is an egg-producing female or a sperm-producing male (fig. 42.3). When animals reproduce sexually, an offspring inherits half its genes from one parent and the other half from the other parent. Therefore, an offspring often has a different combination of genes than either parent. In this way, variation may be introduced and maintained. Such variation is an advantage to the species if the environment is changing, because an offspring might be better adapted to the new environment than is either parent.

Reproductive Timing

The majority of sexually reproducing animals have a single breeding period each year. Reproduction at this time of year is most favorable to survival of offspring. For example, breeding in wild mammals is timed so that the end of the period of internal development (gestation) occurs when conditions are favorable for the growth of the newborn animals. Thus, many temperate-zone mammals mate in the fall, and deliver young in the spring.

Figure 42.2
Stem covered with aphids. In the summer, many generations of aphid females produce up to one hundred young, without prior fertilization. In the autumn, however, aphids reproduce sexually and the zygotes overwinter.

Figure 42.1
Asexual reproductive pattern versus sexual reproductive pattern. (*left*) In asexual reproduction, there is only one parent and the offspring tend to be identical to each other and to the parent. Asexual reproduction occurs in various ways: by regeneration, by budding, or by parthenogenesis. (*right*) In sexual reproduction, there are most often two parents. Each parent contributes one-half of the genes to the offspring either by way of the sperm or the egg. Therefore, the offspring tend to be genetically and phenotypically different from either parent.

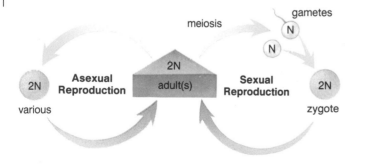

Just as we observed with plants, the reproductive cycles of a large percentage of animals, both vertebrates and invertebrates, are related to day-length changes. Day-length changes are the most reliable indicators that it is the proper time for reproductive behavior, including migration to distant places. Researchers have found that in many species, there is a daily rhythm in the production of the hormone *melatonin* by the *pineal gland*. Production of melatonin is stimulated by darkness and inhibited by light. Apparently, the level of melatonin at certain times of the day provides a photoperiodic signal that controls the activity of the reproductive system in seasonally breeding mammals.

External versus Internal Fertilization

Many aquatic animals practice external fertilization. They shed a large number of gametes into the water at the same time and place (fig. 42.4). External fertilization is possible because the water protects the zygote and embryo from drying out. Often the embryo is a swimming larva that is capable of acquiring its own food.

Internal fertilization is a mechanism that is practiced by animals that lay a shelled egg or in which the embryo develops for a time within the female parent. Even some aquatic animals practice internal fertilization. In certain sharks, skates, and rays, the pelvic fins are specialized to pass sperm to the female, and

Figure 42.3
Male and female vertebrates of the same species are biologically distinguishable. The male sex has testes that produce sperm, and the female sex has ovaries that produce eggs. In addition, the sex hormones often bring about the secondary sex characteristics that cause the sexes to differ in appearance. Note that the female lion in the foreground lacks the mane of the male lion in the background.

Figure 42.4
When frogs mate they shed their eggs and sperm right in the water where fertilization takes place. The watery environment protects the gametes and zygotes from drying out.

Figure 42.5

Side view of the male reproductive system. Notice that the urethra in males carries either urine from the bladder or semen from the testes.

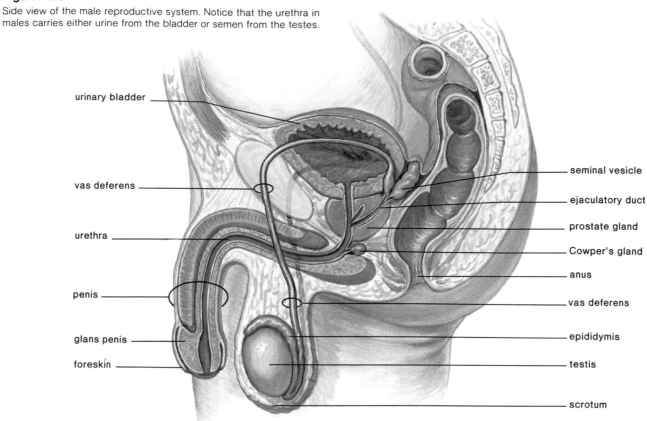

urinary bladder

vas deferens

urethra

penis

glans penis

foreskin

seminal vesicle

ejaculatory duct

prostate gland

Cowper's gland

anus

vas deferens

epididymis

testis

scrotum

in most of these animals, the young develop internally and are born alive. On land, internal fertilization is a necessity because sperm and eggs dry quickly when exposed to air. Often the male has a copulatory organ to transfer sperm to the female. Reptiles and birds lay shelled eggs (fig. 28.10). In placental mammals, the fetus develops within the uterus, where it receives nourishment by way of the placenta. Humans can be used as an example of reproduction in placental mammals.

Some animals practice asexual reproduction, especially when environmental conditions are favorable. Most practice sexual reproduction, timed so that the young are born when food is plentiful. Aquatic animals are apt to practice external fertilization and internal fertilization is common in land animals.

Human Reproduction

The manner in which humans reproduce is similar to that of other mammals and represents an adaptation to the land environment. When the male penis passes sperm to the female, desiccation not only of the male and female gametes, but also of the subsequent embryo is prevented.

Male Reproductive System

Figure 42.5 shows the reproductive system of the male, and table 42.1 lists the organs and their functions. The male gonads are paired **testes,** suspended in a saclike structure, the *scrotum.* The

Table 42.1 Male Reproductive System

Organ	Function
Testes	Produce sperm and sex hormones
Epididymis	Maturation and some storage of sperm
Vas deferens	Conducts and stores sperm
Seminal vesicles	Contribute to seminal fluid
Prostate gland	Contributes to seminal fluid
Urethra	Conducts sperm
Cowper's glands	Contribute to seminal fluid
Penis	Organ of copulation

testes begin their development inside the abdominal cavity, but descend into the scrotum as development proceeds. If the testes do not descend, and the male is not treated by administration of hormones, or operated on to place the testes in the scrotum, *sterility* (the inability to produce offspring) results. The reason for sterility in this case is that normal sperm production does not occur at body temperature; a cooler temperature is required.

A longitudinal cut (fig. 42.6) shows that each testis is composed of compartments called *lobules,* each of which contains one to three tightly coiled **seminiferous tubules.** Altogether, these tubules have a combined length of approximately

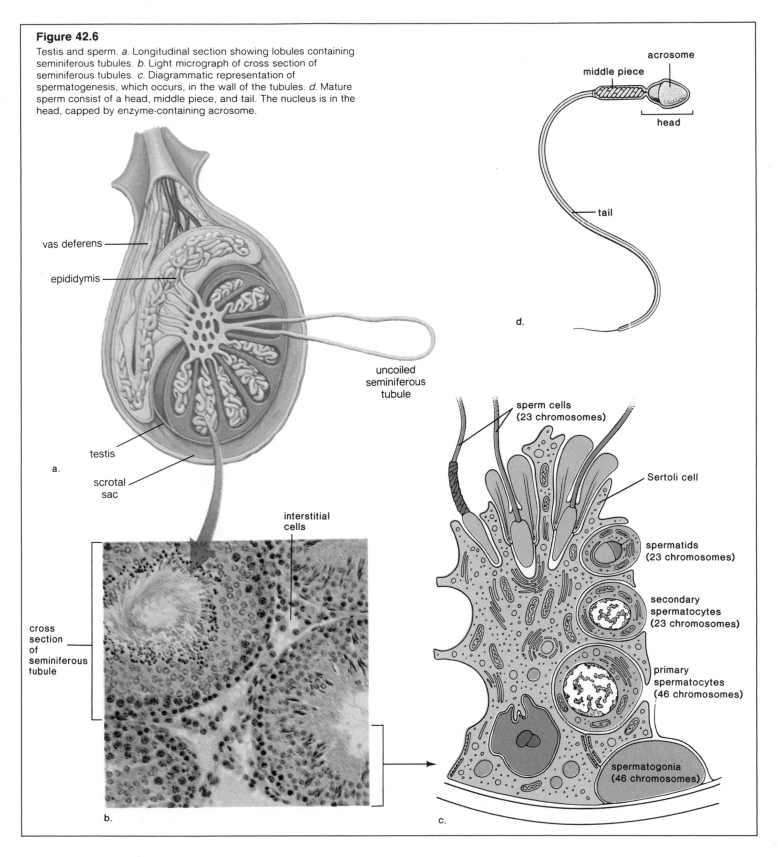

Figure 42.6
Testis and sperm. *a.* Longitudinal section showing lobules containing seminiferous tubules. *b.* Light micrograph of cross section of seminiferous tubules. *c.* Diagrammatic representation of spermatogenesis, which occurs, in the wall of the tubules. *d.* Mature sperm consist of a head, middle piece, and tail. The nucleus is in the head, capped by enzyme-containing acrosome.

250 meters. A microscopic cross section of a tubule (fig. 42.6*b*) shows that it is packed with cells undergoing *spermatogenesis*. These cells are derived from undifferentiated germ cells, called **spermatogonia** (sing., **spermatogonium**), that lie just inside the outer wall, and divide mitotically, producing new spermato-

gonia. (Germ cell is a general term for those cells in males and females that give rise to the sex cells.) Some spermatogonia move away from the outer wall, increase in size, and become primary **spermatocytes** that undergo meiosis. Although these cells have 46 chromosomes, they divide to give secondary spermatocytes, each with 23 chromosomes. Secondary spermatocytes divide to give spermatids that also have 23 chromosomes. Spermatids then differentiate into spermatozoa, or mature sperm. Also present in the tubules are the Sertoli or nurse cells that support, nourish, and regulate the spermatogenic cells.

Spermatozoa (fig. 42.6*d*) have three distinct parts: a head, a middle piece, and a tail. The middle piece and tail contain microtubules usually in the characteristic 9 + 2 pattern of cilia and flagella. In the middle piece, mitochondria are wrapped around the microtubules and provide the energy for movement. The head contains a nucleus covered by a cap called the *acrosome,* which is believed to store enzymes needed for fertilization. The human egg is surrounded by several layers of cells and a thick membrane. These acrosome enzymes play a role in allowing the sperm to reach the surface of the egg, so that one sperm can enter.

Each acrosome may contain so little enzyme that it requires many sperm for one to actually enter the egg. This could explain why so many sperm are required for the process of fertilization. The normal human male usually produces several hundred million sperm per day, assuring an adequate number for fertilization to take place.

Male sex hormones (chiefly **testosterone**) are produced by the **interstitial cells** scattered in the spaces between the seminiferous tubules (fig. 42.6). Sex hormones bring about maturation of the sex organs and are largely responsible for the secondary sex characteristics that appear as the male undergoes puberty. In humans, the male secondary sex characteristics include a beard, deep voice, and increased muscle strength.

Hormone Regulation

The anterior pituitary produces two gonadotropic hormones, *LH* and *FSH,* in both males and females. LH in males is sometimes given the name interstitial cell stimulating hormone (*ICSH*) because it controls the production of testosterone by the interstitial cells. FSH promotes spermatogenesis in the seminiferous tubules, which also release the hormone inhibin. The hypothalamus has ultimate control of the testes' sexual functions, because it secretes a releasing hormone (GnRH = gonadotropin releasing hormone) that stimulates the anterior pituitary to produce the gonadotropic hormones. All of these hormones are involved in a feedback relationship that maintains the fairly constant production of sperm and testosterone (fig. 42.7).

The hypothalamus, anterior pituitary, and testes are engaged in a feedback system that promotes spermatogenesis and maintains the level of testosterone in the male. Testosterone stimulates development of the male sex organs and secondary sex characteristics.

Path of Sperm

Sperm are produced within the seminiferous tubules of the testes, but mature within the **epididymides** (sing., **epididymis**), which are tightly coiled tubules lying just outside the testes. Maturation seems to be required for the sperm to swim to the egg. Once the sperm have matured, they are propelled into the **vas deferentia** (sing., **vas deferens**) by muscular contractions. Sperm

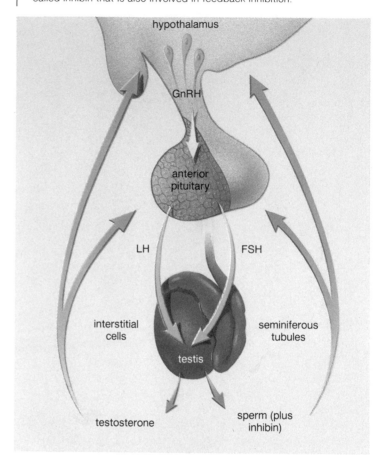

Figure 42.7

The hypothalamus-pituitary-testis control relationship. Testosterone acts on various body tissues and also regulates the amount of hypothalamic GnRH being sent to the pituitary. GnRH affects gonadotropic hormone production by the pituitary. LH regulates the amount of testosterone produced, and FSH controls spermatogenesis. The seminiferous tubules release a substance called inhibin that is also involved in feedback inhibition.

are stored in both the epididymides and the vas deferentia. When a male becomes sexually aroused, sperm enter the urethra, part of which is located within the **penis.**

The penis is a cylindrical organ that usually hangs in front of the scrotum. Spongy erectile tissue containing distensible blood spaces extends through the shaft of the penis (fig. 42.8). During sexual arousal, nervous reflexes cause an increase in arterial blood flow to the penis. This increased blood flow fills the blood spaces in the erectile tissue, and the penis, which is normally limp (flaccid), stiffens and increases in size. These changes are called *erection.* If the penis fails to become erect, the condition is called *impotency,* and may be due to medical or psychological problems. In some male mammals, a bone in the penis gives it a permanent partial rigidity, but the copulatory function of the human penis depends entirely on erection by filling of the blood spaces in the erectile tissue.

Orgasm in Males

Orgasm in both sexes is characterized by a release of neuromuscular tension, particularly in the genital area but also in the body as a whole. In males, orgasm is obvious, because rhythmical muscle contractions compress the urethra and expel semen.

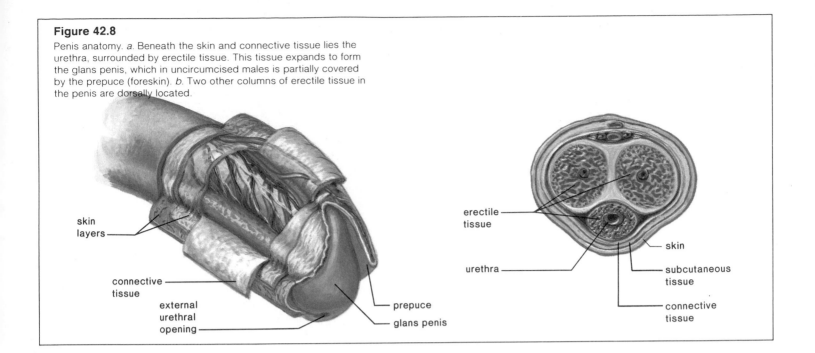

Figure 42.8

Penis anatomy. *a.* Beneath the skin and connective tissue lies the urethra, surrounded by erectile tissue. This tissue expands to form the glans penis, which in uncircumcised males is partially covered by the prepuce (foreskin). *b.* Two other columns of erectile tissue in the penis are dorsally located.

skin layers

connective tissue

external urethral opening

prepuce

glans penis

erectile tissue

urethra

skin

subcutaneous tissue

connective tissue

This is termed **ejaculation,** after which a male typically experiences a *refractory period* when stimulation does not bring about an erection.

Semen

Semen, a thick whitish fluid, contains sperm and seminal fluid. As table 42.1 indicates, three types of glands contribute to the production of seminal fluid. The **seminal vesicles** lie between the bladder and the rectum. Each joins a vas deferens to form an ejaculatory duct that enters the urethra. As sperm pass from the vas deferentia, these vesicles secrete a thick, viscous fluid containing nutrients for possible use by the sperm. Just below the bladder is the **prostate gland,** which secretes a milky alkaline fluid believed to activate or increase the motility of the sperm. In older men, the prostate gland frequently becomes enlarged, thus constricting the urethra and making urination difficult. Slightly below the prostate gland, on either side of the urethra, is a pair of small glands called **Cowper's glands,** which have mucous secretions with a lubricating effect. Notice from figure 42.5 that at different times, the urethra carries urine from the bladder and semen from the vas deferentia.

Sperm produced by the testes mature in the epididymis and pass from the vas deferens to the urethra, where certain glands add seminal fluid prior to ejaculation.

Female Reproductive System

Table 42.2 lists the female reproductive organs and their functions, and figure 42.9 shows the female reproductive system. The **ovaries,** the gonads in the female, lie in shallow depressions on the lateral wall of the pelvis, attached to the uterus and pelvic wall by ligaments. A longitudinal section of an ovary shows that it is made up of an outer cortex and inner medulla. The cortex

Table 42.2 Female Reproductive System

Organ	Function
Ovaries	Produce egg and sex hormones
Oviducts (fallopian tubes)	Conduct egg; location of fertilization
Uterus (womb)	Location of developing fetus
Cervix	Contains opening to uterus
Vagina	Organ of copulation and birth canal

contains **follicles** at various stages of maturation (fig. 42.10). A female is born with a large number of immature follicles (400,000 in both ovaries), each containing a germ cell (oocyte) surrounded by a layer of nongerminal cells. Only a small number of these follicles (about 400) ever mature, because a female produces only one egg each month during her reproductive years. Because follicles are present at birth, they age as the woman ages. This has been given as a reason why an older woman is more apt to produce a child with a genetic defect.

As a follicle undergoes maturation, it develops from a primary to a secondary to a Graafian follicle. In a *primary follicle,* an **oocyte** divides meiotically into two cells, each having 23 chromosomes. One of these cells, termed the secondary oocyte, receives almost all the cytoplasm, nutrients, and enzymes. The other is a *polar body* that disintegrates. A *secondary follicle* contains the secondary oocyte pushed to one side of a fluid-filled cavity. In a *Graafian follicle,* the fluid-filled cavity increases to the point that the follicle wall balloons out on the surface of the ovary and bursts, releasing the secondary oocyte surrounded by

Figure 42.9
Side view of female reproductive system. The ovaries produce one egg per month. Fertilization occurs in the oviduct and development occurs in the uterus. The vagina is the birth canal and organ of copulation.

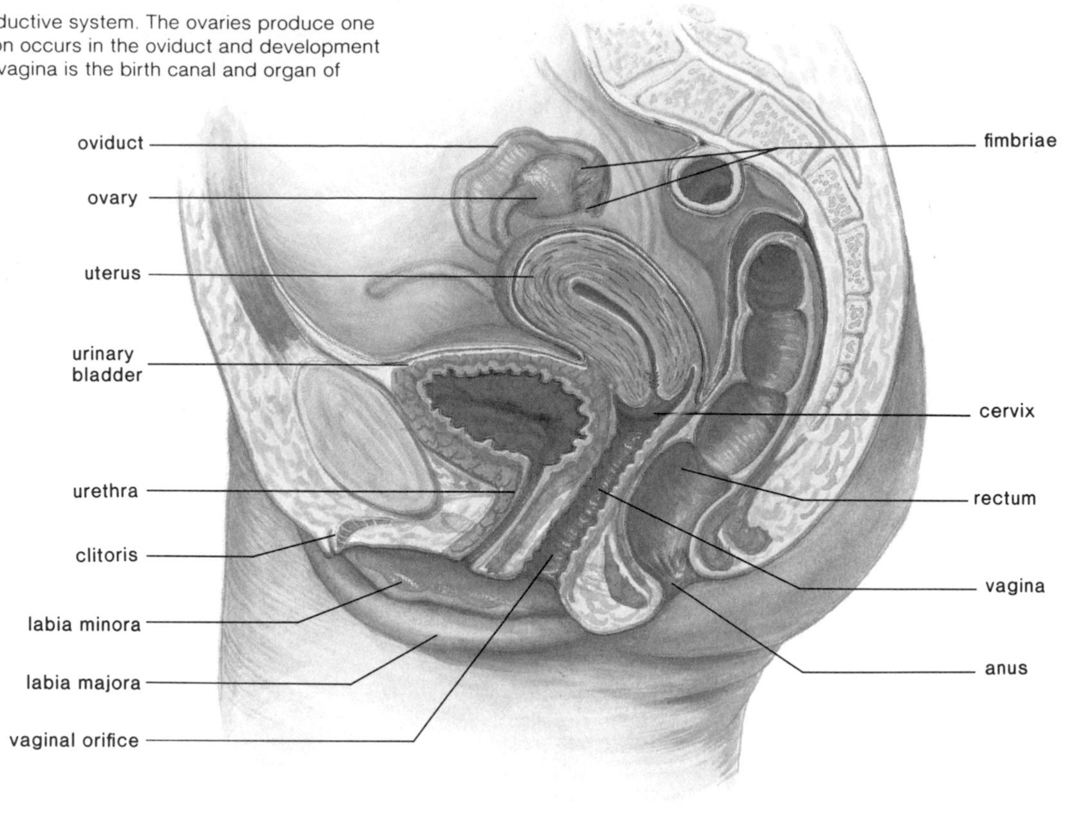

Labels: oviduct, ovary, uterus, urinary bladder, urethra, clitoris, labia minora, labia majora, vaginal orifice, fimbriae, cervix, rectum, vagina, anus

a clear membrane and a few cells. This is referred to as **ovulation,** and for the sake of convenience, the released oocyte is often called an **ovum,** or egg. Actually, the second meiotic division does not take place unless fertilization occurs.

Besides producing eggs, the ovary also secretes the female sex hormones, **estrogen** and **progesterone.** Prior to ovulation, a follicle secretes primarily estrogen, but after ovulation the follicle becomes the **corpus luteum** and secretes increasing amounts of progesterone. If fertilization does not take place, the corpus luteum degenerates into the *corpus albicans,* a whitish scar. The series of stages during which the follicle matures and then degenerates is called the **ovarian cycle.**

In the ovarian cycle, one follicle per month produces a secondary oocyte. Following ovulation, the follicle develops into a corpus luteum. These structures produce the hormones estrogen and progesterone.

Accessory Organs

The oviducts, uterus, and vagina are the other internal organs of the female reproductive system. The **oviducts** (fallopian tubes) are about four inches long and extend from the uterus to the ovaries. They end in fingerlike projections called *fimbria* that sweep over the ovary at the time of ovulation. When the egg bursts from the ovary during ovulation, it is usually sucked up into an oviduct where it may be fertilized, if sperm are present (fig. 42.9). Whether or not fertilization occurs, cilia lining the tubes and tubular contractions propel the egg toward the uterus. The **uterus** is about the size and shape of an inverted pear and lies between the bladder and the rectum. It is a muscular organ with three parts: the fundus, the body, and the **cervix.** The oviducts join the uterus just below the fundus, and the cervix enters into the vagina at nearly a right angle. The opening of the cervix, called the os or cervical canal, leads to the **vagina,** a small muscular tube that makes a 45° angle with the small of the back. The lining of the vagina lies in folds capable of extension as the muscular wall stretches. This capacity to extend is especially important when the vagina serves as the birth canal, and it may facilitate intercourse, when the vagina receives the penis.

Orgasm in Females

Orgasm in the female is a release of neuromuscular tension in the muscles of the genital area, vagina, and uterus. Prior to orgasm, the vagina is lubricated by mucous secretions and expands for reception of the penis. The **clitoris,** which has a limited amount of erectile tissue within a small shaft, and a pealike head or glans, also expands. It is a highly sensitive organ that is located anterior to the opening of the urethra. The urethral and vaginal external openings in the female lie between two folds of skin called **labia minora** and **labia majora** (fig. 42.11). At birth, the vaginal opening is partially or wholly occluded by a membrane called the hymen. All types of physical activities, sexual intercourse in particular, can rupture the hymen.

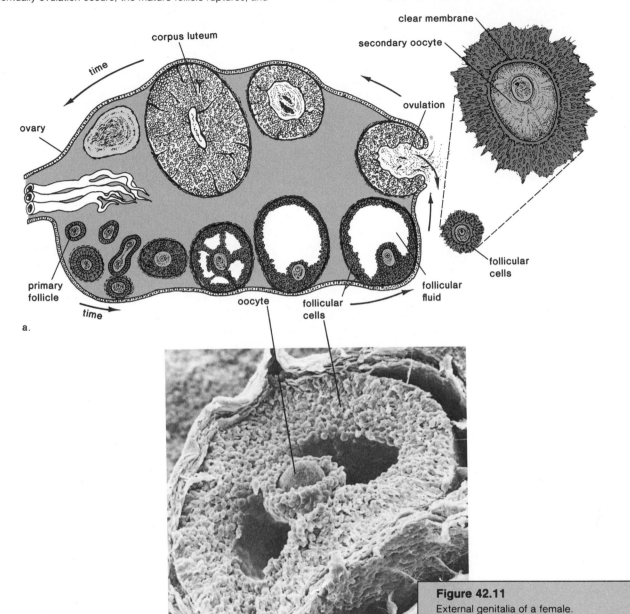

Figure 42.10
Anatomy of ovary and follicle. *a.* As a follicle matures, the oocyte enlarges and become surrounded by a mantle of follicular cells and fluid. Eventually ovulation occurs, the mature follicle ruptures, and the secondary oocyte is released. A single follicle actually goes through all stages in one place within the ovary. *b.* Scanning electron micrograph of a follicle.

corpus luteum

time

ovary

primary follicle

time

a.

oocyte

follicular cells

follicular fluid

ovulation

clear membrane

secondary oocyte

follicular cells

b.

Figure 42.11
External genitalia of a female.

mons pubis
labia majora
glans clitoris
labia minora
urethra
hymen
vagina

anus

Uterine Cycle

There is a **uterine cycle** (fig. 42.12) that runs concurrently with the ovarian cycle, described on previous page. This coordination allows the uterus to be prepared to receive the developing zygote as it implants in the uterine lining. It is customary to describe the uterine cycle as if it begins with **menstruation,** which is the periodic shedding of tissue and blood from the **endometrium,** the inner lining of the uterus. The first day of menstruation is called "day one" of the uterine cycle. During the early days of the cycle, the hypothalamus is secreting moderate amounts of

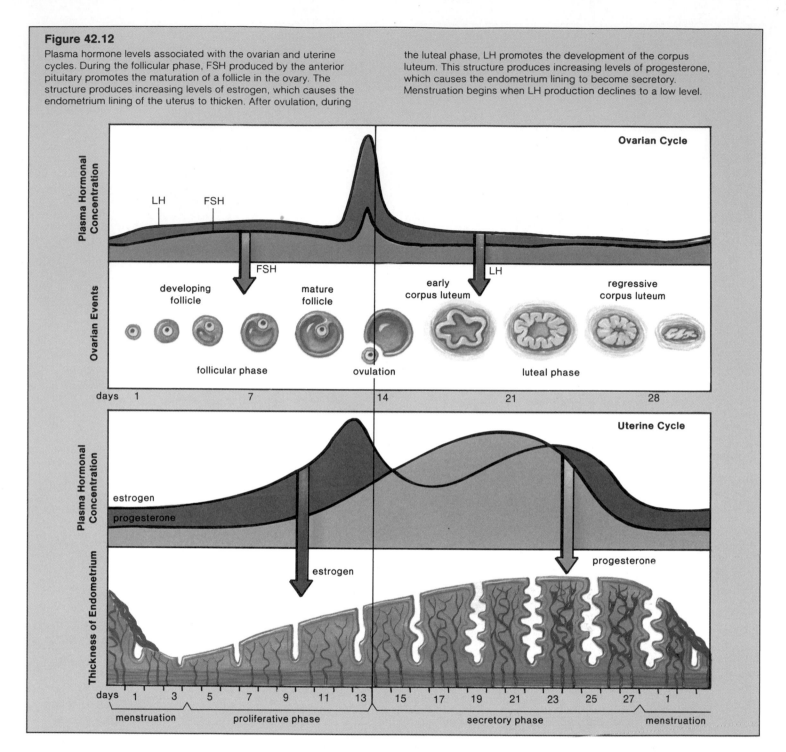

Figure 42.12

Plasma hormone levels associated with the ovarian and uterine cycles. During the follicular phase, FSH produced by the anterior pituitary promotes the maturation of a follicle in the ovary. The structure produces increasing levels of estrogen, which causes the endometrium lining of the uterus to thicken. After ovulation, during the luteal phase, LH promotes the development of the corpus luteum. This structure produces increasing levels of progesterone, which causes the endometrium lining to become secretory. Menstruation begins when LH production declines to a low level.

GnRH, and the anterior pituitary is secreting *FSH (follicle stimulating hormone)*. This hormone causes a follicle to begin to mature and produce an egg. The growing ovarian follicle secretes increasing amounts of *estrogen,* which stimulates repair of the endometrium. This so-called proliferation phase (table 42.3) lasts for 8 to 10 days following the end of menstruation.

Presumably the high level of estrogen in the blood causes the hypothalamus to suddenly secrete a large amount of GnRH at about the midpoint of a 28-day menstrual cycle. This causes a surge of *LH (luteinizing hormone)* production by the pituitary. This LH surge brings about ovulation and eventually influences the follicle to develop into the corpus luteum, which continues hormone production. The maturing corpus luteum secretes both estrogen and increasing quantities of *progesterone*. Progesterone stimulates glycogen accumulation, vascularization, and glandular development in the endometrium of the uterus. All these progesterone-induced changes, which occur during the so-called **secretory phase,** allow an embryo to implant itself easily in the endometrial lining.

Table 42.3 Ovarian and Uterine Cycles (Simplified)			
Ovarian Cycle Phases	**Events**	**Uterine Cycle Phases**	**Events**
Follicular— Days 1–13	FSH	Menstruation— Days 1–5	Endometrium breaks down
	Follicle maturation Estrogen	Proliferation— Days 6–13	Endometrium rebuilds
Ovulation— Day 14*			
Luteal— Days 15–28	LH Corpus luteum Progesterone	Secretory— Days 15–28	Endometrium thickens and glands secrete glycogen

*Assuming a 28-day cycle

Nonpregnant Female   In the usual uterine cycle, when a pregnancy has not begun, the hypothalamus and then the pituitary respond to the high levels of sex hormones secreted by the active corpus luteum (fig. 42.13). This feedback response causes a decrease in LH production, and the resultant fall in LH levels cause the corpus luteum to begin degenerating about the twenty-second or twenty-third day of a 28-day cycle. As the corpus luteum degenerates, its hormone output decreases.

The fully developed endometrium of the uterus is maintained by the high progesterone levels present when the corpus luteum is relatively most active. The uterine lining starts to degenerate and slough off, and menstruation begins, when circulating progesterone levels decrease because of corpus luteum regression. However, before long, the hypothalamus and pituitary respond to decreased circulating estrogen. The whole complex cycle starts over as FSH production rises; follicle growth and maturation are stimulated.

Menstruation marks the end of the time when the uterine endometrium could accept an embryo. It is a clear signal that pregnancy has not begun during that particular menstrual cycle.

Around age 45 to 50, the ovary gradually ceases to respond to the anterior pituitary hormones, and eventually, no more follicles are produced. Following this occurrence, called **menopause,** menstruation ceases entirely.

A negative feedback system involving the hypothalamus, anterior pituitary, and ovaries causes a monthly fluctuation of estrogen and progesterone in the female. These hormones regulate the uterine cycle, in which the endometrium first builds up and then is shed during menstruation.

Pregnant Female   If fertilization has occurred, an embryo begins development even as it travels down the oviduct to the uterus. The endometrium is now prepared to receive the developing embryo, which becomes embedded in the lining several days following fertilization. This process, called **implantation,**

Figure 42.13
Hypothalamic-pituitary-gonad system (simplified) as it functions in the female. GnRH is a hypothalamic-releasing hormone that stimulates the anterior pituitary to secrete FSH and LH. These gonadotropic hormones act on the ovaries. FSH promotes the development of the follicle that later, under the influence of LH, becomes the corpus luteum. Negative feedback controls the level of all hormones involved.

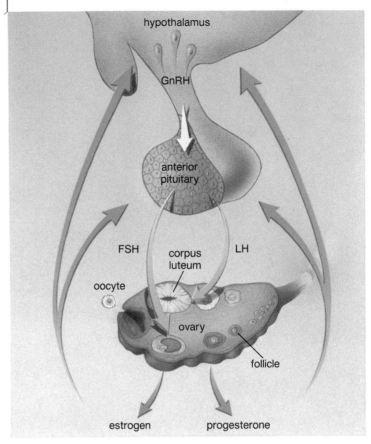

causes the female to become pregnant. During implantation, an outer layer of cells surrounding the embryo produces a gonadotropic hormone (**HCG,** or human chorionic gonadotropin) that prevents degeneration of the corpus luteum, and instead causes it to secrete even larger quantities of progesterone. The corpus luteum may be maintained for as long as six months, even after the placenta is fully developed.

The **placenta** (fig. 43.15) originates from both maternal and fetal tissue. It is the region of exchange of molecules between fetal and maternal blood, although there is rarely any mixing of the two types of blood. After its formation, the placenta continues production of **HCG** and begins production of progesterone and estrogen. The latter hormones have two effects. They shut down the anterior pituitary so no new follicles mature, and they maintain the lining of the uterus so the corpus luteum is not needed. There is no menstruation during pregnancy.

HCG is produced in such large quantities that pregnant women excrete considerable amounts of it in their urine. The chemical tests for pregnancy are based on the detection of this

The Reasons for Orgasm

Why does orgasm exist? A glance at the reproduction of primitive forms of animal life shows that such a fancy thing far exceeds what's necessary to pass genes from one generation to another, which even in mammals can be attained by a no-frills male ejaculation and mere passive receptivity on the part of the female. Yet it's unlikely that nature, as economical as it is, would create something so baroque without good reason.

The simpler the creature, the more difficult it is to determine whether its sexual relations are pleasurable, much less orgasmic. However, all mammals show a marked interest in sex, and will even work for it—a good indication that it's rewarding. Scientists prefer low-key terms like "ejaculatory reflex" and "estrus behavior" when discussing animal sexuality; they're reluctant to apply the term "orgasm," because it can't be verified in animals. The best gauge of orgasm is uniquely human: Did you or didn't you?

Male sexual responses, particularly erection and ejaculation, have been better studied than female processes. Even the humble rat, that scrupulously observed mammal, exhibits the basic criteria associated with male orgasm: ejaculation followed by refraction, accompanied by characteristic movements. "Males of most mammalian species have what at least looks like a precursor to human orgasm, demonstrated by skeletal and facial patterns," says Benjamin Sachs, a reproductive behavior researcher at the University of Connecticut. "You may as well call it orgasm. I'd be more cautious about females." Moreover, in addition to appearances, males have an excellent motive for ejaculation: progeny.

Alas for researchers, human females don't always show characteristic muscular movements during orgasm. Even more vexing, they need not have orgasms to conceive—or even

to seek and enjoy mating. The best evidence for female orgasm is subjective and anecdotal—the yea or nay of the woman in question. Until there's a way to acquire such information from animals, subhuman female orgasm can only be inferred. Does this mean that for female animals mating is simply a selfless matter of lie still and think of England?* Ronald Nadler, a primatologist at the Yerkes Regional Primate Research Center in Atlanta, believes it's unlikely that women are alone among female primates in having orgasms, but the real controversy is not whether, but why, females have orgasms at all.

There are two schools of thought on this. The first maintains that the female has orgasms because the male does, though hers serve no essential reproductive function (an analogy is the male nipple). The female genitals are poorly designed for easy stimulation to orgasm, particularly during intercourse, and if orgasm were really important to procreation, anatomy would have evolved differently to facilitate it. This idea of female orgasm as vestigial is supported by the fact that the inability to have orgasms is common in women yet rare in men, and that orgasm sometimes requires considerable education and practice for women to achieve.

The second school says that female orgasm is designed for a reason or reasons. Sarah Blaffer Hardy, a primatologist at the University of California at Davis, notes that a female could accumulate sufficient stimulation for orgasm as well as increase her chances of conception via the repeated matings typical of primates during estrus. Furthermore, if the motions of the vagina and uterus during orgasm can be proved to enhance the mobility of sperm, as some researchers speculate, orgasm might be a way to enable a female to be somewhat selective about when and even by whom she becomes pregnant. Although neither camp can prove its

thesis, a majority holds that female orgasm has a purpose—without knowing what the purpose is.

While scholars debate, couples unwittingly contend with evolutionary traits that have more practical application. As sex therapists are wont to put it, "Men heat up like light bulbs and women like irons." Trying as this difference in timing can be, it makes sociobiological sense: a female motivated to seek repeated sexual encounters increases her opportunities for pregnancy and assures male benignity to a greater extent than if she were quickly aroused and satisfied. On the other hand, a male, always on the defensive, can't afford lengthy dalliances with his back to potential enemies.

Though many theories about the evolution of orgasm are specific to males or females, the response also makes sense in social, as well as reproductive, terms for both. While the sex lives of other mammals are regulated by hormones—totally so in nonprimates like rats, partially so in nonhuman primates—humans appear to be freed from such controls, and even castrated men and women can enjoy sex and have orgasms. There are reliable reports of children seven years old having orgasms, and there's no limit at the other end of the age spectrum—further indications that there's more to sex than reproduction. Roger Short, a reproductive biologist at Monash University in Melbourne, Australia, thinks that because human females, unlike those of other species, will accept intercourse at any time, whether or not they are in a fertile period, most couplings are performed for social purposes such as strengthening pair-bonding and releasing sexual tension, so that the more mundane business of life can be attended to.

*Traditional advice to Victorian brides on their wedding nights.
Winifred Gallagher, © 1986 DISCOVER PUBLICATIONS.

---

hormone in the urine. HCG is so readily available that it is routinely used in many teaching and research laboratories. For example, it is even used to induce ovulation in female frogs to obtain eggs for embryological studies.

Illness of the mother or a failure of the placenta to continue adequate HCG production may occasionally cause corpus luteum activity to decrease. Due to a resultant drop in progesterone levels, menstruationlike breakdown of the endometrium and loss of the implanted embryo is initiated. Although the blood flow is somewhat heavier and longer than normal as miscarriage occurs, the terminated pregnancy may be mistaken for a somewhat delayed menstrual period.

Female Sex Hormones

The female sex hormones, *estrogen* and *progesterone,* have many other affects in the body, in addition to their affect on the uterus.

Estrogen is largely responsible for the secondary sex characteristics, including female body hair and fat distribution. In general, females have a more rounded appearance than males because of a greater accumulation of fat beneath the skin. Also, the pelvic girdle enlarges in females, and the pelvic cavity has a larger relative size compared to males. This means that females have wider hips. Both estrogen and progesterone are also required for breast development.

Breasts A female breast contains 15 to 25 lobules (fig. 42.14), each with its own milk duct that begins at the nipple and divides into numerous other ducts that end in blind sacs called *alveoli.* In a nonlactating breast, the ducts far outnumber the alveoli, because alveoli are made up of cells that can produce milk.

Milk is not produced during pregnancy. Prolactin is needed for lactation (milk production) to begin, and production of this hormone is suppressed by the feedback inhibition estrogen and progesterone have on the pituitary during pregnancy. It takes a

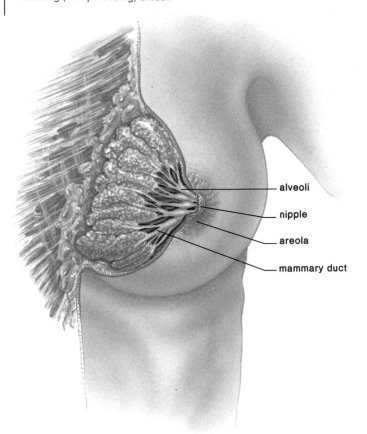

Figure 42.14

Anatomy of breast. The female breast contains lobules consisting of ducts and alveoli. The alveoli are lined by milk-producing cells in the lactating (milk-producing) breast.

— alveoli

— nipple

— areola

— mammary duct

couple of days after delivery of the baby for milk production to begin. In the meantime, the breasts produce a watery, yellowish-white fluid called **colostrum,** which has a similar composition to milk, but contains more protein and less fat.

Control of Reproduction

Birth Control

Several means of birth control have been available for quite some time (table 42.4). The use of contraceptive methods decreases the probability of pregnancy. A common way to discuss pregnancy rate is to indicate the number of pregnancies expected per 100 women per year. For example, it is expected that 80 out of 100 young women, or 80%, who are regularly engaging in unprotected intercourse will be pregnant within a year. Another way to discuss birth control methods is to indicate their effectiveness, in which case the emphasis is placed on the number of women who will not get pregnant. For example, with the least effective method given in table 42.4, we expect that 70 out of 100, or 70%, sexually active women will not get pregnant while 30 women will get pregnant within a year.

Infertility

Sometimes couples do not need to prevent pregnancy; conception or fertilization does not occur despite frequent intercourse. The American Medical Association estimates that 15% of all

Table 42.4 Common Birth Control Methods

Name	Procedure	Methodology	Effectiveness*	Action Needed	Risk
Vasectomy	Vas deferentia are cut and tied	No sperm in semen	Almost 100%	Sexual freedom	Irreversible sterility
Tubal ligation	Oviducts are cut and tied	No eggs in oviduct	Almost 100%	Sexual freedom	Irreversible sterility
Pill	Must take medication daily	Shuts down pituitary	Almost 100%	Sexual freedom	Thromboembolism
IUD	Must be inserted into uterus by physician	Prevents implantation	More than 90%	Sexual freedom	Infection
Diaphragm	Plastic cup inserted into vagina to cover cervix	Blocks entrance of sperm into uterus	With jelly about 90%	Must be inserted each time before intercourse	—
Condom	Sheath that fits over erect penis	Traps sperm	About 85%	Must be placed on penis at time of intercourse	—
Coitus interruptus (withdrawal)	Male withdraws penis before ejaculation	Prevents sperm from entering vagina	About 80%	Intercourse must be interrupted before ejaculation	—
Jellies, creams, foams	Contain spermicidal chemicals	Kill a large number of sperm	About 75%	Must be inserted before intercourse	—
Rhythm method	Determine day of ovulation by record keeping; testing by various methods	Avoid day of ovulation	About 70%	Limits sexual activity	—
Douche	Cleanses vagina and uterus after intercourse	Washes out sperm	Less than 70%	Must be done *immediately* after intercourse	—

*Effectiveness is the average percentage of women who did not become pregnant in a population of 100 sexually active women using the technique for one year.

couples in this country are unable to have any children, and are therefore properly termed sterile. Another 10% have fewer children than they wish and, are therefore termed *infertile*.

Infertility can be due to a number of factors. It is possible that fertilization takes place, but the embryo dies before implantation. One area of concern is that radiation, chemical mutagens, and the use of psychoactive drugs may contribute to sterility, possibly by causing chromosomal mutations that prevent development from proceeding normally. The lack of progesterone can also prevent implantation, and therefore administration of this hormone is sometimes helpful.

It is also possible that fertilization never takes place. There may be a congenital malformation of the reproductive tract, or an obstruction of the oviduct or vas deferens. Sometimes these physical defects can be corrected surgically. If no obstruction is apparent, it is possible to give females a substance rich in FSH and LH that is extracted from the urine of postmenopausal women. This treatment causes multiple ovulations and sometimes multiple pregnancies.

When reproduction does not occur in the usual manner, couples today are seeking alternative reproductive methods that include artificial insemination (sperm are placed in vagina by a physician); in vitro fertilization (fertilization takes place in laboratory glassware and the zygote is inserted into the uterus of the woman), and surrogate motherhood (a woman is paid to have another's child).

Sexually Transmitted Diseases

There are many diseases that are transmitted by sexual contact. We will consider only five of the most troublesome, **AIDS** (acquired immune deficiency syndrome), genital herpes, gonorrhea, chlamydia, and syphilis (fig. 42.15). The first two are viral diseases and, therefore, are difficult to treat because the traditional antibiotics are not helpful. The drug AZT (azidothymidine) is now being used in patients with AIDS (p. 540) and both ointment and oral ACV (acyclovir) are available for herpes. Although gonorrhea and chlamydia are treatable with appropriate antibiotic therapy, they are not always promptly diagnosed. Unfortunately, as yet, there are no vaccines available for these infections.

AIDS

The cause of acquired immune deficiency syndrome (**AIDS**) is a retrovirus that attacks helper T cells and macrophages—the very cells that protect us from disease. The virus, now called **HIV** (human immunodeficiency virus), usually stays hidden inside macrophages, but on occasion buds from T cells.

AIDS has three stages of infection. During the first stage, the individual usually has detectable antibodies in the bloodstream and may exhibit swollen lymph nodes. During the second stage, called AIDS related complex (ARC), symptoms may include weight loss, night sweats, fatigue, fever, and diarrhea. Finally, the person may develop full-blown AIDS, characterized by the development of pneumonia, skin cancer, and also neuromuscular and psychological disturbances.

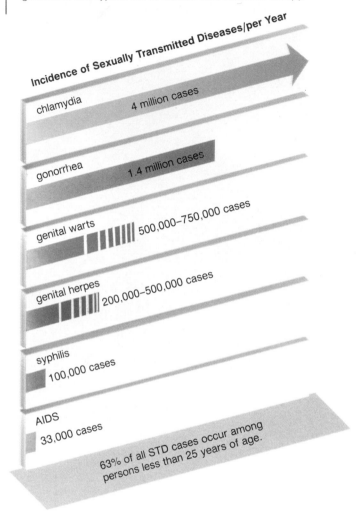

Figure 42.15

Statistics for the most troublesome sexually transmitted diseases show that chlamydia, gonorrhea, and genital warts are all much more common than herpes, syphilis, and AIDS. Chlamydia, gonorrhea, and syphilis are all curable with antibiotic therapy.

Incidence of Sexually Transmitted Diseases/per Year

chlamydia — 4 million cases

gonorrhea — 1.4 million cases

genital warts — 500,000–750,000 cases

genital herpes — 200,000–500,000 cases

syphilis — 100,000 cases

AIDS — 33,000 cases

63% of all STD cases occur among persons less than 25 years of age.

It takes several years before an individual passes through the first two stages and develops AIDS itself. Although a few individuals have had AIDS as long as six years, most die within this time period.

Transmission

AIDS is transmitted by blood, semen, and vaginal fluids. In the United States, the two main affected groups are homosexual men and intravenous drug users (and their sexual partners). In Africa and some parts of South America though, AIDS is transmitted chiefly through heterosexual intercourse, and an equal number of men and women are infected.

Certain portions of the country have been harder hit than others (fig. 42.16). Even in New York City, which reports the highest number of affected individuals, there are regions that have more cases than others. Here, the number of AIDS deaths among needle-using addicts is higher than that of homosexual men. The AIDS virus can cross the placenta, and in the part of New York City known as the Bronx, 1 in 43 babies is currently

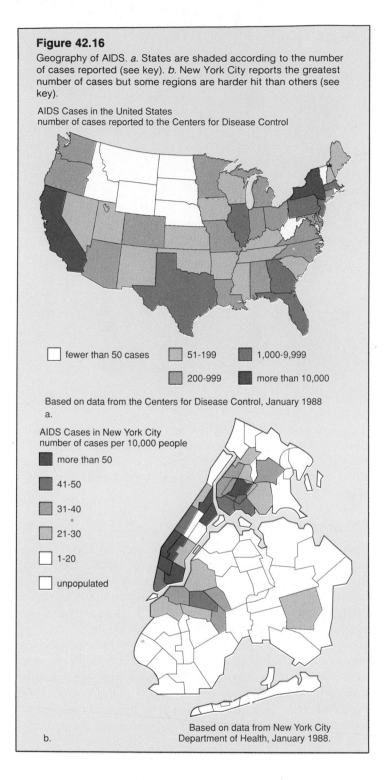

Figure 42.16

Geography of AIDS. *a*. States are shaded according to the number of cases reported (see key). *b*. New York City reports the greatest number of cases but some regions are harder hit than others (see key).

AIDS Cases in the United States
number of cases reported to the Centers for Disease Control

☐ fewer than 50 cases   ☐ 51–199   ☐ 1,000–9,999

☐ 200–999   ☐ more than 10,000

Based on data from the Centers for Disease Control, January 1988

a.

AIDS Cases in New York City
number of cases per 10,000 people

☐ more than 50

☐ 41–50

☐ 31–40

☐ 21–30

☐ 1–20

☐ unpopulated

Based on data from New York City
Department of Health, January 1988.

b.

who is not infected. Casual contact with someone who is infected, such as shaking hands, eating at the same table, or swimming in the same pool, does not transmit the virus.

Treatment

The drug AZT has been found to be helpful in prolonging the lives of AIDS patients. When RNA from the virus is transcribed to DNA, the cell tends to use the drug instead of thymidine, resulting in nonfunctional DNA. Other drugs that interfere with the life cycle of the virus are also being developed.

As discussed on page 540, researchers are trying to develop a vaccine for AIDS. The AIDS virus mutates frequently, but researchers have identified portions of the coat protein that they believe are relatively stable. When these are injected into the bloodstream, antibodies develop, but it is not yet known whether such antibodies will offer protection against infection. After all, persons with AIDS do have antibodies, but for some reason, the antibodies do not halt the course of the disease. The virus rarely enters the bloodstream; it causes T cells to fuse and may spread from cell to cell in this manner.

Genital Herpes

Genital herpes is caused by herpes simplex virus (fig. 42.17), of which there are two types: type 1 usually causes cold sores and fever blisters, and type 2 usually causes genital herpes.

Genital herpes is one of the more common sexually transmitted diseases today (fig. 42.15). An estimated 20.5 million persons in the United States have it, with as many as 500,000 new cases appearing each year. Immediately after infection, there are no symptoms, but the individual may experience a tingling or itching sensation before blisters appear at the infected site in 2 to 20 days. Once the blisters rupture, they leave painful ulcers that may take as long as 3 weeks, or as little as 5 days to heal. These symptoms may be accompanied by fever, pain upon urination, and swollen lymph nodes.

After the ulcers heal, the disease is dormant. But blisters can reoccur repeatedly at variable intervals. Sunlight, sex, menstruation, and stress seem to cause the symptoms of genital herpes to reoccur. When the virus is latent, it resides in nerve cells. Type 1 occasionally travels via a nerve fiber to the eye and causes an eye infection that can lead to blindness. Type 2 has been known to cause a form of meningitis, and has been associated with a form of cervical cancer.

Infection of the newborn can occur if the child comes in contact with a lesion in the birth canal. In 1 to 3 weeks, the infant becomes gravely ill and may become blind, or have neurological disorders, including brain damage, or may die. Birth by caesarean section prevents infection.

Gonorrhea

Gonorrhea is caused by the bacterium *Neisseria gonorrheae*, which is a nonmotile (lacks flagella), nonspore-forming diplococcus, meaning that two cells generally stay together (fig. 42.18).

born with HIV antibodies in the blood. Some of these newborns may have the antibodies without having the virus, but 30%–50% are probably infected.

Although intravenous drug users are liable to spread the disease to the general heterosexual population, to date, infection among the general population is still less than 1%. Health officials emphasize that unprotected intercourse with multiple partners or a single infected partner increases the chance of infection. The use of a condom reduces the risk, but the best preventive measure at this time is monogamy with a sexual partner

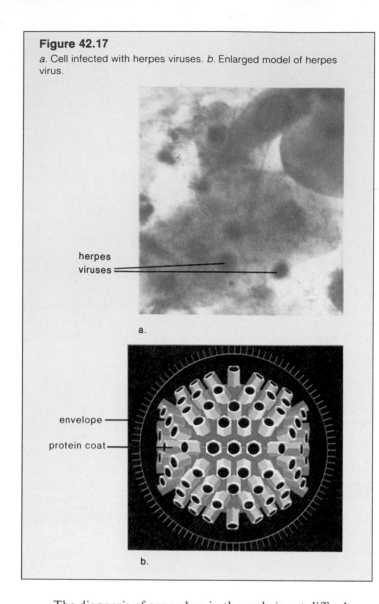

Figure 42.17

a. Cell infected with herpes viruses. *b.* Enlarged model of herpes virus.

herpes viruses

envelope

protein coat

a.

b.

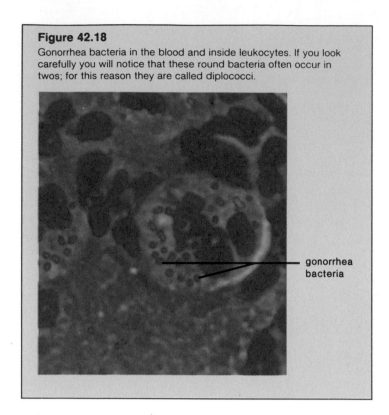

Figure 42.18

Gonorrhea bacteria in the blood and inside leukocytes. If you look carefully you will notice that these round bacteria often occur in twos; for this reason they are called diplococci.

gonorrhea bacteria

The diagnosis of gonorrhea in the male is not difficult, as long as he displays typical symptoms (as many as 40% of males may be asymptomatic). The patient complains of pain on urination and has a thick, greenish-yellow, urethral discharge 3 to 5 days after contact. In the female, the bacteria may first settle within the urethra or near the cervix, from which they may spread to the oviducts, causing **pelvic inflammatory disease (PID).** As the inflamed tubes heal, they may become partially or completely blocked by scar tissue. As a result, the female is sterile or, at best, subject to **ectopic pregnancy,** a pregnancy that begins at a location other than the uterus. Unfortunately, 60% to 80% of females are asymptomatic until they develop severe pain in the abdominal region due to PID.

Homosexual males develop gonorrhea proctitis, or infection of the anus, with symptoms including pain in the anus and blood or pus in the feces. Oral sex can cause infection of the throat and tonsillitis. Gonorrhea may also spread to other parts of the body, causing heart damage or arthritis. If, by chance, the person touches infected genitals and then his or her eyes, a severe eye infection can result.

Eye infection leading to blindness can occur as a baby passes through the birth canal. Because of this, all newborn infants receive eye drops containing antibacterial agents, such as silver nitrate, tetracycline, or penicillin, as a protective measure.

Chlamydia

This sexually transmitted disease is named for the tiny bacterium that causes it *(Chlamydia trachomatis).* New chlamydia infections occur more frequently than gonorrhea infections (fig. 42.15). They are the most common cause of NGU, nongonococcal urethritis. About 8–21 days after exposure, men experience a mild burning sensation upon urinating and a mucoid discharge. Women may have a vaginal discharge, along with the symptoms of a urinary tract infection. Unfortunately, a physician may mistakenly diagnose a gonorrheal or urinary infection and prescribe the wrong type of antibiotic. Or the person may never seek medical help. In either case, the infection may eventually cause PID and sterility or ectopic pregnancy.

If a newborn comes in contact with chlamydia during delivery, inflammation of the eyes or pneumonia may result. There are also those who believe that chlamydia infections increase the possibility of premature and stillborn births.

Syphilis

Syphilis is caused by the bacteria, *Treponema pallidum.* Syphilis has three stages, which may be separated by latent stages when the bacteria are resting before multiplying again. During the *primary stage,* a hard chancre (ulcerated sore with hard edges) indicates the site of infection. The chancre may go unnoticed, especially because it usually heals spontaneously, leaving little scarring. During the *secondary stage,* proof that bacteria have invaded and spread throughout the body is evident when the victim breaks out in a rash. Curiously, the rash does not itch and is seen even on the palms of the hands and soles of the feet.

There may be hair loss and gray patches on the mucous membranes, including the mouth. These symptoms disappear of their own accord. During a *tertiary stage,* which lasts until the patient dies, gummas (large, destructive ulcers) appear on the skin and within the internal organs, especially the small arteries (cardiovascular syphilis) and brain (neurosyphilis). Because syphilis can be cured even if it reaches the tertiary stage, a person should never feel that it is too late for treatment. Penicillin has been used as an effective antibiotic to cure syphilis.

Congenital syphilis is caused by syphilis bacteria crossing the placenta. The newborn is born blind and/or with numerous anatomical malformations.

The sexually transmitted diseases, AIDS, herpes, gonorrhea, chlamydia, and syphilis are troublesome at this time. AIDS often results in death within a few years, recurrent attacks of herpes occur throughout life, and gonorrhea and chlamydia often lead to sterility. The best preventive measure against these diseases is monogamy with a partner who is free of them.

Summary

1. Asexual reproduction may quickly produce a large number of offspring exactly like the single parent, and is advantageous when environmental conditions are relatively unchanging.
2. Sexual reproduction involves the use of gametes and produces offspring that are slightly different from the parents. This may be advantageous if the environment is changing.
3. Drying out of gametes and zygotes is not a threat for aquatic animals and they tend to practice external fertilization. Drying out is a problem for land animals and they tend to practice internal fertilization.
4. In human males, spermatogenesis occurs in the seminiferous tubules of the testes, which also produce testosterone in the interstitial cells. Testosterone maintains the secondary sex characteristics of males, such as the low voice, facial hair, and increased muscle strength.
5. FSH (also called ICSH) from the anterior pituitary stimulates spermatogenesis, and LH stimulates testosterone production. A releasing hormone from the hypothalamus, GnRH, controls anterior pituitary production and release of FSH and LH.

The level of testosterone in the blood controls the secretion of GnRH and the anterior pituitary hormones by a negative feedback system.
6. Sperm mature in the epididymis and may be stored in the vas deferens before entering the urethra, along with seminal fluid (produced by the seminal vesicles, prostate gland, and Cowper's gland), prior to ejaculation during male orgasm, when the penis becomes erect.
7. In females, oogenesis occurs in the ovaries, where one follicle matures, produces a secondary oocyte, and becomes a corpus luteum each month. This is called the ovarian cycle. The follicle and corpus luteum produce estrogen and progesterone, the female sex hormones.
8. The secondary oocyte enters an oviduct, which leads to the uterus. The uterus opens into the vagina. The genital area of women includes the vaginal opening, clitoris, and labia minora and labia majora.
9. A uterine cycle occurs concurrently with the ovarian cycle. In the first half of these cycles (days 1–13 before ovulation), the anterior pituitary is producing FSH and the follicle is producing estrogen. Estrogen causes the uterine lining to increase in thickness. In the second half of these cycles (days 15–28 after ovulation), the anterior pituitary is producing LH and the follicle is producing progesterone. Progesterone causes the uterine lining to become secretory. Feedback control of the hypothalamus and anterior pituitary causes the level of estrogen and progesterone to fluctuate. When they are at a low level, menstruation begins.
10. If fertilization occurs, the corpus luteum is maintained because of HCG production. Progesterone production does not cease, and the zygote implants itself in the thick uterine lining.
11. Estrogen and progesterone maintain the secondary sex characteristics of females, including less body hair than males, a wider pelvic girdle, more rounded appearance, and development of breasts.
12. Numerous birth control methods and devices are available for those who wish to prevent pregnancy. Infertile couples are increasingly resorting to alternative methods of reproduction.
13. The sexually transmitted diseases, AIDS, genital herpes, and chlamydia are of great concern, as are gonorrhea and syphilis, which are still common.

Objective Questions

1. Which of these is a requirement for sexual reproduction?
   a. male and female parents
   b. production of gametes
   c. optimal environmental conditions
   d. aquatic habitat

2. Internal fertilization
   a. prevents the drying out of gametes and zygotes.
   b. must take place on land.
   c. is practiced by humans.
   d. Both (a) and (c).

3. Which of these is mismatched?
   a. interstitial cells—testosterone
   b. seminiferous tubules—sperm production
   c. vas deferens—seminal fluid production
   d. penis—erection

4. FSH
   a. occurs in females but not males.
   b. stimulates the seminiferous tubules to produce sperm.
   c. secretion is controlled by GnRH.
   d. Both (b) and (c).
5. Which of these combinations is most likely to be present before ovulation occurs?
   a. FSH, corpus luteum, estrogen, secretory uterine lining
   b. LH, follicle, progesterone, thick uterine lining
   c. FSH, follicle, estrogen, uterine lining becoming thick
   d. LH, corpus luteum, progesterone, secretory uterine lining
6. In tracing the path of sperm, you would mention vas deferens before
   a. testes.
   b. epididymis.
   c. urethra.
   d. uterus.
7. An oocyte is fertilized in the
   a. vagina.
   b. uterus.
   c. oviduct.
   d. ovary.
8. During pregnancy,
   a. the ovarian cycle and uterine cycle occur more quickly than before.
   b. GnRH is produced at a higher level than before.
   c. the ovarian cycle and uterine cycle do not occur.
   d. the female secondary sex characteristics are not maintained.
9. Which of the following means of birth control is most effective in preventing pregnancy?
   a. condom
   b. pill
   c. diaphragm
   d. spermicidal jelly
10. Which of these sexually transmitted diseases is mismatched with its cause?
   a. AIDS—bacterial infection of red blood cells
   b. gonorrhea—bacterial infection of genital tract
   c. chlamydia—bacterial infection of genital tract
   d. syphilis—systemic bacterial infection

Study Questions

1. Give examples of asexual and sexual reproduction among animals. Relate these practices to environmental conditions.
2. Discuss the human reproductive system as an adaptation to life on land.
3. Discuss the anatomy and physiology of the testes. Describe the structure of sperm.
4. Give the path of sperm. What glands produce seminal fluid?
5. Name the endocrine glands involved in maintaining the sex characteristics of males and the hormones produced by each.
6. Discuss the anatomy and physiology of the ovaries. Describe ovulation.
7. Give the path of the egg. Where do fertilization and implantation occur? Name two functions of the vagina.
8. Discuss hormonal regulation in the female by giving the events of the menstrual cycle and relating these to the ovarian cycle. In what way is menstruation prevented if pregnancy occurs?
9. State the various means of birth control and tell their relative effectiveness.
10. Describe at least three common sexually transmitted diseases.

Thought Questions

1. Construct a diagram to explain the role of melatonin in reproductive timing.
2. Most animals follow the diplontic life cycle (p. 357). How might the diplontic life cycle have arisen from the alternation of generation life cycle? from the haplontic life cycle?
3. All organisms have a reproductive strategy that maximizes the chances of producing offspring that will survive to reproduce. Outline and discuss the human reproductive strategy.

Selected Key Terms

regeneration (re-jen″er-a′shun) 660
budding (bud′ing) 660
parthenogenesis (par″thĕ-no-jen′-ĕ-sis) 660
seminiferous tubule (se″mĭ-nif′er-us tu′bul) 662
spermatozoa (sper″mah-to-zo′ah) 664

testosterone (tes-tos′tĕ-rōn) 664
penis (pe′nis) 664
ovaries (o′vah-rez) 665
follicle (fol′ĭ-k'l) 665
ovulation (o″vu-la′shun) 666
ovum (o′vum) 666

corpus luteum (kor′pus lu′te-um) 666
oviduct (o′vĭ-dukt) 666
uterus (u′ter-us) 666
vagina (vah-ji′nah) 666
uterine cycle (u′ter-ĭn si′k'l) 667
menstruation (men″stroo-a′shun) 667
endometrium (en-do-me′tre-um) 667

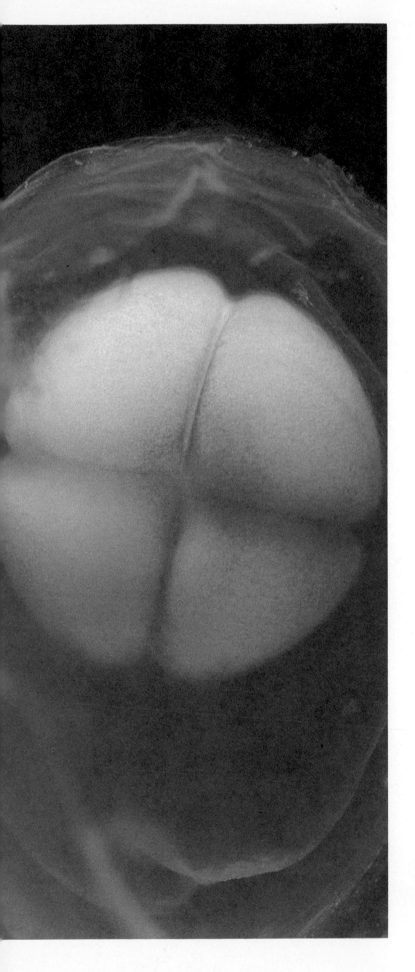

CHAPTER 43

Development in Animals

Your study of this chapter will be complete when you can:

1. Name and define three processes that occur whenever there is a developmental change.
2. Compare the early developmental stages of the lancelet, frog, and chick.
3. Explain the germ layer theory of development and give examples.
4. Draw and label a cross section of a typical vertebrate embryo at the neurula stage of development.
5. Give evidence that differentiation probably begins soon after formation of the zygote.
6. Give evidence that morphogenesis can be accounted for by interactions between tissues.
7. State the stages of embryonic development in humans, and compare these to the development of the chick.
8. List the extraembryonic membranes and give their function in the chick and in humans.
9. Outline briefly the developmental changes in humans, from the fetus to adulthood.
10. Define aging and discuss three theories of the aging process.

Development begins with cleavage, the division of the zygote and subsequent cells. During cleavage, each daughter cell receives a full set of chromosomes but the cytoplasm is proportioned between them. We now realize that the cytoplasm has a profound influence on the developmental potential of each cell.

The study of development concerns the events and processes that occur as a single cell becomes a complex organism. These same processes are also seen as the newly born or hatched organism matures, as lost parts regenerate, as a wound heals, and even during aging. Therefore, it is customary to stress that the study of development encompasses not only embryology (development of the embryo), but these other events as well.

Development requires growth, differentiation, and morphogenesis. When an organism increases in size, we say that it has grown. During **growth,** cells divide, get larger, and divide once again. **Differentiation** occurs when cells become specialized with regard to structure and function. A muscle cell looks and acts quite differently than a nerve cell, for example. **Morphogenesis** goes one step beyond growth and differentiation. It occurs when body parts become shaped and patterned into a certain form. There is a great deal of difference between your arm and leg for example, even though they contain the same types of tissues.

We will be discussing these processes as they apply to development of the embryo, but keep in mind that they also occur whenever an organism goes through any developmental change.

> Growth, differentiation, and morphogenesis are three processes that are seen whenever a developmental change occurs.

Early Developmental Stages

Embryological development begins when the sperm fertilizes the egg (fig. 43.1). Each gamete has a haploid number of chromosomes, therefore the resulting zygote has the diploid number. Gene expression is required for development to proceed normally and we will be stressing this again in later sections.

All chordate embryos go through the same early developmental stages of cleavage, blastulation, gastrulation, and neurulation. However, the presence of **yolk,** which is dense, nutrient material, affects the manner in which embryonic cells complete the first three stages and hence the appearance of the embryo at the end of each stage. Varying amounts of yolk will result in embryos with different appearances. Table 43.1 indicates the amount of yolk in the four embryos discussed in this chapter and relates the amount of yolk to the environment in which the animal develops. The two animals (lancelet and frog)

Figure 43.1
Fertilization occurs when a sperm fuses with an egg. The head of a human sperm (fig. 42.6d) is capped by an organelle called an acrosome. When a sperm meets a secondary oocyte in the oviduct, the acrosome releases enzymes that allow the sperm to penetrate through the follicular cells surrounding the oocyte. As the first sperm penetrates the oocyte, a chemical change occurs in the clear membrane and prevents other sperm from entering. The oocyte now undergoes its second meiotic division and the nucleus of the sperm and egg unite. Like the first meiotic division, there is a polar body that will disintegrate. The ovum has received all of the cytoplasm. a. Sea urchin egg surrounded by sperm. b. One sperm enters the egg.

a.

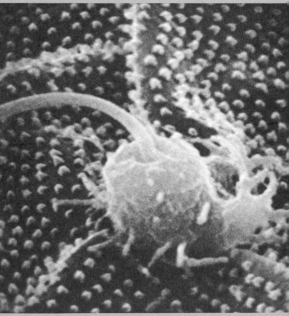

b.

that develop in water have less yolk than the chick, because development in these two animals proceeds quickly to a swimming larval stage that can feed itself. But the chick is representative of animals that have solved the problem of reproduction on land, in part, by providing a great deal of yolk within a hard shell, where development continues to a more mature individual, capable of land existence.

Early stages of human development resemble those of the chick embryo, yet this resemblance cannot be related to the amount of yolk, because the human egg contains little yolk. But the evolutionary history of these two animals can provide an answer for this similarity. Both birds (for example, chicks) and mammals (for example, humans) are related to the reptiles, and this explains why all three groups develop similarly despite a difference in the amount of yolk in the eggs.

Table 43.1 Amount of Yolk in Eggs

Animal	Yolk	Location of Development
Lancelet	Little	External in water
Frog	Some	External in water
Chick	Much	Within hard shell
Human	Little	Inside mother

The amount of yolk affects the manner in which lancelets, frogs, and chicks complete the first three stages of development.

The first three developmental stages of lancelets, frogs, and chicks are contrasted in figure 43.2.

Figure 43.2

a. Comparative morula stages for a lancelet, a frog, and a chick. *b.* Comparative blastula stages. *c.* Comparative early gastrula stages. *d.* Comparative late gastrula stages.

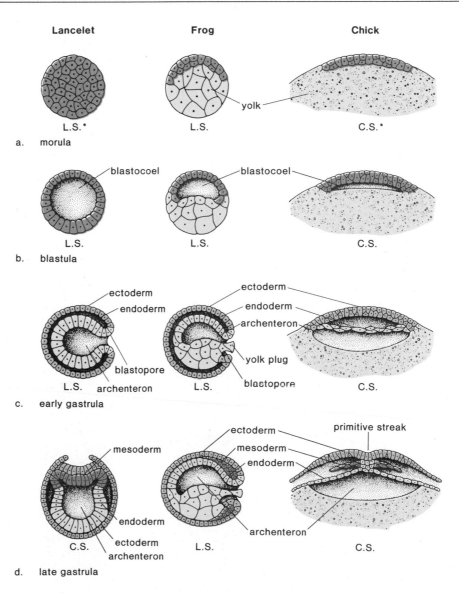

*L.S. = longitudinal section; C.S. = cross section

Cleavage

During **cleavage,** there is cell division without growth. DNA replication and mitosis occur repeatedly, and the cells get smaller with each division. Chordates, being deuterostomes, have a pattern of cleavage that is radial and indeterminate (see fig. 27.2). The term radial means that each new cell lies in the same plane as the parent cell, and the term indeterminate means that the cleavage cells have not differentiated, and therefore their developmental fate is not yet set.

In a lancelet, the cell divisions are equal, and the cells are of uniform size, whereas in a frog, the upper cells are smaller than the lower cells. This difference in size occurs because the upper cells at the animal pole contain little yolk, whereas the lower cells at the vegetal pole contain a large amount of yolk. Cells containing yolk cleave more slowly than those without yolk. The presence of extensive yolk in the chick egg causes cleavage to be incomplete, and only those cells lying on top of the yolk cleave. This means that although cleavage in a lancelet and a frog results in a ball of cells called the morula (fig. 43.2*a*), no such ball is seen in a chick. Instead, the cells spread out on a portion of the yolk.

Blastula

At first there is a solid mass of cells, then a cavity, called the **blastocoel,** develops. This "hollow-ball stage" of development, called the **blastula,** is best exhibited by the lancelet (fig. 43.2*b*). In the frog, the blastocoel is formed at the animal pole only. The heavily laden yolk cells of the vegetal pole do not participate in this step. In a chick, the blastocoel is created when the cells lift up from the yolk and leave a space between the cells and the yolk.

Cleavage results in a ball of cells, which becomes the blastula when an internal cavity develops. In lancelets and frogs, the blastula is a hollow ball. In the chick, the cavity is found beneath cells that lie flat atop the great mass of yolk.

Gastrula

The **gastrula** stage is evident in a lancelet when certain cells begin to push, or invaginate into the blastocoel, creating a double layer of cells (fig. 43.2*c*). The outer layer is called the **ectoderm,** and the inner layer is called the **endoderm.** The space created by invagination will become the gut, and is called either the primitive gut or the archenteron. The pore, or hole created by invagination, is called the blastopore, and in a lancelet, as well as in the other animals discussed here, it eventually becomes the anus (fig. 27.2).

In the frog, the cells containing yolk do not participate in gastrulation, and therefore do not invaginate. Instead, a slitlike blastopore is formed when the animal pole cells begin to invaginate from above. Following this, other animal pole cells move down over the yolk, and the blastopore becomes rounded when

Table 43.2 Organs Developed from the Three Primary Germ Layers

Ectoderm	Mesoderm	Endoderm
Skin epidermis, including hair, nails, and sweat glands	All muscles	Lining of digestive tract, trachea, bronchi, lungs, gallbladder, and urethra
Nervous system, including brain, spinal cord, ganglia, nerves, and sense receptors	Dermis of skin	Liver
	All connective tissue, including bone, cartilage, and blood	Pancreas
Lens and cornea of eye	Blood vessels	Thyroid, parathyroid, and thymus glands
Lining of nose, mouth, and anus	Kidneys	Urinary bladder
Teeth enamel	Reproductive organs	

these cells also invaginate from below. At this stage, there are some yolk cells temporarily left in the region of the pore, and they are called the yolk plug.

In the chick, there is so much yolk that endoderm formation does not occur by invagination. Instead, an upper layer of cells differentiates into ectoderm, and a lower layer differentiates into endoderm.

Mesoderm Formation

Gastrulation is not complete until three layers of cells have formed (fig. 43.2*d*). The third, or middle, layer of cells is called the **mesoderm.** In a lancelet, this layer begins as outpocketings from the primitive gut (fig. 27.2). These outpocketings grow in size until they meet and fuse. In effect, then, two layers of mesoderm are formed, and the space between them is the coelom.

In the frog, cells from the dorsal lip of the blastopore migrate between the ectoderm and endoderm, forming the mesoderm. Later, a splitting of the mesoderm creates the coelom. In the chick, the mesoderm layer arises by an invagination of cells along the edges of a longitudinal furrow in the midline of the embryo. Because of its appearance this furrow is called the *primitive streak.* Later, the newly formed mesoderm will split to give a coelomic cavity.

Germ Layers   Ectoderm, mesoderm, and endoderm are called the primary **germ layers** of the embryo, and no matter how gastrulation takes place, the end result is the same: three germ layers are formed. It is possible to relate the development of future organs to these germ layers, as is done in table 43.2. Von Baer, the nineteenth-century embryologist, first related later development to the early formation of germ layers. This is called the *germ layer theory.*

During gastrulation, the three embryonic germ layers arise when cells invaginate into the blastocoel. The development of organs can be related to the three germ layers: ectoderm, mesoderm, and endoderm.

Figure 43.3

Development of neural tube and coelom in a frog embryo. *a.* Ectoderm cells that lie above the future notochord (called presumptive notochord) thicken to form a neural plate. *b.* The neural groove and folds are noticeable as the neural tube begins to form. *c.* A splitting of the mesoderm produces a coelom that is completely lined by mesoderm. *d.* A neural tube and coelom have now developed.

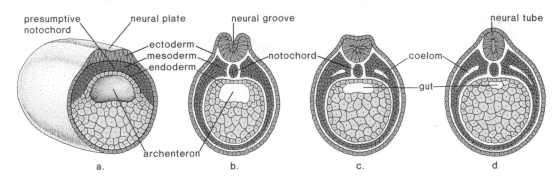

Development of the coelom and nervous system in a frog embryo.

Neurula

In chordate animals, newly formed mesodermal cells that lie along the main longitudinal axis of the animal coalesce to form a dorsal supporting rod called the **notochord.** The notochord persists in lancelets (fig. 27.16), but in the frogs, chicks, and humans it is later replaced by the vertebral column.

The nervous system develops from ectoderm located just above the notochord. At first, a thickening of cells, called the *neural plate,* is seen along the dorsal surface of the embryo. Then, *neural folds* develop on either side of a neural groove that will become the *neural tube* when they fuse. Figure 43.3 shows cross sections of frog development to illustrate the formation of the neural tube. At this point, the embryo is called a **neurula.** Later, the anterior end of the neural tube will develop into the brain.

Midline mesoderm that did not contribute to the formation of the notochord now becomes two longitudinal masses of tissue. These become blocked off as the somites, and will give rise to segmental muscles in all chordates. In vertebrates, the somites also produce the vertebral bones.

With the formation of the nervous system, it is possible to show a generalized diagram (fig. 43.4) of a chordate embryo to illustrate the location of parts. Consideration of this figure with table 43.2 will help you relate the formation of chordate structures and organs to the three embryonic layers of cells: the ectoderm, mesoderm, and endoderm.

> During neurulation, the neural tube develops just above the notochord. At the neurula stage of development, a cross section of all chordate embryos is similar in appearance.

Differentiation and Morphogenesis

These two developmental processes account for the specialization of tissues and the formation of organs that have an overall pattern of shape and form.

Differentiation

Differentiation is apparent when cells are specialized as to structure and function. The process of differentiation must have started long before we can recognize different types of cells.

Figure 43.4

Typical cross section of a chordate embryo at the neurula stage. Each of the germ layers indicated by color (see key) can be associated with the later development of particular parts (see table 43.2). The somites will give rise to the muscles of each segment and the vertebrae that will replace the notochord.

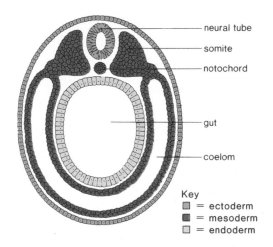

Key
- ☐ = ectoderm
- ■ = mesoderm
- ☐ = endoderm

After all, the cells that make up ectoderm, endoderm, and mesoderm in the gastrula look quite similar, but they are destined to develop into different organs. What causes differentiation to occur and when does it begin?

The contents of the cytoplasm of an egg are not distributed uniformly. Following cleavage, therefore, embryonic cells will differ as to the cytoplasmic contents they have received:

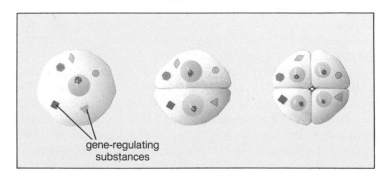

gene-regulating substances

Figure 43.5

Importance of the gray crescent in frog development. *a.* The frog's egg is polar; there is an animal pole and vegetal pole (contains the yolk). The position of the gray crescent can be correlated with the anterior/posterior and dorsal/ventral axes of the body. *b.* The first cleavage normally divides the gray crescent in half and each daughter cell is capable of developing into a complete tadpole. But if only one daughter cell should receive the gray crescent, then only that cell can become a complete embryo.

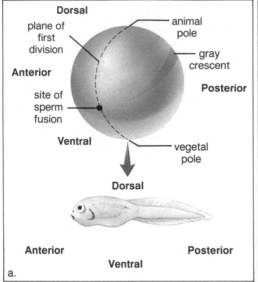

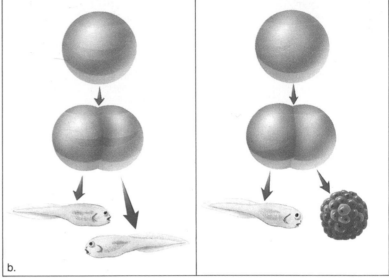

These varying contents probably contain substances that play a role in regulating which genes will be active in a particular cell.

Chapter 17 discusses the ways in which regulation of gene expression can occur. The reading in that chapter, page 249, describes an experiment in which a nucleus from an intestinal tadpole cell is transplanted into an egg, whose own nucleus was destroyed. Development proceeds normally, showing that embryonic nuclei are totipotent—they contain all the genetic information required to bring about complete development of the organism. Therefore differentiation cannot be due to a parceling out of genes into the various embryonic cells. Instead, it must be due to the expression of particular genes, controlled at first by *ooplasmic segregation,* which is the distribution of maternal cytoplasmic contents to the cells of the morula.

The fact that the cytoplasm of the egg does indeed play a role in regulating development can be demonstrated by another experiment with a frog embryo. After the sperm fuses with the egg, some contents of the egg shift position and a *gray crescent* area appears on the egg, opposite the point where the sperm entered (fig. 43.5). Normally, the first cleavage gives each daughter cell half of the gray crescent. In this case, each of the daughter cells has the potential to become a complete embryo. However, if the researcher causes the egg to divide so that only one daughter cell receives the gray crescent, only that cell can become a complete embryo.

Even though the first two cells of a frog embryo are normally able to develop into a complete embryo, this ability will eventually disappear. At some point in all embryos, some earlier than others, the developmental potential of a cell is restricted to a particular fate. Only a transplantation experiment that places the nucleus in the cytoplasmic contents of the complete egg shows that all the genes are still present in embryonic nuclei. This too, indicates the importance of cellular cytoplasmic contents in early development.

Cytoplasmic substances unequally distributed in the egg are parceled out during cleavage. These substances initially influence which genes are active, and how a cell will differentiate.

Morphogenesis

As development proceeds, a cell's differentiation is not only influenced by its cytoplasmic contents, but also by signals given off by neighboring cells. Migration of cells occurs during gastrulation, and there is evidence that one set of cells can influence the migratory path taken by another set of cells. Some cells produce an extracellular matrix that contains fibrils, and in the laboratory it can be shown that the orientation of these fibrils influences migratory cells. The cytoskeleton of the migrating cells is orientated in the same direction as the fibrils. Although this may not be an exact mechanism at work during gastrulation, it suggests that germ layer formation is probably influenced by environmental factors.

More specific information is known about neurulation. In 1924, the German, Hans Spemann, performed a series of experiments on frog embryos showing that presumptive (potential) notochord tissue induces the formation of the nervous system (fig. 43.6). If presumptive nervous system tissue, located just above the notochord, is cut out and transplanted to the belly region of the embryo, it will not form a neural tube. On the other hand, if presumptive notochord tissue is cut out and transplanted beneath what would be belly ectoderm, this ectoderm differentiates into neural tissue.

Figure 43.7 shows how the neural tube, an internal structure, forms from ectoderm. Microfilament contraction brings about invagination of the ectoderm and a pinching off of the neural tube.

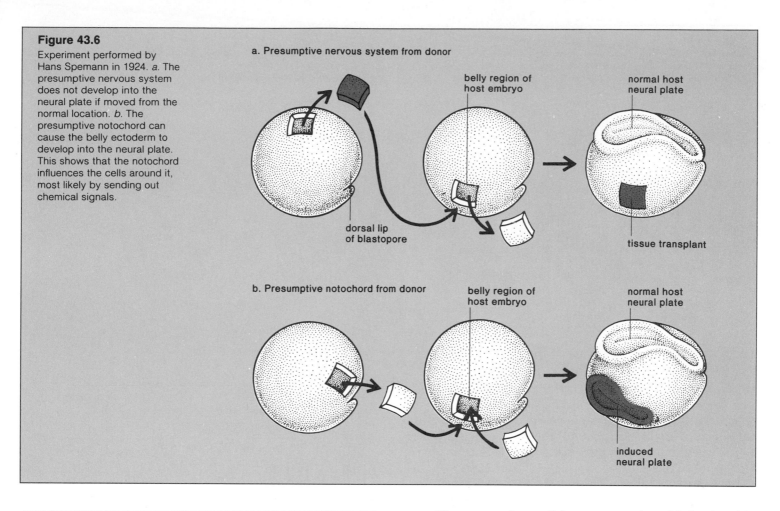

Figure 43.6

Experiment performed by Hans Spemann in 1924. *a.* The presumptive nervous system does not develop into the neural plate if moved from the normal location. *b.* The presumptive notochord can cause the belly ectoderm to develop into the neural plate. This shows that the notochord influences the cells around it, most likely by sending out chemical signals.

a. Presumptive nervous system from donor

belly region of host embryo

normal host neural plate

dorsal lip of blastopore

tissue transplant

b. Presumptive notochord from donor

belly region of host embryo

normal host neural plate

induced neural plate

Figure 43.7

a. Neural plate is an ectoderm layer that will give rise to the neural tube. *b.* As microfilaments contract, cells begin to invaginate. *c.* Continued constriction causes the neural tube to pinch off from the outer ectoderm layer.

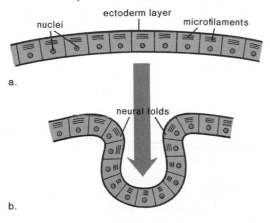

nuclei

ectoderm layer

microfilaments

a.

neural folds

b.

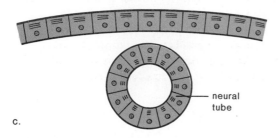

neural tube

c.

There are other well-known examples of **induction** (the ability of one tissue to influence the development of another tissue) besides the neural tube. In 1905, Warren Lewis studied the formation of the eye in frog embryos (fig. 43.8). He found that the optic vesicles, which are lateral outgrowths from developing brain tissue, induced the overlying ectoderm to thicken and become a lens. The developing lens in turn induced the optic vesicle to form the optic cup where the retina develops.

Spemann called the dorsal lip of the blastopore an organizer, because it gave rise to notochord tissue, which in turn induced neural tube formation. He envisioned that morphogenesis was dependent upon a series of organizers that came into being sequentially and directed future development. Today, we see the process of induction going on continuously—neighboring cells are always influencing one another. Either direct contact, or the production of a chemical acts as a signal that activates certain genes and brings about morphogenesis. This diagram shows how morphogenesis can be a sequential process:

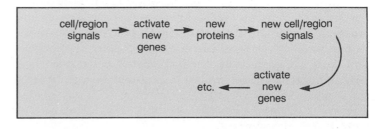

cell/region signals → activate new genes → new proteins → new cell/region signals

etc. ← activate new genes ←

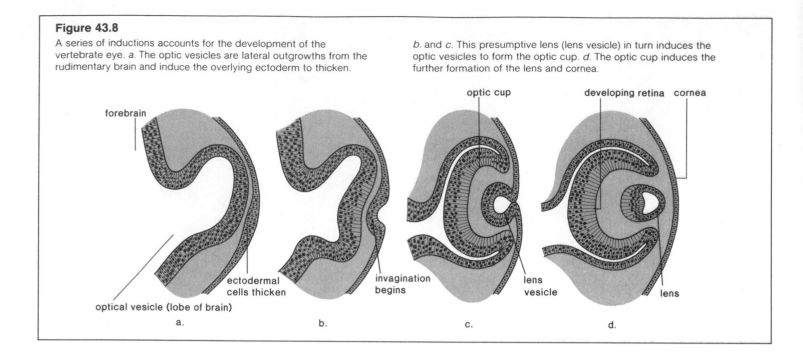

Figure 43.8

A series of inductions accounts for the development of the vertebrate eye. *a*. The optic vesicles are lateral outgrowths from the rudimentary brain and induce the overlying ectoderm to thicken.

b. and *c*. This presumptive lens (lens vesicle) in turn induces the optic vesicles to form the optic cup. *d*. The optic cup induces the further formation of the lens and cornea.

The diagram points out that there are both genetic and chemical aspects to morphogenesis. When the notochord induces the ectoderm to form the neural tube, a chemical signal is involved. The neural tube will form even if the two tissues are separated by a filter that allows only molecules to pass through.

Homeotic Genes and Pattern Formation

Regulatory genes that are specifically involved in development have been discussed. A homeotic gene is a regulatory gene that can turn on a whole series of genes involved in the development of a body part. Evidence for the existence of homeotic genes comes from experiments with *Drosophila,* in which a single gene mutation can have bizarre consequences. Some homeotic mutants have legs on their heads where antennae are supposed to be! Less severe homeotic mutations are known, in which only part of an antenna develops into a leg part instead. It is of interest that if the tip of the antenna is affected, it develops into a foot; if the base of the antenna is affected it develops into the proximal part of a leg. In other words, the structural genes controlled by a homeotic (regulatory) gene code not only for an organ but also for its morphogenic pattern. Homeotic genes have been discovered in other segmented animals with body parts that repeat.

Pattern formation has also been studied in chick embryos. Lewis Wolpert of Middlesex Hospital Medical School in London, after studying wing development, suggested that environmental signals help direct pattern formation. As the wing develops, there is an ectodermal ridge that pushes outward from the body. Cells that are in the ridge for the shortest length of time, develop into proximal parts of the wing, and those that have been there the longest, develop into distal parts. Wolpert further hypothesizes that there is a gradient of chemicals, called morphogens, in a wing bud, and they determine whether posterior or anterior parts will form.

Morphogenesis is dependent upon signals (either contact or chemical) from neighboring cells. These signals activate particular genes.

Human Development

Human development is often divided into embryonic development (first two months) and fetal development (three to nine months) (fig. 43.9). The embryonic period consists of early formation of the major organs, and fetal development is a refinement of these structures.

Of the three animals we have previously discussed, human development best resembles that of the chick, no doubt because reptiles are a common ancestor for both birds and mammals. In the chick, the germ layers extend over the yolk and develop into the extraembryonic membranes. Their placement and function are described in figure 43.10. In humans, the *extraembryonic membranes* (so-called because they lie outside the embryo), develop earlier than in the chick. In contrast to the chick, the human egg has little yolk, and the nutritional needs of the embryo are supplied by the membranes. This is the reason for their earlier development.

Embryonic Development

First Month

Cleavage Following fertilization in the oviduct, the human zygote begins to cleave as it travels down the oviduct to the uterus (fig. 43.11). The first cleavage is completed about a day after fertilization, and subsequent divisions occur at intervals of 8–10 hours. Passage of the embryo through the oviduct takes from 3–3½ days. While the embryo lies free in the uterine cavity,

Figure 43.9

Human development is divided into the embryonic period (first 2 months) and fetal development (third to ninth month). *a.* Embryo is not recognizably human. *b.* Fetus is recognizably human.

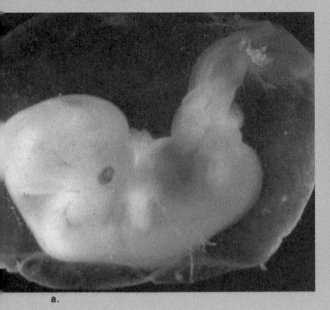

a.

b.

Figure 43.10

Evolution of the extraembryonic membranes in the reptile made development on land possible. If an embryo develops in water, the water supplies oxygen, takes away waste, prevents dessication and provides a protective cushion. On land some of these functions are performed by the extraembryonic membranes, which must be protected from drying out by an outer shell or by internal development. *a.* Chick within its hard shell. The chorion lies just beneath the shell and carries on gas exchange. The allantois collects nitrogenous wastes. The yolk sac provides nourishment, and the amnion provides a watery environment. *b.* In humans, only the chorion and amnion have comparable functions. The chorion forms the fetal half of the placenta, where exchange with the mother's blood occurs, and the amnion provides a watery environment. The yolk sac and allantois are vestigial and disappear as the umbilical cord forms. The allantoic blood vessels become the umbilical blood vessels.

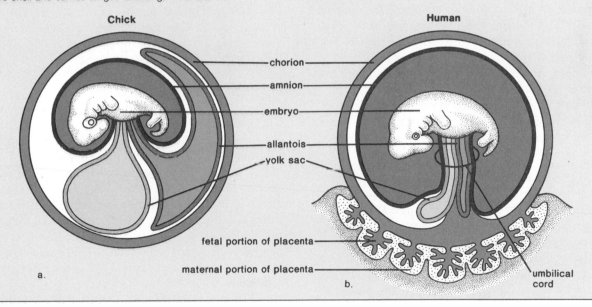

Chick

Human

chorion
amnion
embryo
allantois
yolk sac

fetal portion of placenta
maternal portion of placenta

a.

b.

umbilical cord

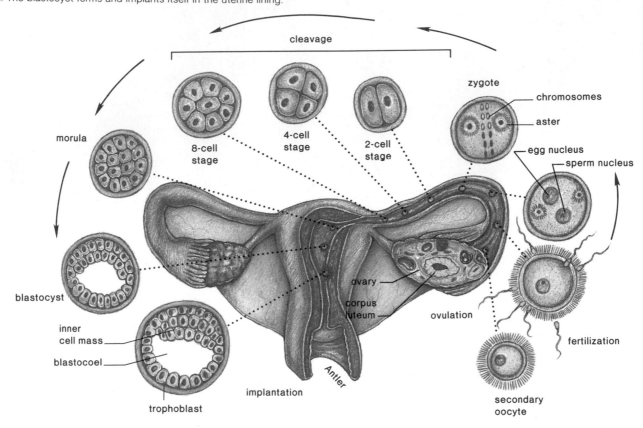

Figure 43.11

Human development before implantation. Structures and events proceed counterclockwise. At ovulation, the secondary oocyte leaves the ovary. Fertilization occurs in the oviduct. As the zygote moves along the oviduct, it undergoes cleavage to produce a morula. The blastocyst forms and implants itself in the uterine lining.

cleavage

zygote

chromosomes

aster

egg nucleus

sperm nucleus

morula

8-cell stage

4-cell stage

2-cell stage

blastocyst

ovary

corpus luteum

ovulation

inner cell mass

blastocoel

fertilization

Antler

implantation

secondary oocyte

trophoblast

the morula becomes the **blastocyst,** consisting of the *inner cell mass* and a single layer of surrounding cells known as the *trophoblast* (fig. 43.11). The trophoblast will be part of the **chorion,** an extraembryonic membrane. The chorion, in turn, will become the fetal part of the placenta.

At the end of the first week, the embryo begins the process of implanting itself in the wall of the uterus. During this process, the trophoblast secretes enzymes to digest away some of the tissue and blood vessels of the uterine wall (fig. 43.11). After the blastocyst has entered the endometrium, uterine tissue grows over the surface and heals the implantation site.

Gastrulation As implantation occurs, the embryo undergoes gastrulation. During the first stage of gastrulation, the inner cell mass becomes the *embryonic disc,* consisting of a layer of endodermal cells beneath an ectodermal layer of cells (fig. 43.12a). The **amnion,** an extraembryonic membrane that contains a cavity filled with amniotic fluid, is now visible above the embryo. The amniotic fluid cushions the embryo and keeps it from drying out. At this time we also see the **yolk sac,** an extraembryonic membrane that is largely vestigial.

As gastrulation continues, mesoderm forms by invagination of cells between the ectoderm and endoderm, along the length of a primitive streak. When the edges of the embryonic disc fold under, the embryo is converted into a tubular, three-layered body. The ectoderm is on the outside, the endoderm is on the inside, and the mesoderm is between these two.

During gastrulation, the placenta is developing between the embryo and its mother. A series of villi (fingerlike outgrowths) develop on the surface of the chorion. Only villi on one side will become a part of the placenta; the rest regress and later disappear. Within the body stalk, the **allantois,** which is the last of the embryonic membranes (fig. 43.10), contributes to the development of the umbilical blood vessels, which take blood to and from the placenta within the umbilical cord.

The fully formed **placenta** has a fetal side (the chorion) and a maternal side. Their blood supplies lie in close proximity so wastes and carbon dioxide can move from the fetal to the maternal side of the placenta, and nutrients and oxygen can move from the maternal to the fetal side. The chorionic villi are surrounded by maternal blood sinuses, yet the blood of mother and fetus do not mix. Instead, exchange takes place across plasma membranes.

The immature placenta begins producing HCG (human chorionic gonadotropic hormone) right away so that the corpus luteum in the ovary does not degenerate for as long as six months. The mature placenta not only produces HCG, but also progesterone and estrogen to maintain the supply of these hormones

Figure 43.12

Stages showing the early appearance of the extraembryonic membranes and the formation of the umbilical cord in the human embryo. *a.* At the first week, as gastrulation occurs, the amniotic cavity appears. *b.* At the second week the chorion and yolk sac are now apparent. *c.* At the third week, the body stalk and allantois form. *d.* and *e.* At weeks four and five, the embryo begins to take shape as the umbilical cord forms.

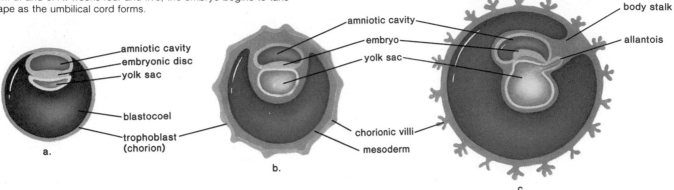

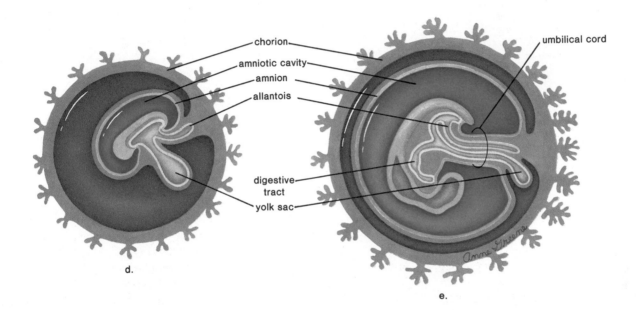

after the corpus luteum does degenerate. The high level of progesterone and estrogen in the pregnant woman's body assures, through the feedback mechanism discussed in the previous chapter, that no new follicles will mature, and that the uterine lining is maintained so embryonic growth can continue.

Organ Formation By the middle of the third week, the mesoderm cells of the central axis have formed a notochord. The central nervous system begins to develop above the notochord. Neural folds arise from the ectoderm and fuse to give the neural tube that swells anteriorly to form the brain (fig. 43.13). The *somites* appear on both sides of the neural tube. They will become the dermis of the skin, the muscle tissue, and the vertebrae and other bones.

The heart and blood vessels also arise from the mesoderm. At first, the heart is two tiny tubes. They fuse to form a single chamber as the third week ends. This primitive heart bulges and is easily observed as it begins to pump blood through simple

Figure 43.13

Human embryo at 21 days. The neural folds still need to close at the anterior and posterior ends of the embryo. The pericardial area contains the primitive heart, and the somites are the precursors of the muscles.

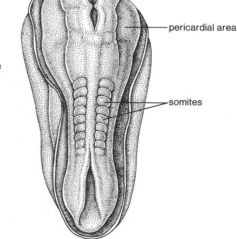

Detecting Birth Defects

A new method called chorionic villi sampling (fig. *a*) allows physicians to collect embryonic cells as early as the fifth week. The doctor inserts a long thin tube through the vagina into the uterus. With the help of ultrasound, which gives a picture of the uterine contents, the tube is placed between the lining of the uterus and the chorion. Then suction is used to remove a sampling of the chorionic villi cells. Chromosomal analysis and biochemical tests to detect some genetic defects can be done immediately on these cells. Before the development of chorionic villi sampling, physicians had to wait until about the sixteenth week of pregnancy to perform amniocentesis. In amniocentesis (fig. *b*), a long needle is passed through the abdominal wall to withdraw a small amount of amniotic fluid along with fetal cells. Since there are only a few cells in the amniotic fluid, testing must be delayed until cell culture has caused them to grow and multiply. Therefore it may be another two to four weeks until the prospective parents are told whether their child has a genetic defect.

In fetoscopy, another possible procedure, the physician uses an endoscope to view the fetus and can withdraw blood for prenatal diagnosis.

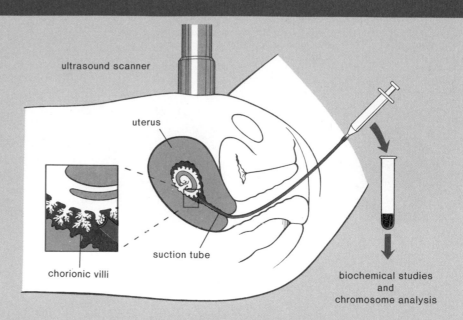

a.

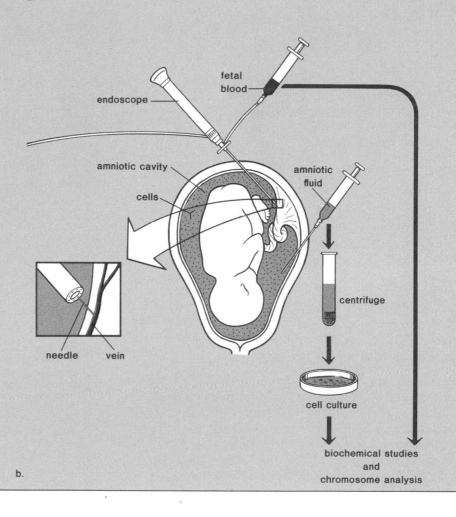

b.

Figure 43.14
Human embryo at beginning of fifth week. *a.* Scanning electron micrograph. *b.* Drawing. The embryo is curled so that the head touches the heart, two organs whose development is further along than the rest of the body. The organs of the gastrointestinal tract are forming. The presence of the tail is an evolutionary remnant; it will later regress. The arms and legs will develop from the bulges that are called limb buds.

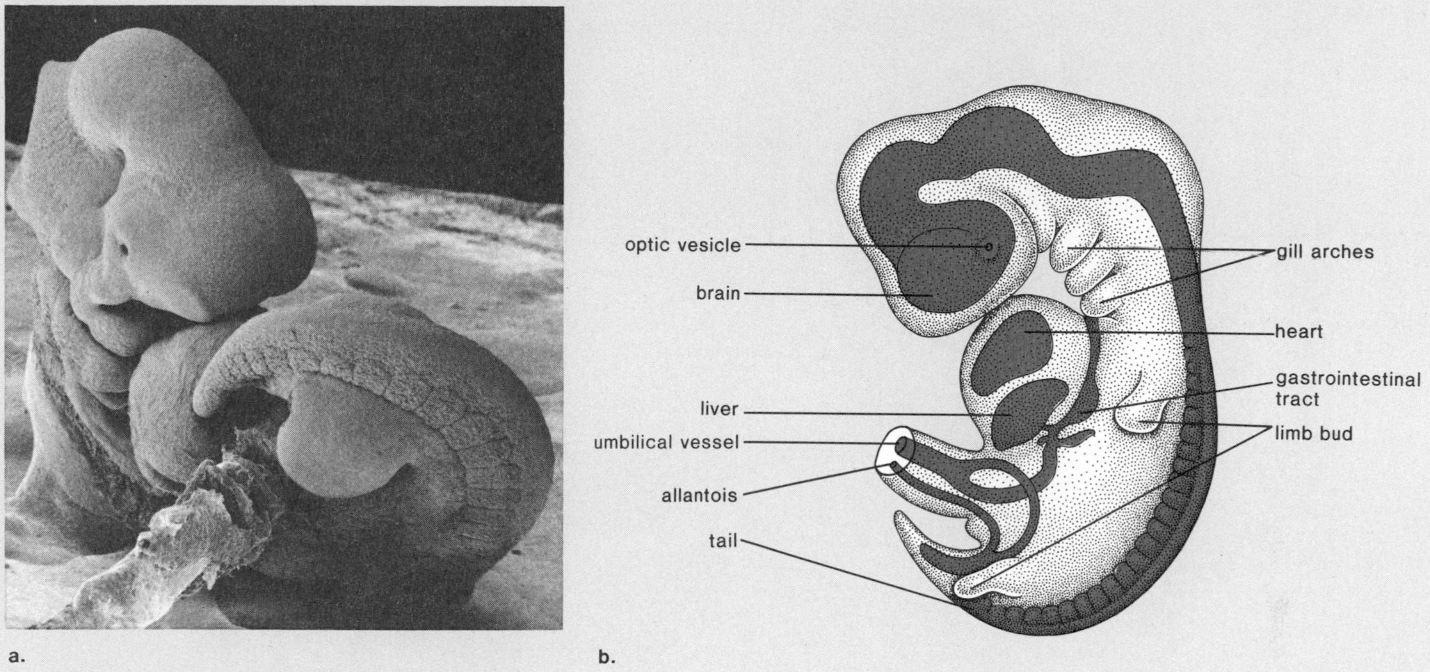

arteries and veins. These vessels take blood to the developing esophagus, stomach, liver, pancreas, and intestine, all emerging from the endoderm.

During the fourth week, eyes, ears, and nose appear as pits on an oversized head that is flexed, and appears to rest on the heart bulging beneath it. The embryo is still quite tiny, only 5–6 mm. Only now is the mother beginning to suspect she is pregnant, even though the embryo has many organs already forming.

Second Month

During the second month, the arms and then legs begin as **limb buds** (fig. 43.14). At first the buds look like paddles. As they grow, hands, forearms, upper arms, and shoulders appear. The newly formed arms bend at the elbow, and the legs bend at the knee to allow these appendages to reach across the body, fingers and toes nearly touching.

At first the fetus' skeleton is cartilage, but it begins to be replaced by bone during the sixth week. Just as the growth of the arms is faster than the growth of the legs, so the growth of the head outstrips the growth of the trunk. At the end of the second month, the embryo measures roughly 2.5 cm from crown to rump; about half of this is head. The eyes, nose, mouth, and ears are all recognizable, and the embryo now has a distinctly human appearance.

Most birth defects are either genetic in origin or they are caused by insults suffered during embryonic development. Two methods are available for detecting birth defects and these are described in the reading on the opposite page.

Fetal Development

Fetal development is marked by an extreme increase in size. Weight multiplies 600 times, going from less than 1 ounce to 7 pounds. In this time, too, the fetus grows to about 50 cm in length. The genitalia finally make their appearance, and it is possible to tell if the fetus is male or female. With this accelerated growth, proportions change. The head is now only one-half to one-fourth of the body's total length.

Soon, hair, eyebrows, and eyelashes add finishing touches to the face and head. In the same way, fingernails, and toenails complete the hands and feet. A fine downy hair (lanugo) covers the limbs and trunk, only to later disappear. The fetus looks like an old man because the skin is growing so fast that it wrinkles. A waxy, almost cheeselike substance (vernix caseasa) protects the wrinkly skin from the watery amniotic fluid.

The fetus at first only flexes its limbs and nods its head, but later it can move its limbs vigorously to avoid discomfort. The mother will feel these movements from about the fourth month on. The other systems of the body also begin to function (fig. 43.15). The fetus begins to suck its thumb, swallow amniotic fluid, and urinate.

After 16 weeks, the fetal heartbeat is heard through a stethoscope. A fetus born at 24 weeks has a chance of surviving, although the lungs are still immature and often cannot capture oxygen adequately. Weight gain during the last couple of months increases the likelihood of survival.

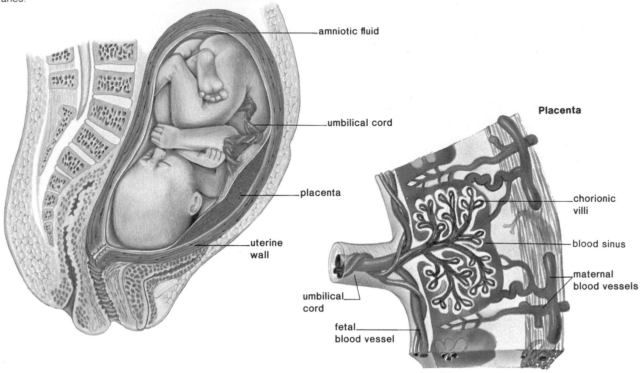

Figure 43.15
This fetus is 6–7 months old and the placenta is well developed. The placenta is the region of exchange between fetus and mother. Here, fetal circulation is separated from maternal circulation only by thin membranes.

amniotic fluid

umbilical cord

placenta

uterine wall

Placenta

chorionic villi

blood sinus

maternal blood vessels

umbilical cord

fetal blood vessel

Birth

The time of birth is usually calculated at 280 days from the start of the last menstruation. As pregnancy progresses, the level of estrogen in the bloodstream exceeds the level of progesterone. This may help bring on birth of the fetus, because estrogen promotes uterine irritability. It is also possible that the fetus itself initiates the birth process by releasing a chemical messenger. Researchers have found in studies with sheep that if the fetal hypothalamus or pituitary is destroyed, birth can be postponed indefinitely.

The process of birth includes three stages: dilation of the cervix, expulsion of the fetus, and delivery of the afterbirth (the placenta and the extraembryonic membranes).

Development after Birth

Development does not cease once birth has occurred, but continues throughout the stages of life: infancy, childhood, adolescence, and adulthood.

Infancy lasts until about two years of age. It is characterized by tremendous growth and sensorimotor development. During *childhood*, the individual grows, and the body proportions change (fig. 43.16). *Adolescence* begins with **puberty**, when the secondary sex characteristics appear, and the sexual organs become functional. At this time, there is an acceleration of growth leading to changes in height, weight, fat distribution, and body proportions. Males commonly experience a growth

spurt later than females; therefore, they grow for a longer period of time. Males are generally taller than females and have broader shoulders and longer legs relative to their trunk length.

Young adults are at their physical peak in muscle strength, reaction time, and sensory perception. The organ systems at this time are best able to respond to altered circumstances in a homeostatic manner. From now on, there will be an almost imperceptible, gradual loss in certain of the body's abilities.

Aging

Aging encompasses the progressive changes that contribute to an increased risk of infirmity, disease, and death. Figure 43.17 compares the percentage of function of various organs in a 75–80 year old man compared to that of a 20 year old, whose organs are assumed to function at 100% capacity. When making this comparison, we may note that the body has a vast functional reserve, so that it can still perform well, even when not at 100% capacity.

There are many theories about what causes aging, and we will consider three of them.

Genetic Origin Several lines of evidence indicate that aging has a genetic basis. For example, the children of long-lived parents tend to live longer than those of short-lived parents. Perhaps the genes are programmed to control aging and time of death. The maximum life span of animals is species specific; for humans it is about 110 years. The number of times a cell will divide is also species specific. The maximum number of times

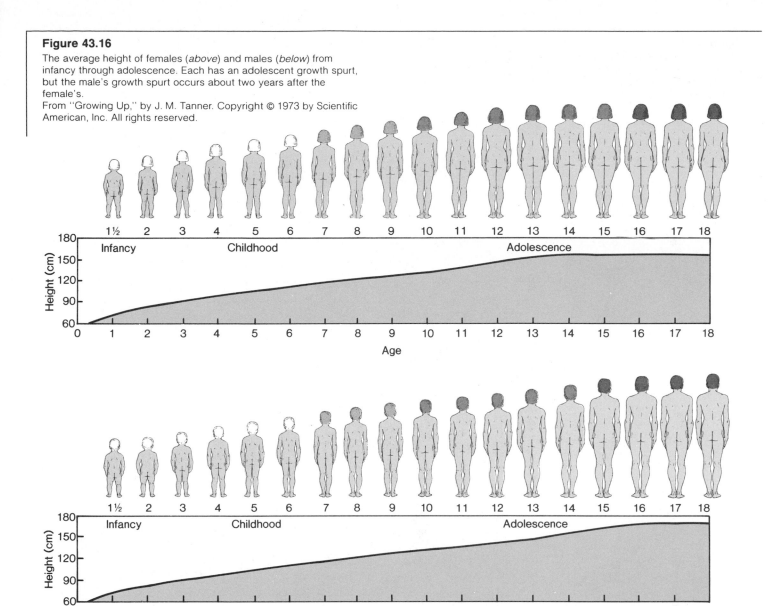

Figure 43.16

The average height of females (*above*) and males (*below*) from infancy through adolescence. Each has an adolescent growth spurt, but the male's growth spurt occurs about two years after the female's.

From "Growing Up," by J. M. Tanner. Copyright © 1973 by Scientific American, Inc. All rights reserved.

for human cells is about fifty. Perhaps as we grow older, more and more cells are unable to divide and instead undergo degenerative changes and die. Some cell lines may become nonfunctional long before the maximum number of divisions has occurred. Whenever DNA replicates, mutations can occur and may lead to production of nonfunctional proteins. Eventually, the number of inadequately functioning cells may build up and contribute to an aging process.

Whole Body Processes A decline in the hormonal system may affect many different organs of the body. For example, diabetes type II is a common occurrence in older individuals. The pancreas makes insulin, but the cells lack the receptors that enable them to respond. Menopause in women occurs for a similar reason. There is plenty of FSH in the bloodstream, but the ovaries do not respond. Perhaps aging results from the loss of hormonal activities, and a decline in the functions they control.

The immune system, too, no longer performs as it once did, and this may affect the body as a whole. The thymus gland gradually decreases in size, and eventually, most of it is replaced by fat and connective tissue. The incidence of cancer increases among the elderly, which may signify that the immune system is no longer functioning as it should. This idea is further substantiated by the increased incidence of autoimmune diseases in older individuals.

It's possible, though, that aging is not due to the failure of a particular system that affects the body as a whole, but to a specific type of tissue change that affects all organs and even the genes. It has been noticed for some time that proteins, such as collagen, which makes up the white fibers and is present in many support tissues, becomes increasingly cross-linked as people age. Undoubtedly, this cross-linking contributes to the

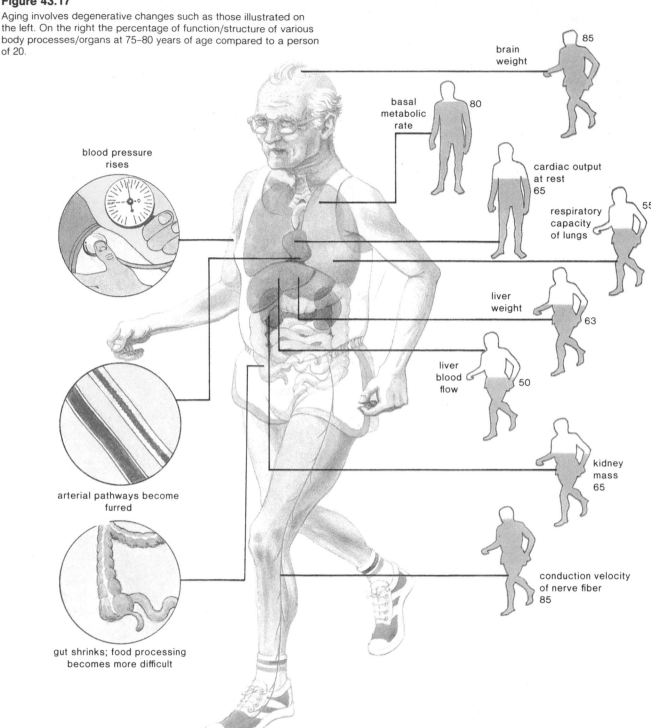

Figure 43.17

Aging involves degenerative changes such as those illustrated on the left. On the right the percentage of function/structure of various body processes/organs at 75–80 years of age compared to a person of 20.

brain weight 85

basal metabolic rate 80

blood pressure rises

cardiac output at rest 65

respiratory capacity of lungs 55

liver weight 63

liver blood flow 50

arterial pathways become furred

kidney mass 65

conduction velocity of nerve fiber 85

gut shrinks; food processing becomes more difficult

stiffening and loss of elasticity characteristic of aging tendons and ligaments. It may also account for the inability of organs, such as the blood vessels, heart, and lungs, to function as they once did. Researchers have found that glucose has a tendency to attach to any type of protein. This attachment is the first step in a cross-linking process that ends with the formation of "advanced glycosylation end products (AGEs)." The presence of AGE-derived cross-links not only explains why cataracts develop, it may also help explain the development of atherosclerosis, and the inefficiency of the kidneys in diabetics and older individuals. Even DNA seems capable of forming AGE-derived cross-links, and perhaps this increases the rate of mutations as we age. These researchers are presently experimenting with the drug aminoguanidine, which can prevent the development of AGEs.

Extrinsic Factors The present data on the effects of aging are often based on comparing the characteristics of the elderly to younger age groups. But today's elderly perhaps were not as aware of the importance of diet and exercise for general health. It's possible, then, that much of what we attribute to aging is instead due to years of poor health habits.

For example, osteoporosis is associated with a progressive decline in bone density in both males and females. Fractures are more likely to occur after only minimal trauma. Osteoporosis is common in the elderly—by age 65 one-third of women will have vertebral fractures and by age 81, one-third of women and one-sixth of men will have suffered a hip fracture. While there is no denying that a decline in bone mass results from aging, certain extrinsic factors are also important. The occurrence of osteoporosis itself is associated with cigarette smoking, heavy alcohol intake, and inadequate calcium intake. Not only is it possible to eliminate these negative factors by personal choice, it is also possible to add a positive factor. A moderate exercise program has been found to slow down the progressive loss of bone mass.

Rather than collecting data on the average changes observed between different age groups, it might be more useful to note the differences within any particular age group so that extrinsic factors that contribute to a decline, or that promote the health of an organ can be identified.

Summary

1. Any developmental change requires three processes: growth, differentiation, and morphogenesis.

2. The development of three types of animals (lancelet, frog, and chick) is compared. The first three stages (cleavage, blastulation, and gastrulation) differ in the amount of yolk in the egg.

3. During cleavage, the zygote divides, but there is no overall growth. The result is a morula, which becomes the blastula when an internal cavity (the blastocoel) appears. During the gastrula stage, invagination of cells into the blastocoel results in formation of the germ layers: ectoderm, mesoderm, and endoderm. Later development of organs can be related to these layers.

4. At the neurula stage, the nervous system develops from mid-line mesoderm, just above the notochord. At this point, it is possible to draw a typical cross section of a vertebrate embryo (fig. 43.4).

5. Differentiation begins with cleavage, as the egg's cytoplasm is partitioned among the numerous cells. The cytoplasm is not uniform in its content, and presumably each of the first few cells will differ as to their cytoplasmic contents. Some probably contain substances that can influence gene activity—turning some genes on and others off.

6. After the first cleavage of a frog embryo, only a daughter cell that receives a portion of the gray crescent is able to develop into a complete embryo. This illustrates the importance of cytoplasmic inheritance to early development.

7. Morphogenesis involves the process of induction. Hans Spemann performed a series of experiments showing that the notochord induces the formation of the neural tube in frog embryos. The reciprocal induction that occurs between the lens and the optic vesicle is another good example of this phenomenon.

8. Today we envision induction as always present, because cells are believed to constantly give off signals that influence the genetic activity of neighboring cells. This is the cause of morphogenesis.

9. The studies of homeotic genes and pattern formation further indicate that morphogenesis involves genetic and environmental influences.

10. Human development can be divided into embryonic development (first two months) and fetal development (three to nine months). The early stages in human development resemble those of the chick. The similarities are probably due to their evolutionary relationship, not the amount of yolk the eggs contain, because the human egg has little yolk.

11. Fertilization occurs in the oviduct, and cleavage occurs as the embryo moves toward the uterus. The morula becomes the blastocyst before implanting in the uterine lining. Gastrulation occurs as in the chick, and the three germ layers appear.

12. The extraembryonic membranes appear early in human development. The trophoblast of the blastocyst signals the early development of the chorion, which goes on to become the fetal part of the placenta, where exchange occurs between fetal and maternal blood. During gastrulation, the amnion appears, as do the vestigial yolk sac and allantois. The amnion contains the amniotic fluid, which cushions and protects the embryo. The other two membranes are largely vestigial, except that the allantoic blood vessels become the umbilical blood vessels.

13. Organ development begins with the neural tube and heart formation. There follows a steady progression of organ formation during embryonic development. During fetal development, refinement of features occurs, and the fetus adds weight. Birth occurs about 280 days after the start of the last menstruation.

14. Development after birth consists of infancy, childhood, adolescence, and adulthood. Young adults are at their prime, and then the aging process begins. Perhaps aging is genetic in origin, due to a change that affects the whole body, or due to extrinsic factors.

Objective Questions

1. Which of these stages is the first one out of sequence?
   a. cleavage
   b. blastula
   c. morula
   d. gastrula

2. Which of these stages is mismatched?
   a. cleavage—cell division
   b. blastula—gut formation
   c. gastrula—three germ layers
   d. neurula—nervous system

3. Which of the germ layers is best associated with development of the heart?
   a. ectoderm
   b. mesoderm
   c. endoderm
   d. All of these.

4. Differentiation begins at what stage?
   a. cleavage
   b. blastula
   c. gastrula
   d. neurula
5. Morphogenesis is best associated with
   a. overall growth.
   b. induction of one tissue by another.
   c. genetic mutations.
   d. All of these.
6. In humans, the placenta develops from the chorion. This indicates that human development
   a. resembles that of the chick.
   b. is dependent upon extraembryonic membranes.

   c. cannot be compared to lower animals.
   d. only begins upon implantation.
7. In humans, the fetus
   a. is surrounded by four extraembryonic membranes.
   b. has developed organs and is recognizably human.
   c. is dependent upon the placenta for excretion of wastes and acquisition of nutrients.
   d. Both (b) and (c).
8. Developmental changes
   a. require growth, differentiation, and morphogenesis.
   b. stop occurring as soon as one is grown.

   c. are dependent upon a parceling out of genes into daughter cells.
   d. Both (a) and (c).
9. Mesoderm forms by invagination of cells in
   a. the lancelet.
   b. the frog.
   c. the chick.
   d. All of these.
10. Which of these is mismatched?
   a. brain—ectoderm
   b. gut—endoderm
   c. bone—mesoderm
   d. lens—endoderm

Study Questions

1. State, define, and give examples of the three processes that occur whenever a developmental change occurs.
2. Compare the process of cleavage and the formation of the blastula and gastrula in lancelets, frogs, and chicks.
3. State the germ layer theory, and name organs derived from each of the germ layers.
4. Draw a cross section of a typical chordate embryo at the neurula stage and label your drawing. Compare to figure 43.4 and correct if necessary.

5. Give reasons for suggesting that differentiation begins with the embryonic stage of cleavage.
6. Explain an experiment performed by Spemann that suggests that the notochord induces formation of the neural tube. Give another well-known example of induction between tissues.
7. Give reasons for suggesting that morphogenesis is dependent upon signals given off by neighboring cells. What do the signals bring about in the receiving cells?

8. List the human extraembryonic membranes, give a function for each, and compare this function to that in the chick.
9. Tell where the stages of fertilization, cleavage, morula, and blastula occur in humans. What happens to the blastula in the uterus?
10. Describe the structure and function of the placenta in humans.
11. Outline the developmental changes that occur in humans from fetal development to adulthood.
12. Discuss three theories of aging.

Thought Questions

1. Support the custom of relying on embryological evidence to determine which animals are closely related.

2. The female contributes most of the cytoplasm to the zygote. Therefore, argue that inheritance from the female parent is more important than from the male parent.

3. Develop an overall theory of aging that would draw from the three different theories presented in this chapter.

Selected Key Terms

differentiation (dif″er-en″she-a′shun) 678
morphogenesis (mor″fo-jen′ĕ-sis) 678
cleavage (klēv′ij) 680
blastocoel (blas′to-sēl) 680
blastula (blas′tu-lah) 680

gastrula (gas′troo-lah) 680
ectoderm (ek′to-derm) 680
endoderm (en′do-derm) 680
mesoderm (mes′o-derm) 680
germ layers (jerm la′erz) 680

chorion (kor′e-on) 686
amnion (am′ne-on) 686
yolk sac (yōk sak) 686
allantois (ah-lan′to-is) 686
placenta (plah-sen′tah) 686

Suggested Readings for Part 5

Brown, M. S., and Goldstein, J. L. November 1984. How LDL receptors influence cholesterol and atherosclerosis. *Scientific American.*

Cantin, M., and Genest, J. February 1986. The heart as an endocrine gland. *Scientific American.*

Cerami, A. et al. May 1987. Glucose and aging. *Scientific American.*

Cohen, I. R. April 1988. The self, the world and autoimmunity. *Scientific American.*

Crapo, L. 1985. *Hormones, the messengers of life.* New York: W. H. Freeman and Co.

Eckert, R., and Randall, D. 1983. *Animal physiology: Mechanisms and adaptations.* 2d ed: New York: W. H. Freeman and Co.

Feder, M. E., and Burggren, W. W. November 1985. Skin breathing in vertebrates. *Scientific American.*

Gehring, W. J. October 1985. The molecular basis of development. *Scientific American.*

Guyton, A. C. 1984. *Physiology of the human body.* 6th ed. Philadelphia: W. B. Saunders.

Hadley, N. F. July 1986. The arthropod cuticle. *Scientific American.*

Hegarty, V. 1988. *Decisions in nutrition.* St. Louis, Missouri: Times Mirror/Mosby College Publishing.

Hole, J. W. 1987. *Human anatomy and physiology.* 4th ed. Dubuque, IA: Wm. C. Brown Publishers.

Koretz, J. F., and Handelman, G. H. July 1988. How the human eye focuses. *Scientific American.*

Mishkin, M., and Appenzeller, T. June 1987. The anatomy of memory. *Scientific American.*

Orci, L. et al. September 1988. The insulin factor. *Scientific American.*

Robinson, T. F. et al. June 1986. The heart as a suction pump. *Scientific American.*

Schnapt, J. L., and Baylor, D. A. April 1987. How photoreceptor cells respond to light. *Scientific American.*

Scientific American. October 1988. Entire issue is devoted to articles on AIDS.

Snyder, S. H. October 1985. The molecular basis of communication between cells. *Scientific American.*

Stryer, L. July 1987. The molecules of visual excitation. *Scientific American.*

Tonegawa, S. October 1985. The molecules of the immune system. *Scientific American.*

Wassarman, P. M. December 1988. Fertilization in mammals. *Scientific American.*

Young, J., and Cohn, Z. January 1988. How killer cells kill. *Scientific American.*

CRITICAL THINKING
CASE STUDY

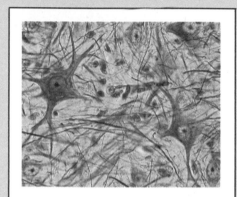

PART 5
Animal Structure and Function

When animals are exposed to foreign substances called antigens, they produce antibodies against them. Early investigators were aware of two hypotheses (*see* figure) to explain the sudden appearance of antibodies in animals immunized to specific antigens:

Template hypothesis—Any antigen (and more than one type at a time) can enter any lymphocyte and once inside, an antigen acts as a *template* causing a newly forming antibody to become specific. Then the antibody is secreted.

Clonal selection hypothesis—Each lymphocyte has one type of membrane-bound antibody that is already specific. When an antigen combines with a membrane-bound antibody, this lymphocyte is *selected* to give rise to a *clone* of plasma cells that secrete many specific antibodies.

Does each lymphocyte have the potential to produce more than one kind of antibody according to the template hypothesis? according to the clonal selection hypothesis? Is it possible for each lymphocyte to produce two or more kinds of antigens according to the template theory? according to the clonal selection hypothesis? Will there, with time, be many more cells to react to a specific antigen according to the template hypothesis? according to the clonal selection hypothesis?

Many experiments were done to test which of these two theories is supported by data. For example, in one experiment mice were immunized to two different types of antigens. The antigens were flagella proteins from two different motile bacteria. After a few days, the experimenter removed the lymph nodes from the mice, and one lymph node cell (lymph node cells are capable of producing antibodies) was placed per depression on a microscope slide. If a lymph node cell is producing antibody to a bacterial flagellum, movement of the bacterium ceases immediately. What do you predict would happen when both types of bacteria are added, one at a time to a depression containing a single lymph node cell according to the template hypothesis? according to the clonal selection hypothesis?

Prediction 1 If the template hypothesis is correct, both types of bacteria would be immobilized by each lymph node cell. If the clonal selection hypothesis is correct, only one type of bacterium would be immobilized by each lymph node cell.

Result 1 Only one type of bacterium was immobilized for each lymph node cell. Therefore, the clonal selection hypothesis was supported.

If there had been more than one cell per depression what might have happened? Would it have been possible to distinguish which hypothesis was supported? Assuming that the clonal selection hypothesis is correct, what would have been the chance of an antibody-antigen reaction if lymph node cells had been taken from nonimmunized mice?

In another series of experiments, the spleen cells (where lymphocytes are also produced) of nonimmunized mice were exposed to a radioactively labeled antigen. The spleen cells were then washed free of any antigens that had not reacted to them. Which hypothesis—the template or clonal selection—would predict that very few spleen cells from nonimmunized animals would combine with the labeled antigen?

Prediction 2 Very few spleen cells from a nonimmunized animal would be found with bound antigen according to the clonal selection hypothesis.

Result 2 Subsequent measurement and analysis showed that, as predicted, only a few cells bound the antigen. These results are consistent with the clonal selection hypothesis.

Which hypothesis—the template or clonal selection—would predict that many cells will combine with the labeled antigen when spleen cells are taken from immunized mice?

Prediction 3 Many spleen cells from an immunized animal would bind to labeled antigen according to the clonal selection theory.

Result 3 Experiments showed that immunized animals have many spleen cells capable of binding with the labeled antigen.

Summarize the data from Results 1 through 3. How do the data fail to support the template hypothesis? On the other hand how do the data support the clonal selection hypothesis?

According to the clonal selection hypothesis, a small number of antigen-binding cells, such as observed in Result 2, produce by division a larger number of antigen-binding cells, such as observed in Result 3.

Suppose lymphocytes that have bound to a specific antigen are destroyed soon thereafter. Would you still be able to observe antibody-producing cells for this antigen according to the template hypothesis? according to the clonal selection hypothesis?

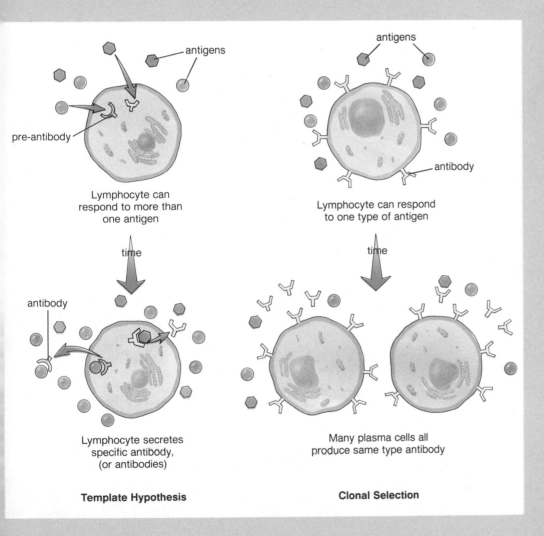

antigens

pre-antibody

Lymphocyte can
respond to more than
one antigen

time

antibody

Lymphocyte secretes
specific antibody,
(or antibodies)

Template Hypothesis

antigens

antibody

Lymphocyte can respond
to one type of antigen

time

Many plasma cells all
produce same type antibody

Clonal Selection

To observe this consequence, this experiment was performed: (1) Mice were injected with radioisotope-labeled antigen that destroyed any lymphocyte bound to the antigen. (2) All remaining lymphocytes were removed from these animals and (3) placed in mice whose immune systems had been destroyed by exposure to X rays. (4) These mice were injected with two types of antigens, a new one and the same one used previously, only this time neither antigen was labeled.

Predict which type antibodies the mice would produce if the clonal selection hypothesis is supported?

Prediction 4 If the clonal selection hypothesis is supported, the animals would produce antibody in response to the new antigen, but not to the antigen used previously.

Result 4 As predicted, the mice could respond to the new antigen but not to the antigen that was used previously.

Why were mice used for this experiment— why not just laboratory glassware? Why do you need two groups of mice–one that can produce new lymphocytes and one that no longer can produce new lymphocytes? Why not just inject the antigen both times into the same animal? Why do you need to use two different antigens? What was the role of the new antigen?

This series of investigations shows how a variety of experiments provide evidence to support one hypothesis over the other. Since experiments, such as those described here, gave increasing support to the clonal selection hypothesis of lymphocyte and antibody production, it is now called the clonal selection theory.

Gordon, L. Ada and Nossal, Sir Gustav. August 1987. The clonal-selection theory. *Scientific American*.

P A R T 6

Behavior and Ecology

It's been surprising to discover that all sorts of mammals live in highly organized societies. Prairie dogs living in burrows beneath the ground form social groups and several groups may make up a town that covers over a hundred acres. When prairie dogs meet they touch noses. This greeting most likely allows them to identify the members of their group.

The burrows provide the prairie dogs with a refuge from predators, such as hawks, and give them a place to rear their young. A dominant male maintains the group's boundaries by challenging (with chases and calls) other prairie dogs from neighboring groups. In a breeding colony there is usually one adult male for every four adult females. Both sexes interact with young pups allowing them to follow behind and even playing with them.

CHAPTER 44

Animal Behavior

Your study of this chapter will be complete when you can:

1. Give two main avenues of research in the field of behavior. Explain the purpose of each.
2. Give evidence that behavior is inherited and adaptive.
3. Distinguish between and give examples of innate and learned behaviors.
4. Give examples of fixed-action patterns and how it is possible to detect a sign stimulus.
5. Give examples of external stimuli that help animals orient themselves and navigate long distances.
6. Describe an experiment that suggests animals are motivated by their internal state.
7. Name three types of learned behavior and give examples of each.
8. List ways in which members of a society communicate.
9. Tell how competitive interactions are controlled among members of a society.
10. Give an example of a courtship pattern that shows how aggression is controlled.
11. Explain the main tenets of sociobiology giving examples of fitness, altruism, kin selection, and inclusive fitness.

These kangaroos are trying to push each other to the ground in a battle over females. Aggression between members of a society is ritualized and neither party in the struggle is usually harmed by the conflict.

Figure 44.1

A diagram that emphasizes current interests in regard to behavior. On the left-hand side of the diagram we note that selected phenotypes contribute genes to the gene pool of the population and in this way behavior can evolve. On the right-hand side of the diagram we note that environmental stimuli cause organisms to display behavioral patterns which differ as to the degree they are innate or learned. In the latter case, the capacity to learn is a part of the phenotype.

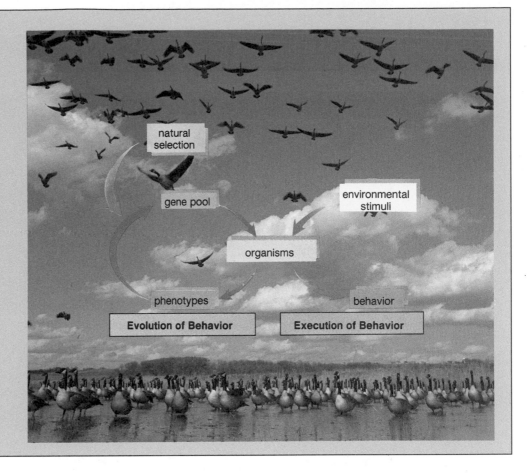

natural selection

gene pool

environmental stimuli

organisms

phenotypes

behavior

Evolution of Behavior

Execution of Behavior

E thologists (scientists who study behavior) ask two types of questions about a behavior, such as the way birds build a nest, bees find food, or elephant seals defend a territory. First, they are interested in how the behavior evolved or in what way it is adaptive. For example, most pet owners have observed that cats tend to be loners, and dogs tend to prefer companionship. This can be traced to the fact that cats were originally adapted to living in forests, where there was no advantage to hunting or living in groups, whereas dogs were originally adapted to living in grasslands, where they learned to hunt cooperatively. Second, ethologists are also interested in the degree to which behavior is instinctive (innate) or learned. For example, the courtship pattern of stickleback fish always occurs in exactly the same way, and digger wasps always lay their eggs in exactly the same way. Therefore, these behaviors seem to be innate.

The left-hand side of figure 44.1 assumes that an animal's behavior is adaptive—that it increases the possibility that the individual's genes will be passed on to the next generation. In other words, behavior patterns are subject to the process of natural selection just as other aspects of the phenotype are. Those individuals with the most adaptive behavior patterns are those that contribute most to the gene pool of the population.

The right-hand side of figure 44.1 concerns the execution of the behavior pattern. The phenotype of the organism includes a repertoire of behavior patterns, and a particular stimulus determines which pattern will be executed. The behavior pattern can range from being primarily innate to being primarily learned. In either instance, both genetic inheritance and the environment have played a role in shaping the behavior pattern. If it is largely innate, then the genetic component most likely plays the dominant role, and if it is learned, then the environment most likely plays the dominant role.

Evolution of Behavior

There are two ways to approach the study of the evolution of behavior—to show that the behavior pattern is adaptive and/or to show that it is inherited.

Adaptation of Behavior

Nikolaas Tinbergen, who wrote *Herring Gull's World* in 1960, emphasized the importance of making field observations of animals under natural conditions. For example, he observed that black-headed gulls always remove broken eggshells from their nests after the young have hatched. In a series of experiments, he determined that nests in which broken shells remained were subject to more predator attacks by crows than those from which the shells had been removed. It can be reasoned that the gulls that remove broken eggshells from their nests will successfully raise more young than the gulls that do not. This behavior pattern has been selected, then, through the evolutionary process because it improves reproductive success.

Figure 44.2

An *Agapornis roseicollis* parrot cuts strips of nesting paper with its beak and then tucks them in the rump feathers. Birds of this species have been bred with another species that carry the strips in their beaks. The resulting hybrids tried to tuck the strips but were unable to perform the behavior. Later, they learned to carry the strips in their beaks, but still turned their heads toward the rump before flying off.

Inheritance of Behavior

Another example of behavioral evolution is nest-building behavior of parrot *Agapornis* (fig. 44.2). Three species of this genus have different methods of building nests. Females of the most primitive species use their sharp bills for cutting bits of bark, which they carry in their feathers to a site where they make a padlike nest. Females of a more advanced species cut long, regular strips of material, which are carried only in the rump feathers. These strips are used to make elaborate nests with a special section for the eggs. Females of the most advanced species carry stronger materials such as sticks, in their beaks. They construct roofed nests with two chambers and a passageway. The steady progression here from primitive to advanced nest-building suggests that evolution has taken place.

The fact that genes control these traits can be supported by the following experiment. The species that carries material in its rump feathers is mated to the species that carries material in the beak. The resulting hybrid birds cut strips and try to tuck them in their rump feathers, but are unsuccessful. It is as if the genes of one parent require them to tuck the strips, but the genes of the other species prevent them from succeeding. With time, however, these birds learn to carry the cut strips in their bills, and only briefly turn the head toward the rump before flying off.

Recently, researchers have reported on even more in-depth work with the snail *Aplysia* (fig. 44.3), which lays long strings of more than a million eggs. The snail winds the string into an irregular mass and attaches it to a solid object like a rock. Several years ago, a hormone that caused the snail to lay eggs, even though it had not mated, was isolated and analyzed. This egg-laying hormone (ELH) was found to be a small protein of 36 amino acids. It acts locally as an excitatory transmitter, augmenting the firing of a particular neuron in the snail's body. It also diffuses into the circulatory system and excites the smooth-muscle cells of the reproductive duct, causing them to contract and expel the egg string. Using recombinant DNA techniques, the investigators isolated the ELH gene. Surprisingly, the gene's product turned out to be a protein with 271 amino acids. The protein could be cleaved into as many as 11 possible products, and ELH is one of these known to be active. The investigators speculate that all the components of the egg-laying behavior could be mediated by this one gene. This work is quite remarkable, because it definitely shows that genes control behavior and further shows the manner in which they do so in a simple organism.

Behavior patterns are inherited and evolve in the same manner as other animal traits. Breeding experiments with the parrot *Agapornis,* and recombinant DNA experimentation with the snail *Aplysia* support this contention.

Execution of Behavior

Figure 44.1 tells us that behavior occurs as a response to a stimulus. Therefore, it is dependent on the sense organs, nervous system, and the musculoskeletal system of the organism. The complexity of behavior increases with the complexity of the nervous system. Animals with simple nervous systems tend to respond automatically to stimuli in a programmed way. This type of response is called instinctive or innate. Animals with complex nervous systems tend to be able to modify their behavior as they perform a particular act. These animals are said to be capable of learning new behavior patterns (fig. 44.4).

Figure 44.3

The snail *Aplysia* lays long strings of eggs that are wound into an irregular mass for deposition on a nearby object. Using modern recombinant DNA techniques, investigators have isolated a gene whose product contains 271 amino acids, 36 of which make up the ELH (egg-laying hormone) peptide. This peptide is partially responsible for the egg-laying behavior. Analysis of the gene shows that there are enough signals for cleavage (vertical lines) to permit the product to include ten other peptides, each of which may play some role in the animal's behavior.

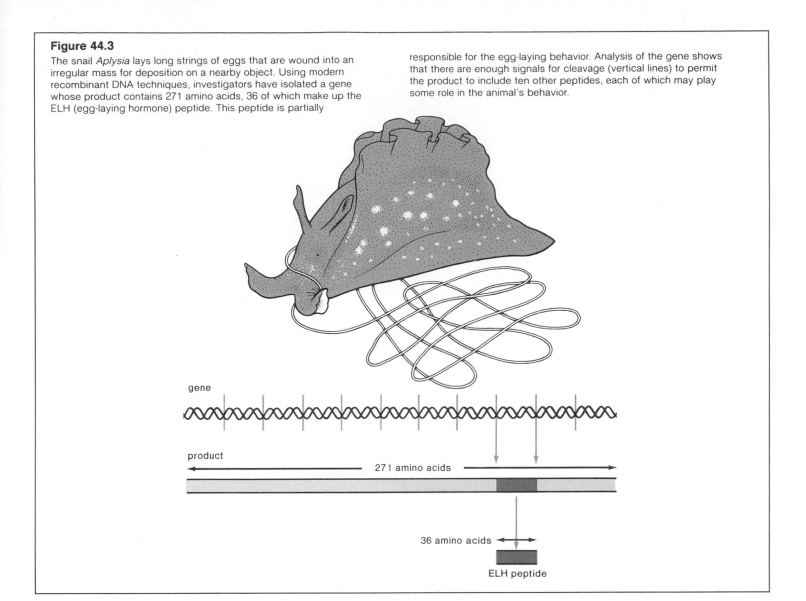

gene

product

271 amino acids

36 amino acids

ELH peptide

Innate Behavior

In **innate behavior,** a stimulus appears to trigger a fixed response that does not vary, even if the circumstances should change.

Fixed-Action Patterns

Reflexes are simply automatic responses to a stimulus, over which the individual appears to have little or no control. When a human knee is hit by a mallet, the lower leg jerks in a characteristic manner. Because reflexes are clearly innate, Tinbergen and other investigators believed that it might be possible to view an innate pattern of behavior as a series of automatic reflexes. The term **fixed-action pattern (FAP)** was given to a behavior pattern that occurs automatically, as if it were a composite of reflex actions. The behavior pattern is stereotyped— it always occurs the same way. A stimulus that initiates a fixed-action pattern is called a **sign stimulus.** For example, robins will attack a red tuft of feathers rather than an exact replica of a male robin without a red breast (fig. 44.5). The color red, not the entire bird, acts as a stimulus that releases the attacking behavior. Presumably, sign stimuli have been chosen through the process of natural selection.

Animals performing FAPs seem to act in a purposeful manner, but research shows that this may not be the case. For example, certain solitary digger wasps dig a hole and seek out a caterpillar, paralyzing it with a series of stings along the undersurface. The wasp carries the prey to the hole (fig. 44.6) and pulls it in. After laying her egg on the side of the caterpillar, she begins to close the hole. If, at this point the researcher removes the caterpillar and puts it on the ground nearby, the wasp will continue to cover the hole, even though the caterpillar is in full view.

Other features of FAPs have been discovered. Animals prefer supernormal stimuli, which are exaggerations of normal stimuli. When gulls are nesting, they will retrieve the most

Figure 44.4

Comparative frequency of modes of adaptive behavior among invertebrates and vertebrates. Innate types of behavior (taxes and reflexes) are more frequently found among invertebrates, and learned types are more frequently found among vertebrates. Data from V. G. Dethier and Eliot Stellar, *Animal Behavior*, Third Edition, © 1970, p. 91. Adapted by permission of Prentice-Hall, Inc., Englewood Cliffs, NJ.

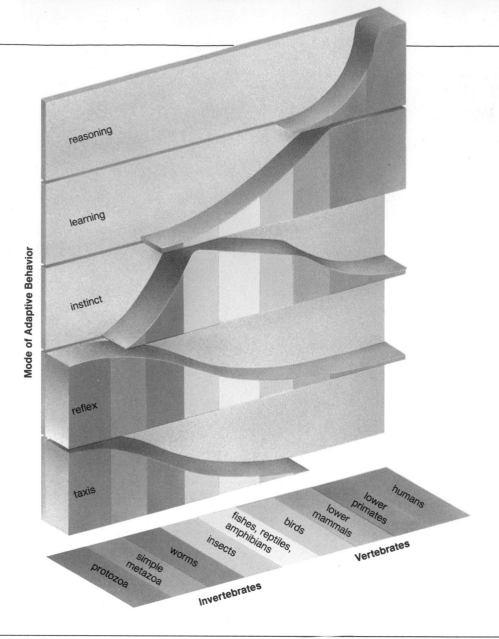

Figure 44.5

Sign stimulus determination. *a.* A male robin is presented with an exact model of a robin without a red breast beside a tuft of red feathers. *b.* It attacks the tuft. This shows that the color red is a sign stimulus that initiates a fixed-action pattern.

Figure 44.6

A digger wasp is about to pull this paralyzed caterpillar into a hole. The prey will serve as food for an offspring that will hatch from an egg she lays on the caterpillar. This is an example of a fixed action pattern because, if by chance, an experimenter removes its contents, the wasp will still close the hole.

Figure 44.7

Feeding behavior among gulls. *a.* Chicks peck at the red spot on a parent's bill when they want to be fed. *b.* In an experiment with laughing gull chicks, it was shown that with time, chicks peck at a more exact replica of parents' bill (first model). In each example, the top bar represents the pecking frequency of newly hatched chicks; the bottom bar represents the pecking frequency of chicks 2–5 days old.

a.

b.

Pecks/30 Seconds

speckled of the models that resemble their own eggs, even if it is more speckled than their eggs. They also prefer the largest model. Gulls will preferably retrieve eggs that are so large, they are unable to brood them!

Still another feature of FAPs is that they are sometimes subject to modification. For example, baby gulls peck at the parent's beak to induce the parent to feed them. Research with various models has shown that the chicks are not very discrim-inatory at first and choose a model that has a large red spot (the sign stimulus), even if it does not resemble the parent. With time, however, chicks become progressively more selective and choose a model that more nearly resembles the parent (fig. 44.7). This shows that there is probably a thin line between what is innate and learned behavior. Probably many types of behavior contain both components.

Figure 44.8
Bats use echolocation (sonar soundings) to locate their prey in the dark. Many moths are able to detect these soundings, however, and begin evasive action that often prevents the bat from finding them. This is an example of coevolution.

Figure 44.9
A land bird that breeds in North America clover fields, the bobolink migrates to the grasslands of Argentina by island hopping through the Caribbean. The small bobolink colonies in the Northwest do not use the most direct route to the south through Mexico; instead, they make a detour east to fly with other bobolinks along an ancestral flyway.

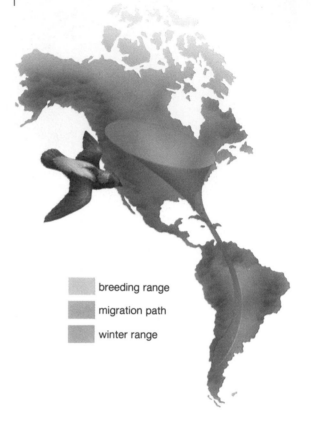

breeding range

migration path

winter range

Orientation and Navigation

Animals use a variety of external stimuli to orientate and navigate as they go about their daily lives.

Orientation of the body toward or away from a stimulus is termed **taxis** in animals. Animals exhibit a number of different types of taxes; *phototaxis* is movement in relation to light and *chemotaxis* is movement in relation to a chemical. Some insects, for example moths and flies, will fly directly toward a light. Often they orient themselves by shifting the body until the light falls equally on both eyes. If one eye is blind, the animal will move in a spiral, forever trying to find the direction in which the light will be balanced between the two eyes. Chemotaxes are quite common. Insects are attracted to minute quantities of chemicals, called **pheromones,** given off by members of their species. Army ants are so set on following a pheromone trail that if, by chance, it leads them into a circle, they will continue to circle until they die of exhaustion. Vertebrates, too, are sometimes highly responsive to chemicals. A few whiffs of a piece of clothing, and bloodhounds are capable of tracking down a single individual.

Only a few animals are able to orientate themselves by means of *sonar* or *echolocation*. Bats send out series of sound pulses and listen for the echoes that come back. The time it takes

for an echo to return indicates the location of both inanimate and animate objects, and enables a bat to find its way through dark caves and to locate food at night (fig. 44.8). Some moths have evolved the ability to hear the sounds of a bat and they begin evasive tactics when they sense that a bat is near.

Some animals have the ability to return home or migrate long distances (fig. 44.9). This ability also requires that they orientate their bodies in a particular direction. Studies have been done to show that they navigate by using the sun in the day and the stars at night as compasses to determine direction (north, south, east, or west). They can even allow for the east-to-west movement of the sun during the course of the day. Bees offered a food source can return to their hive no matter where the sun is located in the sky. Therefore, it is believed that they have an internal or biological clock (see reading on the next page) that tells them the location of the sun according to the time of day.

Although birds use the sun and stars as compasses, they cannot rely on these to tell them that home is in a particular direction. In other words, suppose you were blindfolded and then transported away from home. Upon being set free, you are given a compass to tell direction; how would you know which direction to select? Some investigators now believe that birds and other

Rhythmic or Cyclic Behavior

Certain behaviors of animals reoccur at regular intervals. Behaviors that occur on a daily basis are said to have a circadian rhythm. For example, some animals, like humans, are usually active during the day and sleep at night. Others, such as bats, sleep during the day and hunt at night. There are also behaviors that occur on a yearly basis. For example, birds migrate south in the fall, and the young of many animals are born in the spring.

Originally, it was assumed that environmental changes, such as the coming of night or the length of the day, directly controlled cyclical behavior in animals. But it is now known that cyclical behaviors will occur even when the associated stimulus (daylight or darkness) is lacking. For example, fiddler crabs are dark in color during the day and light in color at night, even when kept in a constant environment. But if the crabs are kept in the constant environment indefinitely, the timing of the daily change tends to drift and become out of synchronization with the natural cycle. For this reason, it has been suggested that rhythmic behavior is under the control of an innate, internal biological clock that runs on its own, but is reset by external stimuli. Aside from keeping time, a biological clock must also be able to bring about the change in behavior. In the fiddler crab, for example, it must be able to stimulate the processes that cause the shell to change color.

Therefore, a biological clock system needs (*see* also figure):

1. a time-keeping mechanism that *keeps time* independently of external stimuli (i.e., a minute is always a minute);
2. a receptor that is sensitive to light/dark periods and *can reset* the clock for circadian rhythms or indicate a change in the length of the day/night for circannual rhythms;
3. a communication mechanism by which the clock *induces* the appropriate behavior.

A review of the discussion concerning flowering in plants shows that the last two components of a biological clock system have been tentatively identified. Phytochrome is believed to be the receptor sensitive to light and dark periods, and plant hormones are believed to be the means by which flowering is induced. The time-keeping mechanism has not been identified, however.

In animals, we know that melatonin is produced at night by the pineal gland; therefore the amount produced begins to increase in the fall and decrease in the spring. Melatonin is believed to inhibit development of the reproductive organs, which would account for why some animals reproduce in the spring. For animals in which the pineal gland is the "third eye" not only the clock, but also the receptor is presumed to be in the pineal gland.

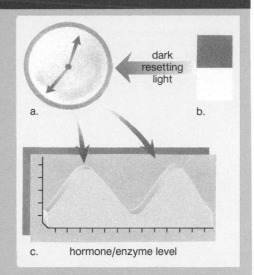

Biological clock system. Biological clock systems have three components: (*a*) an internal timekeeper, (*b*) a means of detecting light/dark periods, (*c*) a method of communication that eventually brings about a behavioral response.

For some behaviors, it would appear that only the time-keeping mechanism is required. For example, bees and some birds use the sun and stars as compasses, even though they change their position throughout the day and night.

animals are sensitive to magnetic lines of force, which are dependent on the earth's magnetic field, and this sensitivity allows them to determine the direction of home.

Salmon use chemotaxis to pinpoint their exact home. Salmon are hatched in a tributary of a river but grow to maturity in the open sea. At spawning time, mature salmon travel back up the river to the same spot where they hatched. Experiments have shown that the fish find the spawning ground by following the chemical scent of their first home.

Motivation

Innate behavior patterns, like the "urge" to migrate, seem to require that the animal be motivated to perform them. **Motivation** includes three stages:

1. An appetitive stage, during which the animal searches for the goal.
2. A consummatory stage or series of responses directed at the goal.
3. A quiescent stage, when the animal no longer seeks the goal.

A good example of motivated behavior concerns the need for food. A hungry animal looks for food; once food is found, it eats the food; and then, being satisfied, it no longer seeks food.

Motivated behavior requires that the animal is readied to perform the behavior by its internal state. A study of reproduction (fig. 44.10) in ringdoves has indeed shown this to be the case. Ringdoves reproduce in the spring, but when male and female ringdoves are separated, neither shows any tendency toward reproductive behavior. In contrast, when a pair are put together in a cage, the male begins courting by repeatedly bowing and cooing. Because castrated males do not do this, it can be reasoned that the hormone testosterone readies the male for this behavior. The sight of the male courting causes the pituitary gland in the female to release FSH and LH. These in turn cause her ovaries to produce eggs and release estrogen into the bloodstream. Now both male and female are ready to construct a nest, during which time copulation takes place. The hormone progesterone is believed to cause the birds to incubate the eggs and, while they are incubating the eggs, the hormone prolactin causes crop growth so that both parents are capable of feeding their young crop milk.

Reproductive behavior in the ringdove can be explained on the basis of both *external* and *internal stimuli*. The external stimuli are processed by the central nervous system, which then

Figure 44.10
Ringdove mating behavior. Successful reproductive behavior requires these steps.
a. Male and female are courting.
b. Copulation takes place. *c.* Creating a nest. *d.* Chicks are fed "crop milk." It can be shown that internal hormonal levels motivate the birds to perform these behavior patterns. Similarly, in all animals, behavior patterns are most likely to occur if the animal is ready to perform them.

directs the secretion of hormones. In animals, the nervous and endocrine systems work together to produce physiological changes that lead to appropriate behavior patterns.

Innate behavior patterns tend to be fixed and unlearned responses that are triggered by external stimuli. Sometimes the response appears to be a series of reflexes called a fixed-action pattern. An innate behavior pattern may require orientation of the animal toward a stimulus and/or that it be motivated to perform the behavior.

Learned Behavior

Learned behavior is characterized by a change in behavior as a result of experience. The capacity to learn is inherited and allows an organism to change its behavior to suit the environment. The organism alters its behavior to respond to a specific stimulus in ways that promote its own survival or that of its offspring and/or near relatives.

Imprinting

Konrad Lorenz, a famous ethologist, observed that birds become attached to, and follow the first moving object they are exposed to. He termed this behavior **imprinting** and suggested that it was a means by which organisms learned to recognize their own species. Ordinarily, the object followed is the mother, however, Lorenz caused goslings to become imprinted on him (fig. 44.11), and showed that they would choose him over their own mother when given the opportunity.

Critical Period   Lorenz found that if goslings were totally isolated for the first 2 days after hatching, they failed to undergo the process of imprinting. Apparently, there was a critical period of time, during which imprinting was possible. Since Lorenz' time, it has been discovered that there are critical periods for other types of learning in birds. A male white-crowned sparrow reared in isolation will sing a song, but it is less complicated than the normal song. If the bird is permitted to hear another adult white-crowned sparrow sing during a critical period of 10–50 days after hatching, however, it will learn to sing the normal song. White-crowned sparrows never learn to sing the song of other species of birds. This shows that an animal's genes sometimes influence what type of learning is possible or what can be learned.

Today, the term imprinting is used in a larger context to refer to any type of learning that has a **critical period.** Also, this phenomenon has been observed in older animals. For example, adult birds are imprinted on their young after hatching, and will accept as their own any chick, even of another species, if it is put in the nest during the critical period.

Figure 44.11

a. Normally, goslings are imprinted on their mother. *b.* Goslings are following Konrad Lorenz, who, like Niko Tinbergen, performed experiments in the field. Lorenz found that goslings learn to follow the first moving object they see, and could be imprinted on an improper object.

a.

b.

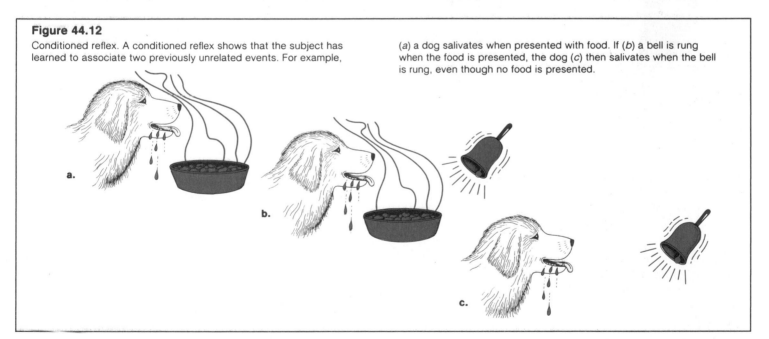

Figure 44.12

Conditioned reflex. A conditioned reflex shows that the subject has learned to associate two previously unrelated events. For example,

(*a*) a dog salivates when presented with food. If (*b*) a bell is rung when the food is presented, the dog (*c*) then salivates when the bell is rung, even though no food is presented.

a.

b.

c.

Conditioned Learning

In a type of conditioned learning called **associative learning,** an animal learns to give a response to an irrelevant stimulus. Pavlov's dogs learned to expect food and began to salivate when a bell rang, because they had learned to associate the ringing of the bell with food (fig. 44.12). Associative learning can explain behavior that seems out of place. For example, the ringing of a bell in and of itself does not have anything to do with food. Even so, associative learning can be useful; for example, it has been suggested that mothers can instill love of learning by holding their children in their laps while reading to them.

Operant Conditioning   **Operant conditioning** is trial and error learning. An animal faced with several alternatives is *rewarded* (positively reinforced) for making the proper choice and learns to make this response without hesitation. This type of learning, employed by animal trainers, can be used successfully to make animals learn all sorts of tricks. Using *positive reinforcement,* B. F. Skinner, a well-known Harvard University behaviorist, has even taught pigeons to play a form of Ping-Pong. Skinner believes that humans too, are controlled primarily by positive reinforcement mechanisms, although many disagree with this idea.

Figure 44.13

Apes are capable of insight learning—reasoning out a solution to a problem without trial and error. In (a), a chimpanzee is unable to reach a banana. In (b), it stands on boxes to reach the banana. If an animal can reason, does it have a cognitive awareness of itself and its surroundings? Researchers are now wrestling with this question.

a.

b.

Extinction Characteristically, conditioned learning is maintained only as long as a reward is given for the behavior. If the reward is withdrawn, the behavior will also disappear. This is called extinction. Extinction is also a type of learning.

Insight Learning

Insight, or reasoning, is the ability to solve a problem using previous experiences to think through to a solution. To the observer, it seems as if the animal employing insight needs no practice to successfully reach a goal. For example, apes can devise the means to get to bananas placed out of their arms' reach. They will pile up boxes or use a pole in order to reach the food (fig. 44.13).

Perhaps the ultimate in reasoning ability is learning and using language. When humans use language, they are able to put words together in various ways to convey new and creative thoughts. For the past several years, investigators have been trying to determine the degree to which other animals can learn to use language in this way. Most of this work has been done with chimpanzees. Investigators have overcome the biological inability of chimpanzees to articulate many sounds by having them learn a sign language. Chimpanzees have learned as many as four hundred signs of an artificial visual language. Thus far, however, it has not been possible to demonstrate unequivocally that these animals are capable of putting the signs together to

Figure 44.14

Example of chemical communication between members of the same species. Antennae of male moths are capable of detecting female pheromone from several miles away.

create new sentences and meanings. Some have suggested that the animals are merely copying their trainers and have no true understanding of language use.

Of course, it can be questioned whether cognition, or the ability to think, requires the use of language at all. Perhaps animals only differ in the degree to which they are aware of themselves and their environment. Research is being conducted to see if even invertebrates are capable of analyzing a situation.

Learned behavior includes imprinting, conditioned learning, and insight learning. The possibility exists that all animals in varying degrees are aware of their surroundings and that nonhuman vertebrates, especially apes, can be taught to use a language.

Societies

A **society** is a group of individuals belonging to the same species that are organized in a cooperative manner. In order to cooperate, evidence suggests that members of a society need a means of reciprocal communication and a means of overcoming aggression.

Communication

Communication by chemical, visual, auditory, or tactile (touch) stimuli often includes a *social releaser,* a sign stimulus used between members of the same species that causes the receiver to respond in a certain way.

Chemical Communication

The term *pheromone* is used to designate chemical signals that are passed between members of the same species. Pheromones can have either releaser effects or primer effects. A pheromone

Figure 44.15

An example of visual communication and ritualized behavior among humans. Eyebrow flashing is a common human trait displayed by (a) a French woman, (b) a Balinese man, and (c) a member of the Woitapmin tribe of New Guinea.

a.

b.

c.

with a *releaser effect* evokes an immediate behavioral response, while a pheromone with a *primer effect* alters the physiology of the recipient, leading to a change in behavior.

Sex attractants are good examples of pheromones with releaser effects. For example, female moths secrete chemicals from special abdominal glands. These chemicals are detected downwind by receptors on male antennae (fig. 44.14). This signaling method is extremely efficient because it has been estimated that only 40 of the 40,000 receptors on the male antennae need to be activated in order for the male to respond.

Ants and honeybees provide an example of a pheromone with a *primer effect*. The queen produces a substance that is passed from worker to worker by regurgitation. This substance prevents the workers from raising other queens, and it also prevents the ovaries of the workers from maturing. Primer effects have also been seen in mice. Male mice produce a substance that can alter the reproductive cycle of females. When a new male and female are placed together, this substance can cause the female to abort her present pregnancy so that she can then be impregnated by her new mate. Similarly, crowding of female mice causes disturbances or even blockage of their estrus (those times when they are sexually receptive). In both instances, removal of the olfactory lobes prevents these occurrences.

It is well known that male dogs and cats are attracted to the opposite sex by means of scent. Research suggests that apes, and even humans, as discussed in the reading on page 713, may also be influenced by pheromones.

Visual Communication

Visual communication includes many social releasers. This term is used for a sign stimulus that triggers an FAP in another member of the species. For example, male birds and fish sometimes undergo a color change that indicates they are ready to mate. Female baboons show that they are in estrus by a reddening of the sex flesh on the buttocks.

Visual communication is also involved in complex defense and courtship behaviors that are dependent upon a series of social releasers and FAPs. Defense and courtship behaviors are discussed on page 714, but in the meantime, we can point out that they are called ritualized behaviors. Their movements are exaggerated and always performed in the same way so that the sign stimuli are clear. **Ritualized behavior** is believed to be derived from body movements associated with such activities as locomotion, feeding, and caring for the young. Facial expressions of humans (fig. 44.15) seem to be universal, and it has been suggested that they, too, are social releasers and ritualizations of movements that were first used for biological processes.

Tactile Communication

For example, baby gulls peck at the parent's beak in order to induce the parent to feed them. In primates, grooming occurs when one animal cleans the coat of another. Grooming helps cement social bonds within a group. It's been found that monkeys who lack a mother can still develop normally if they have the opportunity to play with their peers.

The communication of honeybees is remarkable, because the so-called language of the bees uses a variety of stimuli to impart information about the environment. Karl von Frisch[1], another famous ethologist, carried out many detailed bee experiments in the 1940s and was able to determine that when a foraging bee returns to the hive, it performs a **waggle dance** (fig. 44.16). The dance, which indicates the distance and direction of a food source, has a figure-eight pattern. As the bee moves between the two loops of the figure eight, it buzzes noisily and shakes its entire body in so-called waggles. *Distance* to the food source is believed to be indicated by the number of waggles and/or the amount of time taken to complete the straight run. The straight run also indicates the *location* of the food. When the dance is performed outside the hive, the straightaway indicates the exact direction of the food, but when it is done inside the

---

[1]Karl von Frisch, Konrad Lorenz, and Nikolaas Tinbergen, sometimes called the fathers of modern ethology, jointly received a Nobel Prize in 1973 for their enterprising studies in animal behavior.

Figure 44.16
Tactile and auditory communication among bees relates information not about the bee but about the environment. Honeybees do a waggle dance to indicate the direction of food. *a.* If the dance is done outside the hive on a horizontal surface, the straight run of the dance will point to the food source. *b.* If the dance is done inside the hive on a vertical surface, the angle of the straight run to that of the direction of gravity is the same as the angle of the food source to the sun.

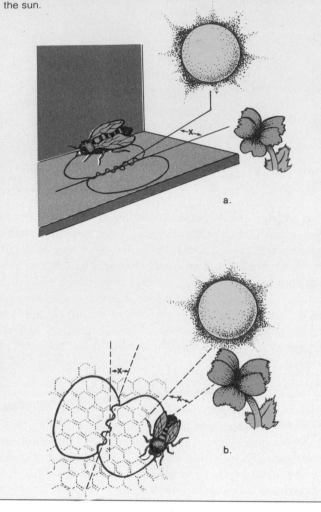

Male crickets have calls and male birds have songs for a number of different occasions. For example, birds may have one song for distress, another for courting, and still another for marking territories. Sound stimuli have been shown to be more important than visual stimuli for birds that live in dense woods, where vision is obstructed. In experiments, a male wood thrush attacked models of an unrelated species, as long as they were silent. But if the unrelated species' song was played via a loudspeaker, the wood thrush paid the model no heed. The intensity with which a wood thrush attacked a model of its own species was directly proportional to the loudness of the song that was played.

One advantage of auditory communication is that the message can be modified by the sound's intensity, duration, and repetition. In an experiment with rats, a researcher discovered that an intruder could avoid attack by increasing the frequency with which it made an appeasement sound. Rats isolated from birth could make the sound but had not learned to increase the frequency.

The fact that organisms with good vision often rely on sound communication has been demonstrated in chickens. Hens will react vigorously when they hear a chick peep, even if they cannot see the chick. However, they will ignore a chick that is peeping within a soundproof glass container.

Language is the ultimate auditory communication, but only humans have the biological ability to produce a large number of different sounds and to put them together in many different ways. Nonhuman primates have at most only forty different vocalizations, each one having a definite meaning, such as the one meaning "baby on the ground" that is uttered by a baboon when a baby baboon falls out of a tree. As discussed on page 710, although chimpanzees can be taught to use an artificial language (fig. 44.17), they never progress beyond the capability level of a two-year-old child. It has also been difficult

hive, the angle of the straightaway to that of the direction of gravity is the same as the angle of the food to the sun. In other words, a 40° angle to the left of vertical means that food is 40° to the left of the sun. The bees can use the sun as a compass to locate food because their biological clocks (see reading on page 707) allow them to compensate for the movement of the sun in the sky.

At first it was not apparent how the bees were communicating in the dark hive. But now it appears that a combination of tactile and auditory communication is being used. Through touch, the bees can determine the direction and waggles of the dance and the buzzing noises of the dancer are also related to the distance to the food source.

Auditory Communication

Because auditory (sound) signals are able to reach a larger audience and can be sent even in the dark, they are sometimes favored even by animals with good vision.

The Chemistry between People

While the idea of human pheromones is intriguing, the dozen or so studies that addressed the possibility in the past 10 years were disappointing; no one established beyond a doubt that human pheromones exist. Now two new studies are stirring up the pheromone debate with the boldest claims yet. Researchers at the University of Pennsylvania and the Monell Chemical Senses Center, a nonprofit research institute in Philadelphia, say that people produce underarm pheromones that can influence menstrual cycles.

The studies, done by chemist George Preti and biologist Winnifred B. Cutler, are not the first of their kind, but they are the first ones rigorous enough to be published in a respected scientific journal, *Hormones and Behavior*. In one study the researchers collected underarm secretions from men who wore a pad in each armpit. This "male essence" was then swabbed, three times a week, on the upper lips of seven women whose cycles typically lasted less than 26 days or more than 33. By the third month of such treatment, the average length of the women's cycles began to approach the optimum 29.5 days—the cycle length associated with highest fertility. Cutler's conclusion: "Male essence" contains at least one pheromone that "helps promote reproductive health."

The experiment was more rigorous than earlier ones for two reasons. It employed a control group—eight women who were swabbed with alcohol showed no effect—and it was performed in "double-blind" fashion; neither the subjects nor the researchers knew whether alcohol or male essence dissolved in alcohol was being applied until after the study. In spite of such safeguards, skeptics charge that the study was flawed. Richard Doty, a psychologist at the Clinical Smell and Taste Research Center of the University of Pennsylvania, says Cutler and Preti tested too few women for too brief a time to be certain that the observed effects are real. "If they're right, it's fascinating," says Doty, "but I'm not convinced yet."

Do pheromones affect our behavior toward the opposite sex?

In Cutler and Preti's second experiment they studied menstrual synchrony—the phenomenon that women who live in close quarters tend to have cycles that coincide. And because this experiment's results were more dramatic, several scientists found it more persuasive. This time Cutler and Preti exposed 10 women with normal cycles to *female* underarm sweat. After three months of the same sweat-on-the-lip treatment, the women's cycles were starting roughly in synchrony with those of the women who had donated the sweat. (Again, women exposed to alcohol showed no change.) Menstrual synchrony was first documented in 1970 when psychologist Martha McClintock studied women living in a college dormitory. But this new study is the first to offer solid evidence that pheromones are what mediate the effect. "Pheromone effects are real in human beings," concludes Preti, "and the anecdotal evidence suggests they even occur here in the United States, where we're all deodorized and perfumized."

The researchers don't know how the pheromone exerts its effect on those brain centers that control reproductive hormones, or whether the women inhale it or absorb it through the skin. Cutler and Preti also aren't sure which of the many compounds in human sweat is the active ingredient—but they're focusing now on substances that are already well known for their role in pig sex.

One of the targeted compounds, a steroid known as androstenol, is also found in boar saliva; when a sow in heat detects the musky smell, it promptly assumes the mating stance. A decade ago researchers found that androstenol was also produced in human underarms by bacteria residing there. One firm specializing in mail-order fragrances decided that if the compound was a sex attractant in pigs, it must be one in people, too: SCIENTISTS DISCOVER MYSTERY CHEMICAL THAT SEEMS TO DRIVE WOMEN WILD! it trumpeted in an ad for an androstenol-containing cologne.

Contrary to that claim, repeated studies have shown that androstenol doesn't work as an aphrodisiac in people. Still, Preti and Cutler say it may have other physiological effects. In a third study they found that the amount of androstenol in female underarms changed predictably during the menstrual cycle, peaking just before ovulation. This may help explain how menstrual synchrony occurs: the changes in a woman's androstenol may be what communicate her place in the cycle to other women, the researchers say, and perhaps set off changes in their cycles as well.

If scientists ever do get their hands on an actual human pheromone, it will probably work gradually like the one that Cutler and Preti describe, says McClintock, who is now at the University of Chicago. "It is very unlikely that you'll find a simple push-button effect in people, where a pheromone *compels* somebody to do something," she says. "At the most it will affect how you feel a little bit." No doubt that is why it is so difficult to identify human pheromones. Their messages are not urgent, like a moth's, but subliminal. In people it appears that pheromones are a most subtle body language.

By Terence Monmaney with Susan Katz, *Newsweek*, January 12, 1987.

---

to prove that chimps understand the concept of grammar or can use their language to reason. It still seems as if humans possess a communication ability unparalleled by other animals.

Members of a species communicate with one another. This communication may be chemical, visual, tactile, and/or auditory.

Competitive Interactions

Members of the same species compete with one another for resources, including food and mates. **Aggression** is belligerent behavior that helps an animal compete. The members of a society have ways of dealing with competition and aggression so that they can live orderly lives and perform courtship and reproduction.

Territoriality and Dominance

Territoriality means that an animal defends a certain area, preventing certain other members of the same species from utilizing it (fig. 44.18). Territoriality spaces animals and thereby reduces aggression. It also avoids overcrowding, ensuring that the young will have enough to eat. In territorial species, such behavior may also have the effect of regulating population density to some extent, because animals that do not have a territory will not mate.

A **dominance hierarchy** exists when animals within a society form a relationship in which a higher ranking animal has access to resources before a lower ranking animal. Dominance hierarchies help keep order among members of a group.

Figure 44.18
Two male elephant seals battle over a piece of beachfront property. This will be the winner's territory. Here he will collect a harem made up of females that will willingly mate with him. Territoriality partitions resources according to the supposed fitness of the animal.

Ritualization of Aggression As mentioned previously on page 711, when animals defend a territory or themselves, the behavior pattern includes many social releasers. Seemingly, the animals are communicating with each other in such a way that one will back down and no blood will be shed. The display often includes postures that make the body appear larger, and color changes that make the animal more conspicuous. Natural selection most likely favors a contest decided in this manner because in the end both animals survive and are capable of reproducing.

When the animals are facing one another, two opposing responses—approach and avoidance—are simultaneously present. The animals are often in conflict as to whether to fight or to escape, and this conflict causes them to threaten one another, rather than to fight outright. When an aggressor is in conflict, redirection or displacement of aggression may occur. As an example of redirection, a bird might peck at the ground instead of his opponent. Displacement of aggression is recognized by the fact that the animal performs an irrelevant activity. In the midst of a territorial encounter, for example, a bird might suddenly decide to preen its feathers.

In the end, one of the animals backs down. **Appeasement,** or submission, occurs when an animal actually exposes itself to the attack of another in a gesture that prevents further attack.

For example, a subordinate baboon of either sex turns away from an aggressor and crouches in the sexual presentation posture. Tinbergen found that when gulls face each other, they have a whole range of postures from actual fighting to appeasement. Food-begging behavior in gull chicks may also be an appeasement gesture. Appeasement behavior may cause even greater conflict in the aggressor so that inactivity results.

Courtship

Courtship is a time during which aggression must be at least temporarily suspended in order for mating to take place. The courtship pattern also includes behaviors that range from aggression to appeasement. As an example, consider the courtship pattern of stickleback fish (fig. 44.19).

In the spring, the male stickleback stakes out a territory and builds a nest. At this time, his body becomes highly colored, including a red belly. The red belly is a sign stimulus that causes a defending male to dart toward and nip an intruder in his territory. On the other hand, the owner entices a female to enter the territory by first darting toward her, and then away in a so-called zigzag dance. Finally, he leads her to the nest where she deposits her eggs. Investigators have pointed out that the zigzag

Figure 44.19
Reproductive behavior of three-spined stickleback fish is based on social releasers (sign stimuli) and fixed action patterns. The belly of the male has turned red and is a sign stimulus that he is ready to mate. If a female approaches, her swollen belly is a sign stimulus for the male and he performs zigzag swimming toward her. She then swims toward him and he leads her to the nest. When he puts his snout inside, she swims into the nest and lays her eggs as she is prodded by the male. After the male enters the nest and fertilizes the eggs, he chases the female away.

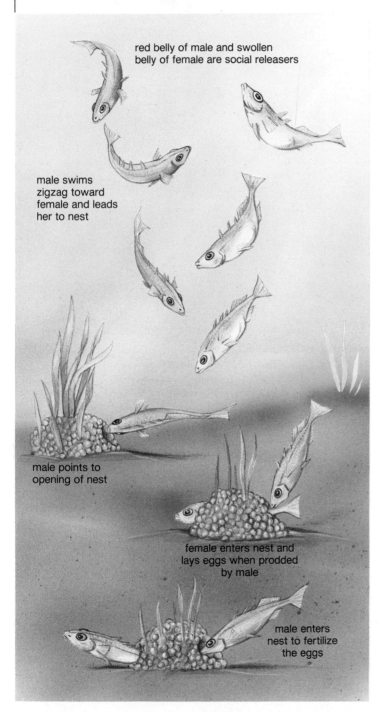

red belly of male and swollen belly of female are social releasers

male swims zigzag toward female and leads her to nest

male points to opening of nest

female enters nest and lays eggs when prodded by male

male enters nest to fertilize the eggs

dance of the male actually contains the same aggressive movements used when the male darts toward and attacks a trespassing male. But aggression is curtailed, and the sign stimuli now signify a readiness to mate.

The members of a society must overcome aggression so that cooperation can occur. Territoriality and dominance are two mechanisms by which aggression is controlled. During courtship, aggression is minimized, although aggressive actions are still detectable.

Sociobiology

It can be shown that the degree to which members of a species cooperate with one another is dependent on their physical characteristics and the nature of the environment. For example, an investigator reports that in the Kibale Forest in eastern Africa, chimpanzees usually forage in groups of three or four. The group spreads out and when one spots a tree in fruit, it calls to the others. Because they are able to travel on the ground, chimpanzees are able to forage in a group, despite the fact that trees in fruit are widely dispersed. On the other hand, orangutans are solitary feeders. They literally have four hands that are adapted to gripping tree limbs. Therefore, orangutans do not travel well on the ground. Under these circumstances, it is best for each one to hoard for itself any tree that is in fruit. We might take another example from reproductive behavior. Song birds are often monogamous, whereas mammals tend to be polygamous. In the case of birds, incubating the eggs and/or gathering enough food requires the effort of both parents. If they did not cooperate, neither would have any offspring. In the case of mammals, the female, not the male, is anatomically and physiologically adapted to care for the offspring. Under these circumstances, it is more adaptive for the male to impregnate as many females as possible.

The basic tenet of **sociobiology,** the biological study of social behavior, is that cooperative behavior among members of a society only exists because such behavior increases the fitness of the individual. *Fitness,* of course, is judged by the frequency that an animal's genes are passed on to the next generation.

Altruism

At times, it may seem as if an animal is acting **altruistically;** that is, for the good of others rather than from self-interest. For example, in ants and bees, only the queen lays eggs and most of the hive consists of female workers that do not have any offspring at all. How can such a society increase the fitness of the workers? The answer may lie in the fact that the male parent is haploid; therefore, siblings have on the average three-fourths of their genes in common (fig. 44.20). Because offspring have only one-half of their genes in common with their parents, it may actually increase the fitness of the worker ants to raise siblings rather than offspring.

Altruism, exemplified by the willingness of daughters to help the queen rather than to have their own offspring, can be explained by noting that **kin selection** increases the animal's **inclusive fitness.** In other words, survival of close relatives (*kin*) increases the frequency of an animal's genes in the next generation. Therefore the animal's overall (*inclusive*) fitness is also increased.

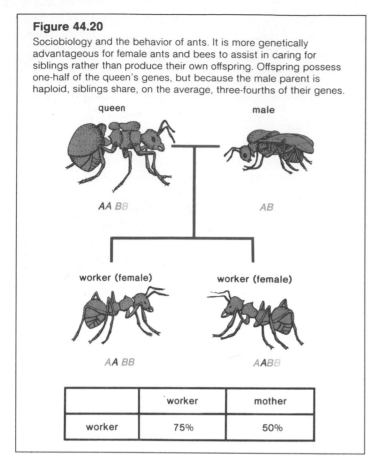

Figure 44.20
Sociobiology and the behavior of ants. It is more genetically advantageous for female ants and bees to assist in caring for siblings rather than produce their own offspring. Offspring possess one-half of the queen's genes, but because the male parent is haploid, siblings share, on the average, three-fourths of their genes.

queen   male

AA BB   AB

worker (female)   worker (female)

AA BB   AABB

	worker	mother
worker	75%	50%

Communal Groups The concept of kin selection is exemplified in acorn woodpeckers (fig. 44.21). These birds received their name because they store acorns in holes drilled in trees. In California and New Mexico, acorn woodpeckers usually live in social groups of about fifteen birds. Each group defends its storage tree from birds that are not members of the group.

Researchers have noted that only certain members of a group are *breeders* and they share mates. Mate sharing, which is rarely seen in birds, in this instance increases the inclusive fitness of the individual bird. Groups have more offspring per adult than do single pairs. For a group of three, investigators found there were 1.16 young per adult male on the average; for pairs there were only 0.92 young per male. Secondly, among the sharers, the males are usually brothers and the females are usually sisters.

In a large group, there are several *nonbreeders,* birds who only help the breeders raise their young. Because nonbreeders are offspring of the breeders, it is possible that they are also increasing their inclusive fitness by their behavior. However, the researchers also note that the adjoining territories are already occupied and birds have a very difficult time establishing themselves as breeders. Under these circumstances, it may simply be more advantageous for the nonbreeders to remain in a group where they can feed on the stored acorns, and wait until there is a vacancy among the breeding birds.

Alarm Callers Sociobiologists suggest that all animal behavior patterns should be interpreted as selfish acts. For example, some animals that live in groups give an alarm call when

Figure 44.21
Acorn woodpeckers live and mate in groups and defend their store of acorns against birds of other groups. It can be shown that group living increases their individual fitness.

a predator approaches. Careful analysis may reveal that the alarm callers are not putting themselves in danger and may, instead, be protecting *themselves* by this action. Sociobiologists maintain that there is always an explanation for social behavior that is consistent with and supports the concept of organic evolution by natural selection.

Sociobiology is the study of social behavior according to the tenets of evolutionary theory. While it may seem as if members of a group are altruistic, it can be shown that their motives are most likely selfish and their behavior serves to increase their own fitness. Sometimes it is necessary to utilize the concept of inclusive fitness when interpreting an animal's behavior.

Humans Sociobiologists interpret human behavior according to these same principles. For example, parental love is clearly selfish in that it promotes the likelihood that an individual's genes will be present in the next generation's gene pool. People tend to keep their aggression under control with those they recognize as blood relatives. Also, we would expect small towns to have less crime than large cities because small towns have a greater percentage of related residents.

Human reproductive behavior can be similarly examined. Human infants are born helpless and have a much better chance of developing properly if both parents contribute to the effort. Perhaps this explains why the human female has evolved to be continuously receptive to sexual intercourse. (Other mammalian females are receptive only during estrus, a physiological state that recurs periodically; sometimes only once a year.) The fact that sex is continuously available may help assure that the human male will remain and help the female raise the young.

Human beings are usually monogamous (one mate at a time); however, there are exceptions that appear to be adaptations to the environment. Among African tribes, for example, one man may have several wives. This is reproductively advantageous to the male, but is also advantageous to the woman. By this arrangement, she has fewer children and they are thereby assured a more nutritious diet. In Africa, sources of protein are scarce and early weaning poses a threat to the health of the child.

In contrast, among the Bihari hillpeople of India, brothers have the same wife. Here the environment is hostile and it takes two men to provide the everyday necessities for one family. Because the men are brothers, they are actually helping each other look after common genes.

It should be stated that not all scientists and laypeople are in favor of applying sociobiological principles to human behavior. Many are concerned that it may lead to the perception that human reproductive behavior is fixed and cannot be changed. Particularly, today, when people are exploring new avenues of behavior, some do not wish to unduly stress any potential obstacles to bringing about a new social order.

Summary

1. Behavior patterns, to varying degrees, are inherited and evolve in the same manner as other animal traits. Therefore behavior patterns are adaptive.
2. Behavior occurs as a response to a stimulus and is often divided into two categories: innate (instinctive) and learned. In innate behaviors, the genetic component is believed to be more influential, and in learned behaviors, the environmental component is believed to be more influential.
3. Observation of eggshell removal by black-headed gulls, breeding experiments with the parrot *Agapornis,* and recombinant DNA experimentation with the snail *Aplysia* support the contention that behaviors are adaptive and inherited.
4. Fixed-action patterns, which appear to be a series of reflexes, occur as a response to a sign stimulus. FAPs are largely inherited, although they often increase in efficiency with practice.
5. External stimuli help organisms orientate themselves. Phototaxis and chemotaxis are both seen. Bats orientate themselves by echolocation. Birds and bees are able to navigate using the location of the sun during the day, and some birds use the stars at night. These animals probably have a biological clock that compensates for the movement of these bodies with time.
6. Sometimes animals seem to be motivated to perform a particular behavior. Research with ringdoves suggests that they are motivated to perform reproductive acts because of the hormone levels in their blood.
7. Learned behavior includes imprinting, which occurs during a critical period. Ducks and geese follow the first object they see, as long as they see it within a short time after hatching.
8. Conditioned learning includes associative learning. Pavlov's dogs learned to associate the ringing of a bell with food. Operant learning occurs when a behavior is constantly reinforced. Conditioned learning is subject to extinction when the association or reward is no longer presented.
9. Insight learning occurs when an animal solves a problem or is able to reason. Use of a complex language implies that the animal can reason.
10. Societies have two main characteristics. Its members communicate with one another and there are ways of coping with competition between members.
11. Communication is classified into chemical (e.g., pheromones); tactile (e.g., the waggle dance of bees); visual (e.g., color changes); auditory (e.g., calls and language).
12. The control of competition between members of a society often takes the form of territoriality and dominance hierarchies. During the times when animals face-off with one another, they often use ritualized behavior, which consists of a series of social releasers (sign stimuli between members of a species).
13. During courtship, aggressive behaviors have been modified to serve as sign stimuli to induce mating. Courtship of the stickleback fish is an example of this.
14. Social behavior can be studied from an evolutionary perspective. According to sociobiologists, although it may seem as if members of a group are altruistic, it has been argued that their motives are probably selfish, and that their behavior serves to increase their own fitness or that of the group as a whole.

Objective Questions

For questions 1 to 3, match the items below with those in the key:
Key:
    a. innate behavior
    b. learned behavior

1. operant conditioning
2. fixed-action pattern
3. Stickleback fish attack any model having the color red.

4. The fact that behavior is subject to evolution is best exemplified by
    a. giving examples of behavior that is believed to be innate.
    b. showing that patterns of behavior are inherited.
    c. showing that blood hormone levels motivate behavior.
    d. All of these.

5. How can you tell if a behavior has a sign stimulus?
    a. The behavior only occurs between members of a society.
    b. It is provoked by some particular aspect of the animal rather than the whole animal.
    c. The behavior involves orientation of the animal.
    d. All of these.

6. *Aplysia* has a gene that codes for the hormones necessary to bring about reproductive behavior. This shows that behavior
   a. is often innate.
   b. requires motivation.
   c. is controlled by genes.
   d. All of these.
7. Which of these best describes imprinting?
   a. learning that occurs during a critical period
   b. learning that requires a sign stimulus
   c. innate behavior that requires practice
   d. All of these.
8. Territoriality and dominance hierarchies
   a. help control aggression among members of a society.
   b. are forms of communication between members of a society.
   c. are evidence of sociobiology.
   d. All of these.
9. Which of these forms of communication is mismatched?
   a. pheromones—chemical
   b. waggle dance—visual
   c. alarm calls—auditory
   d. language—auditory
10. Altruism
   a. seems to be self-sacrifice for the benefit of the group.
   b. may increase inclusive fitness.
   c. fosters kin selection.
   d. All of these.

Study Questions

1. Draw a diagram showing that behavior evolves and occurs as a result of an environmental stimulus.
2. Describe the breeding experiments with the parrot *Agapornis*, and the results of genetic studies with the snail *Aplysia*.
3. Define a fixed-action pattern. Give several examples, and discuss one that is improved by practice.
4. Give examples of orientation and navigation in animals. What kinds of stimuli are utilized for such behavior?
5. Describe an experiment that indicates how ringdoves are motivated to perform reproductive behavior.
6. Name three types of learning and give an example of each type.
7. Define society and name two aspects of behavior that are usually found within a society.
8. Give examples of four common means of communication between members of a society.
9. Name and discuss mechanisms that help reduce overt aggression among members of a society.
10. What is the basic tenet of sociobiology? How do sociobiologists explain altruistic and human reproductive behavior?

Thought Questions

1. What evidence in this chapter suggests that behavior is fixed and cannot be modified.
2. What evidence can you offer to suggest that human behavior can be modified and it is not fixed?
3. Give examples to show that aggression is controlled between members of a human society in the same manner as other animals.

Selected Key Terms

innate behavior (in-nāt′ be-hāv′yor) 703
fixed-action pattern (fikst ak′shun pat′ern) 703
sign stimulus (sīn stim′u-lus) 703
taxis (tak′sis) 706
pheromones (fer′o-mōnz) 706

imprinting (im′print-ing) 708
operant conditioning (op′ē-rant kon-dish′un-ing) 709
insight learning (in′sīt lern′ing) 710
society (so-si′ē-te) 710
aggression (ah-gresh′un) 713

territoriality (ter″ĭ-tor″ĭ-al′ĭ-te) 713
dominance hierarchy (dom′ĭ-nans hi′er-ar″ke) 713
sociobiology (so″se-o-bi-ol′ŏ-je) 715
kin selection (kin sē-lek′shun) 715

C H A P T E R   45

Ecology of Populations

Your study of this chapter will be complete when you can:

1. Calculate the rate of natural increase for a population, when provided with the number of individuals in the population, the birthrate, and the death rate.

2. Contrast a J-shaped growth curve with an S-shaped growth curve.

3. Indicate which parts of an S-shaped growth curve represent biotic potential, environmental resistance, and carrying capacity.

4. Give examples of density-independent and density-dependent effects of environmental resistance.

5. Contrast the three most likely types of survivorship curves for a population.

6. Contrast three common age structure diagrams for a population.

7. Contrast an r-strategist population with a K-strategist population with regard to growth curves, survivorship curves, and age structure diagrams.

8. Describe the growth curve for the world's human population.

9. Contrast the demography of developed and developing countries. Tell why both types of countries contribute heavily to resource consumption.

10. Explain why replacement reproduction does not automatically bring about a zero population growth.

These poppies are members of a population whose size is determined by the carrying capacity of the environment.

Figure 45.1

Temperate deciduous forest and members of various populations; (a) general appearance; (b) chipmunk whose cheeks are bulging with stored nuts; (c) round-lobed hepatica.

a.

b.

c.

The German zoologist Ernst Haeckel coined the word "ecology" from two Greek roots: *oikos,* meaning "household" and *ology,* meaning "study of." He recognized that organisms do not exist alone; rather, they are part of a household, a functional unit that includes other types of organisms and their physical surroundings. Today, we define **ecology** as the study of the interactions of organisms with both their biotic (living) environment and their abiotic (physical) environment.

Ecologists may study these interactions at a particular locale (fig. 45.1). Various ecological factors can affect the size of a **population,** all the organisms of the same species at the locale, and we will begin by taking a look at some of these in this and the next chapter.

Ecology is the study of interactions of organisms with each other and with the physical environment.

Population Growth

Populations increase in size whenever the number of births exceeds the number of deaths. Suppose, for example, a population has 1,000 individuals ($N = 1,000$), a **birthrate** (*b*) of 30 per year, and a **death rate** (*d*) of 10 per year. The annual **rate of natural increase** (*r*) would be:

$$r = \frac{b - d}{N} = \frac{30 - 10}{1,000} = 0.02 = 2\% \text{ per year.}$$

The rate of natural increase does not include any change in population size from either immigration or emigration, which for the purpose of our discussion can be assumed to be equivalent.

Populations with a positive rate of natural increase grow larger each year. The expected increase (*I*) can be calculated by multiplying the rate of natural increase (*r*) by the current population size (*N*)

$$I = rN$$

This formula indicates that population growth is **exponential:** at the end of each year, *N* is larger, and therefore, *I* will also be larger, as demonstrated in table 45.1. This means population size increases by an ever larger amount each year. This is apparent when population size is plotted against time. When an arithmetic scale is used for population size, the resulting growth curve resembles that in figure 45.2. It starts off slowly and then increases dramatically, giving a *J-shaped curve.*

Table 45.1 Exponential Growth ($r = 2.0\%$)					
Unit Time	N	I	Unit Time	N	I
0	10,000.00	200.00	7	11,486.86	229.74
1	10,200.00	204.00	8	11,716.60	234.33
2	10,404.00	208.08	9	11,950.93	239.02
3	10,612.08	212.24	10	12,189.95	243.80
4	10,824.32	216.49	11	12,433.75	248.68
5	11,040.81	220.82	12	12,682.43	253.65
6	11,261.63	225.23			

Figure 45.2

The exponential growth curve is J shaped, indicating that the population is experiencing a steady rise in population size.

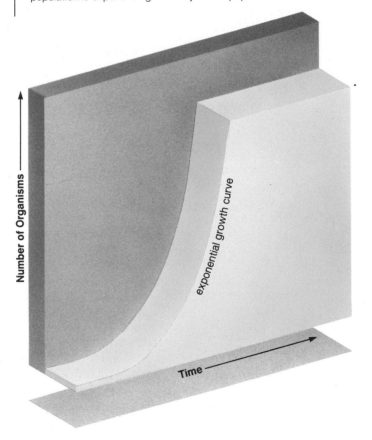

Biotic Potential

The **biotic potential** of a population is the maximum rate of natural increase (r_{max}) that can possibly occur under ideal circumstances. These include plenty of room for each member of the population, unlimited resources, and no hindrances, such as predators or parasites. Because these conditions rarely occur, it is very difficult to determine the r_{max} for most species. A few approximations, however, are given in table 45.2.

Table 45.2 Biotic Potential	
Kind of Organism	Approximate Biotic Potential r_{max} (per year)
Large mammals	0.02–0.5
Birds	0.05–1.5
Small mammals	0.3–8
Larger invertebrates	10–30
Insects	4–50
Small invertebrates (including large protozoans)	30–800
Protozoa and unicellular algae	600–2,000
Bacteria	3,000–20,000

Table 3.3 from *Principles of Ecology,* by Richard Brewer, copyright © 1979 by Saunders College Publishing, a division of Holt, Rinehart and Winston, Inc., reprinted by permission of the publisher.

Whether the biotic potential is high or low depends on such factors as the following:

1. Usual number of offspring per reproduction
2. Chances of survival until age of reproduction
3. How often each individual reproduces
4. Age at which reproduction begins

Maximum growth rate is not experienced by any population for any length of time. This is readily apparent when considering that if a single female pig had her first litter at nine months, and produced two litters a year, each of which contained an average of four females (which, in turn, reproduced at the same rate), there would be 2,220 pigs by the end of three years. Because of exponential growth, even species with a low biotic potential would soon cover the face of the earth with their own kind. For example, as Charles Darwin calculated, a single pair of elephants would have over 19 million live descendants after 750 years.

Environmental Resistance and Carrying Capacity

Populations do not undergo exponential growth for very long. Even in the laboratory setting, where crowding may be the only limiting factor, growth of populations eventually levels off. Figure 45.3 gives the actual growth curve for a fruit fly population reared in a culture bottle. This type of growth curve is often called a logistic curve, or an *S-shaped* (sigmoidal) curve. During an acceleration phase, when births exceed deaths, r is positive; during a deceleration phase, when births and deaths are more nearly equal, r is declining. Finally, when births exactly equal deaths, r is zero, and the population maintains a steady state (fig. 45.4).

Figure 45.3

The number of fruit flies in a colony was counted every other day. When these numbers were plotted, the result was a sigmoidal (S-shaped) growth curve.
Data from *Laboratory Studies in Biology: Observations and Their Implications*, Chester A. Lawson, Ralph W. Lewis, Mary Alice Burmester, and Garrett Hardin. Copyright © 1955 W. H. Freeman and Company, New York. Reprinted by permission.

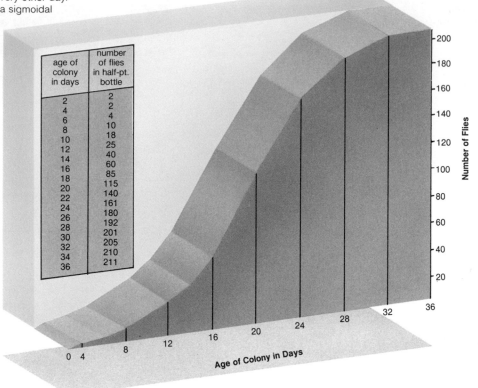

age of colony in days	number of flies in half-pt. bottle
2	2
4	4
6	10
8	18
10	25
12	40
14	60
16	85
18	115
20	140
22	161
24	180
26	192
28	201
30	205
32	210
34	211
36	

Figure 45.4

The logistic or S-shaped growth curve. The population initially grows exponentially, but begins to slow down as resources become limited. Population size remains stable when the carrying capacity of the environment is reached.

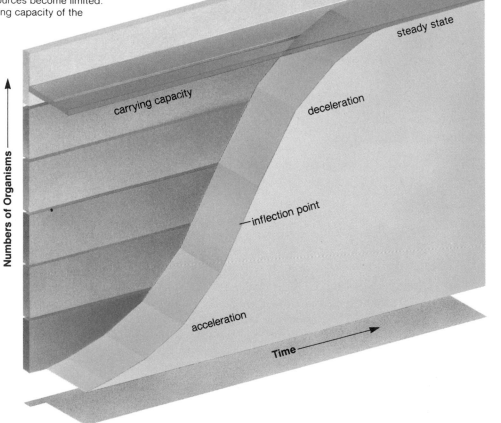

Population size is believed to level off at the **carrying capacity** (K) of the environment. In other words, the environment is capable of sustaining a population of a limited size. As the population increases in size, there will be more and more competition for available space and food. Eventually, this will affect population size. It is said that the environment increasingly resists the biotic potential of the population. **Environmental resistance** contains the effects given in table 45.3. The expression:

$$\frac{K - N}{K}$$

pertains to density-dependent resistance. It also tells us what proportion of the carrying capacity still remains. If N is small, the expression approximates 1, and most of the carrying capacity still remains. If N is large, the expression approaches 0, which indicates that the carrying capacity has almost been reached.

If our argument to this point is sound, then insertion of the expression

$$\frac{(K - N)}{K}$$

into the equation for exponential growth should result in an S-shaped curve.

$$I = r \, \frac{(K - N)}{(K)} \, N$$

At low population densities, when the expression approximates 1, growth would be nearly equal to rN; therefore, growth would be exponential. But as the density increases, the expression approaches 0, and growth levels off. Therefore, it seems that the concept of carrying capacity explains the logistic growth curve. Further, we can predict that if a population happens to overshoot the carrying capacity, it would have to decline once again, because the expression

$$\frac{(K - N)}{K}$$

would then have a negative value.

Fluctuations are often seen in population sizes. In figure 45.5, this fluctuation is due to an abrupt change in carrying capacity from season to season. During the summer, when resources are more than adequate, the bobwhite population increases in size. In winter, when the environment is less favorable, the bobwhite population declines. As a result, population size can fluctuate from year to year. The mean of the increases and decreases over time represents the average carrying capacity of the environment.

Table 45.3 Environmental Resistance to r_{max}

Density-Independent Effects
Climate and weather
Natural disasters
Requirements for growth
} Effects that are independent of population density.

Density-Dependent Effects
Competition
Emigration
Predation
Parasitism
} Effects that intensify with increased population density.

Figure 45.5

Population fluctuations in a bobwhite quail population. Gold columns represent spring population sizes, and orange columns represent fall population sizes. Note that quail populations increase from spring to fall and decrease from fall to spring.

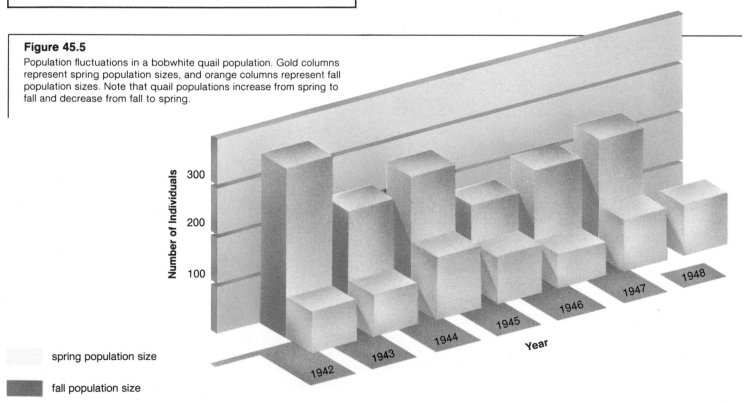

spring population size

fall population size

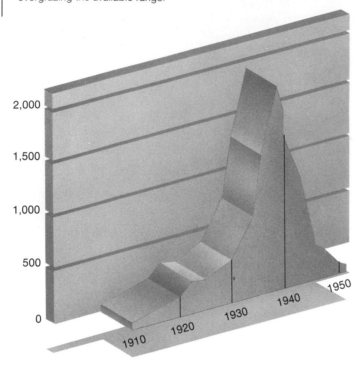

Among some species, notably insects and certain plants, populations grow rapidly for a limited period of time and then quickly die out. The growth curve for these populations is at first J-shaped, but the exponential growth is followed by a sharp decline. For example, a reindeer herd introduced on St. Matthew Island, Alaska, underwent a rapid expansion, and then declined just as quickly when food resources were depleted (fig. 45.6) by the overbrowsing of the herd that had grown too large.

Every population has a biotic potential for increase in size. Biotic potential is normally held in check by environmental resistance so that the growth curve is S-shaped and levels off at the carrying capacity of the environment. J-shaped curves are also known.

Practical Applications

The concepts of biotic potential and environmental resistance can be used to devise the means of regulating population growth. When a population is a human food source, it should be harvested only to the inflection point (fig. 45.4), where the greatest amount of population growth occurs. This amount of cropping results in a *maximum sustained yield,* as long as enough mature individuals capable of reproduction remain. Unfortunately, many marine populations, such as whales, have been overexploited by commercial fisheries, to the point where it will be many years before a maximum sustained yield will again be possible.

For pests, such as rats and mosquitoes, it is far better to reduce breeding sites and food supply or to increase predators than it is to try to kill off the animals. Killing the animals does not restrict their growth potential, which may come into full force as soon as the population size is reduced. On the other hand, when trying to preserve an endangered species, it is best to place them in a protective environment, such as a reserve, which provides for all their needs. In this case, reduction in environmental resistance will permit population growth to occur.

Mortality and Survivorship

Members of a population have a limited life span and death is the means by which most individuals are lost from a population. (The other means is emigration.) The number of members of a population that are dying per unit of population per unit time is the death rate. The death rate is usually determined by dividing the number of individuals that have died during a given time period by the number alive at the beginning of that period of time.

Rates of natural increase in population size are a function of additions by birth and losses by death. The rate of increase sometimes becomes greater, not because of growing birthrate, but because of a decline in the death rate. This decline in deaths is one reason for the explosive growth of the human population in certain countries of the world.

Because living members of the population are more important to the population than are dead ones, another way of looking at mortality is **survivorship,** which is the number of individuals in a given population alive at the beginning of each age interval. A survivorship curve can be drawn by plotting the number of survivors against time.

In theory, there are three types of survivorship curves (fig. 45.7). A type-I curve shows that most individuals within a designated group live out their allotted life span and then die of old age. Under very favorable environmental conditions, certain mammals, such as humans, and annual plants that live one season tend to approach this curve. The type-II curve results

Figure 45.7

The three basic types of survivorship curves. Type-I curve occurs when the members of a species usually live the full physiological life span. Type-II curve occurs when the rate of mortality is fairly constant at all age levels. Type-III curve occurs when there is a high mortality early in life.

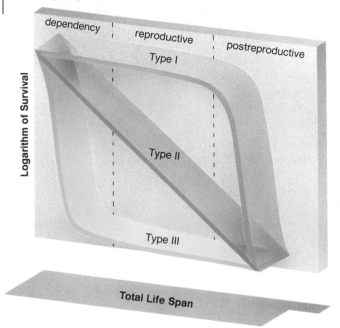

when organisms die at a constant rate through time. Such curves are typical of birds, rodents, and some perennial plants that live more than one season. Type-III curves indicate that many organisms tend to die early in life. This curve is typical of fishes, many invertebrates, and many perennial plants.

Age Structure

Populations have three age groups: *dependency, reproductive,* and *postreproductive.* One way of characterizing populations is by age groups. This is best visualized when the proportion of individuals in each group is plotted on a bar graph, thus producing an **age structure diagram.** If the birthrate is high, and the population is undergoing exponential growth, a *pyramid-shaped* (fig. 45.8a) figure results, because each successive generation is larger than the previous one. When the birthrate equals the death rate, a *bell-shaped* (fig. 45.8b) age diagram is expected, because the dependency and reproductive age groups become more nearly equal. The postreproductive group is still the smallest, however, because of mortality. If the birthrate falls below the death rate, the dependency group will become smaller than the reproductive group. The age structure diagram will be *urn-shaped* (fig. 45.8c) because the postreproductive group is now the largest.

Figure 45.8

Age structure diagrams for three different populations. *a.* Age structure diagram is a pyramid shape, indicating that the population is undergoing exponential growth. *b.* Age structure diagram is tending to be bell shaped, indicating that population size is fairly stationary. *c.* Age structure diagram is tending toward an urn shape, indicating that the population is on the decline.
Data from Ronald Freedman and Bernard Berelson, "The Human Population." Copyright © 1974 by Scientific American, Inc. All rights reserved.

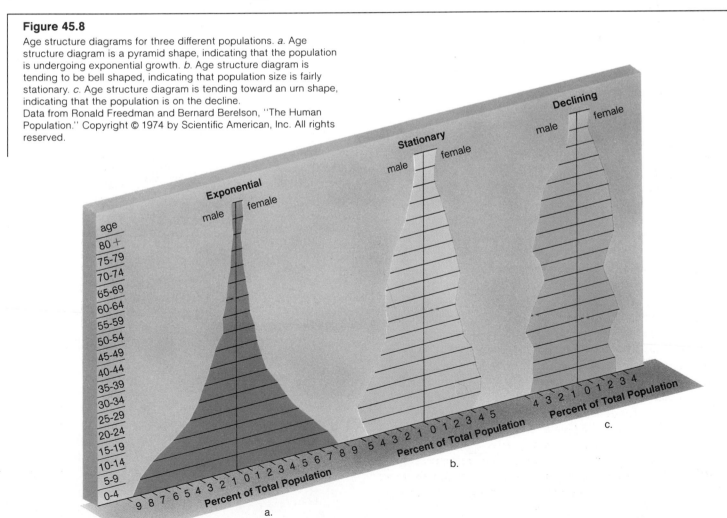

In a study of a honeybee hive, these changes in the age structure diagram were observed in a single season. At the beginning of the season, when the population was expanding, the age structure diagram was pyramid-shaped; in midseason, when the population size was stationary, the age structure diagram was bell-shaped; and finally, as the season came to a close and the population was dying off, the age structure diagram was urn-shaped.

Strategies for Existence

Populations do not increase in size year after year because environmental resistance, including both *density-independent* and *density-dependent* effects (table 45.3), regulates number. Some populations are regulated primarily by density-independent effects. These tend to have J-shaped growth curves until some environmental change causes them to decline, usually within a short time. Most likely these populations are *opportunistic species* that move in and occupy new environments. Ragweeds in gardens and along roadsides, and tent caterpillars in cherry and apple trees are opportunistic species. During a short period of time, they produce many offspring that require little care. Therefore, these populations usually have a survivorship curve similar to type III in figure 45.7. But chances are, some of these offspring will quickly disperse to colonize new habitats, because opportunistic species usually have efficient dispersal mechanisms. From an evolutionary point of view, such species have undergone selection to maximize their rates of natural increase, and for this reason, are said to be *r*-selected or to be **r-strategists.**

Other populations are regulated primarily by density-dependent effects. These tend to have sigmoidal (S-shaped) growth curves. Such populations are termed *equilibrium species*. For example, adult oak trees and whales are relatively large, long-lived, and slow to mature. They produce a small number of offspring and expend considerable energy in producing each individual because resources are limited and competitive superiority is a necessity for survival. These organisms tend to have a survivorship curve like type I in figure 45.7. Equilibrium species are strong competitors and, once established, can dominate or exclude opportunistic species. They are specialists rather than colonizers, and tend to become extinct when their normal way of life is destroyed. When undisturbed, the population size remains stable at or near the carrying capacity of the environment. Equilibrium species are said to be *K*-selected or to be **K-strategists.**

Most populations cannot be characterized as being either *r*- or *K*-strategists because they have intermediate characteristics. Such populations tend to have a type-II survivorship curve.

Populations of *r*-strategists tend to have J-shaped growth curves, pyramid-shaped age structure diagrams, and a type-III survivorship curve. *K*-strategists tend to have S-shaped growth curves, urn-shaped age structure diagrams, and a type-I survivorship curve.

Human Population Growth

The rate of natural increase, or growth rate, of the human population of the world is now 1.8%. The growth curve is J-shaped and exponential (fig. 45.9), but may eventually level off.

The human population is believed to have undergone three phases of exponential growth. Tool making may have been the first technological advance that resulted in a phase of exponential growth. Cultivation of plants and animal husbandry may have resulted in a second phase of growth; and the Industrial Revolution, which occurred about 1850, promoted the third phase.

Developed and Developing Countries

Europe, North America, Russia, and Japan were the first nations to become industrialized. These nations, often referred to collectively as the **developed countries,** doubled their populations between 1850 and 1950. This was largely due to a decline in the death rate caused by the rise of modern medicine and improved socioeconomic conditions. The death rate decline was followed shortly thereafter by a decline in the birthrate, so the populations in the developed countries showed only modest growth between 1950 and 1975. This sequence of events is termed a **demographic transition.**

The growth rate for the developed countries as a whole has now stabilized at about 0.6%. A few developed countries—Italy, Denmark, East Germany, Hungary, Sweden, West Germany—are not growing, or are actually losing population. The United States has a higher growth rate (0.7%) than average, because many people immigrate to the United States each year, and a baby boom between 1947 and 1964 has resulted in an unusually large number of women of reproductive age.

Most countries in Africa, Asia, and Latin America are collectively known as **developing countries,** because they have not yet become industrialized. Although mortality began to decline steeply in these countries following World War II with the importation of modern medicine from the developed countries, the birthrate did not decline to the same extent. Therefore, the populations of the developing countries began to increase dramatically. The developing countries were unable to cope adequately with such rapid population expansion, and many people in these countries are underfed, ill-housed, unschooled, and living in abject poverty.

The growth rate of the developing countries peaked at 2.4% in 1960–1965. Since that time, the mortality decline has slowed, and the birthrate has fallen. The growth is expected to be 1.8% by the end of the century. At that time, however, more than two-thirds of the world population will be in the developing countries.

The reason for a decline in the growth rate in the developing countries is not clear. Previously, it was argued that decline would only be seen when socioeconomic conditions had improved. It was believed that as soon as the populace enjoyed the benefits of an industrialized society, the birthrate would decline. It is now apparent, however, that countries with the

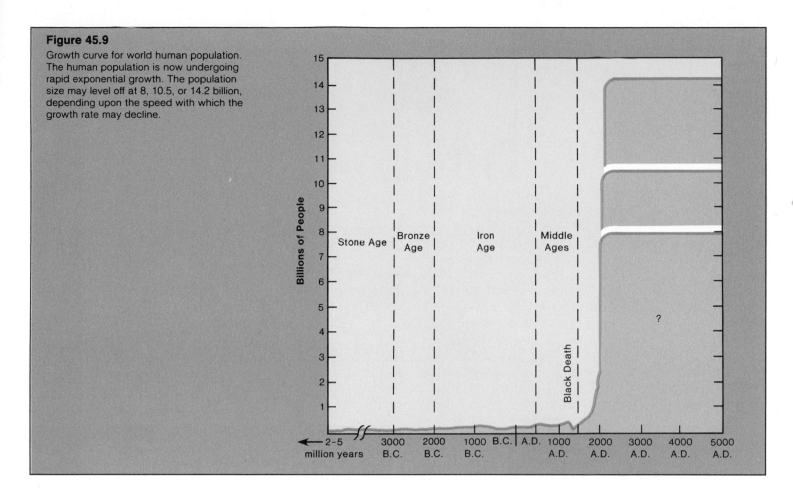

Figure 45.9
Growth curve for world human population. The human population is now undergoing rapid exponential growth. The population size may level off at 8, 10.5, or 14.2 billion, depending upon the speed with which the growth rate may decline.

greatest decline are those with good family-planning programs, supported by community leaders. Such programs may bring about stable population sizes in the developing countries even before socioeconomic conditions have improved. Nevertheless, certain socioeconomic factors have also probably contributed to the decline in the developing countries' growth rate. Relatively high Gross National Product (GNP), urbanization, low infant mortality, increased life expectancy, literacy, and education, especially of women, have all had a dampening effect on the *fertility rate*—the average number of children born to each woman.

The history of the world population shows that the developed countries underwent a demographic transition between 1950 and 1975; the developing countries are just now undergoing demographic transition.

Age Structure Comparison

Lay people are sometimes under the impression that if each couple had two children, **zero population growth** would immediately take place. However, *replacement reproduction,* as it is called, would still cause most countries to continue growth because of the age structure of the population. If there are more young women entering the reproductive years than there are older women leaving them, replacement reproduction will still result in a positive growth rate.

Reproduction is at or below replacement level in some twenty developed countries, including the United States. Even so, some of these countries will continue to grow modestly, in part because of the baby boom after World War II. These young women are now in their reproductive years, and even if each one has fewer than two children, the population will still grow. It should also be kept in mind that even the smallest growth rate can add a considerable number of individuals to a large country. For example, a rate of natural increase of 0.7% added over 1.6 million people to the United States population in 1982.

Whereas many developed countries are tending toward a stabilized age structure (fig. 45.10), most developing countries have a youthful profile—a large proportion of the population is below the age of fifteen. Because there are so many young women entering their reproductive years, the population will still greatly expand, even after replacement reproduction is attained. The more quickly replacement reproduction is achieved, however, the sooner zero population growth will result.

Consequences of Population Growth

Increasing world population is putting extreme pressure on the earth's resources, environment, and social organization. Although population increases in developing countries might seem to be of the gravest concern, it is not necessarily the case. Each

Figure 45.10

Contrasting age pyramids illustrate that the developed countries are approaching stabilization, whereas the developing countries will expand rapidly due to the shape of their age diagrams.

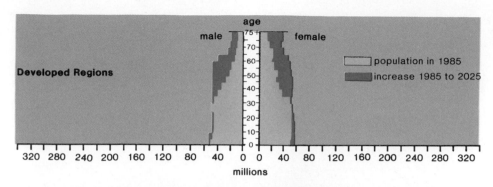

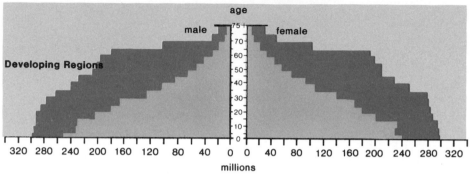

United States Population

The United States now has a population of over 240 million. The geographic center (the point where there are just as many persons in each direction) has moved steadily westward and recently has also moved southward. Another interesting trend is the shifting emphasis from metropolitan areas to nonmetropolitan areas. In the 1970s cities increased by 9.8%, but rural small towns increased by 15.8%.

The population size of the United States is not expected to level off any time soon for two primary reasons.

1. A baby boom between 1947 and 1964 has resulted in an unusually large number of reproductive women at this time (see figure). Thus, although each of these women is having on the average only 1.9 children, the total number of births increased from 3.1 million a year in the mid 1970s to 3.6 million in 1980–1981.
2. Many people immigrate to the United States each year. In 1981 immigration accounted for 43% of the annual population growth. The number of legal immigrants was about 700,000. Even though ordinarily only 20,000 legal

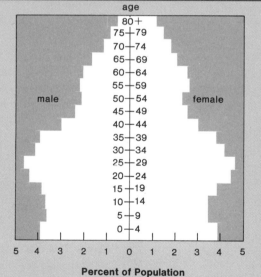

Percent of Population

U.S. population age diagram, 1985.
Source: United States Bureau of the Census.

immigrants can come from any one country, we give special permission for large numbers of political refugees to enter the United States.

In recent years, the majority of refugees have come from Latin America (e.g., Cuba) and Asia (e.g., Indonesia).

There is also substantial illegal immigration into the United States, although the exact number is not known. Estimates range from 100,000 to 500,000 a year or more. About 50% to 60% of these illegal immigrants come from Mexico, according to most estimates. There has been an effort to stem the tide of illegal immigration to the United States.

Whether the United States can ever achieve a stable population size depends on the fertility rate (the average number of children each woman bears) and the net annual immigration. If the fertility rate were kept at 1.8 and immigration limited to 500,000, the population would peak at 274 million in 2050, after which it would decline. Those who favor a curtailment of our population size point out that a fertility rate of 1.8 allows couples a great deal of freedom in deciding the number of children they will have. For example, it means that 50% of all couples can have two children; 30% can have three children; 10% can have one child; 5% can have no children; and 5% can have four children.

person in a developed country consumes more resources, and is therefore responsible for a greater amount of pollution. Environmental impact (*E.I.*) is measured not only in terms of population size, but also in terms of the resources used and the pollution caused by each person in the population.

E.I. = population size × resource use per person

= pollution per unit of resource used

Therefore, there are two possible types of overpopulation. The first type is due to increased population and the second to increased resource consumption. The first type of overpopulation is more obvious in developing countries, and the second type is more obvious in developed countries.

Resource consumption is dependent on population size and on consumption per individual. The developing countries are responsible for the first type of environmental impact and the developed countries for the second type.

Summary

1. Ecology is the study of how organisms interact with the physical environment and with each other. Organisms of the same species living in one locale form a population.
2. Populations with a positive rate of natural increase (*r*) have an increase in size expressed as *I = rN,* where (*N*) is the number of individuals in the population. Because (*N*) is larger each year, so is (*I*). This means the population is increasing in size exponentially, and the growth curve would be a J-shaped, or exponential one.
3. The maximum rate of natural increase (biotic potential) for a population (r_{max}) occurs only when there is no environmental resistance, which includes density-independent and density-dependent effects.
4. An r_{max} might be observed for a short time during exponential growth, but eventually environmental resistance causes the growth curve to become sigmoidal as the population size levels off at the carrying capacity of the environment.
5. Sometimes populations exhibit a J-shaped curve followed by a sharp decline. Opportunistic species often colonize an area and then are cut back by the first frost, for example. Or unrestricted growth is often followed by a crash, as the population greatly exceeds the environment's carrying capacity.
6. Populations tend to have one of three types of survivorship curves depending on whether most individuals live out the normal life span, die at a constant rate regardless of age, or die early.
7. Each population also has a particular type of age structure diagram that indicates the number of individuals in the dependency, reproductive, and postreproductive age groups. The shape of the diagram tells whether the population is a young, expanding one, a stable one, or a dying one.
8. Populations seem to have a particular, overall strategy for continued existence. Those that are opportunists produce many young within a short period of time and rely on rapid dispersal to new, unoccupied environments. These populations tend to have a J-shaped growth curve and to be regulated by density-independent effects. Such populations belong to species that are *r*-selected.
9. Other populations produce a limited number of young that they nurture for a long time. These populations tend to be regulated by density-dependent effects and belong to species that are *K*-selected.
10. The human population is now in an exponential part of its growth curve. Developed countries underwent a demographic transition between 1950 and 1975, with the result that their growth rate is now only about 0.6%. Developing countries are just now undergoing demographic transition. Because it was delayed, their growth rate went as high as 2.4%, but is declining slowly now.
11. Replacement reproduction will not bring about zero population growth even in developed countries if the age structure is biased toward those in their early reproductive years. In developing countries, where the average age is less than fifteen, it will be many years before replacement reproduction will mean zero population growth.
12. Resource consumption depends on population size and on consumption per individual. Developing countries are responsible for the first type of environmental impact and developed countries for the second type.

Objective Questions

1. A population contains all the
   a. organisms that live at one locale.
   b. organisms that are members of the same species at one locale.
   c. similar types of organisms at one locale.
   d. organisms that are interacting with each other at one locale.

2. In a population *N* = 1,500, if the birthrate is 42 per year and the death rate is 27 per year, the rate of natural increase will be
   a. 42%.
   b. 27%.
   c. 1%.
   d. 0.1%.

3. Whether the biotic potential for a species is high or low depends on such factors as
   a. usual number of offspring per reproduction.
   b. chances of survival until age of reproduction.
   c. age at which reproduction begins.
   d. All of these.

4. Environmental resistance
   a. prevents exponential growth from occurring.
   b. includes density-independent effects.
   c. helps determine the carrying capacity.
   d. Both (b) and (c).
5. Populations termed *r*-strategists
   a. have J-shaped growth curves.
   b. have type-III survivorship curves.
   c. are usually pioneer species.
   d. All of these.
6. Replacement reproduction cannot bring about zero population growth unless
   a. all diseases such as cholera, typhus, and diphtheria are brought under control.
   b. there are many fatalities due to a world war.
   c. there are as many young women entering their reproductive years as older women leaving them.
   d. Any one of these.
7. Once the demographic transition has occurred
   a. both the death rate and birthrate are high.
   b. both the death rate and birthrate are low.
   c. the death rate is high but the birthrate is low.
   d. the death rate is low but the birthrate is high.
8. Which shape best describes the age structure for a country whose population has stabilized?
   a. pyramid-shaped
   b. urn-shaped
   c. bell-shaped
   d. exponential

Study Questions

1. Draw a sigmoidal growth curve and relate the concepts of biotic potential, environmental resistance, and carrying capacity to the curve.
2. Draw a J-shaped curve and discuss why populations characterized by such curves may suddenly decline in size.
3. Draw three different survivorship curves and tell what each curve indicates about the expected time of death.
4. Draw three different age structure diagrams and tell what each one indicates about a population.
5. What is an "*r*-strategist"? a "*K*-strategist"?
6. Name four types of density-dependent effects, and explain what is meant by this term.
7. Show that a population of 10,000 persons will add more individuals with each generation, even if the growth rate remains constant. Is it possible that even with a declining growth rate the same number of individuals could be added each generation?
8. Define demographic transition. When did developed countries undergo demographic transition? When did developing countries undergo demographic transition?
9. Give at least three differences between developed countries and developing countries.

Thought Questions

1. Compare and contrast the manner in which carrying capacity and environmental resistance regulate population size.
2. Give added examples of animals from your own knowledge that are *r*-strategists, that are *K*-strategists.
3. Provide arguments for and against the United States allowing immigration into this country.

Selected Key Terms

ecology (e-kol'o-je) 720
population (pop''u-la'shun) 720
birthrate (birth'rāt) 720
death rate (deth rāt) 720
exponential (ex''po-nen'shal) 720
biotic potential (bi-ot'ik po-ten'shal) 721
carrying capacity (kar'e-ing kah-pas'ĭ-te) 723

environmental resistance (en-vi''ron-men'tal re-zis'tans) 723
survivorship (sur-vi'vor-ship) 724
age structure diagram (āj struk'tūr di'ah-gram) 725
r-strategist (ar strat'ĕ-jist) 726
K-strategist (ka strat'e-jist) 726

developed country (de-vel'opt kun'tre) 726
demographic transition (dĕ''mo-graf'ik tran-zish'un) 726
developing country (de-vel'op-ing kun'tre) 726
zero population growth (ze'ro pop''u-la'shun grōth) 727

CHAPTER 46

Interactions within Communities

Your study of this chapter will be complete when you can:

1. *State several characteristics of a niche and distinguish between the niche and the habitat of an organism.*
2. *Give examples of scramble and contest competition, and explain how each regulates population size.*
3. *Discuss the effect that interspecific competition can have on population size.*
4. *State the competitive exclusion principle and relate this principle to the diversity of organisms.*
5. *Discuss the effect that predation has on the size of the prey population.*
6. *Give examples of plant and animal defenses against predators.*
7. *Explain the principle of mimicry and give examples of two kinds of mimicry.*
8. *Give examples of the three types of symbiotic relationships, and tell what effect they can have on population size.*

Monarch butterflies are brightly colored. This serves as a warning to predators that they are to be avoided. Monarch butterflies are poisonous and after a bird has experienced one, it leaves the others alone.

A population contains members of the same species. A **community** contains populations of many species that interact with one another. This chapter will primarily consider interactions involving competition, predation, and symbiosis. In chapter 48, we will see that the community and the physical environment function together to form a system, termed an **ecosystem.**

Before we begin our discussion of community ecology, let's review the important concepts of habitat and niche.

Habitat and Niche

The **habitat** of an organism is where it lives, its mailing address, so to speak. When describing an organism's habitat, it is best to be specific and to describe in as much detail as possible the exact location. For example, if an organism lives in the soil, it would be good to say what type of soil, and give the exact depth it prefers. Or if an organism lives in trees, the type of tree, and the organism's exact height in the tree should be specified.

The **niche** of an organism refers to the role it plays in an ecosystem. It includes physical factors, such as its temperature, light, moisture, and pH requirements. It also includes biological factors, such as how it acquires its food, what season of the year it reproduces, and how it affects other organisms in the community. Table 46.1 lists some factors that should be included when describing the niche of a particular plant or animal species.

The concept of niche complements the concept of character displacement we discussed on page 325. Figure 21.19 shows that three species of finches living on the same island have slightly different beak depths, allowing them to feed on different sizes of seeds. The concept of niche suggests that similar species that differ with respect to niche are able to coexist in one ecosystem because they partition resources. For example, Robert MacArthur recorded the length of time each of five species of warblers spent in different spruce tree zones to determine where each species did most of its feeding. He discovered that different species used different parts of the tree canopy, although some overlap did occur (fig. 46.1).

Resource partitioning is also observed in plants. Three species of annual plants, found in an abandoned field one year after cultivation ceased, exploit different parts of the soil. Bristly foxtail has shallow, fibrous roots that take water only from the top part of the soil. Indian mallow has a sparsely branched taproot that extends to an intermediate depth. Smartweed has a taproot that is moderately branched in the upper soil layer, but develops primarily below the rooting zone of the other species (fig. 46.2).

Table 46.1 Aspects of Niche	
Plants	**Animals**
Season of year for growth and reproduction	Time of day for feeding and season of year for reproduction
Sunlight, water, and soil requirements	Habitat and food requirements
Relationships with other organisms	Relationships with other organisms
Effect on abiotic environment	Effect on abiotic environment

Figure 46.1

Resource partitioning among five species of coexisting warblers. The diagrams that appear beside the birds represent spruce trees. The time each species spent in various portions of the trees was determined, and it was discovered that each species spent more than half its time in the blue regions.

Cape May Warbler   Black-throated Green Warbler   Yellow-rumped Warbler   Bay-breasted Warbler   Blackburnian Warbler

Competition

Even though five species of warblers can coexist in a northern spruce forest, resources, such as food, water, light, and space determine the carrying capacity of the environment and set a limit on population sizes. When competition for these resources occurs between members of the same species, it is called intraspecific competition. When competition occurs between populations belonging to different species, it is called interspecific competition.

Intraspecific Competition

When members of the same species compete, two types of competition can occur: scramble competition and contest competition.

Scramble Competition

In **scramble competition,** each member tries to acquire as much of a particular resource as possible. A. H. Nicholson, an English animal ecologist, showed that this type of competition could lead to population size fluctuations. He chose blowflies, which have the usual insect life cycle, consisting of egg, larva, pupa, and adult, as his experimental animal. Because the larvae live on decaying meat, and the adults feed on sugar-rich foods, he was able to regulate the amount of food available for each life history stage. In one experiment, he rationed food to the larvae, but provided an unlimited amount to the adults. When larval density was high, only a few grew large enough to pupate, and the few adults they produced were small. These small adults laid a small number of eggs, which, in turn, resulted in few larvae. But more members of the small, new larval population now had access to sufficient food, and in the end, there was a larger adult population than before. Thus, the adult population size fluctuated (fig. 46.3) from one generation to the next, its size affected by competition for food among the larvae.

Figure 46.2
Resource partitioning by three species of annual plants in an abandoned field one year after cultivation. The A horizon is the upper layer of soil and contains more organic material and variable amounts of moisture. The B horizon contains inorganic material and a more constant amount of moisture.

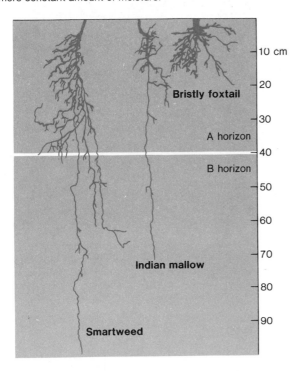

Figure 46.3
Data from a scramble competition experiment. Because food to larvae was limited, scramble competition resulted in few adults whenever there were many larvae.

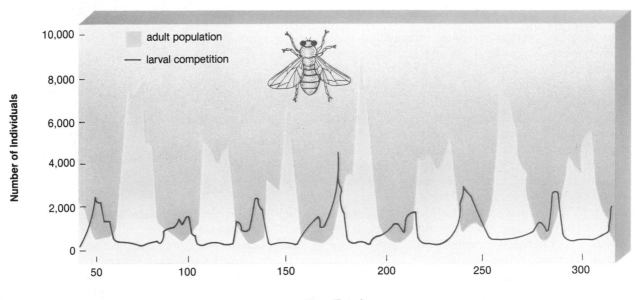

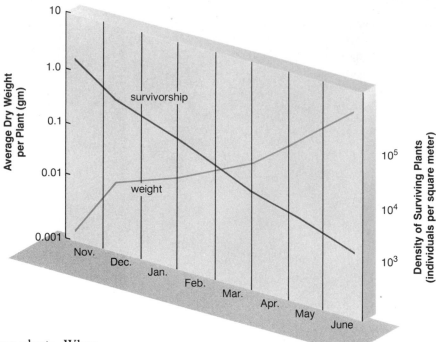

Figure 46.4

A self-thinning curve for an experimental planting. Horseweed was sown at a density of 100,000 seeds per square meter. As the density (number) of the surviving plants decreases, the weight per plant increases. Multiplying these values gives the weight (biomass) that the environment can support—the environment's carrying capacity.
Data from Moran, et. al., *Introduction to Environmental Science*. Copyright W. H. Freeman and Company, New York. Reprinted by permission of W. H. Freeman and Company, and Blackwell Scientific Publications.

Scramble competition is also seen among plants. When horseweeds were sown at low densities, each plant tended to grow vigorously. When they were sown at high densities, many did not survive, and those that did, were smaller than usual. Data from a number of plants produced the curves shown in figure 46.4. These curves indicate the carrying capacity of the environment.

Scramble competition is apt to cause animals to **emigrate**—to move away from centers of great density. Emigration must not be confused with migration, which is an annual movement of animals to feeding or reproductive areas, as when birds go south for the winter. Some animals, notably lemmings in the Arctic, undergo emigrations about every four years, when they reach a maximum population size that exceeds the carrying capacity of the area. Bees and grasshoppers swarm and move on when the colony becomes crowded.

Contest Competition

In contrast to scramble competition, **contest competition** tends to keep the animal population within the capacity of the environment without extreme oscillations in population size. As discussed on page 713, some animals live in groups with *dominance hierarchies* that determine the social position and importance of each animal. Dominance is determined by a contest—either actual fighting or simply a ritual contest of who can frighten whom (fig. 46.5). The dominant animals eat first and may have a chance to mate before lower ranking animals. The other animals either assist the dominant individuals or leave the group. In either case, they may not produce offspring.

Another social interaction that may affect population size is *territoriality,* in which successful competitors establish territories that they defend against other animals (fig. 44.18). Many animals mate and produce offspring within the territory. The

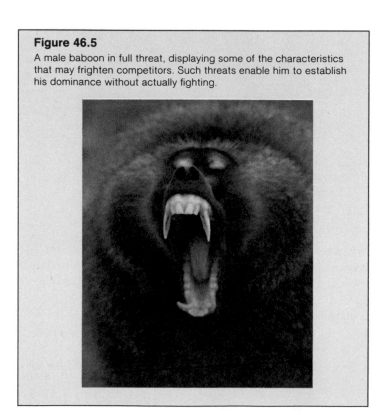

Figure 46.5

A male baboon in full threat, displaying some of the characteristics that may frighten competitors. Such threats enable him to establish his dominance without actually fighting.

best territory defenders often have the best territories—those with the best food resources and, among birds, the best nesting sites. The less successful defenders must make do with less desirable territories, and some have no territories at all. Therefore, these animals produce fewer offspring than those in better territories, or they produce no offspring at all.

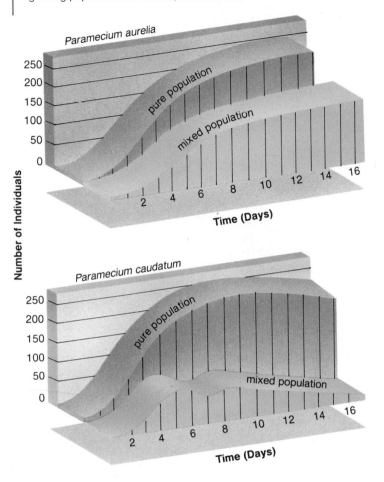

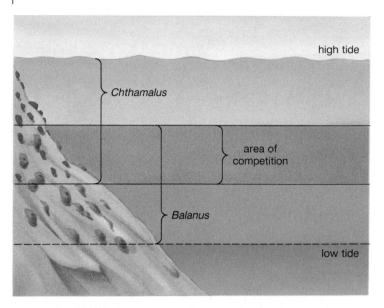

Interspecific Competition

When populations of different species compete with each other, it is possible that one population could be so successful that the other would die out. This was demonstrated in the laboratory by G. F. Gause, who grew two species of paramecia in one vial containing a fixed amount of bacterial food. Although each population survived when grown separately, only one survived when they were grown together (fig. 46.6). The successful paramecium population had a higher rate of natural increase than the unsuccessful population. This is an example of local extinction resulting from interspecific competition.

Instead of extinction in natural ecosystems, there may be restriction in the range of one of the species, as observed by Joseph Connell, when he studied the distribution of barnacles on the Scottish coast. *Chthamalus stellatus* lives on the high part of the intertidal zone, and *Balanus balanoides* lives on the lower part (fig. 46.7). Free-swimming larvae of both species attach themselves to rocks at any point in the intertidal zone, where they develop into the sessile adult form. In the lower zone,

the faster-growing *Balanus* individuals either force *Chthamalus* individuals off the rocks or grow over them. *Balanus*, however, is not as resistant to drying out as is *Chthamalus* and does not survive well in the upper intertidal zone, thus allowing *Chthamalus* to grow there.

Human Intervention

The effects of competition have also been observed after humans have inadvertently introduced a species into a new location. The carp, a fish imported to the United States from Asia, is better able to tolerate polluted water than are native fishes. Therefore, this fish is often more common than native species. An ornamental tree, the melaleuca, was introduced into Florida and has since invaded the Everglades, where it is drying up the cypress swamps and reducing the survival rate of natural vegetation. In Australia, the prickly pear cactus introduced from the United States overran 60 million acres before it was brought under control (see the reading later in this chapter, page 738). In each of these examples, a new species introduced into an area where its natural enemies do not occur was able to outcompete the native species in the area. Control of nonnative species is usually obtained only by introducing natural pest species from the introduced species' original area of distribution.

Exclusion Principle

Competition between species does not always lead to expansion of one population and restriction of another. Gause showed in another laboratory experiment that when two different species of paramecia occupied the same tube, both survived, because one species fed on bacteria on the bottom, and the other fed on bacteria suspended in solution. This experiment illustrated that

Figure 46.8

Size of prey population versus rate of predation. The risk of bird predation was found to be smaller than expected when an insect's population was low, and greater than expected when the population was increasing. This indicates that perhaps as a particular prey becomes more abundant, birds develop a search image that helps them find that prey.

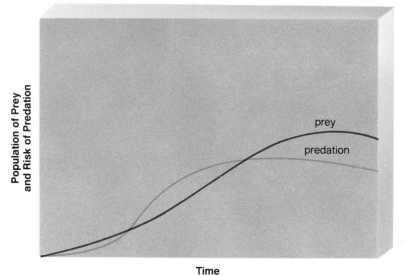

two or more similar species can live together when they occupy different niches (fig. 46.1). Only if the two species are attempting to occupy the same niche will one tend to die out. The fact that no two species can occupy the same niche is called the **competitive exclusion principle.**

We have already seen that competition causes species to specialize and competition can be reduced by occupying different niches. Therefore, competition may lead to increased diversity of living things.

> When two species compete for the same resources, one usually eliminates or restricts the range of the other. This is in keeping with the competitive exclusion principle, which states that no two species can occupy the same niche.

Predation

Predation, simply defined, is one organism feeding on another. Examples include monkeys feeding on flower bulbs, a bear feeding on salmon, or a lion feeding on a gazelle.

Predator Adaptations

Through the evolutionary process, predators become adapted to capturing their prey. The anatomy of browsers enables them to reach high into trees—giraffes have long necks and elephants have long trunks. Birds of prey have talons and sharp beaks that enable them to capture, hold, and kill their prey. Carnivores also need keen eyesight. Hawk eyes have a resolving power that is eight times better than that of the human eye, due to the concentration of cones in two foveas, instead of one.

The efficiency of predation is related to the amount of prey available, but the correlation is not as simple as might be predicted. In an experiment with European titmice, which feed on insect larvae, predation was less than expected when the prey species was rare. Once the prey increased in number, however, predation was greater than expected. Finally, when the prey species was most numerous, predation was again less than expected (fig. 46.8). Because of these results, it has been suggested that predation requires a *search image,* which is a mental picture of the prey. Supposedly, when the prey is rare, the predator is not prompted to form a search image but when the prey numbers increase, the predator does develop a search image and begins to actively seek a particular food item. When the predator has achieved its maximum potential for capturing the prey, there is no further increase in predation, regardless of the amount of prey.

Control of Prey Population Size

Other data suggest that predators might overkill their prey population, eventually leading to a reduction in the size of the predator population. For example, the population size of the lynx in Canada decreases approximately every ten years (fig. 46.9). Lynx population size fluctuations closely follow cyclical changes in the population size of their principal prey, the snowshoe hare. Ecologists are not convinced, however, that these oscillations are due to the predator–prey relationship. These changes may be due to a periodic decline in the availability of food for the hares. As their population size declines, the lynx population also declines. This explanation seems reasonable, because the size of hare populations has fluctuated even in the absence of lynxes.

One other consideration in the predator–prey relationship was illustrated by other experiments performed by G. F. Gause. In one experiment, Gause included a species of paramecia as

Figure 46.9

Changes in abundance of lynx and hare as determined from the number of pelts received by the Hudson's Bay Company for the years indicated.

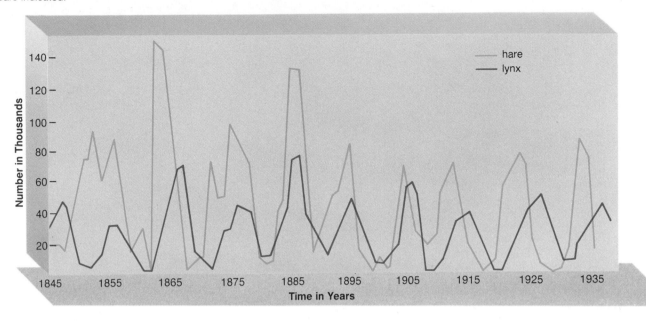

Figure 46.10

a. Scanning electron micrograph of a didinium engulfing a paramecium. b. In an experimental situation, a predatory population of didinia will sometimes kill all the paramecia prey and then die off itself.

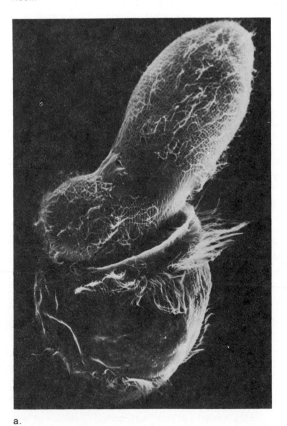

a.

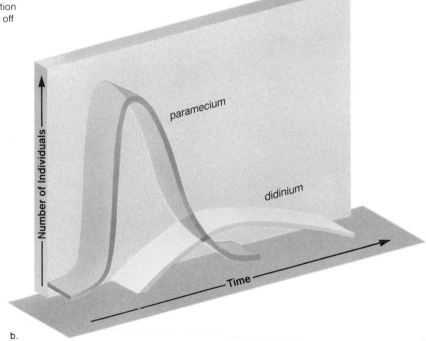

b.

the prey and a species of didinia as the predator. In the experimental test-tube environment, the didinia actually captured all the paramecia and then died out from starvation (fig. 46.10). However, if Gause provided refuge for the prey, in this case sediment on the bottom of the laboratory glassware, part of the prey population survived. This suggests that predator–prey relationships can stabilize when predator pressure eases as the prey

Integrated Pest Management

The use of chemical pesticides to kill agricultural pests has often created more problems than it has solved. Indeed, the percentage of crop losses to insects is now over three times what it was in the 1940s, before the modern pesticides became available. Pesticides invariably reduce the numbers of predators and parasites that naturally control the pests more than they reduce the numbers of pests themselves. Then, when the pests become resistant to the pesticide, there are no natural enemies to hold the pest population in check. The population size then increases manyfold, as does that of other plant-eating insects whose natural enemies were also destroyed by the pesticide.

Pesticides also find their way into the water supply of humans, and/or they concentrate in food chains causing humans to receive larger doses than were applied in the first place. In Hawaii, cows were fed pineapple leaves that had been sprayed, and the cows' milk then passed the pesticide on to human infants.

These problems with pesticide use have caused many to become interested in integrated pest management, which utilizes resistant plants, natural enemies of the pest, and environmental modification to attempt to reduce a pest population. This system seeks to control the pest rather than to eradicate it.

The use of natural predators, parasites, and pathogens can control agricultural pests. The vedalia ladybug was imported from Australia in 1888 to save citrus trees that were literally covered with cottony-cushion scale. The ladybug proved to be such a good predator that the pest was nearly exterminated within two years of the initial release of just 28 beetles. The praying mantis and lacewing are predators that feed on many types of pests. When lacewings were released in cotton fields, they reduced the boll weevil population by 96% and increased cotton yield threefold. A predatory moth was used to reclaim 60 million acres in Australia overrun by the prickly pear cactus (see figures a and b), and another type of moth is now used to control alligator weed in the southern United States. Similarly, in this country the Chrysolina beetle controls Klamath weed, except in shady places where a root-boring beetle is more effective.

Parasites are also sometimes helpful. Many parasitic wasps lay their eggs in caterpillars, which are then used as food for the wasp larvae. Cereal leaf beetles, which attack grain crops, and olive scale, which attacks olive trees,

a.   b.

Figure 1

Example of biological control. a. Prickly pear cactus infestation in Australia made land unfit for human use. b. The same area after the introduction of Argentine moths (*Castoblastis cactorum*).

can both be controlled by specific parasitic wasps. Bacteria and viruses can also successfully control certain pest populations. For example, milky spore disease, caused by a bacterium, kills the larvae of Japanese beetles, and a virus kills the cabbage looper, which damages the leaves of cotton, cabbage, and several other crops.

The use of pheromones is also a type of biological control. Traps are baited with synthetic pheromones to collect male and/or female insects. Outbreaks of gypsy moths have been curtailed in this way. Recently, scientists in Sweden and Norway synthesized the sex pheromone of the Ips bark beetle, which attacks spruce trees. Almost 100,000 baited traps collected females that are normally attracted to males emitting this pheromone. The use of compounds that interfere with insect development has an even more direct effect than do pheromones. Two such compounds isolated from *Ajuga remota*, an East African plant, cause the fall armyworm caterpillar, a common cotton field pest, to develop three heads, burying its functional mouthparts.

Sterile males have also been used to reduce pest populations. The screwworm fly parasitizes cattle in the United States. Flies were raised in the laboratory and made sterile by exposure to radiation. The entire southeast was freed from this parasite when female flies mated with sterile mates and then laid eggs that did not hatch.

Various farming techniques can modify the environment to control pests. When farmers use strip farming and crop rotation, pests do not have a continuous food source. Maintaining hedgerows, weedy patches, and certain trees can provide habitats and food for predators and parasites that help control pests. For example, cottonwood and willow trees are normally inhabited by three species of flies that will kill the cotton bollworm, and wild blackberry bushes are the alternate host for wasps that parasitize grape leafhoppers.

Biological control, which requires in-depth knowledge of pests and their life cycles, is more sophisticated than chemical control of pests. Its great advantage is that it avoids the problems associated with the use of pesticides.

---

population decreases in size. In nature, this can be the result of prey being able to hide, or of the predators switching to a more abundant, and hence more economical, prey species.

Human Intervention

Sometimes humans have neglected to take into consideration that predators can keep prey populations in check. Coyotes have been killed off in the West, without regard to the fact that they tend to keep the mice, rabbit, and prairie dog populations under control. Without them, these rodents and their relatives can increase to the point where grazing land is damaged, and the carrying capacity for cattle and sheep is affected. Similarly, when the dingo, a wild dog in Australia, was killed off because it attacked sheep, the rabbit and wallaby populations increased.

Some farmers now recognize the benefits of maintaining populations of owls, because they capture rodents. In other instances, predators have been used to control pest populations. The prickly pear cactus in Australia was finally brought under control by the introduction of moth borers, and a weevil has been used to reduce the amount of water hyacinth in southern Florida. The cottony-cushion scale, a small, sap-sucking insect named for the large white egg mass deposited by the female on tree limbs, threatened the California citrus industry. It was brought under control by importing its natural predator, a ladybird beetle called the vedalia. These examples indicate that agricultural pests could be biologically rather than chemically controlled. Control of agricultural pests is discussed in the reading on the opposite page.

Predators help control prey population size. Whether oscillations in predator and prey populations will occur is still being investigated.

Prey Defense Adaptations

While predators have evolved strategies to secure the maximum amount of food with minimal expenditure of energy, prey organisms have evolved strategies to escape predation.

Plants

Some plants use mechanical means to discourage predation. The sharp spines of the cactus, the pointed leaves of the holly, and the tough, leathery leaves of oak trees all discourage predation by insects. *Acacia* trees and shrubs occur in tropical and subtropical regions. In Africa and tropical America, where browsers are numerous, the acacia species are protected by thorns that are often well developed. In Australia, where there are no browsers, most species of acacia lack thorns.

Plants also produce noxious and/or poisonous chemicals, such as cardiac glycosides in milkweeds, and tannins in oak leaves. Some of the chemicals produced by plants interfere with the normal metabolism of the adult insect, and others act as hormone analogues that interfere with insect development. Humans often make use of plant chemicals; they are the active ingredients in spices, medicines (digitalis, morphine, quinine), stimulants (caffeine), and other drugs (nicotine, marijuana, and opium).

Animals

Animals have varied defenses. Among these are:

Concealment. Some animals attempt to blend in with their background. Insects look like twigs, katydids may look like sprouting green leaves, and moths may resemble the bark of trees. Larger animals, too, try to conceal themselves by looking like their environment. Decorator crabs cover themselves with debris, and green herons, even young ones, attempt to look like straight tree branches (fig. 46.11).

Fright. Other types of prey try to startle or frighten predators (fig. 46.12). Some make themselves appear much larger than they are, such as the frilled lizard of Australia that

Figure 46.11
Concealment as an antipredator defense. Young green heron birds strike a pose that makes them resemble tree branches.

Figure 46.12
Startle display as an antipredator defense. The South American lantern fly has a large false head that resembles an alligator. This may frighten a predator into thinking it is facing a dangerous animal.

can open up folds of skin around its neck. Many moths have eyelike spots on their underwings that they can flash to startle a bird, long enough to allow their escape.

Warning coloration. Poisonous animals tend to be brightly colored, as a warning to those that may want to prey upon them. Yellow and black-striped bees and wasps,

Warning coloration as an antipredator defense. Poison arrow frogs are so poisonous that they were used by natives to make their arrows instant lethal weapons. The coloration of these frogs warns others to beware.

Figure 46.14
The monarch butterfly (*a*) and its mimic, the viceroy butterfly (*b*). Birds avoid both types of butterflies even though only the monarch is distasteful to them.

a.

b.

yellow or red caterpillars or plant bugs, and ladybird beetles are all advertising that they are not good to eat. The poison-arrow frog (fig. 46.13) found in tropical regions is so poisonous that South American Indians have long used it to poison their arrow tips. When the arrow strikes, it instantly paralyzes the victim.

Vigilance. Flocks of birds, schools of fish, and herds of mammals stick together as protection against predators. Grazing herbivores are constantly on the alert; if one begins to dart away, they all run. Baboons that detect predators visually, and antelope that detect predators by smell sometimes forage together, giving double protection against stealthy predators. The gazellelike springboks of southern Africa jump stiff legged straight up and down (8–12 feet), a number of times as a warning. The jumble of shapes and motion may confuse an attacking lion, and the herd can escape.

Mimicry

Mimicry occurs when one species resembles another that possesses an antipredator defense. For example, the brightly colored monarch butterfly (fig. 46.14*a*) causes predaceous birds to become sick and sometimes vomit. Monarch butterfly larvae feed on milkweed plants. These plants contain digitalislike compounds, a form of which becomes incorporated into the adult butterfly tissue. These compounds activate nerve centers controlling vomiting. Birds that have eaten a poisonous monarch butterfly avoid monarch butterflies in the future.

The viceroy butterfly (fig. 46.14*b*) mimics the monarch butterfly's coloration, but is not toxic. Birds eagerly eat viceroy butterflies, unless they have had previous experience with a poisonous monarch. Then, because the butterflies closely resemble each other, birds avoid both types of butterflies. The queen butterfly also mimics the monarch, but the queen is also poisonous.

A mimic that lacks the defense of the organism it resembles is called a Batesian mimic. The viceroy butterfly is a Batesian mimic of the monarch. Therefore, Batesian mimicry involves the resemblance of a harmless species (the mimic) to a harmful species (the model). A mimic that possesses the same defense as its model is called a Müllerian mimic. The queen butterfly is a Müllerian mimic of the monarch butterfly. In Müllerian mimicry, therefore, both animal species are harmful and resemble each other. In such cases, predators learn more rapidly to avoid contact with any animal that has that particular pattern.

Prey defenses include concealment, fright, warning coloration, and vigilance. Sometimes an animal with a successful antipredator defense is mimicked by other animals that have the same defense (Müllerian mimic) or that lack the defense (Batesian mimic).

Symbiosis

Symbiotic relationships (table 46.2) are close relationships that may occur when two different species live together. As with predation, coevolution occurs, and the species become closely adapted to one another. In parasitism, the parasite benefits, but the host is harmed; in commensalism, one species benefits, but the other is unaffected; and in mutualism, both species benefit.

Table 46.2 Symbiosis		
	Species 1	**Species 2**
Parasitism	+	−
Commensalism	+	0
Mutualism	+	+

Key: + = benefits
  − = harmed
  0 = no effect

Figure 46.15
Slave-making Amazon ants invade the colony of a slave species and carry off cocoons. When ants emerge from the cocoons, they serve as slaves and in so doing they help increase the population size of their captors. This is a form of social parasitism in which one species benefits and the other is harmed.

Parasitism

Parasitism is similar to predation in that the parasite derives nourishment from the host (just as the predator derives nourishment from its prey). Usually, however, the host is larger than the parasite, and an efficient parasite does not kill the host, at least until its own life cycle is complete. Viruses are always parasites, as are a number of bacteria, protists, plants, and animals. The smaller parasites tend to be endoparasites, which live in the bodies of the hosts. Larger parasites tend to be ectoparasites and remain attached to the exterior of the host by means of specialized organs and appendages.

Just as predators can dramatically reduce the size of a prey population that lacks a suitable defense, so parasites can reduce the size of a host population that has no defense. Thousands of elm trees have died in this country from the inadvertent introduction of Dutch elm disease, which is caused by a parasitic fungus. A new method of treating Dutch elm disease utilizes bacteria that produce a fungus-killing antibiotic when injected into the tree. Another means of curbing this fungal infection of trees is to control the bark beetle that spreads the disease.

Many other parasites use a secondary host for dispersal or completion of the early stages of development. For example, sheep liver flukes that mature in sheep use snails and sometimes even fish as intermediate hosts to complete the early stages of their life cycle. The hookworm, judged to be the most troublesome parasitic worm in the United States, does not require a secondary host. New World hookworm males are from 5 to 9 mm in length, and females are usually about 1 cm long. The head of the worm is sharply bent in relation to the rest of the body, accounting for its characteristic hooklike appearance. Adult hookworms attach themselves to the intestinal wall of their host, and pass eggs out with the feces. When deposited on moist, sandy soil, the larvae develop and hatch within 24 to 48 hours. After a period of growth and development, the worms extend their bodies into the air and wave about in this position until they come into contact with the skin of a suitable host, such as a human. Penetration usually occurs through the feet. Once in the blood vessels, the worms are passively carried to the lungs, where they invade the alveoli. The larvae migrate from the lungs to the trachea to be swallowed and passed along to the small intestine where they mature. They attach to the intestinal wall by means of their stout mouthparts and suck blood and tissue juices from the host. Symptoms of hookworm infection include abdominal pains, nausea, diarrhea, and finally iron deficiency anemia.

Social parasitism occurs when one species exploits another species. The cuckoo lays eggs in nests of songbirds, and the newly hatched cuckoo ejects its nestmates, so the songbird parents attend only to it. Slave-making ants of the species *Polyergus rufescens* raid the ant colonies of a slave species *Formica fusca* (fig. 46.15). They destroy any resisting defenders with their mandibles, which are shaped like miniature sabers. *Polyergus* ants are so specialized that they can only groom themselves. To eat, they must beg slave workers for food. The slave workers not only provide food for the slave-making ants but also care for the eggs, larvae, and pupae of their captors.

In parasitism, the host species is harmed. Parasites, which help control the size of their host population, may require a secondary host for dispersal.

Commensalism

In a **commensal** relationship, one species benefits, and the other is neither benefited nor harmed. Often, the host species provides a home and/or transportation for the other species. Barnacles

Figure 46.16
Clownfish often live among a sea anemone's tentacles but are not
seized and eaten as prey. This is a form of commensalism in which
the clownfish species gets a place to live and the anemone species
is not harmed.

that attach themselves to the backs of whales and the shells of horseshoe crabs are provided with both a home and transportation. Remoras are fish that attach themselves to the bellies of sharks by means of a modified dorsal fin acting as a suction cup. The remoras obtain a free ride and also feed on the remains of the shark's meals. Epiphytes (fig. 30.13) grow in the branches of trees, where they receive light, but they take no nourishment from the trees. Instead, their roots obtain nutrients and water from the air. Clownfish (fig. 46.16) live within the tentacles and gut of sea anemones. Because most fish avoid the poisonous tentacles of the anemones, the clownfish are protected from predators. Perhaps this relationship borders on mutualism, because the clownfish may actually attract other fish on which the anemone can feed. The sea anemone's tentacles quickly paralyze and seize other fish as prey.

In commensalism, one species is benefited and the other is unaffected. Often the host simply provides a home and/or transportation.

Mutualism

Mutualism is a symbiotic relationship in which both members of the association benefit. Mutualistic relationships often allow organisms to obtain food or avoid predation.

Bacteria that reside in the human intestinal tract are provided with food, but they also provide humans with vitamins, which are molecules we are unable to synthesize for ourselves. Termites would not even be able to digest wood if it were not for the protozoans that inhabit their intestinal tract. The bacteria in the protozoans digest cellulose, which termites cannot digest. Mycorrhizae, also called "fungal roots," are symbiotic associations between the roots of plants and fungal hyphae. Mycorrhizal hyphae improve the uptake of nutrients for the plant, protect the plant's roots against pathogens, and produce plant growth hormones. In return, the fungus obtains carbohydrates from the plant. As we discussed on page 488, flowers and their pollinators have coevolved and are dependent upon one another. The flower is benefited when the pollinator carries pollen to another flower, assuring cross-fertilization, and the flower provides food for the pollinator.

Figure 46.17

The bullhorn acacia is adapted so as to provide a home for ants of the species *Pseudomyrmex ferruginea*. *a*. The thorns are hollow and the ants live inside. *b*. The base of leaves have nectaries (openings) where ants can feed. *c*. Leaves have bodies at the tips that ants harvest for larval food.

a.

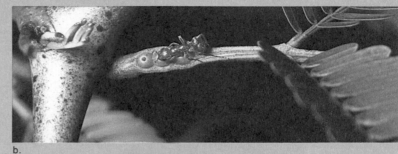

b.

c.

Figure 46.18

Cleaner wrasse in the mouth of a spotted sweetlip. This is a mutualistic relationship. The cleaner wrasse is feeding off parasites and the spotted sweetlip is being cleaned.

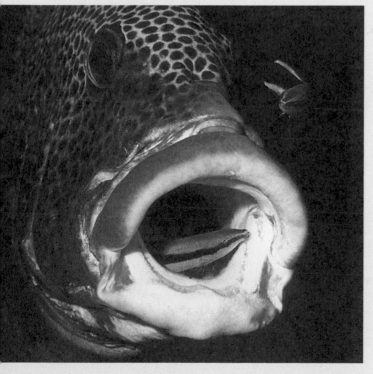

A number of mutualistic relationships involve ants and other organisms. Leaf cutter ants keep fungal gardens. These ants were so named because they gather flowers and leaves, cut them into pieces, and transport them to underground nests. After preparing the leaves, they implant them with fungal mycelia, which then grow in profusion. This relationship gives the fungi a competitive edge over other fungal species and provides the ants with a source of food.

In tropical America, the bullhorn acacia (fig. 46.17) is adapted to provide a home for ants of the species *Pseudomyrmex ferruginea*. Unlike other acacias, this species has swollen thorns with a hollow interior, where ant larvae can grow and develop. In addition to housing the ants, the acacias provide them with food. The ants feed from nectaries at the base of the leaves and eat fat- and protein-containing nodules, called Beltian bodies, which are found at the tips of some of the leaves. Bullhorn acacias have leaves throughout the year, while related acacia species lose their leaves during the dry season. The ants constantly protect the plant from herbivorous insects because, unlike other ants, they are active 24 hours a day.

Cleaning symbiosis (fig. 46.18) is a phenomenon believed to be quite common among marine organisms. There are species of small fish and shrimp that specialize in removing parasites from larger fish. The large fish line up at the "cleaning stations" and wait their turn. The small fish feel so secure they clean even the mouths of the larger fish. Not everyone plays fair, however, because there are small fish that mimic the cleaners and take a bite out of the larger fish, and cleaner fish are sometimes found in the stomachs of the fish they clean.

In mutualism, both species benefit and the two species are often closely adapted to one another.

Summary

1. A community is a group of populations that are interacting with one another. The community and the physical environment together form an ecosystem.

2. Two very important concepts in ecology are habitat and niche. Habitat is where a population lives and niche is the role that it plays in an ecosystem. We find that when there are several similar species living in the same environment, they occupy different niches.

3. Competition may be intraspecific or interspecific. Intraspecific competition may be of the scramble or contest type. Scramble competition often results in fluctuations in population size, and contest competition results in maintenance of population size at or near carrying capacity.

4. Contest competition is common in populations that either establish territories or have dominance hierarchies.

5. When populations increase in size, members sometimes emigrate, and intraspecific competition is reduced.

6. Interspecific competition occurs only for a short period of time, because either one population replaces the other or the two species evolve to occupy different niches. Although it may seem that similar species are occupying the same niche, actually there are slight differences.

7. Evidence suggests that predation tends to control the size of the prey population, but the relationship may involve other factors as well. For example, the availability of food affects population sizes.

8. Just as predators are adapted to capture prey, prey are adapted to escape predators. Plants have both physical and chemical defenses. Animals have defenses that include concealment, fright of the predator, warning coloration, and vigilance.

9. Some animals mimic other animals that have a defense, especially those others that are poisonous to eat.

10. Symbiotic relationships include parasitism, commensalism, and mutualism. Parasitism may control the size of the host populations. Although many parasites require secondary hosts, the hookworm, an animal parasite, does not. In social parasitism, one society exploits another.

11. In commensalism, one species is benefited, and the other is unaffected. Often, the host simply provides a home and/or transportation.

12. In mutualism, both species benefit. The two species are often closely adapted to one another, as are flowers and their pollinators. There are also examples of mutualistic species that live together in the same locale, such as ants that keep fungal gardens.

Objective Questions

1. Six species of monkeys are found in a tropical forest. Most likely, they
   a. occupy the same niche.
   b. eat different foods and occupy different ranges.
   c. spend much time fighting each other.
   d. All of these.

2. Leaf cutter ants keep fungal gardens. The ants provide food for the fungus on which they feed. This is an example of
   a. competition.
   b. predation.
   c. commensalism.
   d. mutualism.

3. When blowfly larvae are given a restricted amount of food, few adults result. This is an example of
   a. predation.
   b. scramble competition.
   c. contest competition.
   d. interspecific competition.

4. Clownfish live among sea anemone tentacles where they are protected. If the clownfish provides no service to the anemone, this is an example of
   a. competition.
   b. predation.
   c. commensalism.
   d. mutualism.

5. Two species of barnacles vie for space in the intertidal zone. The one that remains is
   a. the better competitor.
   b. better adapted to the area.
   c. the better predator on the other.
   d. Both (a) and (b).

6. Territoriality occurs as a result of
   a. predation.
   b. parasitism.
   c. competition.
   d. cooperation.

7. Bullhorn acacia provides a home for ants. Which statement below is correct?
   a. The plant is under the control of pheromones produced by the ants.
   b. The ants protect the plant.
   c. They have coevolved to occupy different niches.
   d. All of these.

8. The niche of an organism
   a. is the same as its habitat.
   b. includes how it competes and acquires food.
   c. is specific to the organism.
   d. Both (b) and (c).

9. The frilled lizard of Australia suddenly opened its mouth wide and unfurled folds of skin around its neck. Most likely this was a way to
   a. conceal itself.
   b. warn that it was noxious to eat.
   c. scare a predator.
   d. All of these.

10. Birds that have eaten monarch butterflies
    a. usually develop a search image for them.
    b. do not eat viceroy butterflies.
    c. hunt out queen butterflies instead.
    d. have the same niche as these butterflies.

Study Questions

1. Explain what is meant by the niche of an organism, and why it is possible for similar types of organisms to be found in the same general area.
2. Give examples of scramble and contest competition and explain why they limit population size.
3. Why doesn't interspecific competition between similar populations always lead to the extinction of one of them?
4. What is the competitive exclusion principle, and how is it related to diversity of organisms?
5. What evidence is there that predators might overkill their prey? that they do not do so?
6. Why would you recommend that humans not kill animals such as coyotes, that might sometimes prey on farm animals?
7. Describe some prey defenses of plants and animals.
8. What are the two types of mimics? What is the benefit of each type?
9. Give examples of parasitism, and state the effects on the host organism.
10. Give examples of commensalism, and state the effects on the host organism.
11. Give examples of mutualism, and state the benefits to each organism involved in the relationship.

Thought Questions

1. Contrast the effects of competition and predation on population size and community structure.
2. Cattle egrets feed on insects flushed from vegetation by grazing cattle. Why is this relationship termed commensalism? What changes in the relationship would make it a mutualistic one?
3. Biologists have presented evidence that fungi feed on algae in a lichen. If the algae still derive benefit from the relationship, how should it be classified?

Selected Key Terms

community (kŏ-mu′ni-te) 732
ecosystem (ek″o-sis′tem) 732
habitat (hab′ĭ-tat) 732
niche (nich) 732
scramble competition (skram′b'l kom″pĕ-tish′un) 733
emigrate (em′ĭ-grāt) 734
contest competition (kon′test kom″pĕ-tish′un) 734
competitive exclusion principle (kom-pet′ĭ-tiv eks-kloo′zhun prin′sĭ-p'l) 736
mimicry (mim′ik-re) 740
symbiosis (sim″bi-o′sis) 740
parasitism (par′ah-si″tizm) 741
commensalism (kŏ-men′sal-izm″) 741
mutualism (mu′tu-al-izm″) 742

CHAPTER 47

The Biosphere

Your study of this chapter will be complete when you can:

1. Explain how the water cycle provides a renewable source of fresh water.

2. List and describe the freshwater communities of the biosphere.

3. Describe the events and causes of the spring and fall turnover of lakes in the temperate zone.

4. Describe the life zones of a lake and the types of organisms that reside in these zones.

5. Relate eutrophication to the process of aquatic succession.

6. Explain why estuaries and the coastal zone are more productive than the open ocean.

7. Describe the life zones of the ocean and the types of organisms that reside in these zones.

8. Describe the process of primary and secondary succession on land. Contrast the productivity of the early stages to the climax stage of succession.

9. Distinguish between deserts, grasslands, and forests on the basis of rainfall and temperature.

10. Describe the characteristics of the tundra, the savanna, and the prairie.

11. Describe the characteristics of the taiga, the temperate deciduous forest, and the tropical rain forest.

12. Contrast the productivity of the various terrestrial biomes, and explain why the tropical rain forest is the most productive of these.

Coral reefs located in shallow tropical waters are among the most productive of the biological communities. Some fish feed on plankton brought by the tides but the reef also contains photosynthesizers that produce food. Dead coral is coated by algae and live coral contains symbiotic algae within its cells. The carrying capacity of a coral reef is quite high.

Figure 47.1
The water (hydrologic) cycle. Evaporation by solar energy,
precipitation in the form of rain and snow, and surface runoff are the
three processes that maintain a stable water balance on earth. A
large portion of the ocean and some stored water in aquifers do not
immediately participate in the cycle.

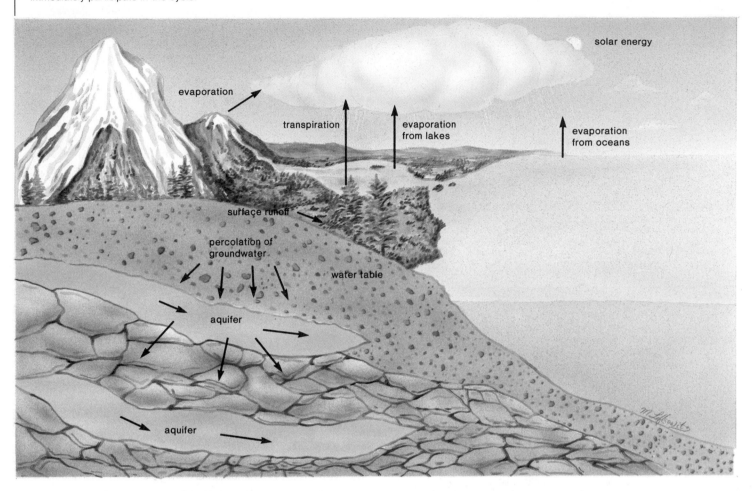

The earth is enveloped by the **biosphere**—a thin realm composed of water, land, and air—where organisms may be found. The biosphere contains all the communities found on our planet, and the largest of these on *land* are called the **biomes.** There is no equivalent term for large *aquatic* communities. This is unfortunate because both the aquatic and the land environment contain large communities that have unique, defined characteristics.

Aquatic Communities

Aquatic communities can be divided into two types: (1) *freshwater* (inland) communities, and (2) *marine* (oceanic or saltwater) communities. In aquatic ecosystems, organisms vary according to whether they are adapted to fresh or salt water, warm or cold water, quiet or turbulent water, and the presence or absence of light. In both fresh and salt water, floating, microscopic organisms called **plankton** are important components of the ecosystem. *Phytoplankton* are photosynthesizing algae that only become noticeable when they reproduce to the extent that a green scum or a "red tide" appears on the water. *Zooplankton* are microscopic animals that feed on the phytoplankton.

Freshwater Communities

Fresh water is distilled from salt water, a fact which is readily apparent from looking at the **water** (hydrologic) **cycle** (fig. 47.1): As the sun's rays cause fresh water to evaporate from seawater, the salts are left behind. The vaporized fresh water rises into the atmosphere, cools, and falls as rain over the oceans and the land. A lesser amount of water also evaporates from bodies of fresh water. Since land lies above sea level, gravity eventually returns all fresh water to the sea, but in the meantime, it is contained within standing waters (lakes and ponds), flowing waters (streams and rivers), and groundwater.

When rain falls, some of the water sinks or percolates into the ground and saturates the earth to a certain level. The top of the saturation zone is called the *groundwater table,* or simply, the water table (fig. 47.1). Sometimes groundwater is also located in a porous layer, called an **aquifer,** that lies between two sloping layers of impervious rock. Wells are used to remove some of this water.

Table 47.1 indicates that fresh water makes up only about 3% of the world's supply of water.

Table 47.1 Water Resources and Their Average Rate of Renewal

Location of Water	Percentage of World Supply	Average Rate of Renewal
Oceans	97	3,100 years
Atmosphere	0.001	9–12 days
On land		
Ice caps and glaciers	2.240	16,000 years
Other	0.759	
Groundwater:		
Upper levels		300 years
Lower levels		4,600 years
Lakes		10–100 years
Rivers and streams		12–120 days

Most of the world's water is in the oceans where the renewal rate is slow. Very little water at any one time is in the atmosphere, but the renewal rate there is rapid. Most of the water on land is frozen in ice caps and glaciers where the renewal rate is slow.

Lakes and Ponds

Lakes and ponds, which are characterized by fresh and still water, tend to have fairly stable abiotic and biotic regions. In the summer, large lakes (but not ponds) in the temperate zone have three layers of water that differ in temperature (fig. 47.2). The surface layer, the *epilimnion,* is warm from solar radiation; the middle layer, the *thermocline,* experiences an abrupt drop in temperature; and the bottom layer, the *hypolimnion,* is cold. These differences in temperature prevent mixing: the warmer, less dense water of the epilimnion "floats" on top of the colder, more dense water of the hypolimnion. (Recall from figure 3.14 that water is most dense at 4° C.)

As the summer progresses, the epilimnion becomes nutrient poor, and the hypolimnion loses oxygen because of the following processes. The phytoplankton located in the sunlit epilimnion use up inorganic nutrients like nitrates and phosphates as they photosynthesize. Photosynthesis releases oxygen, giving this layer a ready supply. Detritus, or debris, naturally falls by gravity to the bottom of the lake, and here oxygen is used up as decomposition occurs. Decomposition releases inorganic nutrients, however.

Figure 47.2
Temperature profiles of a large lake in a temperate region vary with the season. During spring and fall turnover, the deep waters receive oxygen while the shallow waters receive nutrients.

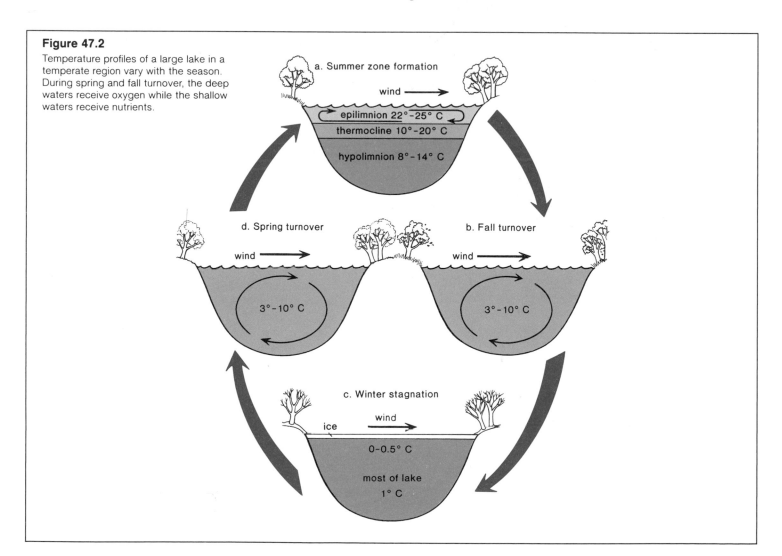

a. Summer zone formation
wind
epilimnion 22°–25° C
thermocline 10°–20° C
hypolimnion 8°–14° C

d. Spring turnover
wind
3°–10° C

b. Fall turnover
wind
3°–10° C

c. Winter stagnation
wind
ice
0–0.5° C
most of lake
1° C

In the autumn, as the surface waters cool, they become denser and sink below the warmer and less dense water on the bottom. Once the temperature is uniform throughout the lake, winds cause a circulation of water so that mixing occurs. This is called the *fall turnover*. Eventually, oxygen and inorganic nutrients become evenly distributed.

As winter approaches, the surface waters cool. Ice formation begins, and the ice remains on the surface because ice is less dense than cool water. Ice has an insulating effect, preventing further cooling of the water below. This permits aquatic organisms to live through the winter in the water beneath the surface of the ice.

In the spring, as the surface waters warm from 0° C, they become densest at 4° C and sink below the colder and relatively less dense water on the bottom. Once the temperature is uniform throughout the lake, winds cause circulation of water to the bottom. This is called the *spring turnover*. When the surface waters warm further, the lake again becomes thermally stratified, with formation of the epilimnion, thermocline, and hypolimnion.

This vertical stratification and seasonal change of temperatures in ponds and lakes influence the seasonal distribution of fishes and other aquatic life. For example, coldwater fishes move to the deeper water in summer where it is cooler, but inhabit the upper water in winter because these oxygenated waters have now cooled off. In the fall and spring, just after mixing occurs, phytoplankton growth is most abundant, because the sunlit surface waters now contain nutrients.

Zones of a Deep Lake   Deep lakes are composed of three zones: the shallow *littoral zone* is closest to the shore; the *limnetic zone* is the sunlit main body of the lake; and the *profundal zone* is the deep part where light does not penetrate. Aquatic plants are rooted in the littoral zone of a lake. Other organisms in lakes are also found in one of these zones, as shown in figure 47.3.

Classification by Nutrient Status   Lakes and ponds are often classified by their nutrient status. *Oligotrophic* (nutrient-poor) lakes are characterized by low organic matter, low nutrient release from bottom sediments, and low levels of phytoplankton productivity. Such a lake is usually situated in a nutrient-poor *watershed*—the area surrounding the lake that drains water into it. *Eutrophic* (nutrient-rich) lakes are characterized by high organic matter, high levels of nutrient release from bottom sediments, and high phytoplankton productivity. They often have a well-developed littoral zone. Such lakes are usually situated in naturally nutrient-rich watersheds or in watersheds affected by agriculture or by urban and suburban development. Oligotrophic lakes can become eutrophic through large inputs of nutrients. This process, called **cultural eutrophication,** accelerates the process of *aquatic succession,* a series of stages by which a pond—or even a lake—can fill in and disappear (fig. 47.4).

Marine Communities

Marine communities begin along the shoreline of oceans. These communities include estuaries and seashores in addition to the great diversity of shallow and deep-water marine communities in the oceans themselves.

Figure 47.3

Zones of a deep lake. In the littoral zone, there are microscopic organisms that cling to abiotic and biotic surfaces. In the limnetic zone there are organisms that live on the surface where atmospheric oxygen is available; zooplankton and phytoplankton which are floating organisms; and free-swimming organisms such as fishes. In the profundal zone, there are mollusks and crustaceans which feed on debris that falls from above.

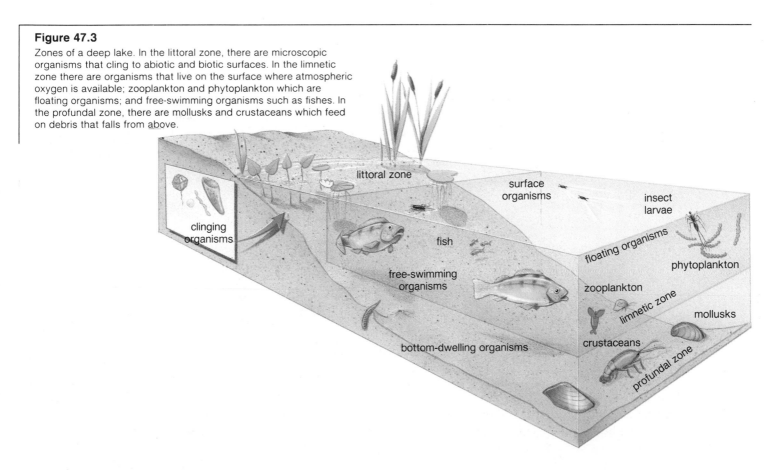

Figure 47.4

Aquatic succession on Presque Isle, Pennsylvania. *a.* As succession progressed in this pond, sediments began filling the edges and the body of the water became shallower. Water lilies began rooting in the bottom sediment. Cattails, rushes, and sedges began to grow at the edges when the water depth was reduced to only a few inches. *b.* As the land surrounding the receding pond dried out, trees began to crowd in, and the remaining wet, marshy areas were invaded by meadow grasses.

a.

b.

Figure 47.5

Estuary structure and function. Since an estuary is located where a river flows into the ocean, it receives nutrients from land and also from the sea by way of the tides. Decaying grasses also provide nutrients. Estuaries serve as a nursery for the spawning and rearing of the young for many species of fishes, shrimp, mollusks, and crustacea.

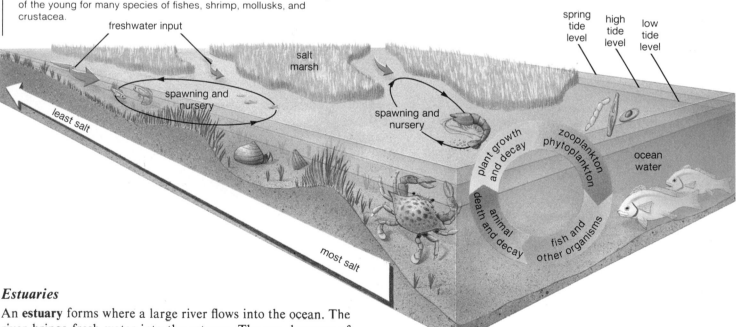

Estuaries

An **estuary** forms where a large river flows into the ocean. The river brings fresh water into the estuary. The sea, because of the tides, brings salt water into this same area. Therefore, an estuary is characterized by a mixture of fresh and salt water, called *brackish* water. Most estuaries are rather shallow, which enhances sunlight penetration, and the temperature of the estuarine waters is generally higher than that of the surrounding marine or freshwater environments. Further, an estuary acts as a nutrient trap; the tides bring nutrients from the sea and at the same time prevent the seaward escape of nutrients brought by the river. Therefore, photosynthesis occurs at a high rate in the marine phytoplankton and sea grass beds of estuaries, making estuaries very productive (fig. 47.21).

Only a few types of small fishes live permanently in an estuary, but many types develop there; thus, both larval and immature fishes are present in large numbers. It has been estimated that well over half of all marine fishes develop in the

Figure 47.6
Some of the animals that typically are found along a rocky coast in the intertidal zone.

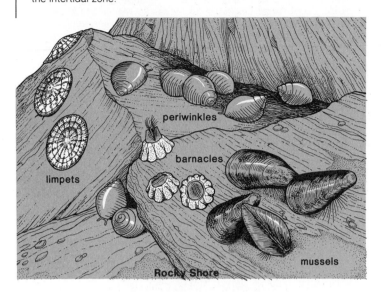

Figure 47.7
Representative animals of a sandy shore.

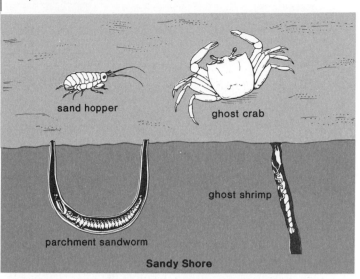

protective environment of an estuary; this is why estuaries are called the *nurseries of the sea.* Shrimp and mollusks, too, use the estuary as a nursery (fig. 47.5).

Salt marshes dominated by salt marsh cordgrass are often adjacent to estuaries. Salt marshes contribute nutrients and detrital material to estuaries, in addition to being an important habitat for wildlife.

Seashores

Both rocky and sandy shores are constantly bombarded by the sea as the tides roll in and out. The maximum elevation of the tide is *high tide* and the minimum elevation is *low tide.* Between these two is the *intertidal* (or littoral) *zone,* where some organisms live. The *rocky shore* (fig. 47.6) offers a firm substratum to which organisms can attach themselves in the intertidal zone and not be swept out to sea. Macroscopic seaweeds, which are the main photosynthesizers, anchor themselves to the rocks by holdfasts. Barnacles are glued so tightly to stones by their own secretions that their calcareous outer plates remain in place even after the enclosed, shrimplike animal dies. Oysters and mussels attach themselves to the rocks by filaments called byssus threads. Limpets are snails, just as are periwinkles. But whereas periwinkles have a coiled shell and secure themselves by hiding in crevices or under seaweeds, limpets press their single, flattened cone tightly to a rock.

The shifting, unstable sands on a *sandy beach* (fig. 47.7) do not provide a suitable substratum for the attachment of organisms; therefore, nearly all the permanent residents dwell underneath the sand. They either burrow during the day and surface to feed at night, or they remain permanently within their burrows and tubes. Ghost crabs and sand hoppers (amphipods) burrow above high tide and feed at night whenever the tide is out. Sandworms and sand (ghost) shrimp remain within their burrows in the intertidal zone and feed on detritus whenever possible. Still lower on the beach, clams, cockles, and sand dollars are found.

Figure 47.8
A close-up of a coral reef. A coral reef is a unique community of marine organisms. The diversity and richness of form and color are the result of the optimal environmental conditions of tropical seas.

Coral Reefs

Coral reefs (fig. 47.8) are areas of biological abundance found in shallow, warm tropical waters that have a minimum temperature of 70° F. Coral reefs begin to form on a rocky substratum beneath the surface of the water. Their chief constituents are stony corals (phylum Cnidaria), which have a

Table 47.2 Zones of the Ocean

Life Zones	Location
Neritic	Shallow coastal waters
Pelagic	Open sea
Epipelagic (sunlight waters)	
Mesopelagic (twilight waters)	
Bathypelagic (dark waters)	
Benthic	Seafloor
Littoral (intertidal) and sublittoral	Continental shelf
Bathyal zone	Continental slope
Abyssal zone	Abyssal plain

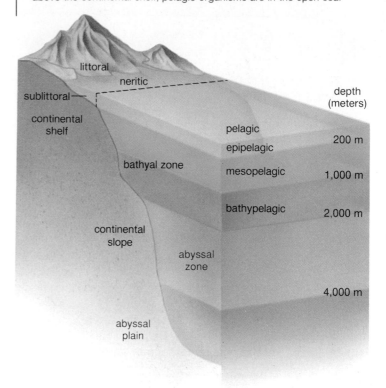

Figure 47.9
The floor of the ocean is divided into the continental shelf, the continental slope, and the abyssal plain. Organisms reside in these zones: benthic organisms are on the ocean floor in the littoral, sublittoral, bathyal, and abyssal zones; neritic organisms are found above the continental shelf; pelagic organisms are in the open sea.

calcium carbonate (limestone) exoskeleton, and calcareous red and green algae. Corals do not usually occur individually; rather, they form colonies derived from an individual coral that has reproduced by means of budding. Most of the solid part of a coral reef is composed of the skeletons of dead coral. Only the outer layer contains living organisms. Many corals provide a home for microscopic algae called zooxanthellae. The corals, which feed at night, and the algae, which photosynthesize during the day, are mutualistic and share materials and nutrients.

A reef is densely populated with animal life. There are many types of small fishes (butterfly, damsel, clown, and sturgeon), all beautifully colored. In addition, the large number of crevices and caves provide shelter for filter feeders (sponges, sea squirts, and fan worms) and for scavengers (crabs and sea urchins). The barracuda and moray eel prey on these animals. Some fishes feed on the coral, but the most deadly coral predator of Pacific reefs is the crown-of-thorns starfish, which grows as large as two feet across and has from 9 to 21 arms. Along the northeastern coast of Australia, the very existence of the Great Barrier Reef is threatened by these animals. A snail called the giant triton is their primary natural foe, and some believe that the crown-of-thorns starfish is proliferating because humans have killed off the triton for its handsome spiral shell.

Oceans

The geographic areas and zones of an ocean (table 47.2) are shown in figure 47.9. The oceans cover approximately three-quarters of our planet. This figure shows that the sea is shallow at first, and then becomes deep as the **continental shelf** gives way to the *continental slope* that leads to the *abyssal plain.* The flat plains are interrupted by enormous underwater mountain chains called oceanic ridges. Along the ridges scientists have discovered ecosystems that seem to depend on chemosynthetic bacteria acting as producers.

Oceans can be divided into two areas: the coastal zone and the open sea.

Coastal Zone   The *coastal zone* includes the continental shelf and the waters above the shelf. Therefore it includes the marine communities we have just been discussing. In reference to table 47.2, it includes the neritic zone and part of the benthic zone.

Seaweed, after it is partially decomposed by bacteria, is the primary source of food for the benthic clams, worms, and sea urchins found in the coastal zone. These are preyed on by starfish, crabs, and brittle stars, all of which are eaten by bottom-dwelling fish. Also, these shallow, sunlit waters receive nutrients from estuaries and even from the sea, especially in regions of upwelling (fig. 47.10), where surface waters are blown offshore and then replaced by cold, nutrient-laden water from the deep. This greatly increases the supply of phytoplankton that grow and provide food not only for zooplankton but also for small fishes. These, in turn, are food for the commercial fishes—herring, cod, and flounder. Because of these factors, the coastal zone is more productive than the open sea (fig. 47.21).

The coastal zone including marshes, swamps, and estuaries, is extremely important to the productivity of the ocean. Coral reefs, areas of biological abundance, occur in tropical seas.

Open Sea   The open sea, or **pelagic zone,** is divided into the epipelagic, mesopelagic, bathypelagic zones (fig. 47.9). Only the epipelagic zone is brightly lit, or euphotic; the mesopelagic zone is in semidarkness; and the bathypelagic zone is in complete darkness.

The open sea, which includes 90% of ocean waters, is not very productive because it is low in nutrients. This lack of nutrients limits phytoplankton growth so much that productivity of the open sea is equivalent to that of a terrestrial desert (fig.

Figure 47.10

Coastal upwelling in the Northern Hemisphere. During upwelling, nutrient-rich water from below rises over nutrient-poor water on the surface. The nutrients allow producers and then consumers to dramatically increase in number. Therefore, fishing is particularly good in regions of upwelling.

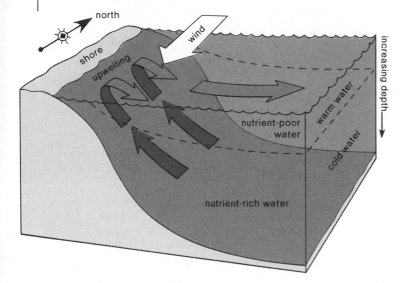

47.21). Then, too, food chains tend to be longer in the ocean than on land. The extremely small size of phytoplankton and herbivorous zooplankton may account for this difference in food-chain length. Among the phytoplankton are diatoms, dinoflagellates, and even smaller microorganisms. The herbivorous zooplankton, such as copepods, krill, foraminifera, and radiolarians, feed on these and in turn are food for the carnivorous zooplankton, such as jellyfishes, comb jellies, wing-footed snails, sea squirts, and arrow worms.

The free-swimming organisms (fig. 47.11) of the *epipelagic* zone include herring and bluefish, which are food for the larger mackerel, tuna, and sharks. Flying fish, which glide above the surface, are preyed upon by dolphins, not to be confused with the mammalian porpoises, which are also present. Whales are other mammals found in this zone. Baleen whales strain krill from the water, and the toothed sperm whales feed primarily on the common squid.

Animals in the *mesopelagic zone* are adapted to the absence of light and tend to be translucent, red-colored, or even luminescent. In addition to luminescent jellyfishes, sea squirts, copepods, shrimp, and squid, there are also luminescent carnivorous fishes, such as lantern and hatchet fishes.

Figure 47.11

Free-swimming fishes of the epipelagic, mesopelagic, bathypelagic, and abyssal zones. In the epipelagic sunlit zone, many types of fish populations exist because the producers are located here. Consumers are found in increased numbers in the dimly lit mesopelagic zone. Inhabitants of the dark bathypelagic zone are adapted to lack of light and to infrequent meals. The organisms in the abyssal zone are scavengers of various types.

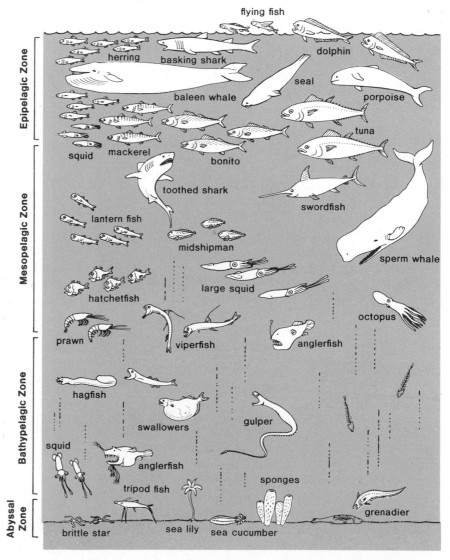

The *bathypelagic zone* is the largest and is in complete darkness except for an occasional flash of bioluminescent light. Strange-looking fishes with distensible mouths and abdomens and small, tubular eyes feed on infrequent prey.

The bathypelagic and the abyssal life zones are inhabited by those animals that live above and in the cold, dark abyssal plain. Because of the cold temperature (averaging 2° C) and the intense pressure (300 to 500 atmospheres), it was once thought that only a few specialized animals would live in the abyssal life zone. Yet a diverse assemblage of organisms has been found. Debris from the mesopelagic zone is taken in by filter feeders, such as the sea lilies that rise above the seafloor, and the clams and tube worms that lie burrowed in the mud. Other animals, such as sea cucumbers and sea urchins, crawl around on the sea bottom, living on detritus and bacteria of decay. They, in turn, are food for predaceous brittle stars and other larger animals found in the abyssal zone.

Terrestrial Biomes

The locations of terrestrial biomes studied in this text are indicated in figure 47.12b. We will be discussing ten of these biomes and will group them in two categories: the treeless biomes (deserts, grasslands, tundra, savanna, and prairie) and the forests of the world (taiga, other coniferous forests, temperate deciduous forests, tropical rain forests, and tropical deciduous forests). Before we begin our discussion, however, we will consider how these biomes arose.

Succession

Succession is an orderly sequence of communities that lead up to a climax community. A biome is a climax community. The complete sequence of communities is called a *sere,* and each transitory community is called a *seral stage.*

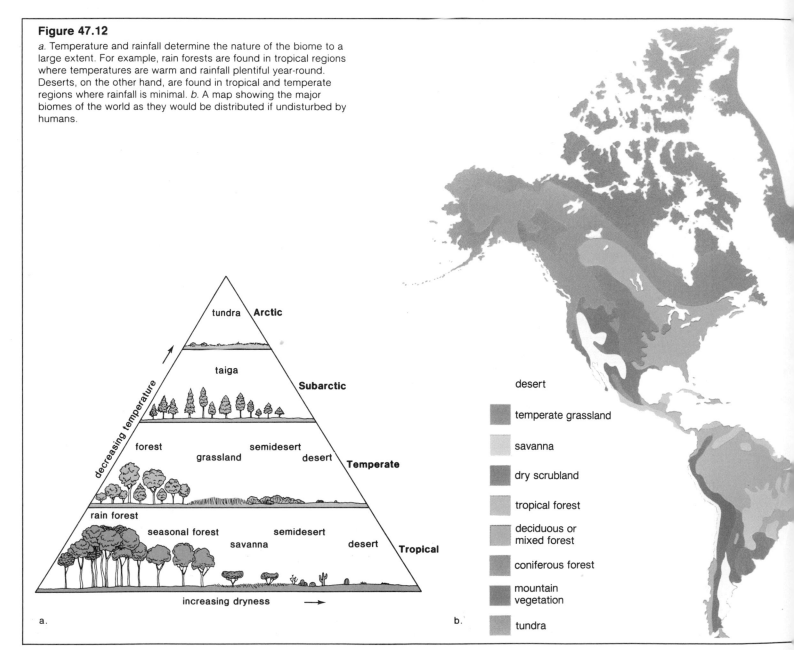

Figure 47.12
a. Temperature and rainfall determine the nature of the biome to a large extent. For example, rain forests are found in tropical regions where temperatures are warm and rainfall plentiful year-round. Deserts, on the other hand, are found in tropical and temperate regions where rainfall is minimal. *b.* A map showing the major biomes of the world as they would be distributed if undisturbed by humans.

Primary succession (fig. 47.13) on land begins with bare rock. At first the rock is subjected to weathering as wind and rain act on it. Then lichens begin growing on the rock. Their acid secretions further break down the rock, and mosses and ferns take hold after a while. As soil begins to build up, longer-growing native plants begin to grow in the area, and these are eventually followed by shrubs and trees.

Secondary succession (fig. 47.14) occurs in regions that have been disturbed. The stages vary from site to site, but figure 47.14 shows one possible succession sequence for abandoned farmland in the eastern United States. The first community, called the pioneer community, includes plants such as weeds, that are able to colonize disturbed areas because of their ability to survive under such harsh conditions as limited soil moisture and direct sunlight. This colonization prepares the way for native plants of the area that most likely have a longer growing period.

Notice in figure 47.14 how one plant community replaces another until a **climax community** has been achieved. A different mix of animals is associated with each of these communities. Whereas previous communities are replaced, the final stage, the climax community, is able to sustain itself indefinitely as long as it is not seriously stressed.

Productivity of Climax Communities versus Early Seral Stages

Researchers find that the early stages of succession, also called pioneer communities, show the most growth and are therefore the most productive, but the inputs into the system in terms of energy and nutrients are high. However, the outputs (heat and loss of nutrients) are also high, because the community lacks the diversity to make full use of all the inputs. Not surprisingly, these communities have a CP/CR ratio greater than 1, where

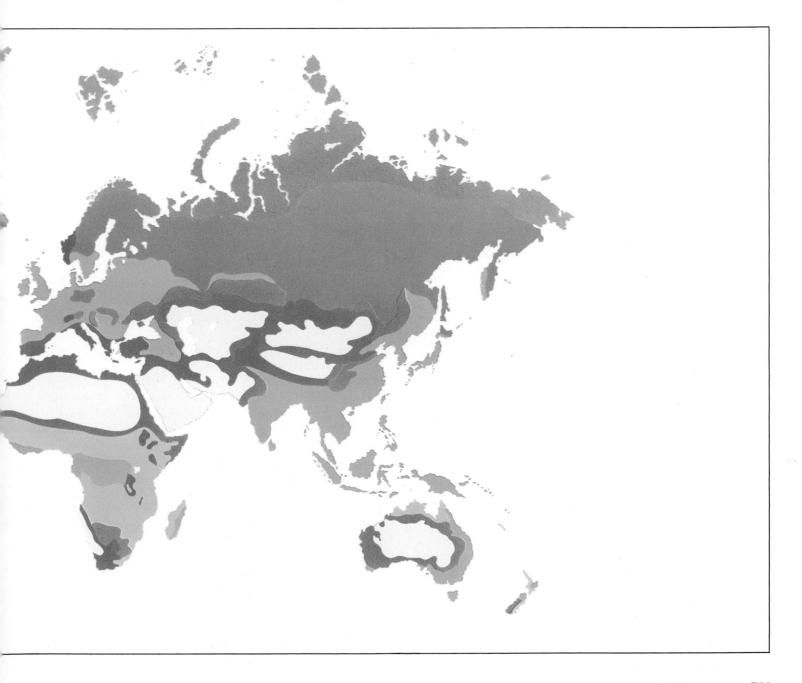

Figure 47.13

Primary succession includes the stages by which bare rock becomes a climax community within a particular biome. These photographs show a possible sequence of early events: (a) lichens growing on bare rock; (b) individual plants taking hold; (c) perennial herbs spread out over the area.

a.

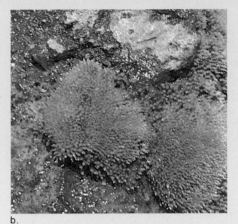
b.

c.

Figure 47.14

Secondary succession includes the stages by which defoliated land areas, such as abandoned farmland or strip-mined land, become a climax community again. During secondary succession, grass and weeds are followed by shrubs and trees and finally by a mature climax forest.

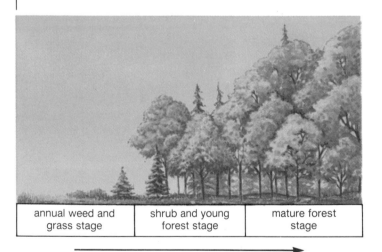

| annual weed and grass stage | shrub and young forest stage | mature forest stage |

CP = community productivity and CR = community respiratory loss. This indicates that some of the energy derived from photosynthesis is not going into growth and increased biomass but is being dissipated as heat.

As succession proceeds, diversity increases, nutrients recycle more, and there is more efficient utilization of energy within the community. Figure 47.21 compares the productivity of the producers found within the climax communities studied in this chapter. Although their productivity varies, *all* climax communities have a CP/CR ratio that approximates unity. This indicates that they are making full use of inputs to maintain their stability.

Table 47.3 Properties of Communities

Pioneer Community	Climax Community
Harsh environment	Most favorable environment
Biomass increasing	Biomass stable
Energy consumption inefficient	Energy consumption efficient
Some nutrient loss	Nutrient cycling
Low species diversity	High species diversity
Fluctuations common	Fluctuations do not usually occur
Little stability	High degree of stability

Humans often replace climax communities, which tend to have the characteristics listed in table 47.3, with simpler communities, which tend to have the same characteristics as those of the pioneer communities listed in table 47.3. Agricultural fields and managed forests show an increase in biomass and have high yields because they are provided with large inputs, such as fertilizer applications. The price paid for this growth potential, however, is a certain amount of waste and loss of stability. We should keep in mind that stable climax communities aid in many ecological services, such as purification of the air and water, that are necessary to our well-being. For this reason, it would be wise to make sure that a large portion of every biome is kept in its natural state.

The term *succession* is applied to the series of stages by which rocks or abandoned land become a climax community. A climax community is the most complex and stable, but the early stages of succession are the most productive in terms of new growth. When humans alter biomes, they should balance productivity against stability, never neglecting the latter because only a stable biosphere ensures human existence.

Figure 47.15

Soil profiles. *a*. Soil has horizontal layers called horizons. The top layer, the A horizon, contains most of the organic matter consisting of litter and humus. It also has a zone of leaching through which dissolved materials move downward. The middle layer, the B horizon, accumulates minerals and particles as water moves from the A to B horizon. The last layer, the C horizon, consists of weathered rock. This layer determines the pH of the soil and its ability to absorb and retain moisture. *b*. Grassland profile. Soils formed in grasslands have a deep A horizon built up from decaying grasses over many years. The shallow B horizon does not have sufficient nutrients to support root growth because the low amount of rainfall in the grassland areas limits the amount of leaching from the topsoil. *c*. Forest profile. In forest soils, both the A and B horizons possess enough nutrients to allow for root growth. In tropical rain forests, the A horizon is more shallow than this generalized profile and the B horizon is deeper, signifying that leaching is more extensive. Since the topsoil of rain forests lacks nutrients, it only has a limited ability to support crops.

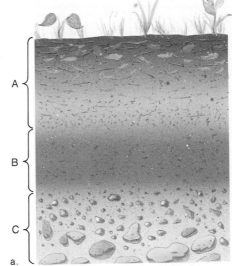

Litter: leaves and other debris

Topsoil: humus plus living organisms

Leaching: removal of nutrients

Subsoil: accumulation of minerals and organic materials

Parent material: weathered rock

a.

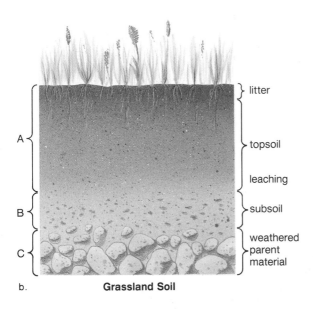

litter

topsoil

leaching

subsoil

weathered parent material

b. **Grassland Soil**

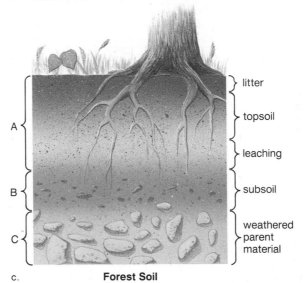

litter

topsoil

leaching

subsoil

weathered parent material

c. **Forest Soil**

Climate and Soil

Physical conditions, particularly climate, determine the biome of an area. Figure 47.12*a* shows how the major biomes are related to temperature and rainfall. Temperature decreases both with increasing latitude and increasing altitude. Therefore, when traveling from the equator to the Arctic or from the bottom of a mountain to the top, it would be possible to observe first a tropical rain forest, followed by a temperate deciduous forest, and then a taiga and a tundra, in that order.

The amount of rainfall is another important aspect of biome location; deserts and the tundra are the biomes with the least amount of rainfall while the tropical forests have the greatest amount.

Temperature and rainfall are very important to the distribution of organisms, but soil type is also a contributing factor. Soil formation requires the physical breakdown of rocks and the addition of dead organic matter called **humus.** Both of these supply the inorganic nutrients that plants need and, in addition, humus helps retain moisture. Figure 47.15 shows a typical soil profile for a grassland and for a forest.

Deserts

Deserts are vast arid areas with less than 25 cm of rainfall a year. There are few deserts with less than 2 cm of rainfall annually, the Sahara in Africa being the largest. Deserts of all types comprise about one-third of all land areas. In deserts, the days are hot because a lack of cloud cover allows the sun's rays to penetrate easily, but the nights are cold because heat easily escapes into the atmosphere.

Desert vegetation is extremely varied (fig. 47.16); annuals, herbaceous and woody perennials, deciduous and evergreen shrubs and trees all grow here. The annuals exist most of the year as seeds and burst into flower during a limited period of time when moisture and temperature are favorable. The entire life cycle takes place within a few weeks, as foliage and then flowers develop in quick succession.

Figure 47.16

Desert vegetation. These plants are adapted to retain water so that they can live in areas with infrequent rain.

Desert perennials are of two types: (1) nonsucculent shrubs and trees, and (2) succulent, herbaceous plants. Sagebrush, a densely branched, evergreen shrub with wedge-shaped leaves, is common in the "cold" northern desert, and the creosote bush—with small, well-protected leaves that may turn brown and drop off—is common to the "hot" southern desert of the United States. The deciduous mesquite tree with compound leaves has very deep roots. The best-known desert plants are the succulent, leafless cacti, which have stems that store water and also carry on photosynthesis. All cacti have widespreading root systems that can absorb great quantities of water during brief periods of rainfall. Their spines provide a means of defense against desert herbivores.

Like the plants that live in deserts, the animals also must have features that allow them to survive heat and lack of water. Millipedes, centipedes, scorpions, spiders, and numerous insects

are found here. Like plants, many desert insects have a compressed life cycle, going from pupa to pupa within a very short time, since the pupa can remain inactive until it rains. Other insects burrow in the soil where a detritus food chain allows survival.

Reptiles, especially lizards and snakes, are perhaps the most characteristic group of vertebrates found in desert regions. Running birds, such as the ostrich in southwest Africa, the emu in Australia, and the roadrunner in the southwest United States, are numerous. Such birds as the hawk and eagle may fly in and out of the desert at will. Bats, rodents, and rabbits are also common desert mammals. Some, such as the kangaroo rat, survive on metabolic water produced by respiration. Ungulates (antelopes, goats, and sheep) are able to roam far afield in search of food and water. The African camel is the best adapted of the ungulates; broad feet enable it to move over sand, and the reservoir of fat within its hump allows it to drink and eat infrequently. Dogs, such as the coyote and kit fox in North America and the dingo in Australia, are prominent carnivores. Lions in Africa and Asia and smaller cats, such as the mountain lion in North America, also hunt in the desert.

Most deserts are located in high-temperature areas, which receive less than 20 centimeters of rainfall a year. Plants and animals living in deserts need adaptations that protect them from the sun's rays and shortage of water.

Grasslands

Grasslands occur where rainfall is greater than 20 cm but is generally insufficient to support trees. The extensive root system of grasses allows them to recover quickly from drought, cold, fire, and grazing. Furthermore, their matted roots, which absorb surface water efficiently, prevent invasion by trees. Most of the grass, both above and below ground, dies each year. Thus, the turnover of biomass and nutrients is rapid; a complete turnover occurs in approximately three years. Because of relatively low rainfall and high evaporation rates, nutrients tend to accumulate in the lower part of the soil. At the same time, organic matter accumulates in the upper soil. The result is very fertile soil that has been exploited worldwide for agriculture.

Grasslands, which occur on all continents (fig. 47.12b), are known by various names, as indicated in table 47.4. We will discuss the tundra, an arctic grassland; the savanna, a tropical grassland; and the prairie, a temperate grassland found in the United States.

Tundra

The **tundra** (fig. 47.17) is an arctic grassland that occurs north of the Arctic Circle in Asia, Europe, and North America, and above timberline on higher mountain ranges to the south.

The arctic tundra is cold and dark much of the year. Since precipitation is only about 20 cm a year, it could possibly be considered a desert, but water that is frozen in the winter is plentiful in the summer because so little of it evaporates. Only the topmost layer of the earth thaws, and the permafrost beneath this is always frozen. Trees are not found in the tundra

Table 47.4 Types of Grassland

Location	Name
Arctic Grassland	
Arctic region of all continents	Tundra
Temperate Grasslands	
North America	Prairie
South America	Pampa
Africa	Veld
Eurasia	Steppe
Tropical Grasslands	
Africa and Australia	Savanna

Figure 47.17
Vegetation of the tundra. Notice the lack of trees, which are unable to root successfully because the ground is always frozen beneath the surface.

Figure 47.18
Vegetation and animal life of the African savanna. There are more types of grazers in the savanna than in any other biome.

small animals—for example, the arctic fox; the snowshoe hare; the lemming, which resembles a rat; and a bird called the ptarmigan—live in the tundra the year-round, but many birds migrate there in summer. At one time, before its virtual extermination by human hunters, musk-ox were a year-round resident. Now, among large animals, only caribou and reindeer migrate to the tundra in the summer, and wolves follow to prey upon them. Polar bears are common near the coast.

Plants and animals need adaptations to survive the extreme cold in the tundra. Plants have a short life cycle and are also low lying due to the limited growing season, high winds, extreme cold, and winter snowpack. To escape the cold, smaller animals burrow, while the large musk-ox has a thick coat and a short, squat body that conserves heat.

Savanna

The **savanna** (fig. 47.18) is a tropical grassland that has scattered trees and therefore supports populations of both browsers and grazers. It occurs in Africa, southern Mexico, Central America, and also Australia. The temperature is warm and there is adequate yearly rainfall, but a severe dry season limits the number of different types of plants. Perhaps the best known of the trees is the flat-topped acacia tree that sheds its leaves during a drought and never grows very large because its water requirements increase as it grows.

The large, always warm African savanna supports the greatest number of different types of large herbivores of all the biomes. Elephants and giraffes are browsers, animals that feed on tree vegetation. Antelopes, zebras, wildebeest, water buffalo, and rhinoceros are grazers, animals that feed on grass. These are preyed upon by cheetahs and lions, whose kill is, at times, scavenged by hyenas and vultures.

for three reasons: the growing season is too short, their roots cannot penetrate the permafrost, and they cannot become anchored in the boggy soil during the summer.

While the ground is covered with sedges, shortgrasses, and forbs during the summer months, there are also dwarf woody shrubs and numerous patches of lichens and mosses. Only a few

These large animals can be active all year because they can migrate in search of new pastures. Some other animals are more active in either the wet or the dry season, but not both. Most insects and birds reproduce during the wet season, while reptiles tend to reproduce during the dry season.

Prairie

A **prairie** is a temperate grassland. When traveling from east to west across the United States, the tallgrass prairie gradually gives way to a shortgrass prairie. Although grasses dominate both types of prairie, they are interspersed by other herbaceous plants called forbs. These may catch the eye because they have colorful flowers, whereas grasses do not.

In contrast to the savanna, the mammalian herbivores of the prairie are all grazers because there are too few trees to support populations of browsers. Small mammals, such as mice, prairie dogs, and rabbits, typically burrow in the ground but usually feed above ground. Hawks, snakes, badgers, coyotes, and kit foxes capture and feed off the small mammals. Large herds of buffalo, in the hundreds of thousands, once roamed the prairies and plains, along with pronghorn antelope, deer, and elk.

Grasslands occur where rainfall is greater than 20 cm but is insufficient to support many trees. The tundra supports only a limited variety of living things. The savanna supports the greatest number of different types of herbivores that all serve as food for carnivores. Much of the Midwest in the United States was once a prairie.

Forests

Forests (table 47.5) are located all over the world, except the Arctic. Coniferous evergreen forests are generally associated with the subarctic zone; broad-leaved deciduous forests with the temperate and the tropical zone; and broad-leaved evergreen forests with the tropical zone.

Taiga

The **taiga** is a coniferous forest extending in a broad belt across northern Eurasia and North America. The climate is characterized by cool summers, cold winters, and a growing season of about 130 days. Precipitation ranges between 40 cm and 100 cm per year, with much of it coming as heavy snow. The great stands of spruce, fir, and pine trees (fig. 47.19) are interrupted by many lakes and swamps. (Taiga is Russian for "swamp land.") The thickness of the forest allows little sunshine to filter through to the forest floor. As a result, there is no shrub understory; instead, there are only lichens, mosses, ferns, and some types of flowers found close to the ground.

Productivity is good (fig. 47.21) because the cone-bearing trees have needlelike evergreen leaves that retain moisture and carry on photosynthesis whenever weather conditions permit. The trees acquire adequate nutrients from the soil even though it is acidic and relatively infertile. Much of the precipitation moves through the soil, carrying with it important nutrients. However, the tree roots have a symbiotic association with fungi collectively called mycorrhizae. Fungal hyphae extend into the soil and speed the transfer of nutrients from it to the plant.

Table 47.5 Types of Forests

Biome	Plants
Coniferous forest (taiga)	Cone-bearing evergreen trees, such as pine and spruce No understory
Temperate deciduous forest	Broad-leaved trees, such as oak and maple Understories
Tropical rain forest	Broad-leaved evergreen trees Multilevel canopy No understory
Tropical deciduous forest	Mixed broad-leaved evergreen and deciduous Rich understories produce jungle

Figure 47.19
A coniferous forest encircles the globe in the northern temperate zone.

Compared to other forests, the taiga has relatively few consumer species. Insects inundate the area in the summer, attack the trees, and are eaten by birds, such as woodpeckers and chickadees. Other birds, such as the crossbills and grosbeaks, extract seeds from the cones. Hawks and eagles prey on small animals, such as the plentiful rodents, rabbits, and squirrels. The large herbivores—moose and bear—are apt to be found in clearings or near the water's edge where small trees, such as willows, aspens, and birches, and berry-bearing shrubs, such as blueberry and raspberry, are found. The carnivores—weasel, lynx, fox, and wolf—are especially noticeable in the summer.

Active signs of life all but disappear in the winter. Insect larvae survive in the dormant pupa stage, and many birds go south. Small animals seek shelter beneath the snow, within the ground, or in tree hollows. The snowshoe rabbit and short-tailed weasel change to their winter coats of white. Only the strongest animals survive the winter to again take advantage of the short summer season of plenty.

Temperate Deciduous Forests

Temperate forests (fig. 45.1) are found around the world just south of the taiga. They are also found in other areas, such as parts of Japan, Australia, and South America, where there is a moderate climate with well-defined winter and summer seasons and relatively high precipitation (75 cm to 150 cm per year). The growing season, which lasts from the last frost of spring to the first frost of fall, ranges between 140 and 300 days. The soil is relatively fertile and mildly acidic, and litter decomposes quite rapidly.

The dominant trees are broad-leaved deciduous trees, such as oak and hickory, or beech, hemlock, and maple, which lose their leaves in the fall and regain them in the spring. The tallest of these trees form a canopy, an upper layer of leaves that are the first to receive sunlight. Even so, enough sunlight penetrates to provide energy for another layer of trees called understory trees. Beneath these trees are shrubs that may flower in the spring before the trees have put forth their leaves. Still another layer of plant growth—mosses, lichens, and ferns—resides beneath the shrub layer. This stratification provides a variety of habitats for insects and birds. Ground life is also plentiful; squirrels, cottontail rabbits, shrews, skunks, woodchucks, and chipmunks are common small herbivores. These and the ground birds—turkeys, pheasants, and grouse—provide food for the carnivores. Mountain lions, bobcats, and timber wolves used to be found throughout the biome, but now only red foxes remain. The white-tail deer, a large herbivore, has increased in number of late, while the black bear, an omnivore, is all but extinct.

Autumn fruits, nuts, and berries provide many animals with a supply of food for the winter, and the leaves, after turning brilliant colors, fall to the ground and contribute to the rich layer of organic matter. The minerals within the soil are washed far into the ground by the spring rains, but the deep tree roots capture these and bring them back up into the forest system again.

Tropical Rain Forests

The largest **tropical rain forest** (fig. 47.20) is found in the Amazon basin of South America, but such a forest also occurs

Figure 47.20
The trees of a tropical rain forest form a canopy in which organisms can spend their entire lives.

at the equator in Africa and in the Indo-Malayan region where it is always warm (between 20° C and 25° C) and rain is plentiful (in excess of 200 cm per year). Therefore, due to the favorable climate, these forests are very productive (fig. 47.21).

Diversity of species and luxurious growth characterize the plants of this biome. There is a multi-layered canopy; some of the broad-leaved evergreen trees grow to 150 feet or taller, some to 100 feet, and some to 50 feet. These tall trees often have trunks buttressed at ground level to prevent their toppling over. Lianas, or woody vines, which encircle the trees as they grow from tree bottom to tree top, also help strengthen the trunk. Sometimes, little light penetrates the canopy. Then there is no understory except in open areas or clearings where the presence of light may produce a thick jungle.

Although there is animal life, including many types of insects and unique small animals, such as the paca, agouti, peccary, armadillo, and coatimundi on the ground floor, most animals have an arboreal existence. Birds, such as hummingbirds, parakeets, parrots, and toucans, are often beautifully colored. Amphibians and reptiles are also well represented by many types of snakes, lizards, and frogs. Monkeys are well-known primates in tropical forests. New-World monkeys such as the capuchin have a prehensile tail that helps them stay aloft; Old-World monkeys such as the rhesus do not. Monkeys feed off the fruit of the trees and are rarely carnivorous. The largest carnivores of the tropical forests are the big cats—the jaguars in South America and the leopards in the Old World.

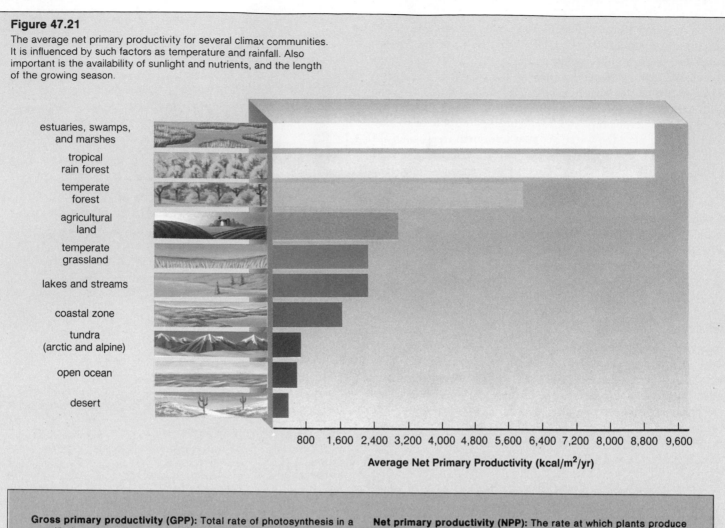

Figure 47.21
The average net primary productivity for several climax communities. It is influenced by such factors as temperature and rainfall. Also important is the availability of sunlight and nutrients, and the length of the growing season.

estuaries, swamps, and marshes
tropical rain forest
temperate forest
agricultural land
temperate grassland
lakes and streams
coastal zone
tundra (arctic and alpine)
open ocean
desert

800   1,600   2,400   3,200   4,000   4,800   5,600   6,400   7,200   8,000   8,800   9,600

Average Net Primary Productivity (kcal/m²/yr)

Gross primary productivity (GPP): Total rate of photosynthesis in a specified area usually expressed as kilocalories of energy produced per square meter per year (kcal/m²/yr).
Autotrophs (mostly photosynthetic plants and algae) absorb the energy of the sun and produce food for all organisms in a community.

Net primary productivity (NPP): The rate at which plants produce usable food or chemical energy (also expressed as kilocalories per square meters per year).

$$NPP = GPP - RS$$

where RS is the energy used by the autotrophs for respiration.

Since plants respire they use some of the food they produce (GPP) for respiration, and what is left (NPP) is available for growth and storage and can be used as food for heterotrophs. Usually, between 50% and 90% of GPP remains as NPP.

Not only do many animals spend their entire lives in the canopy, some plants do also. *Epiphytes* are plants that grow on tall trees but do not parasitize them. Instead, they have roots that take moisture and minerals from the air, so they are sometimes known as "air plants." The most common types belong to the pineapple, orchid, and fern families.

Tropical Deciduous Forests

While we usually think of all tropical forests as being nonseasonal rain forests, there are tropical forests with wet and dry seasons in India, Southeast Asia, West Africa, South and Central America, the West Indies, and northern Australia. Here deciduous trees often allow enough light to pass through so that

several layers of growth are possible. In addition to the animals just mentioned, certain forests also contain elephants, tigers, and hippopotami.

Whereas the soil of a temperate deciduous forest is rich enough for agricultural purposes, the soil of a tropical rain forest is not. Nutrients are cycled directly from the litter to the plants by way of mycorrhizae. Productivity is high (fig. 47.21) because of high temperatures, a year-long growing season, and the rapid recycling of nutrients from the litter. Of the minerals, only aluminum and iron sometimes remain near the surface, producing a red-colored soil known as laterite. When the trees are cleared, laterite bakes in the hot sun to a bricklike consistency that will not support crops. Slash-and-burn agriculture is the only type that has thus far been successful in the tropics. Trees are felled and burned, and the ashes provide enough nutrients for several harvests. Thereafter, the forest must be allowed to regrow, and a new section must be utilized for agriculture. The need for tropical rain forest preservation is discussed on page 783.

Forests require adequate rainfall. The taiga has the least amount of rainfall. The temperate deciduous forest has trees that gain and lose their leaves because of the alternating seasons of summer and winter. The tropical forests are the least studied and the most complex of all biomes.

Summary

1. Aquatic ecosystems, divided into freshwater and saltwater systems nearly always have a population of phytoplankton and zooplankton.
2. Fresh water, a distillate of salt water, falls on the land and saturates the earth to the groundwater table.
3. Lakes in temperate regions may form layers in the summer depending on temperature. Eutrophic lakes tend to fill in and disappear as succession occurs.
4. Marine ecosystems include estuaries and seashores, both rocky and sandy, in addition to the oceans themselves.
5. The rocky seashore, with organisms that cling to rocks, has more obvious signs of life than the sandy seashore, where animals burrow in the sand. Coral reefs are extremely productive communities found in shallow tropical waters.
6. The coastal zone, particularly wherever upwelling occurs, is more productive than the open ocean. In the ocean, the epipelagic zone receives adequate sunlight to support photosynthesis, and the mesopelagic and bathypelagic zones depend on detritus from the first. Each of these zones contains organisms adapted to the conditions found there.
7. Terrestrial ecosystems have communities called biomes that are adapted to climate—that is, temperature and rainfall.
8. Complex biomes arise by a process called succession. The first stage of succession, called the pioneer community, produces new growth but tends to be unstable. The final stage, called the climax community, produces little new growth but tends to be stable. Climax communities provide many beneficial services for humans, and their preservation is necessary for our well-being.
9. Most deserts are located in high-temperature areas which receive less than 20 cm of rainfall a year, and the organisms found here are adapted to these conditions in diverse ways.
10. Grasslands occur where rainfall is greater than 20 cm but is generally insufficient to support trees.
11. The tundra, being a northernmost biome, supports only a limited amount of life. In contrast, the savanna, a tropical grassland, supports the greatest number of different types of large herbivores. The prairie, found in the United States, has a limited variety of vegetation and animal life.
12. Forests require adequate rainfall. The taiga, a coniferous forest, has the least amount of rainfall. The temperate deciduous forest has trees that gain and lose their leaves because of the alternating seasons of summer and winter. Tropical forests, which include deciduous ones that experience dry and wet seasons, are the most complex and productive of all biomes.

Objective Questions

1. Phytoplankton are more likely to be found in which zone of the ocean?
   a. epipelagic zone
   b. mesopelagic zone
   c. bathypelagic zone
   d. continental slope
2. When would you expect the temperature of a temperate lake to be least uniform?
   a. winter
   b. spring
   c. summer
   d. fall
3. In which area of a lake would the temperature be the lowest?
   a. hypolimnion
   b. thermocline
   c. epilimnion
   d. limnetic zone
4. An estuary acts as a nutrient trap because of the
   a. action of rivers and tides.
   b. depth at which photosynthesis can occur.

   c. amount of rainfall received.
   d. the height of the water table.
5. Which area of an ocean has the greatest concentration of nutrients?
   a. epipelagic zone and benthic zone
   b. epipelagic zone only
   c. benthic zone only
   d. neritic zone
6. The forest with a multilevel understory is the
   a. tropical rain forest.
   b. coniferous forest.
   c. tundra.
   d. temperate deciduous forest.
7. All of these phrases describe a tropical rain forest, except
   a. nutrient-rich soil.
   b. many arboreal plants and animals.
   c. canopy composed of many layers.
   d. broad-leaved evergreen trees.

8. Which type of animal would you be least likely to find in a grassland biome?
   a. hoofed herbivore
   b. active carnivore
   c. arboreal primate
   d. flying insect
9. All of these phrases are true of the tundra, except
   a. low-lying vegetation.
   b. northernmost biome.
   c. short growing season.
   d. many different types of species.
10. Which of these is mismatched?
   a. tundra—permafrost
   b. savanna—acacia trees
   c. prairie—epiphytes
   d. coniferous forest—evergreen trees

Study Questions

1. Every year the land receives a quantity of fresh water, and yet no new supply of water is added to the biosphere. Explain.
2. Describe the temperature stratification of a lake in summer.
3. What organisms live in littoral, limnetic, and profundal zones of a lake?
4. Compare a rocky seashore to a sandy seashore.
5. Describe the coastline biomes (including coral reefs), and discuss their importance to the productivity of the ocean.
6. Describe the life zones of the ocean and the organisms you would expect to find in each zone.
7. Arrange the terrestrial biomes discussed in this text in a diagram according to temperature and rainfall.
8. Describe the stages of succession by which an abandoned field becomes a forest.
9. Describe the location, climate, and populations of the treeless biomes and the forests.

Thought Questions

1. Compare and contrast the main abiotic (nonliving) influences on aquatic ecosystems in general with terrestrial ecosystems in general.
2. Contrast the ways in which the productivity of an agricultural field and the productivity of a natural forest are maintained.

Selected Key Terms

biosphere (bi′o-sfēr) 747
biome (bi′ōm) 747
plankton (plank′ton) 747
water cycle (wah′ter si′k′l) 747
aquifer (ah′kwĭ-fer) 747
cultural eutrophication (kul′tūr-al u″tro-fĭ-ka′shun) 749

estuary (es′tu-ar″e) 750
coral reef (kor′al rēf) 751
continental shelf (kon″tĭ-nent′al shelf) 752
pelagic zone (pe-laj′ik zōn) 752
succession (suk-sesh′un) 754

climax community (kli′maks kŏ-mu′nĭ-te) 755
humus (hu′mus) 757
temperate forest (tem′per-it for′ests) 761
tropical rain forest (trop′ĭ-kal rān for′est) 761

CHAPTER 48

Ecosystems: The Flow of Energy and the Cycling of Materials

Your study of this chapter will be complete when you can:

1. *Give examples of biotic components within an ecosystem.*
2. *Name two types of food chains and give an example of each type.*
3. *Give an example of a food web and define trophic level.*
4. *Contrast the human food chain with a natural food web.*
5. *Construct a generalized pyramid of energy and use it as a basis to explain why energy flows through an ecosystem.*
6. *Use a pyramid of biomass to explain biological magnification.*
7. *Name and give a function for each part of a generalized biogeochemical cycle.*
8. *Describe the carbon, nitrogen, and phosphorus biogeochemical cycles.*
9. *Describe the human influence on each of these cycles.*
10. *List ways in which it would be possible for humans to reduce the current level of pollution.*

Large birds like this stork eating a frog are top predators whose population size is usually limited by the quantity of available food. When top predators die, the chemicals making up their bodies are utilized by photosynthesizers who produce food for themselves and indirectly for all other populations within the community.

In previous chapters we have studied the ecology of individual populations and communities. Now we wish to consider the interactions that take place between the community and the environment. Table 48.1 lists various ecological terms that have already been introduced.

All of the populations in a given area, together with the physical environment, comprise an ecosystem. Therefore, an ecosystem possesses both living (biotic) and nonliving (abiotic) components. The nonliving components include soil, water, light, inorganic nutrients, and weather variables. The living components of the ecosystem may be categorized as either producers or consumers. **Producers** are autotrophic organisms with the capability of carrying on photosynthesis and making food for themselves (and indirectly for the other populations as well). In terrestrial ecosystems, the producers are predominantly green plants, while in freshwater and marine ecosystems the dominant producers are various species of algae.

Consumers are heterotrophic organisms that use preformed food. It is possible to distinguish four types of consumers, depending on their food source. **Herbivores** feed directly on green plants; they are termed primary consumers. **Carnivores** feed only on other animals and are thus secondary or tertiary consumers. **Omnivores** feed on both plants and animals. Thus, a caterpillar feeding on a leaf is a herbivore; a green heron feeding on a fish is a carnivore; a human being eating both leafy green vegetables and beef is an omnivore.

Detritus is the remains of plants and animals following their death. The fourth type of consumer is decomposers that feed on detritus. The bacteria and fungi of decay are important detritus feeders, but so are other soil organisms, such as earthworms and various small arthropods. The importance of the latter can be demonstrated by placing leaf litter in bags with mesh too fine to allow soil animals to enter; the leaf litter will not decompose well even though bacteria and fungi are present. Small soil organisms precondition the detritus so that bacteria and fungi can break it down to inorganic matter that producers can use again.

When we diagram all the biotic components of an ecosystem as in figure 48.1, it is possible to illustrate that every ecosystem is characterized by two fundamental phenomena: energy flow and chemical cycling. Energy flow begins when producers absorb solar energy, and chemical cycling begins when producers take in inorganic nutrients from the physical environment. Energy flow occurs because all the energy content of organic food is eventually lost to the environment as heat. However, the original inorganic nutrients—the individual chemical elements that make up organisms—are cycled through the abiotic environment and eventually cycled back to the producers.

Energy Flow

Energy flow in an ecosystem is a consequence of the two laws of thermodynamics we discussed in chapter 7. The first law states that energy cannot be created or destroyed, and the second law states that when energy is transformed from one form to another, there is always a loss of some usable energy as heat. This

Table 48.1 Ecological Terms

Term	Definition
Ecology	Study of the interactions of organisms with the physical environment and with each other.
Population	All the members of the same species that inhabit a particular area.
Community	All the populations that are found in a particular area.
Ecosystem	A community along with its physical environment. Has a living (biotic) and a nonliving (abiotic) component.
Biosphere	That portion of the surface of the earth (air, water, and land) where living things exist.

Figure 48.1

A diagram illustrating energy flow and chemical cycling through an ecosystem. Energy does not cycle because all the energy that is derived from the sun eventually dissipates as heat.

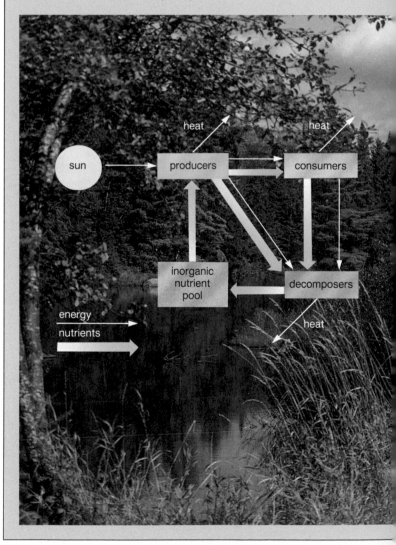

Figure 48.2

Examples of food chains. *a.* Terrestrial. *b.* Aquatic.

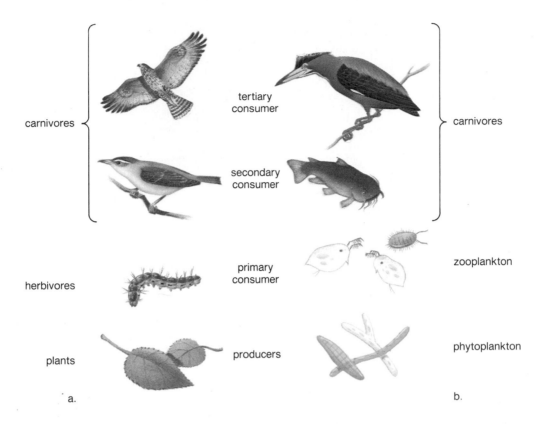

carnivores

tertiary
consumer

carnivores

secondary
consumer

herbivores

primary
consumer

zooplankton

plants

producers

phytoplankton

a.

b.

means that ecosystems are unable to function unless there is a constant energy input from an external source, such as the sun, the ultimate source of energy for our planet.

Food Chains and Food Webs

Energy flows through an ecosystem as the individuals of one population feed on those of another. A **food chain** indicates who eats whom in an ecosystem. Figure 48.2 depicts a possible terrestrial food chain and a possible aquatic food chain. It is important to realize that each represents just one path of energy flow through an ecosystem. Natural ecosystems have numerous food chains, each linked to others to form a complex food web. For example, figure 48.3 shows a forest **food web** in which plants are eaten by a variety of insects, and these, in turn, are eaten by several different birds, while any one of the latter may be eaten by a larger bird such as a hawk. Therefore energy flow is better described in terms of **trophic (feeding) levels,** each one further removed from the producer population, the first (photosynthetic) level. All animals acting as primary consumers are part of a second trophic level; all animals acting as secondary consumers are part of the third level, and so on.

The populations in an ecosystem form food chains in which the producers produce food for the other populations, which are consumers. While it is convenient to study food chains, the populations in an ecosystem actually form a food web in which food chains join and overlap with one another.

Figure 48.3

This drawing depicts populations typical of a deciduous forest ecosystem. The arrows indicate the flow of energy.

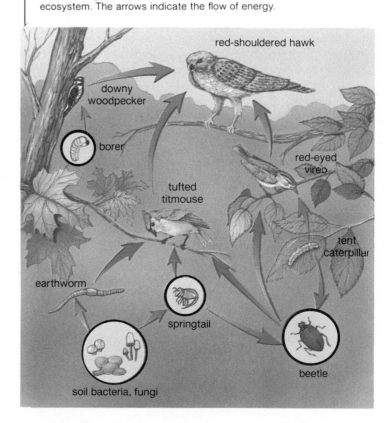

red-shouldered hawk

downy
woodpecker

borer

red-eyed
vireo

tufted
titmouse

tent
caterpillar

earthworm

springtail

beetle

soil bacteria, fungi

One of the food chains depicted in figure 48.2 was taken from the forest food web shown in figure 48.3 and the other was taken from the aquatic food web shown in figure 48.4. Both these food chains are called *grazing food chains* because the primary consumer feeds on a photosynthesizer. In some ecosystems (forests, rivers, and marshes) the primary consumer feeds mostly on detritus (dead organisms). This *detritus food chain* accounts for more energy flow than does the grazing food chain because most organisms die without having been eaten. In the forest, an example of a detritus food chain is:

detritus ⟶ soil bacteria ⟶ earthworms

A detritus food chain is often connected to a grazing food chain, as when earthworms are eaten by a tufted titmouse. Eventually, however, as dead organisms decompose, all the solar energy that was taken up by the producer populations dissipates as heat. Therefore energy does not cycle.

Pyramids of Energy and Biomass

The trophic structure of an ecosystem can be summarized in the form of an **ecological pyramid.** The base of the pyramid represents the producer trophic level; the apex is the tertiary or some higher level consumer, and the other consumer trophic levels are in between. There are three possible kinds of pyramids. One is a *pyramid of numbers,* based on the number of organisms at each trophic level. A second is the *pyramid of biomass.* Biomass is the weight of living material at some particular time. To calculate the biomass for each trophic level, one first determines an average weight for the organisms at each level and then estimates the number of organisms at each level. Now, multiplying the average weight by the estimated number gives the approximate biomass for each trophic level. A third pyramid, the *pyramid of energy* (fig. 48.5), illustrates the fact that each succeeding trophic level is smaller than the previous level. Less energy is found in each succeeding trophic level because of the following:

1. Of the food available, only a certain amount is captured and eaten by the next trophic level. After all, prey are adapted in many ways to escape their predators, as discussed on page 739.
2. Some of the food that is eaten cannot be digested and exits the digestive tract as waste.
3. Only a portion of the food that is digested becomes part of the organisms's body. The rest is used as a source of energy.

In regard to the last point, consider that even when food molecules are used for growth and repair, energy is needed to cause these molecules to react. Synthetic reactions will not occur spontaneously, and when the body builds proteins, carbohydrates, and fats, energy in the form of ATP must be supplied. ATP is also needed by the body for such activities as muscle contraction and nerve conduction. Altogether, a significant portion of the food molecules supply the energy needed to form ATP by means of cellular respiration within mitochondria. We must also remember, as discussed on page 95, that one form of

Figure 48.4
This drawing depicts populations typical of a freshwater pond ecosystem. The arrows indicate the flow of energy.

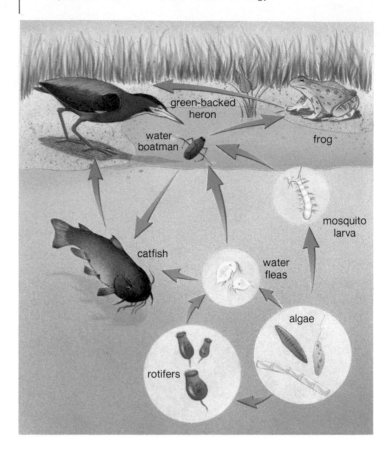

energy can never be transformed completely into another form. There is always some loss of usable energy between trophic levels for this same reason.

The light areas in figure 48.5 stand for the amount of energy that is unavailable to the next trophic level. The unavoidable loss of usable energy between feeding levels explains why food chains are relatively short—at most, four or five links—and why mice (herbivores) are more common than weasels, foxes, or hawks (carnivores). The population sizes are appropriate to the amount of energy available at each trophic level, until finally there is an insufficient amount of energy to support another level. The largest populations in an ecological pyramid are the producer populations at the base of the pyramid, and the smallest populations are the top predators. As with all organisms, the population sizes of top predators are controlled by the amount of food energy available to them.

The energy considerations associated with ecological pyramids have implications for the human population. It is generally stated that only about 10% of the energy available at a particular trophic level is incorporated into the tissues of the next level. This being the case, it can be estimated that 100 pounds of grain could, if consumed directly, result in 10 human pounds, but if fed to cattle, it would result in only 1 human pound. Thus, a larger human population can be sustained by eating grain than by eating grain-consuming animals. Humans generally need some meat in their diet, however, because that is the commonest way to acquire the essential amino acids, as discussed in chapter 35.

Figure 48.5

Energy flow in an ecological pyramid. At each step, an appreciable portion of energy originally trapped by the producer is lost as heat. Accordingly, organisms in each trophic level pass on less energy than they received.

energy retained in the living system   energy lost from the living system

---

In a food web, each successive trophic level has less total energy content. This is due to the fact that when energy is transferred from one level to the next, some energy is lost to the environment as heat.

Biological Magnification

Humans add all sorts of pollutants to the environment: some of these are hazardous substances that cannot be degraded by decomposers. These include heavy metals such as lead and mercury, synthetic organic chemicals such as PCB and DDT[1], and radioactive materials. Since decomposers are unable to break down these substances, they remain in the body and are not excreted. Once they enter a food chain, they become more concentrated at each trophic level. Notice in figure 48.6 that the number of dots representing DDT become more concentrated as the chemical is passed along from producer to tertiary consumer. **Biological magnification** is most apt to occur in aquatic food chains; there are more trophic levels in aquatic food chains

[1]PCB = *p*olychlorinated *b*iphenyl, a chemical used in manufacturing plastics. DDT = *d*ichloro*d*iphenyl*t*richloroethane, a pesticide no longer in use in the United States.

than there are in terrestrial food chains. Humans are the final consumers in both types of food chains, and in some areas, human milk contains detectable amounts of DDT and PCB.

The use of DDT as a pesticide is now banned for most uses in this country. The first indication that the substance was harmful to living things came when birds of prey became reduced in numbers. DDT interfered with the deposition of calcium in their egg shells, which were then too fragile to allow proper development of the enclosed chick.

Human Food Chain

Complex food webs tend to be stable because species diversity provides alternate pathways by which individuals in the community can obtain energy and nutrients. To put this another way, if a large number of species is present, then when one species declines or disappears, its place and function can be assumed in part by another. If the red-eyed vireo population shown in figure 48.3 declined in size, the tufted titmouse population would most likely increase in size, and in this way the hawk population would be maintained. If the hawk population should disappear, competition between its prey populations would keep their numbers in check.

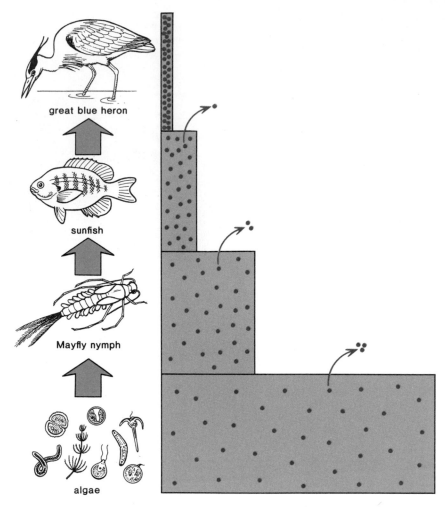

Figure 48.6

A diagram illustrating the biological magnification of a poison (dots), such as DDT, due to minimal excretion (arrows) by members of a food chain.

great blue heron

sunfish

Mayfly nymph

algae

Generally, population sizes remain constant in mature natural communities primarily due to biotic factors, such as competition and predation. Under normal circumstances, most of the solar energy utilized goes into maintenance of the community rather than into the growth of certain populations. The same amount of energy and materials are required by the system, and it is said that the inputs are constant. Also, because of nutrient cycling, there is little output except heat (fig. 48.7).

When humans were hunters and gatherers, human populations, like a few native tribes today, were members of food webs, such as the one depicted in figure 48.3. With the advent of the Agricultural Revolution (about 8000 B.C.) and the Industrial Revolution (about 1650 A.D.), human populations were able to become largely independent of natural food webs. Instead, humans created their own captive populations of producers and herbivores, which they have kept separate from natural food chains even to the present time. In order to maintain this independence from nature, the domesticated plants or animals have to be kept in highly unnatural conditions, with a minimum of species diversity and a high degree of energy input in the form of human care and fossil fuel.

Whereas natural food webs are remarkable for their inner stability, the human food chain is highly unstable. First, it may be noted that the human food chain is very short and usually has at most three links; for example:

grain ──────→ cattle ──────→ humans

Adding to the lack of diversity is the tendency of farmers to plant the same hybrid variety of wheat or corn year after year because it gives the highest yield. This so-called *monoculture farming* contributes to instability since any one environmental factor, either biological or physical, can cause a crop failure in all regions. One frequently cited example is the failure of the potato crop in Ireland in the 1800s due to a fungal infection. A more recent example occurred in 1970 when 15% of the corn crop in the United States was damaged by a corn blight. The damage could have been far worse because 80% of the crop planted that year was of the same susceptible hybrid variety. The fungus causing this particular corn blight is endemic to the South, but it happened to undergo a mutation that made it more lethal than before, and it spread to the cornbelt states where warm, moist weather helped it flourish. The next year resistant hybrid seed was available for planting, but this occurrence still shows that planting the same type of crop over a wide area produces a potentially unstable ecological situation.

Figure 48.7

Natural food chain versus human food chain. *a.* Diagram illustrating the cyclic nature of natural food chains and, thus, of natural ecosystems. Notice, too, that the inputs and outputs remain constant, as does the size of the populations. *b.* Diagram illustrating the noncyclic nature of the human food chain and, thus, the human ecosystem. Notice that the use of supplemental fossil fuel energy and of substances that do not cycle results in pollution. The ever-increasing size of the human food chain and the concomitant use of more artificial supplements cause an ever greater amount of pollution.

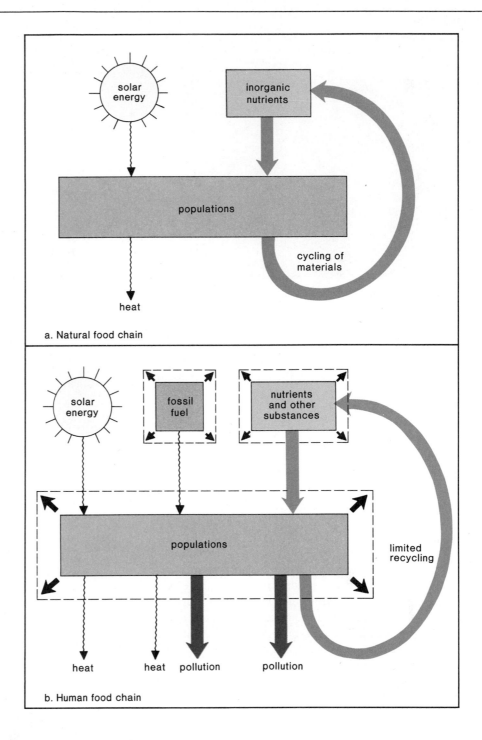

a. Natural food chain

b. Human food chain

The size of the populations in the human food chain are not constant. The human population has tripled since 1900 and the overall size of the population continues to increase dramatically (fig. 45.9). More food must be produced to sustain the ever-growing human population, although this population encroaches on agricultural land. A half-million hectares (one hectare = 2.471 acres) of farmland are taken over each year for other human uses, such as highways and housing. Nevertheless, food production (but not necessarily food distribution) has kept up with population increases. The newly developed hybrid plants are designed to give a high yield as long as they receive adequate supplements of fertilizers, pesticides, and water. The production of fertilizers and pesticides utilizes energy, and the application of these substances as well as water requires the use of mechanized farm machinery run on fossil fuel energy (fig. 48.8).

Fertilizers, pesticides, water, and fossil fuels are visible supplements to the U.S. agricultural system, but there is another supplement as well. Many farmers are not using methods to minimize soil erosion, such as contour farming (fig. 48.9).

Figure 48.8
Intensive mechanized monoculture farming is commonly practiced in the United States today. Here a farmer is harvesting sorghum for silage.

Figure 48.9
In contour farming, crops are planted according to the lay of the land in order to reduce soil erosion. This farmer has planted alfalfa in between the strips of corn in order to replenish the nitrogen content of the soil. Alfalfa, a legume, has root nodules that contain nitrogen-fixing bacteria.

The U.S. Department of Agriculture estimates that erosion is causing a steady drop in the productivity of land that is equivalent to the loss of 1.25 million acres of tillable land per year. Ever more fertilizers, pesticides, and energy supplements will be required to maintain yield.

Supplemental fossil fuel energy also contributes to animal husbandry yields because at least 50% of all cattle are kept in feedlots (fig. 48.10) where they are fed grain. Production of the grain is enhanced by great expenditures of fossil fuel energy, and more fossil fuel energy must then be used to make the grain available to the cattle. Chickens are raised in a completely artificial environment where the climate is controlled and each one has its own cage to which food is delivered on a conveyor belt.

From this discussion it is evident that the human food chain is largely noncyclic (fig. 48.7b). There is a large input of fossil fuel energy (in addition to solar energy) and other substances that do not cycle back to the producers. Unfortunately, this means there is also a large output of wastes. Excess fertilizers and pesticides from farmlands and home lawns are often transported by water and wind to aquatic ecosystems. Animal and human sewage are also often added to natural waters, where the bacteria of decay are often unable to immediately break down the quantity that is disposed. Therefore, it is customary for each community to pretreat sewage in sewage-treatment plants before the effluent is added to nearby bodies of water.

It is clear that the human food chain cannot be isolated completely from the natural communities. In fact, it is dependent on them not only because they by themselves lend stability to the biosphere, but also because they help stabilize the human food chain by absorbing by-products, such as excess fertilizer, pesticides, and fuel breakdown products. Unfortunately, because natural communities have decreased in size as the human population has increased, the natural communities are often

Figure 48.10
In the United States, most cattle are kept in feedlots and fed grain. This means of caring for cattle requires an input of fossil fuel energy. First, supplemental fossil fuel energy must be used to grow the grain, and then fossil fuel energy must be used to make the grain available to the cattle.

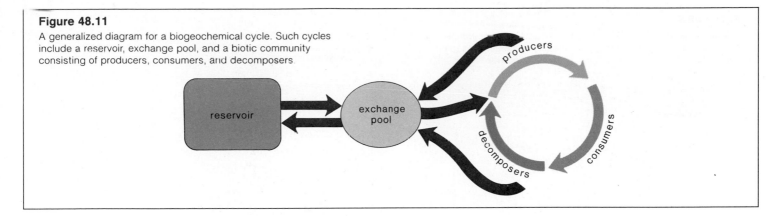

Figure 48.11

A generalized diagram for a biogeochemical cycle. Such cycles include a reservoir, exchange pool, and a biotic community consisting of producers, consumers, and decomposers.

overwhelmed. It is then that humans recognize that something is amiss and realize that pollutants[2] have been added to the environment.

To control pollution it is necessary to control the amount of inputs and/or to control the amount and treatment of the outputs. For example, a few farmers have stopped using artificial additives, such as fertilizers and pesticides, to increase crop yields. A study indicates that this method of farming, labeled organic farming by the investigators, does have reduced yields but can be equally profitable to the farmer because of the reduced costs. Other citizens are attempting to reduce pollution by making the human food chain more cyclical. For example, a new sewage treatment system is now functioning outside Atlanta, Georgia: wastewater sprinkled on a forest area provides nutrients for tree growth and purifies the water at the same time. Not only is the water then fit for drinking, but the trees can be harvested for human use on a four-year cycle.

These examples demonstrate that it is possible to find more natural ways to fulfill our needs. This is often called "working with nature rather than against nature." On the other hand, if we choose to continue practices that create environmental problems, we must ask ourselves if we are willing to bear the cost. As ecologists are fond of saying, "There is no free lunch." By this they mean that for every artificial benefit, there is a price. For example, if we kill off coyotes because they prey on chickens, we then have to realize that the mouse population (mice are a common food for coyotes) will probably increase. In an ecosystem, as we have seen, "everything is connected to everything else."

The human food chain is shorter and less stable than those that are found in natural food webs. As the human population increases in size, a greater quantity of materials and fossil fuel energy is required to sustain the population, and a greater quantity of outputs (e.g., pollutants) results.

Chemical Cycles

In contrast to energy, inorganic nutrients do cycle through large natural ecosystems. Because there is minimal input from the

[2]Pollutant: Material added to the environment by human or natural activities that produces undesirable effects.

outside, the various elements essential to life are used over and over. Since the pathways by which chemicals circulate through ecosystems involve both the living (biosphere) and nonliving (geological) areas, they are known as **biogeochemical cycles.** For each element, the cycling process (fig. 48.11) involves (1) a reservoir—that portion of the earth that acts as a storehouse for the element; (2) an exchange pool—that portion of the environment from which the producers take their nutrients; and (3) the biotic community—through which chemicals move along food chains to and from the exchange pool.

Carbon Cycle

Organisms in both terrestrial and aquatic ecosystems (fig. 48.12) exchange carbon dioxide with the atmosphere. On land, plants take up carbon dioxide from the air, and through photosynthesis they incorporate carbon into food that is used by autotrophs and heterotrophs alike. Then, when all organisms respire, a portion of this carbon is returned to the atmosphere as carbon dioxide.

In aquatic ecosystems, the exchange of carbon dioxide with the atmosphere is indirect. Carbon dioxide from the air combines with water to give bicarbonate ions that are a source of carbon for algae that produce food for themselves and for heterotrophs. Similarly, when aquatic organisms respire, the carbon dioxide they give off becomes the bicarbonate ion. The amount of bicarbonate in the water is in equilibrium with the amount of carbon dioxide in the air.

Carbon Reservoirs

Living and dead organisms contain organic carbon and serve as one of the reservoirs for the carbon cycle. The world's biota, particularly trees, contain 800 billion tons of organic carbon, and an additional 1,000 to 3,000 billion tons are estimated to be held in the remains of plants and animals in the soil. Before decomposition can occur, some of these remains are subjected to physical processes that transform them into coal, oil, and natural gas. We call these materials the fossil fuels. Most of the fossil fuels were formed during the Carboniferous Period, 280 to 350 million years ago, when an exceptionally large amount of organic matter was buried before decomposing. Another reservoir is the inorganic carbonate that accumulates in limestone

Figure 48.12

Carbon cycle. Photosynthesizers take up carbon dioxide from the air or bicarbonate from the water. They and all other organisms return carbon dioxide to the environment. The carbon dioxide level is also increased when volcanoes erupt and fossil fuels are burned. Presently, the oceans are a primary reservoir for carbon in the form of limestone and carbonaceous shells.

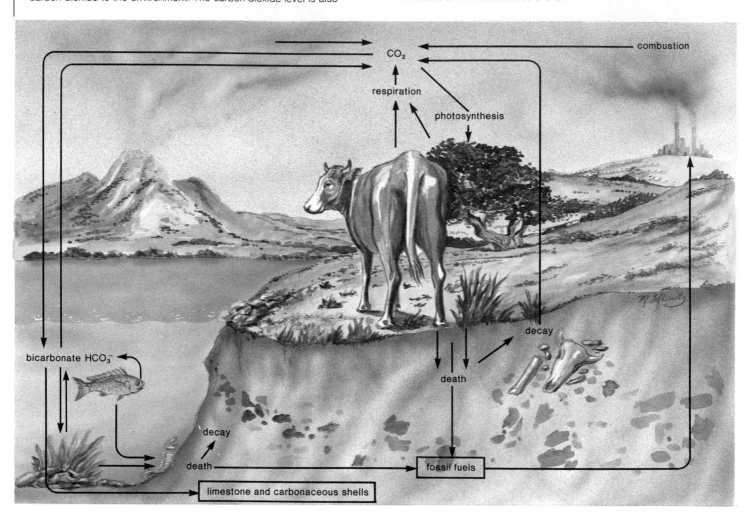

and in carbonaceous shells. The ocean abounds in organisms, some microscopic, that are composed of calcium carbonate shells and accumulate in ocean bottom sediments. Limestone is formed from these sediments by geological transformation.

Human Influence on the Carbon Cycle

The activities of human beings have increased the amount of carbon dioxide and other gases in the atmosphere. Data from monitoring stations record an increase of 20 ppm (parts per million) in carbon dioxide in only 22 years. (This is equivalent to 42 billion tons of carbon.) This buildup is primarily attributed to the burning of fossil fuels and the destruction of the world's forests (fig. 48.13). When we do away with forests, we are reducing a reservoir that takes up excess carbon dioxide. At this time, the oceans are believed to be taking up most of the excess carbon dioxide. The burning of fossil fuels in the last 22 years has probably released 78 billion tons of carbon, yet the atmosphere registers an increase of "only" 42 billion tons.

As discussed on page 790, there is much concern that an increased amount of carbon dioxide (and other gases) in the atmosphere is causing a global warming. These gases allow the sun's rays to pass through but absorb and reradiate heat back to the earth, a phenomenon called the *greenhouse effect*.

In the carbon cycle, carbon dioxide is removed from the atmosphere by photosynthesis but is returned by respiration. Living things and dead matter are carbon reservoirs. The ocean, because it abounds with calcareous shelled animals and accumulates limestone, is also a major carbon reservoir.

Nitrogen Cycle

Nitrogen is an abundant element in the atmosphere. Nitrogen gas (N_2) makes up about 78% of the atmosphere by volume, yet nitrogen deficiency commonly limits plant growth. Plants cannot incorporate N_2 into organic compounds and must absorb nitrogen from the soil, mainly as nitrate (NO_3^-) ions (fig. 48.14). Processes that reduce nitrogen gas and convert it to usable compounds for organisms are called **nitrogen-fixation** processes.

Nitrogen Fixation

Atmospheric nitrogen, the reservoir for the nitrogen cycle, can be converted to ammonia by some cyanobacteria in aquatic ecosystems and by nitrogen-fixing bacteria in terrestrial ecosystems. Some *nitrogen-fixing bacteria* are free-living in the soil, but others infect and live in nodules on the roots of legumes (fig. 30.12*a*). There are at least two ways by which nitrogen can be converted to NO_3^- for use by plants. Nitrogen is converted to NO_3^- in the atmosphere when cosmic radiation, meteor trails, and lightning provide the high energy needed for nitrogen to react with oxygen. Also, humans make a most significant contribution to the nitrogen cycle when they convert gaseous nitrogen to nitrates for use in fertilizers. This industrial process requires an energy input that at least equals that of the eventual increase in crop yield. The application of fertilizers also contributes to water pollution. Since nitrogen-fixing bacteria do not require fossil fuel energy and do not cause pollution, research is now directed toward finding a way to make all plants capable of forming nodules or, even better, through recombinant DNA techniques, to possess the biochemical ability to fix nitrogen themselves.

Figure 48.14

Nitrogen cycle. Several types of bacteria are at work: nitrogen-fixing bacteria reduce gaseous nitrogen, nitrifying bacteria, which include both nitrite-producing and nitrate-producing bacteria, convert ammonia to nitrate; and the denitrifying bacteria convert nitrate back to gaseous nitrogen. Humans contribute to the cycle by using gaseous nitrogen to produce nitrate for fertilizers.

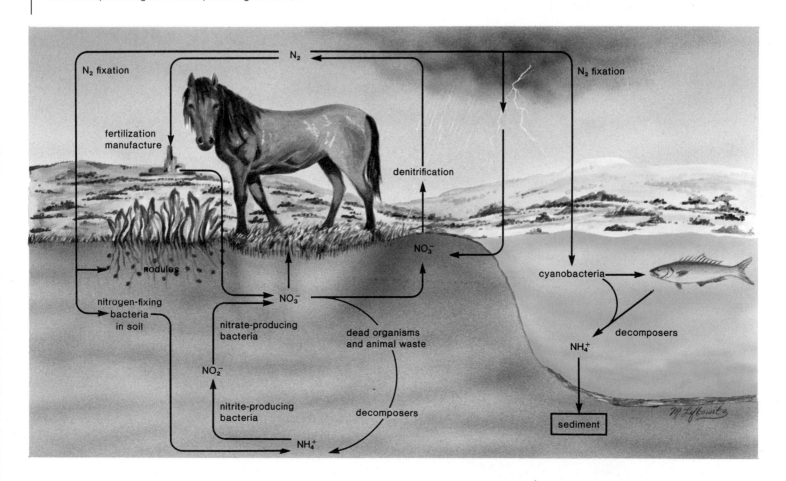

Figure 48.15
Phosphorus cycle. Normally, phosphate is in short supply. However,
humans mine phosphate deposits, including guano, for inclusion in
fertilizers; thus, they increase the amount of available phosphorus.

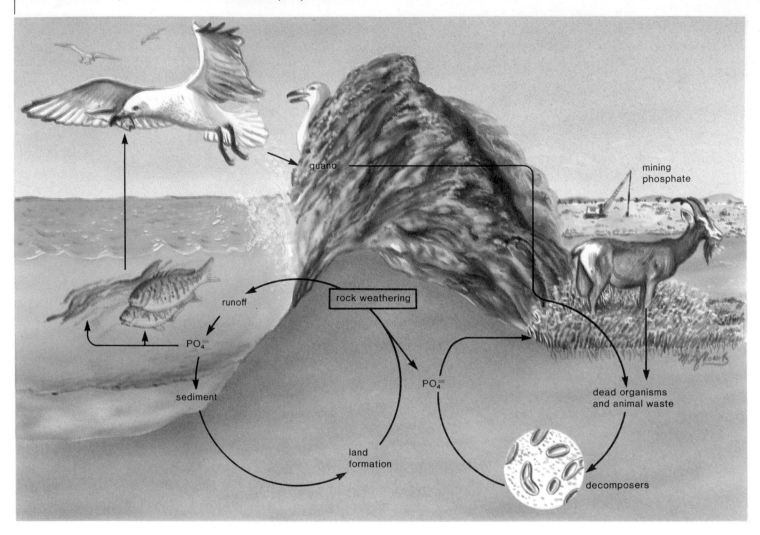

Nitrification and Denitrification

Once plants absorb nitrogen in a usable form they can synthesize proteins and nucleic acids. Consumers require the resulting supply of organic nitrogen to form their own proteins.
Some of this organic nitrogen quickly returns to the soil as fallen
leaves, etc., or as animal wastes. The remaining organic nitrogen returns only when plants and animals die. Decomposers
then break down these organic nitrogen compounds and release
ammonia. Often the ammonia is in turn oxidized to nitrate by
nitrifying bacteria in a process called **nitrification.** Thus, there
is a subcycle among members of food chains that does not involve atmospheric nitrogen (N_2) at all.

   Denitrification is the conversion of nitrate to atmospheric
nitrogen, largely by the action of denitrifying bacteria in both
aquatic and terrestrial ecosystems. Denitrification counterbal-
ances the process of nitrogen fixation, but not entirely. There is
more nitrogen fixation, especially due to fertilizer production.

In the nitrogen cycle:

nitrogen-fixing bacteria (in nodules and in the soil)—convert
   gaseous nitrogen to ammonia
nitrifying bacteria—convert ammonia to nitrate (overall reaction)
denitrifying bacteria—convert nitrates back to gaseous nitrogen

Phosphorus Cycle

The phosphorus that is available for organisms comes from rocks,
on land, and from sediments, in lakes and oceans.

   On land, the weathering of rocks makes phosphate ($PO_4^=$)
available to organisms (fig. 48.15). In water, floating algae have
to take up phosphate from the water because when phosphate
becomes deposited in sediments it is not available to them. When
spring or fall turnover mixes lake waters and sediments or when
a geological upheaval lowers the sea level or raises the ocean
floor, the phosphorus in sediments (or in sedimentary rock) is
available once more to organisms.

The phosphate taken up by producers is incorporated into a variety of molecules, including lipids and nucleic acids. Animals eat producers and incorporate their phosphate into teeth, bones, and shells that do not decompose for very long periods of time. Decomposition of phosphate-containing organic molecules takes place readily, however, most often by the action of decomposers such as bacteria and fungi. These organisms utilize the molecules as a carbon source, releasing inorganic phosphate that can be taken up by producers once again. Phosphate is generally taken up very quickly, and is a limiting nutrient in most ecosystems.

In the phosphorus cycle, weathering of rocks makes phosphate available to producers and from them to consumers. Following death and decay, this phosphate is returned to producers. Phosphate that accumulates in sediment may be lost to the biosphere for a long time.

Human Influence on the Phosphorous Cycle

For the most part, the phosphorus cycle on the land is separate from the one in the ocean, but there is one well-known exception. Seabirds, which feed on fish, deposit excrement near sea cliffs. Decomposition of this material results in *guano,* which contains phosphorus and other nutrients and can be harvested by humans for incorporation into fertilizers. The most famous guano deposits are along the coasts of Ecuador and Peru, where shorebirds feed on anchovies. Overfishing of the anchovies may eventually lead to a decline in the number of seabirds and thus in guano.

Human beings also mine rock high in phosphate to make phosphorus available for fertilizer production, animal feed supplements, pesticides, medicines, and numerous other products. Phosphate used to be added to detergents, but this led to eutrophication of natural waters, so the practice has largely been discontinued.

One well-known region where phosphorus has been mined since the late 1800s lies east of Tampa, Florida. Here, the so-called Bone Valley formation was made by fossilized remains of marine animals laid down 10 to 15 million years ago. Some 150,000 acres of this land have been strip-mined to remove phosphorus ore, but only 30,000 of these acres have been reclaimed or restored to approximate their original condition. When land is strip-mined, the vegetation and soil above the deposit are removed and piled up in a manner convenient to the operation. When the land is reclaimed, the soil must be put back, preferably with the top soil uppermost, and the land must be seeded or replanted. Unless this is done, the land is subject to severe soil erosion.

Another environmental concern is that the phosphate ore is slightly radioactive; therefore, phosphate mining poses a health threat to organisms in the area. There is also concern that these deposits, the best-known phosphate reservoir in the United States, will not last more than another few decades. It would seem, then, that the best course of action would be to take all possible steps to reduce the need for phosphorus. For example, better farming practices to control soil erosion and increase the efficiency of phosphate fertilizer use would help. Also, the phosphate industry could improve their processing procedures and try to extract more phosphorus from the ore that is mined. Up to 35% or more of the phosphate in the original ore is lost during processing.

Summary

1. Ecosystems contain both abiotic (physical) and biotic (living) components. The latter are either producers or consumers. Consumers may be herbivores, carnivores, omnivores, or decomposers.

2. Energy flow and chemical cycling are two phenomena observed in ecosystems.

3. Food chains are paths of energy flow through an ecosystem. Grazing food chains always begin with a producer population that is capable of producing organic food, followed by a series of consumer populations. Detritus food chains begin with dead organic matter that is consumed by decomposers, including bacteria and fungi, as well as various soil animals. Eventually, all members of the food chains die and decompose. Now the very same chemicals are made available to the producer population again, but the energy has been dissipated as heat. Therefore, energy does not cycle through an ecosystem.

4. If we consider the various food chains together, we realize that they form an intricate food web in which there are various trophic (feeding) levels. All producers are on the first level, all primary consumers are on the second level, and so forth. To illustrate that energy does not cycle, it is customary to arrange the various trophic levels to form an energy pyramid. Then we see that each level contains less energy than the previous level. It is estimated that only 10% of the available energy is actually incorporated into the body tissues of the next level.

5. In mature natural food webs, usually the same constant amount of energy is required and additional material inputs are minimal because matter cycles. In contrast, the human food chain tends to be unstable. The human food chain is quite short, containing at most only three links. Adding to this lack of diversity is the tendency of farmers to use only certain high-yield varieties of agricultural plants, which require supplements, such as fertilizers, pesticides, and water. The human food chain is noncyclic, and worse yet, more fossil fuel energy and material supplements are needed each year to produce more food for a human population that continues to increase in size. Pollution is a side effect; it would be beneficial for us to find ways to fulfill our needs in a more natural way.

6. Chemical cycling through an ecosystem involves not only the biotic components of an ecosystem but also the physical environment. Each cycle involves a reservoir where the element is stored, an exchange pool from which the populations take and return nutrients, and the populations themselves.

7. In the carbon cycle, the reservoir is organic matter, carbonaceous shells, and limestone. The exchange pool is the atmosphere: photosynthesis removes carbon dioxide, and respiration and combustion add carbon dioxide.

8. In the nitrogen cycle, the reservoir is the atmosphere, but atmospheric N_2 must be converted to NO_3^- for use by producers. Nitrogen-fixing bacteria, particularly in root nodules, make organic nitrogen available to plants. Other bacteria active in the nitrogen cycle are the nitrifying bacteria, which convert ammonia to nitrates and the denitrifying bacteria, which convert nitrate to gaseous N_2 again.

9. The reservoir for the phosphorus cycle is rocks, which weather and release phosphate. When producers and consumers die and decay, organic phosphate is eventually converted back to inorganic phosphate for use by organisms. Any phosphate that accumulates in oceanic sediment is made available to the biosphere only when a geological upheaval lowers sea level or raises the ocean floor.

10. In general, humans affect the carbon, nitrogen, and phosphorus cycles by withdrawing these substances from their reservoirs. For example, when we burn fossil fuels, we are removing organic carbon from a reservoir. We can also convert gaseous nitrogen to ammonia, which removes nitrogen from the air, and make phosphorus available by mining phosphate and guano.

Objective Questions

1. Compare this food chain: algae → water fleas → catfish → green herons to this food chain: trees → tent caterpillars → red-eyed vireos → hawks. Both water fleas and tent caterpillars are
   a. carnivores.
   b. primary consumers.
   c. detritus feeders.
   d. Both (a) and (b).

2. Consider the components of a food chain: producers → herbivores → carnivores → top carnivores. Which level contains the most energy?
   a. primary producers
   b. herbivores
   c. carnivores
   d. top carnivores

3. Consider the components of a food chain: producers → herbivores → carnivores → top carnivores. Eventually what happens to all the energy passed from one element to the next?
   a. It recycles back to the producers.
   b. It results in a much larger decomposer population.
   c. It is dissipated into the environment.
   d. It is recaptured by another food chain.

4. Which of the following contribute to the carbon cycle?
   a. respiration
   b. photosynthesis
   c. fossil fuel combustion
   d. All of these.

5. A natural food web
   a. contains only grazing food chains.
   b. contains several trophic levels.
   c. is usually unstable.
   d. All of these.

6. How do plants contribute to the carbon cycle?
   a. When they respire they release CO_2 into the atmosphere.
   b. When they photosynthesize they consume carbon dioxide from the atmosphere.
   c. They do not contribute to the carbon cycle.
   d. Both (a) and (b).

7. How do nitrogen-fixing bacteria contribute to the nitrogen cycle?
   a. They return N_2 to the atmosphere.
   b. They change ammonia to nitrate.
   c. They change N_2 to ammonia.
   d. They withdraw nitrate from the soil.

8. In what way are decomposers like producers?
   a. Either one may be the first member of a grazing food chain.
   b. Both produce oxygen for other forms of life.
   c. Both require a source of nutrient molecules and energy.
   d. Both supply organic food for the biosphere.

9. How is a first-order consumer like a second-order consumer?
   a. Both pass on less energy to the next trophic level than they received.
   b. Both pass on the same amount of energy to the next trophic level.
   c. Both tend to be herbivores that produce nutrients for plants.
   d. Both are able to convert organic compounds to ATP without the loss of energy.

10. Which statement is true concerning this food chain: grass → rabbits → snakes → hawks?
    a. Each predator population has a greater biomass than its prey population.
    b. Each prey population has a greater biomass than its predator population.
    c. Each population is omnivorous.
    d. Both (a) and (c).

Study Questions

1. Name four different types of consumers found in natural ecosystems.
2. Give an example of a grazing and a detritus food chain for a terrestrial and for an aquatic food chain.
3. Draw an energy pyramid and explain why such a pyramid can be used to verify that energy does not cycle.
4. Give an example of biological magnification and tell why it occurs.
5. Compare and contrast the basic characteristics of a natural food web to the characteristics of the human food chain.
6. Discuss in detail the carbon, nitrogen, and phosphorus biogeochemical cycles. Indicate the manner in which humans disturb the equilibrium of these cycles.

Thought Questions

1. Argue against the suggestion that natural food webs are stable.

2. Some people think that all bacteria are harmful. What evidence can you give from this chapter to counter this belief?

3. Suggest ways by which it would be possible to make the human food chain more cyclical so pollution might be better controlled.

Selected Key Terms

producer (pro-dūs'er) 766
consumer (kon-sūm'er) 766
carnivore (kar'nĭ-vōr) 766
herbivore (her'bĭ-vōr) 766
omnivore (om'nĭ-vōr) 766
detritus (de-tri'tus) 766

food chain (fōōd chān) 767
food web (fōōd web) 767
trophic level (trof'ik lev'el) 767
ecological pyramid (e''ko-log'ĭ-kal pir'ah-mid) 768

biological magnification (bi''o-loj'ĭ-kal mag''nĭ-fi-ka'shun) 769
biogeochemical cycle (bi''o-je''o-kem'ĭ-kal si'k'l) 773
nitrogen fixation (ni'tro-jen fik-sa'shun) 774
denitrification (de-ni''tri-fi-ka'shun) 776

CHAPTER 49

Environmental Concerns

Your study of this chapter will be complete when you can:

1. Give three general reasons why the quality of the land is being degraded today.

2. Give three reasons why tropical rain forests are being destroyed, and explain why this destruction should be halted immediately.

3. List three types of hazardous wastes that contribute to pollution on land.

4. Tell the difference between primary, secondary, and tertiary sewage treatment plants.

5. Give reasons why underground water supplies are being polluted and tell how the ocean receives pollutants.

6. List the substances that contribute to air pollution; tell their sources and associate each one with either development of photochemical smog, acid deposition, destruction of the ozone shield, or the greenhouse effect.

7. Explain how smog develops, how acids develop in the atmosphere, how chlorofluorocarbons break down the ozone shield, and how the greenhouse gases bring about global warming.

8. List the environmental effects of global warming and tell what can be done to prevent global warming.

9. Review the energy sources available to humans and indicate which of these would best prevent the occurrence of the greenhouse effect.

10. Describe the characteristics of a human society that is in a steady state.

Many agricultural fields are irrigated as shown here. Although water is a renewable resource, the supply during a particular time period is limited. Often water is removed from underground aquifers and if withdrawal exceeds recharging of the aquifer, it can run dry.

As the human population increases in size, the space allotted to natural ecosystems is reduced in size. Natural ecosystems are then no longer able to process and rid the biosphere of wastes, which accumulate and are called pollutants. **Pollutants** are substances added to the environment, particularly by human activities that lead to undesirable effects for all living things. Human beings add pollutants to all parts of the biosphere—land, water, and air.

Land

Whereas 20% of the world's population lived in cities in 1950, it is predicted that 50% will live in cities by the year 2000. Near cities, the development of new housing areas has led to urban sprawl (fig. 49.1). Such areas tend to take over agricultural land or simply further degrade the land in the area.

Soil Erosion and Desertification

Soil erosion causes the productivity of agricultural lands to decline. It occurs when wind blows and rain washes away the topsoil to leave the land exposed without adequate cover. The U.S. Department of Agriculture estimates that erosion is causing a steady drop in the productivity of farmland equivalent to the loss of 1.25 million acres per year. To maintain the productivity of eroding land, more fertilizers, more pesticides, and more energy must be used.

One answer to the problem of erosion is to adopt soil conservation measures. Unfortunately, at the present time many farmers are failing to use any form of soil conservation, and it is estimated that soil erosion is removing topsoil at three times the rate at which it can be re-formed by natural means. However, there are both new and old measures that farmers can adopt. Some farmers have turned to no-tillage farming as a means to prevent soil erosion. In no-tillage farming, the remains of the previous crop are left in place rather than being plowed under. At planting time a disc is used to cut furrows for planting new seeds. There is one drawback to this method: it requires the use of potent herbicides to prevent weed growth, and it may also require a more liberal application of pesticides, since pests thrive in the undisturbed soil and litter. For this reason, some experts prefer the older methods of preventing soil erosion, such as strip and contour farming (fig. 48.9).

The seriousness of land degradation is made clear by some recent trends. People are abandoning marginal cropland and rangeland worldwide because of severe soil erosion. Another process, **desertification,** leads to the transformation of marginal lands to desert conditions. Desertification has been particularly evident along the southern edge of the Sahara desert in Africa, where it is estimated that 350,000 square miles of once-productive grazing land has become desert in the last fifty years. However, desertification also occurs in this country. The U.S. Bureau of Land Management, which opens up federal lands for

Figure 49.1
Urban sprawl now occurs worldwide as cities expand due to newly arrived residents. *a.* Urban sprawl in U.S. *b.* Urban sprawl in Mexico.

a.

b.

Figure 49.2

Tropical rain forest ecology. *a.* In its natural state a tropical rain forest is immensely rich in vegetation and breathtakingly beautiful. *b.* The biological diversity of tropical rain forests is unequaled, and destroying them will do away with this diversity. Already threatened are the green iguana, South American ocelot, and the blue/yellow macaw. *c.* Slash-and-burn agriculture is the first step toward destruction of a portion of the forest. Then cattle ranchers often take over any land that has been cleared. But even if they do not, agriculture cannot be sustained. The infertile soil is subject to leaching, which further removes any nutrients present (see fig. 47.15).

a.

b.

c.

grazing, reports that much of the rangeland it manages is in poor or bad condition, with much of its topsoil gone and with greatly reduced ability to support forage plants.

Humans often use land for their cities and for agriculture. Agricultural land quality is threatened by soil erosion, which can lead to desertification.

Tropical Rain Forest Destruction

In developed countries much of the hardwood forests were long ago converted to rapidly growing softwood forest plantations to provide humans with a source of wood. However, there are still virgin rain forests (fig. 49.2) in Southeast Asia and Oceania, Central and South America, and Africa. These forests are severely threatened by human exploitation. The people living in developed countries want all sorts of things made from beautiful and costly tropical woods and provide a market for such lumber. Most of the loss due to logging is in Southeast Asia, and in Malaysia and the Philippines, for example, the best commercial forests are already gone. Indonesia is cutting heavily to feed its new wood-exporting business and Brazil is racing to catch up. Much of this wood goes to Japan, the United States, and Europe.

Another reason that tropical rain forests are undergoing destruction is a result of the needs of the people that live there. For example, in Brazil there are large numbers of persons who have no means to support their families. To ease social unrest, the government allows citizens to own any land they clear in the Amazon forest (occurs along the Amazon River). In tropical rain forests, it is the custom to practice what is called **slash-and-burn agriculture,** in which trees are cut down and burned to provide space to raise crops. However, unfortunately the land is fertile for only a few years. In tropical forests the inorganic nutrients cycle immediately back to producers (fig. 47.15) and do not accumulate in the soil. Therefore, despite the massive amount of growth above ground, the soil itself is nutrient-poor. Once the cleared land is unable to sustain crops, the farmer moves on to another part of the forest to slash and burn again.

Cattle ranchers are the greatest beneficiaries of deforestation and increased ranching is therefore another reason for tropical rain forest destruction. The ranchers in Brazil are so aggressive that they actually force colonists to sell them newly cleared land at gun point. The cattle that are raised provide beef for export, and much of it is bought by United States fast-food chains. Because the imported beef is cheaper, hamburgers can be sold for five cents less than if the beef were purchased in the United States. The destruction of tropical forests on this account has been termed the "hamburger connection."

A pig-iron industry has also begun in the Amazon and indirectly results in further exploitation of the rain forest. Deposits of iron ore have been found in the Carajas Mountains and are processed before being exported. The pig-iron smelters are fired by charcoal bought from small, private producers. The largest pig-iron company acknowledges having paid for construction of 1,500 small, make-shift ovens used by peasants who burn trees from the forests to produce the charcoal. Conservative estimates are that pig-iron production in Brazil could consume more than half a million acres of rain forest annually.

Altogether it is estimated that 28 million acres of tropical forest have been destroyed worldwide through the combined action of land clearing for crop production, fuel wood gathering, and cattle ranching. Commercial timber harvesting has degraded at least an additional 11 million acres worldwide. Considering the rate of destruction that is presently occurring in Brazil, this estimate may be too low. Satellite pictures show that 2.5 million acres of Amazon rain forest were cleared in 1984 and 1985 alone.

There are currently three primary reasons for tropical rain forest destruction: (1) logging to provide hardwoods for export, (2) slash-and-burn agriculture, and (3) cattle ranching. Industrialization in countries having extensive tropical rain forests will no doubt become an important future reason also.

Loss of Biological Diversity

Tropical rain forests are much more biologically diverse than are temperate forests. For example, temperate forests across the entire United States contain about 400 tree species. In the rain forest, a typical 4-square-mile area holds as many as 750 types of trees. The fresh waters of South America are inhabited by an estimated 5,000 fish species; on the eastern slopes of the Andes there are 80 or more species of frogs and toads, and in Ecuador there are more than 1,200 species of birds—roughly twice as many as those inhabiting all of the United States and Canada. Therefore, a very serious side effect from deforestation in tropical countries is the loss of biological diversity.

Altogether, about half of the world's species are believed to live in tropical forests. A National Academy of Sciences study estimated that a million species of plants and animals are in danger of disappearing within 20 years as a result of deforestation in tropical countries. Many of these life forms have never been studied and yet many could possibly be useful. At present, our entire domesticated crop production around the world relies on less than 30 species of plants and animals that have been domesticated during the last 10,000 years! It's quite possible that many additional species of wild plants and animals now living in tropical forests could be domesticated. As matters now stand, the clearing of tropical forests will very likely prevent humans from ever having the possibility of utilizing more than a tiny fraction of the earth's biological diversity.

There is much worldwide concern about the loss of biological diversity due to the destruction of tropical rain forests. The myriad of plants and animals that live there could possibly benefit human beings.

Waste Disposal

Every year, the U.S. population discards billions of tons of solid wastes, much of it on land. **Solid wastes** include not only household trash but also sewage sludge, agricultural residues, mining

Figure 49.3

Dump sites not only cause land and air pollution, they also allow chemicals to enter groundwater, which may later become part of human drinking water.

refure, and industrial wastes. Some of these solid wastes contain substances that cause human illness and sometimes even death; they are called **hazardous wastes.**

In 1920 the per capita production of waste was about 2.75 pounds per day; in 1970 about 5 pounds per day; and in 1980 about 8 pounds per day. The rapid growth of consumer products with "planned obsolescence" and the use of packaging materials to gain a competitive edge account for much of this increase.

Open dumping (fig. 49.3), sanitary landfills, or incineration have been the most common practices of disposing trash. These disposal methods have become increasingly expensive and also cause pollution problems. It would be far more satisfactory to recycle materials as much as possible and/or to use organic substances as a fuel to generate electricity. One study showed that it was possible to achieve 70% to 90% public participation in recycling by spending only thirty cents per household. A city the size of Washington, D.C., where 500,000 tons of waste are generated each year, could have an increase of 1,300 jobs if the community utilized solid wastes as a resource instead of throwing them away.

Hazardous Wastes

The U.S. Environmental Protection Agency, charged with the task of keeping the environment safe, estimates that in 1980 at least 57 million metric tons of the nation's total wasteload could be classified as hazardous. Hazardous wastes fall into three general categories.

1. *Heavy metals,* such as lead, mercury, cadmium, nickel, and beryllium, contaminate many wastes. These metals can accumulate in various organs, interfering with normal enzymatic actions and causing illness—including cancer.

2. *Chlorinated hydrocarbons,* also called *organochlorides,* include pesticides and numerous organic compounds, such as PCBs (polychlorinated biphenyls), in which chlorine atoms are present in place of hydrogen atoms. Research has often found organochlorides to be cancer-producing in laboratory animals.

3. *Nuclear wastes* include radioactive elements that will be dangerous for thousands of years. ^{239}Plutonium takes 200,000 years before it loses its radioactivity.

Hazardous wastes are subject to biological magnification, as discussed on page 769. The public has become aware of hazardous dump sites that have polluted nearby water supplies. Chemical wastes buried over a quarter century ago in Love Canal, near Niagara Falls, have seriously damaged the health of some residents there. Similarly, the town of Times Beach, Missouri, is abandoned because workmen spread an organochloride (dioxin)-laced oil on the city streets, leading to a myriad of illnesses among its citizens. In other places, such as in Holbrook, Massachusetts, manufacturers have left thousands of drums in abandoned or uncontrolled sites where toxic chemicals are oozing out into the ground and contaminating the water supply. Illnesses, especially forms of cancer, are quite common not only in Holbrook but also in adjoining towns.

Land pollution in the United States is often caused by the dumping of trash, which may contain hazardous materials that can leach into a nearby water supply.

Water

Pollution of surface water, groundwater, and the oceans is of major concern today.

Surface Water Pollution

All sorts of pollutants enter surface waters, as depicted in figure 49.4 and listed in table 49.1. If sewage and other pollutants enter a river directly, biological diversity of the river is apt to suffer because bacteria and fungi of decay use up oxygen while they decompose the material. Such substances are said to have a high **BOD** (biological oxygen demand). Downstream from the point of discharge, the amount of oxygen decreases, giving a characteristic oxygen sag curve. Figure 49.5 illustrates this curve and also shows the effects upon the natural ecosystem. The portion of the river before the discharge contains a variety of bottom-dwelling organisms and fishes, such as trout, pike, or bass. As oxygen level decreases, only certain fishes—carp, catfish, or gar—are present. When the oxygen level approaches zero, sludge worms and bloodworms are typical of the only forms of animal life that can survive. As long as minimal oxygen is available, aerobic breakdown will occur; but should oxygen disappear entirely, anaerobic breakdown gives off methane, ammonia, and hydrogen sulfide. As recovery takes place, the warm-water fishes and then the cold-water fishes return.

To keep high BOD organic materials from polluting rivers, municipalities have built sewage treatment plants. A *primary treatment plant* removes solids, grease, and scum before the effluent (discharge) is chlorinated to kill any harmful bacteria. A

Figure 49.4

Water pollution is caused in all the ways shown here. Many bodies of water are dying due to the introduction of sediments and nutrients.

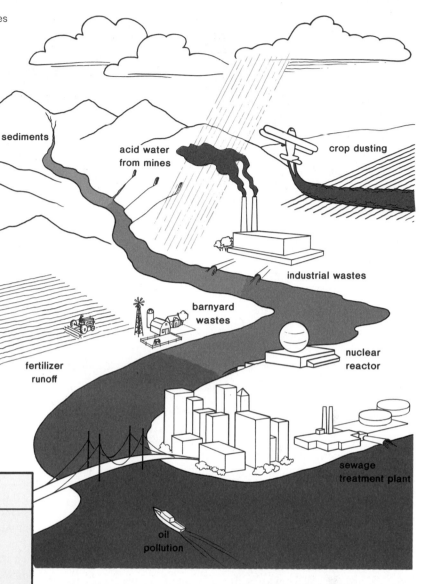

Table 49.1 Sources of Water Pollution

Leading to Cultural Eutrophication	
Oxygen-demanding wastes	Biodegradable organic compounds (e.g., sewage, wastes from food processing plants, paper mills, and tanneries)
Plant nutrients	Nitrates and phosphates from detergents, fertilizers, and sewage treatment plants
Sediments	Enriched soil in water due to soil erosion
Thermal discharges	Heated water from power plants
Health Hazards	
Disease-causing agents	Bacteria and viruses from sewage (e.g., food poisoning and hepatitis)
Synthetic organic compounds	Pesticides, industrial chemicals (e.g., PCBs)
Inorganic chemicals and minerals	Acids from mines and air pollution; dissolved salts; heavy metals (e.g., mercury) from industry
Radioactive substances	From nuclear power plants, medical and research facilities, and nuclear weapons testing

secondary treatment plant uses specially provided bacterial decomposers to break down any remaining organic material after primary treatment before the effluent is chlorinated and discharged. This means that instead of organic wastes, inorganic nutrients like nitrates and phosphates remain in the effluent. However, these nutrients also lead to water pollution, and they hasten the process of *eutrophication* that accompanies succession, or the steps by which a pond or lake becomes filled in and disappears (fig. 47.4). Therefore, these nutrients are said to contribute to *cultural eutrophication,* that portion of eutrophication caused by human activities. A *tertiary treatment plant,* which removes inorganic nutrient molecules, can be used to prevent eutrophication. However, since tertiary treatment plants cost twice as much as secondary plants, other ways to remove excess nutrients are being explored. For example, it is sometimes possible to pass the water through a swampy area before it drains into a nearby waterway. Or the water can be used to irrigate crops and/or grow algae and aquatic plants in an artificial shallow pond. Since the latter can be used as food for animals, this represents a cyclical use of chemicals.

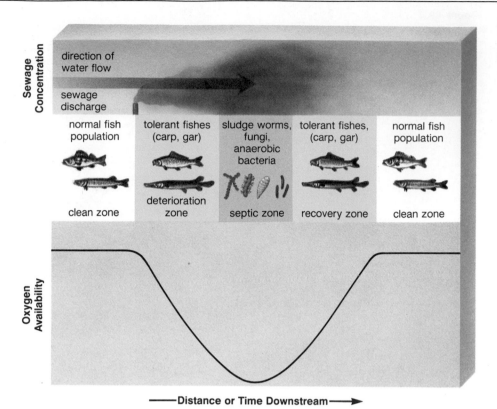

Figure 49.5

Oxygen sag curve. When raw sewage (brown curve) is added to a river, the water is deprived of oxygen and the normal distribution of animals is affected in the manner shown here. Notice that a plentiful amount of oxygen exists in the clean zones. After sewage is discharged, the oxygen level drops in the degradation zone because the decay bacteria use oxygen when they break down the sewage. The septic zone contains almost no oxygen, and mostly anaerobic bacteria are at work. Once the sewage has been decomposed, the oxygen level again rises in the recovery zone.

Groundwater Pollution

Ordinarily, one would expect underground water to be free of pollutants because bacteria and fungi found in the soil can remove most conventional contaminants before water reaches an aquifer. But it has been found that underground water is sometimes polluted with nonbiodegradable pollutants, such as organochlorides and heavy metals, and also with inorganic nitrates and chlorides. Figure 49.6 shows the ways in which pollutants can reach aquifers. Previously, industry was accustomed to running wastewater into a pit. The pollutants could then seep into the ground. Wastewaters and chemical wastes have also been injected into deep wells from which the pollutants constantly discharged. Both of these customs have been or are in the process of being phased out. However, it is very difficult for industry to find any other way to dispose of wastes. More adequately managed and controlled waste-treatment plants are needed, but because citizens do not wish to live near such plants, towns are often successful in preventing their construction.

Oceanic Pollution

Coastal regions are not only the immediate receptors for local pollutants (fig. 49.7), they are also the final receptors for pollutants carried by rivers that empty at the coast. Millions of tons of each of the pollutants listed in table 49.1 are added to ocean water each month. It now appears that nutrients and sediments, together with physical alteration from dredging and the like, will cause eutrophication of coastal ecosystems. Much of the coral reefs along the shores of Oahu, Hawaii, are covered by a putrid slime because of sewage disposal from the city of Honolulu. On this continent, marshes are filling in with sediments and disappearing.

Coastal areas are subject to more petroleum contamination than inland waters. Large oil spills from blowouts and tanker accidents can have disastrous local effects. Oil contaminants remain in estuarine wetland sediments for as long as eight years after an oil spill, and a marsh community is unable to reestablish itself even after three years. Although it is less dramatic, some believe that routine discharges from everyday use of oil and gas represent a longer and even greater threat to the marine environment.

The buildup of wastes in the oceans could eventually lead to the destruction of the marine ecosystem, since the ocean's capacity to dilute wastes is not infinite. However, it is possible to reverse the present trend. For example, a decade ago, off the coast of southern California, pelicans were unable to breed on nearby islands, beds of giant kelp off Palos Verdes had disappeared, and animal populations in large areas of the ocean bottom were substantially altered. California instigated a water quality control plan that set limits on the type and amount of wastes that could be discharged. Now pelicans and kelp are back, and in most areas marine life seems to be in good condition.

Adequate sewage treatment and waste disposal is necessary to prevent the polluting of rivers and the ocean. New methods are also needed to prevent pollution of underground water supplies.

Figure 49.6

Groundwater pollution is caused in all the ways shown here. Discontinuance of these means of disposal for industrial wastes has been difficult to achieve because citizens do not wish to have waste disposal plants located near them.

Source: Adapted from U.S. Environmental Protection Agency, Office of Water Supply and Solid Waste Management Programs, "Waste Disposal Practices and Their Effects on Ground Water" in *Executive Summary* (Washington, D.C., U.S. Government Printing Office, 1977).

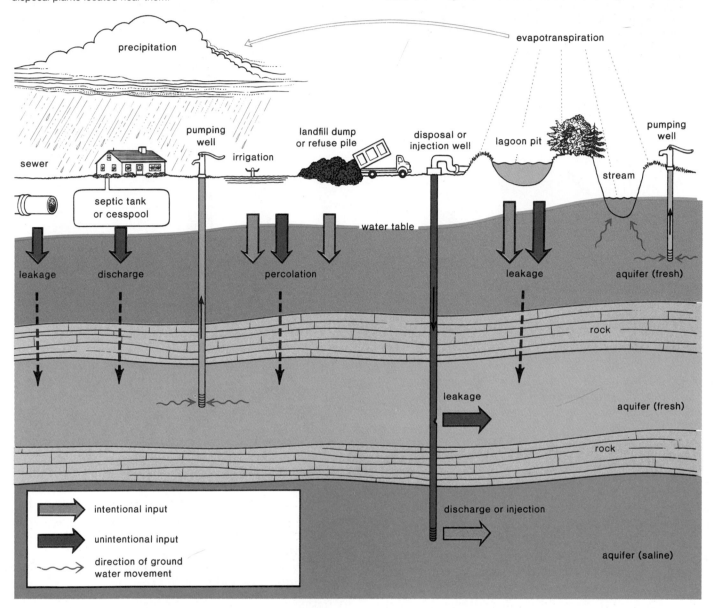

Figure 49.7

The possibility of oceanic pollution was brought home to the public in the summer of 1988 when medical wastes were washed back onto the beaches along the east coast.

Air

There are two layers of the atmosphere that will be of concern to us:

Stratosphere. This layer lies 15 km to 50 km above the surface of the earth. Here the energy of the sun splits oxygen molecules and then these individual oxygen atoms combine with molecular oxygen to give ozone. This ozone layer is called a shield because it absorbs the ultraviolet rays of the sun so that they do not strike the earth. If they did penetrate the atmosphere, life on earth would not be possible because living things cannot tolerate heavy doses of ultraviolet radiation.

Figure 49.8

Air pollutants. These are the gases, along with their sources, that contribute to four environmental effects of major concern: photochemical smog, acid deposition, the greenhouse effect, and destruction of the ozone shield. An examination of the sources of these gases shows that vehicle exhaust and fossil fuel burning are the chief contributors.

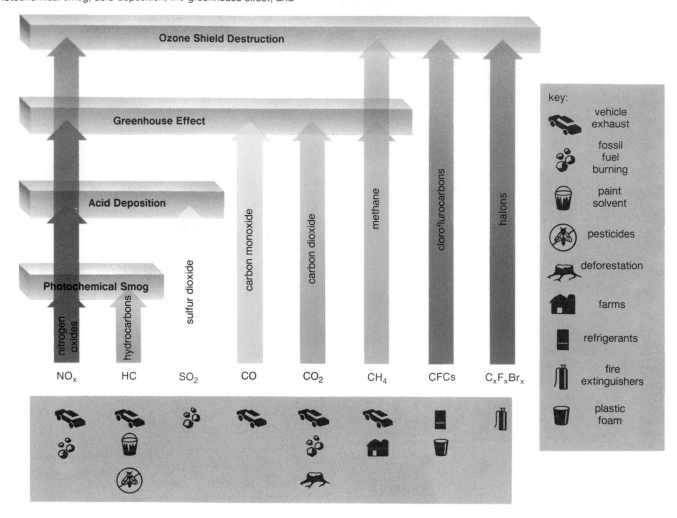

Troposphere. This layer is closest to the earth's surface: it ordinarily contains the gases nitrogen, 78%, oxygen, 21%, and carbon dioxide, 0.03%. Ozone (O_3) may be present in varying amounts as a pollutant in the troposphere.

Outdoor Air Pollution

Four major concerns (photochemical smog, acid deposition, the greenhouse effect, and destruction of the ozone shield) are associated with the air pollutants listed in figure 49.8. You can see that fossil fuel burning and vehicle exhaust are primary causes of air pollution gases. These two are related because gasoline is derived from petroleum (oil), a fossil fuel. The fossil fuels (oil, coal, and natural gas) are burned in the home to provide heat and in power plants to generate electricity. In other words, civilization's need for energy creates the major ecological problems we will be discussing.

Photochemical Smog

Photochemical smog contains two air pollutants—nitrogen oxides (NO_x) and hydrocarbons—which react with one another in the presence of sunlight to produce ozone (O_3) and PAN (peroxylacetyl nitrate). Both nitrogen oxides and hydrocarbons come from fossil fuel combustion, but additional hydrocarbons come from various other sources as well, including industrial solvents.

Ozone and PAN are commonly referred to as oxidants. Breathing ozone affects the respiratory and nervous systems, resulting in respiratory distress, headache, and exhaustion. These symptoms are particularly apt to appear in young people; therefore, in Los Angeles, where ozone levels are often high, schoolchildren must remain inside the school building whenever the ozone level reaches 0.35 ppm (parts per million by weight). Ozone is especially damaging to plants, resulting in leaf mottling and reduced growth (fig. 49.9).

Carbon monoxide is an air pollutant that comes primarily from vehicle exhaust and is also found in photochemical smog. Carbon monoxide is of concern because it combines preferentially with hemoglobin and thereby prevents hemoglobin from carrying oxygen. In fact, breathing large quantities of automobile exhaust results in death because of carbon monoxide poisoning.

a.

b.

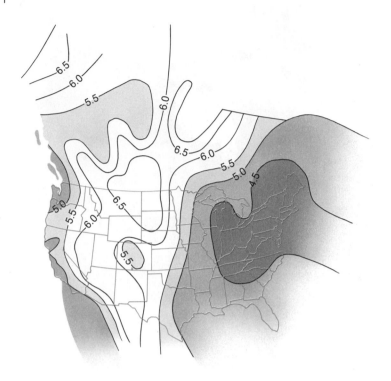

Normally, warm air near the ground is able to escape into the atmosphere. Sometimes, however, air pollutants, including smog and soot, are trapped near the earth due to a long-lasting thermal inversion. During a **thermal inversion,** there is the cold air at ground level beneath a layer of warm stagnant air above. This often occurs at sunset, but turbulence usually mixes these layers during the day. Some areas surrounded by hills are particularly susceptible to the effects of a temperature inversion because the air tends to stagnate and there is little turbulent mixing.

Acid Deposition

The coal and oil burned by power plants releases sulfur dioxide, and automobile exhaust contains nitrogen oxides; both of these are converted to acids when they combine with water vapor in the atmosphere, a reaction that is promoted by ozone in smog. These acids return to earth as either wet deposition (acid rain or snow) or dry deposition (sulfate and nitrate salts).

As discussed earlier in this text in the reading on page 34, **acid deposition** is now associated with dead or dying lakes and forests particularly in North America (fig. 49.10) and Europe. Acid deposition also corrodes marble, metal, and stonework, an effect that is noticeable in cities. It can also degrade our water supply by leaching heavy metals from the soil into drinking-water supplies. Similarly, acid water dissolves copper from pipes and from lead solder that is used to join pipes.

Ozone Shield Destruction

Chlorofluorocarbons, or CFCs, of which the most important is Freon, are heat transfer agents used in refrigerators and air conditioners. They are also used as foaming agents in such products as styrofoam cups and egg cartons. Formerly, they were used as propellants in spray cans, but this application is now banned in the United States.

It was known that CFCs would drift up into the stratosphere, but it was believed that they would be nonreactive there during their 150-year life span. However, it is now apparent that when the temperature drops, these compounds react chemically

Environmental Concerns **789**

Figure 49.11

Ozone depletion has been confirmed by observations of a drastic decrease in ozone over Antarctica each spring during the late 1970s and 1980s. The so-called ozone hole appears as a large white area at the center of this picture taken by the total ozone mapping spectrometer aboard NASA's Nimbus 7 satellite on 5 October 1987, when the ozone loss reached nearly 60%.

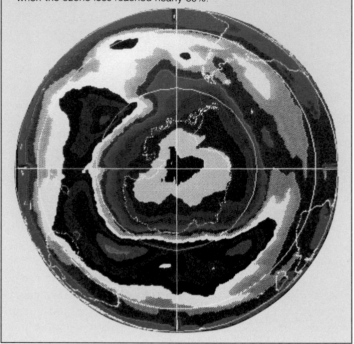

on frozen particle surfaces within stratospheric clouds and release active chlorine that can then react with the ozone in the **ozone shield.** Once freed, a single atom of chlorine destroys about 100,000 molecules of ozone before finally settling to the earth's surface as chloride years later. Measurements suggest that 3% and perhaps up to 5% of the global ozone layer has already been destroyed by CFCs. An even more drastic effect has been found in the Antarctic and Arctic. At the South Pole, up to 50% of the ozone is destroyed each spring over an area the size of North America (fig. 49.11). There is a similar reduction, called an "ozone hole," but not as severe, over the Arctic as well.

Depletion of the ozone layer will allow more ultraviolet rays to enter the troposphere. The incidence of human cancer, especially skin cancer, can be expected to increase, and plants and animals living in the top microlayer of the ocean will begin to die. Increased ultraviolet radiation will also hasten the rate at which smog is formed.

Greenhouse Effect

It is predicted that certain air pollutants will cause the temperature of the earth to rise, perhaps by as much as 0.8° C/decade (1.36° F/decade). These gases let the sun's rays pass through but absorb and re-radiate their heat back toward the earth (fig. 49.12a). This is called the **greenhouse effect** because the glass of a greenhouse allows sunlight to pass through, then traps the resulting heat inside the structure. The pollutants that cause global warming include:

- carbon dioxide (from fossil fuel and wood burning)
- nitrous oxide (NO_2 primarily from fertilizer use and animal wastes)

Figure 49.12

Greenhouse effect. *a.* The greenhouse effect is caused by the accumulation of certain gases such as CO_2 in the atmosphere, which allow the rays of the sun to pass through but absorb and reradiate heat back to the earth. *b.* The greenhouse gases. This graph shows the fraction of warming caused by the gases CO_2, CFCs, methane, and nitrous oxide for the decades from 1950 to 2020. There was no accumulation of CFCs in the 1950s because they were not being manufactured to any degree as yet. By 2020, the CO_2 and all of the other gases taken together will each contribute about 50% to the projected global warming.

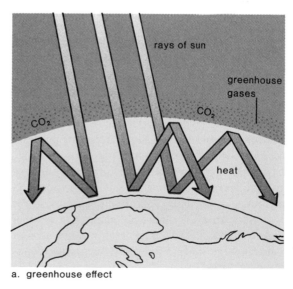

a. greenhouse effect

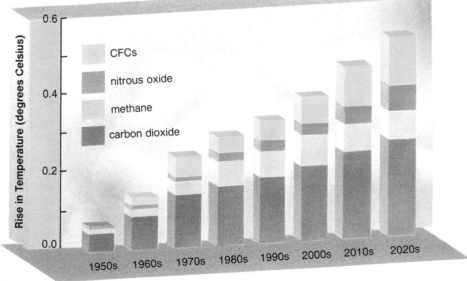

b.

- methane (biogas, bacterial decomposition particularly in the guts of animals, sediments, and flooded rice paddies)
- CFCs (the Freons mentioned earlier)

It so happens that nitrous oxide, methane, and CFCs cause an even more severe greenhouse effect than carbon dioxide, and their combined effect is expected to equal the effect from carbon dioxide by the year 2020 (fig. 49.12b). Just now carbon dioxide levels are 25% higher than they were in 1860. Most of this excess is due to industrialization but tropical **deforestation** is now a major contributor as well. Burning one acre of primary forest puts 200,000 kg of carbon dioxide into the air; moreover, the trees are no longer available to act as a sink to take up carbon dioxide during photosynthesis.

It would appear that global warming has already begun: the four hottest years on record have occurred during the 1980s; there has been greater warming in the winters than summers; there has been greater warming at high altitudes than near the equator; the stratosphere is cooler and the lower atmosphere is warmer than formerly. All of these effects have been predicted by computer models of the greenhouse effect.

The ecological effects of global warming are expected to be severe. First of all, the sea level will rise—melting of the polar ice caps will add more water to the sea, and water expands when it heats, as well. This will cause flooding of many coastal regions and the possible loss of many cities, like New York, Boston, Miami, and Galveston in the United States Coastal ecosystems such as marshes, swamps, and bayous would normally move inland to higher ground as the sea level rises. But many of these ecosystems are blocked in by artificial structures and may be unable to move inland. If so, the loss of fertility will be immense.

There will also be food loss because of regional climate changes. Not only will there be greater heat, there will be drought in the midwestern United States, and the suitable climate for growing wheat and corn will shift to Canada, where the soil is not as suitable.

Drastic measures are recommended to hold the quantity of greenhouse gases to their present level. A 50% decrease in consumption of fossil fuels is recommended and we must find more efficient ways to acquire energy from cleaner fuels, such as natural gas. We should more aggressively use alternative energy sources such as solar and geothermal energy and even perhaps nuclear power. The section below explores all the various sources of energy available to us.

Not only should tropical rain forest deforestation be halted, extensive reforesting all over the globe should take place. These forests could be used as a source of wood as fuel, and at the same time the carbon dioxide given off by the burning of the fuel would be absorbed by the new trees coming along.

The manufacturing and use of CFCs should be completely eliminated. The United States and European countries have already agreed to reduce CFC production by 85% as soon as possible and to try to ban the chemicals altogether by the end of the century.

Outdoor air pollutants are involved in causing four major environmental effects: petrochemical smog, acid deposition, ozone shield destruction, and the greenhouse effect. Each pollutant may be involved in more than one of these.

While each of these environmental effects is bad enough when considered separately, they actually feed on one another, making the total effect much worse than is predicted for the total of each separately.

Indoor Air Pollution

While much attention has been given to the quality of outdoor air, it has been pointed out that indoor air is often more polluted. NO_2 traced to gas combustion in stoves has been found indoors at twice the outdoor level; CO in offices, repair garages, and hockey rinks is routinely in excess of the governmental standard and hydrocarbons from myriad sources appear in high concentrations. Radioactive radon gas, emitted naturally from radioactive decay in uranium-bearing materials in the earth, as well as those incorporated into building materials such as concrete blocks, has been detected indoors at levels that exceed those outdoors by factors of 2 to 20.

Because of these findings, some have questioned the advisability of tightening up residential buildings to save energy. Reduced ventilation unquestionably increases concentrations of pollutants. The EPA is now recommending that ventilation rates not be reduced below one complete air change per hour.

Energy Choices

The threat of global warming due to build up of gases that bring about the greenhouse effect has renewed the desire to use sources of energy that will not add carbon dioxide to the atmosphere. The energy sources currently available or being investigated are listed in table 49.2. **Nonrenewable resources** are those with a supply that can be used up or exhausted. **Renewable resources** are those that can be replenished by either physical or biological means.

Nonrenewable Energy Sources

Although it is difficult to realize it, supplies of the cleaner-burning petroleum and natural gas are not expected to last more than another thirty years. Petroleum became the favored fuel in the United States during the present generation (fig. 49.13) not only because it is cleaner burning than coal, but also because it is easily transportable and can serve as the raw material for gasoline and many organic compounds.

Although coal has been projected to be the primary energy source in the United States in the future, this is now doubtful because it contributes more than other fossil fuels to sulfur dioxide pollution. Also, in order to supply coal, thousands of acres of land are strip mined and this results in huge piles of residue having the potential to leach acids and hazardous chemicals into underground aquifers.

Table 49.2 Renewable and Nonrenewable Energy Sources

	Advantages	Disadvantages
Nonrenewable	*Technology Well Established*	*Finite Fuel Supply*
Fossil fuels		
Coal	Plentiful supply	Surface mining Air and water pollution
Petroleum	Cleaner burning	Limited supply
Natural gas	Cleanest burning	Limited supply
Nuclear		
Fission	Fuel availability	Radiation pollution
Renewable	*Fuel Supply*	*Technology under Development*
Nuclear		
Breeder	Fuel availability	Radiation pollution Nuclear weapons proliferation
Fusion	Fuel availability	Radiation pollution
Geothermal	Less pollution	Availability limited
Solar and wind	Nonpolluting Large and small scale possible	Expensive
Ocean	Nonpolluting	Applicable only in certain areas
Biomass	Utilizes wastes	Air and water pollution

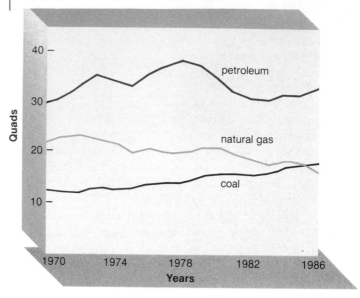

Figure 49.13

Present energy sources in the United States. Oil became the favored fuel during the 1900s because it is cleaner burning than coal and because it can be converted to gasoline. In recent years, use of coal has increased because it is more plentiful than oil and natural gas in the U.S. but it may fall into disfavor once again because CO_2 from coal contributes most heavily to air pollution acid deposition. There is also much concern about the possibility of global warming because of the CO_2 produced by burning coal.

While all fossil fuels release carbon dioxide in addition to other pollutants, **nuclear power** does not contribute to air pollution. Thus far, only fission power plants, which split ^{235}uranium, have been utilized in the United States. In recent years the nuclear power industry has declined in favor because coal-fired power plants are cheaper to build and the public is very much concerned about nuclear power dangers, such as meltdown that occurred in 1986 at the Chernobyl nuclear power plant in Russia. At a proposed plant to be built in Idaho, designers plan to use a series of four small-scale, modular reactors that use fuel in such small quantities their cores could not achieve meltdown temperatures. The fuel would be packed inside tiny heat-resistant ceramic spheres and cooled by inert helium gas.

Even if the possibility of a meltdown can be eliminated, there is still the need to store radioactive wastes that have accumulated since the start of the nuclear age. Another source of radioactive waste, not often mentioned, occurs when uranium is mined. Uranium mill tailings are the sand that remains after uranium has been removed from mined material. The tailings contain ^{226}radium, which has a half-life[1] of 1,620 years. As ^{226}radium decays, it gives off radon gas, which has been associated with the development of lung cancer.

Breeder fission reactors do not have the same environmental problems as conventional fission plants. They need less raw uranium, and they have little waste because they use ^{239}plutonium, a fuel that is actually generated from what would

[1]Length of time required for one-half of the radiation to dissipate.

be reactor waste. Plutonium, however, is a very toxic element that readily causes lung cancer; it is also the element used to make nuclear weapons. (The chances of a nuclear explosion are much greater with breeder than with conventional reactors.) The fear of nuclear weapons proliferation, in particular, has thus far prevented the start-up of a breeder reactor in this country.

Nuclear fusion requires that two atoms, usually deuterium and tritium, be fused, and the energy needed for this is so great that there is no conventional container for the reaction. Most scientists have been perfecting laser-beam ignition and magnetic containment for the reaction, which thus far uses up more energy than it produces. Since the fusion reaction gives off neutrons that can change uranium to plutonium, the best use of fusion plants may be to provide fuel for breeder reactors. One idea is to have hybrid fusion plants that would combine both fusion and breeder fission plants. But, in any case, the fusion process is still experimental and is not expected to be ready for production any time soon.

Renewable Energy Sources

Solar energy is diffuse energy that must be collected and concentrated before utilization is possible. Solar heating has been used primarily in homes and offices. Passive solar systems consist of specially constructed glass for the south wall of a building, and building materials that can collect the sun's energy during the day and release it at night. Solar collectors placed on rooftops absorb radiant energy but do not release the resulting heat. A fluid within the solar collector heats up and is pumped to other parts of the building for space heating, cooling, or the generation of electricity.

Another possibility is the use of photovoltaic (solar) cells that produce electricity directly from sunlight. It has even been suggested that cells might be placed in orbit around the earth where they would collect intense solar energy, generate electricity, and send it back to earth via microwaves.

Solar energy is clean energy. It does not produce air or water pollution, nor does it add additional heat to the atmosphere, since radiant energy is eventually converted to heat anyway. The problem of storage can be overcome in a number of ways, including the use of solar energy to produce hydrogen by means of hydrolysis of water. Hydrogen as either a gas or a liquid can be piped into existing pipelines and used as fuel for automobiles and airplanes. When it is burned, it forms fog, not smog:

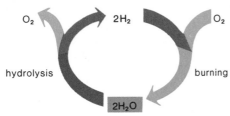

Certain types of renewable energy sources have been utilized for quite some time. Falling water is used to produce electricity in hydroelectric plants. Geothermal energy is derived from the heat of the magma in the earth's core. Water, converted to steam by this heat, may be pumped up and used to heat buildings or to generate electricity. Wind power provides enough force to turn vanes, blades, or propellers attached to shafts which, in turn, spin generator motors that produce electricity. The government has allocated a small amount of money for wind research, particularly the promotion of large windmills, such as those now in operation in central and southern California. Many others believe it would be best to build numerous small windmills, such as those found in the first wind farm now operating in New Hampshire.

The threat of global warming makes it imperative to use a source of energy that does not pollute the atmosphere. Nuclear power and several of the renewable sources such as solar and wind power are possibilities.

Presently, the world's population increases in size each year and as a result an ever greater supply of energy and food is needed each year. Does this pattern have to continue in order to ensure economic growth and our well-being? On the contrary, it seems unlikely that this exponential increase can long continue. The pollution caused by energy consumption and the production of goods has increased so much that it can no longer be ignored, and the once-hidden cost of pollution has now become apparent. The human population, like other populations of the biosphere, could exist in a steady state, with no increase in number of people or in resource consumption.

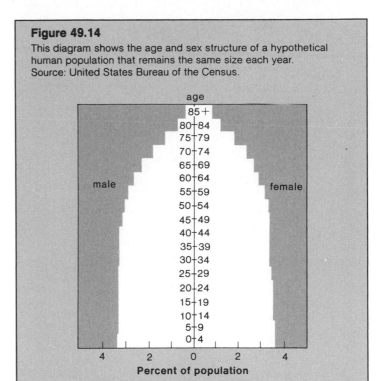

Figure 49.14
This diagram shows the age and sex structure of a hypothetical human population that remains the same size each year.
Source: United States Bureau of the Census.

A stable population size would be a new experience for humans. Figure 49.14 shows the age structure for a hypothetical stable population. As you can see,

1. Over 40% of the people would be fairly youthful, with only about 15% being in the senior citizen category. Further, a combination of good health habits and advances in medical science could well mean that people would remain more youthful and more productive for a longer time than is common today.
2. There would be proportionately fewer children and teenagers than in a rapidly expanding population. This might mean a reduction in automobile accidents and certain types of crime statistically related to the teenage years.
3. There might be increased employment opportunities for women and a generally less competitive workplace, since newly qualified workers would enter the job market at a more moderate rate.
4. Creativity need not be impaired. A study of Nobel Prize winners showed that the average age at which the prizewinning work was done was over 30.
5. The quality of life for children might increase substantially, since fewer unwanted babies might be born, and the opportunity would exist for each child to receive the loving attention it needs.

The economy, like the population, should show no growth. For example, if the population size were to increase, the resource consumption per person should decline. In this way the amount of resource consumption would remain constant. In order

Figure 49.15
In the steady state, environmental preservation will be an important consideration.

to ensure that people have the goods and services they need for a comfortable but not necessarily luxurious life, goods should have a long lifetime expectancy. Frugality is envisioned as absolutely necessary to the steady-state economy. Also, materials must be recycled so that they can be used over and over again. In order to provide jobs, technology could sometimes be labor intensive instead of energy intensive. The steady state would have need of very sophisticated technology, but it is hoped that the technology will not exploit the environment; instead, it is hoped that the technology will work with the environment.

Environmental preservation would be the most important consideration in a steady-state world. Renewable energy sources, such as solar energy, would play a greater role in providing energy needs. Pollution would be minimized. Ecological diversity would be maintained and overexploitation would cease (fig. 49.15). Ecological principles would serve as guidelines for specialists in all fields, creating a unified approach to the environment. In a steady-state world all people would strive to be aware of the environmental consequences of their actions, consciously working toward achieving balance in the ecosystems of their planet.

In a **steady-state society,** there would be no yearly increase in population nor resource consumption. It is forecast that under these circumstances the quality of life would improve.

What would our culture be like if we had steady-state manufacturing and a steady-state population? Perhaps it would be greatly improved. Certainly there are no limits to growth in knowledge, education, art, music, scientific research, human rights, justice, and cooperative human interactions. In a steady-state world the general sense of fearful competition among peoples might diminish, allowing human compassion and creativity to prosper as never before.

Summary

1. An increased human population is causing land, water, and air pollution.
2. Most people of the world now live in cities that are increasing in size and taking over land formerly used for agriculture and other purposes.
3. Soil erosion reduces the quality of land and leads to desertification.
4. The tropical rain forests in Southeast Asia and Oceania, Central and South America, and Africa are being reduced in size for several reasons: (1) to provide wood for export, (2) slash-and-burn agriculture, and (3) cattle ranching. Industrialization may soon be a fourth reason.
5. The loss of biological diversity due to the destruction of tropical rain forests will be immense. Many of these organisms could possibly be of benefit to humans if we had time to study and domesticate them.
6. Solid wastes, including hazardous wastes, are deposited on land. The latter, including metals, organochlorides, and nuclear wastes, may contaminate water supplies.
7. Sewage-treatment plants were designed to degrade sewage and not handle hazardous wastes. Primary treatment plants merely remove solids; secondary treatment plants break down solids but the nutrients can still cause cultural eutrophication. Only tertiary treatment removes the nutrients, but at high cost; however, it is possible to find alternative means of treatment such as using the nutrient water to grow plants.
8. Heavy metals and organochlorides are those hazardous wastes that are apt to pollute both surface and underground water.
9. The ocean is the final recipient for wastes deposited in rivers and along the coasts. Its capacity is not infinite, however, and oceanic pollution is a real threat.
10. Two layers of the atmosphere are of interest: (1) the troposphere, where ozone is a pollutant; and (2) the stratosphere, where ozone forms a shield against ultraviolet radiation.
11. Various substances are associated with air pollution, such as sulfur dioxide, hydrocarbons, nitrogen oxides, methane, CFCs, carbon dioxide, and carbon monoxide. These have various sources but most come from vehicle exhaust and fossil fuel burning.
12. Hydrocarbons and nitrogen oxide (NO_x) react to form smog, which contains ozone and PAN. These oxidants are harmful to animal and plant life.

13. Sulfur dioxide (SO$_2$) and nitrogen oxide react with water vapor to form acids that contribute to acid deposition. Acid deposition is killing lakes and forests and also corrodes marble, metal, and stonework.

14. Ozone destruction is particularly associated with CFCs, which are used as refrigerator coolants and foaming agents. Formerly they were also used in spray cans but this use is now banned. CFCs rise into the stratosphere and react on frozen particles within clouds to release chlorine. Chlorine causes ozone to break down. A 50% reduction in ozone was found above the Antarctic and dubbed an "ozone hole." Since ozone prevents harmful ultraviolet radiation from reaching the surface of the earth, reduction in the amount of ozone will lead to skin cancer in humans and decreased productivity of the oceans.

15. Several major gases of natural or human origin (carbon dioxide, nitrous oxide (N$_2$O), methane, and CFCs) trap solar heat just like the panes of a greenhouse, leading to the so-called greenhouse effect. It is predicted that a buildup in these "greenhouse gases" will lead to a global warming. The effects of global warming will be a rise in sea level and a change in climate patterns. Food shortages will follow.

16. To control global warming it is recommended that we reduce our consumption of fossil fuel energy and halt tropical rain forest destruction. Also, we should increase the amount of land devoted to forests. A ban on the manufacture of CFCs has already begun.

17. A review of the energy sources available indicates that the consumption of all fossil fuels leads to a release of carbon dioxide, but coal, the most abundant, also contributes SO$_2$ to the atmosphere. Cleaner burning petroleum and natural gas are not expected to be available many more years. It has been suggested that nuclear power and renewable sources such as solar and wind power are the best possibilities to prevent global warming.

Objective Questions

For questions 1 to 4, match the terms below with those in the key:

Key:

    a. sulfur dioxide
    b. ozone
    c. carbon dioxide
    d. CFCs

1. acid deposition
2. ozone shield destruction
3. greenhouse effect
4. photochemical smog
5. Which of these is a true statement?
   a. Developed countries do not contribute to the destruction of tropical rain forests.
   b. Carbon dioxide from fossil fuel combustion is the primary greenhouse gas.
   c. Water from underground sources is never subject to contamination.
   d. Sulfur dioxide is given off to the same degree by the burning of all fossil fuels.

6. Which of these is a true statement?
   a. The greenhouse effect is of no immediate concern.
   b. The greenhouse effect is so imminent that nothing can be done.
   c. Reduction in fossil fuel burning will lessen the greenhouse effect.
   d. Since gases not derived from fossil fuel combustion are involved, reduction in fossil fuel burning will not help the greenhouse effect.
7. Tropical rain forest destruction is extremely serious because
   a. it will lead to a severe reduction in biological diversity.
   b. tropical soils cannot support agriculture for long.
   c. large tracts of forest absorb carbon dioxide, reducing the threat of global warming.
   d. All of these.

8. Which of these is mismatched?
   a. fossil fuel burning—carbon dioxide given off
   b. nuclear power—radioactive wastes
   c. solar energy—greenhouse effect
   d. biomass burning—carbon dioxide given off
9. Acid deposition causes
   a. lakes and forests to die.
   b. acid indigestion in humans.
   c. the greenhouse effect to lessen.
   d. All of these.
10. Water is a renewable resource and
   a. there will always be a plentiful supply.
   b. the oceans can never become polluted.
   c. it is still subject to pollution.
   d. primary sewage treatment plants assure clean drinking water.

Study Questions

1. Give three reasons why the quality of the land is being degraded today. What is desertification?
2. Give three reasons why tropical rain forests are being destroyed. What is another possible reason in the future?
3. What are the primary ecological concerns associated with the destruction of rain forests?
4. What are the three types of hazardous wastes that contribute to pollution on land?

5. What is the difference between a primary, secondary, and tertiary sewage treatment plant?
6. What are several ways in which underground water supplies can be polluted?
7. What substances contribute to air pollution? What are their sources? Which ones are associated with photochemical smog, acid deposition, destruction of the ozone shield, or the greenhouse effect?

8. How does smog develop? How do acids develop in the atmosphere? How do CFCs break down the ozone shield? How do the greenhouse gases bring about global warming?
9. What are the environmental effects of global warming, and what can be done to prevent global warming?
10. What are the various energy sources available to humans? What are the drawbacks of each one? Which of these would not contribute to the greenhouse effect?

Thought Questions

1. It is conceivable that biotechnology might be able to help with which of the environmental problems discussed in this chapter?

2. Give reasons why the tropical rain forests should be saved. What could the citizens of the United States and the government do to help save the tropical rain forests?

3. In order for you to have the use of a product, the raw materials must be made available, transported to the factory where the product is made, transported to the store where the product is sold, and transported to your home where the product is used until disposed of. What types of pollution may be associated with your desire to have and use this product?

Selected Key Terms

pollutants (pŏ-lūt′ants) 781
desertification (dez″ert-i-fi-ka′shun) 781
slash-and-burn agriculture (slash and burn ag′ri-kul″tūr) 783
photochemical smog (fo″to-kem′i-kal smog) 788

thermal inversion (ther′mal in-ver′zhun) 789
acid deposition (as′id dep″o-zish′un) 789
ozone shield (o′zōn shēld) 790
greenhouse effect (grēn′hows ĕ-fekt) 790
deforestation (de″for-es-ta′shun) 791
nonrenewable resources (non″re-nu′ah-b′l re-sors′ez) 791

renewable resources (re-nu′ah-b′l re-sors′ez) 791
nuclear power (nu′kle-ar pow′er) 792
solar energy (so′lar en′er-je) 792
steady-state society (sted′e stāt so-si′e-te) 794

Suggested Readings for Part 6

Alcock, J. 1984. *Animal behavior: An evolutionary approach.* 3rd ed. Sunderland, MA: Sinauer, Associates.

Batie, S. S., and Healy, R. G. February 1983. The future of American agriculture. *Scientific American.*

Begon, J., Harper, J., and Townsend, C. 1986. *Ecology: individuals, populations, and communities.* Sunderland, MA: Sinauer, Associates.

Berner, R. A., and Lasaga, A. C. March 1989. Modeling the geochemical carbon cycle. *Scientific American.*

Blaustein, A. R., and O'Hara, R. K. January 1986. Kin recognition in tadpoles. *Scientific American.*

Brown, L. R. et al. 1986. *State of the world: 1990.* New York: W. W. Norton and Co.

Bushbacher, R. J. January 1986. Tropical deforestation and pasture development. *BioScience.*

Ghiglieri, M. P. June 1985. The social ecology of chimpanzees. *Scientific American.*

Hamakawa, Y. April 1987. Photovoltaic power. *Scientific American.*

Horn, M. H., and Gibson, R. N. January 1988. Intertidal fishes. *Scientific American.*

Houghton, R. A., and Woodwell, G. M. April 1989. Global climatic change. *Scientific American.*

Miller, J. T. 1988. *Living in the environment.* 5th ed. Belmont, CA: Wadsworth.

Mohnen, V. A. August 1988. The challenge of acid rain. *Scientific American.*

Nebel, B. J. 1987. *Environmental science: The way the world works.* 2d ed. Englewood Cliffs, NJ: Prentice Hall.

Odum, H. T. 1983. *Systems ecology: An introduction.* New York: Wiley.

O'Leary, P. R., Walsh, P. W., and Ham, R. H. December 1988. Managing solid waste. *Scientific American.*

Power, J. F., and Follett, R. F. March 1987. Monoculture. *Scientific American.*

Ricklefs, R. E. 1986. *Ecology.* 3rd ed. New York: Chiron Press.

Schneider, S. H. May 1987. Climate modeling. *Scientific American.*

Shaw, R. W. August 1987. Air pollution by particles. *Scientific American.*

Smith, R. L. 1985. *Ecology and field biology.* 3rd ed. New York: Harper & Row.

Sumich, J. 1988. *Biology of marine life.* 4th ed. Dubuque, IA: Wm. C. Brown Publishers.

Wursig, B. April 1988. The behavior of baleen whales. *Scientific American.*

CRITICAL THINKING

CASE STUDY

PART 6
Behavior and Ecology

igure 48.1 suggests that all ecosystems have these elements:

- an outside energy source
- producers
- consumers
- decomposers

This case studies the hypothesis that deep-sea ecosystems, recently found 2,600 meters below the surface of the ocean, also have these elements. Packed around occasional thermal vents (deep-sea hot springs) there is a great diversity of organisms including giant tube worms (*see* fig. 23.8), clams, mussels, shrimp, and crabs.

What type of organisms are usually the producers in ecosystems? Would you expect to find any photosynthetic organisms 2,600 meters below the surface of the ocean? Why not?

Investigation revealed that thermal vents release large amounts of hydrogen sulfide (H_2S). It's well known that some chemosynthetic bacteria have the capability of oxidizing hydrogen

sulfide and using the released energy to form ATP molecules. These ATP molecules are then used to drive the Calvin-Benson cycle so that carbohydrates are produced.

Since the vents are a rich source of H_2S, would you predict that bacteria would be found in the vacinity that can use this molecule as an energy source?

Prediction 1 Bacteria should be found that can use H_2S as an energy source.

Result 1 Bacteria that can utilize H_2S as an energy source have been found in the water near the vents.

What does the term producer mean? Is it possible that these sulfur-oxidizing bacteria are the producers in deep-sea ecosystems? Is H_2S equivalent to the sun in deep-sea ecosystems? Why or why not?

If deep-sea ecosystems have the elements of other known ecosystems would you predict that consumers will be present in deep-sea ecosystems?

Prediction 2 There should be consumers that feed off the bacteria in deep-sea ecosystems.

Result 2 Contrary to the prediction, few of the organisms appeared to utilize the sulfur-oxidizing bacteria as a major food source.

Even the clams, which are filter feeders and normally feed on bacteria, did not appear to feed on sulfur-oxidizing bacteria. Also, it was found that the large, 2-meter-long tube worms don't have a mouth or any other sign of a digestive system!

Microscopic examination of the tube worm body, however, indicated that the internal cells contain enormous numbers of a type of sulfur-oxidizing bacteria. A similar examination of the clams revealed that another type of sulfur-oxidizing bacteria had colonized the gills of these organisms. This suggested that there was a mutualistic relationship between the bacteria and their hosts.

In a mutualistic relationship, each organism contributes to the livelihood of the other. What could the bacteria be contributing to their hosts? Design an experiment that would test the hypothesis that tube worms and clams are dependent on sulfur-oxidizing bacteria to supply their nutritional needs.

What could the hosts be contributing to the bacteria? Since the bacteria are deep inside the tissues of the worms, what system in the body of the tube worms would be transporting H_2S to the bacteria?

What do you predict about the circulatory system of the tube worms?

Prediction 3 The circulatory systems of the tube worms must be able to transport H_2S to the bacteria.

Result 3 It was discovered that the hemoglobin of tube worms, which is in plasma and not in red blood cells, is a very special type—it contains a binding site for H_2S. In the clam, where hemoglobin is in red blood cells, there is a special transport protein that transports H_2S to the bacteria.

What does the term consumer mean? Should the tube worm and clams, which harbor sulfur-oxidizing bacteria, be considered consumers? Why or why not? Can we assume that decomposers are present at thermal vents? Does it seem, then, that our title—deep-sea ecosystem— is suitable?

The hemoglobin must not only be able to transport H_2S, it must also carry oxygen in such a way that H_2S is not oxidized before it reaches the bacteria. Also, H_2S is a powerful inhibitor of cytochrome oxidase, equivalent to that of cyanide; therefore, hemoglobin must keep H_2S from diffusing into mitochondria where it would poison cytochrome oxidase.

Considering the fact that the worms flourish in their H_2S environment, would you predict that the hemoglobin must be able to protect the H_2S from oxidation and keep it from diffusing into mitochondria?

Prediction 4 The hemoglobin must be able to protect the H_2S from oxidation and keep it from diffusing into mitochondria.

Result 4 Experimentation showed that the worm's hemoglobin can bind both H_2S and oxygen simultaneously and it appears that the presence of these two different binding sites protects the H_2S from oxidation. Also, the H_2S binding site has a higher affinity for H_2S than does cytochrome oxidase. Extracted cytochrome oxidase is inhibited by H_2S, but the addition of a small amount of tube worm hemoglobin quickly restores its activity.

How would the hemoglobin's "higher affinity for H_2S" protect cytochrome oxidase from Inhibition by H_2S? What do you predict about the bacteria's affinity for H_2S? Rank order from highest to lowest the expected affinity of hemoglobin, cytochrome oxidase, and bacteria for H_2S in the body of the worm.

These observations and experiments show how H_2S provides usable energy for deep-sea ecosystems. In spite of a markedly different environment, the ecosystem is characterized by energetic relationships and a biotic structure similar to those found in other ecosystems.

Childress, J. J., Felbeck, H., and Somero, G. N. May 1987. Symbiosis in the deep sea. *Scientific American*.

Jannasch, H. W., and Mottl, M. J. August 1985. Geomicrobiology of deep-sea hydrothermal vents. *Science*.

Classification of Organisms

The classification system given here is a simplified one, containing all the major kingdoms, as well as the major divisions (called phyla in the kingdom Prostista and the kingdom Animalia). The text does not discuss all the divisions and phyla listed here.

Kingdom Monera
Prokaryotic, unicellular organisms. Nutrition principally by absorption, but some are photosynthetic or chemosynthetic.
   Division Archaebacteria: methanogens, halophiles, and
      thermoacidophiles
   Division Eubacteria: all other bacteria, including cyanobacteria
      (formerly called blue-green algae)

Kingdom Protista
Eukaryotic, unicellular organisms (and the most closely related multicellular forms). Nutrition by photosynthesis, absorption, or ingestion.
   Phylum Sarcodina: amoeboid protozoans
   Phylum Ciliophora: ciliated protozoans
   Phylum Zoomastigina: flagellated protozoans
   Phylum Sporozoa: parasitic protozoans
   Phylum Chlorophyta: green algae
   Phylum Pyrrophyta: dinoflagellates
   Phylum Euglenophyta: *Euglena* and relatives
   Phylum Chrysophyta: diatoms
   Phylum Rhodophyta: red algae
   Phylum Phaeophyta: brown algae
   Phylum Myxomycota: slime molds
   Phylum Oomycota: water molds

Kingdom Fungi
Eukaryotic organisms, usually having haploid or multinucleated hyphal filaments. Spore formation during both asexual and sexual reproduction. Nutrition principally by absorption.
   Division Zygomycota: black bread molds
   Division Ascomycota: sac fungi
   Division Basidiomycota: club fungi
   Division Deuteromycota: imperfect fungi, i.e., means of sexual
      reproduction not known.

Kingdom Plantae
Eukaryotic, terrestrial, multicellular organisms with rigid cellulose cell walls and chlorophyll a and b. Nutrition principally by photosynthesis. Starch is the food reserve.
   Division Bryophyta: mosses and liverworts
   Division Psilophyta: whisk ferns
   Division Lycophyta: club mosses
   Division Sphenophyta: horsetails
   Division Pterophyta: ferns
   Division Cycadophyta: cycads
   Division Ginkgophyta: ginkgoes
   Division Gnetophyta: gnetae
   Division Coniferophyta: conifers
   Division Anthophyta: flowering plants
      Class Dicotyledonae: dicots
      Class Monocotyledonae: monocots

Kindgom Animalia
Eukaryotic, usually motile, multicellular organisms without cell walls or chlorophyll. Nutrition principally ingestive, with digestion in an internal cavity.
   Phylum Porifera: sponges
   Phylum Cnidaria: radially symmetrical marine animals
      Class Hydrozoa: *Hydra,* Portuguese man-of-war
      Class Scyphozoa: jellyfish
      Class Anthozoa: sea anemones and corals
   Phylum Platyhelminthes: flatworms
      Class Turbellaria: free-living flatworms
      Class Trematoda: parasitic flukes
      Class Cestoda: parasitic tapeworms
   Phylum Nematoda: roundworms
   Phylum Rotifera: rotifers
   Phylum Mollusca: soft-bodied, unsegmented animals
      Class Polyplacophora: chitons
      Class Monoplacophora: *Neopilina*
      Class Gastropoda: snails and slugs
      Class Bivalvia: clams and mussels
      Class Cephalopoda: squids and octopuses
   Phylum Annelida: segmented worms
      Class Polychaeta: sandworms
      Class Oligochaeta: earthworms
      Class Hirudinea: leeches

Phylum Arthropoda: joint-legged animals; exoskeleton
    Class Crustacea: lobsters, crabs, barnacles
    Class Arachnida: spiders, scorpions, ticks
    Class Chilopoda: centipedes
    Class Diplopoda: millipedes
    Class Insecta: grasshoppers, termites, beetles
Phylum Echinodermata: marine; spiny, radially symmetrical animals
    Class Crinoidea: sea lilies and feather stars
    Class Asteroidea: sea stars
    Class Ophiuroidea: brittle stars
    Class Echinoidea: sea urchins and sand dollars
    Class Holothuroidea: sea cucumbers
Phylum Chordata: dorsal supporting rod (notochord) at some stage; dorsal hollow nerve cord; pharyngeal gill pouches or slits
    Subphylum Urochordata: tunicates
    Subphylum Cephalochordata: lancelets
    Subphylum Vertebrata: vertebrates
        Class Agnatha: jawless fishes (lampreys, hagfishes)
        Class Chondrichthyes: cartilaginous fishes (sharks, rays)
        Class Osteichthyes: bony fishes
            Subclass Sarcopterygii: lobe-finned fishes
            Subclass Actinopterygii: ray-finned fishes
        Class Amphibia: frogs, toads, salamanders
        Class Reptilia: snakes, lizards, turtles
        Class Aves: birds
        Class Mammalia: mammals
            Subclass Prototheria: egg-laying mammals
                Order Monotremata: duckbilled platypus, spiny anteater

            Subclass Metatheria: marsupial mammals
                Order Marsupialia: opossums, kangaroos
            Subclass Eutheria: placental mammals
                Order Insectivora: shrews, moles
                Order Chiroptera: bats
                Order Edentata: anteaters, armadillos
                Order Rodentia: rats, mice, squirrels
                Order Lagomorpha: rabbits and hares
                Order Cetacea: whales, dolphins, porpoises
                Order Carnivora: dogs, bears, weasels, cats, skunks
                Order Proboscidea: elephants
                Order Sirenia: manatees
                Order Perissodactyla: horse, hippopotamus, zebra
                Order Artiodactyla: pigs, deer, cattle
                Order Primates: lemurs, monkeys, apes, humans
                    Suborder Prosimii: lemurs, tree shrews, tarsiers, lorises, pottos
                    Suborder Anthropoidea: monkeys, apes, humans
                        Superfamily Ceboidea: New world monkeys
                        Superfamily Cercopithecoidea: Old world monkeys
                        Superfamily Hominoidea: apes and humans
                            Family Hylobatidae: gibbons
                            Family Pongidae: chimpanzee, gorilla, orangutan
                            Family Hominidae: *Australopithecus,** *Homo erectus,** *Homo sapiens* neapiens
                        *extinct

A P P E N D I X   B

Table of Chemical Elements

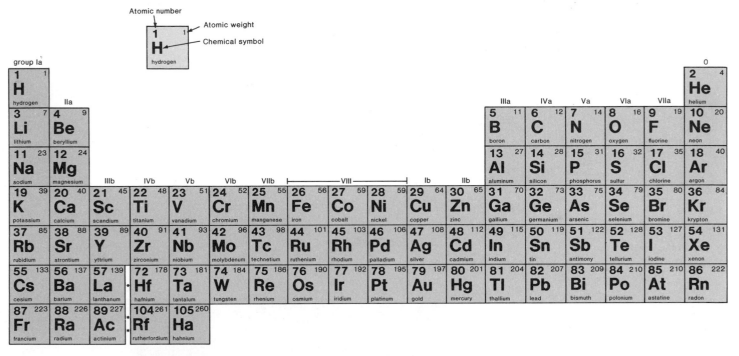

A P P E N D I X   C

Metric System

The Metric System		
Standard Metric Units		**Abbreviations**
Standard unit of mass	gram	g
Standard unit of length	meter	m
Standard unit of volume	liter	l
Common Prefixes		**Examples**
kilo	1,000	a kilogram is 1,000 grams
centi	0.01	a centimeter is 0.01 meter
milli	0.001	a milliliter is 0.001 liter
micro (μ)	one-millionth	a micrometer is 0.000001 (one-millionth) of a meter
nano (n)	one-billionth	a nanogram is 10^{-9} (one-billionth) of a gram
pico (p)	one-trillionth	a picogram is 10^{-12} (one-trillionth) of a gram

Think Metric
Length

1. The speed of a car is 60 miles/hr or 100 km/hr.
2. A man who is six feet tall is 180 cm.
3. A six inch ruler is 15 cm.
4. One yard is almost a meter (0.9 m).

Units of Length		
Unit	**Abbreviation**	**Equivalent**
meter	m	approximately 39 in
centimeter	cm	10^{-2} m
millimeter	mm	10^{-3} m
micrometer	μm	10^{-6} m
nanometer	nm	10^{-9} m
angstrom	Å	10^{-10} m

Length Conversions			
1 in = 2.5 cm	1 mm = 0.039 in		
1 ft = 30 cm	1 cm = 0.39 in		
1 yd = 0.9 m	1 m = 39 in		
1 mi = 1.6 km	1 m = 1.094 yd		
	1 km = 0.6 mi		

To Convert	**Multiply by**	**To Obtain**
inches	2.54	centimeters
feet	30	centimeters
centimeters	0.39	inches
millimeters	0.039	inches

Think Metric
Volume

1. One can of beer (12 oz) contains 360 ml.
2. The average human body contains between 10–12 pints of blood or between 4.7–5.6 liters.
3. One cubic foot of water (7.48 gallons) is 28.426 liters.
4. If a gallon of unleaded gasoline costs $1.00, a liter would cost 38¢.

Units of Volume

Unit	Abbreviation	Equivalent
liter	l	approximately 1.06 qt
milliliter	ml	10^{-3} l (1 ml = 1 cm^3 = 1 cc)
microliter	μl	10^{-6} l

Volume Conversions

1 tsp	=	5 ml	1 pt	=	0.47 l	1 ml	=	0.03 fl oz
1 tbsp	=	15 ml	1 qt	=	0.95 l	1 l	=	2.1 pt
1 fl oz	=	30 ml	1 gal	=	3.8 l	1 l	=	1.06 qt
1 cup	=	0.24 l				1 l	=	0.26 gal

To Convert	Multiply by	To Obtain
fluid ounces	30	milliliters
quarts	0.95	liters
milliliters	0.03	fluid ounces
liters	1.06	quarts

Think Metric
Weight

1. One pound of hamburger is 448 grams.
2. The average human male brain weighs 1.4 kg (3 lb 1.7 oz).
3. A person who weighs 154 lbs weighs 70 kg.
4. Lucia Zarate weighed 5.85 kg (13 lbs) at age 20.

Units of Weight

Unit	Abbreviation	Equivalent
kilogram	kg	10^3 g (approximately 2.2 lb)
gram	g	approximately 0.035 oz
milligram	mg	10^{-3} g
microgram	μg	10^{-6} g
nanogram	ng	10^{-9} g
picogram	pg	10^{-12} g

Weight Conversions

1 oz	=	28.3 g	1 g	=	0.035 oz
1 lb	=	453.6 g	1 kg	=	2.2 lb
1 lb	=	0.45 kg			

To Convert	Multiply by	To Obtain
ounces	28.3	grams
pounds	453.6	grams
pounds	0.45	kilograms
grams	0.035	ounces
kilograms	2.2	pounds

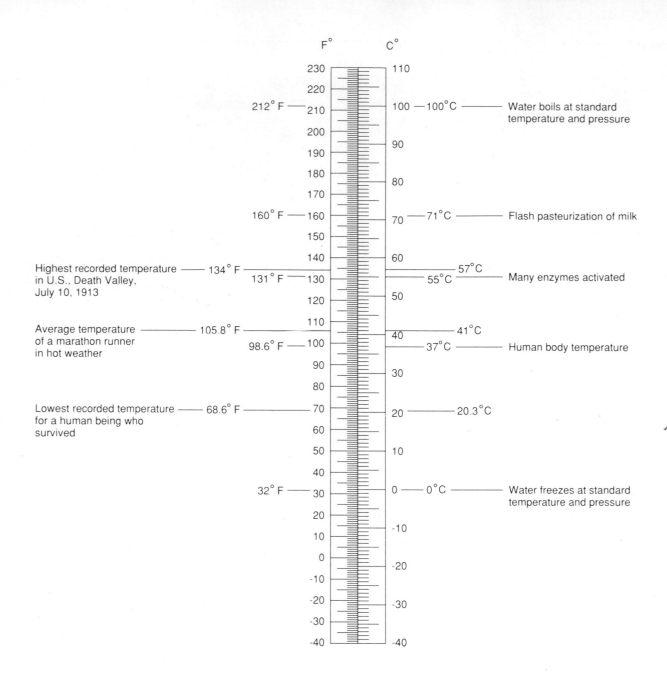

F°   C°

230   110

220

212° F — 210   100 — 100°C ———— Water boils at standard temperature and pressure

200   90

190

180   80

170

160° F — 160   70 — 71°C ———— Flash pasteurization of milk

150

140   60   57°C

Highest recorded temperature — 134° F
in U.S., Death Valley,   131° F — 130   55°C ———— Many enzymes activated
July 10, 1913

120   50

110

Average temperature ———— 105.8° F   40 ———— 41°C
of a marathon runner

98.6° F — 100   37°C ———— Human body temperature
in hot weather

90   30

80

Lowest recorded temperature ——— 68.6° F ——— 70   20 ———— 20.3°C
for a human being who
survived   60

50   10

40

32° F — 30   0 — 0°C ———— Water freezes at standard temperature and pressure

20

10   -10

0

-10   -20

-20   -30

-30

-40   -40

To convert temperature scales:

Fahrenheit to Celsius $°C = \frac{5}{9} (°F - 32)$

Celsius to Fahrenheit $°F = \frac{9}{5} (°C) + 32$

Answers to the Objective Questions

Chapter 1
  1. d, 2. c, 3. a, 4. b, 5. e, 6. c, 7. b, 8. c

Chapter 2
  1. d, 2. a, 3. d, 4. a, 5. b, 6. c, 7. b, 8. d

Chapter 3
  1. d, 2. c, 3. b, 4. c, 5. d, 6. d, 7. d, 8. b, 9. c, 10. a

Chapter 4
  1. d, 2. a, 3. c, 4. b, 5. c, 6. b, 7. d, 8. a, 9. c, 10. a, 11. d, 12. b, 13. c, 14. b, 15. c, 16. c, 17. d, 18. c, 19. a, 20. a

Chapter 5
  1. c, 2. c, 3. d, 4. a, 5. b, 6. c, 7. c, 8. a, 9. d, 10. d, 11. d, 12. b

Chapter 6
  1. b, 2. b, 3. a, 4. c, 5. c, 6. d, 7. c, 8. d, 9. b, 10. b

Chapter 7
  1. b, 2. a, 3. d, 4. c, 5. b, 6. a, 7. d, 8. c, 9. d, 10. c

Chapter 8
  1. d, 2. c, 3. d, 4. a, 5. d, 6. d, 7. c, 8. d, 9. c, 10. d

Chapter 9
  1. b, 2. c, 3. a, 4. c, 5. c, 6. c, 7. a, 8. b, 9. c, 10. d, 11. c, 12. a, 13. b, 14. d, 15. b, 16. a, 17. d, 18. a, 19. c, 20. b

Chapter 10
  1. c, 2. b, 3. d, 4. d, 5. a, 6. c, 7. b, 8. c, 9. b, 10. b

Chapter 11
  1. b, 2. c, 3. a, 4. c, 5. b, 6. a, 7. c, 8. d, 9. d, 10. c

Chapter 12
  1. b, 2. a, 3. c, 4. d, 5. c, 6. d, 7. a, 8. b, 9. b, 10. b, 11. c, 12. d

Chapter 12 Additional Genetics Problems
  1. 100% chance for widow's peak and 0% chance for continuous hairline
  2. *Ee*
  3. 50%
  4. 210 gray body and 70 black body; 140 = heterozygous; cross fly with recessive
  5. F_1 = all brown with short hair; F_2 = 9:3:3:1; offspring would be 1 brown long: 1 brown short: 1 black long: 1 black short
  6. *Bbtt* × *bbTt* and *bbtt*
  7. *GGL1*
  8. 25%

Chapter 12 Answers for Practice Problems 1
  1. a. 100% *W*; b. 50% *W*, 50% *w*; c. 50% *T*, 50% *t*; d. 100% *T*
  2. a. gamete; b. genotype; c. gamete

Chapter 12 Answers for Practice Problems 2
  1. *bb*
  2. *Tt* × *tt*, *tt*
  3. 3/4 or 75%
  4. *Yy* and *yy*

Chapter 12 Answers for Practice Problems 3
  1. a. 100% *tG*; b. 50% *TG*, 50% *tG*; c. 25% *TG*, 25% *Tg*, 25% *tG*, 25% *tg*; d. 50% *TG*, 50% *Tg*
  2. a. genotype; b. gamete; c. genotype; d. gamete

Chapter 12 Answers for Practice Problems 4
  1. *BbTt* only
  2. a. *L1Gg* × *11gg*
     b. *L1Gg* × *L1Gg*
  3. 9/16

Chapter 13
  1. b, 2. c, 3. e, 4. a, 5. b, 6. b, 7. d, 8. b, 9. c, 10. d

Chapter 13 Additional Genetics Problems
  1. pleiotropy
  2. codominance, $F^B F^B$, $F^W F^W$, $F^B F^W$
  3. 50%
  4. man $I^A i$; woman $I^B i$; child *ii*; possible offspring A, AB, B, O
  5. linkage, 10
  6. Both males and females are 1:1.
  7. $X^B X^B \times X^b Y$ = both males and females bar-eyed; $X^b X^b \times X^B Y$ = females bar-eyed; males all normal
  8. mother $X^b X^b ww$; father $X^B Y$ *Ww*; girl $X^b X^b$ *Ww*; cannot be the father because a man with normal vision cannot have a color-blind daughter
  9. females—1 long-haired tortoise shell: 1 long-haired yellow; males—1 long-haired black: 1 long-haired yellow

Chapter 13 Answers for Practice Problems 1
  1. Yes, he would be $T^A i$
  2. 1 pink:1 white
  3. seven
  4. pleiotropy, epistasis

Chapter 13 Answers for Practice Problems 2

1. females: $X^R X^R$, $X^R X^r$, $X^r X^r$
   males: $X^R Y$(gametes X^R, Y); $X^r Y$ (gametes X^r, Y)
2. b. 1:1
3. 100%, none, 100%
4. mother $X^B X^b$ Rr, father $X^B Y$ Rr, son $X^b Y$ rr

Chapter 13 Answers for Practice Problems 3

1. 9:3:3:1, linkage
2. 54.5–13.00 = 41.5%
3. 12
4. bar eye, scalloped wings, garnet eye

Chapter 14

1. c, 2. a, 3. b, 4. c, 5. a, 6. b, 7. d, 8. c, 9. a, 10. autosomal dominant

Chapter 14 Answers to Additional Genetics Problems

1. 0%, 0%, 100%
2. 25%
3. 50%, 50%
4. a. autosomal recessive, *Aa
   b. autosomal dominant, *Aa
   c. sex-linked recessive, *$X^A X^a$

Chapter 15

1. b, 2. d, 3. a, 4. c, 5. d, 6. c, 7. a, 8. a, 9. b, 10. d

Chapter 16

1. b, 2. d, 3. a, 4. c, 5. a, 6. a, 7. c, 8. b, 9. c, 10. d

Chapter 16 Answers to Practice Problems

1. ACU CCU GAA UGC AAA
2. UGA GGA CUU ACG UUU
3. Threonine-Proline-Glutamic acid-Cysteine-Lysine

Chapter 17

1. c, 2. a, 3. d, 4. a, 5. b, 6. b, 7. b, 8. d, 9. d, 10. d

Chapter 18

1. a, 2. c, 3. d, 4. a, 5. b, 6. d, 7. b, 8. c, 9. d, 10. a

Chapter 19

1. d, 2. b, 3. d, 4. b, 5. c, 6. a, 7. b, 8. d, 9. d, 10. b

Chapter 20

1. b, 2. a, 3. b, 4. d, 5. d, 6. d, 7. a, 8. b, 9. d, 10. c

Chapter 21

1. d, 2. b, 3. a, 4. b, 5. d, 6. d, 7. c, 8. d, 9. a, 10. b

Chapter 21 Answers to Population Genetics Problems

1. 29%
2. 99%
3. recessive allele = 0.2; dominant allele = 0.8; homozygous recessive = 0.04; homozygous dominant = 0.64; heterozygous = 0.32.

Chapter 22

1. a, 2. c, 3. d, 4. b, 5. c, 6. b, 7. b, 8. a, 9. c, 10. c

Chapter 23

1. b, 2. b, 3. c, 4. b, 5. c, 6. c, 7. d, 8. c, 9. b, 10. d, 11. d, 12. d

Chapter 24

1. d, 2. c, 3. a, 4. b, 5. d, 6. c, 7. d, 8. a, 9. a, 10. d

Chapter 25

1. d, 2. c, 3. c, 4. c, 5. c, 6. d, 7. b, 8. b, 9. b, 10. d

Chapter 26

1. d, 2. c, 3. c, 4. d, 5. a, 6. d, 7. b, 8. b, 9. b, 10. a

Chapter 27

1. a, 2. a, 3. c, 4. c, 5. a, 6. d, 7. c, 8. d, 9. b, 10. d

Chapter 28

1. a, 2. c, 3. d, 4. a, 5. b, 6. d, 7. b, 8. d, 9. d, 10. b, 11. b, 12. c

Chapter 29

1. c, 2. b, 3. c, 4. c, 5. b, 6. b, 7. b, 8. c, 9. c, 10. b

Chapter 30

1. d, 2. a, 3. c, 4. d, 5. b, 6. c, 7. c, 8. c, 9. d, 10. d

Chapter 31

1. c, 2. a, 3. b, 4. a, 5. a, 6. b, 7. a, 8. d, 9. c, 10. c

Chapter 32

1. d, 2. b, 3. e, 4. a, 5. c, 6. d, 7. b, 8. d, 9. d, 10. a

Chapter 33

1. b, 2. b, 3. a, 4. d, 5. d, 6. c, 7. d, 8. b, 9. b, 10. b, 11. d

Chapter 34

1. d, 2. d, 3. a, 4. b, 5. c, 6. b, 7. a, 8. a, 9. b, 10. d

Chapter 35

1. d, 2. b, 3. d, 4. b, 5. c, 6. a, 7. c, 8. c, 9. d, 10. c, 11. d, 12. c

Chapter 36

1. a, 2. b, 3. b, 4. d, 5. c, 6. b, 7. c, 8. b, 9. d, 10. b

Chapter 37

1. d, 2. a, 3. c, 4. b, 5. a, 6. c, 7. b, 8. d, 9. a, 10. b

Chapter 38

1. c, 2. b, 3. d, 4. c, 5. a, 6. a, 7. b, 8. d, 9. b, 10. b

Chapter 39

1. d, 2. c, 3. c, 4. d, 5. d, 6. c, 7. c, 8. b, 9. d, 10. c

Chapter 40

1. b, 2. f, 3. c, 4. e, 5. b, 6. d, 7. b, 8. b, 9. b, 10. c, 11. d, 12. d

Chapter 41

1. f, 2. b, 3. c, 4. a, 5. e, 6. c, 7. a, 8. d, 9. b, 10. d, 11. b, 12. b

Chapter 42

1. b, 2. d, 3. c, 4. d, 5. c, 6. c, 7. c, 8. c, 9. b, 10. a

Chapter 43

1. b, 2. b, 3. b, 4. a, 5. b, 6. b, 7. d, 8. a, 9. d, 10. d

Chapter 44

1. b, 2. a, 3. a, 4. b, 5. b, 6. d, 7. a, 8. a, 9. b, 10. d

Chapter 45

1. b, 2. c, 3. d, 4. d, 5. d, 6. c, 7. b, 8. c

Chapter 46

1. b, 2. d, 3. b, 4. c, 5. d, 6. c, 7. b, 8. d, 9. c, 10. b

Chapter 47

1. a, 2. c, 3. a, 4. a, 5. c, 6. d, 7. a, 8. c, 9. d, 10. c

Chapter 48

1. b, 2. a, 3. c, 4. d, 5. b, 6. d, 7. c, 8. c, 9. a, 10. b

Chapter 49

1. a, 2. d, 3. c, 4. b, 5. b, 6. c, 7. d, 8. c, 9. a, 10. c

GLOSSARY

A

abscisic acid (ab-sis′ik as′id) a plant hormone causing stomata closing and initiates and maintains dormancy 498

acetylcholine (as″ ĕ-til-ko′lēn) a neurotransmitter utilized within both the peripheral and central nervous systems 598

acetyl CoA (as″ ĕ-til-ko-en′zīm A) molecule made up of a two-carbon acetyl group attached to coenzyme A, that enters the Krebs cycle for further oxidation 132

acid (as′id) a compound tending to yield hydrogen ions in a solution and to lower its pH numerically 33

acid deposition (as′id dep″o-zish′un) the return to earth as rain or snow, the sulfate or nitrate salts of acids produced by commercial and industrial activities on earth 789

actin (ak′tin) a muscle protein making up the thin filaments in a sarcomere; its movement shortens the sarcomere, yielding muscle contraction 636

action potential (ak′shun po-ten′shal) change in the polarity of a neuron caused by opening and closing of ion channels in the axon plasma membrane 595

active site (ak′tiv sīt) that part of an enzyme molecule where the substrate fits and the chemical reaction occurs 98

active transport (ak′tiv trans′port) use of a plasma membrane carrier molecule to move particles from a region of lower to higher concentration; it opposes an equilibrium and requires energy 85

adaptation (ad″ap-ta′shun) an organism's modification in structure or function to suit the environmental conditions and increase the likelihood of continued existence 5

adaptive radiation (ah-dap′tiv ra-de-a′shun) formation of a large number of species from a common ancestor 303

adenine (ad′ĕ-nīn) one of four organic bases in the structure of DNA and RNA 222

adenosine diphosphate (ADP) (ah-den′o-sēn di-fos′fāt) one of the products of the hydrolysis of ATP, a process which liberates energy 103

adenosine triphosphate (ATP) (ah-den′o-sēn tri-fos′fāt) a compound containing adenine, ribose, and three phosphates, two of which are high-energy phosphates. The breakdown of ATP to ADP and makes energy available for energy-requiring processes in cells 103

ADP *see* adenosine diphosphate

adrenal gland (ah-dre′nal gland) lies atop a kidney and secretes the stress hormones epinephrine, norepinephrine, and the corticord hormones 651

age structure diagram (āj struk′tūr di′ah-gram) a representation of the number of individuals in each age group in a population 725

aggression (ah-gresh′un) belligerent behavior shown in competitive situations 713

aldosterone (al″do-ster′ōn) hormone secreted by the adrenal gland that maintains the sodium and potassium balance of the blood 587

allantois (ah-lan′to-is) an extraembryonic membrane that accumulates nitrogenous wastes in the embryo of birds and reptiles, and contributes to the formation of umbilical blood vessels in mammals 686

allele (ah-lēl) alternative forms of a gene, e.g., an allele for long wings and an allele for short wings in fruit flies 175

allopatric speciation (al″o-pat′rik spe″se-a′shun) origin of new species in populations that are separated geographically 325

alternation of generations (aw″ter-na′shun uv jen″ĕ-ra′shunz) life cycle in which a haploid generation alternates with a diploid generation 356

alveoli (sing., **alveolus**) (al-ve′o-li) terminal, microscopic, grapelike air sacs found in vertebrate lungs 569

amino acid (ah-me′no as′id) organic subunits each having an amino group and an acid group, that covalently bond to produce protein molecules 46

amnion (am′ne-on) an extraembryonic membrane of the bird, reptile, and mammal embryo that forms an enclosing fluid-filled sac 686

amoeboid (ah-me′boid) like an amoeba, particularly showing the slow creeping motility of this protozoan 353

amphibian (am-fib′e-an) includes frogs, toads, and salamanders; terrestrial vertebrate still tied to a watery environment for reproduction 421

analogous structure (ah-nal′o-gus struk′tūr) a structure that functions similarly but is not derived from a common ancestor 299

angiosperm (an′je-o-sperm) flowering plant, having double fertilization that results in development of seed-bearing fruit 381

annelid (ah′nel-id) a phylum of segmented worms characterized by a tube-within-a-tube body plan, bilateral symmetry, and segmentation of some organ systems 405

anther (an′ther) part of stamen where pollen grains develop 480

antheridia (sing., **antheridium**) (an″ther-id′ī-ah) male structures in nonseed plants where swimming sperm are produced 372

antibody (an′tī-bod″e) a protein molecule usually formed naturally in an organism to combat antigens or foreign substances 533

anticodon (an′tī-ko′don) three nucleotides on a tRNA molecule, attracted to a complementary codon on mRNA 239

antidiuretic hormone (ADH) (an″tī-di″u-ret′ik hor′mōn) hormone secreted by the posterior pituitary that increases the permeability of the collecting duct in a nephron 587

antigen (an′tī-jen) substances, usually proteins, that are not normally found in the body 532

aorta (a-or′tah) large arterial trunk that supplies the systemic circuit 515

appendicular skeleton (ap″en-dik′u-lar skel′ē-ton) part of the skeleton forming the upper appendages, shoulder girdle, lower appendages, and hip girdle 627

aquifer (ah'kwi-fer) a porous layer that sometimes contains water and lies between two nonporous layers of rock 747

archaebacteria (ar"ke-bak-te're-ah) probably the earliest prokaryotes; their cell wall and plasma membrane differ from other bacteria, and many live in extreme environments 346

archegonia (ar"ke-go'ne-ah) female structures in nonseed plants where eggs are produced 372

artery (ar'ter-e) a blood vessel that transports blood away from the heart toward arterioles 512

arthropod (ar'thro-pod) diverse phylum of animals that have jointed appendages; includes lobsters, insects, and spiders 406

artificial selection (ar"ti-fish'al se-lek'shun) human selection of which organisms will reproduce 289

asexual reproduction (a-seks'u-al re"pro-duk'shun) reproduction involving only mitosis that produces offspring genetically identical to the parent 154

aster (as'ter) short, radiating fibers produced by the centrioles; important during mitosis and meiosis 150

atom (at'om) the smallest particle of an element that displays its properties 24

atomic number (ah-tom'ik num'ber) the number of protons in an atom 25

atomic weight (ah-tom'ik wat) combined weight of protons and neutrons in an atom 25

ATP *see* adenosine triphosphate

australopithecine (aw"strah-lo-pith'e-sin) early hominids, fossils of which have been found dating 3 to 4 million years ago 431

autonomic nervous system (aw"to-nom'ik ner'vus sis'tem) a branch of the peripheral nervous system administering motor control over internal organs 600

autosome (aw'to-som) any chromosome other than the sex-determining pair 191

autotroph (aw'to-trof) self-nourishing organism; referring to producers starting food chains that make organic molecules from inorganic nutrients 331

auxin (awk'sin) plant hormone regulating growth, particularly cell elongation; also called indolacetic acid 493

axial skeleton (ak'se-al skel'e-ton) part of the skeleton forming the vertical support or axis, including the skull, rib cage, and vertebral column 627

axon (ak'son) the part of a neuron that conducts the impulse from cell body to synapse 593

B

B lymphocyte (be lim'fo-sit) white blood cell that produces and secretes antibodies that can combine with antigens 533

bacteriophage (bak-te're-o-faj") a virus that parasitizes a bacterial cell as its host, destroying it by lytic action 221

bacterium (bak-te're-um) organism that lacks a nucleus and cytoplasmic organelles other than ribosomes; reproduces by binary fission; and occurs in one of three shapes (rod, sphere, spiral) 342

Barr body (bahr bod'e) X chromosome in females that is inactive 251

base (bas) a compound tending to lower the hydrogen ion concentration in a solution and raise its pH numerically 33

bicarbonate ion (bi-kar'bo-nat i'on) form in which most of the carbon dioxide is transported in the bloodstream 571

bilaterally symmetrical (bi-lat'er-al-e si-met're-kal) a body having two corresponding or complementary halves 390

bile (bil) a substance released from the gallbladder to small intestine to emulsify fats prior to chemical digestion 553

binary fission (bi'na-re fish'un) splitting of a parent cell into two daughter cells; serves as an asexual form of reproduction in single-celled organisms 146

biogeochemical cycle (bi"o-je"o-kem'i-kal si'k'l) circulating pathway of an element through the biotic and abiotic components of an ecosystem 773

biogeography (bi"o-je-og'rah-fe) the study of the geographical distribution of organisms 285

biological clock (bi-o-loj'i-kal klok) internal mechanism that maintains a biological rhythm in the absence of environmental stimuli 501

biological magnification (bi"o-loj'i-kal mag"ni-fi-ka'shun) process by which substances become more concentrated in organisms in the higher trophic levels of the food chain 769

biome (bi'om) a major biotic community having well-recognized life forms and a typical climax species 747

biosphere (bi'o-sfer) a thin shell of air, land, and water around the earth supporting life 7

biotechnology (bi"o-tek-nol'o-je) the use of a natural biological system to produce a commercial product 261

biotic potential (bi-ot'ik po-ten'shal) maximum rate of natural increase of a population that can occur under ideal circumstances 721

birthrate (birth'rat) the number of births during a specified time period 720

bivalent (tetrad) (bi-va'lent)(tet'rad) homologous chromosomes each having sister chromatids that are joined by a nucleoprotein lattice during meiosis 159

blade (blad) part of a plant leaf that may be single or compound leaflets 458

blastocoel (blas'to-sel) the fluid-filled cavity of a blastula in the early embryo 680

blastula (blas'tu-lah) a hollow, fluid-filled ball of cells prior to gastrula formation of the early embryo 680

Bowman's capsule (bo'manz kap'sul) a cuplike structure that is the initial portion of a nephron; where pressure filtration occurs 583

bronchi (sing., **bronchus**) (brong'ki) two main branches of the trachea in vertebrates that have lungs 569

bronchiole (brong'ke-ol) small tube that conducts air from a bronchus to the alveoli 569

bryophyte (bri'o-fit) a plant group that includes mosses and liverworts 371

budding (bud'ing) in animals an asexual form of reproduction whereby a new organism develops as an outgrowth of the body of the parent 660

buffer (buf'er) a substance or group of substances that tend to resist pH changes in a solution, thus stabilizing its relative acidity 34

bundle sheath (bun'd'l sheth) cells that surround the xylem and phloem of leaf veins 459

C

C₃ plant plant that directly uses the Calvin-Benson cycle and produces an immediate three-carbon molecule during photosynthesis 119

C₄ plant plant that does not directly use the Calvin-Benson cycle and produces an immediate four-carbon molecule during photosynthesis 119

Calvin-Benson cycle (kal'vin ben'sun si'k'l) series of photosynthetic reactions in which carbon dioxide is reduced in the chloroplast 117

capillary (kap'i-lar"e) microscopic blood vessel for gas and nutrient exchange with body cells 512

carbohydrate (kar"bo-hi'drat) a family of organic compounds consisting of carbon, hydrogen, and oxygen atoms; includes subfamilies monosaccharides, disaccharides, and polysaccharides 40

carbon dioxide fixation (kar'bon di-ok'sid fix-sa'shun) photosynthetic reaction in which carbon dioxide is attached to an organic compound 117

carbonic anhydrase (kar-bon'ik an-hi'dras) an enzyme in red blood cells that speeds up the formation of carbonic acid from water and carbon dioxide 571

carcinogen (kar-sin'o-jen) agent that contributes to the development of cancer 255

carnivore (kar'ni-vor) a secondary consumer in a food chain, that eats other animals 546

carrier (ka'e-er) an individual that is capable of transmitting an infectious or genetic disease; a protein that combines with and transports a molecule across the plasma membrane 209

carrying capacity (kar'e-ing kah-pas'i-te) the maximum size of a population that can be supported by the environment in a particular locale 723

cartilage (kar'ti-lij) a type of connective tissue having a matrix of protein and proteinaceous fibers that is found in animal skeletal systems 630

Casparian strip a waxy layer bordering four sides of root endodermal cells; prevents water and solute transport between adjacent cells 451

catastrophism (kah-tas'tro-fizm) belief espoused by Cuvier that periods of catastrophic extinctions occurred, after which repopulation of surviving species took place and gave the appearance of change through time 283

cell (sel) the smallest unit that displays properties of life; composed of cytoplasmic regions, possibly organelles, and surrounded by a plasma membrane 2

cell cycle (sel si'k'l) the sequence of events in a cell that includes interphase and cell division 149

cell-first hypothesis (sel ferst hi-poth'ĕ-sis) proposal that the protocell came before true proteins and nucleic acids 331

cell plate (sel plāt) structure across a dividing plant cell that signals the location of new plasma membranes and cell walls 154

cellular respiration (sel-u-lar res''pi-ra'shun) metabolic reactions that provide energy to cells by the oxidation of carbohydrates 101

cell wall (sel wawl) a relatively rigid structure composed mostly of polysaccharides that surrounds the plasma membrane of plants, fungi, and bacteria 57

central dogma (sen'tral dog'mah) belief that genetic information passes only from DNA to RNA to proteins 235

central nervous system (CNS) (sen'tral ner'vus sis'tem) the brain and nerve (spinal) cord in animals 592

centriole (sen'tri-ōl) cell organelle, existing in pairs, that possibly organizes a mitotic spindle for chromosome movement during cell division 70

centromere (sen'tro-mēr) a constriction where duplicates (chromatids) of chromosomes are held together 147

cephalization (sef'al-i-za'shun) having a well-recognized anterior head with concentrated nerve masses and receptors 392

chemical evolution (kem'i-kal ev''o-lu'shun) an increase in the complexity of chemicals that could have led to first cells 329

chemiosmosis (kem''i-os-mos'is) a process by which an electron transport system pumps hydrogen ions across a membrane resulting in an electrochemical gradient that can be used for ATP synthesis 104

chemiosmotic phosphorylation (kem''i-os-mot'ik fos''for-i-la'shun) production of ATP by utilizing the energy released when H⁺ ions flow through an F_1 or CF_1 particle in mitochondria and chloroplasts 114, 133

chemoreceptor (ke''mo-re-sep'tor) a receptor that is sensitive to chemical stimulation, e.g., receptors for taste and smell 613

chemosynthetic (ke''-mo-sin-thet'ik) capable of oxidizing inorganic compounds, such as ammonia, nitrites, and sulfides, to gain energy to synthesize carbohydrates 345

chiasmata (sing., **chiasma**) (ki-as'mah-tah) regions where nonsister chromatids are joined and crossing-over occurs 162

chitin (ki'tin) a strong but flexible, nitrogenous polysaccharide that is found in the exoskeleton of arthropods 406

chlorophyll (klo'ro-fil) green pigment that absorbs sunlight energy and is important in photosynthesis 112

chloroplast (klo'ro-plast) a membrane-bounded organelle with membranous grana that contain chlorophyll and where photosynthesis takes place 111

cholesterol (ko-les'ter-ol) one of the major lipids found in animal cell membranes; makes the membrane impermeable to many molecules 80

chordate (kor'dāt) phylum of animals that includes lancelets, tunicates, fishes, amphibians, reptiles, birds, and mammals; characterized by a notochord, dorsal nerve cord, and gill pouches 412

chorion (kor'e-on) an extraembryonic membrane functioning for respiratory exchange in birds and reptiles; contributes to placenta formation in mammals 686

chromatin (kro'ma-tin) the mass of DNA and associated proteins observed within a nucleus that is not dividing 64

chromosome (kro'mo-sōm) association of DNA and proteins arranged linearly in genes and observed during cell division 57

chromosome theory of inheritance (kro'mo-sōm the'o-re uv in-her'i-tans) theory stating that the genes are on chromosomes accounting for their similar behavior 191

circadian rhythm (ser''kah-de'an rith'm) a biological rhythm with a 24-hour cycle 501

citric acid cycle (Krebs cycle) (si-trik as'id si'k'l) a cycle of reactions in mitochondria, which begins and ends with citric acid; produces CO_2, ATP, NADH, and $FADH_2$ 132

cleavage (klēv'ij) earliest division of the zygote developmentally, without cytoplasmic addition or enlargement 680

cleavage furrowing (clē-vaj fur'o-ing) indenting of the plasma membrane that leads to division of the cytoplasm during cell division 153

climax community (kli'maks kŏ-mu'ni-te) mature stage of a community that can sustain itself indefinitely as long as it is not stressed 755

cloned (klōnd) the production of identical copies; in genetic engineering, the production of many, identical copies of a gene 261

cnidarian (ni-dah're-an) phylum of animals with a gastrovascular cavity; includes sea anemones, corals, hydras, and jellyfishes 390

cochlea (kok'le-ah) spiral-shaped structure of the inner ear containing the receptor hair cells for hearing 622

codominance (ko-dom'i-nans) a condition in heredity in which both alleles are expressed 185

codon (ko'don) three nucleotides of DNA or mRNA that codes for a particular amino acid 236

coelomate (sēl'o-māt) characterizes animals with a true coelom that is completely lined with mesoderm 400

coenzyme (ko-en'zīm) a nonprotein organic part of an enzyme structure, often with a vitamin as a subpart 100

coevolution (ko-ev''o-lu'shun) evolutionary process where two groups act as selective agents for each other 307

cohesion-tension model (ko-he'zhun ten'shun mod'el) explanation for the transport of water to great heights in a plant by water molecules clinging together as transcription occurs 465

collecting duct (kŏ-lekt'ing dukt) the final portion of a nephron where reabsorption of water can be controlled 583

colonial (ko-lo'ne-al) a life form with a loose association of cells that remain independent for most functions 358

commensalism (kŏ-men'sal-izm'') a symbiotic relationship in which one species profits and the other is neither harmed nor benefited 741

common ancestor (kom'un an'ses''tor) organism related to two different descendant groups; located at points of divergence in an evolutionary tree 300

community (kŏ-mu'ni-te) many different populations of organisms that interact with each other 732

compact bone (kom'pakt bōn) type of bone that is made up of structural units called Haversian canals, through which blood vessels and nerve fibers run 630

competitive exclusion principle (kom-pet'i-tiv eks-kloo'zhun prin'si-p'l) theory that no two species can occupy the same niche 736

complementary base pairing (kom''plĕ-men'tā-re bās pār'ing) bonding between particular purines and pyrimidines in DNA 223

compound (kom'pownd) a chemical substance having two or more different elements in fixed ratio 27

compound eye (kom'pound i) a type of eye found in arthropods composed of many independent visual units 615

cone (kōn) photoreceptors in vertebrate eyes that respond to bright light and allow color vision 616

conifer (ko'ni-fer) one of the four groups of gymnosperm plants; cone-bearing trees that include pine, cedar, and spruce 379

conjugation (kon''ju-ga'shun) the transfer of genetic material from one cell to another through a cytoplasmic bridge 358

consumer (kon-sūm'er) an organism that feeds on another organism in a food chain; primary consumers eat plants, secondary consumers eat animals 766

contest competition (kon'test kom''pē-tish'un) competition in which some members of a population gain access to resources and others do not; outcome of dominance hierarchies, territoriality; tends to keep populations at carrying capacity of the environment 734

continental drift (kon''tĭ-nen'tal drift) large movements of the continents over geologic time, proposed by Alfred Wegener 296

continental shelf (kon''tĭ-nent'al shelf) that part of the ocean floor that projects from the continents to the slope leading to the abyss 752

control group (kon-trōl' grōōp) a sample that goes through all the steps of an experiment except the one being tested; a standard against which results of an experiment are checked 15

convergent evolution (kon-ver'jent ev''o-lu'shun) the appearance of similar structures in different groups of organisms exposed to similar types of environments 307

coral reef (kor'al rēf) area of biological abundance found in shallow, warm tropical waters and centered on coral formations 751

corepressor (ko''re-pres'or) molecule that binds to a repressor, allowing the repressor to bind to an operator in a repressible operon 250

cork (kork) outer covering of bark of trees; made up of dead cells that may be sloughed off 447

corpus luteum (kor'pus lu'te-um) a follicle that has released an egg and increases its secretion of progesterone 666

covalent (ko-va'lent) a chemical bond in which the atoms share electrons 28

Cro-Magnon (kro-mag'non) hominid who lived 40,000 years ago; accomplished hunters, made compound stone tools, and possibly had language 436

crossing-over (kros'ing o'ver) an exchange of segments between nonsister chromatids of a bivalent during meiosis 162

crustacean (krus-ta'shun) a class of arthropods that includes the lobsters, shrimp, crabs, copepods, and barnacles 406

cultural eutrophication (kul'tūr-al u''tro-fĭ-ka'shun) an acceleration of the process by which nutrient poor lakes and ponds become nutrient rich; speeds up aquatic succession, in which a pond or lake eventually fills in and disappears 749

cyanobacterium (si''ah-no-bak-te're-ah) formerly called blue-green alga; photosynthetic prokaryotes that contain chlorophyll and release oxygen 347

cyclic AMP (sik'lik AMP) an ATP related compound that promotes chemical reactions in the body; the 'second messenger' in peptide hormone activity, it initiates activity of the metabolic machinery 643

cyclic photophosphorylation (sik'lik fo''to-fos''for-i-la'shun) photosynthetic pathway in which electrons from photosystem P700 pass through the electron transport system, produce ATP, and eventually return to P700 115

cystic fibrosis (sis'tik fĭ-bro'sis) a lethal, genetic disease affecting the mucus in the lungs and digestive tract 210

cytochrome (si'to-krōme) an electron carrier in the electron transport system of mitochondria and chloroplasts 133

cytokinesis (si''to-ki-ne'sis) the division of the cytoplasm into two halves following mitosis 147

cytokinin (si''to-ki'nin) plant hormone that promotes cell division; development, often works in combination with auxin during organ development 496

cytoplasm (si'to-plazm) contents of a cell between the nucleus (nucleoid) and the plasma membrane 57

cytosine (si'to-sin) one of four bases in the structure of DNA and RNA 222

cytoskeleton (si''to-skel'ĕ-ton) internal framework of the cell that contains microtubules, filaments, and microfilaments 70

cytosol (si'to-sol) fluid medium that bathes the contents of the cytoplasm 57

D

data (da'tah) facts or information collected through observation and/or experimentation 13

death rate (deth rāt) the number of deaths during a specified time period 720

deductive reasoning (de-duk'tiv re'zun-ing) a process of logic and reasoning, using "if . . . then" statements 13

deforestation (de''for-es-ta'shun) the replacement of forests with other types of biomes often due to human activities 791

dehydration synthesis (de''hi-dra'shun sin'thĕ-sis) a joining of two monomers by removing a hydrogen from one monomer and a hydroxyl group from the other so that water results 40

demographic transition (dĕ''mo-graf'ik tran-zish'un) a decline in death rate, followed shortly by a decline in birthrate, resulting in a slower population growth 726

denatured (de-na'tūrd) condition of an enzyme when its shape is changed so that its active site cannot bind substrate molecules 99

dendrite (den'drit) the part of a neuron that sends impulses toward the cell body 593

denitrification (de-ni''trĭ-fĭ-ka'shun) the process of converting nitrogen compounds to atmospheric nitrogen; is a part of the nitrogen cycle 776

deoxyribonucleic acid (DNA) (de-ok''se-ri''bo-nu-kle'ik as'id) an organic molecule produced from covalent bonding of nucleotide subunits; the chemical composition of genes on chromosomes 219

dependent variable (de-pen'dent va're-ah-b'l) result or change that occurs when the experimental variable is varied 14

desertification (dez''ert-i-fi-ka'shun) transformation of marginal lands to desert conditions 781

detritus (de-tri'tus) falling, settled remains of plants and animals on the land-floor or water bed 766

deuterostome (du'ter-o-stōm'') a group of coelomate animals in which the second embryonic opening becomes the mouth; the first embryonic opening, the blastopore, becomes the anus 400

developed country (de-vel'opt kun'tre) highly industrialized countries, e.g., United States, West Germany, Japan 726

developing country (de-vel'op-ing kun'tre) those countries that are not yet fully industrialized, e.g., in Africa, Latin America, Asia 726

diabetes mellitus (di''ah-be'tēz mĕ-li'tus) a disease caused by a lack of insulin production by the pancreas 653

diaphragm (di'ah-fram) a dome-shaped muscle separating the thoracic cavity from the abdominal cavity in the animal body and involved in respiration 568

diastole (di-as'to-le) relaxation period of a heart during the cardiac cycle 518

diatom (di'ah-tom) golden brown alga with a cell wall in two parts, or valves; an important component of the marine food chain 362

dicotyledon (di-kot''i-le'don) a flowering plant group; members show two embryonic leaves, netted-venation leaves, cylindrical arrangement of vascular bundles, and other characteristics 381

differentiation (dif''er-en'she-a'shun) specialization of early embryonic cells with regard to structure and function 678

diffusion (di-fu'zhun) the movement of molecules from a region of high to low concentration, tending toward an equal distribution 81

dinoflagellate (di-no-flaj'ē-lāt) algal phytoplankton with two flagella that is an important component of the marine food chain 362

diploid (2N) (dip'loid) the condition in which cells have two of each type of chromosome 147

diplontic cycle (dip-lon'tik si'k'l) life cycle in which the adult is diploid 356

directional selection (di-rek'shun-al sĕ-lek'shun) evolutionary process that occurs when one atypical phenotype is favored, usually in a changing environment 320

disruptive selection (dis-rup'tiv sĕ-lek'shun) extreme phenotypes are favored over the average phenotype and can lead to polymorphic forms 320

distal convoluted tubule (dis'tal kon'vo-lūt-ed tu'būl) the portion of a nephron between the loop of Henle and the collecting duct where tubular secretion occurs 583

divergent evolution (di-ver'jent ev''o-lu'shun) the formation of new forms in a single lineage 307

DNA see deoxyribonucleic acid

DNA ligase (de'en-a li'gās) an enzyme that links DNA fragments; used in genetic engineering to join foreign DNA to the vector DNA 263

DNA polymerase (de'en-a pol-im'er-ās) enzyme that joins complementary nucleotides to form double-stranded DNA 225

DNA probe (de'en-a prōb) known sequences of DNA that are used to find complementary DNA strands; can be used diagnostically to determine the presence of particular genes 266

dominance hierarchy (dom'i-nans hi'er-ar''ke) organization of animals in a group that determines the order in which the animals have access to resources 713

dominant allele (dom'i-nant ah-lēl') allele of a given gene that hides the effect of the recessive allele in a heterozygous condition 175

Down syndrome (down sin'drōm) a genetic disease caused by trisomy 21 that is characterized by mental retardation and a specific set of physical features 205

Duchenne muscular dystrophy (du-shen' mus'ku-lar dis'tro-fe) an X-linked, recessive disorder that causes muscular weakness and eventually death at a young age 214

dryopithecine (dri''o-pith'ē-sin) a possible hominoid ancestor; a forest-dwelling primate with characteristics somewhat like those of living apes 429

E

echinoderm (e-kin'o-derm) a phylum of marine animals that includes sea stars, sea urchins, and sand dollars; characterized by radial symmetry and a water vascular system 411

ecological pyramid (e''ko-log'i-kal pir'ah-mid) pictorial graph representing the biomass, organism number, or energy content of each trophic level in a food web from the producer to the final consumer populations 768

ecology (e-kol'o-je) the study of the interactions of organisms with their living and physical environment 720

ecosystem (ek''o-sis'tem) a biological community together with the associated abiotic environment 7

ectoderm (ek'to-derm) outermost of an animal's primary germ layers 680

electroencephalogram (EEG) (e-lek''tro-en-sef'ah-lo-gram) a recording of the electrical activity of the brain obtained by taping electrodes to different parts of the scalp 605

electromagnetic spectrum (e-lek''tro-mag-net'ik spek'trum) solar radiation divided on the basis of wavelength with gamma rays having the shortest wavelength and radio waves having the longest wavelength 110

electron transport system (e-lek'tron trans'port sis'tem) a mechanism whereby electrons are passed along a series of carrier molecules to produce energy for the synthesis of ATP 101

element (el'ē-ment) the simplest of substances consisting of only one type of atom; e.g., carbon, hydrogen, oxygen 24

emigrate (em'i-grāt) to move from a geographical source of reference 734

endocrine system (en'do-krin sis'tem) one of the major systems involved in the coordination of bodily activities; uses messengers called hormones that are secreted into the bloodstream 642

endocytosis (en''do-si-to'sis) moving particles or debris into the cell from the environment by phagocytosis (cellular eating) or pinocytosis (cellular drinking) 86

endoderm (en'do-derm) innermost of an animal's primary germ layers 680

endodermis (en'do-der'mis) internal plant root tissue, forming a boundary between the cortex and vascular cylinder 451

endometrium (en-do-me'tre-um) a mucous membrane lining the inside free surface of the uterus 667

endoplasmic reticulum (en''do-plas'mik rē-tik'u-lum) a system of membranous saccules and channels in the cytoplasm 65

endoskeleton (en'do-skel'ē-ton) a protective, internal skeleton, as in vertebrates 627

endosperm (en'do-sperm) nutritive tissue in a seed; often triploid from a fusion of sperm cell with two polar nuclei 483

endospore (en'do-spōr) a bacterium that has shrunk its cell, rounded up within the former plasma membrane, and secreted a new and thicker cell wall in the face of unfavorable environmental conditions 343

energy level (en'er-je lev'el) a designation of the amount of energy of electrons; energy level increases with distance from the nucleus 26

environment (en-vi'ron-ment) surroundings with which a cell or organism interacts 4

environmental resistance (en-vi''ron-men'tal re-zis'tans) the opposing force of the environment on the biotic potential of a population 723

enzyme (en'zīm) protein that acts as an organic catalyst, speeding up reaction rates in living systems 46

epicotyl (ep''i-kot'il) plant embryo portion above the cotyledon that contributes to shoot development 486

epidermis (ep''i-der'mis) covering tissue of roots and leaves of plants, plus stems of nonwoody organisms 446

epiglottis (ep''i-glot'is) a flaplike covering, hinged to the back of the larynx and capable of covering the glottis or air-tract opening 551

epiphyte (ep'i-fīt) a plant that takes its nourishment from the air because its attachment to other plants gives it an aerial position 472

epistatic gene (epistasis) (ēp''i-stat'ik jēn) a gene that interferes with the expression of alleles that are at a different locus 186

erythrocyte (ē-rith'ro-sīt) red blood cell; important for oxygen transport in blood 521

esophagus (ē-sof'ah-gus) a muscular tube for moving swallowed food from pharynx to stomach 551

essential amino acid (ē-sen'shal ah-me'no as'id) an amino acid that is required by the body, but must be obtained in the diet because the body is unable to produce it 557

estuary (es'tu-ar''e) the end of a river where fresh water and salt water mix as they meet 750

ethylene (eth'i-lēn) plant hormone that causes ripening of fruit and is also involved in abscission 497

eubacteria (u''bak-te're-ah) most common types of bacteria; including the photosynthetic bacteria 346

euchromatin (u-kro'mah-tin) diffuse chromatin, which is being transcribed 251

euglenoid (u-gle'noid) freshwater organism with both animal- and plantlike characteristics 360

eukaryotic cell (u''kar-e-ot'ik sel) typical of organisms, except bacteria and cyanobacteria, having organelles and a well-defined nucleus 57

evolution (ev''o-lu'shun) genetic and phenotypic changes that occur in populations of organisms with the passage of time, often resulting in increased adaptation of organisms to the prevailing environment 5

evolutionary tree (ev''o-lu'shun-ar''e tre) a branching diagram linking ancestors and progeny through time 300

excretion (eks-kre'shun) process of removing nitrogenous waste, water, and salts from the body 578

exhalation (eks''hah-la'shun) stage during respiration when air is pushed out of the lungs 568

exocytosis (ex'' o-si-to'sis) a process by which vesicles expel particles or debris through a plasma membrane to the extracellular environment 86

exon (eks'on) the coding portion of genes that are expressed 237

exoskeleton (ek''so-skel'ē-ton) a protective, external skeleton, as in arthropods 627

experimental variable (ek-sper''i-men'tal var're-ah-b'l) condition that is tested in an experiment by varying it and observing the results 14

exponential (ex''po-nen'shal) referring to a geometrically multiplying, rapid population growth rate 720

F

F₁ generation (jen''ē-ra'shun) the first generation offspring of a genetic cross 173

F₁ particle (par'tĕ-k'l) located in the inner membrane of the mitochondria and contains a channel through which hydrogen ions flow during chemiosmotic phosphorylation 133

F₂ generation (jen''e-ra'shun) the second generation offspring of a genetic cross 173

FAD (flavin adenine dinucleotide) (fla'vin ad'ē-nin di-nu'kle-o-tīd) coenzyme that functions as an electron acceptor in cellular oxidation-reduction reactions 101

feedback inhibition (fēd''bak in''hi-bish'un) process by which a substance, often an end product of a metabolic pathway, controls its own continued production 100

fermentation (fer''men-ta'shun) anaerobic breakdown of carbohydrates that results in products such as alcohol and lactic acid 126

fertilization (fer''ti-li-za'shun) joining of haploid egg and sperm to form a diploid zygote 159

fixed-action pattern (fikst ak'shun pat'ern) a stereotyped behavior pattern that occurs automatically 703

fluid-mosaic model (floo'id mo-za'ik mod'el) model for the plasma membrane based on the changing location and pattern of protein molecules in a fluid lipid bilayer 78

follicle (fol'i-k'l) structures in the ovary of animals that contain oocytes; site of egg production 665

food chain (food chān) a succession of organisms in an ecosystem that are linked by an energy flow and the order of who eats whom 767

food web (food web) a complex pattern of interlocking and crisscrossing food chains 767

fossil (fos'il) remains of an organism, usually plant or animal, of a past geological period; usually a petrified part, but sometimes a cast or impression 283

founder effect (fown'der ĕ-fekt') the tendency for a new, small population to experience genetic drift after it has separated from an original, larger population 317

fruit (froot) flowering plant structure consisting of one or more ripened ovaries that usually contain seeds 381

fruiting body (froot'ing bod'e) a spore-bearing structure found in certain types of fungi, such as mushrooms 365

fungus (fun'gus) saprophytic decomposer; body is made up of filaments that are called hyphae and form a mass called a mycelium 363

G

gallbladder (gawl'-blad-der) attached to the liver; serves as a storage organ for bile 553

gamete (gam'ēt) haploid sex cell 159

ganglion (gang'gle-on) a knot or bundle of neuron cell bodies outside the central nervous system 598

gastrovascular cavity (gas''tro-vas'ku-lar kav'i-te) a blind, branched digestive cavity that also serves a circulatory (transport) function in animals that lack a circulatory system 390

gastrula (gas'troo-lah) cup-shaped early embryo with two primary germ layers, ectoderm and endoderm, enclosing a primitive digestive tract 680

gene (jēn) the unit of heredity occupying a particular locus on the chromosome and passed on to offspring 5

gene frequency (jēn fre'kwen-sē) the percentage occurrence of each allele of a given gene in the gene pool of a population 311

gene locus (jēn lo'kus) the specific location of a particular gene on a chromosome 175

gene mutation (jē mu-ta'shun) a change in the base sequence of DNA such that the sequence of amino acids is changed in a protein 243

gene pool (jēn pool) the total of all the genes of all the individuals in a population 311

gene sequencing (jēn se'kwens-ing) determination of the base sequence of a gene 272

gene therapy (jēn thĕr-ah-pe) use of transplanted genes to overcome an inborn error in metabolism 267

genetic drift (je-net'ik drift) evolution by chance processes, as when a population becomes homozygous for one allele of a pair due to nonrandom mating 317

genetic engineering (je-net'ik en''ji-nēr'ing) alteration of the genome of an organism by technological processes 261

genetic recombination (je-net'ik re''kom-by-na'shun) process whereby chromosomes in daughter cells contain genes from both parents 165

genomic library (je-nom'ik li'brer-e) a collection of engineered bacteriophages that together carry all of the genes of the species 262

genotype (jen'o-tīp) the genes of an organism for a particular trait or traits; ie., *BB* or *Aa* 175

germ layer (jerm la'er) developmental layer of body; i.e., ectoderm, mesoderm, and endoderm 680

gibberellin (gib-ber-el'in) plant hormone producing increased stem growth by cell division and enlargement; also involved in flowering and seed germination 494

girdling (ger'd'l-ing) removing a strip of bark from around a tree 473

glomerulus (glo-mer'u-lus) a capillary network within a Bowman's capsule of a nephron 584

glottis (glot'is) opening for airflow in the larynx 569

glycolysis (gli-kol'i-sis) pathway of metabolism converting a sugar, usually glucose, to pyruvate; resulting in a net gain of 2 ATP and 2 NADH molecules 125

goiter (goi'ter) an enlargement of the thyroid gland caused by a lack of iodine in the diet 650

Golgi complex (gol'je kom'pleks) an organelle, consisting of a central region of saccules and vesicles, that modifies and/or activates proteins and produces lysosomes 66

grana (gra'nah) the membrane of the stroma in chloroplasts; made up of a stack of flattened disks that contain chlorophyll 111

gravitropism (grav'i-tro'pizm) growth in response to the earth's gravity; roots demonstrate positive gravitropism 499

greenhouse effect (grēn'hows ĕ-fekt) the reradiation of heat toward the earth caused by gases in the atmosphere 790

guanine (gwan'in) one of four bases in the structure of DNA and RNA 222

guttation (gut-ta'shun) liberation of water droplets from the edges and tips of leaves caused by guttation 465

gymnosperm (jim'no-sperm) a vascular plant producing naked seeds, as in conifers 378

H

habitat (hab'i-tat) region of an ecosystem occupied by an organism 732

haploid (N) (hap'loid) a cell condition in which only one of each type of chromosome is present 147

haplontic cycle (hap-lon'tik si'kl) life cycle in which the adult is haploid 356

Hardy-Weinberg law (har'de vīn'berg law) a law stating that the frequency of an allele in a population remains stable under certain assumptions, such as random mating; therefore, no change or evolution occurs 311

helper T cell (hel'per te sel) secretes lymphokines, which stimulate all kinds of immune cells 537

hemoglobin (he''mo-glo'bin) red respiratory pigment of erythrocytes for transport of oxygen 512

hemophilia (he''mo-fil'e-ah) a disease in which the blood fails to clot; caused by insufficient clotting factor VIII 213

herbaceous (her-ba'shus) characterizes a plant that lacks persistent woody tissue 381

herbivore (her'bi-vōr) a primary consumer in a food chain; a plant eater 546

hermaphroditic (her'maf''ro-dit'ik) characterizes an animal having both male and female sex organs 392

heterochromatin (het''er-o-kro'mah-tin) highly compacted chromatin during interphase 251

heterogamete (het''er-o-gam'ēt) a nonidentical gamete, e.g., large, nonmotile egg and small, flagellated sperm 358

heterotroph (het'er-o-trōf'') an organism that cannot synthesize organic compounds from inorganic substances and therefore must take in preformed food 331

heterozygote superiority (het''er-o-zi'gōt su-pe''re-or'i-te) increased fitness of individuals having the heterozygote genotype over those with either homozygous genotype 316

heterozygous (het''er-o-zi'gus) possessing two different alleles for a particular trait 175

homeostasis (ho''me-o-sta'sis) the maintenance of internal conditions in cell or organisms, e.g., relatively constant temperature, pH, blood sugar 4

hominid (hom'ĭ-nid) referring to humans; ancestors of man who had diverged from the ape lineage 431

hominoid (hom'ĭ-noid) term for humans and apes 429

Homo erectus (ho'mo ē-rek'tus) hominid who lived during the Pleistocene; with a posture and locomotion similar to modern humans 435

Homo habilis (ho'mo hah'bĭ-lis) fossil hominid of 2 million years ago; possible first direct ancestor of modern humans 433

homologous chromosome (ho-mo'o-gus kro'mo-sōm) a pair of chromosomes that carry genes for the same traits, and synapse during prophase of the first meiotic division 159

homologous structures (ho-mol'o-gus struk'tūr) in evolution, structures derived from a common ancestor 299

homologue (hom'o-log) member of homologous pair of chromosomes 159

homozygous (ho''mo-zi'gus) processing two identical alleles for a particular trait 175

hormone (hor'mōn) a chemical secreted into the bloodstream in one part of the body that controls the activity of other parts 493

humus (hu'mus) decomposing organic matter in the soil 757

Huntington disease (HD) (hunt'ing-tun dĭ-zēz) a lethal, genetic disease affecting 1 in 10,000 people in the U.S.; affects neurological functioning and generally does not manifest itself until middle age 212

hybrid (hi'brid) organism that results from a cross of two different species, or strains 314

hybrid vigor (hi'brid vig'or) a phenomenon whereby a heterozygote resulting from the crossing of two different strains is more fit than either homozygous parent 314

hydra (hi'drah) a freshwater cnidarian with a dominant polyp stage that reproduces both sexually and asexually 390

hydrogen bond (hi'dro-jen bond) a weak bond that arises between a partially positive hydrogen and a partially negative oxygen, often on different molecules or separated by some distance 30

hydrolysis (hi-drol'ĭ-sis) splitting of a compound into parts in a reaction that involves addition of water, with the H⁺ ion being incorporated in one fragment and the OH⁻ ion in the other 40

hydrophilic (hi''dro-fil'ik) polar; has a tendency to interact with water 39

hydrophobic (hi''dro-fo'bik) nonpolar; without a tendency to interact with water 39

hydroponics (hi''dro-pon'iks) a water culture method of growing plants that allows an experimenter to vary the nutrients and minerals provided so as to determine specific growth requirements 470

hydrostatic skeleton (hi''dro-stat'ik skel'ē-ton) characteristic of animals in which muscular contractions are applied to fluid-filled body compartments 405

hypertonic (hi''per-ton'ik) lesser water concentration, greater solute concentration than the cytoplasm of a particular cell; situation in which cell tends to lose water 83

hypha (hi'fa) a filament of the vegetative body of a fungus 363

hypocotyl (hi''po-kot'il) plant embryo portion below the cotyledon that contributes to development of stem 486

hypothalamus (hi''po-thal'ah-mus) portion of the brain that regulates the internal environment of the body; involved in control of heart rate, body temperature, water balance, and glandular secretions of the stomach and pituitary gland 602

hypothesis (hi-poth'ē-sis) a supposition that is established by reasoning after consideration of available evidence and that can be tested by obtaining more data, often by experimentation 13

hypotonic (hi''po-ton'ik) greater water concentration, lesser solute concentration than the cytoplasm of a particular cell; situation in which a cell tends to gain water 83

I

immune system (i-mūn sis'tem) includes many leukocytes and various organs and provides specific defense against foreign antigens 532

imprinting (im'print-ing) a behavior pattern with both learned and innate components that is acquired during a specific and limited period, usually when very young 708

incomplete dominance (in''kom-plēt' dom'ĭ-nans) a pattern of inheritance in which the offspring show characteristics intermediate between the parental characteristics, e.g., a red and a white flower producing pink offspring 185

independent assortment (in''de-pend'ent ah-sort'ment) in heredity, referring to the independent segregation of the allelic pair relative to another pair during sex cell formation; i.e., *AaBb* genotype yields *AB,Ab,aB,ab* in gametes 178

inducer (in-dūs'er) a molecule that brings about activity of an operon by joining with a repressor and preventing it from binding to the operator 250

inducible operon (in-dūs'ĭ-b'l op'er-on) an operon that is normally inactive but is turned on by an inducer 250

inductive reasoning (in-duk'tiv re'zun-ing) a process of logic and reasoning, using specific observations to arrive at a hypothesis 13

inhalation (in''hah-la'shun) stage during respiration when air is drawn into the lungs 568

inheritance of acquired characteristics (in-her'ĭ-tans uv ah-kwīrd' kar''ak-ter-is'tiks) Lamarckian belief that organisms become adapted to their environment during their lifetime and passed on these adaptations to their offspring 283

innate behavior (in-nāt' be-hāv'yor) instinctive; as in behavior patterns with nonlearned components to them 703

insight learning (in'sīt lern'ing) ability to respond correctly to a new, different situation the first time it is encountered by using reason 710

interferon (in''ter-fer'on) an antiviral agent produced by an infected cell that blocks the infection of another cell 532

interphase (in'ter-fāz) stage of cell cycle during which DNA synthesis occurs and the nucleus is not actively dividing 149

intron (in'tron) noncoding segments of DNA that are transcribed but removed before mRNA leaves nucleus 237

invertebrate (in-ver'tĕ-brāt) referring to an animal without a serial arrangement of vertebrae or a backbone 387

ionic (i-on'ik) an attraction between charged particles (ions) through their opposite charge that involves a transfer of electrons from one atom to another 28

isomer (i'so-mer) molecules with the same molecular formula but different structure and, therefore, shape 39

isotonic (i''so-ton'ik) a solution that is equal in solute and water concentration to that of the cytoplasm of a particular cell; situation in which cell neither loses nor gains water 83

isotope (i'so-tōp) forms of an element having the same atomic number but a different atomic weight due to a different number of neutrons 25

J

jointed appendage (joint'ed ah-pen'dij) freely-moving appendages of arthropods 406

K

K-strategist (ka strat'e-jist) a species that has evolved characteristics that keep its population size near carrying capacity, e.g., few offspring, longer generation time 726

karyotype (kar'e-o-tīp) an organized display of chromosomes cut from a photograph of the nucleus just prior to cell division 203

killer T cell (kil'er te sel) cell that attacks and destroys cells that bear a foreign antigen 537

kin selection (kin sĕ-lek'shun) an explanation for the evolution of altruistic behavior that benefits one's relatives 715

kingdom (king'dum) a taxonomic category grouping related phyla (animals) or divisions (plants) 2

Klinefelter syndrome (klīn'fel-ter sin'drōm) a condition caused by the inheritance of a chromosome abnormality in number; an XXY (trisomy) condition where normally a chromosome pair exists 207

Krebs cycle (krebz si'k'l) *see* citric acid cycle

L

lacteal (lak′te-al) a lymphatic vessel in an intestinal villus; aids in the absorption of fats 554

larynx (lar′inks) voicebox, cartilage-made box for airflow between the glottis and trachea 569

leukocyte (lu′ko-sīt) white blood cell; important for immune response 521

lichen (li′ken) a symbiotic relationship between certain fungi and algae that has long been thought to be mutualistic, the fungi providing inorganic and the algae providing organic food 348

light-dependent reaction (līt de-pen′dent re-ak′shun) a reaction of photosynthesis that requires light energy to proceed; involved in the production of ATP and NADH 112

light-independent reaction (līt in′′de-pen′dent re-ak′shun) photosynthetic reaction that does not require light energy; involved primarily in the conversion of carbon dioxide to carbohydrate 112

limbic system (lim′bic sis′tem) part of the brain beneath the cortex that contains pathways that connect various portions of the brain; allows the experience of many emotions 605

linkage group (lingk′ij groop) genes that are located on the same chromosome and tend to be inherited together 193

lipase (lip′ās) an enzyme that digests fat molecules in the small intestine 553

lipid (lip′id) a family of organic compounds that tend to be soluble in nonpolar solvents such as alcohol; includes fats and oils 44

liposome (lip′o-sōm) droplet of phospholipid molecules formed in a liquid environment 331

loop of Henle (loop uv hen′le) the portion of a nephron between the proximal and distal convoluted tubules where some water reabsorption occurs 583

lymph (limf) tissue fluid collected by the lymphatic system and returned to the general systemic circuit 520

lymphokine (lim′fo-kīn) chemicals, secreted by T cells, that stimulate immune cells 537

lysogenic cycle (li-so-jen′ik si′kl) one of the bacteriophage life cycles in which the virus incorporates its DNA into that of the bacterium; only later does it begin a lytic cycle that ends with the destruction of the bacterium 340

lysosome (li′so-sōm) membrane-bound vesicles that contain hydrolytic enzymes to digest macromolecules 67

lytic cycle (lit′ik si′kl) one of the bacteriophage life cycles in which the virus takes over the operation of the bacterium immediately upon entering it and subsequently destroys the bacterium 339

M

macroevolution (mak′′ro-ev′′o-lu′shun) large-scale evolutionary change, e.g., the formation of new species 303

macrophage (mak′ro-fāj) large, phagocytotic white blood cell 524

MHC protein (**major histocompatibility complex**) (em āch se pro′tein) protein on the plasma membrane of a macrophage that aids in the stimulation of T cells 336

Malpighian tubules (mal-pig′i-an tu′būl) blind, threadlike excretory tubules near anterior end of insect hindgut 408

mammal (mam′al) a class of vertebrates characterized especially by the presence of hair and mammary glands 425

marsupial (mar-su′pe-al) a mammal bearing immature young nursed in a marsupium, or pouch; i.e., kangaroo, opossum 426

mass extinction (mas eks-ting′shun) recorded in the fossil record as a very large number of extinctions over a relatively short geologic time 298

mechanoreceptor (mek′′ah-no-re-sep′tor) a cell that is sensitive to mechanical stimulation, such as that from pressure, sound waves, and gravity 618

medulla oblongata (mě-dul′ah ob′′long-ga′tah) a brain region at the base of the brainstem controlling heartbeat, blood pressure, breathing and other vital functions 602

meiosis (mi-o′sis) type of nuclear division that occurs during the production of gametes, in which the daughter cells receive the haploid number of chromosomes 147

meiosis I (mi-o′sis wun) first division of meiosis that ends when duplicated homologous chromosomes separate and move into different cells 162

meiosis II (mi-o′sis too) second division of meiosis in which sister chromatids separate and move into different cells, resulting in the formation of gametes 165

memory B cell (mem′o-re be sel) a B lymphocyte that remains in the bloodstream after an immune response ceases and is capable of producing antibodies to an antigen 534

memory T cell (mem′o-re te sel) a T lymphocyte that remains in the bloodstream after an immune response ceases 537

menstruation (men′′stroo-a′shun) periodic shedding of tissue and blood from the inner lining of the uterus 667

meristem (mer′i-stem) undifferentiated, embryonic tissue in the active growth regions of plants 445

mesoderm (mes′o-derm) middle layer of an animal's primary germ layers 680

mesoglea (mes′′-o-gle′ah) a jellylike layer between the epidermis and endodermis of cnidarians 390

messenger RNA (mRNA) (mes′en-jer ar′en-a) a nucleic acid (ribonucleic acid) complementary to genetic DNA and bearing a message to direct cell protein synthesis at the ribosome 237

metabolic pool (mě-ta-bo′lik pool) metabolites that are the products of and/or the substrates for key reactions in cells allowing one type of molecule to be changed into another type, such as the conversion of carbohydrates to fats 137

metabolism (mě-tab′o-lizm) all of the chemical reactions that occur in a cell during growth and repair 4

metafemale (met′′ah-fe′māl) a female with three X chromosomes; most show no obvious physical abnormalities 207

metamorphosis (met′′ah-mor′fo-sis) change in shape and form that some animals, such as insects, undergo during development 409

microbody (mi′kro-bod′ē) membrane-bound vesicle containing specific enzymes involved in lipid and alcohol metabolism, photosynthesis, and germination 68

microfilament (mi′′kro-fil′ah-ment) long, thin fibers that occur in bundles, e.g., actin microfilaments that are involved in muscle contraction 73

microsphere (mi-kro-sfēr) formed from proteinoids exposed to water; has properties similar to today's cells 331

microtubule (mi′′kro-tu′būl) small cylindrical organelle that is believed to be involved in maintaining the shape of the cell and directing various cytoplasmic structures 70

mimicry (mim′ik-re) superficial resemblance of an organism to one of another species; often used to avoid predation 740

mitochondrion (mi′′to-kon′dre-an) membrane-bound organelle of the cell known as the powerhouse because it transforms the energy of carbohydrates and fats to ATP energy 68

mitosis (mi-to′sis) a process in which a parent nucleus reproduces two daughter nuclei, each identical to the parent nucleus; this division is necessary for growth and development 150

model (mod′el) a suggested explanation for experimental results that can help direct future research 16

molecule (mol′ē-kūl) a unit of a chemical substance formed by the union of two or more atoms by covalent bonding 27

mollusk (mol′usk) a vertebrate phylum that includes squids, clams, snails, and chitons and is characterized by a visceral mass, a mantle, and a foot 401

molt (mōlt) periodic shedding of the exoskeleton in arthropods 406

monocotyledon (mon′′o-kot′′i-le′don) a flowering plant group; members show one embryonic leaf, parallel leaf venation, scattered vascular bundles, and other characteristics 381

monotreme (mon′o-trēm) an egg-laying mammal; i.e., duckbill platypus or spiny anteater 426

morphogenesis (mor″fo-jen′ĕ-sis) movement of early embryonic cells to establish body outline and form 678

multiple allele (mul′ti-p'l ah-lēl) more than two alleles for a particular trait 185

muscle fiber (mus′el fi′ber) a cell with myofibrils containing actin and myosin filaments arranged within sarcomeres; a group of muscle fibers is a muscle 634

mutation (mu-ta′shun) a change in the composition of DNA, due to either a chromosomal or a genetic alteration 5

mutualism (mu′tu-al-izm″) a symbiotic relationship in which both species benefit 742

mycelium (mi-se′le-um) a tangled mass of hyphal filaments composing the vegetative body of a fungus 363

mycorrhiza (mi″ko-ri′zah) a 'fungal root' composed of a fungus growing around a plant root that assists the growth and development of the plant 472

myelin sheath (mi′ĕ-lin shēth) covering on neurons that is formed from Schwann cell membrane and aids in transmission of nervous impulses 594

myocardium (mi″o-kar′de-um) muscular wall of the heart 517

myofibril (mi″o-fi′bril) a specific muscle cell organelle containing a linear arrangement of sarcomeres that shorten to produce muscle contraction 635

myosin (mi′o-sin) a muscle protein making up the thick filaments in a sarcomere; it pulls actin to shorten the sarcomere, yielding muscle contraction 636

N

NAD⁺ (nicotinamide-adenine dinucleotide) (nik″o-tin′ah-mīd ad′ĕ-nīn di-nu′kle-o-tīd) a coenzyme that functions as an electron and hydrogen ion carrier in cellular oxidation-reduction reactions of glycolysis and cellular respiration 101

NADP⁺ (nicotinamide-adenine dinucleotide phosphate) (nik″o-tin′ah-mīd ad′ĕ-nīn di-nu′kle-o-tīd fos′fāt) a coenzyme that functions as an electron and hydrogen ion carrier in cellular oxidation-reduction reactions of photosynthesis 102

natural selection (nat′u-ral sĕ-lek′shun) the guiding force of evolution caused by environmental selection of organisms most fit to reproduce, resulting in adaptation 287

Neanderthal (ne-an′der-thawl) hominid who lived during the last ice age in Europe and the Middle East; made stone tools, hunted large game, and lived together in a kind of society 435

nematocyst (nem′ah-to-sist) threadlike fiber enclosed within the capsule of a stinging cell; when released, aids in the capture of prey 390

nephridia (sing., **nephridium**) (nĕ-frid′e-ah) for invertebrates, a tubular structure for excretion; its contents are released outside through a nephridiopore 405

nephron (nef′ron) microscopic kidney unit that regulates blood composition by filtration and reabsorption; over one million per human kidney 583

nerve (nerv) bundle of axons and/or dendrites 598

neuromuscular junction (nu″ro-mus′ku-lar junk′shun) region where a nerve end comes into contact with a muscle fiber; contains a presynaptic membrane, synaptic cleft, and postsynaptic membrane 637

neuron (nu′ron) a nerve cell; composed of dendrite(s), a cell body, and axon 593

neurotransmitter (nu″ro-trans-mit′er) a chemical made at the ends of axons that is responsible for transmission across a synapse or neuromuscular junction 597

neutrophil (nu′tro-fil) white blood cell with granules 523

niche (nich) total description of an organism's functional role in an ecosystem, from activities to reproduction 732

nitrogen fixation (ni″tro-jen fik-sa′shun) process whereby free atmospheric nitrogen is converted into compounds, such as ammonia and nitrates, usually by soil bacteria 774

nitrogen-fixing (ni″tro-jen fiks-ing) an organism that can convert atmospheric nitrogen into compounds such as ammonia and nitrates, usually carried out by soil bacteria associated with the organism 346

nodules (nod′ūlz) structures on plant roots that contain nitrogen-fixing bacteria 472

noncyclic photophosphorylation (non-sik′lik fo″to-fos″for-i-la′shun) photosynthetic pathway in which electrons from photosystem P680 pass through the electron transport system and move on to another photosystem 114

nondisjunction (non″dis-junk′shun) failure of homologous chromosomes (or chromatid pair) to separate during meiosis, yielding an abnormal chromosome number in the produced sex cell 197

nonrenewable resources (non″re-nu′ah-b'l resors′ez) resources found in limited supplies that can be used up 791

norepinephrine (nor″ep-ah-nef′ren) a hormone released by the adrenal medulla as a reaction to stress; also a neurotransmitter released by neurons of the central and peripheral nervous systems 598

notochord (no′to-kord) a dorsal, elongated, supporting structure in chordates beneath the neural tube or spinal cord; present in the embryo of all vertebrates 413

nuclear envelope (nu′kle-ar en′vĕ-lōp) double membrane that separates the nucleus from the cytoplasm 64

nuclear power (nu′kle-ar pow′er) energy source generation by atomic reactions 792

nucleic acid (nu-kle′ik as′id) polymer of nucleotides; both DNA and RNA are nucleic acids 50

nucleoid (nu′kle-oid) area in prokaryotic cell where DNA is found 57

nucleoli (sing., **nucleolus**) (nu′kle′o-li) dark-staining, spherical bodies in the cell nucleus that contains ribosomal RNA 64

nucleotide (nu′kle-o-tīd) the building block subunit of DNA and RNA consisting of a five-carbon sugar bonded to a nitrogen base and phosphorus group 50

nucleus (nu′kle-us) control center of a eukaryotic cell that oversees the metabolic function of the cell 62

O

olfactory cell (ol-fak′to-re sel) modified neuron that is a receptor for the sense of smell 613

omnivore (om′ni-vōr) a food chain organism feeding on both plants and animals 546

oncogene (ong′ko-jēn) a cancer-causing gene 256

oocyte (o′o-sīt) the largest, function product of meiosis that becomes the egg 165

oogenesis (o″o-jen′ĕ-sis) production of eggs in females by the process of meiosis and maturation 165

operant conditioning (op′ĕ-rant kon-dish′un-ing) trial and error learning in which behavior is reinforced with a reward or a punishment 709

operator (op′er-a-tor) the sequence of DNA in an operon to which the repressor protein binds 250

operon (op′er-on) a group of structural and regulating genes that functions as a single unit 250

orbital (or′bi-tal) volume of space around a nucleus where electrons can be found most of the time 26

organelle (or″gan-el′) small, membranous structure in the cytoplasm having a specific function 57

organic (or-gan′ik) containing carbon and hydrogen; such molecules usually also have oxygen attached to the carbon(s) 38

organic soup (or-gan′ik sōōp) accumulation of simple organic compounds in the early oceans 331

organ of Corti (or′gan uv kor′ti) specialized region of the cochlea containing the hair cells for sound detection and discrimination 623

osmosis (oz-mo′sis) the diffusion of water through a selectively permeable membrane 82

osmotic pressure (os-mot′ik presh′ur) measure of the tendency of water to move across a selectively permeable membrane into a solution; visible as an increase in liquid on the side of the membrane with higher solute concentration 82

ovary (o′vah-re) in flowering plants enlarged, base portion of the pistil that eventually develops into the fruit; in animals an egg producing organ 480

oviduct (o′vi-dukt) tubular portion of the female reproductive tract from ovary to the uterus 666

ovulation (o″vu-la′shun) bursting of a follicle at the release of an egg from the ovary 666

ovule (o′vūl) in seed plants, a structure that contains the megasporangium where meiosis occurs and the female gametophyte is produced; develops into the seed 480

ovum (o′vum) a released oocyte; egg 666

oxidation (ok″si-da′shun) a chemical reaction that results in removal of one or more electrons from an atom, ion, or compound; oxidation of one substance occurs simultaneously with reduction of another 31

oxidative phosphorylation (ok″si-da′tiv fos″for-i-la′shun) the process of building ATP by an electron transport system that uses oxygen as the final receptor; occurs in mitochondria 133

oxidizing atmosphere (ok″si-dī′zing at′mos-fēr) an atmosphere with free oxygen, e.g., the atmosphere of today 329

oxygen debt (ok′si-jen det) use of oxygen to convert lactate, which builds up due to anaerobic conditions, to pyruvate 130

ozone shield (o′zōn shēld) formed from oxygen in the upper atmosphere; protects the earth from ultraviolet radiation 333

P

P generation (jen″ĕ-ra′shun) the parent organisms of a genetic cross 173

palisade layer (pal-a-sād la′er) layer in a plant leaf containing elongated cells with many chloroplasts; along with the spongy layer, it is the site of most of photosynthesis 459

pancreas (pan′kre-as) an abdominal organ that produces digestive enzymes and the hormones insulin and glucagon 653

pancreatic amylase (pan″kre-at′ik am′i-lās) enzyme that digests starch to maltose 553

parallel evolution (par′ah-lel ev″o-lu′shun) evolutionary pattern seen when related species change similarly when subjected to the same adaptive pressures, e.g., African ostrich and South American rhea 307

parasitism (par′ah-si″tizm) a symbiotic relationship in which one species (parasite) benefits for growth and reproduction to the harm of the other species (host) 741

parasympathetic nervous system (par″ah-sim″path-thet′ik ner′vus sis′tem) a division of the autonomic nervous system that is involved in normal activities; uses acetylcholine as a neurotransmitter 600

parathyroid gland (par″ah-thi′roid gland) embedded in the posterior surface of the thyroid gland; produces parathyroid hormone, which is involved in the regulation of calcium and phosphorus balance 651

parthenogenesis (par″thĕ-no-jen′-ĕ-sis) development of an egg cell into a whole organism without fertilization 660

pelagic (pe-laj′ik) the open portion of the sea 752

penis (pe′nis) male copulatory organ 664

pepsin (pep′sin) enzyme that digests proteins into peptides 552

peptide (pep′tīd) two or more amino acids joined together 46

peripheral nervous system (PNS) (pĕ-rif′er-al ner′vus sis′tem) made up of the nerves that branch off of the central nervous system 592

peristalsis (per″i-stal′sis) the rhythmic, wavelike contraction that moves food through the digestive tract 551

petiole (pet′ē-ōl) part of a plant leaf that connects the blade to the stem 458

pH scale (pe-āch skāl) measurement scale for the relative concentrations of hydrogen (H^+) and hydroxyl (OH^-) ions in a solution 33

pharynx (far′inks) a common passageway (throat) for both food intake and air movement 569

phenylketonuria (PKU) (fen″il-ke″to-nu′re-ah) a genetic disorder characterized by the absence of an enzyme that metabolizes phenylalanine; results in severe mental retardation 211

phenotype (fe′no-tīp) the visible expression of a genotype; i.e., brown eyes, height 175

pheromone (fer′o-mōn) substance produced and discharged into the environment by an organism that influences the behavior of another organism 706

phloem (flo′em) vascular tissue conducting organic solutes in plants; contains sieve-tube cells and companion cells 374

phospholipid (fos″fo-lip′id) a molecule having the same structure as a neutral fat except one of the bonded fatty acids is replaced by a phosphorous-containing group; an important component of cell membranes 45

photochemical smog (fo″to-kem′i-kal smog) air pollution that contains nitrogen oxides and hydrocarbons that react to produce ozone and peroxyacetyl nitrate 788

photon (fo′ton) discrete packet of light energy; the amount of energy in a photon is inversely related to the wavelength of light in the photon 110

photoperiod (fo″to-pe′re-od) relative lengths of alternating light and dark periods 501

photoperiodism (fo″to-pe′re-od-izm) relative lengths of daylight and darkness that affect the physiology and behavior of an organism 501

photoreceptor (fo″to-re-sep′tor) light sensitive receptor cell 614

photosynthesis (fo″to-sin′thĕ-sis) a process whereby chlorophyll-containing organisms trap sunlight energy to build a sugar from carbon dioxide and water 102

photosystem (fo″to-sis′tem) a photosynthetic unit where light is absorbed; contains an antenna (photosynthetic pigments) and an electron acceptor 113

phototropism (fo-tot′ro-pizm) movement in response to light; in plants, growth toward or away from light 499

phyletic evolution (fi-let′ik ev″o-lu′shun) gradual change over time in a lineage 303

phytochrome (fī′to-krōm) a plant photoreversible pigment; this reversion rate of two molecular forms is believed to measure the photoperiod 501

pinocytosis (pi″no-si-to′sis) process by which vesicles form around and bring large-sized molecules into the cell 86

pistil (pis′til) a female flower structure consisting of an ovary, style, and stigma 480

pituitary gland (pi-tu′i-tār″ĕ gland) a small gland that lies just below the hypothalamus and is important for its hormone storage and production activities 644

placenta (plah-sen′tah) a structure formed from an extraembryonic membrane, the chorion and uterine tissue through which nutrient and waste exchange occur for the embryo and fetus 686

placental mammal (plah-sen′tal mam′al) a mammalian subclass that is characterized by a placenta, which is an organ of exchange between maternal and fetal blood that supplies nutrients to the growing offspring 426

plankton (plank′ton) fresh- and saltwater organisms that float on or near the surface of the water 747

plasma (plaz′mah) liquid portion of blood; contains nutrients, wastes, salts, and proteins 521

plasma cell (plaz′mah sel) an activated B cell that is currently secreting antibodies 534

plasma membrane (plaz′mah mem′brān) membrane that separates the contents of a cell from the environment and regulates the passage of molecules into and out of the cell 56

plasmid (plaz′mid) a self-duplicating ring of accessory DNA in the cytoplasm of bacteria 261

platelet (plāt′let) cell-like disks formed from fragmentation of megakaryocytes that are involved in blood clotting in vertebrate blood 521

pleiotropy (pli-ot′ro-pe) a gene that affects several different traits 187

plumule (ploo′mūl) embryonic plant shoot that bears young leaves 486

polar body (po′lar bod′e) a nonfunctional product of oogenesis; three of the four meiotic products are of this type 165

polar covalent (po′lar ko-va′lent) a bond in which the sharing of electrons between atoms is unequal 29

pollen grain (pol′en grān) a male gametophyte in vascular plants 377

pollination (pol″i-na′shun) transfer of pollen from an anther to a stigma 378

pollutants (pŏ-lūt′ants) substances that are added to the environment and lead to undesirable effects for living organisms 781

polygenic inheritance (pol''ĕ-jēn'ik in-her'ĭ-tans) a pattern of inheritance in which a trait is controlled by several genes that segregate independently 186

polymerization (pol''ĭ-mer''ĭ-za'shun) process by which larger molecules are formed by the joining of smaller molecules 331

polymorphism (pol''e-mor'fizm) the occurrence in the same habitat of two or more distinct forms of a species 316

polyploid (polyploidy) (pol'e-ploid) a condition in which an organism has more than two sets of chromosomes 196

polysome (pol'e-sōm) a string of ribosomes, simultaneously translating different regions of the same mRNA strand during protein synthesis 240

population (pop''u-la'shun) a group of organisms of the same species occupying a certain area and sharing a common gene pool 7

pressure-flow model (presh'ur flo mod'el) model explaining transport through sieve-tube cells of phloem, with leaves serving as a "source" and roots as a "sink" in the summer, and vice versa in the spring 474

primate (pri'māt) an animal order of mammals having hands and feet with five distinct digits, and some having an opposable thumb 427

primitive atmosphere (prim'ĭ-tiv at'mos-fēr) earliest atmosphere of the earth, after the planet's formation 329

producer (pro-dūs'er) organisms in a food chain that make their own foods; they start the chain (e.g., green plants on land and algae in water) 766

prokaryotic cell (pro''kar-e-ot'ik sel) the first, primitive cells on earth, exemplified today by bacteria, which lack a defined nucleus and most organelles 57

promoter (pro-mo'ter) a sequence of DNA in an operon where RNA polymerase begins transcription 250

protein (pro'te-in) macromolecule whose primary structure is produced by peptide bonding of amino acids 46

proteinoid (pro'te-in-oid'') abiotically polymerized amino acids that are joined in a preferred manner; possible early step in cell evolution 331

protocell (pro'to sel) a cell forerunner developed from cell-like microspheres 331

protoplast (pro'to-plast) 'naked' cells that have had their cell walls removed; used in genetic engineering to allow a plasmid to enter a plant cell 267

protostome (pro'to-stōm) a group of coelomate animals in which the blastopore (the first embryonic opening) becomes the mouth 400

protozoan (pro-ti-zo'en) animal-like, heterotrophic, unicellular organisms 353

proximal convoluted tubule (prok'sĭ-mal kon'vo-lūt-ed tu'būl) the portion of a nephron following the Bowman's capsule where selective reabsorption of filtrate occurs 583

pseudocoelom (su''do-se'lom) a coelom that is not completely lined by mesoderm 394

pulmonary (pul'mo-ner''e) referring to lung 514

Punnett square (pun'et skwār) a gridlike device that enables one to calculate the results of simple genetic crosses by lining gametic genotypes of two parents on the outside margin and their recombination in boxes inside the grid 176

purine (pu'rin) a type of nitrogenous base, such as adenine and guanine, having a double-ring structure 222

pyrimidine (pi-rim'ĭ-din) a type of nitrogenous base, such as cytosine, thymine, and uracil 222

pyruvate (pi'roo-vāt) the end product of glycolysis; its further fate involving fermentation or Krebs cycle incorporation depending on oxygen availability 125

R

r-strategist (ar strat'ĕ-jist) a species that has evolved characteristics that maximize its rate of natural increase, e.g., high birthrate 726

radially symmetrical (ra'de-al-e sĭ-met're-kal) body plan in which similar parts are arranged around a central axis like spokes of a wheel 390

radicle (rad'ĭ-k'l) part of the plant embryo that contains the root apical meristem and becomes the first root of the seedling 486

receptor (re-sep'tor) a cell, often in groups, that detects change or stimulus and initiates a nerve impulse 613

recessive allele (re-ses'iv ah-lēl') form of a gene whose effect is hidden by the dominant allele in a heterozyote 175

recombinant DNA (re-kom'bĭ-nant de'en-a) DNA that contains genes from more than one source 263

reducing atmosphere (re-dūs'ing at'mos-fēr) an atmosphere with little if any free oxygen, e.g., the primitive atmosphere 329

reduction (re-duk'shun) a chemical reaction that results in addition of one or more electrons to an atom, ion, or compound. Reduction of one substance occurs simultaneously with oxidation of another 31

reflex (re'fleks) an automatic, involuntary response of an organism to a stimulus 600

regeneration (re-jen''er-a'shun) reforming a lost body part from the remaining body mass by active cell division and growth 660

regulator gene (reg'u-la''tor jēn) gene that codes for a repressor protein 250

renewable resources (re-nu'ah-b'l re-sors'ez) resources that can be replenished by physical or biological means 791

replication (rĕ''pli-ka'shun) the duplication of DNA; occurs when the cell is not dividing 224

repressible operon (re-pres'ĭ-b'l op'er-on) an operon that is normally active because the repressor must combine with a corepressor before the complex can bind to the operator 250

repressor (re-pres'or) protein molecules that binds to an operator site, preventing RNA polymerase from binding to the promotor site of an operon 250

reproduce (re''pro-dūs') make a copy similar to oneself; e.g., bacteria dividing to produce more bacteria, egg and sperm joining to produce offspring in more advanced organisms 5

reptile (rep'tĭl) includes snakes, lizards, turtles, and crocodilians; terrestrial vertebrate with internal fertilization, scaly skin, and an egg with a leathery shell 422

resting potential (rest'ing po-ten'shal) state of an axon when the membrane potential is about −65 mV and no impulse is being conducted 595

restriction enzyme (re-strik'shun en'zīm) enzyme that stops viral reproduction by cutting viral DNA; used in genetic engineering to cut DNA at specific points 262

retina (ret'ĭ-nah) innermost layer of the eyeball containing the light receptors—rods and cones 616

retrovirus (ret''ro-vi'rus) RNA viruses containing the enzyme reverse transcriptase that carries out RNA/DNA transcription; include viruses that carry cancer-causing oncogenes and the AIDS virus 340

rhizoid (ri'zoid) rootlike hairs that anchor a plant and absorb minerals and water from the soil 372

rhizome (ri'zōm) a rootlike, underground stem 457

rhodopsin (ro-dop'sin) a light-absorbing molecule in rods and cones that contains a pigment and its attached protein 617

ribonucleic acid (RNA) (ri''bo-nu-kle'ik as'id) a nucleic acid with a sequence of subunits dictated by DNA; involved in protein synthesis 219

ribosomal RNA (rRNA) (ri''bo-sōm''al ar'en-a) type of RNA found in ribosomes; sometimes called structural RNA 239

ribosome (ri'bo-sōm) cytoplasmic organelle; site of protein synthesis 65

RNA *see* ribonucleic acid

RNA polymerase (ar'en-a pol-im'er-ās) enzyme that links ribonucleotides together during transcription 237

rod (rod) photoreceptors in vertebrate eyes that respond to dim light 616

root pressure (root presh'ur) a force generated by an osmotic gradient that serves to elevate sap through xylem a short distance 465

RuBP five-carbon compound that combines with carbon dioxide during the Calvin-Benson cycle and is later regenerated by the same cycle 117

S

SA node (es a nōd) mass of specialized tissue in right atrium wall of heart initiates a heartbeat; the "pacemaker" 517

sac body plan (sak bod′e plan) a body with a digestive cavity that has only one opening, as in cnidarians and flatworms 392

salivary amylase (sal′ĭ-ver-e am′ĭ-lās) an enzyme in the saliva that initiates the digestion of starch 551

saltatory conduction (sal′tah-to″re kon-duk′shun) movement of a nervous impulse from one node of Ranvier to another along a myelinated axon 597

saprophytic decomposers (sap″ro-fit′ik de″kom-pōz′er) heterotrophic organisms, such as fungi and bacteria, that live on dead organic material 345

sarcolemma (sar″ko-lem′ah) plasma membrane of a muscle fiber that forms the tubules of the T system involved in muscular contraction 635

sarcomere (sar″ko-mēr) a unit of a myofibril, many of which are arranged linearly along its length that shortens to give muscle contraction 636

sarcoplasmic reticulum (sar″ko-plaz′mik rĕ-tik′u-lum) the modified endoplasmic reticulum of a muscle fiber that stores calcium ions whose release initiates muscular contraction 635

scientific method (si″en-tif′ik meth′ud) a step-by-step process for discovery and generation of knowledge, ranging from observation and hypothesis to theory and principle 13

sclera (skle′rah) outer, white, and fibrous layer of the eye that surrounds the transparent cornea 615

scramble competition (skram′b′l kom″pĕ-tish′un) competition in which each member of a species try to acquire as much of a particular resource as possible 733

seed (sēd) a mature ovule that contains an embryo with stored food, enclosed in a protective coat 378

segmented (seg-ment′ed) having repeating body units as is seen in the earthworm 405

segregation (seg″re-ga′shun) in heredity, the separation of alleles and movement into different gametes 174

semicircular canal (sem″e-ser′ku-lar kah-nal′) one of three half-circle-shaped canals of the inner ear that are fluid filled for registering changes in motion 622

semiconservative replication (sem″e-kon-ser′vah-tiv rĕ′pli-ka′shun) duplication of DNA resulting in a double helix having one parental and one new strand 225

seminiferous tubule (se″mĭ-nif′er-us tu′bul) a long, coiled structure contained within chambers of the testis where sperm are produced 662

sessile filter feeder (ses′il fil′ter fēd′er) an organism that stays in one place and filters its food from the water 387

sex chromosome (seks kro′mo-sōm) the chromosome that determines sex 191

sex influenced trait (seks in-flu′enst trāt) a genetic trait that is expressed differently in the two sexes but is not controlled by alleles on the sex chromosomes, e.g., pattern baldness 215

sexual reproduction (seks′u-al re″pro-duk′shun) reproduction involving gamete formation in meiosis and that produces offspring with genes inherited from each parent 155

sickle-cell anemia (sik′l sel ah-ne′me-ah) a genetic disorder common in black people in which 'sickling' of red blood cells clogs blood vessels; individuals who are heterozygous for the allele have a greater resistance to malaria 212

sign stimulus (sīn stim′u-lus) a stimulus that initiates a fixed-action pattern 703

sister chromatid (sis′ter kro′mah-tid) genetically identical chromosomes that are the result of DNA replication and are attached to each other at the centromere 147

slash-and-burn agriculture (slash and burn ag′rĭ-kul″tūr) practice of cutting and burning forest to use the land for agriculture 783

society (so-si′ĕ-te) a group of individuals of the same species that show cooperative behavior 710

sociobiology (so″se-o-bi-ol′ŏ-je) the biological study of social behavior 715

sodium-potassium pump (so′de-um po-tas′e-um pump) a transport protein in the plasma membrane that moves sodium into and potassium out of animal cells; important in nerve and muscle cells 85

solar energy (so′lar en′er-je) energy from the sun; light energy 792

speciation (spe″se-a′shun) the process whereby a new species is produced or originates 322

species (spe′shēz) a taxonomic category that is the subdivision of a genus; its members can breed successfully with each other but not with organisms of another species 5

spermatogenesis (sper″mah-to-jen′ĕ-sis) production of sperm in males by the process of meiosis and maturation 165

spermatozoa (sper″mah-to-zo′ah) male sex cell or gamete; sperm 664

spicule (spik′ūl) skeletal structures of sponges composed of calcium carbonate or silicate 388

spindle apparatus (spin′d′l ap″ah-ra′tus) microtubule structure that brings about chromosome movement during cell division 150

spongy bone (spun′je bōn) type of bone made up of numerous bars and plates; somewhat lighter than compact bone 631

spongy layer (spun′je la′er) layer of loosely packed cells in a plant leaf that increases the amount of surface area for gas exchange; along with the palisade layer, it is the site of most of photosynthesis 459

spore (spōr) usually a haploid reproductive structure that develops into a haploid generation 356

stabilizing selection (sta′bil-i-zing sĕ-lek′shun) process whereby atypical phenotypes are eliminated and the typical phenotype is conserved 320

stamen (sta′men) a pollen-producing flower structure consisting of an anther on a filament tip 480

steady-state society (ste′e stāt so-si′e-te) a society with no yearly increase in population or resource consumption 794

steroid (ste′roid) biologically active lipid molecule having four interlocking rings; examples are cholesterol, progesterone, testosterone 46

stoma (pl., **stomata**) (stō-ma) small opening between two guard cells on the underside of leaf epidermis; their regulated size controls the rate of gas exchange 446

stroma (stro′mah) large, central space in a chloroplast that is fluid filled and contains enzymes used in photosynthesis 111

structural gene (struk′tur-al jēn) gene that codes for proteins in metabolic pathways 250

subspecies (sub′spe-sēz) population of a species that is somewhat isolated from other populations of the same species; race 314

substrate (sub′strāt) the reactant in an enzymatic reaction; each enzyme has a specific substrate 97

substrate-level phosphorylation (sub′strāt lĕv-al fos″fōr-i-la′shun) process in which ATP is formed by transferring a phosphate from a metabolic substrate to ADP 128

succession (suk-sesh′un) an orderly sequence of community replacement, one following the other, to an eventual climax community 754

suppressor T cell (sŭ-pres′or te sel) T lymphocyte that increases in number more slowly than other T cells and eventually brings about an end to the immune response 537

survivorship (sur-vi′vor-ship) usually shown graphically to depict death rates or percentage of remaining survivors of a population over time 724

symbiosis (sim″bi-o′sis) a relationship that occurs when two different species live together in a unique way; may be beneficial, neutral, or detrimental to one and/or the other species 740

symbiotic (sim″bi-ot′ik) one species having a close relationship with another species; includes parasitism, mutualism, and commensalism 346

sympathetic nervous system (sim″pah-thet′ik ner′vus sis′tem) a division of the autonomic nervous system that is involved in fight or flight responses; uses norepinephrine as a neurotransmitter 600

sympatric speciation (sim-pat′rik spe″se-a′shun) origin of new species in populations that overlap geographically 325

synapse (sin′aps) a junction between neurons consisting of presynaptic (axon) membrane, the synaptic cleft, and the postsynaptic (usually dendrite) membrane 597

synapsis (sĭ-nap′sis) pairing of homologous chromosomes during meiosis I 162

synovial joint (sĭ-no′ve-al joint) a freely moving joint in which two bones are separated by a cavity 631

systemic (sis-tem′ik) referring to general circulation serving overall body regions 514

systole (sis′to-le) contraction period of a heart during the cardiac cycle 518

T

taste bud (tāst bud) an elongated cell that functions as a taste receptor 614

taxis (tak'sis) orientation of an organism in response to a stimulus 706

taxonomic categories (tak''so-nom'ik kat'ē-go''rez) the groups on which our system of classifying plants and animals is based 300

taxonomy (tak-son'o-me) a system to meaningfully classify organisms into groups, based on similarities and differences, using morphology and evolution 282

Tay-Sachs disease (ta saks' dī-zēz) a lethal, genetic lysosomal storage disease, best known among the U.S. Jewish population, that results from the lack of a particular enzyme; nervous system damage leads to early death 211

teleological (te''le-o-loj'ĭ-kal) assuming that a process is directed toward a final goal or purpose 286

temperate forest (tem'per-it for'est) forest found in moderate climates with well-defined winter and summer seasons 761

template (tem'plāt) a parental strand of DNA that serves as a mold for the complementary daughter strand produced during DNA replication 224

territoriality (ter''i-tor''i-al'i-te) behavior used to guarantee exclusive use of a given space for reproduction, feeding, etc. 713

testcross (test'kros) a genetic mating in which a possible heterozygote is crossed with an organism homozygous recessive for the characteristic(s) in question in order to determine its genotype 177

testosterone (tes-tos'tĕ-rōn) male sex hormone produced from interstitial cells in the testis 664

thalamus (thal'ah-mus) a lower forebrain region in vertebrates involved in crude sensory perception and screening messages intended for the higher forebrain or cerebrum 602

theory (the'o-re) a conceptual scheme arrived at by the scientific method and supported by innumerable observations and experimentations 16

thermal inversion (ther'mal in-ver'zhun) temperature inversion that traps cold air and its pollutants near the earth with the warm air above it 789

thigmotropism (thig-mot'ro-pizm) movement in response to touch; in plants, the coiling of tendrils 500

thoracic cavity (tho-ras'ik kav'i-te) internal body space of some animals that contains the lungs, protecting them from desiccation 568

thylakoid (thi'lah-koid) flattened disk of the grana; its membrane contain the photosynthetic pigments 111

thymine (thi'min) one of four bases in the structure of DNA 222

thyroid gland (thi'roid gland) large gland in the neck that produces several important hormones, e.g., thyroxin, calcitonin 649

T lymphocyte (te lim'fo-sīt) white blood cell that directly attacks antigen-bearing cells 533

trachea (trak'ke-a) 1. an air tube (windpipe) of the respiratory tract in vertebrates; 568; 2. an air tube in insects 408

tracheophyte (trā'ke-a-fīt) vascular plants, including ferns and seed plants that have a dominant sporophyte generation 373

transcription (trans-krip'shun) the process whereby the DNA code determines (is transcribed into) the sequence of codons in mRNA 235

transfer RNA (tRNA) (trans'fer ar'en-a) active in protein synthesis that transfers a particular amino acid to a ribosome; at one end it binds to the amino acid and at the other end it has an anticodon that binds to an mRNA codon 239

transformed (trans-formd') cell that has been altered by genetic engineering and is capable of producing new protein 261

transgenic organism (trans-jen'ik orgah-nizm) a multicellular organism that contains a transplanted gene 267

transition reaction (tran-zish'un re-ak'shun) a reaction involving the removal of CO_2 from pyruvate that connects glycolysis to the Krebs cycle 132

translation (trans''lā'shun) the process whereby the sequence of codons in mRNA determines (is translated into) the sequence of amino acids in a polypeptide 236

transpiration (tran''spi-ra'shun) the plant loss of water to the atmosphere, mainly through evaporation at leaf stomata 465

transposon (trans-po'zun) movable segments of DNA 228

triplet code (trip'let kōd) code with three base unit coding for an amino acid 236

trophic level (trof'ik lev'el) feeding level of one or more populations in a food web 767

tropical rain forest (trop'i-kal rān for'est) forest found in warm climates with plentiful rainfall 761

tropism (tro'pizm) movement toward or away from an external stimulus 499

trypsin (trip'sin) an enzyme that digests protein molecules to peptides in the small intestine 553

tube-within-a-tube body plan (tūb with-in' ah tūb bod'e plan) a body with a digestive tract that has both a mouth and an anus 392

turgor pressure (tur'gor presh'ur) the pressure of the plasma membrane against the cell wall, determined by the water content of the plant cell vacuole; gives internal support to the cell 83

Turner syndrome (tur'ner sin'drōm) a condition that results from the inheritance of a single chromosome 206

tympanic membrane (tim-pan'ik mem'brān) eardrum; membranous region that receives air vibrations in an auditory organ 620

U

uniformitarian theory (u''ni-for''mi-tār'e-an the'o-re) belief espoused by Hutton that geological forces act at a continuous, uniform rate 285

urea (u-re'ah) main nitrogenous waste of terrestrial amphibians and mammals 579

ureter (u-re'ter) a tubular structure conducting urine from kidney to urinary bladder 583

urethra (u-re'thrah) tubular structure that receives urine from the bladder and carries it to the outside of the body 583

uric acid (u'rik as'id) main nitrogenous waste of insects, reptiles, birds, and some dogs 579

urinary bladder (u'ri-ner''e blad'der) organ where urine is stored 583

uterine cycle (u'ter-in si'k'l) a cycle that runs concurrently with the ovarian cycle; and prepares the uterus to receive a developing zygote 667

uterus (u'ter-us) pear-shaped portion of female reproductive tract that lies between the oviducts and the vagina; site of embryo development 666

V

vaccine (vak'sēn) a substance that wakes the immune response without causing illness 539

vagina (vah-ji'nah) muscular tube leading from the uterus; the female copulatory organ and the birth canal 666

vascular cambium (vas'ku-lar kam'be-um) embryonic tissue in the root and stem of a vascular plant that produces secondary xylem and phloem each year 454

vector (vek'tor) in genetic engineering, a means to transfer foreign genetic material into a cell, e.g., a plasmid 261

vein (vān) a blood vessel that arises from venules and transports blood toward the heart 512

vena cava (ve'nah ka'vah) one of two largest veins in the body; returns blood to the right atrium in a four-chambered heart 517

vertebrate (ver'tĕ-brāt) referring to a chordate animal with a serial arrangement of vertebrae, or backbone 387

vestigial structure (ves-tij'e-al struk'tūr) the remains of a structure that was functional in some ancestor but is no longer functional in the organism in question 299

villi (vil'i) small, fingerlike projections of the inner small intestine wall 552

viroid (vi'roid) unusually infectious particle consisting of a short chain of naked RNA 340

virus (vi'rus) small organism made up of an inner core of nucleic acid and an outer protein coat 338

visible light (viz'i-b'l līt) a portion of the electromagnetic spectrum of light that is visible to the human eye 111

vitamin (vi'tah-min) an organic molecule that is required in small quantities for various biological processes and must be in an organism's diet because it cannot be synthesized by the organism 559

W

water cycle (wah'ter si'k'l) the movement of water through the physical and living components of the earth 747

woody (wood'e) characterizes angiosperm trees with hardwood 381

X

X-linked gene (eks linkt jēn) gene located on the X-chromosome that does not control a sexual feature of the organism 191

xylem (zi'lem) a vascular tissue that transports water and mineral solutes upward through the plant body 374

XYY male (eks wi wi māl) a genetic condition in which affected males have two Y chromosomes, are taller than average, and have low intelligence 207

Y

yolk sac (yōk sak) one of four extraembryonic membranes in vertebrates; important in the development of fishes, amphibians, reptiles, and birds, it is largely vestigial in mammals 686

Z

zero population growth (ze'ro pop''u-la'shun grōth) no growth in population size 727

zygote (zi'gōt) the diploid (2N) cell formed by the union of two gametes, the product of fertilization 483

CREDITS

Photographs

Part Openers
Part 1: © K. G. Murti/Visuals Unlimited; **Part 2:** © Biophoto Associates/Photo Researchers, Inc.; **Part 3:** © Fred Bavendam/Valan Photos; **Part 4:** © Richard Kolar/Earth Scenes; **Part 5:** © Biophoto Associates/Science Source/Photo Researchers, Inc. **Part 6:** © Dennis W. Schmidt/Valan Photos.

Chapter 1
Opener: © Pam Hickman/Valan Photos; **1.1a:** © Rod Allin/Tom Stack and Associates; **b,c:** © Kjell B. Sanved; page 2 **a,b:** © Eric Grave/Photo Researchers, Inc., © John D. Cunningham; page 3 **c–e:** © Rod Planck/Tom Stack and Associates, © Farrell Grehan/Photo Researchers, Inc.; © Leonard Rue Enterprises; **1.2:** © John Gerlach/Animals Animals; **1.3a–c:** © Stephen Kraseman/Photo Researchers, Inc., © Mitch Reardon/Photo Researchers, Inc., © David Fritts/Animals Animals; **1.4a–d:** © Biophoto Associates/Photo Researchers, Inc., © Philip M. Zito, © Breck Kent/Animals Animals, © Breck Kent/Animals Animals; **1.6a–e:** © Ed Robinson/Tom Stack and Associates, © Dave Fleetham/Tom Stack and Associates, © F. Stuart Westmorland/Tom Stack and Associates, © Tom Stack, © C. Ressler/Animals Animals; **1.7a–f:** © Richard Thom/Tom Stack and Associates, © C. W. Schwartz/Animals Animals, © Wendy Neefus/Animals Animals, © Zig Lesczynski/Animals Animals, © Tom McHugh/Photo Researchers, Inc., © Oxford Scientific Films/Animals Animals.

Chapter 2
Opener: © Hank Morgan/Photo Researchers, Inc.; **2.4:** © Dr. Gardner.

Chapter 3
Opener: © Arthur Strange/Valan Photos; **3.4:** © Journalism Services, Inc.; **3.8c:** © Charles Kingery/Micro/Macro Stock Photography; **3.9:** © Douglas Faulkner/Sally Faulkner Collection; **3.12a,b,c:** © Clyde Smith/Peter Arnold, Inc., © JoAnne Kalish/Joe DiMaggio/Peter Arnold, Inc., © Holt Confer/Grant Heilman Photography, Inc.

Chapter 4
Opener: © Dr. Gopal Murti/Science Photo Library/Photo Researchers, Inc.; **4.1a–c:** © John Gerlach/Tom Stack and Associates, © Leonard Lee Rue III/Photo Researchers, Inc., © Science Photo Library/Photo Researchers, Inc.; **4.8b,c:** © Stanley Flegler/Visuals Unlimited, © Don Fawcett/Photo Researchers, Inc.; **4.9:** © Biophoto Associates/Photo Researchers, Inc.; **4.11:** © S. Maslowski/Visuals Unlimited.

Chapter 5
Opener: © Biocosmos/Science Photo Library/Photo Researchers, Inc.; **5.1a–d:** © Bob Coyle, Carolina Biological Supply, © B. Miller/Biological Photo Service, © Dr. Ed Reschke; page 56 **a–c:** © M. Abbey/Visuals Unlimited, © M. Schliwa/Visuals Unlimited, © K. W. Jeon/Visuals Unlimited; page 59 **(all):** © Keith Roberts and James Barnett; **5.3b:** © Henry Aldrich/Visuals Unlimited; **5.4b:** © Richard Rodewald/Biological Photo Service; **5.5b:** © W. P. Wergin, University of Wisconsin, Courtesy E. A. Newcomb/Biological Photo Service; **5.6a:** © L. L. Sims/Visuals Unlimited; **5.7b:** © D. W. Fawcett/Photo Researchers, Inc.; **5.8a,c:** © W. Rosenberg/Biological Photo Service, © Dr. James A. Lake, *Journal of Molecular Biology* 105, 131–59, 1976 © Academic Press, Inc. (London) LTD; **5.10a:** © David M. Phillips/Visuals Unlimited; **5.11:** © K. G. Murti/Visuals Unlimited; **5.12:** © S. E. Frederick, courtesy E. H. Newcomb, University of Wisconsin/Biological Photo Service; **5.13a:** © Dr. Herbert Israel, Cornell University; **5.14a:** © Dr. Keith R. Porter; **5.16:** © L. E. Roth/Biological Photo Service; **5.17:** © John Walsh/Science Photo Library/Photo Researchers, Inc.; **5.18c:** © Dr. Keith Porter.

Chapter 6
Opener: © Hank Morgan/Photo Researchers, Inc.; **6.1:** © Steven C. Kaufman/Peter Arnold, Inc.; **6.3a:** © W. Rosenberg/Biological Photo Service, **d:** © Don Fawcett/Photo Researchers, Inc.; **6.10a:** © Eric Grave/Photo Researchers, Inc., **b,c,** © Thomas Eisner; **6.13:** © Francis Leroy; **6.14:** © Mark S. Bretscher; **6.17a:** © Ray F. Evert.

Chapter 7
Opener: © Marty Cooper/Peter Arnold, Inc.; **7.1:** © Ray Coleman/Photo Researchers, Inc.; **7.5:** © Courtesy of Dr. Alfonso Tramontano; **7.8:** © Doolittle, R. F., University of California, from *Scientific American,* "Proteins", Oct. 1985, page 95.

Chapter 8
Opener: © Breck P. Kent/Earth Scenes; **8.1:** © Louise K. Broman/Root Resources; **8.3b:** © Gordon Leedale/Biophoto Associates; **8.4c:** © Dr. Kenneth Miller; **8.8a:** © Dr. Melvin Calvin; **8.14:** © Pat Armstrong/Visuals Unlimited.

Chapter 9
Opener: © Fritz Polking; **9.1a:** © N. Myers/Bruce Coleman, Inc.; page 131 **(all):** © Bob Coyle. End of Part 1: **1a:** © Gordon Leedale/Biophoto Associates.

Chapter 10
Opener: © John D. Cunningham/Visuals Unlimited; **10.1a,b:** © Dave Davidson/Tom Stack and Associates, © John Cunningham/Visuals Unlimited; **10.2f:** © John J. Cardamone, Jr./University of Pittsburg, Medical Microbiology/Biological Photo Service; page 148: © Dr. Stephen Wolfe; **10.3a:** © Biophoto Associates/Science Source/Photo Researchers, Inc.; **10.5, 10.6, 10.7:** © Dr. Ed Reschke; **10.8a:** © Francis Leroy; **10.9:** © Dr. Ed Reschke; **10.10 (all):** © R. G. Kessel and C. Y. Shih: *Scanning Electron Microscopy* © 1976, Springer Verlag, Berlin; **10.11a–d:** © Dr. Andrew Bajer; **10.12a:** © B. A. Palevity and E. H. Newcomb, University of Wisconsin/Biological Photo Service.

Chapter 11
Opener: © G. I. Bernard/Earth Scenes; **11.1:** © Francis Leroy; **11.5a:** © Dr. D. von Wettstein; **11.6:** © Prof. B. John.

Chapter 12
Opener: © Marion Patterson/Photo Researchers, Inc.; **12.1:** Bettmann Archive.

Chapter 13
Opener: © Manfred Kage/Peter Arnold, Inc.; **13.2:** © Sunquist/Tom Stack and Associates; **13.7:** © Leonard McCombe *Life Magazine,* © Time, Inc.; **13.14 (all):** © Bob Coyle; **13.16:** © John D. Cunningham/Visuals Unlimited.

Chapter 14
Opener: © Brian Parker/Tom Stack and Associates; **14.1b:** © Dr. Digamer Borgoankar, Ph.D., Cytogenetics Lab; **14.3a:** © Jill Cannefax/EKM Nepenthe; **14.6a,b:** © F. A. Davis Company and Dr. R. H. Kampmeier; page 210 **a,b:** © A. Robinson, National Jewish Hospital and Research Center, © Danny Brass/Photo Researchers, Inc.; **14.11c:** © J. Callahan, M.D./IMS/University of Toronto/Custom Medical Stock Photography; **14.12b:** © Bill Longcore/Photo Researchers, Inc.

Chapter 15
Opener: © Dr. Gopal Murti/Science Photo Library/Photo Researchers, Inc.; **15.2b:** © Lee D. Simon/Photo Researchers, Inc.; **15.4b:** © Courtesy of the Biophysics Dept., King's College, London; **15.5b:** © Dr. Nelson Max; **15.10a,b:** Wide World Photos, Inc., © Biological Photo Service.

Chapter 16
Opener: © Breck P. Kent/Earth Scenes; **16.3a:** © Bill Longcore/Photo Researchers, Inc.; **16.9a:** © J. J. Head/Carolina Biological Supply; **16.11b:** © Dr. Nelson Max; **16.13b:** © Professor Alexander Rich.

Chapter 17
Opener: © David Phillips/Visuals Unlimited; **17.3:** © Dr. Monroe Yoder/Peter Arnold, Inc.; **17.6:** © Dr. G. Stephen Martin.

Chapter 18
Opener: © Gerbis Keriman/Peter Arnold, Inc.; **18.1 (all):** Courtesy of Genentech, Inc.; **18.2 (left):** © Lee Simon/Photo Researchers, Inc., **(right):** © CNRI/Science

Chapter 33

33.7: From Kent M. Van De Graaff, *Human Anatomy,* 2d ed. Copyright © 1988 Wm. C. Brown Publishers, Dubuque, Iowa. All Rights Reserved. Reprinted by permission. **33.9** and **33.14a:** From John W. Hole, Jr., *Human Anatomy and Physiology,* 4th ed. Copyright © 1987 Wm. C. Brown Publishers, Dubuque, Iowa. All Rights Reserved. Reprinted by permission.

Chapter 34

34.4: From Stuart Ira Fox, *Human Physiology,* 2d ed. Copyright © 1987 Wm. C. Brown Publishers, Dubuque, Iowa. All Rights Reserved. Reprinted by permission.

Chapter 35

35.14: From John W. Hole, Jr., *Human Anatomy and Physiology,* 4th ed. Copyright © 1987 Wm. C. Brown Publishers, Dubuque, Iowa. All Rights Reserved. Reprinted by permission.

Chapter 36

36.10: From John W. Hole, Jr., *Human Anatomy and Physiology,* 4th ed. Copyright © 1987 Wm. C. Brown Publishers, Dubuque, Iowa. All Rights Reserved. Reprinted by permission. **36.11:** From *Human Physiology* by A. J. Vander, et al. Copyright © 1970 McGraw-Hill Book Company, New York, NY. Reprinted by permission.

Chapter 37

37.4: Figure 233 from *An Introduction to Embryology,* Fifth Edition, by B. I. Balinsky and B. C. Fabian, copyright © 1981 by Holt, Rinehart and Winston, Inc., reprinted by permission of the publisher.

Chapter 38

38.12: From John W. Hole, Jr., *Human Anatomy and Physiology,* 4th ed. Copyright © 1987 Wm. C. Brown Publishers, Dubuque, Iowa. All Rights Reserved. Reprinted by permission.

Chapter 39

39.2b and **39.11a:** From John W. Hole, Jr., *Human Anatomy and Physiology,* 4th ed. Copyright © 1987 Wm. C. Brown Publishers, Dubuque, Iowa. All Rights Reserved. Reprinted by permission.

Chapter 40

40.6a, 40.7, 40.11, and **40.14b,c:** From John W. Hole, Jr., *Human Anatomy and Physiology,* 4th ed. Copyright © 1987 Wm. C. Brown Publishers, Dubuque, Iowa. All Rights Reserved. Reprinted by permission.

Chapter 41

41.12: From Kent M. Van De Graaff and Stuart Ira Fox, *Concepts of Human Anatomy and Physiology,* 2d ed. Copyright © 1989 Wm. C. Brown Publishers, Dubuque, Iowa. All Rights Reserved. Reprinted by permission.

Chapter 42

42.5, 42.9, 42.10a, and **42.12:** From John W. Hole, Jr., *Human Anatomy and Physiology,* 4th ed. Copyright © 1987 Wm. C. Brown Publishers, Dubuque, Iowa. All Rights Reserved. Reprinted by permission. **42.6c, 42.8,** and **42.14:** From Kent M. Van De Graaff and Stuart Ira Fox, *Concepts of Human Anatomy and Physiology,* 2d ed. Copyright © 1989 Wm. C. Brown Publishers, Dubuque, Iowa. All Rights Reserved. Reprinted by permission. **42.16:** Nina Wallace © 1988 DISCOVER PUBLICATIONS.

Chapter 43

43.11 and **43.15:** From John W. Hole, Jr., *Human Anatomy and Physiology,* 4th ed. Copyright © 1987 Wm. C. Brown Publishers, Dubuque, Iowa. All Rights Reserved. Reprinted by permission. **43.17:** From *The Human Body: Growth and Development.* Torstar Books. Reprinted by permission.

Chapter 44

44.7b: From J. P. Hailman, *Behavior Supplement #15.* Copyright © E. J. Brill, Leiden, Holland. Reprinted by permission.

Chapter 45

45.5: Data based on C. Kabal and D. R. Thompson, "Wisconsin Quail 1834–1962 Population Dynamics and Habitat Management" in *Technical Bulletin #30,* 1963. Wisconsin Conservation Department. **45.6 (left):** From V. B. Scheffer, "The Rise and Fall of a Reindeer Herd" in *Scientific Monthly,* Vol. 73, pp. 356–362, Fig. 1, December 1951. Copyright © 1951 American Association for the Advancement of Science, Washington, D.C. Reprinted by permission. Fig 1 (Case Study) From Stuart Ira Fox, *Human Physiology,* 2d ed. Copyright © 1987 Wm. C. Brown Publishers, Dubuque, Iowa. All Rights Reserved. Reprinted by permission. **45.10:** Marshall Green and Robert A. Feary, "World Population: The Silent Explosion." *Department of State Bulletin,* Publication 8956 (October 1978) based on U.S. Bureau of Census Illustrative Projections of World Population to the 21st Century. Current Population Reports Special Studies.

Chapter 46

46.2: From "Physiological Ecology of Three Codominant Successional Annuals" by N. K. Wieland and F. A. Bazzaz, *Ecology* 56:681–88. Copyright © by the Ecological Society of America, Tempe, AZ. Reprinted by permission. **46.3:** Figure from *Ecology and Field Biology* by Robert Leo Smith. Copyright © 1980 by Robert Leo Smith. Reprinted by permission of Harper & Row, Publishers, Inc. **46.6:** Data from G. F. Gause, *The Struggle for Existence.* Copyright © 1934 by Williams and Wilkins Company, Baltimore, MD. Reprinted by permission. **46.9b:** Figure 7.16 from *Fundamentals of Ecology,* Third Edition, by Eugene P. Odum, copyright © 1971 by Saunders College Publishing, a division of Holt, Rinehart and Winston, Inc., reprinted by permission of the publisher.

Chapter 47

47.5: From David L. Correll, "Estaurine Productivity" in *BioScience,* Vol. 281, p. 649. Copyright © 1978 by the American Institute of Biological Sciences. Reprinted by permission.

Illustrators

Kathleen Hagelston

Provided many illustrations throughout the text and also provided sketches for the following illustrators and figures: Marj Leggitt 14.3b, 18.6a, 35.16. Laurie O'Keefe 14.7a, 16.1, 18.2, 18.3a, 34.02, 34.06b, 34.07, 34.08c.

Mark Lefkowitz

The following figures copyright © 1989 Mark Lefkowitz. All Rights Reserved. 2.02, 3.01, Chapter 10 reading art, figure 1, page 148, 10.08b, 10.12b, 12.02, 13.04, 13.06, 15.02a, 17.04, 19.06, 19.08, 19.13, 20.15, 20.16, 21.05, 23.01c, 24.02, 24.04, 24.12, 24.14, 26.02, 26.03, 26.05, 26.07, 27.01, 27.05, 27.12, 28.21, 28.22, 28.23, 28.24, 31.02, 31.08, 33.02, 35.07, 35.08, 36.05, 36.06, 36.08, 36.09, 37.07, 37.08, 37.09, 37.11, 38.01, 38.07, 38.08, 38.09, 38.11, 38.15, 39.01, 39.13, 41.01, 41.03, 47.01, 48.05, 48.12, 48.14, 48.15.

Marj Leggitt

28.16.

Steve Moon

39.02, 39.11a, 42.14.

Diane Nelson

33.07, 33.09, Chapter 33 reading art, b, page 520, 35.14, 36.10, 43.15.

Laurie O'Keefe

5.09, 14.11, 17.7.

Mike Shenk

41.13, 42.06c.

Tom Waldrop

33.13, 34.03, 38.02a, 38.10, 40.11, 40.14, Chapter 41 reading art, page 654, 42.05, 42.08, 42.09.

Rolin Graphics

Chapter 1: Reading text art, page 2. **Chapter 2:** Text art, on top, left and center right of page 14; 2.3; text art, page 16; 2.5. **Chapter 3:** 3.2, 3.5, 3.6, 3.7, 3.8a,b, 3.10, 3.11a,b; text art, pages 31 & 32; 3.13, 3.14; text art, pages 33, 34; 3.15. **Chapter 4:** Text art, pages 38, 39; 4.2, 4.3, 4.4a,b, 4.5, 4.6, 4.7, 4.9, 4.10, text art, pages 44, 45, 46; 4.14, 4.15, 4.16, 4.17, 4.18a, 4.19; text art, pages 50, 53. **Chapter 5:** 5.2a,b; reading art, figure 1, page 58, figure 2, page 59; reading text art, page 59; text art, page 60; 5.3, 5.4b, 5.5a, 5.7a,c, 5.8b,c,d; text art, page 66; 5.10b, text art, page 68; 5.13b, 5.14b, 5.15. **Chapter 6:** Text art, page 77; 6.2; text art, page 78; 6.3b,c, 6.4; text art, pages 79, 80; 6.5, 6.6, 6.9, 6.14a; reading art, page 88; 6.17b. **Chapter 7:** Text art, pages 94, 95, 96; 7.3; text art, pages 98, 99; 7.6, 7.7; text art, page 101, 7.9; text art, page 102; 7.10; text art, page 104. **Chapter 8:** Text art, at top left, page 109, at top right, page 110, page 112; 8.6; text art, page 114; 8.7, 8.8b, 8.11; text art, pages 118 and 119; 8.12, 8.13. **Chapter 9:** Text art, page 125; 9.2; text art, page 126; 9.3; text art, page 127; 9.4, 9.5, 9.6; text art, page 130; 9.7, 9.8; text art, page 132; 9.9, 9.10; text art, pages 133, 134, 136. **Critical Thinking Case Study, Part 1,** figure b, page 141. **Chapter 10:** Text art, pages 145 and 147; 10.4; text art, page 150; 10.5a, 10.6a, 10.7a, 10.9a. **Chapter 11:** Text art, page 159; 11.2, 11.3, 11.4; text art, page 163; 11.5, 11.8. **Chapter 12:** Table 12.1; 12.3, 12.4; text art, pages 175 and 176; 12.5; text art, page 177; 12.6, 12.7; text art, page 180; 12.8, 12.9. **Chapter 13:** Text art, page 185; 13.1; text art, page 186; 13.3a,b; 13.5, 13.9; text art, page 191; 13.10a,b, 13.12, 13.13; text art, pages 195, 197, 199. **Chapter 14:** Text art, page 203; 14.4, 14.7b, 14.9, 14.10, 14.13. **Chapter 15:** Text art, page 219; 15.3a,b; text art, page 221; 15.4a; text art, page 224; reading art a–c, page 226; text art, page 227. **Chapter 16:** Text art, page 233; 16.3c; text art, page 235; 16.7, 16.9b, 16.10, 16.11a; text art, page 240; 16.15; text art, page 244. **Chapter 17:** 17.1; reading art, figures 1, 2, page 249 text art, page 250; 17.2, 17.5. **Chapter 18:** Reading art a–c, page 265; 18.9b. **Chapter 19:** 19.01a. **Chapter 20:** Text art, page 294; 20.2, 20.3, 20.6, reading art a,b, page 301, 20.10, 20.12. **Chapter 21:** Text art, page 311; 21.2, 21.7a,b, 21.8a, 21.10; text art, page 320. **Chapter 22:** 22.1, 22.5. **Chapter 23:** 23.3, 23.5a. **Chapter 24:** 24.8, 24.9, 24.13b, 24.16, 24.19, 24.22. **Chapter 25:** 25.3, 25.4; table 25.3; 25.5, 25.7; text art, page 378; 25.12. **Chapter 27:** Text art, page 400; 27.2, 27.4b, 27.6b, 27.11b. **Chapter 28:** Table 28.1, 28.20a. **Chapter 29:** 29.1, 29.3a,b,c, 29.6, 29.8c, 29.13b, 29.15, 29.17a,b, 29.18, 29.19a,c, 29.20. **Chapter 30:** 30.1, 30.2, 30.3a,b; text art, page 467; 30.6, 30.7, 30.8b,c; text art, page 470; 30.11, 30.14a,b, 30.15a, 30.16; text art, page 475. **Chapter 31:** 31.1; text art, page 480; 31.3. **Chapter 32:** 32.2a,b, 32.3, 32.4, 32.9a,b; text art, page 499; 32.13, 32.14. **Critical Thinking Case Study, Part 4,** page 507. **Chapter 33:** 33.6, 33.14c; text art, page 524. **Chapter 34:** 34.5c. **Chapter 35:** Text art, pages 551, 552, 553, 554, 555. **Chapter 36:** Text art, pages 570, 571; 36.12b. **Chapter 37:** 37.1b, 37.2b; text art, page 579; 37.5. **Chapter 38:** 38.3d, 38.4, 38.5. **Chapter 39:** 39.5, 39.6. **Chapter 40:** 40.8a,b,c; text art, pages 636, 637, 640. **Chapter 41:** 41.2; reading art, page 646; 41.4, 41.5, 41.6, 41.8, 41.9, 41.12, 41.14. **Chapter 42:** 42.1, 42.6a, 42.7, 42.13, 42.15. **Chapter 43:** Text art, page 681; 43.5; text art, page 683. **Critical Thinking Case Study, Part 5,** page 697. **Chapter 44:** 44.1a, 44.4, 44.5, 44.7b, 44.9; reading text art, page 707; 44.13, 44.19. **Chapter 45:** 45.2, 45.3, 45.4, 45.5, 45.6a, 45.7, 45.8. **Chapter 46:** 46.1, 46.3, 46.4, 46.6, 46.7, 46.8, 46.9, 46.10b. **Chapter 47:** 47.3, 47.5, 47.9, 47.12b, 47.14, 47.15, 47.21. **Chapter 48:** Art in 48.1, 48.2, 48.3, 48.4. **Chapter 49:** 49.5, 49.8, 49.10, 49.12b, 49.13.

Other artwork throughout the text were rendered by *Laurel Antler, Anne Greene, Ruth Krabach* and *Mildred Rinehart.*

INDEX

Aneuploidy, 197
   sex chromosome, 206–7
Angiosperm, 371, 377, 381–83
   adaptations of, 382
   extinction of, 382
   importance of, 382
   See also Flowering plant
Angiotensin, 587, 652
Animal, 2–3
   characteristics of, 387
   chromosome number in, 147
   defense adaptations of, 739–40
   evolution of, 294–95, 387–88
   seed and fruit dispersal by, 483
   transgenic, 269–70
Animal cell, 61
   junctions between, 89–90
   mitosis in, 150–53
Animal husbandry, 726, 772
Animal virus, 340–41
Ankle, 629
Annelid, 405–6, 410, 414
Annual ring, 455
Annulus, of fern, 376
Anopheles, 322, 355–56
Ant, 655, 711, 715
   army, 706
   leaf cutter, 743
   slave-making, 741
Anteater, spiny, 426
Antelope, 740, 757, 759
   pronghorn, 760
Antennae
   light-harvesting, 113
   sensory, 408
Anther, 381, 480–81
Antheridia, 372–73, 376
Anthophyta, 371
Anthropoid, 428–29
Antibiotic, 17, 342
Antibiotic resistance, 264, 322, 342
Antibody, 524, 533, 535
   antigen-antibody complex, 535
   constant region of, 535
   fluorescent, 81
   membrane-bound, 534
   monoclonal, 266, 542
   variable region of, 535
Antibody-mediated immunity, 534
Anticodon, 241, 243
Antidiuretic hormone. *See* ADH
Antigen, 524, 532–33, 696–97
Antigen-antibody complex, 535
Antipodal, 482
alpha-Antitrypsin, 266
Antiviral drug, 342
Anus, 547, 551, 556, 680
Anvil, 620–21, 623
Aorta, 515–16
Aortic body, 569, 571
Ape, 427–29, 710
Aphid, 473–74
Aphrodisiac, 713
Apical dominance, 494, 498
Apical meristem, 448–50, 485–86
Aplysia, 702–3
Appeasement, 714
Appendage, arthropod, 406–7
Appendicitis, 556

Appendicular skeleton, 627–30
Appendix, 556
Apple tress, 381, 483–84, 486, 496
   Macintosh, 196
Aquatic community, 747–54
Aquatic succession, 749–50
Aquifer, 747
Arachnid, 409
ARAS, 602, 604–5, 607
Arboreal life, 427
ARC, 672
Archaebacteria, 342, 346
Archegonia, 372, 376–77
Archenteron, 680
Archeopteryx, 301, 424
Archeozoic Era, 294
Aristotle's lantern, 412
Arm, 628–30
Armadillo, 761
   giant, 285
Armyworm, fall, 738
Arousal, 605
Arrow worm, 753
Arsenic, 560
Arteriole, 512
Artery, 512, 516
Arthritis, 652
   rheumatoid, 539, 631
Arthropod, 406–11, 414
   chemoreceptors of, 613
   eye of, 615
Artificial insemination, 672
Artificial selection, 288–89
Artiodactyla, 426
Ascaris, 159, 394–95
Ascending reticular activating
       system. *See* ARAS
Ascomycota, 363
Ascospore, 365
Asexual reproduction, 154–55, 168,
   660
   in plants, 486–88
Asparagus, 267
Aspartic acid, 101
Aspen, 761
Aspergillus nidulans, 147
Associative learning, 709
Aster, 150, 152, 501
Asteroidea, 411
Atherosclerosis, 266, 520–21, 692
Athlete, anabolic steroid use by, 654
Athletes foot, 365
Atmosphere
   oxidizing, 329
   primitive, 329
   reducing, 329
Atom, 23–27
Atomic energy level, 26
Atomic mass unit, 25
Atomic nucleus, 24
Atomic number, 25–26
Atomic symbol, 24–25
Atomic weight, 25
ATP, 50, 52, 68–69
   in coupled reactions, 104
   functions of, 104
   high-energy content of, 103
   production of, 104–6
      in cellular respiration, 102,
      126–27, 132–34, 136

      in fermentation, 126, 130–31
      in glycolysis, 126–30, 136
      in photosynthesis, 102,
      112–16
      in transition reaction, 136
   structure of, 103–4
   use of
      in active transport, 85
      in glycolysis, 116–21
      in muscle contraction, 636–38
      in photosynthesis, 116–21
ATP synthetase, 114, 133
Atrial natriuretic factor, 265–66,
   655
Atrioventricular node. *See* AV node
Atrium, 512
Attacking behavior, 703–4, 712
Auditory canal, 620
Auditory communication, 712
Auditory nerve, 623
Australopithecine, 431–33
Australopithecus, 428
Australopithecus afarensis, 431–32
Australopithecus africanus, 431,
   433
Australopithecus robustus, 431, 433
Autoantibody, 539
Autodigestion, 67
Autoimmune disease, 539, 691
Autonomic nervous system, 593,
   600–602
Autosome, 191
Autotroph, evolution of, 329, 331
Autotrophic bacteria, 345
Auxin, 493–96, 498–500
Avery, Oswald, 219
Aves, 424
AV node, 518
Axial skeleton, 627–30
Axillary bud, 452–53, 488, 494
Axon, 594
Axoplasm, 595
Azidothymidine. *See* AZT
AZT, 342, 540, 672–73

B

Baboon, 428, 711–12, 714, 740
Baby boom, 726–28
Bacillus, 343
Bacillus subtilis, 222
Background mutation, 302
Bacteria, 38, 60, 342
   autotrophic, 345
   chemosynthetic, 345
   classification of, 346–48
   denitrifying, 775–76
   genetic mapping in, 264
   genetic recombination in,
      264–65, 343
   heterotrophic, 345–46
   metabolism in, 345–46
   natural selection in, 322
   nitrifying, 775–76
   nitrogen-fixing, 775
   photosynthetic, 115, 345, 347–48
   reproduction in, 343
   shape of, 343
Bacteriophage, 339–40

Badger, 760
Balance, 603, 620–22
Balanced polymorphism, 316,
   440–41
Baldness, pattern, 215
Baleen, 547
Ball-and-socket joint, 631
Banana, 196, 496
Barash, David P., 17–18
Bark, 455, 473
Barley, 483–84, 495, 501
Barnacle, 408, 735, 741–42, 751
Barr, M. L., 210
Barracuda, 8, 752
Barr body, 210, 251–52
Basal body, 71–72
Basal nucleus, 605–6
Base, 33–35
Basidia, 365–66
Basidiomycota, 365
Basidiospore, 365–66
Basilar membrane, 622–23
Basophils, 523
Bass, 784
Bat, 426, 706–7, 757
Batesian mimic, 740
Bateson, William, 186
Bathypelagic zone, 752–54
Bat-pollinated flower, 489
B cells, 533
   action of, 534–36
   memory, 534
B-DNA, 227–28
Beadle, G. W., 232
Beagle (Darwin's ship), 281, 286
Beagle (dog), 288
Bean, 483–84, 486–87, 558
   broad, 147, 149, 469
   bush, 495
   garden, 500
Bear, 426, 761
   black, 761
   polar, 759
Beaumont, William, 552
Beaver, 426
Bee, 479, 706–7, 715, 734
Beech tree, 761
Bee-pollinated flower, 488
Beer, 131, 587
Beet, 452
Beetle, 479
Behavior, 5
   evolution of, 701–2
   execution of, 702–10
   inheritance of, 702
   innate, 703–8
   learned, 704, 708–10
Behavioral isolation, 323
Beltian body, 743
Beneden, Pierre-Joseph van, 159
Benign tumor, 254
Bentgrass, creeping, 120
Benthic zone, 752
Beriberi, 561
Bernard, Claude, 17, 511
Bernstein, Julius, 594
Berry, 484
Beryllium, 784
Beta sheet, 48–49
Beta wave, 605

Lynx, 2, 736–37, 761
Lyon, Mary, 210
Lyon hypothesis, 210
Lysine, 47
Lysogenic cycle, 339–40
Lysosome, 63, 66–68, 73
Lytic cycle, 339–40

M

MacArthur, Robert, 732
McClintock, Barbara, 228–29
McClintock, Martha, 713
McHenry, Henry, 434
Mackerel, 753
Macrocystis, 361
Macroelement, in plants, 469
Macroevolution, 303–7
Macromolecule, 39
    passage into and out of cells, 82
    on primitive earth, 329, 331
Macronucleus, 354
Macrophages, 86–87, 524, 531, 535, 672
Magnesium, 24, 100
    in humans, 560
    in plants, 469–70
Major histocompatibility protein.
    See MHC protein
Malaria, 213, 266, 316, 322, 355–56
"Male essence," 713
Malic acid, 120
Malignant tumor, 254
    See also Cancer
Mallow, Indian, 732–33
Malpighi, Marcello, 473
Malpighian tubule, 408–9, 583
Malt, 131
Maltase, 97, 554
Malthus, Thomas, 287
Maltose, 41
Malt wort, 131
Mammal, 3, 425–27
    adaptive radiation of, 304–5
    nitrogenous wastes of, 578
    osmotic regulation in, 582
    placental, 426–27
    respiratory system of, 567–68
Mammary gland, 425
Mammoth, 437
Manatee, 289–90
Manganese, 469, 560
Mangrove, 452
Manic depressive psychosis, 213
Mantle, of mollusk, 401, 548
Maple tree, 322–23, 381, 483, 760–61
Marfan syndrome, 187, 211
Marijuana, 382, 607–9, 739
Marine community, 747, 749–54
Marsh, 786
Marsh gas, 346
Marsupial, 296, 299, 426
Martin, Robert, 434
Masked messenger, 254
Mass extinction, 283–84, 294–95, 298
Mast cells, 537
Master gland, 649

Mastoiditis, 627
Mastoid sinus, 627
Mathematical data, 15
Mating, random, 312
Mating behavior, 17–18, 713
Matrix, mitochondrial, 127
Matter, 24
Matthei, J. Heinrich, 236
Maximum sustained yield, 724
Mayfly, 770
Mealworm, 583
Measles, 342
Measurement, biological, 57
Mechanical isolation, 323–24
Mechanical work, 104
Mechanoreceptor, 618–20
Medical diagnostic, 266, 272, 542
Medical genetics, 203
Medulla, of kidney, 583
Medulla oblongata, 602
Medullary cavity, 630
Medusa, 390, 392
Megakaryocyte, 524
Megasporangium, 381
Megaspore, 377, 380, 479–80
Megaspore mother cell, 480–81
Megavitamin therapy, 559
Meiosis, 147, 149, 155, 159
    comparison to mitosis, 166–68
    in humans, 165–66
    importance of, 168
    overview of, 159–62
    in plants, 481
Meiosis I, 159, 161
Meiosis II, 159, 161, 163, 165
Meischer, Friedrich, 219
Meissner's corpuscle, 619
Melaleuca, 735
Melanism, industrial, 321–22
Melanocyte-stimulating hormone
    (MSH), 645, 649
Melatonin, 645, 655, 661, 707
Melon, honeydew, 496
Membrane
    fluid-mosaic model of, 78–80
    movement of molecules across, 81–88
    selectively permeable, 82–83
    structure of, 77–79
    unit membrane model, 78
    *See also specific types of
        membranes*
Membrane channel, 78
Memory, 605–7
    long-term, 606
    short-term, 606
Memory B cells, 534
Memory T cells, 537
Mendel, Gregor, 17, 172–75, 189
Mendelian inheritance, 172–81
Mendel's law of independent
        assortment, 178–80
Mendel's law of segregation, 174, 176
Meninges, 599, 602
Meningitis, 599, 602
Meningococci, 266
Meniscus, 631
Menopause, 669, 691

Menstrual cycle, 667–70, 713
Menstrual synchrony, 713
Menstruation, 667, 669
Mercury, 769, 784–85
Meristem, 154, 448, 454–55
    apical, 448–50, 485–86
    ground, 448–49
    shoot-tip, 452–53
Meselson, Matthew, 225
Mesoderm, 392–94, 680–81
Mesoglea, 390–92
Mesopelagic zone, 752–53
Mesophyll cell, 119–21
Mesophyll tissue, 459
Mesozoic Era, 294, 298, 377, 381, 422, 426
Mesquite tree, 757
Messenger RNA (mRNA), 64, 235–37
    capping of, 238–39
    in differentiated cells, 249
    masked, 254
    persistence of, 254
    poly A tail of, 238–39
    processing of, 237–39, 243, 248, 252–53
    transport into cytoplasm, 252
    See also Transcription;
        Translation
Metabolic pathway, 96–97
Metabolic pool, 136–37
Metabolic rate, 651
Metabolic water, 126–27, 582
Metabolism, 4, 6, 96–103
Metacarpal, 628–29
Metafemale, 205, 207, 210
Metamorphosis
    in amphibian, 421–22, 650
    in insect, 409, 646
Metaphase
    meiosis I, 162–63
    meiosis II, 163, 165
    mitosis, 150–51
Metastasis, 254–55
Metatarsal, 629
Metatheria, 426
Methane, 29, 790–91
Methanogen, 346
Methyl mercury, 34
MHC protein, 536, 540
Microbody, 63, 68, 73
Microelement, in plants, 469
Microfilament, 73
Micronucleus, 354
Micropyle, 482, 486–87
Microscope
    electron, 57
    light, 57
Microsphere, 331–32
Microspore, 377, 379–81, 479, 481–82
Microspore mother cell, 481–82
Microtubule, 70–71, 150, 152, 154
    9 + 2 pattern of, 71
Microtubule organizing center, 70
Microvilli
    of intestine, 73, 552–54
    of taste bud, 614
Midbrain, 602

Middle ear, 620–21
Midge, 253
Mid-ocean ridge system, 345
Migration, 314
    of birds, 45, 425, 706–7
Milkweed, 484, 739–40
Milky spore disease, 738
Miller, Stanley, 329, 331
Millet, 459
Millipede, 409, 757
Mimicry, 740
Mineral
    availability to plants, 470–71
    requirement of humans, 560–62
    requirement of plants, 469–71
    uptake by roots, 464, 471–73
Mineralocorticoid, 645, 651–53
Miocene Epoch, 429
Miracidia, 394
Miscarriage, 670
Mistletoe, 473
Mitchell, Peter, 105, 134
Mite, 409
Mitochondria, 63, 68–70, 73, 105–6, 126–27
    evolution of, 70, 352
    genes of, 237
    pyruvate metabolism in, 132–36
    structure of, 69
Mitochondrial matrix, 69
Mitosis, 149–50
    in animal cells, 150–53
    comparison to meiosis, 166–68
    evolution of, 154
    importance of, 154–55
    in plant cells, 154–55
M line, 635–36
Model, 16
Molar, 549, 551
Mold, 3, 147
    black bread, 363–64
    blue-green, 365
    red bread, 365
Mole, 426
Molecular clock, 302
Molecule, 23, 27–30
    inorganic, 38
    organic, 38–39
    polar, 29–30
Mollusk, 401–4, 410, 414, 749–51
    eye of, 615
    feeding systems of, 547–48
Molt and maturation hormone, 646
Molting, 406, 627, 646
Molybdenum, 469, 560
Monera, 2–3, 342–48
Moniliasis, 365
Monkey, 426–28, 606, 711
    mantled howling, 15–16
    New World, 427–28, 761
    Old World, 427–28, 761
    rhesus, 428
    spider, 428
Monoamine oxidase, 598
Monoclonal antibody, 266, 542
Monocot, 381, 448–49
    embryo of, 485
    flower of, 480
    genetic engineering in, 268

Voluntary muscle, 632
Volvox, 358
Vomiting, 602
von Recklinghausen disease. *See*
    Neurofibromatosis
Vorticella, 354
Vulture, 759

W

Waggle dance, 711–12
Walker, Alan, 434
Walking worm, 410
Wallaby, 738
Wallace, Alfred Russel, 289, 295
Wang, Andrew, 227
Warbler, 732–33
Warning coloration, 739–40
Wart, genital, 672
Wasp, 6
    digger, 701, 703–5
    parasitic, 738
Waste
    disposal of, 783–84
    hazardous, 784
    nuclear, 784
    radioactive, 792
    solid, 783–84
Water
    absorption in large intestine, 556
    availability to plants, 464–65
    cohesive property of, 465–67
    conservation in animals, 578
    control of excretion of, 586–88
    ionization of, 33–35
    metabolic, 126–27, 582
    passage into and out of cells,
        81–84
    on primitive earth, 329
    properties of, 30–33
    reabsorption in kidney, 586–88,
        647
    seed and fruit dispersal by, 483
    soil, 464–65
    structure of, 29–30
    transport in plants, 465–68
    uptake by roots, 464–65
Water balance, 580, 602, 644, 655
Water buffalo, 759
Water cycle, 747
Water flea, 314
Water hyacinth, 739
Watermelon, 196
Water mold, 352, 363
Water pollution, 348, 775, 784–87
Water power, 793
Watershed, 749
Water-soluble vitamin, 561
Water vascular system, 411
Watson, James, 17, 223
Wax, 45, 52
Weasel, 761
    short-tailed, 761
Weather, 723
Weathering, 755, 776
Weeds, 755
Wegener, Alfred, 296
Weinberg, R. A., 255
Weinberg, W., 311–12

Welwitschia, 378
Whale, 426, 582, 724, 726, 742
    baleen, 547, 753
    sperm, 753
Wheat, 42, 119, 196, 267, 314, 381,
        459, 483, 496, 501
Wheat germ, 42
Whelk, 402
Whiskey, 131
White blood cells. *See* Leukocytes
White matter, 599, 602
Wiggleworth, V. B., 646
Wildebeest, 759
Wildt, David, 320
Wilkins, M. H. F., 222
Willow tree, 738, 761
Wilson, H. V., 389
Wilting, 83, 469
Wilting point, permanent, 465
Wind
    pollination by, 489
    seed and fruit dispersal by, 483
Windpipe. *See* Trachea
Wind power, 792–93
Wine, 131
Wing
    of bird, 299, 424, 684
    of insect, 299, 408
Winter bud, 498
Wisdom teeth, 551
Wobble effect, 240
Wolf, 9, 759, 761
    Tasmanian, 426
    timber, 288, 761
Wollman, Elie, 264
Wolpert, Lewis, 684
Womb. *See* Uterus
Wood, 381–82, 454, 783
Woodchuck, 761
Woodpecker, 761
Woody plant, 89
Woody stem, 448; 454–56
Worm, 3
    segmented, 405–6
Worm cast, 406
Wort, 131
Wrangham, Richard, 434
Wrist, 628–29

X

X chromosome, 191, 206–7
    inactive, 210, 251–52
Xenopus, 252
Xeroderma pigmentosum, 210, 245
X-linked gene, 191–92
X-ray, cancer and, 255
X-ray diffraction study, of DNA,
        222
Xylem, 374, 381, 447–48, 472
    cohesion-tension model of
        transport in, 465–68
    secondary, 454–56
XYY male, 205, 207

Y

Yawning, 620
Y chromosome, 191, 206–7, 210
Yeast, 130–31, 147, 365
Yellow marrow, 631
Yolk, 678–80
Yolk plug, 680
Yolk sac, 686–87

Z

Z-DNA, 227–28
Zeatin, 496
Zebra, 426, 759
Zero population growth, 727
Zigzag dance, 714–15
Zinc, 24, 469, 560
Zinder, Norton, 264
Z line, 635–36
Zone of cell division, 450
Zone of elongation, 450
Zone of maturation, 450
Zooflagellate, 353–55
Zoomastigina, 354
Zooplankton, 747, 749, 753
Zoospore, 357–58
Zooxanthellae, 752
Zygomycota, 363
Zygospore, 357–58, 363
Zygote, 147, 166, 479, 481, 483
Zygote mortality, 323–24

History of Biology *continued*

Year	Name	Country	Contribution
1905	Paul Ehrlich	Germany	Discovers first antibiotic, named Salvarsan, to cure syphilis.
1910	Thomas H. Morgan	United States	States that each gene has a locus on a particular chromosome, based on experiments with *Drosophila*.
1914	Robert Feulgen	Germany	Devises a stain for DNA and shows that chromosomes contain DNA.
1922	Sir Frederick Banting Charles Best	Canada	Isolate insulin from the pancreas.
1924	Hans Spemann Hilde Mangold	Germany	Show that induction occurs during development, based on experiments with frog embryos.
1927	Hermann J. Muller	United States	Proves that X rays cause mutations.
1929	Sir Alexander Fleming	Britain	Discovers the toxic effect of a mold product he called penicillin on certain bacteria.
1931	Fred Griffith	Britain	Discovers that nonvirulent bacteria can be transformed and become virulent if exposed to dead virulent bacteria within mice.
1932	Albert Szent-Györgyi von Nagyrapolt	Hungary	Isolates ascorbic acid from tissues and proves it is the anti-scurvy vitamin C.
1937	Konrad Z. Lorenz	Austria	Founds the study of ethology and shows the importance of imprinting as a form of early learning.
1937	Sir Hans A. Krebs	Britain	Discovers the reactions of a cycle that produces carbon dioxide during cellular respiration.
1940	George Beadle Edward Tatum	United States	Develop the one gene-one enzyme theory, based on red bread mold studies.
1944	O. T. Avery Maclyn McCarty Colin MacLeod	United States	Demonstrate that DNA alone from virulent bacteria can transform nonvirulent bacteria.
1945	Melvin Calvin Andrew A. Benson	United States	Discover the individual reactions of a cycle that reduces carbon dioxide during photosynthesis.
1950	Barbara McClintock	United States	Discovers transposons (jumping genes) while doing experiments with corn.
1950	A. L. Hodgkin A. F. Huxley	Britain	Present a model to explain the movement of the nerve impulse along a neuron.
1952	Alfred D. Hershey Martha Chase	United States	Find that only DNA from viruses enters cells and directs the reproduction of new viruses.